LANDOLT-BÖRNSTEIN

NUMERICAL DATA AND FUNCTIONAL RELATIONSHIPS IN SCIENCE AND TECHNOLOGY

NEW SERIES

EDITOR IN CHIEF

K.-H. HELLWEGE

GROUP I: NUCLEAR PHYSICS AND TECHNOLOGY

VOLUME 4

NUMERICAL TABLES FOR BETA-DECAY AND ELECTRON CAPTURE

BY

H. BEHRENS · J. JÄNECKE

EDITOR

H. SCHOPPER

SPRINGER-VERLAG BERLIN HEIDELBERG GMBH
1969

LANDOLT-BÖRNSTEIN

ZAHLENWERTE UND FUNKTIONEN AUS NATURWISSENSCHAFTEN UND TECHNIK

NEUE SERIE

GESAMTHERAUSGABE

K.-H. HELLWEGE

GRUPPE I: KERNPHYSIK UND KERNTECHNIK

BAND 4

NUMERISCHE TABELLEN FÜR BETA-ZERFALL UND ELEKTRONEN-EINFANG

VON

H. BEHRENS · J. JÄNECKE

HERAUSGEBER

H. SCHOPPER

SPRINGER-VERLAG BERLIN HEIDELBERG GMBH
1969

URSPRÜNGLICH ERSCHIENEN BEI SPRINGER-VERLAG, BERLIN/HEIDELBERG 1969
SOFTCOVER REPRINT OF THE HARDCOVER 1ST EDITION 1969

LIBRARY OF CONGRESS CATALOG CARD NUMBER 62—53136

ISBN 978-3-662-37716-1
DOI 10.1007/978-3-662-38531-9

ISBN 978-3-662-38531-9 (eBook)

Titel-Nummer 6185

Vorwort

Die beim β-Zerfall und beim Elektroneneinfang experimentell beobachtbaren Größen hängen nicht nur vom Kern selbst ab, sondern sie werden auch durch die Vorgänge in der Atomhülle bestimmt. Diese Effekte werden mit Hilfe von Funktionen berücksichtigt, für die eine Reihe von Tabellen vorliegen. Diese weisen aber zum Teil erhebliche Widersprüche auf. Ein großer Teil der früher publizierten Meßergebnisse wurde infolgedessen mit unzureichenden Tabellen ausgewertet.

In den letzten Jahren gelang es, durch eine neue Formulierung der Theorie des β-Zerfalls die kernphysikalischen Effekte sauber von den Hülleneinflüssen abzutrennen. Auch wurde die Behandlung verschiedener Korrekturen dadurch transparenter. Da außerdem der Einfluß verschiedener Näherungsverfahren auf die Berechnung der Elektronenwellenfunktionen jetzt besser zu verstehen ist, können verschiedene Rechenverfahren nunmehr zu übereinstimmenden Ergebnissen führen.

Aufgrund dieser neuen Erkenntnisse ergab sich ein dringender Bedarf an verbesserten Funktionen für den β-Zerfall und den Elektroneneinfang. Diese Lücke wird durch die hier publizierten Tabellen geschlossen.

Da die Bezeichnungsweise in der Literatur sehr stark variiert, und in einigen Fällen die Formeln nur unvollständig angegeben wurden, ist den Tabellen eine Zusammenstellung von Formeln vorangestellt, die alle Korrekturen höherer Ordnung enthalten.

Die Ausgabe-Daten der Rechenmaschine wurden unmittelbar als Druckvorlage benutzt. Dieses Verfahren, das sich bereits früher bewährt hat, wurde auch in diesem Fall vom Verlag mit Sachkunde und Verständnis gehandhabt.

Karlsruhe, Dezember 1968

Der Herausgeber

Preface

The measurable quantities in β-decay and electron capture do not depend on the nucleus alone, they are also determined by processes within the atomic shell. These effects have been described by functions presented in a number of tables which, however, contain serious discrepancies. Consequently, many of the experimental results published in the past have been evaluated on the basis of inadequate tables.

Thanks to a new formulation of the theory of β-decay, it has in recent years become possible to distinguish the physical effects in the nucleus from the influence exerted by the atomic shell. This also made the treatment of various corrections much easier to follow. Moreover, since the influence of various approximation procedures for calculating the electron wave functions is now better understood, coincident results can be obtained by different methods of computation.

This changed situation has generated an urgent need for improved β-decay and electron-capture functions, a gap which is closed by the tables published here.

Since there is considerable variation in the nomenclature used in the literature and in some cases the formulae have not been published completely, the tables are preceded by a collection of formulae which contain all higher-order corrections.

The book was reproduced directly from the computer output. This procedure has already proved its worth and was handled here by the publishers with technical competence and good judgement.

Karlsruhe, December 1968

The Editor

Inhaltsverzeichnis

Numerische Tabellen für Beta-Zerfall und Elektronen-Einfang

Von H. Behrens, Institut für experimentelle Kernphysik der Universität und des Kernforschungszentrums Karlsruhe, und J. Jänecke, The University of Michigan, Department of Physics, Ann Arbor, Michigan/USA

Table of Contents

Numerical Tables for Beta-Decay and Electron Capture

By H. Behrens, Institut für experimentelle Kernphysik der Universität und des Kernforschungszentrums Karlsruhe, and J. Jänecke, The University of Michigan, Department of Physics, Ann Arbor, Michigan/USA

1 Introduction

1.1 Effect of the Coulomb field

The main object of experiments in the field of β-decay is to obtain information about form factors. Form factors can be expressed in terms of matrix elements if it is assumed that the weak interaction causes the nucleons inside the nucleus to decay like free particles. Precise values for the above quantities, however, can be extracted only if all effects due to the electromagnetic interaction between the β-particles and the nucleus are properly taken into account. Of major importance is the distortion of the wave function of the emitted β-particle by the electrostatic Coulomb field of the nucleus. The most obvious effect of this distortion of the wave function is that it strongly affects the shape of all β-decay spectra and makes it necessary to introduce the so-called Fermi-function. To obtain the distortions of the wave functions, one has to solve the Dirac equation for an unbound electron in the field of a nucleus. For the exact solution of this problem one has to consider the finite size of the nucleus as well as the screening of the nuclear electrostatic field by the orbital electrons. Such a solution can only be obtained numerically.

1.2 Previous tabulations for β-decay

There are a number of published tables of the Fermi-function and of electron radial wave functions along with the phase shifts. However, in most cases screening corrections were not considered or were applied only in an approximate way. The Fermi-function of a point nucleus without screening corrections is tabulated in the NBS-tables [*1*]. Rose, Perry and Dismuke [*2*] introduced a finite de Broglie wave length correction by including higher terms in the confluent hypergeometric function. These authors [*2*] give tables of Coulomb functions for allowed and forbidden β-transitions. Dzhelepov and Zyrianova [*3*] published tables for the Fermi-function, for the integrated Fermi-function, and for forbidden β-decay Coulomb functions which included corrections for finite size and for screening. These tables, however, were not obtained from an exact solution of the problem of a nucleus of finite size screened by orbital electrons but by adding corrections to the functions which were obtained for a point nucleus. Exact solutions of the Dirac equation for a nucleus with finite size but without screening were calculated by Sliv and Volchok [*4*] and more extensively by Bhalla and Rose [*5*]. The former tabulated Coulomb phase shifts and amplitudes. The latter tabulated Coulomb phase shifts and radial wave functions evaluated at the nuclear radius. It has been pointed out [*6, 7*] that the tables of Bhalla and Rose [*5*] are not entirely correct for positrons of higher momentum.

Until recently it was difficult to treat the screening effect of the orbital electrons properly because the numerical integration of the Dirac equation is particularly complicated and lengthy for this case. Rose [*8*] showed that for higher electron momenta, the calculation of screening corrections for the Fermi-function is simplified by using the WKB method (see also ref. [*9*]). A compilation of the parameters needed for the WKB screening correction, i.e., the potential at the nuclear radius caused by the orbital electrons, is given by Garret and Bhalla [*10*]. A certain discrepancy exists between the values given by these authors [*10*] and the values given by Matese and Johnson [*11*].

The first detailed calculations of screening corrections to the Fermi-function were carried out by Reitz [*12*] who numerically integrated the Dirac equation. These corrections have been used in the above mentioned tables of Dzhelepov and Zyrianova [*3*]. Several authors [*11, 13, 14*] have noted that the corrections of Reitz are not quite correct for higher values of the electron momentum. An exact numerical integration of the Dirac equation which takes into consideration the finite size of the nucleus as well as screening was carried out by Bühring [*16*]. For several values of Z, his tables give certain combinations of the radial wave functions and phase shifts for allowed and first forbidden β-transitions. These are needed in the formulation of the β-decay theory as discussed below. For a few values of Z, Matese and Johnson [*11*] calculated Fermi-function screening corrections by integrating the Dirac equation numerically with a self-consistent atomic potential. These numerical values agree with the values given by Bühring [*16*]. Fermi-function screening corrections can also be obtained by solving the Klein-Gordon equation analytically using a Hulthén model for the atomic field [*13*]. This procedure was applied by Eman [*15*]. His screening corrections, given for several values of Z, are also in satisfactory agreement with the results of Bühring [*16*].

1.3 The present tables for β-decay

The following β-decay tables represent a tabulation of certain combinations of the electron radial wave functions and phase shifts for the cases with and without screening. These combinations are needed for the formulation of the β-decay theory as given by Bühring, Stech and Schülke [17···21]. In a recent book by Schopper [22], this formulation is discussed in detail for allowed and forbidden β-transitions. The tabulated solutions of the Dirac equation including the finite size of the nucleus were obtained numerically, but otherwise exactly. For the calculation of unscreened and screened β-decay functions, a uniform charge distribution potential was taken inside the nucleus. For the screened functions, Hartree-Fock potentials (for parent nuclei with $Z \leqq 36$) or Thomas-Fermi-Dirac potentials (for $Z > 36$) were used outside the nucleus.

The functions without screening are tabulated for all values of Z, and over a wide range of electron momenta. The functions with screening are not given directly. Instead, ratios of the function with screening to the function without screening are listed. These ratios η are tabulated for all values of Z, but for a more limited number of electron momenta for which $|1 - \eta| \geqq 0.02$. These restrictions became necessary because of the computer time required to calculate the screened functions. The text (chapter 2) for the present tables (chapter 4) includes a compilation of the formulae relevant to the various measurable quantities in β-decay and electron capture [17···22]. In addition to the tables, a graphical representation of the integrated Fermi-function as a function of the maximum kinetic energy of the electrons or positrons is included for several values of Z.

1.4 Previous tabulations for electron capture

For the case of electron capture, the electron radial wave functions of the bound orbital electrons at the nuclear radius are needed for theoretical calculations as well as for the analysis of experimental data. Nomograms for these functions and for the Fermi-function were first obtained by using hydrogen like Coulomb functions with screening constants according to Slater [23, 24]. Alternatively, one may obtain the electron radial wave function by numerically integrating the Dirac equation for the Thomas-Fermi-Dirac atom [12] or by employing the Hartree-Fock self-consistent field method [25]. A compilation of all previous calculations and nomograms is given by Dzhelepov and Zyrianova [3]. Brysk and Rose [26] were the first to derive detailed graphical representations of the K and L shell electron radial wave functions at the nuclear radius which took into account both screening and finite size of the nucleus (see also ref. [24]). Extensive tables of functions needed for determining the probability of allowed and forbidden K and L capture were published by Band et al. [27, 28] for a large number of Z-values. The electron radial wave functions were obtained by integrating the Dirac equation with a uniform charge distribution potential inside the nucleus and with a Thomas-Fermi-Dirac potential outside the nucleus. Using the same potential, Brewer, Harmer and Hay [29] calculated, for several values of Z, the Mshell electron radial wave functions at the nuclear radius. Electron capture ratios L_{I}/K for the region $3 \leqq Z \leqq 42$ were calculated by Winter [30] using analytical Hartree-Fock wave functions. His results agree with ours and those of Band et al. [27, 28] as well as with other new calculations [31] but are not in as good agreement with those of Brysk and Rose [26].

Odiot and Daudel [32] showed that for the exact treatment of electron capture one has to take into consideration not only the radial wave function at the nuclear radius of the captured electron, but also the corrections due to an exchange interaction between this particular electron and the electrons in the other shells. There are cases where the latter effect changes the transition probability by almost 20%. Bahcall [33] calculated such exchange corrections, and also corrections for imperfect atomic overlap, for a variety of Z-values. Recently Vatai [34] has pointed out, that these corrections may be not entirely correct.

1.5 The present tables for electron capture

The present tables for electron capture give the amplitudes and the values of the electron radial wave function at the nuclear radius for the K, L and M shell for all values of Z (L and M shell excluded for the very light nuclei). These tabulated quantities were calculated by integrating the Dirac equation with the same potentials as were used for the β-decay functions, but modified to include exchange terms in the potential. As before, a brief compilation of all relevant formulae is given. The notation follows closely the one introduced by Stech, Schülke and Bühring for β-decay [17···21]. As in β-decay, only the amplitudes of the bound electron radial wave functions are needed and not the electron radial wave functions at the nuclear radius. Nevertheless, both quantities are listed in the present tables to facilitate the discussion and the comparison with other representations [26, 35] and existing electron capture tables. Possibly both functions will be useful in connection with other applications. The exchange corrections given in chapter 2.5 were taken from BAHCALL [33].

1.6 Recent reviews of β-decay and electron capture

To close this introduction, we refer the reader to several recent reviews of β-decay and electron capture. Books by KONOPINSKI [36], WU and MOSZKOWSKI [37], PRESTON [38] and DE BENEDETTI [39] treat both topics as does the previously mentioned monograph by SCHOPPER [22]. Review articles have been written by KONOPINSKI and ROSE [40], WEIDENMÜLLER [41], TOLHOEK [42], and BLIN STOYLE and NAIR [43].

For electron capture, the most detailed reviews are the articles of BRYSK and ROSE [26], BOUCHEZ and DEPOMMIER [35] and BERÉNYI [44]. A description of more special aspects of β-decay is also given in a number of articles in ref. [45].

Arrangement, Acknowledgments

In chapter 2, a brief compilation of all the relevant formulae used in the treatment of β-decay and electron capture is given.

In chapter 3, we explain the details of the computations and describe the tables presented in chapter 4. The references are listed in chapter 5.

Sincere thanks are due to Professor H. SCHOPPER for initiating this work and for his continued interest. The authors are particularly grateful to Dr. W. BÜHRING for making available to them his computer programms which provided the starting point for our calculations, for numerous valuable discussions, and for the careful reading of the manuscript. Furthermore, we are indebted to Dr. H. APPEL for constructive comments and for reading the manuscript. The close and excellent cooperation with the editors and the Springer-Verlag is greatly appreciated.

2 Formulae for β-decay and electron capture

2.1 Classification of β-decays

The various types of β-transitions are defined by the following selection rules:

$L = 0, 1$	$\pi_i \cdot \pi_f = +1$	allowed transitions
$L = 0, 1$	$\pi_i \cdot \pi_f = -1$	first non unique forbidden transitions
$L > 1$	$\pi_i \cdot \pi_f = (-1)^L$	L^{th} non unique forbidden transitions
	$\pi_i \cdot \pi_f = (-1)^{L-1}$	$(L-1)^{\text{th}}$ unique forbidden transitions

with $L = \Delta J = |J - J'|$.

Here (J, π_i) and (J', π_f) denote the spins and parities of the initial and final nuclear states, respectively.

2.2 General formulae for β-decay

2.2.1 Shapes of beta spectra

The number of β-particles with momentum p in the interval between p and $p + dp$ emitted per unit time is:

$$N(p_e)\,dp_e = \frac{g^2}{2\pi^3} p_e^2 \cdot p_\nu^2 \cdot F(Z, W_e) \cdot C(W_e)\,dp_e, \tag{1}$$

where

g = weak interaction coupling constant,
p_e = electron momentum in units of $m_0 c$,
$W_e = \sqrt{p_e^2 + 1}$ = total electron energy in units of $m_0 c^2$,
W_0 = maximum value* of W_e,
$p_\nu = W_0 - W_e$ = energy or momentum of the neutrino in units of $m_0 c^2$ or $m_0 c$, respectively,
Z = atomic number of the daughter nucleus.

* For β^--decay the maximum kinetic energy $W_0 - 1$ is equal to the difference of the atomic masses, for β^+-decay $2m_0c^2$ must be substracted from this difference in order to obtain $W_0 - 1$.

The Fermi-function $F(Z, W_e)$ takes into account the distortion of the electron wave function by the nuclear charge. It has the following form [22]:

$$F(Z, W_e) = F_0 L_0,$$

with

$$F_0 = 4 \cdot (2 p_e R)^{-2(1-\gamma_1)} \cdot \frac{|\Gamma(\gamma_1 + i y)|^2}{[\Gamma(2\gamma_1 + 1)]^2} \cdot e^{\pi \cdot y}, \tag{2}$$

where

$$\gamma_{k_e} = \sqrt{k_e^2 - (\alpha Z)^2}, \qquad y = \alpha Z \cdot \frac{W_e}{p_e}$$

$\alpha = 1/137.0388$ (fine structure constant),
R = nuclear radius in units of $\lambda\!\!\!^{-} = \hbar/(m_0 c)$,

and

$$L_0 = \frac{\alpha_{-1}^2 + \alpha_{+1}^2}{2 p_e^2 F_0}. \tag{3}$$

This definition of L_0 differs from the usual one (see footnote on p. 10 and 16).

The quantities α_{k_e} are the so-called Coulomb amplitudes, which must be calculated numerically.

The spectrum shape factor $C(W_e)$ of the allowed, $(L-1)^{\text{th}}$ unique forbidden and L^{th} non unique forbidden transitions has the following form [17, 18, 21, 22] (correction terms are included):

$$\left.\begin{array}{lll} L = \Delta J & \text{if} & \Delta J > 0 \\ L = 1 & \text{if} & \Delta J = 0 \end{array}\right\} \pi_i \cdot \pi_f = (-1)^L \quad \text{or} \quad (-1)^{L-1}.$$

$$\begin{aligned} C(W_e) = & \sum_{\substack{k_e k_\nu \\ k_e + k_\nu = L+1}} \lambda_{k_e} \left\{ M_L^2(k_e, k_\nu) + m_L^2(k_e, k_\nu) - \frac{2\mu_{k_e}\gamma_{k_e}}{k_e W_e} M_L(k_e, k_\nu) m_L(k_e, k_\nu) \right\} + \\ & + \sum_{\substack{k_e, k_\nu \\ k_e + k_\nu = L+2}} \lambda_{k_e} \left\{ M_L^2(k_e, k_\nu) - \frac{2\mu_{k_e}\gamma_{k_e}}{k_e W_e} M_L(k_e, k_\nu) m_L(k_e, k_\nu) + M_{L+1}^2(k_e, k_\nu) \right. \\ & \left. - \frac{2\mu_{k_e}\gamma_{k_e}}{k_e W_e} M_{L+1}(k_e, k_\nu) m_{L+1}(k_e, k_\nu) \right\} + \\ & + \delta_{\Delta J, 0} \left\{ M_0^2(1,1) + m_0^2(1,1) - \frac{2\mu_1\gamma_1}{W_e} M_0(1,1) m_0(1,1) \right\}, \end{aligned} \tag{4}$$

with

$$\lambda_{k_e} = \frac{\alpha_{-k_e}^2 + \alpha_{k_e}^2}{\alpha_{-1}^2 + \alpha_1^2}, \tag{5}$$

$$\mu_{k_e} = \frac{k_e W_e}{\gamma_{k_e}} \frac{\alpha_{-k_e}^2 - \alpha_{k_e}^2}{\alpha_{-k_e}^2 + \alpha_{k_e}^2}, \tag{6}$$

$$\Lambda_{k_e} = \frac{W_e}{p_e} \frac{2\alpha_{-k_e}\alpha_{k_e}\cos(\Delta_{k_e} - \Delta_{-k_e})}{\alpha_{-k_e}^2 + \alpha_{k_e}^2}. \tag{7}$$

(see footnote on p. 10),

$$k = |\varkappa| \qquad \begin{array}{ll} \varkappa = l & \text{for } j = l - \frac{1}{2}, \\ \varkappa = -(l+1) & \text{for } j = l + \frac{1}{2}. \end{array}$$

Here j and l denote the total angular momentum and orbital angular momentum, respectively, of the emitted electron or neutrino. The quantity Δ_{k_e} is the Coulomb phase [16, 17, 21] which must be computed numerically.

The quantities M_L depend on the form factor coefficients and on the energy dependent functions $H_{2\varrho}$ and $D_{2\varrho+1}$ [21, 22]. If we restrict ourselves to the leading terms [17, 18, 22] we obtain the following expressions*:

* In eqs. (8), (9), and (10) only the first three terms N_0, N_1, and N_2 of the neutrino wave function have been kept since the remaining terms are smaller by at least a factor $\frac{(p_\nu R)^3}{30}$ and therefore negligible.

if $\qquad k_e + k_\nu = L + 1, \quad L = \Delta J, \quad \pi_i \pi_f = (-1)^L:$

$$\begin{aligned} M_L(k_e, k_\nu) = {} & K\,(p_e R)^{k_e-1}\,(p_\nu R)^{k_\nu-1} \Big\{ -\sqrt{\frac{2L+1}{L}} \sum_{n=0}^{\infty} {}^{\mathrm{V}}F^{(n)}_{L,L-1,1} \Big(H_{2n} + \frac{1}{2L+1} N_1 D_{2n-1} + \\ & + N_2 H_{2n-2}\Big) - \sum_{n=0}^{\infty} {}^{\mathrm{V}}F^{(n)}_{LL0} \left(D_{2n+1} + N_1 H_{2n} + N_2 D_{2n-1}\right) - \sqrt{\frac{L+1}{L}} \times \\ & \times \sum_{n=0}^{\infty} {}^{\mathrm{A}}F^{(n)}_{LL1} \left(D_{2n+1} - N_1 H_{2n} + N_2 D_{2n-1}\right) + 2\sqrt{\frac{L+1}{2L+1}} \sum_{n=0}^{\infty} {}^{\mathrm{V}}F^{(n)}_{L,L+1,1} N_1 D_{2n+1} \Big\}, \end{aligned} \tag{8}$$

if $\qquad k_e + k_\nu = L + 2, \quad L = \Delta J, \quad \pi_i \pi_f = (-1)^L:$

$$\begin{aligned} M_L(k_e, k_\nu) = {} & K\,(p_e R)^{k_e-1}\,(p_\nu R)^{k_\nu-1} \sqrt{\frac{L+1}{(2k_e-1)(2k_\nu-1)}} \times \\ & \times \Big\{ \sum_{n=0}^{\infty} {}^{\mathrm{V}}F^{(n)}_{LL0} \left(H_{2n} - N_1 D_{2n-1} + N_2 H_{2n-2}\right) - \frac{k_e - k_\nu}{L+1} \sqrt{\frac{L+1}{L}} \sum_{n=0}^{\infty} {}^{\mathrm{A}}F^{(n)}_{LL1} \times \\ & \times \left(H_{2n} + N_1 D_{2n-1} + N_2 H_{2n-2}\right) - \sqrt{\frac{1}{L(2L+1)}} \sum_{n=0}^{\infty} {}^{\mathrm{V}}F^{(n)}_{L,L-1,1} \times \\ & \times \left[2(k_\nu - 1)(D_{2n-1} + N_2 D_{2n-3}) + 2(k_e - 1) N_1 H_{2n-2}\right] - \sqrt{\frac{1}{(L+1)(2L+1)}} \times \\ & \times \sum_{n=0}^{\infty} {}^{\mathrm{V}}F^{(n)}_{L,L+1,1} \left[(2k_e - 1)(D_{2n+1} + N_2 D_{2n-1}) + (2k_\nu - 1) N_1 H_{2n}\right] \Big\}, \end{aligned} \tag{9}$$

and

$$\begin{aligned} M_{L+1}(k_e, k_\nu) = {} & K \cdot (p_e R)^{k_e-1}\,(p_\nu R)^{k_\nu-1} \Big\{ -\sum_{n=0}^{\infty} {}^{\mathrm{A}}F^{(n)}_{L+1,L,1} \Big(H_{2n} + \\ & + \frac{1}{2L+3} N_1 D_{2n-1} + N_2 H_{2n-2}\Big) - \sqrt{\frac{L+1}{2L+3}} \sum_{n=0}^{\infty} {}^{\mathrm{A}}F^{(n)}_{L+1,L+1,0} \times \\ & \times \left(D_{2n+1} + N_1 H_{2n} + N_2 D_{2n-1}\right) - \sqrt{\frac{L+2}{2L+3}} \sum_{n=0}^{\infty} \times \\ & \times {}^{\mathrm{V}}F^{(n)}_{L+1,L+1,1} \left(D_{2n+1} - N_1 H_{2n} + N_2 D_{2n-1}\right) + \\ & + 2\,\frac{\sqrt{(L+1)(L+2)}}{2L+3} \sum_{n=0}^{\infty} {}^{\mathrm{A}}F^{(n)}_{L+1,L+2,1} N_1 D_{2n+1} \Big\}, \end{aligned} \tag{10}$$

with $\quad K = \sqrt{\dfrac{L!}{(2L+1)!!}} \cdot \dfrac{1}{\sqrt{(2k_e-1)!!\,(2k_\nu-1)!!\,(k_e-1)!\,(k_\nu-1)!}}.$

If $\pi_i \pi_f = (-1)^{L-1}$, ${}^{\mathrm{V}}F_{LLs}$ in eqs. (8, 9), and (10) must be replaced by ${}^{\mathrm{A}}F_{LLs}$ and ${}^{\mathrm{A}}F_{LLs}$ by ${}^{\mathrm{V}}F_{LLs}$. The formulae for $m_L(k_e, k_\nu)$ are the same as for $M_L(k_e, k_\nu)$ except that $H_{2\varrho}$ and $D_{2\varrho+1}$ must be re-

Table 1. The coefficients $H_{2\varrho}$, $D_{2\varrho+1}$, $h_{2\varrho}$ and $d_{2\varrho+1}$

$\overline{W}_e = W_e \pm 3\alpha\lvert Z\rvert/(2R)$	$\overline{p}_e^2 = \overline{W}_e^2 - 1$
$H_{2\varrho} = D_{2\varrho+1} = 0$ for $\varrho < 0$	$H_{2\varrho+1} = D_{2\varrho} = 0$ for all ϱ
$H_0(k_e) = 1$	$h_0(k_e) = 0$
$H_2(k_e) = -\dfrac{(\overline{p}_e R)^2}{2(2k_e+1)}$	$h_2(k_e) = 0$
	$h_4(k_e) = \dfrac{\pm\alpha\lvert Z\rvert \cdot R}{4(2k_e+1)(2k_e+3)}$
$D_1(k_e) = \dfrac{\overline{W}_e R}{2k_e+1}$	$d_1(k_e) = \dfrac{R}{2k_e+1}$
$D_3(k_e) = -\dfrac{\overline{W}_e R(\overline{p}_e R)^2}{2(2k_e+1)(2k_e+3)} \mp \dfrac{\alpha\lvert Z\rvert}{2(2k_e+3)}$	$d_3(k_e) = -\dfrac{R(\overline{p}_e R)^2}{2(2k_e+1)(2k_e+3)}$
$N_0(k_\nu) = 1, \quad N_1(k_\nu) = \dfrac{p_\nu R}{2k_\nu+1}, \quad N_2(k_\nu) = -\dfrac{(p_\nu R)^2}{2(2k_\nu+1)}$	

Z is the atomic number of the daughter nucleus for β-decay and of the parent nucleus for electron capture. The upper sign applies for β^--decay and electron capture, the lower sign for β^+-decay. In some cases (expecially for $k_e = k_\nu = 1$) we have omitted the indices k_e and k_ν in the quantities $H_{2\varrho}$, $D_{2\varrho+1}$ and N_ϱ.

placed by $h_{2\varrho}$ and $d_{2\varrho+1}$, respectively. These functions can be computed with the help of recursion relations [*17, 21*]. Since in all practical cases only the first few functions are needed, we give these functions explicitly in table 1.

The so-called form factor coefficients $F^{(n)}_{Lls}$ contain all the relevant information about the nuclear structure. The form factor coefficients $F^{(n)}_{Lls}$ are the only quantities which can be determined from a β-decay experiment.

If the usual assumption that the nucleons inside the nucleus decay like free nucleons is made, the form factor coefficients can be related to the ordinary nuclear matrix elements [*19, 20, 22*] as given in table 2.

Table 2. Connection between form factor coefficients and reduced nuclear matrix elements, neglecting induced interactions (according to [*18, 19, 20, 22*])

Form factor coefficient	Matrix element in cartesian notation	Selection rules			Type of transition
		ΔJ	$\pi_i \pi_f$	forbidden	
${}^VF^{(0)}_{000}$	$C_V \int 1$	0	+1	—	Allowed.
${}^AF^{(0)}_{101}$	$\pm C_A \int \boldsymbol{\sigma}$	0,1	+1	0—0	
${}^VF^{(1)}_{000}$	$C_V \int \left(\frac{r}{R}\right)^2$	0	+1	—	Main corrections to allowed transitions.
${}^VF^{(0)}_{011}$	$C_V \int i\frac{\boldsymbol{\alpha}\cdot\boldsymbol{r}}{R}$	0	+1	—	
${}^AF^{(1)}_{101}$	$\pm C_A \int \boldsymbol{\sigma}\left(\frac{r}{R}\right)^2$	0,1	+1	0—0	
${}^VF^{(0)}_{111}$	$-C_V\sqrt{\frac{3}{2}}\int\frac{\boldsymbol{\alpha}\times\boldsymbol{r}}{R}$	0,1	+1	0—0	
${}^AF^{(0)}_{110}$	$\mp C_A\sqrt{3}\int\gamma_5\frac{i\boldsymbol{r}}{R}$	0,1	+1	0—0	
${}^AF^{(0)}_{121}$	$\pm C_A\frac{3}{\sqrt{2}}\int\frac{(\boldsymbol{\sigma}\cdot\boldsymbol{r})\boldsymbol{r}-\frac{1}{3}\boldsymbol{\sigma}\cdot r^2}{R^2}$	0,1	+1	0—0	
${}^VF^{(0)}_{220}$	$-C_V\sqrt{\frac{15}{2}}\int\frac{R_{ij}}{R^2}$	0, 1, 2	+1	0—0; 0—1; $\frac{1}{2}$—$\frac{1}{2}$	
${}^AF^{(0)}_{221}$	$\pm C_A\frac{\sqrt{5}}{2}\int\frac{iT_{ij}}{R^2}$	0, 1, 2	+1	0—0; 0—1; $\frac{1}{2}$—$\frac{1}{2}$	
${}^VF^{(0)}_{211}$	$-C_V\frac{\sqrt{3}}{2}\int\frac{iA_{ij}}{R^2}$	0, 1, 2	+1	0—0; 0—1; $\frac{1}{2}$—$\frac{1}{2}$	
${}^AF^{(0)}_{000}$	$\mp C_A\int\gamma_5$	0	−1	—	First non unique forbidden.
${}^AF^{(0)}_{011}$	$\mp C_A\int i\frac{\boldsymbol{\sigma}\cdot\boldsymbol{r}}{R}$	0	−1	—	
${}^VF^{(0)}_{101}$	$-C_V\int\boldsymbol{\alpha}$	0, 1	−1	0—0	
${}^VF^{(0)}_{110}$	$C_V\sqrt{3}\int i\frac{\boldsymbol{r}}{R}$	0, 1	−1	0—0	
${}^AF^{(0)}_{111}$	$\pm C_A\sqrt{\frac{3}{2}}\int\frac{\boldsymbol{\sigma}\times\boldsymbol{r}}{R}$	0, 1	−1	0—0	
${}^AF^{(0)}_{211}$	$\pm C_A\frac{\sqrt{3}}{2}\int\frac{iB_{ij}}{R}$	0, 1, 2	−1	0—0; 0—1; $\frac{1}{2}$—$\frac{1}{2}$	
${}^AF^{(1)}_{000}$	$\mp C_A\int\gamma_5\left(\frac{r}{R}\right)^2$	0	−1	—	Main corrections to first non unique forbidden transitions.
${}^AF^{(1)}_{011}$	$\mp C_A\int i\frac{\boldsymbol{\sigma}\cdot\boldsymbol{r}}{R}\cdot\left(\frac{r}{R}\right)^2$	0	−1	—	
${}^VF^{(1)}_{101}$	$-C_V\int\boldsymbol{\alpha}\left(\frac{r}{R}\right)^2$	0, 1	−1	0—0	
${}^VF^{(1)}_{110}$	$C_V\sqrt{3}\int i\frac{\boldsymbol{r}}{R}\left(\frac{r}{R}\right)^2$	0, 1	−1	0—0	
${}^VF^{(0)}_{121}$	$-C_V\frac{3}{\sqrt{2}}\int\frac{(\boldsymbol{\alpha}\cdot\boldsymbol{r})\cdot\boldsymbol{r}-\frac{1}{3}\boldsymbol{\alpha}\cdot r^2}{R^2}$	0, 1	−1	0—0	

Table 2 (continued)

Form factor coefficient	Matrix element in cartesian notation	Selection rules ΔJ	$\pi_i \pi_f$	forbidden	Type of transition
${}^{A}F^{(1)}_{111}$	$\pm C_A \sqrt{\frac{3}{2}} \int \frac{\boldsymbol{\sigma} \times \boldsymbol{r}}{R} \left(\frac{r}{R}\right)^2$	0, 1	-1	0—0	Main corrections to first non unique forbidden transitions. (cont.)
${}^{A}F^{(1)}_{211}$	$\pm C_A \frac{\sqrt{3}}{2} \int \frac{i B_{ij}}{R} \left(\frac{r}{R}\right)^2$	0, 1, 2	-1	0—0; 0—1; $\frac{1}{2}$—$\frac{1}{2}$	
${}^{A}F^{(0)}_{220}$	$\pm C_A \sqrt{\frac{15}{2}} \int \gamma_5 \frac{R_{ij}}{R^2}$	0, 1, 2	-1	0—0; 0—1; $\frac{1}{2}$—$\frac{1}{2}$	
${}^{V}F^{(0)}_{221}$	$-C_V \frac{\sqrt{5}}{2} \int \gamma_5 \frac{i T_{ij}}{R^2}$	0, 1, 2	-1	0—0; 0—1; $\frac{1}{2}$—$\frac{1}{2}$	
${}^{V}F^{(0)}_{LL0}$	$C_V \sqrt{\frac{(2L+1)!!}{L!}} \int \frac{i^L R_{i_1 \cdots i_L}}{R^L}$	$0, 1, \dots L$	$(-1)^L$	$J + J' < L$	
${}^{V}F^{(0)}_{L, L-1, 1}$	$-C_V \frac{1}{L!} \sqrt{\frac{(2L-1)!!}{(L-1)!}} \int \frac{i^{L-1} A_{i_1 \cdots i_L}}{R^{L-1}}$	$0, 1, \dots L$	$(-1)^L$	$J + J' < L$	
${}^{A}F^{(0)}_{LL1}$	$\mp C_A \frac{1}{L!} \sqrt{\frac{(2L+1)!!}{L!}} \sqrt{\frac{L}{L+1}} \times \int \frac{i^{L+1} T_{i_1 \cdots i_L}}{R^L}$	$0, 1, \dots L$	$(-1)^L$	$J + J' < L$	Lth non unique forbidden.
${}^{A}F^{(0)}_{L+1, L, 1}$	$\pm C_A \frac{1}{(L+1)!} \sqrt{\frac{(2L+1)!!}{L!}} \times \int \frac{i^L B_{i_1 \cdots i_{L+1}}}{R^L}$	$0, 1, \dots, L+1$	$(-1)^L$	$J + J' < L + 1$	
${}^{A}F^{(0)}_{211}$	$\pm C_A \frac{\sqrt{3}}{2} \int i \frac{B_{ij}}{R}$	0, 1, 2	-1	0—0; 0—1; $\frac{1}{2}$—$\frac{1}{2}$	First unique forbidden.
${}^{A}F^{(0)}_{L, L-1, 1}$	$\pm C_A \frac{1}{L!} \sqrt{\frac{(2L-1)!!}{(L-1)!}} \int \frac{i^{L-1} B_{i_1 \cdots i_L}}{R^{L-1}}$	$0, 1, \dots L$	$(-1)^{L-1}$	$J + J' < L$	$(L-1)$ th unique forbidden.

$$R_{ij} = r_i r_j - \frac{1}{3} \delta_{ij} r^2 \qquad A_{ij} = \alpha_i r_j + \alpha_j r_i - \frac{2}{3} \delta_{ij} (\boldsymbol{\alpha} \cdot \boldsymbol{r})$$

$$B_{ij} = \sigma_i r_j + \sigma_j r_i - \frac{2}{3} \delta_{ij} (\boldsymbol{\sigma} \cdot \boldsymbol{r}) \qquad T_{ij} = (\boldsymbol{\sigma} \times \boldsymbol{r})_i r_j + (\boldsymbol{\sigma} \times \boldsymbol{r})_j r_i .$$

The upper sign applies for β^--decay and electron capture, the lower sign for β^+-decay. $F^{(n)}_{Lls}$ is obtained from the matrix elements associated with $F^{(0)}_{Lls}$ by multiplying the expression under the integral by $(r/R)^{2n}$. All matrix elements have been expressed as real quantities, thus i's appear in several of our terms. Usually $\int 1$ is denoted as the Fermi-matrix element M_F and $\int \sigma$ as the Gamow-Teller matrix element M_{GT}.

2.2.2. Electron polarization

The longitudinal polarization P of the electrons emitted in β-decay for allowed, L^{th} non unique forbidden and $(L-1)^{\text{th}}$ unique forbidden transitions is

$$P = \mp \frac{p_e}{W_e} \cdot \frac{1}{C(W_e)} \left\{ \sum_{\substack{k_e, k_\nu \\ k_e + k_\nu = L+1}} \Lambda_{k_e} \lambda_{k_e} [M_L^2(k_e, k_\nu) - m_L^2(k_e, k_\nu)] + \right. \tag{11}$$

$$\left. + \sum_{\substack{k_e, k_\nu \\ k_e + k_\nu = L+2}} \Lambda_{k_e} \lambda_{k_e} [M_L^2(k_e, k_\nu) + M_{L+1}^2(k_e, k_\nu)] + \delta_{\Delta J, 0} \Lambda_1 [M_0^2(1,1) - m_0^2(1,1)] \right\},$$

(upper sign for β^--, lower sign for β^+-decay).

The quantities used in this formula have been defined in subsection 2.2.1.

2.2.3 β—γ directional and circular polarization correlation

For the spin sequence $J \xrightarrow{\beta} J' \xrightarrow{\gamma} J''$, the correlation between the β-particle and the γ-rays emitted subsequently has the form

$$\omega(W_e, \theta) = \frac{1}{L_0 \cdot C(W_e)} \left\{ \sum_{m \text{ even}} B^{(m)} \cdot A_m(\gamma) P_m(\cos\theta) - \tau \sum_{m \text{ odd}} B^{(m)} \cdot A_m(\gamma) P_m(\cos\theta) \right\}. \quad (12)$$

Here, θ is the angle between β-particle and γ-quantum. The $P_m(\cos\theta)$ are the Legendre polynomials. The quantity τ, which determines the sign of the circular polarization of the γ quantum, is equal to ± 1. If only the directional correlation is observed, the terms with odd m in eq. (12) must be omitted.

The coefficients $B^{(m)}$ are defined by [22]

$$B^{(m)} = \sum_{L \leqq L'} b_{LL'}^{(m)} \Gamma_{LL'}(m, J, J'). \quad (13)$$

The indices L, L' and m in eq. (13) must satisfy the following relations:

$$|J - J'| \leqq L, L' \leqq J + J', \quad |L - L'| \leqq m \leqq \min(2J', L + L').$$

The functions $b_{LL'}^{(m)}$, called the particle parameters*, contain all the information about the β-transition. The general expressions for these parameters are given by SCHOPPER [22]. For the special cases of the allowed and first forbidden decays, explicit expressions will be given in section 2.3.

For the $\Gamma_{LL'}(m, J, J')$, we have the relation**

$$\Gamma_{LL'}(m, J, J') = (-1)^{J'-J} \cdot W(J'J'\; LL', mJ) \cdot \sqrt{2J'+1}. \quad (14)$$

Here, W is a Racah coefficient. The $\Gamma_{LL'}(m, J, J')$ are tabulated by APPEL in the New Series of Landolt-Börnstein [46].

The $A_m(\gamma)$, which contain all the information about the γ-transition, have the form

$$A_m(\gamma) = \sum_{\lambda\lambda'} F_m(\lambda\lambda' J'' J') \cdot \delta_\lambda \cdot \delta_{\lambda'}, \quad (15)$$

with

$$F_m(\lambda\lambda' J'' J') = (-1)^{J'-J''-1} \cdot \sqrt{(2J'+1)(2\lambda+1)(2\lambda'+1)}\, C(\lambda\lambda' m;\, 1-1)\, W(J'J'\lambda\lambda';\, mJ'').$$

C is a Clebsch-Gordon coefficient. The F-coefficients needed in the theory of γ—γ-correlations are (apart from many other tabulations) tabulated extensively by APPEL in the New Series of Landolt-Börnstein [46]. The δ_λ are reduced matrix elements (usual notaiton, see [46]) for the 2^λ-pole γ-transition and are normalized according to $\sum \delta_\lambda^2 = 1$.

The sum in eq. (12) runs from 0 to $\min(2J', \lambda + \lambda')$. For practical applications, however, we have $m \leqq 2$ in the case of allowed and $m \leqq 4$ for first forbidden transitions, since the terms with larger values of L and L' are neglegible.

2.2.4 Electron emission from oriented nuclei

For the distribution of the emitted electrons from oriented nuclei [22], one has the correlation function

$$\omega(W_e, \theta) = \frac{\sqrt{2J+1}}{C(W_e) \cdot L_0} \sum_n \tilde{B}^{(n)} G_n(J) P_n(\cos\theta), \quad (16)$$

which corresponds to eq. (12). The main difference is that the coefficients $A_m(\gamma)$ must be replaced by the parameters $G_n(J)$. Here, θ is the angle between the axis of orientation and the direction of the

* The shape factor can be expressed in terms of the particle parameters in the following way:

$$C(W_e) = L_0^{-1} \sum_L (-1)^L (2L+1)^{-1/2} b_{LL}^{(0)}.$$

** Between the $\Gamma_{LL'}(m, J, J')$ as defined in the text and the $\Gamma_{LL'}(m, J, J', J'')$ as defined by SCHOPPER [22], we have the relation $\Gamma_{LL'}(m, J, J, 'J'') = A_m(\gamma) \cdot \Gamma_{LL'}(m, J, J')$.

emitted β-rays. The quantity $\tilde{B}^{(n)}$ is defined by

$$\tilde{B}^{(n)} = \sum_{L \leq L'} (-1)^{L+L'+n} \cdot \Gamma(n, J', J) \cdot b_{LL'}^{(n)}. \tag{17}$$

This corresponds essentially to eq. (13), but the spins J and J' in the $\Gamma_{LL'}(n, J, J')$ coefficient must be interchanged. The $G_n(J)$ are special cases of statistical tensors introduced by FANO [47]:

$$G_n(J) = \sum_m (-1)^{J-m} \cdot C(JJn;\, m\, -m) \cdot a_m. \tag{18}$$

C is again a Clebsch-Gordon coefficient. The a_m are the relative populations of the initial magnetic substates. They are normalized such that

$$\sum_m a_m = 1.$$

For the first four terms of $G_n(J)$, one has

$$G_0(J) = \frac{1}{[2J+1]^{1/2}}$$

$$G_1(J) = \frac{\sqrt{3}}{[J(J+1)(2J+1)]^{1/2}} \sum_m m\, a_m, \tag{19}$$

$$G_2(J) = \frac{3 \cdot \sqrt{5}\left[\sum_m m^2 a_m - \frac{1}{3} J(J+1)\right]}{[J(J+1)(2J+3)(2J+1)(2J-1)]^{1/2}},$$

$$G_3(J) = \frac{5 \cdot \sqrt{7}\left[\sum_m m^3 a_m - \frac{1}{5}(3J^2+3J-1)\sum_m m\, a_m\right]}{[J(J+1)(2J+3)(2J+1)(2J-1)(J-1)(J+2)]^{1/2}}.$$

The orientation parameters $f_n(J)$ which are usually used and the parameters $G_n(J)$ are related by [48]

$$\frac{f_n(J)}{G_n(J)} = \binom{2n}{n}^{-1} J^{-n} \left[\frac{(2J+n+1)!}{(2n+1)(2J-n)!}\right]^{1/2}. \tag{20}$$

2.3 Special formulae for allowed, first forbidden, L^{th} non unique forbidden, and $(L-1)^{\text{th}}$ unique forbidden β-decays

2.3.1 Particle parameters

For allowed and first forbidden transitions, the particle parameters $b_{LL'}^{(m)}$ have the form shown in table 3 (according to BÜHRING [17, 22]).

Table 3. Particle parameters $b_{LL'}^{(m)}$ for allowed and first forbidden decays

$$b_{00}^{(0)}/L_0 = M_0^2(1,1) + m_0^2(1,1) - 2\mu_1\gamma_1 \frac{1}{W_e} M_0(1,1)\, m_0(1,1)$$

$$b_{11}^{(0)}/L_0 = -\sqrt{3}\{M_1^2(1,1) + m_1^2(1,1) - 2\mu_1\gamma_1 \frac{1}{W_e} M_1(1,1)\, m_1(1,1) + M_1^2(1,2) -$$
$$- 2\mu_1\gamma_1 \frac{1}{W_e} M_1(1,2)\, m_1(1,2) + \lambda_2[M_1^2(2,1) - \mu_2\gamma_2 \frac{1}{W_e} M_1(2,1)\, m_1(2,1)]\}$$

$$b_{22}^{(0)}/L_0 = \sqrt{5}\{M_2^2(1,2) - 2\mu_1\gamma_1 \frac{1}{W_e} M_2(1,2)\, m_2(1,2) + \lambda_2[M_2^2(2,1) -$$
$$- \mu_2\gamma_2 \frac{1}{W_e} M_2(2,1)\, m_2(2,1)]\}$$

$$b_{01}^{(1)}/L_0 = \pm\{2 \frac{p_e}{W_e} \Lambda_1[M_0(1,1)\, M_1(1,1) - m_0(1,1)\, m_1(1,1)] -$$
$$- 2\sqrt{2}\,[\eta_{12} M_0(1,1)\, M_1(2,1) - \hat{\eta}_{12} \frac{1}{W_e}(M_1(2,1)\, m_0(1,1) + M_0(1,1)\, m_1(2,1))]\}$$

Table 3 (continued)

$$b_{11}^{(1)}/L_0 = \pm \{\sqrt{2}\frac{p_e}{W_e}\Lambda_1[M_1^2(1,1) - m_1^2(1,1)] + 2[\eta_{12}M_1(1,1)M_1(2,1) - \hat{\eta}_{12}\frac{1}{W_e}(M_1(2,1)m_1(1,1) + M_1(1,1)m_1(2,1))] - \frac{1}{\sqrt{2}}\frac{p_e}{W_e} \times [\Lambda_1 M_1^2(1,2) - \lambda_2\Lambda_2 M_1^2(2,1)]\}$$

$$b_{12}^{(1)}/L_0 = \pm \{2\sqrt{5}[\eta_{12}M_1(1,1)M_2(2,1) - \hat{\eta}_{12}\frac{1}{W_e}(M_2(2,1)m_1(1,1) + M_1(1,1)m_2(2,1))] - \sqrt{10}\frac{p_e}{W_e}[\Lambda_1 M_1(1,2)M_2(1,2) + \frac{1}{5}\lambda_2\Lambda_2 M_1(2,1)M_2(2,1)]\}$$

$$b_{22}^{(1)}/L_0 = \mp\left\{\frac{\sqrt{5}}{\sqrt{2}}\frac{p_e}{W_e}[\Lambda_1 M_2^2(1,2) + \frac{3}{5}\lambda_2\Lambda_2 M_2^2(2,1)]\right\}$$

$$b_{11}^{(2)}/L_0 = -2\sqrt{3}[\nu_{12}\frac{p_e}{W_e}M_1(1,1)M_1(2,1) + \hat{\nu}_{12}\alpha^2Z^2\frac{1}{p_e}(M_1(2,1)m_1(1,1) - M_1(1,1)m_1(2,1))] + \frac{\sqrt{3}}{\sqrt{2}}\lambda_2[M_1^2(2,1) - \mu_2\gamma_2\frac{1}{W_e}M_1(2,1)m_1(2,1)]$$

$$b_{02}^{(2)}/L_0 = -2\sqrt{2}[\nu_{12}\frac{p_e}{W_e}M_0(1,1)M_2(2,1) + \hat{\nu}_{12}\alpha^2Z^2\frac{1}{p_e} \times (M_2(2,1)m_0(1,1) - M_0(1,1)m_2(2,1))]$$

$$b_{12}^{(2)}/L_0 = -2\sqrt{3}[\nu_{12}\frac{p_e}{W_e}M_1(1,1)M_2(2,1) + \hat{\nu}_{12}\alpha^2Z^2\frac{1}{p_e} \times (M_2(2,1)m_1(1,1) - M_1(1,1)m_2(2,1))] - \sqrt{6}\lambda_2[M_1(2,1)M_2(2,1) - \mu_2\gamma_2\frac{1}{2W_e}(M_1(2,1)m_2(2,1) + M_2(2,1)m_1(2,1))]$$

$$b_{22}^{(2)}/L_0 = -\frac{\sqrt{7}}{\sqrt{2}}\lambda_2[M_2^2(2,1) - \mu_2\gamma_2\frac{1}{W_e}M_2(2,1)m_2(2,1)]$$

$$b_{12}^{(3)}/L_0 = \pm\frac{6\sqrt{3}}{\sqrt{5}}\lambda_2\Lambda_2\frac{p_e}{W_e}M_1(2,1)M_2(2,1)$$

$$b_{22}^{(3)}/L_0 = \pm\frac{3\sqrt{2}}{\sqrt{5}}\lambda_2\Lambda_2\frac{p_e}{W_e}M_2^2(2,1).$$

Upper (lower) sign for β^--decay (β^+-decay).

General formulae for the linear combinations $M_L(k_e, k_\nu)$ and $m_L(k_e, k_\nu)$ are given in chapter 2.2.1 (eqs. (8, 9), and (10)). Special formulae for allowed and first forbidden β-decays will be given in the next section*.

* Note that the Coulomb functions used here are the same as those of Schopper [22]. They differ from those defined by Bühring [17] in the following way:

$$L_0(\text{Bü}) = H^2(1)\cdot L_0,$$

$$\lambda_{k_e}(\text{Bü}) = \frac{H^2(k_e)}{H^2(1)}\cdot\lambda_{k_e},$$

$$\Lambda_{k_e}(\text{Bü}) = \Lambda_{k_e},$$

$$\mu_{k_e}(\text{Bü}) = \mu_{k_e},$$

$$\nu_{k_e k'_e}(\text{Bü}) = \frac{H(k_e)H(k'_e)}{H^2(1)}\cdot\nu_{k_e k'_e},$$

$$\hat{\nu}_{k_e k'_e}(\text{Bü}) = \frac{W_e}{p_e}\cdot\frac{H(k_e)H(k'_e)}{H^2(1)}\hat{\nu}_{k_e k'_e}$$

$$\eta_{k_e k'_e}(\text{Bü}) = \frac{H(k_e)H(k'_e)}{H^2(1)}\cdot\eta_{k_e k'_e},$$

$$\hat{\eta}_{k_e k'_e}(\text{Bü}) = \frac{H(k_e)H(k'_e)}{H^2(1)}\cdot\hat{\eta}_{k_e k'_e};$$

with

$$H(k_e) = \sum_{n=0}^{\infty} H_{2n}(k_e).$$

The quantities $\nu_{k_e k'_e}$, $\hat{\nu}_{k_e k'_e}$, $\eta_{k_e k'_e}$ and $\hat{\eta}_{k_e k'_e}$ are functions of W_e and Z and are defined as follows [17, 22],

$$\nu_{k_e k'_e} = \frac{W_e}{p_e} \cdot \frac{\alpha_{k_e}\alpha_{-k'_e}\cos(\Delta_{k_e} - \Delta_{-k'_e}) + \alpha_{-k_e}\alpha_{k'_e}\cos(\Delta_{-k_e} - \Delta_{k'_e})}{\alpha_{-1}^2 + \alpha_1^2},$$
$$\hat{\nu}_{k_e k'_e} = \frac{p_e}{(\alpha Z)^2} \cdot \frac{\alpha_{k_e}\alpha_{-k'_e}\cos(\Delta_{k_e} - \Delta_{-k'_e}) - \alpha_{-k_e}\alpha_{k'_e}\cos(\Delta_{-k_e} - \Delta_{k'_e})}{\alpha_{-1}^2 + \alpha_1^2},$$
$$\eta_{k_e k'_e} = \frac{\alpha_{-k_e}\alpha_{-k'_e}\cos(\Delta_{-k_e} - \Delta_{-k'_e}) + \alpha_{k_e}\alpha_{k'_e}\cos(\Delta_{k_e} - \Delta_{k'_e})}{\alpha_{-1}^2 + \alpha_1^2}, \tag{21}$$
$$\hat{\eta}_{k_e k'_e} = W_e \cdot \frac{\alpha_{-k_e}\alpha_{-k'_e}\cos(\Delta_{-k_e} - \Delta_{-k'_e}) - \alpha_{k_e}\alpha_{k'_e}\cos(\Delta_{k_e} - \Delta_{k'_e})}{\alpha_{-1}^2 + \alpha_1^2}.$$

These functions are needed only for $k_e < k'_e$ (for the α_{k_e} and Δ_{k_e} see chapter 2.2.1).

2.3.2 Linear combinations $M_L(k_e, k_\nu)$ and $m_L(k_e, k_\nu)$ of nuclear form factor coefficients

In table 4 we list the linear combinations [17, 22] of form factor coefficients which occur in the dominant terms for allowed transitions and also in the correction terms which must be added to the dominant terms. In table 5, the same is done for the first non unique forbidden transitions.

Table 4. Linear combinations $M_L(k_e, k_\nu)$ and $m_L(k_e, k_\nu)$ of form factor coefficients in allowed transitions*

Dominant terms

$$M_0(1,1) = {}^{V}F_{000}^{(0)}$$
$$M_1(1,1) = -{}^{A}F_{101}^{(0)}$$

Main correction terms

$$M_0(1,1){:}\ -{}^{V}F_{011}^{(0)}(D_1 + N_1) + {}^{V}F_{000}^{(1)}(H_2 - N_1 D_1 + N_2)$$
$$m_0(1,1) = -{}^{V}F_{011}^{(0)} d_1 - {}^{V}F_{000}^{(1)} N_1 d_1$$
$$M_1(1,1){:}\ -{}^{A}F_{101}^{(1)}\left(H_2 + \frac{1}{3}N_1 D_1 + N_2\right) - \sqrt{\frac{1}{3}}\,{}^{A}F_{110}^{(0)}(D_1 + N_1) - $$
$$- \sqrt{\frac{2}{3}}\,{}^{V}F_{111}^{(0)}(D_1 - N_1) + \frac{2}{3}\sqrt{2}\,{}^{A}F_{121}^{(0)} N_1 D_1$$
$$m_1(1,1) = -\frac{1}{3}{}^{A}F_{101}^{(1)} N_1 d_1 - \sqrt{\frac{1}{3}}\,{}^{A}F_{110}^{(0)} d_1 - \sqrt{\frac{2}{3}}\,{}^{V}F_{111}^{(0)} d_1 + \frac{2}{3}\sqrt{2}\cdot{}^{A}F_{121}^{(0)} N_1 d_1$$
$$M_1(1,2) = \frac{p_\nu R}{3}\left[\sqrt{\frac{2}{3}}\,{}^{A}F_{110}^{(0)} + \sqrt{\frac{1}{3}}\,{}^{V}F_{111}^{(0)} - \frac{2}{3}\sqrt{2}\,{}^{A}F_{101}^{(1)} D_1 - \frac{1}{3}{}^{A}F_{121}^{(0)}(D_1 + 3N_1(2))\right]$$
$$m_1(1,2) = -\frac{p_\nu R}{3}\left[\frac{1}{3}{}^{A}F_{121}^{(0)} + \frac{2}{3}\sqrt{2}\,{}^{A}F_{101}^{(1)}\right] d_1$$
$$M_1(2,1) = \frac{p_e R}{3}\left[\sqrt{\frac{2}{3}}\,{}^{A}F_{110}^{(0)} - \sqrt{\frac{1}{3}}\,{}^{V}F_{111}^{(0)} - \frac{2}{3}\sqrt{2}\,{}^{A}F_{101}^{(1)} N_1 - \frac{1}{3}{}^{A}F_{121}^{(0)}(3D_1(2) + N_1)\right]$$
$$m_1(2,1) = -\frac{p_e R}{3}{}^{A}F_{121}^{(0)} d_1(2)$$
$$M_2(1,2) = -\frac{p_\nu R}{3}\left[{}^{V}F_{211}^{(0)} + \sqrt{\frac{2}{5}}\,{}^{V}F_{220}^{(0)}(D_1 + N_1(2)) + \sqrt{\frac{3}{5}}\,{}^{A}F_{221}^{(0)}(D_1 - N_1(2))\right]$$
$$m_2(1,2) = -\frac{p_\nu R}{3}\left[\sqrt{\frac{2}{5}}\,{}^{V}F_{220}^{(0)} + \sqrt{\frac{3}{5}}\,{}^{A}F_{221}^{(0)}\right] d_1$$
$$M_2(2,1) = -\frac{p_e R}{3}\left[{}^{V}F_{211}^{(0)} + \sqrt{\frac{2}{5}}\,{}^{V}F_{220}^{(0)}(D_1(2) + N_1) + \sqrt{\frac{3}{5}}\,{}^{A}F_{221}^{(0)}(D_1(2) - N_1)\right]$$
$$m_2(2,1) = -\frac{p_e R}{3}\left[\sqrt{\frac{2}{5}}\,{}^{V}F_{220}^{(0)} + \sqrt{\frac{3}{5}}\,{}^{A}F_{221}^{(0)}\right] d_1(2)$$

* Note that the often occurring quantity $D_1(1) + N_1(1) = \alpha Z/2 + W_0 R/3$ does not depend on W_e.

Table 5. Linear combinations $M_L(k_e, k_\nu)$ and $m_L(k_e, k_\nu)$ of form factor coefficients in first non unique forbidden transitions

Dominant terms

$$M_0(1,1) = {}^{A}F^{(0)}_{000} - {}^{A}F^{(0)}_{011}\,(D_1 + N_1)$$

$$m_0(1,1) = -{}^{A}F^{(0)}_{011}\,d_1$$

$$M_1(1,1) = -{}^{V}F^{(0)}_{101} - \sqrt{\tfrac{1}{3}}\,{}^{V}F^{(0)}_{110}(D_1 + N_1) - \sqrt{\tfrac{2}{3}}\,{}^{A}F^{(0)}_{111}(D_1 - N_1)$$

$$m_1(1,1) = -\sqrt{\tfrac{1}{3}}\,{}^{V}F^{(0)}_{110}\,d_1 - \sqrt{\tfrac{2}{3}}\,{}^{A}F^{(0)}_{111}\,d_1$$

$$M_1(1,2) = \frac{p_\nu R}{3}\left(\sqrt{\tfrac{2}{3}}\,{}^{V}F^{(0)}_{110} + \sqrt{\tfrac{1}{3}}\,{}^{A}F^{(0)}_{111}\right)$$

$$M_1(2,1) = \frac{p_e R}{3}\left(\sqrt{\tfrac{2}{3}}\,{}^{V}F^{(0)}_{110} - \sqrt{\tfrac{1}{3}}\,{}^{A}F^{(0)}_{111}\right)$$

$$M_2(1,2) = \frac{p_\nu R}{3}\left(-{}^{A}F^{(0)}_{211}\right)$$

$$M_2(2,1) = \frac{p_e R}{3}\left(-{}^{A}F^{(0)}_{211}\right)$$

Main correction terms

$$M_0(1,1):\ {}^{A}F^{(1)}_{000}(H_2 - N_1 D_1 + N_2) - {}^{A}F^{(1)}_{011}(D_3 + N_1 H_2 + N_2 D_1)$$

$$m_0(1,1):\ -{}^{A}F^{(1)}_{000}\,N_1 d_1 - {}^{A}F^{(1)}_{011}(d_3 + N_2 d_1)$$

$$M_1(1,1):\ -{}^{V}F^{(1)}_{101}\left(H_2 + \tfrac{1}{3}N_1 D_1 + N_2\right) + \tfrac{2}{3}\sqrt{2}\,{}^{V}F^{(0)}_{121}\,N_1 D_1 - \\ - \sqrt{\tfrac{1}{3}}\,{}^{V}F^{(1)}_{110}(D_3 + N_1 H_2 + N_2 D_1) - \sqrt{\tfrac{2}{3}}\,{}^{A}F^{(1)}_{111}(D_3 - N_1 H_2 + N_2 D_1)$$

$$m_1(1,1):\ -\tfrac{1}{3}\,{}^{V}F^{(1)}_{101}\,N_1 d_1 + \tfrac{2}{3}\sqrt{2}\,{}^{V}F^{(0)}_{121}\,N_1 d_1 - \sqrt{\tfrac{1}{3}}\,{}^{V}F^{(1)}_{110}(d_3 + N_2 d_1) - \\ - \sqrt{\tfrac{2}{3}}\,{}^{A}F^{(1)}_{111}(d_3 + N_2 d_1)$$

$$M_1(1,2):\ \frac{p_\nu R}{3}\left[\sqrt{\tfrac{2}{3}}\,{}^{V}F^{(1)}_{110}(H_2 - N_1(2)\,D_1 + N_2(2)) + \sqrt{\tfrac{1}{3}}\,{}^{A}F^{(1)}_{111} \times \right. \\ \left. \times (H_2 + N_1(2)\,D_1 + N_2(2)) - \tfrac{2}{3}\sqrt{2}\,{}^{V}F^{(1)}_{101}\,D_1 - \tfrac{1}{3}\,{}^{V}F^{(0)}_{121}(D_1 + 3N_1(2))\right]$$

$$m_1(1,2) = \frac{p_\nu R}{3}\left[-\sqrt{\tfrac{2}{3}}\,{}^{V}F^{(1)}_{110}\,N_1(2)\,d_1 + \sqrt{\tfrac{1}{3}}\,{}^{A}F^{(1)}_{111}\,N_1(2)\,d_1 - \tfrac{2}{3}\sqrt{2}\,{}^{V}F^{(1)}_{101}\,d_1 - \tfrac{1}{3}\,{}^{V}F^{(0)}_{121}\,d_1\right]$$

$$M_1(2,1):\ \frac{p_e R}{3}\left[\sqrt{\tfrac{2}{3}}\,{}^{V}F^{(1)}_{110}(H_2(2) - N_1 D_1(2) + N_2) - \sqrt{\tfrac{1}{3}}\,{}^{A}F^{(1)}_{111} \times \right. \\ \left. \times (H_2(2) + N_1 D_1(2) + N_2) - \tfrac{2}{3}\sqrt{2}\,{}^{V}F^{(1)}_{101}\,N_1 - \tfrac{1}{3}\,{}^{V}F^{(0)}_{121}(3D_1(2) + N_1)\right]$$

$$m_1(2,1) = -\frac{p_e R}{3}\left[\sqrt{\tfrac{2}{3}}\,{}^{V}F^{(1)}_{110}\,N_1 d_1(2) + \sqrt{\tfrac{1}{3}}\,{}^{A}F^{(1)}_{111}\,N_1 d_1(2) + {}^{V}F^{(0)}_{121}\,d_1(2)\right]$$

$$M_2(1,2):\ -\frac{p_\nu R}{3}\left[{}^{A}F^{(1)}_{211}\left(H_2 + \tfrac{1}{5}N_1(2)\,D_1 + N_2(2)\right) + \sqrt{\tfrac{2}{5}} \times \right. \\ \left. \times\,{}^{A}F^{(0)}_{220}(D_1 + N_1(2)) + \sqrt{\tfrac{3}{5}}\,{}^{V}F^{(0)}_{221}(D_1 - N_1(2))\right]$$

Table 5 (continued)

$$m_2(1,2) = -\frac{p_\nu R}{3}\left[\frac{1}{5}\,{}^{\mathrm{A}}F^{(1)}_{211}\,N_1(2) + \sqrt{\frac{2}{5}}\,{}^{\mathrm{A}}F^{(0)}_{220} + \sqrt{\frac{3}{5}}\,{}^{\mathrm{V}}F^{(0)}_{221}\right] d_1$$

$$M_2(2,1){:}\ -\frac{p_\mathrm{e} R}{3}\left[{}^{\mathrm{A}}F^{(1)}_{211}\left(H_2(2) + \frac{1}{5}N_1 D_1(2) + N_2\right) + \sqrt{\frac{2}{5}}\,{}^{\mathrm{A}}F^{(0)}_{220} \times \right.$$
$$\left. \times \left(D_1(2) + N_1\right) + \sqrt{\frac{3}{5}}\,{}^{\mathrm{V}}F^{(0)}_{221}\left(D_1(2) - N_1\right)\right]$$

$$m_2(2,1) = -\frac{p_\mathrm{e} R}{3}\left[\frac{1}{5}\,{}^{\mathrm{A}}F^{(1)}_{211}\,N_1 + \sqrt{\frac{2}{5}}\,{}^{\mathrm{A}}F^{(0)}_{220} + \sqrt{\frac{3}{5}}\,{}^{\mathrm{V}}F^{(0)}_{221}\right] d_1(2).$$

2.3.3 Spectrum shape, electron polarization, β—γ angular and circular polarization correlation, and distribution of electrons emitted from oriented nuclei

The following formulae are given without correction terms. To obtain the formulae with correction terms one must insert the quantities of sections 2.3.1 and 2.3.2 in the formulae (4, 11, 12, 16) of the preceding chapters and of table 3.

2·3.3.1 Allowed decays

Shape factor

$$C(W_\mathrm{e}) = \left({}^{\mathrm{V}}F^{(0)}_{000}\right)^2 + \left({}^{\mathrm{A}}F^{(0)}_{101}\right)^2 \tag{22}$$

Electron polarization

$$P = \mp \Lambda_1 \frac{p_\mathrm{e}}{W_\mathrm{e}} = \mp \Lambda_1 \frac{v}{c}. \tag{23}$$

The upper sign holds for β^- and the lower for β^+ decay. The quantities v and c denote the velocities of the electrons and of light.

β—γ circular polarization correlation

$$\omega(W_\mathrm{e}, \theta) = 1 + \tau \cdot \frac{v}{c}\,\Lambda_1 \tilde{A} \cos\theta, \tag{24}$$

with

$$\tilde{A} = \mp \frac{\sqrt{3}}{6} \cdot \frac{1}{1+y^2}\left[\frac{J'(J'+1) - J(J+1) + 2}{\sqrt{J'(J'+1)}} - 4y\right] \times$$
$$\times \frac{F_1(\lambda,\lambda,J'',J') + 2\delta F_1(\lambda,\lambda+1,J'',J') + \delta^2 F_1(\lambda+1,\lambda+1,J'',J')}{1+\delta^2}, \tag{25}$$

and

$$y = \frac{F^{(0)}_{000}}{F^{(0)}_{101}} = \pm \frac{C_V M_F}{C_A M_{GT}}, \qquad \delta = \frac{\delta_{\lambda+1}}{\delta_\lambda},$$

where the upper (lower) sign in formula (25) applies for β^--decay (β^+-decay). For the other quantities, see chapter 2.2.3.

Distribution of electrons emitted from oriented nuclei

$$\omega(\theta, W_\mathrm{e}) = 1 + f_1(J) \cdot \frac{v}{c} \cdot \Lambda_1 \cdot A \cdot \cos\theta, \tag{26}$$

with

$$A = \mp \frac{1}{1+y^2}\left[\frac{J(J+1) - J'(J'+1) + 2}{2(J+1)} + 2\sqrt{\frac{J}{J+1}} \cdot y\right]. \tag{27}$$

The orientation parameter is given here in the usual form (see eq. (20))

$$f_1(J) = \frac{1}{J}\sum_m m\, a_m. \tag{28}$$

2.3.3.2 First non unique forbidden decays

Shape factor

$$C(W_e) = M_0^2(1,1) + m_0^2(1,1) - 2\frac{\mu_1\gamma_1}{W_e} M_0(1,1)\,m_0(1,1) + M_1^2(1,1) + \\ + m_1^2(1,1) - 2\frac{\mu_1\gamma_1}{W_e} M_1(1,1)\,m_1(1,1) + M_1^2(1,2) + \lambda_2 M_1^2(2,1) + M_2^2(1,2) + \lambda_2 M_2^2(2,1). \quad (29)$$

Electron polarization

$$P = \mp\frac{v}{c} A_1\left[1 + \frac{1}{C(W_e)}\left\{\frac{2\mu_1\gamma_1}{W_e}[M_1(1,1)\,m_1(1,1) + M_0(1,1)\,m_0(1,1)] - \\ - 2(m_1^2(1,1) + m_0^2(1,1)) + \left(\frac{A_2}{A_1} - 1\right)\cdot\lambda_2\cdot(M_1^2(2,1) + M_2^2(2,1))\right\}\right]. \quad (30)$$

β—γ angular correlation

$$\omega(W_e, \theta) = 1 + \varepsilon\left(\frac{3}{2}\cos^2\theta - \frac{1}{2}\right), \quad (31)$$

with

$$\varepsilon = \frac{B^{(2)}\cdot A_2(\gamma)}{C(W_e)\cdot L_0} = \frac{b_{11}^{(2)}\Gamma_{11}(2) + b_{12}^{(2)}\Gamma_{12}(2) + b_{22}^{(2)}\Gamma_{22}(2) + b_{02}^{(2)}\Gamma_{02}(2)}{C(W_e)\cdot L_0}\cdot A_2(\gamma). \quad (32)$$

β—γ circular polarization correlation

$$\omega(W_e, \theta) = 1 + \tau\cdot\frac{v}{c}\cdot\tilde{A}\cdot\cos\theta, \quad (33)$$

with

$$\tilde{A} = -\frac{W_e}{p_e}\cdot\frac{B^{(1)}\cdot A_1(\gamma) + B^{(3)}\cdot A_3(\gamma)\left(\frac{5}{2}\cos^2\theta - \frac{3}{2}\right)}{B^{(0)} + B^{(2)}\cdot A_2(\gamma)\left(\frac{3}{2}\cos^2\theta - \frac{1}{2}\right)}, \quad (34)$$

$$B^{(1)} = b_{01}^{(1)}\Gamma_{01}(1) + b_{11}^{(1)}\Gamma_{11}(1) + b_{12}^{(1)}\Gamma_{12}(1) + b_{22}^{(1)}\Gamma_{22}(1),$$

$$B^{(3)} = b_{12}^{(3)}\Gamma_{12}(3) + b_{22}^{(3)}\Gamma_{22}(3).$$

For $B^{(2)}$, see eq. (32). $B^{(0)}$ is equal to $C(W_e)\cdot L_0$. For 0—0 transitions, one obtains $\tilde{A} = 0$.

Distribution of electrons emitted from oriented nuclei

$$\omega(\theta, W_e) = 1 + A_1 P_1(\cos\theta) + A_2 P_2(\cos\theta) + A_3 P_3(\cos\theta), \quad (35)$$

with

$$A_1 = \frac{\sqrt{2J+1}\,G_1(J)}{C(W_e)\cdot L_0}[b_{01}^{(1)}\Gamma_{01}(1) - b_{11}^{(1)}\Gamma_{11}(1) + b_{12}^{(1)}\Gamma_{12}(1) - b_{22}^{(1)}\Gamma_{22}(1)],$$

$$A_2 = \frac{\sqrt{2J+1}\,G_2(J)}{C(W_e)\cdot L_0}[b_{11}^{(2)}\Gamma_{11}(2) - b_{12}^{(2)}\Gamma_{12}(2) + b_{22}^{(2)}\Gamma_{22}(2) + b_{02}^{(2)}\Gamma_{02}(2)], \quad (36)$$

$$A_3 = \frac{\sqrt{2J+1}\,G_3(J)}{C(W_e)\cdot L_0}[b_{12}^{(3)}\Gamma_{12}(3) - b_{22}^{(3)}\Gamma_{22}(3)].$$

In the Γ-coefficients of the eqs. (36) the spins J and J' must be interchanged with respect to eqs. (31 ··· 34). Therefore, we have $\Gamma_{LL'}(n, J', J)$ instead of $\Gamma_{LL'}(n, J, J')$.

2.3.3.3 First unique forbidden decays

Shape factor

$$C(W_e) = \frac{R^2}{9}\left({}^{A}F_{211}^{(0)}\right)^2 (p_\nu^2 + \lambda_2 p_e^2). \quad (37)$$

Electron polarization

$$P = \mp\frac{v}{c} A_1\left\{1 + \left(\frac{A_2}{A_1} - 1\right)\cdot\lambda_2\cdot\frac{p_e^2}{p_\nu^2 + \lambda_2 p_e^2}\right\}. \quad (38)$$

β—γ angular correlation

Here the anisotropy coefficient is (see eq. (31))

$$\varepsilon = -\sqrt{\frac{7}{2}}\,\Gamma_{22}(2)\cdot A_2(\gamma)\cdot\frac{\lambda_2 p_e^2}{p_\nu^2+\lambda_2 p_e^2}. \tag{39}$$

β—γ circular polarization correlation

$$\tilde{A} = \pm\frac{(5\Lambda_1 p_\nu^2 + 3\Lambda_2\lambda_2 p_e^2)\,\Gamma_{22}(1)\cdot A_1(\gamma) - 6\Lambda_2\lambda_2 p_e^2\cdot(\frac{5}{3}\cos^2\theta - \frac{3}{5})\,\Gamma_{22}(3)\,A_3(\gamma)}{\sqrt{10}\,(p_\nu^2+\lambda_2 p_e^2) - \sqrt{35}\cdot\lambda_2 p_e^2(\frac{3}{2}\cos^2\theta - \frac{1}{2})\,\Gamma_{22}(2)\cdot A_2(\gamma)}. \tag{40}$$

Distribution of electrons emitted from oriented nuclei

$$\omega(\theta, W_e) = 1 + \frac{p_e}{W_e}A_1 P_1(\cos\theta) + A_2 P_2(\cos\theta) + \frac{p_e}{W_e}A_3 P_3(\cos\theta), \tag{41}$$

with

$$A_1 = \pm\sqrt{2J+1}\cdot G_1(J)\cdot\frac{\sqrt{5}}{\sqrt{2}}\cdot\frac{\Lambda_1 p_\nu^2 + \frac{3}{5}\lambda_2\Lambda_2 p_e^2}{p_\nu^2+\lambda_2 p_e^2}\cdot\Gamma_{22}(1),$$

$$A_2 = -\sqrt{2J+1}\cdot G_2(J)\cdot\frac{\sqrt{7}}{\sqrt{2}}\frac{\lambda_2 p_e^2}{p_\nu^2+\lambda_2 p_e^2}\cdot\Gamma_{22}(2), \tag{42}$$

$$A_3 = \mp\sqrt{2J+1}\cdot G_3(J)\frac{3\cdot\sqrt{2}}{\sqrt{5}}\frac{\lambda_2\Lambda_2 p_e^2}{p_\nu^2+\lambda_2 p_e^2}\cdot\Gamma_{22}(3)$$

(see eqs. (13, 14, 16), and (17)). The parameters $G_n(J)$ are listed in eq. (19). The upper (lower) sign is for β^--decay (β^+-decay).

2.3.3..4 L^{th} non unique forbidden decays

Shape factor

$$\begin{aligned} C(W_e) = {} & \frac{L!}{(2L+1)!!}\Bigg[\sum_{n=1}^{L}\frac{\lambda_n(p_e R)^{2(n-1)}(p_\nu R)^{2(L-n)}}{(2n-1)!!\,(n-1)!\,(2\{L-n\}+1)!!\,(L-n)!}\times \\ & \times\Bigg\{\Bigg[\sqrt{\frac{2L+1}{L}}\,{}^{\text{V}}F^{(0)}_{L,L-1,1} + {}^{\text{V}}F^{(0)}_{LL0}[D_1(n)+N_1(L-n+1)] + \sqrt{\frac{L+1}{L}}\times \\ & \times{}^{\text{A}}F^{(0)}_{LL1}[D_1(n) - N_1(L-n+1)]\Bigg]^2 + \Bigg[{}^{\text{V}}F^{(0)}_{LL0} + \sqrt{\frac{L+1}{L}}\,{}^{\text{A}}F^{(0)}_{LL1}\Bigg]^2 d_1^2(n) - \\ & - \frac{2\mu_n\gamma_n}{nW_e}\Bigg[\sqrt{\frac{2L+1}{L}}\,{}^{\text{V}}F^{(0)}_{L,L-1,1} + {}^{\text{V}}F^{(0)}_{LL0}[D_1(n)+N_1(L-n+1)] + \\ & + \sqrt{\frac{L+1}{L}}\,{}^{\text{A}}F^{(0)}_{LL1}[D_1(n) - N_1(L-n+1)]\Bigg]\cdot\Bigg[{}^{\text{V}}F^{(0)}_{LL0} + \sqrt{\frac{L+1}{L}}\,{}^{\text{A}}F^{(0)}_{LL1}\Bigg]\cdot d_1(n)\Bigg\} + \\ & + \sum_{n=1}^{L+1}\frac{\lambda_n(p_e R)^{2(n-1)}(p_\nu R)^{2(L-n+1)}}{(2n-1)!!\,(n-1)!\,(2\{L-n\}+3)!!\,(L-n+1)!}\cdot\Bigg\{\frac{L+1}{(2n-1)(2\{L-n\}+3)}\times \\ & \times[{}^{\text{V}}F^{(0)}_{LL0} - \frac{2n-L-2}{L+1}\sqrt{\frac{L+1}{L}}\,{}^{\text{A}}F^{(0)}_{LL1}]^2 + \left({}^{\text{A}}F^{(0)}_{L+1,L,1}\right)^2\Bigg\}\Bigg]. \end{aligned} \tag{43}$$

For $\Delta J = 0$ see eq. (22) and (29).

2.3.3.5 $(L-1)^{\text{th}}$ unique forbidden decays

Shape factor

$$C(W_e) = \frac{(L-1)!}{(2L-1)!!}\left({}^{\text{A}}F^{(0)}_{L,L-1,1}\right)^2\sum_{n=1}^{L}\frac{\lambda_n\cdot(p_e R)^{2(n-1)}(p_\nu R)^{2(L-n)}}{(2n-1)!!\,(n-1)!\,(2\{L-n\}+1)!!\,(L-n)!}. \tag{44}$$

Electron polarization

$$P = \mp \frac{p_e}{W_e} \frac{\sum_{n=1}^{L} \Lambda_n \lambda_n \frac{(p_e R)^{2(n-1)} (p_\nu R)^{2(L-n)}}{(2n-1)!!\,(n-1)!\,(2\{L-n\}+1)!!\,(L-n)!}}{\sum_{n=1}^{L} \lambda_n \frac{(p_e R)^{2(n-1)} (p_\nu R)^{2(L-n)}}{(2n-1)!!\,(n-1)!\,(2\{L-n\}+1)!!\,(L-n)!}} \tag{45}$$

If the electron energy is not too small so that $\Lambda_n \approx \Lambda_1$ for $n > 1$, eq. (45) reduces to

$$P = \mp \Lambda_1 \frac{p_e}{W_e} = \mp \Lambda_1 \frac{v}{c}.$$

The upper (lower) sign applies for β^--decay (β^+-decay).

2.4 *ft*-value and integrated Fermi-function

The decay constant λ in β-decay is given by (see eq. (1))

$$\lambda = \frac{g^2}{2\pi^3} \int_0^{p_0} p_e^2 (W_0 - W_e)^2 \cdot F(Z, W_e) \cdot C(W_e)\, dp_e, \tag{46}$$

with

$$p_0 = \sqrt{W_0^2 - 1}.$$

W_0 is the maximum total energy of the β-particle in units of $m_0 c^2$.

In the general case of Lth forbidden decays, the shape factor $C(W_e)$ depends on p_e (see eqs. (4) and (43)). The magnitudes of the form factor coefficients $F_{Lls}^{(n)}$ are unknown in most cases. Therefore, the integral in eq. (46) depends on the unknown ratios of form factor coefficients. The energy dependence of $C(W_e)$ is known only in the case of unique forbidden and allowed decays (see eqs. (22) and (44)). For allowed decays, the quantity $C(W_e)$ is independent of the energy of the β-particles (see eq. (22)).

For allowed decays we have

$$\lambda = \frac{g^2}{2\pi^3} \{({}^{\mathrm{V}}F_{000}^{(0)})^2 + ({}^{\mathrm{A}}F_{101}^{(0)})^2\} \cdot f(Z, W_0), \tag{47}$$

with

$$f(Z, W_0) = \int_0^{p_0} F(Z, W_e)\, p_e^2 (W_0 - W_e)^2\, dp_e. \tag{48}$$

The decay constant λ is not very characteristic for an allowed β-decay, because it depends on the function f which is determined by the statistical factor and by Coulomb effects. Therefore one introduces the so-called reduced half life or ft-value which is defined by

$$ft = 2\pi^3 \frac{\ln 2}{g^2 \{{}^{\mathrm{V}}F_{000}^{(0)\,2} + {}^{\mathrm{A}}F_{101}^{(0)\,2}\}}, \tag{49}$$

with the half life

$$t = \frac{\ln 2}{\lambda}.$$

Usually*, the ft-values are quoted in seconds (the half life t is given in seconds, f is dimensionless), but all the other equations are given in natural units.

Since the f-function depends very strongly on the energy W_0, it is better for the tabulation of f to use the ratio

$$S = \frac{f}{f_0}, \tag{50}$$

* It should be noted that the magnitude of the f-function and the ft-values given here are somewhat different from the ones usually given in other tables. Our Fermi-function →

with

$$f_0 = \int_0^{p_0} p_e^2 (W_0 - W_e)^2 \, dp_e .$$

To obtain the most accurate intermediate f values, the functions S must be interpolated rather than f. The quantity f_0 can be expressed analytically in the form

$$f_0 = \frac{1}{60}\left[(2W_0^4 - 9W_0^2 - 8)p_0 + 15W_0 \ln(W_0 + p_0)\right], \tag{51}$$

For $p_0 < 1$, one has

$$f_0 = \frac{2}{105} p_0^7 \left(1 - \frac{2}{3}p_0^2 + \frac{16}{33}p_0^4 - \frac{160}{429}p_0^6 + \cdots\right). \tag{52}$$

2.5 General formulae for electron capture

Electron capture can be treated in complete analogy to β-decay [*22, 26, 35*]. The differences between the two decay modes of a nucleus are as follows:

In electron capture, the electron wave function is a bound state wave function instead of a free particle wave function as in β-decay. This means that for the capture from a given shell, contrary to β-decay, the electron energy is discrete and the angular momentum has a fixed value. From this, it follows that the neutrino energy is also discrete.

In analogy to the eqs. (4) and (46) in β-decay, the decay constant for electron capture has the following form

$$\lambda_c = \frac{\ln 2}{t_c} = \frac{g^2}{2\pi^3} \sum_x n_x \cdot C_x \cdot f_x . \tag{53}$$

$$F(Z, W_e) = \frac{\alpha_{-1}^2 + \alpha_1^2}{2p_e^2} = \frac{g_{-1}^2(0) + f_{+1}^2(0)}{2p_e^2}$$

is defined differently from the usual Fermi-function

$$F(Z, W_e) = \frac{g_{-1}^2(R) + f_1^2(R)}{2p_e^2}.$$

We have

$$f_1(R) = \alpha_1 \sum_{n=0}^{\infty} [H_{2n}(1) + h_{2n}(1)] \approx \alpha_1 \sum_{n=0}^{\infty} H_{2n}(1),$$

$$g_{-1}(R) = \alpha_{-1} \sum_{n=0}^{\infty} [H_{2n}(1) - h_{2n}(1)] \approx \alpha_{-1} \sum_{n=0}^{\infty} H_{2n}(1).$$

See refs. [*17, 18, 21*] for the general form of the functions $H_{2n}(k_e)$ and $h_{2n}(k_e)$. The leading terms of these functions are listed in table 1. In general, higher order terms can be neglected. The disadvantage of the old definitions is that the approximation

$$F_{Lls}^{(0)} = F_{Lls}^{(1)} = F_{Lls}^{(2)} = F_{Lls}^{(3)} = \cdots F_{Lls}^{(n)}$$

is made for the form factor coefficients. For example, in terms of matrix elements, this approximation means for allowed decays (Lls = 000 and 101).

$$\int 1 = \int \left(\frac{r}{R}\right)^2 = \int \left(\frac{r}{R}\right)^4 = \cdots \int \left(\frac{r}{R}\right)^n,$$

$$\int \boldsymbol{\sigma} = \int \boldsymbol{\sigma}\left(\frac{r}{R}\right)^2 = \int \boldsymbol{\sigma}\left(\frac{r}{R}\right)^4 = \cdots \int \boldsymbol{\sigma}\left(\frac{r}{R}\right)^n.$$

In many cases, this approximation is not valid. If one wants to take into account the contributions from the matrix elements $\int (r/R)^2$ and $\int \boldsymbol{\sigma}(r/R)^2$ etc., one must use the equations for the quantities $M_L(k_e, k_\nu)$ and $m_L(k_e, k_\nu)$ with correction terms from the tables 4 and 5. For the higher order corrections, one must use the general eqs. (8, 9), and (10) in combination with the general eq. (4) for $C(W_e)$.

Corresponding to the expression for the shape factor in β-decay (see eq. (4)), one has for the quantitiy C_x:

$$C_x = [M_L(k_x, k_\nu^{(1)}) + \frac{\varkappa_x}{k_x} m_L(k_x, k_\nu^{(1)})]^2 + [M_L(k_x, k_\nu^{(2)}) + \frac{\varkappa_x}{k_x} m_L(k_x, k_\nu^{(2)})]^2 + \\ + [M_{L+1}(k_x, k_\nu^{(2)}) + \frac{\varkappa_x}{k_x} m_{L+1}(k_x, k_\nu^{(2)})]^2 + \delta_{\Delta J,0} [M_0(1,1) + \frac{\varkappa_x}{k_x} m_0(1,1)]^2 . \quad (54)$$

The classification of electron capture decays is the same as in β-decay (see section 2.1 and eq. (43)). Therefore, one has in eq. (54)

$$\begin{aligned} L &= \Delta J \quad \text{for} \quad \Delta J > 0 , \\ L &= 1 \quad \text{for} \quad \Delta J = 0 , \\ k_\nu^{(1)} &= L - k_x + 1 , \\ k_\nu^{(2)} &= L - k_x + 2 . \end{aligned} \quad (55)$$

The quantities $\varkappa_\nu$ and $k_\nu = |\varkappa_\nu|$ for the neutrino are used in the same way as in β-decay to determine the j_ν and l_ν of the continuum wave functions (see p. 4). Similarly, $\varkappa_x$ and $k_x = |\varkappa_x|$ are related to j_x and l_x of the bound electrons in a given subshell.

In table 6 we list the values of $\varkappa_x$ for the electrons bound in different subshells. Here, x stands for the different shells and subshells, i.e., for K, L_I, L_II, L_III, M_I, etc. For closed shells, we have $n_x = 1$. For partly filled shells, n_x is equal to the relative occupation number of the electrons in this shell.

Table 6. $\varkappa_x$-values for the different subshells

Subshell	$\varkappa_x$	Subshell	$\varkappa_x$
K	−1	M_I	−1
L_I	−1	M_II	+1
L_II	+1	M_III	−2
L_III	−2	M_IV	+2
		M_V	−3

The $M_L(k_x, k_\nu)$ and $m_L(k_x, k_\nu)$ can be taken from eqs. (8, 9), and (10). There are two differences from β^--decay. The first* is that we have to reverse the sign of the neutrino momentum q_x and to replace the form factor coefficient $F_{Lls}^{(n)}$ by $(-1)^{L-s} \cdot F_{Lls}^{(n)}$. The second is that the electron momentum p_e must be replaced by $p_x = \sqrt{1 - W_x^2}$ and the energy W_e by W_x. The total energy W_x (in units of m_0c^2) of the bound electron is given by $W_x = 1 - |E_x|$, where E_x is the binding energy of the electron. For the neutrino momentum, the relation

$$q_x = W_0 + W_x$$

holds. Here, W_0 is the nuclear transition energy** in units of m_0c^2. The recoil energy of the nucleus is neglected. If we insert the $M_L(k_x, k_\nu)$ and $m_L(k_x, k_\nu)$ from eqs. (8, 9), and (10) in eq. (54) and take into account only dominant terms, we can give the following explicit expressions for C_x both for Lth non unique forbidden and for $(L-1)$th unique forbidden transitions.

Lth non unique forbidden transitions

For $L \geqq k_x$ ($L = 1$ if $\Delta J = 0$), we have

$$C_x = \frac{L!}{(2L+1)!!} \cdot \frac{(p_x R)^{2(k_x-1)} (q_x R)^{2(L-k_x)}}{(2k_x-1)!! (k_x-1)! (2\{L-k_x\}+1)!! (L-k_x)!} \times \\ \times \{({}^\mathrm{V}F_{L,L-1,1}^{(0)})^2 \cdot \frac{2L+1}{L} + ({}^\mathrm{V}F_{LL0}^{(0)})^2 \cdot f_1 + ({}^\mathrm{A}F_{LL1}^{(0)})^2 \cdot f_2 + ({}^\mathrm{A}F_{L+1,L,1}^{(0)})^2 \times \\ \times f_3 + {}^\mathrm{V}F_{L,L-1,1}^{(0)} \, {}^\mathrm{V}F_{LL0}^{(0)} \cdot f_4 + {}^\mathrm{V}F_{L,L-1,1}^{(0)} \, {}^\mathrm{A}F_{LL1}^{(0)} \cdot f_5 + {}^\mathrm{V}F_{LL0}^{(0)} \, {}^\mathrm{A}F_{LL1}^{(0)} \cdot f_6\} + \\ + \delta_{\Delta J,0} \{({}^\mathrm{A}F_{000}^{(0)})^2 + ({}^\mathrm{A}F_{011}^{(0)})^2 \cdot f_7 + {}^\mathrm{A}F_{000}^{(0)} \, {}^\mathrm{A}F_{011}^{(0)} \cdot f_8\} , \quad (56)$$

* Alternatively we could start from the formulae for β^+-decay. The prescription then is to change the signs of the quantities D_{2n-1} and d_{2n-1}, and to replace $-Z$ by Z and ${}^\mathrm{A}F_{Lls}^{(n)}$ by $-{}^\mathrm{A}F_{Lls}^{(n)}$.

** W_0 has the same value as in the β^+-decay (difference of the atomic masses minus m_0c^2).

$$f_1 = \left\{ \frac{\frac{3}{2}\alpha Z + R\left(W_x + \frac{\varkappa_x}{k_x}\right)}{2k_x+1} - \frac{q_x R}{2(L-k_x)+3} \right\}^2 + \frac{(L+1)\,(q_x R)^2}{\{2(L-k_x)+3\}^2\,(L-k_x+1)\,(2k_x-1)}$$

$$f_2 = \frac{L+1}{L}\left\{ \frac{\frac{3}{2}\alpha Z + R\left(W_x + \frac{\varkappa_x}{k_x}\right)}{2k_x+1} + \frac{q_x R}{2(L-k_x)+3} \right\}^2 + \\ + \frac{(2k_x-L-2)^2\,(q_x R)^2}{L\{2(L-k_x)+3\}^2\,(L-k_x+1)\,(2k_x-1)},$$

$$f_3 = \frac{(q_x R)^2}{\{2(L-k_x)+3\}\,(L-k_x+1)},$$

$$f_4 = -2\sqrt{\frac{2L+1}{L}}\left\{ \frac{\frac{3}{2}\alpha Z + R\left(W_x + \frac{\varkappa_x}{k_x}\right)}{2k_x+1} - \frac{q_x R}{2(L-k_x)+3} \right\},$$

$$f_5 = \frac{2}{L}\sqrt{(2L+1)\,(L+1)}\left\{ \frac{\frac{3}{2}\alpha Z + R\left(W_x + \frac{\varkappa_x}{k_x}\right)}{2k_x+1} + \frac{q_x R}{2(L-k_x)+3} \right\},$$

$$f_6 = -2\sqrt{\frac{L+1}{L}}\left\{ \left(\frac{\frac{3}{2}\alpha Z + R\left(W_x + \frac{\varkappa_x}{k_x}\right)}{2k_x+1} \right)^2 - \frac{\{k_x[2(L-k_x)+5]-(2L+3)\}\,(q_x R)^2}{(2k_x-1)\,(L-k_x+1)\,\{2(L-k_x)+3\}^2} \right\},$$

$$f_7 = \frac{1}{9}\left\{ \frac{3}{2}\alpha Z + R\left(W_x + \frac{\varkappa_x}{k_x}\right) - q_x R \right\}^2,$$

$$f_8 = \frac{2}{3}\left\{ \frac{3}{2}\alpha Z + R\left(W_x + \frac{\varkappa_x}{k_x}\right) - q_x R \right\}.$$

Here, Z is the atomic number of the parent nucleus.

For $L = k_x - 1$ we have

$$C_x = \frac{(p_x R)^{2L}}{\{(2L+1)!!\}^2}\left\{ \frac{1}{2L+1}\left[\sqrt{L+1}\,{}^{\mathrm{V}}F^{(0)}_{LL0} + \sqrt{L}\,{}^{\mathrm{A}}F^{(0)}_{LL1}\right]^2 + \left({}^{\mathrm{A}}F^{(0)}_{L+1,L,1}\right)^2 \right\}. \tag{57}$$

For $L < k_x - 1$, we have $C_x = 0$ if only dominant terms are taken into account.

$(L-1)$th *unique forbidden transitions*:

For $L \geqq k_x$, we have

$$C_x = \frac{(L-1)!}{(2L-1)!!}\,\frac{(p_x R)^{2(k_x-1)}\,(q_x R)^{2(L-k_x)}}{(2k_x-1)!!\,(k_x-1)!\,\{2(L-k_x)+1\}!!\,(L-k_x)!}\left({}^{\mathrm{A}}F^{(0)}_{L,L-1,1}\right)^2. \tag{58}$$

For $L < k_x$, we have $C_x = 0$ if only dominant terms are taken into account.

In complete analogy to the f-function in β-decay (see eq. (48)), the function f_x has the form

$$f_x = \frac{\pi}{2}\,q_x^2\,\beta_x^2 \cdot B_x\,. \tag{59}$$

The quantities* β_x, the Coulomb amplitudes for the bound electron wave function, are similar to the

---

* The commonly used bound Coulomb functions and the quantities β_x are related as follows

$$\varkappa_x > 0 \quad f_{k_x}(R) = \beta_x \frac{(p_x R)^{k_x-1}}{(2k_x-1)!!}\sum_{n=0}^{\infty}[H_{2n}(k_x) + h_{2n}(k_x)],$$

$$g_{k_x}(R) = \beta_x \frac{(p_x R)^{k_x-1}}{(2k_x-1)!!}\sum_{n=0}^{\infty}[D_{2n+1}(k_x) + d_{2n+1}(k_x)].$$

$$\varkappa_x < 0 \quad f_{-k_x}(R) = -\beta_x \frac{(p_x R)^{k_x-1}}{(2k_x-1)!!}\sum_{n=0}^{\infty}[D_{2n+1}(k_x) - d_{2n+1}(k_x)],$$

$$g_{-k_x}(R) = \beta_x \frac{(p_x R)^{k_x-1}}{(2k_x-1)!!}\sum_{n=0}^{\infty}[H_{2n}(k_x) - h_{2n}(k_x)].$$

The Coulomb amplitudes β_x must be computed numerically.

amplitudes α_{ke} for the continuum wave functions (see section 2). The factor B_x takes into account the effects of the electron exchange as well as the imperfect atomic overlap [33]. In table 7, the B_x values computed by Bahcall [33] for some Z-values are listed. As Vatai [34] has noted, these values may be incorrect.

Table 7. Exchange and imperfect atomic overlap correction factors (according to Bahcall [33])

Z	B_K	B_{L_I}	B_{M_I}	Z	B_K	B_{L_I}	B_{M_I}
14	0.924	1.199	1.804	27	0.978	1.103	1.302
15	0.939	1.193	1.711	28	0.980	1.099	1.288
16	0.947	1.181	1.639	29	0.981	1.096	1.275
17	0.954	1.172	1.579	30	0.981*	1.090*	1.266*
18	0.959	1.162	1.530	31	0.981	1.087	1.256
19	0.963*	1.153*	1.489*	32	0.982	1.083	1.247
20	0.966*	1.145*	1.454*	33	0.982	1.080	1.238
21	0.969*	1.137*	1.423*	34	0.983	1.078	1.230
22	0.970	1.128	1.399	35	0.983	1.075	1.222
23	0.973	1.122	1.375	36	0.984	1.072	1.215
24	0.974	1.117	1.354	37	0.984	1.070	1.208
25	0.976	1.112	1.335				
26	0.977	1.107	1.317				

Values marked by an asterisk were interpolated

2.6 Special formulae for allowed and first forbidden electron captures

2.6.1 Allowed transitions

In allowed transitions, electrons can only be captured from orbits with $\varkappa = \pm 1$ (K, L_{I}, L_{II}, M_{I}, M_{II}, etc.). This result is obtained with the approximation that corrections from higher order terms (see table 4) are not taken into account. For the decay constants, we have

$$\lambda_c = \frac{g^2}{4\pi^2} C_0 \, [n_K q_K^2 \beta_K^2 B_K + n_{L_{\mathrm{I}}} q_{L_{\mathrm{I}}}^2 \beta_{L_{\mathrm{I}}}^2 B_{L_{\mathrm{I}}} + n_{L_{\mathrm{II}}} q_{L_{\mathrm{II}}}^2 \beta_{L_{\mathrm{II}}}^2 B_{L_{\mathrm{II}}} + n_{M_{\mathrm{I}}} q_{M_{\mathrm{I}}}^2 \beta_{M_{\mathrm{I}}}^2 B_{M_{\mathrm{I}}} + n_{M_{\mathrm{II}}} q_{M_{\mathrm{II}}}^2 \beta_{M_{\mathrm{II}}}^2 B_{M_{\mathrm{II}}}^2], \quad (60)$$

with

$$C_0 = ({}^{\mathrm{V}}F_{000}^{(0)})^2 + ({}^{\mathrm{A}}F_{101}^{(0)})^2 .$$

From this formula, we can derive the ratios of the capture probabilities in allowed decays from different subshells. As an example, we give the L_{I}/K ratio

$$\frac{\lambda_{L_{\mathrm{I}}}}{\lambda_K} = \frac{n_{L_{\mathrm{I}}} q_{L_{\mathrm{I}}}^2 \beta_{L_{\mathrm{I}}}^2 B_{L_{\mathrm{I}}}}{n_K q_K^2 \beta_K^2 B_K} . \quad (61)$$

For other ratios, the relations are completely analogous.

2.6.2 First non unique forbidden transitions

In the case of non unique forbidden transitions, only electrons with the quantum numbers $\varkappa_x = \pm 1, \pm 2$ can be captured. This result is obtained if the dominant terms only are considered. For K, L_{I}, L_{II}, M_{I}, and M_{II} capture, we have

$$\begin{aligned} C_x = {} & ({}^{\mathrm{V}}F_{101}^{(0)})^2 + \frac{1}{27} ({}^{\mathrm{V}}F_{110}^{(0)})^2 [\{\tfrac{3}{2} \alpha Z + R(W_x \mp 1) - q_x R\}^2 + 2(q_x R)^2] + \\ & + \frac{2}{27} ({}^{\mathrm{A}}F_{111}^{(0)})^2 \cdot [\{\tfrac{3}{2} \alpha Z + R(W_x \mp 1) + q_x R\}^2 + (q_x R)^2] + \frac{1}{9} ({}^{\mathrm{A}}F_{211}^{(0)})^2 \times \\ & \times (q_x R)^2 - \frac{2}{3\sqrt{3}} {}^{\mathrm{V}}F_{101}^{(0)} \, {}^{\mathrm{V}}F_{110}^{(0)} [\tfrac{3}{2} \alpha Z + R(W_x \mp 1) - q_x R] + \frac{2\sqrt{2}}{3\sqrt{3}} \times \\ & \times {}^{\mathrm{V}}F_{101}^{(0)} \, {}^{\mathrm{A}}F_{111}^{(0)} [\tfrac{3}{2} \alpha Z + R(W_x \mp 1) + q_x R] - \frac{2\sqrt{2}}{27} {}^{\mathrm{V}}F_{110}^{(0)} \, {}^{\mathrm{A}}F_{111}^{(0)} \times \\ & \times [\tfrac{3}{2} \alpha Z + R(W_x \mp 1)]^2 + \delta_{\Delta J,0} \{({}^{\mathrm{A}}F_{000}^{(0)})^2 + \frac{1}{9} ({}^{\mathrm{A}}F_{011}^{(0)})^2 \times \\ & \times [\tfrac{3}{2} \alpha Z + R \cdot (W_x \mp 1) - q_x R]^2 + \frac{2}{3} {}^{\mathrm{A}}F_{000}^{(0)} \, {}^{\mathrm{A}}F_{011}^{(0)} [\tfrac{3}{2} \alpha Z + R(W_x \mp 1) - q_x R]\} . \end{aligned} \quad (62)$$

The upper sign is for K, L_{I}, M_{I}, the lower for L_{II}, M_{II} capture.
For L_{III}, M_{III} and M_{IV} capture, one has

$$C_x = \frac{(p_x R)^2}{9}\left\{\left[\sqrt{\frac{2}{3}}\,{}^{\text{V}}F_{110}^{(0)} + \sqrt{\frac{1}{3}}\,{}^{\text{A}}F_{111}^{(0)}\right]^2 + ({}^{\text{A}}F_{211}^{(0)})^2\right\}. \tag{63}$$

2.6.3 First unique forbidden transitions

If we again consider only the dominant terms for these types of transitions, subshells with $\varkappa = \pm 1$, ± 2 can contribute. For K, L_{I}, L_{II}, M_{I} and M_{II} capture, we have

$$\lambda_c = \frac{g^2}{4\pi^2}({}^{\text{A}}F_{211}^{(0)})^2\frac{R^2}{9}\times$$
$$\times\left[n_K q_K^4 \beta_K^2 B_K + n_{L_{\text{I}}} q_{L_{\text{I}}}^4 \beta_{L_{\text{I}}}^2 B_{L_{\text{I}}} + n_{L_{\text{II}}} q_{L_{\text{II}}}^4 \beta_{L_{\text{II}}}^2 B_{L_{\text{II}}} + n_{M_{\text{I}}} q_{M_{\text{I}}}^4 \beta_{M_{\text{I}}}^2 B_{M_{\text{I}}} + n_{M_{\text{I}}} q_{M_{\text{II}}}^4 \beta_{M_{\text{II}}}^2 B_{M_{\text{II}}}\right]. \tag{64}$$

For L_{III}, M_{III} and M_{IV} capture, we have

$$\lambda_c = \frac{g^2}{4\pi^2}({}^{\text{A}}F_{211}^{(0)})^2\frac{R^2}{9}\times$$
$$\times\left[n_{L_{\text{III}}} p_{L_{\text{III}}}^2 q_{L_{\text{III}}}^2 \beta_{L_{\text{III}}}^2 B_{L_{\text{III}}} + n_{M_{\text{III}}} p_{M_{\text{III}}}^2 q_{M_{\text{III}}}^2 \beta_{M_{\text{III}}}^2 B_{M_{\text{III}}} + n_{M_{\text{IV}}} p_{M_{\text{IV}}}^2 q_{M_{\text{IV}}}^2 \beta_{M_{\text{IV}}}^2 B_{M_{\text{IV}}}\right]. \tag{65}$$

Eqs. (64) and (65) enable us to write expressions for the different capture ratios such as

$$\frac{\lambda_{L\text{I}}}{\lambda_K} = \frac{n_{L_{\text{I}}} q_{L_{\text{I}}}^4 \beta_{L_{\text{I}}}^2 \cdot B_{L_{\text{I}}}}{n_K q_K^4 \beta_K^2 \cdot B_K}. \tag{66}$$

The relations for the ratios $\lambda_{L_{\text{II}}}/\lambda_K$, $\lambda_{M_{\text{I}}}/\lambda_K$, $\lambda_{M_{\text{II}}}/\lambda_K$, $\lambda_{L_{\text{II}}}/\lambda_{L_{\text{I}}}$, $\lambda_{M_{\text{I}}}/\lambda_{L_{\text{I}}}$, etc., are completely analogous.
As another example we have, for the $\lambda_{L_{\text{III}}}/\lambda_{L_{\text{I}}}$ ratio,

$$\frac{\lambda_{L_{\text{III}}}}{\lambda_{L_{\text{I}}}} = \frac{n_{L_{\text{III}}} p_{L_{\text{III}}}^2 q_{L_{\text{III}}}^2 \beta_{L_{\text{III}}}^2 B_{L_{\text{III}}}}{n_{L_{\text{I}}} q_{L_{\text{I}}}^4 \beta_{L_{\text{I}}}^2 B_{L_{\text{I}}}}. \tag{67}$$

2.6.4 Ratio of electron capture to positron decay

From the foregoing, eqs. (46⋯67), the capture to β^+-decay ratios can easily be computed. As an example, we give here the K/β^+ ratio for allowed decays:

$$\frac{\lambda_K}{\lambda_{\beta^+}} = \frac{\mathfrak{f}_K}{f_{\beta^+}}. \tag{68}$$

3 Description of the tables including the calculation methods

The numerical calculations and the subsequent printing of all the tables were carried out on the IBM 7070/74 at the Kernforschungszentrum Karlsruhe. With the exception of the functions λ_2, λ_3 and λ_4 for $p = 0.1\cdots0.3$ and $\hat{\nu}_{12}$ for $|Z| < 10$ listed in table II, the computational accuracy* is ± 1 in the last digit. For the unscreened functions λ_2, λ_3 and λ_4, the figures are accurate to 5 digits. For the unscreened function ν_{12} the accuracy decreases from 4 digits after the decimal point at $|Z| = 10$ to 2 digits after the decimal point at $|Z| = 1$.

* Here, only the accuracy of the numerical calculation is meant, and not the uncertainty arising from the fact that for the potential, taking into account the screening by orbital electrons, we have used only an approximation to the true potential. The uncertainty caused by this approximation is much higher than the calculational accuracy and is discussed in detail in ref. [*49*].

Table I

In this table, the total energies, the kinetic energies and the $B\varrho$-values are listed for the electron momentum values used in the following tables II and III.

The momentum p is given in all tables in units of $m_0 c$ (m_0 = electron rest mass, c = velocity of light), the total energy W in units of $m_0 c^2$, the kinetic energy in keV and the magnetic rigidity $B\varrho$ in Gauss cm.

The relations between the various quantities are

$$W = \sqrt{p^2 + 1},$$
$$E = (W - 1)\, m_0 c^2,$$
$$B\varrho = \frac{m_0 c}{e}\, p$$

(e = elementary charge).

The following values of the fundamental constants were used for the computation:

$$m_0 c^2 = 511.006 \text{ keV},$$
$$e/m_0 = 1.7587970 \cdot 10^7 \text{ e.m.u.}\, \text{g}^{-1},$$
$$c = 2.997925 \cdot 10^{10} \text{ cm sec}^{-1}.$$

Table II

The following functions (see eqs. (2, 3, 5, 6, 7, 21)) described in chapter 2. and in the other literature [*17*, *18*, *21*, *22*] are tabulated in table II as a function of the electron momentum p_e (in units of $m_0 c$).

$$F_0 L_0 = \frac{\alpha_{-1}^2 + \alpha_1^2}{2 p_e^2},$$

$$L_0 = \frac{\alpha_{-1}^2 + \alpha_1^2}{2 p_e^2 F_0},$$

with

$$F_0 = 4 (2 p_e R)^{-2(1-\gamma_1)} \cdot \frac{|\Gamma(\gamma_1 + i y)|^2}{[\Gamma(2\gamma_1 + 1)]^2} \cdot e^{\pi \cdot y},$$

$$\lambda_k = \frac{\alpha_{-k}^2 + \alpha_k^2}{\alpha_{-1}^2 + \alpha_1^2} \qquad \text{for } k = 2, 3, 4,$$

$$\mu_k = \frac{k W_e}{\gamma_k} \cdot \frac{\alpha_{-k}^2 - \alpha_k^2}{\alpha_{-k}^2 + \alpha_k^2} \qquad \text{for } k = 1, 2,$$

$$\Lambda_k = \frac{W_e}{p_e} \frac{2\alpha_{-k}\alpha_k \cos(\Delta_k - \Delta_{-k})}{\alpha_{-k}^2 + \alpha_k^2} \qquad \text{for } k = 1, 2,$$

$$\nu_{12} = \frac{W_e}{p_e} \cdot \frac{\alpha_1 \alpha_{-2} \cos(\Delta_1 - \Delta_{-2}) + \alpha_{-1}\alpha_2 \cos(\Delta_{-1} - \Delta_2)}{\alpha_{-1}^2 + \alpha_1^2},$$

$$\hat{\nu}_{12} = \frac{p_e}{(\alpha Z)^2} \cdot \frac{\alpha_1 \alpha_{-2} \cos(\Delta_1 - \Delta_{-2}) - \alpha_{-1}\alpha_2 \cos(\Delta_{-1} - \Delta_2)}{\alpha_{-1}^2 + \alpha_1^2},$$

$$\eta_{12} = \frac{\alpha_{-1} \alpha_{-2} \cos(\Delta_{-1} - \Delta_{-2}) + \alpha_1\alpha_2 \cos(\Delta_1 - \Delta_2)}{\alpha_{-1}^2 + \alpha_1^2},$$

$$\hat{\eta}_{12} = W_e \frac{\alpha_{-1} \alpha_{-2} \cos(\Delta_{-1} - \Delta_{-2}) - \alpha_1\alpha_2 \cos(\Delta_1 - \Delta_2)}{\alpha_{-1}^2 + \alpha_1^2}.$$

The so-called Coulomb amplitudes α_k and Coulomb phases Δ_k were computed by solving the Dirac equation for a nucleus of finite size and with a uniform charge distribution inside but without taking into account screening. The fundamental equations for the calculation of these tabulated functions may be found in the book of ROSE [50]. The special methods [49] used for the computation of these tables are similar to those applied by BHALLA and ROSE [5] who have given a detailed description of the necessary equations. The functions defined above are listed for electrons and positrons (Z is positive for electrons and negative for positrons). Values for the atomic number Z of the daughter nucleus and the mass number A are given at the top of each page.

Throughout this and the subsequent tables, we use natural units ($\hbar = m_0 = c = 1$) and the following values for the fine structure constant α and the nuclear radius R:

$$\alpha = \frac{1}{137.0388},$$

$$R = 0.42587 \cdot \alpha \cdot A^{1/3} \triangleq 1.2 \cdot A^{1/3}\, 10^{-13}\,\text{cm}.$$

It is through this equation that the results depend slightly on the mass number A.

To indicate to what extent the Coulomb functions listed in table II are dependent on the nuclear radius R ($R = r_0 A^{1/3}$), we present a short tabulation of these functions for values of r_0 other than 1.2 fm.

Table 8. The unscreened Coulomb functions listed in table II as a function of r_0 ($R = r_0 A^{1/3}$)

r_0 fm	$Z=20$ $p=1.0$	$Z=20$ $p=20.0$	$Z=50$ $p=1.0$	$Z=50$ $p=20.0$	$Z=80$ $p=1.0$	$Z=80$ $p=20.0$	$Z=-20$ $p=1.0$	$Z=-20$ $p=20.0$	$Z=-50$ $p=1.0$	$Z=-50$ $p=20.0$	$Z=-80$ $p=1.0$	$Z=-80$ $p=20.0$
						F_0L_0						
1.0	2.0505	1.6064	7.6404	3.5087	38.835	7.7753	0.56321	0.67717	0.30447	0.43727	0.22770	0.32645
1.1	2.0458	1.5987	7.5333	3.4247	37.371	7.3303	0.56220	0.67757	0.30077	0.43562	0.22018	0.32137
1.2	2.0415	1.5913	7.4361	3.3469	36.075	6.9374	0.56129	0.67808	0.29745	0.43443	0.21357	0.31726
1.3	2.0375	1.5841	7.3473	3.2743	34.915	6.5873	0.56046	0.67868	0.29446	0.43364	0.20770	0.31395
1.4	2.0338	1.5772	7.2654	3.2062	33.868	6.2730	0.55970	0.67937	0.29173	0.43318	0.20244	0.31128
1.5	2.0303	1.5705	7.1896	3.1421	32.916	5.9891	0.55901	0.68013	0.28923	0.43300	0.19769	0.30916
						L_0						
1.0	1.0021	0.9771	1.0154	0.9155	1.0158	0.7936	1.0070	1.0317	1.0349	1.1322	1.0674	1.3128
1.1	1.0019	0.9743	1.0144	0.9054	1.0132	0.7755	1.0072	1.0344	1.0358	1.1428	1.0698	1.3396
1.2	1.0016	0.9716	1.0134	0.8955	1.0106	0.7583	1.0074	1.0372	1.0368	1.1535	1.0722	1.3664
1.3	1.0014	0.9689	1.0124	0.8858	1.0080	0.7421	1.0077	1.0399	1.0377	1.1642	1.0746	1.3935
1.4	1.0012	0.9662	1.0114	0.8762	1.0054	0.7266	1.0079	1.0426	1.0387	1.1749	1.0770	1.4207
1.5	1.0009	0.9636	1.0104	0.8669	1.0029	0.7120	1.0082	1.0453	1.0396	1.1856	1.0794	1.4481
						λ_2						
1.0	0.9781	1.0045	0.8755	1.0206	0.6914	1.1102	0.9752	0.9720	0.8640	0.8905	0.6664	0.7831
1.1	0.9793	1.0073	0.8820	1.0349	0.7062	1.1488	0.9760	0.9714	0.8693	0.8915	0.6782	0.7882
1.2	0.9804	1.0099	0.8880	1.0486	0.7201	1.1859	0.9768	0.9709	0.8741	0.8921	0.6891	0.7922
1.3	0.9814	1.0125	0.8937	1.0619	0.7333	1.2215	0.9775	0.9702	0.8785	0.8923	0.6992	0.7953
1.4	0.9823	1.0150	0.8990	1.0748	0.7458	1.2557	0.9781	0.9695	0.8826	0.8921	0.7086	0.7977
1.5	0.9832	1.0175	0.9040	1.0873	0.7578	1.2887	0.9787	0.9687	0.8864	0.8917	0.7174	0.7994
						λ_3						
1.0	0.9664	1.0014	0.8163	0.9985	0.5796	1.0536	0.9629	0.9615	0.8036	0.8469	0.5561	0.7002
1.1	0.9680	1.0049	0.8242	1.0161	0.5956	1.1001	0.9641	0.9610	0.8102	0.8485	0.5690	0.7068
1.2	0.9693	1.0082	0.8316	1.0332	0.6106	1.1454	0.9651	0.9603	0.8162	0.8496	0.5810	0.7121
1.3	0.9706	1.0115	0.8385	1.0498	0.6249	1.1895	0.9660	0.9596	0.8218	0.8501	0.5922	0.7164
1.4	0.9718	1.0147	0.8450	1.0660	0.6386	1.2326	0.9669	0.9588	0.8269	0.8503	0.6027	0.7198
1.5	0.9730	1.0178	0.8512	1.0817	0.6516	1.2747	0.9676	0.9579	0.8316	0.8500	0.6125	0.7223
						λ_4						
1.0	0.9597	0.9990	0.7835	0.9823	0.5235	1.0118	0.9559	0.9558	0.7706	0.8227	0.5012	0.6556
1.1	0.9613	1.0028	0.7920	1.0014	0.5394	1.0608	0.9572	0.9552	0.7777	0.8247	0.5143	0.6628
1.2	0.9629	1.0065	0.7999	1.0199	0.5545	1.1087	0.9583	0.9546	0.7841	0.8260	0.5264	0.6687
1.3	0.9643	1.0101	0.8073	1.0379	0.5689	1.1556	0.9594	0.9538	0.7901	0.8267	0.5377	0.6734
1.4	0.9657	1.0136	0.8143	1.0555	0.5826	1.2017	0.9603	0.9529	0.7956	0.8270	0.5483	0.6772
1.5	0.9669	1.0170	0.8209	1.0726	0.5958	1.2469	0.9612	0.9520	0.8008	0.8269	0.5584	0.6801

Table 8 (continued)

r_0 fm	$Z=20$, $p=1.0$	$Z=20$, $p=20.0$	$Z=50$, $p=1.0$	$Z=50$, $p=20.0$	$Z=80$, $p=1.0$	$Z=80$, $p=20.0$	$Z=-20$, $p=1.0$	$Z=-20$, $p=20.0$	$Z=-50$, $p=1.0$	$Z=-50$, $p=20.0$	$Z=-80$, $p=1.0$	$Z=-80$, $p=20.0$
						μ_1						
1.0	0.9995	0.9860	0.9976	0.9404	0.9911	0.8560	1.0005	1.0132	1.0023	1.0502	1.0078	1.1248
1.1	0.9995	0.9846	0.9973	0.9350	0.9902	0.8464	1.0005	1.0145	1.0025	1.0548	1.0085	1.1352
1.2	0.9994	0.9832	0.9971	0.9297	0.9893	0.8376	1.0006	1.0158	1.0027	1.0593	1.0093	1.1454
1.3	0.9993	0.9818	0.9968	0.9245	0.9884	0.8295	1.0006	1.0171	1.0029	1.0639	1.0100	1.1554
1.4	0.9993	0.9804	0.9966	0.9195	0.9875	0.8220	1.0007	1.0183	1.0032	1.0683	1.0107	1.1653
1.5	0.9992	0.9790	0.9963	0.9146	0.9866	0.8152	1.0007	1.0196	1.0034	1.0728	1.0114	1.1750
						μ_2						
1.0	0.9999	0.9973	0.9996	0.9900	0.9992	0.9789	1.0001	1.0027	1.0004	1.0095	1.0008	1.0193
1.1	0.9999	0.9970	0.9996	0.9890	0.9992	0.9766	1.0001	1.0029	1.0004	1.0104	1.0008	1.0212
1.2	0.9999	0.9968	0.9996	0.9880	0.9991	0.9743	1.0001	1.0032	1.0004	1.0114	1.0009	1.0231
1.3	0.9999	0.9965	0.9995	0.9869	0.9990	0.9719	1.0001	1.0035	1.0005	1.0123	1.0010	1.0249
1.4	0.9999	0.9962	0.9995	0.9858	0.9989	0.9695	1.0001	1.0038	1.0005	1.0133	1.0011	1.0268
1.5	0.9999	0.9959	0.9995	0.9847	0.9988	0.9671	1.0001	1.0040	1.0005	1.0142	1.0011	1.0287
						Λ_1						
1.0	1.0005	1.0000	1.0019	1.0001	1.0049	1.0003	0.9995	1.0000	0.9983	0.9999	0.9961	0.9998
1.1	1.0005	1.0000	1.0021	1.0002	1.0055	1.0003	0.9995	1.0000	0.9981	0.9999	0.9958	0.9998
1.2	1.0006	1.0000	1.0023	1.0002	1.0060	1.0004	0.9994	1.0000	0.9979	0.9999	0.9954	0.9998
1.3	1.0006	1.0000	1.0025	1.0002	1.0065	1.0004	0.9994	1.0000	0.9978	0.9999	0.9951	0.9997
1.4	1.0007	1.0000	1.0027	1.0002	1.0071	1.0004	0.9993	1.0000	0.9976	0.9999	0.9947	0.9997
1.5	1.0007	1.0001	1.0029	1.0002	1.0076	1.0004	0.9993	1.0000	0.9974	0.9998	0.9944	0.9997
						Λ_2						
1.0	1.0001	1.0000	1.0003	1.0000	1.0006	1.0000	0.9999	1.0000	0.9997	1.0000	0.9994	1.0000
1.1	1.0001	1.0000	1.0004	1.0000	1.0007	1.0001	0.9999	1.0000	0.9996	1.0000	0.9993	1.0000
1.2	1.0001	1.0000	1.0004	1.0000	1.0008	1.0001	0.9999	1.0000	0.9996	1.0000	0.9992	0.9999
1.3	1.0001	1.0000	1.0004	1.0000	1.0008	1.0001	0.9999	1.0000	0.9996	1.0000	0.9992	0.9999
1.4	1.0001	1.0000	1.0005	1.0000	1.0009	1.0001	0.9999	1.0000	0.9995	1.0000	0.9991	0.9999
1.5	1.0001	1.0000	1.0005	1.0000	1.0010	1.0001	0.9999	1.0000	0.9995	1.0000	0.9990	0.9999
						ν_{12}						
1.0	0.9768	0.9953	0.8517	0.9545	0.6040	0.8654	0.9774	0.9808	0.8828	0.9186	0.7390	0.8344
1.1	0.9775	0.9967	0.8550	0.9617	0.6108	0.8848	0.9778	0.9805	0.8854	0.9192	0.7453	0.8372
1.2	0.9780	0.9980	0.8580	0.9686	0.6172	0.9037	0.9781	0.9802	0.8877	0.9195	0.7510	0.8394
1.3	0.9785	0.9993	0.8609	0.9754	0.6231	0.9219	0.9784	0.9799	0.8899	0.9197	0.7563	0.8412
1.4	0.9790	1.0006	0.8636	0.9819	0.6288	0.9396	0.9787	0.9795	0.8919	0.9196	0.7612	0.8425
1.5	0.9795	1.0018	0.8661	0.9883	0.6342	0.9569	0.9790	0.9792	0.8937	0.9195	0.7656	0.8436
						$\hat{\nu}_{12}$						
1.0	0.7758	1.0515	0.7470	1.0114	0.6515	1.0131	0.6971	0.4573	0.5995	0.4922	0.4722	0.4520
1.1	0.7772	1.0773	0.7504	1.0267	0.6589	1.0229	0.6965	0.4334	0.6008	0.4801	0.4759	0.4447
1.2	0.7785	1.1029	0.7535	1.0410	0.6658	1.0307	0.6959	0.4094	0.6019	0.4680	0.4792	0.4371
1.3	0.7798	1.1281	0.7565	1.0544	0.6723	1.0365	0.6953	0.3857	0.6029	0.4559	0.4822	0.4293
1.4	0.7811	1.1530	0.7593	1.0667	0.6784	1.0405	0.6946	0.3621	0.6037	0.4437	0.4849	0.4214
1.5	0.7824	1.1775	0.7620	1.0781	0.6842	1.0428	0.6939	0.3384	0.6045	0.4314	0.4874	0.4134
						η_{12}						
1.0	0.9711	0.9953	0.8166	0.9543	0.5208	0.8650	0.9726	0.9808	0.8560	0.9186	0.6816	0.8343
1.1	0.9717	0.9966	0.8196	0.9615	0.5265	0.8844	0.9730	0.9805	0.8586	0.9192	0.6876	0.8371
1.2	0.9722	0.9980	0.8224	0.9684	0.5318	0.9032	0.9734	0.9802	0.8610	0.9195	0.6931	0.8394
1.3	0.9727	0.9993	0.9251	0.9752	0.5368	0.9215	0.9738	0.9799	0.8632	0.9197	0.6982	0.8412
1.4	0.9732	1.0006	0.8275	0.9817	0.5415	0.9392	0.9741	0.9796	0.8652	0.9197	0.7028	0.8425
1.5	0.9736	1.0018	0.8298	0.9881	0.5460	0.9564	0.9744	0.9792	0.8670	0.9195	0.7072	0.8436
						$\hat{\eta}_{12}$						
1.0	0.9602	0.9761	0.7547	0.8547	0.3976	0.6398	0.9627	0.9783	0.8010	0.8856	0.5672	0.7452
1.1	0.9607	0.9767	0.7574	0.8588	0.4018	0.6538	0.9631	0.9788	0.8035	0.8886	0.5725	0.7522
1.2	0.9612	0.9772	0.7599	0.8627	0.4057	0.6675	0.9635	0.9793	0.8059	0.8913	0.5773	0.7586
1.3	0.9617	0.9777	0.7623	0.8665	0.4094	0.6809	0.9639	0.9798	0.8080	0.8939	0.5817	0.7645
1.4	0.9621	0.9781	0.7644	C.8702	0.4128	0.6941	0.9642	0.9802	0.8100	0.8963	0.5858	0.7701
1.5	0.9625	0.9786	0.7664	0.8738	0.4161	0.7070	0.9646	0.9805	0.8118	0.8986	0.5897	0.7754

Table III

In table III the ratios of the screened (with asterisk) to the unscreened (without asterisk) functions listed in table II are given (except for λ_3 and λ_4). The β-decay functions, which contain the effect of the nuclear charge and the influence of screening from the atomic electron cloud, were obtained by solving the Dirac equation with a potential which corresponds to an extended uniform nuclear charge distribution and an atomic electron charge distribution of opposite sign outside the nucleus. The methods and computer programs employed in the present work are very similar to those developed by Bühring [16]. A detailed description of this method is given in ref. [16] and [49]. The constants a_i and b_i which appear in the expression for the potentials outside the nucleus

$$V(r) = -\frac{\alpha Z}{r} \sum_i a_i \mathrm{e}^{-b_i r}$$

are taken from the papers of Tietz and Strand [51, 52] for $Z \leqq 36$ and of Bonham and Strand [53] for $Z > 36$. These constants were determined by fitting the above equation to Hartree-Fock potentials for $Z \leqq 36$ and to Thomas-Fermi-Dirac potentials for $Z > 36$ by the method of least squares. Due to the extensive computer time required for this tabulation, the ratios of the screened to the unscreened functions are listed for a much smaller number of momentum values than for the unscreened functions in table II. These ratios η are given only for p-values for which $|1 - \eta| \geqq 0.02$.

In tables II and III the same Z (atomic number of the daughter nucleus) and A (mass number) values were used. The ratio of screened to unscreened functions is nearly independent of the nuclear radius R.

Table IV

In table IV, the integrated statistical factor (in natural units)

$$f_0(E_0) = \int_0^{p_0} p_e^2 (W_0 - W_e)^2 \, dp_e$$

is listed as a function of the maximum kinetic energy E_0 in keV of the electrons or positrons (see eqs. (50, 51) and (52) in chapter 2.). Additional values can easily be computed with the aid of eqs. (51) and (52). The relation between kinetic energy in keV and the total energy in natural units is given in table I.

Diagram I and II

In these diagrams, the ratio

$$S(E_0) = \frac{\int_0^{p_0} F_0 L_0 p_e^2 (W_0 - W_e)^2 \, dp_e}{\int_0^{p_0} p_e^2 (W_0 - W_e)^2 \, dp_e}$$

of the integrated unscreened Fermi function to the integrated statistical factor is shown as a function of the maximum kinetic energy of the β-particles (in keV) for different atomic numbers Z. The diagrams I and II correspond to electrons and positrons, respectively. The same computer program was used as for the computation of table II. The curves are identified by the Z-values. The associated mass numbers are the same as those in table II.

Diagram III and IV

In these diagrams the ratio

$$\varepsilon_f(E_0) = \frac{\int_0^{p_0} F_0 L_0^* p_e^2 (W_0 - W_e)^2 \, dp_e}{\int_0^{p_0} F_0 L_0 p_e^2 (W_0 - W_e)^2 \, dp_e}$$

of the integrated Fermi-function with screening to the same function without screening is displayed as a function of the maximum kinetic energies of the β-particles. For practical reasons, instead of ε_f

the quantities $1 - \varepsilon_f$ or $\varepsilon_f - 1$ are displayed for electrons and positrons, respectively. The Z-values of the daughter nuclei are shown with the curves. They are identical to the ones of diagrams I and II. The same computer programs as were used for tables II and III were employed to obtain the values for these diagrams. For the higher values of E_0, we have used for diagrams III and IV the Rose-WKB screening formula* [8, 9, 10] combined with the computer program for the unscreened case (see table II) to calculate $F_0 L_0$.

Table V

In table V the Coulomb functions and other quantities of the bound orbital electron used in electron capture calculations are tabulated. These quantities are the Coulomb amplitude β_x, the radial electron wave functions $f_x(R)$ and $g_x(R)$ at the nuclear radius R, and both the calculated (one-electron eigenvalues) and experimental electron binding energies E_B (in eV). The notation used in table V is the following:

$Z \triangleq$ atomic number of the parent nucleus

$A \triangleq$ mass number

BETA $\triangleq \beta_x$	EB(CO) $\triangleq E_B$ (computed)
$F(R) \triangleq f_x(R)$	EB(EX) $\triangleq E_B$ (experimental).
$G(R) \triangleq g_x(R)$	

Because the accuracy of the potentials used as input data for the calculation is not sufficient for the lower Z-values, we have restricted the tabulation for $Z \leqq 11$ to the K shell and for $Z < 30$ to the K and L shells. In all the other cases the tabulation includes the K, L and M shells. The experimental values** of the electron binding energies are taken from the work of Hagström et al. [54] and Fahlman et al. [55].

The tabulated quantities were obtained by solving the Dirac equation for the case of the bound orbital electron. The method used for the numerical integration is the same as that used in the case of the β-decay [16] (Table III) except for some special modifications which became necessary due to the difference between the bound and continuum case.

The essential difference from β-decay is that we have to take into account exchange terms in the potential. These exchange terms in the potential are, of course, not existent in the case of β^+-decay. But, as Matese and Johnson [11] have shown, these terms can also be neglected in the case of β^--decay. In electron capture, however, they must be included. The electrostatic potential used both inside and outside the nucleus was exactly the same as the one which we had employed for the calculation of the β-decay functions in table III. Inside the nucleus we have neglected the exchange terms. Outside the nucleus the additional exchange interaction terms needed in the potential are included in the approximation of Slater [56]:

$$V_{exch} = C \varrho^{1/3},$$

with

ϱ = number of electrons per unit of volume (in natural units),

$C = (81/8\pi)^{1/3} \cdot \alpha$.

---

* The parameter V_0 needed in the Rose-WKB screening formula [8, 9, 10] is the potential energy shift at the surface of the nucleus due to the atomic electrons. For the potential outside the nucleus

$$V(r) = -\frac{\alpha Z}{r} \sum_i a_i e^{-b_i r}$$

used to calculate the screened functions (see description of table III) V_0 has the form

$$V_0 \approx \alpha Z \sum_i a_i b_i.$$

The ratios of screened to unscreened $F_0 L_0$ calculated with the Rose-WKB-formula and the above values for V_0 are for higher electron momenta in excellent agreement with the exact ratios from table III. It may also be noted that these V_0's are in good agreement with the V_0's given by Garret and Bhalla [10] using self consistent Hartree-Fock wave functions.

** Not all of the binding energies listed by Hagström et al. [54] and Fahlman et al. [55] were measured experimentally. A number of these values were interpolated or extrapolated. Apart from a few other interpolated values, this is the case for the L_I-binding energies between $Z = 13$ and $Z = 24$ and the binding energies between $Z = 83$ and $Z = 90$ and also those above $Z = 94$.

Therefore the total potential outside the nucleus has the form

$$V(r) = -\frac{\alpha Z}{r}\left(\sum_i a_i\,\mathrm{e}^{-b_i r} + \left(\frac{81}{32\pi^2}\right)^{1/3} \frac{1}{Z^{2/3}}\left\{\sum_i a_i\,(b_i r)^2\,\mathrm{e}^{-b_i r}\right\}^{1/3}\right).$$

This analytical form of the exchange term is not very appropriate when the Dirac equation is solved using BÜHRING's [16] method. For this reason we have replaced this expression by a simpler function of r. Thus the potential outside the nucleus used in the calculation has the form

$$V(r) = -\frac{\alpha Z}{r}\left(\sum_i a_i\,\mathrm{e}^{-b_i r} + \sum_{i=1}^{3} \alpha_i\, r\,\mathrm{e}^{-\beta_i r}\right).$$

The constants α_i and β_i were determined by fitting the above analytical form of the exchange term to the more exact version using the method of least squares. Similar to the calculation of the β-decay functions by BÜHRING [16], the above expression for the potential (apart from a small modification due to the finite extension of the nucleus) is used only in the interval from the nuclear radius R to a distance R_3. From R_3 to a distance R_2 the sum over exponentials is replaced by a polynomial of degree four, and in the range from R_2 to infinity the potential has the form $-\alpha/r$. The coefficients of the polynomial (see [16]) are determined by the condition that the potential itself and its first and second derivatives are continous at R_3 as is the potential and its first derivative at R_2. The same values are used for R_2 and R_3 as in the case of the β-decay functions (table III). A more detailed description of the method used here to solve the Dirac equation for the case of the bound electron is given in [49].

To show the R-dependence of the Coulomb-amplitude β_x, this quantity is listed in table 9 for r_0 values ($R = r_0 A^{1/3}$) from 1.0 to 1.5 fm in steps of 0.1 fm and for different Z-values. The R-dependence of the other quantities is either negligible, as in the case of the binding energies, or can be calculated analytically with aid of the given β_x (see footnote on p. 19 and table 1) as in the case of $f_x(R)$ and $g_x(R)$.

Table 9. β_x as a function of r_0 ($R = r_0 A^{1/3}$)

sub-shell	r_0 fm	β_x $Z = 20$	$Z = 50$	$Z = 80$
K	1.0	$1.173 \cdot 10^{-1}$	$6.470 \cdot 10^{-1}$	2.248
	1.1	$1.172 \cdot 10^{-1}$	$6.425 \cdot 10^{-1}$	2.206
	1.2	$1.171 \cdot 10^{-1}$	$6.383 \cdot 10^{-1}$	2.168
	1.3	$1.170 \cdot 10^{-1}$	$6.346 \cdot 10^{-1}$	2.133
	1.4	$1.168 \cdot 10^{-1}$	$6.311 \cdot 10^{-1}$	2.102
	1.5	$1.167 \cdot 10^{-1}$	$6.278 \cdot 10^{-1}$	2.072
L_{I}	1.0	$3.433 \cdot 10^{-2}$	$2.225 \cdot 10^{-1}$	$8.815 \cdot 10^{-1}$
	1.1	$3.429 \cdot 10^{-2}$	$2.210 \cdot 10^{-1}$	$8.649 \cdot 10^{-1}$
	1.2	$3.426 \cdot 10^{-2}$	$2.196 \cdot 10^{-1}$	$8.498 \cdot 10^{-1}$
	1.3	$3.422 \cdot 10^{-2}$	$2.183 \cdot 10^{-1}$	$8.362 \cdot 10^{-1}$
	1.4	$3.419 \cdot 10^{-2}$	$2.171 \cdot 10^{-1}$	$8.236 \cdot 10^{-1}$
	1.5	$3.416 \cdot 10^{-2}$	$2.159 \cdot 10^{-1}$	$8.121 \cdot 10^{-1}$
L_{II}	1.0	$1.937 \cdot 10^{-3}$	$3.494 \cdot 10^{-2}$	$2.444 \cdot 10^{-1}$
	1.1	$1.935 \cdot 10^{-3}$	$3.471 \cdot 10^{-2}$	$2.400 \cdot 10^{-1}$
	1.2	$1.933 \cdot 10^{-3}$	$3.450 \cdot 10^{-2}$	$2.360 \cdot 10^{-1}$
	1.3	$1.931 \cdot 10^{-3}$	$3.430 \cdot 10^{-2}$	$2.324 \cdot 10^{-1}$
	1.4	$1.929 \cdot 10^{-3}$	$3.412 \cdot 10^{-2}$	$2.291 \cdot 10^{-1}$
	1.5	$1.928 \cdot 10^{-3}$	$3.395 \cdot 10^{-2}$	$2.260 \cdot 10^{-1}$
L_{III}	1.0	$9.848 \cdot 10^{-2}$	$4.331 \cdot 10^{-1}$	1.169
	1.1	$9.843 \cdot 10^{-2}$	$4.316 \cdot 10^{-1}$	1.160
	1.2	$9.838 \cdot 10^{-2}$	$4.303 \cdot 10^{-1}$	1.150
	1.3	$9.833 \cdot 10^{-2}$	$4.291 \cdot 10^{-1}$	1.142
	1.4	$9.829 \cdot 10^{-2}$	$4.280 \cdot 10^{-1}$	1.135
	1.5	$9.825 \cdot 10^{-2}$	$4.270 \cdot 10^{-1}$	1.128
M_{I}	1.0		$9.982 \cdot 10^{-2}$	$4.251 \cdot 10^{-1}$
	1.1		$9.912 \cdot 10^{-2}$	$4.170 \cdot 10^{-1}$
	1.2		$9.848 \cdot 10^{-2}$	$4.098 \cdot 10^{-1}$
	1.3		$9.790 \cdot 10^{-2}$	$4.032 \cdot 10^{-1}$
	1.4		$9.736 \cdot 10^{-2}$	$3.971 \cdot 10^{-1}$
	1.5		$9.685 \cdot 10^{-2}$	$3.916 \cdot 10^{-1}$
M_{II}	1.0		$1.634 \cdot 10^{-2}$	$1.248 \cdot 10^{-1}$
	1.1		$1.623 \cdot 10^{-2}$	$1.225 \cdot 10^{-1}$
	1.2		$1.613 \cdot 10^{-2}$	$1.205 \cdot 10^{-1}$
	1.3		$1.604 \cdot 10^{-2}$	$1.186 \cdot 10^{-1}$
	1.4		$1.596 \cdot 10^{-2}$	$1.169 \cdot 10^{-1}$
	1.5		$1.588 \cdot 10^{-2}$	$1.154 \cdot 10^{-1}$
M_{III}	1.0		$4.748 \cdot 10^{-1}$	1.252
	1.1		$4.732 \cdot 10^{-1}$	1.242
	1.2		$4.718 \cdot 10^{-1}$	1.232
	1.3		$4.705 \cdot 10^{-1}$	1.223
	1.4		$4.692 \cdot 10^{-1}$	1.215
	1.5		$4.681 \cdot 10^{-1}$	1.207
M_{IV}	1.0		$2.740 \cdot 10^{-2}$	$1.243 \cdot 10^{-1}$
	1.1		$2.731 \cdot 10^{-2}$	$1.232 \cdot 10^{-1}$
	1.2		$2.723 \cdot 10^{-2}$	$1.223 \cdot 10^{-1}$
	1.3		$2.715 \cdot 10^{-2}$	$1.214 \cdot 10^{-1}$
	1.4		$2.708 \cdot 10^{-2}$	$1.206 \cdot 10^{-1}$
	1.5		$2.702 \cdot 10^{-2}$	$1.199 \cdot 10^{-1}$
M_{V}	1.0		1.104	2.032
	1.1		1.102	2.020
	1.2		1.100	2.010
	1.3		1.098	2.000
	1.4		1.096	1.992
	1.5		1.094	1.984

4 Tables and diagrams

Contents*)

*) Detailed table of contents see p. VI, VII

Table I. Relations between electron momenta, total energies, kinetic energies, and Bϱ-values used in the following tables

p	W	E [keV]	$B \cdot \varrho$ [Gauss·cm]
0.1	1.00499	2.54867E 00	1.70453E 02
0.2	1.01980	1.01199E 01	3.40907E 02
0.3	1.04403	2.24999E 01	5.11360E 02
0.4	1.07703	3.93643E 01	6.81813E 02
0.5	1.11803	6.03161E 01	8.52266E 02
0.6	1.16619	8.49243E 01	1.02272E 03
0.7	1.22066	1.12756E 02	1.19317E 03
0.8	1.28062	1.43401E 02	1.36363E 03
0.9	1.34536	1.76482E 02	1.53408E 03
1.0	1.41421	2.11666E 02	1.70453E 03
1.2	1.56205	2.87211E 02	2.04544E 03
1.4	1.72047	3.68162E 02	2.38635E 03
1.6	1.88680	4.53158E 02	2.72725E 03
1.8	2.05913	5.41220E 02	3.06816E 03
2.0	2.23607	6.31638E 02	3.40907E 03
2.2	2.41661	7.23896E 02	3.74997E 03
2.4	2.60000	8.17610E 02	4.09088E 03
2.6	2.78568	9.12492E 02	4.43178E 03
2.8	2.97321	1.00832E 03	4.77269E 03
3.0	3.16228	1.10494E 03	5.11360E 03
3.2	3.35261	1.20220E 03	5.45450E 03
3.4	3.54401	1.30000E 03	5.79541E 03
3.6	3.73631	1.39827E 03	6.13632E 03
3.8	3.92938	1.49693E 03	6.47722E 03
4.0	4.12311	1.59593E 03	6.81813E 03
4.5	4.60977	1.84462E 03	7.67040E 03
5.0	5.09902	2.09462E 03	8.52266E 03
5.5	5.59017	2.34560E 03	9.37493E 03
6.0	6.08276	2.59732E 03	1.02272E 04
6.5	6.57647	2.84961E 03	1.10795E 04
7.0	7.07107	3.10235E 03	1.19317E 04
7.5	7.56637	3.35546E 03	1.27840E 04
8.0	8.06226	3.60886E 03	1.36363E 04
9.0	9.05539	4.11635E 03	1.53408E 04
10.0	10.04988	4.62454E 03	1.70453E 04
11.0	11.04536	5.13324E 03	1.87499E 04
12.0	12.04159	5.64232E 03	2.04544E 04
13.0	13.03840	6.15170E 03	2.21589E 04
14.0	14.03567	6.66130E 03	2.38635E 04
16.0	16.03122	7.68104E 03	2.72725E 04
18.0	18.02776	8.70129E 03	3.06816E 04
20.0	20.02498	9.72188E 03	3.40907E 04
25.0	25.01999	1.22744E 04	4.26133E 04
30.0	30.01666	1.48277E 04	5.11360E 04
35.0	35.01428	1.73815E 04	5.96586E 04
40.0	40.01250	1.99356E 04	6.81813E 04
45.0	45.01111	2.24899E 04	7.67040E 04
50.0	50.01000	2.50444E 04	8.52266E 04

p in units of m_0c

W in units of m_0c^2

Table II. Unscreened β-decay Coulomb functions

ELECTRON Z= 1 A= 1

p	F_0L_0	L_0	λ_2	λ_3	λ_4	μ_1	μ_2	Λ_1	Λ_2	ν_{12}	$\hat{\nu}_{12}$	η_{12}	$\hat{\eta}_{12}$
0.1	1.2487E 00	1.0000	1.0051	1.0064	1.0070	1.0000	1.0000	1.0000	1.0000	1.0012	0.7503	0.9999	0.9999
0.2	1.1220E 00	1.0000	1.0012	1.0014	1.0016	1.0000	1.0000	1.0000	1.0000	1.0002	0.7530	0.9999	0.9999
0.3	1.0824E 00	1.0000	1.0004	1.0005	1.0006	1.0000	1.0000	1.0000	1.0000	1.0001	0.7572	0.9999	0.9999
0.4	1.0634E 00	1.0000	1.0002	1.0002	1.0002	1.0000	1.0000	1.0000	1.0000	1.0000	0.7628	0.9999	0.9999
0.5	1.0526E 00	1.0000	1.0001	1.0001	1.0001	1.0000	1.0000	1.0000	1.0000	1.0000	0.7697	0.9999	0.9999
0.6	1.0456E 00	1.0000	1.0000	1.0000	1.0000	1.0000	1.0000	1.0000	1.0000	1.0000	0.7763	0.9999	0.9999
0.7	1.0409E 00	1.0000	1.0000	1.0000	0.9999	1.0000	1.0000	1.0000	1.0000	1.0000	0.7847	0.9999	0.9999
0.8	1.0375E 00	1.0000	1.0000	0.9999	0.9999	1.0000	1.0000	1.0000	1.0000	1.0000	0.7938	0.9999	0.9999
0.9	1.0350E 00	1.0000	0.9999	0.9999	0.9999	1.0000	1.0000	1.0000	1.0000	1.0000	0.8030	0.9999	0.9999
1.0	1.0331E 00	1.0000	0.9999	0.9999	0.9999	1.0000	1.0000	1.0000	1.0000	0.9999	0.8126	0.9999	0.9999
1.2	1.0305E 00	1.0000	0.9999	0.9999	0.9999	1.0000	1.0000	1.0000	1.0000	0.9999	0.8299	0.9999	0.9999
1.4	1.0288E 00	1.0000	0.9999	0.9999	0.9999	1.0000	1.0000	1.0000	1.0000	0.9999	0.8521	0.9999	0.9999
1.6	1.0276E 00	1.0000	0.9999	0.9999	0.9999	1.0000	1.0000	1.0000	1.0000	0.9999	0.8698	0.9999	0.9999
1.8	1.0268E 00	1.0000	0.9999	0.9999	0.9999	1.0000	1.0000	1.0000	1.0000	0.9999	0.8881	0.9999	0.9999
2.0	1.0262E 00	1.0000	0.9999	0.9999	0.9999	1.0000	1.0000	1.0000	1.0000	0.9999	0.9054	0.9999	0.9999
2.2	1.0257E 00	1.0000	0.9999	0.9999	0.9999	1.0000	1.0000	1.0000	1.0000	1.0000	0.9250	0.9999	0.9999
2.4	1.0253E 00	0.9999	0.9999	0.9999	0.9999	1.0000	1.0000	1.0000	1.0000	1.0000	0.9402	0.9999	0.9999
2.6	1.0251E 00	0.9999	0.9999	0.9999	0.9999	1.0000	1.0000	1.0000	1.0000	1.0000	0.9555	0.9999	0.9999
2.8	1.0248E 00	0.9999	0.9999	0.9999	0.9999	1.0000	1.0000	1.0000	1.0000	1.0000	0.9742	1.0000	0.9999
3.0	1.0246E 00	0.9999	0.9999	0.9999	0.9999	1.0000	1.0000	1.0000	1.0000	1.0000	0.9946	1.0000	0.9999
3.2	1.0245E 00	0.9999	0.9999	0.9999	0.9999	1.0000	1.0000	1.0000	1.0000	1.0000	1.0111	1.0000	0.9999
3.4	1.0243E 00	0.9999	0.9999	0.9999	0.9999	1.0000	1.0000	1.0000	1.0000	1.0000	1.0272	1.0000	0.9999
3.6	1.0242E 00	0.9999	1.0000	0.9999	0.9999	1.0000	1.0000	1.0000	1.0000	1.0000	1.0492	1.0000	0.9999
3.8	1.0241E 00	0.9999	1.0000	0.9999	0.9999	1.0000	1.0000	1.0000	1.0000	1.0000	1.0616	1.0000	0.9999
4.0	1.0241E 00	0.9999	1.0000	0.9999	0.9999	1.0000	1.0000	1.0000	1.0000	1.0000	1.0682	1.0000	0.9999
4.5	1.0239E 00	0.9999	1.0000	0.9999	0.9999	1.0000	1.0000	1.0000	1.0000	1.0000	1.1249	1.0000	0.9999
5.0	1.0238E 00	0.9999	1.0000	1.0000	0.9999	0.9999	1.0000	1.0000	1.0000	1.0000	1.1630	1.0000	0.9999
5.5	1.0237E 00	0.9999	1.0000	1.0000	1.0000	0.9999	1.0000	1.0000	1.0000	1.0000	1.2125	1.0000	0.9999
6.0	1.0236E 00	0.9999	1.0000	1.0000	1.0000	0.9999	1.0000	1.0000	1.0000	1.0000	1.2552	1.0000	0.9999
6.5	1.0235E 00	0.9999	1.0000	1.0000	1.0000	0.9999	1.0000	1.0000	1.0000	1.0000	1.3027	1.0000	0.9999
7.0	1.0235E 00	0.9998	1.0000	1.0000	1.0000	0.9999	1.0000	1.0000	1.0000	1.0000	1.3486	1.0000	0.9999
7.5	1.0234E 00	0.9998	1.0000	1.0000	1.0000	0.9999	1.0000	1.0000	1.0000	1.0000	1.3872	1.0000	0.9999
8.0	1.0234E 00	0.9998	1.0000	1.0000	1.0000	0.9999	1.0000	1.0000	1.0000	1.0000	1.4348	1.0000	0.9999
9.0	1.0233E 00	0.9998	1.0000	1.0000	1.0000	0.9999	1.0000	1.0000	1.0000	1.0000	1.5191	1.0000	0.9999
10.0	1.0233E 00	0.9998	1.0001	1.0001	1.0001	0.9999	1.0000	1.0000	1.0000	1.0000	1.6150	1.0000	0.9999
11.0	1.0232E 00	0.9998	1.0001	1.0001	1.0001	0.9999	1.0000	1.0000	1.0000	1.0000	1.6768	1.0000	0.9999
12.0	1.0232E 00	0.9997	1.0001	1.0001	1.0001	0.9999	1.0000	1.0000	1.0000	1.0000	1.7742	1.0000	0.9999
13.0	1.0231E 00	0.9997	1.0001	1.0001	1.0001	0.9999	1.0000	1.0000	1.0000	1.0000	1.8637	1.0000	0.9999
14.0	1.0231E 00	0.9997	1.0001	1.0001	1.0001	0.9998	1.0000	1.0000	1.0000	1.0000	1.9208	1.0000	0.9999
16.0	1.0230E 00	0.9996	1.0002	1.0002	1.0002	0.9998	1.0000	1.0000	1.0000	1.0001	2.1282	1.0001	0.9999
18.0	1.0230E 00	0.9996	1.0002	1.0002	1.0002	0.9998	1.0000	1.0000	1.0000	1.0001	2.2896	1.0001	0.9999
20.0	1.0229E 00	0.9996	1.0002	1.0003	1.0003	0.9998	1.0000	1.0000	1.0000	1.0001	2.4739	1.0001	0.9999
25.0	1.0228E 00	0.9994	1.0003	1.0003	1.0004	0.9997	0.9999	1.0000	1.0000	1.0001	2.8681	1.0001	0.9999
30.0	1.0226E 00	0.9993	1.0004	1.0004	1.0005	0.9997	0.9999	1.0000	1.0000	1.0002	3.3208	1.0002	0.9999
35.0	1.0225E 00	0.9992	1.0004	1.0005	1.0006	0.9996	0.9999	1.0000	1.0000	1.0002	3.8044	1.0002	0.9999
40.0	1.0224E 00	0.9991	1.0005	1.0006	1.0007	0.9995	0.9999	1.0000	1.0000	1.0002	4.1788	1.0002	0.9999
45.0	1.0223E 00	0.9990	1.0006	1.0007	1.0008	0.9995	0.9999	1.0000	1.0000	1.0003	4.6335	1.0003	0.9999
50.0	1.0221E 00	0.9989	1.0007	1.0008	1.0009	0.9994	0.9999	1.0000	1.0000	1.0003	4.9607	1.0003	0.9999

Instr. see p. 22

ELECTRON Z= 2 A= 3

p	F_0L_0	L_0	λ_2	λ_3	λ_4	μ_1	μ_2	Λ_1	Λ_2	ν_{12}	$\hat{\nu}_{12}$	η_{12}	$\hat{\eta}_{12}$
0.1	1.5336E 00	1.0000	1.0206	1.0257	1.0280	1.0000	1.0000	1.0000	1.0000	1.0049	0.7507	0.9996	0.9995
0.2	1.2542E 00	0.9999	1.0047	1.0058	1.0063	1.0000	1.0000	1.0000	1.0000	1.0010	0.7530	0.9996	0.9995
0.3	1.1700E 00	0.9999	1.0018	1.0021	1.0023	1.0000	1.0000	1.0000	1.0000	1.0003	0.7563	0.9996	0.9995
0.4	1.1304E 00	0.9999	1.0008	1.0009	1.0009	1.0000	1.0000	1.0000	1.0000	1.0000	0.7605	0.9997	0.9996
0.5	1.1078E 00	0.9999	1.0003	1.0003	1.0003	1.0000	1.0000	1.0000	1.0000	0.9999	0.7655	0.9997	0.9996
0.6	1.0935E 00	0.9999	1.0001	1.0000	1.0000	1.0000	1.0000	1.0000	1.0000	0.9998	0.7709	0.9997	0.9996
0.7	1.0837E 00	0.9999	0.9999	0.9999	0.9998	1.0000	1.0000	1.0000	1.0000	0.9998	0.7768	0.9997	0.9996
0.8	1.0768E 00	0.9999	0.9999	0.9997	0.9997	1.0000	1.0000	1.0000	1.0000	0.9998	0.7832	0.9997	0.9996
0.9	1.0717E 00	0.9999	0.9998	0.9997	0.9996	1.0000	1.0000	1.0000	1.0000	0.9998	0.7899	0.9997	0.9996
1.0	1.0678E 00	0.9999	0.9998	0.9996	0.9996	1.0000	1.0000	1.0000	1.0000	0.9998	0.7962	0.9997	0.9996
1.2	1.0624E 00	0.9999	0.9997	0.9996	0.9995	1.0000	1.0000	1.0000	1.0000	0.9998	0.8098	0.9997	0.9996
1.4	1.0588E 00	0.9999	0.9997	0.9996	0.9995	1.0000	1.0000	1.0000	1.0000	0.9998	0.8239	0.9997	0.9996
1.6	1.0564E 00	0.9999	0.9997	0.9996	0.9995	1.0000	1.0000	1.0000	1.0000	0.9998	0.8367	0.9997	0.9996
1.8	1.0547E 00	0.9999	0.9997	0.9996	0.9995	0.9999	1.0000	1.0000	1.0000	0.9998	0.8509	0.9998	0.9996
2.0	1.0535E 00	0.9999	0.9997	0.9996	0.9995	0.9999	1.0000	1.0000	1.0000	0.9998	0.8640	0.9998	0.9996
2.2	1.0525E 00	0.9999	0.9997	0.9996	0.9995	0.9999	1.0000	1.0000	1.0000	0.9998	0.8770	0.9998	0.9996
2.4	1.0517E 00	0.9999	0.9997	0.9996	0.9995	0.9999	1.0000	1.0000	1.0000	0.9998	0.8907	0.9998	0.9996
2.6	1.0511E 00	0.9999	0.9997	0.9996	0.9995	0.9999	1.0000	1.0000	1.0000	0.9998	0.9024	0.9998	0.9996
2.8	1.0507E 00	0.9998	0.9997	0.9996	0.9995	0.9999	1.0000	1.0000	1.0000	0.9998	0.9164	0.9998	0.9996
3.0	1.0503E 00	0.9998	0.9998	0.9996	0.9996	0.9999	1.0000	1.0000	1.0000	0.9998	0.9290	0.9998	0.9996
3.2	1.0499E 00	0.9998	0.9998	0.9996	0.9996	0.9999	1.0000	1.0000	1.0000	0.9998	0.9404	0.9998	0.9996
3.4	1.0497E 00	0.9998	0.9998	0.9997	0.9996	0.9999	1.0000	1.0000	1.0000	0.9998	0.9531	0.9998	0.9996
3.6	1.0494E 00	0.9998	0.9998	0.9997	0.9996	0.9999	1.0000	1.0000	1.0000	0.9998	0.9681	0.9998	0.9996
3.8	1.0492E 00	0.9998	0.9998	0.9997	0.9996	0.9999	1.0000	1.0000	1.0000	0.9998	0.9815	0.9998	0.9997
4.0	1.0490E 00	0.9998	0.9998	0.9997	0.9996	0.9999	1.0000	1.0000	1.0000	0.9998	0.9907	0.9998	0.9997
4.5	1.0487E 00	0.9997	0.9998	0.9997	0.9997	0.9999	1.0000	1.0000	1.0000	0.9999	1.0227	0.9999	0.9997
5.0	1.0484E 00	0.9997	0.9999	0.9998	0.9997	0.9998	1.0000	1.0000	1.0000	0.9999	1.0550	0.9999	0.9997
5.5	1.0482E 00	0.9997	0.9999	0.9998	0.9998	0.9998	1.0000	1.0000	1.0000	0.9999	1.0860	0.9999	0.9997
6.0	1.0480E 00	0.9996	0.9999	0.9998	0.9998	0.9998	1.0000	1.0000	1.0000	0.9999	1.1181	0.9999	0.9997
6.5	1.0479E 00	0.9996	0.9999	0.9999	0.9998	0.9998	1.0000	1.0000	1.0000	0.9999	1.1494	0.9999	0.9997
7.0	1.0477E 00	0.9996	1.0000	0.9999	0.9999	0.9998	1.0000	1.0000	1.0000	0.9999	1.1830	0.9999	0.9997
7.5	1.0476E 00	0.9995	1.0000	0.9999	0.9999	0.9998	1.0000	1.0000	1.0000	0.9999	1.2121	0.9999	0.9997
8.0	1.0475E 00	0.9995	1.0000	1.0000	0.9999	0.9997	0.9999	1.0000	1.0000	1.0000	1.2437	1.0000	0.9997
9.0	1.0474E 00	0.9994	1.0001	1.0000	1.0000	0.9997	0.9999	1.0000	1.0000	1.0000	1.3010	1.0000	0.9997
10.0	1.0472E 00	0.9994	1.0001	1.0001	1.0001	0.9997	0.9999	1.0000	1.0000	1.0000	1.3651	1.0000	0.9997
11.0	1.0471E 00	0.9993	1.0002	1.0002	1.0002	0.9996	0.9999	1.0000	1.0000	1.0000	1.4287	1.0000	0.9997
12.0	1.0470E 00	0.9993	1.0002	1.0002	1.0002	0.9996	0.9999	1.0000	1.0000	1.0001	1.4929	1.0001	0.9997
13.0	1.0469E 00	0.9992	1.0003	1.0003	1.0003	0.9996	0.9999	1.0000	1.0000	1.0001	1.5455	1.0001	0.9997
14.0	1.0468E 00	0.9991	1.0003	1.0003	1.0003	0.9995	0.9999	1.0000	1.0000	1.0001	1.6163	1.0001	0.9997
16.0	1.0466E 00	0.9990	1.0004	1.0005	1.0005	0.9995	0.9999	1.0000	1.0000	1.0001	1.7411	1.0001	0.9997
18.0	1.0464E 00	0.9989	1.0005	1.0006	1.0006	0.9994	0.9999	1.0000	1.0000	1.0002	1.8620	1.0002	0.9997
20.0	1.0462E 00	0.9987	1.0006	1.0007	1.0007	0.9993	0.9999	1.0000	1.0000	1.0002	1.9858	1.0002	0.9997
25.0	1.0458E 00	0.9984	1.0008	1.0010	1.0010	0.9992	0.9998	1.0000	1.0000	1.0003	2.2914	1.0003	0.9997
30.0	1.0454E 00	0.9981	1.0010	1.0012	1.0013	0.9990	0.9998	1.0000	1.0000	1.0005	2.5928	1.0005	0.9998
35.0	1.0450E 00	0.9978	1.0012	1.0015	1.0016	0.9989	0.9998	1.0000	1.0000	1.0006	2.9074	1.0006	0.9998
40.0	1.0446E 00	0.9974	1.0015	1.0018	1.0019	0.9987	0.9997	1.0000	1.0000	1.0007	3.2247	1.0007	0.9998
45.0	1.0443E 00	0.9971	1.0017	1.0020	1.0022	0.9985	0.9997	1.0000	1.0000	1.0008	3.5018	1.0008	0.9998
50.0	1.0439E 00	0.9968	1.0019	1.0023	1.0025	0.9984	0.9997	1.0000	1.0000	1.0009	3.8139	1.0009	0.9998

Instr. see p. 22

ELECTRON Z= 3 A= 6

p	F_0L_0	L_0	λ_2	λ_3	λ_4	μ_1	μ_2	Λ_1	Λ_2	ν_{12}	$\hat{\nu}_{12}$	η_{12}	$\hat{\eta}_{12}$
0.1	1.8536E 00	0.9999	1.0463	1.0582	1.0635	1.0000	1.0000	1.0001	1.0000	1.0111	0.7506	0.9990	0.9988
0.2	1.3969E 00	0.9999	1.0106	1.0131	1.0141	1.0000	1.0000	1.0001	1.0000	1.0022	0.7528	0.9992	0.9989
0.3	1.2632E 00	0.9999	1.0040	1.0049	1.0052	1.0000	1.0000	1.0001	1.0000	1.0006	0.7559	0.9992	0.9990
0.4	1.2010E 00	0.9999	1.0018	1.0021	1.0021	1.0000	1.0000	1.0001	1.0000	1.0000	0.7597	0.9993	0.9990
0.5	1.1658E 00	0.9999	1.0008	1.0008	1.0007	1.0000	1.0000	1.0001	1.0000	0.9998	0.7640	0.9993	0.9990
0.6	1.1436E 00	0.9999	1.0002	1.0001	1.0000	1.0000	1.0000	1.0001	1.0000	0.9997	0.7689	0.9993	0.9991
0.7	1.1285E 00	0.9999	0.9999	0.9997	0.9996	1.0000	1.0000	1.0001	1.0000	0.9996	0.7740	0.9993	0.9991
0.8	1.1178E 00	0.9999	0.9997	0.9995	0.9993	1.0000	1.0000	1.0000	1.0000	0.9996	0.7795	0.9993	0.9991
0.9	1.1099E 00	0.9999	0.9996	0.9993	0.9992	1.0000	1.0000	1.0000	1.0000	0.9995	0.7850	0.9994	0.9991
1.0	1.1039E 00	0.9999	0.9995	0.9992	0.9991	1.0000	1.0000	1.0000	1.0000	0.9995	0.7908	0.9994	0.9991
1.2	1.0956E 00	0.9999	0.9994	0.9991	0.9989	0.9999	1.0000	1.0000	1.0000	0.9995	0.8024	0.9994	0.9991
1.4	1.0902E 00	0.9999	0.9994	0.9991	0.9989	0.9999	1.0000	1.0000	1.0000	0.9995	0.8141	0.9994	0.9991
1.6	1.0865E 00	0.9998	0.9994	0.9990	0.9989	0.9999	1.0000	1.0000	1.0000	0.9995	0.8252	0.9994	0.9992
1.8	1.0838E 00	0.9998	0.9994	0.9990	0.9989	0.9999	1.0000	1.0000	1.0000	0.9995	0.8370	0.9995	0.9992
2.0	1.0819E 00	0.9998	0.9994	0.9991	0.9989	0.9999	1.0000	1.0000	1.0000	0.9995	0.8480	0.9995	0.9992
2.2	1.0804E 00	0.9998	0.9994	0.9991	0.9989	0.9999	1.0000	1.0000	1.0000	0.9995	0.8591	0.9995	0.9992
2.4	1.0792E 00	0.9998	0.9994	0.9991	0.9989	0.9999	1.0000	1.0000	1.0000	0.9996	0.8703	0.9995	0.9992
2.6	1.0783E 00	0.9997	0.9994	0.9991	0.9990	0.9999	1.0000	1.0000	1.0000	0.9996	0.8806	0.9995	0.9992
2.8	1.0775E 00	0.9997	0.9994	0.9992	0.9990	0.9998	1.0000	1.0000	1.0000	0.9996	0.8922	0.9995	0.9992
3.0	1.0769E 00	0.9997	0.9995	0.9992	0.9990	0.9998	1.0000	1.0000	1.0000	0.9996	0.9030	0.9996	0.9992
3.2	1.0764E 00	0.9997	0.9995	0.9992	0.9991	0.9998	1.0000	1.0000	1.0000	0.9996	0.9135	0.9996	0.9992
3.4	1.0759E 00	0.9996	0.9995	0.9992	0.9991	0.9998	1.0000	1.0000	1.0000	0.9996	0.9249	0.9996	0.9992
3.6	1.0756E 00	0.9996	0.9995	0.9993	0.9991	0.9998	1.0000	1.0000	1.0000	0.9996	0.9344	0.9996	0.9992
3.8	1.0752E 00	0.9996	0.9995	0.9993	0.9992	0.9998	1.0000	1.0000	1.0000	0.9996	0.9458	0.9996	0.9992
4.0	1.0750E 00	0.9996	0.9996	0.9993	0.9992	0.9998	1.0000	1.0000	1.0000	0.9996	0.9553	0.9996	0.9992
4.5	1.0744E 00	0.9995	0.9996	0.9994	0.9993	0.9997	0.9999	1.0000	1.0000	0.9997	0.9821	0.9997	0.9993
5.0	1.0739E 00	0.9995	0.9997	0.9995	0.9993	0.9997	0.9999	1.0000	1.0000	0.9997	1.0090	0.9997	0.9993
5.5	1.0736E 00	0.9994	0.9997	0.9995	0.9994	0.9997	0.9999	1.0000	1.0000	0.9997	1.0372	0.9997	0.9993
6.0	1.0733E 00	0.9993	0.9998	0.9996	0.9995	0.9996	0.9999	1.0000	1.0000	0.9998	1.0636	0.9997	0.9993
6.5	1.0730E 00	0.9993	0.9998	0.9997	0.9996	0.9996	0.9999	1.0000	1.0000	0.9998	1.0904	0.9998	0.9993
7.0	1.0728E 00	0.9992	0.9999	0.9997	0.9996	0.9996	0.9999	1.0000	1.0000	0.9998	1.1159	0.9998	0.9993
7.5	1.0726E 00	0.9992	0.9999	0.9998	0.9997	0.9995	0.9999	1.0000	1.0000	0.9998	1.1413	0.9998	0.9993
8.0	1.0724E 00	0.9991	1.0000	0.9999	0.9998	0.9995	0.9999	1.0000	1.0000	0.9999	1.1676	0.9999	0.9993
9.0	1.0721E 00	0.9990	1.0001	1.0000	0.9999	0.9994	0.9999	1.0000	1.0000	0.9999	1.2201	0.9999	0.9993
10.0	1.0718E 00	0.9989	1.0002	1.0001	1.0001	0.9994	0.9999	1.0000	1.0000	1.0000	1.2722	1.0000	0.9993
11.0	1.0716E 00	0.9987	1.0003	1.0002	1.0002	0.9993	0.9999	1.0000	1.0000	1.0000	1.3206	1.0000	0.9994
12.0	1.0714E 00	0.9986	1.0004	1.0004	1.0003	0.9993	0.9999	1.0000	1.0000	1.0001	1.3787	1.0001	0.9994
13.0	1.0712E 00	0.9985	1.0005	1.0005	1.0005	0.9992	0.9998	1.0000	1.0000	1.0001	1.4250	1.0001	0.9994
14.0	1.0710E 00	0.9984	1.0006	1.0006	1.0006	0.9991	0.9998	1.0000	1.0000	1.0001	1.4780	1.0001	0.9994
16.0	1.0706E 00	0.9981	1.0007	1.0008	1.0008	0.9990	0.9998	1.0000	1.0000	1.0002	1.5799	1.0002	0.9994
18.0	1.0702E 00	0.9979	1.0009	1.0010	1.0011	0.9989	0.9998	1.0000	1.0000	1.0003	1.6821	1.0003	0.9994
20.0	1.0699E 00	0.9976	1.0011	1.0013	1.0013	0.9988	0.9998	1.0000	1.0000	1.0004	1.7904	1.0004	0.9994
25.0	1.0691E 00	0.9970	1.0015	1.0018	1.0019	0.9985	0.9997	1.0000	1.0000	1.0006	2.0459	1.0006	0.9995
30.0	1.0683E 00	0.9964	1.0019	1.0023	1.0025	0.9981	0.9996	1.0000	1.0000	1.0008	2.3067	1.0008	0.9995
35.0	1.0676E 00	0.9958	1.0023	1.0028	1.0030	0.9978	0.9996	1.0000	1.0000	1.0010	2.5549	1.0010	0.9995
40.0	1.0668E 00	0.9952	1.0028	1.0033	1.0036	0.9975	0.9995	1.0000	1.0000	1.0012	2.8116	1.0012	0.9995
45.0	1.0661E 00	0.9945	1.0031	1.0038	1.0041	0.9972	0.9994	1.0000	1.0000	1.0014	3.0555	1.0014	0.9995
50.0	1.0654E 00	0.9939	1.0035	1.0043	1.0047	0.9969	0.9994	1.0000	1.0000	1.0016	3.3139	1.0016	0.9995

Instr. see p. 22

ELECTRON Z= 4 A= 8

p	F_0L_0	L_0	λ_2	λ_3	λ_4	μ_1	μ_2	Λ_1	Λ_2	ν_{12}	$\hat{\nu}_{12}$	η_{12}	$\hat{\eta}_{12}$
0.1	2.2066E 00	0.9999	1.0822	1.1041	1.1140	1.0000	1.0000	1.0001	1.0000	1.0196	0.7504	0.9983	0.9979
0.2	1.5503E 00	0.9999	1.0188	1.0233	1.0252	1.0000	1.0000	1.0001	1.0000	1.0039	0.7526	0.9985	0.9981
0.3	1.3620E 00	0.9999	1.0072	1.0087	1.0092	1.0000	1.0000	1.0001	1.0000	1.0010	0.7554	0.9986	0.9982
0.4	1.2754E 00	0.9999	1.0032	1.0036	1.0038	1.0000	1.0000	1.0001	1.0000	1.0000	0.7588	0.9987	0.9982
0.5	1.2267E 00	0.9999	1.0014	1.0014	1.0013	1.0000	1.0000	1.0001	1.0000	0.9996	0.7627	0.9987	0.9983
0.6	1.1961E 00	0.9999	1.0004	1.0002	1.0000	1.0000	1.0000	1.0001	1.0000	0.9994	0.7669	0.9988	0.9983
0.7	1.1754E 00	0.9999	0.9998	0.9995	0.9993	1.0000	1.0000	1.0001	1.0000	0.9993	0.7714	0.9988	0.9983
0.8	1.1608E 00	0.9999	0.9995	0.9991	0.9988	1.0000	1.0000	1.0001	1.0000	0.9992	0.7761	0.9988	0.9984
0.9	1.1499E 00	0.9999	0.9993	0.9988	0.9985	0.9999	1.0000	1.0001	1.0000	0.9992	0.7807	0.9989	0.9984
1.0	1.1417E 00	0.9999	0.9991	0.9986	0.9983	0.9999	1.0000	1.0001	1.0000	0.9991	0.7855	0.9989	0.9984
1.2	1.1303E 00	0.9998	0.9989	0.9984	0.9981	0.9999	1.0000	1.0001	1.0000	0.9991	0.7953	0.9989	0.9984
1.4	1.1229E 00	0.9998	0.9989	0.9983	0.9980	0.9999	1.0000	1.0001	1.0000	0.9991	0.8050	0.9990	0.9985
1.6	1.1178E 00	0.9998	0.9988	0.9983	0.9980	0.9999	1.0000	1.0000	1.0000	0.9991	0.8146	0.9990	0.9985
1.8	1.1142E 00	0.9998	0.9988	0.9983	0.9980	0.9999	1.0000	1.0000	1.0000	0.9991	0.8240	0.9990	0.9985
2.0	1.1115E 00	0.9997	0.9988	0.9983	0.9980	0.9998	1.0000	1.0000	1.0000	0.9991	0.8335	0.9991	0.9985
2.2	1.1095E 00	0.9997	0.9989	0.9983	0.9980	0.9998	1.0000	1.0000	1.0000	0.9992	0.8426	0.9991	0.9986
2.4	1.1079E 00	0.9997	0.9989	0.9984	0.9981	0.9998	1.0000	1.0000	1.0000	0.9992	0.8520	0.9991	0.9986
2.6	1.1066E 00	0.9996	0.9989	0.9984	0.9981	0.9998	1.0000	1.0000	1.0000	0.9992	0.8609	0.9991	0.9986
2.8	1.1055E 00	0.9996	0.9990	0.9985	0.9982	0.9998	1.0000	1.0000	1.0000	0.9992	0.8700	0.9992	0.9986
3.0	1.1047E 00	0.9996	0.9990	0.9985	0.9982	0.9997	0.9999	1.0000	1.0000	0.9992	0.8789	0.9992	0.9986
3.2	1.1039E 00	0.9995	0.9990	0.9985	0.9983	0.9997	0.9999	1.0000	1.0000	0.9993	0.8881	0.9992	0.9986
3.4	1.1033E 00	0.9995	0.9991	0.9986	0.9983	0.9997	0.9999	1.0000	1.0000	0.9993	0.8969	0.9992	0.9986
3.6	1.1028E 00	0.9995	0.9991	0.9986	0.9984	0.9997	0.9999	1.0000	1.0000	0.9993	0.9056	0.9993	0.9987
3.8	1.1023E 00	0.9994	0.9991	0.9987	0.9984	0.9997	0.9999	1.0000	1.0000	0.9993	0.9143	0.9993	0.9987
4.0	1.1019E 00	0.9994	0.9992	0.9987	0.9985	0.9996	0.9999	1.0000	1.0000	0.9993	0.9231	0.9993	0.9987
4.5	1.1011E 00	0.9993	0.9993	0.9989	0.9986	0.9996	0.9999	1.0000	1.0000	0.9994	0.9448	0.9994	0.9987
5.0	1.1005E 00	0.9992	0.9994	0.9990	0.9987	0.9996	0.9999	1.0000	1.0000	0.9994	0.9675	0.9994	0.9987
5.5	1.0999E 00	0.9992	0.9994	0.9991	0.9989	0.9995	0.9999	1.0000	1.0000	0.9995	0.9898	0.9995	0.9987
6.0	1.0995E 00	0.9991	0.9995	0.9992	0.9990	0.9995	0.9999	1.0000	1.0000	0.9995	1.0110	0.9995	0.9988
6.5	1.0991E 00	0.9990	0.9996	0.9993	0.9991	0.9994	0.9999	1.0000	1.0000	0.9996	1.0329	0.9995	0.9988
7.0	1.0988E 00	0.9989	0.9997	0.9994	0.9992	0.9994	0.9999	1.0000	1.0000	0.9996	1.0549	0.9996	0.9988
7.5	1.0985E 00	0.9988	0.9998	0.9995	0.9993	0.9993	0.9999	1.0000	1.0000	0.9996	1.0765	0.9996	0.9988
8.0	1.0983E 00	0.9987	0.9999	0.9996	0.9994	0.9993	0.9999	1.0000	1.0000	0.9997	1.0980	0.9997	0.9988
9.0	1.0978E 00	0.9985	1.0000	0.9998	0.9997	0.9992	0.9998	1.0000	1.0000	0.9998	1.1402	0.9998	0.9988
10.0	1.0974E 00	0.9983	1.0002	1.0000	0.9999	0.9991	0.9998	1.0000	1.0000	0.9998	1.1835	0.9998	0.9989
11.0	1.0970E 00	0.9982	1.0003	1.0002	1.0001	0.9990	0.9998	1.0000	1.0000	0.9999	1.2275	0.9999	0.9989
12.0	1.0967E 00	0.9980	1.0004	1.0004	1.0003	0.9989	0.9998	1.0000	1.0000	1.0000	1.2692	1.0000	0.9989
13.0	1.0963E 00	0.9978	1.0006	1.0005	1.0005	0.9988	0.9998	1.0000	1.0000	1.0001	1.3127	1.0000	0.9989
14.0	1.0960E 00	0.9976	1.0007	1.0007	1.0007	0.9987	0.9997	1.0000	1.0000	1.0001	1.3554	1.0001	0.9989
16.0	1.0955E 00	0.9973	1.0010	1.0011	1.0010	0.9985	0.9997	1.0000	1.0000	1.0003	1.4408	1.0003	0.9990
18.0	1.0949E 00	0.9969	1.0013	1.0014	1.0014	0.9984	0.9997	1.0000	1.0000	1.0004	1.5273	1.0004	0.9990
20.0	1.0944E 00	0.9965	1.0015	1.0017	1.0018	0.9982	0.9996	1.0000	1.0000	1.0005	1.6117	1.0005	0.9990
25.0	1.0931E 00	0.9956	1.0022	1.0025	1.0027	0.9977	0.9995	1.0000	1.0000	1.0008	1.8234	1.0008	0.9990
30.0	1.0919E 00	0.9947	1.0028	1.0033	1.0035	0.9973	0.9995	1.0000	1.0000	1.0012	2.0344	1.0012	0.9991
35.0	1.0908E 00	0.9938	1.0034	1.0041	1.0043	0.9968	0.9994	1.0000	1.0000	1.0015	2.2397	1.0015	0.9991
40.0	1.0897E 00	0.9929	1.0040	1.0048	1.0052	0.9964	0.9993	1.0000	1.0000	1.0018	2.4559	1.0018	0.9992
45.0	1.0886E 00	0.9920	1.0046	1.0056	1.0060	0.9959	0.9992	1.0000	1.0000	1.0021	2.6584	1.0021	0.9992
50.0	1.0875E 00	0.9911	1.0052	1.0063	1.0068	0.9955	0.9991	1.0000	1.0000	1.0024	2.8569	1.0023	0.9992

ELECTRON Z= 5 A= 10

p	F_0L_0	L_0	λ_2	λ_3	λ_4	μ_1	μ_2	Λ_1	Λ_2	ν_{12}	$\hat{\nu}_{12}$	η_{12}	$\hat{\eta}_{12}$
0.1	2.5899E 00	0.9999	1.1283	1.1640	1.1801	1.0000	1.0000	1.0001	1.0000	1.0306	0.7501	0.9973	0.9966
0.2	1.7145E 00	0.9999	1.0294	1.0364	1.0394	1.0000	1.0000	1.0001	1.0000	1.0060	0.7524	0.9976	0.9970
0.3	1.4667E 00	0.9999	1.0112	1.0135	1.0145	1.0000	1.0000	1.0001	1.0000	1.0016	0.7552	0.9978	0.9971
0.4	1.3538E 00	0.9999	1.0050	1.0057	1.0059	1.0000	1.0000	1.0001	1.0000	1.0001	0.7584	0.9979	0.9972
0.5	1.2907E 00	0.9999	1.0021	1.0022	1.0021	1.0000	1.0000	1.0001	1.0000	0.9994	0.7620	0.9980	0.9973
0.6	1.2512E 00	0.9999	1.0006	1.0003	1.0001	1.0000	1.0000	1.0001	1.0000	0.9991	0.7659	0.9981	0.9974
0.7	1.2245E 00	0.9999	0.9998	0.9992	0.9989	1.0000	1.0000	1.0001	1.0000	0.9989	0.7700	0.9981	0.9974
0.8	1.2057E 00	0.9999	0.9992	0.9986	0.9982	0.9999	1.0000	1.0001	1.0000	0.9987	0.7742	0.9982	0.9975
0.9	1.1918E 00	0.9999	0.9989	0.9981	0.9977	0.9999	1.0000	1.0001	1.0000	0.9987	0.7785	0.9982	0.9975
1.0	1.1812E 00	0.9999	0.9986	0.9979	0.9974	0.9999	1.0000	1.0001	1.0000	0.9986	0.7827	0.9983	0.9975
1.2	1.1665E 00	0.9998	0.9984	0.9975	0.9971	0.9999	1.0000	1.0001	1.0000	0.9986	0.7913	0.9983	0.9976
1.4	1.1570E 00	0.9998	0.9982	0.9974	0.9969	0.9999	1.0000	1.0001	1.0000	0.9986	0.7998	0.9984	0.9976
1.6	1.1505E 00	0.9998	0.9982	0.9973	0.9968	0.9998	1.0000	1.0001	1.0000	0.9986	0.8082	0.9984	0.9977
1.8	1.1459E 00	0.9997	0.9982	0.9973	0.9968	0.9998	1.0000	1.0001	1.0000	0.9986	0.8166	0.9985	0.9977
2.0	1.1424E 00	0.9997	0.9982	0.9973	0.9969	0.9998	1.0000	1.0001	1.0000	0.9987	0.8244	0.9985	0.9977
2.2	1.1398E 00	0.9996	0.9982	0.9974	0.9969	0.9998	1.0000	1.0001	1.0000	0.9987	0.8328	0.9986	0.9978
2.4	1.1377E 00	0.9996	0.9983	0.9974	0.9970	0.9997	0.9999	1.0000	1.0000	0.9987	0.8408	0.9986	0.9978
2.6	1.1360E 00	0.9996	0.9983	0.9975	0.9970	0.9997	0.9999	1.0000	1.0000	0.9987	0.8486	0.9987	0.9978
2.8	1.1347E 00	0.9995	0.9983	0.9976	0.9971	0.9997	0.9999	1.0000	1.0000	0.9988	0.8563	0.9987	0.9978
3.0	1.1336E 00	0.9995	0.9984	0.9976	0.9972	0.9997	0.9999	1.0000	1.0000	0.9988	0.8642	0.9987	0.9979
3.2	1.1326E 00	0.9994	0.9984	0.9977	0.9973	0.9996	0.9999	1.0000	1.0000	0.9988	0.8720	0.9988	0.9979
3.4	1.1318E 00	0.9994	0.9985	0.9978	0.9973	0.9996	0.9999	1.0000	1.0000	0.9989	0.8797	0.9988	0.9979
3.6	1.1311E 00	0.9993	0.9985	0.9978	0.9974	0.9996	0.9999	1.0000	1.0000	0.9989	0.8873	0.9988	0.9979
3.8	1.1305E 00	0.9993	0.9986	0.9979	0.9975	0.9995	0.9999	1.0000	1.0000	0.9989	0.8950	0.9989	0.9979
4.0	1.1300E 00	0.9992	0.9987	0.9980	0.9976	0.9995	0.9999	1.0000	1.0000	0.9989	0.9024	0.9989	0.9979
4.5	1.1289E 00	0.9991	0.9988	0.9981	0.9978	0.9995	0.9999	1.0000	1.0000	0.9990	0.9215	0.9990	0.9980
5.0	1.1280E 00	0.9990	0.9989	0.9983	0.9979	0.9994	0.9999	1.0000	1.0000	0.9991	0.9405	0.9990	0.9980
5.5	1.1273E 00	0.9989	0.9990	0.9985	0.9981	0.9993	0.9999	1.0000	1.0000	0.9991	0.9599	0.9991	0.9980
6.0	1.1267E 00	0.9988	0.9992	0.9986	0.9983	0.9993	0.9999	1.0000	1.0000	0.9992	0.9792	0.9992	0.9981
6.5	1.1262E 00	0.9987	0.9993	0.9988	0.9985	0.9992	0.9998	1.0000	1.0000	0.9993	0.9971	0.9992	0.9981
7.0	1.1258E 00	0.9985	0.9994	0.9989	0.9986	0.9991	0.9998	1.0000	1.0000	0.9993	1.0157	0.9993	0.9981
7.5	1.1254E 00	0.9984	0.9995	0.9991	0.9988	0.9991	0.9998	1.0000	1.0000	0.9994	1.0349	0.9994	0.9981
8.0	1.1250E 00	0.9983	0.9996	0.9992	0.9989	0.9990	0.9998	1.0000	1.0000	0.9994	1.0537	0.9994	0.9982
9.0	1.1244E 00	0.9981	0.9998	0.9995	0.9992	0.9989	0.9998	1.0000	1.0000	0.9995	1.0906	0.9995	0.9982
10.0	1.1238E 00	0.9978	1.0001	0.9998	0.9995	0.9988	0.9998	1.0000	1.0000	0.9996	1.1270	0.9996	0.9982
11.0	1.1233E 00	0.9976	1.0003	1.0000	0.9998	0.9987	0.9997	1.0000	1.0000	0.9997	1.1646	0.9997	0.9983
12.0	1.1228E 00	0.9973	1.0005	1.0003	1.0001	0.9985	0.9997	1.0000	1.0000	0.9998	1.2021	0.9998	0.9983
13.0	1.1223E 00	0.9971	1.0007	1.0005	1.0004	0.9984	0.9997	1.0000	1.0000	0.9999	1.2393	0.9999	0.9983
14.0	1.1219E 00	0.9968	1.0009	1.0008	1.0007	0.9983	0.9997	1.0000	1.0000	1.0000	1.2763	1.0000	0.9983
16.0	1.1211E 00	0.9964	1.0012	1.0013	1.0012	0.9980	0.9996	1.0000	1.0000	1.0002	1.3506	1.0002	0.9984
18.0	1.1203E 00	0.9959	1.0016	1.0017	1.0017	0.9978	0.9996	1.0000	1.0000	1.0004	1.4230	1.0004	0.9984
20.0	1.1195E 00	0.9954	1.0020	1.0022	1.0022	0.9975	0.9995	1.0000	1.0000	1.0006	1.4954	1.0006	0.9985
25.0	1.1178E 00	0.9942	1.0029	1.0033	1.0034	0.9969	0.9994	1.0000	1.0000	1.0010	1.6826	1.0010	0.9985
30.0	1.1161E 00	0.9929	1.0037	1.0044	1.0046	0.9963	0.9993	1.0000	1.0000	1.0015	1.8621	1.0015	0.9986
35.0	1.1145E 00	0.9917	1.0046	1.0054	1.0058	0.9957	0.9991	1.0000	1.0000	1.0019	2.0413	1.0019	0.9987
40.0	1.1129E 00	0.9905	1.0054	1.0065	1.0069	0.9951	0.9990	1.0000	1.0000	1.0023	2.2198	1.0023	0.9987
45.0	1.1114E 00	0.9893	1.0062	1.0075	1.0080	0.9945	0.9989	1.0000	1.0000	1.0027	2.3957	1.0027	0.9987
50.0	1.1099E 00	0.9881	1.0070	1.0085	1.0091	0.9939	0.9988	1.0000	1.0000	1.0031	2.5686	1.0031	0.9988

Instr. see p. 22

ELECTRON Z= 6 A= 12

p	F_0L_0	L_0	λ_2	λ_3	λ_4	μ_1	μ_2	Λ_1	Λ_2	ν_{12}	$\hat{\nu}_{12}$	η_{12}	$\hat{\eta}_{12}$
0.1	3.0006E 00	1.0000	1.1844	1.2383	1.2631	1.0000	1.0000	1.0001	1.0000	1.0440	0.7496	0.9960	0.9951
0.2	1.8897E 00	0.9999	1.0422	1.0525	1.0569	1.0000	1.0000	1.0002	1.0000	1.0086	0.7521	0.9966	0.9956
0.3	1.5774E 00	0.9999	1.0161	1.0195	1.0208	1.0000	1.0000	1.0001	1.0000	1.0022	0.7549	0.9968	0.9958
0.4	1.4363E 00	0.9999	1.0071	1.0082	1.0086	1.0000	1.0000	1.0001	1.0000	1.0001	0.7581	0.9970	0.9960
0.5	1.3578E 00	0.9999	1.0031	1.0032	1.0030	1.0000	1.0000	1.0001	1.0000	0.9991	0.7617	0.9971	0.9961
0.6	1.3088E 00	0.9999	1.0009	1.0005	1.0001	1.0000	1.0000	1.0001	1.0000	0.9986	0.7653	0.9972	0.9962
0.7	1.2759E 00	0.9999	0.9997	0.9989	0.9984	0.9999	1.0000	1.0001	1.0000	0.9983	0.7692	0.9973	0.9963
0.8	1.2526E 00	0.9999	0.9989	0.9980	0.9974	0.9999	1.0000	1.0001	1.0000	0.9982	0.7731	0.9974	0.9963
0.9	1.2355E 00	0.9999	0.9984	0.9973	0.9967	0.9999	1.0000	1.0001	1.0000	0.9981	0.7770	0.9974	0.9964
1.0	1.2225E 00	0.9999	0.9980	0.9969	0.9963	0.9999	1.0000	1.0001	1.0000	0.9980	0.7810	0.9975	0.9964
1.2	1.2044E 00	0.9998	0.9976	0.9965	0.9958	0.9999	1.0000	1.0001	1.0000	0.9980	0.7888	0.9976	0.9965
1.4	1.1927E 00	0.9998	0.9975	0.9962	0.9955	0.9998	1.0000	1.0001	1.0000	0.9980	0.7966	0.9977	0.9966
1.6	1.1847E 00	0.9997	0.9974	0.9962	0.9955	0.9998	1.0000	1.0001	1.0000	0.9980	0.8041	0.9977	0.9967
1.8	1.1789E 00	0.9997	0.9974	0.9961	0.9955	0.9998	1.0000	1.0001	1.0000	0.9980	0.8116	0.9978	0.9967
2.0	1.1746E 00	0.9996	0.9974	0.9962	0.9955	0.9997	0.9999	1.0001	1.0000	0.9980	0.8189	0.9979	0.9968
2.2	1.1714E 00	0.9996	0.9974	0.9962	0.9956	0.9997	0.9999	1.0001	1.0000	0.9981	0.8263	0.9979	0.9968
2.4	1.1688E 00	0.9995	0.9975	0.9963	0.9956	0.9997	0.9999	1.0001	1.0000	0.9981	0.8333	0.9980	0.9968
2.6	1.1667E 00	0.9995	0.9975	0.9964	0.9957	0.9996	0.9999	1.0001	1.0000	0.9982	0.8406	0.9981	0.9969
2.8	1.1650E 00	0.9994	0.9976	0.9965	0.9958	0.9996	0.9999	1.0001	1.0000	0.9982	0.8474	0.9981	0.9969
3.0	1.1636E 00	0.9994	0.9977	0.9966	0.9959	0.9996	0.9999	1.0000	1.0000	0.9982	0.8543	0.9982	0.9969
3.2	1.1625E 00	0.9993	0.9977	0.9966	0.9960	0.9995	0.9999	1.0000	1.0000	0.9983	0.8613	0.9982	0.9970
3.4	1.1614E 00	0.9993	0.9978	0.9967	0.9961	0.9995	0.9999	1.0000	1.0000	0.9983	0.8679	0.9983	0.9970
3.6	1.1606E 00	0.9992	0.9979	0.9968	0.9962	0.9995	0.9999	1.0000	1.0000	0.9984	0.8751	0.9983	0.9970
3.8	1.1598E 00	0.9991	0.9979	0.9969	0.9963	0.9994	0.9999	1.0000	1.0000	0.9984	0.8818	0.9983	0.9970
4.0	1.1591E 00	0.9991	0.9980	0.9970	0.9964	0.9994	0.9999	1.0000	1.0000	0.9984	0.8883	0.9984	0.9971
4.5	1.1577E 00	0.9989	0.9982	0.9972	0.9967	0.9993	0.9999	1.0000	1.0000	0.9985	0.9056	0.9985	0.9971
5.0	1.1566E 00	0.9988	0.9984	0.9975	0.9969	0.9992	0.9998	1.0000	1.0000	0.9986	0.9224	0.9986	0.9972
5.5	1.1557E 00	0.9986	0.9985	0.9977	0.9972	0.9992	0.9998	1.0000	1.0000	0.9987	0.9391	0.9987	0.9972
6.0	1.1549E 00	0.9985	0.9987	0.9979	0.9974	0.9991	0.9998	1.0000	1.0000	0.9988	0.9562	0.9988	0.9972
6.5	1.1543E 00	0.9983	0.9989	0.9981	0.9977	0.9990	0.9998	1.0000	1.0000	0.9989	0.9731	0.9988	0.9973
7.0	1.1537E 00	0.9982	0.9990	0.9983	0.9979	0.9989	0.9998	1.0000	1.0000	0.9989	0.9893	0.9989	0.9973
7.5	1.1532E 00	0.9980	0.9992	0.9985	0.9981	0.9988	0.9998	1.0000	1.0000	0.9990	1.0057	0.9990	0.9973
8.0	1.1527E 00	0.9979	0.9993	0.9987	0.9983	0.9988	0.9998	1.0000	1.0000	0.9991	1.0223	0.9991	0.9974
9.0	1.1518E 00	0.9976	0.9996	0.9991	0.9987	0.9986	0.9997	1.0000	1.0000	0.9992	1.0559	0.9992	0.9974
10.0	1.1511E 00	0.9973	0.9999	0.9994	0.9991	0.9984	0.9997	1.0000	1.0000	0.9994	1.0886	0.9994	0.9975
11.0	1.1504E 00	0.9969	1.0002	0.9998	0.9995	0.9983	0.9997	1.0000	1.0000	0.9995	1.1213	0.9995	0.9975
12.0	1.1497E 00	0.9966	1.0004	1.0001	0.9999	0.9981	0.9996	1.0000	1.0000	0.9997	1.1544	0.9996	0.9976
13.0	1.1491E 00	0.9963	1.0007	1.0004	1.0002	0.9980	0.9996	1.0000	1.0000	0.9998	1.1871	0.9998	0.9976
14.0	1.1485E 00	0.9960	1.0010	1.0008	1.0006	0.9978	0.9996	1.0000	1.0000	0.9999	1.2202	0.9999	0.9976
16.0	1.1474E 00	0.9954	1.0015	1.0014	1.0013	0.9975	0.9995	1.0000	1.0000	1.0002	1.2862	1.0002	0.9977
18.0	1.1464E 00	0.9948	1.0020	1.0020	1.0020	0.9972	0.9994	1.0000	1.0000	1.0004	1.3519	1.0004	0.9978
20.0	1.1454E 00	0.9941	1.0024	1.0026	1.0026	0.9969	0.9994	1.0000	1.0000	1.0007	1.4160	1.0006	0.9978
25.0	1.1430E 00	0.9926	1.0036	1.0041	1.0042	0.9961	0.9992	1.0000	1.0000	1.0012	1.5796	1.0012	0.9979
30.0	1.1408E 00	0.9910	1.0047	1.0055	1.0058	0.9953	0.9991	1.0000	1.0000	1.0018	1.7407	1.0018	0.9980
35.0	1.1387E 00	0.9895	1.0058	1.0068	1.0073	0.9945	0.9989	1.0000	1.0000	1.0023	1.8993	1.0023	0.9981
40.0	1.1366E 00	0.9879	1.0069	1.0082	1.0087	0.9937	0.9987	1.0000	1.0000	1.0029	2.0562	1.0029	0.9982
45.0	1.1345E 00	0.9864	1.0079	1.0095	1.0102	0.9930	0.9986	1.0000	1.0000	1.0034	2.2121	1.0034	0.9982
50.0	1.1325E 00	0.9849	1.0089	1.0108	1.0116	0.9922	0.9984	1.0000	1.0000	1.0039	2.3615	1.0039	0.9983

ELECTRON Z= 7 A= 14

p	F_0L_0	L_0	λ_2	λ_3	λ_4	μ_1	μ_2	Λ_1	Λ_2	ν_{12}	$\hat{\nu}_{12}$	η_{12}	$\hat{\eta}_{12}$
0.1	3.4358E 00	1.0000	1.2505	1.3279	1.3641	1.0000	1.0000	1.0002	1.0000	1.0598	0.7491	0.9945	0.9932
0.2	2.0758E 00	1.0000	1.0574	1.0715	1.0777	1.0000	1.0000	1.0002	1.0000	1.0117	0.7518	0.9953	0.9940
0.3	1.6944E 00	1.0000	1.0219	1.0266	1.0284	1.0000	1.0000	1.0002	1.0000	1.0030	0.7547	0.9956	0.9943
0.4	1.5231E 00	1.0000	1.0097	1.0112	1.0117	1.0000	1.0000	1.0002	1.0000	1.0001	0.7580	0.9959	0.9946
0.5	1.4283E 00	1.0000	1.0042	1.0043	1.0042	1.0000	1.0000	1.0002	1.0000	0.9988	0.7614	0.9961	0.9947
0.6	1.3693E 00	1.0000	1.0013	1.0007	1.0002	0.9999	1.0000	1.0002	1.0000	0.9981	0.7650	0.9962	0.9948
0.7	1.3297E 00	0.9999	0.9996	0.9986	0.9979	0.9999	1.0000	1.0002	1.0000	0.9977	0.7687	0.9963	0.9950
0.8	1.3017E 00	0.9999	0.9985	0.9973	0.9965	0.9999	1.0000	1.0001	1.0000	0.9975	0.7724	0.9964	0.9950
0.9	1.2812E 00	0.9999	0.9978	0.9964	0.9956	0.9999	1.0000	1.0001	1.0000	0.9974	0.7761	0.9965	0.9951
1.0	1.2656E 00	0.9999	0.9973	0.9959	0.9950	0.9999	1.0000	1.0001	1.0000	0.9973	0.7799	0.9966	0.9952
1.2	1.2439E 00	0.9998	0.9968	0.9952	0.9943	0.9998	1.0000	1.0001	1.0000	0.9972	0.7872	0.9967	0.9953
1.4	1.2299E 00	0.9998	0.9966	0.9949	0.9940	0.9998	1.0000	1.0001	1.0000	0.9972	0.7943	0.9968	0.9954
1.6	1.2203E 00	0.9997	0.9965	0.9948	0.9939	0.9997	0.9999	1.0001	1.0000	0.9973	0.8013	0.9969	0.9955
1.8	1.2134E 00	0.9997	0.9964	0.9948	0.9938	0.9997	0.9999	1.0001	1.0000	0.9973	0.8082	0.9970	0.9955
2.0	1.2082E 00	0.9996	0.9964	0.9948	0.9939	0.9997	0.9999	1.0001	1.0000	0.9973	0.8149	0.9971	0.9956
2.2	1.2043E 00	0.9996	0.9965	0.9949	0.9940	0.9996	0.9999	1.0001	1.0000	0.9974	0.8215	0.9972	0.9957
2.4	1.2012E 00	0.9995	0.9966	0.9950	0.9941	0.9996	0.9999	1.0001	1.0000	0.9974	0.8280	0.9973	0.9957
2.6	1.1987E 00	0.9994	0.9966	0.9951	0.9942	0.9995	0.9999	1.0001	1.0000	0.9975	0.8344	0.9973	0.9958
2.8	1.1966E 00	0.9994	0.9967	0.9952	0.9943	0.9995	0.9999	1.0001	1.0000	0.9975	0.8408	0.9974	0.9958
3.0	1.1949E 00	0.9993	0.9968	0.9953	0.9945	0.9995	0.9999	1.0001	1.0000	0.9976	0.8472	0.9975	0.9958
3.2	1.1935E 00	0.9992	0.9969	0.9954	0.9946	0.9994	0.9999	1.0001	1.0000	0.9976	0.8536	0.9975	0.9959
3.4	1.1923E 00	0.9991	0.9970	0.9955	0.9947	0.9994	0.9999	1.0001	1.0000	0.9977	0.8598	0.9976	0.9959
3.6	1.1912E 00	0.9991	0.9971	0.9957	0.9949	0.9993	0.9999	1.0001	1.0000	0.9977	0.8659	0.9977	0.9959
3.8	1.1902E 00	0.9990	0.9972	0.9958	0.9950	0.9993	0.9999	1.0000	1.0000	0.9978	0.8720	0.9977	0.9960
4.0	1.1894E 00	0.9989	0.9973	0.9959	0.9951	0.9993	0.9999	1.0000	1.0000	0.9978	0.8782	0.9978	0.9960
4.5	1.1877E 00	0.9987	0.9975	0.9962	0.9955	0.9992	0.9998	1.0000	1.0000	0.9980	0.8934	0.9979	0.9961
5.0	1.1863E 00	0.9986	0.9977	0.9965	0.9958	0.9991	0.9998	1.0000	1.0000	0.9981	0.9088	0.9980	0.9961
5.5	1.1852E 00	0.9984	0.9979	0.9968	0.9961	0.9990	0.9998	1.0000	1.0000	0.9982	0.9239	0.9981	0.9962
6.0	1.1842E 00	0.9982	0.9981	0.9970	0.9964	0.9989	0.9998	1.0000	1.0000	0.9983	0.9390	0.9983	0.9963
6.5	1.1834E 00	0.9980	0.9984	0.9973	0.9967	0.9988	0.9998	1.0000	1.0000	0.9984	0.9548	0.9984	0.9963
7.0	1.1826E 00	0.9978	0.9986	0.9976	0.9970	0.9987	0.9997	1.0000	1.0000	0.9985	0.9692	0.9985	0.9964
7.5	1.1820E 00	0.9976	0.9987	0.9978	0.9972	0.9986	0.9997	1.0000	1.0000	0.9986	0.9842	0.9986	0.9964
8.0	1.1813E 00	0.9974	0.9989	0.9981	0.9975	0.9985	0.9997	1.0000	1.0000	0.9987	0.9995	0.9987	0.9964
9.0	1.1802E 00	0.9971	0.9993	0.9985	0.9980	0.9983	0.9997	1.0000	1.0000	0.9989	1.0294	0.9989	0.9965
10.0	1.1793E 00	0.9967	0.9997	0.9990	0.9986	0.9981	0.9996	1.0000	1.0000	0.9991	1.0590	0.9991	0.9966
11.0	1.1784E 00	0.9963	1.0000	0.9994	0.9990	0.9979	0.9996	1.0000	1.0000	0.9992	1.0887	0.9992	0.9966
12.0	1.1775E 00	0.9959	1.0004	0.9999	0.9995	0.9977	0.9995	1.0000	1.0000	0.9994	1.1188	0.9994	0.9967
13.0	1.1767E 00	0.9955	1.0007	1.0003	1.0000	0.9975	0.9995	1.0000	1.0000	0.9996	1.1488	0.9996	0.9968
14.0	1.1760E 00	0.9951	1.0010	1.0007	1.0004	0.9973	0.9995	1.0000	1.0000	0.9997	1.1781	0.9997	0.9968
16.0	1.1746E 00	0.9944	1.0017	1.0015	1.0013	0.9969	0.9994	1.0000	1.0000	1.0001	1.2376	1.0001	0.9969
18.0	1.1732E 00	0.9936	1.0023	1.0023	1.0022	0.9965	0.9993	1.0000	1.0000	1.0004	1.2965	1.0004	0.9970
20.0	1.1719E 00	0.9929	1.0029	1.0031	1.0030	0.9961	0.9992	1.0000	1.0000	1.0007	1.3552	1.0007	0.9970
25.0	1.1689E 00	0.9909	1.0043	1.0049	1.0050	0.9952	0.9990	1.0000	1.0000	1.0014	1.5026	1.0014	0.9972
30.0	1.1661E 00	0.9890	1.0057	1.0066	1.0069	0.9942	0.9988	1.0000	1.0000	1.0021	1.6483	1.0021	0.9973
35.0	1.1633E 00	0.9871	1.0071	1.0084	1.0088	0.9933	0.9986	1.0000	1.0000	1.0028	1.7908	1.0028	0.9974
40.0	1.1607E 00	0.9852	1.0084	1.0100	1.0106	0.9923	0.9985	1.0000	1.0000	1.0035	1.9317	1.0035	0.9975
45.0	1.1581E 00	0.9833	1.0097	1.0117	1.0125	0.9914	0.9983	1.0000	1.0000	1.0041	2.0702	1.0041	0.9976
50.0	1.1556E 00	0.9815	1.0110	1.0133	1.0142	0.9905	0.9980	1.0000	1.0000	1.0048	2.2049	1.0047	0.9977

 Instr. see p. 22

ELECTRON Z= 8 A= 16

p	F_0L_0	L_0	λ_2	λ_3	λ_4	μ_1	μ_2	Λ_1	Λ_2	ν_{12}	$\hat{\nu}_{12}$	η_{12}	$\hat{\eta}_{12}$
0.1	3.8933E 00	1.0001	1.3265	1.4336	1.4846	1.0000	1.0000	1.0002	1.0000	1.0780	0.7485	0.9927	0.9910
0.2	2.2729E 00	1.0000	1.0748	1.0936	1.1018	1.0000	1.0000	1.0002	1.0000	1.0153	0.7514	0.9938	0.9921
0.3	1.8179E 00	1.0000	1.0286	1.0347	1.0371	1.0000	1.0000	1.0002	1.0000	1.0039	0.7545	0.9943	0.9926
0.4	1.6144E 00	1.0000	1.0127	1.0147	1.0153	1.0000	1.0000	1.0002	1.0000	1.0001	0.7578	0.9946	0.9929
0.5	1.5023E 00	1.0000	1.0055	1.0057	1.0055	0.9999	1.0000	1.0002	1.0000	0.9984	0.7612	0.9948	0.9931
0.6	1.4326E 00	1.0000	1.0017	1.0009	1.0003	0.9999	1.0000	1.0002	1.0000	0.9975	0.7648	0.9950	0.9933
0.7	1.3860E 00	1.0000	0.9994	0.9982	0.9973	0.9999	1.0000	1.0002	1.0000	0.9970	0.7684	0.9952	0.9934
0.8	1.3531E 00	1.0000	0.9981	0.9965	0.9955	0.9999	1.0000	1.0002	1.0000	0.9968	0.7720	0.9953	0.9935
0.9	1.3289E 00	0.9999	0.9972	0.9954	0.9943	0.9999	1.0000	1.0002	1.0000	0.9966	0.7756	0.9954	0.9936
1.0	1.3106E 00	0.9999	0.9966	0.9946	0.9935	0.9998	1.0000	1.0002	1.0000	0.9965	0.7791	0.9955	0.9937
1.2	1.2852E 00	0.9999	0.9959	0.9938	0.9926	0.9998	1.0000	1.0001	1.0000	0.9964	0.7861	0.9957	0.9939
1.4	1.2687E 00	0.9998	0.9955	0.9934	0.9922	0.9997	0.9999	1.0001	1.0000	0.9964	0.7929	0.9959	0.9940
1.6	1.2574E 00	0.9997	0.9954	0.9932	0.9920	0.9997	0.9999	1.0001	1.0000	0.9964	0.7994	0.9960	0.9941
1.8	1.2493E 00	0.9997	0.9953	0.9932	0.9920	0.9996	0.9999	1.0001	1.0000	0.9964	0.8058	0.9961	0.9942
2.0	1.2432E 00	0.9996	0.9954	0.9932	0.9920	0.9996	0.9999	1.0001	1.0000	0.9965	0.8121	0.9962	0.9943
2.2	1.2386E 00	0.9995	0.9954	0.9933	0.9921	0.9995	0.9999	1.0001	1.0000	0.9966	0.8182	0.9963	0.9943
2.4	1.2349E 00	0.9995	0.9955	0.9934	0.9923	0.9995	0.9999	1.0001	1.0000	0.9966	0.8242	0.9964	0.9944
2.6	1.2320E 00	0.9994	0.9956	0.9936	0.9924	0.9994	0.9999	1.0001	1.0000	0.9967	0.8302	0.9965	0.9945
2.8	1.2295E 00	0.9993	0.9957	0.9937	0.9926	0.9994	0.9999	1.0001	1.0000	0.9968	0.8361	0.9966	0.9945
3.0	1.2275E 00	0.9992	0.9958	0.9939	0.9928	0.9993	0.9999	1.0001	1.0000	0.9968	0.8418	0.9967	0.9946
3.2	1.2258E 00	0.9991	0.9959	0.9940	0.9929	0.9993	0.9999	1.0001	1.0000	0.9969	0.8476	0.9968	0.9946
3.4	1.2243E 00	0.9990	0.9960	0.9942	0.9931	0.9992	0.9999	1.0001	1.0000	0.9970	0.8533	0.9969	0.9947
3.6	1.2230E 00	0.9990	0.9962	0.9943	0.9933	0.9992	0.9998	1.0001	1.0000	0.9970	0.8591	0.9969	0.9947
3.8	1.2219E 00	0.9989	0.9963	0.9945	0.9934	0.9992	0.9998	1.0001	1.0000	0.9971	0.8649	0.9970	0.9948
4.0	1.2209E 00	0.9988	0.9964	0.9946	0.9936	0.9991	0.9998	1.0001	1.0000	0.9972	0.8705	0.9971	0.9948
4.5	1.2188E 00	0.9986	0.9967	0.9950	0.9940	0.9990	0.9998	1.0001	1.0000	0.9973	0.8847	0.9972	0.9949
5.0	1.2171E 00	0.9983	0.9970	0.9953	0.9944	0.9989	0.9998	1.0000	1.0000	0.9975	0.8985	0.9974	0.9950
5.5	1.2157E 00	0.9981	0.9972	0.9957	0.9948	0.9987	0.9998	1.0000	1.0000	0.9976	0.9125	0.9975	0.9951
6.0	1.2146E 00	0.9979	0.9975	0.9960	0.9952	0.9986	0.9997	1.0000	1.0000	0.9977	0.9264	0.9977	0.9951
6.5	1.2135E 00	0.9977	0.9977	0.9964	0.9956	0.9985	0.9997	1.0000	1.0000	0.9979	0.9402	0.9978	0.9952
7.0	1.2126E 00	0.9974	0.9980	0.9967	0.9959	0.9984	0.9997	1.0000	1.0000	0.9980	0.9543	0.9979	0.9953
7.5	1.2118E 00	0.9972	0.9982	0.9970	0.9963	0.9983	0.9997	1.0000	1.0000	0.9981	0.9678	0.9981	0.9953
8.0	1.2110E 00	0.9970	0.9985	0.9973	0.9966	0.9982	0.9996	1.0000	1.0000	0.9982	0.9815	0.9982	0.9954
9.0	1.2096E 00	0.9965	0.9989	0.9979	0.9972	0.9979	0.9996	1.0000	1.0000	0.9985	1.0091	0.9984	0.9955
10.0	1.2084E 00	0.9961	0.9994	0.9985	0.9979	0.9977	0.9995	1.0000	1.0000	0.9987	1.0365	0.9987	0.9956
11.0	1.2073E 00	0.9956	0.9998	0.9990	0.9985	0.9975	0.9995	1.0000	1.0000	0.9989	1.0637	0.9989	0.9956
12.0	1.2062E 00	0.9952	1.0003	0.9996	0.9991	0.9972	0.9995	1.0000	1.0000	0.9991	1.0914	0.9991	0.9957
13.0	1.2052E 00	0.9947	1.0007	1.0001	0.9996	0.9970	0.9994	1.0000	1.0000	0.9993	1.1186	0.9993	0.9958
14.0	1.2043E 00	0.9943	1.0011	1.0006	1.0002	0.9968	0.9994	1.0000	1.0000	0.9995	1.1459	0.9995	0.9959
16.0	1.2025E 00	0.9933	1.0018	1.0016	1.0013	0.9963	0.9993	1.0000	1.0000	0.9999	1.2002	0.9999	0.9960
18.0	1.2008E 00	0.9924	1.0026	1.0025	1.0023	0.9958	0.9992	1.0000	1.0000	1.0003	1.2548	1.0003	0.9961
20.0	1.1992E 00	0.9915	1.0033	1.0035	1.0034	0.9954	0.9991	1.0000	1.0000	1.0007	1.3092	1.0006	0.9962
25.0	1.1955E 00	0.9892	1.0051	1.0057	1.0058	0.9942	0.9989	1.0000	1.0000	1.0015	1.4430	1.0015	0.9964
30.0	1.1919E 00	0.9869	1.0068	1.0078	1.0082	0.9931	0.9986	1.0000	1.0000	1.0024	1.5752	1.0024	0.9965
35.0	1.1885E 00	0.9847	1.0085	1.0099	1.0104	0.9919	0.9984	1.0000	1.0000	1.0032	1.7060	1.0032	0.9967
40.0	1.1852E 00	0.9824	1.0101	1.0120	1.0127	0.9908	0.9981	1.0000	1.0000	1.0040	1.8348	1.0040	0.9968
45.0	1.1820E 00	0.9802	1.0117	1.0140	1.0149	0.9897	0.9979	1.0000	1.0000	1.0048	1.9597	1.0048	0.9969
50.0	1.1789E 00	0.9779	1.0132	1.0159	1.0170	0.9886	0.9977	1.0000	1.0000	1.0056	2.0803	1.0056	0.9970

Instr. see p. 22

ELECTRON Z= 9 A= 19

p	F_0L_0	L_0	λ_2	λ_3	λ_4	μ_1	μ_2	Λ_1	Λ_2	ν_{12}	$\hat{\nu}_{12}$	η_{12}	$\hat{\eta}_{12}$
0.1	4.3706E 00	1.0001	1.4122	1.5564	1.6265	1.0000	1.0000	1.0003	1.0000	1.0985	0.7479	0.9907	0.9885
0.2	2.4809E 00	1.0001	1.0945	1.1188	1.1294	1.0000	1.0000	1.0003	1.0001	1.0193	0.7510	0.9921	0.9900
0.3	1.9478E 00	1.0001	1.0362	1.0440	1.0471	1.0000	1.0000	1.0003	1.0001	1.0050	0.7543	0.9928	0.9906
0.4	1.7103E 00	1.0001	1.0161	1.0187	1.0194	1.0000	1.0000	1.0003	1.0000	1.0001	0.7577	0.9932	0.9910
0.5	1.5797E 00	1.0001	1.0070	1.0073	1.0070	0.9999	1.0000	1.0002	1.0000	0.9980	0.7612	0.9935	0.9913
0.6	1.4989E 00	1.0001	1.0022	1.0013	1.0005	0.9999	1.0000	1.0002	1.0000	0.9969	0.7648	0.9937	0.9915
0.7	1.4448E 00	1.0000	0.9994	0.9978	0.9967	0.9999	1.0000	1.0002	1.0000	0.9963	0.7684	0.9939	0.9917
0.8	1.4067E 00	1.0000	0.9976	0.9956	0.9944	0.9999	1.0000	1.0002	1.0000	0.9959	0.7720	0.9941	0.9918
0.9	1.3787E 00	1.0000	0.9965	0.9942	0.9929	0.9998	1.0000	1.0002	1.0000	0.9957	0.7755	0.9942	0.9920
1.0	1.3576E 00	1.0000	0.9957	0.9933	0.9919	0.9998	1.0000	1.0002	1.0000	0.9956	0.7790	0.9944	0.9921
1.2	1.3282E 00	0.9999	0.9948	0.9922	0.9908	0.9997	0.9999	1.0002	1.0000	0.9955	0.7858	0.9946	0.9923
1.4	1.3091E 00	0.9998	0.9944	0.9917	0.9902	0.9997	0.9999	1.0002	1.0000	0.9954	0.7924	0.9948	0.9924
1.6	1.2960E 00	0.9998	0.9942	0.9915	0.9900	0.9996	0.9999	1.0001	1.0000	0.9955	0.7987	0.9950	0.9926
1.8	1.2866E 00	0.9997	0.9942	0.9915	0.9900	0.9996	0.9999	1.0001	1.0000	0.9955	0.8048	0.9951	0.9927
2.0	1.2796E 00	0.9996	0.9942	0.9916	0.9901	0.9995	0.9999	1.0001	1.0000	0.9956	0.8108	0.9953	0.9928
2.2	1.2742E 00	0.9995	0.9943	0.9917	0.9902	0.9994	0.9999	1.0001	1.0000	0.9957	0.8167	0.9954	0.9929
2.4	1.2699E 00	0.9994	0.9944	0.9918	0.9903	0.9994	0.9999	1.0001	1.0000	0.9958	0.8224	0.9955	0.9930
2.6	1.2665E 00	0.9993	0.9945	0.9920	0.9905	0.9993	0.9999	1.0001	1.0000	0.9958	0.8280	0.9956	0.9930
2.8	1.2636E 00	0.9992	0.9946	0.9921	0.9907	0.9993	0.9999	1.0001	1.0000	0.9959	0.8336	0.9957	0.9931
3.0	1.2613E 00	0.9991	0.9948	0.9923	0.9909	0.9992	0.9998	1.0001	1.0000	0.9960	0.8392	0.9958	0.9932
3.2	1.2592E 00	0.9990	0.9949	0.9925	0.9911	0.9992	0.9998	1.0001	1.0000	0.9961	0.8445	0.9959	0.9933
3.4	1.2575E 00	0.9989	0.9950	0.9927	0.9913	0.9991	0.9998	1.0001	1.0000	0.9962	0.8501	0.9960	0.9933
3.6	1.2560E 00	0.9988	0.9952	0.9929	0.9916	0.9990	0.9998	1.0001	1.0000	0.9963	0.8554	0.9961	0.9934
3.8	1.2546E 00	0.9987	0.9953	0.9931	0.9918	0.9990	0.9998	1.0001	1.0000	0.9963	0.8609	0.9962	0.9934
4.0	1.2534E 00	0.9986	0.9955	0.9932	0.9920	0.9989	0.9998	1.0001	1.0000	0.9964	0.8663	0.9963	0.9935
4.5	1.2510E 00	0.9983	0.9958	0.9937	0.9925	0.9988	0.9998	1.0001	1.0000	0.9966	0.8795	0.9965	0.9936
5.0	1.2490E 00	0.9981	0.9962	0.9942	0.9930	0.9986	0.9997	1.0001	1.0000	0.9968	0.8928	0.9967	0.9937
5.5	1.2473E 00	0.9978	0.9965	0.9946	0.9935	0.9985	0.9997	1.0000	1.0000	0.9969	0.9061	0.9969	0.9938
6.0	1.2459E 00	0.9975	0.9968	0.9950	0.9939	0.9984	0.9997	1.0000	1.0000	0.9971	0.9191	0.9971	0.9939
6.5	1.2446E 00	0.9973	0.9971	0.9954	0.9944	0.9982	0.9996	1.0000	1.0000	0.9973	0.9324	0.9972	0.9940
7.0	1.2435E 00	0.9970	0.9975	0.9958	0.9948	0.9981	0.9996	1.0000	1.0000	0.9974	0.9454	0.9974	0.9941
7.5	1.2425E 00	0.9967	0.9978	0.9962	0.9952	0.9979	0.9996	1.0000	1.0000	0.9976	0.9584	0.9975	0.9941
8.0	1.2415E 00	0.9965	0.9980	0.9966	0.9957	0.9978	0.9996	1.0000	1.0000	0.9977	0.9714	0.9977	0.9942
9.0	1.2398E 00	0.9959	0.9986	0.9973	0.9965	0.9975	0.9995	1.0000	1.0000	0.9980	0.9975	0.9980	0.9943
10.0	1.2383E 00	0.9954	0.9992	0.9980	0.9972	0.9972	0.9995	1.0000	1.0000	0.9983	1.0233	0.9983	0.9944
11.0	1.2369E 00	0.9948	0.9997	0.9987	0.9980	0.9970	0.9994	1.0000	1.0000	0.9986	1.0492	0.9985	0.9945
12.0	1.2356E 00	0.9943	1.0002	0.9993	0.9987	0.9967	0.9993	1.0000	1.0000	0.9988	1.0749	0.9988	0.9946
13.0	1.2344E 00	0.9937	1.0007	0.9999	0.9994	0.9964	0.9993	1.0000	1.0000	0.9991	1.1008	0.9990	0.9947
14.0	1.2332E 00	0.9932	1.0012	1.0006	1.0001	0.9961	0.9992	1.0000	1.0000	0.9993	1.1266	0.9993	0.9948
16.0	1.2310E 00	0.9921	1.0021	1.0018	1.0014	0.9956	0.9991	1.0000	1.0000	0.9998	1.1782	0.9998	0.9949
18.0	1.2290E 00	0.9910	1.0031	1.0029	1.0027	0.9950	0.9990	1.0000	1.0000	1.0002	1.2293	1.0002	0.9951
20.0	1.2270E 00	0.9899	1.0040	1.0041	1.0039	0.9945	0.9989	1.0000	1.0000	1.0007	1.2802	1.0007	0.9952
25.0	1.2223E 00	0.9872	1.0061	1.0068	1.0069	0.9931	0.9986	1.0000	1.0000	1.0018	1.4067	1.0018	0.9954
30.0	1.2179E 00	0.9845	1.0082	1.0094	1.0097	0.9917	0.9983	1.0000	1.0000	1.0028	1.5305	1.0028	0.9957
35.0	1.2137E 00	0.9818	1.0102	1.0119	1.0125	0.9904	0.9981	1.0000	1.0000	1.0038	1.6526	1.0038	0.9959
40.0	1.2097E 00	0.9791	1.0121	1.0143	1.0152	0.9890	0.9978	1.0000	1.0000	1.0048	1.7726	1.0048	0.9960
45.0	1.2058E 00	0.9764	1.0140	1.0167	1.0178	0.9877	0.9975	1.0000	1.0000	1.0057	1.8897	1.0057	0.9962
50.0	1.2019E 00	0.9738	1.0159	1.0191	1.0204	0.9864	0.9972	1.0000	1.0000	1.0067	2.0032	1.0067	0.9963

Instr. see p. 22

ELECTRON Z= 10 A= 20

p	F_0L_0	L_0	λ_2	λ_3	λ_4	μ_1	μ_2	Λ_1	Λ_2	ν_{12}	$\hat{\nu}_{12}$	η_{12}	$\hat{\eta}_{12}$
C.1	4.8675E OC	1.CCC2	1.5073	1.6971	1.7915	1.0000	1.0000	1.0003	1.0001	1.1212	0.7470	0.9883	0.9857
0.2	2.7005E 00	1.0002	1.1163	1.1468	1.1602	1.0000	1.0000	1.0003	1.0001	1.0237	0.7504	0.9902	0.9875
C.3	2.0851E OC	1.CCC2	1.0446	1.0543	1.0582	1.0000	1.0000	1.0003	1.0001	1.0060	0.7539	0.9910	0.9883
0.4	1.8114E 00	1.0002	1.0198	1.0230	1.0239	1.0000	1.0000	1.0003	1.0001	1.0001	0.7574	0.9915	0.9889
0.5	1.6613E 00	1.CCC2	1.0086	1.0089	1.0086	0.9999	1.0000	1.0003	1.0001	0.9975	0.7610	0.9919	0.9892
C.6	1.5685E OC	1.CC02	1.0026	1.0015	1.0006	0.9999	1.0000	1.0003	1.0001	0.9961	0.7645	0.9922	0.9895
0.7	1.5066E OC	1.CCC1	0.9992	0.9973	0.9960	0.9999	1.0000	1.0003	1.0001	0.9954	0.7681	0.9925	0.9897
C.8	1.4631E OC	1.CCC1	0.9970	0.9946	0.9931	0.9998	1.0000	1.0002	1.0000	0.9949	0.7716	0.9927	0.9899
C.9	1.4311E 00	1.0001	0.9956	C.9929	0.9912	0.9998	1.0000	1.0002	1.0000	0.9947	0.7751	0.9928	0.9901
1.0	1.4069E OC	1.0C01	0.9947	0.9917	0.9900	0.9998	1.0000	1.0002	1.0000	0.9945	0.7784	0.9930	0.9902
1.2	1.3733E OC	1.CCC0	0.9936	0.9904	0.9886	0.9997	0.9999	1.0002	1.0000	0.9944	0.7849	0.9933	0.9904
1.4	1.3516E OC	0.9999	0.9931	0.9898	0.9879	0.9996	0.9999	1.0002	1.0000	0.9943	0.7911	0.9935	0.9906
1.6	1.3366E OC	0.9998	0.9929	0.9895	0.9877	0.9996	0.9999	1.0002	1.0000	0.9944	0.7971	0.9937	0.9908
1.8	1.3258E OC	0.9997	0.9928	0.9895	0.9876	0.9995	0.9999	1.0002	1.0000	0.9944	0.8029	0.9939	0.9910
2.0	1.3178E OC	0.9996	0.9928	0.9895	0.9877	0.9994	0.9999	1.C001	1.0000	0.9945	0.8085	0.9941	0.9911
2.2	1.3116E OC	0.9995	0.9929	0.9897	0.9878	0.9994	0.9999	1.0001	1.0000	0.9946	0.8138	0.9943	0.9912
2.4	1.3067E 00	0.9994	0.9930	0.9898	0.9880	0.9993	0.9999	1.0001	1.0000	0.9947	0.8192	0.9944	0.9913
2.6	1.3027E 00	0.9993	0.9932	0.9900	0.9883	0.9992	0.9999	1.0001	1.0000	0.9948	0.8245	0.9946	0.9914
2.8	1.2994E 00	0.9992	0.9933	0.9902	0.9885	0.9992	0.9998	1.0001	1.0000	0.9949	0.8296	0.9947	0.9915
3.0	1.2967E 00	0.9991	0.9935	0.9904	0.9887	0.9991	0.9998	1.0001	1.0000	0.9950	0.8347	0.9948	0.9916
3.2	1.2943E 00	0.9990	0.9936	0.9907	0.9890	0.9990	0.9998	1.0001	1.0000	0.9951	0.8398	0.9949	0.9917
3.4	1.2923E 00	0.9989	0.9938	0.9909	0.9892	0.9990	0.9998	1.0001	1.0000	0.9952	0.8449	0.9951	0.9918
3.6	1.29C5E 00	0.9987	0.9940	0.9911	0.9895	0.9989	0.9998	1.0001	1.0000	0.9953	0.8499	0.9952	0.9918
3.8	1.2890E OC	0.9986	0.9942	0.9913	0.9897	0.9989	0.9998	1.0001	1.0000	0.9954	0.8548	0.9953	0.9919
4.0	1.2876E OC	0.9985	0.9943	0.9916	0.9900	0.9988	0.9998	1.0001	1.0000	0.9955	0.8598	0.9954	0.9920
4.5	1.2847E OC	0.9982	0.9947	0.9921	0.9906	0.9986	0.9997	1.0001	1.0000	0.9957	0.872C	0.9956	0.9921
5.0	1.2823E 00	0.9979	0.9951	0.9926	0.9912	0.9985	0.9997	1.0001	1.0000	0.9959	0.8843	0.9958	0.9922
5.5	1.2804E 00	0.9976	0.9955	0.9931	0.9918	0.9983	0.9997	1.0001	1.0000	0.9961	0.8965	0.9961	0.9924
6.0	1.2787E 00	0.9973	0.9959	0.9936	0.9923	0.9981	0.9996	1.0001	1.0000	0.9963	0.9086	0.9963	0.9925
6.5	1.2772E OC	0.9970	0.9963	0.9941	0.9928	0.9980	0.9996	1.0000	1.0000	0.9965	0.9207	0.9965	0.9926
7.0	1.2759E OC	0.9967	0.9967	0.9946	0.9934	0.9978	0.9996	1.0000	1.0000	0.9967	0.9327	0.9967	0.9927
7.5	1.2747E OC	0.9964	0.9970	0.9950	0.9939	0.9977	0.9995	1.0000	1.0000	0.9969	0.9448	0.9968	0.9928
8.0	1.2735E 00	0.9961	0.9974	0.9955	0.9943	0.9975	0.9995	1.0000	1.0000	0.9971	0.9566	0.9970	0.9928
9.0	1.2715E 00	0.9955	0.9980	0.9963	0.9953	0.9972	0.9994	1.0000	1.0000	0.9974	0.9805	0.9974	0.9930
1C.0	1.2697E 00	0.9949	0.9987	0.9972	0.9962	0.9969	0.9994	1.0000	1.0000	0.9977	1.0045	0.9977	0.9931
11.0	1.2680E 00	0.9942	0.9993	0.9979	0.9971	0.9966	0.9993	1.0000	1.0000	0.9980	1.0283	0.9980	0.9933
12.0	1.2665E 00	0.9936	0.9999	0.9987	0.9979	0.9962	0.9993	1.0000	1.0000	0.9983	1.0521	0.9983	0.9934
13.0	1.2650E OC	0.9930	1.0005	0.9995	0.9987	0.9959	0.9992	1.0000	1.0000	0.9986	1.0759	0.9986	0.9935
14.0	1.2636E 00	0.9924	1.0011	1.0002	0.9995	0.9956	0.9991	1.0000	1.0000	0.9989	1.0993	0.9989	0.9936
16.0	1.2610E OC	0.9912	1.0022	1.0016	1.0011	0.9950	0.9990	1.0000	1.0000	0.9995	1.1464	0.9995	0.9938
18.0	1.2585E 00	0.9899	1.0032	1.0029	1.0025	0.9944	0.9989	1.0000	1.0000	1.0000	1.1932	1.0000	0.9939
20.0	1.2562E 00	0.9887	1.0043	1.0042	1.0040	0.9937	0.9988	1.0000	1.0000	1.0005	1.2406	1.0005	0.9941
25.0	1.2506E 00	0.9856	1.0068	1.0074	1.0074	0.9922	0.9984	1.0000	1.0000	1.0018	1.3557	1.0018	0.9944
30.0	1.2455E 00	0.9825	1.0091	1.0104	1.0107	0.9906	0.9981	1.0000	1.0000	1.0030	1.4692	1.0030	0.9947
35.0	1.2405E 00	0.9795	1.0114	1.0133	1.0139	0.9891	0.9978	1.0000	1.0000	1.0041	1.5818	1.0041	0.9949
40.0	1.2358E OC	0.9764	1.0137	1.0161	1.0170	0.9876	0.9975	1.0000	1.0000	1.0052	1.6914	1.0052	0.9951
45.0	1.2312E OC	0.9734	1.0159	1.0189	1.0200	0.9861	0.9972	1.0000	1.0000	1.0063	1.7976	1.0063	0.9953
50.0	1.2267E OC	0.9704	1.0180	1.0216	1.0230	0.9846	0.9968	1.0000	1.0000	1.0074	1.9000	1.0074	0.9955

ELECTRON Z= 11 A= 23

p	F_0L_0	L_0	λ_2	λ_3	λ_4	μ_1	μ_2	Λ_1	Λ_2	ν_{12}	$\hat{\nu}_{12}$	η_{12}	$\hat{\eta}_{12}$
0.1	5.3813E 00	1.0003	1.6119	1.8573	1.9822	1.0000	1.0000	1.0003	1.0001	1.1464	0.7462	0.9856	0.9825
0.2	2.9308E 00	1.0003	1.1404	1.1782	1.1948	1.0000	1.0000	1.0004	1.0001	1.0286	0.7498	0.9881	0.9849
0.3	2.2290E 00	1.0003	1.0538	1.0658	1.0705	1.0000	1.0000	1.0003	1.0001	1.0073	0.7535	0.9891	0.9859
0.4	1.9171E 00	1.0003	1.0240	1.0279	1.0291	0.9999	1.0000	1.0003	1.0001	1.0001	0.7572	0.9898	0.9865
0.5	1.7465E 00	1.0003	1.0104	1.0109	1.0106	0.9999	1.0000	1.0003	1.0001	0.9969	0.7609	0.9902	0.9870
0.6	1.6412E 00	1.0002	1.0033	1.0020	1.0009	0.9999	1.0000	1.0003	1.0001	0.9953	0.7646	0.9906	0.9873
0.7	1.5711E 00	1.0002	0.9991	0.9968	0.9953	0.9999	1.0000	1.0003	1.0001	0.9944	0.7682	0.9909	0.9876
0.8	1.5217E 00	1.0002	0.9965	0.9936	0.9918	0.9998	1.0000	1.0003	1.0001	0.9939	0.7717	0.9912	0.9878
0.9	1.4856E 00	1.0002	0.9948	0.9915	0.9896	0.9998	1.0000	1.0003	1.0001	0.9936	0.7752	0.9914	0.9880
1.0	1.4582E 00	1.0001	0.9937	0.9901	0.9881	0.9997	0.9999	1.0003	1.0001	0.9934	0.7785	0.9916	0.9882
1.2	1.4202E 00	1.0000	0.9924	0.9885	0.9864	0.9997	0.9999	1.0002	1.0000	0.9932	0.7850	0.9919	0.9885
1.4	1.3956E 00	0.9999	0.9917	0.9878	0.9856	0.9996	0.9999	1.0002	1.0000	0.9932	0.7911	0.9922	0.9887
1.6	1.3786E 00	0.9998	0.9915	0.9875	0.9853	0.9995	0.9999	1.0002	1.0000	0.9932	0.7969	0.9925	0.9889
1.8	1.3664E 00	0.9997	0.9914	0.9874	0.9852	0.9994	0.9999	1.0002	1.0000	0.9933	0.8025	0.9927	0.9891
2.0	1.3573E 00	0.9996	0.9914	0.9875	0.9853	0.9994	0.9999	1.0002	1.0000	0.9934	0.8079	0.9929	0.9893
2.2	1.3502E 00	0.9995	0.9915	0.9876	0.9855	0.9993	0.9999	1.0001	1.0000	0.9935	0.8131	0.9931	0.9894
2.4	1.3447E 00	0.9994	0.9916	0.9878	0.9857	0.9992	0.9998	1.0001	1.0000	0.9936	0.8183	0.9933	0.9895
2.6	1.3401E 00	0.9993	0.9918	0.9880	0.9859	0.9991	0.9998	1.0001	1.0000	0.9938	0.8233	0.9934	0.9897
2.8	1.3364E 00	0.9991	0.9920	0.9883	0.9862	0.9990	0.9998	1.0001	1.0000	0.9939	0.8283	0.9936	0.9898
3.0	1.3332E 00	0.9990	0.9922	0.9886	0.9865	0.9990	0.9998	1.0001	1.0000	0.9940	0.8332	0.9937	0.9899
3.2	1.3305E 00	0.9989	0.9924	0.9888	0.9868	0.9989	0.9998	1.0001	1.0000	0.9941	0.8381	0.9939	0.9900
3.4	1.3282E 00	0.9987	0.9926	0.9891	0.9871	0.9988	0.9998	1.0001	1.0000	0.9942	0.8430	0.9940	0.9901
3.6	1.3262E 00	0.9986	0.9928	0.9893	0.9874	0.9987	0.9998	1.0001	1.0000	0.9943	0.8478	0.9942	0.9902
3.8	1.3244E 00	0.9985	0.9930	0.9896	0.9877	0.9987	0.9997	1.0001	1.0000	0.9945	0.8526	0.9943	0.9902
4.0	1.3228E 00	0.9983	0.9932	0.9899	0.9880	0.9986	0.9997	1.0001	1.0000	0.9946	0.8573	0.9944	0.9903
4.5	1.3194E 00	0.9980	0.9937	0.9905	0.9887	0.9984	0.9997	1.0001	1.0000	0.9948	0.8690	0.9947	0.9905
5.0	1.3166E 00	0.9977	0.9942	0.9912	0.9894	0.9982	0.9997	1.0001	1.0000	0.9951	0.8808	0.9950	0.9907
5.5	1.3143E 00	0.9973	0.9946	0.9918	0.9901	0.9980	0.9996	1.0001	1.0000	0.9953	0.8925	0.9952	0.9908
6.0	1.3123E 00	0.9970	0.9951	0.9923	0.9907	0.9979	0.9996	1.0001	1.0000	0.9956	0.9042	0.9955	0.9909
6.5	1.3106E 00	0.9966	0.9955	0.9929	0.9914	0.9977	0.9995	1.0001	1.0000	0.9958	0.9157	0.9957	0.9911
7.0	1.3090E 00	0.9963	0.9960	0.9935	0.9920	0.9975	0.9995	1.0001	1.0000	0.9960	0.9272	0.9959	0.9912
7.5	1.3076E 00	0.9959	0.9964	0.9940	0.9926	0.9973	0.9995	1.0000	1.0000	0.9962	0.9387	0.9962	0.9913
8.0	1.3062E 00	0.9956	0.9968	0.9945	0.9932	0.9971	0.9994	1.0000	1.0000	0.9964	0.9502	0.9964	0.9914
9.0	1.3038E 00	0.9948	0.9976	0.9955	0.9943	0.9968	0.9994	1.0000	1.0000	0.9968	0.9729	0.9968	0.9916
10.0	1.3017E 00	0.9941	0.9983	0.9965	0.9953	0.9964	0.9993	1.0000	1.0000	0.9972	0.9957	0.9972	0.9918
11.0	1.2997E 00	0.9934	0.9991	0.9974	0.9964	0.9960	0.9992	1.0000	1.0000	0.9976	1.0185	0.9975	0.9919
12.0	1.2978E 00	0.9927	0.9998	0.9983	0.9973	0.9957	0.9991	1.0000	1.0000	0.9979	1.0413	0.9979	0.9920
13.0	1.2960E 00	0.9920	1.0005	0.9992	0.9983	0.9953	0.9991	1.0000	1.0000	0.9983	1.0639	0.9982	0.9922
14.0	1.2944E 00	0.9913	1.0011	1.0000	0.9992	0.9949	0.9990	1.0000	1.0000	0.9986	1.0866	0.9986	0.9923
16.0	1.2912E 00	0.9899	1.0024	1.0017	1.0010	0.9942	0.9989	1.0000	1.0000	0.9993	1.1315	0.9992	0.9925
18.0	1.2883E 00	0.9884	1.0037	1.0033	1.0028	0.9935	0.9987	1.0000	1.0000	0.9999	1.1762	0.9999	0.9927
20.0	1.2854E 00	0.9870	1.0049	1.0048	1.0045	0.9927	0.9986	1.0000	1.0000	1.0005	1.2207	1.0005	0.9929
25.0	1.2788E 00	0.9834	1.0078	1.0085	1.0085	0.9909	0.9982	1.0000	1.0000	1.0020	1.3316	1.0019	0.9933
30.0	1.2726E 00	0.9799	1.0106	1.0120	1.0124	0.9892	0.9978	1.0000	1.0000	1.0034	1.4393	1.0033	0.9936
35.0	1.2668E 00	0.9764	1.0133	1.0154	1.0161	0.9874	0.9975	1.0000	1.0000	1.0047	1.5443	1.0047	0.9939
40.0	1.2611E 00	0.9729	1.0159	1.0187	1.0197	0.9856	0.9971	1.0000	1.0000	1.0060	1.6486	1.0060	0.9942
45.0	1.2557E 00	0.9694	1.0184	1.0219	1.0232	0.9839	0.9967	1.0000	1.0000	1.0073	1.7480	1.0073	0.9944
50.0	1.2503E 00	0.9660	1.0209	1.0251	1.0267	0.9822	0.9964	1.0000	1.0000	1.0085	1.8459	1.0085	0.9946

ELECTRON Z= 12 A= 24

p	F_0L_0	L_0	λ_2	λ_3	λ_4	μ_1	μ_2	Λ_1	Λ_2	ν_{12}	$\hat{\nu}_{12}$	η_{12}	$\hat{\eta}_{12}$
0.1	5.9138E 00	1.0005	1.7255	2.0377	2.2007	1.0000	1.0000	1.0004	1.0001	1.1737	0.7452	0.9826	0.9788
0.2	3.1730E 00	1.0005	1.1665	1.2124	1.2328	1.0000	1.0000	1.0004	1.0001	1.0338	0.7491	0.9857	0.9819
0.3	2.3808E 00	1.0004	1.0639	1.0781	1.0839	1.0000	1.0000	1.0004	1.0001	1.0086	0.7530	0.9870	0.9831
0.4	2.0287E 00	1.0004	1.0284	1.0331	1.0345	0.9999	1.0000	1.0004	1.0001	1.0000	0.7568	0.9878	0.9839
0.5	1.8363E 00	1.0004	1.0124	1.0129	1.0125	0.9999	1.0000	1.0004	1.0001	0.9963	0.7606	0.9883	0.9844
0.6	1.7178E 00	1.0004	1.0039	1.0023	1.0010	0.9999	1.0000	1.0003	1.0001	0.9944	0.7643	0.9888	0.9849
0.7	1.6389E 00	1.0004	0.9989	0.9962	0.9943	0.9998	1.0000	1.0003	1.0001	0.9933	0.7680	0.9891	0.9852
0.8	1.5835E 00	1.0003	0.9958	0.9924	0.9902	0.9998	1.0000	1.0003	1.0001	0.9927	0.7715	0.9894	0.9855
0.9	1.5429E 00	1.0003	0.9938	0.9899	0.9876	0.9998	1.0000	1.0003	1.0001	0.9923	0.7749	0.9897	0.9857
1.0	1.5122E 00	1.0002	0.9925	0.9883	0.9858	0.9997	0.9999	1.0003	1.0001	0.9921	0.7782	0.9899	0.9859
1.2	1.4696E 00	1.0002	0.9909	0.9864	0.9838	0.9996	0.9999	1.0003	1.0001	0.9919	0.7845	0.9903	0.9863
1.4	1.4419E 00	1.0001	0.9902	0.9855	0.9828	0.9995	0.9999	1.0002	1.0000	0.9918	0.7904	0.9907	0.9866
1.6	1.4229E 00	0.9999	0.9898	0.9851	0.9825	0.9995	0.9999	1.0002	1.0000	0.9919	0.7960	0.9910	0.9868
1.8	1.4091E 00	0.9998	0.9897	0.9850	0.9824	0.9994	0.9999	1.0002	1.0000	0.9920	0.8013	0.9913	0.9870
2.0	1.3988E 00	0.9997	0.9897	0.9851	0.9825	0.9993	0.9999	1.0002	1.0000	0.9921	0.8064	0.9915	0.9872
2.2	1.3909E 00	0.9996	0.9899	0.9852	0.9827	0.9992	0.9998	1.0002	1.0000	0.9922	0.8114	0.9917	0.9874
2.4	1.3846E 00	0.9994	0.9900	0.9855	0.9829	0.9991	0.9998	1.0002	1.0000	0.9924	0.8163	0.9920	0.9876
2.6	1.3794E 00	0.9993	0.9902	0.9857	0.9832	0.9990	0.9998	1.0001	1.0000	0.9925	0.8211	0.9921	0.9877
2.8	1.3752E 00	0.9992	0.9904	0.9860	0.9835	0.9989	0.9998	1.0001	1.0000	0.9927	0.8258	0.9923	0.9878
3.0	1.3716E 00	0.9990	0.9906	0.9863	0.9839	0.9989	0.9998	1.0001	1.0000	0.9928	0.8305	0.9925	0.9880
3.2	1.3686E 00	0.9989	0.9909	0.9866	0.9842	0.9988	0.9998	1.0001	1.0000	0.9929	0.8350	0.9927	0.9881
3.4	1.3659E 00	0.9987	0.9911	0.9869	0.9846	0.9987	0.9997	1.0001	1.0000	0.9931	0.8396	0.9928	0.9882
3.6	1.3636E 00	0.9986	0.9913	0.9872	0.9849	0.9986	0.9997	1.0001	1.0000	0.9932	0.8442	0.9930	0.9883
3.8	1.3615E 00	0.9984	0.9916	0.9875	0.9853	0.9985	0.9997	1.0001	1.0000	0.9933	0.8486	0.9931	0.9884
4.0	1.3597E 00	0.9983	0.9918	0.9878	0.9856	0.9984	0.9997	1.0001	1.0000	0.9935	0.8530	0.9933	0.9885
4.5	1.3558E 00	0.9979	0.9924	0.9886	0.9864	0.9982	0.9997	1.0001	1.0000	0.9938	0.8641	0.9936	0.9887
5.0	1.3526E 00	0.9975	0.9929	0.9893	0.9872	0.9980	0.9996	1.0001	1.0000	0.9941	0.8751	0.9939	0.9889
5.5	1.3500E 00	0.9971	0.9935	0.9900	0.9880	0.9978	0.9996	1.0001	1.0000	0.9944	0.8860	0.9942	0.9891
6.0	1.3477E 00	0.9967	0.9940	0.9907	0.9888	0.9976	0.9995	1.0001	1.0000	0.9946	0.8969	0.9945	0.9892
6.5	1.3456E 00	0.9964	0.9945	0.9914	0.9895	0.9974	0.9995	1.0001	1.0000	0.9949	0.9077	0.9948	0.9894
7.0	1.3438E 00	0.9960	0.9950	0.9920	0.9902	0.9972	0.9995	1.0001	1.0000	0.9951	0.9186	0.9951	0.9895
7.5	1.3421E 00	0.9956	0.9955	0.9926	0.9909	0.9970	0.9994	1.0001	1.0000	0.9954	0.9293	0.9953	0.9897
8.0	1.3406E 00	0.9952	0.9960	0.9932	0.9916	0.9968	0.9994	1.0000	1.0000	0.9956	0.9399	0.9956	0.9898
9.0	1.3377E 00	0.9944	0.9969	0.9944	0.9928	0.9964	0.9993	1.0000	1.0000	0.9961	0.9612	0.9960	0.9900
10.0	1.3352E 00	0.9936	0.9977	0.9955	0.9941	0.9960	0.9992	1.0000	1.0000	0.9965	0.9825	0.9965	0.9902
11.0	1.3329E 00	0.9928	0.9986	0.9965	0.9952	0.9956	0.9991	1.0000	1.0000	0.9969	1.0039	0.9969	0.9904
12.0	1.3307E 00	0.9920	0.9994	0.9976	0.9964	0.9952	0.9991	1.0000	1.0000	0.9973	1.0251	0.9973	0.9905
13.0	1.3287E 00	0.9912	1.0001	0.9986	0.9975	0.9948	0.9990	1.0000	1.0000	0.9977	1.0463	0.9977	0.9907
14.0	1.3267E 00	0.9904	1.0009	0.9995	0.9985	0.9944	0.9989	1.0000	1.0000	0.9981	1.0674	0.9981	0.9908
16.0	1.3230E 00	0.9889	1.0024	1.0014	1.0006	0.9936	0.9987	1.0000	1.0000	0.9988	1.1091	0.9988	0.9911
18.0	1.3196E 00	0.9873	1.0038	1.0032	1.0026	0.9928	0.9986	1.0000	1.0000	0.9996	1.1512	0.9995	0.9913
20.0	1.3163E 00	0.9857	1.0052	1.0049	1.0045	0.9919	0.9984	1.0000	1.0000	1.0002	1.1927	1.0002	0.9916
25.0	1.3087E 00	0.9818	1.0085	1.0091	1.0090	0.9899	0.9980	1.0000	1.0000	1.0019	1.2951	1.0019	0.9920
30.0	1.3015E 00	0.9778	1.0116	1.0131	1.0134	0.9880	0.9976	1.0000	1.0000	1.0035	1.3958	1.0035	0.9924
35.0	1.2948E 00	0.9739	1.0147	1.0169	1.0176	0.9860	0.9972	1.0000	1.0000	1.0050	1.4940	1.0050	0.9927
40.0	1.2883E 00	0.9701	1.0176	1.0206	1.0216	0.9841	0.9968	1.0000	1.0000	1.0065	1.5902	1.0065	0.9931
45.0	1.2820E 00	0.9662	1.0204	1.0242	1.0256	0.9822	0.9964	1.0000	1.0000	1.0079	1.6823	1.0079	0.9933
50.0	1.2759E 00	0.9625	1.0232	1.0278	1.0295	0.9803	0.9960	1.0000	1.0000	1.0093	1.7722	1.0093	0.9936

Instr. see p. 22

ELECTRON Z= 13 A= 27

p	F_0L_0	L_0	λ_2	λ_3	λ_4	μ_1	μ_2	Λ_1	Λ_2	ν_{12}	$\hat{\nu}_{12}$	η_{12}	$\hat{\eta}_{12}$
0.1	6.4625E 00	1.0006	1.8483	2.2403	2.4505	1.0000	1.0000	1.0004	1.0001	1.2033	0.7441	0.9793	0.9749
0.2	3.4260E 00	1.0006	1.1948	1.2500	1.2747	1.0000	1.0000	1.0004	1.0001	1.0396	0.7484	0.9831	0.9786
0.3	2.5396E 00	1.0006	1.0748	1.0918	1.0986	1.0000	1.0000	1.0004	1.0001	1.0100	0.7525	0.9847	0.9802
0.4	2.1453E 00	1.0006	1.0334	1.0389	1.0406	0.9999	1.0000	1.0004	1.0001	1.0000	0.7565	0.9856	0.9811
0.5	1.9300E 00	1.0005	1.0146	1.0153	1.0148	0.9999	1.0000	1.0004	1.0001	0.9956	0.7605	0.9863	0.9818
0.6	1.7976E 00	1.0005	1.0046	1.0029	1.0013	0.9999	1.0000	1.0004	1.0001	0.9934	0.7643	0.9868	0.9823
0.7	1.7096E 00	1.0005	0.9988	0.9957	0.9935	0.9998	1.0000	1.0004	1.0001	0.9922	0.7680	0.9873	0.9827
0.8	1.6477E 00	1.0004	0.9952	0.9912	0.9887	0.9998	1.0000	1.0003	1.0001	0.9914	0.7716	0.9876	0.9830
0.9	1.6025E 00	1.0004	0.9929	0.9883	0.9856	0.9997	0.9999	1.0003	1.0001	0.9910	0.7751	0.9880	0.9833
1.0	1.5683E 00	1.0004	0.9913	0.9864	0.9835	0.9997	0.9999	1.0003	1.0001	0.9907	0.7784	0.9882	0.9835
1.2	1.5207E 00	1.0002	0.9895	0.9842	0.9812	0.9996	0.9999	1.0003	1.0001	0.9905	0.7847	0.9887	0.9839
1.4	1.4899E 00	1.0001	0.9886	0.9832	0.9801	0.9995	0.9999	1.0003	1.0001	0.9904	0.7905	0.9891	0.9843
1.6	1.4686E 00	1.0000	0.9882	0.9827	0.9796	0.9994	0.9999	1.0002	1.0000	0.9905	0.7961	0.9895	0.9846
1.8	1.4533E 00	0.9999	0.9881	0.9826	0.9795	0.9993	0.9999	1.0002	1.0000	0.9906	0.8013	0.9898	0.9848
2.0	1.4418E 00	0.9997	0.9881	0.9827	0.9796	0.9992	0.9998	1.0002	1.0000	0.9908	0.8063	0.9901	0.9851
2.2	1.4329E 00	0.9996	0.9882	0.9829	0.9799	0.9991	0.9998	1.0002	1.0000	0.9909	0.8112	0.9903	0.9853
2.4	1.4258E 00	0.9994	0.9884	0.9831	0.9802	0.9990	0.9998	1.0002	1.0000	0.9911	0.8160	0.9906	0.9854
2.6	1.4200E 00	0.9993	0.9886	0.9834	0.9805	0.9989	0.9998	1.0002	1.0000	0.9913	0.8206	0.9908	0.9856
2.8	1.4153E 00	0.9991	0.9889	0.9838	0.9809	0.9988	0.9998	1.0002	1.0000	0.9914	0.8252	0.9910	0.9858
3.0	1.4112E 00	0.9990	0.9891	0.9841	0.9813	0.9987	0.9997	1.0001	1.0000	0.9916	0.8297	0.9912	0.9859
3.2	1.4078E 00	0.9988	0.9894	0.9844	0.9817	0.9986	0.9997	1.0001	1.0000	0.9917	0.8342	0.9914	0.9861
3.4	1.4048E 00	0.9986	0.9897	0.9848	0.9821	0.9985	0.9997	1.0001	1.0000	0.9919	0.8386	0.9916	0.9862
3.6	1.4021E 00	0.9985	0.9899	0.9852	0.9825	0.9984	0.9997	1.0001	1.0000	0.9921	0.8430	0.9918	0.9863
3.8	1.3998E 00	0.9983	0.9902	0.9855	0.9829	0.9983	0.9997	1.0001	1.0000	0.9922	0.8474	0.9920	0.9864
4.0	1.3977E 00	0.9981	0.9905	0.9859	0.9833	0.9982	0.9997	1.0001	1.0000	0.9924	0.8517	0.9921	0.9865
4.5	1.3932E 00	0.9977	0.9911	0.9867	0.9842	0.9980	0.9996	1.0001	1.0000	0.9927	0.8624	0.9925	0.9868
5.0	1.3896E 00	0.9973	0.9918	0.9876	0.9852	0.9978	0.9996	1.0001	1.0000	0.9930	0.8730	0.9929	0.9870
5.5	1.3865E 00	0.9968	0.9924	0.9884	0.9861	0.9975	0.9995	1.0001	1.0000	0.9934	0.8835	0.9932	0.9872
6.0	1.3838E 00	0.9964	0.9930	0.9892	0.9869	0.9973	0.9995	1.0001	1.0000	0.9937	0.8940	0.9936	0.9874
6.5	1.3815E 00	0.9960	0.9936	0.9899	0.9878	0.9971	0.9994	1.0001	1.0000	0.9940	0.9045	0.9939	0.9876
7.0	1.3793E 00	0.9955	0.9942	0.9907	0.9886	0.9969	0.9994	1.0001	1.0000	0.9943	0.9150	0.9942	0.9878
7.5	1.3774E 00	0.9951	0.9947	0.9914	0.9894	0.9966	0.9993	1.0001	1.0000	0.9946	0.9252	0.9945	0.9879
8.0	1.3756E 00	0.9946	0.9953	0.9921	0.9901	0.9964	0.9993	1.0001	1.0000	0.9948	0.9356	0.9948	0.9881
9.0	1.3723E 00	0.9937	0.9963	0.9934	0.9916	0.9959	0.9992	1.0001	1.0000	0.9954	0.9563	0.9953	0.9883
10.0	1.3693E 00	0.9929	0.9973	0.9947	0.9930	0.9955	0.9991	1.0000	1.0000	0.9959	0.9767	0.9958	0.9886
11.0	1.3666E 00	0.9920	0.9983	0.9959	0.9943	0.9950	0.9990	1.0000	1.0000	0.9963	0.9973	0.9963	0.9888
12.0	1.3641E 00	0.9911	0.9992	0.9971	0.9956	0.9946	0.9989	1.0000	1.0000	0.9968	1.0176	0.9968	0.9890
13.0	1.3617E 00	0.9902	1.0001	0.9982	0.9969	0.9941	0.9988	1.0000	1.0000	0.9973	1.0380	0.9972	0.9892
14.0	1.3594E 00	0.9893	1.0010	0.9993	0.9981	0.9936	0.9988	1.0000	1.0000	0.9977	1.0582	0.9977	0.9893
16.0	1.3551E 00	0.9875	1.0026	1.0015	1.0005	0.9927	0.9986	1.0000	1.0000	0.9985	1.0987	0.9985	0.9896
18.0	1.3511E 00	0.9857	1.0043	1.0035	1.0027	0.9918	0.9984	1.0000	1.0000	0.9994	1.1390	0.9993	0.9899
20.0	1.3472E 00	0.9839	1.0059	1.0055	1.0049	0.9909	0.9982	1.0000	1.0000	1.0002	1.1788	1.0001	0.9902
25.0	1.3383E 00	0.9795	1.0096	1.0103	1.0102	0.9886	0.9978	1.0000	1.0000	1.0020	1.2772	1.0020	0.9907
30.0	1.3300E 00	0.9751	1.0132	1.0148	1.0151	0.9864	0.9973	1.0000	1.0000	1.0038	1.3736	1.0038	0.9912
35.0	1.3221E 00	0.9707	1.0167	1.0192	1.0199	0.9842	0.9968	1.0000	1.0000	1.0056	1.4675	1.0056	0.9916
40.0	1.3146E 00	0.9663	1.0200	1.0234	1.0245	0.9820	0.9964	1.0000	1.0000	1.0073	1.5589	1.0072	0.9919
45.0	1.3073E 00	0.9620	1.0233	1.0275	1.0291	0.9799	0.9959	1.0000	1.0000	1.0089	1.6465	1.0089	0.9923
50.0	1.3003E 00	0.9578	1.0264	1.0315	1.0335	0.9778	0.9954	1.0000	1.0000	1.0105	1.7306	1.0105	0.9926

Instr. see p. 22

ELECTRON Z= 14 A= 29

p	F_0L_0	L_0	λ_2	λ_3	λ_4	μ_1	μ_2	Λ_1	Λ_2	ν_{12}	$\hat{\nu}_{12}$	η_{12}	$\hat{\eta}_{12}$
0.1	7.0298E 00	1.0008	1.9798	2.4661	2.7342	1.0000	1.0000	1.0005	1.0001	1.2351	0.7429	0.9756	0.9705
0.2	3.6910E 00	1.0007	1.2251	1.2906	1.3202	1.0000	1.0000	1.0005	1.0001	1.0457	0.7475	0.9803	0.9751
0.3	2.7066E 00	1.0007	1.0865	1.1064	1.1144	1.0000	1.0000	1.0005	1.0001	1.0115	0.7519	0.9822	0.9770
0.4	2.2679E 00	1.0007	1.0386	1.0451	1.0471	0.9999	1.0000	1.0005	1.0001	1.0000	0.7561	0.9833	0.9781
0.5	2.0286E 00	1.0007	1.0169	1.0177	1.0172	0.9999	1.0000	1.0004	1.0001	0.9949	0.7602	0.9841	0.9788
0.6	1.8815E 00	1.0007	1.0054	1.0034	1.0016	0.9998	1.0000	1.0004	1.0001	0.9923	0.7641	0.9847	0.9794
0.7	1.7838E 00	1.0006	0.9987	0.9950	0.9926	0.9998	1.0000	1.0004	1.0001	0.9909	0.7679	0.9852	0.9799
0.8	1.7152E 00	1.0006	0.9945	0.9899	0.9870	0.9997	1.0000	1.0004	1.0001	0.9900	0.7716	0.9857	0.9803
0.9	1.6650E 00	1.0005	0.9918	0.9866	0.9834	0.9997	0.9999	1.0004	1.0001	0.9895	0.7751	0.9860	0.9806
1.0	1.6271E 00	1.0005	0.9900	0.9843	0.9810	0.9996	0.9999	1.0003	1.0001	0.9892	0.7784	0.9864	0.9809
1.2	1.5744E 00	1.0004	0.9878	0.9818	0.9783	0.9995	0.9999	1.0003	1.0001	0.9890	0.7846	0.9869	0.9814
1.4	1.5402E 00	1.0002	0.9868	0.9806	0.9770	0.9994	0.9999	1.0003	1.0001	0.9889	0.7904	0.9874	0.9818
1.6	1.5166E 00	1.0001	0.9864	0.9801	0.9765	0.9993	0.9999	1.0003	1.0001	0.9890	0.7958	0.9878	0.9821
1.8	1.4996E 00	1.0000	0.9862	0.9799	0.9764	0.9992	0.9998	1.0002	1.0000	0.9891	0.8010	0.9882	0.9824
2.0	1.4868E 00	0.9998	0.9863	0.9800	0.9765	0.9991	0.9998	1.0002	1.0000	0.9893	0.8059	0.9885	0.9827
2.2	1.4769E 00	0.9996	0.9864	0.9802	0.9768	0.9990	0.9998	1.0002	1.0000	0.9895	0.8107	0.9888	0.9829
2.4	1.4690E 00	0.9995	0.9866	0.9805	0.9771	0.9989	0.9998	1.0002	1.0000	0.9897	0.8153	0.9891	0.9832
2.6	1.4625E 00	0.9993	0.9869	0.9809	0.9775	0.9988	0.9998	1.0002	1.0000	0.9899	0.8198	0.9893	0.9834
2.8	1.4572E 00	0.9991	0.9871	0.9812	0.9779	0.9987	0.9997	1.0002	1.0000	0.9900	0.8242	0.9896	0.9835
3.0	1.4526E 00	0.9989	0.9874	0.9816	0.9784	0.9986	0.9997	1.0002	1.0000	0.9902	0.8285	0.9898	0.9837
3.2	1.4487E 00	0.9988	0.9877	0.9820	0.9788	0.9985	0.9997	1.0001	1.0000	0.9904	0.8329	0.9901	0.9839
3.4	1.4454E 00	0.9986	0.9880	0.9824	0.9793	0.9984	0.9997	1.0001	1.0000	0.9906	0.8371	0.9903	0.9840
3.6	1.4424E 00	0.9984	0.9883	0.9828	0.9797	0.9983	0.9997	1.0001	1.0000	0.9908	0.8413	0.9905	0.9842
3.8	1.4397E 00	0.9982	0.9886	0.9832	0.9802	0.9982	0.9996	1.0001	1.0000	0.9909	0.8455	0.9907	0.9843
4.0	1.4373E 00	0.9980	0.9890	0.9836	0.9806	0.9981	0.9996	1.0001	1.0000	0.9911	0.8496	0.9909	0.9844
4.5	1.4323E 00	0.9976	0.9897	0.9846	0.9817	0.9978	0.9996	1.0001	1.0000	0.9915	0.8599	0.9913	0.9847
5.0	1.4281E 00	0.9971	0.9904	0.9856	0.9828	0.9975	0.9995	1.0001	1.0000	0.9919	0.8701	0.9917	0.9850
5.5	1.4246E 00	0.9966	0.9912	0.9865	0.9838	0.9973	0.9995	1.0001	1.0000	0.9923	0.8803	0.9921	0.9852
6.0	1.4216E 00	0.9961	0.9918	0.9874	0.9848	0.9970	0.9994	1.0001	1.0000	0.9926	0.8902	0.9925	0.9855
6.5	1.4189E 00	0.9956	0.9925	0.9883	0.9858	0.9968	0.9994	1.0001	1.0000	0.9930	0.9003	0.9929	0.9857
7.0	1.4164E 00	0.9951	0.9932	0.9891	0.9867	0.9965	0.9993	1.0001	1.0000	0.9933	0.9102	0.9932	0.9858
7.5	1.4142E 00	0.9947	0.9938	0.9899	0.9876	0.9963	0.9993	1.0001	1.0000	0.9936	0.9202	0.9935	0.9860
8.0	1.4121E 00	0.9942	0.9944	0.9907	0.9884	0.9960	0.9992	1.0001	1.0000	0.9939	0.9300	0.9938	0.9862
9.0	1.4083E 00	0.9932	0.9956	0.9922	0.9901	0.9955	0.9991	1.0001	1.0000	0.9945	0.9499	0.9945	0.9865
10.0	1.4049E 00	0.9922	0.9967	0.9936	0.9917	0.9950	0.9990	1.0000	1.0000	0.9951	0.9695	0.9950	0.9868
11.0	1.4018E 00	0.9912	0.9978	0.9950	0.9932	0.9945	0.9989	1.0000	1.0000	0.9956	0.9890	0.9956	0.9870
12.0	1.3988E 00	0.9902	0.9988	0.9963	0.9947	0.9940	0.9988	1.0000	1.0000	0.9962	1.0085	0.9961	0.9872
13.0	1.3961E 00	0.9892	0.9998	0.9976	0.9961	0.9935	0.9987	1.0000	1.0000	0.9967	1.0281	0.9966	0.9875
14.0	1.3934E 00	0.9883	1.0008	0.9989	0.9975	0.9930	0.9986	1.0000	1.0000	0.9972	1.0475	0.9971	0.9877
16.0	1.3885E 00	0.9863	1.0027	1.0013	1.0001	0.9919	0.9984	1.0000	1.0000	0.9981	1.0861	0.9981	0.9880
18.0	1.3838E 00	0.9843	1.0046	1.0036	1.0027	0.9909	0.9982	1.0000	1.0000	0.9990	1.1243	0.9990	0.9883
20.0	1.3794E 00	0.9823	1.0063	1.0058	1.0051	0.9899	0.9980	1.0000	1.0000	0.9999	1.1623	0.9999	0.9886
25.0	1.3692E 00	0.9774	1.0106	1.0112	1.0110	0.9874	0.9975	1.0000	1.0000	1.0020	1.2561	1.0020	0.9893
30.0	1.3596E 00	0.9726	1.0146	1.0163	1.0166	0.9849	0.9970	1.0000	1.0000	1.0041	1.3473	1.0041	0.9898
35.0	1.3506E 00	0.9677	1.0185	1.0211	1.0219	0.9825	0.9965	1.0000	1.0000	1.0060	1.4365	1.0060	0.9903
40.0	1.3421E 00	0.9629	1.0222	1.0259	1.0271	0.9801	0.9960	1.0000	1.0000	1.0079	1.5226	1.0079	0.9907
45.0	1.3338E 00	0.9582	1.0258	1.0305	1.0322	0.9778	0.9955	1.0000	1.0000	1.0097	1.6052	1.0097	0.9911
50.0	1.3258E 00	0.9535	1.0293	1.0350	1.0371	0.9755	0.9949	1.0000	1.0000	1.0115	1.6841	1.0115	0.9915

Instr. see p. 22

ELECTRON Z= 15 A= 31

p	F_0L_0	L_0	λ_2	λ_3	λ_4	μ_1	μ_2	Λ_1	Λ_2	ν_{12}	$\hat{\nu}_{12}$	η_{12}	$\hat{\eta}_{12}$
0.1	7.6156E 00	1.0009	2.1198	2.7165	3.0553	1.0000	1.0000	1.0005	1.0001	1.2691	0.7417	0.9716	0.9657
0.2	3.9680E 00	1.0009	1.2574	1.3345	1.3696	1.0000	1.0000	1.0005	1.0001	1.0522	0.7466	0.9772	0.9713
0.3	2.8818E 00	1.0009	1.0989	1.1220	1.1313	0.9999	1.0000	1.0005	1.0001	1.0131	0.7512	0.9794	0.9735
0.4	2.3966E 00	1.0009	1.0442	1.0517	1.0540	0.9999	1.0000	1.0005	1.0001	0.9999	0.7556	0.9808	0.9748
0.5	2.1320E 00	1.0009	1.0193	1.0203	1.0197	0.9999	1.0000	1.0005	1.0001	0.9941	0.7599	0.9817	0.9757
0.6	1.9694E 00	1.0008	1.0062	1.0039	1.0019	0.9998	1.0000	1.0005	1.0001	0.9912	0.7639	0.9824	0.9764
0.7	1.8615E 00	1.0008	0.9985	0.9944	0.9916	0.9998	1.0000	1.0004	1.0001	0.9895	0.7678	0.9830	0.9769
0.8	1.7858E 00	1.0008	0.9937	0.9885	0.9852	0.9997	0.9999	1.0004	1.0001	0.9886	0.7715	0.9835	0.9774
0.9	1.7305E 00	1.0007	0.9906	0.9847	0.9811	0.9997	0.9999	1.0004	1.0001	0.9880	0.7750	0.9840	0.9778
1.0	1.6886E 00	1.0006	0.9885	0.9821	0.9784	0.9996	0.9999	1.0004	1.0001	0.9876	0.7784	0.9843	0.9781
1.2	1.6305E 00	1.0005	0.9861	0.9792	0.9752	0.9995	0.9999	1.0003	1.0001	0.9873	0.7846	0.9850	0.9787
1.4	1.5928E 00	1.0004	0.9849	0.9778	0.9738	0.9994	0.9999	1.0003	1.0001	0.9873	0.7904	0.9855	0.9791
1.6	1.5668E 00	1.0002	0.9844	0.9772	0.9732	0.9993	0.9999	1.0003	1.0001	0.9874	0.7957	0.9860	0.9795
1.8	1.5479E 00	1.0001	0.9842	0.9771	0.9730	0.9991	0.9998	1.0003	1.0001	0.9875	0.8008	0.9864	0.9799
2.0	1.5337E 00	0.9999	0.9843	0.9772	0.9732	0.9990	0.9998	1.0002	1.0000	0.9877	0.8056	0.9868	0.9802
2.2	1.5227E 00	0.9997	0.9844	0.9774	0.9734	0.9989	0.9998	1.0002	1.0000	0.9879	0.8102	0.9871	0.9804
2.4	1.5140E 00	0.9995	0.9847	0.9777	0.9738	0.9988	0.9998	1.0002	1.0000	0.9881	0.8147	0.9875	0.9807
2.6	1.5068E 00	0.9993	0.9850	0.9781	0.9743	0.9987	0.9997	1.0002	1.0000	0.9883	0.8191	0.9878	0.9809
2.8	1.5008E 00	0.9991	0.9853	0.9785	0.9748	0.9985	0.9997	1.0002	1.0000	0.9885	0.8234	0.9880	0.9811
3.0	1.4958E 00	0.9990	0.9856	0.9790	0.9753	0.9984	0.9997	1.0002	1.0000	0.9888	0.8276	0.9883	0.9813
3.2	1.4914E 00	0.9988	0.9859	0.9794	0.9758	0.9983	0.9997	1.0002	1.0000	0.9890	0.8318	0.9886	0.9815
3.4	1.4876E 00	0.9986	0.9863	0.9799	0.9763	0.9982	0.9996	1.0002	1.0000	0.9892	0.8359	0.9888	0.9817
3.6	1.4843E 00	0.9984	0.9866	0.9803	0.9768	0.9981	0.9996	1.0001	1.0000	0.9894	0.8400	0.9890	0.9818
3.8	1.4813E 00	0.9981	0.9870	0.9808	0.9773	0.9980	0.9996	1.0001	1.0000	0.9896	0.8440	0.9893	0.9820
4.0	1.4786E 00	0.9979	0.9873	0.9813	0.9778	0.9979	0.9996	1.0001	1.0000	0.9898	0.8480	0.9895	0.9822
4.5	1.4729E 00	0.9974	0.9882	0.9824	0.9791	0.9976	0.9995	1.0001	1.0000	0.9902	0.8579	0.9900	0.9825
5.0	1.4682E 00	0.9969	0.9890	0.9834	0.9802	0.9973	0.9995	1.0001	1.0000	0.9907	0.8677	0.9905	0.9828
5.5	1.4643E 00	0.9964	0.9898	0.9845	0.9814	0.9970	0.9994	1.0001	1.0000	0.9911	0.8775	0.9909	0.9831
6.0	1.4608E 00	0.9958	0.9906	0.9855	0.9825	0.9967	0.9994	1.0001	1.0000	0.9915	0.8871	0.9913	0.9833
6.5	1.4577E 00	0.9953	0.9913	0.9865	0.9836	0.9964	0.9993	1.0001	1.0000	0.9919	0.8967	0.9917	0.9836
7.0	1.4549E 00	0.9948	0.9921	0.9874	0.9846	0.9962	0.9993	1.0001	1.0000	0.9922	0.9063	0.9921	0.9838
7.5	1.4524E 00	0.9942	0.9928	0.9883	0.9856	0.9959	0.9992	1.0001	1.0000	0.9926	0.9159	0.9925	0.9840
8.0	1.4500E 00	0.9937	0.9935	0.9892	0.9866	0.9956	0.9991	1.0001	1.0000	0.9929	0.9253	0.9929	0.9842
9.0	1.4456E 00	0.9926	0.9948	0.9909	0.9885	0.9950	0.9990	1.0001	1.0000	0.9936	0.9444	0.9935	0.9845
10.0	1.4417E 00	0.9915	0.9961	0.9925	0.9902	0.9945	0.9989	1.0001	1.0000	0.9942	0.9632	0.9942	0.9848
11.0	1.4381E 00	0.9905	0.9973	0.9940	0.9920	0.9939	0.9988	1.0000	1.0000	0.9949	0.9819	0.9948	0.9851
12.0	1.4348E 00	0.9894	0.9984	0.9955	0.9936	0.9934	0.9987	1.0000	1.0000	0.9955	1.0007	0.9954	0.9854
13.0	1.4316E 00	0.9883	0.9996	0.9970	0.9952	0.9928	0.9986	1.0000	1.0000	0.9960	1.0193	0.9960	0.9856
14.0	1.4286E 00	0.9872	1.0007	0.9984	0.9967	0.9923	0.9985	1.0000	1.0000	0.9966	1.0378	0.9965	0.9859
16.0	1.4229E 00	0.9850	1.0028	1.0011	0.9997	0.9911	0.9983	1.0000	1.0000	0.9976	1.0749	0.9976	0.9863
18.0	1.4176E 00	0.9829	1.0049	1.0037	1.0025	0.9900	0.9981	1.0000	1.0000	0.9987	1.1115	0.9986	0.9866
20.0	1.4125E 00	0.9807	1.0068	1.0062	1.0053	0.9889	0.9978	1.0000	1.0000	0.9997	1.1479	0.9996	0.9870
25.0	1.4008E 00	0.9753	1.0115	1.0121	1.0118	0.9862	0.9973	1.0000	1.0000	1.0020	1.2377	1.0020	0.9877
30.0	1.3900E 00	0.9700	1.0160	1.0178	1.0180	0.9835	0.9967	1.0000	1.0000	1.0043	1.3247	1.0043	0.9883
35.0	1.3798E 00	0.9647	1.0203	1.0232	1.0240	0.9808	0.9962	1.0000	1.0000	1.0064	1.4088	1.0064	0.9889
40.0	1.3701E 00	0.9595	1.0245	1.0284	1.0297	0.9782	0.9956	1.0000	1.0000	1.0085	1.4905	1.0085	0.9894
45.0	1.3608E 00	0.9543	1.0285	1.0336	1.0354	0.9756	0.9950	1.0000	1.0000	1.0105	1.5689	1.0105	0.9899
50.0	1.3517E 00	0.9492	1.0324	1.0386	1.0409	0.9731	0.9945	1.0000	1.0000	1.0125	1.6431	1.0125	0.9903

 Instr. see p. 22

ELECTRON Z= 16 A= 34

p	F_0L_0	L_0	λ_2	λ_3	λ_4	μ_1	μ_2	Λ_1	Λ_2	ν_{12}	$\hat{\nu}_{12}$	η_{12}	$\hat{\eta}_{12}$
0.1	8.2196E 00	1.0011	2.2684	2.9937	3.4181	1.0000	1.0000	1.0006	1.0001	1.3052	0.7403	0.9672	0.9605
0.2	4.2564E 00	1.0011	1.2917	1.3817	1.4231	1.0000	1.0000	1.0006	1.0001	1.0592	0.7456	0.9739	0.9672
0.3	3.0649E 00	1.0011	1.1122	1.1388	1.1496	0.9999	1.0000	1.0006	1.0001	1.0148	0.7505	0.9765	0.9698
0.4	2.5312E 00	1.0011	1.0502	1.0588	1.0615	0.9999	1.0000	1.0006	1.0001	0.9999	0.7551	0.9781	0.9713
0.5	2.2400E 00	1.0010	1.0220	1.0232	1.0226	0.9999	1.0000	1.0005	1.0001	0.9933	0.7596	0.9792	0.9723
0.6	2.0613E 00	1.0010	1.0071	1.0046	1.0023	0.9998	1.0000	1.0005	1.0001	0.9900	0.7638	0.9800	0.9731
0.7	1.9427E 00	1.0010	0.9984	0.9938	0.9906	0.9998	1.0000	1.0005	1.0001	0.9881	0.7678	0.9807	0.9738
0.8	1.8595E 00	1.0009	0.9930	0.9871	0.9834	0.9997	0.9999	1.0005	1.0001	0.9870	0.7716	0.9813	0.9743
0.9	1.7987E 00	1.0009	0.9895	0.9828	0.9787	0.9996	0.9999	1.0004	1.0001	0.9863	0.7752	0.9818	0.9747
1.0	1.7528E 00	1.0008	0.9871	0.9799	0.9756	0.9996	0.9999	1.0004	1.0001	0.9860	0.7786	0.9822	0.9751
1.2	1.6889E 00	1.0007	0.9844	0.9766	0.9721	0.9994	0.9999	1.0004	1.0001	0.9856	0.7849	0.9829	0.9758
1.4	1.6475E 00	1.0005	0.9830	0.9750	0.9704	0.9993	0.9999	1.0003	1.0001	0.9856	0.7906	0.9836	0.9763
1.6	1.6188E 00	1.0003	0.9824	0.9743	0.9698	0.9992	0.9998	1.0003	1.0001	0.9857	0.7960	0.9841	0.9767
1.8	1.5981E 00	1.0002	0.9822	0.9741	0.9696	0.9990	0.9998	1.0003	1.0001	0.9858	0.8010	0.9846	0.9771
2.0	1.5825E 00	1.0000	0.9823	0.9742	0.9697	0.9989	0.9998	1.0003	1.0001	0.9861	0.8058	0.9850	0.9775
2.2	1.5703E 00	0.9998	0.9825	0.9745	0.9701	0.9988	0.9998	1.0002	1.0000	0.9863	0.8103	0.9854	0.9778
2.4	1.5606E 00	0.9996	0.9827	0.9749	0.9705	0.9987	0.9997	1.0002	1.0000	0.9865	0.8148	0.9858	0.9781
2.6	1.5527E 00	0.9994	0.9830	0.9753	0.9710	0.9985	0.9997	1.0002	1.0000	0.9868	0.8191	0.9861	0.9784
2.8	1.5460E 00	0.9991	0.9834	0.9758	0.9715	0.9984	0.9997	1.0002	1.0000	0.9870	0.8233	0.9864	0.9786
3.0	1.5404E 00	0.9989	0.9838	0.9763	0.9721	0.9983	0.9997	1.0002	1.0000	0.9872	0.8274	0.9867	0.9788
3.2	1.5356E 00	0.9987	0.9841	0.9768	0.9727	0.9981	0.9996	1.0002	1.0000	0.9875	0.8315	0.9870	0.9790
3.4	1.5314E 00	0.9985	0.9845	0.9773	0.9732	0.9980	0.9996	1.0002	1.0000	0.9877	0.8355	0.9873	0.9792
3.6	1.5276E 00	0.9983	0.9849	0.9778	0.9738	0.9979	0.9996	1.0002	1.0000	0.9879	0.8395	0.9876	0.9794
3.8	1.5243E 00	0.9980	0.9853	0.9784	0.9744	0.9978	0.9996	1.0002	1.0000	0.9882	0.8435	0.9878	0.9796
4.0	1.5213E 00	0.9978	0.9857	0.9789	0.9750	0.9976	0.9995	1.0001	1.0000	0.9884	0.8474	0.9881	0.9798
4.5	1.5148E 00	0.9972	0.9867	0.9801	0.9764	0.9973	0.9995	1.0001	1.0000	0.9889	0.8571	0.9886	0.9801
5.0	1.5096E 00	0.9967	0.9876	0.9813	0.9777	0.9970	0.9994	1.0001	1.0000	0.9894	0.8666	0.9892	0.9805
5.5	1.5051E 00	0.9961	0.9885	0.9825	0.9790	0.9967	0.9994	1.0001	1.0000	0.9899	0.8761	0.9897	0.9808
6.0	1.5011E 00	0.9955	0.9894	0.9836	0.9803	0.9964	0.9993	1.0001	1.0000	0.9903	0.8856	0.9901	0.9811
6.5	1.4976E 00	0.9949	0.9902	0.9847	0.9815	0.9961	0.9992	1.0001	1.0000	0.9907	0.8949	0.9906	0.9814
7.0	1.4945E 00	0.9943	0.9910	0.9857	0.9826	0.9958	0.9992	1.0001	1.0000	0.9912	0.9043	0.9910	0.9816
7.5	1.4915E 00	0.9937	0.9918	0.9868	0.9837	0.9955	0.9991	1.0001	1.0000	0.9916	0.9136	0.9915	0.9818
8.0	1.4888E 00	0.9931	0.9926	0.9878	0.9848	0.9951	0.9991	1.0001	1.0000	0.9920	0.9229	0.9919	0.9821
9.0	1.4839E 00	0.9919	0.9941	0.9896	0.9869	0.9945	0.9989	1.0001	1.0000	0.9927	0.9413	0.9926	0.9825
10.0	1.4794E 00	0.9908	0.9955	0.9915	0.9889	0.9939	0.9988	1.0001	1.0000	0.9934	0.9596	0.9933	0.9828
11.0	1.4753E 00	0.9896	0.9969	0.9932	0.9908	0.9933	0.9987	1.0001	1.0000	0.9941	0.9779	0.9940	0.9831
12.0	1.4714E 00	0.9884	0.9982	0.9949	0.9927	0.9927	0.9986	1.0001	1.0000	0.9948	0.9960	0.9947	0.9834
13.0	1.4678E 00	0.9872	0.9994	0.9965	0.9944	0.9921	0.9985	1.0000	1.0000	0.9954	1.0143	0.9954	0.9837
14.0	1.4643E 00	0.9860	1.0007	0.9980	0.9962	0.9914	0.9983	1.0000	1.0000	0.9960	1.0322	0.9960	0.9840
16.0	1.4579E 00	0.9836	1.0031	1.0011	0.9995	0.9902	0.9981	1.0000	1.0000	0.9972	1.0682	0.9972	0.9845
18.0	1.4518E 00	0.9812	1.0053	1.0039	1.0027	0.9890	0.9979	1.0000	1.0000	0.9984	1.1037	0.9983	0.9849
20.0	1.4460E 00	0.9788	1.0076	1.0067	1.0057	0.9878	0.9976	1.0000	1.0000	0.9995	1.1390	0.9994	0.9853
25.0	1.4327E 00	0.9729	1.0128	1.0134	1.0130	0.9847	0.9970	1.0000	1.0000	1.0021	1.2254	1.0021	0.9861
30.0	1.4204E 00	0.9670	1.0178	1.0197	1.0199	0.9817	0.9964	1.0000	1.0000	1.0046	1.3093	1.0046	0.9868
35.0	1.4088E 00	0.9612	1.0226	1.0257	1.0266	0.9788	0.9958	1.0000	1.0000	1.0070	1.3904	1.0070	0.9875
40.0	1.3978E 00	0.9555	1.0272	1.0316	1.0330	0.9759	0.9951	1.0000	1.0000	1.0093	1.4680	1.0093	0.9881
45.0	1.3872E 00	0.9498	1.0316	1.0372	1.0392	0.9731	0.9945	1.0000	1.0000	1.0116	1.5427	1.0116	0.9886
50.0	1.3771E 00	0.9442	1.0360	1.0428	1.0453	0.9704	0.9939	1.0000	1.0000	1.0138	1.6130	1.0138	0.9891

ELECTRON Z= 17 A= 37

p	F_0L_0	L_0	λ_2	λ_3	λ_4	μ_1	μ_2	Λ_1	Λ_2	ν_{12}	$\hat{\nu}_{12}$	η_{12}	$\hat{\eta}_{12}$
0.1	8.8436E 00	1.0013	2.4250	3.2988	3.8263	1.0000	1.0000	1.0006	1.0001	1.3435	0.7389	0.9624	0.9549
0.2	4.5572E 00	1.0013	1.3280	1.4323	1.4806	1.0000	1.0000	1.0006	1.0001	1.0666	0.7446	0.9704	0.9628
0.3	3.2569E 00	1.0013	1.1262	1.1566	1.1690	0.9999	1.0000	1.0006	1.0001	1.0167	0.7497	0.9734	0.9658
0.4	2.6724E 00	1.0013	1.0565	1.0663	1.0694	0.9999	1.0000	1.0006	1.0001	0.9998	0.7546	0.9753	0.9676
0.5	2.3534E 00	1.0012	1.0249	1.0263	1.0256	0.9998	1.0000	1.0006	1.0001	0.9924	0.7592	0.9765	0.9688
0.6	2.1576E 00	1.0012	1.0081	1.0053	1.0028	0.9998	1.0000	1.0006	1.0001	0.9887	0.7636	0.9775	0.9697
0.7	2.0277E 00	1.0012	0.9983	0.9931	0.9896	0.9997	0.9999	1.0005	1.0001	0.9866	0.7677	0.9783	0.9704
0.8	1.9367E 00	1.0011	0.9923	0.9856	0.9815	0.9997	0.9999	1.0005	1.0001	0.9853	0.7716	0.9789	0.9710
0.9	1.8701E 00	1.0010	0.9883	0.9808	0.9763	0.9996	0.9999	1.0005	1.0001	0.9846	0.7752	0.9795	0.9715
1.0	1.8199E 00	1.0010	0.9856	0.9775	0.9728	0.9995	0.9999	1.0005	1.0001	0.9842	0.7787	0.9800	0.9720
1.2	1.7500E 00	1.0008	0.9825	0.9738	0.9688	0.9994	0.9999	1.0004	1.0001	0.9838	0.7851	0.9808	0.9727
1.4	1.7046E 00	1.0006	0.9810	0.9720	0.9669	0.9992	0.9999	1.0004	1.0001	0.9837	0.7909	0.9815	0.9733
1.6	1.6732E 00	1.0005	0.9803	0.9713	0.9662	0.9991	0.9998	1.0003	1.0001	0.9839	0.7962	0.9821	0.9738
1.8	1.6505E 00	1.0003	0.9801	0.9711	0.9660	0.9989	0.9998	1.0003	1.0001	0.9841	0.8012	0.9826	0.9743
2.0	1.6333E 00	1.0001	0.9802	0.9712	0.9661	0.9988	0.9998	1.0003	1.0001	0.9843	0.8059	0.9831	0.9746
2.2	1.6199E 00	0.9998	0.9804	0.9715	0.9665	0.9987	0.9997	1.0003	1.0001	0.9845	0.8105	0.9836	0.9750
2.4	1.6092E 00	0.9996	0.9806	0.9719	0.9670	0.9985	0.9997	1.0003	1.0000	0.9848	0.8148	0.9840	0.9753
2.6	1.6005E 00	0.9994	0.9810	0.9724	0.9675	0.9984	0.9997	1.0002	1.0000	0.9851	0.8191	0.9844	0.9756
2.8	1.5931E 00	0.9992	0.9814	0.9729	0.9681	0.9982	0.9997	1.0002	1.0000	0.9854	0.8233	0.9847	0.9759
3.0	1.5869E 00	0.9989	0.9818	0.9735	0.9687	0.9981	0.9996	1.0002	1.0000	0.9856	0.8273	0.9851	0.9762
3.2	1.5815E 00	0.9987	0.9822	0.9740	0.9694	0.9980	0.9996	1.0002	1.0000	0.9859	0.8314	0.9854	0.9764
3.4	1.5768E 00	0.9984	0.9827	0.9746	0.9700	0.9978	0.9996	1.0002	1.0000	0.9861	0.8353	0.9857	0.9766
3.6	1.5727E 00	0.9982	0.9831	0.9752	0.9707	0.9977	0.9996	1.0002	1.0000	0.9864	0.8392	0.9860	0.9768
3.8	1.5690E 00	0.9979	0.9836	0.9758	0.9713	0.9975	0.9995	1.0002	1.0000	0.9866	0.8431	0.9863	0.9770
4.0	1.5656E 00	0.9977	0.9840	0.9763	0.9720	0.9974	0.9995	1.0002	1.0000	0.9869	0.8469	0.9865	0.9772
4.5	1.5584E 00	0.9971	0.9851	0.9777	0.9735	0.9971	0.9994	1.0001	1.0000	0.9875	0.8564	0.9872	0.9777
5.0	1.5525E 00	0.9964	0.9861	0.9791	0.9750	0.9967	0.9994	1.0001	1.0000	0.9880	0.8657	0.9878	0.9780
5.5	1.5474E 00	0.9958	0.9871	0.9804	0.9765	0.9964	0.9993	1.0001	1.0000	0.9886	0.8750	0.9883	0.9784
6.0	1.5430E 00	0.9952	0.9881	0.9816	0.9779	0.9960	0.9992	1.0001	1.0000	0.9891	0.8842	0.9889	0.9787
6.5	1.5390E 00	0.9945	0.9890	0.9828	0.9792	0.9957	0.9992	1.0001	1.0000	0.9895	0.8934	0.9894	0.9790
7.0	1.5354E 00	0.9939	0.9899	0.9840	0.9805	0.9954	0.9991	1.0001	1.0000	0.9900	0.9024	0.9899	0.9793
7.5	1.5321E 00	0.9932	0.9908	0.9851	0.9817	0.9950	0.9990	1.0001	1.0000	0.9905	0.9116	0.9903	0.9796
8.0	1.5290E 00	0.9926	0.9917	0.9862	0.9830	0.9947	0.9990	1.0001	1.0000	0.9909	0.9205	0.9908	0.9798
9.0	1.5234E 00	0.9913	0.9933	0.9883	0.9853	0.9940	0.9988	1.0001	1.0000	0.9917	0.9385	0.9916	0.9803
10.0	1.5183E 00	0.9900	0.9949	0.9904	0.9875	0.9933	0.9987	1.0001	1.0000	0.9925	0.9563	0.9924	0.9807
11.0	1.5136E 00	0.9887	0.9964	0.9923	0.9896	0.9926	0.9986	1.0001	1.0000	0.9933	0.9742	0.9932	0.9810
12.0	1.5093E 00	0.9873	0.9979	0.9941	0.9917	0.9920	0.9984	1.0001	1.0000	0.9940	0.9918	0.9940	0.9814
13.0	1.5051E 00	0.9860	0.9993	0.9959	0.9936	0.9913	0.9983	1.0001	1.0000	0.9947	1.0094	0.9947	0.9817
14.0	1.5012E 00	0.9847	1.0007	0.9977	0.9956	0.9906	0.9982	1.0000	1.0000	0.9954	1.0270	0.9954	0.9820
16.0	1.4938E 00	0.9821	1.0033	1.0010	0.9992	0.9893	0.9979	1.0000	1.0000	0.9967	1.0618	0.9967	0.9825
18.0	1.4870E 00	0.9795	1.0058	1.0042	1.0028	0.9879	0.9977	1.0000	1.0000	0.9980	1.0964	0.9980	0.9830
20.0	1.4804E 00	0.9769	1.0083	1.0073	1.0062	0.9866	0.9974	1.0000	1.0000	0.9992	1.1306	0.9992	0.9834
25.0	1.4653E 00	0.9704	1.0141	1.0147	1.0142	0.9832	0.9967	1.0000	1.0000	1.0022	1.2141	1.0022	0.9844
30.0	1.4514E 00	0.9640	1.0196	1.0216	1.0219	0.9800	0.9961	1.0000	1.0000	1.0049	1.2950	1.0049	0.9852
35.0	1.4384E 00	0.9577	1.0249	1.0283	1.0292	0.9768	0.9954	1.0000	1.0000	1.0076	1.3728	1.0076	0.9860
40.0	1.4260E 00	0.9514	1.0300	1.0348	1.0363	0.9736	0.9947	1.0000	1.0000	1.0102	1.4473	1.0101	0.9866
45.0	1.4141E 00	0.9452	1.0349	1.0411	1.0432	0.9706	0.9940	1.0000	1.0000	1.0126	1.5182	1.0126	0.9873
50.0	1.4027E 00	0.9392	1.0397	1.0472	1.0500	0.9676	0.9933	1.0000	1.0000	1.0151	1.5850	1.0150	0.9879

 Instr. see p. 22

ELECTRON Z= 18 A= 39

p	F_0L_0	L_0	λ_2	λ_3	λ_4	μ_1	μ_2	Λ_1	Λ_2	ν_{12}	$\hat{\nu}_{12}$	η_{12}	$\hat{\eta}_{12}$
0.1	9.4907E 00	1.0016	2.5893	3.6332	4.2839	1.0000	1.0000	1.0007	1.0001	1.3838	0.7374	0.9572	0.9488
0.2	4.8716E 00	1.0015	1.3659	1.4860	1.5422	1.0000	1.0000	1.0007	1.0001	1.0742	0.7434	0.9665	0.9580
0.3	3.4584E 00	1.0015	1.1408	1.1753	1.1895	0.9999	1.0000	1.0007	1.0001	1.0185	0.7488	0.9701	0.9616
0.4	2.8209E 00	1.0015	1.0631	1.0742	1.0776	0.9999	1.0000	1.0007	1.0001	0.9997	0.7539	0.9722	0.9636
0.5	2.4726E 00	1.0015	1.0277	1.0293	1.0286	0.9998	1.0000	1.0006	1.0001	0.9914	0.7587	0.9736	0.9650
0.6	2.2589E 00	1.0014	1.0090	1.0059	1.0031	0.9998	1.0000	1.0006	1.0001	0.9872	0.7632	0.9747	0.9660
0.7	2.1172E 00	1.0014	0.9981	0.9924	0.9884	0.9997	0.9999	1.0006	1.0001	0.9849	0.7674	0.9756	0.9668
0.8	2.0178E 00	1.0013	0.9914	0.9840	0.9794	0.9996	0.9999	1.0006	1.0001	0.9835	0.7714	0.9763	0.9675
0.9	1.9452E 00	1.0012	0.9869	0.9786	0.9735	0.9996	0.9999	1.0005	1.0001	0.9827	0.7752	0.9770	0.9681
1.0	1.8904E 00	1.0012	0.9840	0.9749	0.9696	0.9995	0.9999	1.0005	1.0001	0.9822	0.7787	0.9775	0.9686
1.2	1.8142E 00	1.0010	0.9805	0.9707	0.9652	0.9993	0.9999	1.0005	1.0001	0.9818	0.7851	0.9784	0.9694
1.4	1.7646E 00	1.0008	0.9788	0.9688	0.9631	0.9992	0.9998	1.0004	1.0001	0.9818	0.7909	0.9792	0.9701
1.6	1.7303E 00	1.0006	0.9781	0.9679	0.9622	0.9990	0.9998	1.0004	1.0001	0.9819	0.7962	0.9799	0.9707
1.8	1.7054E 00	1.0004	0.9778	0.9677	0.9620	0.9989	0.9998	1.0003	1.0001	0.9821	0.8012	0.9805	0.9712
2.0	1.6866E 00	1.0002	0.9778	0.9678	0.9622	0.9987	0.9998	1.0003	1.0001	0.9824	0.8059	0.9811	0.9716
2.2	1.6719E 00	1.0000	0.9781	0.9681	0.9626	0.9986	0.9997	1.0003	1.0001	0.9827	0.8103	0.9816	0.9720
2.4	1.6602E 00	0.9997	0.9784	0.9686	0.9631	0.9984	0.9997	1.0003	1.0001	0.9830	0.8146	0.9820	0.9724
2.6	1.6506E 00	0.9995	0.9788	0.9691	0.9637	0.9982	0.9997	1.0003	1.0000	0.9833	0.8188	0.9824	0.9727
2.8	1.6425E 00	0.9992	0.9792	0.9697	0.9644	0.9981	0.9996	1.0002	1.0000	0.9836	0.8229	0.9828	0.9730
3.0	1.6356E 00	0.9990	0.9797	0.9703	0.9651	0.9979	0.9996	1.0002	1.0000	0.9839	0.8269	0.9832	0.9733
3.2	1.6297E 00	0.9987	0.9801	0.9710	0.9658	0.9978	0.9996	1.0002	1.0000	0.9842	0.8308	0.9836	0.9736
3.4	1.6245E 00	0.9984	0.9806	0.9716	0.9665	0.9976	0.9995	1.0002	1.0000	0.9844	0.8346	0.9839	0.9738
3.6	1.6199E 00	0.9982	0.9811	0.9722	0.9672	0.9975	0.9995	1.0002	1.0000	0.9847	0.8385	0.9843	0.9740
3.8	1.6157E 00	0.9979	0.9816	0.9729	0.9679	0.9973	0.9995	1.0002	1.0000	0.9850	0.8422	0.9846	0.9743
4.0	1.6120E 00	0.9976	0.9821	0.9735	0.9686	0.9972	0.9995	1.0002	1.0000	0.9853	0.8460	0.9849	0.9745
4.5	1.6040E 00	0.9970	0.9833	0.9751	0.9704	0.9968	0.9994	1.0002	1.0000	0.9859	0.8552	0.9856	0.9750
5.0	1.5974E 00	0.9963	0.9844	0.9765	0.9720	0.9965	0.9993	1.0001	1.0000	0.9865	0.8643	0.9863	0.9754
5.5	1.5917E 00	0.9956	0.9855	0.9780	0.9736	0.9961	0.9992	1.0001	1.0000	0.9871	0.8733	0.9869	0.9758
6.0	1.5867E 00	0.9949	0.9866	0.9794	0.9752	0.9957	0.9992	1.0001	1.0000	0.9877	0.8822	0.9875	0.9762
6.5	1.5823E 00	0.9942	0.9876	0.9807	0.9766	0.9953	0.9991	1.0001	1.0000	0.9882	0.8911	0.9880	0.9765
7.0	1.5782E 00	0.9935	0.9886	0.9820	0.9781	0.9950	0.9990	1.0001	1.0000	0.9887	0.8999	0.9886	0.9768
7.5	1.5745E 00	0.9928	0.9896	0.9832	0.9794	0.9946	0.9990	1.0001	1.0000	0.9892	0.9087	0.9891	0.9771
8.0	1.5710E 00	0.9921	0.9906	0.9845	0.9808	0.9942	0.9989	1.0001	1.0000	0.9897	0.9174	0.9896	0.9774
9.0	1.5647E 00	0.9907	0.9924	0.9868	0.9833	0.9935	0.9988	1.0001	1.0000	0.9906	0.9349	0.9905	0.9779
10.0	1.5590E 00	0.9893	0.9941	0.9890	0.9858	0.9928	0.9986	1.0001	1.0000	0.9915	0.9521	0.9914	0.9784
11.0	1.5537E 00	0.9879	0.9958	0.9911	0.9881	0.9920	0.9985	1.0001	1.0000	0.9923	0.9693	0.9923	0.9788
12.0	1.5488E 00	0.9865	0.9974	0.9932	0.9904	0.9913	0.9983	1.0001	1.0000	0.9931	0.9864	0.9931	0.9792
13.0	1.5441E 00	0.9850	0.9989	0.9951	0.9925	0.9906	0.9982	1.0001	1.0000	0.9939	1.0034	0.9939	0.9795
14.0	1.5397E 00	0.9836	1.0004	0.9970	0.9946	0.9898	0.9980	1.0001	1.0000	0.9947	1.0203	0.9946	0.9798
16.0	1.5314E 00	0.9808	1.0034	1.0007	0.9987	0.9884	0.9978	1.0000	1.0000	0.9961	1.0540	0.9961	0.9804
18.0	1.5237E 00	0.9780	1.0061	1.0042	1.0026	0.9869	0.9975	1.0000	1.0000	0.9975	1.0872	0.9975	0.9810
20.0	1.5164E 00	0.9752	1.0088	1.0076	1.0063	0.9855	0.9972	1.0000	1.0000	0.9989	1.1202	0.9988	0.9815
25.0	1.4995E 00	0.9682	1.0152	1.0157	1.0151	0.9819	0.9965	1.0000	1.0000	1.0021	1.2005	1.0021	0.9826
30.0	1.4839E 00	0.9613	1.0212	1.0233	1.0235	0.9783	0.9957	1.0000	1.0000	1.0051	1.2783	1.0051	0.9835
35.0	1.4694E 00	0.9544	1.0270	1.0306	1.0315	0.9749	0.9950	1.0000	1.0000	1.0080	1.3528	1.0080	0.9843
40.0	1.4556E 00	0.9477	1.0325	1.0377	1.0393	0.9715	0.9943	1.0000	1.0000	1.0108	1.4237	1.0108	0.9851
45.0	1.4424E 00	0.9411	1.0379	1.0445	1.0468	0.9683	0.9935	1.0000	1.0000	1.0135	1.4908	1.0135	0.9858
50.0	1.4298E 00	0.9345	1.0431	1.0512	1.0542	0.9651	0.9927	1.0000	1.0000	1.0161	1.5542	1.0161	0.9865

Instr. see p. 22

ELECTRON Z= 19 A= 41

p	F_0L_0	L_0	λ_2	λ_3	λ_4	μ_1	μ_2	Λ_1	Λ_2	ν_{12}	$\hat{\nu}_{12}$	η_{12}	$\hat{\eta}_{12}$
0.1	1.0161E 01	1.0018	2.7611	3.9988	4.7961	1.0000	1.0000	1.0008	1.0001	1.4260	0.7358	0.9516	0.9423
0.2	5.1994E 00	1.0018	1.4056	1.5431	1.6081	1.0000	1.0000	1.0007	1.0001	1.0823	0.7421	0.9624	0.9530
0.3	3.6695E 00	1.0018	1.1561	1.1951	1.2111	0.9999	1.0000	1.0007	1.0001	1.0204	0.7478	0.9665	0.9570
0.4	2.9766E 00	1.0017	1.0699	1.0824	1.0863	0.9999	1.0000	1.0007	1.0001	0.9995	0.7531	0.9689	0.9593
0.5	2.5977E 00	1.0017	1.0308	1.0326	1.0317	0.9998	1.0000	1.0007	1.0001	0.9904	0.7581	0.9705	0.9609
0.6	2.3651E 00	1.0017	1.0100	1.0066	1.0035	0.9997	1.0000	1.0007	1.0001	0.9857	0.7628	0.9718	0.9621
0.7	2.2109E 00	1.0016	0.9979	0.9915	0.9872	0.9997	0.9999	1.0006	1.0001	0.9831	0.7672	0.9728	0.9630
0.8	2.1028E 00	1.0015	0.9904	0.9822	0.9771	0.9996	0.9999	1.0006	1.0001	0.9816	0.7712	0.9736	0.9638
0.9	2.0238E 00	1.0015	0.9855	0.9762	0.9707	0.9995	0.9999	1.0006	1.0001	0.9807	0.7750	0.9743	0.9644
1.0	1.9642E 00	1.0014	0.9822	0.9722	0.9663	0.9994	0.9999	1.0005	1.0001	0.9802	0.7786	0.9749	0.9650
1.2	1.8813E 00	1.0012	0.9783	0.9676	0.9614	0.9993	0.9999	1.0005	1.0001	0.9797	0.7851	0.9760	0.9659
1.4	1.8273E 00	1.0010	0.9765	0.9654	0.9591	0.9991	0.9998	1.0004	1.0001	0.9797	0.7909	0.9769	0.9667
1.6	1.7900E 00	1.0008	0.9757	0.9644	0.9581	0.9989	0.9998	1.0004	1.0001	0.9798	0.7963	0.9776	0.9673
1.8	1.7628E 00	1.0006	0.9754	0.9642	0.9579	0.9988	0.9998	1.0004	1.0001	0.9800	0.8012	0.9783	0.9679
2.0	1.7423E 00	1.0003	0.9754	0.9643	0.9581	0.9986	0.9997	1.0003	1.0001	0.9803	0.8059	0.9789	0.9684
2.2	1.7262E 00	1.0001	0.9756	0.9647	0.9585	0.9984	0.9997	1.0003	1.0001	0.9807	0.8103	0.9794	0.9688
2.4	1.7134E 00	0.9998	0.9760	0.9652	0.9591	0.9983	0.9997	1.0003	1.0001	0.9810	0.8145	0.9800	0.9692
2.6	1.7028E 00	0.9996	0.9764	0.9658	0.9598	0.9981	0.9996	1.0003	1.0001	0.9813	0.8186	0.9804	0.9696
2.8	1.6939E 00	0.9993	0.9769	0.9664	0.9605	0.9979	0.9996	1.0003	1.0001	0.9817	0.8226	0.9809	0.9699
3.0	1.6864E 00	0.9990	0.9774	0.9671	0.9613	0.9978	0.9996	1.0002	1.0000	0.9820	0.8265	0.9813	0.9703
3.2	1.6798E 00	0.9987	0.9779	0.9678	0.9620	0.9976	0.9995	1.0002	1.0000	0.9823	0.8304	0.9817	0.9706
3.4	1.6741E 00	0.9985	0.9785	0.9685	0.9628	0.9975	0.9995	1.0002	1.0000	0.9826	0.8342	0.9821	0.9709
3.6	1.6690E 00	0.9982	0.9790	0.9692	0.9636	0.9973	0.9995	1.0002	1.0000	0.9829	0.8379	0.9824	0.9711
3.8	1.6644E 00	0.9979	0.9795	0.9699	0.9644	0.9971	0.9994	1.0002	1.0000	0.9832	0.8415	0.9828	0.9714
4.0	1.6603E 00	0.9976	0.9801	0.9706	0.9652	0.9970	0.9994	1.0002	1.0000	0.9835	0.8452	0.9831	0.9716
4.5	1.6514E 00	0.9969	0.9814	0.9723	0.9671	0.9966	0.9993	1.0002	1.0000	0.9843	0.8542	0.9839	0.9721
5.0	1.6440E 00	0.9961	0.9826	0.9739	0.9689	0.9962	0.9993	1.0002	1.0000	0.9849	0.8631	0.9846	0.9726
5.5	1.6377E 00	0.9954	0.9839	0.9755	0.9706	0.9958	0.9992	1.0001	1.0000	0.9856	0.8718	0.9853	0.9731
6.0	1.6321E 00	0.9946	0.9850	0.9770	0.9723	0.9954	0.9991	1.0001	1.0000	0.9862	0.8805	0.9860	0.9735
6.5	1.6271E 00	0.9939	0.9862	0.9785	0.9740	0.9950	0.9990	1.0001	1.0000	0.9868	0.8892	0.9866	0.9739
7.0	1.6226E 00	0.9931	0.9873	0.9799	0.9755	0.9946	0.9990	1.0001	1.0000	0.9873	0.8977	0.9872	0.9742
7.5	1.6184E 00	0.9924	0.9884	0.9813	0.9770	0.9942	0.9989	1.0001	1.0000	0.9879	0.9063	0.9877	0.9745
8.0	1.6145E 00	0.9916	0.9894	0.9826	0.9785	0.9938	0.9988	1.0001	1.0000	0.9884	0.9148	0.9883	0.9749
9.0	1.6074E 00	0.9901	0.9914	0.9851	0.9813	0.9930	0.9987	1.0001	1.0000	0.9894	0.9317	0.9893	0.9754
10.0	1.6010E 00	0.9886	0.9933	0.9876	0.9840	0.9922	0.9985	1.0001	1.0000	0.9904	0.9484	0.9903	0.9759
11.0	1.5951E 00	0.9871	0.9951	0.9899	0.9866	0.9914	0.9984	1.0001	1.0000	0.9913	0.9651	0.9912	0.9764
12.0	1.5896E 00	0.9855	0.9969	0.9921	0.9890	0.9906	0.9982	1.0001	1.0000	0.9922	0.9817	0.9921	0.9768
13.0	1.5844E 00	0.9840	0.9986	0.9943	0.9914	0.9898	0.9980	1.0001	1.0000	0.9930	0.9982	0.9930	0.9772
14.0	1.5794E 00	0.9825	1.0002	0.9964	0.9937	0.9890	0.9979	1.0001	1.0000	0.9939	1.0146	0.9938	0.9776
16.0	1.5701E 00	0.9795	1.0034	1.0004	0.9981	0.9875	0.9976	1.0000	1.0000	0.9955	1.0471	0.9954	0.9782
18.0	1.5615E 00	0.9764	1.0064	1.0042	1.0023	0.9859	0.9973	1.0000	1.0000	0.9970	1.0793	0.9970	0.9788
20.0	1.5533E 00	0.9734	1.0094	1.0079	1.0064	0.9843	0.9970	1.0000	1.0000	0.9985	1.1110	0.9984	0.9794
25.0	1.5344E 00	0.9659	1.0163	1.0167	1.0160	0.9805	0.9962	1.0000	1.0000	1.0020	1.1886	1.0019	0.9806
30.0	1.5172E 00	0.9585	1.0229	1.0250	1.0251	0.9767	0.9954	1.0000	1.0000	1.0053	1.2632	1.0052	0.9817
35.0	1.5010E 00	0.9512	1.0292	1.0330	1.0339	0.9730	0.9946	1.0000	1.0000	1.0084	1.3342	1.0084	0.9826
40.0	1.4857E 00	0.9439	1.0352	1.0407	1.0423	0.9694	0.9938	1.0000	1.0000	1.0115	1.4024	1.0115	0.9835
45.0	1.4712E 00	0.9368	1.0410	1.0481	1.0505	0.9658	0.9930	1.0000	1.0000	1.0144	1.4662	1.0144	0.9843
50.0	1.4572E 00	0.9299	1.0466	1.0554	1.0585	0.9625	0.9922	1.0000	1.0000	1.0173	1.5254	1.0172	0.9851

ELECTRON Z= 20 A= 43

p	F_0L_0	L_0	λ_2	λ_3	λ_4	μ_1	μ_2	Λ_1	Λ_2	ν_{12}	$\hat{\nu}_{12}$	η_{12}	$\hat{\eta}_{12}$
0.1	1.0855E 01	1.0021	2.9402	4.3974	5.3677	1.0000	1.0000	1.0008	1.0002	1.4703	0.7341	0.9455	0.9352
0.2	5.5411E 00	1.0021	1.4470	1.6035	1.6783	1.0000	1.0000	1.0008	1.0002	1.0907	0.7408	0.9580	0.9476
0.3	3.8905E 00	1.0020	1.1721	1.2158	1.2338	0.9999	1.0000	1.0008	1.0002	1.0224	0.7468	0.9627	0.9522
0.4	3.1400E 00	1.0020	1.0771	1.0910	1.0954	0.9999	1.0000	1.0008	1.0001	0.9994	0.7523	0.9654	0.9549
0.5	2.7289E 00	1.0020	1.0339	1.0360	1.0350	0.9998	1.0000	1.0007	1.0001	0.9893	0.7575	0.9673	0.9566
0.6	2.4766E 00	1.0019	1.0111	1.0073	1.0038	0.9997	0.9999	1.0007	1.0001	0.9841	0.7623	0.9687	0.9580
0.7	2.3092E 00	1.0019	0.9977	0.9907	0.9859	0.9997	0.9999	1.0007	1.0001	0.9813	0.7668	0.9698	0.9590
0.8	2.1919E 00	1.0018	0.9894	0.9804	0.9748	0.9996	0.9999	1.0006	1.0001	0.9796	0.7710	0.9707	0.9599
0.9	2.1062E 00	1.0017	0.9840	0.9738	0.9676	0.9995	0.9999	1.0006	1.0001	0.9786	0.7749	0.9715	0.9606
1.0	2.0415E 00	1.0016	0.9804	0.9693	0.9629	0.9994	0.9999	1.0006	1.0001	0.9780	0.7785	0.9722	0.9612
1.2	1.9515E 00	1.0014	0.9761	0.9642	0.9574	0.9992	0.9999	1.0005	1.0001	0.9775	0.7851	0.9734	0.9622
1.4	1.8930E 00	1.0012	0.9741	0.9618	0.9549	0.9990	0.9998	1.0005	1.0001	0.9774	0.7910	0.9743	0.9631
1.6	1.8523E 00	1.0010	0.9731	0.9608	0.9538	0.9989	0.9998	1.0004	1.0001	0.9776	0.7963	0.9752	0.9638
1.8	1.8228E 00	1.0008	0.9728	0.9605	0.9536	0.9987	0.9997	1.0004	1.0001	0.9779	0.8013	0.9759	0.9644
2.0	1.8004E 00	1.0005	0.9728	0.9606	0.9538	0.9985	0.9997	1.0004	1.0001	0.9782	0.8059	0.9766	0.9650
2.2	1.7829E 00	1.0002	0.9731	0.9610	0.9543	0.9983	0.9997	1.0003	1.0001	0.9786	0.8102	0.9772	0.9655
2.4	1.7688E 00	0.9999	0.9735	0.9616	0.9549	0.9981	0.9996	1.0003	1.0001	0.9789	0.8145	0.9778	0.9659
2.6	1.7573E 00	0.9997	0.9740	0.9622	0.9556	0.9980	0.9996	1.0003	1.0001	0.9793	0.8185	0.9783	0.9663
2.8	1.7475E 00	0.9994	0.9745	0.9629	0.9564	0.9978	0.9996	1.0003	1.0001	0.9797	0.8225	0.9788	0.9667
3.0	1.7392E 00	0.9991	0.9750	0.9636	0.9573	0.9976	0.9995	1.0003	1.0001	0.9800	0.8263	0.9792	0.9671
3.2	1.7320E 00	0.9988	0.9756	0.9644	0.9581	0.9974	0.9995	1.0002	1.0000	0.9804	0.8301	0.9797	0.9674
3.4	1.7257E 00	0.9985	0.9762	0.9652	0.9590	0.9973	0.9995	1.0002	1.0000	0.9807	0.8338	0.9801	0.9677
3.6	1.7201E 00	0.9982	0.9768	0.9659	0.9598	0.9971	0.9994	1.0002	1.0000	0.9811	0.8375	0.9805	0.9680
3.8	1.7151E 00	0.9979	0.9774	0.9667	0.9607	0.9969	0.9994	1.0002	1.0000	0.9814	0.8411	0.9809	0.9683
4.0	1.7105E 00	0.9976	0.9780	0.9675	0.9615	0.9968	0.9994	1.0002	1.0000	0.9817	0.8446	0.9813	0.9686
4.5	1.7006E 00	0.9968	0.9794	0.9693	0.9636	0.9963	0.9993	1.0002	1.0000	0.9825	0.8534	0.9821	0.9692
5.0	1.6925E 00	0.9960	0.9808	0.9711	0.9656	0.9959	0.9992	1.0002	1.0000	0.9832	0.8621	0.9829	0.9697
5.5	1.6854E 00	0.9952	0.9821	0.9729	0.9675	0.9955	0.9991	1.0001	1.0000	0.9840	0.8707	0.9837	0.9702
6.0	1.6792E 00	0.9944	0.9834	0.9745	0.9694	0.9950	0.9990	1.0001	1.0000	0.9846	0.8791	0.9844	0.9707
6.5	1.6737E 00	0.9936	0.9846	0.9761	0.9711	0.9946	0.9990	1.0001	1.0000	0.9853	0.8875	0.9851	0.9711
7.0	1.6686E 00	0.9928	0.9859	0.9777	0.9728	0.9942	0.9989	1.0001	1.0000	0.9859	0.8959	0.9857	0.9715
7.5	1.6640E 00	0.9920	0.9870	0.9792	0.9745	0.9937	0.9988	1.0001	1.0000	0.9865	0.9042	0.9863	0.9718
8.0	1.6596E 00	0.9911	0.9882	0.9806	0.9761	0.9933	0.9987	1.0001	1.0000	0.9871	0.9125	0.9869	0.9722
9.0	1.6517E 00	0.9895	0.9903	0.9834	0.9792	0.9925	0.9986	1.0001	1.0000	0.9882	0.9289	0.9880	0.9728
10.0	1.6445E 00	0.9879	0.9924	0.9861	0.9821	0.9916	0.9984	1.0001	1.0000	0.9892	0.9452	0.9891	0.9733
11.0	1.6379E 00	0.9863	0.9944	0.9886	0.9849	0.9908	0.9982	1.0001	1.0000	0.9902	0.9614	0.9901	0.9739
12.0	1.6317E 00	0.9846	0.9963	0.9910	0.9876	0.9899	0.9981	1.0001	1.0000	0.9912	0.9775	0.9911	0.9743
13.0	1.6259E 00	0.9830	0.9982	0.9934	0.9901	0.9891	0.9979	1.0001	1.0000	0.9921	0.9936	0.9920	0.9748
14.0	1.6204E 00	0.9814	1.0000	0.9956	0.9926	0.9882	0.9977	1.0001	1.0000	0.9930	1.0095	0.9930	0.9752
16.0	1.6100E 00	0.9781	1.0034	1.0000	0.9975	0.9865	0.9974	1.0001	1.0000	0.9948	1.0410	0.9947	0.9759
18.0	1.6004E 00	0.9749	1.0067	1.0042	1.0021	0.9848	0.9971	1.0000	1.0000	0.9964	1.0721	0.9964	0.9766
20.0	1.5913E 00	0.9716	1.0099	1.0082	1.0065	0.9832	0.9968	1.0000	1.0000	0.9980	1.1029	0.9980	0.9772
25.0	1.5703E 00	0.9636	1.0175	1.0178	1.0169	0.9790	0.9959	1.0000	1.0000	1.0018	1.1778	1.0018	0.9786
30.0	1.5512E 00	0.9556	1.0246	1.0268	1.0268	0.9750	0.9951	1.0000	1.0000	1.0054	1.2495	1.0054	0.9797
35.0	1.5333E 00	0.9478	1.0314	1.0354	1.0363	0.9710	0.9943	1.0000	1.0000	1.0088	1.3181	1.0088	0.9808
40.0	1.5164E 00	0.9401	1.0379	1.0437	1.0454	0.9671	0.9934	1.0000	1.0000	1.0121	1.3826	1.0121	0.9818
45.0	1.5004E 00	0.9325	1.0442	1.0518	1.0543	0.9634	0.9925	1.0000	1.0000	1.0153	1.4436	1.0153	0.9827
50.0	1.4851E 00	0.9251	1.0503	1.0596	1.0630	0.9598	0.9916	1.0000	1.0000	1.0184	1.5000	1.0184	0.9836

ELECTRON Z= 21 A= 47

p	F_0L_0	L_0	λ_2	λ_3	λ_4	μ_1	μ_2	Λ_1	Λ_2	ν_{12}	$\hat{\nu}_{12}$	η_{12}	$\hat{\eta}_{12}$
0.1	1.1571E 01	1.0023	3.1270	4.8319	6.0062	1.0000	1.0000	1.0009	1.0002	1.5167	0.7324	0.9391	0.9278
0.2	5.8952E 00	1.0023	1.4903	1.6678	1.7535	0.9999	1.0000	1.0009	1.0002	1.0995	0.7394	0.9535	0.9420
0.3	4.1203E 00	1.0023	1.1889	1.2378	1.2581	0.9999	1.0000	1.0008	1.0002	1.0245	0.7457	0.9588	0.9473
0.4	3.3101E 00	1.0023	1.0847	1.1003	1.1052	0.9998	1.0000	1.0008	1.0002	0.9992	0.7516	0.9619	0.9502
0.5	2.8655E 00	1.0022	1.0374	1.0397	1.0388	0.9998	1.0000	1.0008	1.0002	0.9882	0.7570	0.9639	0.9522
0.6	2.5925E 00	1.0022	1.0123	1.0082	1.0045	0.9997	0.9999	1.0008	1.0001	0.9825	0.7620	0.9655	0.9537
0.7	2.4114E 00	1.0021	0.9977	0.9900	0.98;8	0.9996	0.9999	1.0007	1.0001	0.9794	0.7666	0.9667	0.9549
0.8	2.2845E 00	1.0020	0.9886	0.9788	0.9726	0.9995	0.9999	1.0007	1.0001	0.9776	0.7709	0.9678	0.9558
0.9	2.1917E 00	1.0019	0.9826	0.9715	0.9648	0.9994	0.9999	1.0007	1.0001	0.9765	0.7749	0.9687	0.9566
1.0	2.1217E 00	1.0018	0.9786	0.9666	0.9596	0.9993	0.9999	1.0006	1.0001	0.9758	0.7786	0.9694	0.9573
1.2	2.0243E 00	1.0016	0.9740	0.9610	0.9536	0.9992	0.9998	1.0006	1.0001	0.9753	0.7854	0.9707	0.9585
1.4	1.9608E 00	1.0014	0.9717	0.9584	0.9508	0.9990	0.9998	1.0005	1.0001	0.9752	0.7914	0.9718	0.9594
1.6	1.9168E 00	1.0012	0.9707	0.9572	0.9497	0.9988	0.9998	1.0005	1.0001	0.9754	0.7968	0.9727	0.9602
1.8	1.8847E 00	1.0009	0.9703	0.9569	0.9494	0.9986	0.9997	1.0004	1.0001	0.9757	0.8017	0.9736	0.9609
2.0	1.8604E 00	1.0006	0.9704	0.9570	0.9496	0.9984	0.9997	1.0004	1.0001	0.9761	0.8064	0.9743	0.9615
2.2	1.8413E 00	1.0003	0.9707	0.9575	0.9501	0.9982	0.9997	1.0004	1.0001	0.9765	0.8108	0.9750	0.9621
2.4	1.8260E 00	1.0000	0.9711	0.9581	0.9508	0.9980	0.9996	1.0003	1.0001	0.9769	0.8150	0.9756	0.9626
2.6	1.8133E 00	0.9997	0.9716	0.9588	0.9516	0.9978	0.9996	1.0003	1.0001	0.9773	0.8191	0.9762	0.9630
2.8	1.8027E 00	0.9994	0.9722	0.9596	0.9525	0.9976	0.9995	1.0003	1.0001	0.9777	0.8230	0.9767	0.9634
3.0	1.7936E 00	0.9991	0.9728	0.9604	0.9534	0.9974	0.9995	1.0003	1.0001	0.9781	0.8269	0.9772	0.9638
3.2	1.7857E 00	0.9987	0.9734	0.9612	0.9543	0.9972	0.9995	1.0003	1.0001	0.9785	0.8306	0.9777	0.9642
3.4	1.7788E 00	0.9984	0.9741	0.9620	0.9553	0.9970	0.9994	1.0003	1.0000	0.9788	0.8343	0.9782	0.9645
3.6	1.7726E 00	0.9981	0.9747	0.9629	0.9562	0.9968	0.9994	1.0002	1.0000	0.9792	0.8380	0.9786	0.9649
3.8	1.7670E 00	0.9977	0.9753	0.9637	0.9571	0.9967	0.9994	1.0002	1.0000	0.9796	0.8416	0.9790	0.9652
4.0	1.7620E 00	0.9974	0.9760	0.9646	0.9581	0.9965	0.9993	1.0002	1.0000	0.9799	0.8451	0.9794	0.9655
4.5	1.7511E 00	0.9966	0.9776	0.9666	0.9603	0.9960	0.9992	1.0002	1.0000	0.9808	0.8539	0.9804	0.9661
5.0	1.7420E 00	0.9957	0.9791	0.9686	0.9625	0.9955	0.9991	1.0002	1.0000	0.9816	0.8625	0.9813	0.9667
5.5	1.7342E 00	0.9948	0.9805	0.9705	0.9646	0.9951	0.9991	1.0002	1.0000	0.9824	0.8710	0.9821	0.9673
6.0	1.7273E 00	0.9940	0.9820	0.9723	0.9666	0.9946	0.9990	1.0001	1.0000	0.9831	0.8794	0.9829	0.9678
6.5	1.7211E 00	0.9931	0.9833	0.9740	0.9686	0.9941	0.9989	1.0001	1.0000	0.9838	0.8877	0.9836	0.9682
7.0	1.7155E 00	0.9922	0.9846	0.9757	0.9705	0.9937	0.9988	1.0001	1.0000	0.9845	0.8960	0.9843	0.9687
7.5	1.7103E 00	0.9913	0.9859	0.9774	0.9723	0.9932	0.9987	1.0001	1.0000	0.9852	0.9042	0.9850	0.9691
8.0	1.7054E 00	0.9905	0.9872	0.9790	0.9740	0.9927	0.9986	1.0001	1.0000	0.9858	0.9124	0.9856	0.9694
9.0	1.6965E 00	0.9887	0.9896	0.9820	0.9774	0.9918	0.9984	1.0001	1.0000	0.9870	0.9286	0.9869	0.9701
10.0	1.6885E 00	0.9869	0.9918	0.9849	0.9806	0.9909	0.9983	1.0001	1.0000	0.9881	0.9447	0.9880	0.9707
11.0	1.6811E 00	0.9852	0.9940	0.9877	0.9836	0.9900	0.9981	1.0001	1.0000	0.9892	0.9607	0.9891	0.9713
12.0	1.6742E 00	0.9834	0.9961	0.9903	0.9865	0.9890	0.9979	1.0001	1.0000	0.9903	0.9766	0.9902	0.9718
13.0	1.6677E 00	0.9816	0.9981	0.9929	0.9894	0.9881	0.9977	1.0001	1.0000	0.9913	0.9924	0.9912	0.9723
14.0	1.6615E 00	0.9799	1.0001	0.9954	0.9921	0.9872	0.9976	1.0001	1.0000	0.9923	1.0081	0.9922	0.9728
16.0	1.6499E 00	0.9763	1.0039	1.0002	0.9974	0.9854	0.9972	1.0001	1.0000	0.9942	1.0391	0.9942	0.9736
18.0	1.6391E 00	0.9728	1.0075	1.0047	1.0024	0.9835	0.9969	1.0001	1.0000	0.9960	1.0697	0.9960	0.9743
20.0	1.6289E 00	0.9693	1.0110	1.0091	1.0072	0.9817	0.9965	1.0000	1.0000	0.9978	1.0999	0.9977	0.9750
25.0	1.6055E 00	0.9606	1.0193	1.0196	1.0187	0.9772	0.9956	1.0000	1.0000	1.0019	1.1731	1.0019	0.9765
30.0	1.5842E 00	0.9520	1.0270	1.0294	1.0295	0.9728	0.9947	1.0000	1.0000	1.0059	1.2431	1.0058	0.9778
35.0	1.5643E 00	0.9436	1.0344	1.0388	1.0398	0.9685	0.9938	1.0000	1.0000	1.0096	1.3094	1.0096	0.9790
40.0	1.5456E 00	0.9353	1.0416	1.0479	1.0498	0.9644	0.9928	1.0000	1.0000	1.0132	1.3717	1.0132	0.9801
45.0	1.5279E 00	0.9271	1.0484	1.0567	1.0595	0.9604	0.9919	1.0000	1.0000	1.0167	1.4301	1.0166	0.9811
50.0	1.5109E 00	0.9192	1.0550	1.0653	1.0689	0.9565	0.9909	1.0000	1.0000	1.0200	1.4840	1.0200	0.9821

ELECTRON Z= 22 A= 48

p	F_0L_0	L_0	λ_2	λ_3	λ_4	μ_1	μ_2	Λ_1	Λ_2	ν_{12}	$\hat{\nu}_{12}$	η_{12}	$\hat{\eta}_{12}$
0.1	1.2321E 01	1.0026	3.3195	5.3009	6.7129	1.0000	1.0000	1.0009	1.0002	1.5648	0.7305	0.9321	0.9198
0.2	6.2676E 00	1.0026	1.5347	1.7348	1.8326	0.9999	1.0000	1.0009	1.0002	1.1085	0.7379	0.9485	0.9360
0.3	4.3634E 00	1.0026	1.2060	1.2603	1.2830	0.9999	1.0000	1.0009	1.0002	1.0265	0.7445	0.9545	0.9419
0.4	3.4903E 00	1.0026	1.0923	1.1094	1.1149	0.9998	1.0000	1.0009	1.0002	0.9989	0.7506	0.9580	0.9452
0.5	3.0105E 00	1.0025	1.0407	1.0432	1.0422	0.9998	1.0000	1.0008	1.0002	0.9868	0.7561	0.9603	0.9475
0.6	2.7156E 00	1.0025	1.0133	1.0088	1.0047	0.9997	0.9999	1.0008	1.0002	0.9807	0.7613	0.9620	0.9491
0.7	2.5199E 00	1.0024	0.9973	0.9889	0.9832	0.9996	0.9999	1.0008	1.0001	0.9773	0.7661	0.9634	0.9504
0.8	2.3828E 00	1.0023	0.9874	0.9766	0.9699	0.9995	0.9999	1.0007	1.0001	0.9753	0.7705	0.9645	0.9515
0.9	2.2826E 00	1.0022	0.9809	0.9687	0.9614	0.9994	0.9999	1.0007	1.0001	0.9741	0.7746	0.9655	0.9523
1.0	2.2070E 00	1.0021	0.9765	0.9634	0.9557	0.9993	0.9999	1.0007	1.0001	0.9734	0.7784	0.9664	0.9531
1.2	2.1016E 00	1.0019	0.9714	0.9573	0.9492	0.9991	0.9998	1.0006	1.0001	0.9728	0.7852	0.9678	0.9544
1.4	2.0330E 00	1.0017	0.9690	0.9544	0.9461	0.9989	0.9998	1.0005	1.0001	0.9727	0.7912	0.9690	0.9554
1.6	1.9853E 00	1.0014	0.9678	0.9531	0.9449	0.9987	0.9998	1.0005	1.0001	0.9729	0.7966	0.9700	0.9563
1.8	1.9505E 00	1.0011	0.9675	0.9528	0.9445	0.9985	0.9997	1.0005	1.0001	0.9733	0.8016	0.9709	0.9571
2.0	1.9241E 00	1.0008	0.9675	0.9529	0.9448	0.9983	0.9997	1.0004	1.0001	0.9736	0.8062	0.9717	0.9577
2.2	1.9034E 00	1.0005	0.9678	0.9534	0.9453	0.9981	0.9996	1.0004	1.0001	0.9741	0.8106	0.9725	0.9583
2.4	1.8867E 00	1.0002	0.9683	0.9540	0.9461	0.9979	0.9996	1.0004	1.0001	0.9745	0.8147	0.9731	0.9589
2.6	1.8729E 00	0.9999	0.9688	0.9548	0.9470	0.9977	0.9996	1.0003	1.0001	0.9750	0.8188	0.9738	0.9594
2.8	1.8613E 00	0.9995	0.9695	0.9556	0.9479	0.9975	0.9995	1.0003	1.0001	0.9754	0.8226	0.9743	0.9598
3.0	1.8514E 00	0.9992	0.9701	0.9565	0.9489	0.9973	0.9995	1.0003	1.0001	0.9758	0.8264	0.9749	0.9603
3.2	1.8428E 00	0.9989	0.9708	0.9574	0.9499	0.9971	0.9994	1.0003	1.0001	0.9763	0.8301	0.9754	0.9607
3.4	1.8352E 00	0.9985	0.9715	0.9583	0.9509	0.9969	0.9994	1.0003	1.0001	0.9767	0.8337	0.9759	0.9611
3.6	1.8284E 00	0.9982	0.9722	0.9592	0.9519	0.9967	0.9994	1.0003	1.0000	0.9771	0.8373	0.9764	0.9614
3.8	1.8223E 00	0.9978	0.9729	0.9601	0.9530	0.9965	0.9993	1.0002	1.0000	0.9775	0.8408	0.9769	0.9617
4.0	1.8167E 00	0.9974	0.9736	0.9611	0.9540	0.9963	0.9993	1.0002	1.0000	0.9779	0.8443	0.9773	0.9621
4.5	1.8047E 00	0.9965	0.9753	0.9633	0.9564	0.9958	0.9992	1.0002	1.0000	0.9788	0.8528	0.9783	0.9628
5.0	1.7947E 00	0.9956	0.9769	0.9654	0.9588	0.9953	0.9991	1.0002	1.0000	0.9797	0.8612	0.9793	0.9635
5.5	1.7861E 00	0.9947	0.9785	0.9674	0.9611	0.9948	0.9990	1.0002	1.0000	0.9805	0.8695	0.9802	0.9641
6.0	1.7785E 00	0.9938	0.9800	0.9694	0.9632	0.9943	0.9989	1.0002	1.0000	0.9813	0.8777	0.9810	0.9646
6.5	1.7717E 00	0.9929	0.9815	0.9713	0.9653	0.9938	0.9988	1.0001	1.0000	0.9821	0.8858	0.9818	0.9651
7.0	1.7654E 00	0.9919	0.9830	0.9731	0.9674	0.9933	0.9987	1.0001	1.0000	0.9828	0.8938	0.9826	0.9656
7.5	1.7596E 00	0.9910	0.9843	0.9749	0.9693	0.9928	0.9986	1.0001	1.0000	0.9835	0.9018	0.9833	0.9660
8.0	1.7543E 00	0.9901	0.9857	0.9766	0.9712	0.9923	0.9985	1.0001	1.0000	0.9842	0.9098	0.9840	0.9664
9.0	1.7444E 00	0.9882	0.9883	0.9799	0.9748	0.9913	0.9984	1.0001	1.0000	0.9855	0.9255	0.9854	0.9672
10.0	1.7355E 00	0.9863	0.9907	0.9830	0.9783	0.9903	0.9982	1.0001	1.0000	0.9868	0.9411	0.9866	0.9679
11.0	1.7273E 00	0.9845	0.9931	0.9860	0.9816	0.9894	0.9980	1.0001	1.0000	0.9880	0.9567	0.9878	0.9685
12.0	1.7197E 00	0.9826	0.9953	0.9889	0.9847	0.9884	0.9978	1.0001	1.0000	0.9891	0.9720	0.9890	0.9691
13.0	1.7125E 00	0.9807	0.9975	0.9917	0.9878	0.9874	0.9976	1.0001	1.0000	0.9902	0.9873	0.9901	0.9696
14.0	1.7056E 00	0.9789	0.9996	0.9943	0.9907	0.9864	0.9974	1.0001	1.0000	0.9913	1.0025	0.9912	0.9701
16.0	1.6928E 00	0.9751	1.0037	0.9995	0.9964	0.9845	0.9971	1.0001	1.0000	0.9933	1.0325	0.9932	0.9710
18.0	1.6809E 00	0.9714	1.0076	1.0044	1.0018	0.9825	0.9967	1.0001	1.0000	0.9953	1.0620	0.9952	0.9718
20.0	1.6697E 00	0.9677	1.0113	1.0091	1.0069	0.9806	0.9963	1.0000	1.0000	0.9971	1.0911	0.9971	0.9726
25.0	1.6440E 00	0.9585	1.0202	1.0203	1.0192	0.9759	0.9953	1.0000	1.0000	1.0016	1.1616	1.0016	0.9742
30.0	1.6206E 00	0.9494	1.0285	1.0309	1.0308	0.9712	0.9944	1.0000	1.0000	1.0058	1.2287	1.0058	0.9757
35.0	1.5989E 00	0.9405	1.0365	1.0410	1.0419	0.9667	0.9934	1.0000	1.0000	1.0098	1.2923	1.0098	0.9770
40.0	1.5785E 00	0.9318	1.0441	1.0507	1.0526	0.9623	0.9924	1.0000	1.0000	1.0137	1.3519	1.0136	0.9782
45.0	1.5592E 00	0.9232	1.0514	1.0601	1.0629	0.9581	0.9914	1.0000	1.0000	1.0174	1.4071	1.0173	0.9794
50.0	1.5407E 00	0.9148	1.0584	1.0693	1.0730	0.9540	0.9904	1.0000	1.0000	1.0209	1.4580	1.0209	0.9804

ELECTRON Z= 23 A= 51

p	F_0L_0	L_0	λ_2	λ_3	λ_4	μ_1	μ_2	Λ_1	Λ_2	ν_{12}	$\hat{\nu}_{12}$	η_{12}	$\hat{\eta}_{12}$
0.1	1.3096E 01	1.0029	3.5192	5.8095	7.4987	1.0000	1.0000	1.0010	1.0002	1.6148	0.7286	0.9247	0.9113
0.2	6.6535E 00	1.0029	1.5810	1.8058	1.9170	0.9999	1.0000	1.0010	1.0002	1.1180	0.7364	0.9433	0.9297
0.3	4.6160E 00	1.0029	1.2240	1.2841	1.3093	0.9999	1.0000	1.0010	1.0002	1.0287	0.7433	0.9501	0.9364
0.4	3.6780E 00	1.0029	1.1004	1.1193	1.1253	0.9998	1.0000	1.0009	1.0002	0.9987	0.7496	0.9539	0.9401
0.5	3.1614E 00	1.0028	1.0443	1.0471	1.0460	0.9997	1.0000	1.0009	1.0002	0.9855	0.7554	0.9565	0.9426
0.6	2.8437E 00	1.0028	1.0145	1.0097	1.0052	0.9997	0.9999	1.0009	1.0002	0.9788	0.7608	0.9585	0.9444
0.7	2.6328E 00	1.0027	0.9972	0.9881	0.9818	0.9996	0.9999	1.0008	1.0002	0.9751	0.7657	0.9600	0.9458
0.8	2.4850E 00	1.0026	0.9864	0.9747	0.9674	0.9995	0.9999	1.0008	1.0002	0.9730	0.7702	0.9613	0.9470
0.9	2.3770E 00	1.0025	0.9793	0.9661	0.9581	0.9994	0.9999	1.0007	1.0001	0.9717	0.7744	0.9623	0.9480
1.0	2.2954E 00	1.0024	0.9745	0.9603	0.9519	0.9993	0.9999	1.0007	1.0001	0.9709	0.7783	0.9633	0.9488
1.2	2.1818E 00	1.0022	0.9690	0.9536	0.9449	0.9990	0.9998	1.0006	1.0001	0.9703	0.7853	0.9648	0.9502
1.4	2.1077E 00	1.0019	0.9663	0.9505	0.9416	0.9988	0.9998	1.0006	1.0001	0.9702	0.7914	0.9661	0.9514
1.6	2.0561E 00	1.0016	0.9651	0.9491	0.9402	0.9986	0.9997	1.0005	1.0001	0.9704	0.7968	0.9673	0.9523
1.8	2.0185E 00	1.0013	0.9647	0.9487	0.9398	0.9984	0.9997	1.0005	1.0001	0.9708	0.8019	0.9682	0.9531
2.0	1.9899E 00	1.0010	0.9647	0.9489	0.9401	0.9981	0.9996	1.0004	1.0001	0.9712	0.8065	0.9691	0.9539
2.2	1.9674E 00	1.0007	0.9651	0.9494	0.9407	0.9979	0.9996	1.0004	1.0001	0.9717	0.8109	0.9699	0.9545
2.4	1.9493E 00	1.0003	0.9656	0.9501	0.9415	0.9977	0.9996	1.0004	1.0001	0.9722	0.8151	0.9707	0.9551
2.6	1.9343E 00	1.0000	0.9662	0.9510	0.9425	0.9975	0.9995	1.0004	1.0001	0.9727	0.8191	0.9714	0.9557
2.8	1.9217E 00	0.9996	0.9668	0.9519	0.9435	0.9973	0.9995	1.0003	1.0001	0.9731	0.8229	0.9720	0.9562
3.0	1.9109E 00	0.9992	0.9676	0.9528	0.9445	0.9971	0.9994	1.0003	1.0001	0.9736	0.8267	0.9726	0.9567
3.2	1.9014E 00	0.9989	0.9683	0.9538	0.9457	0.9968	0.9994	1.0003	1.0001	0.9741	0.8304	0.9732	0.9571
3.4	1.8931E 00	0.9985	0.9691	0.9548	0.9468	0.9966	0.9994	1.0003	1.0001	0.9745	0.8340	0.9737	0.9575
3.6	1.8857E 00	0.9981	0.9698	0.9558	0.9479	0.9964	0.9993	1.0003	1.0001	0.9750	0.8375	0.9742	0.9579
3.8	1.8790E 00	0.9978	0.9706	0.9568	0.9490	0.9962	0.9993	1.0003	1.0000	0.9754	0.8410	0.9747	0.9583
4.0	1.8729E 00	0.9974	0.9713	0.9577	0.9500	0.9960	0.9992	1.0002	1.0000	0.9758	0.8445	0.9752	0.9586
4.5	1.8597E 00	0.9964	0.9732	0.9601	0.9527	0.9955	0.9991	1.0002	1.0000	0.9768	0.8529	0.9763	0.9594
5.0	1.8487E 00	0.9954	0.9750	0.9624	0.9553	0.9949	0.9990	1.0002	1.0000	0.9778	0.8612	0.9774	0.9601
5.5	1.8391E 00	0.9945	0.9767	0.9647	0.9577	0.9944	0.9989	1.0002	1.0000	0.9787	0.8694	0.9784	0.9608
6.0	1.8307E 00	0.9935	0.9783	0.9668	0.9601	0.9939	0.9988	1.0002	1.0000	0.9796	0.8775	0.9793	0.9614
6.5	1.8231E 00	0.9925	0.9799	0.9688	0.9624	0.9933	0.9987	1.0002	1.0000	0.9804	0.8855	0.9801	0.9619
7.0	1.8162E 00	0.9915	0.9815	0.9708	0.9645	0.9928	0.9986	1.0001	1.0000	0.9812	0.8934	0.9810	0.9624
7.5	1.8098E 00	0.9905	0.9830	0.9728	0.9667	0.9923	0.9985	1.0001	1.0000	0.9820	0.9013	0.9818	0.9629
8.0	1.8039E 00	0.9895	0.9845	0.9746	0.9687	0.9917	0.9984	1.0001	1.0000	0.9827	0.9091	0.9825	0.9634
9.0	1.7929E 00	0.9875	0.9873	0.9782	0.9726	0.9907	0.9982	1.0001	1.0000	0.9841	0.9247	0.9840	0.9642
10.0	1.7831E 00	0.9855	0.9899	0.9816	0.9764	0.9896	0.9980	1.0001	1.0000	0.9855	0.9400	0.9853	0.9649
11.0	1.7740E 00	0.9835	0.9925	0.9848	0.9799	0.9886	0.9978	1.0001	1.0000	0.9868	0.9552	0.9867	0.9656
12.0	1.7655E 00	0.9815	0.9949	0.9879	0.9833	0.9875	0.9977	1.0001	1.0000	0.9880	0.9704	0.9879	0.9663
13.0	1.7575E 00	0.9795	0.9973	0.9909	0.9866	0.9865	0.9974	1.0001	1.0000	0.9892	0.9853	0.9891	0.9668
14.0	1.7499E 00	0.9775	0.9996	0.9938	0.9898	0.9854	0.9973	1.0001	1.0000	0.9903	1.0002	0.9903	0.9674
16.0	1.7357E 00	0.9735	1.0040	0.9993	0.9959	0.9833	0.9968	1.0001	1.0000	0.9926	1.0296	0.9925	0.9684
18.0	1.7226E 00	0.9695	1.0082	1.0047	1.0018	0.9813	0.9964	1.0001	1.0000	0.9947	1.0585	0.9946	0.9693
20.0	1.7102E 00	0.9656	1.0122	1.0098	1.0074	0.9792	0.9960	1.0001	1.0000	0.9967	1.0869	0.9967	0.9701
25.0	1.6818E 00	0.9557	1.0218	1.0219	1.0206	0.9741	0.9950	1.0000	1.0000	1.0015	1.1557	1.0015	0.9719
30.0	1.6560E 00	0.9461	1.0308	1.0333	1.0331	0.9692	0.9940	1.0000	1.0000	1.0061	1.2209	1.0060	0.9735
35.0	1.6322E 00	0.9366	1.0394	1.0442	1.0451	0.9644	0.9930	1.0000	1.0000	1.0104	1.2824	1.0104	0.9750
40.0	1.6098E 00	0.9273	1.0476	1.0546	1.0566	0.9597	0.9919	1.0000	1.0000	1.0145	1.3395	1.0145	0.9763
45.0	1.5886E 00	0.9182	1.0554	1.0648	1.0678	0.9552	0.9908	1.0000	1.0000	1.0185	1.3923	1.0185	0.9776
50.0	1.5685E 00	0.9093	1.0630	1.0746	1.0787	0.9510	0.9897	1.0000	1.0000	1.0224	1.4408	1.0224	0.9789

Instr. see p. 22

ELECTRON Z= 24 A= 53

p	F_0L_0	L_0	λ_2	λ_3	λ_4	μ_1	μ_2	Λ_1	Λ_2	ν_{12}	$\hat{\nu}_{12}$	η_{12}	$\hat{\eta}_{12}$
0.1	1.3904E 01	1.0033	3.7245	6.3570	8.3669	1.0000	1.0000	1.0011	1.0002	1.6667	0.7266	0.9168	0.9023
0.2	7.0574E 00	1.0033	1.6285	1.8801	2.0060	0.9999	1.0000	1.0010	1.0002	1.1277	0.7347	0.9378	0.9230
0.3	4.8815E 00	1.0032	1.2423	1.3086	1.3366	0.9999	1.0000	1.0010	1.0002	1.0309	0.7419	0.9454	0.9305
0.4	3.8755E 00	1.0032	1.1086	1.1293	1.1359	0.9998	1.0000	1.0010	1.0002	0.9984	0.7485	0.9497	0.9346
0.5	3.3205E 00	1.0031	1.0479	1.0510	1.0498	0.9997	1.0000	1.0009	1.0002	0.9841	0.7545	0.9526	0.9374
0.6	2.9787E 00	1.0031	1.0157	1.0104	1.0056	0.9996	0.9999	1.0009	1.0002	0.9769	0.7601	0.9547	0.9394
0.7	2.7518E 00	1.0030	0.9969	0.9870	0.9802	0.9995	0.9999	1.0009	1.0002	0.9729	0.7651	0.9564	0.9410
0.8	2.5927E 00	1.0029	0.9852	0.9725	0.9646	0.9994	0.9999	1.0008	1.0002	0.9705	0.7698	0.9578	0.9423
0.9	2.4764E 00	1.0028	0.9775	0.9632	0.9546	0.9993	0.9999	1.0008	1.0002	0.9691	0.7741	0.9589	0.9433
1.0	2.3885E 00	1.0027	0.9724	0.9569	0.9479	0.9992	0.9999	1.0007	1.0001	0.9683	0.7781	0.9600	0.9443
1.2	2.2662E 00	1.0025	0.9664	0.9497	0.9403	0.9990	0.9998	1.0007	1.0001	0.9676	0.7851	0.9617	0.9458
1.4	2.1863E 00	1.0022	0.9635	0.9463	0.9367	0.9987	0.9998	1.0006	1.0001	0.9676	0.7913	0.9631	0.9470
1.6	2.1307E 00	1.0019	0.9621	0.9449	0.9352	0.9985	0.9997	1.0006	1.0001	0.9678	0.7969	0.9643	0.9481
1.8	2.0900E 00	1.0016	0.9617	0.9444	0.9348	0.9983	0.9997	1.0005	1.0001	0.9682	0.8019	0.9654	0.9490
2.0	2.0591E 00	1.0012	0.9617	0.9446	0.9351	0.9980	0.9996	1.0005	1.0001	0.9687	0.8066	0.9664	0.9498
2.2	2.0348E 00	1.0009	0.9621	0.9452	0.9357	0.9978	0.9996	1.0004	1.0001	0.9692	0.8110	0.9672	0.9505
2.4	2.0151E 00	1.0005	0.9626	0.9459	0.9366	0.9976	0.9995	1.0004	1.0001	0.9697	0.8151	0.9680	0.9512
2.6	1.9988E 00	1.0001	0.9633	0.9468	0.9376	0.9973	0.9995	1.0004	1.0001	0.9702	0.8191	0.9688	0.9518
2.8	1.9851E 00	0.9997	0.9640	0.9478	0.9387	0.9971	0.9995	1.0004	1.0001	0.9707	0.8229	0.9695	0.9523
3.0	1.9733E 00	0.9994	0.9648	0.9488	0.9399	0.9969	0.9994	1.0003	1.0001	0.9712	0.8267	0.9701	0.9528
3.2	1.9630E 00	0.9990	0.9656	0.9499	0.9411	0.9966	0.9994	1.0003	1.0001	0.9717	0.8303	0.9708	0.9533
3.4	1.9539E 00	0.9986	0.9664	0.9509	0.9423	0.9964	0.9993	1.0003	1.0001	0.9722	0.8339	0.9713	0.9538
3.6	1.9458E 00	0.9982	0.9672	0.9520	0.9434	0.9962	0.9993	1.0003	1.0001	0.9727	0.8374	0.9719	0.9542
3.8	1.9384E 00	0.9978	0.9680	0.9531	0.9446	0.9960	0.9992	1.0003	1.0001	0.9732	0.8409	0.9724	0.9546
4.0	1.9318E 00	0.9974	0.9688	0.9541	0.9458	0.9957	0.9992	1.0003	1.0000	0.9736	0.8443	0.9730	0.9550
4.5	1.9173E 00	0.9963	0.9708	0.9567	0.9487	0.9952	0.9991	1.0002	1.0000	0.9747	0.8526	0.9742	0.9558
5.0	1.9052E 00	0.9953	0.9728	0.9592	0.9514	0.9946	0.9990	1.0002	1.0000	0.9758	0.8608	0.9753	0.9566
5.5	1.8947E 00	0.9943	0.9746	0.9616	0.9541	0.9940	0.9989	1.0002	1.0000	0.9767	0.8688	0.9764	0.9573
6.0	1.8854E 00	0.9932	0.9764	0.9639	0.9566	0.9935	0.9988	1.0002	1.0000	0.9777	0.8768	0.9773	0.9580
6.5	1.8771E 00	0.9922	0.9782	0.9661	0.9591	0.9929	0.9987	1.0002	1.0000	0.9786	0.8846	0.9783	0.9586
7.0	1.8694E 00	0.9911	0.9798	0.9682	0.9614	0.9923	0.9986	1.0002	1.0000	0.9794	0.8924	0.9792	0.9591
7.5	1.8624E 00	0.9901	0.9814	0.9703	0.9637	0.9918	0.9985	1.0001	1.0000	0.9803	0.9001	0.9800	0.9597
8.0	1.8558E 00	0.9890	0.9830	0.9723	0.9659	0.9912	0.9984	1.0001	1.0000	0.9811	0.9078	0.9809	0.9601
9.0	1.8437E 00	0.9869	0.9860	0.9761	0.9701	0.9901	0.9981	1.0001	1.0000	0.9826	0.9231	0.9824	0.9610
10.0	1.8328E 00	0.9848	0.9889	0.9798	0.9741	0.9890	0.9979	1.0001	1.0000	0.9840	0.9381	0.9839	0.9619
11.0	1.8228E 00	0.9826	0.9916	0.9832	0.9780	0.9879	0.9977	1.0001	1.0000	0.9854	0.9530	0.9853	0.9626
12.0	1.8134E 00	0.9805	0.9943	0.9866	0.9816	0.9867	0.9975	1.0001	1.0000	0.9868	0.9677	0.9866	0.9633
13.0	1.8046E 00	0.9784	0.9968	0.9898	0.9852	0.9856	0.9973	1.0001	1.0000	0.9880	0.9825	0.9879	0.9639
14.0	1.7963E 00	0.9763	0.9993	0.9929	0.9886	0.9845	0.9971	1.0001	1.0000	0.9893	0.9969	0.9892	0.9645
16.0	1.7806E 00	0.9720	1.0040	0.9989	0.9951	0.9823	0.9967	1.0001	1.0000	0.9917	1.0256	0.9916	0.9656
18.0	1.7661E 00	0.9678	1.0085	1.0046	1.0014	0.9801	0.9962	1.0001	1.0000	0.9939	1.0538	0.9939	0.9666
20.0	1.7525E 00	0.9636	1.0129	1.0101	1.0074	0.9779	0.9958	1.0001	1.0000	0.9961	1.0814	0.9961	0.9675
25.0	1.7214E 00	0.9532	1.0231	1.0231	1.0216	0.9725	0.9947	1.0000	1.0000	1.0013	1.1481	1.0012	0.9695
30.0	1.6932E 00	0.9430	1.0328	1.0353	1.0350	0.9673	0.9936	1.0000	1.0000	1.0061	1.2112	1.0061	0.9713
35.0	1.6671E 00	0.9330	1.0419	1.0469	1.0478	0.9622	0.9925	1.0000	1.0000	1.0108	1.2702	1.0108	0.9729
40.0	1.6427E 00	0.9232	1.0507	1.0582	1.0602	0.9573	0.9914	1.0000	1.0000	1.0152	1.3251	1.0152	0.9743
45.0	1.6197E 00	0.9136	1.0591	1.0690	1.0721	0.9526	0.9903	1.0000	1.0000	1.0195	1.3756	1.0195	0.9758
50.0	1.5979E 00	0.9043	1.0672	1.0795	1.0838	0.9481	0.9891	1.0000	1.0000	1.0236	1.4211	1.0236	0.9771

Instr. see p. 22

ELECTRON Z= 25 A= 55

p	F_0L_0	L_0	λ_2	λ_3	λ_4	μ_1	μ_2	Λ_1	Λ_2	ν_{12}	$\hat{\nu}_{12}$	η_{12}	$\hat{\eta}_{12}$
0.1	1.4746E 01	1.0036	3.9358	6.9459	9.3254	1.0000	1.0000	1.0011	1.0002	1.7203	0.7245	0.9084	0.8927
0.2	7.4789E 00	1.0036	1.6774	1.9579	2.0999	0.9999	1.0000	1.0011	1.0002	1.1377	0.7330	0.9320	0.9160
0.3	5.1594E 00	1.0036	1.2612	1.3340	1.3650	0.9999	1.0000	1.0011	1.0002	1.0331	0.7405	0.9405	0.9243
0.4	4.0828E 00	1.0035	1.1171	1.1396	1.1469	0.9998	1.0000	1.0010	1.0002	0.9980	0.7473	0.9452	0.9289
0.5	3.4874E 00	1.0035	1.0516	1.0550	1.0537	0.9997	0.9999	1.0010	1.0002	0.9826	0.7536	0.9484	0.9320
0.6	3.1204E 00	1.0034	1.0168	1.0111	1.0059	0.9996	0.9999	1.0010	1.0002	0.9748	0.7593	0.9507	0.9342
0.7	2.8767E 00	1.0033	0.9966	0.9859	0.9786	0.9995	0.9999	1.0009	1.0002	0.9705	0.7645	0.9526	0.9359
0.8	2.7057E 00	1.0032	0.9839	0.9703	0.9618	0.9994	0.9999	1.0009	1.0002	0.9680	0.7693	0.9541	0.9373
0.9	2.5807E 00	1.0031	0.9757	0.9602	0.9509	0.9993	0.9999	1.0008	1.0002	0.9665	0.7737	0.9554	0.9385
1.0	2.4862E 00	1.0030	0.9701	0.9535	0.9437	0.9992	0.9998	1.0008	1.0002	0.9656	0.7778	0.9565	0.9395
1.2	2.3546E 00	1.0028	0.9636	0.9457	0.9355	0.9989	0.9998	1.0007	1.0001	0.9648	0.7850	0.9584	0.9412
1.4	2.2686E 00	1.0025	0.9605	0.9420	0.9316	0.9987	0.9998	1.0007	1.0001	0.9648	0.7913	0.9599	0.9426
1.6	2.2087E 00	1.0021	0.9591	0.9404	0.9300	0.9984	0.9997	1.0006	1.0001	0.9650	0.7969	0.9613	0.9437
1.8	2.1648E 00	1.0018	0.9586	0.9400	0.9296	0.9982	0.9997	1.0005	1.0001	0.9655	0.8020	0.9624	0.9447
2.0	2.1314E 00	1.0014	0.9586	0.9402	0.9299	0.9979	0.9996	1.0005	1.0001	0.9660	0.8066	0.9635	0.9456
2.2	2.1051E 00	1.0011	0.9590	0.9407	0.9306	0.9977	0.9996	1.0005	1.0001	0.9665	0.8110	0.9644	0.9463
2.4	2.0838E 00	1.0007	0.9596	0.9416	0.9315	0.9974	0.9995	1.0004	1.0001	0.9671	0.8152	0.9653	0.9470
2.6	2.0661E 00	1.0003	0.9603	0.9425	0.9326	0.9972	0.9995	1.0004	1.0001	0.9676	0.8192	0.9661	0.9477
2.8	2.0512E 00	0.9999	0.9611	0.9436	0.9338	0.9969	0.9994	1.0004	1.0001	0.9682	0.8230	0.9669	0.9483
3.0	2.0384E 00	0.9995	0.9619	0.9447	0.9351	0.9967	0.9994	1.0004	1.0001	0.9687	0.8267	0.9676	0.9488
3.2	2.0272E 00	0.9991	0.9628	0.9458	0.9363	0.9964	0.9993	1.0003	1.0001	0.9693	0.8303	0.9682	0.9494
3.4	2.0173E 00	0.9986	0.9636	0.9470	0.9376	0.9962	0.9993	1.0003	1.0001	0.9698	0.8339	0.9689	0.9498
3.6	2.0084E 00	0.9982	0.9645	0.9481	0.9389	0.9960	0.9992	1.0003	1.0001	0.9703	0.8374	0.9695	0.9503
3.8	2.0004E 00	0.9978	0.9654	0.9493	0.9402	0.9957	0.9992	1.0003	1.0001	0.9708	0.8408	0.9701	0.9507
4.0	1.9931E 00	0.9974	0.9663	0.9504	0.9414	0.9955	0.9992	1.0003	1.0001	0.9713	0.8442	0.9706	0.9511
4.5	1.9773E 00	0.9963	0.9684	0.9532	0.9445	0.9949	0.9990	1.0002	1.0000	0.9725	0.8524	0.9719	0.9521
5.0	1.9640E 00	0.9952	0.9705	0.9558	0.9475	0.9943	0.9989	1.0002	1.0000	0.9736	0.8605	0.9731	0.9529
5.5	1.9525E 00	0.9941	0.9725	0.9584	0.9503	0.9937	0.9988	1.0002	1.0000	0.9747	0.8684	0.9743	0.9537
6.0	1.9423E 00	0.9930	0.9744	0.9609	0.9530	0.9931	0.9987	1.0002	1.0000	0.9757	0.8762	0.9753	0.9544
6.5	1.9331E 00	0.9919	0.9763	0.9633	0.9557	0.9925	0.9986	1.0002	1.0000	0.9767	0.8840	0.9763	0.9551
7.0	1.9247E 00	0.9908	0.9781	0.9656	0.9582	0.9919	0.9985	1.0002	1.0000	0.9776	0.8916	0.9773	0.9557
7.5	1.9169E 00	0.9896	0.9798	0.9678	0.9606	0.9913	0.9984	1.0002	1.0000	0.9785	0.8992	0.9782	0.9562
8.0	1.9096E 00	0.9885	0.9815	0.9699	0.9630	0.9907	0.9983	1.0001	1.0000	0.9793	0.9067	0.9791	0.9568
9.0	1.8963E 00	0.9863	0.9848	0.9740	0.9675	0.9895	0.9980	1.0001	1.0000	0.9810	0.9217	0.9808	0.9577
10.0	1.8843E 00	0.9840	0.9878	0.9779	0.9718	0.9883	0.9978	1.0001	1.0000	0.9825	0.9364	0.9824	0.9586
11.0	1.8733E 00	0.9818	0.9908	0.9817	0.9759	0.9871	0.9976	1.0001	1.0000	0.9840	0.9510	0.9839	0.9594
12.0	1.8630E 00	0.9795	0.9936	0.9852	0.9798	0.9859	0.9974	1.0001	1.0000	0.9854	0.9655	0.9853	0.9602
13.0	1.8533E 00	0.9773	0.9963	0.9887	0.9836	0.9848	0.9971	1.0001	1.0000	0.9868	0.9798	0.9867	0.9609
14.0	1.8441E 00	0.9750	0.9990	0.9920	0.9873	0.9836	0.9969	1.0001	1.0000	0.9882	0.9941	0.9881	0.9615
16.0	1.8269E 00	0.9706	1.0040	0.9984	0.9943	0.9812	0.9965	1.0001	1.0000	0.9907	1.0220	0.9906	0.9627
18.0	1.8110E 00	0.9661	1.0089	1.0045	1.0010	0.9789	0.9960	1.0001	1.0000	0.9931	1.0496	0.9931	0.9638
20.0	1.7961E 00	0.9617	1.0135	1.0104	1.0074	0.9766	0.9956	1.0001	1.0000	0.9955	1.0765	0.9954	0.9648
25.0	1.7620E 00	0.9507	1.0245	1.0243	1.0227	0.9709	0.9944	1.0000	1.0000	1.0010	1.1414	1.0010	0.9670
30.0	1.7312E 00	0.9399	1.0348	1.0374	1.0370	0.9654	0.9933	1.0000	1.0000	1.0062	1.2023	1.0062	0.9689
35.0	1.7028E 00	0.9294	1.0446	1.0498	1.0507	0.9600	0.9921	1.0000	1.0000	1.0112	1.2592	1.0111	0.9706
40.0	1.6763E 00	0.9191	1.0539	1.0618	1.0638	0.9548	0.9909	1.0000	1.0000	1.0159	1.3120	1.0159	0.9723
45.0	1.6513E 00	0.9090	1.0629	1.0734	1.0766	0.9499	0.9897	1.0000	1.0000	1.0205	1.3599	1.0204	0.9738
50.0	1.6277E 00	0.8992	1.0715	1.0846	1.0890	0.9452	0.9885	1.0000	1.0000	1.0248	1.4032	1.0248	0.9753

Instr. see p. 22

ELECTRON Z= 26 A= 56

p	F_0L_0	L_0	λ_2	λ_3	λ_4	μ_1	μ_2	Λ_1	Λ_2	ν_{12}	$\hat{\nu}_{12}$	η_{12}	$\hat{\eta}_{12}$
0.1	1.5627E 01	1.0040	4.1520	7.5765	10.3795	0.9999	1.0000	1.0012	1.0002	1.7755	0.7222	0.8994	0.8825
0.2	7.9209E 00	1.0040	1.7273	2.0388	2.1986	0.9999	1.0000	1.0012	1.0002	1.1479	0.7311	0.9258	0.9086
0.3	5.4518E 00	1.0040	1.2803	1.3601	1.3942	0.9999	1.0000	1.0011	1.0002	1.0353	0.7389	0.9353	0.9179
0.4	4.3013E 00	1.0039	1.1256	1.1501	1.1580	0.9998	1.0000	1.0011	1.0002	0.9975	0.7460	0.9405	0.9229
0.5	3.6636E 00	1.0039	1.0552	1.0589	1.0575	0.9997	0.9999	1.0010	1.0002	0.9809	0.7525	0.9440	0.9263
0.6	3.2700E 00	1.0038	1.0179	1.0117	1.0061	0.9996	0.9999	1.0010	1.0002	0.9725	0.7584	0.9466	0.9287
0.7	3.0085E 00	1.0037	0.9961	0.9845	0.9767	0.9995	0.9999	1.0010	1.0002	0.9679	0.7638	0.9486	0.9306
0.8	2.8250E 00	1.0036	0.9825	0.9678	0.9586	0.9994	0.9999	1.0009	1.0002	0.9652	0.7687	0.9502	0.9321
0.9	2.6908E 00	1.0035	0.9736	0.9569	0.9470	0.9992	0.9999	1.0009	1.0002	0.9636	0.7732	0.9516	0.9334
1.0	2.5893E 00	1.0034	0.9677	0.9497	0.9392	0.9991	0.9998	1.0008	1.0002	0.9627	0.7773	0.9529	0.9345
1.2	2.4479E 00	1.0031	0.9607	0.9414	0.9304	0.9989	0.9998	1.0008	1.0001	0.9618	0.7847	0.9549	0.9363
1.4	2.3554E 00	1.0028	0.9573	0.9374	0.9263	0.9986	0.9997	1.0007	1.0001	0.9618	0.7911	0.9566	0.9378
1.6	2.2909E 00	1.0024	0.9558	0.9357	0.9245	0.9983	0.9997	1.0006	1.0001	0.9621	0.7967	0.9580	0.9391
1.8	2.2436E 00	1.0021	0.9552	0.9352	0.9241	0.9981	0.9996	1.0006	1.0001	0.9625	0.8018	0.9593	0.9401
2.0	2.2076E 00	1.0017	0.9553	0.9354	0.9244	0.9978	0.9996	1.0005	1.0001	0.9631	0.8065	0.9604	0.9411
2.2	2.1791E 00	1.0013	0.9557	0.9360	0.9251	0.9975	0.9995	1.0005	1.0001	0.9637	0.8109	0.9614	0.9419
2.4	2.1561E 00	1.0009	0.9563	0.9369	0.9261	0.9973	0.9995	1.0005	1.0001	0.9643	0.8150	0.9624	0.9427
2.6	2.1370E 00	1.0005	0.9571	0.9379	0.9273	0.9970	0.9994	1.0004	1.0001	0.9649	0.8190	0.9632	0.9434
2.8	2.1208E 00	1.0001	0.9579	0.9391	0.9286	0.9968	0.9994	1.0004	1.0001	0.9655	0.8228	0.9640	0.9440
3.0	2.1069E 00	0.9996	0.9588	0.9402	0.9299	0.9965	0.9993	1.0004	1.0001	0.9661	0.8265	0.9648	0.9446
3.2	2.0947E 00	0.9992	0.9597	0.9415	0.9312	0.9963	0.9993	1.0004	1.0001	0.9667	0.8301	0.9655	0.9452
3.4	2.0839E 00	0.9988	0.9607	0.9427	0.9326	0.9960	0.9993	1.0003	1.0001	0.9672	0.8336	0.9662	0.9457
3.6	2.0742E 00	0.9983	0.9616	0.9439	0.9340	0.9957	0.9992	1.0003	1.0001	0.9678	0.8370	0.9669	0.9462
3.8	2.0655E 00	0.9979	0.9626	0.9451	0.9353	0.9955	0.9992	1.0003	1.0001	0.9683	0.8404	0.9675	0.9467
4.0	2.0575E 00	0.9974	0.9635	0.9464	0.9367	0.9952	0.9991	1.0003	1.0001	0.9689	0.8437	0.9681	0.9471
4.5	2.0402E 00	0.9963	0.9658	0.9493	0.9400	0.9946	0.9990	1.0003	1.0000	0.9701	0.8519	0.9695	0.9482
5.0	2.0257E 00	0.9951	0.9680	0.9522	0.9431	0.9940	0.9989	1.0002	1.0000	0.9713	0.8598	0.9708	0.9491
5.5	2.0131E 00	0.9940	0.9702	0.9549	0.9462	0.9933	0.9988	1.0002	1.0000	0.9725	0.8676	0.9720	0.9499
6.0	2.0019E 00	0.9928	0.9722	0.9576	0.9491	0.9927	0.9986	1.0002	1.0000	0.9735	0.8753	0.9732	0.9507
6.5	1.9918E 00	0.9917	0.9742	0.9601	0.9519	0.9921	0.9985	1.0002	1.0000	0.9746	0.8829	0.9742	0.9514
7.0	1.9826E 00	0.9905	0.9761	0.9626	0.9546	0.9915	0.9984	1.0002	1.0000	0.9756	0.8904	0.9753	0.9520
7.5	1.9741E 00	0.9893	0.9780	0.9649	0.9572	0.9908	0.9983	1.0002	1.0000	0.9765	0.8978	0.9763	0.9526
8.0	1.9661E 00	0.9881	0.9798	0.9672	0.9597	0.9902	0.9982	1.0001	1.0000	0.9774	0.9052	0.9772	0.9532
9.0	1.9515E 00	0.9858	0.9832	0.9716	0.9646	0.9890	0.9979	1.0001	1.0000	0.9792	0.9198	0.9790	0.9543
10.0	1.9383E 00	0.9834	0.9865	0.9758	0.9691	0.9877	0.9977	1.0001	1.0000	0.9809	0.9342	0.9807	0.9552
11.0	1.9262E 00	0.9811	0.9897	0.9798	0.9735	0.9865	0.9975	1.0001	1.0000	0.9824	0.9484	0.9823	0.9561
12.0	1.9149E 00	0.9787	0.9927	0.9836	0.9777	0.9852	0.9972	1.0001	1.0000	0.9840	0.9626	0.9838	0.9569
13.0	1.9043E 00	0.9763	0.9956	0.9872	0.9817	0.9840	0.9970	1.0001	1.0000	0.9854	0.9765	0.9853	0.9576
14.0	1.8942E 00	0.9740	0.9984	0.9908	0.9856	0.9827	0.9968	1.0001	1.0000	0.9869	0.9904	0.9868	0.9583
16.0	1.8754E 00	0.9693	1.0038	0.9976	0.9931	0.9803	0.9963	1.0001	1.0000	0.9896	1.0176	0.9895	0.9596
18.0	1.8580E 00	0.9646	1.0090	1.0041	1.0002	0.9778	0.9958	1.0001	1.0000	0.9922	1.0443	0.9921	0.9608
20.0	1.8418E 00	0.9600	1.0139	1.0104	1.0071	0.9754	0.9954	1.0001	1.0000	0.9947	1.0705	0.9946	0.9619
25.0	1.8046E 00	0.9485	1.0256	1.0251	1.0233	0.9694	0.9942	1.0000	1.0000	1.0005	1.1333	1.0005	0.9642
30.0	1.7712E 00	0.9372	1.0366	1.0390	1.0385	0.9636	0.9930	1.0000	1.0000	1.0061	1.1921	1.0060	0.9663
35.0	1.7404E 00	0.9262	1.0469	1.0522	1.0530	0.9580	0.9917	1.0000	1.0000	1.0113	1.2468	1.0113	0.9683
40.0	1.7118E 00	0.9154	1.0568	1.0649	1.0670	0.9526	0.9905	1.0000	1.0000	1.0163	1.2972	1.0163	0.9700
45.0	1.6848E 00	0.9049	1.0663	1.0772	1.0805	0.9474	0.9892	1.0000	1.0000	1.0212	1.3431	1.0212	0.9718
50.0	1.6594E 00	0.8946	1.0754	1.0891	1.0936	0.9425	0.9879	1.0000	1.0000	1.0258	1.3841	1.0258	0.9734

ELECTRON Z= 27 A= 59

p	F_0L_0	L_0	λ_2	λ_3	λ_4	μ_1	μ_2	Λ_1	Λ_2	ν_{12}	$\hat{\nu}_{12}$	η_{12}	$\hat{\eta}_{12}$
0.1	1.6538E 01	1.0044	4.3743	8.2541	11.5423	0.9999	1.0000	1.0012	1.0002	1.8326	0.7200	0.8900	0.8719
0.2	8.3788E 00	1.0043	1.7788	2.1240	2.3033	0.9999	1.0000	1.0012	1.0002	1.1585	0.7292	0.9194	0.9009
0.3	5.7555E 00	1.0043	1.3003	1.3875	1.4251	0.9998	1.0000	1.0012	1.0002	1.0376	0.7374	0.9299	0.9112
0.4	4.5285E 00	1.0043	1.1346	1.1612	1.1698	0.9998	1.0000	1.0012	1.0002	0.9971	0.7448	0.9357	0.9168
0.5	3.8468E 00	1.0042	1.0592	1.0632	1.0617	0.9997	0.9999	1.0011	1.0002	0.9793	0.7515	0.9395	0.9204
0.6	3.4255E 00	1.0041	1.0192	1.0126	1.0065	0.9996	0.9999	1.0011	1.0002	0.9703	0.7576	0.9423	0.9231
0.7	3.1454E 00	1.0040	0.9958	0.9835	0.9750	0.9994	0.9999	1.0010	1.0002	0.9653	0.7631	0.9445	0.9252
0.8	2.9488E 00	1.0039	0.9813	0.9655	0.9557	0.9993	0.9999	1.0010	1.0002	0.9625	0.7682	0.9463	0.9268
0.9	2.8049E 00	1.0038	0.9718	0.9539	0.9432	0.9992	0.9999	1.0009	1.0002	0.9607	0.7728	0.9479	0.9282
1.0	2.6961E 00	1.0037	0.9654	0.9461	0.9349	0.9991	0.9998	1.0009	1.0002	0.9597	0.7771	0.9492	0.9294
1.2	2.5444E 00	1.0034	0.9579	0.9372	0.9255	0.9988	0.9998	1.0008	1.0002	0.9589	0.7846	0.9514	0.9314
1.4	2.4451E 00	1.0031	0.9543	0.9330	0.9210	0.9985	0.9997	1.0007	1.0001	0.9588	0.7911	0.9532	0.9330
1.6	2.3758E 00	1.0027	0.9526	0.9311	0.9192	0.9982	0.9997	1.0007	1.0001	0.9591	0.7968	0.9548	0.9344
1.8	2.3249E 00	1.0023	0.9520	0.9306	0.9187	0.9979	0.9996	1.0006	1.0001	0.9596	0.8020	0.9561	0.9355
2.0	2.2861E 00	1.0019	0.9521	0.9308	0.9190	0.9977	0.9996	1.0006	1.0001	0.9602	0.8068	0.9573	0.9365
2.2	2.2554E 00	1.0015	0.9525	0.9315	0.9198	0.9974	0.9995	1.0005	1.0001	0.9609	0.8112	0.9584	0.9374
2.4	2.2305E 00	1.0011	0.9532	0.9324	0.9209	0.9971	0.9995	1.0005	1.0001	0.9615	0.8154	0.9594	0.9383
2.6	2.2098E 00	1.0006	0.9540	0.9335	0.9222	0.9968	0.9994	1.0004	1.0001	0.9622	0.8193	0.9604	0.9390
2.8	2.1923E 00	1.0002	0.9549	0.9347	0.9235	0.9966	0.9994	1.0004	1.0001	0.9628	0.8232	0.9613	0.9397
3.0	2.1772E 00	0.9997	0.9559	0.9360	0.9249	0.9963	0.9993	1.0004	1.0001	0.9635	0.8269	0.9621	0.9404
3.2	2.1640E 00	0.9993	0.9569	0.9373	0.9264	0.9960	0.9993	1.0004	1.0001	0.9641	0.8305	0.9629	0.9410
3.4	2.1522E 00	0.9988	0.9579	0.9386	0.9278	0.9957	0.9992	1.0004	1.0001	0.9647	0.8340	0.9636	0.9415
3.6	2.1417E 00	0.9983	0.9589	0.9399	0.9293	0.9955	0.9992	1.0003	1.0001	0.9653	0.8374	0.9643	0.9421
3.8	2.1322E 00	0.9979	0.9599	0.9412	0.9308	0.9952	0.9991	1.0003	1.0001	0.9659	0.8408	0.9650	0.9426
4.0	2.1235E 00	0.9974	0.9609	0.9425	0.9322	0.9949	0.9991	1.0003	1.0001	0.9664	0.8441	0.9656	0.9431
4.5	2.1046E 00	0.9962	0.9634	0.9457	0.9357	0.9943	0.9989	1.0003	1.0001	0.9678	0.8522	0.9671	0.9442
5.0	2.0887E 00	0.9950	0.9657	0.9488	0.9391	0.9936	0.9988	1.0002	1.0000	0.9691	0.8601	0.9685	0.9452
5.5	2.0749E 00	0.9937	0.9680	0.9517	0.9424	0.9929	0.9987	1.0002	1.0000	0.9703	0.8679	0.9698	0.9460
6.0	2.0626E 00	0.9925	0.9703	0.9546	0.9455	0.9923	0.9986	1.0002	1.0000	0.9715	0.8755	0.9710	0.9469
6.5	2.0516E 00	0.9913	0.9724	0.9573	0.9485	0.9916	0.9984	1.0002	1.0000	0.9726	0.8830	0.9722	0.9476
7.0	2.0414E 00	0.9900	0.9744	0.9599	0.9514	0.9909	0.9983	1.0002	1.0000	0.9736	0.8905	0.9733	0.9483
7.5	2.0321E 00	0.9888	0.9764	0.9624	0.9542	0.9903	0.9982	1.0002	1.0000	0.9747	0.8978	0.9744	0.9490
8.0	2.0233E 00	0.9875	0.9784	0.9649	0.9569	0.9896	0.9981	1.0002	1.0000	0.9756	0.9051	0.9754	0.9496
9.0	2.0073E 00	0.9850	0.9821	0.9696	0.9620	0.9883	0.9978	1.0001	1.0000	0.9775	0.9196	0.9773	0.9507
10.0	1.9928E 00	0.9825	0.9856	0.9741	0.9669	0.9869	0.9976	1.0001	1.0000	0.9793	0.9338	0.9791	0.9518
11.0	1.9795E 00	0.9800	0.9890	0.9783	0.9716	0.9856	0.9973	1.0001	1.0000	0.9810	0.9479	0.9809	0.9527
12.0	1.9671E 00	0.9775	0.9922	0.9824	0.9761	0.9843	0.9971	1.0001	1.0000	0.9826	0.9618	0.9825	0.9536
13.0	1.9554E 00	0.9750	0.9953	0.9863	0.9804	0.9830	0.9968	1.0001	1.0000	0.9842	0.9756	0.9841	0.9544
14.0	1.9443E 00	0.9725	0.9983	0.9901	0.9846	0.9816	0.9966	1.0001	1.0000	0.9857	0.9892	0.9856	0.9552
16.0	1.9237E 00	0.9675	1.0041	0.9974	0.9926	0.9790	0.9961	1.0001	1.0000	0.9886	1.0161	0.9885	0.9565
18.0	1.9047E 00	0.9626	1.0096	1.0044	1.0002	0.9764	0.9956	1.0001	1.0000	0.9914	1.0423	0.9913	0.9578
20.0	1.8869E 00	0.9577	1.0149	1.0111	1.0075	0.9738	0.9951	1.0001	1.0000	0.9941	1.0679	0.9940	0.9590
25.0	1.8463E 00	0.9455	1.0274	1.0268	1.0248	0.9675	0.9938	1.0001	1.0000	1.0004	1.1293	1.0003	0.9616
30.0	1.8098E 00	0.9336	1.0391	1.0417	1.0411	0.9614	0.9925	1.0000	1.0000	1.0063	1.1867	1.0062	0.9639
35.0	1.7764E 00	0.9220	1.0502	1.0558	1.0566	0.9555	0.9912	1.0000	1.0000	1.0119	1.2396	1.0119	0.9660
40.0	1.7453E 00	0.9106	1.0608	1.0694	1.0715	0.9498	0.9899	1.0000	1.0000	1.0173	1.2881	1.0172	0.9679
45.0	1.7161E 00	0.8996	1.0709	1.0825	1.0860	0.9444	0.9885	1.0000	1.0000	1.0224	1.3317	1.0224	0.9698
50.0	1.6885E 00	0.8888	1.0806	1.0952	1.1000	0.9392	0.9871	1.0000	1.0000	1.0274	1.3706	1.0273	0.9716

Instr. see p. 22

ELECTRON Z= 28 A= 61

p	F_0L_0	L_0	λ_2	λ_3	λ_4	μ_1	μ_2	Λ_1	Λ_2	ν_{12}	$\hat{\nu}_{12}$	η_{12}	$\hat{\eta}_{12}$
0.1	1.7493E 01	1.0048	4.6010	8.9765	12.8160	0.9999	1.0000	1.0013	1.0002	1.8913	0.7177	0.8800	0.8607
0.2	8.8592E 00	1.0047	1.8313	2.2124	2.4129	0.9999	1.0000	1.0013	1.0002	1.1693	0.7273	0.9126	0.8929
0.3	6.0750E 00	1.0047	1.3205	1.4155	1.4567	0.9998	1.0000	1.0013	1.0002	1.0399	0.7358	0.9242	0.9042
0.4	4.7679E 00	1.0047	1.1436	1.1724	1.1817	0.9997	1.0000	1.0012	1.0002	0.9965	0.7434	0.9306	0.9103
0.5	4.0401E 00	1.0046	1.0631	1.0674	1.0659	0.9996	0.9999	1.0012	1.0002	0.9775	0.7503	0.9348	0.9143
0.6	3.5897E 00	1.0045	1.0204	1.0133	1.0068	0.9995	0.9999	1.0011	1.0002	0.9679	0.7566	0.9378	0.9172
0.7	3.2900E 00	1.0044	0.9954	0.9822	0.9731	0.9994	0.9999	1.0011	1.0002	0.9626	0.7624	0.9402	0.9195
0.8	3.0795E 00	1.0043	0.9798	0.9630	0.9525	0.9993	0.9999	1.0010	1.0002	0.9595	0.7676	0.9422	0.9213
0.9	2.9255E 00	1.0042	0.9697	0.9506	0.9392	0.9991	0.9998	1.0010	1.0002	0.9577	0.7723	0.9439	0.9228
1.0	2.8089E 00	1.0040	0.9628	0.9423	0.9303	0.9990	0.9998	1.0009	1.0002	0.9566	0.7767	0.9453	0.9241
1.2	2.6463E 00	1.0037	0.9549	0.9328	0.9202	0.9987	0.9998	1.0008	1.0002	0.9557	0.7843	0.9477	0.9263
1.4	2.5397E 00	1.0034	0.9510	0.9283	0.9155	0.9984	0.9997	1.0008	1.0001	0.9557	0.7909	0.9497	0.9280
1.6	2.4653E 00	1.0030	0.9492	0.9263	0.9135	0.9981	0.9997	1.0007	1.0001	0.9560	0.7968	0.9513	0.9294
1.8	2.4106E 00	1.0026	0.9486	0.9257	0.9130	0.9978	0.9996	1.0006	1.0001	0.9566	0.8020	0.9528	0.9307
2.0	2.3688E 00	1.0022	0.9487	0.9259	0.9133	0.9975	0.9995	1.0006	1.0001	0.9572	0.8068	0.9541	0.9318
2.2	2.3357E 00	1.0017	0.9491	0.9267	0.9142	0.9972	0.9995	1.0005	1.0001	0.9579	0.8112	0.9553	0.9328
2.4	2.3089E 00	1.0013	0.9498	0.9277	0.9153	0.9970	0.9994	1.0005	1.0001	0.9586	0.8154	0.9564	0.9336
2.6	2.2865E 00	1.0008	0.9507	0.9288	0.9167	0.9967	0.9994	1.0005	1.0001	0.9593	0.8194	0.9574	0.9345
2.8	2.2676E 00	1.0004	0.9517	0.9301	0.9181	0.9964	0.9993	1.0004	1.0001	0.9600	0.8233	0.9583	0.9352
3.0	2.2512E 00	0.9999	0.9527	0.9315	0.9196	0.9961	0.9993	1.0004	1.0001	0.9607	0.8270	0.9592	0.9359
3.2	2.2368E 00	0.9994	0.9538	0.9329	0.9212	0.9958	0.9992	1.0004	1.0001	0.9613	0.8305	0.9600	0.9365
3.4	2.2241E 00	0.9989	0.9548	0.9343	0.9228	0.9955	0.9992	1.0004	1.0001	0.9620	0.8340	0.9608	0.9372
3.6	2.2127E 00	0.9984	0.9559	0.9357	0.9243	0.9952	0.9991	1.0004	1.0001	0.9626	0.8375	0.9615	0.9377
3.8	2.2023E 00	0.9979	0.9570	0.9371	0.9259	0.9949	0.9991	1.0003	1.0001	0.9632	0.8408	0.9623	0.9383
4.0	2.1929E 00	0.9974	0.9581	0.9384	0.9274	0.9947	0.9990	1.0003	1.0001	0.9638	0.8441	0.9630	0.9388
4.5	2.1723E 00	0.9961	0.9607	0.9418	0.9311	0.9940	0.9989	1.0003	1.0001	0.9653	0.8522	0.9646	0.9400
5.0	2.1549E 00	0.9948	0.9632	0.9451	0.9348	0.9932	0.9988	1.0003	1.0000	0.9667	0.8600	0.9661	0.9410
5.5	2.1398E 00	0.9936	0.9657	0.9482	0.9382	0.9925	0.9986	1.0002	1.0000	0.9680	0.8677	0.9675	0.9420
6.0	2.1264E 00	0.9923	0.9681	0.9513	0.9415	0.9918	0.9985	1.0002	1.0000	0.9692	0.8752	0.9688	0.9429
6.5	2.1142E 00	0.9909	0.9703	0.9542	0.9447	0.9911	0.9984	1.0002	1.0000	0.9704	0.8827	0.9700	0.9437
7.0	2.1032E 00	0.9896	0.9725	0.9570	0.9478	0.9904	0.9982	1.0002	1.0000	0.9715	0.8901	0.9712	0.9444
7.5	2.0929E 00	0.9883	0.9747	0.9597	0.9508	0.9897	0.9981	1.0002	1.0000	0.9726	0.8973	0.9723	0.9452
8.0	2.0833E 00	0.9870	0.9767	0.9623	0.9537	0.9890	0.9980	1.0002	1.0000	0.9737	0.9045	0.9734	0.9458
9.0	2.0657E 00	0.9844	0.9807	0.9673	0.9592	0.9876	0.9977	1.0001	1.0000	0.9757	0.9188	0.9755	0.9470
10.0	2.0499E 00	0.9817	0.9844	0.9720	0.9644	0.9862	0.9975	1.0001	1.0000	0.9776	0.9328	0.9774	0.9482
11.0	2.0353E 00	0.9791	0.9880	0.9766	0.9693	0.9848	0.9972	1.0001	1.0000	0.9794	0.9467	0.9792	0.9492
12.0	2.0217E 00	0.9765	0.9914	0.9809	0.9741	0.9834	0.9969	1.0001	1.0000	0.9811	0.9604	0.9810	0.9501
13.0	2.0089E 00	0.9738	0.9948	0.9851	0.9787	0.9820	0.9967	1.0001	1.0000	0.9828	0.9739	0.9827	0.9510
14.0	1.9969E 00	0.9712	0.9980	0.9891	0.9831	0.9806	0.9964	1.0001	1.0000	0.9844	0.9873	0.9843	0.9518
16.0	1.9744E 00	0.9660	1.0041	0.9969	0.9917	0.9779	0.9959	1.0001	1.0000	0.9875	1.0136	0.9874	0.9533
18.0	1.9536E 00	0.9608	1.0100	1.0043	0.9998	0.9751	0.9954	1.0001	1.0000	0.9905	1.0392	0.9904	0.9547
20.0	1.9342E 00	0.9556	1.0156	1.0114	1.0075	0.9724	0.9948	1.0001	1.0000	0.9933	1.0643	0.9933	0.9559
25.0	1.8900E 00	0.9429	1.0289	1.0282	1.0259	0.9658	0.9935	1.0001	1.0000	1.0000	1.1241	1.0000	0.9587
30.0	1.8505E 00	0.9304	1.0414	1.0439	1.0432	0.9593	0.9922	1.0000	1.0000	1.0063	1.1798	1.0063	0.9612
35.0	1.8143E 00	0.9182	1.0531	1.0589	1.0597	0.9531	0.9908	1.0000	1.0000	1.0123	1.2310	1.0122	0.9635
40.0	1.7807E 00	0.9063	1.0643	1.0734	1.0755	0.9472	0.9894	1.0000	1.0000	1.0180	1.2775	1.0179	0.9656
45.0	1.7492E 00	0.8948	1.0750	1.0872	1.0909	0.9415	0.9879	1.0000	1.0000	1.0234	1.3191	1.0234	0.9677
50.0	1.7196E 00	0.8836	1.0853	1.1007	1.1057	0.9362	0.9865	1.0000	1.0000	1.0287	1.3557	1.0286	0.9696

ELECTRON Z= 29 A= 65

p	F_0L_0	L_0	λ_2	λ_3	λ_4	μ_1	μ_2	Λ_1	Λ_2	ν_{12}	$\hat{\nu}_{12}$	η_{12}	$\hat{\eta}_{12}$
0.1	1.8482E 01	1.0051	4.8336	9.7496	14.2163	0.9999	1.0000	1.0014	1.0003	1.9519	0.7154	0.8696	0.8491
0.2	9.3567E 00	1.0051	1.8853	2.3051	2.5290	0.9999	1.0000	1.0014	1.0003	1.1805	0.7253	0.9057	0.8846
0.3	6.4066E 00	1.0051	1.3415	1.4449	1.4899	0.9998	1.0000	1.0013	1.0003	1.0424	0.7342	0.9185	0.8971
0.4	5.0168E 00	1.0050	1.1531	1.1842	1.1944	0.9997	1.0000	1.0013	1.0002	0.9961	0.7421	0.9254	0.9038
0.5	4.2410E 00	1.0049	1.0673	1.0721	1.0705	0.9996	0.9999	1.0012	1.0002	0.9758	0.7493	0.9300	0.9081
0.6	3.7603E 00	1.0049	1.0218	1.0143	1.0074	0.9995	0.9999	1.0012	1.0002	0.9656	0.7558	0.9333	0.9113
0.7	3.4401E 00	1.0048	0.9952	0.9811	0.9715	0.9994	0.9999	1.0011	1.0002	0.9599	0.7617	0.9359	0.9137
0.8	3.2152E 00	1.0046	0.9786	0.9607	0.9495	0.9992	0.9999	1.0011	1.0002	0.9566	0.7670	0.9380	0.9157
0.9	3.0504E 00	1.0045	0.9678	0.9475	0.9354	0.9991	0.9998	1.0010	1.0002	0.9547	0.7719	0.9398	0.9173
1.0	2.9258E 00	1.0044	0.9605	0.9387	0.9260	0.9989	0.9998	1.0010	1.0002	0.9536	0.7764	0.9414	0.9187
1.2	2.7517E 00	1.0040	0.9520	0.9285	0.9152	0.9986	0.9998	1.0009	1.0002	0.9526	0.7842	0.9440	0.9210
1.4	2.6375E 00	1.0037	0.9479	0.9237	0.9102	0.9983	0.9997	1.0008	1.0002	0.9525	0.7910	0.9461	0.9229
1.6	2.5577E 00	1.0033	0.9460	0.9216	0.9081	0.9980	0.9996	1.0007	1.0001	0.9529	0.7969	0.9479	0.9244
1.8	2.4990E 00	1.0028	0.9453	0.9210	0.9075	0.9977	0.9996	1.0007	1.0001	0.9535	0.8023	0.9495	0.9258
2.0	2.4540E 00	1.0024	0.9454	0.9213	0.9079	0.9974	0.9995	1.0006	1.0001	0.9542	0.8071	0.9509	0.9270
2.2	2.4184E 00	1.0019	0.9459	0.9220	0.9088	0.9971	0.9995	1.0006	1.0001	0.9549	0.8117	0.9521	0.9280
2.4	2.3894E 00	1.0014	0.9467	0.9231	0.9100	0.9968	0.9994	1.0005	1.0001	0.9557	0.8159	0.9533	0.9290
2.6	2.3653E 00	1.0009	0.9476	0.9243	0.9114	0.9964	0.9993	1.0005	1.0001	0.9564	0.8200	0.9544	0.9298
2.8	2.3448E 00	1.0004	0.9486	0.9257	0.9130	0.9961	0.9993	1.0005	1.0001	0.9572	0.8238	0.9554	0.9306
3.0	2.3271E 00	0.9999	0.9497	0.9272	0.9146	0.9958	0.9992	1.0004	1.0001	0.9579	0.8275	0.9563	0.9314
3.2	2.3115E 00	0.9994	0.9509	0.9286	0.9163	0.9955	0.9992	1.0004	1.0001	0.9586	0.8312	0.9572	0.9321
3.4	2.2977E 00	0.9989	0.9520	0.9301	0.9179	0.9952	0.9991	1.0004	1.0001	0.9593	0.8347	0.9580	0.9327
3.6	2.2853E 00	0.9984	0.9532	0.9316	0.9196	0.9949	0.9991	1.0004	1.0001	0.9600	0.8381	0.9588	0.9333
3.8	2.2740E 00	0.9978	0.9543	0.9331	0.9212	0.9946	0.9990	1.0004	1.0001	0.9607	0.8415	0.9596	0.9339
4.0	2.2638E 00	0.9973	0.9555	0.9346	0.9229	0.9943	0.9990	1.0003	1.0001	0.9613	0.8448	0.9604	0.9345
4.5	2.2413E 00	0.9959	0.9583	0.9382	0.9269	0.9936	0.9988	1.0003	1.0001	0.9629	0.8529	0.9621	0.9357
5.0	2.2223E 00	0.9946	0.9610	0.9417	0.9307	0.9928	0.9987	1.0003	1.0001	0.9643	0.8608	0.9637	0.9369
5.5	2.2058E 00	0.9932	0.9636	0.9451	0.9344	0.9921	0.9985	1.0003	1.0000	0.9657	0.8685	0.9652	0.9379
6.0	2.1912E 00	0.9918	0.9661	0.9483	0.9380	0.9913	0.9984	1.0002	1.0000	0.9670	0.8760	0.9666	0.9389
6.5	2.1779E 00	0.9904	0.9686	0.9514	0.9414	0.9906	0.9983	1.0002	1.0000	0.9683	0.8834	0.9679	0.9397
7.0	2.1657E 00	0.9891	0.9709	0.9544	0.9447	0.9898	0.9981	1.0002	1.0000	0.9695	0.8908	0.9692	0.9405
7.5	2.1545E 00	0.9877	0.9732	0.9572	0.9478	0.9891	0.9980	1.0002	1.0000	0.9707	0.8981	0.9704	0.9413
8.0	2.1440E 00	0.9863	0.9754	0.9600	0.9509	0.9883	0.9979	1.0002	1.0000	0.9718	0.9053	0.9715	0.9420
9.0	2.1248E 00	0.9835	0.9796	0.9654	0.9568	0.9868	0.9976	1.0002	1.0000	0.9740	0.9194	0.9737	0.9433
10.0	2.1074E 00	0.9807	0.9836	0.9705	0.9623	0.9853	0.9973	1.0001	1.0000	0.9760	0.9334	0.9758	0.9445
11.0	2.0914E 00	0.9779	0.9875	0.9753	0.9676	0.9838	0.9970	1.0001	1.0000	0.9779	0.9472	0.9778	0.9456
12.0	2.0765E 00	0.9751	0.9911	0.9799	0.9727	0.9823	0.9968	1.0001	1.0000	0.9798	0.9608	0.9796	0.9466
13.0	2.0625E 00	0.9723	0.9947	0.9844	0.9776	0.9809	0.9965	1.0001	1.0000	0.9816	0.9742	0.9815	0.9476
14.0	2.0493E 00	0.9695	0.9981	0.9887	0.9824	0.9794	0.9962	1.0001	1.0000	0.9833	0.9875	0.9832	0.9484
16.0	2.0246E 00	0.9640	1.0047	0.9970	0.9914	0.9765	0.9957	1.0001	1.0000	0.9866	1.0135	0.9865	0.9501
18.0	2.0019E 00	0.9584	1.0110	1.0049	1.0001	0.9736	0.9951	1.0001	1.0000	0.9898	1.0388	0.9897	0.9515
20.0	1.9807E 00	0.9530	1.0170	1.0125	1.0084	0.9707	0.9945	1.0001	1.0000	0.9928	1.0635	0.9928	0.9529
25.0	1.9326E 00	0.9395	1.0312	1.0305	1.0281	0.9636	0.9931	1.0001	1.0000	1.0000	1.1223	0.9999	0.9559
30.0	1.8895E 00	0.9263	1.0445	1.0473	1.0465	0.9568	0.9917	1.0000	1.0000	1.0067	1.1766	1.0067	0.9586
35.0	1.8502E 00	0.9134	1.0571	1.0633	1.0641	0.9503	0.9902	1.0000	1.0000	1.0131	1.2262	1.0130	0.9611
40.0	1.8138E 00	0.9009	1.0690	1.0787	1.0810	0.9440	0.9887	1.0000	1.0000	1.0191	1.2709	1.0191	0.9635
45.0	1.7798E 00	0.8888	1.0804	1.0935	1.0974	0.9381	0.9872	1.0000	1.0000	1.0250	1.3106	1.0249	0.9657
50.0	1.7478E 00	0.8770	1.0913	1.1079	1.1132	0.9326	0.9856	1.0000	1.0000	1.0305	1.3451	1.0305	0.9678

 Instr. see p. 22

ELECTRON Z= 30 A= 67

p	F_0L_0	L_0	λ_2	λ_3	λ_4	μ_1	μ_2	Λ_1	Λ_2	ν_{12}	$\hat{\nu}_{12}$	η_{12}	$\hat{\eta}_{12}$
0.1	1.9524E 01	1.0055	5.0691	10.5685	15.7410	0.9999	1.0000	1.0015	1.0003	2.0137	0.7129	0.8585	0.8367
0.2	9.8818E 00	1.0055	1.9398	2.4006	2.6498	0.9999	1.0000	1.0014	1.0003	1.1917	0.7232	0.8983	0.8758
0.3	6.7573E 00	1.0055	1.3625	1.4745	1.5237	0.9998	1.0000	1.0014	1.0003	1.0447	0.7324	0.9123	0.8895
0.4	5.2805E 00	1.0054	1.1624	1.1959	1.2070	0.9997	1.0000	1.0014	1.0003	0.9954	0.7406	0.9199	0.8968
0.5	4.4542E 00	1.0053	1.0713	1.0764	1.0747	0.9996	0.9999	1.0013	1.0002	0.9739	0.7480	0.9249	0.9016
0.6	3.9414E 00	1.0053	1.0230	1.0150	1.0076	0.9995	0.9999	1.0012	1.0002	0.9630	0.7547	0.9285	0.9050
0.7	3.5996E 00	1.0052	0.9947	0.9797	0.9695	0.9993	0.9999	1.0012	1.0002	0.9570	0.7608	0.9313	0.9076
0.8	3.3593E 00	1.0050	0.9771	0.9580	0.9461	0.9992	0.9999	1.0011	1.0002	0.9535	0.7663	0.9336	0.9097
0.9	3.1832E 00	1.0049	0.9656	0.9440	0.9311	0.9990	0.9998	1.0011	1.0002	0.9514	0.7713	0.9355	0.9115
1.0	3.0499E 00	1.0047	0.9578	0.9346	0.9211	0.9989	0.9998	1.0010	1.0002	0.9502	0.7758	0.9372	0.9130
1.2	2.8636E 00	1.0044	0.9488	0.9238	0.9097	0.9985	0.9997	1.0009	1.0002	0.9492	0.7839	0.9400	0.9155
1.4	2.7414E 00	1.0040	0.9444	0.9187	0.9044	0.9982	0.9997	1.0008	1.0002	0.9492	0.7908	0.9423	0.9175
1.6	2.6558E 00	1.0036	0.9424	0.9165	0.9021	0.9979	0.9996	1.0008	1.0001	0.9496	0.7968	0.9442	0.9192
1.8	2.5928E 00	1.0031	0.9417	0.9159	0.9015	0.9976	0.9996	1.0007	1.0001	0.9502	0.8022	0.9459	0.9206
2.0	2.5445E 00	1.0027	0.9418	0.9161	0.9019	0.9972	0.9995	1.0007	1.0001	0.9509	0.8071	0.9474	0.9218
2.2	2.5062E 00	1.0022	0.9423	0.9169	0.9029	0.9969	0.9994	1.0006	1.0001	0.9517	0.8117	0.9487	0.9230
2.4	2.4750E 00	1.0017	0.9431	0.9180	0.9042	0.9966	0.9994	1.0006	1.0001	0.9525	0.8160	0.9500	0.9240
2.6	2.4489E 00	1.0012	0.9441	0.9194	0.9057	0.9963	0.9993	1.0005	1.0001	0.9533	0.8200	0.9511	0.9249
2.8	2.4268E 00	1.0006	0.9452	0.9208	0.9073	0.9959	0.9993	1.0005	1.0001	0.9541	0.8239	0.9522	0.9258
3.0	2.4076E 00	1.0001	0.9464	0.9224	0.9090	0.9956	0.9992	1.0005	1.0001	0.9549	0.8276	0.9532	0.9266
3.2	2.3908E 00	0.9995	0.9476	0.9240	0.9108	0.9953	0.9991	1.0004	1.0001	0.9556	0.8313	0.9541	0.9273
3.4	2.3758E 00	0.9990	0.9488	0.9255	0.9125	0.9950	0.9991	1.0004	1.0001	0.9564	0.8348	0.9550	0.9280
3.6	2.3623E 00	0.9984	0.9500	0.9271	0.9143	0.9947	0.9990	1.0004	1.0001	0.9571	0.8382	0.9559	0.9287
3.8	2.3501E 00	0.9979	0.9513	0.9287	0.9161	0.9943	0.9990	1.0004	1.0001	0.9578	0.8416	0.9567	0.9293
4.0	2.3390E 00	0.9973	0.9525	0.9303	0.9178	0.9940	0.9989	1.0004	1.0001	0.9585	0.8449	0.9575	0.9299
4.5	2.3145E 00	0.9959	0.9555	0.9341	0.9220	0.9932	0.9988	1.0003	1.0001	0.9602	0.8530	0.9594	0.9312
5.0	2.2939E 00	0.9945	0.9584	0.9378	0.9261	0.9924	0.9986	1.0003	1.0001	0.9617	0.8608	0.9611	0.9325
5.5	2.2759E 00	0.9930	0.9612	0.9414	0.9300	0.9916	0.9985	1.0003	1.0000	0.9632	0.8684	0.9626	0.9336
6.0	2.2599E 00	0.9916	0.9638	0.9448	0.9338	0.9909	0.9983	1.0002	1.0000	0.9646	0.8759	0.9641	0.9346
6.5	2.2454E 00	0.9901	0.9664	0.9481	0.9374	0.9901	0.9982	1.0002	1.0000	0.9660	0.8833	0.9655	0.9355
7.0	2.2322E 00	0.9887	0.9689	0.9512	0.9409	0.9893	0.9980	1.0002	1.0000	0.9673	0.8906	0.9669	0.9364
7.5	2.2199E 00	0.9872	0.9713	0.9543	0.9443	0.9885	0.9979	1.0002	1.0000	0.9685	0.8977	0.9682	0.9372
8.0	2.2084E 00	0.9857	0.9737	0.9573	0.9475	0.9877	0.9978	1.0002	1.0000	0.9697	0.9048	0.9694	0.9380
9.0	2.1874E 00	0.9828	0.9782	0.9629	0.9537	0.9861	0.9975	1.0002	1.0000	0.9720	0.9189	0.9717	0.9394
10.0	2.1684E 00	0.9799	0.9824	0.9683	0.9596	0.9846	0.9972	1.0001	1.0000	0.9741	0.9327	0.9739	0.9406
11.0	2.1510E 00	0.9769	0.9865	0.9735	0.9653	0.9830	0.9969	1.0001	1.0000	0.9762	0.9463	0.9760	0.9418
12.0	2.1347E 00	0.9740	0.9904	0.9784	0.9707	0.9814	0.9966	1.0001	1.0000	0.9782	0.9597	0.9780	0.9429
13.0	2.1195E 00	0.9711	0.9942	0.9831	0.9758	0.9799	0.9963	1.0001	1.0000	0.9801	0.9729	0.9799	0.9439
14.0	2.1051E 00	0.9682	0.9978	0.9877	0.9809	0.9783	0.9960	1.0001	1.0000	0.9819	0.9859	0.9818	0.9448
16.0	2.0783E 00	0.9623	1.0048	0.9965	0.9905	0.9753	0.9955	1.0001	1.0000	0.9854	1.0114	0.9853	0.9466
18.0	2.0536E 00	0.9566	1.0114	1.0049	0.9997	0.9722	0.9949	1.0001	1.0000	0.9888	1.0363	0.9887	0.9482
20.0	2.0306E 00	0.9508	1.0178	1.0129	1.0085	0.9692	0.9943	1.0001	1.0000	0.9920	1.0605	0.9919	0.9496
25.0	1.9784E 00	0.9367	1.0329	1.0319	1.0293	0.9618	0.9928	1.0001	1.0000	0.9996	1.1178	0.9995	0.9529
30.0	1.9319E 00	0.9229	1.0469	1.0497	1.0488	0.9547	0.9913	1.0000	1.0000	1.0067	1.1705	1.0066	0.9558
35.0	1.8895E 00	0.9095	1.0602	1.0667	1.0674	0.9478	0.9897	1.0000	1.0000	1.0134	1.2182	1.0134	0.9585
40.0	1.8503E 00	0.8965	1.0728	1.0829	1.0853	0.9413	0.9882	1.0000	1.0000	1.0199	1.2612	1.0198	0.9611
45.0	1.8138E 00	0.8839	1.0848	1.0986	1.1026	0.9352	0.9866	1.0000	1.0000	1.0260	1.2990	1.0260	0.9635
50.0	1.7796E 00	0.8717	1.0963	1.1138	1.1194	0.9294	0.9849	1.0000	1.0000	1.0319	1.3315	1.0319	0.9658

ELECTRON Z= 31 A= 71

p	F_0L_0	L_0	λ_2	λ_3	λ_4	μ_1	μ_2	Λ_1	Λ_2	ν_{12}	$\hat{\nu}_{12}$	η_{12}	$\hat{\eta}_{12}$
0.1	2.0603E 01	1.0059	5.3100	11.4417	17.4112	0.9999	1.0000	1.0016	1.0003	2.0774	0.7105	0.8470	0.8239
0.2	1.0426E 01	1.0059	1.9957	2.5006	2.7775	0.9999	1.0000	1.0015	1.0003	1.2034	0.7211	0.8907	0.8669
0.3	7.1213E 00	1.0059	1.3842	1.5055	1.5591	0.9998	1.0000	1.0015	1.0003	1.0472	0.7306	0.9060	0.8818
0.4	5.5545E 00	1.0058	1.1722	1.2083	1.2203	0.9997	1.0000	1.0014	1.0003	0.9948	0.7391	0.9143	0.8897
0.5	4.6757E 00	1.0057	1.0756	1.0812	1.0794	0.9996	0.9999	1.0014	1.0003	0.9720	0.7467	0.9197	0.8949
0.6	4.1296E 00	1.0056	1.0244	1.0160	1.0082	0.9994	0.9999	1.0013	1.0003	0.9604	0.7537	0.9236	0.8986
0.7	3.7652E 00	1.0055	0.9945	0.9786	0.9677	0.9993	0.9999	1.0013	1.0002	0.9540	0.7599	0.9267	0.9014
0.8	3.5088E 00	1.0054	0.9758	0.9555	0.9430	0.9991	0.9998	1.0012	1.0002	0.9504	0.7656	0.9291	0.9037
0.9	3.3209E 00	1.0052	0.9636	0.9407	0.9270	0.9990	0.9998	1.0011	1.0002	0.9482	0.7708	0.9312	0.9056
1.0	3.1785E 00	1.0051	0.9553	0.9308	0.9165	0.9988	0.9998	1.0011	1.0002	0.9469	0.7755	0.9330	0.9072
1.2	2.9795E 00	1.0047	0.9458	0.9193	0.9044	0.9985	0.9997	1.0010	1.0002	0.9458	0.7837	0.9360	0.9099
1.4	2.8488E 00	1.0043	0.9411	0.9139	0.8987	0.9981	0.9997	1.0009	1.0002	0.9458	0.7907	0.9384	0.9120
1.6	2.7571E 00	1.0039	0.9390	0.9116	0.8964	0.9978	0.9996	1.0008	1.0002	0.9462	0.7969	0.9405	0.9138
1.8	2.6895E 00	1.0034	0.9383	0.9109	0.8958	0.9974	0.9995	1.0007	1.0001	0.9469	0.8024	0.9423	0.9153
2.0	2.6377E 00	1.0029	0.9383	0.9112	0.8962	0.9971	0.9995	1.0007	1.0001	0.9477	0.8074	0.9439	0.9167
2.2	2.5965E 00	1.0024	0.9389	0.9120	0.8972	0.9967	0.9994	1.0006	1.0001	0.9485	0.8120	0.9453	0.9179
2.4	2.5629E 00	1.0018	0.9398	0.9132	0.8986	0.9964	0.9993	1.0006	1.0001	0.9494	0.8164	0.9467	0.9190
2.6	2.5348E 00	1.0013	0.9408	0.9147	0.9002	0.9960	0.9993	1.0006	1.0001	0.9502	0.8205	0.9479	0.9200
2.8	2.5109E 00	1.0007	0.9420	0.9162	0.9019	0.9957	0.9992	1.0005	1.0001	0.9511	0.8244	0.9490	0.9209
3.0	2.4902E 00	1.0002	0.9432	0.9178	0.9037	0.9953	0.9992	1.0005	1.0001	0.9519	0.8282	0.9501	0.9217
3.2	2.4720E 00	0.9996	0.9445	0.9195	0.9056	0.9950	0.9991	1.0005	1.0001	0.9527	0.8318	0.9511	0.9225
3.4	2.4558E 00	0.9990	0.9458	0.9212	0.9075	0.9947	0.9990	1.0004	1.0001	0.9535	0.8354	0.9521	0.9233
3.6	2.4412E 00	0.9984	0.9471	0.9229	0.9093	0.9943	0.9990	1.0004	1.0001	0.9543	0.8389	0.9530	0.9240
3.8	2.4280E 00	0.9978	0.9484	0.9246	0.9112	0.9940	0.9989	1.0004	1.0001	0.9550	0.8423	0.9539	0.9246
4.0	2.4159E 00	0.9972	0.9497	0.9262	0.9130	0.9937	0.9988	1.0004	1.0001	0.9558	0.8456	0.9547	0.9253
4.5	2.3893E 00	0.9957	0.9529	0.9303	0.9175	0.9928	0.9987	1.0003	1.0001	0.9576	0.8537	0.9567	0.9267
5.0	2.3668E 00	0.9942	0.9560	0.9343	0.9219	0.9920	0.9985	1.0003	1.0001	0.9592	0.8615	0.9585	0.9280
5.5	2.3472E 00	0.9927	0.9590	0.9380	0.9260	0.9911	0.9984	1.0003	1.0001	0.9608	0.8692	0.9602	0.9292
6.0	2.3298E 00	0.9911	0.9618	0.9417	0.9300	0.9903	0.9982	1.0003	1.0000	0.9623	0.8767	0.9618	0.9303
6.5	2.3140E 00	0.9896	0.9646	0.9452	0.9339	0.9895	0.9981	1.0002	1.0000	0.9637	0.8841	0.9633	0.9313
7.0	2.2995E 00	0.9881	0.9672	0.9485	0.9376	0.9886	0.9979	1.0002	1.0000	0.9651	0.8913	0.9647	0.9322
7.5	2.2861E 00	0.9865	0.9698	0.9518	0.9411	0.9878	0.9978	1.0002	1.0000	0.9664	0.8985	0.9661	0.9331
8.0	2.2736E 00	0.9850	0.9723	0.9549	0.9446	0.9870	0.9976	1.0002	1.0000	0.9677	0.9056	0.9674	0.9339
9.0	2.2506E 00	0.9819	0.9771	0.9610	0.9512	0.9853	0.9973	1.0002	1.0000	0.9701	0.9196	0.9699	0.9354
10.0	2.2299E 00	0.9788	0.9816	0.9667	0.9575	0.9836	0.9970	1.0002	1.0000	0.9724	0.9333	0.9722	0.9367
11.0	2.2108E 00	0.9756	0.9859	0.9721	0.9635	0.9820	0.9967	1.0001	1.0000	0.9746	0.9468	0.9744	0.9380
12.0	2.1931E 00	0.9725	0.9901	0.9774	0.9692	0.9803	0.9964	1.0001	1.0000	0.9767	0.9601	0.9765	0.9391
13.0	2.1765E 00	0.9695	0.9941	0.9824	0.9747	0.9787	0.9961	1.0001	1.0000	0.9787	0.9733	0.9786	0.9402
14.0	2.1607E 00	0.9664	0.9980	0.9873	0.9800	0.9770	0.9958	1.0001	1.0000	0.9807	0.9862	0.9806	0.9412
16.0	2.1315E 00	0.9602	1.0054	0.9967	0.9903	0.9738	0.9952	1.0001	1.0000	0.9845	1.0115	0.9843	0.9431
18.0	2.1046E 00	0.9541	1.0125	1.0056	1.0000	0.9705	0.9946	1.0001	1.0000	0.9880	1.0360	0.9879	0.9448
20.0	2.0796E 00	0.9481	1.0193	1.0141	1.0094	0.9673	0.9940	1.0001	1.0000	0.9914	1.0598	0.9914	0.9464
25.0	2.0229E 00	0.9332	1.0353	1.0343	1.0315	0.9595	0.9924	1.0001	1.0000	0.9995	1.1162	0.9995	0.9499
30.0	1.9725E 00	0.9187	1.0503	1.0533	1.0523	0.9521	0.9908	1.0001	1.0000	1.0071	1.1675	1.0070	0.9531
35.0	1.9266E 00	0.9046	1.0644	1.0713	1.0721	0.9449	0.9892	1.0000	1.0000	1.0142	1.2139	1.0142	0.9560
40.0	1.8844E 00	0.8910	1.0778	1.0886	1.0911	0.9381	0.9875	1.0000	1.0000	1.0211	1.2551	1.0210	0.9588
45.0	1.8451E 00	0.8778	1.0905	1.1053	1.1095	0.9317	0.9858	1.0000	1.0000	1.0276	1.2909	1.0276	0.9614
50.0	1.8084E 00	0.8651	1.1027	1.1213	1.1273	0.9257	0.9841	1.0000	1.0000	1.0339	1.3213	1.0338	0.9639

Instr. see p. 22

ELECTRON Z= 32 A= 72

p	F_0L_0	L_0	λ_2	λ_3	λ_4	μ_1	μ_2	Λ_1	Λ_2	ν_{12}	$\hat{\nu}_{12}$	η_{12}	$\hat{\eta}_{12}$
0.1	2.1749E 01	1.0064	5.5523	12.3599	19.2182	0.9999	1.0000	1.0016	1.0003	2.1420	0.7078	0.8348	0.8104
0.2	1.1004E 01	1.0064	2.0516	2.6027	2.9093	0.9998	1.0000	1.0016	1.0003	1.2150	0.7188	0.8826	0.8573
0.3	7.5089E 00	1.0063	1.4056	1.5364	1.5946	0.9998	1.0000	1.0015	1.0003	1.0494	0.7286	0.8993	0.8736
0.4	5.8468E 00	1.0063	1.1816	1.2202	1.2330	0.9997	1.0000	1.0015	1.0003	0.9940	0.7374	0.9083	0.8822
0.5	4.9124E 00	1.0062	1.0796	1.0854	1.0836	0.9995	0.9999	1.0014	1.0003	0.9698	0.7453	0.9141	0.8878
0.6	4.3308E 00	1.0061	1.0254	1.0164	1.0081	0.9994	0.9999	1.0014	1.0003	0.9575	0.7524	0.9184	0.8918
0.7	3.9424E 00	1.0060	0.9937	0.9768	0.9653	0.9992	0.9999	1.0013	1.0002	0.9508	0.7588	0.9216	0.8949
0.8	3.6689E 00	1.0058	0.9740	0.9525	0.9391	0.9991	0.9998	1.0012	1.0002	0.9469	0.7646	0.9243	0.8973
0.9	3.4683E 00	1.0057	0.9611	0.9368	0.9223	0.9989	0.9998	1.0012	1.0002	0.9446	0.7699	0.9265	0.8993
1.0	3.3163E 00	1.0055	0.9524	0.9263	0.9111	0.9987	0.9998	1.0011	1.0002	0.9433	0.7747	0.9285	0.9011
1.2	3.1036E 00	1.0051	0.9422	0.9142	0.8984	0.9984	0.9997	1.0010	1.0002	0.9421	0.7831	0.9317	0.9039
1.4	2.9637E 00	1.0047	0.9373	0.9085	0.8924	0.9980	0.9996	1.0009	1.0002	0.9421	0.7903	0.9343	0.9062
1.6	2.8656E 00	1.0042	0.9351	0.9060	0.8899	0.9977	0.9996	1.0008	1.0002	0.9426	0.7965	0.9365	0.9081
1.8	2.7932E 00	1.0037	0.9343	0.9053	0.8893	0.9973	0.9995	1.0008	1.0001	0.9433	0.8021	0.9384	0.9097
2.0	2.7375E 00	1.0032	0.9344	0.9056	0.8897	0.9969	0.9994	1.0007	1.0001	0.9441	0.8072	0.9401	0.9112
2.2	2.6932E 00	1.0027	0.9350	0.9065	0.8908	0.9966	0.9994	1.0007	1.0001	0.9450	0.8119	0.9416	0.9125
2.4	2.6571E 00	1.0021	0.9359	0.9078	0.8922	0.9962	0.9993	1.0006	1.0001	0.9459	0.8162	0.9430	0.9136
2.6	2.6269E 00	1.0015	0.9370	0.9093	0.8939	0.9959	0.9993	1.0006	1.0001	0.9468	0.8203	0.9443	0.9147
2.8	2.6011E 00	1.0010	0.9383	0.9109	0.8958	0.9955	0.9992	1.0005	1.0001	0.9477	0.8243	0.9455	0.9157
3.0	2.5788E 00	1.0004	0.9396	0.9126	0.8977	0.9951	0.9991	1.0005	1.0001	0.9486	0.8281	0.9467	0.9166
3.2	2.5591E 00	0.9998	0.9409	0.9144	0.8996	0.9948	0.9991	1.0005	1.0001	0.9495	0.8317	0.9478	0.9174
3.4	2.5416E 00	0.9992	0.9423	0.9162	0.9016	0.9944	0.9990	1.0005	1.0001	0.9503	0.8353	0.9488	0.9182
3.6	2.5258E 00	0.9986	0.9437	0.9180	0.9036	0.9941	0.9989	1.0004	1.0001	0.9511	0.8387	0.9498	0.9189
3.8	2.5114E 00	0.9979	0.9451	0.9197	0.9056	0.9937	0.9989	1.0004	1.0001	0.9519	0.8421	0.9507	0.9196
4.0	2.4983E 00	0.9973	0.9464	0.9215	0.9075	0.9934	0.9988	1.0004	1.0001	0.9527	0.8454	0.9516	0.9203
4.5	2.4695E 00	0.9958	0.9498	0.9258	0.9123	0.9925	0.9986	1.0004	1.0001	0.9546	0.8534	0.9537	0.9219
5.0	2.4451E 00	0.9942	0.9531	0.9300	0.9168	0.9916	0.9985	1.0003	1.0001	0.9564	0.8613	0.9556	0.9232
5.5	2.4238E 00	0.9926	0.9562	0.9339	0.9212	0.9907	0.9983	1.0003	1.0001	0.9580	0.8688	0.9574	0.9245
6.0	2.4048E 00	0.9910	0.9592	0.9378	0.9254	0.9899	0.9982	1.0003	1.0001	0.9596	0.8763	0.9590	0.9256
6.5	2.3876E 00	0.9894	0.9621	0.9415	0.9295	0.9890	0.9980	1.0002	1.0000	0.9611	0.8836	0.9606	0.9267
7.0	2.3719E 00	0.9878	0.9649	0.9450	0.9334	0.9881	0.9979	1.0002	1.0000	0.9626	0.8908	0.9621	0.9277
7.5	2.3573E 00	0.9862	0.9677	0.9485	0.9371	0.9872	0.9977	1.0002	1.0000	0.9640	0.8979	0.9636	0.9286
8.0	2.3437E 00	0.9845	0.9703	0.9518	0.9408	0.9864	0.9975	1.0002	1.0000	0.9653	0.9049	0.9650	0.9295
9.0	2.3187E 00	0.9813	0.9753	0.9581	0.9478	0.9846	0.9972	1.0002	1.0000	0.9679	0.9187	0.9676	0.9311
10.0	2.2962E 00	0.9781	0.9801	0.9642	0.9544	0.9829	0.9969	1.0002	1.0000	0.9703	0.9322	0.9701	0.9325
11.0	2.2755E 00	0.9748	0.9847	0.9699	0.9607	0.9812	0.9966	1.0001	1.0000	0.9726	0.9455	0.9724	0.9339
12.0	2.2562E 00	0.9716	0.9891	0.9754	0.9667	0.9795	0.9963	1.0001	1.0000	0.9749	0.9585	0.9747	0.9351
13.0	2.2382E 00	0.9684	0.9933	0.9808	0.9725	0.9777	0.9960	1.0001	1.0000	0.9770	0.9714	0.9768	0.9362
14.0	2.2212E 00	0.9652	0.9974	0.9859	0.9781	0.9760	0.9957	1.0001	1.0000	0.9790	0.9841	0.9789	0.9373
16.0	2.1895E 00	0.9588	1.0052	0.9958	0.9889	0.9726	0.9950	1.0001	1.0000	0.9830	1.0089	0.9829	0.9393
18.0	2.1604E 00	0.9524	1.0127	1.0052	0.9992	0.9693	0.9944	1.0001	1.0000	0.9868	1.0329	0.9867	0.9411
20.0	2.1333E 00	0.9462	1.0199	1.0142	1.0090	0.9659	0.9937	1.0001	1.0000	0.9904	1.0561	0.9903	0.9428
25.0	2.0723E 00	0.9307	1.0367	1.0354	1.0323	0.9578	0.9921	1.0001	1.0000	0.9989	1.1110	0.9988	0.9466
30.0	2.0181E 00	0.9157	1.0525	1.0554	1.0542	0.9500	0.9904	1.0001	1.0000	1.0068	1.1608	1.0068	0.9500
35.0	1.9689E 00	0.9011	1.0673	1.0743	1.0750	0.9426	0.9887	1.0000	1.0000	1.0144	1.2055	1.0143	0.9532
40.0	1.9237E 00	0.8870	1.0813	1.0925	1.0950	0.9356	0.9870	1.0000	1.0000	1.0215	1.2449	1.0215	0.9561
45.0	1.8817E 00	0.8733	1.0947	1.1100	1.1143	0.9290	0.9852	1.0000	1.0000	1.0284	1.2790	1.0284	0.9590
50.0	1.8426E 00	0.8602	1.1074	1.1269	1.1330	0.9228	0.9834	1.0000	1.0000	1.0350	1.3076	1.0349	0.9617

ELECTRON Z= 33 A= 77

p	F_0L_0	L_0	λ_2	λ_3	λ_4	μ_1	μ_2	Λ_1	Λ_2	ν_{12}	$\hat{\nu}_{12}$	η_{12}	$\hat{\eta}_{12}$
0.1	2.2924E 01	1.0068	5.8010	13.3403	21.1988	0.9999	1.0000	1.0017	1.0003	2.2088	0.7052	0.8221	0.7964
0.2	1.1597E 01	1.0067	2.1095	2.7104	3.0495	0.9998	1.0000	1.0017	1.0003	1.2272	0.7166	0.8744	0.8477
0.3	7.9069E 00	1.0067	1.4282	1.5692	1.6324	0.9997	1.0000	1.0016	1.0003	1.0519	0.7268	0.8926	0.8654
0.4	6.1472E 00	1.0066	1.1918	1.2332	1.2471	0.9996	0.9999	1.0016	1.0003	0.9933	0.7358	0.9023	0.8747
0.5	5.1556E 00	1.0066	1.0841	1.0905	1.0886	0.9995	0.9999	1.0015	1.0003	0.9677	0.7440	0.9086	0.8807
0.6	4.5374E 00	1.0064	1.0270	1.0175	1.0088	0.9994	0.9999	1.0015	1.0003	0.9548	0.7513	0.9132	0.8850
0.7	4.1241E 00	1.0063	0.9935	0.9757	0.9636	0.9992	0.9999	1.0014	1.0003	0.9477	0.7580	0.9167	0.8883
0.8	3.8328E 00	1.0062	0.9727	0.9500	0.9360	0.9990	0.9998	1.0013	1.0003	0.9436	0.7639	0.9196	0.8909
0.9	3.6191E 00	1.0060	0.9590	0.9334	0.9182	0.9988	0.9998	1.0013	1.0002	0.9412	0.7694	0.9220	0.8931
1.0	3.4571E 00	1.0058	0.9499	0.9224	0.9065	0.9987	0.9998	1.0012	1.0002	0.9398	0.7743	0.9241	0.8950
1.2	3.2302E 00	1.0054	0.9392	0.9097	0.8930	0.9983	0.9997	1.0011	1.0002	0.9386	0.7829	0.9275	0.8980
1.4	3.0809E 00	1.0050	0.9340	0.9036	0.8868	0.9979	0.9996	1.0010	1.0002	0.9386	0.7903	0.9302	0.9005
1.6	2.9759E 00	1.0045	0.9316	0.9010	0.8841	0.9975	0.9996	1.0009	1.0002	0.9391	0.7967	0.9326	0.9025
1.8	2.8984E 00	1.0040	0.9308	0.9003	0.8834	0.9971	0.9995	1.0008	1.0002	0.9398	0.8024	0.9346	0.9042
2.0	2.8387E 00	1.0034	0.9309	0.9006	0.8839	0.9967	0.9994	1.0008	1.0001	0.9407	0.8075	0.9364	0.9058
2.2	2.7912E 00	1.0028	0.9315	0.9016	0.8850	0.9964	0.9993	1.0007	1.0001	0.9417	0.8123	0.9381	0.9071
2.4	2.7524E 00	1.0022	0.9325	0.9029	0.8866	0.9960	0.9993	1.0007	1.0001	0.9426	0.8167	0.9396	0.9083
2.6	2.7198E 00	1.0016	0.9337	0.9045	0.8884	0.9956	0.9992	1.0006	1.0001	0.9436	0.8209	0.9410	0.9095
2.8	2.6921E 00	1.0010	0.9350	0.9062	0.8903	0.9952	0.9991	1.0006	1.0001	0.9446	0.8249	0.9423	0.9105
3.0	2.6680E 00	1.0004	0.9364	0.9081	0.8924	0.9948	0.9991	1.0005	1.0001	0.9455	0.8288	0.9435	0.9115
3.2	2.6467E 00	0.9998	0.9378	0.9099	0.8944	0.9945	0.9990	1.0005	1.0001	0.9464	0.8325	0.9446	0.9124
3.4	2.6278E 00	0.9991	0.9393	0.9118	0.8965	0.9941	0.9989	1.0005	1.0001	0.9473	0.8361	0.9457	0.9132
3.6	2.6107E 00	0.9985	0.9408	0.9137	0.8986	0.9937	0.9989	1.0005	1.0001	0.9482	0.8396	0.9467	0.9140
3.8	2.5952E 00	0.9978	0.9422	0.9156	0.9007	0.9933	0.9988	1.0004	1.0001	0.9490	0.8430	0.9477	0.9147
4.0	2.5809E 00	0.9972	0.9437	0.9175	0.9027	0.9930	0.9987	1.0004	1.0001	0.9499	0.8463	0.9487	0.9155
4.5	2.5497E 00	0.9955	0.9473	0.9220	0.9078	0.9920	0.9986	1.0004	1.0001	0.9519	0.8545	0.9509	0.9171
5.0	2.5232E 00	0.9938	0.9507	0.9264	0.9126	0.9911	0.9984	1.0003	1.0001	0.9537	0.8623	0.9529	0.9186
5.5	2.5000E 00	0.9921	0.9541	0.9307	0.9173	0.9902	0.9982	1.0003	1.0001	0.9555	0.8700	0.9548	0.9199
6.0	2.4794E 00	0.9904	0.9573	0.9347	0.9217	0.9892	0.9981	1.0003	1.0001	0.9572	0.8775	0.9566	0.9211
6.5	2.4607E 00	0.9887	0.9604	0.9386	0.9260	0.9883	0.9979	1.0003	1.0000	0.9588	0.8848	0.9583	0.9222
7.0	2.4435E 00	0.9870	0.9634	0.9424	0.9301	0.9874	0.9977	1.0002	1.0000	0.9604	0.8920	0.9599	0.9233
7.5	2.4276E 00	0.9853	0.9662	0.9460	0.9341	0.9865	0.9976	1.0002	1.0000	0.9618	0.8991	0.9614	0.9243
8.0	2.4127E 00	0.9836	0.9690	0.9496	0.9380	0.9855	0.9974	1.0002	1.0000	0.9633	0.9062	0.9629	0.9252
9.0	2.3855E 00	0.9802	0.9744	0.9563	0.9454	0.9837	0.9971	1.0002	1.0000	0.9660	0.9200	0.9657	0.9269
10.0	2.3610E 00	0.9768	0.9795	0.9627	0.9524	0.9818	0.9967	1.0002	1.0000	0.9686	0.9335	0.9683	0.9284
11.0	2.3384E 00	0.9733	0.9844	0.9688	0.9591	0.9800	0.9964	1.0002	1.0000	0.9710	0.9468	0.9708	0.9299
12.0	2.3174E 00	0.9699	0.9890	0.9747	0.9655	0.9782	0.9961	1.0001	1.0000	0.9734	0.9599	0.9732	0.9312
13.0	2.2978E 00	0.9665	0.9935	0.9804	0.9717	0.9764	0.9957	1.0001	1.0000	0.9757	0.9727	0.9755	0.9324
14.0	2.2792E 00	0.9631	0.9979	0.9858	0.9776	0.9745	0.9954	1.0001	1.0000	0.9779	0.9854	0.9777	0.9336
16.0	2.2448E 00	0.9564	1.0062	0.9963	0.9891	0.9709	0.9947	1.0001	1.0000	0.9821	1.0100	0.9819	0.9357
18.0	2.2131E 00	0.9497	1.0142	1.0063	1.0000	0.9674	0.9940	1.0001	1.0000	0.9861	1.0338	0.9860	0.9376
20.0	2.1837E 00	0.9430	1.0218	1.0159	1.0105	0.9639	0.9934	1.0001	1.0000	0.9899	1.0568	0.9898	0.9395
25.0	2.1175E 00	0.9268	1.0397	1.0385	1.0352	0.9553	0.9916	1.0001	1.0000	0.9990	1.1107	0.9989	0.9435
30.0	2.0588E 00	0.9109	1.0564	1.0597	1.0585	0.9471	0.9899	1.0001	1.0000	1.0074	1.1594	1.0074	0.9472
35.0	2.0058E 00	0.8956	1.0722	1.0799	1.0806	0.9394	0.9881	1.0000	1.0000	1.0154	1.2027	1.0154	0.9506
40.0	1.9571E 00	0.8808	1.0871	1.0992	1.1019	0.9320	0.9862	1.0000	1.0000	1.0231	1.2403	1.0230	0.9539
45.0	1.9121E 00	0.8666	1.1012	1.1177	1.1224	0.9252	0.9844	1.0000	1.0000	1.0303	1.2726	1.0303	0.9570
50.0	1.8702E 00	0.8529	1.1147	1.1357	1.1423	0.9188	0.9824	1.0000	1.0000	1.0373	1.2989	1.0373	0.9599

 Instr. see p. 22

ELECTRON Z= 34 A= 78

p	F_0L_0	L_0	λ_2	λ_3	λ_4	μ_1	μ_2	Λ_1	Λ_2	ν_{12}	$\hat{\nu}_{12}$	η_{12}	$\hat{\eta}_{12}$
0.1	2.4182E 01	1.0073	6.0494	14.3640	23.3281	0.9999	1.0000	1.0018	1.0003	2.2762	0.7024	0.8087	0.7816
0.2	1.2232E 01	1.0072	2.1668	2.8196	3.1933	0.9998	1.0000	1.0017	1.0003	1.2391	0.7142	0.8656	0.8374
0.3	8.3335E 00	1.0072	1.4501	1.6015	1.6698	0.9997	1.0000	1.0017	1.0003	1.0542	0.7246	0.8853	0.8566
0.4	6.4698E 00	1.0071	1.2014	1.2455	1.2604	0.9996	0.9999	1.0016	1.0003	0.9923	0.7339	0.8959	0.8667
0.5	5.4171E 00	1.0070	1.0881	1.0948	1.0927	0.9995	0.9999	1.0016	1.0003	0.9653	0.7423	0.9027	0.8732
0.6	4.7598E 00	1.0069	1.0279	1.0178	1.0086	0.9993	0.9999	1.0015	1.0003	0.9517	0.7499	0.9076	0.8778
0.7	4.3199E 00	1.0068	0.9927	0.9738	0.9610	0.9992	0.9999	1.0014	1.0003	0.9442	0.7567	0.9114	0.8813
0.8	4.0097E 00	1.0066	0.9707	0.9468	0.9319	0.9990	0.9998	1.0014	1.0003	0.9399	0.7629	0.9144	0.8841
0.9	3.7819E 00	1.0065	0.9564	0.9293	0.9132	0.9988	0.9998	1.0013	1.0002	0.9374	0.7684	0.9170	0.8865
1.0	3.6090E 00	1.0063	0.9467	0.9177	0.9009	0.9986	0.9998	1.0012	1.0002	0.9359	0.7735	0.9192	0.8885
1.2	3.3669E 00	1.0059	0.9354	0.9043	0.8867	0.9982	0.9997	1.0011	1.0002	0.9347	0.7823	0.9228	0.8917
1.4	3.2074E 00	1.0054	0.9300	0.8980	0.8802	0.9978	0.9996	1.0010	1.0002	0.9347	0.7897	0.9258	0.8943
1.6	3.0951E 00	1.0049	0.9275	0.8952	0.8774	0.9974	0.9995	1.0009	1.0002	0.9352	0.7962	0.9283	0.8964
1.8	3.0121E 00	1.0043	0.9266	0.8944	0.8767	0.9970	0.9995	1.0009	1.0002	0.9360	0.8020	0.9305	0.8983
2.0	2.9481E 00	1.0038	0.9267	0.8948	0.8772	0.9966	0.9994	1.0008	1.0001	0.9369	0.8072	0.9324	0.8999
2.2	2.8972E 00	1.0032	0.9274	0.8958	0.8784	0.9962	0.9993	1.0007	1.0001	0.9379	0.8120	0.9341	0.9014
2.4	2.8554E 00	1.0025	0.9285	0.8972	0.8800	0.9958	0.9993	1.0007	1.0001	0.9390	0.8165	0.9357	0.9027
2.6	2.8204E 00	1.0019	0.9297	0.8989	0.8819	0.9954	0.9992	1.0006	1.0001	0.9400	0.8208	0.9372	0.9039
2.8	2.7905E 00	1.0013	0.9311	0.9007	0.8839	0.9950	0.9991	1.0006	1.0001	0.9410	0.8248	0.9385	0.9049
3.0	2.7645E 00	1.0006	0.9326	0.9026	0.8861	0.9946	0.9990	1.0006	1.0001	0.9420	0.8286	0.9398	0.9060
3.2	2.7416E 00	1.0000	0.9341	0.9046	0.8882	0.9942	0.9990	1.0005	1.0001	0.9430	0.8323	0.9410	0.9069
3.4	2.7212E 00	0.9993	0.9356	0.9066	0.8904	0.9938	0.9989	1.0005	1.0001	0.9439	0.8359	0.9422	0.9078
3.6	2.7027E 00	0.9986	0.9372	0.9086	0.8926	0.9934	0.9988	1.0005	1.0001	0.9448	0.8394	0.9433	0.9086
3.8	2.6859E 00	0.9979	0.9387	0.9105	0.8948	0.9931	0.9988	1.0005	1.0001	0.9457	0.8429	0.9443	0.9094
4.0	2.6705E 00	0.9973	0.9402	0.9125	0.8970	0.9927	0.9987	1.0004	1.0001	0.9466	0.8462	0.9453	0.9102
4.5	2.6367E 00	0.9955	0.9440	0.9173	0.9023	0.9917	0.9985	1.0004	1.0001	0.9487	0.8543	0.9477	0.9119
5.0	2.6079E 00	0.9938	0.9477	0.9219	0.9073	0.9907	0.9983	1.0004	1.0001	0.9507	0.8621	0.9498	0.9135
5.5	2.5828E 00	0.9920	0.9512	0.9264	0.9122	0.9897	0.9982	1.0003	1.0001	0.9526	0.8697	0.9518	0.9149
6.0	2.5604E 00	0.9903	0.9546	0.9307	0.9169	0.9888	0.9980	1.0003	1.0001	0.9543	0.8772	0.9537	0.9162
6.5	2.5401E 00	0.9885	0.9578	0.9348	0.9214	0.9878	0.9978	1.0003	1.0001	0.9560	0.8844	0.9555	0.9174
7.0	2.5215E 00	0.9867	0.9610	0.9388	0.9258	0.9868	0.9977	1.0003	1.0000	0.9577	0.8916	0.9572	0.9185
7.5	2.5042E 00	0.9849	0.9640	0.9426	0.9300	0.9859	0.9975	1.0002	1.0000	0.9592	0.8987	0.9588	0.9195
8.0	2.4881E 00	0.9832	0.9670	0.9463	0.9340	0.9849	0.9973	1.0002	1.0000	0.9607	0.9056	0.9603	0.9205
9.0	2.4586E 00	0.9796	0.9726	0.9534	0.9418	0.9830	0.9970	1.0002	1.0000	0.9636	0.9193	0.9633	0.9223
10.0	2.4320E 00	0.9760	0.9780	0.9601	0.9492	0.9811	0.9966	1.0002	1.0000	0.9663	0.9326	0.9661	0.9240
11.0	2.4075E 00	0.9725	0.9831	0.9665	0.9562	0.9792	0.9963	1.0002	1.0000	0.9689	0.9457	0.9687	0.9255
12.0	2.3848E 00	0.9689	0.9880	0.9727	0.9629	0.9773	0.9959	1.0002	1.0000	0.9714	0.9586	0.9712	0.9269
13.0	2.3635E 00	0.9654	0.9927	0.9786	0.9694	0.9754	0.9956	1.0001	1.0000	0.9738	0.9712	0.9736	0.9282
14.0	2.3435E 00	0.9619	0.9973	0.9844	0.9757	0.9735	0.9952	1.0001	1.0000	0.9761	0.9836	0.9759	0.9294
16.0	2.3062E 00	0.9549	1.0061	0.9954	0.9877	0.9697	0.9945	1.0001	1.0000	0.9805	1.0078	0.9804	0.9317
18.0	2.2721E 00	0.9479	1.0144	1.0059	0.9991	0.9660	0.9938	1.0001	1.0000	0.9847	1.0311	0.9846	0.9337
20.0	2.2404E 00	0.9410	1.0224	1.0159	1.0101	0.9624	0.9931	1.0001	1.0000	0.9888	1.0536	0.9887	0.9357
25.0	2.1692E 00	0.9242	1.0413	1.0397	1.0361	0.9535	0.9913	1.0001	1.0000	0.9983	1.1061	0.9982	0.9401
30.0	2.1063E 00	0.9078	1.0588	1.0620	1.0605	0.9450	0.9895	1.0001	1.0000	1.0072	1.1534	1.0071	0.9440
35.0	2.0496E 00	0.8920	1.0753	1.0831	1.0837	0.9370	0.9876	1.0001	1.0000	1.0156	1.1950	1.0155	0.9476
40.0	1.9977E 00	0.8767	1.0909	1.1033	1.1060	0.9294	0.9857	1.0000	1.0000	1.0236	1.2311	1.0235	0.9511
45.0	1.9497E 00	0.8620	1.1058	1.1228	1.1275	0.9224	0.9838	1.0000	1.0000	1.0312	1.2616	1.0311	0.9544
50.0	1.9051E 00	0.8479	1.1198	1.1416	1.1483	0.9158	0.9818	1.0000	1.0000	1.0384	1.2863	1.0384	0.9576

Instr. see p. 22

ELECTRON Z= 35 A= 79

p	F_0L_0	L_0	λ_2	λ_3	λ_4	μ_1	μ_2	Λ_1	Λ_2	ν_{12}	$\hat{\nu}_{12}$	η_{12}	$\hat{\eta}_{12}$
0.1	2.5502E 01	1.0078	6.3000	15.4412	25.6317	0.9999	1.0000	1.0019	1.0003	2.3449	0.6995	0.7946	0.7663
0.2	1.2898E 01	1.0077	2.2246	2.9321	3.3431	0.9998	1.0000	1.0018	1.0003	1.2513	0.7116	0.8565	0.8268
0.3	8.7818E 00	1.0077	1.4722	1.6344	1.7082	0.9997	1.0000	1.0018	1.0003	1.0563	0.7224	0.8778	0.8475
0.4	6.8091E 00	1.0076	1.2111	1.2579	1.2738	0.9996	0.9999	1.0017	1.0003	0.9912	0.7320	0.8892	0.8584
0.5	5.6924E 00	1.0075	1.0920	1.0991	1.0969	0.9995	0.9999	1.0016	1.0003	0.9628	0.7406	0.8965	0.8654
0.6	4.9941E 00	1.0074	1.0288	1.0181	1.0083	0.9993	0.9999	1.0016	1.0003	0.9485	0.7484	0.9018	0.8703
0.7	4.5260E 00	1.0073	0.9918	0.9718	0.9582	0.9991	0.9999	1.0015	1.0003	0.9406	0.7554	0.9058	0.8741
0.8	4.1958E 00	1.0071	0.9687	0.9434	0.9277	0.9989	0.9998	1.0014	1.0003	0.9361	0.7617	0.9091	0.8771
0.9	3.9531E 00	1.0069	0.9536	0.9251	0.9081	0.9987	0.9998	1.0014	1.0003	0.9334	0.7674	0.9119	0.8796
1.0	3.7689E 00	1.0067	0.9435	0.9128	0.8951	0.9985	0.9997	1.0013	1.0002	0.9319	0.7725	0.9142	0.8817
1.2	3.5106E 00	1.0063	0.9316	0.8988	0.8803	0.9981	0.9997	1.0012	1.0002	0.9306	0.7815	0.9181	0.8852
1.4	3.3402E 00	1.0058	0.9259	0.8921	0.8734	0.9977	0.9996	1.0011	1.0002	0.9306	0.7892	0.9212	0.8879
1.6	3.2203E 00	1.0053	0.9233	0.8893	0.8705	0.9973	0.9995	1.0010	1.0002	0.9312	0.7958	0.9239	0.8902
1.8	3.1314E 00	1.0047	0.9224	0.8884	0.8698	0.9969	0.9994	1.0009	1.0002	0.9320	0.8016	0.9262	0.8922
2.0	3.0628E 00	1.0041	0.9225	0.8888	0.8703	0.9965	0.9994	1.0008	1.0002	0.9330	0.8069	0.9282	0.8939
2.2	3.0081E 00	1.0035	0.9232	0.8899	0.8715	0.9960	0.9993	1.0008	1.0001	0.9341	0.8118	0.9300	0.8954
2.4	2.9633E 00	1.0028	0.9243	0.8914	0.8733	0.9956	0.9992	1.0007	1.0001	0.9352	0.8163	0.9317	0.8968
2.6	2.9257E 00	1.0022	0.9256	0.8931	0.8752	0.9952	0.9992	1.0007	1.0001	0.9362	0.8206	0.9333	0.8981
2.8	2.8935E 00	1.0015	0.9271	0.8950	0.8774	0.9948	0.9991	1.0006	1.0001	0.9373	0.8246	0.9347	0.8992
3.0	2.8655E 00	1.0008	0.9286	0.8970	0.8796	0.9944	0.9990	1.0006	1.0001	0.9384	0.8285	0.9361	0.9003
3.2	2.8408E 00	1.0002	0.9302	0.8991	0.8819	0.9940	0.9989	1.0006	1.0001	0.9394	0.8322	0.9373	0.9013
3.4	2.8187E 00	0.9995	0.9318	0.9012	0.8842	0.9936	0.9989	1.0005	1.0001	0.9404	0.8358	0.9386	0.9022
3.6	2.7988E 00	0.9988	0.9335	0.9033	0.8865	0.9932	0.9988	1.0005	1.0001	0.9414	0.8393	0.9397	0.9031
3.8	2.7807E 00	0.9981	0.9351	0.9054	0.8888	0.9928	0.9987	1.0005	1.0001	0.9423	0.8427	0.9408	0.9040
4.0	2.7640E 00	0.9973	0.9367	0.9074	0.8911	0.9924	0.9986	1.0004	1.0001	0.9432	0.8461	0.9419	0.9048
4.5	2.7274E 00	0.9956	0.9407	0.9125	0.8966	0.9913	0.9985	1.0004	1.0001	0.9454	0.8542	0.9443	0.9066
5.0	2.6963E 00	0.9937	0.9445	0.9174	0.9020	0.9903	0.9983	1.0004	1.0001	0.9475	0.8620	0.9466	0.9082
5.5	2.6690E 00	0.9919	0.9482	0.9220	0.9071	0.9893	0.9981	1.0003	1.0001	0.9495	0.8695	0.9487	0.9097
6.0	2.6448E 00	0.9901	0.9518	0.9265	0.9120	0.9883	0.9979	1.0003	1.0001	0.9514	0.8769	0.9507	0.9111
6.5	2.6228E 00	0.9882	0.9552	0.9308	0.9167	0.9873	0.9977	1.0003	1.0001	0.9532	0.8842	0.9526	0.9124
7.0	2.6026E 00	0.9864	0.9585	0.9350	0.9213	0.9863	0.9976	1.0003	1.0000	0.9549	0.8912	0.9543	0.9135
7.5	2.5839E 00	0.9846	0.9617	0.9390	0.9257	0.9853	0.9974	1.0002	1.0000	0.9565	0.8982	0.9561	0.9146
8.0	2.5664E 00	0.9827	0.9648	0.9429	0.9300	0.9843	0.9972	1.0002	1.0000	0.9581	0.9051	0.9577	0.9157
9.0	2.5344E 00	0.9790	0.9707	0.9504	0.9381	0.9823	0.9969	1.0002	1.0000	0.9611	0.9186	0.9608	0.9176
10.0	2.5056E 00	0.9753	0.9764	0.9575	0.9458	0.9803	0.9965	1.0002	1.0000	0.9640	0.9319	0.9637	0.9193
11.0	2.4791E 00	0.9716	0.9818	0.9642	0.9532	0.9783	0.9961	1.0002	1.0000	0.9667	0.9448	0.9665	0.9209
12.0	2.4545E 00	0.9679	0.9869	0.9707	0.9603	0.9763	0.9958	1.0002	1.0000	0.9693	0.9575	0.9691	0.9224
13.0	2.4315E 00	0.9643	0.9919	0.9769	0.9671	0.9744	0.9954	1.0001	1.0000	0.9718	0.9699	0.9716	0.9238
14.0	2.4099E 00	0.9606	0.9967	0.9829	0.9736	0.9724	0.9951	1.0001	1.0000	0.9743	0.9821	0.9741	0.9251
16.0	2.3697E 00	0.9533	1.0059	0.9945	0.9863	0.9685	0.9943	1.0001	1.0000	0.9789	1.0058	0.9788	0.9275
18.0	2.3329E 00	0.9461	1.0147	1.0055	0.9983	0.9647	0.9936	1.0001	1.0000	0.9833	1.0287	0.9832	0.9297
20.0	2.2988E 00	0.9390	1.0231	1.0160	1.0098	0.9609	0.9929	1.0001	1.0000	0.9876	1.0507	0.9875	0.9318
25.0	2.2223E 00	0.9215	1.0429	1.0409	1.0370	0.9517	0.9910	1.0001	1.0000	0.9975	1.1020	0.9975	0.9364
30.0	2.1550E 00	0.9046	1.0612	1.0643	1.0626	0.9429	0.9891	1.0001	1.0000	1.0069	1.1477	1.0068	0.9406
35.0	2.0943E 00	0.8883	1.0785	1.0864	1.0869	0.9346	0.9872	1.0001	1.0000	1.0157	1.1880	1.0156	0.9446
40.0	2.0390E 00	0.8725	1.0949	1.1076	1.1102	0.9268	0.9852	1.0000	1.0000	1.0240	1.2225	1.0240	0.9483
45.0	1.9879E 00	0.8574	1.1104	1.1280	1.1328	0.9195	0.9832	1.0000	1.0000	1.0320	1.2514	1.0320	0.9518
50.0	1.9406E 00	0.8429	1.1251	1.1476	1.1545	0.9128	0.9811	1.0000	1.0000	1.0396	1.2746	1.0396	0.9552

ELECTRON Z= 36 A= 84

p	F_0L_0	L_0	λ_2	λ_3	λ_4	μ_1	μ_2	Λ_1	Λ_2	ν_{12}	$\hat{\nu}_{12}$	η_{12}	$\hat{\eta}_{12}$
0.1	2.6854E 01	1.0082	6.5570	16.5863	28.1454	0.9998	1.0000	1.0020	1.0004	2.4157	0.6968	0.7802	0.7505
0.2	1.3581E 01	1.0081	2.2843	3.0505	3.5023	0.9998	1.0000	1.0019	1.0004	1.2640	0.7092	0.8472	0.8160
0.3	9.2417E 00	1.0081	1.4954	1.6693	1.7490	0.9997	1.0000	1.0019	1.0003	1.0589	0.7203	0.8703	0.8384
0.4	7.1574E 00	1.0080	1.2215	1.2715	1.2885	0.9996	0.9999	1.0018	1.0003	0.9904	0.7302	0.8826	0.8502
0.5	5.9749E 00	1.0079	1.0966	1.1043	1.1020	0.9994	0.9999	1.0017	1.0003	0.9605	0.7391	0.8904	0.8576
0.6	5.2342E 00	1.0078	1.0303	1.0191	1.0088	0.9992	0.9999	1.0016	1.0003	0.9455	0.7471	0.8961	0.8629
0.7	4.7373E 00	1.0076	0.9915	0.9705	0.9563	0.9991	0.9998	1.0016	1.0003	0.9372	0.7543	0.9004	0.8670
0.8	4.3863E 00	1.0075	0.9673	0.9407	0.9243	0.9989	0.9998	1.0015	1.0003	0.9324	0.7608	0.9039	0.8702
0.9	4.1282E 00	1.0073	0.9514	0.9215	0.9037	0.9987	0.9998	1.0014	1.0003	0.9297	0.7667	0.9069	0.8728
1.0	3.9322E 00	1.0071	0.9408	0.9087	0.8901	0.9984	0.9997	1.0014	1.0003	0.9280	0.7720	0.9093	0.8751
1.2	3.6571E 00	1.0066	0.9283	0.8939	0.8746	0.9980	0.9997	1.0012	1.0002	0.9267	0.7812	0.9135	0.8787
1.4	3.4755E 00	1.0061	0.9223	0.8870	0.8674	0.9976	0.9996	1.0011	1.0002	0.9267	0.7890	0.9168	0.8817
1.6	3.3474E 00	1.0056	0.9196	0.8840	0.8643	0.9971	0.9995	1.0010	1.0002	0.9273	0.7958	0.9196	0.8841
1.8	3.2524E 00	1.0049	0.9186	0.8831	0.8636	0.9967	0.9994	1.0009	1.0002	0.9282	0.8018	0.9220	0.8861
2.0	3.1790E 00	1.0043	0.9188	0.8835	0.8641	0.9962	0.9993	1.0009	1.0002	0.9293	0.8072	0.9242	0.8880
2.2	3.1204E 00	1.0037	0.9195	0.8846	0.8655	0.9958	0.9993	1.0008	1.0001	0.9304	0.8121	0.9261	0.8896
2.4	3.0723E 00	1.0030	0.9207	0.8862	0.8672	0.9954	0.9992	1.0007	1.0001	0.9316	0.8167	0.9279	0.8910
2.6	3.0319E 00	1.0023	0.9221	0.8880	0.8693	0.9949	0.9991	1.0007	1.0001	0.9327	0.8211	0.9296	0.8924
2.8	2.9973E 00	1.0016	0.9236	0.8901	0.8716	0.9945	0.9990	1.0007	1.0001	0.9338	0.8252	0.9311	0.8936
3.0	2.9672E 00	1.0009	0.9252	0.8922	0.8739	0.9941	0.9990	1.0006	1.0001	0.9349	0.8291	0.9325	0.8947
3.2	2.9405E 00	1.0001	0.9269	0.8944	0.8763	0.9936	0.9989	1.0006	1.0001	0.9360	0.8329	0.9339	0.8958
3.4	2.9167E 00	0.9994	0.9286	0.8966	0.8788	0.9932	0.9988	1.0005	1.0001	0.9371	0.8365	0.9352	0.8968
3.6	2.8952E 00	0.9987	0.9304	0.8988	0.8812	0.9928	0.9987	1.0005	1.0001	0.9381	0.8401	0.9364	0.8977
3.8	2.8756E 00	0.9979	0.9321	0.9010	0.8836	0.9923	0.9986	1.0005	1.0001	0.9391	0.8436	0.9375	0.8986
4.0	2.8576E 00	0.9972	0.9338	0.9031	0.8860	0.9919	0.9986	1.0005	1.0001	0.9401	0.8470	0.9387	0.8995
4.5	2.8180E 00	0.9953	0.9380	0.9085	0.8919	0.9908	0.9984	1.0004	1.0001	0.9424	0.8551	0.9413	0.9014
5.0	2.7843E 00	0.9934	0.9420	0.9136	0.8975	0.9898	0.9982	1.0004	1.0001	0.9446	0.8630	0.9437	0.9031
5.5	2.7548E 00	0.9914	0.9460	0.9185	0.9029	0.9887	0.9980	1.0004	1.0001	0.9467	0.8707	0.9459	0.9047
6.0	2.7284E 00	0.9895	0.9497	0.9233	0.9081	0.9876	0.9978	1.0003	1.0001	0.9487	0.8781	0.9480	0.9061
6.5	2.7045E 00	0.9876	0.9533	0.9278	0.9131	0.9866	0.9976	1.0003	1.0001	0.9506	0.8854	0.9500	0.9075
7.0	2.6826E 00	0.9856	0.9568	0.9322	0.9179	0.9855	0.9974	1.0003	1.0001	0.9524	0.8926	0.9519	0.9087
7.5	2.6623E 00	0.9837	0.9602	0.9365	0.9225	0.9844	0.9973	1.0003	1.0000	0.9541	0.8996	0.9537	0.9099
8.0	2.6433E 00	0.9817	0.9635	0.9406	0.9270	0.9834	0.9971	1.0002	1.0000	0.9558	0.9065	0.9554	0.9110
9.0	2.6086E 00	0.9778	0.9698	0.9485	0.9356	0.9813	0.9967	1.0002	1.0000	0.9590	0.9201	0.9587	0.9130
10.0	2.5772E 00	0.9739	0.9757	0.9559	0.9438	0.9792	0.9963	1.0002	1.0000	0.9621	0.9333	0.9618	0.9149
11.0	2.5485E 00	0.9700	0.9814	0.9631	0.9515	0.9771	0.9959	1.0002	1.0000	0.9649	0.9462	0.9647	0.9166
12.0	2.5219E 00	0.9662	0.9869	0.9699	0.9590	0.9750	0.9956	1.0002	1.0000	0.9677	0.9589	0.9675	0.9181
13.0	2.4969E 00	0.9623	0.9922	0.9765	0.9662	0.9729	0.9952	1.0002	1.0000	0.9704	0.9713	0.9702	0.9196
14.0	2.4734E 00	0.9584	0.9973	0.9829	0.9731	0.9708	0.9948	1.0001	1.0000	0.9729	0.9835	0.9727	0.9210
16.0	2.4299E 00	0.9508	1.0070	0.9951	0.9865	0.9667	0.9940	1.0001	1.0000	0.9778	1.0071	0.9777	0.9236
18.0	2.3901E 00	0.9432	1.0163	1.0067	0.9992	0.9627	0.9933	1.0001	1.0000	0.9825	1.0298	0.9824	0.9259
20.0	2.3532E 00	0.9357	1.0252	1.0179	1.0114	0.9587	0.9925	1.0001	1.0000	0.9870	1.0515	0.9869	0.9281
25.0	2.2706E 00	0.9174	1.0461	1.0443	1.0402	0.9490	0.9905	1.0001	1.0000	0.9976	1.1020	0.9975	0.9331
30.0	2.1981E 00	0.8997	1.0656	1.0690	1.0672	0.9399	0.9885	1.0001	1.0000	1.0075	1.1465	1.0074	0.9376
35.0	2.1330E 00	0.8826	1.0839	1.0924	1.0930	0.9312	0.9865	1.0001	1.0000	1.0168	1.1854	1.0167	0.9419
40.0	2.0737E 00	0.8662	1.1011	1.1149	1.1177	0.9231	0.9844	1.0000	1.0000	1.0256	1.2183	1.0256	0.9459
45.0	2.0192E 00	0.8505	1.1175	1.1364	1.1415	0.9156	0.9823	1.0000	1.0000	1.0340	1.2453	1.0340	0.9497
50.0	1.9688E 00	0.8354	1.1329	1.1571	1.1646	0.9087	0.9801	1.0000	1.0000	1.0421	1.2664	1.0420	0.9534

Instr. see p. 22

ELECTRON Z= 37 A= 88

p	F_0L_0	L_0	λ_2	λ_3	λ_4	μ_1	μ_2	Λ_1	Λ_2	ν_{12}	$\hat{\nu}_{12}$	η_{12}	$\hat{\eta}_{12}$
0.1	2.8281E 01	1.0086	6.8150	17.7845	30.8507	0.9998	1.0000	1.0021	1.0004	2.4875	0.6939	0.7651	0.7340
0.2	1.4301E 01	1.0086	2.3441	3.1718	3.6672	0.9998	1.0000	1.0020	1.0004	1.2767	0.7067	0.8376	0.8048
0.3	9.7273E 00	1.0085	1.5185	1.7045	1.7904	0.9997	1.0000	1.0019	1.0004	1.0612	0.7181	0.8625	0.8290
0.4	7.5256E 00	1.0084	1.2318	1.2850	1.3032	0.9995	0.9999	1.0019	1.0003	0.9894	0.7284	0.8757	0.8416
0.5	6.2738E 00	1.0083	1.1011	1.1092	1.1069	0.9994	0.9999	1.0018	1.0003	0.9580	0.7375	0.8841	0.8496
0.6	5.4884E 00	1.0082	1.0316	1.0199	1.0091	0.9992	0.9999	1.0017	1.0003	0.9422	0.7457	0.8901	0.8553
0.7	4.9609E 00	1.0081	0.9910	0.9689	0.9540	0.9990	0.9998	1.0017	1.0003	0.9336	0.7531	0.8948	0.8596
0.8	4.5880E 00	1.0079	0.9656	0.9377	0.9205	0.9988	0.9998	1.0016	1.0003	0.9286	0.7598	0.8985	0.8630
0.9	4.3136E 00	1.0077	0.9490	0.9176	0.8990	0.9986	0.9998	1.0015	1.0003	0.9257	0.7658	0.9016	0.8658
1.0	4.1050E 00	1.0075	0.9378	0.9042	0.8848	0.9984	0.9997	1.0014	1.0003	0.9240	0.7712	0.9043	0.8682
1.2	3.8121E 00	1.0070	0.9248	0.8887	0.8685	0.9979	0.9996	1.0013	1.0002	0.9226	0.7807	0.9086	0.8721
1.4	3.6184E 00	1.0065	0.9185	0.8815	0.8610	0.9974	0.9996	1.0012	1.0002	0.9226	0.7887	0.9122	0.8751
1.6	3.4817E 00	1.0059	0.9156	0.8783	0.8578	0.9970	0.9995	1.0011	1.0002	0.9233	0.7956	0.9151	0.8777
1.8	3.3802E 00	1.0052	0.9146	0.8774	0.8570	0.9965	0.9994	1.0010	1.0002	0.9243	0.8017	0.9177	0.8799
2.0	3.3017E 00	1.0046	0.9148	0.8779	0.8576	0.9960	0.9993	1.0009	1.0002	0.9254	0.8072	0.9200	0.8818
2.2	3.2389E 00	1.0039	0.9156	0.8790	0.8590	0.9956	0.9992	1.0008	1.0002	0.9266	0.8122	0.9221	0.8835
2.4	3.1874E 00	1.0032	0.9168	0.8807	0.8609	0.9951	0.9991	1.0008	1.0001	0.9278	0.8169	0.9239	0.8851
2.6	3.1440E 00	1.0024	0.9183	0.8827	0.8631	0.9947	0.9991	1.0007	1.0001	0.9290	0.8213	0.9257	0.8865
2.8	3.1067E 00	1.0017	0.9199	0.8848	0.8655	0.9942	0.9990	1.0007	1.0001	0.9302	0.8255	0.9273	0.8878
3.0	3.0743E 00	1.0009	0.9216	0.8870	0.8679	0.9938	0.9989	1.0006	1.0001	0.9313	0.8295	0.9288	0.8889
3.2	3.0456E 00	1.0002	0.9234	0.8893	0.8705	0.9933	0.9988	1.0006	1.0001	0.9325	0.8333	0.9302	0.8901
3.4	3.0200E 00	0.9994	0.9252	0.8916	0.8730	0.9928	0.9987	1.0006	1.0001	0.9336	0.8370	0.9316	0.8911
3.6	2.9968E 00	0.9986	0.9270	0.8939	0.8756	0.9924	0.9987	1.0005	1.0001	0.9347	0.8406	0.9328	0.8921
3.8	2.9756E 00	0.9979	0.9288	0.8962	0.8781	0.9919	0.9986	1.0005	1.0001	0.9357	0.8441	0.9341	0.8930
4.0	2.9562E 00	0.9971	0.9306	0.8985	0.8806	0.9915	0.9985	1.0005	1.0001	0.9368	0.8476	0.9352	0.8939
4.5	2.9134E 00	0.9951	0.9351	0.9041	0.8867	0.9904	0.9983	1.0004	1.0001	0.9392	0.8558	0.9380	0.8960
5.0	2.8769E 00	0.9931	0.9393	0.9095	0.8927	0.9892	0.9981	1.0004	1.0001	0.9416	0.8638	0.9405	0.8978
5.5	2.8449E 00	0.9910	0.9434	0.9147	0.8983	0.9881	0.9979	1.0004	1.0001	0.9438	0.8714	0.9429	0.8994
6.0	2.8164E 00	0.9890	0.9474	0.9197	0.9038	0.9870	0.9977	1.0003	1.0001	0.9458	0.8789	0.9451	0.9010
6.5	2.7905E 00	0.9870	0.9512	0.9245	0.9090	0.9859	0.9975	1.0003	1.0001	0.9478	0.8862	0.9472	0.9024
7.0	2.7667E 00	0.9849	0.9549	0.9291	0.9141	0.9847	0.9973	1.0003	1.0001	0.9497	0.8934	0.9492	0.9037
7.5	2.7447E 00	0.9829	0.9585	0.9336	0.9190	0.9836	0.9971	1.0003	1.0000	0.9516	0.9005	0.9511	0.9049
8.0	2.7241E 00	0.9809	0.9619	0.9379	0.9237	0.9825	0.9969	1.0003	1.0000	0.9534	0.9074	0.9529	0.9061
9.0	2.6865E 00	0.9768	0.9686	0.9462	0.9327	0.9803	0.9965	1.0002	1.0000	0.9567	0.9210	0.9564	0.9082
10.0	2.6525E 00	0.9727	0.9749	0.9541	0.9413	0.9781	0.9962	1.0002	1.0000	0.9599	0.9342	0.9596	0.9102
11.0	2.6214E 00	0.9686	0.9809	0.9616	0.9495	0.9759	0.9958	1.0002	1.0000	0.9630	0.9471	0.9627	0.9120
12.0	2.5925E 00	0.9645	0.9866	0.9688	0.9573	0.9737	0.9954	1.0002	1.0000	0.9659	0.9597	0.9656	0.9136
13.0	2.5655E 00	0.9605	0.9922	0.9757	0.9649	0.9715	0.9950	1.0002	1.0000	0.9687	0.9720	0.9685	0.9152
14.0	2.5401E 00	0.9564	0.9976	0.9825	0.9722	0.9693	0.9946	1.0001	1.0000	0.9714	0.9841	0.9712	0.9167
16.0	2.4931E 00	0.9484	1.0079	0.9953	0.9863	0.9650	0.9938	1.0001	1.0000	0.9766	1.0075	0.9764	0.9194
18.0	2.4501E 00	0.9405	1.0177	1.0076	0.9997	0.9608	0.9930	1.0001	1.0000	0.9815	1.0300	0.9814	0.9219
20.0	2.4104E 00	0.9327	1.0270	1.0194	1.0125	0.9566	0.9922	1.0001	1.0000	0.9863	1.0514	0.9861	0.9243
25.0	2.3215E 00	0.9136	1.0491	1.0472	1.0429	0.9465	0.9901	1.0001	1.0000	0.9974	1.1009	0.9973	0.9296
30.0	2.2437E 00	0.8951	1.0696	1.0732	1.0714	0.9370	0.9880	1.0001	1.0000	1.0078	1.1443	1.0078	0.9345
35.0	2.1740E 00	0.8774	1.0888	1.0980	1.0986	0.9280	0.9859	1.0001	1.0000	1.0176	1.1817	1.0176	0.9390
40.0	2.1107E 00	0.8604	1.1070	1.1216	1.1246	0.9196	0.9837	1.0000	1.0000	1.0270	1.2130	1.0269	0.9433
45.0	2.0527E 00	0.8441	1.1241	1.1443	1.1497	0.9119	0.9814	1.0000	1.0000	1.0358	1.2382	1.0358	0.9474
50.0	1.9991E 00	0.8286	1.1403	1.1661	1.1739	0.9048	0.9791	1.0000	1.0000	1.0442	1.2573	1.0442	0.9514

 Instr. see p. 22

ELECTRON Z= 38 A= 89

p	F_0L_0	L_0	λ_2	λ_3	λ_4	μ_1	μ_2	Λ_1	Λ_2	ν_{12}	$\hat{\nu}_{12}$	η_{12}	$\hat{\eta}_{12}$
0.1	2.9807E 01	1.0091	7.0713	19.0279	33.7404	0.9998	1.0000	1.0021	1.0004	2.5598	0.6908	0.7492	0.7168
0.2	1.5072E 01	1.0091	2.4031	3.2944	3.8361	0.9997	1.0000	1.0021	1.0004	1.2892	0.7040	0.8274	0.7931
0.3	1.0247E 01	1.0090	1.5410	1.7391	1.8314	0.9996	1.0000	1.0020	1.0004	1.0633	0.7157	0.8542	0.8191
0.4	7.9199E 00	1.0090	1.2415	1.2977	1.3171	0.9995	0.9999	1.0019	1.0004	0.9880	0.7262	0.8683	0.8326
0.5	6.5943E 00	1.0089	1.1050	1.1134	1.1109	0.9993	0.9999	1.0019	1.0003	0.9552	0.7356	0.8773	0.8411
0.6	5.7612E 00	1.0087	1.0324	1.0199	1.0085	0.9992	0.9999	1.0018	1.0003	0.9387	0.7440	0.8838	0.8472
0.7	5.2010E 00	1.0086	0.9899	0.9667	0.9510	0.9990	0.9998	1.0017	1.0003	0.9296	0.7516	0.8887	0.8517
0.8	4.8046E 00	1.0084	0.9633	0.9340	0.9159	0.9987	0.9998	1.0016	1.0003	0.9244	0.7584	0.8927	0.8554
0.9	4.5127E 00	1.0082	0.9460	0.9130	0.8935	0.9985	0.9998	1.0016	1.0003	0.9214	0.7646	0.8960	0.8583
1.0	4.2907E 00	1.0080	0.9343	0.8990	0.8787	0.9983	0.9997	1.0015	1.0003	0.9196	0.7701	0.8988	0.8609
1.2	3.9787E 00	1.0075	0.9207	0.8829	0.8617	0.9978	0.9996	1.0013	1.0002	0.9182	0.7798	0.9034	0.8650
1.4	3.7721E 00	1.0069	0.9141	0.8753	0.8539	0.9973	0.9995	1.0012	1.0002	0.9182	0.7879	0.9072	0.8682
1.6	3.6261E 00	1.0063	0.9111	0.8720	0.8506	0.9969	0.9995	1.0011	1.0002	0.9189	0.7950	0.9103	0.8709
1.8	3.5176E 00	1.0056	0.9101	0.8711	0.8498	0.9964	0.9994	1.0010	1.0002	0.9199	0.8012	0.9130	0.8733
2.0	3.4336E 00	1.0049	0.9102	0.8715	0.8504	0.9959	0.9993	1.0009	1.0002	0.9211	0.8068	0.9154	0.8753
2.2	3.3663E 00	1.0042	0.9111	0.8728	0.8518	0.9954	0.9992	1.0009	1.0002	0.9224	0.8119	0.9176	0.8771
2.4	3.3110E 00	1.0035	0.9124	0.8745	0.8538	0.9949	0.9991	1.0008	1.0001	0.9236	0.8166	0.9196	0.8787
2.6	3.2644E 00	1.0027	0.9139	0.8766	0.8561	0.9945	0.9990	1.0008	1.0001	0.9249	0.8211	0.9214	0.8802
2.8	3.2244E 00	1.0020	0.9156	0.8788	0.8586	0.9940	0.9990	1.0007	1.0001	0.9262	0.8253	0.9231	0.8815
3.0	3.1895E 00	1.0012	0.9174	0.8811	0.8611	0.9935	0.9989	1.0007	1.0001	0.9274	0.8293	0.9247	0.8828
3.2	3.1586E 00	1.0004	0.9193	0.8835	0.8638	0.9930	0.9988	1.0006	1.0001	0.9286	0.8331	0.9262	0.8840
3.4	3.1310E 00	0.9996	0.9212	0.8859	0.8665	0.9926	0.9987	1.0006	1.0001	0.9298	0.8369	0.9276	0.8851
3.6	3.1060E 00	0.9988	0.9231	0.8884	0.8691	0.9921	0.9986	1.0006	1.0001	0.9309	0.8405	0.9290	0.8861
3.8	3.0832E 00	0.9980	0.9250	0.8908	0.8718	0.9916	0.9985	1.0005	1.0001	0.9320	0.8440	0.9302	0.8871
4.0	3.0622E 00	0.9972	0.9269	0.8932	0.8744	0.9911	0.9985	1.0005	1.0001	0.9331	0.8474	0.9315	0.8880
4.5	3.0160E 00	0.9951	0.9315	0.8990	0.8808	0.9900	0.9983	1.0005	1.0001	0.9357	0.8557	0.9344	0.8902
5.0	2.9765E 00	0.9930	0.9360	0.9047	0.8870	0.9888	0.9981	1.0004	1.0001	0.9381	0.8636	0.9370	0.8921
5.5	2.9420E 00	0.9909	0.9403	0.9101	0.8929	0.9876	0.9978	1.0004	1.0001	0.9404	0.8713	0.9395	0.8939
6.0	2.9111E 00	0.9888	0.9445	0.9153	0.8986	0.9865	0.9976	1.0004	1.0001	0.9426	0.8788	0.9418	0.8955
6.5	2.8831E 00	0.9867	0.9485	0.9203	0.9041	0.9853	0.9974	1.0003	1.0001	0.9447	0.8861	0.9440	0.8969
7.0	2.8574E 00	0.9846	0.9523	0.9252	0.9094	0.9841	0.9972	1.0003	1.0001	0.9467	0.8932	0.9461	0.8983
7.5	2.8336E 00	0.9825	0.9561	0.9299	0.9145	0.9830	0.9970	1.0003	1.0001	0.9486	0.9002	0.9481	0.8996
8.0	2.8114E 00	0.9803	0.9597	0.9344	0.9194	0.9818	0.9968	1.0003	1.0000	0.9505	0.9071	0.9500	0.9008
9.0	2.7707E 00	0.9761	0.9666	0.9431	0.9289	0.9795	0.9964	1.0002	1.0000	0.9540	0.9206	0.9536	0.9031
10.0	2.7340E 00	0.9719	0.9732	0.9513	0.9378	0.9772	0.9960	1.0002	1.0000	0.9574	0.9336	0.9570	0.9052
11.0	2.7004E 00	0.9676	0.9795	0.9591	0.9464	0.9749	0.9956	1.0002	1.0000	0.9606	0.9464	0.9603	0.9071
12.0	2.6693E 00	0.9634	0.9856	0.9667	0.9546	0.9727	0.9952	1.0002	1.0000	0.9636	0.9589	0.9634	0.9088
13.0	2.6402E 00	0.9592	0.9914	0.9739	0.9625	0.9704	0.9948	1.0002	1.0000	0.9665	0.9710	0.9663	0.9105
14.0	2.6129E 00	0.9551	0.9970	0.9810	0.9702	0.9681	0.9944	1.0002	1.0000	0.9694	0.9830	0.9692	0.9120
16.0	2.5623E 00	0.9468	1.0078	0.9945	0.9848	0.9637	0.9936	1.0001	1.0000	0.9748	1.0060	0.9747	0.9149
18.0	2.5161E 00	0.9386	1.0180	1.0073	0.9988	0.9593	0.9927	1.0001	1.0000	0.9800	1.0280	0.9798	0.9176
20.0	2.4735E 00	0.9305	1.0279	1.0196	1.0123	0.9550	0.9919	1.0001	1.0000	0.9849	1.0491	0.9848	0.9201
25.0	2.3783E 00	0.9108	1.0509	1.0487	1.0440	0.9446	0.9898	1.0001	1.0000	0.9966	1.0974	0.9965	0.9257
30.0	2.2953E 00	0.8918	1.0723	1.0759	1.0738	0.9347	0.9876	1.0001	1.0000	1.0075	1.1395	1.0074	0.9309
35.0	2.2211E 00	0.8735	1.0924	1.1017	1.1021	0.9255	0.9854	1.0001	1.0000	1.0177	1.1755	1.0177	0.9357
40.0	2.1538E 00	0.8560	1.1114	1.1264	1.1293	0.9169	0.9831	1.0001	1.0000	1.0275	1.2053	1.0274	0.9403
45.0	2.0923E 00	0.8393	1.1292	1.1501	1.1555	0.9090	0.9808	1.0000	1.0000	1.0367	1.2290	1.0367	0.9447
50.0	2.0357E 00	0.8234	1.1461	1.1728	1.1808	0.9017	0.9784	1.0000	1.0000	1.0455	1.2466	1.0455	0.9489

Instr. see p. 22

ELECTRON Z= 39 A= 91

p	F_0L_0	L_0	λ_2	λ_3	λ_4	μ_1	μ_2	Λ_1	Λ_2	ν_{12}	$\hat{\nu}_{12}$	η_{12}	$\hat{\eta}_{12}$
0.1	3.1401E 01	1.0096	7.3299	20.3309	36.8525	0.9998	1.0000	1.0022	1.0004	2.6335	0.6877	0.7327	0.6990
0.2	1.5877E 01	1.0096	2.4627	3.4210	4.0122	0.9997	1.0000	1.0022	1.0004	1.3020	0.7012	0.8169	0.7810
0.3	1.0790E 01	1.0095	1.5637	1.7747	1.8737	0.9996	1.0000	1.0021	1.0004	1.0654	0.7132	0.8457	0.8089
0.4	8.3326E 00	1.0095	1.2515	1.3108	1.3314	0.9995	0.9999	1.0020	1.0004	0.9866	0.7240	0.8608	0.8234
0.5	6.9298E 00	1.0093	1.1090	1.1178	1.1152	0.9993	0.9999	1.0019	1.0004	0.9523	0.7337	0.8704	0.8325
0.6	6.0468E 00	1.0092	1.0332	1.0201	1.0081	0.9991	0.9999	1.0019	1.0003	0.9350	0.7423	0.8773	0.8389
0.7	5.4522E 00	1.0090	0.9889	0.9645	0.9480	0.9989	0.9998	1.0018	1.0003	0.9256	0.7501	0.8826	0.8438
0.8	5.0312E 00	1.0089	0.9612	0.9304	0.9115	0.9987	0.9998	1.0017	1.0003	0.9202	0.7571	0.8868	0.8476
0.9	4.7209E 00	1.0086	0.9431	0.9085	0.8881	0.9985	0.9997	1.0016	1.0003	0.9170	0.7634	0.8903	0.8508
1.0	4.4847E 00	1.0084	0.9309	0.8939	0.8727	0.9982	0.9997	1.0015	1.0003	0.9152	0.7691	0.8933	0.8535
1.2	4.1525E 00	1.0079	0.9167	0.8771	0.8550	0.9977	0.9996	1.0014	1.0003	0.9137	0.7789	0.8982	0.8578
1.4	3.9324E 00	1.0073	0.9098	0.8692	0.8468	0.9972	0.9995	1.0013	1.0002	0.9137	0.7873	0.9021	0.8612
1.6	3.7766E 00	1.0067	0.9067	0.8658	0.8434	0.9967	0.9994	1.0012	1.0002	0.9145	0.7944	0.9054	0.8641
1.8	3.6607E 00	1.0060	0.9056	0.8648	0.8426	0.9962	0.9994	1.0011	1.0002	0.9156	0.8007	0.9083	0.8665
2.0	3.5708E 00	1.0053	0.9058	0.8653	0.8432	0.9957	0.9993	1.0010	1.0002	0.9168	0.8064	0.9108	0.8686
2.2	3.4987E 00	1.0045	0.9067	0.8666	0.8447	0.9952	0.9992	1.0009	1.0002	0.9181	0.8116	0.9131	0.8705
2.4	3.4394E 00	1.0038	0.9080	0.8684	0.8468	0.9947	0.9991	1.0008	1.0002	0.9195	0.8164	0.9152	0.8722
2.6	3.3894E 00	1.0030	0.9096	0.8706	0.8492	0.9942	0.9990	1.0008	1.0001	0.9208	0.8209	0.9171	0.8738
2.8	3.3465E 00	1.0022	0.9114	0.8729	0.8518	0.9937	0.9989	1.0007	1.0001	0.9221	0.8252	0.9189	0.8752
3.0	3.3089E 00	1.0014	0.9133	0.8753	0.8545	0.9932	0.9988	1.0007	1.0001	0.9234	0.8292	0.9206	0.8766
3.2	3.2757E 00	1.0005	0.9153	0.8778	0.8572	0.9927	0.9987	1.0007	1.0001	0.9247	0.8331	0.9221	0.8778
3.4	3.2459E 00	0.9997	0.9173	0.8804	0.8600	0.9922	0.9987	1.0006	1.0001	0.9259	0.8369	0.9236	0.8790
3.6	3.2190E 00	0.9989	0.9193	0.8829	0.8628	0.9917	0.9986	1.0006	1.0001	0.9271	0.8405	0.9250	0.8800
3.8	3.1944E 00	0.9980	0.9213	0.8854	0.8656	0.9913	0.9985	1.0006	1.0001	0.9283	0.8441	0.9264	0.8811
4.0	3.1718E 00	0.9972	0.9232	0.8879	0.8683	0.9908	0.9984	1.0005	1.0001	0.9294	0.8475	0.9277	0.8821
4.5	3.1219E 00	0.9950	0.9281	0.8941	0.8750	0.9895	0.9982	1.0005	1.0001	0.9321	0.8559	0.9307	0.8843
5.0	3.0793E 00	0.9929	0.9328	0.9000	0.8815	0.9883	0.9980	1.0004	1.0001	0.9347	0.8638	0.9335	0.8863
5.5	3.0419E 00	0.9907	0.9373	0.9056	0.8877	0.9871	0.9978	1.0004	1.0001	0.9371	0.8715	0.9361	0.8882
6.0	3.0086E 00	0.9885	0.9417	0.9111	0.8936	0.9859	0.9976	1.0004	1.0001	0.9394	0.8790	0.9385	0.8899
6.5	2.9783E 00	0.9863	0.9459	0.9164	0.8993	0.9847	0.9974	1.0003	1.0001	0.9416	0.8863	0.9408	0.8914
7.0	2.9506E 00	0.9841	0.9499	0.9214	0.9049	0.9835	0.9971	1.0003	1.0001	0.9437	0.8934	0.9430	0.8929
7.5	2.9248E 00	0.9819	0.9539	0.9263	0.9102	0.9823	0.9969	1.0003	1.0001	0.9457	0.9004	0.9451	0.8942
8.0	2.9008E 00	0.9797	0.9577	0.9311	0.9154	0.9811	0.9967	1.0003	1.0001	0.9476	0.9072	0.9471	0.8955
9.0	2.8569E 00	0.9753	0.9649	0.9401	0.9253	0.9787	0.9963	1.0002	1.0000	0.9513	0.9206	0.9509	0.8979
10.0	2.8172E 00	0.9709	0.9718	0.9488	0.9346	0.9763	0.9959	1.0002	1.0000	0.9549	0.9337	0.9545	0.9001
11.0	2.7810E 00	0.9665	0.9784	0.9570	0.9436	0.9739	0.9955	1.0002	1.0000	0.9582	0.9464	0.9579	0.9021
12.0	2.7474E 00	0.9621	0.9848	0.9649	0.9522	0.9715	0.9950	1.0002	1.0000	0.9614	0.9587	0.9611	0.9039
13.0	2.7161E 00	0.9578	0.9909	0.9725	0.9604	0.9692	0.9946	1.0002	1.0000	0.9645	0.9708	0.9642	0.9057
14.0	2.6866E 00	0.9534	0.9968	0.9798	0.9684	0.9668	0.9942	1.0002	1.0000	0.9674	0.9825	0.9672	0.9073
16.0	2.6322E 00	0.9449	1.0081	0.9939	0.9838	0.9622	0.9933	1.0001	1.0000	0.9731	1.0053	0.9730	0.9104
18.0	2.5826E 00	0.9364	1.0188	1.0074	0.9984	0.9576	0.9925	1.0001	1.0000	0.9786	1.0270	0.9784	0.9132
20.0	2.5368E 00	0.9280	1.0291	1.0203	1.0125	0.9532	0.9916	1.0001	1.0000	0.9837	1.0477	0.9836	0.9158
25.0	2.4349E 00	0.9076	1.0533	1.0507	1.0457	0.9424	0.9894	1.0001	1.0000	0.9960	1.0949	0.9959	0.9218
30.0	2.3462E 00	0.8879	1.0757	1.0792	1.0769	0.9322	0.9871	1.0001	1.0000	1.0074	1.1358	1.0073	0.9273
35.0	2.2672E 00	0.8691	1.0967	1.1062	1.1065	0.9227	0.9848	1.0001	1.0000	1.0181	1.1704	1.0180	0.9325
40.0	2.1958E 00	0.8511	1.1165	1.1320	1.1350	0.9138	0.9825	1.0001	1.0000	1.0283	1.1988	1.0282	0.9374
45.0	2.1306E 00	0.8340	1.1351	1.1568	1.1624	0.9057	0.9800	1.0000	1.0000	1.0379	1.2210	1.0379	0.9420
50.0	2.0707E 00	0.8176	1.1527	1.1805	1.1888	0.8983	0.9775	1.0000	1.0000	1.0471	1.2370	1.0471	0.9465

Instr. see p. 22

ELECTRON Z= 40 A= 93

p	F_0L_0	L_0	λ_2	λ_3	λ_4	μ_1	μ_2	Λ_1	Λ_2	ν_{12}	$\hat{\nu}_{12}$	η_{12}	$\hat{\eta}_{12}$
0.1	3.3077E 01	1.0102	7.5894	21.6901	40.1892	0.9998	1.0000	1.0023	1.0004	2.7081	0.6845	0.7155	0.6805
0.2	1.6723E 01	1.0101	2.5225	3.5507	4.1948	0.9997	1.0000	1.0022	1.0004	1.3148	0.6983	0.8060	0.7686
0.3	1.1362E 01	1.0100	1.5865	1.8107	1.9167	0.9996	1.0000	1.0022	1.0004	1.0675	0.7107	0.8369	0.7984
0.4	8.7671E 00	1.0100	1.2613	1.3240	1.3458	0.9994	0.9999	1.0021	1.0004	0.9851	0.7218	0.8531	0.8139
0.5	7.2832E 00	1.0098	1.1129	1.1222	1.1194	0.9993	0.9999	1.0020	1.0004	0.9493	0.7317	0.8634	0.8236
0.6	6.3477E 00	1.0097	1.0340	1.0202	1.0076	0.9991	0.9999	1.0019	1.0004	0.9313	0.7406	0.8707	0.8304
0.7	5.7170E 00	1.0095	0.9878	0.9622	0.9450	0.9989	0.9998	1.0018	1.0003	0.9214	0.7485	0.8762	0.8356
0.8	5.2699E 00	1.0093	0.9589	0.9268	0.9069	0.9986	0.9998	1.0018	1.0003	0.9158	0.7557	0.8807	0.8397
0.9	4.9401E 00	1.0091	0.9401	0.9039	0.8826	0.9984	0.9997	1.0017	1.0003	0.9125	0.7621	0.8844	0.8430
1.0	4.6890E 00	1.0089	0.9274	0.8887	0.8665	0.9981	0.9997	1.0016	1.0003	0.9106	0.7679	0.8876	0.8458
1.2	4.3355E 00	1.0083	0.9125	0.8712	0.8482	0.9976	0.9996	1.0014	1.0003	0.9091	0.7780	0.8928	0.8504
1.4	4.1008E 00	1.0077	0.9054	0.8630	0.8397	0.9971	0.9995	1.0013	1.0002	0.9091	0.7865	0.8969	0.8541
1.6	3.9346E 00	1.0071	0.9021	0.8595	0.8361	0.9966	0.9994	1.0012	1.0002	0.9099	0.7938	0.9004	0.8571
1.8	3.8108E 00	1.0063	0.9010	0.8585	0.8353	0.9960	0.9993	1.0011	1.0002	0.9111	0.8003	0.9034	0.8596
2.0	3.7147E 00	1.0056	0.9012	0.8590	0.8359	0.9955	0.9992	1.0010	1.0002	0.9124	0.8061	0.9061	0.8618
2.2	3.6376E 00	1.0048	0.9022	0.8603	0.8375	0.9950	0.9991	1.0009	1.0002	0.9138	0.8113	0.9085	0.8638
2.4	3.5740E 00	1.0040	0.9036	0.8622	0.8397	0.9945	0.9991	1.0009	1.0002	0.9152	0.8162	0.9107	0.8656
2.6	3.5203E 00	1.0032	0.9053	0.8645	0.8422	0.9940	0.9990	1.0008	1.0001	0.9166	0.8208	0.9127	0.8673
2.8	3.4742E 00	1.0024	0.9071	0.8669	0.8449	0.9935	0.9989	1.0008	1.0001	0.9179	0.8251	0.9146	0.8688
3.0	3.4339E 00	1.0015	0.9091	0.8695	0.8477	0.9929	0.9988	1.0007	1.0001	0.9193	0.8292	0.9163	0.8702
3.2	3.3982E 00	1.0007	0.9112	0.8721	0.8506	0.9924	0.9987	1.0007	1.0001	0.9206	0.8331	0.9180	0.8715
3.4	3.3661E 00	0.9998	0.9133	0.8747	0.8535	0.9919	0.9986	1.0006	1.0001	0.9219	0.8369	0.9195	0.8727
3.6	3.3371E 00	0.9989	0.9154	0.8774	0.8564	0.9914	0.9985	1.0006	1.0001	0.9232	0.8406	0.9210	0.8738
3.8	3.3106E 00	0.9981	0.9174	0.8800	0.8593	0.9909	0.9984	1.0006	1.0001	0.9244	0.8442	0.9224	0.8749
4.0	3.2862E 00	0.9972	0.9195	0.8826	0.8621	0.9904	0.9983	1.0006	1.0001	0.9256	0.8476	0.9238	0.8759
4.5	3.2324E 00	0.9950	0.9246	0.8890	0.8691	0.9891	0.9981	1.0005	1.0001	0.9284	0.8560	0.9270	0.8783
5.0	3.1864E 00	0.9927	0.9295	0.8952	0.8758	0.9878	0.9979	1.0005	1.0001	0.9311	0.8640	0.9299	0.8804
5.5	3.1461E 00	0.9904	0.9343	0.9011	0.8823	0.9866	0.9977	1.0004	1.0001	0.9336	0.8717	0.9326	0.8823
6.0	3.1100E 00	0.9882	0.9388	0.9068	0.8885	0.9853	0.9975	1.0004	1.0001	0.9360	0.8792	0.9352	0.8841
6.5	3.0774E 00	0.9859	0.9432	0.9123	0.8945	0.9840	0.9973	1.0004	1.0001	0.9383	0.8865	0.9376	0.8857
7.0	3.0474E 00	0.9836	0.9475	0.9176	0.9003	0.9828	0.9970	1.0003	1.0001	0.9405	0.8936	0.9399	0.8873
7.5	3.0196E 00	0.9813	0.9516	0.9227	0.9058	0.9815	0.9968	1.0003	1.0001	0.9427	0.9006	0.9421	0.8887
8.0	2.9936E 00	0.9790	0.9556	0.9277	0.9112	0.9803	0.9966	1.0003	1.0001	0.9447	0.9074	0.9442	0.8901
9.0	2.9462E 00	0.9744	0.9632	0.9372	0.9216	0.9778	0.9962	1.0003	1.0000	0.9486	0.9208	0.9481	0.8926
10.0	2.9034E 00	0.9699	0.9704	0.9462	0.9314	0.9753	0.9957	1.0002	1.0000	0.9523	0.9338	0.9519	0.8948
11.0	2.8643E 00	0.9653	0.9773	0.9548	0.9407	0.9728	0.9953	1.0002	1.0000	0.9558	0.9464	0.9554	0.8969
12.0	2.8282E 00	0.9608	0.9840	0.9630	0.9497	0.9703	0.9949	1.0002	1.0000	0.9591	0.9586	0.9588	0.8989
13.0	2.7944E 00	0.9563	0.9903	0.9710	0.9583	0.9679	0.9944	1.0002	1.0000	0.9623	0.9706	0.9621	0.9007
14.0	2.7627E 00	0.9518	0.9965	0.9787	0.9667	0.9655	0.9940	1.0002	1.0000	0.9654	0.9822	0.9652	0.9025
16.0	2.7042E 00	0.9429	1.0084	0.9935	0.9828	0.9606	0.9931	1.0001	1.0000	0.9714	1.0047	0.9712	0.9057
18.0	2.6509E 00	0.9341	1.0196	1.0075	0.9981	0.9559	0.9922	1.0001	1.0000	0.9771	1.0261	0.9769	0.9087
20.0	2.6019E 00	0.9254	1.0304	1.0210	1.0128	0.9513	0.9913	1.0001	1.0000	0.9825	1.0464	0.9824	0.9115
25.0	2.4928E 00	0.9043	1.0557	1.0528	1.0475	0.9401	0.9890	1.0001	1.0000	0.9953	1.0926	0.9952	0.9179
30.0	2.3982E 00	0.8840	1.0791	1.0826	1.0801	0.9296	0.9867	1.0001	1.0000	1.0072	1.1323	1.0071	0.9237
35.0	2.3142E 00	0.8647	1.1010	1.1109	1.1111	0.9198	0.9843	1.0001	1.0000	1.0184	1.1657	1.0184	0.9292
40.0	2.2383E 00	0.8462	1.1217	1.1378	1.1408	0.9108	0.9818	1.0001	1.0000	1.0291	1.1926	1.0290	0.9344
45.0	2.1693E 00	0.8286	1.1411	1.1637	1.1694	0.9025	0.9793	1.0000	1.0000	1.0392	1.2132	1.0391	0.9394
50.0	2.1060E 00	0.8118	1.1594	1.1885	1.1970	0.8949	0.9767	1.0000	1.0000	1.0487	1.2277	1.0487	0.9442

Instr. see p. 22

ELECTRON Z= 41 A= 97

p	F_0L_0	L_0	λ_2	λ_3	λ_4	μ_1	μ_2	Λ_1	Λ_2	ν_{12}	$\hat{\nu}_{12}$	η_{12}	$\hat{\eta}_{12}$
0.1	3.4817E 01	1.0106	7.8521	23.1160	43.7837	0.9997	1.0000	1.0024	1.0004	2.7844	0.6814	0.6979	0.6616
0.2	1.7602E 01	1.0106	2.5833	3.6853	4.3863	0.9997	1.0000	1.0023	1.0004	1.3280	0.6955	0.7949	0.7559
0.3	1.1955E 01	1.0105	1.6099	1.8481	1.9616	0.9996	1.0000	1.0023	1.0004	1.0697	0.7083	0.8280	0.7878
0.4	9.2185E 00	1.0104	1.2716	1.3378	1.3610	0.9994	0.9999	1.0022	1.0004	0.9838	0.7196	0.8452	0.8043
0.5	7.6504E 00	1.0103	1.1172	1.1270	1.1242	0.9992	0.9999	1.0021	1.0004	0.9464	0.7298	0.8562	0.8147
0.6	6.6602E 00	1.0101	1.0351	1.0207	1.0076	0.9990	0.9999	1.0020	1.0004	0.9276	0.7389	0.8640	0.8219
0.7	5.9918E 00	1.0100	0.9870	0.9603	0.9423	0.9988	0.9998	1.0019	1.0004	0.9173	0.7471	0.8699	0.8274
0.8	5.5175E 00	1.0098	0.9570	0.9234	0.9027	0.9985	0.9998	1.0018	1.0003	0.9115	0.7544	0.8746	0.8317
0.9	5.1675E 00	1.0095	0.9373	0.8996	0.8775	0.9983	0.9997	1.0017	1.0003	0.9080	0.7610	0.8786	0.8352
1.0	4.9007E 00	1.0093	0.9241	0.8838	0.8607	0.9980	0.9997	1.0017	1.0003	0.9060	0.7670	0.8819	0.8382
1.2	4.5248E 00	1.0087	0.9086	0.8656	0.8417	0.9975	0.9996	1.0015	1.0003	0.9045	0.7773	0.8874	0.8430
1.4	4.2750E 00	1.0081	0.9012	0.8571	0.8329	0.9969	0.9995	1.0014	1.0003	0.9046	0.7860	0.8918	0.8468
1.6	4.0978E 00	1.0074	0.8978	0.8534	0.8292	0.9964	0.9994	1.0012	1.0002	0.9054	0.7934	0.8954	0.8500
1.8	3.9656E 00	1.0066	0.8967	0.8524	0.8283	0.9959	0.9993	1.0011	1.0002	0.9066	0.8000	0.8986	0.8527
2.0	3.8629E 00	1.0059	0.8969	0.8529	0.8290	0.9953	0.9992	1.0011	1.0002	0.9080	0.8059	0.9014	0.8550
2.2	3.7804E 00	1.0050	0.8979	0.8544	0.8307	0.9948	0.9991	1.0010	1.0002	0.9095	0.8113	0.9039	0.8571
2.4	3.7123E 00	1.0042	0.8994	0.8564	0.8329	0.9942	0.9990	1.0009	1.0002	0.9109	0.8163	0.9062	0.8590
2.6	3.6548E 00	1.0034	0.9012	0.8587	0.8355	0.9937	0.9989	1.0009	1.0002	0.9124	0.8209	0.9083	0.8607
2.8	3.6053E 00	1.0025	0.9031	0.8612	0.8383	0.9931	0.9988	1.0008	1.0001	0.9138	0.8253	0.9103	0.8623
3.0	3.5620E 00	1.0016	0.9052	0.8639	0.8413	0.9926	0.9987	1.0008	1.0001	0.9153	0.8295	0.9121	0.8637
3.2	3.5236E 00	1.0007	0.9074	0.8667	0.8443	0.9920	0.9986	1.0007	1.0001	0.9167	0.8335	0.9139	0.8651
3.4	3.4891E 00	0.9998	0.9096	0.8694	0.8473	0.9915	0.9986	1.0007	1.0001	0.9180	0.8373	0.9155	0.8664
3.6	3.4579E 00	0.9989	0.9117	0.8722	0.8504	0.9910	0.9985	1.0006	1.0001	0.9193	0.8410	0.9171	0.8676
3.8	3.4293E 00	0.9980	0.9139	0.8749	0.8534	0.9904	0.9984	1.0006	1.0001	0.9206	0.8446	0.9186	0.8687
4.0	3.4030E 00	0.9970	0.9161	0.8777	0.8564	0.9899	0.9983	1.0006	1.0001	0.9219	0.8482	0.9200	0.8698
4.5	3.3450E 00	0.9947	0.9215	0.8844	0.8637	0.9886	0.9981	1.0005	1.0001	0.9248	0.8566	0.9233	0.8723
5.0	3.2954E 00	0.9924	0.9266	0.8908	0.8707	0.9872	0.9978	1.0005	1.0001	0.9276	0.8647	0.9264	0.8745
5.5	3.2518E 00	0.9900	0.9316	0.8970	0.8775	0.9859	0.9976	1.0004	1.0001	0.9303	0.8725	0.9292	0.8765
6.0	3.2129E 00	0.9876	0.9364	0.9030	0.8840	0.9846	0.9974	1.0004	1.0001	0.9328	0.8800	0.9319	0.8784
6.5	3.1776E 00	0.9852	0.9410	0.9088	0.8902	0.9833	0.9971	1.0004	1.0001	0.9352	0.8874	0.9344	0.8801
7.0	3.1452E 00	0.9828	0.9454	0.9143	0.8962	0.9819	0.9969	1.0003	1.0001	0.9375	0.8945	0.9368	0.8817
7.5	3.1152E 00	0.9804	0.9498	0.9197	0.9021	0.9806	0.9967	1.0003	1.0001	0.9398	0.9015	0.9391	0.8832
8.0	3.0871E 00	0.9781	0.9539	0.9249	0.9077	0.9793	0.9965	1.0003	1.0001	0.9419	0.9084	0.9413	0.8846
9.0	3.0359E 00	0.9733	0.9619	0.9348	0.9185	0.9767	0.9960	1.0003	1.0000	0.9460	0.9218	0.9455	0.8872
10.0	2.9898E 00	0.9685	0.9695	0.9443	0.9288	0.9741	0.9956	1.0002	1.0000	0.9498	0.9348	0.9495	0.8896
11.0	2.9476E 00	0.9637	0.9768	0.9533	0.9386	0.9715	0.9951	1.0002	1.0000	0.9535	0.9473	0.9532	0.8919
12.0	2.9086E 00	0.9590	0.9837	0.9619	0.9480	0.9689	0.9947	1.0002	1.0000	0.9570	0.9596	0.9567	0.8939
13.0	2.8722E 00	0.9543	0.9904	0.9703	0.9570	0.9664	0.9942	1.0002	1.0000	0.9604	0.9715	0.9602	0.8959
14.0	2.8381E 00	0.9496	0.9969	0.9783	0.9658	0.9638	0.9937	1.0002	1.0000	0.9637	0.9831	0.9635	0.8977
16.0	2.7751E 00	0.9403	1.0094	0.9938	0.9826	0.9588	0.9928	1.0002	1.0000	0.9700	1.0054	0.9698	0.9011
18.0	2.7179E 00	0.9312	1.0212	1.0086	0.9987	0.9539	0.9919	1.0001	1.0000	0.9759	1.0266	0.9758	0.9043
20.0	2.6652E 00	0.9222	1.0325	1.0227	1.0141	0.9490	0.9909	1.0001	1.0000	0.9816	1.0466	0.9815	0.9072
25.0	2.5484E 00	0.9002	1.0590	1.0561	1.0505	0.9374	0.9886	1.0001	1.0000	0.9951	1.0920	0.9950	0.9140
30.0	2.4474E 00	0.8792	1.0836	1.0874	1.0847	0.9266	0.9861	1.0001	1.0000	1.0076	1.1305	1.0075	0.9202
35.0	2.3578E 00	0.8592	1.1066	1.1171	1.1173	0.9165	0.9836	1.0001	1.0000	1.0194	1.1624	1.0193	0.9261
40.0	2.2773E 00	0.8401	1.1282	1.1453	1.1485	0.9071	0.9810	1.0001	1.0000	1.0305	1.1878	1.0304	0.9316
45.0	2.2042E 00	0.8220	1.1485	1.1724	1.1785	0.8987	0.9784	1.0000	1.0000	1.0411	1.2067	1.0410	0.9370
50.0	2.1373E 00	0.8048	1.1676	1.1984	1.2075	0.8910	0.9757	1.0000	1.0000	1.0511	1.2194	1.0510	0.9421

Instr. see p. 22

ELECTRON Z= 42 A= 96

p	F_0L_0	L_0	λ_2	λ_3	λ_4	μ_1	μ_2	Λ_1	Λ_2	ν_{12}	$\hat{\nu}_{12}$	η_{12}	$\hat{\eta}_{12}$
0.1	3.6712E 01	1.0112	8.1078	24.5701	47.5635	0.9997	1.0000	1.0025	1.0004	2.8601	0.6778	0.6792	0.6416
0.2	1.8558E 01	1.0112	2.6417	3.8184	4.5786	0.9997	1.0000	1.0024	1.0004	1.3406	0.6923	0.7830	0.7425
0.3	1.2602E 01	1.0111	1.6317	1.8836	2.0046	0.9995	1.0000	1.0023	1.0004	1.0713	0.7054	0.8184	0.7766
0.4	9.7106E 00	1.0110	1.2807	1.3501	1.3744	0.9994	0.9999	1.0022	1.0004	0.9818	0.7170	0.8368	0.7941
0.5	8.0512E 00	1.0109	1.1204	1.1304	1.1274	0.9992	0.9999	1.0022	1.0004	0.9429	0.7274	0.8485	0.8051
0.6	7.0017E 00	1.0107	1.0352	1.0198	1.0060	0.9990	0.9999	1.0021	1.0004	0.9233	0.7368	0.8567	0.8128
0.7	6.2924E 00	1.0105	0.9852	0.9571	0.9383	0.9987	0.9998	1.0020	1.0004	0.9127	0.7451	0.8630	0.8186
0.8	5.7886E 00	1.0103	0.9540	0.9188	0.8972	0.9985	0.9998	1.0019	1.0003	0.9066	0.7526	0.8680	0.8231
0.9	5.4164E 00	1.0101	0.9336	0.8941	0.8710	0.9982	0.9997	1.0018	1.0003	0.9030	0.7593	0.8721	0.8268
1.0	5.1326E 00	1.0098	0.9199	0.8777	0.8537	0.9980	0.9997	1.0017	1.0003	0.9010	0.7654	0.8757	0.8300
1.2	4.7323E 00	1.0093	0.9038	0.8589	0.8339	0.9974	0.9996	1.0015	1.0003	0.8994	0.7759	0.8814	0.8351
1.4	4.4659E 00	1.0086	0.8961	0.8501	0.8248	0.9969	0.9995	1.0014	1.0003	0.8995	0.7848	0.8860	0.8391
1.6	4.2768E 00	1.0079	0.8926	0.8462	0.8210	0.9963	0.9994	1.0013	1.0002	0.9004	0.7924	0.8899	0.8424
1.8	4.1356E 00	1.0071	0.8915	0.8452	0.8201	0.9957	0.9993	1.0012	1.0002	0.9016	0.7991	0.8932	0.8452
2.0	4.0257E 00	1.0063	0.8917	0.8457	0.8208	0.9952	0.9992	1.0011	1.0002	0.9031	0.8050	0.8962	0.8477
2.2	3.9374E 00	1.0055	0.8927	0.8472	0.8225	0.9946	0.9991	1.0010	1.0002	0.9046	0.8105	0.8988	0.8499
2.4	3.8644E 00	1.0046	0.8943	0.8493	0.8249	0.9940	0.9990	1.0009	1.0002	0.9061	0.8155	0.9012	0.8518
2.6	3.8027E 00	1.0037	0.8961	0.8517	0.8276	0.9935	0.9989	1.0009	1.0002	0.9077	0.8202	0.9034	0.8536
2.8	3.7495E 00	1.0028	0.8982	0.8544	0.8305	0.9929	0.9988	1.0008	1.0001	0.9092	0.8246	0.9055	0.8553
3.0	3.7030E 00	1.0019	0.9003	0.8572	0.8336	0.9924	0.9987	1.0008	1.0001	0.9107	0.8288	0.9074	0.8568
3.2	3.6617E 00	1.0010	0.9026	0.8600	0.8367	0.9918	0.9986	1.0007	1.0001	0.9121	0.8328	0.9092	0.8582
3.4	3.6246E 00	1.0001	0.9048	0.8629	0.8398	0.9913	0.9985	1.0007	1.0001	0.9135	0.8367	0.9109	0.8595
3.6	3.5910E 00	0.9992	0.9071	0.8657	0.8430	0.9907	0.9984	1.0007	1.0001	0.9149	0.8404	0.9125	0.8608
3.8	3.5602E 00	0.9982	0.9094	0.8686	0.8461	0.9902	0.9983	1.0006	1.0001	0.9162	0.8440	0.9141	0.8620
4.0	3.5319E 00	0.9973	0.9117	0.8715	0.8492	0.9896	0.9982	1.0006	1.0001	0.9175	0.8476	0.9156	0.8631
4.5	3.4694E 00	0.9949	0.9172	0.8784	0.8568	0.9882	0.9980	1.0005	1.0001	0.9206	0.8560	0.9191	0.8657
5.0	3.4159E 00	0.9924	0.9226	0.8851	0.8641	0.9869	0.9978	1.0005	1.0001	0.9236	0.8641	0.9223	0.8680
5.5	3.3690E 00	0.9900	0.9278	0.8916	0.8711	0.9855	0.9975	1.0004	1.0001	0.9263	0.8719	0.9252	0.8701
6.0	3.3270E 00	0.9876	0.9328	0.8978	0.8778	0.9841	0.9973	1.0004	1.0001	0.9290	0.8794	0.9280	0.8721
6.5	3.2890E 00	0.9851	0.9376	0.9038	0.8843	0.9828	0.9971	1.0004	1.0001	0.9315	0.8867	0.9306	0.8739
7.0	3.2541E 00	0.9827	0.9422	0.9095	0.8906	0.9814	0.9969	1.0004	1.0001	0.9339	0.8938	0.9331	0.8756
7.5	3.2217E 00	0.9802	0.9467	0.9151	0.8966	0.9800	0.9966	1.0003	1.0001	0.9362	0.9008	0.9355	0.8771
8.0	3.1915E 00	0.9777	0.9511	0.9205	0.9025	0.9787	0.9964	1.0003	1.0001	0.9384	0.9076	0.9378	0.8786
9.0	3.1364E 00	0.9728	0.9594	0.9308	0.9137	0.9760	0.9959	1.0003	1.0000	0.9427	0.9208	0.9422	0.8814
10.0	3.0868E 00	0.9679	0.9673	0.9406	0.9243	0.9733	0.9955	1.0002	1.0000	0.9467	0.9337	0.9463	0.8839
11.0	3.0415E 00	0.9631	0.9748	0.9500	0.9345	0.9706	0.9950	1.0002	1.0000	0.9505	0.9461	0.9502	0.8862
12.0	2.9996E 00	0.9582	0.9820	0.9590	0.9442	0.9680	0.9945	1.0002	1.0000	0.9542	0.9581	0.9539	0.8884
13.0	2.9606E 00	0.9534	0.9890	0.9676	0.9537	0.9654	0.9941	1.0002	1.0000	0.9577	0.9699	0.9574	0.8904
14.0	2.9240E 00	0.9486	0.9958	0.9760	0.9628	0.9628	0.9936	1.0002	1.0000	0.9611	0.9813	0.9608	0.8923
16.0	2.8566E 00	0.9391	1.0087	0.9921	0.9802	0.9576	0.9926	1.0002	1.0000	0.9676	1.0032	0.9674	0.8959
18.0	2.7954E 00	0.9297	1.0210	1.0074	0.9968	0.9526	0.9917	1.0001	1.0000	0.9738	1.0240	0.9736	0.8992
20.0	2.7392E 00	0.9204	1.0327	1.0220	1.0128	0.9476	0.9907	1.0001	1.0000	0.9797	1.0436	0.9796	0.9024
25.0	2.6149E 00	0.8980	1.0603	1.0567	1.0506	0.9357	0.9883	1.0001	1.0000	0.9937	1.0878	0.9936	0.9095
30.0	2.5076E 00	0.8765	1.0858	1.0891	1.0860	0.9246	0.9858	1.0001	1.0000	1.0066	1.1252	1.0066	0.9161
35.0	2.4128E 00	0.8561	1.1096	1.1198	1.1198	0.9142	0.9832	1.0001	1.0000	1.0189	1.1559	1.0188	0.9222
40.0	2.3277E 00	0.8366	1.1320	1.1492	1.1521	0.9047	0.9806	1.0001	1.0000	1.0304	1.1801	1.0304	0.9281
45.0	2.2506E 00	0.8182	1.1530	1.1772	1.1832	0.8961	0.9779	1.0001	1.0000	1.0414	1.1981	1.0413	0.9337
50.0	2.1801E 00	0.8007	1.1728	1.2042	1.2132	0.8883	0.9751	1.0000	1.0000	1.0517	1.2096	1.0517	0.9391

Instr. see p. 22

ELECTRON Z= 43 A= 99

p	F_0L_0	L_0	λ_2	λ_3	λ_4	μ_1	μ_2	Λ_1	Λ_2	ν_{12}	$\hat{\nu}_{12}$	η_{12}	$\hat{\eta}_{12}$
0.1	3.8655E 01	1.0117	8.3692	26.1027	51.6460	0.9997	1.0000	1.0026	1.0005	2.9378	0.6745	0.6601	0.6213
0.2	1.9540E 01	1.0117	2.7019	3.9581	4.7822	0.9996	1.0000	1.0025	1.0004	1.3537	0.6894	0.7711	0.7290
0.3	1.3265E 01	1.0116	1.6547	1.9213	2.0504	0.9995	0.9999	1.0024	1.0004	1.0733	0.7027	0.8088	0.7653
0.4	1.0215E 01	1.0115	1.2907	1.3636	1.3893	0.9993	0.9999	1.0023	1.0004	0.9801	0.7146	0.8284	0.7840
0.5	8.4623E 00	1.0114	1.1244	1.1348	1.1316	0.9991	0.9999	1.0022	1.0004	0.9396	0.7253	0.8408	0.7956
0.6	7.3518E 00	1.0112	1.0359	1.0199	1.0055	0.9989	0.9998	1.0021	1.0004	0.9193	0.7349	0.8496	0.8037
0.7	6.6003E 00	1.0110	0.9841	0.9548	0.9351	0.9987	0.9998	1.0021	1.0004	0.9082	0.7434	0.8562	0.8099
0.8	6.0660E 00	1.0108	0.9517	0.9150	0.8925	0.9984	0.9998	1.0020	1.0004	0.9019	0.7511	0.8615	0.8147
0.9	5.6710E 00	1.0106	0.9305	0.8894	0.8654	0.9981	0.9997	1.0019	1.0003	0.8982	0.7580	0.8659	0.8186
1.0	5.3655E 00	1.0103	0.9163	0.8724	0.8474	0.9979	0.9997	1.0018	1.0003	0.8961	0.7642	0.8696	0.8219
1.2	4.9438E 00	1.0097	0.8996	0.8529	0.8270	0.9973	0.9996	1.0016	1.0003	0.8945	0.7750	0.8757	0.8272
1.4	4.6603E 00	1.0090	0.8916	0.8437	0.8175	0.9967	0.9995	1.0015	1.0003	0.8946	0.7840	0.8805	0.8314
1.6	4.4587E 00	1.0082	0.8880	0.8398	0.8136	0.9961	0.9994	1.0013	1.0002	0.8955	0.7918	0.8845	0.8349
1.8	4.3080E 00	1.0074	0.8868	0.8387	0.8127	0.9955	0.9993	1.0012	1.0002	0.8969	0.7986	0.8880	0.8379
2.0	4.1906E 00	1.0066	0.8870	0.8393	0.8134	0.9949	0.9992	1.0011	1.0002	0.8984	0.8047	0.8911	0.8404
2.2	4.0961E 00	1.0057	0.8881	0.8409	0.8153	0.9944	0.9991	1.0010	1.0002	0.9000	0.8102	0.8939	0.8427
2.4	4.0179E 00	1.0048	0.8897	0.8430	0.8177	0.9938	0.9990	1.0010	1.0002	0.9016	0.8153	0.8964	0.8448
2.6	3.9518E 00	1.0039	0.8917	0.8456	0.8205	0.9932	0.9989	1.0009	1.0002	0.9032	0.8201	0.8987	0.8467
2.8	3.8948E 00	1.0030	0.8938	0.8483	0.8235	0.9926	0.9988	1.0009	1.0002	0.9048	0.8246	0.9009	0.8484
3.0	3.8448E 00	1.0020	0.8961	0.8512	0.8267	0.9920	0.9987	1.0008	1.0001	0.9063	0.8288	0.9029	0.8500
3.2	3.8005E 00	1.0011	0.8984	0.8542	0.8300	0.9914	0.9986	1.0008	1.0001	0.9078	0.8329	0.9048	0.8515
3.4	3.7606E 00	1.0001	0.9008	0.8572	0.8332	0.9909	0.9985	1.0007	1.0001	0.9093	0.8368	0.9066	0.8529
3.6	3.7245E 00	0.9991	0.9032	0.8602	0.8365	0.9903	0.9984	1.0007	1.0001	0.9107	0.8406	0.9083	0.8542
3.8	3.6914E 00	0.9982	0.9056	0.8632	0.8398	0.9897	0.9983	1.0006	1.0001	0.9121	0.8443	0.9099	0.8554
4.0	3.6609E 00	0.9972	0.9080	0.8661	0.8430	0.9891	0.9982	1.0006	1.0001	0.9135	0.8479	0.9114	0.8566
4.5	3.5935E 00	0.9947	0.9138	0.8734	0.8509	0.9877	0.9979	1.0006	1.0001	0.9168	0.8564	0.9151	0.8593
5.0	3.5359E 00	0.9922	0.9194	0.8804	0.8585	0.9863	0.9977	1.0005	1.0001	0.9198	0.8646	0.9184	0.8617
5.5	3.4853E 00	0.9896	0.9248	0.8871	0.8658	0.9848	0.9975	1.0005	1.0001	0.9227	0.8724	0.9215	0.8639
6.0	3.4400E 00	0.9871	0.9300	0.8936	0.8728	0.9834	0.9972	1.0004	1.0001	0.9255	0.8799	0.9245	0.8660
6.5	3.3990E 00	0.9845	0.9350	0.8998	0.8796	0.9820	0.9970	1.0004	1.0001	0.9281	0.8873	0.9272	0.8679
7.0	3.3613E 00	0.9820	0.9399	0.9058	0.8861	0.9806	0.9967	1.0004	1.0001	0.9306	0.8944	0.9298	0.8696
7.5	3.3265E 00	0.9794	0.9446	0.9116	0.8924	0.9792	0.9965	1.0003	1.0001	0.9330	0.9014	0.9323	0.8713
8.0	3.2939E 00	0.9769	0.9491	0.9173	0.8985	0.9778	0.9963	1.0003	1.0001	0.9353	0.9082	0.9347	0.8728
9.0	3.2345E 00	0.9718	0.9578	0.9280	0.9102	0.9749	0.9958	1.0003	1.0001	0.9398	0.9215	0.9393	0.8757
10.0	3.1810E 00	0.9667	0.9661	0.9383	0.9213	0.9722	0.9953	1.0003	1.0000	0.9440	0.9343	0.9436	0.8783
11.0	3.1323E 00	0.9616	0.9740	0.9481	0.9319	0.9694	0.9948	1.0002	1.0000	0.9480	0.9467	0.9476	0.8808
12.0	3.0872E 00	0.9566	0.9816	0.9574	0.9421	0.9666	0.9943	1.0002	1.0000	0.9518	0.9588	0.9515	0.8831
13.0	3.0453E 00	0.9515	0.9889	0.9665	0.9519	0.9639	0.9938	1.0002	1.0000	0.9555	0.9704	0.9552	0.8852
14.0	3.0059E 00	0.9466	0.9959	0.9753	0.9614	0.9612	0.9934	1.0002	1.0000	0.9591	0.9818	0.9588	0.8872
16.0	2.9336E 00	0.9367	1.0094	0.9920	0.9796	0.9558	0.9924	1.0002	1.0000	0.9659	1.0035	0.9657	0.8910
18.0	2.8681E 00	0.9270	1.0223	1.0080	0.9970	0.9506	0.9914	1.0001	1.0000	0.9723	1.0240	0.9722	0.8945
20.0	2.8079E 00	0.9174	1.0346	1.0234	1.0137	0.9454	0.9904	1.0001	1.0000	0.9785	1.0433	0.9784	0.8978
25.0	2.6750E 00	0.8942	1.0634	1.0596	1.0532	0.9331	0.9879	1.0001	1.0000	0.9931	1.0866	0.9930	0.9054
30.0	2.5608E 00	0.8721	1.0900	1.0935	1.0903	0.9217	0.9853	1.0001	1.0000	1.0067	1.1229	1.0066	0.9123
35.0	2.4601E 00	0.8510	1.1149	1.1257	1.1255	0.9110	0.9826	1.0001	1.0000	1.0195	1.1523	1.0194	0.9189
40.0	2.3700E 00	0.8310	1.1383	1.1563	1.1593	0.9013	0.9798	1.0001	1.0000	1.0316	1.1751	1.0315	0.9251
45.0	2.2885E 00	0.8121	1.1602	1.1856	1.1918	0.8925	0.9770	1.0001	1.0000	1.0430	1.1914	1.0429	0.9311
50.0	2.2143E 00	0.7942	1.1808	1.2137	1.2231	0.8846	0.9741	1.0000	1.0000	1.0538	1.2014	1.0538	0.9369

Instr. see p. 22

ELECTRON Z= 44 A= 99

p	F_0L_0	L_0	λ_2	λ_3	λ_4	μ_1	μ_2	Λ_1	Λ_2	ν_{12}	$\hat{\nu}_{12}$	η_{12}	$\hat{\eta}_{12}$
0.1	4.0747E 01	1.0123	8.6252	27.6694	55.9464	0.9997	1.0000	1.0027	1.0005	3.0154	0.6709	0.6402	0.6002
0.2	2.0596E 01	1.0122	2.7603	4.0974	4.9883	0.9996	1.0000	1.0026	1.0005	1.3665	0.6861	0.7586	0.7149
0.3	1.3979E 01	1.0122	1.6765	1.9578	2.0951	0.9995	0.9999	1.0025	1.0004	1.0748	0.6997	0.7987	0.7535
0.4	1.0760E 01	1.0121	1.2997	1.3760	1.4031	0.9993	0.9999	1.0024	1.0004	0.9780	0.7119	0.8196	0.7733
0.5	8.9058E 00	1.0119	1.1276	1.1383	1.1348	0.9991	0.9999	1.0023	1.0004	0.9359	0.7229	0.8327	0.7857
0.6	7.7297E 00	1.0118	1.0360	1.0190	1.0039	0.9989	0.9998	1.0022	1.0004	0.9149	0.7326	0.8420	0.7943
0.7	6.9328E 00	1.0116	0.9823	0.9516	0.9310	0.9986	0.9998	1.0021	1.0004	0.9034	0.7414	0.8490	0.8007
0.8	6.3656E 00	1.0113	0.9487	0.9104	0.8870	0.9984	0.9998	1.0020	1.0004	0.8968	0.7492	0.8546	0.8058
0.9	5.9460E 00	1.0111	0.9268	0.8839	0.8589	0.9981	0.9997	1.0019	1.0003	0.8930	0.7563	0.8592	0.8099
1.0	5.6255E 00	1.0108	0.9120	0.8663	0.8403	0.9978	0.9997	1.0018	1.0003	0.8908	0.7626	0.8631	0.8134
1.2	5.1725E 00	1.0102	0.8948	0.8461	0.8192	0.9972	0.9996	1.0016	1.0003	0.8891	0.7736	0.8695	0.8190
1.4	4.8704E 00	1.0095	0.8865	0.8366	0.8095	0.9966	0.9994	1.0015	1.0003	0.8893	0.7828	0.8746	0.8234
1.6	4.6554E 00	1.0087	0.8827	0.8326	0.8054	0.9960	0.9993	1.0014	1.0002	0.8903	0.7907	0.8788	0.8270
1.8	4.4944E 00	1.0079	0.8815	0.8315	0.8044	0.9954	0.9992	1.0013	1.0002	0.8917	0.7976	0.8825	0.8301
2.0	4.3689E 00	1.0070	0.8818	0.8321	0.8053	0.9948	0.9991	1.0012	1.0002	0.8933	0.8038	0.8857	0.8328
2.2	4.2677E 00	1.0061	0.8829	0.8337	0.8071	0.9942	0.9990	1.0011	1.0002	0.8949	0.8094	0.8886	0.8352
2.4	4.1840E 00	1.0052	0.8846	0.8360	0.8097	0.9936	0.9989	1.0010	1.0002	0.8966	0.8146	0.8912	0.8373
2.6	4.1131E 00	1.0043	0.8866	0.8386	0.8126	0.9930	0.9988	1.0009	1.0002	0.8983	0.8194	0.8936	0.8393
2.8	4.0519E 00	1.0033	0.8889	0.8415	0.8157	0.9924	0.9987	1.0009	1.0002	0.8999	0.8240	0.8959	0.8411
3.0	3.9982E 00	1.0023	0.8912	0.8445	0.8190	0.9918	0.9986	1.0008	1.0001	0.9016	0.8283	0.8980	0.8428
3.2	3.9506E 00	1.0013	0.8937	0.8475	0.8224	0.9912	0.9985	1.0008	1.0001	0.9031	0.8324	0.8999	0.8443
3.4	3.9077E 00	1.0003	0.8961	0.8506	0.8258	0.9906	0.9984	1.0007	1.0001	0.9047	0.8364	0.9018	0.8458
3.6	3.8688E 00	0.9993	0.8986	0.8538	0.8292	0.9900	0.9983	1.0007	1.0001	0.9062	0.8402	0.9036	0.8471
3.8	3.8332E 00	0.9983	0.9011	0.8569	0.8325	0.9894	0.9982	1.0007	1.0001	0.9076	0.8439	0.9053	0.8484
4.0	3.8004E 00	0.9973	0.9036	0.8599	0.8359	0.9888	0.9981	1.0006	1.0001	0.9090	0.8475	0.9069	0.8497
4.5	3.7279E 00	0.9947	0.9096	0.8675	0.8441	0.9873	0.9979	1.0006	1.0001	0.9124	0.8561	0.9107	0.8525
5.0	3.6658E 00	0.9921	0.9155	0.8747	0.8519	0.9858	0.9976	1.0005	1.0001	0.9156	0.8642	0.9142	0.8550
5.5	3.6113E 00	0.9895	0.9211	0.8817	0.8595	0.9843	0.9974	1.0005	1.0001	0.9186	0.8721	0.9174	0.8573
6.0	3.5625E 00	0.9869	0.9265	0.8884	0.8668	0.9829	0.9971	1.0004	1.0001	0.9215	0.8796	0.9204	0.8595
6.5	3.5183E 00	0.9843	0.9317	0.8949	0.8738	0.9814	0.9969	1.0004	1.0001	0.9242	0.8869	0.9233	0.8614
7.0	3.4778E 00	0.9817	0.9368	0.9011	0.8806	0.9799	0.9967	1.0004	1.0001	0.9268	0.8941	0.9260	0.8633
7.5	3.4402E 00	0.9790	0.9417	0.9072	0.8871	0.9785	0.9964	1.0004	1.0001	0.9294	0.9010	0.9286	0.8650
8.0	3.4052E 00	0.9764	0.9464	0.9130	0.8935	0.9770	0.9962	1.0003	1.0001	0.9318	0.9078	0.9311	0.8666
9.0	3.3413E 00	0.9711	0.9555	0.9242	0.9056	0.9741	0.9957	1.0003	1.0001	0.9364	0.9211	0.9359	0.8696
10.0	3.2839E 00	0.9659	0.9641	0.9349	0.9171	0.9713	0.9952	1.0003	1.0000	0.9408	0.9338	0.9403	0.8724
11.0	3.2315E 00	0.9607	0.9723	0.9450	0.9281	0.9684	0.9947	1.0002	1.0000	0.9449	0.9461	0.9446	0.8749
12.0	3.1832E 00	0.9555	0.9801	0.9547	0.9386	0.9656	0.9942	1.0002	1.0000	0.9489	0.9580	0.9486	0.8773
13.0	3.1382E 00	0.9503	0.9877	0.9642	0.9488	0.9627	0.9937	1.0002	1.0000	0.9527	0.9695	0.9525	0.8796
14.0	3.0960E 00	0.9452	0.9951	0.9732	0.9587	0.9599	0.9932	1.0002	1.0000	0.9564	0.9807	0.9562	0.8817
16.0	3.0186E 00	0.9351	1.0091	0.9907	0.9776	0.9544	0.9922	1.0002	1.0000	0.9635	1.0021	0.9633	0.8857
18.0	2.9486E 00	0.9251	1.0225	1.0073	0.9956	0.9490	0.9912	1.0001	1.0000	0.9703	1.0223	0.9701	0.8893
20.0	2.8844E 00	0.9153	1.0352	1.0232	1.0130	0.9438	0.9901	1.0001	1.0000	0.9767	1.0413	0.9766	0.8928
25.0	2.7430E 00	0.8915	1.0652	1.0609	1.0539	0.9311	0.9875	1.0001	1.0000	0.9919	1.0836	0.9918	0.9008
30.0	2.6217E 00	0.8688	1.0929	1.0960	1.0924	0.9194	0.9849	1.0001	1.0000	1.0060	1.1188	1.0059	0.9081
35.0	2.5151E 00	0.8473	1.1187	1.1294	1.1289	0.9085	0.9821	1.0001	1.0000	1.0192	1.1471	1.0192	0.9150
40.0	2.4200E 00	0.8269	1.1429	1.1612	1.1640	0.8986	0.9793	1.0001	1.0000	1.0318	1.1688	1.0317	0.9216
45.0	2.3341E 00	0.8076	1.1656	1.1916	1.1977	0.8896	0.9764	1.0001	1.0000	1.0436	1.1839	1.0436	0.9279
50.0	2.2561E 00	0.7894	1.1869	1.2207	1.2302	0.8816	0.9734	1.0000	1.0000	1.0548	1.1927	1.0548	0.9340

Instr. see p. 22

ELECTRON Z= 45 A= 103

p	F_0L_0	L_0	λ_2	λ_3	λ_4	μ_1	μ_2	Λ_1	Λ_2	ν_{12}	$\hat{\nu}_{12}$	η_{12}	$\hat{\eta}_{12}$
0.1	4.2890E 01	1.0128	8.8871	29.3196	60.5840	0.9997	1.0000	1.0028	1.0005	3.0951	0.6675	0.6198	0.5787
0.2	2.1678E 01	1.0127	2.8207	4.2437	5.2068	0.9996	1.0000	1.0027	1.0005	1.3798	0.6830	0.7459	0.7007
0.3	1.4710E 01	1.0126	1.6996	1.9966	2.1428	0.9994	0.9999	1.0026	1.0005	1.0767	0.6970	0.7887	0.7418
0.4	1.1317E 01	1.0125	1.3097	1.3899	1.4184	0.9993	0.9999	1.0025	1.0004	0.9761	0.7095	0.8108	0.7628
0.5	9.3597E 00	1.0124	1.1316	1.1428	1.1393	0.9990	0.9999	1.0024	1.0004	0.9325	0.7207	0.8248	0.7758
0.6	8.1163E 00	1.0122	1.0368	1.0191	1.0034	0.9988	0.9998	1.0023	1.0004	0.9107	0.7307	0.8345	0.7849
0.7	7.2727E 00	1.0120	0.9812	0.9492	0.9279	0.9985	0.9998	1.0022	1.0004	0.8987	0.7396	0.8420	0.7917
0.8	6.6717E 00	1.0118	0.9465	0.9066	0.8823	0.9983	0.9997	1.0021	1.0004	0.8920	0.7477	0.8478	0.7970
0.9	6.2266E 00	1.0115	0.9237	0.8792	0.8533	0.9980	0.9997	1.0020	1.0004	0.8880	0.7549	0.8527	0.8014
1.0	5.8865E 00	1.0112	0.9084	0.8610	0.8341	0.9977	0.9996	1.0019	1.0003	0.8858	0.7614	0.8568	0.8050
1.2	5.4052E 00	1.0106	0.8906	0.8401	0.8123	0.9971	0.9995	1.0017	1.0003	0.8841	0.7726	0.8635	0.8109
1.4	5.0837E 00	1.0098	0.8820	0.8304	0.8022	0.9964	0.9994	1.0016	1.0003	0.8842	0.7820	0.8688	0.8155
1.6	4.8547E 00	1.0090	0.8781	0.8262	0.7981	0.9958	0.9993	1.0014	1.0003	0.8853	0.7901	0.8733	0.8193
1.8	4.6830E 00	1.0082	0.8769	0.8251	0.7971	0.9952	0.9992	1.0013	1.0002	0.8868	0.7972	0.8771	0.8225
2.0	4.5490E 00	1.0073	0.8771	0.8257	0.7979	0.9945	0.9991	1.0012	1.0002	0.8884	0.8035	0.8805	0.8253
2.2	4.4409E 00	1.0063	0.8783	0.8274	0.7999	0.9939	0.9990	1.0011	1.0002	0.8901	0.8092	0.8835	0.8278
2.4	4.3513E 00	1.0054	0.8801	0.8297	0.8025	0.9933	0.9989	1.0010	1.0002	0.8919	0.8145	0.8862	0.8301
2.6	4.2753E 00	1.0044	0.8822	0.8325	0.8056	0.9926	0.9988	1.0010	1.0002	0.8937	0.8194	0.8888	0.8321
2.8	4.2097E 00	1.0034	0.8846	0.8355	0.8088	0.9920	0.9987	1.0009	1.0002	0.8954	0.8241	0.8911	0.8340
3.0	4.1521E 00	1.0024	0.8870	0.8386	0.8123	0.9914	0.9986	1.0009	1.0002	0.8971	0.8284	0.8933	0.8357
3.2	4.1010E 00	1.0013	0.8896	0.8418	0.8158	0.9907	0.9985	1.0008	1.0001	0.8987	0.8326	0.8954	0.8373
3.4	4.0549E 00	1.0003	0.8922	0.8450	0.8193	0.9901	0.9984	1.0008	1.0001	0.9003	0.8366	0.8973	0.8389
3.6	4.0131E 00	0.9992	0.8948	0.8483	0.8228	0.9895	0.9983	1.0007	1.0001	0.9019	0.8405	0.8992	0.8403
3.8	3.9749E 00	0.9982	0.8974	0.8515	0.8263	0.9889	0.9982	1.0007	1.0001	0.9034	0.8443	0.9010	0.8416
4.0	3.9395E 00	0.9971	0.8999	0.8547	0.8298	0.9883	0.9981	1.0007	1.0001	0.9049	0.8479	0.9027	0.8429
4.5	3.8615E 00	0.9944	0.9063	0.8625	0.8383	0.9867	0.9978	1.0006	1.0001	0.9084	0.8566	0.9066	0.8459
5.0	3.7946E 00	0.9917	0.9124	0.8701	0.8465	0.9852	0.9975	1.0005	1.0001	0.9118	0.8649	0.9103	0.8485
5.5	3.7359E 00	0.9890	0.9182	0.8774	0.8544	0.9836	0.9973	1.0005	1.0001	0.9149	0.8728	0.9136	0.8509
6.0	3.6834E 00	0.9863	0.9239	0.8844	0.8620	0.9821	0.9970	1.0005	1.0001	0.9179	0.8805	0.9168	0.8531
6.5	3.6357E 00	0.9835	0.9294	0.8911	0.8693	0.9806	0.9968	1.0004	1.0001	0.9208	0.8878	0.9198	0.8552
7.0	3.5920E 00	0.9808	0.9347	0.8977	0.8763	0.9790	0.9965	1.0004	1.0001	0.9235	0.8950	0.9227	0.8571
7.5	3.5516E 00	0.9780	0.9398	0.9040	0.8831	0.9775	0.9963	1.0004	1.0001	0.9261	0.9020	0.9254	0.8589
8.0	3.5138E 00	0.9753	0.9447	0.9101	0.8897	0.9760	0.9960	1.0003	1.0001	0.9287	0.9089	0.9280	0.8606
9.0	3.4450E 00	0.9698	0.9542	0.9217	0.9024	0.9730	0.9955	1.0003	1.0001	0.9335	0.9221	0.9330	0.8638
10.0	3.3832E 00	0.9644	0.9632	0.9328	0.9144	0.9699	0.9950	1.0003	1.0000	0.9381	0.9349	0.9376	0.8667
11.0	3.3269E 00	0.9590	0.9717	0.9434	0.9259	0.9670	0.9945	1.0003	1.0000	0.9424	0.9472	0.9420	0.8693
12.0	3.2749E 00	0.9536	0.9800	0.9536	0.9369	0.9640	0.9940	1.0002	1.0000	0.9466	0.9591	0.9462	0.8718
13.0	3.2266E 00	0.9482	0.9879	0.9635	0.9475	0.9611	0.9934	1.0002	1.0000	0.9506	0.9706	0.9503	0.8742
14.0	3.1814E 00	0.9429	0.9956	0.9730	0.9578	0.9582	0.9929	1.0002	1.0000	0.9545	0.9818	0.9542	0.8764
16.0	3.0984E 00	0.9323	1.0103	0.9912	0.9776	0.9524	0.9919	1.0002	1.0000	0.9619	1.0031	0.9617	0.8806
18.0	3.0233E 00	0.9220	1.0243	1.0086	0.9964	0.9468	0.9908	1.0002	1.0000	0.9689	1.0231	0.9687	0.8845
20.0	2.9546E 00	0.9118	1.0377	1.0252	1.0146	0.9413	0.9898	1.0001	1.0000	0.9757	1.0418	0.9755	0.8881
25.0	2.8037E 00	0.8872	1.0690	1.0646	1.0574	0.9283	0.9870	1.0001	1.0000	0.9915	1.0832	0.9914	0.8966
30.0	2.6747E 00	0.8638	1.0979	1.1014	1.0976	0.9162	0.9843	1.0001	1.0000	1.0063	1.1173	1.0062	0.9044
35.0	2.5617E 00	0.8416	1.1249	1.1363	1.1359	0.9050	0.9814	1.0001	1.0000	1.0202	1.1443	1.0201	0.9117
40.0	2.4610E 00	0.8207	1.1502	1.1695	1.1725	0.8949	0.9785	1.0001	1.0000	1.0333	1.1644	1.0332	0.9187
45.0	2.3704E 00	0.8009	1.1739	1.2013	1.2078	0.8857	0.9755	1.0001	1.0000	1.0456	1.1779	1.0456	0.9255
50.0	2.2883E 00	0.7823	1.1960	1.2317	1.2417	0.8776	0.9723	1.0000	1.0000	1.0573	1.1850	1.0573	0.9319

Instr. see p. 22

ELECTRON Z= 46 A= 106

p	F_0L_0	L_0	λ_2	λ_3	λ_4	μ_1	μ_2	Λ_1	Λ_2	ν_{12}	$\hat{\nu}_{12}$	η_{12}	$\hat{\eta}_{12}$
0.1	4.5165E 01	1.0133	9.1463	31.0168	65.4896	0.9996	1.0000	1.0029	1.0005	3.1752	0.6640	0.5987	0.5564
0.2	2.2826E 01	1.0132	2.8802	4.3916	5.4305	0.9995	1.0000	1.0028	1.0005	1.3930	0.6798	0.7329	0.6861
0.3	1.5487E 01	1.0131	1.7221	2.0351	2.1906	0.9994	0.9999	1.0027	1.0005	1.0784	0.6941	0.7783	0.7296
0.4	1.1909E 01	1.0130	1.3194	1.4033	1.4334	0.9992	0.9999	1.0026	1.0005	0.9741	0.7069	0.8018	0.7519
0.5	9.8422E 00	1.0129	1.1353	1.1469	1.1432	0.9990	0.9999	1.0025	1.0004	0.9289	0.7183	0.8165	0.7657
0.6	8.5273E 00	1.0127	1.0372	1.0188	1.0024	0.9987	0.9998	1.0024	1.0004	0.9062	0.7286	0.8268	0.7752
0.7	7.6341E 00	1.0125	0.9798	0.9465	0.9244	0.9985	0.9998	1.0023	1.0004	0.8939	0.7377	0.8346	0.7824
0.8	6.9971E 00	1.0122	0.9439	0.9025	0.8773	0.9982	0.9997	1.0022	1.0004	0.8869	0.7459	0.8408	0.7880
0.9	6.5250E 00	1.0120	0.9204	0.8742	0.8473	0.9979	0.9997	1.0021	1.0004	0.8828	0.7533	0.8459	0.7925
1.0	6.1638E 00	1.0117	0.9045	0.8554	0.8275	0.9976	0.9996	1.0020	1.0004	0.8805	0.7600	0.8503	0.7964
1.2	5.6523E 00	1.0110	0.8861	0.8338	0.8050	0.9969	0.9995	1.0018	1.0003	0.8788	0.7714	0.8573	0.8025
1.4	5.3103E 00	1.0102	0.8772	0.8238	0.7947	0.9963	0.9994	1.0016	1.0003	0.8790	0.7811	0.8629	0.8073
1.6	5.0662E 00	1.0094	0.8732	0.8194	0.7904	0.9956	0.9993	1.0015	1.0003	0.8801	0.7893	0.8675	0.8113
1.8	4.8831E 00	1.0085	0.8720	0.8183	0.7894	0.9949	0.9992	1.0014	1.0002	0.8816	0.7965	0.8715	0.8147
2.0	4.7400E 00	1.0075	0.8723	0.8190	0.7903	0.9943	0.9991	1.0013	1.0002	0.8833	0.8030	0.8751	0.8176
2.2	4.6244E 00	1.0066	0.8735	0.8208	0.7923	0.9936	0.9990	1.0012	1.0002	0.8852	0.8088	0.8782	0.8202
2.4	4.5285E 00	1.0056	0.8754	0.8232	0.7951	0.9930	0.9989	1.0011	1.0002	0.8870	0.8142	0.8811	0.8225
2.6	4.4471E 00	1.0045	0.8776	0.8261	0.7982	0.9923	0.9987	1.0010	1.0002	0.8888	0.8192	0.8837	0.8247
2.8	4.3768E 00	1.0035	0.8800	0.8292	0.8016	0.9916	0.9986	1.0009	1.0002	0.8906	0.8239	0.8862	0.8266
3.0	4.3150E 00	1.0024	0.8826	0.8324	0.8052	0.9910	0.9985	1.0009	1.0002	0.8924	0.8284	0.8885	0.8284
3.2	4.2601E 00	1.0014	0.8853	0.8357	0.8088	0.9903	0.9984	1.0008	1.0001	0.8941	0.8326	0.8906	0.8301
3.4	4.2107E 00	1.0003	0.8879	0.8391	0.8125	0.9897	0.9983	1.0008	1.0001	0.8958	0.8367	0.8926	0.8317
3.6	4.1657E 00	0.9992	0.8907	0.8425	0.8161	0.9890	0.9982	1.0008	1.0001	0.8974	0.8406	0.8946	0.8332
3.8	4.1246E 00	0.9981	0.8934	0.8458	0.8198	0.9884	0.9981	1.0007	1.0001	0.8990	0.8444	0.8964	0.8346
4.0	4.0866E 00	0.9970	0.8960	0.8491	0.8234	0.9878	0.9980	1.0007	1.0001	0.9005	0.8481	0.8982	0.8359
4.5	4.0026E 00	0.9942	0.9026	0.8573	0.8322	0.9861	0.9977	1.0006	1.0001	0.9042	0.8569	0.9023	0.8390
5.0	3.9306E 00	0.9914	0.9090	0.8652	0.8407	0.9845	0.9975	1.0006	1.0001	0.9077	0.8653	0.9061	0.8418
5.5	3.8674E 00	0.9886	0.9151	0.8727	0.8489	0.9829	0.9972	1.0005	1.0001	0.9110	0.8733	0.9097	0.8443
6.0	3.8109E 00	0.9857	0.9210	0.8800	0.8568	0.9813	0.9969	1.0005	1.0001	0.9141	0.8810	0.9130	0.8466
6.5	3.7596E 00	0.9829	0.9267	0.8870	0.8644	0.9797	0.9967	1.0004	1.0001	0.9171	0.8884	0.9161	0.8487
7.0	3.7125E 00	0.9800	0.9322	0.8938	0.8717	0.9781	0.9964	1.0004	1.0001	0.9199	0.8957	0.9191	0.8507
7.5	3.6690E 00	0.9772	0.9376	0.9004	0.8788	0.9766	0.9961	1.0004	1.0001	0.9227	0.9027	0.9219	0.8526
8.0	3.6284E 00	0.9743	0.9427	0.9067	0.8856	0.9750	0.9959	1.0004	1.0001	0.9253	0.9096	0.9246	0.8544
9.0	3.5543E 00	0.9687	0.9526	0.9189	0.8988	0.9718	0.9953	1.0003	1.0001	0.9304	0.9229	0.9298	0.8577
10.0	3.4879E 00	0.9630	0.9620	0.9305	0.9113	0.9687	0.9948	1.0003	1.0001	0.9351	0.9356	0.9347	0.8607
11.0	3.4273E 00	0.9574	0.9709	0.9415	0.9232	0.9656	0.9943	1.0003	1.0000	0.9397	0.9479	0.9393	0.8635
12.0	3.3716E 00	0.9518	0.9795	0.9521	0.9347	0.9625	0.9938	1.0002	1.0000	0.9440	0.9598	0.9437	0.8661
13.0	3.3198E 00	0.9462	0.9878	0.9623	0.9458	0.9595	0.9932	1.0002	1.0000	0.9482	0.9713	0.9479	0.8686
14.0	3.2713E 00	0.9407	0.9959	0.9723	0.9565	0.9565	0.9927	1.0002	1.0000	0.9522	0.9824	0.9520	0.8709
16.0	3.1824E 00	0.9298	1.0112	0.9913	0.9771	0.9505	0.9916	1.0002	1.0000	0.9600	1.0035	0.9597	0.8753
18.0	3.1021E 00	0.9191	1.0258	1.0094	0.9967	0.9447	0.9905	1.0002	1.0000	0.9673	1.0232	0.9671	0.8794
20.0	3.0288E 00	0.9086	1.0398	1.0268	1.0156	0.9391	0.9894	1.0001	1.0000	0.9744	1.0417	0.9742	0.8833
25.0	2.8680E 00	0.8832	1.0725	1.0679	1.0603	0.9256	0.9866	1.0001	1.0000	0.9910	1.0823	0.9908	0.8922
30.0	2.7310E 00	0.8591	1.1026	1.1063	1.1022	0.9131	0.9837	1.0001	1.0000	1.0064	1.1153	1.0063	0.9004
35.0	2.6113E 00	0.8364	1.1308	1.1426	1.1422	0.9017	0.9807	1.0001	1.0000	1.0208	1.1410	1.0208	0.9082
40.0	2.5050E 00	0.8149	1.1570	1.1773	1.1804	0.8914	0.9777	1.0001	1.0000	1.0345	1.1597	1.0344	0.9157
45.0	2.4095E 00	0.7947	1.1816	1.2104	1.2171	0.8821	0.9746	1.0001	1.0000	1.0473	1.1717	1.0473	0.9228
50.0	2.3232E 00	0.7757	1.2046	1.2421	1.2525	0.8739	0.9713	1.0001	1.0000	1.0595	1.1772	1.0594	0.9296

ELECTRON Z= 47 A= 109

p	F_0L_0	L_0	λ_2	λ_3	λ_4	μ_1	μ_2	Λ_1	Λ_2	ν_{12}	$\hat{\nu}_{12}$	η_{12}	$\hat{\eta}_{12}$
0.1	4.7564E 01	1.0138	9.4042	32.7676	70.6888	0.9996	1.0000	1.0031	1.0005	3.2559	0.6604	0.5769	0.5335
0.2	2.4037E 01	1.0137	2.9394	4.5421	5.6611	0.9995	1.0000	1.0029	1.0005	1.4062	0.6765	0.7194	0.6711
0.3	1.6306E 01	1.0136	1.7445	2.0739	2.2390	0.9994	0.9999	1.0028	1.0005	1.0800	0.6911	0.7676	0.7172
0.4	1.2533E 01	1.0135	1.3288	1.4166	1.4483	0.9992	0.9999	1.0027	1.0005	0.9719	0.7042	0.7924	0.7407
0.5	1.0351E 01	1.0134	1.1388	1.1509	1.1471	0.9989	0.9999	1.0026	1.0005	0.9251	0.7159	0.8080	0.7553
0.6	8.9610E 00	1.0132	1.0376	1.0184	1.0014	0.9987	0.9998	1.0025	1.0004	0.9017	0.7264	0.8189	0.7654
0.7	8.0155E 00	1.0129	0.9783	0.9437	0.9207	0.9984	0.9998	1.0024	1.0004	0.8889	0.7358	0.8271	0.7729
0.8	7.3404E 00	1.0127	0.9412	0.8982	0.8722	0.9981	0.9997	1.0023	1.0004	0.8817	0.7442	0.8337	0.7787
0.9	6.8396E 00	1.0124	0.9170	0.8690	0.8412	0.9978	0.9997	1.0021	1.0004	0.8775	0.7517	0.8390	0.7835
1.0	6.4562E 00	1.0121	0.9006	0.8496	0.8208	0.9974	0.9996	1.0020	1.0004	0.8751	0.7585	0.8436	0.7876
1.2	5.9126E 00	1.0114	0.8815	0.8274	0.7977	0.9968	0.9995	1.0019	1.0003	0.8733	0.7702	0.8509	0.7940
1.4	5.5487E 00	1.0106	0.8724	0.8171	0.7870	0.9961	0.9994	1.0017	1.0003	0.8736	0.7800	0.8568	0.7990
1.6	5.2887E 00	1.0097	0.8683	0.8126	0.7826	0.9954	0.9993	1.0015	1.0003	0.8747	0.7885	0.8616	0.8032
1.8	5.0934E 00	1.0088	0.8670	0.8115	0.7816	0.9947	0.9992	1.0014	1.0003	0.8763	0.7958	0.8658	0.8067
2.0	4.9405E 00	1.0078	0.8673	0.8122	0.7826	0.9940	0.9990	1.0013	1.0002	0.8782	0.8024	0.8695	0.8097
2.2	4.8169E 00	1.0068	0.8686	0.8140	0.7847	0.9933	0.9989	1.0012	1.0002	0.8801	0.8084	0.8728	0.8124
2.4	4.7143E 00	1.0058	0.8706	0.8166	0.7875	0.9927	0.9988	1.0011	1.0002	0.8820	0.8138	0.8758	0.8149
2.6	4.6272E 00	1.0047	0.8729	0.8195	0.7908	0.9920	0.9987	1.0010	1.0002	0.8839	0.8189	0.8785	0.8171
2.8	4.5518E 00	1.0036	0.8754	0.8228	0.7943	0.9913	0.9986	1.0010	1.0002	0.8857	0.8237	0.8811	0.8191
3.0	4.4856E 00	1.0025	0.8781	0.8261	0.7980	0.9906	0.9985	1.0009	1.0002	0.8876	0.8282	0.8835	0.8210
3.2	4.4266E 00	1.0014	0.8809	0.8296	0.8018	0.9899	0.9984	1.0009	1.0002	0.8894	0.8326	0.8857	0.8227
3.4	4.3735E 00	1.0003	0.8837	0.8331	0.8056	0.9893	0.9983	1.0008	1.0001	0.8911	0.8367	0.8878	0.8244
3.6	4.3253E 00	0.9991	0.8865	0.8366	0.8093	0.9886	0.9981	1.0008	1.0001	0.8928	0.8407	0.8899	0.8259
3.8	4.2810E 00	0.9980	0.8893	0.8401	0.8131	0.9879	0.9980	1.0007	1.0001	0.8945	0.8445	0.8918	0.8274
4.0	4.2402E 00	0.9968	0.8921	0.8435	0.8169	0.9872	0.9979	1.0007	1.0001	0.8961	0.8483	0.8936	0.8288
4.5	4.1499E 00	0.9939	0.8989	0.8520	0.8260	0.9856	0.9977	1.0006	1.0001	0.8999	0.8572	0.8979	0.8320
5.0	4.0724E 00	0.9910	0.9056	0.8602	0.8348	0.9839	0.9974	1.0006	1.0001	0.9035	0.8657	0.9019	0.8349
5.5	4.0044E 00	0.9881	0.9120	0.8680	0.8433	0.9822	0.9971	1.0005	1.0001	0.9070	0.8737	0.9056	0.8375
6.0	3.9435E 00	0.9851	0.9181	0.8756	0.8515	0.9805	0.9968	1.0005	1.0001	0.9102	0.8815	0.9090	0.8399
6.5	3.8883E 00	0.9822	0.9241	0.8829	0.8594	0.9789	0.9966	1.0005	1.0001	0.9133	0.8890	0.9123	0.8421
7.0	3.8377E 00	0.9792	0.9298	0.8899	0.8670	0.9772	0.9963	1.0004	1.0001	0.9163	0.8963	0.9154	0.8442
7.5	3.7909E 00	0.9763	0.9353	0.8967	0.8743	0.9756	0.9960	1.0004	1.0001	0.9191	0.9034	0.9183	0.8462
8.0	3.7472E 00	0.9733	0.9407	0.9034	0.8815	0.9739	0.9957	1.0004	1.0001	0.9219	0.9103	0.9212	0.8480
9.0	3.6676E 00	0.9674	0.9510	0.9160	0.8952	0.9707	0.9952	1.0003	1.0001	0.9272	0.9236	0.9266	0.8515
10.0	3.5962E 00	0.9616	0.9608	0.9280	0.9081	0.9674	0.9946	1.0003	1.0001	0.9321	0.9364	0.9316	0.8546
11.0	3.5312E 00	0.9557	0.9702	0.9395	0.9206	0.9642	0.9941	1.0003	1.0000	0.9369	0.9487	0.9364	0.8575
12.0	3.4714E 00	0.9500	0.9791	0.9506	0.9325	0.9610	0.9935	1.0003	1.0000	0.9414	0.9605	0.9410	0.8603
13.0	3.4158E 00	0.9442	0.9878	0.9613	0.9440	0.9579	0.9930	1.0002	1.0000	0.9457	0.9719	0.9454	0.8629
14.0	3.3639E 00	0.9385	0.9961	0.9716	0.9552	0.9547	0.9924	1.0002	1.0000	0.9500	0.9830	0.9497	0.8653
16.0	3.2688E 00	0.9272	1.0122	0.9914	0.9766	0.9486	0.9913	1.0002	1.0000	0.9580	1.0039	0.9578	0.8700
18.0	3.1830E 00	0.9162	1.0274	1.0103	0.9971	0.9426	0.9902	1.0002	1.0000	0.9657	1.0234	0.9655	0.8743
20.0	3.1048E 00	0.9053	1.0420	1.0284	1.0167	0.9367	0.9890	1.0001	1.0000	0.9731	1.0416	0.9729	0.8783
25.0	2.9337E 00	0.8792	1.0760	1.0712	1.0633	0.9228	0.9861	1.0001	1.0000	0.9904	1.0813	0.9902	0.8877
30.0	2.7882E 00	0.8544	1.1075	1.1113	1.1070	0.9100	0.9831	1.0001	1.0000	1.0064	1.1133	1.0063	0.8964
35.0	2.6616E 00	0.8311	1.1368	1.1492	1.1486	0.8984	0.9801	1.0001	1.0000	1.0215	1.1377	1.0214	0.9047
40.0	2.5494E 00	0.8091	1.1641	1.1853	1.1885	0.8878	0.9769	1.0001	1.0000	1.0357	1.1551	1.0356	0.9126
45.0	2.4490E 00	0.7885	1.1896	1.2198	1.2267	0.8784	0.9737	1.0001	1.0000	1.0491	1.1655	1.0490	0.9201
50.0	2.3584E 00	0.7691	1.2134	1.2527	1.2636	0.8701	0.9703	1.0001	1.0000	1.0617	1.1695	1.0617	0.9273

ELECTRON Z= 48 A= 111

p	F_0L_0	L_0	λ_2	λ_3	λ_4	μ_1	μ_2	Λ_1	Λ_2	ν_{12}	$\hat{\nu}_{12}$	η_{12}	$\hat{\eta}_{12}$
0.1	5.0116E 01	1.0143	9.6585	34.5616	76.1661	0.9996	1.0000	1.0032	1.0005	3.3369	0.6567	0.5543	0.5100
0.2	2.5326E 01	1.0142	2.9975	4.6936	5.8966	0.9995	1.0000	1.0030	1.0005	1.4193	0.6731	0.7055	0.6556
0.3	1.7177E 01	1.0141	1.7662	2.1123	2.2871	0.9993	0.9999	1.0029	1.0005	1.0813	0.6880	0.7565	0.7044
0.4	1.3198E 01	1.0140	1.3379	1.4295	1.4627	0.9991	0.9999	1.0028	1.0005	0.9695	0.7014	0.7828	0.7292
0.5	1.0893E 01	1.0139	1.1420	1.1544	1.1504	0.9989	0.9999	1.0027	1.0005	0.9211	0.7134	0.7992	0.7446
0.6	9.4226E 00	1.0137	1.0377	1.0176	0.9999	0.9986	0.9998	1.0026	1.0005	0.8969	0.7241	0.8107	0.7552
0.7	8.4214E 00	1.0134	0.9765	0.9406	0.9167	0.9983	0.9998	1.0024	1.0004	0.8837	0.7336	0.8194	0.7631
0.8	7.7058E 00	1.0132	0.9383	0.8937	0.8666	0.9980	0.9997	1.0023	1.0004	0.8762	0.7422	0.8262	0.7693
0.9	7.1744E 00	1.0129	0.9132	0.8636	0.8348	0.9977	0.9997	1.0022	1.0004	0.8719	0.7499	0.8319	0.7743
1.0	6.7673E 00	1.0125	0.8964	0.8436	0.8138	0.9973	0.9996	1.0021	1.0004	0.8695	0.7568	0.8366	0.7785
1.2	6.1896E 00	1.0118	0.8767	0.8207	0.7900	0.9966	0.9995	1.0019	1.0003	0.8677	0.7688	0.8443	0.7852
1.4	5.8022E 00	1.0110	0.8673	0.8100	0.7790	0.9959	0.9994	1.0017	1.0003	0.8680	0.7788	0.8505	0.7905
1.6	5.5251E 00	1.0101	0.8631	0.8055	0.7745	0.9952	0.9992	1.0016	1.0003	0.8692	0.7874	0.8555	0.7948
1.8	5.3167E 00	1.0091	0.8617	0.8043	0.7735	0.9945	0.9991	1.0015	1.0003	0.8709	0.7949	0.8599	0.7984
2.0	5.1534E 00	1.0081	0.8621	0.8051	0.7745	0.9938	0.9990	1.0013	1.0002	0.8727	0.8016	0.8637	0.8016
2.2	5.0213E 00	1.0071	0.8635	0.8070	0.7767	0.9931	0.9989	1.0012	1.0002	0.8747	0.8077	0.8672	0.8044
2.4	4.9115E 00	1.0060	0.8655	0.8096	0.7797	0.9924	0.9988	1.0012	1.0002	0.8767	0.8133	0.8703	0.8069
2.6	4.8181E 00	1.0049	0.8679	0.8127	0.7830	0.9917	0.9987	1.0011	1.0002	0.8787	0.8184	0.8731	0.8092
2.8	4.7373E 00	1.0038	0.8705	0.8161	0.7867	0.9910	0.9985	1.0010	1.0002	0.8806	0.8233	0.8758	0.8114
3.0	4.6663E 00	1.0026	0.8733	0.8195	0.7905	0.9902	0.9984	1.0010	1.0002	0.8826	0.8279	0.8783	0.8133
3.2	4.6030E 00	1.0015	0.8762	0.8231	0.7944	0.9895	0.9983	1.0009	1.0002	0.8844	0.8323	0.8806	0.8151
3.4	4.5460E 00	1.0003	0.8791	0.8267	0.7983	0.9888	0.9982	1.0009	1.0001	0.8862	0.8365	0.8828	0.8168
3.6	4.4942E 00	0.9991	0.8820	0.8304	0.8022	0.9881	0.9981	1.0008	1.0001	0.8880	0.8405	0.8849	0.8184
3.8	4.4466E 00	0.9979	0.8850	0.8340	0.8061	0.9874	0.9980	1.0008	1.0001	0.8897	0.8444	0.8869	0.8200
4.0	4.4027E 00	0.9968	0.8879	0.8376	0.8100	0.9867	0.9979	1.0007	1.0001	0.8914	0.8482	0.8888	0.8214
4.5	4.3056E 00	0.9938	0.8950	0.8463	0.8195	0.9850	0.9976	1.0007	1.0001	0.8954	0.8573	0.8933	0.8247
5.0	4.2223E 00	0.9907	0.9019	0.8548	0.8286	0.9833	0.9973	1.0006	1.0001	0.8992	0.8658	0.8974	0.8277
5.5	4.1491E 00	0.9877	0.9085	0.8630	0.8374	0.9815	0.9970	1.0006	1.0001	0.9027	0.8739	0.9013	0.8304
6.0	4.0837E 00	0.9846	0.9149	0.8708	0.8459	0.9798	0.9967	1.0005	1.0001	0.9061	0.8817	0.9048	0.8329
6.5	4.0243E 00	0.9816	0.9211	0.8784	0.8540	0.9781	0.9965	1.0005	1.0001	0.9093	0.8893	0.9082	0.8353
7.0	3.9699E 00	0.9785	0.9271	0.8857	0.8619	0.9764	0.9962	1.0004	1.0001	0.9124	0.8966	0.9114	0.8374
7.5	3.9195E 00	0.9755	0.9329	0.8928	0.8695	0.9747	0.9959	1.0004	1.0001	0.9154	0.9037	0.9145	0.8395
8.0	3.8726E 00	0.9724	0.9384	0.8996	0.8769	0.9730	0.9956	1.0004	1.0001	0.9182	0.9106	0.9175	0.8414
9.0	3.7871E 00	0.9664	0.9492	0.9128	0.8911	0.9696	0.9951	1.0003	1.0001	0.9237	0.9240	0.9231	0.8450
10.0	3.7105E 00	0.9603	0.9593	0.9253	0.9046	0.9662	0.9945	1.0003	1.0001	0.9289	0.9367	0.9284	0.8483
11.0	3.6408E 00	0.9543	0.9691	0.9372	0.9174	0.9629	0.9939	1.0003	1.0000	0.9338	0.9490	0.9334	0.8514
12.0	3.5767E 00	0.9483	0.9784	0.9487	0.9298	0.9596	0.9933	1.0003	1.0000	0.9385	0.9608	0.9381	0.8542
13.0	3.5172E 00	0.9424	0.9874	0.9598	0.9418	0.9563	0.9928	1.0002	1.0000	0.9431	0.9722	0.9427	0.8569
14.0	3.4616E 00	0.9365	0.9961	0.9705	0.9534	0.9531	0.9922	1.0002	1.0000	0.9474	0.9831	0.9471	0.8595
16.0	3.3599E 00	0.9249	1.0128	0.9911	0.9757	0.9467	0.9910	1.0002	1.0000	0.9558	1.0039	0.9556	0.8643
18.0	3.2685E 00	0.9135	1.0286	1.0107	0.9969	0.9406	0.9899	1.0002	1.0000	0.9638	1.0232	0.9636	0.8689
20.0	3.1851E 00	0.9023	1.0438	1.0296	1.0173	0.9345	0.9887	1.0002	1.0000	0.9715	1.0410	0.9713	0.8731
25.0	3.0032E 00	0.8755	1.0792	1.0741	1.0657	0.9203	0.9857	1.0001	1.0000	0.9895	1.0799	0.9893	0.8830
30.0	2.8491E 00	0.8501	1.1119	1.1157	1.1112	0.9072	0.9826	1.0001	1.0000	1.0062	1.1108	1.0061	0.8922
35.0	2.7152E 00	0.8262	1.1423	1.1551	1.1544	0.8953	0.9794	1.0001	1.0000	1.0218	1.1341	1.0218	0.9009
40.0	2.5970E 00	0.8038	1.1707	1.1926	1.1958	0.8845	0.9762	1.0001	1.0000	1.0366	1.1501	1.0365	0.9092
45.0	2.4915E 00	0.7827	1.1971	1.2284	1.2356	0.8750	0.9728	1.0001	1.0000	1.0505	1.1592	1.0505	0.9172
50.0	2.3965E 00	0.7630	1.2217	1.2626	1.2738	0.8666	0.9694	1.0001	1.0000	1.0636	1.1618	1.0635	0.9248

Instr. see p. 22

ELECTRON Z= 49 A= 113

p	F_0L_0	L_0	λ_2	λ_3	λ_4	μ_1	μ_2	Λ_1	Λ_2	ν_{12}	$\hat{\nu}_{12}$	η_{12}	$\hat{\eta}_{12}$
0.1	5.2812E 01	1.0148	9.9106	36.4052	81.9480	0.9995	1.0000	1.0033	1.0005	3.4183	0.6529	0.5311	0.4857
0.2	2.6686E 01	1.0147	3.0550	4.8473	6.1385	0.9994	1.0000	1.0031	1.0005	1.4322	0.6696	0.6911	0.6398
0.3	1.8097E 01	1.0147	1.7877	2.1507	2.3357	0.9993	0.9999	1.0030	1.0005	1.0826	0.6848	0.7452	0.6913
0.4	1.3899E 01	1.0145	1.3467	1.4422	1.4770	0.9991	0.9999	1.0029	1.0005	0.9669	0.6985	0.7729	0.7175
0.5	1.1465E 01	1.0144	1.1451	1.1578	1.1536	0.9988	0.9999	1.0028	1.0005	0.9169	0.7107	0.7902	0.7337
0.6	9.9104E 00	1.0142	1.0376	1.0166	0.9982	0.9985	0.9998	1.0027	1.0005	0.8919	0.7216	0.8024	0.7448
0.7	8.8502E 00	1.0139	0.9746	0.9372	0.9125	0.9982	0.9998	1.0025	1.0004	0.8784	0.7314	0.8115	0.7531
0.8	8.0917E 00	1.0137	0.9352	0.8889	0.8610	0.9979	0.9997	1.0024	1.0004	0.8707	0.7402	0.8186	0.7596
0.9	7.5280E 00	1.0133	0.9094	0.8580	0.8282	0.9976	0.9996	1.0023	1.0004	0.8662	0.7480	0.8245	0.7648
1.0	7.0957E 00	1.0130	0.8920	0.8374	0.8067	0.9972	0.9996	1.0022	1.0004	0.8637	0.7551	0.8295	0.7692
1.2	6.4816E 00	1.0122	0.8718	0.8139	0.7822	0.9965	0.9995	1.0020	1.0003	0.8619	0.7673	0.8376	0.7762
1.4	6.0693E 00	1.0114	0.8621	0.8029	0.7710	0.9958	0.9993	1.0018	1.0003	0.8622	0.7776	0.8440	0.7817
1.6	5.7741E 00	1.0104	0.8578	0.7983	0.7664	0.9950	0.9992	1.0016	1.0003	0.8635	0.7863	0.8493	0.7862
1.8	5.5517E 00	1.0094	0.8564	0.7971	0.7653	0.9943	0.9991	1.0015	1.0003	0.8652	0.7940	0.8538	0.7900
2.0	5.3773E 00	1.0084	0.8568	0.7979	0.7664	0.9935	0.9990	1.0014	1.0002	0.8672	0.8008	0.8578	0.7933
2.2	5.2360E 00	1.0073	0.8583	0.7999	0.7687	0.9928	0.9989	1.0013	1.0002	0.8693	0.8070	0.8614	0.7962
2.4	5.1185E 00	1.0062	0.8603	0.8026	0.7717	0.9921	0.9987	1.0012	1.0002	0.8713	0.8127	0.8646	0.7988
2.6	5.0185E 00	1.0051	0.8628	0.8058	0.7752	0.9913	0.9986	1.0011	1.0002	0.8734	0.8179	0.8676	0.8012
2.8	4.9319E 00	1.0039	0.8656	0.8093	0.7790	0.9906	0.9985	1.0011	1.0002	0.8754	0.8229	0.8704	0.8034
3.0	4.8557E 00	1.0027	0.8685	0.8129	0.7829	0.9899	0.9984	1.0010	1.0002	0.8774	0.8275	0.8730	0.8055
3.2	4.7878E 00	1.0015	0.8715	0.8166	0.7869	0.9891	0.9983	1.0009	1.0002	0.8794	0.8320	0.8754	0.8074
3.4	4.7267E 00	1.0003	0.8745	0.8203	0.7910	0.9884	0.9982	1.0009	1.0002	0.8812	0.8362	0.8777	0.8091
3.6	4.6710E 00	0.9991	0.8775	0.8241	0.7950	0.9877	0.9980	1.0008	1.0001	0.8831	0.8403	0.8799	0.8108
3.8	4.6159E 00	0.9979	0.8806	0.8278	0.7991	0.9870	0.9979	1.0008	1.0001	0.8849	0.8443	0.8819	0.8124
4.0	4.5727E 00	0.9967	0.8836	0.8315	0.8031	0.9862	0.9978	1.0008	1.0001	0.8866	0.8481	0.8839	0.8139
4.5	4.4684E 00	0.9936	0.8910	0.8406	0.8129	0.9844	0.9975	1.0007	1.0001	0.8907	0.8573	0.8886	0.8173
5.0	4.3788E 00	0.9904	0.8982	0.8494	0.8223	0.9826	0.9972	1.0006	1.0001	0.8947	0.8659	0.8929	0.8204
5.5	4.3001E 00	0.9873	0.9051	0.8578	0.8314	0.9808	0.9969	1.0006	1.0001	0.8984	0.8741	0.8968	0.8232
6.0	4.2297E 00	0.9841	0.9117	0.8660	0.8402	0.9790	0.9966	1.0005	1.0001	0.9019	0.8820	0.9006	0.8258
6.5	4.1659E 00	0.9810	0.9181	0.8738	0.8486	0.9773	0.9964	1.0005	1.0001	0.9052	0.8895	0.9041	0.8283
7.0	4.1074E 00	0.9778	0.9243	0.8814	0.8568	0.9755	0.9961	1.0005	1.0001	0.9084	0.8969	0.9074	0.8305
7.5	4.0532E 00	0.9747	0.9303	0.8887	0.8647	0.9737	0.9958	1.0004	1.0001	0.9115	0.9040	0.9106	0.8327
8.0	4.0028E 00	0.9715	0.9361	0.8958	0.8723	0.9720	0.9955	1.0004	1.0001	0.9145	0.9110	0.9137	0.8347
9.0	3.9110E 00	0.9652	0.9473	0.9095	0.8870	0.9684	0.9949	1.0004	1.0001	0.9202	0.9243	0.9195	0.8384
10.0	3.8288E 00	0.9590	0.9579	0.9224	0.9010	0.9650	0.9943	1.0003	1.0001	0.9256	0.9371	0.9250	0.8418
11.0	3.7541E 00	0.9528	0.9680	0.9348	0.9143	0.9615	0.9937	1.0003	1.0000	0.9307	0.9494	0.9302	0.8450
12.0	3.6855E 00	0.9466	0.9777	0.9468	0.9272	0.9581	0.9932	1.0003	1.0000	0.9356	0.9611	0.9352	0.8480
13.0	3.6218E 00	0.9405	0.9871	0.9583	0.9396	0.9547	0.9926	1.0002	1.0000	0.9403	0.9724	0.9399	0.8509
14.0	3.5624E 00	0.9344	0.9961	0.9694	0.9516	0.9514	0.9920	1.0002	1.0000	0.9448	0.9833	0.9445	0.8536
16.0	3.4538E 00	0.9225	1.0135	0.9908	0.9747	0.9449	0.9908	1.0002	1.0000	0.9536	1.0039	0.9533	0.8586
18.0	3.3563E 00	0.9107	1.0299	1.0112	0.9968	0.9385	0.9896	1.0002	1.0000	0.9619	1.0229	0.9617	0.8634
20.0	3.2675E 00	0.8993	1.0457	1.0308	1.0180	0.9323	0.9884	1.0002	1.0000	0.9698	1.0405	0.9697	0.8678
25.0	3.0742E 00	0.8717	1.0825	1.0770	1.0682	0.9177	0.9853	1.0001	1.0000	0.9885	1.0785	0.9884	0.8783
30.0	2.9110E 00	0.8457	1.1165	1.1203	1.1154	0.9042	0.9821	1.0001	1.0000	1.0059	1.1084	1.0058	0.8880
35.0	2.7697E 00	0.8213	1.1481	1.1612	1.1604	0.8921	0.9788	1.0001	1.0000	1.0222	1.1305	1.0221	0.8971
40.0	2.6453E 00	0.7984	1.1774	1.2002	1.2033	0.8812	0.9754	1.0001	1.0000	1.0375	1.1452	1.0375	0.9059
45.0	2.5344E 00	0.7770	1.2048	1.2374	1.2446	0.8715	0.9720	1.0001	1.0000	1.0519	1.1531	1.0519	0.9143
50.0	2.4349E 00	0.7570	1.2303	1.2728	1.2843	0.8631	0.9684	1.0001	1.0000	1.0655	1.1543	1.0655	0.9223

Instr. see p. 22

ELECTRON Z= 50 A= 117

p	F_0L_0	L_0	λ_2	λ_3	λ_4	μ_1	μ_2	Λ_1	Λ_2	ν_{12}	$\hat{\nu}_{12}$	η_{12}	$\hat{\eta}_{12}$
0.1	5.5613E 01	1.0153	10.1649	38.3194	88.0996	0.9995	1.0000	1.0035	1.0006	3.5011	0.6492	0.5072	0.4610
0.2	2.8100E 01	1.0152	3.1132	5.0059	6.3909	0.9994	1.0000	1.0032	1.0006	1.4455	0.6662	0.6766	0.6237
0.3	1.9053E 01	1.0151	1.8097	2.1905	2.3863	0.9992	0.9999	1.0031	1.0005	1.0839	0.6817	0.7337	0.6781
0.4	1.4629E 01	1.0150	1.3560	1.4556	1.4921	0.9990	0.9999	1.0030	1.0005	0.9645	0.6957	0.7630	0.7057
0.5	1.2060E 01	1.0148	1.1485	1.1617	1.1574	0.9988	0.9999	1.0029	1.0005	0.9128	0.7082	0.7812	0.7227
0.6	1.0417E 01	1.0146	1.0379	1.0161	0.9970	0.9985	0.9998	1.0028	1.0005	0.8871	0.7193	0.7940	0.7344
0.7	9.2956E 00	1.0143	0.9730	0.9343	0.9087	0.9981	0.9998	1.0026	1.0005	0.8731	0.7293	0.8035	0.7431
0.8	8.4923E 00	1.0141	0.9325	0.8846	0.8557	0.9978	0.9997	1.0025	1.0004	0.8652	0.7383	0.8111	0.7499
0.9	7.8947E 00	1.0137	0.9059	0.8528	0.8221	0.9975	0.9996	1.0024	1.0004	0.8606	0.7463	0.8172	0.7554
1.0	7.4361E 00	1.0134	0.8880	0.8316	0.7999	0.9971	0.9996	1.0023	1.0004	0.8580	0.7535	0.8224	0.7599
1.2	6.7839E 00	1.0126	0.8672	0.8075	0.7749	0.9963	0.9995	1.0021	1.0004	0.8562	0.7660	0.8309	0.7673
1.4	6.3455E 00	1.0117	0.8572	0.7962	0.7633	0.9956	0.9993	1.0019	1.0003	0.8565	0.7765	0.8375	0.7730
1.6	6.0310E 00	1.0107	0.8528	0.7914	0.7586	0.9948	0.9992	1.0017	1.0003	0.8579	0.7854	0.8431	0.7776
1.8	5.7939E 00	1.0097	0.8514	0.7903	0.7576	0.9940	0.9991	1.0016	1.0003	0.8597	0.7932	0.8478	0.7816
2.0	5.6078E 00	1.0086	0.8518	0.7911	0.7587	0.9933	0.9989	1.0014	1.0003	0.8618	0.8002	0.8520	0.7850
2.2	5.4569E 00	1.0075	0.8534	0.7932	0.7611	0.9925	0.9988	1.0013	1.0002	0.8639	0.8065	0.8557	0.7880
2.4	5.3312E 00	1.0063	0.8555	0.7960	0.7642	0.9917	0.9987	1.0012	1.0002	0.8661	0.8123	0.8591	0.7908
2.6	5.2242E 00	1.0051	0.8581	0.7993	0.7678	0.9910	0.9986	1.0012	1.0002	0.8682	0.8177	0.8622	0.7933
2.8	5.1314E 00	1.0039	0.8610	0.8029	0.7717	0.9902	0.9985	1.0011	1.0002	0.8703	0.8227	0.8650	0.7955
3.0	5.0498E 00	1.0027	0.8640	0.8067	0.7758	0.9894	0.9983	1.0010	1.0002	0.8724	0.8275	0.8677	0.7976
3.2	4.9769E 00	1.0014	0.8671	0.8105	0.7799	0.9887	0.9982	1.0010	1.0002	0.8744	0.8320	0.8703	0.7996
3.4	4.9113E 00	1.0002	0.8703	0.8144	0.7841	0.9879	0.9981	1.0009	1.0002	0.8763	0.8363	0.8726	0.8014
3.6	4.8515E 00	0.9989	0.8734	0.8183	0.7883	0.9871	0.9980	1.0009	1.0002	0.8783	0.8405	0.8749	0.8032
3.8	4.7967E 00	0.9976	0.8766	0.8221	0.7925	0.9864	0.9978	1.0008	1.0001	0.8801	0.8445	0.8771	0.8048
4.0	4.7460E 00	0.9964	0.8797	0.8260	0.7966	0.9856	0.9977	1.0008	1.0001	0.8819	0.8484	0.8792	0.8064
4.5	4.6339E 00	0.9931	0.8874	0.8354	0.8068	0.9837	0.9974	1.0007	1.0001	0.8862	0.8577	0.8840	0.8099
5.0	4.5376E 00	0.9899	0.8949	0.8445	0.8166	0.9819	0.9971	1.0006	1.0001	0.8903	0.8664	0.8884	0.8132
5.5	4.4530E 00	0.9866	0.9021	0.8533	0.8260	0.9800	0.9968	1.0006	1.0001	0.8941	0.8747	0.8926	0.8161
6.0	4.3773E 00	0.9833	0.9090	0.8617	0.8351	0.9781	0.9965	1.0005	1.0001	0.8978	0.8827	0.8964	0.8188
6.5	4.3087E 00	0.9801	0.9157	0.8699	0.8438	0.9763	0.9962	1.0005	1.0001	0.9013	0.8904	0.9001	0.8213
7.0	4.2458E 00	0.9768	0.9221	0.8777	0.8523	0.9744	0.9959	1.0005	1.0001	0.9046	0.8978	0.9036	0.8237
7.5	4.1877E 00	0.9735	0.9284	0.8854	0.8605	0.9726	0.9956	1.0004	1.0001	0.9078	0.9050	0.9069	0.8259
8.0	4.1336E 00	0.9702	0.9344	0.8928	0.8684	0.9708	0.9953	1.0004	1.0001	0.9109	0.9120	0.9101	0.8280
9.0	4.0350E 00	0.9637	0.9460	0.9069	0.8837	0.9671	0.9947	1.0004	1.0001	0.9169	0.9254	0.9162	0.8319
10.0	3.9468E 00	0.9572	0.9570	0.9204	0.8982	0.9635	0.9941	1.0003	1.0001	0.9225	0.9382	0.9219	0.8354
11.0	3.8668E 00	0.9508	0.9676	0.9333	0.9121	0.9599	0.9935	1.0003	1.0001	0.9278	0.9505	0.9273	0.8388
12.0	3.7932E 00	0.9444	0.9777	0.9457	0.9254	0.9564	0.9929	1.0003	1.0000	0.9329	0.9623	0.9325	0.8419
13.0	3.7251E 00	0.9381	0.9875	0.9577	0.9383	0.9529	0.9923	1.0003	1.0000	0.9378	0.9736	0.9374	0.8449
14.0	3.6615E 00	0.9318	0.9969	0.9693	0.9508	0.9494	0.9917	1.0002	1.0000	0.9426	0.9844	0.9422	0.8477
16.0	3.5455E 00	0.9194	1.0150	0.9916	0.9749	0.9426	0.9905	1.0002	1.0000	0.9517	1.0049	0.9514	0.8530
18.0	3.4415E 00	0.9073	1.0322	1.0128	0.9978	0.9361	0.9892	1.0002	1.0000	0.9603	1.0237	0.9601	0.8580
20.0	3.3469E 00	0.8955	1.0486	1.0332	1.0199	0.9297	0.9880	1.0002	1.0000	0.9686	1.0410	0.9684	0.8627
25.0	3.1415E 00	0.8670	1.0870	1.0814	1.0723	0.9146	0.9847	1.0001	1.0000	0.9881	1.0782	0.9880	0.8737
30.0	2.9687E 00	0.8404	1.1224	1.1265	1.1215	0.9008	0.9814	1.0001	1.0000	1.0063	1.1069	1.0062	0.8840
35.0	2.8195E 00	0.8153	1.1552	1.1692	1.1683	0.8884	0.9780	1.0001	1.0000	1.0232	1.1277	1.0231	0.8937
40.0	2.6885E 00	0.7920	1.1858	1.2098	1.2131	0.8774	0.9745	1.0001	1.0000	1.0392	1.1410	1.0391	0.9029
45.0	2.5721E 00	0.7701	1.2142	1.2485	1.2561	0.8676	0.9709	1.0001	1.0000	1.0542	1.1472	1.0541	0.9118
50.0	2.4679E 00	0.7498	1.2405	1.2854	1.2975	0.8591	0.9672	1.0001	1.0000	1.0682	1.1468	1.0682	0.9202

ELECTRON Z= 51 A= 125

p	F_0L_0	L_0	λ_2	λ_3	λ_4	μ_1	μ_2	Λ_1	Λ_2	ν_{12}	$\hat{\nu}_{12}$	η_{12}	$\hat{\eta}_{12}$
0.1	5.8474E 01	1.0156	10.4259	40.3292	94.7006	0.9994	1.0000	1.0037	1.0006	3.5861	0.6457	0.4829	0.4358
0.2	2.9543E 01	1.0155	3.1737	5.1725	6.6585	0.9993	1.0000	1.0034	1.0006	1.4594	0.6631	0.6619	0.6076
0.3	2.0028E 01	1.0154	1.8329	2.2329	2.4404	0.9992	0.9999	1.0033	1.0006	1.0857	0.6789	0.7222	0.6649
0.4	1.5373E 01	1.0153	1.3663	1.4705	1.5091	0.9989	0.9999	1.0032	1.0005	0.9623	0.6932	0.7531	0.6940
0.5	1.2667E 01	1.0151	1.1528	1.1668	1.1625	0.9987	0.9998	1.0030	1.0005	0.9090	0.7059	0.7723	0.7119
0.6	1.0934E 01	1.0149	1.0390	1.0166	0.9970	0.9984	0.9998	1.0029	1.0005	0.8825	0.7173	0.7857	0.7241
0.7	9.7493E 00	1.0146	0.9722	0.9324	0.9060	0.9980	0.9997	1.0028	1.0005	0.8681	0.7275	0.7958	0.7332
0.8	8.9001E 00	1.0143	0.9305	0.8812	0.8515	0.9977	0.9997	1.0026	1.0005	0.8599	0.7367	0.8037	0.7403
0.9	8.2677E 00	1.0140	0.9031	0.8484	0.8169	0.9973	0.9996	1.0025	1.0004	0.8552	0.7449	0.8101	0.7460
1.0	7.7820E 00	1.0136	0.8847	0.8267	0.7942	0.9969	0.9996	1.0024	1.0004	0.8526	0.7523	0.8156	0.7508
1.2	7.0905E 00	1.0128	0.8633	0.8019	0.7684	0.9961	0.9994	1.0022	1.0004	0.8507	0.7651	0.8244	0.7584
1.4	6.6249E 00	1.0118	0.8530	0.7903	0.7566	0.9953	0.9993	1.0020	1.0003	0.8511	0.7758	0.8314	0.7644
1.6	6.2906E 00	1.0108	0.8485	0.7855	0.7518	0.9945	0.9992	1.0018	1.0003	0.8525	0.7850	0.8371	0.7692
1.8	6.0382E 00	1.0097	0.8471	0.7843	0.7508	0.9937	0.9990	1.0016	1.0003	0.8544	0.7930	0.8421	0.7733
2.0	5.8398E 00	1.0086	0.8476	0.7852	0.7519	0.9929	0.9989	1.0015	1.0003	0.8566	0.8001	0.8464	0.7769
2.2	5.6788E 00	1.0074	0.8492	0.7874	0.7544	0.9921	0.9988	1.0014	1.0002	0.8588	0.8066	0.8503	0.7800
2.4	5.5445E 00	1.0062	0.8515	0.7903	0.7577	0.9913	0.9986	1.0013	1.0002	0.8611	0.8125	0.8538	0.7829
2.6	5.4301E 00	1.0049	0.8542	0.7938	0.7614	0.9904	0.9985	1.0012	1.0002	0.8633	0.8180	0.8570	0.7854
2.8	5.3309E 00	1.0036	0.8572	0.7975	0.7655	0.9896	0.9984	1.0011	1.0002	0.8655	0.8231	0.8600	0.7878
3.0	5.2435E 00	1.0024	0.8603	0.8014	0.7697	0.9888	0.9983	1.0011	1.0002	0.8676	0.8280	0.8628	0.7900
3.2	5.1655E 00	1.0011	0.8636	0.8054	0.7740	0.9880	0.9981	1.0010	1.0002	0.8697	0.8327	0.8654	0.7920
3.4	5.0951E 00	0.9997	0.8668	0.8094	0.7783	0.9872	0.9980	1.0010	1.0002	0.8718	0.8371	0.8679	0.7939
3.6	5.0310E 00	0.9984	0.8701	0.8134	0.7827	0.9864	0.9979	1.0009	1.0002	0.8738	0.8414	0.8703	0.7957
3.8	4.9722E 00	0.9971	0.8734	0.8175	0.7870	0.9856	0.9978	1.0009	1.0001	0.8757	0.8455	0.8725	0.7974
4.0	4.9178E 00	0.9957	0.8767	0.8215	0.7913	0.9848	0.9976	1.0008	1.0001	0.8776	0.8494	0.8747	0.7990
4.5	4.7974E 00	0.9923	0.8847	0.8313	0.8019	0.9829	0.9973	1.0007	1.0001	0.8821	0.8589	0.8797	0.8027
5.0	4.6940E 00	0.9889	0.8925	0.8407	0.8120	0.9809	0.9970	1.0007	1.0001	0.8863	0.8678	0.8844	0.8061
5.5	4.6030E 00	0.9855	0.9000	0.8499	0.8218	0.9789	0.9967	1.0006	1.0001	0.8903	0.8763	0.8887	0.8091
6.0	4.5217E 00	0.9820	0.9072	0.8587	0.8313	0.9770	0.9964	1.0006	1.0001	0.8941	0.8844	0.8927	0.8119
6.5	4.4480E 00	0.9786	0.9142	0.8671	0.8404	0.9750	0.9961	1.0005	1.0001	0.8978	0.8922	0.8966	0.8146
7.0	4.3805E 00	0.9752	0.9210	0.8754	0.8492	0.9731	0.9957	1.0005	1.0001	0.9013	0.8997	0.9002	0.8170
7.5	4.3180E 00	0.9717	0.9275	0.8833	0.8577	0.9711	0.9954	1.0005	1.0001	0.9046	0.9070	0.9037	0.8193
8.0	4.2598E 00	0.9683	0.9338	0.8910	0.8660	0.9692	0.9951	1.0004	1.0001	0.9079	0.9141	0.9070	0.8215
9.0	4.1540E 00	0.9615	0.9460	0.9058	0.8819	0.9654	0.9945	1.0004	1.0001	0.9141	0.9277	0.9134	0.8256
10.0	4.0594E 00	0.9547	0.9575	0.9199	0.8970	0.9616	0.9939	1.0003	1.0001	0.9199	0.9407	0.9193	0.8293
11.0	3.9735E 00	0.9479	0.9685	0.9334	0.9115	0.9578	0.9932	1.0003	1.0001	0.9255	0.9531	0.9250	0.8328
12.0	3.8947E 00	0.9413	0.9791	0.9463	0.9255	0.9541	0.9926	1.0003	1.0000	0.9308	0.9649	0.9304	0.8361
13.0	3.8217E 00	0.9346	0.9894	0.9589	0.9390	0.9505	0.9919	1.0003	1.0000	0.9360	0.9762	0.9356	0.8392
14.0	3.7536E 00	0.9281	0.9993	0.9710	0.9521	0.9468	0.9913	1.0003	1.0000	0.9410	0.9871	0.9406	0.8422
16.0	3.6296E 00	0.9152	1.0182	0.9944	0.9772	0.9398	0.9900	1.0002	1.0000	0.9505	1.0074	0.9502	0.8478
18.0	3.5185E 00	0.9026	1.0362	1.0166	1.0013	0.9329	0.9887	1.0002	1.0000	0.9596	1.0262	0.9594	0.8530
20.0	3.4177E 00	0.8903	1.0535	1.0380	1.0245	0.9263	0.9874	1.0002	1.0000	0.9683	1.0433	0.9681	0.8580
25.0	3.1993E 00	0.8608	1.0938	1.0886	1.0794	0.9107	0.9840	1.0001	1.0000	0.9888	1.0795	0.9887	0.8697
30.0	3.0162E 00	0.8332	1.1308	1.1360	1.1310	0.8965	0.9806	1.0001	1.0000	1.0078	1.1069	1.0077	0.8806
35.0	2.8587E 00	0.8075	1.1651	1.1807	1.1802	0.8838	0.9770	1.0001	1.0000	1.0256	1.1260	1.0255	0.8909
40.0	2.7209E 00	0.7835	1.1970	1.2233	1.2272	0.8726	0.9733	1.0001	1.0000	1.0423	1.1373	1.0422	0.9008
45.0	2.5988E 00	0.7612	1.2265	1.2637	1.2723	0.8628	0.9695	1.0001	1.0000	1.0579	1.1413	1.0578	0.9102
50.0	2.4899E 00	0.7404	1.2538	1.3023	1.3156	0.8543	0.9656	1.0001	1.0000	1.0726	1.1385	1.0725	0.9191

Instr. see p. 22

ELECTRON Z= 52 A= 124

p	F_0L_0	L_0	λ_2	λ_3	λ_4	μ_1	μ_2	Λ_1	Λ_2	ν_{12}	$\hat{\nu}_{12}$	η_{12}	$\hat{\eta}_{12}$
0.1	6.1723E 01	1.0162	10.6636	42.2825	101.3620	0.9994	1.0000	1.0038	1.0006	3.6675	0.6415	0.4573	0.4095
0.2	3.1182E 01	1.0161	3.2271	5.3276	6.9135	0.9993	1.0000	1.0035	1.0006	1.4716	0.6592	0.6462	0.5904
0.3	2.1137E 01	1.0160	1.8522	2.2695	2.4878	0.9991	0.9999	1.0034	1.0006	1.0862	0.6753	0.7097	0.6508
0.4	1.6219E 01	1.0159	1.3736	1.4815	1.5216	0.9989	0.9999	1.0032	1.0006	0.9590	0.6898	0.7423	0.6813
0.5	1.3357E 01	1.0157	1.1546	1.1687	1.1641	0.9986	0.9998	1.0031	1.0005	0.9042	0.7028	0.7625	0.7001
0.6	1.1522E 01	1.0155	1.0379	1.0143	0.9939	0.9983	0.9998	1.0030	1.0005	0.8768	0.7144	0.7765	0.7129
0.7	1.0267E 01	1.0152	0.9694	0.9279	0.9006	0.9980	0.9997	1.0028	1.0005	0.8620	0.7248	0.7870	0.7224
0.8	9.3656E 00	1.0149	0.9265	0.8754	0.8447	0.9976	0.9997	1.0027	1.0005	0.8536	0.7341	0.7953	0.7298
0.9	8.6939E 00	1.0145	0.8985	0.8418	0.8093	0.9972	0.9996	1.0026	1.0004	0.8488	0.7425	0.8020	0.7358
1.0	8.1776E 00	1.0141	0.8796	0.8195	0.7860	0.9968	0.9995	1.0024	1.0004	0.8461	0.7501	0.8077	0.7408
1.2	7.4418E 00	1.0133	0.8576	0.7941	0.7596	0.9960	0.9994	1.0022	1.0004	0.8442	0.7631	0.8169	0.7487
1.4	6.9457E 00	1.0123	0.8470	0.7823	0.7476	0.9952	0.9993	1.0020	1.0003	0.8447	0.7740	0.8242	0.7549
1.6	6.5890E 00	1.0113	0.8424	0.7773	0.7426	0.9943	0.9991	1.0018	1.0003	0.8462	0.7833	0.8302	0.7599
1.8	6.3195E 00	1.0101	0.8410	0.7761	0.7416	0.9935	0.9990	1.0017	1.0003	0.8482	0.7915	0.8353	0.7642
2.0	6.1074E 00	1.0090	0.8415	0.7771	0.7428	0.9927	0.9989	1.0015	1.0003	0.8504	0.7987	0.8398	0.7679
2.2	5.9350E 00	1.0078	0.8432	0.7793	0.7454	0.9918	0.9987	1.0014	1.0002	0.8527	0.8053	0.8438	0.7711
2.4	5.7913E 00	1.0065	0.8455	0.7824	0.7487	0.9910	0.9986	1.0013	1.0002	0.8550	0.8113	0.8475	0.7741
2.6	5.6687E 00	1.0052	0.8484	0.7859	0.7526	0.9902	0.9985	1.0013	1.0002	0.8573	0.8169	0.8508	0.7767
2.8	5.5623E 00	1.0039	0.8514	0.7897	0.7568	0.9893	0.9984	1.0012	1.0002	0.8596	0.8221	0.8539	0.7792
3.0	5.4685E 00	1.0026	0.8547	0.7938	0.7611	0.9885	0.9982	1.0011	1.0002	0.8618	0.8270	0.8568	0.7814
3.2	5.3847E 00	1.0013	0.8580	0.7979	0.7655	0.9877	0.9981	1.0010	1.0002	0.8640	0.8317	0.8595	0.7835
3.4	5.3092E 00	0.9999	0.8614	0.8020	0.7700	0.9869	0.9980	1.0010	1.0002	0.8661	0.8362	0.8621	0.7855
3.6	5.2404E 00	0.9985	0.8648	0.8062	0.7745	0.9860	0.9978	1.0009	1.0002	0.8682	0.8405	0.8645	0.7874
3.8	5.1772E 00	0.9972	0.8682	0.8103	0.7789	0.9852	0.9977	1.0009	1.0002	0.8702	0.8447	0.8669	0.7891
4.0	5.1188E 00	0.9958	0.8716	0.8145	0.7834	0.9844	0.9976	1.0009	1.0001	0.8721	0.8487	0.8691	0.7908
4.5	4.9894E 00	0.9923	0.8799	0.8246	0.7942	0.9824	0.9973	1.0008	1.0001	0.8768	0.8582	0.8743	0.7946
5.0	4.8782E 00	0.9888	0.8880	0.8343	0.8047	0.9803	0.9969	1.0007	1.0001	0.8811	0.8672	0.8791	0.7981
5.5	4.7806E 00	0.9853	0.8957	0.8437	0.8147	0.9783	0.9966	1.0006	1.0001	0.8853	0.8757	0.8836	0.8012
6.0	4.6933E 00	0.9818	0.9032	0.8528	0.8245	0.9763	0.9963	1.0006	1.0001	0.8892	0.8838	0.8878	0.8042
6.5	4.6141E 00	0.9783	0.9104	0.8615	0.8338	0.9743	0.9960	1.0005	1.0001	0.8930	0.8917	0.8917	0.8069
7.0	4.5416E 00	0.9747	0.9174	0.8700	0.8429	0.9723	0.9957	1.0005	1.0001	0.8966	0.8992	0.8955	0.8094
7.5	4.4746E 00	0.9712	0.9241	0.8782	0.8517	0.9703	0.9953	1.0005	1.0001	0.9001	0.9065	0.8991	0.8118
8.0	4.4123E 00	0.9677	0.9307	0.8861	0.8602	0.9683	0.9950	1.0004	1.0001	0.9034	0.9136	0.9025	0.8141
9.0	4.2989E 00	0.9607	0.9432	0.9014	0.8766	0.9644	0.9944	1.0004	1.0001	0.9098	0.9272	0.9091	0.8183
10.0	4.1976E 00	0.9538	0.9551	0.9159	0.8922	0.9605	0.9937	1.0004	1.0001	0.9159	0.9401	0.9153	0.8222
11.0	4.1058E 00	0.9469	0.9666	0.9298	0.9071	0.9567	0.9931	1.0003	1.0001	0.9216	0.9524	0.9211	0.8258
12.0	4.0216E 00	0.9400	0.9775	0.9432	0.9215	0.9529	0.9924	1.0003	1.0001	0.9272	0.9641	0.9267	0.8292
13.0	3.9437E 00	0.9333	0.9881	0.9561	0.9354	0.9491	0.9918	1.0003	1.0000	0.9325	0.9754	0.9321	0.8324
14.0	3.8711E 00	0.9266	0.9983	0.9687	0.9489	0.9454	0.9911	1.0003	1.0000	0.9376	0.9861	0.9373	0.8355
16.0	3.7390E 00	0.9134	1.0179	0.9927	0.9748	0.9382	0.9898	1.0002	1.0000	0.9475	1.0062	0.9472	0.8414
18.0	3.6209E 00	0.9005	1.0365	1.0157	0.9996	0.9312	0.9885	1.0002	1.0000	0.9569	1.0247	0.9566	0.8468
20.0	3.5139E 00	0.8880	1.0543	1.0378	1.0235	0.9244	0.9871	1.0002	1.0000	0.9659	1.0415	0.9657	0.8521
25.0	3.2827E 00	0.8580	1.0959	1.0900	1.0801	0.9085	0.9837	1.0001	1.0000	0.9870	1.0769	0.9869	0.8642
30.0	3.0894E 00	0.8300	1.1342	1.1389	1.1334	0.8941	0.9802	1.0001	1.0000	1.0066	1.1035	1.0065	0.8756
35.0	2.9236E 00	0.8038	1.1696	1.1850	1.1840	0.8813	0.9765	1.0001	1.0000	1.0250	1.1217	1.0249	0.8864
40.0	2.7789E 00	0.7795	1.2024	1.2289	1.2325	0.8699	0.9728	1.0001	1.0000	1.0422	1.1321	1.0421	0.8967
45.0	2.6511E 00	0.7569	1.2328	1.2706	1.2790	0.8600	0.9689	1.0001	1.0000	1.0584	1.1353	1.0583	0.9066
50.0	2.5372E 00	0.7359	1.2609	1.3104	1.3237	0.8515	0.9649	1.0001	1.0000	1.0735	1.1317	1.0734	0.9159

ELECTRON Z= 53 A= 127

p	F_0L_0	L_0	λ_2	λ_3	λ_4	μ_1	μ_2	Λ_1	Λ_2	ν_{12}	$\hat{\nu}_{12}$	η_{12}	$\hat{\eta}_{12}$
0.1	6.5055E 01	1.0166	10.9073	44.3264	108.4802	0.9993	1.0000	1.0039	1.0006	3.7510	0.6375	0.4313	0.3828
0.2	3.2863E 01	1.0166	3.2825	5.4902	7.1833	0.9992	1.0000	1.0036	1.0006	1.4845	0.6555	0.6303	0.5732
0.3	2.2274E 01	1.0165	1.8728	2.3086	2.5387	0.9991	0.9999	1.0035	1.0006	1.0870	0.6719	0.6973	0.6367
0.4	1.7086E 01	1.0163	1.3820	1.4939	1.5359	0.9988	0.9999	1.0033	1.0006	0.9560	0.6867	0.7315	0.6687
0.5	1.4065E 01	1.0161	1.1573	1.1717	1.1669	0.9985	0.9998	1.0032	1.0005	0.8995	0.6999	0.7527	0.6884
0.6	1.2125E 01	1.0159	1.0375	1.0131	0.9919	0.9982	0.9998	1.0031	1.0005	0.8714	0.7118	0.7674	0.7018
0.7	1.0796E 01	1.0156	0.9673	0.9243	0.8961	0.9979	0.9997	1.0029	1.0005	0.8562	0.7224	0.7784	0.7118
0.8	9.8417E 00	1.0153	0.9233	0.8705	0.8388	0.9975	0.9997	1.0028	1.0005	0.8476	0.7319	0.7871	0.7195
0.9	9.1295E 00	1.0149	0.8945	0.8361	0.8026	0.9971	0.9996	1.0026	1.0005	0.8427	0.7405	0.7942	0.7257
1.0	8.5816E 00	1.0145	0.8751	0.8132	0.7787	0.9967	0.9995	1.0025	1.0004	0.8400	0.7482	0.8001	0.7309
1.2	7.7998E 00	1.0136	0.8525	0.7872	0.7518	0.9958	0.9994	1.0023	1.0004	0.8380	0.7615	0.8097	0.7392
1.4	7.2720E 00	1.0126	0.8417	0.7751	0.7394	0.9950	0.9993	1.0021	1.0004	0.8385	0.7726	0.8172	0.7456
1.6	6.8920E 00	1.0115	0.8370	0.7700	0.7344	0.9941	0.9991	1.0019	1.0003	0.8401	0.7821	0.8235	0.7508
1.8	6.6045E 00	1.0104	0.8356	0.7688	0.7334	0.9932	0.9990	1.0017	1.0003	0.8421	0.7904	0.8288	0.7552
2.0	6.3781E 00	1.0092	0.8361	0.7698	0.7347	0.9924	0.9988	1.0016	1.0003	0.8444	0.7978	0.8335	0.7590
2.2	6.1939E 00	1.0079	0.8379	0.7722	0.7373	0.9915	0.9987	1.0015	1.0003	0.8468	0.8045	0.8376	0.7624
2.4	6.0401E 00	1.0066	0.8403	0.7753	0.7408	0.9906	0.9986	1.0014	1.0002	0.8493	0.8106	0.8414	0.7654
2.6	5.9089E 00	1.0053	0.8433	0.7790	0.7448	0.9898	0.9984	1.0013	1.0002	0.8517	0.8163	0.8449	0.7682
2.8	5.7949E 00	1.0039	0.8465	0.7830	0.7491	0.9889	0.9983	1.0012	1.0002	0.8540	0.8216	0.8481	0.7707
3.0	5.6944E 00	1.0025	0.8498	0.7871	0.7535	0.9881	0.9982	1.0011	1.0002	0.8563	0.8266	0.8511	0.7730
3.2	5.6046E 00	1.0012	0.8533	0.7913	0.7581	0.9872	0.9980	1.0011	1.0002	0.8586	0.8314	0.8539	0.7752
3.4	5.5235E 00	0.9998	0.8568	0.7956	0.7627	0.9863	0.9979	1.0010	1.0002	0.8608	0.8360	0.8566	0.7773
3.6	5.4497E 00	0.9983	0.8604	0.7999	0.7673	0.9855	0.9978	1.0010	1.0002	0.8629	0.8404	0.8591	0.7792
3.8	5.3819E 00	0.9969	0.8639	0.8042	0.7719	0.9846	0.9976	1.0009	1.0002	0.8650	0.8446	0.8616	0.7810
4.0	5.3192E 00	0.9955	0.8674	0.8085	0.7765	0.9838	0.9975	1.0009	1.0001	0.8670	0.8486	0.8639	0.7827
4.5	5.1803E 00	0.9919	0.8760	0.8189	0.7877	0.9817	0.9972	1.0008	1.0001	0.8718	0.8583	0.8693	0.7867
5.0	5.0609E 00	0.9883	0.8843	0.8290	0.7985	0.9796	0.9968	1.0007	1.0001	0.8763	0.8674	0.8743	0.7903
5.5	4.9561E 00	0.9846	0.8924	0.8387	0.8089	0.9775	0.9965	1.0007	1.0001	0.8806	0.8760	0.8789	0.7936
6.0	4.8623E 00	0.9810	0.9001	0.8481	0.8189	0.9754	0.9962	1.0006	1.0001	0.8847	0.8842	0.8832	0.7966
6.5	4.7773E 00	0.9773	0.9076	0.8572	0.8286	0.9733	0.9959	1.0006	1.0001	0.8886	0.8921	0.8873	0.7994
7.0	4.6996E 00	0.9737	0.9148	0.8659	0.8380	0.9712	0.9955	1.0005	1.0001	0.8924	0.8997	0.8912	0.8020
7.5	4.6277E 00	0.9701	0.9218	0.8744	0.8471	0.9692	0.9952	1.0005	1.0001	0.8960	0.9071	0.8949	0.8045
8.0	4.5608E 00	0.9664	0.9286	0.8826	0.8559	0.9671	0.9949	1.0005	1.0001	0.8995	0.9142	0.8985	0.8068
9.0	4.4393E 00	0.9592	0.9417	0.8984	0.8728	0.9631	0.9942	1.0004	1.0001	0.9061	0.9279	0.9053	0.8112
10.0	4.3308E 00	0.9520	0.9541	0.9135	0.8889	0.9590	0.9936	1.0004	1.0001	0.9124	0.9408	0.9117	0.8153
11.0	4.2326E 00	0.9449	0.9659	0.9279	0.9044	0.9551	0.9929	1.0003	1.0001	0.9184	0.9531	0.9178	0.8191
12.0	4.1426E 00	0.9379	0.9773	0.9418	0.9193	0.9511	0.9922	1.0003	1.0001	0.9241	0.9649	0.9236	0.8226
13.0	4.0594E 00	0.9309	0.9883	0.9552	0.9337	0.9473	0.9915	1.0003	1.0000	0.9296	0.9761	0.9292	0.8260
14.0	3.9819E 00	0.9240	0.9989	0.9682	0.9477	0.9435	0.9909	1.0003	1.0000	0.9350	0.9868	0.9346	0.8292
16.0	3.8412E 00	0.9104	1.0193	0.9932	0.9746	0.9360	0.9895	1.0002	1.0000	0.9452	1.0068	0.9449	0.8353
18.0	3.7155E 00	0.8972	1.0386	1.0171	1.0003	0.9288	0.9881	1.0002	1.0000	0.9550	1.0250	0.9547	0.8411
20.0	3.6018E 00	0.8843	1.0571	1.0400	1.0251	0.9218	0.9868	1.0002	1.0000	0.9643	1.0415	0.9641	0.8465
25.0	3.3568E 00	0.8536	1.1003	1.0942	1.0839	0.9055	0.9832	1.0001	1.0000	0.9863	1.0761	0.9862	0.8593
30.0	3.1527E 00	0.8249	1.1400	1.1449	1.1392	0.8908	0.9795	1.0001	1.0000	1.0067	1.1015	1.0066	0.8713
35.0	2.9781E 00	0.7982	1.1767	1.1929	1.1918	0.8778	0.9758	1.0001	1.0000	1.0257	1.1185	1.0256	0.8827
40.0	2.8262E 00	0.7735	1.2107	1.2384	1.2421	0.8663	0.9719	1.0001	1.0000	1.0436	1.1276	1.0435	0.8935
45.0	2.6924E 00	0.7505	1.2422	1.2817	1.2904	0.8563	0.9679	1.0001	1.0000	1.0603	1.1294	1.0603	0.9039
50.0	2.5735E 00	0.7293	1.2712	1.3229	1.3367	0.8478	0.9638	1.0001	1.0000	1.0760	1.1243	1.0759	0.9137

Instr. see p. 22

ELECTRON Z= 54 A= 131

p	F_0L_0	L_0	λ_2	λ_3	λ_4	μ_1	μ_2	Λ_1	Λ_2	ν_{12}	$\hat{\nu}_{12}$	η_{12}	$\hat{\eta}_{12}$
0.1	6.8550E 01	1.0170	11.1503	46.4268	115.9788	0.9993	1.0000	1.0041	1.0006	3.8352	0.6335	0.4045	0.3554
0.2	3.4627E 01	1.0170	3.3378	5.6560	7.4618	0.9992	0.9999	1.0038	1.0006	1.4974	0.6518	0.6142	0.5556
0.3	2.3465E 01	1.0169	1.8933	2.3483	2.5907	0.9990	0.9999	1.0036	1.0006	1.0878	0.6685	0.6846	0.6223
0.4	1.7995E 01	1.0167	1.3904	1.5066	1.5505	0.9988	0.9999	1.0035	1.0006	0.9530	0.6836	0.7206	0.6560
0.5	1.4807E 01	1.0165	1.1601	1.1750	1.1700	0.9985	0.9998	1.0033	1.0006	0.8949	0.6971	0.7428	0.6766
0.6	1.2757E 01	1.0163	1.0373	1.0120	0.9901	0.9981	0.9998	1.0032	1.0005	0.8660	0.7092	0.7583	0.6906
0.7	1.1351E 01	1.0160	0.9653	0.9209	0.8919	0.9977	0.9997	1.0030	1.0005	0.8504	0.7200	0.7698	0.7010
0.8	1.0340E 01	1.0156	0.9201	0.8657	0.8332	0.9973	0.9997	1.0029	1.0005	0.8416	0.7297	0.7788	0.7090
0.9	9.5856E 00	1.0153	0.8906	0.8305	0.7960	0.9969	0.9996	1.0027	1.0005	0.8365	0.7384	0.7862	0.7155
1.0	9.0043E 00	1.0148	0.8707	0.8071	0.7716	0.9965	0.9995	1.0026	1.0004	0.8337	0.7463	0.7924	0.7209
1.2	8.1740E 00	1.0139	0.8476	0.7804	0.7441	0.9956	0.9994	1.0024	1.0004	0.8318	0.7599	0.8024	0.7295
1.4	7.6127E 00	1.0129	0.8366	0.7681	0.7315	0.9947	0.9992	1.0022	1.0004	0.8323	0.7712	0.8102	0.7362
1.6	7.2080E 00	1.0117	0.8317	0.7629	0.7264	0.9938	0.9991	1.0020	1.0003	0.8339	0.7809	0.8167	0.7416
1.8	6.9015E 00	1.0105	0.8303	0.7617	0.7254	0.9929	0.9990	1.0018	1.0003	0.8361	0.7894	0.8222	0.7461
2.0	6.6598E 00	1.0093	0.8309	0.7628	0.7267	0.9920	0.9988	1.0017	1.0003	0.8385	0.7970	0.8271	0.7501
2.2	6.4631E 00	1.0080	0.8327	0.7652	0.7295	0.9911	0.9987	1.0015	1.0003	0.8410	0.8038	0.8314	0.7536
2.4	6.2986E 00	1.0066	0.8353	0.7685	0.7330	0.9902	0.9985	1.0014	1.0002	0.8435	0.8100	0.8353	0.7567
2.6	6.1582E 00	1.0052	0.8383	0.7723	0.7372	0.9893	0.9984	1.0013	1.0002	0.8460	0.8158	0.8389	0.7596
2.8	6.0362E 00	1.0038	0.8417	0.7764	0.7416	0.9884	0.9983	1.0013	1.0002	0.8484	0.8213	0.8423	0.7622
3.0	5.9285E 00	1.0024	0.8452	0.7806	0.7462	0.9875	0.9981	1.0012	1.0002	0.8508	0.8264	0.8454	0.7646
3.2	5.8323E 00	1.0010	0.8488	0.7850	0.7509	0.9867	0.9980	1.0011	1.0002	0.8531	0.8312	0.8483	0.7669
3.4	5.7453E 00	0.9995	0.8524	0.7895	0.7557	0.9858	0.9978	1.0011	1.0002	0.8554	0.8359	0.8511	0.7690
3.6	5.6662E 00	0.9980	0.8561	0.7939	0.7604	0.9849	0.9977	1.0010	1.0002	0.8576	0.8403	0.8537	0.7710
3.8	5.5934E 00	0.9966	0.8597	0.7983	0.7652	0.9840	0.9976	1.0010	1.0002	0.8598	0.8446	0.8562	0.7728
4.0	5.5261E 00	0.9951	0.8634	0.8027	0.7699	0.9831	0.9974	1.0009	1.0002	0.8619	0.8488	0.8586	0.7746
4.5	5.3770E 00	0.9914	0.8723	0.8135	0.7814	0.9809	0.9971	1.0008	1.0001	0.8669	0.8586	0.8642	0.7787
5.0	5.2489E 00	0.9876	0.8809	0.8240	0.7925	0.9787	0.9967	1.0007	1.0001	0.8716	0.8678	0.8694	0.7824
5.5	5.1363E 00	0.9838	0.8893	0.8340	0.8033	0.9765	0.9964	1.0007	1.0001	0.8760	0.8765	0.8742	0.7858
6.0	5.0357E 00	0.9800	0.8973	0.8437	0.8137	0.9744	0.9961	1.0006	1.0001	0.8803	0.8849	0.8787	0.7889
6.5	4.9445E 00	0.9763	0.9051	0.8531	0.8237	0.9722	0.9957	1.0006	1.0001	0.8843	0.8928	0.8829	0.7918
7.0	4.8611E 00	0.9725	0.9126	0.8622	0.8334	0.9701	0.9954	1.0005	1.0001	0.8882	0.9005	0.8870	0.7946
7.5	4.7840E 00	0.9687	0.9199	0.8709	0.8428	0.9679	0.9951	1.0005	1.0001	0.8919	0.9079	0.8909	0.7971
8.0	4.7123E 00	0.9650	0.9269	0.8795	0.8519	0.9658	0.9947	1.0005	1.0001	0.8955	0.9151	0.8946	0.7996
9.0	4.5821E 00	0.9575	0.9405	0.8959	0.8694	0.9616	0.9940	1.0004	1.0001	0.9024	0.9289	0.9016	0.8041
10.0	4.4660E 00	0.9501	0.9533	0.9115	0.8861	0.9574	0.9933	1.0004	1.0001	0.9090	0.9419	0.9083	0.8084
11.0	4.3609E 00	0.9427	0.9657	0.9264	0.9022	0.9533	0.9927	1.0003	1.0001	0.9152	0.9542	0.9146	0.8123
12.0	4.2647E 00	0.9354	0.9775	0.9408	0.9176	0.9492	0.9920	1.0003	1.0001	0.9211	0.9659	0.9206	0.8160
13.0	4.1759E 00	0.9282	0.9889	0.9548	0.9325	0.9452	0.9913	1.0003	1.0000	0.9269	0.9771	0.9265	0.8195
14.0	4.0933E 00	0.9211	1.0000	0.9683	0.9470	0.9413	0.9906	1.0003	1.0000	0.9324	0.9878	0.9321	0.8229
16.0	3.9433E 00	0.9071	1.0212	0.9942	0.9750	0.9336	0.9892	1.0002	1.0000	0.9431	1.0076	0.9428	0.8293
18.0	3.8096E 00	0.8935	1.0413	1.0190	1.0017	0.9262	0.9878	1.0002	1.0000	0.9533	1.0256	0.9530	0.8353
20.0	3.6889E 00	0.8802	1.0605	1.0429	1.0274	0.9190	0.9863	1.0002	1.0000	0.9630	1.0419	0.9628	0.8410
25.0	3.4293E 00	0.8487	1.1054	1.0992	1.0885	0.9023	0.9827	1.0001	1.0000	0.9858	1.0756	0.9857	0.8545
30.0	3.2139E 00	0.8193	1.1467	1.1520	1.1460	0.8873	0.9789	1.0001	1.0000	1.0070	1.0999	1.0069	0.8671
35.0	3.0303E 00	0.7921	1.1848	1.2018	1.2007	0.8740	0.9750	1.0001	1.0000	1.0268	1.1156	1.0267	0.8791
40.0	2.8710E 00	0.7669	1.2200	1.2491	1.2530	0.8624	0.9709	1.0001	1.0000	1.0454	1.1232	1.0453	0.8906
45.0	2.7310E 00	0.7436	1.2525	1.2941	1.3032	0.8524	0.9668	1.0001	1.0000	1.0627	1.1233	1.0627	0.9014
50.0	2.6070E 00	0.7221	1.2824	1.3368	1.3513	0.8439	0.9625	1.0001	1.0000	1.0789	1.1167	1.0788	0.9117

Instr. see p. 22

ELECTRON Z= 55 A= 135

p	F_0L_0	L_0	λ_2	λ_3	λ_4	μ_1	μ_2	Λ_1	Λ_2	ν_{12}	$\hat{\nu}_{12}$	η_{12}	$\hat{\eta}_{12}$
0.1	7.2246E 01	1.0174	11.3897	48.5689	123.8273	0.9992	1.0000	1.0043	1.0006	3.9196	0.6295	0.3771	0.3275
0.2	3.6491E 01	1.0174	3.3922	5.8232	7.7466	0.9991	0.9999	1.0039	1.0006	1.5102	0.6481	0.5976	0.5377
0.3	2.4726E 01	1.0172	1.9134	2.3878	2.6429	0.9989	0.9999	1.0038	1.0006	1.0885	0.6651	0.6717	0.6077
0.4	1.8957E 01	1.0171	1.3985	1.5191	1.5649	0.9987	0.9999	1.0036	1.0006	0.9498	0.6804	0.7094	0.6430
0.5	1.5591E 01	1.0169	1.1627	1.1780	1.1729	0.9984	0.9998	1.0035	1.0006	0.8901	0.6942	0.7327	0.6645
0.6	1.3425E 01	1.0166	1.0370	1.0107	0.9882	0.9980	0.9998	1.0033	1.0006	0.8604	0.7065	0.7489	0.6792
0.7	1.1938E 01	1.0163	0.9631	0.9174	0.8875	0.9976	0.9997	1.0031	1.0005	0.8444	0.7176	0.7609	0.6900
0.8	1.0868E 01	1.0160	0.9169	0.8609	0.8274	0.9972	0.9996	1.0030	1.0005	0.8354	0.7275	0.7704	0.6984
0.9	1.0067E 01	1.0156	0.8867	0.8248	0.7894	0.9968	0.9996	1.0028	1.0005	0.8302	0.7364	0.7781	0.7052
1.0	9.4508E 00	1.0152	0.8663	0.8008	0.7645	0.9963	0.9995	1.0027	1.0005	0.8274	0.7444	0.7845	0.7108
1.2	8.5689E 00	1.0142	0.8426	0.7736	0.7363	0.9954	0.9994	1.0025	1.0004	0.8254	0.7582	0.7949	0.7197
1.4	7.9719E 00	1.0131	0.8313	0.7610	0.7235	0.9945	0.9992	1.0022	1.0004	0.8260	0.7698	0.8031	0.7266
1.6	7.5410E 00	1.0119	0.8264	0.7557	0.7183	0.9936	0.9991	1.0020	1.0003	0.8277	0.7797	0.8098	0.7322
1.8	7.2142E 00	1.0107	0.8250	0.7545	0.7173	0.9926	0.9989	1.0019	1.0003	0.8299	0.7884	0.8155	0.7369
2.0	6.9562E 00	1.0094	0.8256	0.7557	0.7187	0.9917	0.9988	1.0017	1.0003	0.8324	0.7961	0.8206	0.7410
2.2	6.7460E 00	1.0080	0.8275	0.7582	0.7216	0.9907	0.9986	1.0016	1.0003	0.8350	0.8030	0.8251	0.7446
2.4	6.5702E 00	1.0066	0.8302	0.7616	0.7253	0.9898	0.9985	1.0015	1.0003	0.8376	0.8094	0.8291	0.7479
2.6	6.4199E 00	1.0052	0.8333	0.7655	0.7295	0.9889	0.9983	1.0014	1.0002	0.8402	0.8153	0.8329	0.7508
2.8	6.2892E 00	1.0037	0.8368	0.7697	0.7341	0.9879	0.9982	1.0013	1.0002	0.8427	0.8209	0.8363	0.7535
3.0	6.1739E 00	1.0022	0.8404	0.7741	0.7388	0.9870	0.9981	1.0012	1.0002	0.8452	0.8261	0.8395	0.7560
3.2	6.0708E 00	1.0007	0.8442	0.7787	0.7437	0.9861	0.9979	1.0012	1.0002	0.8476	0.8310	0.8426	0.7583
3.4	5.9776E 00	0.9992	0.8479	0.7832	0.7486	0.9852	0.9978	1.0011	1.0002	0.8499	0.8357	0.8455	0.7605
3.6	5.8926E 00	0.9977	0.8517	0.7878	0.7535	0.9842	0.9976	1.0010	1.0002	0.8522	0.8403	0.8482	0.7626
3.8	5.8146E 00	0.9962	0.8555	0.7924	0.7584	0.9833	0.9975	1.0010	1.0002	0.8544	0.8446	0.8508	0.7645
4.0	5.7424E 00	0.9946	0.8593	0.7969	0.7632	0.9824	0.9973	1.0010	1.0002	0.8566	0.8488	0.8533	0.7663
4.5	5.5824E 00	0.9908	0.8685	0.8081	0.7751	0.9801	0.9970	1.0009	1.0001	0.8618	0.8588	0.8591	0.7706
5.0	5.4449E 00	0.9869	0.8775	0.8189	0.7866	0.9778	0.9966	1.0008	1.0001	0.8667	0.8682	0.8644	0.7744
5.5	5.3241E 00	0.9830	0.8861	0.8293	0.7977	0.9756	0.9963	1.0007	1.0001	0.8713	0.8770	0.8694	0.7779
6.0	5.2161E 00	0.9790	0.8945	0.8393	0.8084	0.9733	0.9959	1.0007	1.0001	0.8757	0.8854	0.8740	0.7812
6.5	5.1184E 00	0.9751	0.9025	0.8490	0.8188	0.9711	0.9956	1.0006	1.0001	0.8799	0.8935	0.8785	0.7842
7.0	5.0289E 00	0.9712	0.9103	0.8584	0.8288	0.9688	0.9952	1.0006	1.0001	0.8839	0.9013	0.8827	0.7870
7.5	4.9463E 00	0.9673	0.9179	0.8675	0.8385	0.9666	0.9949	1.0005	1.0001	0.8878	0.9088	0.8867	0.7897
8.0	4.8694E 00	0.9634	0.9252	0.8763	0.8479	0.9644	0.9945	1.0005	1.0001	0.8915	0.9160	0.8905	0.7922
9.0	4.7300E 00	0.9557	0.9393	0.8933	0.8660	0.9600	0.9938	1.0004	1.0001	0.8987	0.9298	0.8979	0.7969
10.0	4.6058E 00	0.9480	0.9526	0.9095	0.8833	0.9557	0.9931	1.0004	1.0001	0.9055	0.9429	0.9048	0.8013
11.0	4.4934E 00	0.9404	0.9654	0.9250	0.8999	0.9515	0.9924	1.0004	1.0001	0.9119	0.9552	0.9113	0.8054
12.0	4.3907E 00	0.9329	0.9777	0.9399	0.9159	0.9473	0.9917	1.0003	1.0001	0.9181	0.9670	0.9176	0.8093
13.0	4.2959E 00	0.9255	0.9896	0.9544	0.9314	0.9431	0.9910	1.0003	1.0001	0.9241	0.9781	0.9236	0.8129
14.0	4.2079E 00	0.9181	1.0011	0.9684	0.9465	0.9391	0.9903	1.0003	1.0000	0.9299	0.9888	0.9295	0.8165
16.0	4.0481E 00	0.9037	1.0231	0.9953	0.9754	0.9311	0.9888	1.0002	1.0000	0.9409	1.0085	0.9406	0.8232
18.0	3.9061E 00	0.8897	1.0440	1.0211	1.0032	0.9235	0.9874	1.0002	1.0000	0.9515	1.0263	0.9512	0.8295
20.0	3.7779E 00	0.8761	1.0640	1.0459	1.0299	0.9161	0.9859	1.0002	1.0000	0.9616	1.0423	0.9614	0.8355
25.0	3.5031E 00	0.8437	1.1107	1.1044	1.0933	0.8990	0.9821	1.0002	1.0000	0.9854	1.0751	0.9852	0.8497
30.0	3.2759E 00	0.8137	1.1535	1.1592	1.1531	0.8837	0.9782	1.0001	1.0000	1.0074	1.0982	1.0073	0.8629
35.0	3.0829E 00	0.7860	1.1930	1.2110	1.2099	0.8703	0.9741	1.0001	1.0000	1.0280	1.1125	1.0279	0.8756
40.0	2.9160E 00	0.7603	1.2295	1.2601	1.2643	0.8586	0.9699	1.0001	1.0000	1.0472	1.1187	1.0471	0.8876
45.0	2.7698E 00	0.7367	1.2630	1.3068	1.3163	0.8485	0.9656	1.0001	1.0000	1.0652	1.1173	1.0651	0.8990
50.0	2.6405E 00	0.7149	1.2938	1.3511	1.3662	0.8401	0.9612	1.0001	1.0000	1.0819	1.1090	1.0818	0.9098

Instr. see p. 22

ELECTRON Z= 56 A= 135

p	F_0L_0	L_0	λ_2	λ_3	λ_4	μ_1	μ_2	Λ_1	Λ_2	ν_{12}	$\hat{\nu}_{12}$	η_{12}	$\hat{\eta}_{12}$
0.1	7.6296E 01	1.0179	11.6142	50.6872	131.8478	0.9992	1.0000	1.0044	1.0007	4.0019	0.6251	0.3488	0.2988
0.2	3.8534E 01	1.0179	3.4423	5.9840	8.0262	0.9991	0.9999	1.0041	1.0007	1.5220	0.6440	0.5803	0.5192
0.3	2.6106E 01	1.0177	1.9313	2.4241	2.6915	0.9989	0.9999	1.0039	1.0006	1.0884	0.6612	0.6581	0.5926
0.4	2.0011E 01	1.0176	1.4050	1.5294	1.5768	0.9986	0.9999	1.0037	1.0006	0.9459	0.6768	0.6976	0.6295
0.5	1.6451E 01	1.0174	1.1640	1.1793	1.1739	0.9983	0.9998	1.0035	1.0006	0.8847	0.6909	0.7221	0.6520
0.6	1.4158E 01	1.0171	1.0355	1.0080	0.9846	0.9979	0.9998	1.0034	1.0006	0.8542	0.7034	0.7390	0.6673
0.7	1.2582E 01	1.0168	0.9600	0.9126	0.8817	0.9975	0.9997	1.0032	1.0005	0.8378	0.7147	0.7515	0.6786
0.8	1.1446E 01	1.0164	0.9127	0.8548	0.8203	0.9971	0.9996	1.0031	1.0005	0.8286	0.7247	0.7614	0.6873
0.9	1.0596E 01	1.0160	0.8818	0.8179	0.7816	0.9967	0.9996	1.0029	1.0005	0.8233	0.7338	0.7694	0.6943
1.0	9.9408E 00	1.0156	0.8610	0.7935	0.7561	0.9962	0.9995	1.0028	1.0005	0.8204	0.7419	0.7761	0.7002
1.2	9.0025E 00	1.0146	0.8367	0.7657	0.7275	0.9953	0.9993	1.0025	1.0004	0.8185	0.7560	0.7869	0.7094
1.4	8.3663E 00	1.0135	0.8252	0.7529	0.7144	0.9943	0.9992	1.0023	1.0004	0.8191	0.7678	0.7954	0.7165
1.6	7.9067E 00	1.0123	0.8202	0.7475	0.7092	0.9934	0.9990	1.0021	1.0004	0.8209	0.7779	0.8024	0.7223
1.8	7.5576E 00	1.0110	0.8188	0.7463	0.7082	0.9924	0.9989	1.0019	1.0003	0.8232	0.7867	0.8083	0.7272
2.0	7.2818E 00	1.0096	0.8195	0.7475	0.7096	0.9914	0.9987	1.0018	1.0003	0.8258	0.7946	0.8135	0.7314
2.2	7.0569E 00	1.0082	0.8214	0.7501	0.7126	0.9904	0.9986	1.0016	1.0003	0.8285	0.8016	0.8182	0.7352
2.4	6.8687E 00	1.0068	0.8242	0.7536	0.7164	0.9895	0.9985	1.0015	1.0003	0.8311	0.8081	0.8224	0.7385
2.6	6.7077E 00	1.0053	0.8275	0.7576	0.7207	0.9885	0.9983	1.0014	1.0002	0.8338	0.8141	0.8262	0.7415
2.8	6.5676E 00	1.0038	0.8310	0.7620	0.7254	0.9876	0.9982	1.0013	1.0002	0.8364	0.8198	0.8298	0.7443
3.0	6.4439E 00	1.0023	0.8348	0.7665	0.7303	0.9866	0.9980	1.0013	1.0002	0.8390	0.8250	0.8331	0.7469
3.2	6.3332E 00	1.0008	0.8386	0.7712	0.7352	0.9857	0.9979	1.0012	1.0002	0.8415	0.8301	0.8363	0.7493
3.4	6.2333E 00	0.9992	0.8425	0.7759	0.7403	0.9847	0.9977	1.0011	1.0002	0.8439	0.8349	0.8392	0.7515
3.6	6.1421E 00	0.9977	0.8464	0.7806	0.7453	0.9838	0.9976	1.0011	1.0002	0.8462	0.8394	0.8420	0.7537
3.8	6.0583E 00	0.9961	0.8503	0.7853	0.7503	0.9828	0.9974	1.0010	1.0002	0.8485	0.8439	0.8447	0.7557
4.0	5.9808E 00	0.9945	0.8542	0.7900	0.7553	0.9819	0.9973	1.0010	1.0002	0.8508	0.8481	0.8473	0.7576
4.5	5.8091E 00	0.9905	0.8638	0.8014	0.7675	0.9795	0.9969	1.0009	1.0001	0.8561	0.8582	0.8533	0.7619
5.0	5.6615E 00	0.9865	0.8730	0.8125	0.7793	0.9772	0.9966	1.0008	1.0001	0.8612	0.8677	0.8588	0.7659
5.5	5.5318E 00	0.9825	0.8819	0.8232	0.7907	0.9748	0.9962	1.0007	1.0001	0.8659	0.8766	0.8639	0.7695
6.0	5.4159E 00	0.9785	0.8906	0.8336	0.8017	0.9725	0.9959	1.0007	1.0001	0.8705	0.8851	0.8688	0.7728
6.5	5.3111E 00	0.9745	0.8989	0.8436	0.8124	0.9702	0.9955	1.0006	1.0001	0.8748	0.8932	0.8733	0.7760
7.0	5.2151E 00	0.9705	0.9069	0.8532	0.8227	0.9679	0.9951	1.0006	1.0001	0.8789	0.9010	0.8776	0.7789
7.5	5.1266E 00	0.9665	0.9147	0.8626	0.8326	0.9656	0.9948	1.0005	1.0001	0.8830	0.9085	0.8818	0.7816
8.0	5.0443E 00	0.9625	0.9223	0.8717	0.8423	0.9634	0.9944	1.0005	1.0001	0.8868	0.9157	0.8858	0.7843
9.0	4.8952E 00	0.9546	0.9368	0.8892	0.8610	0.9589	0.9937	1.0005	1.0001	0.8942	0.9296	0.8934	0.7892
10.0	4.7623E 00	0.9467	0.9506	0.9058	0.8788	0.9544	0.9930	1.0004	1.0001	0.9012	0.9426	0.9005	0.7937
11.0	4.6424E 00	0.9389	0.9638	0.9218	0.8959	0.9501	0.9923	1.0004	1.0001	0.9079	0.9550	0.9073	0.7979
12.0	4.5327E 00	0.9312	0.9766	0.9372	0.9124	0.9458	0.9915	1.0003	1.0001	0.9143	0.9667	0.9137	0.8020
13.0	4.4317E 00	0.9236	0.9888	0.9521	0.9283	0.9415	0.9908	1.0003	1.0001	0.9204	0.9778	0.9200	0.8058
14.0	4.3379E 00	0.9161	1.0007	0.9666	0.9438	0.9374	0.9901	1.0003	1.0000	0.9264	0.9883	0.9260	0.8094
16.0	4.1681E 00	0.9014	1.0234	0.9944	0.9736	0.9292	0.9886	1.0003	1.0000	0.9378	1.0078	0.9375	0.8164
18.0	4.0173E 00	0.8871	1.0450	1.0209	1.0022	0.9214	0.9871	1.0002	1.0000	0.9487	1.0254	0.9485	0.8230
20.0	3.8814E 00	0.8732	1.0657	1.0465	1.0298	0.9139	0.9856	1.0002	1.0000	0.9592	1.0412	0.9589	0.8292
25.0	3.5910E 00	0.8403	1.1139	1.1069	1.0952	0.8965	0.9817	1.0002	1.0000	0.9837	1.0732	0.9836	0.8440
30.0	3.3516E 00	0.8098	1.1581	1.1635	1.1568	0.8810	0.9777	1.0001	1.0000	1.0065	1.0955	1.0063	0.8579
35.0	3.1489E 00	0.7816	1.1988	1.2169	1.2155	0.8674	0.9736	1.0001	1.0000	1.0277	1.1089	1.0276	0.8711
40.0	2.9740E 00	0.7557	1.2364	1.2676	1.2715	0.8556	0.9693	1.0001	1.0000	1.0475	1.1141	1.0475	0.8836
45.0	2.8213E 00	0.7318	1.2709	1.3157	1.3252	0.8455	0.9649	1.0001	1.0000	1.0661	1.1118	1.0660	0.8956
50.0	2.6865E 00	0.7098	1.3026	1.3613	1.3767	0.8370	0.9604	1.0001	1.0000	1.0833	1.1026	1.0833	0.9068

ELECTRON Z= 57 A= 139

p	F_0L_0	L_0	λ_2	λ_3	λ_4	μ_1	μ_2	Λ_1	Λ_2	ν_{12}	$\hat{\nu}_{12}$	η_{12}	$\hat{\eta}_{12}$
0.1	8.0445E 01	1.0183	11.8456	52.9028	140.3965	0.9991	1.0000	1.0046	1.0007	4.0864	0.6210	0.3199	0.2696
0.2	4.0626E 01	1.0182	3.4947	6.1532	8.3225	0.9990	0.9999	1.0042	1.0007	1.5344	0.6401	0.5630	0.5006
0.3	2.7520E 01	1.0181	1.9505	2.4632	2.7440	0.9988	0.9999	1.0040	1.0007	1.0888	0.6577	0.6445	0.5774
0.4	2.1089E 01	1.0179	1.4126	1.5413	1.5907	0.9985	0.9999	1.0038	1.0006	0.9424	0.6735	0.6860	0.6160
0.5	1.7331E 01	1.0177	1.1663	1.1819	1.1763	0.9982	0.9998	1.0037	1.0006	0.8796	0.6878	0.7116	0.6395
0.6	1.4907E 01	1.0174	1.0348	1.0064	0.9823	0.9978	0.9998	1.0035	1.0006	0.8484	0.7006	0.7292	0.6555
0.7	1.3240E 01	1.0171	0.9576	0.9088	0.8770	0.9974	0.9997	1.0033	1.0006	0.8316	0.7121	0.7423	0.6673
0.8	1.2037E 01	1.0167	0.9093	0.8497	0.8143	0.9970	0.9996	1.0032	1.0005	0.8221	0.7223	0.7526	0.6764
0.9	1.1136E 01	1.0163	0.8777	0.8120	0.7748	0.9965	0.9995	1.0030	1.0005	0.8167	0.7315	0.7609	0.6837
1.0	1.0440E 01	1.0159	0.8564	0.7871	0.7488	0.9960	0.9995	1.0029	1.0005	0.8138	0.7399	0.7679	0.6897
1.2	9.4436E 00	1.0148	0.8316	0.7587	0.7196	0.9951	0.9993	1.0026	1.0004	0.8118	0.7542	0.7792	0.6993
1.4	8.7669E 00	1.0137	0.8198	0.7456	0.7063	0.9941	0.9992	1.0024	1.0004	0.8125	0.7663	0.7880	0.7067
1.6	8.2773E 00	1.0124	0.8147	0.7402	0.7010	0.9931	0.9990	1.0022	1.0004	0.8144	0.7766	0.7952	0.7127
1.8	7.9051E 00	1.0111	0.8133	0.7390	0.7000	0.9921	0.9989	1.0020	1.0003	0.8168	0.7856	0.8014	0.7177
2.0	7.6107E 00	1.0097	0.8141	0.7403	0.7016	0.9910	0.9987	1.0018	1.0003	0.8195	0.7936	0.8068	0.7221
2.2	7.3704E 00	1.0082	0.8161	0.7430	0.7046	0.9900	0.9986	1.0017	1.0003	0.8222	0.8008	0.8116	0.7259
2.4	7.1691E 00	1.0067	0.8190	0.7466	0.7085	0.9890	0.9984	1.0016	1.0003	0.8250	0.8074	0.8159	0.7293
2.6	6.9968E 00	1.0052	0.8224	0.7507	0.7130	0.9880	0.9983	1.0015	1.0002	0.8278	0.8135	0.8199	0.7325
2.8	6.8469E 00	1.0036	0.8261	0.7552	0.7178	0.9870	0.9981	1.0014	1.0002	0.8305	0.8192	0.8236	0.7353
3.0	6.7143E 00	1.0021	0.8299	0.7599	0.7228	0.9860	0.9980	1.0013	1.0002	0.8331	0.8246	0.8271	0.7380
3.2	6.5958E 00	1.0005	0.8339	0.7647	0.7279	0.9851	0.9978	1.0012	1.0002	0.8357	0.8297	0.8303	0.7405
3.4	6.4886E 00	0.9989	0.8380	0.7696	0.7331	0.9841	0.9977	1.0012	1.0002	0.8382	0.8346	0.8334	0.7428
3.6	6.3909E 00	0.9973	0.8420	0.7744	0.7383	0.9831	0.9975	1.0011	1.0002	0.8406	0.8393	0.8363	0.7450
3.8	6.3010E 00	0.9956	0.8461	0.7793	0.7434	0.9821	0.9974	1.0011	1.0002	0.8430	0.8438	0.8390	0.7470
4.0	6.2179E 00	0.9940	0.8501	0.7841	0.7486	0.9811	0.9972	1.0010	1.0002	0.8453	0.8481	0.8417	0.7490
4.5	6.0337E 00	0.9899	0.8600	0.7959	0.7611	0.9787	0.9968	1.0009	1.0001	0.8508	0.8584	0.8479	0.7535
5.0	5.8753E 00	0.9857	0.8695	0.8074	0.7733	0.9762	0.9965	1.0008	1.0001	0.8560	0.8680	0.8536	0.7576
5.5	5.7362E 00	0.9816	0.8788	0.8184	0.7851	0.9738	0.9961	1.0008	1.0001	0.8610	0.8770	0.8589	0.7613
6.0	5.6120E 00	0.9774	0.8877	0.8291	0.7964	0.9714	0.9957	1.0007	1.0001	0.8657	0.8856	0.8639	0.7648
6.5	5.4996E 00	0.9733	0.8963	0.8394	0.8074	0.9690	0.9954	1.0006	1.0001	0.8702	0.8938	0.8686	0.7680
7.0	5.3968E 00	0.9691	0.9047	0.8494	0.8180	0.9666	0.9950	1.0006	1.0001	0.8745	0.9017	0.8731	0.7710
7.5	5.3020E 00	0.9650	0.9127	0.8591	0.8283	0.9643	0.9946	1.0006	1.0001	0.8786	0.9093	0.8774	0.7739
8.0	5.2139E 00	0.9608	0.9206	0.8685	0.8383	0.9619	0.9943	1.0005	1.0001	0.8826	0.9166	0.8816	0.7766
9.0	5.0543E 00	0.9526	0.9357	0.8866	0.8576	0.9573	0.9935	1.0005	1.0001	0.8903	0.9305	0.8894	0.7817
10.0	4.9122E 00	0.9445	0.9500	0.9039	0.8760	0.9527	0.9928	1.0004	1.0001	0.8975	0.9436	0.8968	0.7864
11.0	4.7841E 00	0.9365	0.9637	0.9204	0.8937	0.9482	0.9920	1.0004	1.0001	0.9044	0.9560	0.9038	0.7908
12.0	4.6671E 00	0.9286	0.9769	0.9364	0.9107	0.9437	0.9913	1.0004	1.0001	0.9111	0.9677	0.9105	0.7950
13.0	4.5595E 00	0.9207	0.9896	0.9518	0.9272	0.9393	0.9905	1.0003	1.0001	0.9175	0.9788	0.9170	0.7990
14.0	4.4596E 00	0.9130	1.0020	0.9668	0.9433	0.9350	0.9898	1.0003	1.0000	0.9237	0.9893	0.9233	0.8028
16.0	4.2789E 00	0.8979	1.0256	0.9956	0.9742	0.9267	0.9882	1.0003	1.0000	0.9356	1.0087	0.9352	0.8101
18.0	4.1188E 00	0.8832	1.0480	1.0232	1.0039	0.9187	0.9867	1.0002	1.0000	0.9469	1.0261	0.9466	0.8169
20.0	3.9747E 00	0.8689	1.0695	1.0497	1.0324	0.9110	0.9851	1.0002	1.0000	0.9577	1.0415	0.9575	0.8235
25.0	3.6677E 00	0.8352	1.1195	1.1125	1.1003	0.8931	0.9811	1.0002	1.0000	0.9832	1.0727	0.9831	0.8390
30.0	3.4156E 00	0.8040	1.1653	1.1712	1.1643	0.8774	0.9770	1.0001	1.0000	1.0068	1.0938	1.0067	0.8536
35.0	3.2029E 00	0.7754	1.2075	1.2266	1.2252	0.8636	0.9727	1.0001	1.0000	1.0289	1.1059	1.0288	0.8675
40.0	3.0199E 00	0.7490	1.2464	1.2792	1.2833	0.8517	0.9683	1.0001	1.0000	1.0494	1.1096	1.0493	0.8807
45.0	2.8605E 00	0.7248	1.2820	1.3290	1.3390	0.8416	0.9637	1.0001	1.0000	1.0686	1.1058	1.0685	0.8932
50.0	2.7202E 00	0.7026	1.3146	1.3763	1.3924	0.8333	0.9590	1.0001	1.0000	1.0864	1.0950	1.0864	0.9050

Instr. see p. 22

ELECTRON Z= 58 A= 140

p	F_0L_0	L_0	λ_2	λ_3	λ_4	μ_1	μ_2	Λ_1	Λ_2	ν_{12}	$\hat{\nu}_{12}$	η_{12}	$\hat{\eta}_{12}$
0.1	8.4959E 01	1.0187	12.0636	55.0988	149.1421	0.9991	1.0000	1.0048	1.0007	4.1691	0.6165	0.2902	0.2398
0.2	4.2902E 01	1.0187	3.5432	6.3169	8.6150	0.9989	0.9999	1.0044	1.0007	1.5461	0.6359	0.5450	0.4815
0.3	2.9058E 01	1.0185	1.9677	2.4995	2.7934	0.9987	0.9999	1.0041	1.0007	1.0885	0.6537	0.6304	0.5618
0.4	2.2263E 01	1.0183	1.4188	1.5514	1.6026	0.9984	0.9999	1.0040	1.0006	0.9384	0.6699	0.6738	0.6021
0.5	1.8288E 01	1.0181	1.1674	1.1831	1.1772	0.9981	0.9998	1.0038	1.0006	0.8739	0.6844	0.7005	0.6267
0.6	1.5723E 01	1.0178	1.0332	1.0037	0.9787	0.9977	0.9997	1.0036	1.0006	0.8420	0.6974	0.7190	0.6433
0.7	1.3956E 01	1.0175	0.9544	0.9039	0.8712	0.9973	0.9997	1.0034	1.0006	0.8248	0.7091	0.7327	0.6555
0.8	1.2680E 01	1.0171	0.9051	0.8437	0.8073	0.9968	0.9996	1.0033	1.0005	0.8152	0.7195	0.7433	0.6650
0.9	1.1723E 01	1.0167	0.8728	0.8052	0.7670	0.9964	0.9995	1.0031	1.0005	0.8097	0.7289	0.7520	0.6725
1.0	1.0984E 01	1.0162	0.8510	0.7798	0.7406	0.9959	0.9995	1.0030	1.0005	0.8067	0.7374	0.7593	0.6788
1.2	9.9239E 00	1.0151	0.8258	0.7509	0.7109	0.9949	0.9993	1.0027	1.0004	0.8047	0.7520	0.7709	0.6887
1.4	9.2030E 00	1.0140	0.8137	0.7376	0.6974	0.9938	0.9992	1.0024	1.0004	0.8054	0.7643	0.7801	0.6964
1.6	8.6808E 00	1.0127	0.8086	0.7321	0.6920	0.9928	0.9990	1.0022	1.0004	0.8074	0.7748	0.7876	0.7026
1.8	8.2834E 00	1.0113	0.8072	0.7309	0.6911	0.9918	0.9988	1.0020	1.0003	0.8099	0.7839	0.7939	0.7078
2.0	7.9687E 00	1.0098	0.8080	0.7323	0.6926	0.9907	0.9987	1.0019	1.0003	0.8127	0.7921	0.7995	0.7123
2.2	7.7117E 00	1.0083	0.8101	0.7350	0.6957	0.9897	0.9985	1.0017	1.0003	0.8155	0.7994	0.8045	0.7162
2.4	7.4962E 00	1.0068	0.8131	0.7387	0.6997	0.9887	0.9984	1.0016	1.0003	0.8184	0.8061	0.8090	0.7198
2.6	7.3117E 00	1.0052	0.8166	0.7430	0.7043	0.9876	0.9982	1.0015	1.0003	0.8212	0.8124	0.8131	0.7230
2.8	7.1510E 00	1.0036	0.8204	0.7476	0.7093	0.9866	0.9981	1.0014	1.0002	0.8240	0.8182	0.8169	0.7259
3.0	7.0089E 00	1.0020	0.8244	0.7525	0.7144	0.9856	0.9979	1.0013	1.0002	0.8267	0.8237	0.8205	0.7287
3.2	6.8817E 00	1.0004	0.8285	0.7574	0.7197	0.9845	0.9978	1.0013	1.0002	0.8294	0.8289	0.8238	0.7312
3.4	6.7667E 00	0.9987	0.8327	0.7624	0.7250	0.9835	0.9976	1.0012	1.0002	0.8320	0.8338	0.8270	0.7336
3.6	6.6619E 00	0.9971	0.8368	0.7674	0.7303	0.9825	0.9975	1.0011	1.0002	0.8345	0.8386	0.8300	0.7359
3.8	6.5655E 00	0.9954	0.8410	0.7724	0.7356	0.9815	0.9973	1.0011	1.0002	0.8370	0.8431	0.8329	0.7380
4.0	6.4763E 00	0.9937	0.8452	0.7773	0.7409	0.9805	0.9971	1.0010	1.0002	0.8393	0.8475	0.8356	0.7400
4.5	6.2785E 00	0.9895	0.8554	0.7895	0.7538	0.9780	0.9968	1.0009	1.0002	0.8450	0.8579	0.8420	0.7447
5.0	6.1086E 00	0.9852	0.8652	0.8013	0.7663	0.9754	0.9964	1.0008	1.0001	0.8504	0.8676	0.8479	0.7489
5.5	5.9594E 00	0.9809	0.8748	0.8126	0.7783	0.9729	0.9960	1.0008	1.0001	0.8555	0.8767	0.8534	0.7527
6.0	5.8262E 00	0.9766	0.8840	0.8236	0.7900	0.9705	0.9956	1.0007	1.0001	0.8604	0.8854	0.8585	0.7563
6.5	5.7057E 00	0.9724	0.8929	0.8342	0.8013	0.9680	0.9953	1.0007	1.0001	0.8650	0.8937	0.8634	0.7596
7.0	5.5955E 00	0.9681	0.9015	0.8445	0.8122	0.9656	0.9949	1.0006	1.0001	0.8694	0.9016	0.8680	0.7627
7.5	5.4940E 00	0.9639	0.9099	0.8545	0.8228	0.9631	0.9945	1.0006	1.0001	0.8737	0.9092	0.8725	0.7657
8.0	5.3997E 00	0.9596	0.9180	0.8642	0.8331	0.9607	0.9941	1.0005	1.0001	0.8779	0.9166	0.8767	0.7685
9.0	5.2290E 00	0.9512	0.9335	0.8828	0.8529	0.9559	0.9934	1.0005	1.0001	0.8858	0.9306	0.8848	0.7737
10.0	5.0772E 00	0.9429	0.9483	0.9006	0.8719	0.9512	0.9926	1.0004	1.0001	0.8932	0.9437	0.8925	0.7786
11.0	4.9405E 00	0.9347	0.9625	0.9177	0.8901	0.9466	0.9918	1.0004	1.0001	0.9004	0.9561	0.8997	0.7832
12.0	4.8157E 00	0.9265	0.9762	0.9342	0.9076	0.9420	0.9911	1.0004	1.0001	0.9072	0.9677	0.9067	0.7875
13.0	4.7010E 00	0.9185	0.9893	0.9501	0.9247	0.9375	0.9903	1.0003	1.0001	0.9139	0.9788	0.9134	0.7917
14.0	4.5947E 00	0.9106	1.0021	0.9655	0.9412	0.9331	0.9895	1.0003	1.0001	0.9203	0.9892	0.9198	0.7956
16.0	4.4028E 00	0.8951	1.0265	0.9953	0.9731	0.9246	0.9880	1.0003	1.0000	0.9325	1.0084	0.9322	0.8032
18.0	4.2328E 00	0.8801	1.0497	1.0238	1.0037	0.9164	0.9864	1.0002	1.0000	0.9442	1.0256	0.9439	0.8104
20.0	4.0803E 00	0.8656	1.0719	1.0512	1.0332	0.9085	0.9848	1.0002	1.0000	0.9555	1.0408	0.9552	0.8172
25.0	3.7561E 00	0.8312	1.1236	1.1160	1.1032	0.8903	0.9807	1.0002	1.0000	0.9818	1.0712	0.9816	0.8334
30.0	3.4907E 00	0.7995	1.1709	1.1766	1.1693	0.8743	0.9764	1.0001	1.0000	1.0062	1.0914	1.0061	0.8487
35.0	3.2675E 00	0.7705	1.2144	1.2339	1.2321	0.8604	0.9720	1.0001	1.0000	1.0290	1.1025	1.0288	0.8632
40.0	3.0761E 00	0.7438	1.2544	1.2881	1.2922	0.8485	0.9675	1.0001	1.0000	1.0502	1.1052	1.0501	0.8769
45.0	2.9097E 00	0.7194	1.2911	1.3396	1.3497	0.8384	0.9628	1.0001	1.0000	1.0700	1.1003	1.0699	0.8900
50.0	2.7637E 00	0.6970	1.3246	1.3883	1.4047	0.8301	0.9580	1.0001	1.0000	1.0883	1.0885	1.0883	0.9023

ELECTRON Z= 59 A= 141

p	F_0L_0	L_0	λ_2	λ_3	λ_4	μ_1	μ_2	Λ_1	Λ_2	ν_{12}	$\hat{\nu}_{12}$	η_{12}	$\hat{\eta}_{12}$
0.1	8.9753E 01	1.0191	12.2764	57.3202	158.2212	0.9990	1.0000	1.0050	1.0007	4.2516	0.6120	0.2597	0.2093
0.2	4.5319E 01	1.0191	3.5905	6.4807	8.9120	0.9989	0.9999	1.0045	1.0007	1.5575	0.6317	0.5266	0.4620
0.3	3.0691E 01	1.0189	1.9843	2.5354	2.8426	0.9987	0.9999	1.0043	1.0007	1.0880	0.6498	0.6160	0.5459
0.4	2.3509E 01	1.0188	1.4247	1.5611	1.6142	0.9984	0.9999	1.0041	1.0007	0.9341	0.6662	0.6614	0.5880
0.5	1.9305E 01	1.0185	1.1683	1.1840	1.1778	0.9980	0.9998	1.0039	1.0006	0.8681	0.6809	0.6893	0.6136
0.6	1.6589E 01	1.0182	1.0315	1.0007	0.9750	0.9976	0.9997	1.0037	1.0006	0.8355	0.6942	0.7085	0.6309
0.7	1.4716E 01	1.0179	0.9511	0.8989	0.8652	0.9972	0.9997	1.0035	1.0006	0.8179	0.7060	0.7228	0.6436
0.8	1.3362E 01	1.0175	0.9007	0.8375	0.8001	0.9967	0.9996	1.0034	1.0006	0.8080	0.7166	0.7339	0.6534
0.9	1.2346E 01	1.0170	0.8678	0.7983	0.7591	0.9962	0.9995	1.0032	1.0005	0.8025	0.7262	0.7430	0.6613
1.0	1.1561E 01	1.0165	0.8456	0.7725	0.7323	0.9957	0.9994	1.0030	1.0005	0.7994	0.7348	0.7505	0.6678
1.2	1.0433E 01	1.0154	0.8198	0.7431	0.7021	0.9947	0.9993	1.0028	1.0005	0.7975	0.7497	0.7626	0.6780
1.4	9.6645E 00	1.0142	0.8076	0.7295	0.6884	0.9936	0.9991	1.0025	1.0004	0.7983	0.7622	0.7720	0.6859
1.6	9.1075E 00	1.0129	0.8023	0.7240	0.6830	0.9926	0.9990	1.0023	1.0004	0.8003	0.7729	0.7798	0.6923
1.8	8.6830E 00	1.0115	0.8009	0.7228	0.6820	0.9915	0.9988	1.0021	1.0003	0.8029	0.7822	0.7864	0.6977
2.0	8.3467E 00	1.0100	0.8018	0.7242	0.6837	0.9904	0.9986	1.0019	1.0003	0.8057	0.7905	0.7922	0.7023
2.2	8.0716E 00	1.0084	0.8040	0.7270	0.6868	0.9893	0.9985	1.0018	1.0003	0.8087	0.7980	0.7973	0.7064
2.4	7.8409E 00	1.0069	0.8071	0.7309	0.6910	0.9883	0.9983	1.0017	1.0003	0.8116	0.8048	0.8019	0.7100
2.6	7.6433E 00	1.0052	0.8107	0.7352	0.6957	0.9872	0.9982	1.0016	1.0003	0.8146	0.8112	0.8062	0.7133
2.8	7.4710E 00	1.0036	0.8146	0.7400	0.7007	0.9861	0.9980	1.0015	1.0002	0.8175	0.8171	0.8101	0.7164
3.0	7.3187E 00	1.0019	0.8188	0.7449	0.7060	0.9851	0.9979	1.0014	1.0002	0.8203	0.8227	0.8138	0.7192
3.2	7.1823E 00	1.0002	0.8230	0.7500	0.7114	0.9840	0.9977	1.0013	1.0002	0.8230	0.8279	0.8173	0.7218
3.4	7.0590E 00	0.9985	0.8273	0.7551	0.7169	0.9830	0.9976	1.0012	1.0002	0.8257	0.8330	0.8205	0.7243
3.6	6.9464E 00	0.9968	0.8316	0.7603	0.7223	0.9819	0.9974	1.0012	1.0002	0.8283	0.8378	0.8236	0.7266
3.8	6.8430E 00	0.9951	0.8359	0.7654	0.7277	0.9809	0.9972	1.0011	1.0002	0.8308	0.8424	0.8266	0.7288
4.0	6.7472E 00	0.9934	0.8402	0.7705	0.7331	0.9798	0.9971	1.0011	1.0002	0.8333	0.8468	0.8294	0.7309
4.5	6.5351E 00	0.9890	0.8507	0.7830	0.7464	0.9772	0.9967	1.0010	1.0002	0.8391	0.8574	0.8360	0.7357
5.0	6.3528E 00	0.9846	0.8609	0.7951	0.7592	0.9746	0.9963	1.0009	1.0001	0.8447	0.8672	0.8421	0.7400
5.5	6.1927E 00	0.9802	0.8708	0.8068	0.7716	0.9721	0.9959	1.0008	1.0001	0.8499	0.8764	0.8477	0.7439
6.0	6.0499E 00	0.9758	0.8803	0.8181	0.7835	0.9695	0.9955	1.0007	1.0001	0.8549	0.8852	0.8530	0.7476
6.5	5.9207E 00	0.9714	0.8895	0.8290	0.7951	0.9670	0.9952	1.0007	1.0001	0.8597	0.8935	0.8581	0.7510
7.0	5.8027E 00	0.9671	0.8983	0.8396	0.8063	0.9644	0.9948	1.0006	1.0001	0.8643	0.9015	0.8628	0.7543
7.5	5.6940E 00	0.9627	0.9070	0.8499	0.8172	0.9619	0.9944	1.0006	1.0001	0.8687	0.9092	0.8674	0.7573
8.0	5.5931E 00	0.9584	0.9153	0.8598	0.8278	0.9595	0.9940	1.0006	1.0001	0.8730	0.9166	0.8718	0.7602
9.0	5.4105E 00	0.9497	0.9314	0.8791	0.8482	0.9545	0.9932	1.0005	1.0001	0.8811	0.9306	0.8802	0.7656
10.0	5.2485E 00	0.9412	0.9467	0.8974	0.8677	0.9497	0.9925	1.0004	1.0001	0.8889	0.9438	0.8881	0.7707
11.0	5.1025E 00	0.9328	0.9614	0.9150	0.8865	0.9449	0.9917	1.0004	1.0001	0.8962	0.9561	0.8956	0.7754
12.0	4.9696E 00	0.9244	0.9755	0.9319	0.9046	0.9403	0.9909	1.0004	1.0001	0.9033	0.9678	0.9027	0.7799
13.0	4.8474E 00	0.9162	0.9891	0.9484	0.9221	0.9357	0.9901	1.0003	1.0001	0.9101	0.9788	0.9096	0.7842
14.0	4.7343E 00	0.9081	1.0022	0.9643	0.9391	0.9311	0.9893	1.0003	1.0001	0.9168	0.9892	0.9163	0.7884
16.0	4.5304E 00	0.8923	1.0275	0.9950	0.9720	0.9224	0.9877	1.0003	1.0000	0.9294	1.0082	0.9291	0.7962
18.0	4.3502E 00	0.8770	1.0514	1.0244	1.0035	0.9140	0.9861	1.0002	1.0000	0.9415	1.0252	0.9412	0.8037
20.0	4.1888E 00	0.8621	1.0743	1.0527	1.0339	0.9060	0.9844	1.0002	1.0000	0.9531	1.0402	0.9529	0.8108
25.0	3.8464E 00	0.8271	1.1278	1.1196	1.1062	0.8875	0.9802	1.0002	1.0000	0.9803	1.0698	0.9802	0.8278
30.0	3.5673E 00	0.7950	1.1766	1.1822	1.1744	0.8713	0.9759	1.0001	1.0000	1.0056	1.0891	1.0054	0.8437
35.0	3.3332E 00	0.7656	1.2215	1.2413	1.2392	0.8572	0.9714	1.0001	1.0000	1.0290	1.0991	1.0289	0.8588
40.0	3.1331E 00	0.7386	1.2627	1.2973	1.3012	0.8452	0.9667	1.0001	1.0000	1.0509	1.1008	1.0509	0.8732
45.0	2.9596E 00	0.7139	1.3005	1.3504	1.3605	0.8352	0.9620	1.0001	1.0000	1.0713	1.0949	1.0713	0.8868
50.0	2.8076E 00	0.6914	1.3349	1.4006	1.4173	0.8269	0.9570	1.0001	1.0000	1.0903	1.0820	1.0902	0.8996

 Instr. see p. 22

ELECTRON Z= 60 A= 142

p	F_0L_0	L_0	λ_2	λ_3	λ_4	μ_1	μ_2	Λ_1	Λ_2	ν_{12}	$\hat{\nu}_{12}$	η_{12}	$\hat{\eta}_{12}$
0.1	9.4846E 01	1.0195	12.4838	59.5636	167.6322	0.9989	1.0000	1.0052	1.0007	4.3339	0.6074	0.2286	0.1783
0.2	4.7887E 01	1.0195	3.6364	6.6442	9.2131	0.9988	0.9999	1.0046	1.0007	1.5686	0.6273	0.5078	0.4422
0.3	3.2426E 01	1.0193	2.0002	2.5706	2.8915	0.9986	0.9999	1.0044	1.0007	1.0873	0.6457	0.6013	0.5298
0.4	2.4832E 01	1.0191	1.4302	1.5704	1.6253	0.9983	0.9999	1.0042	1.0007	0.9297	0.6624	0.6487	0.5737
0.5	2.0384E 01	1.0189	1.1690	1.1847	1.1781	0.9979	0.9998	1.0040	1.0006	0.8622	0.6774	0.6779	0.6003
0.6	1.7508E 01	1.0186	1.0295	0.9975	0.9709	0.9975	0.9997	1.0038	1.0006	0.8288	0.6908	0.6979	0.6183
0.7	1.5522E 01	1.0182	0.9476	0.8937	0.8591	0.9970	0.9997	1.0036	1.0006	0.8108	0.7029	0.7127	0.6315
0.8	1.4086E 01	1.0178	0.8963	0.8312	0.7929	0.9966	0.9996	1.0035	1.0006	0.8008	0.7137	0.7243	0.6417
0.9	1.3007E 01	1.0174	0.8627	0.7913	0.7512	0.9961	0.9995	1.0033	1.0005	0.7951	0.7234	0.7337	0.6498
1.0	1.2172E 01	1.0169	0.8401	0.7650	0.7239	0.9955	0.9994	1.0031	1.0005	0.7920	0.7322	0.7415	0.6565
1.2	1.0972E 01	1.0157	0.8138	0.7351	0.6932	0.9945	0.9993	1.0028	1.0005	0.7901	0.7473	0.7541	0.6671
1.4	1.0153E 01	1.0145	0.8014	0.7213	0.6793	0.9934	0.9991	1.0026	1.0004	0.7909	0.7600	0.7638	0.6753
1.6	9.5587E 00	1.0131	0.7960	0.7157	0.6739	0.9923	0.9989	1.0023	1.0004	0.7930	0.7709	0.7719	0.6819
1.8	9.1053E 00	1.0116	0.7946	0.7146	0.6729	0.9912	0.9988	1.0022	1.0004	0.7957	0.7804	0.7787	0.6874
2.0	8.7457E 00	1.0101	0.7956	0.7160	0.6746	0.9901	0.9986	1.0020	1.0003	0.7987	0.7889	0.7846	0.6922
2.2	8.4514E 00	1.0085	0.7979	0.7190	0.6779	0.9890	0.9985	1.0018	1.0003	0.8017	0.7965	0.7899	0.6964
2.4	8.2044E 00	1.0069	0.8010	0.7229	0.6821	0.9879	0.9983	1.0017	1.0003	0.8048	0.8034	0.7947	0.7001
2.6	7.9926E 00	1.0052	0.8048	0.7274	0.6870	0.9868	0.9981	1.0016	1.0003	0.8078	0.8099	0.7991	0.7036
2.8	7.8080E 00	1.0035	0.8088	0.7323	0.6922	0.9857	0.9980	1.0015	1.0002	0.8108	0.8159	0.8032	0.7067
3.0	7.6446E 00	1.0018	0.8131	0.7374	0.6976	0.9846	0.9978	1.0014	1.0002	0.8137	0.8216	0.8070	0.7096
3.2	7.4984E 00	1.0001	0.8175	0.7426	0.7031	0.9835	0.9977	1.0013	1.0002	0.8165	0.8269	0.8105	0.7123
3.4	7.3661E 00	0.9983	0.8219	0.7478	0.7087	0.9824	0.9975	1.0013	1.0002	0.8192	0.8320	0.8139	0.7148
3.6	7.2454E 00	0.9965	0.8263	0.7531	0.7142	0.9813	0.9973	1.0012	1.0002	0.8219	0.8369	0.8171	0.7172
3.8	7.1343E 00	0.9948	0.8308	0.7584	0.7198	0.9802	0.9972	1.0011	1.0002	0.8245	0.8416	0.8201	0.7194
4.0	7.0316E 00	0.9930	0.8352	0.7636	0.7253	0.9792	0.9970	1.0011	1.0002	0.8270	0.8461	0.8231	0.7216
4.5	6.8040E 00	0.9885	0.8460	0.7764	0.7389	0.9765	0.9966	1.0010	1.0002	0.8331	0.8568	0.8299	0.7265
5.0	6.6084E 00	0.9840	0.8565	0.7889	0.7520	0.9738	0.9962	1.0009	1.0001	0.8388	0.8667	0.8361	0.7309
5.5	6.4367E 00	0.9795	0.8667	0.8009	0.7648	0.9712	0.9958	1.0008	1.0001	0.8442	0.8761	0.8420	0.7350
6.0	6.2836E 00	0.9750	0.8765	0.8125	0.7770	0.9685	0.9954	1.0008	1.0001	0.8494	0.8849	0.8474	0.7388
6.5	6.1452E 00	0.9705	0.8860	0.8237	0.7889	0.9659	0.9951	1.0007	1.0001	0.8543	0.8933	0.8526	0.7423
7.0	6.0188E 00	0.9660	0.8952	0.8346	0.8005	0.9633	0.9947	1.0006	1.0001	0.8590	0.9014	0.8575	0.7457
7.5	5.9024E 00	0.9615	0.9040	0.8452	0.8116	0.9607	0.9943	1.0006	1.0001	0.8636	0.9091	0.8623	0.7488
8.0	5.7944E 00	0.9570	0.9127	0.8555	0.8225	0.9582	0.9939	1.0006	1.0001	0.8680	0.9165	0.8668	0.7518
9.0	5.5953E 00	0.9482	0.9293	0.8753	0.8435	0.9531	0.9931	1.0005	1.0001	0.8764	0.9306	0.8754	0.7574
10.0	5.4262E 00	0.9394	0.9451	0.8941	0.8636	0.9481	0.9923	1.0005	1.0001	0.8844	0.9438	0.8836	0.7626
11.0	5.2705E 00	0.9308	0.9602	0.9123	0.8828	0.9433	0.9915	1.0004	1.0001	0.8920	0.9562	0.8913	0.7676
12.0	5.1289E 00	0.9223	0.9748	0.9297	0.9015	0.9385	0.9907	1.0004	1.0001	0.8993	0.9678	0.8987	0.7722
13.0	4.9988E 00	0.9139	0.9888	0.9467	0.9195	0.9337	0.9899	1.0004	1.0001	0.9063	0.9788	0.9058	0.7767
14.0	4.8785E 00	0.9056	1.0024	0.9631	0.9371	0.9291	0.9891	1.0003	1.0001	0.9132	0.9891	0.9127	0.7810
16.0	4.6620E 00	0.8894	1.0285	0.9948	0.9709	0.9202	0.9874	1.0003	1.0000	0.9263	1.0080	0.9259	0.7891
18.0	4.4710E 00	0.8738	1.0532	1.0251	1.0034	0.9116	0.9858	1.0003	1.0000	0.9388	1.0248	0.9384	0.7969
20.0	4.3002E 00	0.8586	1.0769	1.0543	1.0348	0.9034	0.9841	1.0002	1.0000	0.9507	1.0395	0.9505	0.8044
25.0	3.9389E 00	0.8230	1.1321	1.1233	1.1093	0.8847	0.9798	1.0002	1.0000	0.9789	1.0683	0.9787	0.8220
30.0	3.6454E 00	0.7904	1.1825	1.1880	1.1796	0.8682	0.9753	1.0001	1.0000	1.0049	1.0868	1.0047	0.8386
35.0	3.4000E 00	0.7606	1.2287	1.2489	1.2465	0.8541	0.9707	1.0001	1.0000	1.0291	1.0959	1.0290	0.8544
40.0	3.1908E 00	0.7333	1.2712	1.3067	1.3105	0.8420	0.9660	1.0001	1.0000	1.0517	1.0965	1.0516	0.8694
45.0	3.0100E 00	0.7085	1.3100	1.3614	1.3717	0.8319	0.9610	1.0001	1.0000	1.0727	1.0895	1.0727	0.8836
50.0	2.8520E 00	0.6857	1.3454	1.4132	1.4303	0.8237	0.9560	1.0001	1.0000	1.0922	1.0757	1.0922	0.8969

ELECTRON Z= 61 A= 147

p	F_0L_0	L_0	λ_2	λ_3	λ_4	μ_1	μ_2	Λ_1	Λ_2	ν_{12}	$\hat{\nu}_{12}$	η_{12}	$\hat{\eta}_{12}$
0.1	1.0005E 02	1.0198	12.6992	61.9132	177.6525	0.9989	1.0000	1.0054	1.0007	4.4185	0.6031	0.1969	0.1468
0.2	5.0512E 01	1.0197	3.6848	6.8169	9.5333	0.9987	0.9999	1.0048	1.0007	1.5805	0.6233	0.4889	0.4223
0.3	3.4199E 01	1.0195	2.0177	2.6090	2.9447	0.9985	0.9999	1.0046	1.0007	1.0870	0.6419	0.5867	0.5137
0.4	2.6183E 01	1.0193	1.4368	1.5816	1.6386	0.9982	0.9998	1.0044	1.0007	0.9256	0.6588	0.6362	0.5594
0.5	2.1486E 01	1.0191	1.1706	1.1867	1.1800	0.9978	0.9998	1.0042	1.0007	0.8565	0.6741	0.6666	0.5871
0.6	1.8446E 01	1.0188	1.0285	0.9955	0.9682	0.9974	0.9997	1.0040	1.0006	0.8224	0.6878	0.6875	0.6059
0.7	1.6345E 01	1.0184	0.9450	0.8897	0.8542	0.9969	0.9997	1.0038	1.0006	0.8041	0.7001	0.7029	0.6196
0.8	1.4824E 01	1.0180	0.8926	0.8259	0.7867	0.9964	0.9996	1.0036	1.0006	0.7938	0.7111	0.7149	0.6301
0.9	1.3680E 01	1.0175	0.8584	0.7853	0.7443	0.9959	0.9995	1.0034	1.0006	0.7881	0.7210	0.7247	0.6386
1.0	1.2794E 01	1.0170	0.8354	0.7585	0.7165	0.9953	0.9994	1.0032	1.0005	0.7849	0.7299	0.7328	0.6455
1.2	1.1519E 01	1.0158	0.8086	0.7281	0.6854	0.9942	0.9993	1.0029	1.0005	0.7830	0.7454	0.7458	0.6565
1.4	1.0648E 01	1.0145	0.7959	0.7142	0.6713	0.9931	0.9991	1.0027	1.0004	0.7839	0.7583	0.7559	0.6649
1.6	1.0015E 01	1.0130	0.7905	0.7085	0.6658	0.9919	0.9989	1.0024	1.0004	0.7861	0.7694	0.7642	0.6717
1.8	9.5314E 00	1.0115	0.7892	0.7074	0.6649	0.9908	0.9987	1.0022	1.0004	0.7889	0.7791	0.7713	0.6774
2.0	9.1476E 00	1.0099	0.7902	0.7089	0.6667	0.9896	0.9986	1.0021	1.0003	0.7919	0.7877	0.7774	0.6823
2.2	8.8332E 00	1.0083	0.7926	0.7119	0.6701	0.9885	0.9984	1.0019	1.0003	0.7951	0.7955	0.7829	0.6866
2.4	8.5691E 00	1.0066	0.7959	0.7160	0.6744	0.9873	0.9982	1.0018	1.0003	0.7982	0.8026	0.7878	0.6905
2.6	8.3426E 00	1.0049	0.7997	0.7206	0.6794	0.9862	0.9981	1.0017	1.0003	0.8014	0.8092	0.7924	0.6940
2.8	8.1449E 00	1.0031	0.8039	0.7257	0.6847	0.9850	0.9979	1.0016	1.0003	0.8044	0.8153	0.7966	0.6972
3.0	7.9700E 00	1.0013	0.8083	0.7309	0.6903	0.9839	0.9977	1.0015	1.0002	0.8074	0.8211	0.8005	0.7002
3.2	7.8134E 00	0.9995	0.8128	0.7363	0.6959	0.9828	0.9976	1.0014	1.0002	0.8103	0.8266	0.8042	0.7029
3.4	7.6716E 00	0.9977	0.8174	0.7417	0.7017	0.9816	0.9974	1.0013	1.0002	0.8132	0.8318	0.8076	0.7055
3.6	7.5423E 00	0.9959	0.8220	0.7471	0.7074	0.9805	0.9973	1.0013	1.0002	0.8159	0.8368	0.8109	0.7080
3.8	7.4233E 00	0.9940	0.8266	0.7525	0.7131	0.9794	0.9971	1.0012	1.0002	0.8186	0.8415	0.8141	0.7103
4.0	7.3132E 00	0.9922	0.8312	0.7579	0.7188	0.9783	0.9969	1.0011	1.0002	0.8212	0.8461	0.8171	0.7125
4.5	7.0692E 00	0.9875	0.8424	0.7711	0.7327	0.9755	0.9965	1.0010	1.0002	0.8275	0.8570	0.8241	0.7176
5.0	6.8596E 00	0.9829	0.8533	0.7839	0.7463	0.9727	0.9961	1.0009	1.0001	0.8334	0.8671	0.8306	0.7222
5.5	6.6756E 00	0.9782	0.8638	0.7963	0.7594	0.9699	0.9957	1.0008	1.0001	0.8390	0.8766	0.8366	0.7264
6.0	6.5116E 00	0.9735	0.8739	0.8083	0.7720	0.9672	0.9953	1.0008	1.0001	0.8443	0.8855	0.8423	0.7303
6.5	6.3634E 00	0.9688	0.8837	0.8199	0.7843	0.9645	0.9949	1.0007	1.0001	0.8494	0.8941	0.8476	0.7339
7.0	6.2281E 00	0.9642	0.8932	0.8312	0.7961	0.9618	0.9945	1.0007	1.0001	0.8543	0.9022	0.8527	0.7373
7.5	6.1036E 00	0.9595	0.9024	0.8421	0.8077	0.9591	0.9941	1.0006	1.0001	0.8590	0.9100	0.8576	0.7406
8.0	5.9882E 00	0.9549	0.9114	0.8527	0.8189	0.9565	0.9937	1.0006	1.0001	0.8636	0.9176	0.8623	0.7437
9.0	5.7797E 00	0.9458	0.9286	0.8731	0.8406	0.9512	0.9929	1.0005	1.0001	0.8723	0.9318	0.8713	0.7495
10.0	5.5949E 00	0.9367	0.9450	0.8927	0.8613	0.9461	0.9920	1.0005	1.0001	0.8805	0.9450	0.8797	0.7549
11.0	5.4289E 00	0.9278	0.9607	0.9115	0.8812	0.9410	0.9912	1.0004	1.0001	0.8884	0.9575	0.8877	0.7600
12.0	5.2780E 00	0.9190	0.9758	0.9296	0.9005	0.9361	0.9904	1.0004	1.0001	0.8960	0.9691	0.8954	0.7649
13.0	5.1396E 00	0.9104	0.9904	0.9471	0.9192	0.9312	0.9896	1.0004	1.0001	0.9033	0.9801	0.9028	0.7695
14.0	5.0118E 00	0.9019	1.0045	0.9642	0.9373	0.9265	0.9887	1.0003	1.0001	0.9104	0.9904	0.9099	0.7740
16.0	4.7819E 00	0.8853	1.0315	0.9970	0.9724	0.9173	0.9870	1.0003	1.0000	0.9240	1.0091	0.9236	0.7825
18.0	4.5795E 00	0.8692	1.0572	1.0284	1.0061	0.9085	0.9853	1.0003	1.0000	0.9370	1.0257	0.9366	0.7907
20.0	4.3988E 00	0.8537	1.0818	1.0587	1.0386	0.9001	0.9836	1.0002	1.0000	0.9494	1.0401	0.9491	0.7985
25.0	4.0178E 00	0.8172	1.1390	1.1304	1.1160	0.8809	0.9791	1.0002	1.0000	0.9786	1.0680	0.9784	0.8170
30.0	3.7096E 00	0.7840	1.1913	1.1975	1.1890	0.8643	0.9745	1.0001	1.0000	1.0056	1.0852	1.0055	0.8345
35.0	3.4527E 00	0.7537	1.2391	1.2607	1.2584	0.8500	0.9697	1.0001	1.0000	1.0308	1.0929	1.0306	0.8510
40.0	3.2345E 00	0.7261	1.2830	1.3206	1.3248	0.8379	0.9648	1.0001	1.0000	1.0542	1.0920	1.0541	0.8667
45.0	3.0465E 00	0.7010	1.3230	1.3773	1.3883	0.8280	0.9597	1.0001	1.0000	1.0759	1.0834	1.0758	0.8816
50.0	2.8826E 00	0.6781	1.3593	1.4309	1.4490	0.8199	0.9545	1.0001	1.0000	1.0960	1.0678	1.0959	0.8955

 Instr. see p. 22

ELECTRON Z= 62 A= 150

p	F_0L_0	L_0	λ_2	λ_3	λ_4	μ_1	μ_2	Λ_1	Λ_2	ν_{12}	$\hat{\nu}_{12}$	η_{12}	$\hat{\eta}_{12}$
0.1	1.0568E 02	1.0200	12.9026	64.2424	187.8893	0.9988	1.0000	1.0057	1.0008	4.5014	0.5985	0.1644	0.1147
0.2	5.3349E 01	1.0199	3.7299	6.9848	9.8505	0.9986	0.9999	1.0050	1.0007	1.5916	0.6190	0.4694	0.4019
0.3	3.6114E 01	1.0198	2.0335	2.6450	2.9954	0.9984	0.9999	1.0047	1.0007	1.0862	0.6379	0.5715	0.4972
0.4	2.7644E 01	1.0196	1.4424	1.5912	1.6502	0.9981	0.9998	1.0045	1.0007	0.9211	0.6551	0.6232	0.5448
0.5	2.2677E 01	1.0193	1.1714	1.1876	1.1807	0.9977	0.9998	1.0043	1.0007	0.8505	0.6706	0.6549	0.5736
0.6	1.9459E 01	1.0190	1.0267	0.9926	0.9646	0.9972	0.9997	1.0041	1.0007	0.8156	0.6845	0.6766	0.5931
0.7	1.7234E 01	1.0186	0.9417	0.8848	0.8485	0.9967	0.9996	1.0039	1.0006	0.7969	0.6970	0.6927	0.6073
0.8	1.5621E 01	1.0182	0.8884	0.8199	0.7798	0.9962	0.9996	1.0037	1.0006	0.7865	0.7082	0.7052	0.6183
0.9	1.4407E 01	1.0177	0.8536	0.7787	0.7367	0.9957	0.9995	1.0035	1.0006	0.7806	0.7182	0.7153	0.6270
1.0	1.3466E 01	1.0171	0.8301	0.7514	0.7085	0.9951	0.9994	1.0034	1.0005	0.7775	0.7273	0.7237	0.6342
1.2	1.2110E 01	1.0159	0.8029	0.7205	0.6770	0.9940	0.9992	1.0030	1.0005	0.7755	0.7431	0.7372	0.6455
1.4	1.1182E 01	1.0146	0.7900	0.7064	0.6627	0.9928	0.9991	1.0028	1.0004	0.7765	0.7563	0.7476	0.6541
1.6	1.0507E 01	1.0131	0.7846	0.7007	0.6572	0.9916	0.9989	1.0025	1.0004	0.7788	0.7676	0.7562	0.6611
1.8	9.9912E 00	1.0115	0.7832	0.6996	0.6563	0.9904	0.9987	1.0023	1.0004	0.7817	0.7775	0.7635	0.6670
2.0	9.5810E 00	1.0099	0.7843	0.7012	0.6581	0.9892	0.9985	1.0021	1.0003	0.7848	0.7862	0.7699	0.6720
2.2	9.2449E 00	1.0082	0.7868	0.7043	0.6616	0.9880	0.9984	1.0020	1.0003	0.7881	0.7942	0.7755	0.6765
2.4	8.9622E 00	1.0064	0.7902	0.7085	0.6661	0.9868	0.9982	1.0018	1.0003	0.7913	0.8014	0.7806	0.6805
2.6	8.7197E 00	1.0046	0.7942	0.7133	0.6712	0.9856	0.9980	1.0017	1.0003	0.7946	0.8081	0.7853	0.6841
2.8	8.5080E 00	1.0028	0.7985	0.7184	0.6767	0.9845	0.9979	1.0016	1.0003	0.7977	0.8144	0.7896	0.6874
3.0	8.3206E 00	1.0010	0.8031	0.7238	0.6823	0.9833	0.9977	1.0015	1.0002	0.8008	0.8203	0.7936	0.6904
3.2	8.1527E 00	0.9991	0.8077	0.7293	0.6882	0.9821	0.9975	1.0014	1.0002	0.8038	0.8258	0.7974	0.6933
3.4	8.0007E 00	0.9972	0.8124	0.7349	0.6940	0.9809	0.9974	1.0014	1.0002	0.8067	0.8311	0.8010	0.6959
3.6	7.8621E 00	0.9953	0.8172	0.7405	0.6999	0.9798	0.9972	1.0013	1.0002	0.8096	0.8362	0.8044	0.6984
3.8	7.7345E 00	0.9934	0.8219	0.7460	0.7058	0.9786	0.9970	1.0012	1.0002	0.8123	0.8410	0.8077	0.7008
4.0	7.6164E 00	0.9915	0.8266	0.7516	0.7116	0.9774	0.9968	1.0012	1.0002	0.8150	0.8457	0.8108	0.7031
4.5	7.3548E 00	0.9867	0.8382	0.7652	0.7259	0.9745	0.9964	1.0011	1.0002	0.8215	0.8567	0.8180	0.7083
5.0	7.1301E 00	0.9819	0.8494	0.7783	0.7398	0.9717	0.9960	1.0010	1.0002	0.8276	0.8670	0.8247	0.7130
5.5	6.9330E 00	0.9771	0.8603	0.7911	0.7532	0.9688	0.9956	1.0009	1.0001	0.8333	0.8766	0.8309	0.7173
6.0	6.7572E 00	0.9722	0.8707	0.8034	0.7662	0.9660	0.9952	1.0008	1.0001	0.8388	0.8857	0.8367	0.7214
6.5	6.5985E 00	0.9674	0.8809	0.8154	0.7788	0.9632	0.9948	1.0007	1.0001	0.8441	0.8943	0.8423	0.7251
7.0	6.4538E 00	0.9626	0.8907	0.8269	0.7910	0.9604	0.9943	1.0007	1.0001	0.8491	0.9025	0.8475	0.7287
7.5	6.3206E 00	0.9579	0.9002	0.8382	0.8029	0.9577	0.9939	1.0006	1.0001	0.8540	0.9104	0.8526	0.7320
8.0	6.1972E 00	0.9531	0.9095	0.8491	0.8145	0.9549	0.9935	1.0006	1.0001	0.8587	0.9180	0.8574	0.7352
9.0	5.9744E 00	0.9437	0.9272	0.8702	0.8368	0.9495	0.9927	1.0005	1.0001	0.8677	0.9323	0.8667	0.7412
10.0	5.7773E 00	0.9344	0.9442	0.8904	0.8581	0.9442	0.9918	1.0005	1.0001	0.8762	0.9456	0.8754	0.7469
11.0	5.6003E 00	0.9252	0.9604	0.9097	0.8787	0.9390	0.9910	1.0004	1.0001	0.8844	0.9581	0.8836	0.7521
12.0	5.4396E 00	0.9162	0.9760	0.9284	0.8985	0.9339	0.9901	1.0004	1.0001	0.8922	0.9697	0.8916	0.7572
13.0	5.2925E 00	0.9073	0.9911	0.9466	0.9178	0.9289	0.9893	1.0004	1.0001	0.8998	0.9807	0.8992	0.7620
14.0	5.1566E 00	0.8986	1.0057	0.9642	0.9365	0.9241	0.9884	1.0003	1.0001	0.9071	0.9909	0.9066	0.7667
16.0	4.9127E 00	0.8816	1.0337	0.9981	0.9727	0.9146	0.9867	1.0003	1.0000	0.9212	1.0096	0.9208	0.7756
18.0	4.6984E 00	0.8652	1.0604	1.0306	1.0075	0.9056	0.9849	1.0003	1.0000	0.9346	1.0259	0.9343	0.7840
20.0	4.5074E 00	0.8493	1.0858	1.0619	1.0411	0.8971	0.9831	1.0002	1.0000	0.9475	1.0401	0.9472	0.7922
25.0	4.1058E 00	0.8122	1.1450	1.1361	1.1211	0.8776	0.9785	1.0002	1.0000	0.9777	1.0671	0.9775	0.8115
30.0	3.7821E 00	0.7784	1.1990	1.2055	1.1966	0.8607	0.9738	1.0001	1.0000	1.0057	1.0832	1.0055	0.8298
35.0	3.5134E 00	0.7477	1.2484	1.2709	1.2685	0.8464	0.9689	1.0001	1.0000	1.0316	1.0897	1.0315	0.8471
40.0	3.2858E 00	0.7198	1.2935	1.3328	1.3371	0.8343	0.9638	1.0001	1.0000	1.0558	1.0875	1.0557	0.8636
45.0	3.0902E 00	0.6945	1.3347	1.3914	1.4027	0.8244	0.9586	1.0001	1.0000	1.0782	1.0776	1.0782	0.8790
50.0	2.9202E 00	0.6714	1.3719	1.4467	1.4655	0.8165	0.9532	1.0001	1.0000	1.0989	1.0607	1.0989	0.8935

Instr. see p. 22

ELECTRON Z= 63 A= 152

p	F_0L_0	L_0	λ_2	λ_3	λ_4	μ_1	μ_2	Λ_1	Λ_2	ν_{12}	$\hat{\nu}_{12}$	η_{12}	$\hat{\eta}_{12}$
0.1	1.1172E 02	1.0203	13.0969	66.5648	198.3906	0.9987	0.9999	1.0059	1.0008	4.5832	0.5938	0.1311	0.0820
0.2	5.6391E 01	1.0202	3.7726	7.1497	10.1677	0.9985	0.9999	1.0052	1.0008	1.6022	0.6145	0.4495	0.3812
0.3	3.8168E 01	1.0201	2.0481	2.6794	3.0445	0.9983	0.9999	1.0049	1.0007	1.0850	0.6337	0.5560	0.4804
0.4	2.9210E 01	1.0198	1.4472	1.5999	1.6608	0.9980	0.9998	1.0047	1.0007	0.9162	0.6511	0.6099	0.5299
0.5	2.3954E 01	1.0196	1.1717	1.1878	1.1806	0.9976	0.9998	1.0044	1.0007	0.8441	0.6669	0.6429	0.5599
0.6	2.0546E 01	1.0192	1.0245	0.9892	0.9603	0.9971	0.9997	1.0042	1.0007	0.8085	0.6810	0.6655	0.5801
0.7	1.8187E 01	1.0188	0.9380	0.8795	0.8422	0.9966	0.9996	1.0040	1.0006	0.7895	0.6937	0.6822	0.5948
0.8	1.6475E 01	1.0184	0.8839	0.8136	0.7725	0.9961	0.9996	1.0038	1.0006	0.7789	0.7051	0.6952	0.6061
0.9	1.5186E 01	1.0179	0.8484	0.7716	0.7288	0.9955	0.9995	1.0036	1.0006	0.7730	0.7153	0.7056	0.6151
1.0	1.4186E 01	1.0173	0.8245	0.7440	0.7002	0.9949	0.9994	1.0035	1.0006	0.7698	0.7246	0.7144	0.6226
1.2	1.2742E 01	1.0161	0.7969	0.7127	0.6682	0.9937	0.9992	1.0031	1.0005	0.7678	0.7406	0.7283	0.6342
1.4	1.1754E 01	1.0146	0.7838	0.6983	0.6538	0.9925	0.9990	1.0028	1.0005	0.7689	0.7540	0.7391	0.6432
1.6	1.1034E 01	1.0131	0.7783	0.6926	0.6482	0.9913	0.9989	1.0026	1.0004	0.7713	0.7655	0.7480	0.6504
1.8	1.0482E 01	1.0115	0.7770	0.6915	0.6474	0.9900	0.9987	1.0024	1.0004	0.7742	0.7756	0.7555	0.6564
2.0	1.0044E 01	1.0098	0.7781	0.6931	0.6493	0.9888	0.9985	1.0022	1.0003	0.7775	0.7845	0.7620	0.6616
2.2	9.6841E 00	1.0080	0.7807	0.6964	0.6529	0.9876	0.9983	1.0020	1.0003	0.7808	0.7926	0.7679	0.6661
2.4	9.3815E 00	1.0062	0.7842	0.7007	0.6575	0.9863	0.9982	1.0019	1.0003	0.7842	0.8000	0.7731	0.6702
2.6	9.1217E 00	1.0044	0.7884	0.7056	0.6627	0.9851	0.9980	1.0018	1.0003	0.7875	0.8068	0.7779	0.6739
2.8	8.8948E 00	1.0025	0.7928	0.7109	0.6683	0.9839	0.9978	1.0017	1.0003	0.7907	0.8132	0.7824	0.6773
3.0	8.6939E 00	1.0006	0.7975	0.7164	0.6741	0.9827	0.9976	1.0016	1.0002	0.7939	0.8192	0.7865	0.6804
3.2	8.5139E 00	0.9987	0.8023	0.7221	0.6801	0.9815	0.9975	1.0015	1.0002	0.7970	0.8248	0.7904	0.6834
3.4	8.3510E 00	0.9968	0.8072	0.7278	0.6861	0.9802	0.9973	1.0014	1.0002	0.8000	0.8302	0.7941	0.6861
3.6	8.2022E 00	0.9948	0.8121	0.7335	0.6921	0.9790	0.9971	1.0013	1.0002	0.8029	0.8354	0.7976	0.6887
3.8	8.0654E 00	0.9928	0.8169	0.7392	0.6981	0.9778	0.9969	1.0013	1.0002	0.8058	0.8403	0.8010	0.6911
4.0	7.9388E 00	0.9909	0.8218	0.7449	0.7041	0.9766	0.9968	1.0012	1.0002	0.8086	0.8451	0.8042	0.6934
4.5	7.6582E 00	0.9859	0.8337	0.7588	0.7187	0.9737	0.9963	1.0011	1.0002	0.8152	0.8562	0.8116	0.6988
5.0	7.4172E 00	0.9810	0.8453	0.7724	0.7329	0.9707	0.9959	1.0010	1.0002	0.8215	0.8666	0.8185	0.7036
5.5	7.2060E 00	0.9760	0.8564	0.7855	0.7467	0.9678	0.9955	1.0009	1.0001	0.8274	0.8764	0.8249	0.7081
6.0	7.0177E 00	0.9710	0.8673	0.7982	0.7600	0.9648	0.9951	1.0008	1.0001	0.8331	0.8855	0.8309	0.7122
6.5	6.8478E 00	0.9661	0.8777	0.8104	0.7730	0.9619	0.9946	1.0008	1.0001	0.8385	0.8943	0.8366	0.7161
7.0	6.6929E 00	0.9612	0.8878	0.8223	0.7855	0.9591	0.9942	1.0007	1.0001	0.8437	0.9026	0.8420	0.7197
7.5	6.5504E 00	0.9563	0.8977	0.8339	0.7977	0.9562	0.9938	1.0007	1.0001	0.8487	0.9105	0.8472	0.7232
8.0	6.4185E 00	0.9514	0.9072	0.8452	0.8096	0.9534	0.9934	1.0006	1.0001	0.8536	0.9182	0.8523	0.7265
9.0	6.1806E 00	0.9417	0.9256	0.8669	0.8325	0.9479	0.9925	1.0006	1.0001	0.8628	0.9325	0.8618	0.7327
10.0	5.9703E 00	0.9322	0.9431	0.8876	0.8545	0.9424	0.9916	1.0005	1.0001	0.8716	0.9459	0.8708	0.7385
11.0	5.7817E 00	0.9228	0.9598	0.9076	0.8756	0.9371	0.9908	1.0005	1.0001	0.8801	0.9584	0.8793	0.7440
12.0	5.6107E 00	0.9136	0.9760	0.9269	0.8960	0.9319	0.9899	1.0004	1.0001	0.8882	0.9700	0.8875	0.7492
13.0	5.4542E 00	0.9045	0.9915	0.9456	0.9159	0.9267	0.9890	1.0004	1.0001	0.8960	0.9810	0.8954	0.7542
14.0	5.3099E 00	0.8955	1.0066	0.9637	0.9352	0.9218	0.9881	1.0004	1.0001	0.9035	0.9912	0.9030	0.7591
16.0	5.0512E 00	0.8781	1.0356	0.9987	0.9725	0.9121	0.9864	1.0003	1.0001	0.9181	1.0097	0.9177	0.7683
18.0	4.8243E 00	0.8614	1.0631	1.0323	1.0084	0.9029	0.9845	1.0003	1.0000	0.9319	1.0258	0.9316	0.7771
20.0	4.6225E 00	0.8452	1.0893	1.0646	1.0431	0.8942	0.9827	1.0002	1.0000	0.9452	1.0397	0.9449	0.7856
25.0	4.1994E 00	0.8075	1.1505	1.1412	1.1256	0.8744	0.9780	1.0002	1.0000	0.9765	1.0659	0.9762	0.8058
30.0	3.8597E 00	0.7732	1.2062	1.2128	1.2035	0.8574	0.9731	1.0002	1.0000	1.0053	1.0811	1.0052	0.8249
35.0	3.5786E 00	0.7422	1.2571	1.2804	1.2777	0.8430	0.9681	1.0001	1.0000	1.0322	1.0865	1.0320	0.8430
40.0	3.3413E 00	0.7140	1.3036	1.3442	1.3485	0.8309	0.9629	1.0001	1.0000	1.0571	1.0832	1.0570	0.8601
45.0	3.1379E 00	0.6885	1.3458	1.4046	1.4163	0.8211	0.9575	1.0001	1.0000	1.0802	1.0721	1.0801	0.8762
50.0	2.9615E 00	0.6653	1.3840	1.4617	1.4810	0.8132	0.9520	1.0001	1.0000	1.1015	1.0541	1.1014	0.8912

Instr. see p. 22

ELECTRON Z= 64 A= 155

p	F_0L_0	L_0	λ_2	λ_3	λ_4	μ_1	μ_2	Λ_1	Λ_2	ν_{12}	$\hat{\nu}_{12}$	η_{12}	$\hat{\eta}_{12}$
0.1	1.1807E 02	1.0205	13.2886	68.9236	209.3027	0.9986	0.9999	1.0062	1.0008	4.6652	0.5891	0.0972	0.0488
0.2	5.9594E 01	1.0204	3.8148	7.3162	10.4925	0.9984	0.9999	1.0054	1.0008	1.6128	0.6101	0.4293	0.3602
0.3	4.0330E 01	1.0202	2.0626	2.7142	3.0944	0.9982	0.9999	1.0051	1.0008	1.0838	0.6295	0.5403	0.4634
0.4	3.0858E 01	1.0200	1.4520	1.6087	1.6716	0.9978	0.9998	1.0048	1.0007	0.9113	0.6472	0.5964	0.5149
0.5	2.5297E 01	1.0197	1.1720	1.1882	1.1807	0.9974	0.9998	1.0046	1.0007	0.8377	0.6632	0.6308	0.5460
0.6	2.1689E 01	1.0194	1.0224	0.9859	0.9563	0.9969	0.9997	1.0044	1.0007	0.8014	0.6776	0.6543	0.5670
0.7	1.9188E 01	1.0190	0.9345	0.8744	0.8362	0.9964	0.9996	1.0042	1.0007	0.7821	0.6904	0.6716	0.5822
0.8	1.7373E 01	1.0185	0.8794	0.8074	0.7654	0.9959	0.9995	1.0040	1.0006	0.7713	0.7020	0.6851	0.5939
0.9	1.6004E 01	1.0180	0.8434	0.7648	0.7211	0.9953	0.9995	1.0038	1.0006	0.7653	0.7124	0.6959	0.6032
1.0	1.4941E 01	1.0174	0.8191	0.7367	0.6921	0.9947	0.9994	1.0036	1.0006	0.7620	0.7218	0.7050	0.6109
1.2	1.3405E 01	1.0161	0.7910	0.7050	0.6597	0.9935	0.9992	1.0032	1.0005	0.7601	0.7381	0.7194	0.6229
1.4	1.2352E 01	1.0146	0.7778	0.6905	0.6452	0.9922	0.9990	1.0029	1.0005	0.7612	0.7518	0.7305	0.6321
1.6	1.1584E 01	1.0131	0.7722	0.6847	0.6395	0.9909	0.9988	1.0027	1.0004	0.7637	0.7635	0.7397	0.6395
1.8	1.0995E 01	1.0114	0.7709	0.6836	0.6387	0.9896	0.9986	1.0025	1.0004	0.7668	0.7738	0.7474	0.6457
2.0	1.0527E 01	1.0096	0.7721	0.6854	0.6407	0.9884	0.9985	1.0023	1.0004	0.7701	0.7829	0.7542	0.6510
2.2	1.0142E 01	1.0078	0.7748	0.6887	0.6444	0.9871	0.9983	1.0021	1.0003	0.7736	0.7911	0.7602	0.6557
2.4	9.8180E 00	1.0059	0.7785	0.6931	0.6491	0.9858	0.9981	1.0020	1.0003	0.7770	0.7987	0.7656	0.6599
2.6	9.5398E 00	1.0040	0.7827	0.6982	0.6545	0.9845	0.9979	1.0018	1.0003	0.7804	0.8056	0.7706	0.6637
2.8	9.2968E 00	1.0021	0.7873	0.7036	0.6602	0.9833	0.9978	1.0017	1.0003	0.7838	0.8121	0.7751	0.6672
3.0	9.0815E 00	1.0001	0.7922	0.7093	0.6662	0.9820	0.9976	1.0016	1.0003	0.7870	0.8182	0.7794	0.6704
3.2	8.8885E 00	0.9982	0.7971	0.7151	0.6722	0.9807	0.9974	1.0015	1.0002	0.7902	0.8240	0.7834	0.6734
3.4	8.7139E 00	0.9962	0.8021	0.7209	0.6784	0.9795	0.9972	1.0014	1.0002	0.7933	0.8294	0.7872	0.6762
3.6	8.5544E 00	0.9941	0.8072	0.7268	0.6846	0.9782	0.9970	1.0014	1.0002	0.7963	0.8347	0.7908	0.6788
3.8	8.4077E 00	0.9921	0.8122	0.7327	0.6907	0.9770	0.9969	1.0013	1.0002	0.7993	0.8397	0.7943	0.6813
4.0	8.2720E 00	0.9901	0.8172	0.7385	0.6968	0.9758	0.9967	1.0012	1.0002	0.8021	0.8445	0.7976	0.6837
4.5	7.9711E 00	0.9850	0.8295	0.7528	0.7118	0.9727	0.9963	1.0011	1.0002	0.8090	0.8559	0.8053	0.6892
5.0	7.7129E 00	0.9799	0.8414	0.7667	0.7264	0.9696	0.9958	1.0010	1.0002	0.8154	0.8664	0.8124	0.6942
5.5	7.4865E 00	0.9748	0.8529	0.7802	0.7406	0.9666	0.9954	1.0009	1.0001	0.8215	0.8763	0.8189	0.6988
6.0	7.2849E 00	0.9696	0.8641	0.7932	0.7542	0.9636	0.9949	1.0009	1.0001	0.8274	0.8856	0.8251	0.7030
6.5	7.1031E 00	0.9645	0.8749	0.8059	0.7675	0.9606	0.9945	1.0008	1.0001	0.8329	0.8944	0.8310	0.7070
7.0	6.9373E 00	0.9595	0.8853	0.8181	0.7804	0.9576	0.9941	1.0007	1.0001	0.8383	0.9028	0.8366	0.7108
7.5	6.7850E 00	0.9544	0.8954	0.8300	0.7929	0.9547	0.9936	1.0007	1.0001	0.8435	0.9108	0.8420	0.7144
8.0	6.6441E 00	0.9494	0.9053	0.8416	0.8051	0.9518	0.9932	1.0006	1.0001	0.8485	0.9185	0.8471	0.7178
9.0	6.3901E 00	0.9395	0.9243	0.8640	0.8287	0.9461	0.9923	1.0006	1.0001	0.8581	0.9330	0.8570	0.7242
10.0	6.1658E 00	0.9297	0.9423	0.8854	0.8513	0.9405	0.9914	1.0005	1.0001	0.8672	0.9465	0.8662	0.7302
11.0	5.9649E 00	0.9200	0.9597	0.9060	0.8731	0.9350	0.9905	1.0005	1.0001	0.8759	0.9589	0.8751	0.7359
12.0	5.7830E 00	0.9106	0.9764	0.9258	0.8941	0.9296	0.9896	1.0004	1.0001	0.8842	0.9706	0.8835	0.7413
13.0	5.6166E 00	0.9013	0.9925	0.9451	0.9146	0.9244	0.9887	1.0004	1.0001	0.8923	0.9815	0.8917	0.7465
14.0	5.4634E 00	0.8921	1.0081	0.9639	0.9345	0.9192	0.9878	1.0004	1.0001	0.9001	0.9917	0.8996	0.7515
16.0	5.1892E 00	0.8743	1.0380	1.0001	0.9730	0.9094	0.9860	1.0003	1.0001	0.9151	1.0100	0.9147	0.7611
18.0	4.9492E 00	0.8572	1.0664	1.0347	1.0100	0.9000	0.9841	1.0003	1.0000	0.9295	1.0259	0.9291	0.7703
20.0	4.7361E 00	0.8407	1.0936	1.0681	1.0458	0.8911	0.9823	1.0003	1.0000	0.9432	1.0396	0.9429	0.7792
25.0	4.2906E 00	0.8023	1.1568	1.1473	1.1311	0.8710	0.9774	1.0002	1.0000	0.9755	1.0650	0.9753	0.8003
30.0	3.9344E 00	0.7675	1.2143	1.2213	1.2117	0.8538	0.9724	1.0002	1.0000	1.0054	1.0791	1.0052	0.8202
35.0	3.6406E 00	0.7361	1.2668	1.2911	1.2883	0.8393	0.9672	1.0001	1.0000	1.0331	1.0833	1.0330	0.8391
40.0	3.3934E 00	0.7077	1.3147	1.3571	1.3615	0.8273	0.9619	1.0001	1.0000	1.0588	1.0787	1.0587	0.8569
45.0	3.1820E 00	0.6820	1.3581	1.4194	1.4315	0.8176	0.9563	1.0001	1.0000	1.0826	1.0663	1.0825	0.8737
50.0	2.9953E 00	0.6587	1.3972	1.4783	1.4983	0.8099	0.9507	1.0001	1.0000	1.1045	1.0470	1.1045	0.8893

Instr. see p. 22

ELECTRON Z= 65 A= 159

p	F_0L_0	L_0	λ_2	λ_3	λ_4	μ_1	μ_2	Λ_1	Λ_2	ν_{12}	$\hat{\nu}_{12}$	η_{12}	$\hat{\eta}_{12}$
0.1	1.2476E 02	1.0206	13.4782	71.3195	220.6397	0.9985	0.9999	1.0065	1.0008	4.7474	0.5845	0.0626	0.0150
0.2	6.2963E 01	1.0205	3.8566	7.4846	10.8253	0.9983	0.9999	1.0056	1.0008	1.6233	0.6057	0.4087	0.3390
0.3	4.2604E 01	1.0204	2.0770	2.7492	3.1452	0.9981	0.9999	1.0053	1.0008	1.0825	0.6254	0.5244	0.4462
0.4	3.2590E 01	1.0201	1.4568	1.6177	1.6827	0.9977	0.9998	1.0050	1.0008	0.9063	0.6433	0.5828	0.4998
0.5	2.6709E 01	1.0198	1.1724	1.1887	1.1809	0.9973	0.9998	1.0048	1.0007	0.8312	0.6595	0.6185	0.5320
0.6	2.2889E 01	1.0195	1.0204	0.9828	0.9524	0.9968	0.9997	1.0045	1.0007	0.7943	0.6741	0.6429	0.5538
0.7	2.0241E 01	1.0190	0.9311	0.8694	0.8304	0.9962	0.9996	1.0043	1.0007	0.7746	0.6872	0.6609	0.5696
0.8	1.8315E 01	1.0186	0.8751	0.8014	0.7585	0.9957	0.9995	1.0041	1.0006	0.7636	0.6990	0.6749	0.5816
0.9	1.6862E 01	1.0180	0.8385	0.7581	0.7136	0.9951	0.9994	1.0039	1.0006	0.7575	0.7096	0.6861	0.5913
1.0	1.5733E 01	1.0174	0.8138	0.7297	0.6842	0.9944	0.9994	1.0037	1.0006	0.7542	0.7191	0.6955	0.5992
1.2	1.4100E 01	1.0161	0.7853	0.6975	0.6514	0.9932	0.9992	1.0033	1.0005	0.7524	0.7357	0.7104	0.6116
1.4	1.2979E 01	1.0145	0.7719	0.6828	0.6367	0.9918	0.9990	1.0030	1.0005	0.7536	0.7497	0.7219	0.6210
1.6	1.2159E 01	1.0129	0.7663	0.6770	0.6311	0.9905	0.9988	1.0028	1.0004	0.7561	0.7616	0.7314	0.6286
1.8	1.1531E 01	1.0112	0.7650	0.6760	0.6303	0.9892	0.9986	1.0025	1.0004	0.7593	0.7721	0.7394	0.6350
2.0	1.1030E 01	1.0094	0.7663	0.6778	0.6324	0.9879	0.9984	1.0023	1.0004	0.7627	0.7814	0.7463	0.6405
2.2	1.0619E 01	1.0075	0.7691	0.6813	0.6362	0.9865	0.9982	1.0022	1.0003	0.7663	0.7897	0.7525	0.6453
2.4	1.0272E 01	1.0055	0.7729	0.6858	0.6410	0.9852	0.9981	1.0020	1.0003	0.7698	0.7974	0.7581	0.6496
2.6	9.9743E 00	1.0036	0.7773	0.6910	0.6465	0.9839	0.9979	1.0019	1.0003	0.7734	0.8045	0.7632	0.6534
2.8	9.7141E 00	1.0016	0.7820	0.6966	0.6524	0.9826	0.9977	1.0018	1.0003	0.7768	0.8111	0.7679	0.6570
3.0	9.4834E 00	0.9995	0.7870	0.7024	0.6585	0.9813	0.9975	1.0017	1.0003	0.7802	0.8173	0.7723	0.6603
3.2	9.2767E 00	0.9975	0.7921	0.7084	0.6647	0.9800	0.9973	1.0016	1.0002	0.7834	0.8232	0.7764	0.6634
3.4	9.0895E 00	0.9954	0.7973	0.7144	0.6710	0.9787	0.9971	1.0015	1.0002	0.7866	0.8288	0.7803	0.6663
3.6	8.9186E 00	0.9933	0.8025	0.7204	0.6773	0.9774	0.9970	1.0014	1.0002	0.7897	0.8341	0.7841	0.6690
3.8	8.7613E 00	0.9912	0.8077	0.7264	0.6836	0.9761	0.9968	1.0014	1.0002	0.7927	0.8392	0.7876	0.6715
4.0	8.6158E 00	0.9891	0.8128	0.7324	0.6899	0.9748	0.9966	1.0013	1.0002	0.7957	0.8441	0.7910	0.6740
4.5	8.2935E 00	0.9839	0.8255	0.7471	0.7053	0.9716	0.9962	1.0012	1.0002	0.8027	0.8557	0.7989	0.6796
5.0	8.0168E 00	0.9786	0.8378	0.7614	0.7203	0.9684	0.9957	1.0011	1.0002	0.8094	0.8664	0.8062	0.6848
5.5	7.7744E 00	0.9733	0.8497	0.7753	0.7348	0.9653	0.9953	1.0010	1.0001	0.8157	0.8764	0.8130	0.6895
6.0	7.5586E 00	0.9680	0.8612	0.7887	0.7488	0.9622	0.9948	1.0009	1.0001	0.8217	0.8858	0.8194	0.6938
6.5	7.3640E 00	0.9628	0.8723	0.8017	0.7624	0.9591	0.9944	1.0008	1.0001	0.8274	0.8947	0.8254	0.6979
7.0	7.1868E 00	0.9575	0.8831	0.8143	0.7757	0.9560	0.9939	1.0008	1.0001	0.8330	0.9032	0.8312	0.7018
7.5	7.0241E 00	0.9523	0.8936	0.8266	0.7886	0.9530	0.9935	1.0007	1.0001	0.8383	0.9113	0.8368	0.7055
8.0	6.8735E 00	0.9472	0.9038	0.8385	0.8011	0.9500	0.9930	1.0007	1.0001	0.8435	0.9191	0.8421	0.7090
9.0	6.6024E 00	0.9369	0.9234	0.8616	0.8254	0.9441	0.9921	1.0006	1.0001	0.8534	0.9337	0.8522	0.7156
10.0	6.3634E 00	0.9269	0.9421	0.8836	0.8487	0.9383	0.9912	1.0005	1.0001	0.8628	0.9472	0.8618	0.7219
11.0	6.1496E 00	0.9170	0.9600	0.9049	0.8711	0.9327	0.9903	1.0005	1.0001	0.8718	0.9598	0.8709	0.7277
12.0	5.9561E 00	0.9072	0.9773	0.9254	0.8928	0.9272	0.9894	1.0004	1.0001	0.8804	0.9714	0.8797	0.7334
13.0	5.7794E 00	0.8977	0.9940	0.9453	0.9139	0.9218	0.9884	1.0004	1.0001	0.8887	0.9823	0.8881	0.7388
14.0	5.6168E 00	0.8883	1.0101	0.9647	0.9345	0.9165	0.9875	1.0004	1.0001	0.8968	0.9924	0.8963	0.7440
16.0	5.3263E 00	0.8701	1.0411	1.0021	0.9743	0.9064	0.9856	1.0003	1.0001	0.9124	1.0106	0.9119	0.7540
18.0	5.0725E 00	0.8526	1.0705	1.0379	1.0125	0.8969	0.9837	1.0003	1.0000	0.9272	1.0263	0.9269	0.7636
20.0	4.8476E 00	0.8358	1.0986	1.0725	1.0495	0.8878	0.9818	1.0003	1.0000	0.9415	1.0397	0.9411	0.7729
25.0	4.3789E 00	0.7966	1.1640	1.1544	1.1378	0.8674	0.9768	1.0002	1.0000	0.9749	1.0642	0.9747	0.7949
30.0	4.0057E 00	0.7614	1.2234	1.2310	1.2211	0.8501	0.9716	1.0002	1.0000	1.0058	1.0771	1.0056	0.8157
35.0	3.6991E 00	0.7296	1.2776	1.3033	1.3004	0.8355	0.9663	1.0001	1.0000	1.0344	1.0800	1.0343	0.8355
40.0	3.4418E 00	0.7009	1.3269	1.3715	1.3761	0.8235	0.9607	1.0001	1.0000	1.0610	1.0740	1.0609	0.8541
45.0	3.2225E 00	0.6751	1.3715	1.4358	1.4485	0.8139	0.9550	1.0001	1.0000	1.0856	1.0602	1.0855	0.8716
50.0	3.0335E 00	0.6517	1.4115	1.4965	1.5175	0.8065	0.9492	1.0001	1.0000	1.1081	1.0395	1.1080	0.8878

Instr. see p. 22

ELECTRON Z= 66 A= 161

p	F_0L_0	L_0	λ_2	λ_3	λ_4	μ_1	μ_2	Λ_1	Λ_2	ν_{12}	$\hat{\nu}_{12}$	η_{12}	$\hat{\eta}_{12}$
0.1	1.3201E 02	1.0208	13.6539	73.6692	232.1172	0.9984	0.9999	1.0068	1.0008	4.8274	0.5796	0.0273	-0.0192
0.2	6.6616E 01	1.0207	3.8946	7.6462	11.1518	0.9982	0.9999	1.0058	1.0008	1.6329	0.6010	0.3876	0.3173
0.3	4.5069E 01	1.0205	2.0895	2.7814	3.1928	0.9979	0.9999	1.0055	1.0008	1.0806	0.6210	0.5080	0.4287
0.4	3.4468E 01	1.0203	1.4604	1.6250	1.6919	0.9976	0.9998	1.0052	1.0008	0.9008	0.6391	0.5688	0.4843
0.5	2.8239E 01	1.0200	1.1718	1.1880	1.1799	0.9971	0.9998	1.0049	1.0007	0.8243	0.6556	0.6059	0.5177
0.6	2.4191E 01	1.0196	1.0177	0.9787	0.9475	0.9966	0.9997	1.0047	1.0007	0.7867	0.6704	0.6312	0.5402
0.7	2.1381E 01	1.0191	0.9270	0.8637	0.8237	0.9961	0.9996	1.0044	1.0007	0.7667	0.6837	0.6499	0.5566
0.8	1.9336E 01	1.0186	0.8702	0.7946	0.7509	0.9955	0.9995	1.0042	1.0007	0.7556	0.6956	0.6643	0.5690
0.9	1.7792E 01	1.0181	0.8330	0.7508	0.7054	0.9948	0.9994	1.0040	1.0006	0.7494	0.7064	0.6760	0.5790
1.0	1.6591E 01	1.0174	0.8080	0.7220	0.6757	0.9942	0.9993	1.0038	1.0006	0.7461	0.7161	0.6856	0.5871
1.2	1.4852E 01	1.0161	0.7791	0.6894	0.6426	0.9929	0.9991	1.0034	1.0005	0.7442	0.7330	0.7010	0.5999
1.4	1.3656E 01	1.0145	0.7655	0.6746	0.6278	0.9915	0.9990	1.0031	1.0005	0.7455	0.7472	0.7129	0.6096
1.6	1.2781E 01	1.0128	0.7598	0.6688	0.6221	0.9902	0.9988	1.0029	1.0004	0.7481	0.7593	0.7227	0.6174
1.8	1.2110E 01	1.0110	0.7586	0.6678	0.6214	0.9888	0.9986	1.0026	1.0004	0.7514	0.7700	0.7309	0.6240
2.0	1.1574E 01	1.0091	0.7600	0.6697	0.6235	0.9874	0.9984	1.0024	1.0004	0.7550	0.7794	0.7381	0.6296
2.2	1.1134E 01	1.0072	0.7629	0.6733	0.6274	0.9860	0.9982	1.0022	1.0003	0.7586	0.7880	0.7444	0.6345
2.4	1.0763E 01	1.0052	0.7668	0.6780	0.6324	0.9847	0.9980	1.0021	1.0003	0.7623	0.7958	0.7502	0.6389
2.6	1.0444E 01	1.0032	0.7713	0.6833	0.6380	0.9833	0.9978	1.0019	1.0003	0.7659	0.8030	0.7554	0.6429
2.8	1.0165E 01	1.0011	0.7762	0.6890	0.6440	0.9820	0.9976	1.0018	1.0003	0.7694	0.8097	0.7603	0.6465
3.0	9.9174E 00	0.9990	0.7814	0.6949	0.6502	0.9806	0.9975	1.0017	1.0003	0.7729	0.8161	0.7648	0.6499
3.2	9.6957E 00	0.9969	0.7866	0.7010	0.6566	0.9793	0.9973	1.0016	1.0002	0.7762	0.8220	0.7690	0.6530
3.4	9.4949E 00	0.9948	0.7919	0.7072	0.6630	0.9779	0.9971	1.0015	1.0002	0.7795	0.8277	0.7731	0.6560
3.6	9.3116E 00	0.9926	0.7973	0.7134	0.6695	0.9766	0.9969	1.0015	1.0002	0.7827	0.8331	0.7769	0.6588
3.8	9.1430E 00	0.9905	0.8026	0.7195	0.6759	0.9752	0.9967	1.0014	1.0002	0.7858	0.8383	0.7805	0.6614
4.0	8.9870E 00	0.9883	0.8079	0.7257	0.6823	0.9739	0.9965	1.0013	1.0002	0.7888	0.8433	0.7840	0.6639
4.5	8.6413E 00	0.9829	0.8210	0.7408	0.6981	0.9706	0.9961	1.0012	1.0002	0.7961	0.8550	0.7922	0.6697
5.0	8.3448E 00	0.9775	0.8336	0.7554	0.7134	0.9673	0.9956	1.0011	1.0002	0.8029	0.8659	0.7997	0.6749
5.5	8.0851E 00	0.9720	0.8459	0.7696	0.7282	0.9641	0.9951	1.0010	1.0002	0.8094	0.8760	0.8066	0.6798
6.0	7.8540E 00	0.9666	0.8577	0.7834	0.7426	0.9609	0.9947	1.0009	1.0001	0.8156	0.8856	0.8132	0.6843
6.5	7.6458E 00	0.9612	0.8692	0.7967	0.7566	0.9577	0.9942	1.0008	1.0001	0.8215	0.8946	0.8195	0.6885
7.0	7.4562E 00	0.9558	0.8803	0.8097	0.7702	0.9546	0.9938	1.0008	1.0001	0.8272	0.9032	0.8254	0.6925
7.5	7.2823E 00	0.9505	0.8911	0.8223	0.7834	0.9514	0.9933	1.0007	1.0001	0.8327	0.9113	0.8311	0.6963
8.0	7.1214E 00	0.9452	0.9016	0.8346	0.7963	0.9484	0.9928	1.0007	1.0001	0.8380	0.9192	0.8366	0.6999
9.0	6.8322E 00	0.9347	0.9218	0.8583	0.8212	0.9423	0.9919	1.0006	1.0001	0.8482	0.9338	0.8470	0.7067
10.0	6.5774E 00	0.9244	0.9411	0.8810	0.8451	0.9364	0.9910	1.0006	1.0001	0.8579	0.9474	0.8569	0.7132
11.0	6.3497E 00	0.9142	0.9596	0.9029	0.8681	0.9306	0.9900	1.0005	1.0001	0.8672	0.9600	0.8663	0.7192
12.0	6.1440E 00	0.9043	0.9774	0.9240	0.8905	0.9249	0.9891	1.0005	1.0001	0.8761	0.9717	0.8753	0.7250
13.0	5.9563E 00	0.8945	0.9946	0.9445	0.9122	0.9194	0.9882	1.0004	1.0001	0.8847	0.9825	0.8840	0.7306
14.0	5.7838E 00	0.8849	1.0113	0.9645	0.9333	0.9141	0.9872	1.0004	1.0001	0.8930	0.9926	0.8925	0.7361
16.0	5.4761E 00	0.8664	1.0433	1.0030	0.9743	0.9038	0.9853	1.0003	1.0001	0.9091	1.0106	0.9086	0.7465
18.0	5.2079E 00	0.8485	1.0737	1.0400	1.0137	0.8940	0.9833	1.0003	1.0000	0.9244	1.0261	0.9240	0.7565
20.0	4.9706E 00	0.8314	1.1027	1.0756	1.0518	0.8848	0.9813	1.0003	1.0000	0.9391	1.0393	0.9388	0.7661
25.0	4.4776E 00	0.7917	1.1701	1.1601	1.1428	0.8641	0.9762	1.0002	1.0000	0.9736	1.0629	0.9734	0.7891
30.0	4.0865E 00	0.7560	1.2314	1.2392	1.2287	0.8467	0.9709	1.0002	1.0000	1.0055	1.0749	1.0053	0.8108
35.0	3.7663E 00	0.7239	1.2871	1.3136	1.3105	0.8321	0.9654	1.0001	1.0000	1.0350	1.0768	1.0349	0.8313
40.0	3.4985E 00	0.6950	1.3378	1.3839	1.3886	0.8202	0.9598	1.0001	1.0000	1.0624	1.0697	1.0623	0.8507
45.0	3.2709E 00	0.6691	1.3835	1.4503	1.4632	0.8107	0.9539	1.0001	1.0000	1.0877	1.0547	1.0876	0.8688
50.0	3.0751E 00	0.6456	1.4245	1.5127	1.5343	0.8034	0.9479	1.0001	1.0000	1.1108	1.0329	1.1108	0.8857

ELECTRON Z= 67 A= 165

p	F_0L_0	L_0	λ_2	λ_3	λ_4	μ_1	μ_2	Λ_1	Λ_2	ν_{12}	$\hat{\nu}_{12}$	η_{12}	$\hat{\eta}_{12}$
0.1	1.3958E 02	1.0208	13.8312	76.0778	244.1086	0.9983	0.9999	1.0071	1.0008	4.9084	0.5748	-0.0088	-0.0540
0.2	7.0426E 01	1.0207	3.9332	7.8122	11.4906	0.9981	0.9999	1.0061	1.0008	1.6428	0.5965	0.3662	0.2955
0.3	4.7639E 01	1.0205	2.1025	2.8150	3.2426	0.9978	0.9999	1.0057	1.0008	1.0788	0.6167	0.4915	0.4111
0.4	3.6426E 01	1.0203	1.4644	1.6330	1.7021	0.9974	0.9998	1.0054	1.0008	0.8955	0.6351	0.5547	0.4688
0.5	2.9834E 01	1.0200	1.1717	1.1879	1.1796	0.9970	0.9997	1.0051	1.0008	0.8175	0.6518	0.5932	0.5034
0.6	2.5546E 01	1.0196	1.0153	0.9752	0.9432	0.9964	0.9997	1.0048	1.0007	0.7792	0.6668	0.6195	0.5267
0.7	2.2568E 01	1.0191	0.9233	0.8584	0.8176	0.9959	0.9996	1.0046	1.0007	0.7589	0.6803	0.6388	0.5436
0.8	2.0399E 01	1.0186	0.8657	0.7884	0.7438	0.9952	0.9995	1.0044	1.0007	0.7476	0.6924	0.6538	0.5565
0.9	1.8759E 01	1.0180	0.8280	0.7440	0.6977	0.9946	0.9994	1.0041	1.0006	0.7413	0.7034	0.6658	0.5667
1.0	1.7483E 01	1.0174	0.8026	0.7148	0.6677	0.9939	0.9993	1.0039	1.0006	0.7380	0.7133	0.6759	0.5751
1.2	1.5632E 01	1.0159	0.7732	0.6819	0.6342	0.9926	0.9991	1.0036	1.0005	0.7362	0.7304	0.6917	0.5883
1.4	1.4358E 01	1.0143	0.7595	0.6669	0.6193	0.9912	0.9989	1.0032	1.0005	0.7375	0.7449	0.7040	0.5983
1.6	1.3425E 01	1.0126	0.7538	0.6610	0.6136	0.9897	0.9987	1.0030	1.0005	0.7403	0.7573	0.7141	0.6063
1.8	1.2708E 01	1.0107	0.7526	0.6601	0.6130	0.9883	0.9985	1.0027	1.0004	0.7437	0.7681	0.7225	0.6130
2.0	1.2135E 01	1.0088	0.7541	0.6621	0.6152	0.9869	0.9983	1.0025	1.0004	0.7473	0.7777	0.7299	0.6187
2.2	1.1664E 01	1.0067	0.7571	0.6658	0.6192	0.9855	0.9982	1.0023	1.0004	0.7511	0.7864	0.7364	0.6237
2.4	1.1267E 01	1.0047	0.7612	0.6706	0.6243	0.9840	0.9980	1.0022	1.0003	0.7549	0.7944	0.7423	0.6282
2.6	1.0926E 01	1.0026	0.7658	0.6761	0.6300	0.9826	0.9978	1.0020	1.0003	0.7586	0.8018	0.7477	0.6323
2.8	1.0627E 01	1.0004	0.7709	0.6819	0.6361	0.9812	0.9976	1.0019	1.0003	0.7622	0.8086	0.7527	0.6361
3.0	1.0362E 01	0.9983	0.7762	0.6880	0.6425	0.9798	0.9974	1.0018	1.0003	0.7657	0.8151	0.7574	0.6395
3.2	1.0124E 01	0.9961	0.7816	0.6943	0.6490	0.9784	0.9972	1.0017	1.0003	0.7692	0.8211	0.7618	0.6427
3.4	9.9091E 00	0.9939	0.7871	0.7006	0.6556	0.9770	0.9970	1.0016	1.0002	0.7726	0.8269	0.7659	0.6458
3.6	9.7127E 00	0.9917	0.7926	0.7069	0.6622	0.9756	0.9968	1.0015	1.0002	0.7758	0.8324	0.7698	0.6486
3.8	9.5320E 00	0.9895	0.7981	0.7133	0.6688	0.9743	0.9966	1.0014	1.0002	0.7790	0.8377	0.7736	0.6513
4.0	9.3648E 00	0.9872	0.8035	0.7196	0.6754	0.9729	0.9964	1.0014	1.0002	0.7821	0.8428	0.7772	0.6539
4.5	8.9944E 00	0.9816	0.8170	0.7350	0.6915	0.9695	0.9960	1.0012	1.0002	0.7896	0.8547	0.7856	0.6598
5.0	8.6768E 00	0.9760	0.8300	0.7501	0.7072	0.9661	0.9955	1.0011	1.0002	0.7966	0.8657	0.7933	0.6652
5.5	8.3987E 00	0.9704	0.8427	0.7647	0.7224	0.9627	0.9950	1.0010	1.0002	0.8033	0.8760	0.8005	0.6702
6.0	8.1515E 00	0.9648	0.8549	0.7788	0.7372	0.9594	0.9945	1.0009	1.0001	0.8097	0.8857	0.8072	0.6748
6.5	7.9288E 00	0.9592	0.8667	0.7925	0.7515	0.9561	0.9941	1.0009	1.0001	0.8158	0.8948	0.8137	0.6791
7.0	7.7262E 00	0.9537	0.8782	0.8059	0.7655	0.9529	0.9936	1.0008	1.0001	0.8217	0.9035	0.8198	0.6832
7.5	7.5404E 00	0.9482	0.8893	0.8189	0.7791	0.9497	0.9931	1.0008	1.0001	0.8273	0.9118	0.8257	0.6871
8.0	7.3687E 00	0.9428	0.9002	0.8315	0.7923	0.9465	0.9927	1.0007	1.0001	0.8328	0.9197	0.8313	0.6909
9.0	7.0602E 00	0.9320	0.9211	0.8559	0.8179	0.9402	0.9917	1.0006	1.0001	0.8433	0.9345	0.8421	0.6979
10.0	6.7889E 00	0.9214	0.9410	0.8793	0.8425	0.9341	0.9907	1.0006	1.0001	0.8533	0.9481	0.8523	0.7046
11.0	6.5467E 00	0.9110	0.9601	0.9019	0.8663	0.9282	0.9898	1.0005	1.0001	0.8629	0.9607	0.8620	0.7109
12.0	6.3280E 00	0.9008	0.9785	0.9237	0.8893	0.9224	0.9888	1.0005	1.0001	0.8721	0.9724	0.8713	0.7169
13.0	6.1288E 00	0.8908	0.9963	0.9449	0.9117	0.9167	0.9878	1.0004	1.0001	0.8810	0.9832	0.8803	0.7227
14.0	5.9460E 00	0.8809	1.0136	0.9655	0.9335	0.9112	0.9869	1.0004	1.0001	0.8896	0.9933	0.8890	0.7284
16.0	5.6203E 00	0.8619	1.0467	1.0053	0.9758	0.9007	0.9849	1.0003	1.0001	0.9062	1.0112	0.9057	0.7392
18.0	5.3369E 00	0.8437	1.0781	1.0435	1.0165	0.8908	0.9829	1.0003	1.0000	0.9220	1.0265	0.9217	0.7496
20.0	5.0868E 00	0.8263	1.1081	1.0804	1.0559	0.8814	0.9808	1.0003	1.0000	0.9372	1.0393	0.9369	0.7597
25.0	4.5687E 00	0.7859	1.1778	1.1677	1.1499	0.8605	0.9756	1.0002	1.0000	0.9730	1.0620	0.9727	0.7836
30.0	4.1595E 00	0.7497	1.2410	1.2495	1.2387	0.8429	0.9701	1.0002	1.0000	1.0059	1.0729	1.0057	0.8063
35.0	3.8256E 00	0.7173	1.2985	1.3264	1.3233	0.8283	0.9645	1.0001	1.0000	1.0364	1.0734	1.0363	0.8278
40.0	3.5473E 00	0.6883	1.3506	1.3991	1.4040	0.8165	0.9586	1.0001	1.0000	1.0647	1.0649	1.0646	0.8479
45.0	3.3115E 00	0.6622	1.3975	1.4675	1.4811	0.8072	0.9526	1.0001	1.0000	1.0907	1.0486	1.0906	0.8668
50.0	3.1092E 00	0.6387	1.4395	1.5319	1.5545	0.8002	0.9464	1.0001	1.0000	1.1145	1.0253	1.1145	0.8843

 Instr. see p. 22

ELECTRON Z= 68 A= 166

p	F_0L_0	L_0	λ_2	λ_3	λ_4	μ_1	μ_2	Λ_1	Λ_2	ν_{12}	$\hat{\nu}_{12}$	η_{12}	$\hat{\eta}_{12}$
0.1	1.4787E 02	1.0209	13.9893	78.3968	256.0846	0.9982	0.9999	1.0074	1.0009	4.9859	0.5697	-0.0455	-0.0893
0.2	7.4603E 01	1.0208	3.9667	7.9674	11.8164	0.9980	0.9999	1.0063	1.0009	1.6514	0.5916	0.3443	0.2732
0.3	5.0457E 01	1.0206	2.1129	2.8444	3.2874	0.9977	0.9999	1.0058	1.0008	1.0762	0.6120	0.4745	0.3931
0.4	3.8572E 01	1.0204	1.4667	1.6387	1.7095	0.9973	0.9998	1.0055	1.0008	0.8894	0.6307	0.5401	0.4528
0.5	3.1582E 01	1.0200	1.1703	1.1862	1.1774	0.9968	0.9997	1.0052	1.0008	0.8101	0.6476	0.5801	0.4887
0.6	2.7032E 01	1.0196	1.0118	0.9703	0.9375	0.9963	0.9997	1.0050	1.0007	0.7712	0.6628	0.6073	0.5128
0.7	2.3869E 01	1.0192	0.9187	0.8520	0.8102	0.9957	0.9996	1.0047	1.0007	0.7505	0.6765	0.6273	0.5302
0.8	2.1563E 01	1.0186	0.8603	0.7811	0.7357	0.9950	0.9995	1.0045	1.0007	0.7391	0.6888	0.6428	0.5435
0.9	1.9819E 01	1.0180	0.8221	0.7362	0.6891	0.9944	0.9994	1.0043	1.0006	0.7328	0.6999	0.6553	0.5541
1.0	1.8460E 01	1.0173	0.7964	0.7067	0.6588	0.9937	0.9993	1.0041	1.0006	0.7295	0.7100	0.6656	0.5627
1.2	1.6488E 01	1.0159	0.7666	0.6734	0.6250	0.9923	0.9991	1.0037	1.0006	0.7277	0.7274	0.6820	0.5762
1.4	1.5128E 01	1.0142	0.7528	0.6584	0.6100	0.9908	0.9989	1.0033	1.0005	0.7291	0.7421	0.6946	0.5865
1.6	1.4130E 01	1.0124	0.7471	0.6525	0.6043	0.9894	0.9987	1.0030	1.0005	0.7319	0.7547	0.7050	0.5947
1.8	1.3363E 01	1.0105	0.7459	0.6516	0.6037	0.9879	0.9985	1.0028	1.0004	0.7354	0.7657	0.7137	0.6015
2.0	1.2750E 01	1.0085	0.7475	0.6537	0.6060	0.9864	0.9983	1.0026	1.0004	0.7392	0.7755	0.7212	0.6074
2.2	1.2245E 01	1.0064	0.7506	0.6575	0.6101	0.9850	0.9981	1.0024	1.0004	0.7430	0.7844	0.7280	0.6126
2.4	1.1820E 01	1.0043	0.7548	0.6624	0.6153	0.9835	0.9979	1.0022	1.0003	0.7469	0.7925	0.7340	0.6172
2.6	1.1453E 01	1.0021	0.7596	0.6680	0.6211	0.9820	0.9977	1.0021	1.0003	0.7507	0.8000	0.7396	0.6214
2.8	1.1133E 01	0.9999	0.7648	0.6740	0.6274	0.9806	0.9975	1.0019	1.0003	0.7544	0.8069	0.7447	0.6252
3.0	1.0849E 01	0.9977	0.7702	0.6802	0.6339	0.9791	0.9973	1.0018	1.0003	0.7581	0.8135	0.7495	0.6287
3.2	1.0594E 01	0.9955	0.7758	0.6866	0.6406	0.9777	0.9971	1.0017	1.0003	0.7616	0.8197	0.7540	0.6320
3.4	1.0363E 01	0.9932	0.7814	0.6931	0.6473	0.9763	0.9969	1.0016	1.0002	0.7651	0.8255	0.7582	0.6351
3.6	1.0152E 01	0.9909	0.7870	0.6996	0.6540	0.9748	0.9967	1.0016	1.0002	0.7684	0.8311	0.7622	0.6380
3.8	9.9583E 00	0.9886	0.7927	0.7060	0.6608	0.9734	0.9966	1.0015	1.0002	0.7717	0.8365	0.7661	0.6408
4.0	9.7789E 00	0.9864	0.7983	0.7125	0.6674	0.9720	0.9964	1.0014	1.0002	0.7749	0.8417	0.7698	0.6434
4.5	9.3817E 00	0.9806	0.8121	0.7283	0.6839	0.9685	0.9959	1.0013	1.0002	0.7825	0.8538	0.7784	0.6495
5.0	9.0412E 00	0.9748	0.8255	0.7437	0.6999	0.9650	0.9954	1.0011	1.0002	0.7898	0.8649	0.7863	0.6550
5.5	8.7433E 00	0.9691	0.8385	0.7587	0.7155	0.9615	0.9949	1.0010	1.0002	0.7966	0.8753	0.7937	0.6601
6.0	8.4785E 00	0.9634	0.8511	0.7732	0.7306	0.9581	0.9944	1.0010	1.0001	0.8032	0.8851	0.8007	0.6648
6.5	8.2402E 00	0.9577	0.8633	0.7872	0.7453	0.9547	0.9940	1.0009	1.0001	0.8094	0.8943	0.8073	0.6693
7.0	8.0236E 00	0.9520	0.8751	0.8009	0.7595	0.9514	0.9935	1.0008	1.0001	0.8155	0.9031	0.8136	0.6735
7.5	7.8250E 00	0.9464	0.8865	0.8142	0.7734	0.9481	0.9930	1.0008	1.0001	0.8213	0.9114	0.8196	0.6775
8.0	7.6417E 00	0.9408	0.8977	0.8272	0.7870	0.9449	0.9925	1.0007	1.0001	0.8269	0.9194	0.8254	0.6814
9.0	7.3125E 00	0.9297	0.9192	0.8522	0.8133	0.9385	0.9915	1.0006	1.0001	0.8378	0.9343	0.8365	0.6886
10.0	7.0234E 00	0.9189	0.9397	0.8763	0.8385	0.9322	0.9906	1.0006	1.0001	0.8480	0.9480	0.8470	0.6955
11.0	6.7657E 00	0.9083	0.9594	0.8994	0.8628	0.9261	0.9896	1.0005	1.0001	0.8579	0.9606	0.8570	0.7020
12.0	6.5333E 00	0.8979	0.9784	0.9219	0.8865	0.9202	0.9886	1.0005	1.0001	0.8673	0.9723	0.8666	0.7082
13.0	6.3219E 00	0.8877	0.9967	0.9436	0.9094	0.9144	0.9876	1.0004	1.0001	0.8765	0.9831	0.8758	0.7142
14.0	6.1280E 00	0.8777	1.0145	0.9648	0.9319	0.9088	0.9866	1.0004	1.0001	0.8854	0.9931	0.8848	0.7201
16.0	5.7831E 00	0.8583	1.0486	1.0058	0.9753	0.8981	0.9846	1.0004	1.0001	0.9025	1.0109	0.9020	0.7313
18.0	5.4838E 00	0.8398	1.0810	1.0451	1.0171	0.8880	0.9825	1.0003	1.0001	0.9188	1.0259	0.9184	0.7421
20.0	5.2201E 00	0.8221	1.1119	1.0830	1.0576	0.8785	0.9804	1.0003	1.0000	0.9345	1.0386	0.9341	0.7525
25.0	4.6755E 00	0.7812	1.1837	1.1730	1.1544	0.8574	0.9751	1.0002	1.0000	0.9713	1.0606	0.9710	0.7775
30.0	4.2469E 00	0.7447	1.2487	1.2571	1.2458	0.8397	0.9695	1.0002	1.0000	1.0052	1.0706	1.0050	0.8010
35.0	3.8986E 00	0.7120	1.3078	1.3364	1.3329	0.8251	0.9637	1.0001	1.0000	1.0366	1.0702	1.0365	0.8233
40.0	3.6090E 00	0.6828	1.3613	1.4111	1.4160	0.8133	0.9577	1.0001	1.0000	1.0657	1.0608	1.0656	0.8443
45.0	3.3643E 00	0.6566	1.4094	1.4815	1.4952	0.8042	0.9516	1.0001	1.0000	1.0924	1.0435	1.0924	0.8638
50.0	3.1549E 00	0.6331	1.4523	1.5477	1.5708	0.7973	0.9452	1.0001	1.0000	1.1169	1.0193	1.1168	0.8819

Instr. see p. 22

ELECTRON Z= 69 A= 171

p	F_0L_0	L_0	λ_2	λ_3	λ_4	μ_1	μ_2	Λ_1	Λ_2	ν_{12}	$\hat{\nu}_{12}$	η_{12}	$\hat{\eta}_{12}$
0.1	1.5636E 02	1.0208	14.1575	80.8338	268.8020	0.9980	0.9999	1.0078	1.0009	5.0660	0.5649	0.0829	-0.1250
0.2	7.8878E 01	1.0207	4.0032	8.1330	12.1646	0.9978	0.9999	1.0066	1.0009	1.6608	0.5870	0.3222	0.2508
0.3	5.3339E 01	1.0205	2.1251	2.8773	3.3372	0.9975	0.9999	1.0061	1.0009	1.0740	0.6077	0.4574	0.3751
0.4	4.0766E 01	1.0202	1.4703	1.6464	1.7195	0.9971	0.9998	1.0057	1.0008	0.8838	0.6266	0.5256	0.4370
0.5	3.3368E 01	1.0199	1.1700	1.1859	1.1769	0.9966	0.9997	1.0054	1.0008	0.8030	0.6437	0.5670	0.4741
0.6	2.8549E 01	1.0195	1.0094	0.9668	0.9332	0.9961	0.9997	1.0052	1.0008	0.7635	0.6592	0.5953	0.4990
0.7	2.5197E 01	1.0190	0.9150	0.8468	0.8042	0.9955	0.9996	1.0049	1.0007	0.7425	0.6731	0.6160	0.5170
0.8	2.2751E 01	1.0184	0.8558	0.7750	0.7287	0.9948	0.9995	1.0047	1.0007	0.7309	0.6856	0.6320	0.5307
0.9	2.0859E 01	1.0178	0.8171	0.7295	0.6816	0.9941	0.9994	1.0044	1.0007	0.7245	0.6969	0.6449	0.5416
1.0	1.9454E 01	1.0171	0.7910	0.6997	0.6510	0.9934	0.9993	1.0042	1.0006	0.7212	0.7071	0.6556	0.5505
1.2	1.7356E 01	1.0156	0.7609	0.6660	0.6169	0.9919	0.9991	1.0038	1.0006	0.7195	0.7248	0.6725	0.5644
1.4	1.5907E 01	1.0138	0.7469	0.6509	0.6018	0.9904	0.9989	1.0034	1.0005	0.7210	0.7398	0.6855	0.5749
1.6	1.4843E 01	1.0120	0.7412	0.6450	0.5961	0.9889	0.9987	1.0031	1.0005	0.7239	0.7526	0.6962	0.5833
1.8	1.4024E 01	1.0100	0.7401	0.6442	0.5955	0.9873	0.9985	1.0029	1.0004	0.7275	0.7638	0.7051	0.5904
2.0	1.3369E 01	1.0079	0.7417	0.6463	0.5980	0.9858	0.9983	1.0027	1.0004	0.7314	0.7738	0.7129	0.5964
2.2	1.2829E 01	1.0058	0.7450	0.6502	0.6022	0.9843	0.9981	1.0025	1.0004	0.7354	0.7828	0.7198	0.6017
2.4	1.2374E 01	1.0036	0.7493	0.6553	0.6075	0.9828	0.9979	1.0023	1.0003	0.7393	0.7911	0.7261	0.6064
2.6	1.1982E 01	1.0013	0.7543	0.6610	0.6135	0.9813	0.9977	1.0021	1.0003	0.7432	0.7987	0.7318	0.6106
2.8	1.1638E 01	0.9991	0.7596	0.6672	0.6199	0.9798	0.9975	1.0020	1.0003	0.7471	0.8058	0.7370	0.6145
3.0	1.1334E 01	0.9968	0.7652	0.6736	0.6265	0.9783	0.9973	1.0019	1.0003	0.7508	0.8125	0.7419	0.6182
3.2	1.1061E 01	0.9944	0.7710	0.6801	0.6333	0.9768	0.9971	1.0018	1.0003	0.7544	0.8188	0.7466	0.6215
3.4	1.0814E 01	0.9921	0.7768	0.6868	0.6402	0.9753	0.9969	1.0017	1.0003	0.7580	0.8248	0.7509	0.6247
3.6	1.0588E 01	0.9897	0.7826	0.6934	0.6471	0.9738	0.9967	1.0016	1.0002	0.7614	0.8305	0.7551	0.6277
3.8	1.0380E 01	0.9874	0.7884	0.7000	0.6540	0.9723	0.9965	1.0015	1.0002	0.7648	0.8360	0.7590	0.6305
4.0	1.0188E 01	0.9850	0.7942	0.7066	0.6608	0.9708	0.9963	1.0015	1.0002	0.7681	0.8412	0.7628	0.6332
4.5	9.7626E 00	0.9790	0.8084	0.7229	0.6777	0.9672	0.9958	1.0013	1.0002	0.7760	0.8535	0.7717	0.6394
5.0	9.3981E 00	0.9731	0.8222	0.7387	0.6941	0.9636	0.9953	1.0012	1.0002	0.7834	0.8649	0.7798	0.6451
5.5	9.0794E 00	0.9672	0.8356	0.7541	0.7101	0.9600	0.9948	1.0011	1.0002	0.7904	0.8754	0.7874	0.6503
6.0	8.7962E 00	0.9612	0.8486	0.7690	0.7256	0.9565	0.9943	1.0010	1.0001	0.7972	0.8853	0.7946	0.6552
6.5	8.5415E 00	0.9554	0.8612	0.7835	0.7406	0.9530	0.9938	1.0009	1.0001	0.8036	0.8947	0.8014	0.6597
7.0	8.3102E 00	0.9495	0.8734	0.7975	0.7553	0.9495	0.9933	1.0009	1.0001	0.8099	0.9036	0.8079	0.6641
7.5	8.0982E 00	0.9437	0.8852	0.8112	0.7696	0.9461	0.9928	1.0008	1.0001	0.8159	0.9120	0.8141	0.6682
8.0	7.9027E 00	0.9380	0.8967	0.8246	0.7835	0.9428	0.9923	1.0008	1.0001	0.8217	0.9200	0.8201	0.6722
9.0	7.5519E 00	0.9266	0.9189	0.8504	0.8106	0.9362	0.9913	1.0007	1.0001	0.8328	0.9350	0.8316	0.6797
10.0	7.2442E 00	0.9155	0.9402	0.8752	0.8365	0.9297	0.9903	1.0006	1.0001	0.8435	0.9488	0.8424	0.6868
11.0	6.9703E 00	0.9046	0.9605	0.8991	0.8616	0.9235	0.9893	1.0005	1.0001	0.8536	0.9615	0.8527	0.6935
12.0	6.7236E 00	0.8939	0.9802	0.9223	0.8860	0.9174	0.9883	1.0005	1.0001	0.8634	0.9732	0.8626	0.7000
13.0	6.4994E 00	0.8834	0.9991	0.9448	0.9097	0.9115	0.9872	1.0005	1.0001	0.8729	0.9840	0.8722	0.7063
14.0	6.2941E 00	0.8732	1.0175	0.9667	0.9328	0.9058	0.9862	1.0004	1.0001	0.8821	0.9939	0.8815	0.7123
16.0	5.9296E 00	0.8534	1.0528	1.0090	0.9777	0.8948	0.9841	1.0004	1.0001	0.8998	1.0115	0.8993	0.7240
18.0	5.6138E 00	0.8345	1.0863	1.0496	1.0209	0.8845	0.9820	1.0003	1.0001	0.9167	1.0264	0.9162	0.7353
20.0	5.3362E 00	0.8165	1.1183	1.0889	1.0628	0.8748	0.9798	1.0003	1.0000	0.9329	1.0387	0.9325	0.7462
25.0	4.7647E 00	0.7749	1.1925	1.1820	1.1629	0.8534	0.9743	1.0002	1.0000	0.9710	1.0597	0.9707	0.7723
30.0	4.3169E 00	0.7379	1.2597	1.2690	1.2575	0.8357	0.9686	1.0002	1.0000	1.0061	1.0684	1.0059	0.7969
35.0	3.9544E 00	0.7050	1.3206	1.3510	1.3476	0.8211	0.9626	1.0001	1.0000	1.0386	1.0666	1.0384	0.8201
40.0	3.6541E 00	0.6756	1.3755	1.4282	1.4335	0.8096	0.9564	1.0001	1.0000	1.0685	1.0557	1.0684	0.8419
45.0	3.4010E 00	0.6493	1.4248	1.5009	1.5155	0.8006	0.9501	1.0001	1.0000	1.0961	1.0369	1.0960	0.8622
50.0	3.1851E 00	0.6258	1.4687	1.5691	1.5935	0.7941	0.9436	1.0001	1.0000	1.1212	1.0112	1.1211	0.8809

Instr. see p. 22

ELECTRON Z= 70 A= 170

p	F_0L_0	L_0	λ_2	λ_3	λ_4	μ_1	μ_2	Λ_1	Λ_2	ν_{12}	$\hat{\nu}_{12}$	η_{12}	$\hat{\eta}_{12}$
0.1	1.6596E 02	1.0209	14.2924	83.0676	281.0761	0.9979	0.9999	1.0081	1.0009	5.1397	0.5594	-0.1209	-0.1611
0.2	8.3713E 01	1.0207	4.0306	8.2769	12.4814	0.9977	0.9999	1.0068	1.0009	1.6680	0.5818	0.2994	0.2260
0.3	5.6599E 01	1.0206	2.1325	2.9023	3.3772	0.9974	0.9999	1.0063	1.0009	1.0705	0.6026	0.4397	0.3566
0.4	4.3248E 01	1.0203	1.4708	1.6496	1.7242	0.9970	0.9998	1.0059	1.0008	0.8770	0.6218	0.5103	0.4206
0.5	3.5388E 01	1.0199	1.1672	1.1825	1.1730	0.9965	0.9997	1.0056	1.0008	0.7950	0.6391	0.5533	0.4589
0.6	3.0267E 01	1.0195	1.0049	0.9607	0.9262	0.9959	0.9996	1.0053	1.0008	0.7548	0.6548	0.5825	0.4846
0.7	2.6700E 01	1.0190	0.9094	0.8393	0.7958	0.9953	0.9996	1.0050	1.0007	0.7336	0.6689	0.6039	0.5032
0.8	2.4095E 01	1.0184	0.8496	0.7669	0.7197	0.9946	0.9995	1.0048	1.0007	0.7219	0.6816	0.6205	0.5173
0.9	2.2122E 01	1.0178	0.8104	0.7210	0.6722	0.9939	0.9994	1.0045	1.0007	0.7154	0.6930	0.6338	0.5285
1.0	2.0581E 01	1.0171	0.7841	0.6909	0.6414	0.9931	0.9993	1.0043	1.0006	0.7121	0.7034	0.6448	0.5377
1.2	1.8341E 01	1.0155	0.7537	0.6570	0.6071	0.9916	0.9991	1.0039	1.0006	0.7104	0.7214	0.6622	0.5519
1.4	1.6791E 01	1.0137	0.7396	0.6417	0.5919	0.9901	0.9989	1.0035	1.0005	0.7120	0.7365	0.6756	0.5627
1.6	1.5653E 01	1.0118	0.7338	0.6358	0.5862	0.9885	0.9987	1.0032	1.0005	0.7150	0.7496	0.6865	0.5713
1.8	1.4775E 01	1.0098	0.7328	0.6351	0.5857	0.9869	0.9984	1.0030	1.0004	0.7187	0.7610	0.6957	0.5785
2.0	1.4073E 01	1.0076	0.7345	0.6373	0.5882	0.9854	0.9982	1.0027	1.0004	0.7227	0.7711	0.7037	0.5846
2.2	1.3494E 01	1.0054	0.7379	0.6413	0.5925	0.9838	0.9980	1.0025	1.0004	0.7268	0.7803	0.7108	0.5900
2.4	1.3005E 01	1.0032	0.7423	0.6465	0.5979	0.9822	0.9978	1.0023	1.0003	0.7308	0.7887	0.7172	0.5949
2.6	1.2584E 01	1.0009	0.7474	0.6523	0.6040	0.9807	0.9976	1.0022	1.0003	0.7348	0.7964	0.7231	0.5992
2.8	1.2215E 01	0.9986	0.7529	0.6586	0.6105	0.9791	0.9974	1.0021	1.0003	0.7388	0.8037	0.7285	0.6032
3.0	1.1888E 01	0.9962	0.7587	0.6652	0.6173	0.9776	0.9972	1.0019	1.0003	0.7426	0.8104	0.7335	0.6069
3.2	1.1595E 01	0.9938	0.7645	0.6718	0.6242	0.9761	0.9970	1.0018	1.0003	0.7463	0.8168	0.7382	0.6103
3.4	1.1330E 01	0.9914	0.7705	0.6786	0.6312	0.9745	0.9968	1.0017	1.0003	0.7499	0.8229	0.7427	0.6136
3.6	1.1087E 01	0.9890	0.7764	0.6854	0.6382	0.9730	0.9966	1.0016	1.0002	0.7535	0.8287	0.7469	0.6166
3.8	1.0864E 01	0.9866	0.7824	0.6922	0.6453	0.9715	0.9964	1.0016	1.0002	0.7569	0.8342	0.7510	0.6195
4.0	1.0658E 01	0.9842	0.7883	0.6989	0.6522	0.9700	0.9962	1.0015	1.0002	0.7603	0.8396	0.7549	0.6222
4.5	1.0201E 01	0.9781	0.8029	0.7155	0.6694	0.9662	0.9957	1.0013	1.0002	0.7683	0.8520	0.7639	0.6286
5.0	9.8100E 00	0.9720	0.8171	0.7316	0.6861	0.9625	0.9952	1.0012	1.0002	0.7760	0.8635	0.7723	0.6344
5.5	9.4683E 00	0.9659	0.8308	0.7473	0.7024	0.9589	0.9947	1.0011	1.0002	0.7832	0.8742	0.7801	0.6397
6.0	9.1650E 00	0.9599	0.8441	0.7625	0.7182	0.9553	0.9942	1.0010	1.0002	0.7901	0.8842	0.7874	0.6447
6.5	8.8924E 00	0.9539	0.8570	0.7773	0.7335	0.9517	0.9937	1.0009	1.0001	0.7967	0.8937	0.7944	0.6494
7.0	8.6449E 00	0.9479	0.8696	0.7917	0.7485	0.9482	0.9932	1.0009	1.0001	0.8031	0.9026	0.8010	0.6538
7.5	8.4183E 00	0.9420	0.8817	0.8057	0.7631	0.9447	0.9927	1.0008	1.0001	0.8092	0.9111	0.8074	0.6581
8.0	8.2094E 00	0.9362	0.8936	0.8194	0.7774	0.9413	0.9922	1.0008	1.0001	0.8152	0.9192	0.8136	0.6622
9.0	7.8351E 00	0.9246	0.9164	0.8459	0.8049	0.9345	0.9912	1.0007	1.0001	0.8267	0.9343	0.8253	0.6699
10.0	7.5073E 00	0.9133	0.9381	0.8712	0.8315	0.9280	0.9901	1.0006	1.0001	0.8375	0.9481	0.8364	0.6772
11.0	7.2158E 00	0.9022	0.9591	0.8957	0.8572	0.9216	0.9891	1.0006	1.0001	0.8480	0.9608	0.8470	0.6841
12.0	6.9536E 00	0.8913	0.9792	0.9194	0.8821	0.9154	0.9881	1.0005	1.0001	0.8580	0.9725	0.8572	0.6908
13.0	6.7157E 00	0.8807	0.9987	0.9425	0.9063	0.9094	0.9870	1.0005	1.0001	0.8678	0.9832	0.8670	0.6972
14.0	6.4980E 00	0.8703	1.0176	0.9650	0.9300	0.9036	0.9860	1.0004	1.0001	0.8772	0.9932	0.8766	0.7035
16.0	6.1121E 00	0.8502	1.0539	1.0084	0.9760	0.8924	0.9839	1.0004	1.0001	0.8953	1.0107	0.8948	0.7156
18.0	5.7787E 00	0.8311	1.0883	1.0501	1.0203	0.8820	0.9817	1.0003	1.0001	0.9127	1.0254	0.9123	0.7272
20.0	5.4860E 00	0.8129	1.1211	1.0903	1.0632	0.8722	0.9795	1.0003	1.0000	0.9293	1.0375	0.9290	0.7385
25.0	4.8853E 00	0.7709	1.1974	1.1859	1.1658	0.8507	0.9739	1.0002	1.0000	0.9685	1.0579	0.9682	0.7655
30.0	4.4164E 00	0.7336	1.2664	1.2753	1.2629	0.8328	0.9680	1.0002	1.0000	1.0046	1.0660	1.0044	0.7910
35.0	4.0381E 00	0.7005	1.3288	1.3594	1.3553	0.8183	0.9620	1.0001	1.0000	1.0379	1.0636	1.0378	0.8150
40.0	3.7255E 00	0.6710	1.3852	1.4386	1.4435	0.8068	0.9557	1.0001	1.0000	1.0687	1.0521	1.0686	0.8376
45.0	3.4628E 00	0.6447	1.4357	1.5132	1.5276	0.7980	0.9492	1.0001	1.0000	1.0970	1.0326	1.0969	0.8586
50.0	3.2391E 00	0.6211	1.4805	1.5832	1.6077	0.7916	0.9426	1.0001	1.0000	1.1227	1.0064	1.1227	0.8779

Instr. see p. 22

ELECTRON Z= 71 A= 175

p	F_0L_0	L_0	λ_2	λ_3	λ_4	μ_1	μ_2	Λ_1	Λ_2	ν_{12}	$\hat{\nu}_{12}$	η_{12}	$\hat{\eta}_{12}$
0.1	1.7560E 02	1.0206	14.4470	85.4885	294.3657	0.9978	0.9999	1.0086	1.0009	5.2178	0.5545	-0.1596	-0.1978
0.2	8.8568E 01	1.0205	4.0637	8.4382	12.8325	0.9976	0.9999	1.0071	1.0009	1.6765	0.5771	0.2766	0.2051
0.3	5.9871E 01	1.0203	2.1431	2.9333	3.4254	0.9972	0.9999	1.0065	1.0009	1.0678	0.5982	0.4221	0.3382
0.4	4.5738E 01	1.0200	1.4735	1.6563	1.7331	0.9968	0.9998	1.0061	1.0009	0.8709	0.6175	0.4953	0.4044
0.5	3.7414E 01	1.0197	1.1663	1.1816	1.1719	0.9963	0.9997	1.0058	1.0008	0.7875	0.6351	0.5399	0.4440
0.6	3.1986E 01	1.0192	1.0021	0.9567	0.9215	0.9957	0.9996	1.0055	1.0008	0.7468	0.6510	0.5701	0.4706
0.7	2.8204E 01	1.0187	0.9054	0.8338	0.7895	0.9950	0.9995	1.0052	1.0008	0.7252	0.6653	0.5923	0.4897
0.8	2.5439E 01	1.0181	0.8449	0.7605	0.7125	0.9943	0.9995	1.0050	1.0007	0.7134	0.6781	0.6094	0.5042
0.9	2.3342E 01	1.0174	0.8052	0.7142	0.6646	0.9936	0.9994	1.0047	1.0007	0.7069	0.6898	0.6231	0.5158
1.0	2.1705E 01	1.0167	0.7786	0.6837	0.6335	0.9928	0.9992	1.0045	1.0007	0.7036	0.7003	0.6344	0.5252
1.2	1.9319E 01	1.0150	0.7479	0.6495	0.5989	0.9912	0.9990	1.0040	1.0006	0.7019	0.7186	0.6524	0.5398
1.4	1.7668E 01	1.0132	0.7336	0.6341	0.5837	0.9896	0.9988	1.0037	1.0005	0.7036	0.7340	0.6662	0.5509
1.6	1.6453E 01	1.0112	0.7279	0.6283	0.5780	0.9880	0.9986	1.0033	1.0005	0.7067	0.7473	0.6774	0.5597
1.8	1.5516E 01	1.0091	0.7269	0.6276	0.5776	0.9864	0.9984	1.0031	1.0005	0.7105	0.7589	0.6869	0.5670
2.0	1.4765E 01	1.0069	0.7287	0.6299	0.5802	0.9847	0.9982	1.0028	1.0004	0.7146	0.7693	0.6951	0.5733
2.2	1.4145E 01	1.0046	0.7323	0.6341	0.5846	0.9831	0.9980	1.0026	1.0004	0.7188	0.7786	0.7024	0.5788
2.4	1.3622E 01	1.0023	0.7368	0.6394	0.5901	0.9815	0.9978	1.0024	1.0004	0.7230	0.7872	0.7090	0.5838
2.6	1.3171E 01	0.9999	0.7421	0.6454	0.5963	0.9799	0.9976	1.0023	1.0003	0.7271	0.7951	0.7150	0.5882
2.8	1.2777E 01	0.9975	0.7477	0.6518	0.6030	0.9783	0.9973	1.0021	1.0003	0.7311	0.8024	0.7205	0.5923
3.0	1.2426E 01	0.9951	0.7537	0.6585	0.6099	0.9766	0.9971	1.0020	1.0003	0.7350	0.8093	0.7257	0.5961
3.2	1.2112E 01	0.9926	0.7597	0.6653	0.6170	0.9751	0.9969	1.0019	1.0003	0.7389	0.8159	0.7305	0.5996
3.4	1.1828E 01	0.9901	0.7658	0.6723	0.6242	0.9735	0.9967	1.0018	1.0003	0.7426	0.8220	0.7351	0.6029
3.6	1.1568E 01	0.9876	0.7720	0.6792	0.6313	0.9719	0.9965	1.0017	1.0002	0.7462	0.8279	0.7395	0.6060
3.8	1.1329E 01	0.9851	0.7781	0.6862	0.6385	0.9703	0.9963	1.0016	1.0002	0.7498	0.8336	0.7436	0.6089
4.0	1.1108E 01	0.9826	0.7842	0.6931	0.6456	0.9688	0.9961	1.0016	1.0002	0.7532	0.8390	0.7476	0.6117
4.5	1.0619E 01	0.9764	0.7992	0.7101	0.6632	0.9649	0.9956	1.0014	1.0002	0.7615	0.8517	0.7570	0.6183
5.0	1.0201E 01	0.9701	0.8138	0.7267	0.6803	0.9610	0.9951	1.0013	1.0002	0.7693	0.8634	0.7656	0.6242
5.5	9.8351E 00	0.9638	0.8280	0.7428	0.6970	0.9573	0.9946	1.0012	1.0002	0.7768	0.8742	0.7736	0.6297
6.0	9.5108E 00	0.9576	0.8417	0.7584	0.7132	0.9535	0.9940	1.0011	1.0002	0.7839	0.8844	0.7811	0.6348
6.5	9.2196E 00	0.9514	0.8550	0.7736	0.7290	0.9498	0.9935	1.0010	1.0001	0.7907	0.8939	0.7883	0.6396
7.0	8.9553E 00	0.9452	0.8679	0.7884	0.7443	0.9462	0.9930	1.0009	1.0001	0.7972	0.9030	0.7952	0.6442
7.5	8.7136E 00	0.9392	0.8805	0.8028	0.7593	0.9426	0.9925	1.0009	1.0001	0.8036	0.9116	0.8017	0.6485
8.0	8.4909E 00	0.9331	0.8927	0.8169	0.7739	0.9391	0.9920	1.0008	1.0001	0.8097	0.9197	0.8081	0.6528
9.0	8.0923E 00	0.9212	0.9163	0.8441	0.8023	0.9321	0.9909	1.0007	1.0001	0.8215	0.9349	0.8202	0.6607
10.0	7.7436E 00	0.9096	0.9388	0.8703	0.8297	0.9254	0.9899	1.0006	1.0001	0.8328	0.9488	0.8316	0.6683
11.0	7.4340E 00	0.8982	0.9604	0.8956	0.8561	0.9188	0.9888	1.0006	1.0001	0.8436	0.9616	0.8426	0.6755
12.0	7.1560E 00	0.8871	0.9812	0.9200	0.8818	0.9125	0.9877	1.0005	1.0001	0.8539	0.9733	0.8531	0.6824
13.0	6.9038E 00	0.8762	1.0014	0.9438	0.9068	0.9063	0.9867	1.0005	1.0001	0.8640	0.9840	0.8633	0.6891
14.0	6.6735E 00	0.8656	1.0209	0.9670	0.9312	0.9004	0.9856	1.0004	1.0001	0.8737	0.9939	0.8731	0.6956
16.0	6.2660E 00	0.8451	1.0584	1.0119	0.9786	0.8890	0.9834	1.0004	1.0001	0.8925	1.0112	0.8920	0.7082
18.0	5.9145E 00	0.8256	1.0940	1.0550	1.0244	0.8784	0.9812	1.0003	1.0001	0.9105	1.0257	0.9100	0.7203
20.0	5.6067E 00	0.8070	1.1280	1.0966	1.0687	0.8685	0.9789	1.0003	1.0000	0.9277	1.0375	0.9273	0.7321
25.0	4.9771E 00	0.7644	1.2068	1.1954	1.1748	0.8467	0.9731	1.0002	1.0000	0.9682	1.0569	0.9679	0.7603
30.0	4.4878E 00	0.7268	1.2780	1.2878	1.2752	0.8288	0.9671	1.0002	1.0000	1.0055	1.0638	1.0053	0.7869
35.0	4.0945E 00	0.6934	1.3423	1.3748	1.3708	0.8144	0.9608	1.0001	1.0000	1.0399	1.0600	1.0398	0.8119
40.0	3.7707E 00	0.6638	1.4001	1.4566	1.4620	0.8031	0.9544	1.0001	1.0000	1.0717	1.0470	1.0715	0.8354
45.0	3.4995E 00	0.6374	1.4518	1.5336	1.5488	0.7946	0.9477	1.0001	1.0000	1.1007	1.0260	1.1006	0.8571
50.0	3.2691E 00	0.6139	1.4975	1.6056	1.6315	0.7885	0.9409	1.0001	1.0000	1.1272	0.9982	1.1271	0.8771

Instr. see p. 22

ELECTRON Z= 72 A= 177

p	F_0L_0	L_0	λ_2	λ_3	λ_4	μ_1	μ_2	Λ_1	Λ_2	ν_{12}	$\hat{\nu}_{12}$	η_{12}	$\hat{\eta}_{12}$
0.1	1.8620E 02	1.0205	14.5806	87.7893	307.5135	0.9976	0.9999	1.0090	1.0009	5.2919	0.5492	-0.1990	-0.2348
0.2	9.3903E 01	1.0203	4.0912	8.5865	13.1665	0.9974	0.9999	1.0074	1.0009	1.6837	0.5720	0.2532	0.1818
0.3	6.3467E 01	1.0201	2.1509	2.9597	3.4680	0.9971	0.9998	1.0068	1.0009	1.0642	0.5933	0.4040	0.3195
0.4	4.8473E 01	1.0198	1.4745	1.6603	1.7390	0.9966	0.9998	1.0064	1.0009	0.8641	0.6129	0.4799	0.3878
0.5	3.9640E 01	1.0195	1.1641	1.1789	1.1688	0.9961	0.9997	1.0060	1.0008	0.7795	0.6307	0.5259	0.4288
0.6	3.3876E 01	1.0190	0.9981	0.9513	0.9153	0.9955	0.9996	1.0057	1.0008	0.7382	0.6467	0.5572	0.4561
0.7	2.9856E 01	1.0185	0.9004	0.8271	0.7819	0.9948	0.9995	1.0054	1.0008	0.7164	0.6612	0.5801	0.4758
0.8	2.6915E 01	1.0178	0.8392	0.7531	0.7043	0.9940	0.9994	1.0051	1.0007	0.7044	0.6743	0.5978	0.4908
0.9	2.4683E 01	1.0171	0.7992	0.7063	0.6560	0.9933	0.9993	1.0048	1.0007	0.6979	0.6861	0.6119	0.5027
1.0	2.2938E 01	1.0164	0.7723	0.6756	0.6246	0.9925	0.9992	1.0046	1.0007	0.6945	0.6968	0.6236	0.5123
1.2	2.0395E 01	1.0147	0.7412	0.6412	0.5899	0.9909	0.9990	1.0042	1.0006	0.6929	0.7154	0.6421	0.5273
1.4	1.8631E 01	1.0128	0.7269	0.6257	0.5746	0.9892	0.9988	1.0038	1.0005	0.6947	0.7310	0.6563	0.5387
1.6	1.7332E 01	1.0107	0.7212	0.6199	0.5690	0.9875	0.9986	1.0034	1.0005	0.6979	0.7445	0.6678	0.5477
1.8	1.6328E 01	1.0086	0.7203	0.6193	0.5686	0.9858	0.9984	1.0032	1.0005	0.7018	0.7563	0.6776	0.5552
2.0	1.5524E 01	1.0063	0.7222	0.6217	0.5713	0.9841	0.9981	1.0029	1.0004	0.7060	0.7669	0.6860	0.5616
2.2	1.4861E 01	1.0040	0.7258	0.6260	0.5758	0.9825	0.9979	1.0027	1.0004	0.7103	0.7764	0.6935	0.5672
2.4	1.4300E 01	1.0016	0.7306	0.6314	0.5815	0.9808	0.9977	1.0025	1.0004	0.7146	0.7851	0.7002	0.5723
2.6	1.3816E 01	0.9991	0.7360	0.6375	0.5878	0.9791	0.9975	1.0023	1.0003	0.7188	0.7931	0.7064	0.5768
2.8	1.3393E 01	0.9966	0.7418	0.6441	0.5946	0.9775	0.9973	1.0022	1.0003	0.7229	0.8006	0.7120	0.5810
3.0	1.3017E 01	0.9941	0.7478	0.6509	0.6017	0.9758	0.9971	1.0021	1.0003	0.7270	0.8076	0.7173	0.5848
3.2	1.2680E 01	0.9916	0.7540	0.6579	0.6089	0.9742	0.9969	1.0020	1.0003	0.7309	0.8143	0.7223	0.5884
3.4	1.2375E 01	0.9890	0.7603	0.6650	0.6162	0.9725	0.9967	1.0019	1.0003	0.7347	0.8206	0.7270	0.5918
3.6	1.2096E 01	0.9865	0.7666	0.6721	0.6235	0.9709	0.9964	1.0018	1.0003	0.7384	0.8266	0.7315	0.5949
3.8	1.1840E 01	0.9839	0.7729	0.6792	0.6308	0.9693	0.9962	1.0017	1.0002	0.7421	0.8323	0.7358	0.5979
4.0	1.1603E 01	0.9813	0.7792	0.6863	0.6380	0.9677	0.9960	1.0016	1.0002	0.7456	0.8378	0.7398	0.6008
4.5	1.1079E 01	0.9749	0.7946	0.7037	0.6560	0.9637	0.9955	1.0014	1.0002	0.7541	0.8507	0.7494	0.6075
5.0	1.0630E 01	0.9684	0.8096	0.7206	0.6734	0.9597	0.9950	1.0013	1.0002	0.7621	0.8625	0.7582	0.6135
5.5	1.0239E 01	0.9620	0.8242	0.7371	0.6905	0.9558	0.9944	1.0012	1.0002	0.7697	0.8735	0.7664	0.6191
6.0	9.8920E 00	0.9556	0.8383	0.7531	0.7070	0.9520	0.9939	1.0011	1.0002	0.7770	0.8838	0.7742	0.6244
6.5	9.5805E 00	0.9493	0.8520	0.7687	0.7231	0.9482	0.9934	1.0010	1.0001	0.7840	0.8935	0.7816	0.6293
7.0	9.2982E 00	0.9430	0.8652	0.7839	0.7389	0.9444	0.9929	1.0009	1.0001	0.7908	0.9026	0.7886	0.6340
7.5	9.0402E 00	0.9367	0.8782	0.7987	0.7542	0.9408	0.9923	1.0009	1.0001	0.7973	0.9113	0.7954	0.6385
8.0	8.8026E 00	0.9305	0.8907	0.8131	0.7692	0.9371	0.9918	1.0008	1.0001	0.8036	0.9195	0.8019	0.6428
9.0	8.3778E 00	0.9184	0.9150	0.8411	0.7983	0.9300	0.9907	1.0007	1.0001	0.8157	0.9348	0.8143	0.6511
10.0	8.0068E 00	0.9065	0.9382	0.8679	0.8263	0.9231	0.9896	1.0007	1.0001	0.8273	0.9488	0.8261	0.6588
11.0	7.6777E 00	0.8949	0.9604	0.8939	0.8534	0.9164	0.9886	1.0006	1.0001	0.8384	0.9615	0.8374	0.6663
12.0	7.3826E 00	0.8835	0.9819	0.9190	0.8797	0.9099	0.9875	1.0005	1.0001	0.8491	0.9733	0.8482	0.6734
13.0	7.1154E 00	0.8724	1.0027	0.9435	0.9054	0.9037	0.9864	1.0005	1.0001	0.8594	0.9840	0.8586	0.6803
14.0	6.8716E 00	0.8616	1.0228	0.9673	0.9305	0.8976	0.9853	1.0005	1.0001	0.8694	0.9938	0.8688	0.6871
16.0	6.4409E 00	0.8408	1.0615	1.0135	0.9792	0.8861	0.9830	1.0004	1.0001	0.8888	1.0110	0.8882	0.7001
18.0	6.0703E 00	0.8210	1.0981	1.0579	1.0263	0.8753	0.9808	1.0003	1.0001	0.9073	1.0253	0.9068	0.7127
20.0	5.7464E 00	0.8022	1.1331	1.1007	1.0719	0.8653	0.9785	1.0003	1.0000	0.9250	1.0369	0.9246	0.7249
25.0	5.0860E 00	0.7591	1.2143	1.2025	1.1811	0.8433	0.9725	1.0002	1.0000	0.9667	1.0555	0.9665	0.7542
30.0	4.5750E 00	0.7211	1.2875	1.2977	1.2845	0.8254	0.9664	1.0002	1.0000	1.0052	1.0614	1.0050	0.7818
35.0	4.1657E 00	0.6876	1.3536	1.3872	1.3829	0.8111	0.9599	1.0002	1.0000	1.0406	1.0566	1.0405	0.8079
40.0	3.8297E 00	0.6578	1.4129	1.4714	1.4768	0.7999	0.9533	1.0001	1.0000	1.0733	1.0425	1.0731	0.8322
45.0	3.5490E 00	0.6315	1.4658	1.5505	1.5662	0.7916	0.9465	1.0001	1.0000	1.1031	1.0205	1.1030	0.8547
50.0	3.3112E 00	0.6080	1.5124	1.6246	1.6512	0.7858	0.9395	1.0001	1.0000	1.1303	0.9918	1.1302	0.8753

Instr. see p. 22

ELECTRON Z= 73 A= 181

p	F_0L_0	L_0	λ_2	λ_3	λ_4	μ_1	μ_2	Λ_1	Λ_2	ν_{12}	$\hat{\nu}_{12}$	η_{12}	$\hat{\eta}_{12}$
0.1	1.9728E 02	1.0201	14.7164	90.1447	321.1895	0.9975	0.9999	1.0095	1.0010	5.3667	0.5440	-0.2391	-0.2723
0.2	9.9477E 01	1.0200	4.1194	8.7392	13.5126	0.9972	0.9999	1.0077	1.0009	1.6910	0.5670	0.2295	0.1584
0.3	6.7222E 01	1.0198	2.1593	2.9874	3.5128	0.9969	0.9998	1.0071	1.0009	1.0607	0.5886	0.3858	0.3007
0.4	5.1329E 01	1.0195	1.4759	1.6652	1.7460	0.9964	0.9998	1.0066	1.0009	0.8574	0.6084	0.4643	0.3713
0.5	4.1962E 01	1.0191	1.1623	1.1769	1.1665	0.9959	0.9997	1.0062	1.0009	0.7715	0.6264	0.5120	0.4135
0.6	3.5846E 01	1.0186	0.9946	0.9465	0.9097	0.9952	0.9996	1.0059	1.0008	0.7296	0.6427	0.5444	0.4417
0.7	3.1578E 01	1.0181	0.8959	0.8210	0.7750	0.9945	0.9995	1.0056	1.0008	0.7076	0.6573	0.5680	0.4620
0.8	2.8453E 01	1.0174	0.8340	0.7462	0.6966	0.9937	0.9994	1.0053	1.0008	0.6955	0.6706	0.5862	0.4774
0.9	2.6079E 01	1.0167	0.7936	0.6990	0.6479	0.9930	0.9993	1.0050	1.0007	0.6889	0.6825	0.6008	0.4896
1.0	2.4222E 01	1.0159	0.7664	0.6681	0.6164	0.9921	0.9992	1.0048	1.0007	0.6855	0.6934	0.6128	0.4995
1.2	2.1511E 01	1.0142	0.7350	0.6334	0.5815	0.9904	0.9990	1.0043	1.0006	0.6840	0.7123	0.6319	0.5149
1.4	1.9629E 01	1.0122	0.7206	0.6179	0.5661	0.9887	0.9988	1.0039	1.0006	0.6859	0.7282	0.6465	0.5265
1.6	1.8241E 01	1.0101	0.7149	0.6120	0.5605	0.9870	0.9985	1.0036	1.0005	0.6892	0.7419	0.6583	0.5357
1.8	1.7168E 01	1.0078	0.7141	0.6115	0.5602	0.9852	0.9983	1.0033	1.0005	0.6932	0.7540	0.6683	0.5434
2.0	1.6308E 01	1.0055	0.7161	0.6140	0.5630	0.9835	0.9981	1.0030	1.0004	0.6976	0.7647	0.6770	0.5499
2.2	1.5597E 01	1.0031	0.7199	0.6184	0.5676	0.9817	0.9979	1.0028	1.0004	0.7020	0.7744	0.6846	0.5557
2.4	1.4996E 01	1.0006	0.7248	0.6240	0.5734	0.9800	0.9977	1.0026	1.0004	0.7063	0.7832	0.6915	0.5608
2.6	1.4478E 01	0.9980	0.7303	0.6302	0.5799	0.9783	0.9974	1.0024	1.0003	0.7106	0.7914	0.6978	0.5654
2.8	1.4025E 01	0.9955	0.7363	0.6370	0.5868	0.9765	0.9972	1.0023	1.0003	0.7149	0.7991	0.7037	0.5697
3.0	1.3622E 01	0.9929	0.7425	0.6440	0.5940	0.9748	0.9970	1.0021	1.0003	0.7190	0.8062	0.7091	0.5736
3.2	1.3261E 01	0.9903	0.7489	0.6511	0.6014	0.9731	0.9968	1.0020	1.0003	0.7230	0.8130	0.7142	0.5773
3.4	1.2933E 01	0.9876	0.7554	0.6584	0.6088	0.9714	0.9966	1.0019	1.0003	0.7269	0.8194	0.7190	0.5807
3.6	1.2635E 01	0.9850	0.7619	0.6656	0.6163	0.9697	0.9964	1.0018	1.0003	0.7308	0.8255	0.7236	0.5839
3.8	1.2360E 01	0.9823	0.7683	0.6729	0.6237	0.9681	0.9961	1.0017	1.0002	0.7345	0.8313	0.7280	0.5870
4.0	1.2106E 01	0.9797	0.7748	0.6801	0.6311	0.9664	0.9959	1.0017	1.0002	0.7381	0.8369	0.7322	0.5900
4.5	1.1545E 01	0.9730	0.7906	0.6980	0.6494	0.9623	0.9954	1.0015	1.0002	0.7468	0.8500	0.7420	0.5968
5.0	1.1065E 01	0.9664	0.8061	0.7153	0.6673	0.9582	0.9948	1.0013	1.0002	0.7550	0.8620	0.7511	0.6030
5.5	1.0646E 01	0.9598	0.8211	0.7322	0.6847	0.9542	0.9943	1.0012	1.0002	0.7629	0.8732	0.7595	0.6087
6.0	1.0275E 01	0.9532	0.8356	0.7487	0.7017	0.9502	0.9938	1.0011	1.0002	0.7704	0.8836	0.7675	0.6141
6.5	9.9423E 00	0.9467	0.8497	0.7646	0.7182	0.9463	0.9932	1.0010	1.0001	0.7776	0.8934	0.7750	0.6192
7.0	9.6409E 00	0.9402	0.8633	0.7802	0.7343	0.9424	0.9927	1.0010	1.0001	0.7845	0.9026	0.7823	0.6240
7.5	9.3656E 00	0.9338	0.8767	0.7954	0.7500	0.9387	0.9921	1.0009	1.0001	0.7912	0.9114	0.7892	0.6286
8.0	9.1124E 00	0.9275	0.8896	0.8102	0.7654	0.9349	0.9916	1.0008	1.0001	0.7977	0.9197	0.7959	0.6331
9.0	8.6602E 00	0.9150	0.9146	0.8390	0.7952	0.9276	0.9905	1.0008	1.0001	0.8102	0.9351	0.8088	0.6415
10.0	8.2657E 00	0.9028	0.9385	0.8666	0.8240	0.9205	0.9894	1.0007	1.0001	0.8221	0.9492	0.8209	0.6496
11.0	7.9163E 00	0.8909	0.9615	0.8933	0.8519	0.9137	0.9883	1.0006	1.0001	0.8335	0.9620	0.8325	0.6572
12.0	7.6035E 00	0.8793	0.9837	0.9192	0.8790	0.9071	0.9872	1.0006	1.0001	0.8445	0.9737	0.8437	0.6646
13.0	7.3205E 00	0.8680	1.0051	0.9445	0.9054	0.9006	0.9860	1.0005	1.0001	0.8552	0.9844	0.8544	0.6718
14.0	7.0627E 00	0.8570	1.0259	0.9691	0.9313	0.8945	0.9849	1.0005	1.0001	0.8656	0.9942	0.8649	0.6788
16.0	6.6081E 00	0.8358	1.0658	1.0166	0.9815	0.8827	0.9826	1.0004	1.0001	0.8856	1.0112	0.8850	0.6924
18.0	6.2178E 00	0.8156	1.1037	1.0624	1.0299	0.8718	0.9803	1.0004	1.0001	0.9046	1.0253	0.9042	0.7055
20.0	5.8774E 00	0.7965	1.1398	1.1067	1.0770	0.8617	0.9779	1.0003	1.0001	0.9230	1.0365	0.9226	0.7182
25.0	5.1857E 00	0.7529	1.2235	1.2117	1.1897	0.8396	0.9718	1.0002	1.0000	0.9660	1.0543	0.9658	0.7487
30.0	4.6528E 00	0.7146	1.2989	1.3100	1.2964	0.8217	0.9655	1.0002	1.0000	1.0057	1.0590	1.0055	0.7774
35.0	4.2275E 00	0.6809	1.3669	1.4023	1.3980	0.8075	0.9589	1.0002	1.0000	1.0423	1.0529	1.0421	0.8045
40.0	3.8797E 00	0.6511	1.4277	1.4891	1.4948	0.7965	0.9520	1.0001	1.0000	1.0759	1.0375	1.0757	0.8297
45.0	3.5900E 00	0.6247	1.4818	1.5706	1.5870	0.7885	0.9450	1.0001	1.0000	1.1066	1.0142	1.1065	0.8530
50.0	3.3451E 00	0.6013	1.5293	1.6468	1.6746	0.7830	0.9379	1.0001	1.0000	1.1344	0.9841	1.1343	0.8743

 Instr. see p. 22

ELECTRON Z= 74 A= 182

p	F_0L_0	L_0	λ_2	λ_3	λ_4	μ_1	μ_2	Λ_1	Λ_2	ν_{12}	$\hat{\nu}_{12}$	η_{12}	$\hat{\eta}_{12}$
0.1	2.0947E 02	1.0199	14.8298	92.3589	334.6225	0.9973	0.9999	1.0099	1.0010	5.4370	0.5385	-0.2796	-0.3101
0.2	1.0561E 02	1.0197	4.1417	8.8773	13.8389	0.9971	0.9999	1.0080	1.0010	1.6969	0.5617	0.2054	0.1347
0.3	7.1356E 01	1.0195	2.1647	3.0102	3.5514	0.9967	0.9998	1.0073	1.0009	1.0564	0.5834	0.3671	0.2815
0.4	5.4473E 01	1.0192	1.4754	1.6674	1.7498	0.9962	0.9998	1.0068	1.0009	0.8499	0.6034	0.4483	0.3544
0.5	4.4518E 01	1.0188	1.1590	1.1731	1.1622	0.9956	0.9997	1.0064	1.0009	0.7629	0.6217	0.4976	0.3979
0.6	3.8015E 01	1.0183	0.9898	0.9402	0.9026	0.9950	0.9996	1.0061	1.0008	0.7205	0.6381	0.5310	0.4269
0.7	3.3473E 01	1.0177	0.8903	0.8136	0.7668	0.9942	0.9995	1.0058	1.0008	0.6982	0.6530	0.5554	0.4478
0.8	3.0145E 01	1.0171	0.8279	0.7383	0.6879	0.9935	0.9994	1.0055	1.0008	0.6860	0.6664	0.5742	0.4637
0.9	2.7615E 01	1.0163	0.7870	0.6907	0.6389	0.9927	0.9993	1.0052	1.0007	0.6794	0.6786	0.5892	0.4761
1.0	2.5634E 01	1.0155	0.7596	0.6596	0.6072	0.9918	0.9992	1.0049	1.0007	0.6761	0.6896	0.6016	0.4863
1.2	2.2739E 01	1.0137	0.7281	0.6247	0.5721	0.9901	0.9990	1.0044	1.0006	0.6746	0.7087	0.6212	0.5021
1.4	2.0727E 01	1.0117	0.7135	0.6092	0.5568	0.9883	0.9987	1.0040	1.0006	0.6765	0.7249	0.6362	0.5139
1.6	1.9241E 01	1.0095	0.7078	0.6033	0.5512	0.9865	0.9985	1.0037	1.0005	0.6800	0.7388	0.6483	0.5234
1.8	1.8092E 01	1.0072	0.7071	0.6029	0.5510	0.9847	0.9983	1.0034	1.0005	0.6841	0.7511	0.6586	0.5312
2.0	1.7169E 01	1.0048	0.7092	0.6055	0.5539	0.9829	0.9981	1.0031	1.0004	0.6886	0.7620	0.6674	0.5379
2.2	1.6407E 01	1.0023	0.7131	0.6100	0.5586	0.9811	0.9978	1.0029	1.0004	0.6931	0.7718	0.6753	0.5437
2.4	1.5762E 01	0.9997	0.7182	0.6157	0.5645	0.9793	0.9976	1.0027	1.0004	0.6975	0.7808	0.6824	0.5490
2.6	1.5206E 01	0.9971	0.7239	0.6221	0.5711	0.9775	0.9974	1.0025	1.0004	0.7020	0.7891	0.6888	0.5537
2.8	1.4719E 01	0.9945	0.7300	0.6290	0.5781	0.9757	0.9972	1.0023	1.0003	0.7063	0.7969	0.6948	0.5580
3.0	1.4287E 01	0.9918	0.7364	0.6361	0.5854	0.9740	0.9969	1.0022	1.0003	0.7105	0.8042	0.7003	0.5620
3.2	1.3899E 01	0.9891	0.7429	0.6434	0.5929	0.9722	0.9967	1.0021	1.0003	0.7146	0.8110	0.7056	0.5657
3.4	1.3548E 01	0.9864	0.7496	0.6508	0.6005	0.9705	0.9965	1.0020	1.0003	0.7186	0.8175	0.7105	0.5692
3.6	1.3227E 01	0.9837	0.7562	0.6582	0.6081	0.9687	0.9963	1.0019	1.0003	0.7225	0.8237	0.7152	0.5725
3.8	1.2932E 01	0.9810	0.7629	0.6656	0.6157	0.9670	0.9961	1.0018	1.0003	0.7263	0.8297	0.7197	0.5757
4.0	1.2660E 01	0.9783	0.7695	0.6730	0.6232	0.9653	0.9958	1.0017	1.0002	0.7300	0.8353	0.7240	0.5787
4.5	1.2058E 01	0.9715	0.7857	0.6912	0.6419	0.9610	0.9953	1.0015	1.0002	0.7389	0.8486	0.7340	0.5856
5.0	1.1543E 01	0.9646	0.8016	0.7090	0.6601	0.9568	0.9947	1.0014	1.0002	0.7474	0.8608	0.7433	0.5919
5.5	1.1095E 01	0.9579	0.8169	0.7262	0.6778	0.9527	0.9942	1.0013	1.0002	0.7554	0.8721	0.7519	0.5978
6.0	1.0698E 01	0.9511	0.8319	0.7430	0.6951	0.9486	0.9936	1.0012	1.0002	0.7631	0.8827	0.7601	0.6033
6.5	1.0342E 01	0.9445	0.8463	0.7594	0.7120	0.9446	0.9931	1.0011	1.0002	0.7705	0.8926	0.7679	0.6085
7.0	1.0020E 01	0.9379	0.8604	0.7753	0.7284	0.9407	0.9925	1.0010	1.0001	0.7776	0.9019	0.7753	0.6135
7.5	9.7260E 00	0.9313	0.8740	0.7909	0.7445	0.9368	0.9920	1.0009	1.0001	0.7844	0.9107	0.7824	0.6182
8.0	9.4559E 00	0.9248	0.8874	0.8061	0.7602	0.9330	0.9914	1.0009	1.0001	0.7911	0.9191	0.7893	0.6228
9.0	8.9740E 00	0.9121	0.9131	0.8355	0.7907	0.9255	0.9903	1.0008	1.0001	0.8039	0.9346	0.8024	0.6315
10.0	8.5542E 00	0.8997	0.9377	0.8638	0.8202	0.9183	0.9892	1.0007	1.0001	0.8162	0.9487	0.8149	0.6397
11.0	8.1831E 00	0.8876	0.9613	0.8912	0.8487	0.9113	0.9881	1.0006	1.0001	0.8279	0.9616	0.8268	0.6477
12.0	7.8512E 00	0.8758	0.9841	0.9178	0.8765	0.9046	0.9869	1.0006	1.0001	0.8392	0.9733	0.8383	0.6553
13.0	7.5513E 00	0.8643	1.0062	0.9437	0.9036	0.8980	0.9858	1.0005	1.0001	0.8502	0.9840	0.8494	0.6627
14.0	7.2785E 00	0.8531	1.0275	0.9690	0.9301	0.8917	0.9846	1.0005	1.0001	0.8608	0.9938	0.8601	0.6700
16.0	6.7982E 00	0.8315	1.0686	1.0178	0.9816	0.8798	0.9823	1.0004	1.0001	0.8814	1.0107	0.8808	0.6840
18.0	6.3869E 00	0.8111	1.1076	1.0649	1.0314	0.8688	0.9799	1.0004	1.0001	0.9010	1.0246	0.9005	0.6975
20.0	6.0289E 00	0.7918	1.1447	1.1104	1.0797	0.8586	0.9775	1.0003	1.0001	0.9199	1.0356	0.9195	0.7107
25.0	5.3037E 00	0.7478	1.2308	1.2183	1.1955	0.8363	0.9712	1.0002	1.0000	0.9642	1.0526	0.9639	0.7423
30.0	4.7473E 00	0.7093	1.3083	1.3194	1.3052	0.8184	0.9648	1.0002	1.0000	1.0050	1.0566	1.0048	0.7721
35.0	4.3048E 00	0.6754	1.3780	1.4143	1.4096	0.8043	0.9580	1.0002	1.0000	1.0426	1.0497	1.0424	0.8001
40.0	3.9440E 00	0.6456	1.4403	1.5035	1.5091	0.7936	0.9510	1.0001	1.0000	1.0771	1.0334	1.0769	0.8262
45.0	3.6443E 00	0.6192	1.4956	1.5872	1.6038	0.7858	0.9439	1.0001	1.0000	1.1086	1.0092	1.1085	0.8503
50.0	3.3916E 00	0.5958	1.5441	1.6654	1.6938	0.7806	0.9366	1.0001	1.0000	1.1371	0.9783	1.1370	0.8722

Instr. see p. 22

ELECTRON Z= 75 A= 187

p	F_0L_0	L_0	λ_2	λ_3	λ_4	μ_1	μ_2	Λ_1	Λ_2	ν_{12}	$\hat{\nu}_{12}$	η_{12}	$\hat{\eta}_{12}$
0.1	2.2196E 02	1.0194	14.9557	94.7060	348.9042	0.9971	0.9999	1.0105	1.0010	5.5101	0.5333	-0.3210	-0.3484
0.2	1.1190E 02	1.0192	4.1677	9.0274	14.1906	0.9969	0.9999	1.0084	1.0010	1.7035	0.5567	0.1810	0.1108
0.3	7.5587E 01	1.0190	2.1722	3.0369	3.5957	0.9965	0.9998	1.0076	1.0010	1.0525	0.5786	0.3484	0.2624
0.4	5.7687E 01	1.0187	1.4764	1.6719	1.7564	0.9960	0.9998	1.0071	1.0009	0.8429	0.5989	0.4324	0.3375
0.5	4.7130E 01	1.0182	1.1570	1.1709	1.1597	0.9954	0.9997	1.0067	1.0009	0.7547	0.6173	0.4833	0.3824
0.6	4.0230E 01	1.0177	0.9862	0.9354	0.8971	0.9947	0.9996	1.0063	1.0009	0.7118	0.6340	0.5178	0.4123
0.7	3.5407E 01	1.0171	0.8857	0.8076	0.7599	0.9939	0.9995	1.0060	1.0008	0.6892	0.6491	0.5430	0.4338
0.8	3.1870E 01	1.0164	0.8227	0.7315	0.6804	0.9931	0.9994	1.0057	1.0008	0.6769	0.6627	0.5624	0.4501
0.9	2.9179E 01	1.0157	0.7815	0.6836	0.6310	0.9923	0.9993	1.0054	1.0007	0.6703	0.6750	0.5779	0.4629
1.0	2.7070E 01	1.0148	0.7538	0.6522	0.5992	0.9914	0.9992	1.0051	1.0007	0.6669	0.6861	0.5906	0.4733
1.2	2.3986E 01	1.0129	0.7220	0.6171	0.5639	0.9896	0.9989	1.0046	1.0006	0.6655	0.7056	0.6107	0.4895
1.4	2.1839E 01	1.0108	0.7074	0.6015	0.5486	0.9877	0.9987	1.0042	1.0006	0.6676	0.7221	0.6262	0.5016
1.6	2.0252E 01	1.0086	0.7017	0.5957	0.5430	0.9859	0.9985	1.0038	1.0005	0.6711	0.7362	0.6387	0.5113
1.8	1.9023E 01	1.0062	0.7011	0.5954	0.5429	0.9840	0.9982	1.0035	1.0005	0.6754	0.7487	0.6491	0.5192
2.0	1.8036E 01	1.0037	0.7033	0.5981	0.5459	0.9821	0.9980	1.0032	1.0004	0.6799	0.7598	0.6582	0.5261
2.2	1.7220E 01	1.0011	0.7074	0.6027	0.5507	0.9803	0.9978	1.0030	1.0004	0.6845	0.7698	0.6663	0.5320
2.4	1.6529E 01	0.9985	0.7126	0.6086	0.5567	0.9784	0.9976	1.0028	1.0004	0.6891	0.7790	0.6735	0.5373
2.6	1.5933E 01	0.9958	0.7185	0.6151	0.5634	0.9765	0.9973	1.0026	1.0004	0.6936	0.7875	0.6802	0.5422
2.8	1.5411E 01	0.9931	0.7248	0.6222	0.5706	0.9747	0.9971	1.0024	1.0003	0.6981	0.7954	0.6863	0.5466
3.0	1.4948E 01	0.9903	0.7314	0.6295	0.5781	0.9729	0.9969	1.0023	1.0003	0.7024	0.8028	0.6920	0.5506
3.2	1.4533E 01	0.9875	0.7381	0.6369	0.5858	0.9711	0.9966	1.0022	1.0003	0.7066	0.8097	0.6973	0.5544
3.4	1.4156E 01	0.9847	0.7449	0.6445	0.5935	0.9692	0.9964	1.0020	1.0003	0.7107	0.8164	0.7024	0.5580
3.6	1.3813E 01	0.9819	0.7517	0.6521	0.6012	0.9674	0.9962	1.0019	1.0003	0.7147	0.8227	0.7072	0.5614
3.8	1.3497E 01	0.9791	0.7586	0.6597	0.6090	0.9657	0.9960	1.0018	1.0003	0.7186	0.8287	0.7118	0.5646
4.0	1.3205E 01	0.9763	0.7654	0.6672	0.6167	0.9639	0.9957	1.0018	1.0002	0.7224	0.8345	0.7162	0.5676
4.5	1.2560E 01	0.9693	0.7821	0.6859	0.6357	0.9595	0.9952	1.0016	1.0002	0.7316	0.8480	0.7265	0.5747
5.0	1.2010E 01	0.9623	0.7984	0.7040	0.6544	0.9551	0.9946	1.0014	1.0002	0.7402	0.8604	0.7360	0.5812
5.5	1.1530E 01	0.9553	0.8142	0.7218	0.6725	0.9509	0.9940	1.0013	1.0002	0.7485	0.8718	0.7449	0.5872
6.0	1.1106E 01	0.9483	0.8296	0.7390	0.6902	0.9467	0.9935	1.0012	1.0002	0.7564	0.8825	0.7533	0.5929
6.5	1.0727E 01	0.9415	0.8445	0.7558	0.7075	0.9425	0.9929	1.0011	1.0002	0.7639	0.8926	0.7613	0.5982
7.0	1.0383E 01	0.9347	0.8590	0.7721	0.7244	0.9385	0.9923	1.0010	1.0001	0.7712	0.9020	0.7689	0.6033
7.5	1.0070E 01	0.9280	0.8730	0.7881	0.7408	0.9345	0.9918	1.0010	1.0001	0.7783	0.9109	0.7762	0.6082
8.0	9.7821E 00	0.9213	0.8868	0.8037	0.7570	0.9306	0.9912	1.0009	1.0001	0.7852	0.9194	0.7833	0.6129
9.0	9.2695E 00	0.9083	0.9133	0.8340	0.7883	0.9229	0.9901	1.0008	1.0001	0.7984	0.9351	0.7969	0.6219
10.0	8.8236E 00	0.8955	0.9387	0.8632	0.8186	0.9155	0.9889	1.0007	1.0001	0.8110	0.9492	0.8097	0.6304
11.0	8.4300E 00	0.8831	0.9631	0.8914	0.8480	0.9083	0.9877	1.0006	1.0001	0.8231	0.9621	0.8220	0.6386
12.0	8.0785E 00	0.8711	0.9866	0.9188	0.8765	0.9014	0.9866	1.0006	1.0001	0.8348	0.9739	0.8338	0.6465
13.0	7.7614E 00	0.8593	1.0094	0.9455	0.9044	0.8948	0.9854	1.0005	1.0001	0.8461	0.9845	0.8453	0.6542
14.0	7.4732E 00	0.8479	1.0315	0.9716	0.9318	0.8884	0.9842	1.0005	1.0001	0.8571	0.9942	0.8564	0.6617
16.0	6.9669E 00	0.8260	1.0739	1.0220	0.9848	0.8762	0.9818	1.0004	1.0001	0.8783	1.0110	0.8777	0.6763
18.0	6.5343E 00	0.8053	1.1142	1.0706	1.0362	0.8650	0.9793	1.0004	1.0001	0.8986	1.0246	0.8981	0.6904
20.0	6.1586E 00	0.7857	1.1525	1.1176	1.0861	0.8547	0.9768	1.0003	1.0001	0.9181	1.0353	0.9177	0.7042
25.0	5.4003E 00	0.7412	1.2414	1.2291	1.2056	0.8324	0.9704	1.0003	1.0000	0.9639	1.0513	0.9636	0.7371
30.0	4.8211E 00	0.7023	1.3212	1.3334	1.3189	0.8145	0.9638	1.0002	1.0000	1.0060	1.0541	1.0058	0.7681
35.0	4.3623E 00	0.6683	1.3928	1.4314	1.4268	0.8007	0.9568	1.0002	1.0000	1.0447	1.0457	1.0446	0.7972
40.0	3.9895E 00	0.6384	1.4567	1.5234	1.5295	0.7902	0.9496	1.0001	1.0000	1.0803	1.0280	1.0801	0.8243
45.0	3.6807E 00	0.6121	1.5132	1.6096	1.6273	0.7828	0.9423	1.0001	1.0000	1.1126	1.0023	1.1125	0.8491
50.0	3.4211E 00	0.5889	1.5625	1.6900	1.7200	0.7780	0.9347	1.0001	1.0000	1.1419	0.9700	1.1418	0.8717

Instr. see p. 22

ELECTRON Z= 76 A= 188

p	F_0L_0	L_0	λ_2	λ_3	λ_4	μ_1	μ_2	Λ_1	Λ_2	ν_{12}	$\hat{\nu}_{12}$	η_{12}	$\hat{\eta}_{12}$
0.1	2.3587E 02	1.0189	15.0534	96.8555	362.6927	0.9969	0.9999	1.0110	1.0010	5.5773	0.5277	-0.3629	-0.3868
0.2	1.1889E 02	1.0188	4.1861	9.1581	14.5131	0.9967	0.9999	1.0088	1.0010	1.7083	0.5512	0.1562	0.0867
0.3	8.0299E 01	1.0185	2.1759	3.0571	3.6317	0.9963	0.9998	1.0079	1.0010	1.0474	0.5734	0.3291	0.2429
0.4	6.1268E 01	1.0182	1.4749	1.6728	1.7589	0.9958	0.9998	1.0074	1.0009	0.8350	0.5938	0.4159	0.3204
0.5	5.0040E 01	1.0178	1.1532	1.1663	1.1546	0.9952	0.9997	1.0069	1.0009	0.7456	0.6124	0.4685	0.3666
0.6	4.2697E 01	1.0172	0.9811	0.9287	0.8896	0.9944	0.9996	1.0065	1.0009	0.7023	0.6293	0.5041	0.3973
0.7	3.7561E 01	1.0166	0.8798	0.7999	0.7514	0.9937	0.9995	1.0062	1.0008	0.6795	0.6446	0.5301	0.4194
0.8	3.3791E 01	1.0159	0.8163	0.7234	0.6714	0.9928	0.9994	1.0058	1.0008	0.6671	0.6583	0.5500	0.4361
0.9	3.0921E 01	1.0151	0.7748	0.6752	0.6219	0.9919	0.9993	1.0055	1.0008	0.6605	0.6708	0.5659	0.4492
1.0	2.8670E 01	1.0142	0.7469	0.6437	0.5899	0.9910	0.9992	1.0052	1.0007	0.6571	0.6821	0.5791	0.4599
1.2	2.5375E 01	1.0123	0.7149	0.6084	0.5546	0.9892	0.9989	1.0047	1.0007	0.6558	0.7019	0.5997	0.4764
1.4	2.3077E 01	1.0101	0.7003	0.5928	0.5393	0.9873	0.9987	1.0043	1.0006	0.6579	0.7186	0.6155	0.4888
1.6	2.1378E 01	1.0078	0.6946	0.5871	0.5338	0.9853	0.9985	1.0039	1.0005	0.6616	0.7329	0.6283	0.4987
1.8	2.0060E 01	1.0054	0.6940	0.5867	0.5337	0.9834	0.9982	1.0036	1.0005	0.6660	0.7456	0.6391	0.5068
2.0	1.9001E 01	1.0028	0.6964	0.5896	0.5368	0.9815	0.9980	1.0033	1.0005	0.6706	0.7569	0.6484	0.5138
2.2	1.8125E 01	1.0001	0.7006	0.5943	0.5417	0.9795	0.9977	1.0031	1.0004	0.6753	0.7671	0.6566	0.5198
2.4	1.7384E 01	0.9974	0.7059	0.6003	0.5478	0.9776	0.9975	1.0028	1.0004	0.6800	0.7764	0.6640	0.5252
2.6	1.6744E 01	0.9947	0.7120	0.6070	0.5547	0.9757	0.9973	1.0027	1.0004	0.6847	0.7850	0.6708	0.5301
2.8	1.6184E 01	0.9919	0.7184	0.6142	0.5620	0.9738	0.9970	1.0025	1.0003	0.6892	0.7930	0.6771	0.5346
3.0	1.5687E 01	0.9890	0.7252	0.6216	0.5696	0.9719	0.9968	1.0023	1.0003	0.6936	0.8006	0.6829	0.5388
3.2	1.5240E 01	0.9862	0.7321	0.6292	0.5774	0.9701	0.9966	1.0022	1.0003	0.6979	0.8077	0.6884	0.5426
3.4	1.4836E 01	0.9833	0.7391	0.6369	0.5852	0.9682	0.9963	1.0021	1.0003	0.7021	0.8144	0.6936	0.5463
3.6	1.4467E 01	0.9804	0.7461	0.6447	0.5931	0.9663	0.9961	1.0020	1.0003	0.7062	0.8208	0.6985	0.5497
3.8	1.4129E 01	0.9776	0.7531	0.6524	0.6010	0.9645	0.9959	1.0019	1.0003	0.7102	0.8269	0.7032	0.5530
4.0	1.3815E 01	0.9747	0.7600	0.6601	0.6088	0.9627	0.9957	1.0018	1.0002	0.7141	0.8328	0.7077	0.5561
4.5	1.3124E 01	0.9675	0.7772	0.6791	0.6282	0.9582	0.9951	1.0016	1.0002	0.7234	0.8465	0.7182	0.5633
5.0	1.2534E 01	0.9603	0.7939	0.6977	0.6472	0.9537	0.9945	1.0015	1.0002	0.7323	0.8590	0.7280	0.5699
5.5	1.2020E 01	0.9531	0.8101	0.7158	0.6657	0.9493	0.9939	1.0013	1.0002	0.7407	0.8706	0.7371	0.5761
6.0	1.1567E 01	0.9460	0.8259	0.7334	0.6837	0.9450	0.9934	1.0012	1.0002	0.7488	0.8815	0.7457	0.5818
6.5	1.1161E 01	0.9390	0.8412	0.7506	0.7014	0.9408	0.9928	1.0011	1.0002	0.7566	0.8916	0.7538	0.5873
7.0	1.0794E 01	0.9321	0.8561	0.7673	0.7186	0.9366	0.9922	1.0011	1.0001	0.7640	0.9012	0.7616	0.5925
7.5	1.0459E 01	0.9252	0.8705	0.7836	0.7354	0.9325	0.9916	1.0010	1.0001	0.7713	0.9102	0.7692	0.5975
8.0	1.0152E 01	0.9184	0.8846	0.7996	0.7519	0.9285	0.9910	1.0009	1.0001	0.7783	0.9187	0.7764	0.6023
9.0	9.6064E 00	0.9051	0.9119	0.8307	0.7839	0.9207	0.9899	1.0008	1.0001	0.7919	0.9345	0.7903	0.6116
10.0	9.1323E 00	0.8922	0.9380	0.8605	0.8149	0.9131	0.9887	1.0007	1.0001	0.8048	0.9487	0.8035	0.6204
11.0	8.7143E 00	0.8795	0.9631	0.8895	0.8449	0.9058	0.9875	1.0007	1.0001	0.8172	0.9616	0.8161	0.6288
12.0	8.3415E 00	0.8673	0.9873	0.9176	0.8742	0.8988	0.9863	1.0006	1.0001	0.8292	0.9734	0.8283	0.6370
13.0	8.0058E 00	0.8554	1.0107	0.9450	0.9028	0.8920	0.9851	1.0006	1.0001	0.8409	0.9840	0.8400	0.6449
14.0	7.7010E 00	0.8438	1.0334	0.9717	0.9308	0.8855	0.9839	1.0005	1.0001	0.8522	0.9937	0.8514	0.6527
16.0	7.1666E 00	0.8216	1.0770	1.0235	0.9852	0.8733	0.9814	1.0004	1.0001	0.8740	1.0104	0.8734	0.6678
18.0	6.7110E 00	0.8006	1.1184	1.0734	1.0379	0.8620	0.9789	1.0004	1.0001	0.8948	1.0238	0.8943	0.6824
20.0	6.3162E 00	0.7808	1.1579	1.1217	1.0891	0.8516	0.9764	1.0003	1.0001	0.9149	1.0343	0.9145	0.6966
25.0	5.5218E 00	0.7359	1.2492	1.2363	1.2119	0.8291	0.9699	1.0003	1.0000	0.9620	1.0497	0.9617	0.7307
30.0	4.9176E 00	0.6969	1.3312	1.3435	1.3283	0.8114	0.9630	1.0002	1.0000	1.0053	1.0516	1.0051	0.7628
35.0	4.4408E 00	0.6628	1.4046	1.4442	1.4391	0.7976	0.9559	1.0002	1.0000	1.0451	1.0424	1.0449	0.7929
40.0	4.0545E 00	0.6329	1.4700	1.5387	1.5447	0.7874	0.9486	1.0001	1.0000	1.0816	1.0238	1.0814	0.8209
45.0	3.7353E 00	0.6067	1.5277	1.6272	1.6451	0.7802	0.9411	1.0001	1.0000	1.1148	0.9973	1.1146	0.8465
50.0	3.4677E 00	0.5835	1.5780	1.7096	1.7404	0.7757	0.9334	1.0001	1.0000	1.1447	0.9642	1.1446	0.8698

Instr. see p. 22

ELECTRON Z= 77 A= 191

p	F_0L_0	L_0	λ_2	λ_3	λ_4	μ_1	μ_2	Λ_1	Λ_2	ν_{12}	$\hat{\nu}_{12}$	η_{12}	$\hat{\eta}_{12}$
0.1	2.5043E 02	1.0183	15.1535	99.0527	376.9885	0.9967	0.9999	1.0117	1.0010	5.6451	0.5221	-0.4054	-0.4256
0.2	1.2622E 02	1.0182	4.2053	9.2928	14.8474	0.9965	0.9999	1.0092	1.0010	1.7132	0.5459	0.1310	0.0624
0.3	8.5231E 01	1.0179	2.1801	3.0786	3.6699	0.9961	0.9998	1.0082	1.0010	1.0424	0.5682	0.3097	0.2234
0.4	6.5014E 01	1.0176	1.4739	1.6746	1.7625	0.9955	0.9997	1.0076	1.0010	0.8271	0.5888	0.3994	0.3032
0.5	5.3082E 01	1.0171	1.1498	1.1624	1.1503	0.9949	0.9997	1.0072	1.0009	0.7367	0.6077	0.4537	0.3507
0.6	4.5275E 01	1.0166	0.9764	0.9226	0.8827	0.9941	0.9996	1.0068	1.0009	0.6929	0.6248	0.4904	0.3823
0.7	3.9810E 01	1.0159	0.8743	0.7928	0.7436	0.9933	0.9995	1.0064	1.0009	0.6699	0.6402	0.5171	0.4050
0.8	3.5797E 01	1.0152	0.8104	0.7158	0.6631	0.9925	0.9994	1.0060	1.0008	0.6574	0.6541	0.5376	0.4221
0.9	3.2738E 01	1.0144	0.7685	0.6673	0.6134	0.9916	0.9993	1.0057	1.0008	0.6508	0.6668	0.5540	0.4356
1.0	3.0338E 01	1.0135	0.7405	0.6356	0.5812	0.9906	0.9991	1.0054	1.0007	0.6474	0.6782	0.5675	0.4466
1.2	2.6819E 01	1.0115	0.7082	0.6002	0.5459	0.9887	0.9989	1.0049	1.0007	0.6462	0.6983	0.5888	0.4635
1.4	2.4364E 01	1.0092	0.6936	0.5846	0.5305	0.9867	0.9987	1.0044	1.0006	0.6484	0.7152	0.6050	0.4761
1.6	2.2545E 01	1.0068	0.6880	0.5789	0.5251	0.9847	0.9984	1.0040	1.0006	0.6522	0.7299	0.6181	0.4861
1.8	2.1134E 01	1.0043	0.6875	0.5787	0.5251	0.9827	0.9982	1.0037	1.0005	0.6567	0.7427	0.6291	0.4944
2.0	1.9999E 01	1.0016	0.6900	0.5816	0.5283	0.9807	0.9979	1.0034	1.0005	0.6614	0.7542	0.6387	0.5015
2.2	1.9060E 01	0.9989	0.6943	0.5865	0.5333	0.9787	0.9977	1.0032	1.0004	0.6663	0.7645	0.6471	0.5077
2.4	1.8265E 01	0.9961	0.6998	0.5926	0.5396	0.9767	0.9975	1.0029	1.0004	0.6711	0.7740	0.6547	0.5132
2.6	1.7578E 01	0.9933	0.7060	0.5995	0.5465	0.9748	0.9972	1.0027	1.0004	0.6758	0.7828	0.6616	0.5182
2.8	1.6977E 01	0.9904	0.7127	0.6068	0.5540	0.9728	0.9970	1.0026	1.0004	0.6804	0.7910	0.6680	0.5227
3.0	1.6443E 01	0.9875	0.7196	0.6144	0.5617	0.9709	0.9967	1.0024	1.0003	0.6849	0.7986	0.6740	0.5270
3.2	1.5965E 01	0.9845	0.7267	0.6222	0.5696	0.9689	0.9965	1.0023	1.0003	0.6894	0.8058	0.6796	0.5309
3.4	1.5531E 01	0.9816	0.7338	0.6300	0.5776	0.9670	0.9963	1.0022	1.0003	0.6936	0.8127	0.6849	0.5346
3.6	1.5136E 01	0.9786	0.7410	0.6379	0.5856	0.9651	0.9960	1.0021	1.0003	0.6978	0.8192	0.6899	0.5381
3.8	1.4772E 01	0.9757	0.7482	0.6458	0.5937	0.9632	0.9958	1.0020	1.0003	0.7019	0.8254	0.6947	0.5414
4.0	1.4437E 01	0.9727	0.7553	0.6537	0.6017	0.9613	0.9956	1.0019	1.0003	0.7059	0.8314	0.6993	0.5446
4.5	1.3695E 01	0.9653	0.7730	0.6731	0.6214	0.9566	0.9950	1.0017	1.0002	0.7155	0.8453	0.7101	0.5520
5.0	1.3064E 01	0.9579	0.7901	0.6921	0.6408	0.9521	0.9944	1.0015	1.0002	0.7245	0.8580	0.7201	0.5587
5.5	1.2514E 01	0.9506	0.8068	0.7106	0.6597	0.9475	0.9938	1.0014	1.0002	0.7332	0.8698	0.7294	0.5650
6.0	1.2029E 01	0.9433	0.8230	0.7286	0.6781	0.9431	0.9932	1.0013	1.0002	0.7414	0.8808	0.7382	0.5709
6.5	1.1596E 01	0.9361	0.8387	0.7462	0.6961	0.9388	0.9926	1.0012	1.0002	0.7494	0.8910	0.7466	0.5765
7.0	1.1204E 01	0.9290	0.8540	0.7633	0.7137	0.9345	0.9920	1.0011	1.0001	0.7571	0.9007	0.7546	0.5818
7.5	1.0847E 01	0.9220	0.8689	0.7801	0.7309	0.9303	0.9914	1.0010	1.0001	0.7645	0.9098	0.7624	0.5870
8.0	1.0521E 01	0.9150	0.8834	0.7965	0.7478	0.9262	0.9908	1.0009	1.0001	0.7718	0.9184	0.7698	0.5919
9.0	9.9396E 00	0.9014	0.9114	0.8283	0.7806	0.9182	0.9896	1.0008	1.0001	0.7857	0.9343	0.7841	0.6015
10.0	9.4359E 00	0.8882	0.9383	0.8590	0.8123	0.9104	0.9884	1.0008	1.0001	0.7990	0.9486	0.7976	0.6105
11.0	8.9926E 00	0.8753	0.9641	0.8888	0.8432	0.9030	0.9872	1.0007	1.0001	0.8118	0.9616	0.8106	0.6192
12.0	8.5977E 00	0.8628	0.9891	0.9176	0.8732	0.8958	0.9860	1.0006	1.0001	0.8241	0.9733	0.8231	0.6277
13.0	8.2426E 00	0.8507	1.0132	0.9458	0.9026	0.8890	0.9848	1.0006	1.0001	0.8361	0.9840	0.8352	0.6359
14.0	7.9207E 00	0.8389	1.0366	0.9733	0.9313	0.8824	0.9835	1.0005	1.0001	0.8477	0.9936	0.8470	0.6439
16.0	7.3573E 00	0.8164	1.0815	1.0266	0.9873	0.8699	0.9810	1.0005	1.0001	0.8702	1.0101	0.8696	0.6596
18.0	6.8781E 00	0.7951	1.1242	1.0780	1.0415	0.8585	0.9785	1.0004	1.0001	0.8917	1.0233	0.8912	0.6747
20.0	6.4638E 00	0.7751	1.1649	1.1277	1.0942	0.8480	0.9759	1.0003	1.0001	0.9123	1.0336	0.9119	0.6895
25.0	5.6331E 00	0.7299	1.2589	1.2457	1.2206	0.8255	0.9692	1.0003	1.0000	0.9608	1.0481	0.9606	0.7249
30.0	5.0040E 00	0.6907	1.3431	1.3562	1.3404	0.8079	0.9622	1.0002	1.0000	1.0054	1.0490	1.0052	0.7582
35.0	4.5093E 00	0.6565	1.4184	1.4598	1.4546	0.7943	0.9549	1.0002	1.0000	1.0464	1.0388	1.0463	0.7894
40.0	4.1059E 00	0.6267	1.4854	1.5571	1.5632	0.7844	0.9474	1.0001	1.0000	1.0839	1.0190	1.0838	0.8183
45.0	3.7809E 00	0.6005	1.5443	1.6480	1.6666	0.7775	0.9397	1.0001	1.0000	1.1179	0.9914	1.1178	0.8447
50.0	3.5056E 00	0.5774	1.5954	1.7326	1.7645	0.7734	0.9318	1.0001	1.0000	1.1486	0.9571	1.1485	0.8686

Instr. see p. 22

ELECTRON Z= 78 A= 194

p	F_0L_0	L_0	λ_2	λ_3	λ_4	μ_1	μ_2	Λ_1	Λ_2	ν_{12}	$\hat{\nu}_{12}$	η_{12}	$\hat{\eta}_{12}$
0.1	2.6600E 02	1.0176	15.2459	101.2114	391.4266	0.9965	0.9999	1.0123	1.0011	5.7110	0.5165	-0.4485	-0.4648
0.2	1.3405E 02	1.0174	4.2226	9.4234	15.1789	0.9962	0.9999	1.0096	1.0010	1.7175	0.5405	0.1056	0.0379
0.3	9.0501E 01	1.0172	2.1836	3.0988	3.7067	0.9958	0.9998	1.0086	1.0010	1.0371	0.5630	0.2901	0.2038
0.4	6.9016E 01	1.0168	1.4725	1.6757	1.7654	0.9953	0.9997	1.0079	1.0010	0.8190	0.5838	0.3826	0.2858
0.5	5.6331E 01	1.0164	1.1461	1.1581	1.1456	0.9946	0.9997	1.0074	1.0009	0.7275	0.6028	0.4386	0.3347
0.6	4.8027E 01	1.0158	0.9715	0.9163	0.8757	0.9938	0.9996	1.0070	1.0009	0.6833	0.6201	0.4765	0.3672
0.7	4.2210E 01	1.0151	0.8687	0.7856	0.7356	0.9930	0.9995	1.0066	1.0009	0.6601	0.6357	0.5040	0.3905
0.8	3.7935E 01	1.0144	0.8043	0.7081	0.6547	0.9921	0.9993	1.0063	1.0008	0.6476	0.6499	0.5251	0.4081
0.9	3.4675E 01	1.0135	0.7622	0.6593	0.6047	0.9912	0.9992	1.0059	1.0008	0.6409	0.6627	0.5420	0.4219
1.0	3.2114E 01	1.0126	0.7339	0.6275	0.5725	0.9902	0.9991	1.0056	1.0007	0.6376	0.6743	0.5559	0.4331
1.2	2.8357E 01	1.0105	0.7015	0.5920	0.5371	0.9882	0.9989	1.0051	1.0007	0.6364	0.6946	0.5776	0.4504
1.4	2.5731E 01	1.0082	0.6868	0.5764	0.5218	0.9861	0.9986	1.0046	1.0006	0.6388	0.7118	0.5943	0.4633
1.6	2.3784E 01	1.0057	0.6813	0.5708	0.5164	0.9841	0.9984	1.0042	1.0006	0.6427	0.7267	0.6077	0.4735
1.8	2.2272E 01	1.0031	0.6809	0.5706	0.5165	0.9820	0.9981	1.0038	1.0005	0.6473	0.7397	0.6190	0.4820
2.0	2.1056E 01	1.0004	0.6835	0.5737	0.5198	0.9799	0.9979	1.0035	1.0005	0.6521	0.7514	0.6288	0.4892
2.2	2.0048E 01	0.9976	0.6880	0.5787	0.5250	0.9779	0.9976	1.0033	1.0004	0.6571	0.7619	0.6374	0.4954
2.4	1.9195E 01	0.9947	0.6937	0.5849	0.5313	0.9758	0.9974	1.0030	1.0004	0.6620	0.7716	0.6452	0.5010
2.6	1.8459E 01	0.9918	0.7000	0.5919	0.5384	0.9738	0.9972	1.0028	1.0004	0.6668	0.7805	0.6522	0.5061
2.8	1.7814E 01	0.9888	0.7069	0.5994	0.5460	0.9718	0.9969	1.0027	1.0004	0.6715	0.7888	0.6588	0.5108
3.0	1.7241E 01	0.9858	0.7140	0.6072	0.5539	0.9697	0.9967	1.0025	1.0003	0.6761	0.7966	0.6649	0.5150
3.2	1.6727E 01	0.9828	0.7212	0.6151	0.5619	0.9677	0.9964	1.0024	1.0003	0.6806	0.8040	0.6706	0.5191
3.4	1.6262E 01	0.9797	0.7286	0.6231	0.5700	0.9657	0.9962	1.0022	1.0003	0.6850	0.8109	0.6760	0.5228
3.6	1.5838E 01	0.9767	0.7360	0.6312	0.5782	0.9638	0.9959	1.0021	1.0003	0.6893	0.8175	0.6812	0.5264
3.8	1.5449E 01	0.9736	0.7433	0.6392	0.5864	0.9618	0.9957	1.0020	1.0003	0.6935	0.8239	0.6861	0.5298
4.0	1.5089E 01	0.9706	0.7507	0.6473	0.5945	0.9599	0.9955	1.0019	1.0003	0.6976	0.8299	0.6908	0.5330
4.5	1.4294E 01	0.9630	0.7687	0.6671	0.6147	0.9551	0.9949	1.0017	1.0002	0.7073	0.8440	0.7019	0.5405
5.0	1.3618E 01	0.9554	0.7863	0.6865	0.6344	0.9503	0.9943	1.0016	1.0002	0.7166	0.8570	0.7121	0.5474
5.5	1.3030E 01	0.9479	0.8035	0.7054	0.6537	0.9457	0.9937	1.0014	1.0002	0.7255	0.8689	0.7216	0.5538
6.0	1.2512E 01	0.9404	0.8201	0.7239	0.6725	0.9412	0.9931	1.0013	1.0002	0.7340	0.8800	0.7307	0.5598
6.5	1.2049E 01	0.9331	0.8362	0.7419	0.6909	0.9367	0.9925	1.0012	1.0002	0.7421	0.8904	0.7393	0.5656
7.0	1.1631E 01	0.9258	0.8520	0.7594	0.7088	0.9323	0.9918	1.0011	1.0002	0.7500	0.9001	0.7475	0.5711
7.5	1.1251E 01	0.9186	0.8673	0.7766	0.7264	0.9280	0.9912	1.0010	1.0001	0.7577	0.9093	0.7554	0.5763
8.0	1.0903E 01	0.9115	0.8822	0.7934	0.7437	0.9238	0.9906	1.0010	1.0001	0.7651	0.9180	0.7631	0.5814
9.0	1.0285E 01	0.8976	0.9111	0.8261	0.7773	0.9156	0.9894	1.0009	1.0001	0.7794	0.9340	0.7777	0.5912
10.0	9.7496E 00	0.8841	0.9387	0.8576	0.8098	0.9077	0.9882	1.0008	1.0001	0.7930	0.9484	0.7917	0.6006
11.0	9.2795E 00	0.8710	0.9653	0.8881	0.8415	0.9001	0.9869	1.0007	1.0001	0.8062	0.9614	0.8050	0.6096
12.0	8.8614E 00	0.8583	0.9910	0.9178	0.8723	0.8928	0.9857	1.0006	1.0001	0.8189	0.9732	0.8179	0.6183
13.0	8.4860E 00	0.8459	1.0159	0.9467	0.9024	0.8858	0.9844	1.0006	1.0001	0.8312	0.9838	0.8303	0.6268
14.0	8.1461E 00	0.8340	1.0400	0.9750	0.9319	0.8791	0.9832	1.0005	1.0001	0.8432	0.9934	0.8424	0.6351
16.0	7.5523E 00	0.8111	1.0863	1.0299	0.9895	0.8666	0.9806	1.0005	1.0001	0.8663	1.0098	0.8657	0.6513
18.0	7.0486E 00	0.7896	1.1302	1.0827	1.0452	0.8550	0.9780	1.0004	1.0001	0.8884	1.0228	0.8879	0.6670
20.0	6.6141E 00	0.7694	1.1721	1.1339	1.0994	0.8445	0.9753	1.0004	1.0001	0.9097	1.0328	0.9093	0.6824
25.0	5.7457E 00	0.7238	1.2688	1.2555	1.2295	0.8220	0.9684	1.0003	1.0000	0.9597	1.0464	0.9594	0.7191
30.0	5.0911E 00	0.6844	1.3554	1.3692	1.3529	0.8044	0.9613	1.0002	1.0000	1.0056	1.0464	1.0054	0.7536
35.0	4.5783E 00	0.6502	1.4327	1.4759	1.4704	0.7911	0.9538	1.0002	1.0000	1.0478	1.0350	1.0476	0.7859
40.0	4.1655E 00	0.6204	1.5012	1.5759	1.5823	0.7814	0.9461	1.0001	1.0000	1.0863	1.0142	1.0862	0.8158
45.0	3.8265E 00	0.5943	1.5612	1.6693	1.6886	0.7749	0.9382	1.0001	1.0000	1.1212	0.9854	1.1211	0.8430
50.0	3.5435E 00	0.5714	1.6133	1.7562	1.7893	0.7712	0.9302	1.0001	1.0000	1.1526	0.9501	1.1525	0.8676

ELECTRON Z= 79 A= 197

p	F_0L_0	L_0	λ_2	λ_3	λ_4	μ_1	μ_2	Λ_1	Λ_2	ν_{12}	$\hat{\nu}_{12}$	η_{12}	$\hat{\eta}_{12}$
0.1	2.8265E 02	1.0168	15.3301	103.3270	405.9883	0.9963	0.9999	1.0131	1.0011	5.7750	0.5109	-0.4921	-0.5042
0.2	1.4243E 02	1.0166	4.2380	9.5497	15.5074	0.9960	0.9999	1.0101	1.0011	1.7212	0.5350	0.0797	0.0133
0.3	9.6135E 01	1.0164	2.1861	3.1176	3.7421	0.9956	0.9998	1.0089	1.0010	1.0314	0.5577	0.2702	0.1840
0.4	7.3293E 01	1.0160	1.4706	1.6762	1.7676	0.9950	0.9997	1.0083	1.0010	0.8106	0.5787	0.3657	0.2684
0.5	5.9802E 01	1.0155	1.1421	1.1535	1.1406	0.9943	0.9996	1.0077	1.0010	0.7181	0.5979	0.4234	0.3187
0.6	5.0966E 01	1.0149	0.9664	0.9098	0.8684	0.9935	0.9996	1.0073	1.0009	0.6735	0.6154	0.4624	0.3520
0.7	4.4772E 01	1.0142	0.8629	0.7782	0.7275	0.9926	0.9994	1.0069	1.0009	0.6501	0.6312	0.4907	0.3759
0.8	4.0217E 01	1.0134	0.7981	0.7003	0.6462	0.9917	0.9993	1.0065	1.0008	0.6376	0.6455	0.5125	0.3939
0.9	3.6740E 01	1.0126	0.7558	0.6513	0.5961	0.9907	0.9992	1.0061	1.0008	0.6309	0.6584	0.5298	0.4081
1.0	3.4007E 01	1.0116	0.7273	0.6194	0.5638	0.9897	0.9991	1.0058	1.0008	0.6276	0.6702	0.5440	0.4195
1.2	2.9993E 01	1.0095	0.6948	0.5838	0.5284	0.9877	0.9988	1.0052	1.0007	0.6265	0.6908	0.5664	0.4372
1.4	2.7184E 01	1.0071	0.6800	0.5682	0.5131	0.9855	0.9986	1.0047	1.0006	0.6289	0.7083	0.5835	0.4503
1.6	2.5099E 01	1.0045	0.6746	0.5626	0.5078	0.9834	0.9983	1.0043	1.0006	0.6330	0.7234	0.5972	0.4608
1.8	2.3479E 01	1.0018	0.6743	0.5626	0.5080	0.9813	0.9981	1.0039	1.0005	0.6377	0.7367	0.6088	0.4694
2.0	2.2175E 01	0.9990	0.6771	0.5658	0.5113	0.9791	0.9978	1.0036	1.0005	0.6426	0.7485	0.6187	0.4767
2.2	2.1095E 01	0.9961	0.6817	0.5709	0.5166	0.9770	0.9976	1.0034	1.0004	0.6477	0.7592	0.6275	0.4831
2.4	2.0179E 01	0.9931	0.6875	0.5773	0.5231	0.9749	0.9973	1.0031	1.0004	0.6527	0.7691	0.6355	0.4888
2.6	1.9389E 01	0.9901	0.6941	0.5844	0.5303	0.9727	0.9971	1.0029	1.0004	0.6576	0.7781	0.6427	0.4939
2.8	1.8697E 01	0.9871	0.7011	0.5920	0.5380	0.9707	0.9968	1.0027	1.0004	0.6625	0.7866	0.6494	0.4986
3.0	1.8082E 01	0.9840	0.7084	0.5999	0.5460	0.9686	0.9966	1.0026	1.0003	0.6672	0.7945	0.6556	0.5030
3.2	1.7531E 01	0.9809	0.7158	0.6080	0.5542	0.9665	0.9963	1.0024	1.0003	0.6718	0.8020	0.6615	0.5071
3.4	1.7032E 01	0.9777	0.7233	0.6162	0.5625	0.9645	0.9961	1.0023	1.0003	0.6763	0.8090	0.6670	0.5109
3.6	1.6577E 01	0.9746	0.7309	0.6245	0.5708	0.9624	0.9959	1.0022	1.0003	0.6806	0.8158	0.6723	0.5145
3.8	1.6159E 01	0.9715	0.7385	0.6327	0.5791	0.9604	0.9956	1.0021	1.0003	0.6849	0.8222	0.6773	0.5180
4.0	1.5773E 01	0.9683	0.7460	0.6409	0.5874	0.9584	0.9954	1.0020	1.0003	0.6891	0.8283	0.6822	0.5213
4.5	1.4922E 01	0.9605	0.7645	0.6611	0.6079	0.9534	0.9948	1.0018	1.0002	0.6991	0.8427	0.6935	0.5289
5.0	1.4198E 01	0.9527	0.7826	0.6810	0.6280	0.9486	0.9941	1.0016	1.0002	0.7086	0.8558	0.7039	0.5360
5.5	1.3569E 01	0.9450	0.8002	0.7003	0.6477	0.9438	0.9935	1.0015	1.0002	0.7177	0.8679	0.7137	0.5425
6.0	1.3015E 01	0.9374	0.8173	0.7192	0.6669	0.9391	0.9929	1.0013	1.0002	0.7264	0.8791	0.7230	0.5487
6.5	1.2521E 01	0.9298	0.8339	0.7376	0.6857	0.9346	0.9923	1.0012	1.0002	0.7347	0.8897	0.7318	0.5545
7.0	1.2075E 01	0.9224	0.8500	0.7556	0.7041	0.9301	0.9917	1.0012	1.0002	0.7428	0.8995	0.7402	0.5602
7.5	1.1670E 01	0.9150	0.8658	0.7732	0.7221	0.9257	0.9911	1.0011	1.0001	0.7506	0.9088	0.7483	0.5656
8.0	1.1299E 01	0.9078	0.8811	0.7904	0.7397	0.9213	0.9904	1.0010	1.0001	0.7583	0.9176	0.7562	0.5708
9.0	1.0642E 01	0.8936	0.9108	0.8239	0.7741	0.9129	0.9892	1.0009	1.0001	0.7729	0.9337	0.7713	0.5809
10.0	1.0074E 01	0.8799	0.9392	0.8562	0.8074	0.9049	0.9879	1.0008	1.0001	0.7870	0.9482	0.7856	0.5905
11.0	9.5753E 00	0.8665	0.9666	0.8876	0.8399	0.8972	0.9867	1.0007	1.0001	0.8005	0.9612	0.7993	0.5998
12.0	9.1328E 00	0.8536	0.9931	0.9181	0.8715	0.8898	0.9854	1.0007	1.0001	0.8136	0.9730	0.8125	0.6088
13.0	8.7360E 00	0.8410	1.0187	0.9478	0.9024	0.8827	0.9841	1.0006	1.0001	0.8262	0.9836	0.8253	0.6176
14.0	8.3773E 00	0.8289	1.0435	0.9769	0.9327	0.8759	0.9828	1.0006	1.0001	0.8386	0.9931	0.8378	0.6262
16.0	7.7518E 00	0.8057	1.0912	1.0333	0.9918	0.8631	0.9802	1.0005	1.0001	0.8624	1.0094	0.8617	0.6430
18.0	7.2225E 00	0.7840	1.1365	1.0877	1.0491	0.8515	0.9775	1.0004	1.0001	0.8852	1.0221	0.8846	0.6593
20.0	6.7670E 00	0.7636	1.1796	1.1404	1.1049	0.8409	0.9747	1.0004	1.0001	0.9071	1.0319	0.9067	0.6752
25.0	5.8598E 00	0.7176	1.2791	1.2656	1.2388	0.8184	0.9677	1.0003	1.0000	0.9586	1.0447	0.9583	0.7133
30.0	5.1789E 00	0.6781	1.3680	1.3827	1.3658	0.8010	0.9604	1.0002	1.0000	1.0059	1.0437	1.0056	0.7491
35.0	4.6475E 00	0.6439	1.4473	1.4924	1.4868	0.7879	0.9527	1.0002	1.0000	1.0492	1.0312	1.0490	0.7825
40.0	4.2212E 00	0.6141	1.5173	1.5952	1.6019	0.7786	0.9448	1.0001	1.0000	1.0887	1.0093	1.0886	0.8133
45.0	3.8721E 00	0.5882	1.5786	1.6912	1.7112	0.7725	0.9368	1.0001	1.0000	1.1245	0.9794	1.1244	0.8414
50.0	3.5814E 00	0.5654	1.6315	1.7803	1.8147	0.7691	0.9285	1.0001	1.0000	1.1566	0.9430	1.1565	0.8666

Instr. see p. 22

ELECTRON Z= 80 A= 198

p	F_0L_0	L_0	λ_2	λ_3	λ_4	μ_1	μ_2	Λ_1	Λ_2	ν_{12}	$\hat{\nu}_{12}$	η_{12}	$\hat{\eta}_{12}$
0.1	3.0088E 02	1.0160	15.3950	105.2965	420.2103	0.9960	0.9999	1.0138	1.0011	5.8346	0.5049	-0.5362	-0.5437
0.2	1.5159E 02	1.0158	4.2481	9.6622	15.8154	0.9958	0.9999	1.0105	1.0011	1.7235	0.5292	0.0536	-0.0115
0.3	1.0230E 02	1.0155	2.1862	3.1321	3.7720	0.9953	0.9998	1.0093	1.0010	1.0250	0.5521	0.2499	0.1640
0.4	7.7972E 01	1.0151	1.4671	1.6744	1.7672	0.9947	0.9997	1.0086	1.0010	0.8017	0.5733	0.3483	0.2507
C.5	6.3598E 01	1.0146	1.1370	1.1474	1.1340	0.9940	0.9996	1.0080	1.0010	0.7083	0.5927	0.4078	0.3024
0.6	5.4178E 01	1.0140	0.9604	0.9022	0.8600	0.9932	0.9995	1.0075	1.0009	0.6632	0.6103	0.4479	0.3366
0.7	4.7573E 01	1.0133	0.8564	0.7700	0.7185	0.9923	0.9994	1.0071	1.0009	0.6397	0.6263	0.4771	0.3611
0.8	4.2710E 01	1.0125	0.7913	0.6918	0.6370	0.9913	0.9993	1.0067	1.0009	0.6271	0.6408	0.4994	0.3795
C.9	3.8997E 01	1.0116	0.7487	0.6426	0.5868	0.9903	0.9992	1.0063	1.0008	0.6204	0.6539	0.5172	0.3940
1.0	3.6075E 01	1.0106	0.7201	0.6106	0.5545	0.9893	0.9991	1.0060	1.0008	0.6172	0.6658	0.5318	0.4057
1.2	3.1779E 01	1.CC84	0.6875	0.5750	0.5191	0.9871	0.9988	1.0054	1.0007	0.6161	0.6867	0.5547	0.4237
1.4	2.8770E 01	1.0060	0.6727	0.5594	0.5039	0.9850	0.9986	1.0049	1.0006	0.6187	0.7044	0.5722	0.4371
1.6	2.6534E 01	1.CC33	0.6673	0.5539	0.4987	0.9827	0.9983	1.0044	1.0006	0.6229	0.7197	0.5863	0.4477
1.8	2.4795E 01	1.0006	0.6671	0.5540	0.4989	0.9805	0.9981	1.0041	1.0005	0.6277	0.7332	0.5981	0.4565
2.0	2.3395E 01	C.9977	0.6700	0.5573	0.5024	0.9783	0.9978	1.0037	1.0005	0.6328	0.7452	0.6083	0.4639
2.2	2.2234E 01	0.9947	0.6748	0.5626	0.5078	0.9761	0.9975	1.0035	1.0005	0.6379	0.7561	0.6173	0.4704
2.4	2.1250E 01	0.9916	0.6808	0.5691	0.5143	0.9739	0.9973	1.0032	1.0004	0.6430	0.7661	0.6254	0.4762
2.6	2.04C1E 01	0.9885	0.6875	0.5763	0.5217	0.9718	0.9970	1.0030	1.0004	0.6481	0.7753	0.6328	0.4814
2.8	1.9657E 01	0.9854	0.6947	0.5841	0.5295	0.9696	0.9968	1.0028	1.0004	0.6530	0.7839	0.6396	0.4862
3.0	1.8997E 01	0.9822	0.7022	0.5921	0.5376	0.9675	0.9965	1.0026	1.0003	0.6578	0.7919	0.6460	0.4906
3.2	1.84C4E 01	0.9790	0.7098	0.6004	0.5459	0.9653	0.9963	1.0025	1.0003	0.6625	0.7995	0.6520	0.4948
3.4	1.7868E 01	0.9758	0.7175	0.6087	0.5543	0.9632	0.9960	1.0024	1.0003	0.6671	0.8067	0.6576	0.4987
3.6	1.7380E 01	0.9726	0.7252	0.6171	0.5628	0.9611	0.9958	1.0022	1.0003	0.6715	0.8135	0.6630	0.5024
3.8	1.6931E 01	0.9694	0.7330	0.6255	0.5712	0.9591	0.9955	1.0021	1.0003	0.6759	0.8200	0.6681	0.5059
4.0	1.6517E 01	0.9662	0.7407	0.6338	0.5796	0.9570	0.9953	1.0020	1.0003	0.6801	0.8263	0.6730	0.5092
4.5	1.5604E 01	0.9582	0.7597	0.6545	0.6005	0.9519	0.9947	1.0018	1.0002	0.6904	0.8408	0.6846	0.5170
5.0	1.4828E 01	0.95C2	0.7782	0.6747	0.6210	0.9469	0.9940	1.0016	1.0002	0.7001	0.8541	0.6953	0.5242
5.5	1.4155E 01	0.9423	0.7962	0.6944	0.6410	0.9420	0.9934	1.0015	1.0002	0.7093	0.8664	0.7053	0.5308
6.0	1.3562E 01	0.9345	0.8137	0.7137	0.6605	0.9373	0.9928	1.0014	1.0002	0.7182	0.8778	0.7147	0.5371
6.5	1.3033E 01	0.9268	0.8307	0.7325	0.6797	0.9326	0.9922	1.0013	1.0002	0.7268	0.8884	0.7238	0.5431
7.0	1.2557E 01	0.9192	0.8473	0.7509	0.6984	0.9280	0.9915	1.0012	1.0002	0.7351	0.8984	0.7324	0.5489
7.5	1.2125E 01	0.9117	0.8635	0.7689	0.7168	0.9235	0.9909	1.0011	1.0001	0.7431	0.9077	0.7407	0.5544
8.0	1.1730E 01	0.9C44	0.8792	0.7865	0.7348	0.9190	0.9903	1.0010	1.0001	0.7509	0.9166	0.7488	0.5598
9.0	1.1031E 01	0.8899	0.9097	0.8207	0.7699	0.9105	0.9890	1.0009	1.0001	0.7659	0.9328	0.7642	0.5702
1C.0	1.0427E 01	0.8759	0.9389	0.8539	0.8040	0.9023	0.9877	1.0008	1.0001	0.7803	0.9474	0.7789	0.5800
11.0	9.8984E 0C	0.8624	0.9671	0.8860	0.8371	0.8945	0.9864	1.0007	1.0001	0.7942	0.9605	0.7930	0.5896
12.0	9.4297E 0C	0.8493	0.9942	0.9172	0.8695	0.8869	0.9851	1.0007	1.0001	0.8076	0.9722	0.8065	0.5989
13.0	9.01C1E 0C	0.8366	1.0205	0.9477	0.9011	0.8798	0.9838	1.0006	1.0001	0.8206	0.9828	0.8197	0.6080
14.0	8.6312E 00	0.8243	1.0461	0.9776	0.9322	0.8729	0.9825	1.0006	1.0001	0.8333	0.9924	0.8325	0.6169
16.0	7.9719E 00	0.8CC9	1.0951	1.0355	0.9927	0.8600	0.9798	1.0005	1.0001	0.8577	1.0085	0.8571	0.6342
18.0	7.4154E 00	0.7789	1.1416	1.0913	1.0515	0.8483	0.9771	1.0004	1.0001	0.8811	1.0211	0.8806	0.6511
2C.0	6.9374E C0	0.7583	1.1859	1.1454	1.1087	0.8376	0.9743	1.0004	1.0001	0.9037	1.0307	0.9032	0.6675
25.0	5.9888E 00	0.7122	1.2881	1.2739	1.2461	0.8151	0.9671	1.0003	1.0000	0.9566	1.0429	0.9563	0.7069
30.0	5.2798E 0C	0.6725	1.3793	1.3942	1.3765	0.7979	0.9596	1.C002	1.0000	1.0051	1.0411	1.0049	0.7439
35.0	4.7286E 0C	0.6383	1.4605	1.5069	1.5007	0.7851	0.9518	1.0002	1.0000	1.0497	1.0279	1.0495	0.7784
40.0	4.2877E 00	0.6C87	1.5321	1.6124	1.6189	0.7760	0.9438	1.0001	1.0000	1.0902	1.0051	1.0901	0.8101
45.0	3.9275E 0C	0.5828	1.5946	1.7108	1.7311	0.7702	0.9355	1.0001	1.0000	1.1269	0.9744	1.1268	0.8390
5C.0	3.6283E CC	0.5602	1.6484	1.8021	1.8374	0.7673	0.9271	1.C001	1.000C	1.1597	0.9373	1.1596	0.8650

ELECTRON Z= 81 A= 203

p	F_0L_0	L_0	λ_2	λ_3	λ_4	μ_1	μ_2	Λ_1	Λ_2	ν_{12}	$\hat{\nu}_{12}$	η_{12}	$\hat{\eta}_{12}$
0.1	3.1955E 02	1.0148	15.4747	107.4162	435.3901	0.9958	0.9999	1.0147	1.0011	5.8970	0.4993	-0.5812	-0.5837
0.2	1.6097E 02	1.0147	4.2626	9.7884	16.1525	0.9955	0.9998	1.0111	1.0011	1.7266	0.5238	0.0271	-0.0364
0.3	1.0861E 02	1.0144	2.1886	3.1511	3.8082	0.9950	0.9998	1.0097	1.0011	1.0190	0.5468	0.2296	0.1440
0.4	8.2758E 01	1.0140	1.4653	1.6752	1.7699	0.9944	0.9997	1.0089	1.0010	0.7931	0.5682	0.3311	0.2332
0.5	6.7479E 01	1.0135	1.1333	1.1431	1.1294	0.9936	0.9996	1.0083	1.0010	0.6987	0.5878	0.3923	0.2862
0.6	5.7461E 01	1.0128	0.9556	0.8961	0.8533	0.9928	0.9995	1.0078	1.0010	0.6533	0.6056	0.4336	0.3214
0.7	5.0431E 01	1.0121	0.8510	0.7631	0.7109	0.9919	0.9994	1.0074	1.0009	0.6297	0.6218	0.4636	0.3464
0.8	4.5253E 01	1.0113	0.7855	0.6845	0.6291	0.9909	0.9993	1.0070	1.0009	0.6170	0.6364	0.4866	0.3653
0.9	4.1294E 01	1.0103	0.7427	0.6351	0.5787	0.9898	0.9992	1.0066	1.0008	0.6103	0.6497	0.5049	0.3801
1.0	3.8178E 01	1.0093	0.7139	0.6030	0.5464	0.9887	0.9991	1.0062	1.0008	0.6071	0.6618	0.5199	0.3921
1.2	3.3592E 01	1.0070	0.6812	0.5673	0.5109	0.9865	0.9988	1.0056	1.0007	0.6062	0.6830	0.5434	0.4105
1.4	3.0375E 01	1.0045	0.6664	0.5518	0.4957	0.9842	0.9985	1.0051	1.0007	0.6089	0.7010	0.5613	0.4242
1.6	2.7982E 01	1.0018	0.6611	0.5464	0.4906	0.9820	0.9983	1.0046	1.0006	0.6131	0.7165	0.5757	0.4350
1.8	2.6120E 01	0.9989	0.6610	0.5465	0.4910	0.9797	0.9980	1.0042	1.0005	0.6180	0.7302	0.5878	0.4438
2.0	2.4619E 01	0.9959	0.6641	0.5500	0.4946	0.9774	0.9977	1.0039	1.0005	0.6232	0.7425	0.5982	0.4514
2.2	2.3375E 01	0.9928	0.6690	0.5554	0.5001	0.9751	0.9975	1.0036	1.0005	0.6285	0.7535	0.6074	0.4580
2.4	2.2321E 01	0.9896	0.6752	0.5620	0.5068	0.9728	0.9972	1.0033	1.0004	0.6337	0.7637	0.6157	0.4639
2.6	2.1410E 01	0.9864	0.6821	0.5694	0.5143	0.9706	0.9970	1.0031	1.0004	0.6389	0.7731	0.6232	0.4692
2.8	2.0612E 01	0.9832	0.6895	0.5774	0.5222	0.9683	0.9967	1.0029	1.0004	0.6439	0.7818	0.6302	0.4741
3.0	1.9904E 01	0.9799	0.6971	0.5856	0.5305	0.9661	0.9965	1.0027	1.0004	0.6488	0.7900	0.6367	0.4786
3.2	1.9269E 01	0.9766	0.7050	0.5940	0.5390	0.9639	0.9962	1.0026	1.0003	0.6536	0.7977	0.6428	0.4828
3.4	1.8695E 01	0.9733	0.7129	0.6025	0.5475	0.9617	0.9959	1.0025	1.0003	0.6583	0.8050	0.6486	0.4867
3.6	1.8171E 01	0.9700	0.7208	0.6111	0.5561	0.9596	0.9957	1.0023	1.0003	0.6629	0.8120	0.6541	0.4905
3.8	1.7690E 01	0.9667	0.7288	0.6196	0.5647	0.9574	0.9954	1.0022	1.0003	0.6673	0.8186	0.6594	0.4940
4.0	1.7247E 01	0.9634	0.7367	0.6282	0.5733	0.9553	0.9952	1.0021	1.0003	0.6717	0.8249	0.6644	0.4974
4.5	1.6270E 01	0.9552	0.7562	0.6493	0.5946	0.9500	0.9945	1.0019	1.0002	0.6821	0.8397	0.6762	0.5054
5.0	1.5439E 01	0.9470	0.7752	0.6700	0.6154	0.9449	0.9939	1.0017	1.0002	0.6921	0.8532	0.6872	0.5127
5.5	1.4720E 01	0.9389	0.7937	0.6902	0.6359	0.9399	0.9932	1.0016	1.0002	0.7016	0.8656	0.6974	0.5195
6.0	1.4087E 01	0.9309	0.8117	0.7099	0.6559	0.9349	0.9926	1.0014	1.0002	0.7107	0.8772	0.7071	0.5260
6.5	1.3524E 01	0.9230	0.8293	0.7292	0.6755	0.9301	0.9920	1.0013	1.0002	0.7195	0.8879	0.7164	0.5321
7.0	1.3017E 01	0.9152	0.8463	0.7480	0.6946	0.9254	0.9913	1.0012	1.0002	0.7280	0.8980	0.7252	0.5380
7.5	1.2557E 01	0.9075	0.8629	0.7665	0.7134	0.9208	0.9907	1.0011	1.0002	0.7362	0.9075	0.7338	0.5438
8.0	1.2137E 01	0.9000	0.8792	0.7846	0.7319	0.9162	0.9900	1.0011	1.0001	0.7442	0.9164	0.7421	0.5493
9.0	1.1394E 01	0.8852	0.9106	0.8198	0.7679	0.9075	0.9887	1.0009	1.0001	0.7597	0.9328	0.7579	0.5599
10.0	1.0754E 01	0.8709	0.9407	0.8538	0.8029	0.8991	0.9874	1.0008	1.0001	0.7745	0.9474	0.7730	0.5701
11.0	1.0194E 01	0.8571	0.9697	0.8869	0.8369	0.8911	0.9861	1.0008	1.0001	0.7888	0.9605	0.7875	0.5800
12.0	9.6993E 00	0.8438	0.9977	0.9190	0.8702	0.8835	0.9847	1.0007	1.0001	0.8026	0.9723	0.8015	0.5896
13.0	9.2565E 00	0.8309	1.0249	0.9505	0.9028	0.8762	0.9834	1.0006	1.0001	0.8160	0.9829	0.8151	0.5990
14.0	8.8574E 00	0.8184	1.0512	0.9812	0.9347	0.8692	0.9820	1.0006	1.0001	0.8291	0.9924	0.8282	0.6083
16.0	8.1643E 00	0.7946	1.1018	1.0408	0.9971	0.8562	0.9793	1.0005	1.0001	0.8543	1.0083	0.8536	0.6263
18.0	7.5807E 00	0.7724	1.1497	1.0984	1.0576	0.8444	0.9765	1.0004	1.0001	0.8785	1.0207	0.8779	0.6438
20.0	7.0807E 00	0.7516	1.1954	1.1542	1.1166	0.8337	0.9736	1.0004	1.0001	0.9017	1.0299	0.9012	0.6609
25.0	6.0920E 00	0.7052	1.3007	1.2868	1.2582	0.8113	0.9662	1.0003	1.0000	0.9563	1.0411	0.9560	0.7018
30.0	5.3566E 00	0.6655	1.3944	1.4108	1.3928	0.7942	0.9585	1.0002	1.0000	1.0063	1.0381	1.0061	0.7402
35.0	4.7870E 00	0.6313	1.4776	1.5269	1.5209	0.7818	0.9505	1.0002	1.0000	1.0521	1.0235	1.0520	0.7759
40.0	4.3329E 00	0.6017	1.5508	1.6355	1.6427	0.7732	0.9422	1.0001	1.0000	1.0938	0.9994	1.0937	0.8087
45.0	3.9632E 00	0.5761	1.6145	1.7367	1.7583	0.7679	0.9337	1.0001	1.0000	1.1314	0.9673	1.1313	0.8384
50.0	3.6568E 00	0.5537	1.6690	1.8305	1.8676	0.7654	0.9252	1.0001	1.0000	1.1650	0.9289	1.1649	0.8650

Instr. see p. 22

ELECTRON Z= 82 A= 208

p	F_0L_0	L_0	λ_2	λ_3	λ_4	μ_1	μ_2	Λ_1	Λ_2	ν_{12}	$\hat{\nu}_{12}$	η_{12}	$\hat{\eta}_{12}$
0.1	3.3952E 02	1.0136	15.5468	109.4881	450.6627	0.9955	0.9999	1.0157	1.0011	5.9574	0.4937	-0.6267	-0.6240
0.2	1.7101E 02	1.0134	4.2752	9.9101	16.4858	0.9952	0.9998	1.0117	1.0011	1.7290	0.5182	0.0004	-0.0614
0.3	1.1535E 02	1.0131	2.1902	3.1687	3.8430	0.9947	0.9998	1.0102	1.0011	1.0126	0.5415	0.2091	0.1239
0.4	8.7873E 01	1.0127	1.4631	1.6754	1.7719	0.9940	0.9997	1.0094	1.0011	0.7842	0.5631	0.3136	0.2155
0.5	7.1624E 01	1.0122	1.1292	1.1385	1.1245	0.9933	0.9996	1.0087	1.0010	0.6890	0.5829	0.3767	0.2699
0.6	6.0965E 01	1.0115	0.9506	0.8898	0.8463	0.9924	0.9995	1.0082	1.0010	0.6432	0.6009	0.4192	0.3060
0.7	5.3481E 01	1.0107	0.8454	0.7560	0.7032	0.9914	0.9994	1.0077	1.0009	0.6194	0.6172	0.4500	0.3317
0.8	4.7964E 01	1.0099	0.7796	0.6771	0.6211	0.9904	0.9993	1.0072	1.0009	0.6067	0.6320	0.4736	0.3510
0.9	4.3744E 01	1.0089	0.7365	0.6276	0.5706	0.9893	0.9992	1.0068	1.0009	0.6000	0.6455	0.4924	0.3662
1.0	4.0419E 01	1.0078	0.7077	0.5954	0.5382	0.9882	0.9990	1.0065	1.0008	0.5968	0.6577	0.5078	0.3784
1.2	3.5520E 01	1.0055	0.6748	0.5596	0.5028	0.9859	0.9988	1.0058	1.0007	0.5960	0.6792	0.5318	0.3971
1.4	3.2080E 01	1.0029	0.6601	0.5441	0.4877	0.9835	0.9985	1.0052	1.0007	0.5989	0.6975	0.5502	0.4111
1.6	2.9519E 01	1.0000	0.6548	0.5388	0.4826	0.9811	0.9982	1.0048	1.0006	0.6032	0.7133	0.5649	0.4221
1.8	2.7525E 01	0.9970	0.6549	0.5391	0.4831	0.9787	0.9980	1.0044	1.0006	0.6083	0.7272	0.5773	0.4311
2.0	2.5916E 01	0.9940	0.6581	0.5426	0.4868	0.9764	0.9977	1.0040	1.0005	0.6136	0.7396	0.5879	0.4388
2.2	2.4582E 01	0.9908	0.6632	0.5482	0.4924	0.9740	0.9974	1.0037	1.0005	0.6190	0.7509	0.5973	0.4455
2.4	2.3451E 01	0.9875	0.6696	0.5550	0.4993	0.9716	0.9972	1.0035	1.0004	0.6243	0.7612	0.6058	0.4515
2.6	2.2475E 01	0.9842	0.6767	0.5626	0.5069	0.9693	0.9969	1.0032	1.0004	0.6296	0.7707	0.6135	0.4568
2.8	2.1619E 01	0.9808	0.6842	0.5707	0.5150	0.9670	0.9966	1.0030	1.0004	0.6347	0.7796	0.6206	0.4618
3.0	2.0860E 01	0.9775	0.6921	0.5791	0.5234	0.9647	0.9964	1.0028	1.0004	0.6397	0.7879	0.6273	0.4664
3.2	2.0179E 01	0.9741	0.7001	0.5877	0.5320	0.9624	0.9961	1.0027	1.0003	0.6446	0.7958	0.6335	0.4706
3.4	1.9563E 01	0.9707	0.7083	0.5964	0.5407	0.9602	0.9958	1.0025	1.0003	0.6494	0.8032	0.6395	0.4747
3.6	1.9002E 01	0.9673	0.7164	0.6051	0.5495	0.9579	0.9956	1.0024	1.0003	0.6541	0.8103	0.6451	0.4785
3.8	1.8487E 01	0.9639	0.7246	0.6138	0.5583	0.9557	0.9953	1.0023	1.0003	0.6586	0.8170	0.6504	0.4821
4.0	1.8012E 01	0.9605	0.7327	0.6225	0.5670	0.9535	0.9950	1.0022	1.0003	0.6631	0.8235	0.6556	0.4856
4.5	1.6966E 01	0.9520	0.7527	0.6441	0.5887	0.9481	0.9944	1.0020	1.0003	0.6738	0.8385	0.6677	0.4937
5.0	1.6078E 01	0.9436	0.7723	0.6653	0.6100	0.9428	0.9937	1.0018	1.0002	0.6839	0.8522	0.6789	0.5011
5.5	1.5310E 01	0.9353	0.7913	0.6860	0.6308	0.9376	0.9931	1.0016	1.0002	0.6937	0.8648	0.6894	0.5081
6.0	1.4634E 01	0.9271	0.8098	0.7062	0.6513	0.9325	0.9924	1.0015	1.0002	0.7030	0.8765	0.6994	0.5147
6.5	1.4033E 01	0.9190	0.8279	0.7259	0.6713	0.9276	0.9918	1.0014	1.0002	0.7120	0.8874	0.7088	0.5210
7.0	1.3493E 01	0.9110	0.8454	0.7453	0.6909	0.9227	0.9911	1.0013	1.0002	0.7208	0.8976	0.7180	0.5271
7.5	1.3004E 01	0.9031	0.8625	0.7642	0.7102	0.9180	0.9904	1.0012	1.0002	0.7292	0.9072	0.7267	0.5330
8.0	1.2557E 01	0.8954	0.8793	0.7828	0.7291	0.9134	0.9898	1.0011	1.0001	0.7375	0.9161	0.7352	0.5386
9.0	1.1769E 01	0.8803	0.9116	0.8189	0.7660	0.9044	0.9884	1.0010	1.0001	0.7534	0.9326	0.7515	0.5496
10.0	1.1091E 01	0.8658	0.9426	0.8539	0.8019	0.8959	0.9871	1.0009	1.0001	0.7686	0.9473	0.7671	0.5602
11.0	1.0499E 01	0.8517	0.9726	0.8879	0.8369	0.8877	0.9857	1.0008	1.0001	0.7833	0.9605	0.7820	0.5704
12.0	9.9757E 00	0.8381	1.0014	0.9210	0.8711	0.8800	0.9844	1.0007	1.0001	0.7975	0.9723	0.7964	0.5803
13.0	9.5088E 00	0.8250	1.0294	0.9534	0.9045	0.8726	0.9830	1.0006	1.0001	0.8113	0.9829	0.8104	0.5901
14.0	9.0886E 00	0.8123	1.0566	0.9850	0.9374	0.8655	0.9816	1.0006	1.0001	0.8248	0.9923	0.8239	0.5996
16.0	8.3602E 00	0.7883	1.1087	1.0465	1.0016	0.8524	0.9787	1.0005	1.0001	0.8508	1.0080	0.8501	0.6183
18.0	7.7486E 00	0.7659	1.1582	1.1058	1.0639	0.8405	0.9759	1.0004	1.0001	0.8757	1.0201	0.8752	0.6365
20.0	7.2259E 00	0.7449	1.2052	1.1633	1.1247	0.8298	0.9729	1.0004	1.0001	0.8997	1.0290	0.8992	0.6542
25.0	6.1960E 00	0.6982	1.3136	1.3001	1.2708	0.8074	0.9653	1.0003	1.0001	0.9560	1.0392	0.9557	0.6967
30.0	5.4336E 00	0.6584	1.4100	1.4279	1.4095	0.7907	0.9574	1.0002	1.0000	1.0076	1.0350	1.0073	0.7365
35.0	4.8454E 00	0.6243	1.4952	1.5476	1.5416	0.7786	0.9492	1.0002	1.0000	1.0547	1.0192	1.0545	0.7735
40.0	4.3781E 00	0.5949	1.5700	1.6593	1.6671	0.7704	0.9406	1.0001	1.0000	1.0975	0.9936	1.0973	0.8073
45.0	3.9987E 00	0.5694	1.6347	1.7633	1.7861	0.7656	0.9320	1.0001	1.0000	1.1360	0.9602	1.1359	0.8378
50.0	3.6851E 00	0.5472	1.6899	1.8594	1.8985	0.7637	0.9231	1.0001	1.0000	1.1703	0.9205	1.1702	0.8651

Instr. see p. 22

ELECTRON Z= 83 A= 214

p	F_0L_0	L_0	λ_2	λ_3	λ_4	μ_1	μ_2	Λ_1	Λ_2	ν_{12}	$\hat{\nu}_{12}$	η_{12}	$\hat{\eta}_{12}$
0.1	3.6063E 02	1.0122	15.6169	111.5634	466.2542	0.9951	0.9999	1.0168	1.0012	6.0167	0.4880	-0.6729	-0.6646
0.2	1.8162E 02	1.0120	4.2875	10.0319	16.8237	0.9948	0.9998	1.0124	1.0011	1.7312	0.5128	-0.0267	-0.0866
0.3	1.2248E 02	1.0117	2.1918	3.1865	3.8783	0.9943	0.9998	1.0107	1.0011	1.0060	0.5362	0.1883	0.1038
0.4	9.3275E 01	1.0112	1.4609	1.6758	1.7742	0.9937	0.9997	1.0098	1.0011	0.7752	0.5579	0.2959	0.1978
0.5	7.6000E 01	1.0107	1.1253	1.1342	1.1198	0.9929	0.9996	1.0091	1.0010	0.6791	0.5779	0.3609	0.2537
0.6	6.4662E 01	1.0100	0.9458	0.8837	0.8396	0.9919	0.9995	1.0085	1.0010	0.6330	0.5961	0.4046	0.2906
0.7	5.6697E 01	1.0092	0.8401	0.7493	0.6958	0.9909	0.9994	1.0080	1.0010	0.6091	0.6127	0.4363	0.3169
0.8	5.0821E 01	1.0083	0.7739	0.6700	0.6134	0.9898	0.9993	1.0075	1.0009	0.5964	0.6276	0.4605	0.3367
0.9	4.6323E 01	1.0073	0.7306	0.6203	0.5628	0.9887	0.9991	1.0071	1.0009	0.5897	0.6413	0.4798	0.3522
1.0	4.2776E 01	1.0062	0.7017	0.5880	0.5303	0.9876	0.9990	1.0067	1.0008	0.5866	0.6537	0.4956	0.3647
1.2	3.7546E 01	1.0037	0.6687	0.5522	0.4949	0.9852	0.9987	1.0060	1.0008	0.5859	0.6754	0.5203	0.3838
1.4	3.3869E 01	1.0010	0.6540	0.5367	0.4799	0.9827	0.9985	1.0055	1.0007	0.5888	0.6940	0.5391	0.3980
1.6	3.1128E 01	0.9981	0.6488	0.5315	0.4749	0.9802	0.9982	1.0050	1.0006	0.5933	0.7100	0.5541	0.4092
1.8	2.8993E 01	0.9950	0.6490	0.5319	0.4755	0.9777	0.9979	1.0045	1.0006	0.5985	0.7242	0.5668	0.4183
2.0	2.7269E 01	0.9918	0.6524	0.5356	0.4793	0.9753	0.9976	1.0042	1.0005	0.6039	0.7368	0.5776	0.4261
2.2	2.5840E 01	0.9885	0.6577	0.5413	0.4851	0.9728	0.9974	1.0039	1.0005	0.6094	0.7483	0.5873	0.4330
2.4	2.4628E 01	0.9851	0.6642	0.5483	0.4921	0.9704	0.9971	1.0036	1.0005	0.6148	0.7588	0.5959	0.4390
2.6	2.3581E 01	0.9817	0.6715	0.5560	0.4998	0.9679	0.9968	1.0034	1.0004	0.6202	0.7685	0.6038	0.4445
2.8	2.2663E 01	0.9783	0.6793	0.5643	0.5081	0.9655	0.9965	1.0031	1.0004	0.6255	0.7775	0.6111	0.4495
3.0	2.1849E 01	0.9748	0.6874	0.5729	0.5167	0.9632	0.9963	1.0030	1.0004	0.6306	0.7860	0.6179	0.4541
3.2	2.1120E 01	0.9713	0.6956	0.5817	0.5254	0.9608	0.9960	1.0028	1.0004	0.6356	0.7940	0.6242	0.4585
3.4	2.0460E 01	0.9678	0.7040	0.5905	0.5343	0.9585	0.9957	1.0026	1.0003	0.6405	0.8015	0.6303	0.4626
3.6	1.9859E 01	0.9643	0.7124	0.5995	0.5432	0.9562	0.9955	1.0025	1.0003	0.6453	0.8087	0.6360	0.4665
3.8	1.9308E 01	0.9607	0.7207	0.6084	0.5522	0.9539	0.9952	1.0024	1.0003	0.6499	0.8156	0.6415	0.4701
4.0	1.8759E 01	0.9572	0.7291	0.6173	0.5611	0.9516	0.9949	1.0023	1.0003	0.6545	0.8221	0.6468	0.4737
4.5	1.7680E 01	0.9485	0.7497	0.6394	0.5832	0.9460	0.9942	1.0020	1.0003	0.6654	0.8374	0.6592	0.4819
5.0	1.6731E 01	0.9399	0.7698	0.6610	0.6049	0.9405	0.9936	1.0018	1.0002	0.6758	0.8513	0.6707	0.4896
5.5	1.5911E 01	0.9313	0.7893	0.6822	0.6263	0.9352	0.9929	1.0017	1.0002	0.6858	0.8641	0.6814	0.4967
6.0	1.5191E 01	0.9229	0.8084	0.7029	0.6472	0.9300	0.9922	1.0015	1.0002	0.6954	0.8759	0.6916	0.5035
6.5	1.4551E 01	0.9146	0.8269	0.7232	0.6677	0.9249	0.9915	1.0014	1.0002	0.7046	0.8870	0.7014	0.5099
7.0	1.3976E 01	0.9064	0.8450	0.7430	0.6878	0.9199	0.9909	1.0013	1.0002	0.7136	0.8973	0.7107	0.5162
7.5	1.3456E 01	0.8984	0.8627	0.7625	0.7075	0.9150	0.9902	1.0012	1.0002	0.7223	0.9069	0.7197	0.5222
8.0	1.2981E 01	0.8905	0.8799	0.7816	0.7269	0.9103	0.9895	1.0011	1.0001	0.7308	0.9160	0.7285	0.5281
9.0	1.2144E 01	0.8751	0.9132	0.8187	0.7648	0.9012	0.9881	1.0010	1.0001	0.7471	0.9326	0.7453	0.5394
10.0	1.1427E 01	0.8602	0.9452	0.8547	0.8016	0.8925	0.9867	1.0009	1.0001	0.7628	0.9474	0.7613	0.5503
11.0	1.0801E 01	0.8459	0.9761	0.8897	0.8376	0.8842	0.9853	1.0008	1.0001	0.7779	0.9606	0.7766	0.5608
12.0	1.0249E 01	0.8321	1.0059	0.9238	0.8727	0.8763	0.9839	1.0007	1.0001	0.7926	0.9724	0.7915	0.5711
13.0	9.7574E 00	0.8188	1.0348	0.9571	0.9072	0.8688	0.9825	1.0007	1.0001	0.8069	0.9829	0.8059	0.5812
14.0	9.3154E 00	0.8059	1.0628	0.9898	0.9410	0.8616	0.9811	1.0006	1.0001	0.8208	0.9923	0.8199	0.5912
16.0	8.5510E 00	0.7816	1.1166	1.0531	1.0072	0.8484	0.9782	1.0005	1.0001	0.8476	1.0078	0.8469	0.6106
18.0	7.9108E 00	0.7589	1.1676	1.1144	1.0714	0.8364	0.9752	1.0005	1.0001	0.8733	1.0196	0.8727	0.6295
20.0	7.3650E 00	0.7378	1.2162	1.1737	1.1341	0.8257	0.9722	1.0004	1.0001	0.8981	1.0282	0.8976	0.6479
25.0	6.2937E 00	0.6908	1.3278	1.3148	1.2849	0.8035	0.9644	1.0003	1.0001	0.9562	1.0373	0.9558	0.6921
30.0	5.5043E 00	0.6510	1.4268	1.4467	1.4281	0.7871	0.9562	1.0002	1.0000	1.0093	1.0318	1.0091	0.7334
35.0	4.8979E 00	0.6170	1.5142	1.5700	1.5643	0.7754	0.9477	1.0002	1.0000	1.0578	1.0144	1.0576	0.7715
40.0	4.4178E 00	0.5878	1.5905	1.6851	1.6937	0.7678	0.9390	1.0002	1.0000	1.1018	0.9874	1.1016	0.8064
45.0	4.0291E 00	0.5625	1.6563	1.7919	1.8163	0.7635	0.9300	1.0001	1.0000	1.1412	0.9525	1.1411	0.8378
50.0	3.7087E 00	0.5406	1.7120	1.8906	1.9320	0.7622	0.9210	1.0001	1.0000	1.1762	0.9113	1.1761	0.8657

Instr. see p. 22

ELECTRON Z= 84 A= 210

p	F_0L_0	L_0	λ_2	λ_3	λ_4	μ_1	μ_2	Λ_1	Λ_2	ν_{12}	$\hat{\nu}_{12}$	η_{12}	$\hat{\eta}_{12}$
0.1	3.8588E 02	1.0111	15.6186	113.0252	479.2841	0.9949	0.9999	1.0176	1.0012	6.0602	0.4813	-0.7184	-0.7042
0.2	1.9431E 02	1.0109	4.2812	10.0989	17.0639	0.9945	0.9998	1.0128	1.0012	1.7287	0.5062	-0.0541	-0.1118
0.3	1.3101E 02	1.0106	2.1840	3.1870	3.8907	0.9940	0.9998	1.0111	1.0011	0.9968	0.5297	0.1669	0.0833
0.4	9.9743E 01	1.0101	1.4526	1.6672	1.7661	0.9934	0.9997	1.0101	1.0011	0.7643	0.5516	0.2775	0.1796
0.5	8.1241E 01	1.0095	1.1168	1.1239	1.1087	0.9925	0.9996	1.0094	1.0011	0.6676	0.5717	0.3442	0.2368
0.6	6.9092E 01	1.0088	0.9371	0.8731	0.8281	0.9916	0.9995	1.0088	1.0010	0.6213	0.5901	0.3890	0.2746
0.7	6.0552E 01	1.0080	0.8313	0.7386	0.6844	0.9906	0.9994	1.0082	1.0010	0.5974	0.6067	0.4216	0.3015
0.8	5.4248E 01	1.0071	0.7650	0.6594	0.6023	0.9894	0.9993	1.0077	1.0009	0.5846	0.6219	0.4464	0.3217
0.9	4.9419E 01	1.0061	0.7218	0.6099	0.5519	0.9883	0.9991	1.0073	1.0009	0.5780	0.6356	0.4661	0.3374
1.0	4.5608E 01	1.0049	0.6928	0.5776	0.5196	0.9871	0.9990	1.0069	1.0008	0.5749	0.6482	0.4823	0.3502
1.2	3.9984E 01	1.0024	0.6599	0.5420	0.4844	0.9846	0.9987	1.0062	1.0008	0.5743	0.6702	0.5075	0.3696
1.4	3.6025E 01	0.9997	0.6452	0.5267	0.4695	0.9821	0.9984	1.0056	1.0007	0.5774	0.6889	0.5267	0.3840
1.6	3.3073E 01	0.9967	0.6402	0.5216	0.4646	0.9796	0.9982	1.0051	1.0006	0.5820	0.7052	0.5420	0.3954
1.8	3.0770E 01	0.9935	0.6405	0.5221	0.4653	0.9770	0.9979	1.0046	1.0006	0.5872	0.7195	0.5549	0.4047
2.0	2.8911E 01	0.9902	0.6440	0.5258	0.4692	0.9745	0.9976	1.0043	1.0005	0.5928	0.7323	0.5660	0.4126
2.2	2.7369E 01	0.9869	0.6494	0.5317	0.4750	0.9720	0.9973	1.0039	1.0005	0.5983	0.7439	0.5758	0.4195
2.4	2.6061E 01	0.9834	0.6561	0.5387	0.4821	0.9695	0.9971	1.0037	1.0005	0.6039	0.7545	0.5845	0.4256
2.6	2.4932E 01	0.9800	0.6636	0.5466	0.4899	0.9670	0.9968	1.0034	1.0004	0.6093	0.7644	0.5926	0.4311
2.8	2.3942E 01	0.9764	0.6715	0.5550	0.4983	0.9646	0.9965	1.0032	1.0004	0.6147	0.7735	0.6000	0.4362
3.0	2.3064E 01	0.9729	0.6798	0.5636	0.5069	0.9621	0.9962	1.0030	1.0004	0.6199	0.7821	0.6069	0.4409
3.2	2.2278E 01	0.9693	0.6882	0.5725	0.5158	0.9597	0.9960	1.0028	1.0004	0.6250	0.7902	0.6134	0.4453
3.4	2.1567E 01	0.9658	0.6967	0.5815	0.5247	0.9573	0.9957	1.0027	1.0003	0.6299	0.7979	0.6195	0.4494
3.6	2.0919E 01	0.9622	0.7052	0.5906	0.5337	0.9550	0.9954	1.0025	1.0003	0.6348	0.8051	0.6254	0.4533
3.8	2.0325E 01	0.9586	0.7137	0.5996	0.5428	0.9526	0.9951	1.0024	1.0003	0.6395	0.8121	0.6309	0.4571
4.0	1.9777E 01	0.9551	0.7222	0.6086	0.5518	0.9503	0.9949	1.0023	1.0003	0.6441	0.8187	0.6363	0.4606
4.5	1.8574E 01	0.9462	0.7432	0.6310	0.5741	0.9446	0.9942	1.0021	1.0003	0.6552	0.8342	0.6489	0.4690
5.0	1.7554E 01	0.9374	0.7637	0.6530	0.5961	0.9390	0.9935	1.0019	1.0002	0.6658	0.8482	0.6606	0.4767
5.5	1.6673E 01	0.9288	0.7837	0.6745	0.6177	0.9336	0.9928	1.0017	1.0002	0.6760	0.8612	0.6715	0.4840
6.0	1.5901E 01	0.9202	0.8031	0.6955	0.6389	0.9283	0.9922	1.0016	1.0002	0.6857	0.8731	0.6819	0.4909
6.5	1.5215E 01	0.9118	0.8221	0.7161	0.6596	0.9231	0.9915	1.0014	1.0002	0.6952	0.8843	0.6918	0.4975
7.0	1.4600E 01	0.9035	0.8405	0.7362	0.6800	0.9181	0.9908	1.0013	1.0002	0.7043	0.8947	0.7013	0.5038
7.5	1.4044E 01	0.8954	0.8585	0.7560	0.7000	0.9132	0.9901	1.0012	1.0002	0.7131	0.9044	0.7105	0.5099
8.0	1.3538E 01	0.8874	0.8761	0.7754	0.7197	0.9084	0.9894	1.0012	1.0001	0.7218	0.9136	0.7194	0.5159
9.0	1.2645E 01	0.8718	0.9102	0.8132	0.7581	0.8991	0.9880	1.0010	1.0001	0.7384	0.9303	0.7365	0.5275
10.0	1.1881E 01	0.8569	0.9429	0.8498	0.7955	0.8903	0.9866	1.0009	1.0001	0.7544	0.9452	0.7528	0.5386
11.0	1.1217E 01	0.8424	0.9744	0.8855	0.8321	0.8819	0.9852	1.0008	1.0001	0.7699	0.9585	0.7685	0.5494
12.0	1.0631E 01	0.8285	1.0049	0.9202	0.8678	0.8740	0.9838	1.0007	1.0001	0.7848	0.9703	0.7837	0.5600
13.0	1.0110E 01	0.8151	1.0344	0.9542	0.9029	0.8664	0.9824	1.0007	1.0001	0.7994	0.9809	0.7983	0.5703
14.0	9.6427E 00	0.8022	1.0631	0.9874	0.9373	0.8592	0.9809	1.0006	1.0001	0.8135	0.9902	0.8126	0.5805
16.0	8.8357E 00	0.7777	1.1180	1.0521	1.0046	0.8459	0.9780	1.0005	1.0001	0.8409	1.0058	0.8402	0.6004
18.0	8.1613E 00	0.7550	1.1702	1.1145	1.0701	0.8340	0.9750	1.0005	1.0001	0.8672	1.0176	0.8666	0.6198
20.0	7.5877E 00	0.7338	1.2198	1.1751	1.1339	0.8232	0.9719	1.0004	1.0001	0.8925	1.0262	0.8920	0.6387
25.0	6.4652E 00	0.6868	1.3340	1.3191	1.2876	0.8011	0.9640	1.0003	1.0001	0.9519	1.0352	0.9516	0.6840
30.0	5.6414E 00	0.6471	1.4352	1.4538	1.4338	0.7848	0.9558	1.0002	1.0000	1.0063	1.0296	1.0060	0.7264
35.0	5.0105E 00	0.6132	1.5244	1.5798	1.5728	0.7734	0.9472	1.0002	1.0000	1.0559	1.0122	1.0557	0.7656
40.0	4.5123E 00	0.5841	1.6024	1.6973	1.7050	0.7660	0.9384	1.0002	1.0000	1.1008	0.9851	1.1007	0.8014
45.0	4.1098E 00	0.5590	1.6695	1.8065	1.8302	0.7619	0.9293	1.0001	1.0000	1.1412	0.9502	1.1410	0.8336
50.0	3.7787E 00	0.5372	1.7264	1.9074	1.9484	0.7608	0.9202	1.0001	1.0000	1.1770	0.9090	1.1769	0.8622

Instr. see p. 22

ELECTRON Z= 85 A= 217

p	F_0L_0	L_0	λ_2	λ_3	λ_4	μ_1	μ_2	Λ_1	Λ_2	ν_{12}	$\hat{\nu}_{12}$	η_{12}	$\hat{\eta}_{12}$
0.1	4.0992E 02	1.0093	15.6792	115.0407	495.1610	0.9945	0.9999	1.0189	1.0012	6.1160	0.4756	-0.7657	-0.7451
0.2	2.0638E 02	1.0091	4.2913	10.2155	17.4002	0.9942	0.9998	1.0136	1.0012	1.7297	0.5006	-0.0818	-0.1372
0.3	1.3912E 02	1.0088	2.1847	3.2035	3.9248	0.9936	0.9998	1.0117	1.0012	0.9897	0.5243	0.1457	0.0630
0.4	1.0588E 02	1.0083	1.4501	1.6671	1.7680	0.9929	0.9997	1.0106	1.0011	0.7549	0.5464	0.2595	0.1618
0.5	8.6209E 01	1.0077	1.1128	1.1194	1.1040	0.9921	0.9996	1.0098	1.0011	0.6574	0.5667	0.3281	0.2204
0.6	7.3286E 01	1.0070	0.9323	0.8671	0.8215	0.9911	0.9995	1.0092	1.0010	0.6108	0.5853	0.3742	0.2591
0.7	6.4195E 01	1.0061	0.8260	0.7320	0.6772	0.9900	0.9994	1.0086	1.0010	0.5868	0.6021	0.4076	0.2866
0.8	5.7480E 01	1.0052	0.7595	0.6525	0.5948	0.9888	0.9992	1.0081	1.0009	0.5741	0.6174	0.4331	0.3073
0.9	5.2333E 01	1.0041	0.7160	0.6028	0.5443	0.9876	0.9991	1.0076	1.0009	0.5675	0.6313	0.4533	0.3234
1.0	4.8268E 01	1.0029	0.6870	0.5705	0.5120	0.9864	0.9990	1.0072	1.0009	0.5644	0.6440	0.4699	0.3364
1.2	4.2263E 01	1.0003	0.6539	0.5349	0.4768	0.9838	0.9987	1.0065	1.0008	0.5640	0.6664	0.4957	0.3562
1.4	3.8031E 01	0.9974	0.6393	0.5196	0.4620	0.9812	0.9984	1.0058	1.0007	0.5671	0.6854	0.5153	0.3709
1.6	3.4872E 01	0.9943	0.6344	0.5146	0.4573	0.9786	0.9981	1.0053	1.0006	0.5719	0.7019	0.5310	0.3824
1.8	3.2406E 01	0.9911	0.6349	0.5152	0.4581	0.9759	0.9978	1.0048	1.0006	0.5773	0.7164	0.5442	0.3918
2.0	3.0415E 01	0.9877	0.6385	0.5192	0.4621	0.9733	0.9975	1.0044	1.0005	0.5829	0.7294	0.5555	0.3998
2.2	2.8763E 01	0.9842	0.6442	0.5251	0.4681	0.9707	0.9973	1.0041	1.0005	0.5886	0.7412	0.5655	0.4068
2.4	2.7361E 01	0.9807	0.6511	0.5324	0.4753	0.9681	0.9970	1.0038	1.0005	0.5943	0.7521	0.5745	0.4130
2.6	2.6151E 01	0.9771	0.6588	0.5404	0.4833	0.9655	0.9967	1.0036	1.0004	0.5998	0.7621	0.5827	0.4186
2.8	2.5090E 01	0.9734	0.6669	0.5490	0.4918	0.9630	0.9964	1.0033	1.0004	0.6053	0.7714	0.5902	0.4238
3.0	2.4150E 01	0.9697	0.6754	0.5578	0.5006	0.9604	0.9961	1.0031	1.0004	0.6106	0.7801	0.5973	0.4286
3.2	2.3307E 01	0.9661	0.6840	0.5669	0.5096	0.9579	0.9959	1.0030	1.0004	0.6158	0.7884	0.6039	0.4330
3.4	2.2546E 01	0.9624	0.6928	0.5761	0.5187	0.9555	0.9956	1.0028	1.0003	0.6209	0.7962	0.6102	0.4372
3.6	2.1852E 01	0.9587	0.7015	0.5854	0.5279	0.9530	0.9953	1.0026	1.0003	0.6258	0.8036	0.6162	0.4412
3.8	2.1217E 01	0.9550	0.7103	0.5946	0.5371	0.9506	0.9950	1.0025	1.0003	0.6306	0.8106	0.6219	0.4450
4.0	2.0631E 01	0.9513	0.7190	0.6038	0.5463	0.9482	0.9947	1.0024	1.0003	0.6354	0.8174	0.6274	0.4487
4.5	1.9344E 01	0.9422	0.7406	0.6267	0.5691	0.9423	0.9940	1.0021	1.0003	0.6468	0.8331	0.6403	0.4572
5.0	1.8256E 01	0.9332	0.7617	0.6492	0.5916	0.9366	0.9933	1.0019	1.0002	0.6576	0.8473	0.6522	0.4651
5.5	1.7316E 01	0.9243	0.7822	0.6712	0.6136	0.9310	0.9926	1.0018	1.0002	0.6680	0.8605	0.6634	0.4725
6.0	1.6493E 01	0.9155	0.8022	0.6928	0.6353	0.9255	0.9919	1.0016	1.0002	0.6780	0.8726	0.6741	0.4795
6.5	1.5763E 01	0.9069	0.8217	0.7139	0.6565	0.9202	0.9912	1.0015	1.0002	0.6877	0.8839	0.6843	0.4863
7.0	1.5109E 01	0.8984	0.8408	0.7346	0.6774	0.9150	0.9905	1.0014	1.0002	0.6971	0.8944	0.6940	0.4928
7.5	1.4519E 01	0.8901	0.8593	0.7549	0.6979	0.9100	0.9898	1.0013	1.0002	0.7062	0.9042	0.7035	0.4991
8.0	1.3981E 01	0.8819	0.8775	0.7749	0.7181	0.9050	0.9891	1.0012	1.0002	0.7150	0.9135	0.7126	0.5053
9.0	1.3036E 01	0.8660	0.9126	0.8138	0.7576	0.8956	0.9877	1.0011	1.0001	0.7322	0.9303	0.7302	0.5172
10.0	1.2228E 01	0.8507	0.9463	0.8515	0.7961	0.8866	0.9863	1.0009	1.0001	0.7487	0.9453	0.7470	0.5287
11.0	1.1527E 01	0.8360	0.9789	0.8882	0.8337	0.8781	0.9848	1.0008	1.0001	0.7646	0.9586	0.7632	0.5399
12.0	1.0910E 01	0.8219	1.0103	0.9240	0.8705	0.8700	0.9833	1.0008	1.0001	0.7800	0.9704	0.7788	0.5509
13.0	1.0362E 01	0.8083	1.0408	0.9590	0.9066	0.8624	0.9819	1.0007	1.0001	0.7950	0.9809	0.7940	0.5616
14.0	9.8706E 00	0.7952	1.0704	0.9933	0.9421	0.8551	0.9804	1.0006	1.0001	0.8097	0.9902	0.8087	0.5722
16.0	9.0250E 00	0.7705	1.1272	1.0601	1.0115	0.8417	0.9773	1.0005	1.0001	0.8379	1.0056	0.8372	0.5929
18.0	8.3203E 00	0.7475	1.1810	1.1246	1.0790	0.8297	0.9743	1.0005	1.0001	0.8651	1.0171	0.8645	0.6130
20.0	7.7225E 00	0.7262	1.2322	1.1871	1.1450	0.8190	0.9711	1.0004	1.0001	0.8912	1.0253	0.8907	0.6327
25.0	6.5570E 00	0.6790	1.3498	1.3359	1.3037	0.7971	0.9630	1.0003	1.0001	0.9525	1.0331	0.9522	0.6798
30.0	5.7058E 00	0.6393	1.4538	1.4749	1.4547	0.7813	0.9545	1.0002	1.0000	1.0086	1.0261	1.0084	0.7238
35.0	5.0566E 00	0.6056	1.5452	1.6048	1.5982	0.7703	0.9456	1.0002	1.0000	1.0597	1.0071	1.0595	0.7643
40.0	4.5459E 00	0.5767	1.6247	1.7258	1.7346	0.7635	0.9365	1.0002	1.0000	1.1058	0.9784	1.1056	0.8012
45.0	4.1345E 00	0.5519	1.6928	1.8381	1.8636	0.7601	0.9272	1.0001	1.0000	1.1471	0.9419	1.1470	0.8342
50.0	3.7969E 00	0.5304	1.7502	1.9415	1.9852	0.7596	0.9178	1.0001	1.0000	1.1836	0.8992	1.1835	0.8634

ELECTRON Z= 86 A= 219

p	F_0L_0	L_0	λ_2	λ_3	λ_4	μ_1	μ_2	Λ_1	Λ_2	ν_{12}	$\hat{\nu}_{12}$	η_{12}	$\hat{\eta}_{12}$
0.1	4.3719E 02	1.0076	15.7007	116.7024	509.6402	0.9941	0.9999	1.0201	1.0012	6.1622	0.4694	-0.8129	-0.7855
0.2	2.2008E 02	1.0074	4.2909	10.3010	17.6817	0.9938	0.9998	1.0143	1.0012	1.7279	0.4945	-0.1097	-0.1625
0.3	1.4832E 02	1.0071	2.1803	3.2103	3.9461	0.9932	0.9998	1.0123	1.0012	0.9810	0.5183	0.1241	0.0426
0.4	1.1285E 02	1.0066	1.4442	1.6622	1.7642	0.9925	0.9997	1.0111	1.0011	0.7443	0.5406	0.2410	0.1438
0.5	9.1846E 01	1.0060	1.1062	1.1117	1.0958	0.9916	0.9996	1.0102	1.0011	0.6463	0.5610	0.3115	0.2037
0.6	7.8043E 01	1.0052	0.9254	0.8587	0.8124	0.9906	0.9995	1.0095	1.0011	0.5995	0.5797	0.3588	0.2433
0.7	6.8328E 01	1.0044	0.8189	0.7234	0.6680	0.9895	0.9993	1.0089	1.0010	0.5754	0.5967	0.3931	0.2714
0.8	6.1148E 01	1.0034	0.7523	0.6438	0.5856	0.9883	0.9992	1.0084	1.0010	0.5627	0.6122	0.4192	0.2925
0.9	5.5639E 01	1.0022	0.7088	0.5941	0.5352	0.9870	0.9991	1.0079	1.0009	0.5562	0.6263	0.4399	0.3089
1.0	5.1287E 01	1.0010	0.6797	0.5619	0.5030	0.9858	0.9989	1.0075	1.0009	0.5532	0.6391	0.4568	0.3221
1.2	4.4850E 01	0.9984	0.6466	0.5264	0.4680	0.9831	0.9987	1.0067	1.0008	0.5528	0.6617	0.4833	0.3423
1.4	4.0309E 01	0.9954	0.6321	0.5112	0.4533	0.9804	0.9984	1.0060	1.0007	0.5561	0.6810	0.5033	0.3572
1.6	3.6916E 01	0.9922	0.6273	0.5063	0.4487	0.9777	0.9981	1.0055	1.0007	0.5610	0.6977	0.5193	0.3689
1.8	3.4267E 01	0.9888	0.6279	0.5071	0.4496	0.9749	0.9978	1.0050	1.0006	0.5665	0.7124	0.5327	0.3785
2.0	3.2126E 01	0.9853	0.6318	0.5111	0.4537	0.9722	0.9975	1.0046	1.0005	0.5722	0.7257	0.5443	0.3866
2.2	3.0349E 01	0.9818	0.6376	0.5172	0.4598	0.9695	0.9972	1.0042	1.0005	0.5781	0.7376	0.5544	0.3937
2.4	2.8842E 01	0.9781	0.6447	0.5246	0.4671	0.9668	0.9969	1.0039	1.0005	0.5838	0.7486	0.5636	0.3999
2.6	2.7541E 01	0.9744	0.6525	0.5328	0.4752	0.9642	0.9966	1.0037	1.0004	0.5895	0.7588	0.5719	0.4056
2.8	2.6401E 01	0.9707	0.6609	0.5415	0.4838	0.9616	0.9963	1.0034	1.0004	0.5950	0.7683	0.5796	0.4108
3.0	2.5390E 01	0.9669	0.6696	0.5505	0.4928	0.9590	0.9961	1.0032	1.0004	0.6004	0.7771	0.5868	0.4157
3.2	2.4484E 01	0.9632	0.6784	0.5598	0.5019	0.9564	0.9958	1.0030	1.0004	0.6057	0.7855	0.5936	0.4202
3.4	2.3666E 01	0.9594	0.6873	0.5691	0.5112	0.9538	0.9955	1.0029	1.0003	0.6109	0.7934	0.6000	0.4245
3.6	2.2921E 01	0.9556	0.6963	0.5785	0.5205	0.9513	0.9952	1.0027	1.0003	0.6159	0.8009	0.6061	0.4285
3.8	2.2239E 01	0.9518	0.7053	0.5879	0.5298	0.9488	0.9949	1.0026	1.0003	0.6208	0.8081	0.6119	0.4323
4.0	2.1611E 01	0.9481	0.7142	0.5973	0.5392	0.9464	0.9946	1.0025	1.0003	0.6257	0.8149	0.6175	0.4360
4.5	2.0231E 01	0.9387	0.7363	0.6206	0.5623	0.9403	0.9939	1.0022	1.0003	0.6373	0.8308	0.6306	0.4447
5.0	1.9065E 01	0.9295	0.7579	0.6436	0.5852	0.9344	0.9932	1.0020	1.0002	0.6484	0.8453	0.6428	0.4527
5.5	1.8060E 01	0.9204	0.7790	0.6660	0.6076	0.9287	0.9925	1.0018	1.0002	0.6590	0.8586	0.6543	0.4603
6.0	1.7181E 01	0.9115	0.7995	0.6881	0.6297	0.9231	0.9918	1.0017	1.0002	0.6692	0.8709	0.6652	0.4675
6.5	1.6402E 01	0.9027	0.8195	0.7096	0.6513	0.9177	0.9911	1.0015	1.0002	0.6791	0.8823	0.6756	0.4744
7.0	1.5705E 01	0.8940	0.8390	0.7308	0.6726	0.9124	0.9904	1.0014	1.0002	0.6887	0.8929	0.6856	0.4811
7.5	1.5076E 01	0.8856	0.8580	0.7515	0.6935	0.9073	0.9896	1.0013	1.0002	0.6980	0.9029	0.6952	0.4875
8.0	1.4504E 01	0.8773	0.8766	0.7719	0.7141	0.9023	0.9889	1.0012	1.0002	0.7071	0.9122	0.7046	0.4938
9.0	1.3500E 01	0.8611	0.9127	0.8117	0.7544	0.8926	0.9875	1.0011	1.0001	0.7246	0.9292	0.7226	0.5061
10.0	1.2643E 01	0.8456	0.9474	0.8503	0.7938	0.8835	0.9860	1.0010	1.0001	0.7415	0.9442	0.7398	0.5179
11.0	1.1901E 01	0.8308	0.9808	0.8879	0.8322	0.8749	0.9845	1.0009	1.0001	0.7578	0.9576	0.7564	0.5294
12.0	1.1249E 01	0.8164	1.0131	0.9246	0.8698	0.8668	0.9830	1.0008	1.0001	0.7737	0.9694	0.7724	0.5407
13.0	1.0671E 01	0.8027	1.0444	0.9606	0.9068	0.8590	0.9815	1.0007	1.0001	0.7891	0.9799	0.7880	0.5518
14.0	1.0154E 01	0.7895	1.0749	0.9958	0.9432	0.8517	0.9800	1.0007	1.0001	0.8041	0.9892	0.8032	0.5627
16.0	9.2655E 00	0.7646	1.1332	1.0642	1.0143	0.8383	0.9769	1.0006	1.0001	0.8332	1.0045	0.8324	0.5841
18.0	8.5269E 00	0.7415	1.1885	1.1305	1.0836	0.8263	0.9737	1.0005	1.0001	0.8611	1.0158	0.8604	0.6049
20.0	7.9018E 00	0.7201	1.2411	1.1947	1.1513	0.8156	0.9705	1.0004	1.0001	0.8879	1.0237	0.8874	0.6252
25.0	6.6875E 00	0.6729	1.3618	1.3475	1.3143	0.7939	0.9622	1.0003	1.0001	0.9509	1.0309	0.9506	0.6739
30.0	5.8046E 00	0.6333	1.4684	1.4903	1.4692	0.7784	0.9535	1.0002	1.0000	1.0085	1.0231	1.0082	0.7193
35.0	5.1338E 00	0.5997	1.5618	1.6236	1.6167	0.7678	0.9445	1.0002	1.0000	1.0609	1.0032	1.0607	0.7610
40.0	4.6075E 00	0.5711	1.6429	1.7478	1.7566	0.7614	0.9352	1.0002	1.0000	1.1082	0.9737	1.1080	0.7989
45.0	4.1848E 00	0.5465	1.7121	1.8628	1.8890	0.7585	0.9257	1.0001	1.0000	1.1504	0.9363	1.1503	0.8328
50.0	3.8387E 00	0.5252	1.7702	1.9687	2.0138	0.7586	0.9162	1.0001	1.0000	1.1877	0.8928	1.1876	0.8626

ELECTRON Z= 87 A= 221

p	F_0L_0	L_0	λ_2	λ_3	λ_4	μ_1	μ_2	Λ_1	Λ_2	ν_{12}	$\hat{\nu}_{12}$	η_{12}	$\hat{\eta}_{12}$
0.1	4.6650E 02	1.0058	15.7139	118.2903	524.0117	0.9937	0.9998	1.0214	1.0012	6.2053	0.4630	-0.8604	-0.8260
0.2	2.3480E 02	1.0056	4.2884	10.3804	17.9552	0.9934	0.9998	1.0151	1.0012	1.7253	0.4883	-0.1379	-0.1879
0.3	1.5820E 02	1.0052	2.1749	3.2155	3.9656	0.9928	0.9997	1.0129	1.0012	0.9718	0.5123	0.1023	0.0221
0.4	1.2033E 02	1.0047	1.4378	1.6565	1.7597	0.9921	0.9997	1.0116	1.0012	0.7335	0.5346	0.2224	0.1257
0.5	9.7895E 01	1.0041	1.0994	1.1037	1.0873	0.9911	0.9996	1.0106	1.0011	0.6349	0.5553	0.2947	0.1870
0.6	8.3146E 01	1.0033	0.9183	0.8501	0.8031	0.9901	0.9995	1.0099	1.0011	0.5880	0.5741	0.3432	0.2274
0.7	7.2759E 01	1.0024	0.8117	0.7146	0.6586	0.9889	0.9993	1.0093	1.0010	0.5639	0.5913	0.3783	0.2562
0.8	6.5076E 01	1.0014	0.7450	0.6351	0.5764	0.9877	0.9992	1.0087	1.0010	0.5512	0.6069	0.4051	0.2777
0.9	5.9179E 01	1.0002	0.7014	0.5854	0.5261	0.9864	0.9991	1.0082	1.0009	0.5447	0.6211	0.4263	0.2944
1.0	5.4516E 01	0.9990	0.6723	0.5533	0.4940	0.9851	0.9989	1.0077	1.0009	0.5418	0.6341	0.4436	0.3079
1.2	4.7614E 01	0.9962	0.6393	0.5179	0.4592	0.9823	0.9986	1.0069	1.0008	0.5416	0.6569	0.4706	0.3284
1.4	4.2740E 01	0.9931	0.6249	0.5028	0.4446	0.9795	0.9983	1.0062	1.0007	0.5450	0.6764	0.4911	0.3435
1.6	3.9095E 01	0.9899	0.6202	0.4981	0.4402	0.9767	0.9980	1.0057	1.0007	0.5499	0.6934	0.5075	0.3553
1.8	3.6247E 01	0.9864	0.6210	0.4989	0.4412	0.9739	0.9977	1.0052	1.0006	0.5556	0.7084	0.5211	0.3650
2.0	3.3945E 01	0.9828	0.6250	0.5031	0.4454	0.9711	0.9974	1.0047	1.0006	0.5614	0.7218	0.5329	0.3733
2.2	3.2034E 01	0.9791	0.6310	0.5094	0.4516	0.9683	0.9972	1.0044	1.0005	0.5673	0.7339	0.5432	0.3804
2.4	3.0412E 01	0.9754	0.6383	0.5169	0.4590	0.9655	0.9969	1.0041	1.0005	0.5732	0.7451	0.5525	0.3868
2.6	2.9012E 01	0.9716	0.6463	0.5252	0.4672	0.9628	0.9966	1.0038	1.0004	0.5790	0.7554	0.5610	0.3925
2.8	2.7786E 01	0.9678	0.6549	0.5341	0.4760	0.9601	0.9963	1.0036	1.0004	0.5846	0.7650	0.5689	0.3978
3.0	2.6700E 01	0.9639	0.6637	0.5433	0.4850	0.9574	0.9960	1.0033	1.0004	0.5901	0.7740	0.5762	0.4027
3.2	2.5726E 01	0.9600	0.6728	0.5527	0.4943	0.9548	0.9957	1.0031	1.0004	0.5955	0.7825	0.5831	0.4073
3.4	2.4848E 01	0.9562	0.6819	0.5622	0.5037	0.9522	0.9954	1.0030	1.0004	0.6007	0.7905	0.5896	0.4116
3.6	2.4048E 01	0.9523	0.6911	0.5717	0.5131	0.9496	0.9951	1.0028	1.0003	0.6059	0.7981	0.5958	0.4157
3.8	2.3316E 01	0.9484	0.7003	0.5813	0.5226	0.9470	0.9948	1.0027	1.0003	0.6109	0.8054	0.6017	0.4196
4.0	2.2641E 01	0.9446	0.7094	0.5909	0.5321	0.9445	0.9945	1.0025	1.0003	0.6158	0.8124	0.6074	0.4233
4.5	2.1162E 01	0.9350	0.7320	0.6146	0.5556	0.9383	0.9938	1.0023	1.0003	0.6276	0.8285	0.6208	0.4321
5.0	1.9913E 01	0.9256	0.7542	0.6380	0.5788	0.9322	0.9931	1.0020	1.0002	0.6389	0.8432	0.6333	0.4403
5.5	1.8838E 01	0.9163	0.7757	0.6609	0.6016	0.9264	0.9924	1.0019	1.0002	0.6498	0.8566	0.6450	0.4480
6.0	1.7899E 01	0.9072	0.7968	0.6833	0.6241	0.9207	0.9916	1.0017	1.0002	0.6602	0.8690	0.6561	0.4553
6.5	1.7068E 01	0.8983	0.8173	0.7054	0.6461	0.9151	0.9909	1.0016	1.0002	0.6703	0.8806	0.6667	0.4624
7.0	1.6324E 01	0.8895	0.8373	0.7270	0.6678	0.9098	0.9902	1.0015	1.0002	0.6801	0.8913	0.6770	0.4692
7.5	1.5655E 01	0.8809	0.8568	0.7482	0.6891	0.9045	0.9894	1.0013	1.0002	0.6896	0.9014	0.6868	0.4758
8.0	1.5046E 01	0.8724	0.8759	0.7690	0.7102	0.8994	0.9887	1.0013	1.0002	0.6989	0.9108	0.6964	0.4823
9.0	1.3980E 01	0.8561	0.9129	0.8097	0.7514	0.8897	0.9872	1.0011	1.0001	0.7169	0.9279	0.7149	0.4949
10.0	1.3072E 01	0.8404	0.9485	0.8493	0.7915	0.8804	0.9857	1.0010	1.0001	0.7342	0.9430	0.7325	0.5070
11.0	1.2287E 01	0.8253	0.9829	0.8878	0.8308	0.8717	0.9842	1.0009	1.0001	0.7510	0.9564	0.7495	0.5189
12.0	1.1599E 01	0.8109	1.0161	0.9254	0.8694	0.8635	0.9827	1.0008	1.0001	0.7672	0.9683	0.7660	0.5305
13.0	1.0989E 01	0.7970	1.0483	0.9623	0.9072	0.8557	0.9812	1.0007	1.0001	0.7830	0.9788	0.7819	0.5420
14.0	1.0445E 01	0.7837	1.0795	0.9984	0.9445	0.8483	0.9796	1.0007	1.0001	0.7985	0.9881	0.7975	0.5532
16.0	9.5113E 00	0.7586	1.1395	1.0686	1.0174	0.8348	0.9764	1.0006	1.0001	0.8283	1.0032	0.8275	0.5753
18.0	8.7376E 00	0.7354	1.1963	1.1366	1.0884	0.8228	0.9732	1.0005	1.0001	0.8570	1.0144	0.8563	0.5968
20.0	8.0842E 00	0.7139	1.2503	1.2026	1.1579	0.8122	0.9699	1.0004	1.0001	0.8845	1.0222	0.8840	0.6178
25.0	6.8196E 00	0.6667	1.3742	1.3595	1.3252	0.7908	0.9615	1.0003	1.0001	0.9493	1.0287	0.9489	0.6680
30.0	5.9043E 00	0.6272	1.4833	1.5061	1.4843	0.7756	0.9526	1.0002	1.0000	1.0084	1.0201	1.0082	0.7148
35.0	5.2113E 00	0.5939	1.5789	1.6430	1.6357	0.7655	0.9434	1.0002	1.0000	1.0622	0.9994	1.0620	0.7578
40.0	4.6694E 00	0.5655	1.6615	1.7703	1.7793	0.7595	0.9339	1.0002	1.0000	1.1106	0.9690	1.1104	0.7967
45.0	4.2352E 00	0.5411	1.7319	1.8881	1.9152	0.7571	0.9242	1.0001	1.0000	1.1538	0.9307	1.1537	0.8314
50.0	3.8805E 00	0.5202	1.7907	1.9965	2.0431	0.7577	0.9145	1.0001	1.0000	1.1919	0.8864	1.1917	0.8619

Instr. see p. 22

ELECTRON Z= 88 A= 222

p	F_0L_0	L_0	λ_2	λ_3	λ_4	μ_1	μ_2	Λ_1	Λ_2	ν_{12}	$\hat{\nu}_{12}$	η_{12}	$\hat{\eta}_{12}$
0.1	4.9838E 02	1.0038	15.7122	119.7384	537.9316	0.9933	0.9998	1.0228	1.0013	6.2439	0.4565	-0.9081	-0.8662
0.2	2.5080E 02	1.0036	4.2821	10.4480	18.2091	0.9930	0.9998	1.0159	1.0012	1.7214	0.4819	-0.1664	-0.2133
0.3	1.6894E 02	1.0032	2.1678	3.2173	3.9805	0.9924	0.9997	1.0135	1.0012	0.9619	0.5060	0.0803	0.0016
0.4	1.2846E 02	1.0027	1.4304	1.6493	1.7533	0.9916	0.9997	1.0121	1.0012	0.7221	0.5285	0.2035	0.1075
0.5	1.0447E 02	1.0021	1.0918	1.0947	1.0778	0.9906	0.9996	1.0111	1.0011	0.6231	0.5493	0.2776	0.1702
0.6	8.8688E 01	1.0013	0.9106	0.8408	0.7932	0.9896	0.9994	1.0103	1.0011	0.5761	0.5683	0.3274	0.2115
0.7	7.7569E 01	1.0003	0.8040	0.7054	0.6488	0.9884	0.9993	1.0096	1.0010	0.5520	0.5856	0.3634	0.2408
0.8	6.9340E 01	0.9993	0.7372	0.6259	0.5667	0.9871	0.9992	1.0090	1.0010	0.5394	0.6013	0.3908	0.2627
0.9	6.3019E 01	0.9981	0.6937	0.5764	0.5167	0.9858	0.9990	1.0085	1.0009	0.5329	0.6157	0.4125	0.2798
1.0	5.8017E 01	0.9968	0.6646	0.5443	0.4847	0.9844	0.9989	1.0080	1.0009	0.5301	0.6288	0.4302	0.2935
1.2	5.0608E 01	0.9939	0.6317	0.5091	0.4501	0.9816	0.9986	1.0072	1.0008	0.5300	0.6519	0.4578	0.3143
1.4	4.5369E 01	0.9908	0.6174	0.4942	0.4358	0.9787	0.9983	1.0064	1.0007	0.5335	0.6716	0.4786	0.3297
1.6	4.1448E 01	0.9874	0.6128	0.4896	0.4314	0.9758	0.9980	1.0058	1.0007	0.5386	0.6888	0.4953	0.3417
1.8	3.8384E 01	0.9839	0.6138	0.4906	0.4325	0.9728	0.9977	1.0053	1.0006	0.5444	0.7040	0.5092	0.3515
2.0	3.5906E 01	0.9802	0.6179	0.4949	0.4369	0.9700	0.9974	1.0049	1.0006	0.5503	0.7176	0.5212	0.3598
2.2	3.3848E 01	0.9764	0.6241	0.5013	0.4432	0.9671	0.9971	1.0045	1.0005	0.5563	0.7299	0.5317	0.3670
2.4	3.2102E 01	0.9726	0.6316	0.5089	0.4507	0.9642	0.9968	1.0042	1.0005	0.5623	0.7412	0.5412	0.3734
2.6	3.0596E 01	0.9687	0.6398	0.5174	0.4590	0.9614	0.9965	1.0039	1.0005	0.5681	0.7517	0.5499	0.3793
2.8	2.9276E 01	0.9647	0.6486	0.5264	0.4679	0.9586	0.9962	1.0037	1.0004	0.5739	0.7614	0.5578	0.3846
3.0	2.8107E 01	0.9608	0.6576	0.5357	0.4770	0.9559	0.9959	1.0034	1.0004	0.5795	0.7706	0.5653	0.3895
3.2	2.7060E 01	0.9568	0.6669	0.5453	0.4864	0.9532	0.9956	1.0032	1.0004	0.5849	0.7792	0.5723	0.3941
3.4	2.6115E 01	0.9529	0.6762	0.5549	0.4959	0.9505	0.9953	1.0031	1.0004	0.5903	0.7873	0.5789	0.3985
3.6	2.5256E 01	0.9489	0.6856	0.5646	0.5055	0.9478	0.9950	1.0029	1.0003	0.5955	0.7951	0.5852	0.4027
3.8	2.4469E 01	0.9450	0.6950	0.5744	0.5151	0.9452	0.9947	1.0028	1.0003	0.6006	0.8024	0.5913	0.4066
4.0	2.3745E 01	0.9410	0.7044	0.5841	0.5247	0.9426	0.9944	1.0026	1.0003	0.6056	0.8095	0.5971	0.4104
4.5	2.2158E 01	0.9313	0.7275	0.6083	0.5486	0.9363	0.9937	1.0023	1.0003	0.6177	0.8258	0.6107	0.4193
5.0	2.0819E 01	0.9217	0.7501	0.6320	0.5721	0.9301	0.9930	1.0021	1.0003	0.6292	0.8407	0.6234	0.4277
5.5	1.9669E 01	0.9122	0.7722	0.6554	0.5953	0.9241	0.9922	1.0019	1.0002	0.6402	0.8543	0.6353	0.4355
6.0	1.8665E 01	0.9030	0.7937	0.6783	0.6181	0.9183	0.9915	1.0017	1.0002	0.6509	0.8669	0.6467	0.4430
6.5	1.7777E 01	0.8938	0.8147	0.7007	0.6405	0.9126	0.9907	1.0016	1.0002	0.6612	0.8786	0.6575	0.4502
7.0	1.6985E 01	0.8849	0.8352	0.7228	0.6626	0.9071	0.9900	1.0015	1.0002	0.6712	0.8894	0.6680	0.4571
7.5	1.6271E 01	0.8762	0.8552	0.7444	0.6843	0.9018	0.9893	1.0014	1.0002	0.6809	0.8995	0.6781	0.4639
8.0	1.5624E 01	0.8676	0.8748	0.7657	0.7058	0.8966	0.9885	1.0013	1.0002	0.6904	0.9090	0.6879	0.4705
9.0	1.4491E 01	0.8510	0.9128	0.8073	0.7478	0.8867	0.9870	1.0011	1.0001	0.7088	0.9263	0.7067	0.4834
10.0	1.3529E 01	0.8352	0.9493	0.8477	0.7888	0.8774	0.9855	1.0010	1.0001	0.7265	0.9415	0.7248	0.4959
11.0	1.2697E 01	0.8199	0.9845	0.8872	0.8289	0.8686	0.9839	1.0009	1.0001	0.7437	0.9550	0.7422	0.5081
12.0	1.1970E 01	0.8054	1.0186	0.9257	0.8683	0.8603	0.9824	1.0008	1.0001	0.7603	0.9669	0.7591	0.5201
13.0	1.1327E 01	0.7914	1.0517	0.9634	0.9070	0.8524	0.9808	1.0007	1.0001	0.7765	0.9774	0.7754	0.5318
14.0	1.0754E 01	0.7780	1.0837	1.0004	0.9451	0.8450	0.9792	1.0007	1.0001	0.7924	0.9867	0.7914	0.5434
16.0	9.7732E 00	0.7527	1.1453	1.0724	1.0197	0.8315	0.9760	1.0006	1.0001	0.8230	1.0018	0.8222	0.5662
18.0	8.9622E 00	0.7295	1.2036	1.1421	1.0925	0.8195	0.9727	1.0005	1.0001	0.8524	1.0128	0.8518	0.5883
20.0	8.2791E 00	0.7080	1.2590	1.2097	1.1636	0.8090	0.9694	1.0004	1.0001	0.8807	1.0204	0.8802	0.6100
25.0	6.9615E 00	0.6609	1.3860	1.3707	1.3351	0.7878	0.9608	1.0003	1.0001	0.9472	1.0264	0.9468	0.6618
30.0	6.0120E 00	0.6215	1.4978	1.5211	1.4983	0.7730	0.9517	1.0003	1.0000	1.0078	1.0171	1.0076	0.7100
35.0	5.2957E 00	0.5884	1.5954	1.6615	1.6536	0.7633	0.9423	1.0002	1.0000	1.0629	0.9958	1.0627	0.7542
40.0	4.7373E 00	0.5602	1.6796	1.7919	1.8009	0.7578	0.9326	1.0002	1.0000	1.1125	0.9646	1.1123	0.7941
45.0	4.2910E 00	0.5361	1.7512	1.9126	1.9401	0.7559	0.9228	1.0001	1.0000	1.1567	0.9257	1.1565	0.8297
50.0	3.9272E 00	0.5154	1.8108	2.0234	2.0712	0.7569	0.9129	1.0001	1.0000	1.1955	0.8808	1.1954	0.8608

ELECTRON Z= 89 A= 225

p	F_0L_0	L_0	λ_2	λ_3	λ_4	μ_1	μ_2	Λ_1	Λ_2	ν_{12}	$\hat{\nu}_{12}$	η_{12}	$\hat{\eta}_{12}$
0.1	5.3189E 02	1.0016	15.7153	121.2346	552.3224	0.9929	0.9998	1.0245	1.0013	6.2821	0.4502	-0.9566	-0.9067
0.2	2.6763E 02	1.0013	4.2774	10.5203	18.4753	0.9925	0.9998	1.0168	1.0013	1.7174	0.4756	-0.1951	-0.2387
0.3	1.8023E 02	1.0010	2.1615	3.2209	3.9981	0.9919	0.9997	1.0142	1.0012	0.9520	0.4998	0.0581	-0.0189
0.4	1.3699E 02	1.0004	1.4236	1.6432	1.7482	0.9911	0.9996	1.0126	1.0012	0.7108	0.5225	0.1845	0.0894
0.5	1.1137E 02	0.9997	1.0848	1.0866	1.0692	0.9901	0.9995	1.0116	1.0011	0.6114	0.5434	0.2606	0.1534
0.6	9.4499E 01	0.9989	0.9035	0.8323	0.7840	0.9890	0.9994	1.0107	1.0011	0.5643	0.5626	0.3116	0.1956
0.7	8.2609E 01	0.9979	0.7968	0.6968	0.6397	0.9877	0.9993	1.0100	1.0011	0.5402	0.5800	0.3485	0.2255
0.8	7.3803E 01	0.9968	0.7300	0.6174	0.5578	0.9864	0.9992	1.0094	1.0010	0.5276	0.5959	0.3765	0.2479
0.9	6.7034E 01	0.9956	0.6865	0.5680	0.5079	0.9850	0.9990	1.0088	1.0010	0.5212	0.6104	0.3987	0.2652
1.0	6.1675E 01	0.9943	0.6574	0.5360	0.4760	0.9836	0.9989	1.0083	1.0009	0.5185	0.6237	0.4168	0.2792
1.2	5.3729E 01	0.9914	0.6246	0.5009	0.4417	0.9807	0.9986	1.0074	1.0008	0.5185	0.6471	0.4450	0.3003
1.4	4.8105E 01	0.9881	0.6104	0.4862	0.4275	0.9777	0.9983	1.0067	1.0007	0.5222	0.6670	0.4662	0.3159
1.6	4.3892E 01	0.9846	0.6060	0.4817	0.4232	0.9747	0.9980	1.0061	1.0007	0.5274	0.6844	0.4832	0.3280
1.8	4.0598E 01	0.9810	0.6071	0.4828	0.4245	0.9717	0.9977	1.0055	1.0006	0.5332	0.6998	0.4974	0.3380
2.0	3.7933E 01	0.9772	0.6114	0.4873	0.4289	0.9687	0.9974	1.0051	1.0006	0.5393	0.7136	0.5096	0.3464
2.2	3.5720E 01	0.9733	0.6178	0.4938	0.4353	0.9657	0.9970	1.0047	1.0005	0.5454	0.7261	0.5203	0.3537
2.4	3.3843E 01	0.9693	0.6255	0.5016	0.4430	0.9628	0.9967	1.0043	1.0005	0.5515	0.7376	0.5300	0.3602
2.6	3.2223E 01	0.9653	0.6339	0.5102	0.4514	0.9599	0.9964	1.0040	1.0005	0.5574	0.7482	0.5388	0.3661
2.8	3.0804E 01	0.9613	0.6429	0.5193	0.4604	0.9570	0.9961	1.0038	1.0004	0.5633	0.7581	0.5469	0.3715
3.0	2.9547E 01	0.9573	0.6522	0.5288	0.4697	0.9542	0.9958	1.0036	1.0004	0.5690	0.7674	0.5545	0.3765
3.2	2.8423E 01	0.9532	0.6616	0.5386	0.4792	0.9514	0.9955	1.0033	1.0004	0.5745	0.7761	0.5616	0.3811
3.4	2.7408E 01	0.9491	0.6712	0.5484	0.4889	0.9486	0.9952	1.0032	1.0004	0.5800	0.7844	0.5683	0.3856
3.6	2.6485E 01	0.9451	0.6808	0.5583	0.4986	0.9459	0.9949	1.0030	1.0003	0.5853	0.7922	0.5748	0.3898
3.8	2.5641E 01	0.9410	0.6904	0.5682	0.5083	0.9432	0.9946	1.0028	1.0003	0.5905	0.7997	0.5809	0.3938
4.0	2.4865E 01	0.9370	0.7000	0.5781	0.5181	0.9405	0.9943	1.0027	1.0003	0.5956	0.8069	0.5868	0.3976
4.5	2.3164E 01	0.9271	0.7237	0.6027	0.5423	0.9340	0.9936	1.0024	1.0003	0.6079	0.8235	0.6008	0.4067
5.0	2.1732E 01	0.9172	0.7469	0.6269	0.5663	0.9277	0.9928	1.0022	1.0003	0.6196	0.8385	0.6137	0.4151
5.5	2.0502E 01	0.9076	0.7695	0.6508	0.5899	0.9215	0.9921	1.0020	1.0002	0.6309	0.8523	0.6259	0.4231
6.0	1.9430E 01	0.8981	0.7916	0.6741	0.6131	0.9156	0.9913	1.0018	1.0002	0.6418	0.8651	0.6375	0.4308
6.5	1.8484E 01	0.8889	0.8131	0.6971	0.6360	0.9098	0.9906	1.0017	1.0002	0.6523	0.8768	0.6486	0.4381
7.0	1.7640E 01	0.8798	0.8341	0.7196	0.6585	0.9042	0.9898	1.0015	1.0002	0.6625	0.8878	0.6592	0.4453
7.5	1.6881E 01	0.8709	0.8547	0.7417	0.6806	0.8988	0.9890	1.0014	1.0002	0.6725	0.8981	0.6696	0.4522
8.0	1.6193E 01	0.8622	0.8748	0.7635	0.7025	0.8935	0.9883	1.0013	1.0002	0.6822	0.9076	0.6796	0.4590
9.0	1.4992E 01	0.8454	0.9138	0.8061	0.7454	0.8835	0.9867	1.0012	1.0001	0.7011	0.9250	0.6990	0.4722
10.0	1.3973E 01	0.8293	0.9513	0.8476	0.7874	0.8740	0.9852	1.0010	1.0001	0.7193	0.9403	0.7175	0.4851
11.0	1.3094E 01	0.8139	0.9876	0.8880	0.8284	0.8651	0.9836	1.0009	1.0001	0.7368	0.9539	0.7353	0.4977
12.0	1.2327E 01	0.7992	1.0226	0.9275	0.8688	0.8567	0.9820	1.0008	1.0001	0.7539	0.9658	0.7526	0.5100
13.0	1.1651E 01	0.7851	1.0566	0.9662	0.9084	0.8489	0.9804	1.0008	1.0001	0.7706	0.9763	0.7694	0.5221
14.0	1.1048E 01	0.7716	1.0896	1.0042	0.9475	0.8414	0.9788	1.0007	1.0001	0.7868	0.9855	0.7858	0.5341
16.0	1.0019E 01	0.7462	1.1528	1.0781	1.0241	0.8279	0.9755	1.0006	1.0001	0.8183	1.0005	0.8175	0.5576
18.0	9.1709E 00	0.7229	1.2128	1.1497	1.0988	0.8160	0.9722	1.0005	1.0001	0.8485	1.0114	0.8479	0.5805
20.0	8.4582E 00	0.7013	1.2697	1.2192	1.1718	0.8055	0.9688	1.0004	1.0001	0.8776	1.0188	0.8771	0.6029
25.0	7.0884E 00	0.6543	1.4000	1.3847	1.3480	0.7846	0.9599	1.0003	1.0001	0.9460	1.0240	0.9456	0.6564
30.0	6.1059E 00	0.6151	1.5145	1.5392	1.5157	0.7702	0.9507	1.0003	1.0000	1.0083	1.0139	1.0080	0.7061
35.0	5.3674E 00	0.5822	1.6143	1.6834	1.6752	0.7610	0.9411	1.0002	1.0000	1.0648	0.9916	1.0646	0.7516
40.0	4.7936E 00	0.5544	1.7001	1.8172	1.8265	0.7561	0.9312	1.0002	1.0000	1.1156	0.9594	1.1154	0.7925
45.0	4.3361E 00	0.5306	1.7727	1.9408	1.9694	0.7548	0.9211	1.0001	1.0000	1.1607	0.9195	1.1606	0.8289
50.0	3.9641E 00	0.5102	1.8329	2.0543	2.1039	0.7563	0.9110	1.0001	1.0000	1.2004	0.8737	1.2003	0.8607

Instr. see p. 22

ELECTRON Z= 90 A= 228

p	F_0L_0	L_0	λ_2	λ_3	λ_4	μ_1	μ_2	Λ_1	Λ_2	ν_{12}	$\hat{\nu}_{12}$	η_{12}	$\hat{\eta}_{12}$
0.1	5.6792E 02	0.9991	15.7105	122.6543	566.5530	0.9924	0.9998	1.0262	1.0013	6.3170	0.4437	-1.0054	-0.9471
0.2	2.8571E 02	0.9989	4.2706	10.5865	18.7326	0.9920	0.9998	1.0178	1.0013	1.7125	0.4693	-0.2240	-0.2641
0.3	1.9236E 02	0.9985	2.1545	3.2228	4.0137	0.9914	0.9997	1.0149	1.0013	0.9417	0.4936	0.0357	-0.0394
0.4	1.4616E 02	0.9979	1.4164	1.6364	1.7425	0.9905	0.9996	1.0133	1.0012	0.6991	0.5164	0.1654	0.0713
0.5	1.1877E 02	0.9972	1.0776	1.0782	1.0602	0.9895	0.9995	1.0121	1.0012	0.5994	0.5374	0.2434	0.1367
0.6	1.0073E 02	0.9964	0.8963	0.8235	0.7747	0.9884	0.9994	1.0112	1.0011	0.5522	0.5567	0.2957	0.1797
0.7	8.8012E 01	0.9954	0.7896	0.6881	0.6305	0.9871	0.9993	1.0104	1.0011	0.5282	0.5744	0.3334	0.2102
0.8	7.8585E 01	0.9942	0.7228	0.6089	0.5488	0.9857	0.9991	1.0098	1.0010	0.5157	0.5904	0.3620	0.2330
0.9	7.1334E 01	0.9930	0.6793	0.5595	0.4991	0.9843	0.9990	1.0092	1.0010	0.5094	0.6051	0.3847	0.2506
1.0	6.5589E 01	0.9916	0.6502	0.5276	0.4674	0.9828	0.9989	1.0086	1.0009	0.5067	0.6184	0.4032	0.2648
1.2	5.7063E 01	0.9886	0.6175	0.4928	0.4333	0.9798	0.9985	1.0077	1.0008	0.5069	0.6421	0.4320	0.2863
1.4	5.1023E 01	0.9852	0.6034	0.4782	0.4192	0.9767	0.9982	1.0069	1.0008	0.5107	0.6623	0.4537	0.3021
1.6	4.6495E 01	0.9816	0.5992	0.4738	0.4151	0.9735	0.9979	1.0063	1.0007	0.5160	0.6799	0.4710	0.3144
1.8	4.2953E 01	0.9779	0.6004	0.4751	0.4165	0.9704	0.9976	1.0057	1.0006	0.5220	0.6955	0.4854	0.3244
2.0	4.0086E 01	0.9740	0.6050	0.4797	0.4211	0.9673	0.9973	1.0053	1.0006	0.5282	0.7095	0.4978	0.3329
2.2	3.7706E 01	0.9700	0.6115	0.4863	0.4276	0.9643	0.9970	1.0048	1.0005	0.5344	0.7222	0.5088	0.3403
2.4	3.5687E 01	0.9659	0.6194	0.4943	0.4354	0.9613	0.9967	1.0045	1.0005	0.5405	0.7339	0.5186	0.3469
2.6	3.3944E 01	0.9618	0.6281	0.5031	0.4439	0.9583	0.9964	1.0042	1.0005	0.5466	0.7446	0.5275	0.3528
2.8	3.2419E 01	0.9576	0.6372	0.5124	0.4530	0.9553	0.9961	1.0039	1.0004	0.5525	0.7547	0.5358	0.3583
3.0	3.1068E 01	0.9535	0.6467	0.5220	0.4625	0.9524	0.9958	1.0037	1.0004	0.5583	0.7641	0.5435	0.3633
3.2	2.9860E 01	0.9493	0.6564	0.5319	0.4721	0.9495	0.9954	1.0035	1.0004	0.5640	0.7730	0.5508	0.3680
3.4	2.8770E 01	0.9452	0.6662	0.5419	0.4819	0.9467	0.9951	1.0033	1.0004	0.5695	0.7813	0.5576	0.3725
3.6	2.7779E 01	0.9410	0.6761	0.5520	0.4917	0.9438	0.9948	1.0031	1.0004	0.5749	0.7893	0.5642	0.3768
3.8	2.6874E 01	0.9369	0.6859	0.5621	0.5016	0.9411	0.9945	1.0029	1.0003	0.5802	0.7969	0.5704	0.3808
4.0	2.6041E 01	0.9328	0.6957	0.5722	0.5115	0.9383	0.9942	1.0028	1.0003	0.5854	0.8041	0.5765	0.3847
4.5	2.4218E 01	0.9226	0.7199	0.5972	0.5361	0.9316	0.9935	1.0025	1.0003	0.5979	0.8210	0.5906	0.3940
5.0	2.2686E 01	0.9126	0.7437	0.6219	0.5605	0.9252	0.9927	1.0022	1.0003	0.6099	0.8362	0.6038	0.4026
5.5	2.1372E 01	0.9027	0.7669	0.6462	0.5845	0.9189	0.9919	1.0020	1.0002	0.6214	0.8502	0.6163	0.4107
6.0	2.0228E 01	0.8931	0.7895	0.6701	0.6082	0.9128	0.9911	1.0019	1.0002	0.6325	0.8631	0.6281	0.4185
6.5	1.9219E 01	0.8837	0.8116	0.6935	0.6315	0.9069	0.9904	1.0017	1.0002	0.6433	0.8750	0.6394	0.4260
7.0	1.8320E 01	0.8744	0.8332	0.7165	0.6544	0.9013	0.9896	1.0016	1.0002	0.6537	0.8861	0.6504	0.4333
7.5	1.7513E 01	0.8654	0.8543	0.7392	0.6770	0.8957	0.9888	1.0015	1.0002	0.6639	0.8965	0.6609	0.4404
8.0	1.6783E 01	0.8566	0.8750	0.7615	0.6994	0.8904	0.9880	1.0014	1.0002	0.6739	0.9061	0.6712	0.4474
9.0	1.5508E 01	0.8395	0.9150	0.8051	0.7432	0.8802	0.9865	1.0012	1.0001	0.6932	0.9236	0.6910	0.4610
10.0	1.4430E 01	0.8233	0.9536	0.8475	0.7861	0.8706	0.9849	1.0011	1.0001	0.7118	0.9390	0.7100	0.4742
11.0	1.3502E 01	0.8077	0.9908	0.8890	0.8281	0.8616	0.9833	1.0010	1.0001	0.7299	0.9526	0.7283	0.4872
12.0	1.2694E 01	0.7929	1.0268	0.9295	0.8694	0.8532	0.9817	1.0009	1.0001	0.7474	0.9646	0.7461	0.4999
13.0	1.1982E 01	0.7787	1.0618	0.9692	0.9101	0.8453	0.9800	1.0008	1.0001	0.7645	0.9751	0.7634	0.5124
14.0	1.1348E 01	0.7651	1.0957	1.0082	0.9501	0.8378	0.9784	1.0007	1.0001	0.7812	0.9843	0.7802	0.5248
16.0	1.0270E 01	0.7396	1.1607	1.0841	1.0287	0.8243	0.9750	1.0006	1.0001	0.8136	0.9992	0.8127	0.5490
18.0	9.3829E 00	0.7162	1.2223	1.1577	1.1053	0.8124	0.9716	1.0005	1.0001	0.8446	1.0099	0.8440	0.5727
20.0	8.6396E 00	0.6946	1.2808	1.2291	1.1803	0.8020	0.9681	1.0005	1.0001	0.8746	1.0171	0.8740	0.5958
25.0	7.2163E 00	0.6477	1.4145	1.3992	1.3613	0.7815	0.9590	1.0003	1.0001	0.9448	1.0216	0.9444	0.6511
30.0	6.2001E 00	0.6088	1.5318	1.5579	1.5336	0.7676	0.9496	1.0003	1.0000	1.0088	1.0106	1.0085	0.7023
35.0	5.4392E 00	0.5762	1.6337	1.7059	1.6975	0.7590	0.9398	1.0002	1.0000	1.0668	0.9873	1.0666	0.7490
40.0	4.8497E 00	0.5486	1.7210	1.8432	1.8528	0.7546	0.9297	1.0002	1.0000	1.1188	0.9542	1.1186	0.7910
45.0	4.3810E 00	0.5252	1.7946	1.9698	1.9995	0.7538	0.9194	1.0001	1.0000	1.1649	0.9134	1.1648	0.8282
50.0	4.0007E 00	0.5051	1.8554	2.0859	2.1374	0.7560	0.9091	1.0001	1.0000	1.2053	0.8666	1.2052	0.8607

ELECTRON Z= 91 A= 234

p	F_0L_0	L_0	λ_2	λ_3	λ_4	μ_1	μ_2	Λ_1	Λ_2	ν_{12}	$\hat{\nu}_{12}$	η_{12}	$\hat{\eta}_{12}$
0.1	6.0527E 02	0.9963	15.7184	124.1993	581.6581	0.9919	0.9998	1.0283	1.0013	6.3532	0.4375	-1.0552	-0.9880
0.2	3.0445E 02	0.9961	4.2676	10.6642	19.0152	0.9915	0.9998	1.0190	1.0013	1.7078	0.4631	-0.2533	-0.2896
0.3	2.0492E 02	0.9956	2.1494	3.2285	4.0346	0.9908	0.9997	1.0158	1.0013	0.9315	0.4876	0.0132	-0.0599
0.4	1.5565E 02	0.9951	1.4106	1.6316	1.7391	0.9899	0.9996	1.0140	1.0012	0.6877	0.5105	0.1462	0.0532
0.5	1.2643E 02	0.9943	1.0715	1.0712	1.0529	0.9889	0.9995	1.0127	1.0012	0.5876	0.5318	0.2262	0.1200
0.6	1.0718E 02	0.9934	0.8900	0.8160	0.7667	0.9876	0.9994	1.0118	1.0011	0.5404	0.5512	0.2797	0.1640
0.7	9.3589E 01	0.9924	0.7832	0.6805	0.6224	0.9863	0.9993	1.0109	1.0011	0.5164	0.5690	0.3184	0.1950
0.8	8.3515E 01	0.9912	0.7164	0.6012	0.5408	0.9849	0.9991	1.0102	1.0010	0.5039	0.5852	0.3477	0.2182
0.9	7.5761E 01	0.9899	0.6729	0.5520	0.4912	0.9834	0.9990	1.0096	1.0010	0.4977	0.6000	0.3709	0.2362
1.0	6.9614E 01	0.9885	0.6438	0.5202	0.4596	0.9819	0.9988	1.0090	1.0009	0.4951	0.6135	0.3898	0.2506
1.2	6.0484E 01	0.9853	0.6112	0.4854	0.4257	0.9787	0.9985	1.0080	1.0009	0.4955	0.6374	0.4192	0.2724
1.4	5.4009E 01	0.9819	0.5972	0.4710	0.4118	0.9755	0.9982	1.0072	1.0008	0.4994	0.6579	0.4413	0.2884
1.6	4.9151E 01	0.9782	0.5931	0.4668	0.4079	0.9722	0.9979	1.0065	1.0007	0.5049	0.6758	0.4590	0.3009
1.8	4.5349E 01	0.9743	0.5946	0.4682	0.4094	0.9690	0.9976	1.0060	1.0006	0.5109	0.6916	0.4737	0.3110
2.0	4.2272E 01	0.9702	0.5993	0.4730	0.4141	0.9658	0.9972	1.0055	1.0006	0.5172	0.7058	0.4863	0.3196
2.2	3.9716E 01	0.9661	0.6061	0.4798	0.4207	0.9626	0.9969	1.0050	1.0006	0.5236	0.7187	0.4974	0.3271
2.4	3.7548E 01	0.9619	0.6142	0.4879	0.4286	0.9595	0.9966	1.0047	1.0005	0.5298	0.7305	0.5074	0.3337
2.6	3.5677E 01	0.9577	0.6231	0.4969	0.4374	0.9564	0.9963	1.0044	1.0005	0.5360	0.7414	0.5165	0.3397
2.8	3.4041E 01	0.9534	0.6325	0.5064	0.4466	0.9534	0.9960	1.0041	1.0005	0.5420	0.7516	0.5249	0.3452
3.0	3.2591E 01	0.9492	0.6423	0.5162	0.4562	0.9503	0.9956	1.0038	1.0004	0.5479	0.7612	0.5328	0.3504
3.2	3.1296E 01	0.9449	0.6522	0.5263	0.4660	0.9474	0.9953	1.0036	1.0004	0.5537	0.7702	0.5402	0.3552
3.4	3.0128E 01	0.9406	0.6623	0.5365	0.4759	0.9444	0.9950	1.0034	1.0004	0.5593	0.7787	0.5472	0.3597
3.6	2.9066E 01	0.9364	0.6724	0.5467	0.4860	0.9415	0.9947	1.0032	1.0004	0.5648	0.7868	0.5539	0.3640
3.8	2.8097E 01	0.9321	0.6825	0.5570	0.4960	0.9387	0.9944	1.0030	1.0003	0.5702	0.7945	0.5602	0.3681
4.0	2.7205E 01	0.9279	0.6925	0.5673	0.5061	0.9358	0.9941	1.0029	1.0003	0.5755	0.8019	0.5664	0.3721
4.5	2.5257E 01	0.9175	0.7174	0.5929	0.5312	0.9290	0.9933	1.0026	1.0003	0.5883	0.8190	0.5809	0.3815
5.0	2.3620E 01	0.9072	0.7418	0.6182	0.5560	0.9223	0.9925	1.0023	1.0003	0.6005	0.8344	0.5943	0.3902
5.5	2.2219E 01	0.8971	0.7656	0.6430	0.5805	0.9159	0.9917	1.0021	1.0002	0.6123	0.8486	0.6071	0.3985
6.0	2.1000E 01	0.8873	0.7889	0.6675	0.6047	0.9097	0.9909	1.0019	1.0002	0.6237	0.8616	0.6192	0.4065
6.5	1.9927E 01	0.8777	0.8116	0.6915	0.6285	0.9037	0.9901	1.0018	1.0002	0.6347	0.8737	0.6308	0.4142
7.0	1.8973E 01	0.8683	0.8338	0.7151	0.6519	0.8979	0.9893	1.0016	1.0002	0.6454	0.8849	0.6420	0.4217
7.5	1.8116E 01	0.8591	0.8555	0.7383	0.6751	0.8922	0.9885	1.0015	1.0002	0.6559	0.8954	0.6528	0.4290
8.0	1.7342E 01	0.8501	0.8768	0.7612	0.6980	0.8868	0.9877	1.0014	1.0002	0.6661	0.9051	0.6634	0.4362
9.0	1.5994E 01	0.8328	0.9180	0.8059	0.7429	0.8765	0.9861	1.0012	1.0002	0.6859	0.9228	0.6837	0.4502
10.0	1.4856E 01	0.8163	0.9577	0.8496	0.7869	0.8668	0.9845	1.0011	1.0001	0.7051	0.9383	0.7032	0.4639
11.0	1.3879E 01	0.8006	0.9961	0.8922	0.8301	0.8577	0.9829	1.0010	1.0001	0.7237	0.9519	0.7221	0.4772
12.0	1.3029E 01	0.7856	1.0332	0.9338	0.8725	0.8492	0.9812	1.0009	1.0001	0.7417	0.9638	0.7404	0.4904
13.0	1.2281E 01	0.7713	1.0692	0.9747	0.9143	0.8412	0.9795	1.0008	1.0001	0.7593	0.9743	0.7581	0.5034
14.0	1.1618E 01	0.7576	1.1041	1.0148	0.9555	0.8338	0.9778	1.0007	1.0001	0.7765	0.9835	0.7755	0.5162
16.0	1.0490E 01	0.7320	1.1711	1.0931	1.0363	0.8203	0.9744	1.0006	1.0001	0.8099	0.9982	0.8090	0.5413
18.0	9.5656E 00	0.7085	1.2346	1.1689	1.1152	0.8085	0.9709	1.0005	1.0001	0.8419	1.0087	0.8412	0.5658
20.0	8.7929E 00	0.6870	1.2948	1.2425	1.1925	0.7982	0.9673	1.0005	1.0001	0.8727	1.0155	0.8722	0.5898
25.0	7.3190E 00	0.6401	1.4322	1.4177	1.3790	0.7782	0.9580	1.0003	1.0001	0.9451	1.0190	0.9447	0.6470
30.0	6.2716E 00	0.6015	1.5524	1.5812	1.5566	0.7648	0.9482	1.0003	1.0001	1.0110	1.0068	1.0107	0.6998
35.0	5.4906E 00	0.5692	1.6564	1.7334	1.7254	0.7568	0.9381	1.0002	1.0000	1.0705	0.9821	1.0703	0.7479
40.0	4.8874E 00	0.5420	1.7452	1.8745	1.8853	0.7532	0.9278	1.0002	1.0000	1.1238	0.9476	1.1236	0.7911
45.0	4.4092E 00	0.5190	1.8197	2.0044	2.0361	0.7531	0.9173	1.0001	1.0000	1.1709	0.9054	1.1707	0.8291
50.0	4.0222E 00	0.4993	1.8806	2.1231	2.1776	0.7560	0.9067	1.0001	1.0000	1.2120	0.8573	1.2119	0.8621

Instr. see p. 22

ELECTRON Z= 92 A= 233

p	F_0L_0	L_0	λ_2	λ_3	λ_4	μ_1	μ_2	Λ_1	Λ_2	ν_{12}	$\hat{\nu}_{12}$	η_{12}	$\hat{\eta}_{12}$
0.1	6.4889E 02	0.9936	15.6700	125.1771	594.0369	0.9914	0.9998	1.0302	1.0013	6.3744	0.4305	-1.1038	-1.0274
0.2	3.2634E 02	0.9934	4.2494	10.6937	19.2057	0.9909	0.9998	1.0201	1.0013	1.6991	0.4562	-0.2825	-0.3147
0.3	2.1959E 02	0.9930	2.1368	3.2200	4.0361	0.9903	0.9997	1.0165	1.0013	0.9192	0.4808	-0.0096	-0.0803
0.4	1.6674E 02	0.9924	1.4000	1.6199	1.7275	0.9894	0.9996	1.0146	1.0013	0.6746	0.5038	0.1266	0.0350
0.5	1.3537E 02	0.9916	1.0618	1.0597	1.0407	0.9883	0.9995	1.0133	1.0012	0.5745	0.5251	0.2084	0.1031
0.6	1.1470E 02	0.9907	0.8808	0.8051	0.7551	0.9870	0.9994	1.0122	1.0012	0.5273	0.5447	0.2632	0.1479
0.7	1.0011E 02	0.9896	0.7743	0.6701	0.6115	0.9856	0.9992	1.0114	1.0011	0.5035	0.5626	0.3027	0.1795
0.8	8.9280E 01	0.9884	0.7077	0.5912	0.5304	0.9842	0.9991	1.0106	1.0011	0.4911	0.5789	0.3326	0.2031
0.9	8.0940E 01	0.9871	0.6643	0.5422	0.4812	0.9826	0.9990	1.0099	1.0010	0.4850	0.5938	0.3563	0.2213
1.0	7.4324E 01	0.9856	0.6354	0.5107	0.4499	0.9811	0.9988	1.0093	1.0010	0.4825	0.6075	0.3756	0.2359
1.2	6.4490E 01	0.9824	0.6029	0.4762	0.4164	0.9778	0.9985	1.0083	1.0009	0.4830	0.6316	0.4055	0.2581
1.4	5.7509E 01	0.9788	0.5892	0.4620	0.4027	0.9745	0.9982	1.0075	1.0008	0.4871	0.6523	0.4280	0.2742
1.6	5.2269E 01	0.9750	0.5852	0.4579	0.3989	0.9712	0.9978	1.0068	1.0007	0.4927	0.6704	0.4460	0.2868
1.8	4.8165E 01	0.9710	0.5869	0.4595	0.4005	0.9678	0.9975	1.0062	1.0007	0.4988	0.6864	0.4609	0.2971
2.0	4.4844E 01	0.9669	0.5918	0.4644	0.4053	0.9645	0.9972	1.0056	1.0006	0.5052	0.7007	0.4737	0.3057
2.2	4.2084E 01	0.9627	0.5988	0.4713	0.4121	0.9613	0.9969	1.0052	1.0006	0.5116	0.7138	0.4850	0.3133
2.4	3.9745E 01	0.9584	0.6070	0.4796	0.4200	0.9581	0.9966	1.0048	1.0005	0.5180	0.7258	0.4952	0.3200
2.6	3.7726E 01	0.9541	0.6161	0.4886	0.4288	0.9549	0.9962	1.0045	1.0005	0.5242	0.7369	0.5044	0.3260
2.8	3.5961E 01	0.9497	0.6257	0.4983	0.4382	0.9517	0.9959	1.0042	1.0005	0.5303	0.7472	0.5130	0.3316
3.0	3.4398E 01	0.9454	0.6357	0.5082	0.4479	0.9487	0.9956	1.0039	1.0004	0.5363	0.7569	0.5209	0.3367
3.2	3.3002E 01	0.9410	0.6458	0.5185	0.4578	0.9456	0.9953	1.0037	1.0004	0.5422	0.7660	0.5284	0.3415
3.4	3.1744E 01	0.9367	0.6561	0.5288	0.4678	0.9426	0.9950	1.0035	1.0004	0.5479	0.7746	0.5355	0.3461
3.6	3.0601E 01	0.9323	0.6664	0.5392	0.4779	0.9396	0.9946	1.0033	1.0004	0.5535	0.7828	0.5423	0.3505
3.8	2.9557E 01	0.9280	0.6767	0.5496	0.4881	0.9367	0.9943	1.0031	1.0003	0.5589	0.7906	0.5488	0.3546
4.0	2.8599E 01	0.9237	0.6869	0.5601	0.4983	0.9338	0.9940	1.0030	1.0003	0.5643	0.7981	0.5550	0.3587
4.5	2.6505E 01	0.9131	0.7123	0.5860	0.5236	0.9268	0.9932	1.0026	1.0003	0.5773	0.8154	0.5697	0.3681
5.0	2.4749E 01	0.9027	0.7372	0.6117	0.5488	0.9201	0.9924	1.0024	1.0003	0.5897	0.8311	0.5834	0.3770
5.5	2.3248E 01	0.8925	0.7615	0.6369	0.5736	0.9135	0.9916	1.0022	1.0002	0.6017	0.8454	0.5964	0.3855
6.0	2.1944E 01	0.8825	0.7853	0.6618	0.5981	0.9072	0.9908	1.0020	1.0002	0.6133	0.8586	0.6087	0.3936
6.5	2.0797E 01	0.8728	0.8085	0.6862	0.6222	0.9011	0.9900	1.0018	1.0002	0.6245	0.8708	0.6205	0.4014
7.0	1.9779E 01	0.8633	0.8312	0.7102	0.6461	0.8952	0.9892	1.0017	1.0002	0.6354	0.8821	0.6319	0.4090
7.5	1.8866E 01	0.8540	0.8535	0.7338	0.6696	0.8895	0.9884	1.0015	1.0002	0.6461	0.8926	0.6430	0.4165
8.0	1.8042E 01	0.8449	0.8752	0.7571	0.6928	0.8840	0.9876	1.0014	1.0002	0.6565	0.9025	0.6537	0.4238
9.0	1.6609E 01	0.8275	0.9174	0.8028	0.7385	0.8736	0.9859	1.0013	1.0002	0.6767	0.9203	0.6745	0.4381
10.0	1.5402E 01	0.8109	0.9581	0.8473	0.7833	0.8638	0.9843	1.0011	1.0001	0.6963	0.9359	0.6944	0.4521
11.0	1.4367E 01	0.7951	0.9974	0.8908	0.8273	0.8547	0.9826	1.0010	1.0001	0.7153	0.9496	0.7137	0.4658
12.0	1.3469E 01	0.7800	1.0354	0.9333	0.8705	0.8462	0.9809	1.0009	1.0001	0.7337	0.9617	0.7323	0.4793
13.0	1.2681E 01	0.7656	1.0722	0.9751	0.9131	0.8382	0.9792	1.0008	1.0001	0.7517	0.9722	0.7505	0.4926
14.0	1.1982E 01	0.7519	1.1080	1.0161	0.9552	0.8307	0.9775	1.0008	1.0001	0.7693	0.9814	0.7683	0.5058
16.0	1.0797E 01	0.7263	1.1766	1.0961	1.0377	0.8173	0.9740	1.0006	1.0001	0.8034	0.9961	0.8026	0.5316
18.0	9.8281E 00	0.7028	1.2416	1.1736	1.1183	0.8055	0.9704	1.0005	1.0001	0.8362	1.0065	0.8355	0.5568
20.0	9.0201E 00	0.6813	1.3033	1.2489	1.1973	0.7954	0.9668	1.0005	1.0001	0.8678	1.0133	0.8672	0.5814
25.0	7.4844E 00	0.6347	1.4439	1.4282	1.3880	0.7757	0.9574	1.0003	1.0001	0.9419	1.0166	0.9415	0.6402
30.0	6.3977E 00	0.5964	1.5668	1.5956	1.5696	0.7628	0.9475	1.0003	1.0001	1.0094	1.0040	1.0091	0.6944
35.0	5.5901E 00	0.5644	1.6729	1.7513	1.7423	0.7553	0.9372	1.0002	1.0000	1.0703	0.9790	1.0701	0.7438
40.0	4.9681E 00	0.5375	1.7634	1.8956	1.9059	0.7521	0.9267	1.0002	1.0000	1.1248	0.9441	1.1246	0.7879
45.0	4.4762E 00	0.5148	1.8390	2.0284	2.0602	0.7525	0.9160	1.0001	1.0000	1.1729	0.9016	1.1728	0.8268
50.0	4.0787E 00	0.4954	1.9008	2.1497	2.2050	0.7559	0.9053	1.0001	1.0000	1.2149	0.8531	1.2148	0.8604

Instr. see p. 22

ELECTRON Z= 93 A= 237

p	F_0L_0	L_0	λ_2	λ_3	λ_4	μ_1	μ_2	Λ_1	Λ_2	ν_{12}	$\hat{\nu}_{12}$	η_{12}	$\hat{\eta}_{12}$
0.1	6.9327E 02	0.9905	15.6487	126.4209	607.9870	0.9908	0.9998	1.0325	1.0014	6.3998	0.4239	-1.1537	-1.0676
0.2	3.4860E 02	0.9902	4.2388	10.7468	19.4451	0.9903	0.9998	1.0214	1.0013	1.6916	0.4497	-0.3121	-0.3399
0.3	2.3451E 02	0.9858	2.1282	3.2189	4.0479	0.9896	0.9997	1.0175	1.0013	0.9075	0.4744	-0.0325	-0.1007
0.4	1.7800E 02	0.9891	1.3921	1.6120	1.7206	0.9887	0.9996	1.0154	1.0013	0.6621	0.4975	0.1070	0.0170
0.5	1.4445E 02	0.9884	1.0542	1.0509	1.0315	0.9876	0.9995	1.0139	1.0012	0.5619	0.5190	0.1908	0.0864
0.6	1.2233E 02	0.9874	0.8734	0.7963	0.7458	0.9863	0.9994	1.0128	1.0012	0.5148	0.5387	0.2468	0.1320
0.7	1.0671E 02	0.9863	0.7670	0.6615	0.6025	0.9848	0.9992	1.0119	1.0011	0.4910	0.5567	0.2872	0.1642
0.8	9.5106E 01	0.9850	0.7005	0.5829	0.5217	0.9833	0.9991	1.0111	1.0011	0.4788	0.5731	0.3178	0.1882
0.9	8.6166E 01	0.9837	0.6572	0.5341	0.4727	0.9817	0.9989	1.0104	1.0010	0.4728	0.5882	0.3420	0.2067
1.0	7.9070E 01	0.9821	0.6283	0.5026	0.4417	0.9801	0.9988	1.0097	1.0010	0.4703	0.6020	0.3617	0.2216
1.2	6.8513E 01	0.9788	0.5960	0.4684	0.4084	0.9767	0.9985	1.0087	1.0009	0.4710	0.6264	0.3922	0.2440
1.4	6.1012E 01	0.9751	0.5825	0.4543	0.3949	0.9733	0.9981	1.0078	1.0008	0.4752	0.6473	0.4152	0.2604
1.6	5.5378E 01	0.9712	0.5787	0.4504	0.3913	0.9698	0.9978	1.0070	1.0007	0.4809	0.6656	0.4334	0.2731
1.8	5.0964E 01	0.9671	0.5805	0.4522	0.3930	0.9664	0.9975	1.0064	1.0007	0.4872	0.6818	0.4486	0.2834
2.0	4.7391E 01	0.9629	0.5857	0.4572	0.3979	0.9630	0.9971	1.0059	1.0006	0.4937	0.6964	0.4616	0.2922
2.2	4.4423E 01	0.9585	0.5928	0.4643	0.4048	0.9596	0.9968	1.0054	1.0006	0.5002	0.7096	0.4731	0.2998
2.4	4.1906E 01	0.9541	0.6013	0.4727	0.4129	0.9563	0.9965	1.0050	1.0005	0.5067	0.7218	0.4834	0.3065
2.6	3.9736E 01	0.9497	0.6107	0.4820	0.4219	0.9530	0.9962	1.0046	1.0005	0.5130	0.7331	0.4928	0.3126
2.8	3.7838E 01	0.9452	0.6205	0.4918	0.4313	0.9498	0.9958	1.0043	1.0005	0.5192	0.7435	0.5015	0.3182
3.0	3.6159E 01	0.9408	0.6307	0.5019	0.4411	0.9466	0.9955	1.0041	1.0004	0.5253	0.7534	0.5096	0.3234
3.2	3.4659E 01	0.9363	0.6411	0.5123	0.4512	0.9434	0.9952	1.0038	1.0004	0.5312	0.7626	0.5172	0.3283
3.4	3.3308E 01	0.9318	0.6516	0.5228	0.4614	0.9404	0.9948	1.0036	1.0004	0.5370	0.7714	0.5244	0.3330
3.6	3.2082E 01	0.9274	0.6621	0.5334	0.4716	0.9373	0.9945	1.0034	1.0004	0.5427	0.7797	0.5313	0.3374
3.8	3.0963E 01	0.9230	0.6727	0.5441	0.4820	0.9343	0.9942	1.0032	1.0004	0.5483	0.7876	0.5379	0.3416
4.0	2.9936E 01	0.9186	0.6832	0.5547	0.4923	0.9314	0.9939	1.0031	1.0003	0.5538	0.7952	0.5443	0.3457
4.5	2.7695E 01	0.9078	0.7092	0.5811	0.5181	0.9242	0.9931	1.0027	1.0003	0.5670	0.8127	0.5593	0.3553
5.0	2.5817E 01	0.8971	0.7347	0.6073	0.5436	0.9172	0.9922	1.0025	1.0003	0.5797	0.8286	0.5732	0.3643
5.5	2.4214E 01	0.8867	0.7597	0.6331	0.5689	0.9106	0.9914	1.0022	1.0002	0.5919	0.8431	0.5864	0.3729
6.0	2.2824E 01	0.8766	0.7840	0.6584	0.5939	0.9041	0.9906	1.0020	1.0002	0.6037	0.8565	0.5990	0.3812
6.5	2.1604E 01	0.8667	0.8079	0.6834	0.6185	0.8979	0.9898	1.0019	1.0002	0.6152	0.8688	0.6111	0.3892
7.0	2.0521E 01	0.8570	0.8312	0.7080	0.6428	0.8919	0.9890	1.0017	1.0002	0.6263	0.8802	0.6228	0.3970
7.5	1.9551E 01	0.8476	0.8540	0.7322	0.6668	0.8862	0.9881	1.0016	1.0002	0.6373	0.8909	0.6341	0.4047
8.0	1.8677E 01	0.8385	0.8764	0.7560	0.6906	0.8806	0.9873	1.0015	1.0002	0.6479	0.9008	0.6451	0.4122
9.0	1.7160E 01	0.8208	0.9197	0.8028	0.7373	0.8700	0.9856	1.0013	1.0002	0.6687	0.9188	0.6664	0.4269
10.0	1.5885E 01	0.8041	0.9615	0.8484	0.7831	0.8602	0.9839	1.0012	1.0001	0.6887	0.9345	0.6868	0.4413
11.0	1.4794E 01	0.7881	1.0019	0.8930	0.8281	0.8510	0.9822	1.0010	1.0001	0.7082	0.9482	0.7066	0.4554
12.0	1.3849E 01	0.7729	1.0410	0.9367	0.8725	0.8424	0.9805	1.0009	1.0001	0.7272	0.9603	0.7258	0.4694
13.0	1.3021E 01	0.7585	1.0789	0.9796	0.9162	0.8344	0.9788	1.0008	1.0001	0.7457	0.9708	0.7444	0.4831
14.0	1.2289E 01	0.7447	1.1157	1.0217	0.9593	0.8270	0.9770	1.0008	1.0001	0.7638	0.9800	0.7627	0.4967
16.0	1.1049E 01	0.7190	1.1863	1.1039	1.0440	0.8136	0.9734	1.0006	1.0001	0.7988	0.9946	0.7980	0.5233
18.0	1.0039E 01	0.6956	1.2531	1.1836	1.1268	0.8020	0.9698	1.0006	1.0001	0.8326	1.0049	0.8319	0.5494
20.0	9.1980E 00	0.6741	1.3165	1.2610	1.2079	0.7919	0.9660	1.0005	1.0001	0.8651	1.0114	0.8645	0.5748
25.0	7.6062E 00	0.6278	1.4608	1.4454	1.4040	0.7728	0.9564	1.0004	1.0001	0.9413	1.0139	0.9409	0.6355
30.0	6.4851E 00	0.5897	1.5865	1.6174	1.5907	0.7605	0.9463	1.0003	1.0001	1.0105	1.0004	1.0102	0.6914
35.0	5.6550E 00	0.5582	1.6948	1.7773	1.7682	0.7536	0.9358	1.0002	1.0000	1.0730	0.9743	1.0728	0.7420
40.0	5.0178E 00	0.5317	1.7867	1.9253	1.9363	0.7511	0.9250	1.0002	1.0000	1.1288	0.9384	1.1286	0.7872
45.0	4.5150E 00	0.5093	1.8633	2.0613	2.0946	0.7522	0.9141	1.0001	1.0000	1.1779	0.8947	1.1778	0.8269
50.0	4.1098E 00	0.4903	1.9254	2.1853	2.2431	0.7562	0.9032	1.0001	1.0000	1.2206	0.8454	1.2205	0.8611

 Instr. see p. 22

ELECTRON Z= 94 A= 238

p	F_0L_0	L_0	λ_2	λ_3	λ_4	μ_1	μ_2	Λ_1	Λ_2	ν_{12}	$\hat{\nu}_{12}$	η_{12}	$\hat{\eta}_{12}$
0.1	7.4284E 02	0.9872	15.5980	127.3562	620.4876	0.9902	0.9998	1.0348	1.0014	6.4161	0.4169	-1.2032	-1.1069
0.2	3.7345E 02	0.9869	4.2204	10.7745	19.6361	0.9897	0.9998	1.0227	1.0014	1.6816	0.4428	-0.3417	-0.3649
0.3	2.5116E 02	0.9865	2.1158	3.2105	4.0498	0.9890	0.9997	1.0184	1.0013	0.8945	0.4675	-0.0555	-0.1210
0.4	1.9056E 02	0.9858	1.3818	1.6007	1.7097	0.9880	0.9996	1.0161	1.0013	0.6487	0.4908	0.0872	-0.0011
0.5	1.5458E 02	0.9850	1.0449	1.0399	1.0200	0.9868	0.9995	1.0146	1.0012	0.5485	0.5123	0.1728	0.0696
0.6	1.3084E 02	0.9841	0.8646	0.7860	0.7350	0.9855	0.9994	1.0134	1.0012	0.5016	0.5321	0.2302	0.1160
0.7	1.1406E 02	0.9829	0.7585	0.6517	0.5923	0.9840	0.9992	1.0124	1.0011	0.4780	0.5503	0.2714	0.1488
0.8	1.0160E 02	0.9816	0.6922	0.5735	0.5120	0.9825	0.9991	1.0115	1.0011	0.4659	0.5669	0.3026	0.1731
0.9	9.1991E 01	0.9802	0.6491	0.5249	0.4634	0.9808	0.9989	1.0108	1.0010	0.4600	0.5821	0.3273	0.1919
1.0	8.4358E 01	0.9786	0.6204	0.4937	0.4326	0.9791	0.9988	1.0101	1.0010	0.4576	0.5960	0.3474	0.2070
1.2	7.2994E 01	0.9752	0.5883	0.4598	0.3997	0.9756	0.9984	1.0090	1.0009	0.4585	0.6206	0.3784	0.2297
1.4	6.4912E 01	0.9714	0.5749	0.4460	0.3864	0.9721	0.9981	1.0081	1.0008	0.4629	0.6418	0.4018	0.2463
1.6	5.8838E 01	0.9674	0.5714	0.4422	0.3830	0.9685	0.9978	1.0073	1.0007	0.4687	0.6603	0.4204	0.2591
1.8	5.4078E 01	0.9632	0.5734	0.4441	0.3848	0.9650	0.9974	1.0066	1.0007	0.4750	0.6767	0.4358	0.2695
2.0	5.0224E 01	0.9588	0.5787	0.4493	0.3899	0.9615	0.9971	1.0061	1.0006	0.4816	0.6914	0.4490	0.2784
2.2	4.7023E 01	0.9544	0.5861	0.4565	0.3968	0.9580	0.9968	1.0056	1.0006	0.4882	0.7048	0.4607	0.2860
2.4	4.4309E 01	0.9499	0.5948	0.4651	0.4050	0.9546	0.9964	1.0052	1.0005	0.4948	0.7171	0.4711	0.2928
2.6	4.1969E 01	0.9454	0.6043	0.4745	0.4141	0.9512	0.9961	1.0048	1.0005	0.5012	0.7285	0.4806	0.2989
2.8	3.9924E 01	0.9408	0.6144	0.4844	0.4236	0.9479	0.9958	1.0045	1.0005	0.5075	0.7392	0.4894	0.3046
3.0	3.8116E 01	0.9362	0.6248	0.4947	0.4336	0.9446	0.9954	1.0042	1.0004	0.5137	0.7491	0.4977	0.3098
3.2	3.6502E 01	0.9317	0.6354	0.5052	0.4437	0.9414	0.9951	1.0040	1.0004	0.5197	0.7585	0.5054	0.3148
3.4	3.5048E 01	0.9271	0.6461	0.5159	0.4540	0.9382	0.9948	1.0037	1.0004	0.5256	0.7674	0.5127	0.3194
3.6	3.3730E 01	0.9226	0.6569	0.5267	0.4644	0.9351	0.9944	1.0035	1.0004	0.5313	0.7758	0.5197	0.3239
3.8	3.2527E 01	0.9181	0.6677	0.5375	0.4749	0.9321	0.9941	1.0033	1.0004	0.5370	0.7838	0.5264	0.3282
4.0	3.1424E 01	0.9136	0.6784	0.5483	0.4853	0.9290	0.9938	1.0032	1.0003	0.5425	0.7915	0.5329	0.3323
4.5	2.9019E 01	0.9026	0.7050	0.5752	0.5114	0.9217	0.9929	1.0028	1.0003	0.5560	0.8093	0.5481	0.3420
5.0	2.7008E 01	0.8918	0.7311	0.6018	0.5374	0.9146	0.9921	1.0025	1.0003	0.5689	0.8253	0.5623	0.3512
5.5	2.5293E 01	0.8813	0.7566	0.6280	0.5630	0.9078	0.9913	1.0023	1.0003	0.5813	0.8400	0.5758	0.3599
6.0	2.3808E 01	0.8710	0.7816	0.6538	0.5884	0.9013	0.9905	1.0021	1.0002	0.5933	0.8535	0.5886	0.3684
6.5	2.2506E 01	0.8609	0.8060	0.6792	0.6134	0.8950	0.9896	1.0019	1.0002	0.6050	0.8660	0.6009	0.3765
7.0	2.1352E 01	0.8512	0.8298	0.7043	0.6381	0.8889	0.9888	1.0018	1.0002	0.6164	0.8776	0.6128	0.3845
7.5	2.0321E 01	0.8417	0.8532	0.7290	0.6626	0.8831	0.9879	1.0016	1.0002	0.6276	0.8883	0.6243	0.3923
8.0	1.9392E 01	0.8324	0.8761	0.7533	0.6867	0.8774	0.9871	1.0015	1.0002	0.6385	0.8984	0.6356	0.4000
9.0	1.7783E 01	0.8146	0.9206	0.8011	0.7344	0.8668	0.9854	1.0013	1.0002	0.6597	0.9165	0.6573	0.4151
10.0	1.6433E 01	0.7977	0.9634	0.8477	0.7811	0.8569	0.9837	1.0012	1.0001	0.6802	0.9323	0.6782	0.4299
11.0	1.5281E 01	0.7817	1.0049	0.8933	0.8271	0.8476	0.9819	1.0011	1.0001	0.7001	0.9461	0.6985	0.4444
12.0	1.4284E 01	0.7665	1.0450	0.9380	0.8723	0.8391	0.9802	1.0009	1.0001	0.7195	0.9583	0.7181	0.4587
13.0	1.3413E 01	0.7520	1.0838	0.9819	0.9170	0.8311	0.9784	1.0009	1.0001	0.7385	0.9688	0.7372	0.4728
14.0	1.2643E 01	0.7382	1.1216	1.0251	0.9611	0.8237	0.9766	1.0008	1.0001	0.7570	0.9780	0.7559	0.4868
16.0	1.1344E 01	0.7125	1.1940	1.1092	1.0477	0.8103	0.9730	1.0007	1.0001	0.7930	0.9926	0.7921	0.5142
18.0	1.0288E 01	0.6891	1.2625	1.1909	1.1325	0.7988	0.9693	1.0006	1.0001	0.8276	1.0028	0.8269	0.5410
20.0	9.4110E 00	0.6678	1.3274	1.2703	1.2155	0.7890	0.9655	1.0005	1.0001	0.8609	1.0092	0.8603	0.5672
25.0	7.7569E 00	0.6217	1.4752	1.4593	1.4164	0.7703	0.9556	1.0004	1.0001	0.9390	1.0113	0.9386	0.6296
30.0	6.5972E 00	0.5840	1.6037	1.6356	1.6078	0.7585	0.9453	1.0003	1.0001	1.0101	0.9972	1.0098	0.6870
35.0	5.7415E 00	0.5528	1.7142	1.7994	1.7897	0.7522	0.9346	1.0002	1.0000	1.0740	0.9706	1.0738	0.7390
40.0	5.0864E 00	0.5267	1.8078	1.9509	1.9619	0.7503	0.9237	1.0002	1.0000	1.1311	0.9340	1.1309	0.7852
45.0	4.5709E 00	0.5047	1.8853	2.0900	2.1241	0.7520	0.9126	1.0001	1.0000	1.1813	0.8897	1.1811	0.8257
50.0	4.1562E 00	0.4861	1.9480	2.2167	2.2761	0.7566	0.9015	1.0001	1.0000	1.2248	0.8397	1.2247	0.8605

ELECTRON Z= 95 A= 243

p	F_0L_0	L_0	λ_2	λ_3	λ_4	μ_1	μ_2	Λ_1	Λ_2	ν_{12}	$\hat{\nu}_{12}$	η_{12}	$\hat{\eta}_{12}$
0.1	7.9374E 02	0.9835	15.5690	128.5032	634.2911	0.9895	0.9998	1.0377	1.0014	6.4349	0.4103	-1.2538	-1.1468
0.2	3.9897E 02	0.9832	4.2081	10.8209	19.8666	0.9890	0.9997	1.0242	1.0014	1.6722	0.4362	-0.3717	-0.3900
0.3	2.6824E 02	0.9827	2.1066	3.2080	4.0600	0.9882	0.9997	1.0196	1.0014	0.8820	0.4611	-0.0787	-0.1413
0.4	2.0345E 02	0.9821	1.3737	1.5926	1.7025	0.9872	0.9996	1.0170	1.0013	0.6357	0.4844	0.0673	-0.0191
0.5	1.6495E 02	0.9812	1.0373	1.0311	1.0108	0.9860	0.9995	1.0153	1.0013	0.5355	0.5061	0.1550	0.0530
0.6	1.3955E 02	0.9802	0.8573	0.7774	0.7259	0.9846	0.9993	1.0141	1.0012	0.4887	0.5260	0.2136	0.1002
0.7	1.2158E 02	0.9790	0.7514	0.6434	0.5836	0.9831	0.9992	1.0130	1.0012	0.4652	0.5443	0.2557	0.1335
0.8	1.0823E 02	0.9777	0.6852	0.5654	0.5037	0.9815	0.9990	1.0121	1.0011	0.4533	0.5610	0.2876	0.1583
0.9	9.7927E 01	0.9762	0.6422	0.5171	0.4553	0.9797	0.9989	1.0113	1.0011	0.4475	0.5764	0.3128	0.1774
1.0	8.9739E 01	0.9746	0.6136	0.4860	0.4247	0.9780	0.9987	1.0106	1.0010	0.4453	0.5904	0.3333	0.1926
1.2	7.7538E 01	0.9710	0.5817	0.4524	0.3921	0.9744	0.9984	1.0094	1.0009	0.4463	0.6153	0.3649	0.2157
1.4	6.8855E 01	0.9671	0.5685	0.4387	0.3791	0.9707	0.9980	1.0084	1.0008	0.4508	0.6367	0.3887	0.2324
1.6	6.2325E 01	0.9630	0.5651	0.4352	0.3757	0.9670	0.9977	1.0076	1.0008	0.4567	0.6554	0.4076	0.2454
1.8	5.7205E 01	0.9587	0.5674	0.4372	0.3778	0.9633	0.9974	1.0069	1.0007	0.4632	0.6720	0.4233	0.2559
2.0	5.3060E 01	0.9542	0.5729	0.4425	0.3829	0.9597	0.9970	1.0063	1.0006	0.4699	0.6870	0.4367	0.2648
2.2	4.9617E 01	0.9496	0.5806	0.4500	0.3900	0.9561	0.9967	1.0058	1.0006	0.4766	0.7006	0.4485	0.2725
2.4	4.6699E 01	0.9450	0.5895	0.4587	0.3984	0.9526	0.9963	1.0054	1.0005	0.4833	0.7131	0.4591	0.2794
2.6	4.4184E 01	0.9403	0.5993	0.4682	0.4075	0.9491	0.9960	1.0050	1.0005	0.4898	0.7246	0.4688	0.2856
2.8	4.1987E 01	0.9356	0.6096	0.4783	0.4172	0.9457	0.9957	1.0047	1.0005	0.4962	0.7354	0.4778	0.2913
3.0	4.0045E 01	0.9310	0.6203	0.4888	0.4273	0.9423	0.9953	1.0044	1.0005	0.5024	0.7455	0.4861	0.2966
3.2	3.8312E 01	0.9263	0.6311	0.4996	0.4376	0.9390	0.9950	1.0041	1.0004	0.5086	0.7550	0.4940	0.3015
3.4	3.6753E 01	0.9216	0.6421	0.5104	0.4481	0.9358	0.9946	1.0039	1.0004	0.5145	0.7640	0.5014	0.3063
3.6	3.5339E 01	0.9170	0.6531	0.5214	0.4586	0.9326	0.9943	1.0037	1.0004	0.5204	0.7726	0.5086	0.3108
3.8	3.4051E 01	0.9124	0.6642	0.5324	0.4692	0.9294	0.9940	1.0035	1.0004	0.5262	0.7807	0.5154	0.3151
4.0	3.2869E 01	0.9078	0.6752	0.5434	0.4799	0.9263	0.9936	1.0033	1.0003	0.5318	0.7885	0.5220	0.3193
4.5	3.0296E 01	0.8966	0.7024	0.5708	0.5064	0.9188	0.9928	1.0029	1.0003	0.5455	0.8065	0.5375	0.3292
5.0	2.8148E 01	0.8856	0.7292	0.5980	0.5328	0.9116	0.9919	1.0026	1.0003	0.5587	0.8228	0.5520	0.3385
5.5	2.6319E 01	0.8748	0.7554	0.6247	0.5589	0.9046	0.9911	1.0024	1.0003	0.5713	0.8377	0.5657	0.3474
6.0	2.4738E 01	0.8644	0.7810	0.6511	0.5848	0.8980	0.9902	1.0022	1.0002	0.5836	0.8513	0.5788	0.3560
6.5	2.3354E 01	0.8542	0.8061	0.6772	0.6103	0.8916	0.9894	1.0020	1.0002	0.5956	0.8640	0.5914	0.3644
7.0	2.2128E 01	0.8443	0.8306	0.7028	0.6356	0.8854	0.9885	1.0018	1.0002	0.6073	0.8756	0.6036	0.3726
7.5	2.1035E 01	0.8347	0.8546	0.7281	0.6606	0.8795	0.9877	1.0017	1.0002	0.6187	0.8865	0.6154	0.3806
8.0	2.0051E 01	0.8253	0.8781	0.7530	0.6853	0.8738	0.9868	1.0016	1.0002	0.6299	0.8966	0.6269	0.3885
9.0	1.8350E 01	0.8073	0.9239	0.8020	0.7340	0.8630	0.9851	1.0014	1.0002	0.6516	0.9149	0.6492	0.4040
10.0	1.6926E 01	0.7903	0.9680	0.8498	0.7819	0.8530	0.9833	1.0012	1.0001	0.6727	0.9308	0.6707	0.4192
11.0	1.5714E 01	0.7742	1.0106	0.8967	0.8290	0.8438	0.9815	1.0011	1.0001	0.6931	0.9447	0.6914	0.4342
12.0	1.4667E 01	0.7589	1.0519	0.9426	0.8754	0.8352	0.9797	1.0010	1.0001	0.7131	0.9569	0.7116	0.4490
13.0	1.3753E 01	0.7443	1.0919	0.9877	0.9213	0.8272	0.9779	1.0009	1.0001	0.7326	0.9674	0.7313	0.4636
14.0	1.2948E 01	0.7305	1.1307	1.0321	0.9665	0.8198	0.9761	1.0008	1.0001	0.7516	0.9766	0.7505	0.4780
16.0	1.1591E 01	0.7048	1.2052	1.1187	1.0555	0.8066	0.9723	1.0007	1.0001	0.7886	0.9911	0.7877	0.5063
18.0	1.0492E 01	0.6815	1.2757	1.2027	1.1427	0.7953	0.9685	1.0006	1.0001	0.8242	1.0011	0.8235	0.5341
20.0	9.5814E 00	0.6602	1.3424	1.2843	1.2282	0.7856	0.9646	1.0005	1.0001	0.8585	1.0072	0.8579	0.5611
25.0	7.8706E 00	0.6144	1.4940	1.4788	1.4348	0.7675	0.9545	1.0004	1.0001	0.9389	1.0084	0.9385	0.6255
30.0	6.6766E 00	0.5772	1.6255	1.6601	1.6317	0.7564	0.9439	1.0003	1.0001	1.0119	0.9933	1.0116	0.6846
35.0	5.7990E 00	0.5464	1.7382	1.8284	1.8188	0.7509	0.9330	1.0002	1.0000	1.0775	0.9655	1.0773	0.7379
40.0	5.1293E 00	0.5208	1.8331	1.9838	1.9958	0.7497	0.9218	1.0002	1.0000	1.1359	0.9277	1.1357	0.7852
45.0	4.6036E 00	0.4992	1.9114	2.1262	2.1623	0.7522	0.9105	1.0001	1.0000	1.1870	0.8823	1.1869	0.8265
50.0	4.1817E 00	0.4810	1.9742	2.2557	2.3181	0.7576	0.8992	1.0001	1.0000	1.2313	0.8312	1.2312	0.8619

Instr. see p. 22

ELECTRON Z= 96 A= 245

p	F_0L_0	L_0	λ_2	λ_3	λ_4	μ_1	μ_2	Λ_1	Λ_2	ν_{12}	$\hat{\nu}_{12}$	η_{12}	$\hat{\eta}_{12}$
0.1	8.5063E 02	0.9796	15.5104	129.3331	646.5395	0.9888	0.9998	1.0406	1.0014	6.4440	0.4033	-1.3039	-1.1856
0.2	4.2749E 02	0.9793	4.1880	10.8415	20.0467	0.9883	0.9997	1.0258	1.0014	1.6603	0.4293	-0.4016	-0.4147
0.3	2.8733E 02	0.9788	2.0936	3.1982	4.0602	0.9875	0.9997	1.0207	1.0014	0.8681	0.4542	-0.1021	-0.1614
0.4	2.1784E 02	0.9782	1.3632	1.5810	1.6912	0.9865	0.9996	1.0180	1.0013	0.6217	0.4776	0.0473	-0.0370
0.5	1.7654E 02	0.9773	1.0279	1.0202	0.9993	0.9852	0.9995	1.0161	1.0013	0.5217	0.4994	0.1368	0.0363
0.6	1.4927E 02	0.9762	0.8486	0.7672	0.7153	0.9837	0.9993	1.0147	1.0012	0.4751	0.5194	0.1967	0.0844
0.7	1.2997E 02	0.9750	0.7431	0.6338	0.5737	0.9822	0.9992	1.0136	1.0012	0.4519	0.5378	0.2397	0.1182
0.8	1.1562E 02	0.9736	0.6772	0.5562	0.4943	0.9805	0.9990	1.0126	1.0011	0.4401	0.5547	0.2723	0.1433
0.9	1.0455E 02	0.9721	0.6344	0.5082	0.4463	0.9787	0.9989	1.0118	1.0011	0.4345	0.5701	0.2979	0.1627
1.0	9.5738E 01	0.9704	0.6059	0.4774	0.4160	0.9769	0.9987	1.0110	1.0010	0.4324	0.5843	0.3188	0.1781
1.2	8.2603E 01	0.9668	0.5743	0.4441	0.3838	0.9731	0.9984	1.0098	1.0009	0.4336	0.6094	0.3510	0.2014
1.4	7.3247E 01	0.9628	0.5613	0.4307	0.3710	0.9693	0.9980	1.0087	1.0008	0.4382	0.6310	0.3751	0.2183
1.6	6.6207E 01	0.9585	0.5581	0.4273	0.3679	0.9655	0.9977	1.0079	1.0008	0.4443	0.6499	0.3943	0.2314
1.8	6.0686E 01	0.9541	0.5605	0.4295	0.3700	0.9617	0.9973	1.0072	1.0007	0.4509	0.6667	0.4102	0.2420
2.0	5.6216E 01	0.9495	0.5663	0.4350	0.3753	0.9580	0.9970	1.0065	1.0006	0.4576	0.6819	0.4239	0.2510
2.2	5.2504E 01	0.9448	0.5742	0.4426	0.3825	0.9543	0.9966	1.0060	1.0006	0.4644	0.6956	0.4359	0.2587
2.4	4.9358E 01	0.9401	0.5833	0.4514	0.3909	0.9507	0.9963	1.0056	1.0006	0.4712	0.7083	0.4466	0.2656
2.6	4.6648E 01	0.9353	0.5933	0.4612	0.4002	0.9471	0.9959	1.0052	1.0005	0.4778	0.7200	0.4564	0.2719
2.8	4.4281E 01	0.9305	0.6039	0.4714	0.4100	0.9436	0.9956	1.0048	1.0005	0.4843	0.7309	0.4655	0.2776
3.0	4.2191E 01	0.9257	0.6148	0.4821	0.4202	0.9401	0.9952	1.0045	1.0005	0.4906	0.7412	0.4740	0.2830
3.2	4.0326E 01	0.9210	0.6259	0.4930	0.4306	0.9367	0.9949	1.0042	1.0004	0.4968	0.7508	0.4820	0.2880
3.4	3.8649E 01	0.9162	0.6371	0.5040	0.4412	0.9334	0.9945	1.0040	1.0004	0.5029	0.7599	0.4895	0.2927
3.6	3.7131E 01	0.9115	0.6484	0.5151	0.4519	0.9301	0.9942	1.0038	1.0004	0.5088	0.7686	0.4968	0.2973
3.8	3.5747E 01	0.9068	0.6597	0.5263	0.4627	0.9269	0.9939	1.0036	1.0004	0.5147	0.7768	0.5037	0.3017
4.0	3.4478E 01	0.9022	0.6709	0.5375	0.4734	0.9238	0.9935	1.0034	1.0004	0.5204	0.7847	0.5104	0.3059
4.5	3.1720E 01	0.8907	0.6988	0.5654	0.5003	0.9161	0.9927	1.0030	1.0003	0.5343	0.8030	0.5261	0.3159
5.0	2.9421E 01	0.8795	0.7262	0.5930	0.5271	0.9087	0.9918	1.0027	1.0003	0.5477	0.8195	0.5409	0.3254
5.5	2.7466E 01	0.8687	0.7530	0.6203	0.5537	0.9017	0.9909	1.0024	1.0003	0.5606	0.8345	0.5549	0.3345
6.0	2.5779E 01	0.8581	0.7792	0.6472	0.5799	0.8949	0.9901	1.0022	1.0002	0.5732	0.8484	0.5682	0.3432
6.5	2.4304E 01	0.8478	0.8049	0.6737	0.6059	0.8884	0.9892	1.0020	1.0002	0.5854	0.8611	0.5810	0.3518
7.0	2.3000E 01	0.8378	0.8300	0.6999	0.6316	0.8822	0.9883	1.0019	1.0002	0.5973	0.8729	0.5935	0.3601
7.5	2.1838E 01	0.8280	0.8547	0.7257	0.6571	0.8762	0.9874	1.0017	1.0002	0.6089	0.8839	0.6055	0.3683
8.0	2.0794E 01	0.8186	0.8788	0.7512	0.6823	0.8704	0.9866	1.0016	1.0002	0.6203	0.8941	0.6173	0.3764
9.0	1.8992E 01	0.8005	0.9257	0.8012	0.7320	0.8596	0.9848	1.0014	1.0002	0.6426	0.9125	0.6401	0.3923
10.0	1.7487E 01	0.7834	0.9709	0.8502	0.7808	0.8495	0.9830	1.0012	1.0001	0.6641	0.9286	0.6621	0.4079
11.0	1.6209E 01	0.7672	1.0147	0.8981	0.8290	0.8403	0.9812	1.0011	1.0001	0.6851	0.9426	0.6833	0.4233
12.0	1.5107E 01	0.7518	1.0570	0.9451	0.8764	0.8317	0.9794	1.0010	1.0001	0.7055	0.9548	0.7040	0.4385
13.0	1.4147E 01	0.7373	1.0981	0.9914	0.9233	0.8237	0.9775	1.0009	1.0001	0.7255	0.9654	0.7242	0.4535
14.0	1.3302E 01	0.7234	1.1380	1.0368	0.9696	0.8164	0.9756	1.0008	1.0001	0.7450	0.9746	0.7439	0.4684
16.0	1.1883E 01	0.6978	1.2144	1.1256	1.0608	0.8033	0.9718	1.0007	1.0001	0.7830	0.9890	0.7821	0.4976
18.0	1.0735E 01	0.6746	1.2867	1.2118	1.1501	0.7921	0.9679	1.0006	1.0001	0.8195	0.9985	0.8188	0.5261
20.0	9.7880E 00	0.6534	1.3551	1.2955	1.2377	0.7826	0.9640	1.0005	1.0001	0.8547	1.0049	0.8541	0.5540
25.0	8.0138E 00	0.6080	1.5105	1.4950	1.4496	0.7651	0.9537	1.0004	1.0001	0.9373	1.0057	0.9368	0.6202
30.0	6.7812E 00	0.5712	1.6449	1.6809	1.6515	0.7547	0.9429	1.0003	1.0001	1.0121	0.9899	1.0118	0.6809
35.0	5.8783E 00	0.5409	1.7597	1.8534	1.8434	0.7499	0.9317	1.0002	1.0000	1.0793	0.9614	1.0790	0.7355
40.0	5.1914E 00	0.5157	1.8561	2.0125	2.0248	0.7494	0.9203	1.0002	1.0000	1.1389	0.9228	1.1387	0.7839
45.0	4.6535E 00	0.4945	1.9353	2.1582	2.1954	0.7525	0.9088	1.0001	1.0000	1.1912	0.8766	1.1910	0.8260
50.0	4.2227E 00	0.4767	1.9985	2.2904	2.3550	0.7585	0.8973	1.0001	1.0000	1.2362	0.8249	1.2361	0.8619

Instr. see p. 22

ELECTRON Z= 97 A= 249

p	F_0L_0	L_0	λ_2	λ_3	λ_4	μ_1	μ_2	Λ_1	Λ_2	ν_{12}	$\hat{\nu}_{12}$	η_{12}	$\hat{\eta}_{12}$
0.1	9.1053E 02	0.9753	15.4594	130.2285	659.2914	0.9880	0.9998	1.0439	1.0015	6.4519	0.3964	-1.3547	-1.2246
0.2	4.5750E 02	0.9750	4.1701	10.8685	20.2414	0.9875	0.9997	1.0276	1.0014	1.6481	0.4224	-0.4319	-0.4395
0.3	3.0741E 02	0.9745	2.0819	3.1907	4.0635	0.9867	0.9997	1.0220	1.0014	0.8541	0.4474	-0.1255	-0.1815
0.4	2.3297E 02	0.9738	1.3536	1.5707	1.6815	0.9856	0.9996	1.0190	1.0014	0.6077	0.4709	0.0271	-0.0549
0.5	1.8871E 02	0.9729	1.0194	1.0103	0.9889	0.9843	0.9994	1.0170	1.0013	0.5080	0.4928	0.1186	0.0197
0.6	1.5947E 02	0.9719	0.8405	0.7579	0.7055	0.9828	0.9993	1.0155	1.0013	0.4616	0.5130	0.1798	0.0686
0.7	1.3877E 02	0.9706	0.7354	0.6250	0.5645	0.9811	0.9992	1.0143	1.0012	0.4386	0.5315	0.2237	0.1030
0.8	1.2337E 02	0.9692	0.6697	0.5478	0.4856	0.9794	0.9990	1.0132	1.0011	0.4270	0.5484	0.2569	0.1284
0.9	1.1147E 02	0.9676	0.6271	0.5001	0.4380	0.9775	0.9988	1.0123	1.0011	0.4215	0.5640	0.2831	0.1481
1.0	1.0201E 02	0.9659	0.5988	0.4695	0.4080	0.9756	0.9987	1.0115	1.0010	0.4195	0.5783	0.3043	0.1638
1.2	8.7881E 01	0.9621	0.5674	0.4365	0.3761	0.9718	0.9983	1.0102	1.0009	0.4209	0.6036	0.3371	0.1873
1.4	7.7812E 01	0.9580	0.5546	0.4233	0.3635	0.9678	0.9980	1.0091	1.0009	0.4257	0.6255	0.3616	0.2044
1.6	7.0232E 01	0.9536	0.5517	0.4201	0.3606	0.9639	0.9976	1.0082	1.0008	0.4319	0.6446	0.3811	0.2176
1.8	6.4286E 01	0.9490	0.5543	0.4225	0.3628	0.9600	0.9973	1.0074	1.0007	0.4386	0.6616	0.3973	0.2283
2.0	5.9471E 01	0.9443	0.5603	0.4281	0.3682	0.9561	0.9969	1.0068	1.0007	0.4455	0.6770	0.4111	0.2373
2.2	5.5474E 01	0.9395	0.5684	0.4359	0.3756	0.9523	0.9965	1.0063	1.0006	0.4524	0.6909	0.4233	0.2451
2.4	5.2087E 01	0.9347	0.5778	0.4449	0.3842	0.9486	0.9962	1.0058	1.0006	0.4592	0.7037	0.4342	0.2520
2.6	4.9171E 01	0.9298	0.5881	0.4548	0.3936	0.9449	0.9958	1.0054	1.0005	0.4659	0.7156	0.4442	0.2583
2.8	4.6625E 01	0.9249	0.5989	0.4652	0.4035	0.9413	0.9955	1.0050	1.0005	0.4724	0.7267	0.4534	0.2641
3.0	4.4378E 01	0.9200	0.6101	0.4760	0.4138	0.9377	0.9951	1.0047	1.0005	0.4789	0.7371	0.4620	0.2695
3.2	4.2375E 01	0.9151	0.6214	0.4871	0.4244	0.9343	0.9948	1.0044	1.0004	0.4852	0.7468	0.4701	0.2746
3.4	4.0575E 01	0.9103	0.6329	0.4983	0.4351	0.9308	0.9944	1.0041	1.0004	0.4913	0.7561	0.4778	0.2794
3.6	3.8945E 01	0.9054	0.6444	0.5097	0.4460	0.9275	0.9941	1.0039	1.0004	0.4974	0.7649	0.4851	0.2840
3.8	3.7461E 01	0.9007	0.6560	0.5210	0.4568	0.9242	0.9937	1.0037	1.0004	0.5033	0.7732	0.4921	0.2884
4.0	3.6102E 01	0.8959	0.6675	0.5324	0.4678	0.9209	0.9934	1.0035	1.0004	0.5092	0.7812	0.4989	0.2927
4.5	3.3150E 01	0.8843	0.6960	0.5608	0.4951	0.9131	0.9925	1.0031	1.0003	0.5233	0.7997	0.5150	0.3029
5.0	3.0693E 01	0.8729	0.7241	0.5890	0.5223	0.9056	0.9916	1.0028	1.0003	0.5369	0.8164	0.5300	0.3125
5.5	2.8608E 01	0.8619	0.7516	0.6168	0.5494	0.8984	0.9907	1.0025	1.0003	0.5501	0.8316	0.5442	0.3217
6.0	2.6810E 01	0.8511	0.7785	0.6443	0.5761	0.8915	0.9899	1.0023	1.0002	0.5629	0.8456	0.5578	0.3307
6.5	2.5241E 01	0.8407	0.8048	0.6714	0.6026	0.8850	0.9890	1.0021	1.0002	0.5754	0.8585	0.5710	0.3394
7.0	2.3857E 01	0.8306	0.8306	0.6981	0.6288	0.8786	0.9881	1.0019	1.0002	0.5875	0.8705	0.5837	0.3479
7.5	2.2624E 01	0.8208	0.8559	0.7246	0.6548	0.8726	0.9872	1.0018	1.0002	0.5994	0.8816	0.5960	0.3564
8.0	2.1519E 01	0.8112	0.8807	0.7506	0.6305	0.8668	0.9863	1.0017	1.0002	0.6111	0.8919	0.6081	0.3647
9.0	1.9613E 01	0.7930	0.9289	0.8019	0.7313	0.8559	0.9845	1.0014	1.0002	0.6339	0.9104	0.6314	0.3810
10.0	1.8026E 01	0.7758	0.9754	0.8520	0.7813	0.8458	0.9826	1.0013	1.0002	0.6560	0.9266	0.6539	0.3971
11.0	1.6680E 01	0.7595	1.0204	0.9012	0.8306	0.8365	0.9808	1.0011	1.0001	0.6775	0.9407	0.6757	0.4129
12.0	1.5524E 01	0.7441	1.0639	0.9495	0.8792	0.8279	0.9789	1.0010	1.0001	0.6985	0.9530	0.6970	0.4286
13.0	1.4517E 01	0.7296	1.1061	0.9969	0.9273	0.8200	0.9770	1.0009	1.0001	0.7190	0.9636	0.7177	0.4440
14.0	1.3633E 01	0.7158	1.1471	1.0436	0.9748	0.8127	0.9751	1.0008	1.0001	0.7391	0.9727	0.7379	0.4594
16.0	1.2151E 01	0.6902	1.2257	1.1348	1.0683	0.7998	0.9712	1.0007	1.0001	0.7781	0.9871	0.7772	0.4895
18.0	1.0957E 01	0.6671	1.2999	1.2233	1.1600	0.7888	0.9672	1.0006	1.0001	0.8157	0.9969	0.8149	0.5190
20.0	9.9732E 00	0.6460	1.3701	1.3094	1.2500	0.7796	0.9632	1.0005	1.0001	0.8519	1.0027	0.8512	0.5477
25.0	8.1381E 00	0.6011	1.5294	1.5144	1.4677	0.7627	0.9526	1.0004	1.0001	0.9367	1.0027	0.9362	0.6159
30.0	6.8690E 00	0.5648	1.6668	1.7054	1.6752	0.7531	0.9416	1.0003	1.0001	1.0134	0.9861	1.0131	0.6783
35.0	5.9428E 00	0.5349	1.7837	1.8824	1.8723	0.7490	0.9302	1.0002	1.0000	1.0823	0.9565	1.0821	0.7343
40.0	5.2403E 00	0.5102	1.8815	2.0454	2.0585	0.7493	0.9185	1.0002	1.0000	1.1433	0.9169	1.1431	0.7837
45.0	4.6916E 00	0.4895	1.9614	2.1945	2.2335	0.7532	0.9067	1.0001	1.0000	1.1965	0.8698	1.1964	0.8265
50.0	4.2530E 00	0.4721	2.0247	2.3295	2.3970	0.7600	0.8950	1.0001	1.0000	1.2423	0.8171	1.2422	0.8630

 Instr. see p. 22

ELECTRON Z= 98 A= 249

p	F_0L_0	L_0	λ_2	λ_3	λ_4	μ_1	μ_2	Λ_1	Λ_2	ν_{12}	$\hat{\nu}_{12}$	η_{12}	$\hat{\eta}_{12}$
0.1	9.7844E 02	0.9710	15.3709	130.7103	669.8683	0.9872	0.9998	1.0472	1.0015	6.4475	0.3890	-1.4043	-1.2621
0.2	4.9153E 02	0.9707	4.1423	10.8618	20.3686	0.9867	0.9997	1.0294	1.0014	1.6327	0.4150	-0.4619	-0.4636
0.3	3.3017E 02	0.9702	2.0653	3.1737	4.0536	0.9858	0.9996	1.0233	1.0014	0.8385	0.4400	-0.1491	-0.2013
0.4	2.5012E 02	0.9695	1.3410	1.5559	1.6665	0.9847	0.9995	1.0200	1.0014	0.5926	0.4636	0.0068	-0.0726
0.5	2.0250E 02	0.9685	1.0085	0.9975	0.9756	0.9834	0.9994	1.0178	1.0013	0.4934	0.4856	0.1002	0.0032
0.6	1.7103E 02	0.9674	0.8307	0.7465	0.6937	0.9818	0.9993	1.0162	1.0013	0.4474	0.5058	0.1625	0.0528
0.7	1.4874E 02	0.9661	0.7262	0.6145	0.5538	0.9801	0.9991	1.0149	1.0012	0.4246	0.5244	0.2073	0.0876
0.8	1.3214E 02	0.9646	0.6609	0.5379	0.4756	0.9783	0.9990	1.0138	1.0012	0.4132	0.5415	0.2411	0.1135
0.9	1.1931E 02	0.9630	0.6185	0.4907	0.4285	0.9764	0.9988	1.0128	1.0011	0.4079	0.5572	0.2677	0.1334
1.0	1.0910E 02	0.9613	0.5905	0.4604	0.3989	0.9745	0.9986	1.0120	1.0010	0.4060	0.5716	0.2893	0.1492
1.2	9.3855E 01	0.9574	0.5594	0.4278	0.3674	0.9705	0.9983	1.0106	1.0009	0.4076	0.5971	0.3227	0.1730
1.4	8.2979E 01	0.9532	0.5469	0.4149	0.3551	0.9664	0.9979	1.0095	1.0009	0.4125	0.6192	0.3476	0.1902
1.6	7.4787E 01	0.9487	0.5441	0.4119	0.3523	0.9624	0.9976	1.0085	1.0008	0.4188	0.6385	0.3673	0.2035
1.8	6.8360E 01	0.9440	0.5470	0.4144	0.3548	0.9583	0.9972	1.0077	1.0007	0.4256	0.6557	0.3837	0.2142
2.0	6.3156E 01	0.9392	0.5532	0.4202	0.3603	0.9544	0.9969	1.0070	1.0007	0.4326	0.6712	0.3977	0.2233
2.2	5.8836E 01	0.9343	0.5615	0.4281	0.3677	0.9505	0.9965	1.0065	1.0006	0.4396	0.6853	0.4100	0.2311
2.4	5.5178E 01	0.9293	0.5712	0.4372	0.3764	0.9466	0.9961	1.0060	1.0006	0.4465	0.6983	0.4211	0.2381
2.6	5.2029E 01	0.9243	0.5816	0.4472	0.3858	0.9429	0.9958	1.0055	1.0005	0.4532	0.7103	0.4312	0.2444
2.8	4.9282E 01	0.9193	0.5927	0.4578	0.3959	0.9392	0.9954	1.0052	1.0005	0.4599	0.7215	0.4405	0.2502
3.0	4.6858E 01	0.9144	0.6041	0.4688	0.4063	0.9355	0.9951	1.0048	1.0005	0.4664	0.7320	0.4492	0.2556
3.2	4.4699E 01	0.9094	0.6156	0.4800	0.4170	0.9320	0.9947	1.0045	1.0004	0.4727	0.7419	0.4574	0.2607
3.4	4.2760E 01	0.9045	0.6274	0.4914	0.4278	0.9285	0.9944	1.0043	1.0004	0.4790	0.7513	0.4652	0.2656
3.6	4.1006E 01	0.8996	0.6391	0.5029	0.4387	0.9251	0.9940	1.0040	1.0004	0.4851	0.7602	0.4726	0.2702
3.8	3.9410E 01	0.8947	0.6509	0.5144	0.4498	0.9217	0.9936	1.0038	1.0004	0.4911	0.7686	0.4798	0.2747
4.0	3.7949E 01	0.8899	0.6627	0.5259	0.4608	0.9184	0.9933	1.0036	1.0004	0.4970	0.7767	0.4866	0.2790
4.5	3.4779E 01	0.8781	0.6918	0.5548	0.4885	0.9104	0.9924	1.0032	1.0003	0.5114	0.7954	0.5029	0.2893
5.0	3.2146E 01	0.8666	0.7204	0.5834	0.5160	0.9028	0.9915	1.0029	1.0003	0.5252	0.8123	0.5182	0.2990
5.5	2.9914E 01	0.8555	0.7485	0.6117	0.5434	0.8955	0.9906	1.0026	1.0003	0.5386	0.8278	0.5326	0.3084
6.0	2.7994E 01	0.8446	0.7761	0.6396	0.5705	0.8886	0.9897	1.0023	1.0002	0.5516	0.8419	0.5465	0.3175
6.5	2.6320E 01	0.8341	0.8030	0.6672	0.5974	0.8819	0.9888	1.0021	1.0002	0.5643	0.8550	0.5598	0.3264
7.0	2.4845E 01	0.8239	0.8294	0.6944	0.6240	0.8755	0.9879	1.0020	1.0002	0.5767	0.8670	0.5728	0.3351
7.5	2.3534E 01	0.8141	0.8553	0.7213	0.6504	0.8694	0.9870	1.0018	1.0002	0.5888	0.8782	0.5853	0.3437
8.0	2.2359E 01	0.8045	0.8806	0.7479	0.6766	0.8636	0.9861	1.0017	1.0002	0.6007	0.8886	0.5976	0.3522
9.0	2.0339E 01	0.7862	0.9300	0.8002	0.7283	0.8526	0.9842	1.0015	1.0002	0.6240	0.9074	0.6214	0.3689
10.0	1.8660E 01	0.7689	0.9776	0.8514	0.7792	0.8426	0.9824	1.0013	1.0002	0.6465	0.9237	0.6444	0.3853
11.0	1.7239E 01	0.7526	1.0236	0.9016	0.8294	0.8333	0.9805	1.0012	1.0001	0.6685	0.9379	0.6667	0.4016
12.0	1.6020E 01	0.7373	1.0683	0.9509	0.8790	0.8247	0.9786	1.0010	1.0001	0.6900	0.9503	0.6884	0.4176
13.0	1.4962E 01	0.7227	1.1115	0.9994	0.9280	0.8169	0.9767	1.0009	1.0001	0.7110	0.9610	0.7096	0.4335
14.0	1.4034E 01	0.7090	1.1535	1.0471	0.9766	0.8097	0.9747	1.0009	1.0001	0.7315	0.9702	0.7303	0.4492
16.0	1.2482E 01	0.6835	1.2340	1.1404	1.0721	0.7969	0.9708	1.0007	1.0001	0.7715	0.9846	0.7705	0.4802
18.0	1.1234E 01	0.6605	1.3100	1.2310	1.1658	0.7861	0.9667	1.0006	1.0001	0.8099	0.9944	0.8092	0.5104
20.0	1.0209E 01	0.6397	1.3820	1.3192	1.2578	0.7771	0.9626	1.0005	1.0001	0.8470	1.0002	0.8464	0.5399
25.0	8.3039E 00	0.5952	1.5448	1.5290	1.4806	0.7609	0.9519	1.0004	1.0001	0.9339	1.0000	0.9334	0.6099
30.0	6.9918E 00	0.5594	1.6851	1.7245	1.6930	0.7518	0.9406	1.0003	1.0001	1.0125	0.9830	1.0122	0.6738
35.0	6.0376E 00	0.5301	1.8043	1.9056	1.8947	0.7484	0.9290	1.0002	1.0001	1.0829	0.9530	1.0827	0.7311
40.0	5.3157E 00	0.5058	1.9037	2.0724	2.0852	0.7493	0.9172	1.0002	1.0000	1.1453	0.9130	1.1451	0.7816
45.0	4.7532E 00	0.4855	1.9846	2.2247	2.2643	0.7538	0.9052	1.0002	1.0000	1.1996	0.8653	1.1995	0.8252
50.0	4.3043E 00	0.4684	2.0484	2.3626	2.4315	0.7612	0.8934	1.0001	1.0000	1.2463	0.8122	1.2462	0.8623

Instr. see p. 22

ELECTRON Z= 99 A= 253

p	F_0L_0	L_0	λ_2	λ_3	λ_4	μ_1	μ_2	Λ_1	Λ_2	ν_{12}	$\hat{\nu}_{12}$	η_{12}	$\hat{\eta}_{12}$
0.1	1.0484E 03	0.9661	15.3057	131.4220	681.8516	0.9863	0.9998	1.0512	1.0015	6.4453	0.3819	-1.4553	-1.3002
0.2	5.2654E 02	0.9658	4.1210	10.8755	20.5385	0.9858	0.9997	1.0315	1.0015	1.6178	0.4080	-0.4923	-0.4879
0.3	3.5358E 02	0.9652	2.0521	3.1630	4.0526	0.9849	0.9996	1.0248	1.0014	0.8232	0.4331	-0.1727	-0.2211
0.4	2.6773E 02	0.9645	1.3306	1.5445	1.6554	0.9837	0.9995	1.0212	1.0014	0.5778	0.4567	-0.0136	-0.0902
0.5	2.1665E 02	0.9635	0.9995	0.9871	0.9647	0.9823	0.9994	1.0188	1.0013	0.4790	0.4788	0.0818	-0.0132
0.6	1.8288E 02	0.9624	0.8224	0.7370	0.6837	0.9807	0.9993	1.0171	1.0013	0.4334	0.4991	0.1454	0.0371
0.7	1.5894E 02	0.9610	0.7183	0.6056	0.5446	0.9790	0.9991	1.0157	1.0012	0.4109	0.5178	0.1910	0.0725
0.8	1.4110E 02	0.9595	0.6534	0.5295	0.4670	0.9771	0.9990	1.0145	1.0012	0.3997	0.5350	0.2255	0.0987
0.9	1.2732E 02	0.9578	0.6112	0.4825	0.4203	0.9751	0.9988	1.0135	1.0011	0.3946	0.5508	0.2526	0.1188
1.0	1.1633E 02	0.9560	0.5833	0.4525	0.3909	0.9731	0.9986	1.0126	1.0011	0.3928	0.5653	0.2746	0.1349
1.2	9.9923E 01	0.9520	0.5525	0.4203	0.3599	0.9690	0.9983	1.0111	1.0010	0.3946	0.5911	0.3085	0.1589
1.4	8.8208E 01	0.9477	0.5402	0.4076	0.3478	0.9648	0.9979	1.0099	1.0009	0.3997	0.6133	0.3338	0.1762
1.6	7.9380E 01	0.9431	0.5378	0.4048	0.3452	0.9606	0.9975	1.0089	1.0008	0.4061	0.6329	0.3538	0.1896
1.8	7.2454E 01	0.9383	0.5409	0.4075	0.3478	0.9564	0.9972	1.0080	1.0007	0.4130	0.6503	0.3704	0.2004
2.0	6.6846E 01	0.9333	0.5473	0.4135	0.3534	0.9523	0.9968	1.0073	1.0007	0.4201	0.6659	0.3846	0.2095
2.2	6.2192E 01	0.9283	0.5559	0.4215	0.3610	0.9483	0.9964	1.0067	1.0006	0.4271	0.6802	0.3971	0.2174
2.4	5.8253E 01	0.9232	0.5658	0.4308	0.3698	0.9443	0.9961	1.0062	1.0006	0.4341	0.6934	0.4083	0.2244
2.6	5.4863E 01	0.9181	0.5765	0.4410	0.3794	0.9405	0.9957	1.0058	1.0005	0.4410	0.7055	0.4185	0.2308
2.8	5.1908E 01	0.9130	0.5878	0.4518	0.3895	0.9367	0.9953	1.0054	1.0005	0.4477	0.7169	0.4280	0.2367
3.0	4.9302E 01	0.9080	0.5995	0.4629	0.4001	0.9330	0.9950	1.0050	1.0005	0.4543	0.7276	0.4368	0.2421
3.2	4.6982E 01	0.9029	0.6113	0.4743	0.4109	0.9293	0.9946	1.0047	1.0005	0.4608	0.7376	0.4451	0.2473
3.4	4.4900E 01	0.8979	0.6233	0.4859	0.4219	0.9257	0.9942	1.0044	1.0004	0.4671	0.7471	0.4530	0.2521
3.6	4.3018E 01	0.8929	0.6353	0.4976	0.4330	0.9222	0.9939	1.0042	1.0004	0.4733	0.7561	0.4606	0.2568
3.8	4.1306E 01	0.8879	0.6474	0.5093	0.4442	0.9188	0.9935	1.0039	1.0004	0.4794	0.7647	0.4678	0.2613
4.0	3.9741E 01	0.8831	0.6594	0.5211	0.4554	0.9154	0.9931	1.0037	1.0004	0.4854	0.7728	0.4748	0.2657
4.5	3.6348E 01	0.8711	0.6893	0.5504	0.4834	0.9073	0.9922	1.0033	1.0003	0.5000	0.7918	0.4914	0.2762
5.0	3.3535E 01	0.8594	0.7186	0.5796	0.5115	0.8995	0.9913	1.0030	1.0003	0.5141	0.8089	0.5069	0.2861
5.5	3.1155E 01	0.8481	0.7474	0.6084	0.5393	0.8921	0.9904	1.0027	1.0003	0.5277	0.8245	0.5216	0.2956
6.0	2.9111E 01	0.8371	0.7756	0.6369	0.5669	0.8851	0.9895	1.0024	1.0003	0.5410	0.8389	0.5358	0.3049
6.5	2.7331E 01	0.8265	0.8033	0.6651	0.5943	0.8784	0.9886	1.0022	1.0002	0.5539	0.8521	0.5494	0.3140
7.0	2.5766E 01	0.8162	0.8304	0.6930	0.6215	0.8719	0.9877	1.0020	1.0002	0.5666	0.8643	0.5626	0.3229
7.5	2.4376E 01	0.8063	0.8569	0.7205	0.6484	0.8658	0.9867	1.0019	1.0002	0.5790	0.8756	0.5755	0.3317
8.0	2.3133E 01	0.7966	0.8830	0.7477	0.6751	0.8599	0.9858	1.0017	1.0002	0.5912	0.8861	0.5880	0.3404
9.0	2.0998E 01	0.7782	0.9337	0.8012	0.7279	0.8489	0.9839	1.0015	1.0002	0.6150	0.9050	0.6124	0.3575
10.0	1.9229E 01	0.7609	0.9826	0.8537	0.7800	0.8388	0.9820	1.0013	1.0002	0.6381	0.9215	0.6360	0.3745
11.0	1.7735E 01	0.7446	1.0299	0.9052	0.8314	0.8295	0.9801	1.0012	1.0001	0.6607	0.9358	0.6588	0.3912
12.0	1.6456E 01	0.7293	1.0758	0.9558	0.8822	0.8211	0.9782	1.0011	1.0001	0.6827	0.9482	0.6811	0.4077
13.0	1.5348E 01	0.7148	1.1203	1.0055	0.9325	0.8133	0.9762	1.0010	1.0001	0.7042	0.9589	0.7029	0.4241
14.0	1.4377E 01	0.7010	1.1634	1.0545	0.9822	0.8061	0.9742	1.0009	1.0001	0.7254	0.9681	0.7242	0.4403
16.0	1.2758E 01	0.6757	1.2461	1.1503	1.0803	0.7936	0.9701	1.0007	1.0001	0.7664	0.9825	0.7655	0.4722
18.0	1.1461E 01	0.6529	1.3242	1.2435	1.1765	0.7830	0.9660	1.0006	1.0001	0.8060	0.9922	0.8052	0.5034
20.0	1.0398E 01	0.6322	1.3980	1.3340	1.2709	0.7743	0.9618	1.0005	1.0001	0.8441	0.9978	0.8434	0.5338
25.0	8.4294E 00	0.5883	1.5649	1.5497	1.4998	0.7588	0.9508	1.0004	1.0001	0.9333	0.9969	0.9329	0.6058
30.0	7.0799E 00	0.5530	1.7082	1.7503	1.7180	0.7506	0.9393	1.0003	1.0001	1.0140	0.9791	1.0137	0.6714
35.0	6.1019E 00	0.5242	1.8295	1.9361	1.9251	0.7480	0.9275	1.0002	1.0001	1.0861	0.9481	1.0859	0.7301
40.0	5.3643E 00	0.5005	1.9302	2.1070	2.1207	0.7498	0.9153	1.0002	1.0000	1.1498	0.9070	1.1496	0.7816
45.0	4.7908E 00	0.4807	2.0117	2.2627	2.3043	0.7551	0.9032	1.0002	1.0000	1.2052	0.8584	1.2050	0.8259
50.0	4.3342E 00	0.4641	2.0755	2.4035	2.4755	0.7632	0.8911	1.0001	1.0000	1.2526	0.8044	1.2525	0.8635

Instr. see p. 22

ELECTRON Z= 100 A= 254

p	F_0L_0	L_0	λ_2	λ_3	λ_4	μ_1	μ_2	Λ_1	Λ_2	ν_{12}	$\hat{\nu}_{12}$	η_{12}	$\hat{\eta}_{12}$
0.1	1.1268E 03	0.9610	15.2103	131.7898	691.9707	0.9854	0.9997	1.0553	1.0015	6.4318	0.3744	-1.5052	-1.3369
0.2	5.6582E 02	0.9607	4.0917	10.8616	20.6520	0.9848	0.9997	1.0337	1.0015	1.6001	0.4005	-0.5225	-0.5116
0.3	3.7983E 02	0.9601	2.0350	3.1447	4.0409	0.9839	0.9996	1.0263	1.0015	0.8065	0.4256	-0.1964	-0.2406
0.4	2.8748E 02	0.9594	1.3179	1.5295	1.6402	0.9827	0.9995	1.0224	1.0014	0.5621	0.4493	-0.0341	-0.1077
0.5	2.3251E 02	0.9584	0.9887	0.9745	0.9515	0.9813	0.9994	1.0198	1.0014	0.4639	0.4714	0.0631	-0.0296
0.6	1.9615E 02	0.9572	0.8127	0.7258	0.6722	0.9796	0.9993	1.0179	1.0013	0.4188	0.4919	0.1280	0.0215
0.7	1.7036E 02	0.9558	0.7093	0.5954	0.5343	0.9778	0.9991	1.0164	1.0012	0.3966	0.5107	0.1744	0.0574
0.8	1.5114E 02	0.9542	0.6448	0.5200	0.4574	0.9759	0.9989	1.0152	1.0012	0.3857	0.5279	0.2095	0.0839
0.9	1.3627E 02	0.9525	0.6029	0.4735	0.4112	0.9738	0.9988	1.0141	1.0011	0.3807	0.5438	0.2371	0.1042
1.0	1.2442E 02	0.9506	0.5753	0.4438	0.3822	0.9717	0.9986	1.0131	1.0011	0.3791	0.5584	0.2595	0.1204
1.2	1.0671E 02	0.9466	0.5448	0.4120	0.3517	0.9675	0.9982	1.0115	1.0010	0.3811	0.5844	0.2939	0.1447
1.4	9.4050E 01	0.9421	0.5328	0.3996	0.3399	0.9631	0.9979	1.0103	1.0009	0.3863	0.6069	0.3195	0.1622
1.6	8.4511E 01	0.9374	0.5306	0.3970	0.3374	0.9588	0.9975	1.0092	1.0008	0.3928	0.6266	0.3398	0.1756
1.8	7.7025E 01	0.9325	0.5339	0.3999	0.3401	0.9546	0.9971	1.0083	1.0007	0.3998	0.6442	0.3567	0.1865
2.0	7.0965E 01	0.9274	0.5406	0.4060	0.3459	0.9504	0.9967	1.0076	1.0007	0.4070	0.6600	0.3710	0.1956
2.2	6.5936E 01	0.9223	0.5494	0.4141	0.3535	0.9462	0.9964	1.0070	1.0006	0.4141	0.6745	0.3837	0.2035
2.4	6.1682E 01	0.9171	0.5595	0.4236	0.3624	0.9422	0.9960	1.0064	1.0006	0.4212	0.6877	0.3950	0.2106
2.6	5.8024E 01	0.9119	0.5705	0.4340	0.3721	0.9382	0.9956	1.0060	1.0005	0.4281	0.7001	0.4053	0.2169
2.8	5.4836E 01	0.9067	0.5820	0.4449	0.3824	0.9343	0.9953	1.0055	1.0005	0.4349	0.7116	0.4149	0.2228
3.0	5.2027E 01	0.9016	0.5939	0.4562	0.3931	0.9305	0.9949	1.0052	1.0005	0.4416	0.7223	0.4238	0.2283
3.2	4.9528E 01	0.8964	0.6060	0.4677	0.4040	0.9268	0.9945	1.0049	1.0005	0.4481	0.7325	0.4322	0.2335
3.4	4.7287E 01	0.8913	0.6183	0.4795	0.4151	0.9231	0.9941	1.0046	1.0004	0.4545	0.7421	0.4402	0.2384
3.6	4.5262E 01	0.8862	0.6306	0.4913	0.4263	0.9195	0.9938	1.0043	1.0004	0.4608	0.7512	0.4479	0.2431
3.8	4.3422E 01	0.8812	0.6429	0.5032	0.4376	0.9160	0.9934	1.0041	1.0004	0.4670	0.7599	0.4552	0.2477
4.0	4.1741E 01	0.8763	0.6552	0.5151	0.4489	0.9126	0.9930	1.0039	1.0004	0.4731	0.7682	0.4623	0.2521
4.5	3.8102E 01	0.8641	0.6856	0.5449	0.4774	0.9044	0.9921	1.0034	1.0003	0.4879	0.7874	0.4791	0.2626
5.0	3.5089E 01	0.8523	0.7156	0.5746	0.5057	0.8965	0.9912	1.0030	1.0003	0.5022	0.8047	0.4949	0.2727
5.5	3.2546E 01	0.8409	0.7451	0.6039	0.5340	0.8890	0.9903	1.0027	1.0003	0.5160	0.8205	0.5098	0.2824
6.0	3.0364E 01	0.8299	0.7740	0.6330	0.5620	0.8819	0.9893	1.0025	1.0003	0.5295	0.8350	0.5242	0.2918
6.5	2.8468E 01	0.8192	0.8023	0.6617	0.5898	0.8751	0.9884	1.0023	1.0002	0.5427	0.8484	0.5381	0.3011
7.0	2.6802E 01	0.8089	0.8300	0.6900	0.6174	0.8686	0.9875	1.0021	1.0002	0.5556	0.8607	0.5515	0.3102
7.5	2.5326E 01	0.7989	0.8572	0.7181	0.6448	0.8624	0.9865	1.0019	1.0002	0.5683	0.8722	0.5647	0.3191
8.0	2.4007E 01	0.7892	0.8839	0.7458	0.6720	0.8565	0.9856	1.0018	1.0002	0.5807	0.8828	0.5775	0.3280
9.0	2.1747E 01	0.7707	0.9358	0.8005	0.7258	0.8455	0.9837	1.0016	1.0002	0.6050	0.9019	0.6024	0.3456
10.0	1.9878E 01	0.7534	0.9859	0.8541	0.7789	0.8354	0.9817	1.0014	1.0002	0.6286	0.9185	0.6265	0.3629
11.0	1.8304E 01	0.7372	1.0345	0.9067	0.8314	0.8262	0.9798	1.0012	1.0001	0.6517	0.9329	0.6498	0.3801
12.0	1.6958E 01	0.7218	1.0815	0.9585	0.8833	0.8178	0.9778	1.0011	1.0001	0.6742	0.9454	0.6726	0.3970
13.0	1.5795E 01	0.7074	1.1271	1.0094	0.9346	0.8101	0.9758	1.0010	1.0001	0.6963	0.9562	0.6949	0.4138
14.0	1.4778E 01	0.6937	1.1713	1.0595	0.9854	0.8030	0.9738	1.0009	1.0001	0.7179	0.9655	0.7167	0.4305
16.0	1.3085E 01	0.6686	1.2561	1.1577	1.0856	0.7907	0.9696	1.0007	1.0001	0.7600	0.9799	0.7590	0.4633
18.0	1.1733E 01	0.6460	1.3361	1.2530	1.1841	0.7804	0.9654	1.0006	1.0001	0.8005	0.9896	0.7997	0.4953
20.0	1.0628E 01	0.6255	1.4117	1.3458	1.2808	0.7719	0.9611	1.0005	1.0001	0.8396	0.9952	0.8389	0.5265
25.0	8.5875E 00	0.5822	1.5824	1.5667	1.5151	0.7572	0.9499	1.0004	1.0001	0.9311	0.9940	0.9306	0.6004
30.0	7.1951E 00	0.5475	1.7288	1.7723	1.7386	0.7497	0.9382	1.0003	1.0001	1.0137	0.9757	1.0134	0.6677
35.0	6.1895E 00	0.5192	1.8522	1.9625	1.9508	0.7478	0.9262	1.0002	1.0001	1.0875	0.9442	1.0873	0.7277
40.0	5.4331E 00	0.4959	1.9544	2.1372	2.1510	0.7503	0.9139	1.0002	1.0000	1.1526	0.9026	1.1524	0.7802
45.0	4.8464E 00	0.4766	2.0367	2.2964	2.3389	0.7564	0.9015	1.0002	1.0000	1.2090	0.8534	1.2089	0.8253
50.0	4.3802E 00	0.4605	2.1009	2.4400	2.5141	0.7652	0.8893	1.0001	1.0000	1.2573	0.7989	1.2571	0.8635

Instr. see p. 22

ELECTRON Z= 101 A= 254

p	F_0L_0	L_0	λ_2	λ_3	λ_4	μ_1	μ_2	Λ_1	Λ_2	ν_{12}	$\hat{\nu}_{12}$	η_{12}	$\hat{\eta}_{12}$
0.1	1.2128E 03	0.9556	15.0998	131.9685	701.0734	0.9845	0.9997	1.0597	1.0015	6.4104	0.3668	-1.5547	-1.3726
0.2	6.0891E 02	0.9553	4.0585	10.8333	20.7360	0.9839	0.9997	1.0360	1.0015	1.5803	0.3929	-0.5526	-0.5348
0.3	4.0862E 02	0.9547	2.0162	3.1226	4.0239	0.9829	0.9996	1.0279	1.0015	0.7889	0.4180	-0.2201	-0.2598
0.4	3.0914E 02	0.9539	1.3040	1.5129	1.6231	0.9817	0.9995	1.0237	1.0014	0.5457	0.4417	-0.0546	-0.1250
0.5	2.4989E 02	0.9529	0.9772	0.9610	0.9374	0.9802	0.9994	1.0209	1.0014	0.4484	0.4639	0.0444	-0.0457
0.6	2.1068E 02	0.9517	0.8025	0.7141	0.6602	0.9785	0.9992	1.0188	1.0013	0.4037	0.4844	0.1104	0.0060
0.7	1.8286E 02	0.9503	0.6998	0.5849	0.5235	0.9766	0.9991	1.0172	1.0013	0.3820	0.5033	0.1577	0.0423
0.8	1.6212E 02	0.9487	0.6358	0.5101	0.4475	0.9746	0.9989	1.0159	1.0012	0.3713	0.5206	0.1934	0.0691
0.9	1.4606E 02	0.9469	0.5943	0.4641	0.4019	0.9725	0.9987	1.0147	1.0011	0.3665	0.5366	0.2214	0.0897
1.0	1.3326E 02	0.9450	0.5669	0.4348	0.3733	0.9704	0.9986	1.0137	1.0011	0.3651	0.5512	0.2441	0.1060
1.2	1.1410E 02	0.9408	0.5368	0.4035	0.3433	0.9659	0.9982	1.0120	1.0010	0.3673	0.5774	0.2790	0.1305
1.4	1.0041E 02	0.9362	0.5251	0.3914	0.3318	0.9615	0.9978	1.0107	1.0009	0.3727	0.6001	0.3050	0.1480
1.6	9.0089E 01	0.9314	0.5231	0.3890	0.3295	0.9571	0.9974	1.0096	1.0008	0.3793	0.6200	0.3256	0.1616
1.8	8.1988E 01	0.9264	0.5267	0.3920	0.3324	0.9527	0.9971	1.0087	1.0007	0.3864	0.6377	0.3426	0.1725
2.0	7.5431E 01	0.9212	0.5336	0.3983	0.3382	0.9484	0.9967	1.0079	1.0007	0.3936	0.6537	0.3572	0.1816
2.2	6.9992E 01	0.9160	0.5427	0.4066	0.3460	0.9441	0.9963	1.0072	1.0006	0.4008	0.6683	0.3699	0.1896
2.4	6.5393E 01	0.9107	0.5530	0.4162	0.3549	0.9400	0.9959	1.0067	1.0006	0.4079	0.6818	0.3814	0.1966
2.6	6.1440E 01	0.9054	0.5642	0.4267	0.3647	0.9359	0.9956	1.0062	1.0006	0.4149	0.6942	0.3918	0.2030
2.8	5.7998E 01	0.9002	0.5760	0.4377	0.3751	0.9319	0.9952	1.0057	1.0005	0.4218	0.7058	0.4015	0.2089
3.0	5.4967E 01	0.8949	0.5881	0.4492	0.3858	0.9280	0.9948	1.0054	1.0005	0.4285	0.7168	0.4105	0.2144
3.2	5.2272E 01	0.8897	0.6005	0.4609	0.3969	0.9242	0.9944	1.0050	1.0005	0.4351	0.7270	0.4190	0.2196
3.4	4.9858E 01	0.8845	0.6129	0.4728	0.4081	0.9205	0.9941	1.0047	1.0004	0.4416	0.7368	0.4271	0.2246
3.6	4.7678E 01	0.8794	0.6255	0.4848	0.4194	0.9169	0.9937	1.0044	1.0004	0.4480	0.7460	0.4348	0.2293
3.8	4.5699E 01	0.8743	0.6380	0.4968	0.4308	0.9133	0.9933	1.0042	1.0004	0.4542	0.7548	0.4423	0.2339
4.0	4.3891E 01	0.8693	0.6506	0.5089	0.4422	0.9098	0.9929	1.0040	1.0004	0.4604	0.7632	0.4495	0.2384
4.5	3.9984E 01	0.8570	0.6817	0.5392	0.4710	0.9015	0.9920	1.0035	1.0003	0.4754	0.7826	0.4665	0.2490
5.0	3.6756E 01	0.8451	0.7123	0.5692	0.4997	0.8935	0.9911	1.0031	1.0003	0.4899	0.8001	0.4824	0.2592
5.5	3.4035E 01	0.8336	0.7424	0.5991	0.5283	0.8859	0.9901	1.0028	1.0003	0.5039	0.8161	0.4977	0.2690
6.0	3.1704E 01	0.8225	0.7719	0.6286	0.5567	0.8787	0.9892	1.0026	1.0003	0.5177	0.8308	0.5123	0.2786
6.5	2.9683E 01	0.8118	0.8009	0.6578	0.5850	0.8719	0.9882	1.0023	1.0002	0.5311	0.8443	0.5264	0.2880
7.0	2.7910E 01	0.8014	0.8292	0.6867	0.6130	0.8654	0.9873	1.0021	1.0002	0.5442	0.8568	0.5401	0.2973
7.5	2.6340E 01	0.7914	0.8570	0.7153	0.6408	0.8592	0.9863	1.0020	1.0002	0.5571	0.8683	0.5534	0.3064
8.0	2.4940E 01	0.7817	0.8844	0.7435	0.6685	0.8533	0.9853	1.0018	1.0002	0.5698	0.8791	0.5665	0.3155
9.0	2.2546E 01	0.7632	0.9375	0.7993	0.7232	0.8423	0.9834	1.0016	1.0002	0.5945	0.8984	0.5919	0.3334
10.0	2.0570E 01	0.7459	0.9889	0.8540	0.7773	0.8322	0.9814	1.0014	1.0002	0.6187	0.9152	0.6164	0.3512
11.0	1.8910E 01	0.7297	1.0386	0.9077	0.8307	0.8231	0.9795	1.0012	1.0001	0.6422	0.9297	0.6403	0.3687
12.0	1.7494E 01	0.7145	1.0867	0.9606	0.8836	0.8147	0.9774	1.0011	1.0001	0.6653	0.9423	0.6636	0.3861
13.0	1.6272E 01	0.7001	1.1334	1.0126	0.9360	0.8071	0.9754	1.0010	1.0001	0.6878	0.9532	0.6864	0.4033
14.0	1.5206E 01	0.6865	1.1788	1.0639	0.9879	0.8001	0.9734	1.0009	1.0001	0.7100	0.9626	0.7087	0.4204
16.0	1.3435E 01	0.6616	1.2656	1.1642	1.0902	0.7880	0.9692	1.0008	1.0001	0.7530	0.9771	0.7520	0.4540
18.0	1.2024E 01	0.6392	1.3475	1.2618	1.1908	0.7780	0.9649	1.0006	1.0001	0.7945	0.9869	0.7937	0.4869
20.0	1.0874E 01	0.6190	1.4249	1.3568	1.2897	0.7698	0.9605	1.0006	1.0001	0.8345	0.9924	0.8339	0.5189
25.0	8.7584E 00	0.5763	1.5994	1.5829	1.5294	0.7558	0.9491	1.0004	1.0001	0.9282	0.9911	0.9278	0.5947
30.0	7.3206E 00	0.5422	1.7488	1.7933	1.7581	0.7491	0.9372	1.0003	1.0001	1.0128	0.9726	1.0125	0.6635
35.0	6.2857E 00	0.5145	1.8745	1.9879	1.9752	0.7479	0.9250	1.0002	1.0001	1.0883	0.9407	1.0880	0.7248
40.0	5.5093E 00	0.4917	1.9782	2.1665	2.1801	0.7511	0.9125	1.0002	1.0000	1.1547	0.8986	1.1545	0.7784
45.0	4.9084E 00	0.4729	2.0614	2.3291	2.3722	0.7578	0.9000	1.0002	1.0000	1.2123	0.8490	1.2122	0.8243
50.0	4.4317E 00	0.4571	2.1260	2.4756	2.5513	0.7674	0.8876	1.0001	1.0000	1.2614	0.7940	1.2613	0.8630

Instr. see p. 22

ELECTRON Z= 102 A= 256

p	F_0L_0	L_0	λ_2	λ_3	λ_4	μ_1	μ_2	Λ_1	Λ_2	ν_{12}	$\hat{\nu}_{12}$	η_{12}	$\hat{\eta}_{12}$
0.1	1.3038E 03	0.9498	14.9985	132.2233	710.6668	0.9834	0.9997	1.0647	1.0016	6.3868	0.3593	-1.6049	-1.4083
0.2	6.5445E 02	0.9494	4.0280	10.8125	20.8354	0.9828	0.9997	1.0387	1.0015	1.5600	0.3853	-0.5829	-0.5580
0.3	4.3903E 02	0.9488	1.9988	3.1032	4.0104	0.9818	0.9996	1.0298	1.0015	0.7710	0.4104	-0.2440	-0.2789
0.4	3.3199E 02	0.9480	1.2912	1.4978	1.6078	0.9805	0.9995	1.0251	1.0014	0.5293	0.4342	-0.0753	-0.1421
0.5	2.6822E 02	0.9469	0.9665	0.9485	0.9244	0.9790	0.9994	1.0221	1.0014	0.4328	0.4564	0.0257	-0.0618
0.6	2.2599E 02	0.9457	0.7930	0.7033	0.6490	0.9772	0.9992	1.0198	1.0013	0.3887	0.4770	0.0929	-0.0094
0.7	1.9602E 02	0.9442	0.6910	0.5751	0.5135	0.9753	0.9991	1.0181	1.0013	0.3673	0.4960	0.1410	0.0274
0.8	1.7365E 02	0.9425	0.6274	0.5010	0.4383	0.9732	0.9989	1.0166	1.0012	0.3570	0.5134	0.1773	0.0545
0.9	1.5633E 02	0.9407	0.5863	0.4555	0.3932	0.9710	0.9987	1.0154	1.0012	0.3524	0.5294	0.2058	0.0753
1.0	1.4252E 02	0.9388	0.5591	0.4265	0.3650	0.9688	0.9985	1.0143	1.0011	0.3511	0.5442	0.2289	0.0918
1.2	1.2183E 02	0.9345	0.5295	0.3956	0.3355	0.9643	0.9982	1.0126	1.0010	0.3536	0.5706	0.2643	0.1164
1.4	1.0704E 02	0.9298	0.5181	0.3838	0.3243	0.9597	0.9978	1.0111	1.0009	0.3591	0.5934	0.2906	0.1341
1.6	9.5887E 01	0.9248	0.5163	0.3816	0.3222	0.9551	0.9974	1.0100	1.0008	0.3658	0.6135	0.3114	0.1477
1.8	8.7133E 01	0.9197	0.5202	0.3848	0.3252	0.9506	0.9970	1.0090	1.0008	0.3730	0.6314	0.3286	0.1586
2.0	8.0048E 01	0.9144	0.5273	0.3913	0.3312	0.9461	0.9966	1.0082	1.0007	0.3803	0.6476	0.3433	0.1678
2.2	7.4174E 01	0.9091	0.5366	0.3997	0.3390	0.9418	0.9962	1.0075	1.0006	0.3876	0.6624	0.3563	0.1758
2.4	6.9209E 01	0.9037	0.5472	0.4095	0.3481	0.9375	0.9959	1.0069	1.0006	0.3948	0.6760	0.3678	0.1828
2.6	6.4945E 01	0.8984	0.5586	0.4201	0.3580	0.9334	0.9955	1.0064	1.0006	0.4018	0.6886	0.3784	0.1892
2.8	6.1234E 01	0.8930	0.5707	0.4313	0.3685	0.9293	0.9951	1.0059	1.0005	0.4088	0.7003	0.3882	0.1952
3.0	5.7969E 01	0.8877	0.5831	0.4429	0.3793	0.9254	0.9947	1.0055	1.0005	0.4156	0.7114	0.3973	0.2007
3.2	5.5068E 01	0.8824	0.5957	0.4548	0.3905	0.9215	0.9943	1.0052	1.0005	0.4223	0.7218	0.4059	0.2059
3.4	5.2470E 01	0.8771	0.6084	0.4669	0.4018	0.9177	0.9940	1.0049	1.0004	0.4288	0.7316	0.4141	0.2109
3.6	5.0128E 01	0.8719	0.6213	0.4790	0.4132	0.9140	0.9936	1.0046	1.0004	0.4353	0.7410	0.4219	0.2157
3.8	4.8002E 01	0.8668	0.6341	0.4913	0.4248	0.9103	0.9932	1.0043	1.0004	0.4416	0.7499	0.4295	0.2203
4.0	4.6062E 01	0.8617	0.6469	0.5036	0.4364	0.9068	0.9928	1.0041	1.0004	0.4478	0.7584	0.4367	0.2248
4.5	4.1875E 01	0.8493	0.6787	0.5343	0.4655	0.8983	0.9919	1.0036	1.0003	0.4630	0.7780	0.4540	0.2356
5.0	3.8422E 01	0.8373	0.7100	0.5649	0.4946	0.8902	0.9909	1.0032	1.0003	0.4778	0.7958	0.4702	0.2459
5.5	3.5516E 01	0.8257	0.7408	0.5953	0.5237	0.8826	0.9899	1.0029	1.0003	0.4921	0.8120	0.4857	0.2559
6.0	3.3033E 01	0.8145	0.7710	0.6253	0.5525	0.8753	0.9890	1.0026	1.0003	0.5060	0.8268	0.5005	0.2656
6.5	3.0882E 01	0.8037	0.8007	0.6551	0.5812	0.8685	0.9880	1.0024	1.0002	0.5197	0.8405	0.5149	0.2752
7.0	2.8998E 01	0.7933	0.8297	0.6846	0.6097	0.8619	0.9870	1.0022	1.0002	0.5331	0.8531	0.5289	0.2847
7.5	2.7333E 01	0.7832	0.8582	0.7137	0.6381	0.8557	0.9861	1.0020	1.0002	0.5462	0.8648	0.5425	0.2940
8.0	2.5850E 01	0.7735	0.8862	0.7426	0.6662	0.8498	0.9851	1.0019	1.0002	0.5592	0.8757	0.5558	0.3033
9.0	2.3319E 01	0.7551	0.9407	0.7996	0.7220	0.8388	0.9831	1.0016	1.0002	0.5845	0.8952	0.5818	0.3217
10.0	2.1235E 01	0.7378	0.9934	0.8555	0.7772	0.8288	0.9811	1.0014	1.0002	0.6091	0.9121	0.6069	0.3399
11.0	1.9489E 01	0.7216	1.0444	0.9105	0.8318	0.8197	0.9791	1.0013	1.0001	0.6333	0.9268	0.6313	0.3579
12.0	1.8002E 01	0.7065	1.0938	0.9646	0.8859	0.8114	0.9770	1.0011	1.0001	0.6568	0.9395	0.6552	0.3757
13.0	1.6722E 01	0.6922	1.1417	1.0179	0.9394	0.8039	0.9750	1.0010	1.0001	0.6800	0.9505	0.6786	0.3934
14.0	1.5606E 01	0.6787	1.1882	1.0704	0.9925	0.7971	0.9729	1.0009	1.0001	0.7027	0.9599	0.7014	0.4110
16.0	1.3758E 01	0.6540	1.2773	1.1733	1.0973	0.7852	0.9686	1.0008	1.0001	0.7468	0.9745	0.7458	0.4455
18.0	1.2290E 01	0.6319	1.3613	1.2733	1.2003	0.7755	0.9642	1.0007	1.0001	0.7894	0.9842	0.7886	0.4793
20.0	1.1096E 01	0.6120	1.4405	1.3706	1.3016	0.7676	0.9598	1.0006	1.0001	0.8304	0.9897	0.8298	0.5122
25.0	8.9084E 00	0.5699	1.6191	1.6025	1.5473	0.7544	0.9482	1.0004	1.0001	0.9265	0.9881	0.9260	0.5899
30.0	7.4280E 00	0.5364	1.7715	1.8182	1.7817	0.7486	0.9361	1.0003	1.0001	1.0132	0.9689	1.0128	0.6605
35.0	6.3660E 00	0.5094	1.8994	2.0174	2.0042	0.7483	0.9236	1.0002	1.0001	1.0904	0.9364	1.0901	0.7231
40.0	5.5715E 00	0.4872	2.0045	2.2002	2.2141	0.7523	0.9109	1.0002	1.0000	1.1582	0.8936	1.1580	0.7777
45.0	4.9579E 00	0.4688	2.0884	2.3663	2.4108	0.7598	0.8981	1.0002	1.0000	1.2169	0.8433	1.2168	0.8244
50.0	4.4722E 00	0.4535	2.1531	2.5157	2.5940	0.7701	0.8855	1.0001	1.0000	1.2669	0.7878	1.2667	0.8636

Instr. see p. 22

POSITRON Z= -1 A= 1

p	F_0L_0	L_0	λ_2	λ_3	λ_4	μ_1	μ_2	Λ_1	Λ_2	ν_{12}	$\hat{\nu}_{12}$	η_{12}	$\hat{\eta}_{12}$
0.1	7.8770E-01	1.0001	1.0051	1.0064	1.0069	1.0000	1.0000	1.0000	1.0000	1.0012	0.7481	0.9999	0.9998
0.2	8.8812E-01	1.0000	1.0011	1.0014	1.0015	1.0000	1.0000	1.0000	1.0000	1.0002	0.7454	0.9999	0.9999
0.3	9.2281E-01	1.0000	1.0004	1.0005	1.0005	1.0000	1.0000	1.0000	1.0000	1.0000	0.7412	0.9999	0.9999
0.4	9.4001E-01	1.0000	1.0001	1.0002	1.0002	1.0000	1.0000	1.0000	1.0000	1.0000	0.7357	0.9999	0.9999
0.5	9.5007E-01	1.0000	1.0000	1.0000	1.0000	1.0000	1.0000	1.0000	1.0000	0.9999	0.7290	0.9999	0.9999
0.6	9.5656E-01	1.0000	1.0000	0.9999	0.9999	1.0000	1.0000	1.0000	1.0000	0.9999	0.7218	0.9999	0.9999
0.7	9.6100E-01	1.0000	0.9999	0.9999	0.9999	1.0000	1.0000	1.0000	1.0000	0.9999	0.7129	0.9999	0.9999
0.8	9.6419E-01	1.0000	0.9999	0.9999	0.9999	1.0000	1.0000	1.0000	1.0000	0.9999	0.7064	0.9999	0.9999
0.9	9.6656E-01	1.0000	0.9999	0.9999	0.9998	1.0000	1.0000	1.0000	1.0000	0.9999	0.6981	0.9999	0.9999
1.0	9.6836E-01	1.0001	0.9999	0.9998	0.9998	1.0000	1.0000	1.0000	1.0000	0.9999	0.6871	0.9999	0.9999
1.2	9.7088E-01	1.0001	0.9999	0.9998	0.9998	1.0000	1.0000	1.0000	1.0000	0.9999	0.6714	0.9999	0.9999
1.4	9.7251E-01	1.0001	0.9999	0.9998	0.9998	1.0000	1.0000	1.0000	1.0000	0.9999	0.6517	0.9999	0.9999
1.6	9.7363E-01	1.0001	0.9999	0.9998	0.9998	1.0000	1.0000	1.0000	1.0000	0.9999	0.6305	0.9999	0.9999
1.8	9.7442E-01	1.0001	0.9999	0.9998	0.9998	1.0000	1.0000	1.0000	1.0000	0.9999	0.6162	0.9999	0.9999
2.0	9.7500E-01	1.0001	0.9999	0.9998	0.9998	1.0000	1.0000	1.0000	1.0000	0.9999	0.5990	0.9999	0.9999
2.2	9.7544E-01	1.0001	0.9999	0.9998	0.9998	1.0000	1.0000	1.0000	1.0000	0.9999	0.5719	0.9999	0.9999
2.4	9.7578E-01	1.0001	0.9999	0.9998	0.9998	1.0000	1.0000	1.0000	1.0000	0.9999	0.5561	0.9999	0.9999
2.6	9.7605E-01	1.0001	0.9999	0.9998	0.9998	1.0000	1.0000	1.0000	1.0000	0.9999	0.5380	0.9999	0.9999
2.8	9.7627E-01	1.0001	0.9999	0.9998	0.9998	1.0000	1.0000	1.0000	1.0000	0.9999	0.5262	0.9999	0.9999
3.0	9.7644E-01	1.0001	0.9999	0.9998	0.9998	1.0000	1.0000	1.0000	1.0000	0.9999	0.5049	0.9999	0.9999
3.2	9.7659E-01	1.0001	0.9998	0.9998	0.9998	1.0000	1.0000	1.0000	1.0000	0.9999	0.5012	0.9999	0.9999
3.4	9.7671E-01	1.0001	0.9998	0.9998	0.9998	1.0000	1.0000	1.0000	1.0000	0.9999	0.4750	0.9999	0.9999
3.6	9.7681E-01	1.0001	0.9998	0.9998	0.9998	1.0000	1.0000	1.0000	1.0000	0.9999	0.4513	0.9999	0.9999
3.8	9.7690E-01	1.0001	0.9998	0.9998	0.9998	1.0000	1.0000	1.0000	1.0000	0.9999	0.4300	0.9999	0.9999
4.0	9.7697E-01	1.0001	0.9998	0.9998	0.9998	1.0000	1.0000	1.0000	1.0000	0.9999	0.4157	0.9999	0.9999
4.5	9.7712E-01	1.0001	0.9998	0.9998	0.9998	1.0000	1.0000	1.0000	1.0000	0.9999	0.3737	0.9999	0.9999
5.0	9.7723E-01	1.0001	0.9998	0.9998	0.9998	1.0001	1.0000	1.0000	1.0000	0.9999	0.3305	0.9999	0.9999
5.5	9.7731E-01	1.0001	0.9998	0.9998	0.9997	1.0001	1.0000	1.0000	1.0000	0.9999	0.2830	0.9999	0.9999
6.0	9.7738E-01	1.0002	0.9998	0.9998	0.9997	1.0001	1.0000	1.0000	1.0000	0.9999	0.2434	0.9999	0.9999
6.5	9.7743E-01	1.0002	0.9998	0.9998	0.9997	1.0001	1.0000	1.0000	1.0000	0.9999	0.2084	0.9999	0.9999
7.0	9.7748E-01	1.0002	0.9998	0.9998	0.9997	1.0001	1.0000	1.0000	1.0000	0.9999	0.1523	0.9999	0.9999
7.5	9.7751E-01	1.0002	0.9998	0.9998	0.9997	1.0001	1.0000	1.0000	1.0000	0.9999	0.1217	0.9999	0.9999
8.0	9.7754E-01	1.0002	0.9998	0.9997	0.9997	1.0001	1.0000	1.0000	1.0000	0.9999	0.0684	0.9999	0.9999
9.0	9.7760E-01	1.0002	0.9998	0.9997	0.9997	1.0001	1.0000	1.0000	1.0000	0.9999	-0.0064	0.9999	0.9999
10.0	9.7764E-01	1.0002	0.9998	0.9997	0.9997	1.0001	1.0000	1.0000	1.0000	0.9999	-0.1008	0.9999	0.9999
11.0	9.7768E-01	1.0003	0.9998	0.9997	0.9997	1.0001	1.0000	1.0000	1.0000	0.9999	-0.1834	0.9999	0.9999
12.0	9.7771E-01	1.0003	0.9998	0.9997	0.9997	1.0001	1.0000	1.0000	1.0000	0.9999	-0.2481	0.9999	0.9999
13.0	9.7774E-01	1.0003	0.9998	0.9997	0.9996	1.0001	1.0000	1.0000	1.0000	0.9999	-0.3546	0.9999	0.9999
14.0	9.7776E-01	1.0003	0.9997	0.9997	0.9996	1.0002	1.0000	1.0000	1.0000	0.9999	-0.4253	0.9999	0.9999
16.0	9.7781E-01	1.0004	0.9997	0.9996	0.9996	1.0002	1.0000	1.0000	1.0000	0.9998	-0.5942	0.9998	0.9999
18.0	9.7786E-01	1.0004	0.9997	0.9996	0.9996	1.0002	1.0000	1.0000	1.0000	0.9998	-0.7629	0.9998	0.9999
20.0	9.7791E-01	1.0005	0.9997	0.9996	0.9995	1.0002	1.0000	1.0000	1.0000	0.9998	-0.9458	0.9998	0.9999
25.0	9.7801E-01	1.0006	0.9996	0.9995	0.9995	1.0003	1.0001	1.0000	1.0000	0.9998	-1.3518	0.9998	0.9999
30.0	9.7812E-01	1.0007	0.9995	0.9994	0.9994	1.0003	1.0001	1.0000	1.0000	0.9998	-1.8176	0.9998	0.9999
35.0	9.7823E-01	1.0008	0.9995	0.9994	0.9993	1.0004	1.0001	1.0000	1.0000	0.9997	-2.1666	0.9997	0.9999
40.0	9.7833E-01	1.0009	0.9994	0.9993	0.9992	1.0004	1.0001	1.0000	1.0000	0.9997	-2.5674	0.9997	0.9999
45.0	9.7843E-01	1.0010	0.9994	0.9992	0.9991	1.0005	1.0001	1.0000	1.0000	0.9997	-2.8790	0.9997	0.9999
50.0	9.7854E-01	1.0011	0.9993	0.9991	0.9990	1.0006	1.0001	1.0000	1.0000	0.9996	-3.4929	0.9996	0.9999

POSITRON Z= −2 A= 3

p	F_0L_0	L_0	λ_2	λ_3	λ_4	μ_1	μ_2	Λ_1	Λ_2	ν_{12}	$\hat{\nu}_{12}$	η_{12}	$\hat{\eta}_{12}$
0.1	6.1033E-01	1.0002	1.0204	1.0256	1.0279	1.0000	1.0000	1.0000	1.0000	1.0048	0.7486	0.9995	0.9994
0.2	7.8594E-01	1.0001	1.0046	1.0056	1.0061	1.0000	1.0000	1.0000	1.0000	1.0009	0.7464	0.9996	0.9995
0.3	8.5054E-01	1.0001	1.0016	1.0020	1.0021	1.0000	1.0000	1.0000	1.0000	1.0002	0.7431	0.9996	0.9995
0.4	8.8326E-01	1.0001	1.0006	1.0007	1.0008	1.0000	1.0000	1.0000	1.0000	0.9999	0.7389	0.9996	0.9995
0.5	9.0261E-01	1.0001	1.0002	1.0002	1.0001	1.0000	1.0000	1.0000	1.0000	0.9998	0.7342	0.9996	0.9995
0.6	9.1515E-01	1.0002	0.9999	0.9999	0.9998	1.0000	1.0000	1.0000	1.0000	0.9998	0.7286	0.9996	0.9995
0.7	9.2379E-01	1.0002	0.9998	0.9997	0.9996	1.0000	1.0000	1.0000	1.0000	0.9997	0.7229	0.9996	0.9995
0.8	9.3000E-01	1.0002	0.9997	0.9996	0.9995	1.0000	1.0000	1.0000	1.0000	0.9997	0.7161	0.9996	0.9995
0.9	9.3461E-01	1.0002	0.9997	0.9995	0.9994	1.0000	1.0000	1.0000	1.0000	0.9997	0.7098	0.9996	0.9995
1.0	9.3813E-01	1.0002	0.9996	0.9995	0.9994	1.0000	1.0000	1.0000	1.0000	0.9997	0.7029	0.9996	0.9996
1.2	9.4305E-01	1.0002	0.9996	0.9994	0.9993	1.0000	1.0000	1.0000	1.0000	0.9997	0.6899	0.9997	0.9996
1.4	9.4625E-01	1.0002	0.9995	0.9994	0.9993	1.0000	1.0000	1.0000	1.0000	0.9997	0.6766	0.9997	0.9996
1.6	9.4843E-01	1.0002	0.9995	0.9993	0.9992	1.0000	1.0000	1.0000	1.0000	0.9997	0.6632	0.9997	0.9996
1.8	9.4997E-01	1.0002	0.9995	0.9993	0.9992	1.0001	1.0000	1.0000	1.0000	0.9997	0.6499	0.9997	0.9996
2.0	9.5111E-01	1.0002	0.9995	0.9993	0.9992	1.0001	1.0000	1.0000	1.0000	0.9997	0.6354	0.9997	0.9996
2.2	9.5197E-01	1.0002	0.9995	0.9993	0.9992	1.0001	1.0000	1.0000	1.0000	0.9997	0.6235	0.9997	0.9996
2.4	9.5263E-01	1.0002	0.9995	0.9993	0.9992	1.0001	1.0000	1.0000	1.0000	0.9997	0.6109	0.9997	0.9996
2.6	9.5316E-01	1.0002	0.9995	0.9993	0.9992	1.0001	1.0000	1.0000	1.0000	0.9997	0.5971	0.9997	0.9996
2.8	9.5358E-01	1.0003	0.9995	0.9993	0.9992	1.0001	1.0000	1.0000	1.0000	0.9997	0.5855	0.9997	0.9996
3.0	9.5392E-01	1.0003	0.9995	0.9993	0.9992	1.0001	1.0000	1.0000	1.0000	0.9997	0.5715	0.9997	0.9996
3.2	9.5420E-01	1.0003	0.9995	0.9993	0.9992	1.0001	1.0000	1.0000	1.0000	0.9997	0.5583	0.9997	0.9996
3.4	9.5444E-01	1.0003	0.9995	0.9993	0.9992	1.0001	1.0000	1.0000	1.0000	0.9997	0.5418	0.9997	0.9996
3.6	9.5464E-01	1.0003	0.9995	0.9993	0.9992	1.0001	1.0000	1.0000	1.0000	0.9997	0.5314	0.9997	0.9996
3.8	9.5481E-01	1.0003	0.9995	0.9993	0.9992	1.0001	1.0000	1.0000	1.0000	0.9997	0.5182	0.9997	0.9996
4.0	9.5496E-01	1.0003	0.9995	0.9993	0.9992	1.0001	1.0000	1.0000	1.0000	0.9997	0.5052	0.9997	0.9996
4.5	9.5525E-01	1.0004	0.9995	0.9993	0.9992	1.0001	1.0000	1.0000	1.0000	0.9997	0.4746	0.9997	0.9996
5.0	9.5546E-01	1.0004	0.9995	0.9993	0.9992	1.0002	1.0000	1.0000	1.0000	0.9997	0.4413	0.9997	0.9997
5.5	9.5562E-01	1.0004	0.9994	0.9993	0.9992	1.0002	1.0000	1.0000	1.0000	0.9997	0.4131	0.9997	0.9997
6.0	9.5575E-01	1.0004	0.9994	0.9992	0.9991	1.0002	1.0000	1.0000	1.0000	0.9997	0.3795	0.9997	0.9997
6.5	9.5585E-01	1.0005	0.9994	0.9992	0.9991	1.0002	1.0000	1.0000	1.0000	0.9996	0.3472	0.9997	0.9997
7.0	9.5594E-01	1.0005	0.9994	0.9992	0.9991	1.0002	1.0000	1.0000	1.0000	0.9996	0.3210	0.9996	0.9997
7.5	9.5602E-01	1.0005	0.9994	0.9992	0.9991	1.0002	1.0000	1.0000	1.0000	0.9996	0.2845	0.9996	0.9997
8.0	9.5608E-01	1.0006	0.9994	0.9992	0.9991	1.0003	1.0001	1.0000	1.0000	0.9996	0.2560	0.9996	0.9997
9.0	9.5619E-01	1.0006	0.9994	0.9992	0.9990	1.0003	1.0001	1.0000	1.0000	0.9996	0.1994	0.9996	0.9997
10.0	9.5628E-01	1.0007	0.9993	0.9991	0.9990	1.0003	1.0001	1.0000	1.0000	0.9996	0.1377	0.9996	0.9997
11.0	9.5636E-01	1.0008	0.9993	0.9991	0.9990	1.0004	1.0001	1.0000	1.0000	0.9996	0.0803	0.9996	0.9997
12.0	9.5644E-01	1.0008	0.9993	0.9991	0.9989	1.0004	1.0001	1.0000	1.0000	0.9996	0.0041	0.9996	0.9997
13.0	9.5651E-01	1.0009	0.9992	0.9990	0.9989	1.0004	1.0001	1.0000	1.0000	0.9996	−0.0510	0.9996	0.9997
14.0	9.5657E-01	1.0010	0.9992	0.9990	0.9989	1.0005	1.0001	1.0000	1.0000	0.9995	−0.1034	0.9995	0.9997
16.0	9.5669E-01	1.0011	0.9992	0.9989	0.9988	1.0005	1.0001	1.0000	1.0000	0.9995	−0.2362	0.9995	0.9997
18.0	9.5681E-01	1.0012	0.9991	0.9988	0.9987	1.0006	1.0001	1.0000	1.0000	0.9995	−0.3544	0.9995	0.9997
20.0	9.5693E-01	1.0014	0.9990	0.9987	0.9986	1.0007	1.0001	1.0000	1.0000	0.9994	−0.4747	0.9994	0.9997
25.0	9.5721E-01	1.0017	0.9988	0.9985	0.9984	1.0008	1.0002	1.0000	1.0000	0.9994	−0.7710	0.9994	0.9997
30.0	9.5750E-01	1.0020	0.9987	0.9983	0.9981	1.0010	1.0002	1.0000	1.0000	0.9993	−1.0992	0.9993	0.9998
35.0	9.5779E-01	1.0023	0.9985	0.9981	0.9979	1.0011	1.0002	1.0000	1.0000	0.9992	−1.4005	0.9992	0.9998
40.0	9.5807E-01	1.0027	0.9983	0.9979	0.9977	1.0013	1.0003	1.0000	1.0000	0.9991	−1.6906	0.9991	0.9998
45.0	9.5836E-01	1.0030	0.9981	0.9977	0.9974	1.0015	1.0003	1.0000	1.0000	0.9990	−1.9813	0.9990	0.9998
50.0	9.5865E-01	1.0033	0.9980	0.9974	0.9972	1.0016	1.0003	1.0000	1.0000	0.9989	−2.2773	0.9989	0.9998

POSITRON Z= -3 A= 6

p	F_0L_0	L_0	λ_2	λ_3	λ_4	μ_1	μ_2	Λ_1	Λ_2	ν_{12}	$\hat{\nu}_{12}$	η_{12}	$\hat{\eta}_{12}$
0.1	4.6539E-01	1.0003	1.0460	1.0579	1.0632	1.0000	1.0000	0.9999	1.0000	1.0109	0.7481	0.9990	0.9988
0.2	6.9299E-01	1.0003	1.0103	1.0128	1.0138	1.0000	1.0000	0.9999	1.0000	1.0020	0.7461	0.9991	0.9988
0.3	7.8294E-01	1.0003	1.0038	1.0046	1.0049	1.0000	1.0000	0.9999	1.0000	1.0004	0.7431	0.9991	0.9989
0.4	8.2957E-01	1.0003	1.0015	1.0018	1.0018	1.0000	1.0000	0.9999	1.0000	0.9999	0.7393	0.9991	0.9989
0.5	8.5746E-01	1.0003	1.0005	1.0005	1.0004	1.0000	1.0000	0.9999	1.0000	0.9996	0.7350	0.9992	0.9989
0.6	8.7564E-01	1.0003	1.0000	0.9998	0.9997	1.0000	1.0000	0.9999	1.0000	0.9995	0.7302	0.9992	0.9990
0.7	8.8822E-01	1.0003	0.9996	0.9994	0.9992	1.0000	1.0000	0.9999	1.0000	0.9994	0.7250	0.9992	0.9990
0.8	8.9728E-01	1.0003	0.9994	0.9992	0.9990	1.0000	1.0000	1.0000	1.0000	0.9994	0.7198	0.9992	0.9990
0.9	9.0403E-01	1.0003	0.9993	0.9990	0.9988	1.0000	1.0000	1.0000	1.0000	0.9994	0.7141	0.9992	0.9990
1.0	9.0918E-01	1.0003	0.9992	0.9989	0.9987	1.0000	1.0000	1.0000	1.0000	0.9993	0.7086	0.9992	0.9990
1.2	9.1640E-01	1.0003	0.9991	0.9987	0.9985	1.0001	1.0000	1.0000	1.0000	0.9993	0.6971	0.9993	0.9991
1.4	9.2108E-01	1.0004	0.9991	0.9987	0.9985	1.0001	1.0000	1.0000	1.0000	0.9993	0.6859	0.9993	0.9991
1.6	9.2427E-01	1.0004	0.9990	0.9986	0.9984	1.0001	1.0000	1.0000	1.0000	0.9993	0.6744	0.9993	0.9991
1.8	9.2654E-01	1.0004	0.9990	0.9986	0.9984	1.0001	1.0000	1.0000	1.0000	0.9993	0.6625	0.9993	0.9991
2.0	9.2821E-01	1.0004	0.9990	0.9986	0.9984	1.0001	1.0000	1.0000	1.0000	0.9993	0.6512	0.9993	0.9991
2.2	9.2947E-01	1.0004	0.9990	0.9986	0.9984	1.0001	1.0000	1.0000	1.0000	0.9993	0.6405	0.9993	0.9991
2.4	9.3044E-01	1.0004	0.9990	0.9986	0.9984	1.0001	1.0000	1.0000	1.0000	0.9993	0.6292	0.9993	0.9992
2.6	9.3121E-01	1.0005	0.9990	0.9986	0.9984	1.0002	1.0000	1.0000	1.0000	0.9993	0.6183	0.9993	0.9992
2.8	9.3182E-01	1.0005	0.9990	0.9986	0.9984	1.0002	1.0000	1.0000	1.0000	0.9993	0.6079	0.9993	0.9992
3.0	9.3232E-01	1.0005	0.9990	0.9986	0.9984	1.0002	1.0000	1.0000	1.0000	0.9993	0.5969	0.9993	0.9992
3.2	9.3274E-01	1.0005	0.9989	0.9986	0.9983	1.0002	1.0000	1.0000	1.0000	0.9993	0.5868	0.9993	0.9992
3.4	9.3308E-01	1.0006	0.9989	0.9986	0.9983	1.0002	1.0000	1.0000	1.0000	0.9993	0.5745	0.9993	0.9992
3.6	9.3337E-01	1.0006	0.9989	0.9986	0.9983	1.0002	1.0000	1.0000	1.0000	0.9993	0.5657	0.9993	0.9992
3.8	9.3362E-01	1.0006	0.9989	0.9985	0.9983	1.0002	1.0000	1.0000	1.0000	0.9993	0.5535	0.9993	0.9992
4.0	9.3383E-01	1.0006	0.9989	0.9985	0.9983	1.0002	1.0000	1.0000	1.0000	0.9993	0.5430	0.9993	0.9992
4.5	9.3426E-01	1.0007	0.9989	0.9985	0.9983	1.0003	1.0001	1.0000	1.0000	0.9993	0.5159	0.9993	0.9992
5.0	9.3456E-01	1.0007	0.9989	0.9985	0.9983	1.0003	1.0001	1.0000	1.0000	0.9993	0.4910	0.9993	0.9993
5.5	9.3480E-01	1.0008	0.9989	0.9985	0.9983	1.0003	1.0001	1.0000	1.0000	0.9993	0.4649	0.9993	0.9993
6.0	9.3499E-01	1.0009	0.9989	0.9985	0.9983	1.0004	1.0001	1.0000	1.0000	0.9993	0.4389	0.9993	0.9993
6.5	9.3514E-01	1.0009	0.9989	0.9985	0.9982	1.0004	1.0001	1.0000	1.0000	0.9993	0.4121	0.9993	0.9993
7.0	9.3527E-01	1.0010	0.9988	0.9984	0.9982	1.0004	1.0001	1.0000	1.0000	0.9993	0.3869	0.9993	0.9993
7.5	9.3539E-01	1.0010	0.9988	0.9984	0.9982	1.0005	1.0001	1.0000	1.0000	0.9993	0.3608	0.9993	0.9993
8.0	9.3548E-01	1.0011	0.9988	0.9984	0.9982	1.0005	1.0001	1.0000	1.0000	0.9993	0.3346	0.9993	0.9993
9.0	9.3565E-01	1.0012	0.9987	0.9983	0.9981	1.0006	1.0001	1.0000	1.0000	0.9992	0.2851	0.9992	0.9993
10.0	9.3579E-01	1.0014	0.9987	0.9983	0.9980	1.0006	1.0001	1.0000	1.0000	0.9992	0.2331	0.9992	0.9993
11.0	9.3592E-01	1.0015	0.9986	0.9982	0.9980	1.0007	1.0001	1.0000	1.0000	0.9992	0.1844	0.9992	0.9994
12.0	9.3604E-01	1.0016	0.9986	0.9981	0.9979	1.0007	1.0001	1.0000	1.0000	0.9992	0.1291	0.9992	0.9994
13.0	9.3616E-01	1.0017	0.9985	0.9981	0.9978	1.0008	1.0002	1.0000	1.0000	0.9991	0.0737	0.9991	0.9994
14.0	9.3626E-01	1.0018	0.9985	0.9980	0.9978	1.0009	1.0002	1.0000	1.0000	0.9991	0.0255	0.9991	0.9994
16.0	9.3647E-01	1.0021	0.9984	0.9979	0.9976	1.0010	1.0002	1.0000	1.0000	0.9990	-0.0766	0.9991	0.9994
18.0	9.3668E-01	1.0023	0.9982	0.9977	0.9975	1.0011	1.0002	1.0000	1.0000	0.9990	-0.1813	0.9990	0.9994
20.0	9.3688E-01	1.0026	0.9981	0.9976	0.9973	1.0012	1.0003	1.0000	1.0000	0.9989	-0.2778	0.9989	0.9994
25.0	9.3738E-01	1.0032	0.9978	0.9972	0.9969	1.0015	1.0003	1.0000	1.0000	0.9988	-0.5325	0.9988	0.9995
30.0	9.3789E-01	1.0038	0.9975	0.9968	0.9965	1.0018	1.0004	1.0000	1.0000	0.9986	-0.7891	0.9986	0.9995
35.0	9.3840E-01	1.0044	0.9972	0.9964	0.9960	1.0021	1.0004	1.0000	1.0000	0.9985	-1.0419	0.9985	0.9995
40.0	9.3892E-01	1.0050	0.9968	0.9960	0.9956	1.0024	1.0005	1.0000	1.0000	0.9983	-1.2945	0.9983	0.9995
45.0	9.3944E-01	1.0056	0.9965	0.9956	0.9951	1.0027	1.0006	1.0000	1.0000	0.9981	-1.5350	0.9981	0.9995
50.0	9.3997E-01	1.0063	0.9962	0.9952	0.9947	1.0030	1.0006	1.0000	1.0000	0.9980	-1.7937	0.9980	0.9995

 Instr. see p. 22

POSITRON Z= -4 A= 8

p	F_0L_0	L_0	λ_2	λ_3	λ_4	μ_1	μ_2	Λ_1	Λ_2	ν_{12}	$\hat{\nu}_{12}$	η_{12}	$\hat{\eta}_{12}$
0.1	3.4953E-01	1.0005	1.0818	1.1036	1.1134	1.0000	1.0000	0.9999	1.0000	1.0195	0.7476	0.9983	0.9979
0.2	6.0887E-01	1.0005	1.0184	1.0228	1.0247	1.0000	1.0000	0.9999	1.0000	1.0037	0.7456	0.9984	0.9980
0.3	7.1981E-01	1.0005	1.0068	1.0082	1.0088	1.0000	1.0000	0.9999	1.0000	1.0008	0.7429	0.9985	0.9981
0.4	7.7879E-01	1.0005	1.0028	1.0032	1.0033	1.0000	1.0000	0.9999	1.0000	0.9998	0.7396	0.9985	0.9981
0.5	8.1449E-01	1.0005	1.0010	1.0009	1.0008	1.0000	1.0000	0.9999	1.0000	0.9994	0.7357	0.9986	0.9982
0.6	8.3793E-01	1.0005	1.0000	0.9997	0.9995	1.0000	1.0000	0.9999	1.0000	0.9991	0.7316	0.9986	0.9982
0.7	8.5421E-01	1.0005	0.9995	0.9990	0.9988	1.0000	1.0000	0.9999	1.0000	0.9990	0.7272	0.9986	0.9982
0.8	8.6597E-01	1.0005	0.9991	0.9986	0.9983	1.0000	1.0000	0.9999	1.0000	0.9989	0.7226	0.9987	0.9983
0.9	8.7474E-01	1.0005	0.9989	0.9983	0.9980	1.0001	1.0000	0.9999	1.0000	0.9989	0.7179	0.9987	0.9983
1.0	8.8146E-01	1.0005	0.9987	0.9981	0.9978	1.0001	1.0000	0.9999	1.0000	0.9989	0.7130	0.9987	0.9983
1.2	8.9086E-01	1.0005	0.9985	0.9979	0.9975	1.0001	1.0000	0.9999	1.0000	0.9988	0.7034	0.9987	0.9984
1.4	8.9696E-01	1.0005	0.9984	0.9978	0.9974	1.0001	1.0000	0.9999	1.0000	0.9988	0.6938	0.9988	0.9984
1.6	9.0112E-01	1.0006	0.9984	0.9977	0.9973	1.0001	1.0000	1.0000	1.0000	0.9988	0.6841	0.9988	0.9984
1.8	9.0408E-01	1.0006	0.9983	0.9977	0.9973	1.0001	1.0000	1.0000	1.0000	0.9988	0.6747	0.9988	0.9985
2.0	9.0625E-01	1.0006	0.9983	0.9976	0.9973	1.0002	1.0000	1.0000	1.0000	0.9988	0.6655	0.9988	0.9985
2.2	9.0789E-01	1.0007	0.9983	0.9976	0.9973	1.0002	1.0000	1.0000	1.0000	0.9988	0.6560	0.9988	0.9985
2.4	9.0915E-01	1.0007	0.9983	0.9976	0.9972	1.0002	1.0000	1.0000	1.0000	0.9988	0.6471	0.9988	0.9985
2.6	9.1015E-01	1.0007	0.9983	0.9976	0.9972	1.0002	1.0000	1.0000	1.0000	0.9988	0.6383	0.9988	0.9985
2.8	9.1094E-01	1.0008	0.9983	0.9976	0.9972	1.0002	1.0000	1.0000	1.0000	0.9989	0.6289	0.9988	0.9986
3.0	9.1158E-01	1.0008	0.9983	0.9976	0.9972	1.0003	1.0001	1.0000	1.0000	0.9989	0.6197	0.9988	0.9986
3.2	9.1212E-01	1.0008	0.9983	0.9976	0.9972	1.0003	1.0001	1.0000	1.0000	0.9989	0.6111	0.9989	0.9986
3.4	9.1256E-01	1.0009	0.9982	0.9976	0.9972	1.0003	1.0001	1.0000	1.0000	0.9989	0.6024	0.9989	0.9986
3.6	9.1293E-01	1.0009	0.9982	0.9976	0.9972	1.0003	1.0001	1.0000	1.0000	0.9989	0.5935	0.9989	0.9986
3.8	9.1326E-01	1.0009	0.9982	0.9976	0.9972	1.0003	1.0001	1.0000	1.0000	0.9989	0.5848	0.9989	0.9986
4.0	9.1353E-01	1.0010	0.9982	0.9976	0.9972	1.0004	1.0001	1.0000	1.0000	0.9989	0.5756	0.9989	0.9986
4.5	9.1407E-01	1.0010	0.9982	0.9976	0.9972	1.0004	1.0001	1.0000	1.0000	0.9989	0.5536	0.9989	0.9987
5.0	9.1446E-01	1.0011	0.9982	0.9976	0.9972	1.0004	1.0001	1.0000	1.0000	0.9989	0.5316	0.9989	0.9987
5.5	9.1476E-01	1.0012	0.9982	0.9975	0.9972	1.0005	1.0001	1.0000	1.0000	0.9988	0.5107	0.9989	0.9987
6.0	9.1500E-01	1.0013	0.9982	0.9975	0.9972	1.0005	1.0001	1.0000	1.0000	0.9988	0.4885	0.9988	0.9987
6.5	9.1519E-01	1.0014	0.9982	0.9975	0.9971	1.0006	1.0001	1.0000	1.0000	0.9988	0.4676	0.9988	0.9988
7.0	9.1535E-01	1.0015	0.9981	0.9975	0.9971	1.0006	1.0001	1.0000	1.0000	0.9988	0.4453	0.9988	0.9988
7.5	9.1550E-01	1.0016	0.9981	0.9974	0.9971	1.0007	1.0001	1.0000	1.0000	0.9988	0.4244	0.9988	0.9988
8.0	9.1562E-01	1.0017	0.9981	0.9974	0.9971	1.0007	1.0001	1.0000	1.0000	0.9988	0.4052	0.9988	0.9988
9.0	9.1583E-01	1.0018	0.9980	0.9973	0.9970	1.0008	1.0002	1.0000	1.0000	0.9988	0.3610	0.9988	0.9988
10.0	9.1601E-01	1.0020	0.9980	0.9973	0.9969	1.0009	1.0002	1.0000	1.0000	0.9987	0.3175	0.9987	0.9989
11.0	9.1618E-01	1.0022	0.9979	0.9972	0.9968	1.0010	1.0002	1.0000	1.0000	0.9987	0.2760	0.9987	0.9989
12.0	9.1633E-01	1.0024	0.9978	0.9971	0.9967	1.0011	1.0002	1.0000	1.0000	0.9987	0.2343	0.9987	0.9989
13.0	9.1648E-01	1.0026	0.9977	0.9970	0.9966	1.0012	1.0002	1.0000	1.0000	0.9986	0.1911	0.9986	0.9989
14.0	9.1662E-01	1.0027	0.9977	0.9969	0.9965	1.0013	1.0003	1.0000	1.0000	0.9986	0.1505	0.9986	0.9989
16.0	9.1690E-01	1.0031	0.9975	0.9967	0.9963	1.0014	1.0003	1.0000	1.0000	0.9985	0.0656	0.9985	0.9990
18.0	9.1717E-01	1.0035	0.9973	0.9965	0.9961	1.0016	1.0003	1.0000	1.0000	0.9984	-0.0210	0.9984	0.9990
20.0	9.1744E-01	1.0038	0.9972	0.9963	0.9959	1.0018	1.0004	1.0000	1.0000	0.9983	-0.1017	0.9984	0.9990
25.0	9.1813E-01	1.0047	0.9967	0.9958	0.9953	1.0022	1.0005	1.0000	1.0000	0.9981	-0.3091	0.9981	0.9991
30.0	9.1884E-01	1.0056	0.9963	0.9952	0.9947	1.0027	1.0005	1.0000	1.0000	0.9979	-0.5222	0.9979	0.9991
35.0	9.1955E-01	1.0065	0.9958	0.9947	0.9941	1.0031	1.0006	1.0000	1.0000	0.9977	-0.7239	0.9977	0.9991
40.0	9.2028E-01	1.0074	0.9953	0.9941	0.9934	1.0036	1.0007	1.0000	1.0000	0.9974	-0.9295	0.9974	0.9992
45.0	9.2101E-01	1.0083	0.9949	0.9935	0.9928	1.0040	1.0008	1.0000	1.0000	0.9972	-1.1399	0.9972	0.9992
50.0	9.2175E-01	1.0092	0.9944	0.9929	0.9922	1.0044	1.0009	1.0000	1.0000	0.9970	-1.3383	0.9970	0.9992

POSITRON Z= -5 A= 10

p	F_0L_0	L_0	λ_2	λ_3	λ_4	μ_1	μ_2	Λ_1	Λ_2	ν_{12}	$\hat{\nu}_{12}$	η_{12}	$\hat{\eta}_{12}$
0.1	2.5883E-01	1.0007	1.1278	1.1633	1.1794	1.0000	1.0000	0.9999	1.0000	1.0305	0.7469	0.9974	0.9968
0.2	5.3309E-01	1.0007	1.0288	1.0358	1.0387	1.0000	1.0000	0.9999	1.0000	1.0058	0.7450	0.9976	0.9969
0.3	6.6095E-01	1.0007	1.0107	1.0129	1.0138	1.0000	1.0000	0.9999	1.0000	1.0013	0.7423	0.9977	0.9970
0.4	7.3080E-01	1.0007	1.0045	1.0051	1.0053	1.0000	1.0000	0.9999	1.0000	0.9998	0.7392	0.9978	0.9971
0.5	7.7361E-01	1.0007	1.0016	1.0016	1.0014	1.0000	1.0000	0.9999	1.0000	0.9991	0.7357	0.9978	0.9972
0.6	8.0193E-01	1.0007	1.0001	0.9997	0.9994	1.0000	1.0000	0.9999	1.0000	0.9987	0.7319	0.9979	0.9972
0.7	8.2168E-01	1.0007	0.9992	0.9986	0.9982	1.0000	1.0000	0.9999	1.0000	0.9985	0.7280	0.9979	0.9973
0.8	8.3599E-01	1.0007	0.9987	0.9979	0.9975	1.0001	1.0000	0.9999	1.0000	0.9984	0.7239	0.9980	0.9973
0.9	8.4669E-01	1.0007	0.9983	0.9975	0.9970	1.0001	1.0000	0.9999	1.0000	0.9983	0.7196	0.9980	0.9974
1.0	8.5488E-01	1.0007	0.9981	0.9972	0.9967	1.0001	1.0000	0.9999	1.0000	0.9983	0.7154	0.9980	0.9974
1.2	8.6637E-01	1.0007	0.9978	0.9968	0.9963	1.0001	1.0000	0.9999	1.0000	0.9983	0.7069	0.9981	0.9975
1.4	8.7383E-01	1.0008	0.9976	0.9966	0.9961	1.0001	1.0000	0.9999	1.0000	0.9982	0.6985	0.9981	0.9975
1.6	8.7892E-01	1.0008	0.9975	0.9965	0.9960	1.0002	1.0000	0.9999	1.0000	0.9982	0.6903	0.9981	0.9976
1.8	8.8253E-01	1.0009	0.9975	0.9965	0.9959	1.0002	1.0000	0.9999	1.0000	0.9982	0.6819	0.9982	0.9976
2.0	8.8518E-01	1.0009	0.9974	0.9964	0.9959	1.0002	1.0000	0.9999	1.0000	0.9982	0.6737	0.9982	0.9977
2.2	8.8718E-01	1.0009	0.9974	0.9964	0.9959	1.0002	1.0000	0.9999	1.0000	0.9983	0.6659	0.9982	0.9977
2.4	8.8872E-01	1.0010	0.9974	0.9964	0.9959	1.0003	1.0001	1.0000	1.0000	0.9983	0.6578	0.9982	0.9977
2.6	8.8993E-01	1.0010	0.9974	0.9964	0.9959	1.0003	1.0001	1.0000	1.0000	0.9983	0.6499	0.9982	0.9978
2.8	8.9089E-01	1.0011	0.9974	0.9964	0.9959	1.0003	1.0001	1.0000	1.0000	0.9983	0.6421	0.9983	0.9978
3.0	8.9167E-01	1.0011	0.9974	0.9964	0.9959	1.0003	1.0001	1.0000	1.0000	0.9983	0.6346	0.9983	0.9978
3.2	8.9232E-01	1.0011	0.9974	0.9964	0.9959	1.0004	1.0001	1.0000	1.0000	0.9983	0.6270	0.9983	0.9978
3.4	8.9285E-01	1.0012	0.9974	0.9964	0.9959	1.0004	1.0001	1.0000	1.0000	0.9983	0.6193	0.9983	0.9979
3.6	8.9330E-01	1.0012	0.9974	0.9964	0.9959	1.0004	1.0001	1.0000	1.0000	0.9983	0.6115	0.9983	0.9979
3.8	8.9369E-01	1.0013	0.9974	0.9964	0.9959	1.0004	1.0001	1.0000	1.0000	0.9983	0.6042	0.9983	0.9979
4.0	8.9402E-01	1.0013	0.9974	0.9964	0.9959	1.0005	1.0001	1.0000	1.0000	0.9983	0.5969	0.9983	0.9979
4.5	8.9466E-01	1.0014	0.9974	0.9964	0.9959	1.0005	1.0001	1.0000	1.0000	0.9983	0.5777	0.9983	0.9980
5.0	8.9513E-01	1.0016	0.9974	0.9964	0.9959	1.0006	1.0001	1.0000	1.0000	0.9983	0.5593	0.9983	0.9980
5.5	8.9548E-01	1.0017	0.9974	0.9964	0.9959	1.0007	1.0001	1.0000	1.0000	0.9983	0.5405	0.9983	0.9980
6.0	8.9576E-01	1.0018	0.9973	0.9964	0.9958	1.0007	1.0001	1.0000	1.0000	0.9983	0.5227	0.9983	0.9981
6.5	8.9599E-01	1.0019	0.9973	0.9963	0.9958	1.0008	1.0002	1.0000	1.0000	0.9983	0.5035	0.9983	0.9981
7.0	8.9618E-01	1.0020	0.9973	0.9963	0.9958	1.0008	1.0002	1.0000	1.0000	0.9983	0.4854	0.9983	0.9981
7.5	8.9634E-01	1.0022	0.9973	0.9963	0.9958	1.0009	1.0002	1.0000	1.0000	0.9983	0.4670	0.9983	0.9981
8.0	8.9649E-01	1.0023	0.9972	0.9963	0.9957	1.0010	1.0002	1.0000	1.0000	0.9982	0.4478	0.9982	0.9982
9.0	8.9673E-01	1.0025	0.9972	0.9962	0.9956	1.0011	1.0002	1.0000	1.0000	0.9982	0.4113	0.9982	0.9982
10.0	8.9695E-01	1.0028	0.9971	0.9961	0.9956	1.0012	1.0002	1.0000	1.0000	0.9982	0.3755	0.9982	0.9982
11.0	8.9714E-01	1.0030	0.9970	0.9960	0.9954	1.0013	1.0003	1.0000	1.0000	0.9981	0.3383	0.9981	0.9983
12.0	8.9733E-01	1.0032	0.9969	0.9959	0.9953	1.0015	1.0003	1.0000	1.0000	0.9981	0.3028	0.9981	0.9983
13.0	8.9750E-01	1.0035	0.9968	0.9958	0.9952	1.0016	1.0003	1.0000	1.0000	0.9980	0.2649	0.9980	0.9983
14.0	8.9767E-01	1.0037	0.9967	0.9957	0.9951	1.0017	1.0003	1.0000	1.0000	0.9980	0.2289	0.9980	0.9984
16.0	8.9801E-01	1.0042	0.9965	0.9954	0.9948	1.0019	1.0004	1.0000	1.0000	0.9979	0.1566	0.9979	0.9984
18.0	8.9835E-01	1.0047	0.9963	0.9952	0.9946	1.0022	1.0004	1.0000	1.0000	0.9978	0.0829	0.9978	0.9984
20.0	8.9869E-01	1.0052	0.9961	0.9949	0.9943	1.0024	1.0005	1.0000	1.0000	0.9977	0.0142	0.9977	0.9985
25.0	8.9955E-01	1.0064	0.9955	0.9942	0.9935	1.0030	1.0006	1.0000	1.0000	0.9974	-0.1665	0.9974	0.9986
30.0	9.0045E-01	1.0076	0.9949	0.9935	0.9928	1.0036	1.0007	1.0000	1.0000	0.9971	-0.3463	0.9971	0.9986
35.0	9.0136E-01	1.0088	0.9943	0.9927	0.9919	1.0042	1.0009	1.0000	1.0000	0.9968	-0.5274	0.9968	0.9987
40.0	9.0229E-01	1.0100	0.9937	0.9920	0.9911	1.0048	1.0010	1.0000	1.0000	0.9965	-0.7004	0.9965	0.9987
45.0	9.0324E-01	1.0112	0.9931	0.9912	0.9903	1.0054	1.0011	1.0000	1.0000	0.9962	-0.8723	0.9962	0.9988
50.0	9.0420E-01	1.0124	0.9925	0.9904	0.9894	1.0060	1.0012	1.0000	1.0000	0.9959	-1.0543	0.9959	0.9988

Instr. see p. 22

POSITRON Z= -6 A= 12

p	F_0L_0	L_0	λ_2	λ_3	λ_4	μ_1	μ_2	Λ_1	Λ_2	ν_{12}	$\hat{\nu}_{12}$	η_{12}	$\hat{\eta}_{12}$
0.1	1.8920E-01	1.0009	1.1837	1.2374	1.2621	1.0000	1.0000	0.9998	1.0000	1.0440	0.7460	0.9964	0.9954
0.2	4.6516E-01	1.0009	1.0416	1.0517	1.0560	1.0000	1.0000	0.9998	1.0000	1.0084	0.7441	0.9966	0.9956
0.3	6.0614E-01	1.0009	1.0155	1.0187	1.0200	1.0000	1.0000	0.9998	1.0000	1.0019	0.7415	0.9967	0.9958
0.4	6.8544E-01	1.0009	1.0065	1.0075	1.0077	1.0000	1.0000	0.9999	1.0000	0.9997	0.7385	0.9968	0.9959
0.5	7.3471E-01	1.0009	1.0024	1.0024	1.0022	1.0000	1.0000	0.9999	1.0000	0.9987	0.7352	0.9969	0.9960
0.6	7.6755E-01	1.0009	1.0003	0.9997	0.9993	1.0000	1.0000	0.9999	1.0000	0.9982	0.7316	0.9970	0.9961
0.7	7.9055E-01	1.0009	0.9990	0.9981	0.9976	1.0001	1.0000	0.9999	1.0000	0.9979	0.7280	0.9971	0.9961
0.8	8.0727E-01	1.0009	0.9982	0.9971	0.9965	1.0001	1.0000	0.9999	1.0000	0.9978	0.7242	0.9971	0.9962
0.9	8.1979E-01	1.0009	0.9977	0.9965	0.9958	1.0001	1.0000	0.9999	1.0000	0.9977	0.7203	0.9972	0.9963
1.0	8.2939E-01	1.0010	0.9973	0.9961	0.9954	1.0001	1.0000	0.9999	1.0000	0.9976	0.7164	0.9972	0.9963
1.2	8.4287E-01	1.0010	0.9969	0.9955	0.9948	1.0001	1.0000	0.9999	1.0000	0.9975	0.7088	0.9973	0.9964
1.4	8.5163E-01	1.0010	0.9967	0.9953	0.9945	1.0002	1.0000	0.9999	1.0000	0.9975	0.7012	0.9973	0.9965
1.6	8.5761E-01	1.0011	0.9965	0.9951	0.9944	1.0002	1.0000	0.9999	1.0000	0.9975	0.6938	0.9974	0.9966
1.8	8.6185E-01	1.0011	0.9965	0.9951	0.9943	1.0002	1.0000	0.9999	1.0000	0.9975	0.6865	0.9974	0.9966
2.0	8.6496E-01	1.0012	0.9964	0.9950	0.9942	1.0003	1.0001	0.9999	1.0000	0.9975	0.6792	0.9975	0.9967
2.2	8.6729E-01	1.0012	0.9964	0.9950	0.9942	1.0003	1.0001	0.9999	1.0000	0.9976	0.6720	0.9975	0.9967
2.4	8.6909E-01	1.0013	0.9964	0.9950	0.9942	1.0003	1.0001	0.9999	1.0000	0.9976	0.6650	0.9975	0.9968
2.6	8.7050E-01	1.0013	0.9964	0.9950	0.9942	1.0004	1.0001	0.9999	1.0000	0.9976	0.6581	0.9975	0.9968
2.8	8.7162E-01	1.0014	0.9964	0.9950	0.9942	1.0004	1.0001	0.9999	1.0000	0.9976	0.6513	0.9976	0.9969
3.0	8.7254E-01	1.0014	0.9964	0.9950	0.9942	1.0004	1.0001	1.0000	1.0000	0.9976	0.6445	0.9976	0.9969
3.2	8.7328E-01	1.0015	0.9964	0.9950	0.9943	1.0005	1.0001	1.0000	1.0000	0.9976	0.6375	0.9976	0.9969
3.4	8.7390E-01	1.0016	0.9964	0.9950	0.9943	1.0005	1.0001	1.0000	1.0000	0.9976	0.6306	0.9976	0.9970
3.6	8.7442E-01	1.0016	0.9964	0.9950	0.9943	1.0005	1.0001	1.0000	1.0000	0.9976	0.6242	0.9976	0.9970
3.8	8.7486E-01	1.0017	0.9964	0.9950	0.9943	1.0006	1.0001	1.0000	1.0000	0.9976	0.6173	0.9976	0.9970
4.0	8.7524E-01	1.0017	0.9964	0.9950	0.9943	1.0006	1.0001	1.0000	1.0000	0.9976	0.6106	0.9976	0.9970
4.5	8.7598E-01	1.0019	0.9964	0.9951	0.9943	1.0007	1.0001	1.0000	1.0000	0.9976	0.5941	0.9976	0.9971
5.0	8.7651E-01	1.0020	0.9964	0.9951	0.9943	1.0008	1.0002	1.0000	1.0000	0.9977	0.5772	0.9977	0.9972
5.5	8.7691E-01	1.0022	0.9964	0.9951	0.9943	1.0008	1.0002	1.0000	1.0000	0.9977	0.5611	0.9977	0.9972
6.0	8.7722E-01	1.0023	0.9964	0.9950	0.9943	1.0009	1.0002	1.0000	1.0000	0.9976	0.5447	0.9976	0.9973
6.5	8.7748E-01	1.0025	0.9964	0.9950	0.9943	1.0010	1.0002	1.0000	1.0000	0.9976	0.5285	0.9976	0.9973
7.0	8.7769E-01	1.0026	0.9963	0.9950	0.9943	1.0011	1.0002	1.0000	1.0000	0.9976	0.5122	0.9976	0.9973
7.5	8.7787E-01	1.0028	0.9963	0.9950	0.9942	1.0012	1.0002	1.0000	1.0000	0.9976	0.4963	0.9976	0.9974
8.0	8.7803E-01	1.0029	0.9963	0.9949	0.9942	1.0012	1.0002	1.0000	1.0000	0.9976	0.4802	0.9976	0.9974
9.0	8.7831E-01	1.0033	0.9962	0.9948	0.9941	1.0014	1.0003	1.0000	1.0000	0.9976	0.4479	0.9976	0.9975
10.0	8.7855E-01	1.0036	0.9961	0.9947	0.9940	1.0015	1.0003	1.0000	1.0000	0.9975	0.4155	0.9975	0.9975
11.0	8.7877E-01	1.0039	0.9960	0.9946	0.9939	1.0017	1.0003	1.0000	1.0000	0.9975	0.3826	0.9975	0.9976
12.0	8.7897E-01	1.0042	0.9959	0.9945	0.9938	1.0019	1.0004	1.0000	1.0000	0.9974	0.3499	0.9974	0.9976
13.0	8.7917E-01	1.0045	0.9958	0.9944	0.9936	1.0020	1.0004	1.0000	1.0000	0.9974	0.3188	0.9974	0.9976
14.0	8.7937E-01	1.0048	0.9957	0.9943	0.9935	1.0022	1.0004	1.0000	1.0000	0.9973	0.2864	0.9973	0.9977
16.0	8.7976E-01	1.0054	0.9954	0.9940	0.9932	1.0025	1.0005	1.0000	1.0000	0.9972	0.2214	0.9972	0.9977
18.0	8.8016E-01	1.0060	0.9952	0.9937	0.9929	1.0028	1.0006	1.0000	1.0000	0.9971	0.1567	0.9971	0.9978
20.0	8.8056E-01	1.0067	0.9949	0.9933	0.9925	1.0031	1.0006	1.0000	1.0000	0.9969	0.0960	0.9969	0.9978
25.0	8.8159E-01	1.0082	0.9942	0.9925	0.9916	1.0038	1.0008	1.0000	1.0000	0.9966	-0.0637	0.9966	0.9980
30.0	8.8267E-01	1.0097	0.9935	0.9916	0.9906	1.0046	1.0009	1.0000	1.0000	0.9962	-0.2229	0.9962	0.9981
35.0	8.8378E-01	1.0113	0.9927	0.9907	0.9896	1.0054	1.0011	1.0000	1.0000	0.9959	-0.3803	0.9959	0.9981
40.0	8.8491E-01	1.0128	0.9920	0.9897	0.9886	1.0061	1.0013	1.0000	1.0000	0.9955	-0.5350	0.9955	0.9982
45.0	8.8607E-01	1.0144	0.9912	0.9888	0.9876	1.0068	1.0014	1.0000	1.0000	0.9951	-0.6869	0.9951	0.9983
50.0	8.8724E-01	1.0159	0.9904	0.9878	0.9865	1.0076	1.0016	1.0000	1.0000	0.9947	-0.8414	0.9947	0.9983

POSITRON Z= -7 A= 14

p	F_0L_0	L_0	λ_2	λ_3	λ_4	μ_1	μ_2	Λ_1	Λ_2	ν_{12}	$\hat{\nu}_{12}$	η_{12}	$\hat{\eta}_{12}$
0.1	1.3669E-01	1.0012	1.2496	1.3267	1.3628	1.0000	1.0000	0.9998	1.0000	1.0599	0.7451	0.9952	0.9939
0.2	4.0453E-01	1.0011	1.0566	1.0706	1.0766	1.0000	1.0000	0.9998	1.0000	1.0115	0.7431	0.9954	0.9941
0.3	5.5520E-01	1.0011	1.0211	1.0256	1.0274	1.0000	1.0000	0.9998	1.0000	1.0027	0.7406	0.9956	0.9943
0.4	6.4260E-01	1.0012	1.0089	1.0103	1.0107	1.0000	1.0000	0.9998	1.0000	0.9997	0.7376	0.9958	0.9945
0.5	6.9769E-01	1.0012	1.0034	1.0034	1.0031	1.0000	1.0000	0.9998	1.0000	0.9984	0.7344	0.9959	0.9946
0.6	7.3469E-01	1.0012	1.0005	0.9997	0.9992	1.0001	1.0000	0.9998	1.0000	0.9977	0.7310	0.9960	0.9947
0.7	7.6074E-01	1.0012	0.9987	0.9976	0.9969	1.0001	1.0000	0.9998	1.0000	0.9973	0.7275	0.9961	0.9948
0.8	7.7973E-01	1.0012	0.9977	0.9963	0.9954	1.0001	1.0000	0.9999	1.0000	0.9971	0.7239	0.9961	0.9949
0.9	7.9399E-01	1.0012	0.9970	0.9954	0.9945	1.0001	1.0000	0.9999	1.0000	0.9969	0.7203	0.9962	0.9950
1.0	8.0493E-01	1.0012	0.9965	0.9948	0.9938	1.0001	1.0000	0.9999	1.0000	0.9968	0.7167	0.9963	0.9951
1.2	8.2031E-01	1.0013	0.9959	0.9941	0.9931	1.0002	1.0000	0.9999	1.0000	0.9967	0.7096	0.9964	0.9952
1.4	8.3032E-01	1.0013	0.9956	0.9937	0.9927	1.0002	1.0000	0.9999	1.0000	0.9967	0.7026	0.9965	0.9953
1.6	8.3714E-01	1.0014	0.9954	0.9935	0.9925	1.0003	1.0001	0.9999	1.0000	0.9967	0.6958	0.9965	0.9954
1.8	8.4198E-01	1.0014	0.9953	0.9934	0.9924	1.0003	1.0001	0.9999	1.0000	0.9967	0.6891	0.9966	0.9955
2.0	8.4552E-01	1.0015	0.9953	0.9934	0.9923	1.0003	1.0001	0.9999	1.0000	0.9967	0.6826	0.9966	0.9955
2.2	8.4818E-01	1.0016	0.9953	0.9934	0.9923	1.0004	1.0001	0.9999	1.0000	0.9968	0.6761	0.9967	0.9956
2.4	8.5023E-01	1.0016	0.9952	0.9934	0.9923	1.0004	1.0001	0.9999	1.0000	0.9968	0.6698	0.9967	0.9957
2.6	8.5183E-01	1.0017	0.9952	0.9934	0.9923	1.0005	1.0001	0.9999	1.0000	0.9968	0.6635	0.9967	0.9957
2.8	8.5310E-01	1.0018	0.9953	0.9934	0.9924	1.0005	1.0001	0.9999	1.0000	0.9968	0.6573	0.9968	0.9958
3.0	8.5413E-01	1.0018	0.9953	0.9934	0.9924	1.0005	1.0001	0.9999	1.0000	0.9968	0.6511	0.9968	0.9958
3.2	8.5497E-01	1.0019	0.9953	0.9934	0.9924	1.0006	1.0001	0.9999	1.0000	0.9968	0.6449	0.9968	0.9959
3.4	8.5567E-01	1.0020	0.9953	0.9934	0.9924	1.0006	1.0001	0.9999	1.0000	0.9969	0.6386	0.9968	0.9959
3.6	8.5626E-01	1.0020	0.9953	0.9935	0.9925	1.0007	1.0001	0.9999	1.0000	0.9969	0.6329	0.9968	0.9960
3.8	8.5675E-01	1.0021	0.9953	0.9935	0.9925	1.0007	1.0001	1.0000	1.0000	0.9969	0.6267	0.9969	0.9960
4.0	8.5717E-01	1.0022	0.9953	0.9935	0.9925	1.0007	1.0001	1.0000	1.0000	0.9969	0.6203	0.9969	0.9960
4.5	8.5799E-01	1.0024	0.9953	0.9935	0.9925	1.0008	1.0002	1.0000	1.0000	0.9969	0.6054	0.9969	0.9961
5.0	8.5858E-01	1.0026	0.9953	0.9935	0.9926	1.0009	1.0002	1.0000	1.0000	0.9969	0.5907	0.9969	0.9962
5.5	8.5902E-01	1.0027	0.9953	0.9935	0.9926	1.0010	1.0002	1.0000	1.0000	0.9969	0.5759	0.9969	0.9962
6.0	8.5936E-01	1.0029	0.9953	0.9935	0.9926	1.0011	1.0002	1.0000	1.0000	0.9969	0.5613	0.9969	0.9963
6.5	8.5964E-01	1.0031	0.9953	0.9935	0.9926	1.0012	1.0002	1.0000	1.0000	0.9969	0.5466	0.9969	0.9964
7.0	8.5986E-01	1.0033	0.9953	0.9935	0.9925	1.0013	1.0003	1.0000	1.0000	0.9969	0.5321	0.9969	0.9964
7.5	8.6006E-01	1.0035	0.9952	0.9935	0.9925	1.0014	1.0003	1.0000	1.0000	0.9969	0.5169	0.9969	0.9965
8.0	8.6023E-01	1.0037	0.9952	0.9934	0.9925	1.0015	1.0003	1.0000	1.0000	0.9969	0.5024	0.9969	0.9965
9.0	8.6052E-01	1.0041	0.9951	0.9934	0.9924	1.0017	1.0003	1.0000	1.0000	0.9968	0.4731	0.9968	0.9966
10.0	8.6078E-01	1.0044	0.9950	0.9933	0.9923	1.0019	1.0004	1.0000	1.0000	0.9968	0.4439	0.9968	0.9967
11.0	8.6102E-01	1.0048	0.9949	0.9931	0.9922	1.0021	1.0004	1.0000	1.0000	0.9967	0.4152	0.9968	0.9967
12.0	8.6124E-01	1.0052	0.9948	0.9930	0.9920	1.0023	1.0005	1.0000	1.0000	0.9967	0.3862	0.9967	0.9968
13.0	8.6146E-01	1.0056	0.9947	0.9929	0.9919	1.0025	1.0005	1.0000	1.0000	0.9966	0.3561	0.9966	0.9968
14.0	8.6168E-01	1.0059	0.9946	0.9927	0.9917	1.0027	1.0005	1.0000	1.0000	0.9966	0.3277	0.9966	0.9969
16.0	8.6212E-01	1.0067	0.9943	0.9924	0.9914	1.0030	1.0006	1.0000	1.0000	0.9964	0.2697	0.9964	0.9970
18.0	8.6256E-01	1.0075	0.9940	0.9920	0.9910	1.0034	1.0007	1.0000	1.0000	0.9963	0.2128	0.9963	0.9970
20.0	8.6302E-01	1.0082	0.9937	0.9917	0.9906	1.0038	1.0008	1.0000	1.0000	0.9961	0.1544	0.9961	0.9971
25.0	8.6421E-01	1.0101	0.9928	0.9907	0.9895	1.0047	1.0010	1.0000	1.0000	0.9957	0.0135	0.9957	0.9973
30.0	8.6546E-01	1.0120	0.9920	0.9896	0.9884	1.0056	1.0011	1.0000	1.0000	0.9953	-0.1303	0.9953	0.9974
35.0	8.6676E-01	1.0139	0.9911	0.9885	0.9872	1.0066	1.0013	1.0000	1.0000	0.9948	-0.2710	0.9948	0.9975
40.0	8.6809E-01	1.0158	0.9902	0.9874	0.9859	1.0075	1.0015	1.0000	1.0000	0.9944	-0.4106	0.9944	0.9976
45.0	8.6945E-01	1.0177	0.9892	0.9862	0.9847	1.0084	1.0017	1.0000	1.0000	0.9939	-0.5486	0.9939	0.9977
50.0	8.7084E-01	1.0196	0.9883	0.9850	0.9834	1.0093	1.0019	1.0000	1.0000	0.9935	-0.6861	0.9935	0.9978

POSITRON Z= -8 A= 17

p	F_0L_0	L_0	λ_2	λ_3	λ_4	μ_1	μ_2	Λ_1	Λ_2	ν_{12}	$\hat{\nu}_{12}$	η_{12}	$\hat{\eta}_{12}$
0.1	9.7716E-02	1.0015	1.3253	1.4321	1.4829	1.0000	1.0000	0.9998	0.9999	1.0783	0.7440	0.9938	0.9922
0.2	3.5067E-01	1.0014	1.0738	1.0924	1.1005	1.0000	1.0000	0.9998	1.0000	1.0151	0.7420	0.9941	0.9925
0.3	5.0790E-01	1.0015	1.0277	1.0336	1.0359	1.0000	1.0000	0.9998	1.0000	1.0036	0.7394	0.9944	0.9927
0.4	6.0213E-01	1.0015	1.0118	1.0136	1.0141	1.0000	1.0000	0.9998	1.0000	0.9997	0.7365	0.9946	0.9929
0.5	6.6243E-01	1.0015	1.0045	1.0046	1.0043	1.0001	1.0000	0.9998	1.0000	0.9979	0.7332	0.9947	0.9931
0.6	7.0327E-01	1.0015	1.0007	0.9998	0.9991	1.0001	1.0000	0.9998	1.0000	0.9970	0.7299	0.9948	0.9932
0.7	7.3216E-01	1.0015	0.9985	0.9970	0.9961	1.0001	1.0000	0.9998	1.0000	0.9965	0.7264	0.9949	0.9933
0.8	7.5330E-01	1.0015	0.9971	0.9953	0.9942	1.0001	1.0000	0.9998	1.0000	0.9962	0.7230	0.9950	0.9934
0.9	7.6919E-01	1.0015	0.9962	0.9942	0.9930	1.0001	1.0000	0.9998	1.0000	0.9961	0.7195	0.9951	0.9935
1.0	7.8141E-01	1.0016	0.9955	0.9934	0.9922	1.0002	1.0000	0.9998	1.0000	0.9959	0.7161	0.9952	0.9936
1.2	7.9862E-01	1.0016	0.9948	0.9925	0.9912	1.0002	1.0000	0.9999	1.0000	0.9958	0.7093	0.9953	0.9938
1.4	8.0981E-01	1.0017	0.9944	0.9920	0.9907	1.0003	1.0001	0.9999	1.0000	0.9958	0.7026	0.9954	0.9939
1.6	8.1745E-01	1.0017	0.9942	0.9918	0.9904	1.0003	1.0001	0.9999	1.0000	0.9958	0.6962	0.9955	0.9940
1.8	8.2286E-01	1.0018	0.9941	0.9916	0.9903	1.0004	1.0001	0.9999	1.0000	0.9958	0.6899	0.9956	0.9942
2.0	8.2682E-01	1.0019	0.9940	0.9916	0.9902	1.0004	1.0001	0.9999	1.0000	0.9958	0.6837	0.9957	0.9942
2.2	8.2979E-01	1.0020	0.9940	0.9915	0.9902	1.0005	1.0001	0.9999	1.0000	0.9959	0.6778	0.9957	0.9943
2.4	8.3206E-01	1.0020	0.9940	0.9915	0.9902	1.0005	1.0001	0.9999	1.0000	0.9959	0.6718	0.9958	0.9944
2.6	8.3385E-01	1.0021	0.9940	0.9916	0.9902	1.0006	1.0001	0.9999	1.0000	0.9959	0.6658	0.9958	0.9945
2.8	8.3526E-01	1.0022	0.9940	0.9916	0.9903	1.0006	1.0001	0.9999	1.0000	0.9959	0.6600	0.9959	0.9946
3.0	8.3640E-01	1.0023	0.9940	0.9916	0.9903	1.0007	1.0001	0.9999	1.0000	0.9960	0.6542	0.9959	0.9946
3.2	8.3733E-01	1.0024	0.9940	0.9916	0.9903	1.0007	1.0001	0.9999	1.0000	0.9960	0.6486	0.9959	0.9947
3.4	8.3810E-01	1.0025	0.9940	0.9917	0.9904	1.0008	1.0002	0.9999	1.0000	0.9960	0.6428	0.9960	0.9947
3.6	8.3875E-01	1.0025	0.9940	0.9917	0.9904	1.0008	1.0002	0.9999	1.0000	0.9960	0.6372	0.9960	0.9948
3.8	8.3929E-01	1.0026	0.9940	0.9917	0.9904	1.0009	1.0002	0.9999	1.0000	0.9960	0.6314	0.9960	0.9948
4.0	8.3975E-01	1.0027	0.9940	0.9917	0.9905	1.0009	1.0002	0.9999	1.0000	0.9960	0.6257	0.9960	0.9949
4.5	8.4064E-01	1.0029	0.9941	0.9918	0.9905	1.0010	1.0002	0.9999	1.0000	0.9961	0.6116	0.9961	0.9950
5.0	8.4128E-01	1.0032	0.9941	0.9918	0.9906	1.0011	1.0002	1.0000	1.0000	0.9961	0.5977	0.9961	0.9951
5.5	8.4175E-01	1.0034	0.9941	0.9918	0.9906	1.0013	1.0003	1.0000	1.0000	0.9961	0.5839	0.9961	0.9952
6.0	8.4212E-01	1.0036	0.9941	0.9918	0.9906	1.0014	1.0003	1.0000	1.0000	0.9961	0.5703	0.9961	0.9952
6.5	8.4242E-01	1.0038	0.9941	0.9918	0.9906	1.0015	1.0003	1.0000	1.0000	0.9961	0.5567	0.9961	0.9953
7.0	8.4266E-01	1.0041	0.9941	0.9918	0.9906	1.0016	1.0003	1.0000	1.0000	0.9961	0.5425	0.9961	0.9954
7.5	8.4287E-01	1.0043	0.9940	0.9918	0.9906	1.0017	1.0003	1.0000	1.0000	0.9961	0.5294	0.9961	0.9954
8.0	8.4305E-01	1.0045	0.9940	0.9918	0.9906	1.0018	1.0004	1.0000	1.0000	0.9961	0.5154	0.9961	0.9955
9.0	8.4336E-01	1.0050	0.9939	0.9917	0.9905	1.0021	1.0004	1.0000	1.0000	0.9960	0.4879	0.9960	0.9956
10.0	8.4364E-01	1.0055	0.9938	0.9916	0.9904	1.0023	1.0005	1.0000	1.0000	0.9960	0.4611	0.9960	0.9957
11.0	8.4389E-01	1.0059	0.9937	0.9915	0.9902	1.0025	1.0005	1.0000	1.0000	0.9959	0.4338	0.9959	0.9958
12.0	8.4414E-01	1.0064	0.9936	0.9913	0.9901	1.0028	1.0006	1.0000	1.0000	0.9959	0.4068	0.9959	0.9959
13.0	8.4438E-01	1.0068	0.9934	0.9911	0.9899	1.0030	1.0006	1.0000	1.0000	0.9958	0.3797	0.9958	0.9959
14.0	8.4462E-01	1.0073	0.9933	0.9910	0.9897	1.0032	1.0006	1.0000	1.0000	0.9957	0.3527	0.9957	0.9960
16.0	8.4511E-01	1.0082	0.9930	0.9906	0.9893	1.0037	1.0007	1.0000	1.0000	0.9956	0.2988	0.9956	0.9961
18.0	8.4562E-01	1.0092	0.9926	0.9902	0.9889	1.0041	1.0008	1.0000	1.0000	0.9954	0.2458	0.9954	0.9962
20.0	8.4614E-01	1.0101	0.9922	0.9897	0.9884	1.0046	1.0009	1.0000	1.0000	0.9952	0.1917	0.9952	0.9963
25.0	8.4751E-01	1.0124	0.9913	0.9886	0.9872	1.0057	1.0012	1.0000	1.0000	0.9947	0.0595	0.9947	0.9965
30.0	8.4897E-01	1.0147	0.9902	0.9873	0.9858	1.0069	1.0014	1.0000	1.0000	0.9942	-0.0729	0.9942	0.9967
35.0	8.5048E-01	1.0170	0.9892	0.9860	0.9844	1.0080	1.0016	1.0000	1.0000	0.9937	-0.2045	0.9937	0.9968
40.0	8.5204E-01	1.0194	0.9881	0.9846	0.9829	1.0091	1.0019	1.0000	1.0000	0.9931	-0.3332	0.9931	0.9969
45.0	8.5364E-01	1.0217	0.9870	0.9833	0.9814	1.0102	1.0021	1.0000	1.0000	0.9926	-0.4606	0.9926	0.9971
50.0	8.5528E-01	1.0240	0.9858	0.9819	0.9799	1.0113	1.0024	1.0000	1.0000	0.9920	-0.5894	0.9920	0.9972

POSITRON Z= -9 A= 19

p	F_0L_0	L_0	λ_2	λ_3	λ_4	μ_1	μ_2	Λ_1	Λ_2	ν_{12}	$\hat{\nu}_{12}$	η_{12}	$\hat{\eta}_{12}$
0.1	6.9216E-02	1.0018	1.4106	1.5544	1.6242	1.0000	1.0000	0.9997	0.9999	1.0991	0.7428	0.9924	0.9902
0.2	3.0306E-01	1.0018	1.0933	1.1173	1.1278	1.0000	1.0000	0.9997	0.9999	1.0192	0.7408	0.9927	0.9906
0.3	4.6410E-01	1.0018	1.0351	1.0426	1.0456	1.0000	1.0000	0.9997	0.9999	1.0047	0.7382	0.9930	0.9909
0.4	5.6398E-01	1.0018	1.0150	1.0173	1.0180	1.0000	1.0000	0.9997	0.9999	0.9997	0.7352	0.9932	0.9911
0.5	6.2892E-01	1.0018	1.0058	1.0059	1.0056	1.0001	1.0000	0.9998	1.0000	0.9975	0.7320	0.9934	0.9913
0.6	6.7328E-01	1.0018	1.0010	0.9999	0.9990	1.0001	1.0000	0.9998	1.0000	0.9964	0.7288	0.9936	0.9915
0.7	7.0482E-01	1.0018	0.9982	0.9964	0.9952	1.0001	1.0000	0.9998	1.0000	0.9957	0.7254	0.9937	0.9916
0.8	7.2798E-01	1.0018	0.9964	0.9942	0.9929	1.0001	1.0000	0.9998	1.0000	0.9953	0.7221	0.9938	0.9918
0.9	7.4543E-01	1.0019	0.9953	0.9928	0.9913	1.0002	1.0000	0.9998	1.0000	0.9951	0.7187	0.9939	0.9919
1.0	7.5887E-01	1.0019	0.9945	0.9918	0.9903	1.0002	1.0000	0.9998	1.0000	0.9950	0.7154	0.9940	0.9920
1.2	7.7781E-01	1.0020	0.9935	0.9907	0.9891	1.0003	1.0001	0.9998	1.0000	0.9948	0.7090	0.9942	0.9922
1.4	7.9014E-01	1.0020	0.9930	0.9901	0.9884	1.0003	1.0001	0.9998	1.0000	0.9948	0.7028	0.9943	0.9924
1.6	7.9856E-01	1.0021	0.9928	0.9898	0.9881	1.0004	1.0001	0.9999	1.0000	0.9948	0.6967	0.9944	0.9925
1.8	8.0452E-01	1.0022	0.9926	0.9896	0.9879	1.0004	1.0001	0.9999	1.0000	0.9948	0.6909	0.9945	0.9927
2.0	8.0887E-01	1.0023	0.9926	0.9895	0.9879	1.0005	1.0001	0.9999	1.0000	0.9948	0.6852	0.9946	0.9928
2.2	8.1213E-01	1.0024	0.9925	0.9895	0.9879	1.0005	1.0001	0.9999	1.0000	0.9948	0.6796	0.9947	0.9929
2.4	8.1463E-01	1.0025	0.9925	0.9895	0.9879	1.0006	1.0001	0.9999	1.0000	0.9949	0.6740	0.9948	0.9930
2.6	8.1658E-01	1.0026	0.9925	0.9895	0.9879	1.0007	1.0001	0.9999	1.0000	0.9949	0.6686	0.9948	0.9931
2.8	8.1812E-01	1.0026	0.9925	0.9896	0.9879	1.0007	1.0001	0.9999	1.0000	0.9949	0.6632	0.9949	0.9932
3.0	8.1937E-01	1.0027	0.9926	0.9896	0.9880	1.0008	1.0002	0.9999	1.0000	0.9950	0.6578	0.9949	0.9933
3.2	8.2038E-01	1.0028	0.9926	0.9896	0.9880	1.0008	1.0002	0.9999	1.0000	0.9950	0.6524	0.9950	0.9933
3.4	8.2122E-01	1.0029	0.9926	0.9897	0.9881	1.0009	1.0002	0.9999	1.0000	0.9950	0.6472	0.9950	0.9934
3.6	8.2191E-01	1.0030	0.9926	0.9897	0.9881	1.0009	1.0002	0.9999	1.0000	0.9951	0.6419	0.9950	0.9935
3.8	8.2250E-01	1.0031	0.9927	0.9898	0.9882	1.0010	1.0002	0.9999	1.0000	0.9951	0.6368	0.9950	0.9935
4.0	8.2300E-01	1.0033	0.9927	0.9898	0.9882	1.0011	1.0002	0.9999	1.0000	0.9951	0.6315	0.9951	0.9936
4.5	8.2395E-01	1.0035	0.9927	0.9899	0.9883	1.0012	1.0002	0.9999	1.0000	0.9951	0.6185	0.9951	0.9937
5.0	8.2463E-01	1.0038	0.9927	0.9899	0.9884	1.0013	1.0003	0.9999	1.0000	0.9952	0.6057	0.9952	0.9938
5.5	8.2512E-01	1.0040	0.9928	0.9900	0.9885	1.0015	1.0003	1.0000	1.0000	0.9952	0.5932	0.9952	0.9939
6.0	8.2550E-01	1.0043	0.9928	0.9900	0.9885	1.0016	1.0003	1.0000	1.0000	0.9952	0.5803	0.9952	0.9940
6.5	8.2580E-01	1.0046	0.9928	0.9900	0.9885	1.0017	1.0004	1.0000	1.0000	0.9952	0.5678	0.9952	0.9941
7.0	8.2605E-01	1.0048	0.9928	0.9900	0.9885	1.0019	1.0004	1.0000	1.0000	0.9952	0.5551	0.9952	0.9942
7.5	8.2626E-01	1.0051	0.9927	0.9900	0.9885	1.0020	1.0004	1.0000	1.0000	0.9952	0.5424	0.9952	0.9943
8.0	8.2645E-01	1.0054	0.9927	0.9900	0.9885	1.0022	1.0004	1.0000	1.0000	0.9952	0.5300	0.9952	0.9944
9.0	8.2677E-01	1.0059	0.9926	0.9899	0.9884	1.0024	1.0005	1.0000	1.0000	0.9952	0.5050	0.9952	0.9945
10.0	8.2705E-01	1.0064	0.9925	0.9898	0.9883	1.0027	1.0005	1.0000	1.0000	0.9951	0.4798	0.9951	0.9946
11.0	8.2731E-01	1.0070	0.9924	0.9897	0.9882	1.0030	1.0006	1.0000	1.0000	0.9950	0.4545	0.9951	0.9947
12.0	8.2756E-01	1.0075	0.9923	0.9895	0.9880	1.0032	1.0007	1.0000	1.0000	0.9950	0.4297	0.9950	0.9948
13.0	8.2781E-01	1.0081	0.9921	0.9894	0.9879	1.0035	1.0007	1.0000	1.0000	0.9949	0.4050	0.9949	0.9949
14.0	8.2807E-01	1.0086	0.9920	0.9892	0.9877	1.0038	1.0008	1.0000	1.0000	0.9948	0.3604	0.9948	0.9950
16.0	8.2859E-01	1.0097	0.9916	0.9888	0.9872	1.0043	1.0009	1.0000	1.0000	0.9947	0.3303	0.9947	0.9951
18.0	8.2913E-01	1.0108	0.9912	0.9883	0.9867	1.0048	1.0010	1.0000	1.0000	0.9945	0.2813	0.9945	0.9953
20.0	8.2969E-01	1.0119	0.9908	0.9878	0.9862	1.0054	1.0011	1.0000	1.0000	0.9943	0.2321	0.9943	0.9954
25.0	8.3119E-01	1.0146	0.9897	0.9865	0.9848	1.0067	1.0014	1.0000	1.0000	0.9937	0.1102	0.9937	0.9956
30.0	8.3280E-01	1.0173	0.9886	0.9851	0.9833	1.0080	1.0016	1.0000	1.0000	0.9931	-0.0119	0.9931	0.9958
35.0	8.3449E-01	1.0200	0.9873	0.9836	0.9817	1.0093	1.0019	1.0000	1.0000	0.9925	-0.1322	0.9925	0.9960
40.0	8.3624E-01	1.0227	0.9861	0.9820	0.9800	1.0106	1.0022	1.0000	1.0000	0.9919	-0.2520	0.9919	0.9962
45.0	8.3803E-01	1.0254	0.9848	0.9805	0.9783	1.0119	1.0025	1.0000	1.0000	0.9913	-0.3695	0.9913	0.9963
50.0	8.3986E-01	1.0281	0.9835	0.9789	0.9766	1.0131	1.0028	1.0000	1.0000	0.9907	-0.4859	0.9907	0.9965

Instr. see p. 22

POSITRON Z= -10 A= 21

p	F_0L_0	L_0	λ_2	λ_3	λ_4	μ_1	μ_2	Λ_1	Λ_2	ν_{12}	$\hat{\nu}_{12}$	η_{12}	$\hat{\eta}_{12}$
0.1	4.8632E-02	1.0021	1.5054	1.6946	1.7887	1.0000	1.0000	0.9997	0.9999	1.1223	0.7416	0.9907	0.9881
0.2	2.6115E-01	1.0021	1.1150	1.1452	1.1584	1.0000	1.0000	0.9997	0.9999	1.0238	0.7395	0.9911	0.9885
0.3	4.2358E-01	1.0021	1.0433	1.0528	1.0565	1.0000	1.0000	0.9997	0.9999	1.0059	0.7368	0.9914	0.9888
0.4	5.2801E-01	1.0021	1.0186	1.0215	1.0223	1.0000	1.0000	0.9997	0.9999	0.9997	0.7338	0.9917	0.9891
0.5	5.9704E-01	1.0021	1.0073	1.0074	1.0070	1.0001	1.0000	0.9997	0.9999	0.9970	0.7307	0.9919	0.9894
0.6	6.4461E-01	1.0022	1.0014	1.0000	0.9990	1.0001	1.0000	0.9997	0.9999	0.9956	0.7274	0.9921	0.9896
0.7	6.7862E-01	1.0022	0.9979	0.9957	0.9943	1.0001	1.0000	0.9997	0.9999	0.9948	0.7242	0.9923	0.9898
0.8	7.0368E-01	1.0022	0.9957	0.9930	0.9914	1.0002	1.0000	0.9998	1.0000	0.9943	0.7209	0.9925	0.9899
0.9	7.2261E-01	1.0022	0.9943	0.9913	0.9895	1.0002	1.0000	0.9998	1.0000	0.9941	0.7177	0.9926	0.9901
1.0	7.3721E-01	1.0023	0.9933	0.9901	0.9882	1.0002	1.0000	0.9998	1.0000	0.9939	0.7145	0.9927	0.9902
1.2	7.5781E-01	1.0023	0.9922	0.9887	0.9867	1.0003	1.0001	0.9998	1.0000	0.9937	0.7084	0.9929	0.9905
1.4	7.7124E-01	1.0024	0.9916	0.9879	0.9859	1.0004	1.0001	0.9998	1.0000	0.9936	0.7025	0.9931	0.9907
1.6	7.8040E-01	1.0025	0.9912	0.9876	0.9855	1.0004	1.0001	0.9998	1.0000	0.9936	0.6967	0.9932	0.9909
1.8	7.8688E-01	1.0026	0.9911	0.9874	0.9853	1.0005	1.0001	0.9998	1.0000	0.9937	0.6912	0.9933	0.9910
2.0	7.9161E-01	1.0027	0.9910	0.9873	0.9853	1.0006	1.0001	0.9999	1.0000	0.9937	0.6859	0.9935	0.9912
2.2	7.9515E-01	1.0028	0.9909	0.9873	0.9853	1.0006	1.0001	0.9999	1.0000	0.9937	0.6806	0.9935	0.9913
2.4	7.9786E-01	1.0029	0.9909	0.9873	0.9853	1.0007	1.0001	0.9999	1.0000	0.9938	0.6754	0.9936	0.9914
2.6	7.9997E-01	1.0030	0.9909	0.9873	0.9853	1.0008	1.0002	0.9999	1.0000	0.9938	0.6702	0.9937	0.9915
2.8	8.0163E-01	1.0031	0.9910	0.9874	0.9854	1.0008	1.0002	0.9999	1.0000	0.9939	0.6653	0.9938	0.9916
3.0	8.0297E-01	1.0032	0.9910	0.9874	0.9855	1.0009	1.0002	0.9999	1.0000	0.9939	0.6603	0.9938	0.9917
3.2	8.0406E-01	1.0033	0.9910	0.9875	0.9855	1.0010	1.0002	0.9999	1.0000	0.9939	0.6553	0.9939	0.9918
3.4	8.0495E-01	1.0035	0.9911	0.9875	0.9856	1.0010	1.0002	0.9999	1.0000	0.9940	0.6504	0.9939	0.9919
3.6	8.0570E-01	1.0036	0.9911	0.9876	0.9857	1.0011	1.0002	0.9999	1.0000	0.9940	0.6455	0.9940	0.9920
3.8	8.0632E-01	1.0037	0.9911	0.9876	0.9857	1.0012	1.0002	0.9999	1.0000	0.9940	0.6407	0.9940	0.9921
4.0	8.0685E-01	1.0038	0.9912	0.9877	0.9858	1.0012	1.0002	0.9999	1.0000	0.9941	0.6358	0.9940	0.9921
4.5	8.0785E-01	1.0041	0.9912	0.9878	0.9859	1.0014	1.0003	0.9999	1.0000	0.9941	0.6237	0.9941	0.9923
5.0	8.0856E-01	1.0044	0.9913	0.9879	0.9860	1.0015	1.0003	0.9999	1.0000	0.9942	0.6121	0.9941	0.9924
5.5	8.0907E-01	1.0047	0.9913	0.9880	0.9861	1.0017	1.0003	0.9999	1.0000	0.9942	0.6002	0.9942	0.9926
6.0	8.0946E-01	1.0050	0.9913	0.9880	0.9862	1.0019	1.0004	0.9999	1.0000	0.9942	0.5883	0.9942	0.9927
6.5	8.0977E-01	1.0053	0.9914	0.9880	0.9862	1.0020	1.0004	1.0000	1.0000	0.9942	0.5767	0.9942	0.9928
7.0	8.1002E-01	1.0056	0.9914	0.9881	0.9863	1.0022	1.0004	1.0000	1.0000	0.9942	0.5647	0.9942	0.9929
7.5	8.1023E-01	1.0059	0.9913	0.9881	0.9863	1.0023	1.0005	1.0000	1.0000	0.9942	0.5530	0.9942	0.9930
8.0	8.1041E-01	1.0062	0.9913	0.9880	0.9863	1.0025	1.0005	1.0000	1.0000	0.9942	0.5416	0.9942	0.9931
9.0	8.1072E-01	1.0069	0.9913	0.9880	0.9862	1.0028	1.0006	1.0000	1.0000	0.9942	0.5180	0.9942	0.9933
10.0	8.1100E-01	1.0075	0.9912	0.9879	0.9861	1.0031	1.0006	1.0000	1.0000	0.9942	0.4948	0.9942	0.9934
11.0	8.1126E-01	1.0081	0.9910	0.9878	0.9860	1.0034	1.0007	1.0000	1.0000	0.9941	0.4717	0.9941	0.9935
12.0	8.1152E-01	1.0087	0.9909	0.9876	0.9858	1.0037	1.0007	1.0000	1.0000	0.9940	0.4481	0.9940	0.9936
13.0	8.1177E-01	1.0093	0.9907	0.9874	0.9857	1.0040	1.0008	1.0000	1.0000	0.9940	0.4253	0.9940	0.9938
14.0	8.1203E-01	1.0100	0.9906	0.9872	0.9855	1.0043	1.0009	1.0000	1.0000	0.9939	0.4021	0.9939	0.9939
16.0	8.1257E-01	1.0112	0.9902	0.9868	0.9850	1.0049	1.0010	1.0000	1.0000	0.9937	0.3563	0.9937	0.9940
18.0	8.1315E-01	1.0125	0.9898	0.9863	0.9845	1.0056	1.0011	1.0000	1.0000	0.9935	0.3102	0.9935	0.9942
20.0	8.1375E-01	1.0137	0.9893	0.9858	0.9839	1.0062	1.0013	1.0000	1.0000	0.9933	0.2647	0.9933	0.9944
25.0	8.1537E-01	1.0168	0.9881	0.9843	0.9823	1.0077	1.0016	1.0000	1.0000	0.9927	0.1510	0.9927	0.9947
30.0	8.1712E-01	1.0199	0.9868	0.9827	0.9806	1.0092	1.0019	1.0000	1.0000	0.9920	0.0378	0.9920	0.9949
35.0	8.1897E-01	1.0230	0.9855	0.9811	0.9788	1.0107	1.0022	1.0000	1.0000	0.9913	-0.0734	0.9913	0.9951
40.0	8.2089E-01	1.0262	0.9841	0.9794	0.9770	1.0122	1.0025	1.0000	1.0000	0.9906	-0.1822	0.9906	0.9954
45.0	8.2288E-01	1.0293	0.9826	0.9776	0.9751	1.0136	1.0028	1.0000	1.0000	0.9899	-0.2924	0.9899	0.9955
50.0	8.2490E-01	1.0324	0.9812	0.9758	0.9731	1.0151	1.0032	1.0000	1.0000	0.9892	-0.4011	0.9892	0.9957

POSITRON Z= -11 A= 23

p	F_0L_0	L_0	λ_2	λ_3	λ_4	μ_1	μ_2	Λ_1	Λ_2	ν_{12}	$\hat{\nu}_{12}$	η_{12}	$\hat{\eta}_{12}$
0.1	3.3927E-02	1.0025	1.6096	1.8541	1.9786	1.0000	1.0000	0.9996	0.9999	1.1480	0.7402	0.9890	0.9858
0.2	2.2442E-01	1.0025	1.1388	1.1761	1.1926	1.0000	1.0000	0.9997	0.9999	1.0289	0.7380	0.9894	0.9862
0.3	3.8618E-01	1.0025	1.0523	1.0639	1.0685	1.0000	1.0000	0.9997	0.9999	1.0072	0.7353	0.9898	0.9866
0.4	4.9413E-01	1.0025	1.0225	1.0261	1.0272	1.0001	1.0000	0.9997	0.9999	0.9998	0.7323	0.9901	0.9870
0.5	5.6671E-01	1.0025	1.0090	1.0091	1.0087	1.0001	1.0000	0.9997	0.9999	0.9965	0.7292	0.9904	0.9872
0.6	6.1721E-01	1.0025	1.0018	1.0002	0.9989	1.0001	1.0000	0.9997	0.9999	0.9948	0.7260	0.9906	0.9875
0.7	6.5352E-01	1.0026	0.9976	0.9950	0.9933	1.0001	1.0000	0.9997	0.9999	0.9938	0.7227	0.9908	0.9877
0.8	6.8037E-01	1.0026	0.9950	0.9917	0.9898	1.0002	1.0000	0.9997	0.9999	0.9933	0.7195	0.9910	0.9879
0.9	7.0070E-01	1.0026	0.9932	0.9896	0.9875	1.0002	1.0000	0.9997	0.9999	0.9929	0.7164	0.9911	0.9881
1.0	7.1641E-01	1.0027	0.9921	0.9882	0.9860	1.0003	1.0001	0.9997	0.9999	0.9927	0.7133	0.9913	0.9883
1.2	7.3860E-01	1.0027	0.9907	0.9865	0.9841	1.0003	1.0001	0.9998	1.0000	0.9925	0.7074	0.9915	0.9886
1.4	7.5307E-01	1.0028	0.9899	0.9856	0.9832	1.0004	1.0001	0.9998	1.0000	0.9924	0.7018	0.9917	0.9888
1.6	7.6294E-01	1.0029	0.9896	0.9852	0.9827	1.0005	1.0001	0.9998	1.0000	0.9924	0.6963	0.9919	0.9890
1.8	7.6993E-01	1.0030	0.9894	0.9849	0.9825	1.0006	1.0001	0.9998	1.0000	0.9924	0.6911	0.9920	0.9892
2.0	7.7502E-01	1.0031	0.9892	0.9848	0.9824	1.0006	1.0001	0.9998	1.0000	0.9925	0.6860	0.9922	0.9894
2.2	7.7882E-01	1.0033	0.9892	0.9848	0.9824	1.0007	1.0001	0.9999	1.0000	0.9925	0.6810	0.9923	0.9896
2.4	7.8173E-01	1.0034	0.9892	0.9848	0.9825	1.0008	1.0002	0.9999	1.0000	0.9926	0.6761	0.9924	0.9897
2.6	7.8398E-01	1.0035	0.9892	0.9849	0.9825	1.0009	1.0002	0.9999	1.0000	0.9926	0.6713	0.9925	0.9898
2.8	7.8577E-01	1.0036	0.9893	0.9850	0.9826	1.0009	1.0002	0.9999	1.0000	0.9927	0.6666	0.9926	0.9900
3.0	7.8719E-01	1.0038	0.9893	0.9850	0.9827	1.0010	1.0002	0.9999	1.0000	0.9927	0.6619	0.9926	0.9901
3.2	7.8835E-01	1.0039	0.9894	0.9851	0.9828	1.0011	1.0002	0.9999	1.0000	0.9928	0.6573	0.9927	0.9902
3.4	7.8930E-01	1.0040	0.9894	0.9852	0.9829	1.0012	1.0002	0.9999	1.0000	0.9928	0.6527	0.9928	0.9903
3.6	7.9009E-01	1.0042	0.9894	0.9852	0.9830	1.0012	1.0002	0.9999	1.0000	0.9929	0.6481	0.9928	0.9904
3.8	7.9074E-01	1.0043	0.9895	0.9853	0.9830	1.0013	1.0003	0.9999	1.0000	0.9929	0.6437	0.9929	0.9905
4.0	7.9129E-01	1.0044	0.9895	0.9854	0.9831	1.0014	1.0003	0.9999	1.0000	0.9929	0.6390	0.9929	0.9906
4.5	7.9234E-01	1.0048	0.9896	0.9855	0.9833	1.0016	1.0003	0.9999	1.0000	0.9930	0.6279	0.9930	0.9908
5.0	7.9307E-01	1.0051	0.9897	0.9857	0.9835	1.0017	1.0003	0.9999	1.0000	0.9931	0.6167	0.9931	0.9909
5.5	7.9359E-01	1.0054	0.9898	0.9858	0.9836	1.0019	1.0004	0.9999	1.0000	0.9931	0.6056	0.9931	0.9911
6.0	7.9398E-01	1.0058	0.9898	0.9858	0.9837	1.0021	1.0004	0.9999	1.0000	0.9932	0.5946	0.9931	0.9912
6.5	7.9429E-01	1.0061	0.9898	0.9859	0.9837	1.0023	1.0005	0.9999	1.0000	0.9932	0.5836	0.9932	0.9914
7.0	7.9454E-01	1.0065	0.9898	0.9859	0.9838	1.0025	1.0005	1.0000	1.0000	0.9932	0.5725	0.9932	0.9915
7.5	7.9474E-01	1.0068	0.9898	0.9859	0.9838	1.0026	1.0005	1.0000	1.0000	0.9932	0.5615	0.9932	0.9916
8.0	7.9492E-01	1.0072	0.9898	0.9860	0.9838	1.0028	1.0006	1.0000	1.0000	0.9932	0.5507	0.9932	0.9917
9.0	7.9522E-01	1.0079	0.9898	0.9859	0.9838	1.0032	1.0006	1.0000	1.0000	0.9932	0.5289	0.9932	0.9919
10.0	7.9549E-01	1.0086	0.9897	0.9858	0.9837	1.0035	1.0007	1.0000	1.0000	0.9931	0.5073	0.9931	0.9921
11.0	7.9574E-01	1.0093	0.9896	0.9857	0.9836	1.0039	1.0008	1.0000	1.0000	0.9931	0.4855	0.9931	0.9922
12.0	7.9599E-01	1.0100	0.9894	0.9856	0.9835	1.0042	1.0008	1.0000	1.0000	0.9930	0.4636	0.9930	0.9924
13.0	7.9625E-01	1.0107	0.9893	0.9854	0.9833	1.0046	1.0009	1.0000	1.0000	0.9929	0.4423	0.9929	0.9925
14.0	7.9651E-01	1.0114	0.9891	0.9852	0.9831	1.0049	1.0010	1.0000	1.0000	0.9929	0.4207	0.9929	0.9926
16.0	7.9707E-01	1.0128	0.9887	0.9847	0.9826	1.0056	1.0011	1.0000	1.0000	0.9927	0.3773	0.9927	0.9929
18.0	7.9766E-01	1.0142	0.9882	0.9842	0.9821	1.0063	1.0013	1.0000	1.0000	0.9924	0.3345	0.9924	0.9931
20.0	7.9829E-01	1.0156	0.9878	0.9836	0.9814	1.0070	1.0014	1.0000	1.0000	0.9922	0.2916	0.9922	0.9932
25.0	8.0002E-01	1.0192	0.9864	0.9821	0.9798	1.0087	1.0018	1.0000	1.0000	0.9915	0.1861	0.9915	0.9936
30.0	8.0190E-01	1.0227	0.9850	0.9803	0.9779	1.0104	1.0021	1.0000	1.0000	0.9908	0.0795	0.9908	0.9939
35.0	8.0390E-01	1.0262	0.9835	0.9785	0.9759	1.0121	1.0025	1.0000	1.0000	0.9901	-0.0242	0.9901	0.9942
40.0	8.0599E-01	1.0298	0.9819	0.9766	0.9738	1.0138	1.0029	1.0000	1.0000	0.9893	-0.1276	0.9893	0.9944
45.0	8.0816E-01	1.0333	0.9804	0.9746	0.9717	1.0154	1.0032	1.0000	1.0000	0.9885	-0.2294	0.9885	0.9947
50.0	8.1037E-01	1.0368	0.9788	0.9726	0.9696	1.0170	1.0036	1.0000	1.0000	0.9877	-0.3298	0.9877	0.9949

Instr. see p. 22

POSITRON Z= -12 A= 24

p	F_0L_0	L_0	λ_2	λ_3	λ_4	μ_1	μ_2	Λ_1	Λ_2	ν_{12}	$\hat{\nu}_{12}$	η_{12}	$\hat{\eta}_{12}$
0.1	2.3524E-02	1.0029	1.7228	2.0338	2.1962	1.0000	1.0000	0.9996	0.9999	1.1760	0.7387	0.9871	0.9833
0.2	1.9236E-01	1.0028	1.1647	1.2101	1.2303	1.0000	1.0000	0.9996	0.9999	1.0345	0.7365	0.9875	0.9838
0.3	3.5172E-01	1.0029	1.0622	1.0761	1.0816	1.0000	1.0000	0.9996	0.9999	1.0087	0.7338	0.9880	0.9842
0.4	4.6226E-01	1.0029	1.0268	1.0311	1.0324	1.0001	1.0000	0.9996	0.9999	0.9998	0.7308	0.9883	0.9846
0.5	5.3791E-01	1.0029	1.0107	1.0109	1.0104	1.0001	1.0000	0.9996	0.9999	0.9959	0.7276	0.9886	0.9849
0.6	5.9107E-01	1.0029	1.0022	1.0003	0.9989	1.0001	1.0000	0.9997	0.9999	0.9939	0.7244	0.9889	0.9852
0.7	6.2951E-01	1.0029	0.9972	0.9942	0.9922	1.0002	1.0000	0.9997	0.9999	0.9928	0.7213	0.9891	0.9855
0.8	6.5805E-01	1.0030	0.9941	0.9903	0.9880	1.0002	1.0000	0.9997	0.9999	0.9921	0.7181	0.9893	0.9857
0.9	6.7971E-01	1.0030	0.9921	0.9878	0.9853	1.0002	1.0000	0.9997	0.9999	0.9917	0.7151	0.9895	0.9859
1.0	6.9647E-01	1.0030	0.9907	0.9861	0.9835	1.0003	1.0001	0.9997	0.9999	0.9914	0.7121	0.9897	0.9861
1.2	7.2018E-01	1.0031	0.9890	0.9841	0.9813	1.0004	1.0001	0.9997	0.9999	0.9911	0.7064	0.9900	0.9865
1.4	7.3565E-01	1.0032	0.9882	0.9831	0.9802	1.0005	1.0001	0.9998	1.0000	0.9910	0.7011	0.9902	0.9868
1.6	7.4621E-01	1.0034	0.9877	0.9825	0.9797	1.0005	1.0001	0.9998	1.0000	0.9910	0.6959	0.9904	0.9870
1.8	7.5368E-01	1.0035	0.9875	0.9823	0.9794	1.0006	1.0001	0.9998	1.0000	0.9911	0.6910	0.9906	0.9872
2.0	7.5911E-01	1.0036	0.9874	0.9822	0.9793	1.0007	1.0001	0.9998	1.0000	0.9911	0.6862	0.9908	0.9875
2.2	7.6317E-01	1.0037	0.9873	0.9821	0.9793	1.0008	1.0002	0.9998	1.0000	0.9912	0.6816	0.9909	0.9876
2.4	7.6626E-01	1.0039	0.9873	0.9822	0.9794	1.0009	1.0002	0.9998	1.0000	0.9913	0.6770	0.9910	0.9878
2.6	7.6866E-01	1.0040	0.9874	0.9822	0.9794	1.0010	1.0002	0.9999	1.0000	0.9913	0.6725	0.9912	0.9880
2.8	7.7055E-01	1.0041	0.9874	0.9823	0.9796	1.0010	1.0002	0.9999	1.0000	0.9914	0.6681	0.9913	0.9881
3.0	7.7206E-01	1.0043	0.9875	0.9824	0.9797	1.0011	1.0002	0.9999	1.0000	0.9915	0.6637	0.9913	0.9883
3.2	7.7327E-01	1.0044	0.9875	0.9825	0.9798	1.0012	1.0002	0.9999	1.0000	0.9915	0.6594	0.9914	0.9884
3.4	7.7427E-01	1.0046	0.9876	0.9826	0.9799	1.0013	1.0003	0.9999	1.0000	0.9916	0.6552	0.9915	0.9885
3.6	7.7509E-01	1.0047	0.9876	0.9827	0.9800	1.0014	1.0003	0.9999	1.0000	0.9916	0.6509	0.9916	0.9886
3.8	7.7577E-01	1.0049	0.9877	0.9828	0.9801	1.0014	1.0003	0.9999	1.0000	0.9917	0.6466	0.9916	0.9887
4.0	7.7635E-01	1.0050	0.9878	0.9829	0.9802	1.0015	1.0003	0.9999	1.0000	0.9917	0.6425	0.9917	0.9888
4.5	7.7743E-01	1.0054	0.9879	0.9831	0.9805	1.0017	1.0003	0.9999	1.0000	0.9918	0.6320	0.9918	0.9891
5.0	7.7816E-01	1.0058	0.9880	0.9833	0.9807	1.0019	1.0004	0.9999	1.0000	0.9919	0.6217	0.9919	0.9893
5.5	7.7868E-01	1.0061	0.9881	0.9834	0.9808	1.0021	1.0004	0.9999	1.0000	0.9920	0.6115	0.9919	0.9895
6.0	7.7907E-01	1.0065	0.9881	0.9835	0.9810	1.0023	1.0005	0.9999	1.0000	0.9920	0.6012	0.9920	0.9896
6.5	7.7936E-01	1.0069	0.9882	0.9836	0.9811	1.0025	1.0005	0.9999	1.0000	0.9920	0.5910	0.9920	0.9898
7.0	7.7959E-01	1.0073	0.9882	0.9837	0.9812	1.0027	1.0005	0.9999	1.0000	0.9921	0.5809	0.9921	0.9899
7.5	7.7979E-01	1.0077	0.9882	0.9837	0.9812	1.0029	1.0006	0.9999	1.0000	0.9921	0.5704	0.9921	0.9901
8.0	7.7995E-01	1.0081	0.9883	0.9837	0.9813	1.0031	1.0006	1.0000	1.0000	0.9921	0.5604	0.9921	0.9902
9.0	7.8023E-01	1.0088	0.9882	0.9837	0.9813	1.0035	1.0007	1.0000	1.0000	0.9921	0.5403	0.9921	0.9904
10.0	7.8047E-01	1.0096	0.9881	0.9837	0.9812	1.0039	1.0008	1.0000	1.0000	0.9921	0.5200	0.9921	0.9906
11.0	7.8070E-01	1.0104	0.9880	0.9836	0.9812	1.0043	1.0009	1.0000	1.0000	0.9920	0.4998	0.9920	0.9908
12.0	7.8094E-01	1.0112	0.9879	0.9834	0.9810	1.0047	1.0009	1.0000	1.0000	0.9919	0.4798	0.9920	0.9910
13.0	7.8118E-01	1.0120	0.9878	0.9833	0.9808	1.0051	1.0010	1.0000	1.0000	0.9919	0.4598	0.9919	0.9911
14.0	7.8143E-01	1.0127	0.9876	0.9831	0.9806	1.0054	1.0011	1.0000	1.0000	0.9918	0.4396	0.9918	0.9913
16.0	7.8198E-01	1.0143	0.9872	0.9826	0.9802	1.0062	1.0013	1.0000	1.0000	0.9916	0.3996	0.9916	0.9916
18.0	7.8257E-01	1.0159	0.9867	0.9821	0.9796	1.0070	1.0014	1.0000	1.0000	0.9914	0.3598	0.9914	0.9918
20.0	7.8321E-01	1.0174	0.9862	0.9815	0.9790	1.0077	1.0016	1.0000	1.0000	0.9911	0.3205	0.9911	0.9920
25.0	7.8499E-01	1.0213	0.9848	0.9798	0.9772	1.0096	1.0020	1.0000	1.0000	0.9904	0.2215	0.9904	0.9924
30.0	7.8695E-01	1.0253	0.9833	0.9780	0.9752	1.0115	1.0024	1.0000	1.0000	0.9897	0.1238	0.9897	0.9928
35.0	7.8905E-01	1.0292	0.9817	0.9760	0.9731	1.0134	1.0028	1.0000	1.0000	0.9889	0.0266	0.9889	0.9931
40.0	7.9127E-01	1.0331	0.9800	0.9740	0.9709	1.0152	1.0032	1.0000	1.0000	0.9880	-0.0692	0.9880	0.9934
45.0	7.9356E-01	1.0370	0.9783	0.9719	0.9686	1.0170	1.0036	1.0000	1.0000	0.9872	-0.1632	0.9872	0.9937
50.0	7.9591E-01	1.0409	0.9765	0.9697	0.9663	1.0188	1.0040	1.0000	1.0000	0.9863	-0.2571	0.9863	0.9939

Instr. see p. 22

POSITRON Z= -13 A= 27

p	F_0L_0	L_0	λ_2	λ_3	λ_4	μ_1	μ_2	Λ_1	Λ_2	ν_{12}	$\hat{\nu}_{12}$	η_{12}	$\hat{\eta}_{12}$
0.1	1.6220E-02	1.0033	1.8450	2.2355	2.4448	1.0000	1.0000	0.9996	0.9999	1.2064	0.7371	0.9851	0.9807
0.2	1.6445E-01	1.0033	1.1927	1.2473	1.2717	1.0000	1.0000	0.9996	0.9999	1.0406	0.7349	0.9856	0.9812
0.3	3.1995E-01	1.0033	1.0728	1.0894	1.0960	1.0000	1.0000	0.9996	0.9999	1.0103	0.7321	0.9861	0.9817
0.4	4.3219E-01	1.0033	1.0315	1.0366	1.0382	1.0001	1.0000	0.9996	0.9999	1.0000	0.7291	0.9865	0.9821
0.5	5.1044E-01	1.0033	1.0127	1.0130	1.0124	1.0001	1.0000	0.9996	0.9999	0.9954	0.7259	0.9868	0.9825
0.6	5.6598E-01	1.0034	1.0027	1.0006	0.9989	1.0001	1.0000	0.9996	0.9999	0.9930	0.7226	0.9871	0.9828
0.7	6.0639E-01	1.0034	0.9969	0.9934	0.9911	1.0002	1.0000	0.9996	0.9999	0.9916	0.7194	0.9874	0.9831
0.8	6.3651E-01	1.0034	0.9933	0.9889	0.9862	1.0002	1.0000	0.9997	0.9999	0.9908	0.7163	0.9876	0.9834
0.9	6.5942E-01	1.0035	0.9909	0.9860	0.9831	1.0003	1.0001	0.9997	0.9999	0.9904	0.7133	0.9879	0.9836
1.0	6.7718E-01	1.0035	0.9893	0.9840	0.9809	1.0003	1.0001	0.9997	0.9999	0.9900	0.7104	0.9881	0.9838
1.2	7.0235E-01	1.0036	0.9873	0.9816	0.9784	1.0004	1.0001	0.9997	0.9999	0.9897	0.7048	0.9884	0.9842
1.4	7.1879E-01	1.0037	0.9863	0.9804	0.9772	1.0005	1.0001	0.9997	0.9999	0.9896	0.6995	0.9887	0.9846
1.6	7.3001E-01	1.0039	0.9858	0.9798	0.9765	1.0006	1.0001	0.9998	1.0000	0.9896	0.6945	0.9889	0.9849
1.8	7.3793E-01	1.0040	0.9855	0.9795	0.9762	1.0007	1.0001	0.9998	1.0000	0.9897	0.6897	0.9891	0.9852
2.0	7.4370E-01	1.0041	0.9854	0.9794	0.9761	1.0008	1.0002	0.9998	1.0000	0.9897	0.6851	0.9893	0.9854
2.2	7.4800E-01	1.0043	0.9854	0.9794	0.9761	1.0009	1.0002	0.9998	1.0000	0.9898	0.6806	0.9895	0.9856
2.4	7.5127E-01	1.0044	0.9854	0.9794	0.9762	1.0010	1.0002	0.9998	1.0000	0.9899	0.6762	0.9896	0.9858
2.6	7.5380E-01	1.0046	0.9854	0.9795	0.9763	1.0011	1.0002	0.9998	1.0000	0.9900	0.6719	0.9898	0.9860
2.8	7.5579E-01	1.0047	0.9855	0.9796	0.9764	1.0012	1.0002	0.9999	1.0000	0.9901	0.6676	0.9899	0.9862
3.0	7.5737E-01	1.0049	0.9855	0.9797	0.9765	1.0013	1.0003	0.9999	1.0000	0.9901	0.6635	0.9900	0.9863
3.2	7.5865E-01	1.0051	0.9856	0.9798	0.9767	1.0014	1.0003	0.9999	1.0000	0.9902	0.6593	0.9901	0.9865
3.4	7.5969E-01	1.0052	0.9857	0.9799	0.9768	1.0015	1.0003	0.9999	1.0000	0.9903	0.6552	0.9902	0.9866
3.6	7.6055E-01	1.0054	0.9858	0.9801	0.9769	1.0015	1.0003	0.9999	1.0000	0.9903	0.6511	0.9902	0.9868
3.8	7.6125E-01	1.0056	0.9858	0.9802	0.9771	1.0016	1.0003	0.9999	1.0000	0.9904	0.6472	0.9903	0.9869
4.0	7.6185E-01	1.0057	0.9859	0.9803	0.9772	1.0017	1.0003	0.9999	1.0000	0.9905	0.6431	0.9904	0.9870
4.5	7.6296E-01	1.0061	0.9861	0.9805	0.9775	1.0020	1.0004	0.9999	1.0000	0.9906	0.6330	0.9905	0.9873
5.0	7.6370E-01	1.0066	0.9862	0.9807	0.9777	1.0022	1.0004	0.9999	1.0000	0.9907	0.6232	0.9906	0.9875
5.5	7.6423E-01	1.0070	0.9863	0.9809	0.9780	1.0024	1.0005	0.9999	1.0000	0.9907	0.6133	0.9907	0.9877
6.0	7.6461E-01	1.0074	0.9864	0.9811	0.9781	1.0026	1.0005	0.9999	1.0000	0.9908	0.6035	0.9908	0.9879
6.5	7.6489E-01	1.0079	0.9865	0.9812	0.9783	1.0028	1.0006	0.9999	1.0000	0.9908	0.5937	0.9908	0.9881
7.0	7.6512E-01	1.0083	0.9865	0.9813	0.9784	1.0031	1.0006	0.9999	1.0000	0.9909	0.5839	0.9909	0.9883
7.5	7.6530E-01	1.0087	0.9866	0.9813	0.9785	1.0033	1.0007	0.9999	1.0000	0.9909	0.5741	0.9909	0.9885
8.0	7.6545E-01	1.0092	0.9866	0.9814	0.9785	1.0035	1.0007	0.9999	1.0000	0.9909	0.5644	0.9909	0.9886
9.0	7.6571E-01	1.0100	0.9866	0.9814	0.9786	1.0039	1.0008	1.0000	1.0000	0.9909	0.5450	0.9909	0.9889
10.0	7.6594E-01	1.0109	0.9865	0.9814	0.9786	1.0044	1.0009	1.0000	1.0000	0.9909	0.5256	0.9909	0.9891
11.0	7.6616E-01	1.0118	0.9864	0.9813	0.9785	1.0048	1.0010	1.0000	1.0000	0.9909	0.5064	0.9909	0.9893
12.0	7.6639E-01	1.0127	0.9863	0.9811	0.9784	1.0053	1.0011	1.0000	1.0000	0.9908	0.4870	0.9908	0.9895
13.0	7.6664E-01	1.0136	0.9861	0.9810	0.9782	1.0057	1.0011	1.0000	1.0000	0.9907	0.4675	0.9907	0.9897
14.0	7.6689E-01	1.0144	0.9859	0.9808	0.9780	1.0061	1.0012	1.0000	1.0000	0.9906	0.4488	0.9906	0.9899
16.0	7.6745E-01	1.0162	0.9855	0.9803	0.9775	1.0070	1.0014	1.0000	1.0000	0.9904	0.4101	0.9904	0.9902
18.0	7.6807E-01	1.0180	0.9850	0.9797	0.9769	1.0079	1.0016	1.0000	1.0000	0.9902	0.3720	0.9902	0.9905
20.0	7.6875E-01	1.0197	0.9844	0.9791	0.9762	1.0087	1.0018	1.0000	1.0000	0.9899	0.3341	0.9899	0.9907
25.0	7.7064E-01	1.0242	0.9829	0.9773	0.9743	1.0108	1.0022	1.0000	1.0000	0.9891	0.2396	0.9892	0.9912
30.0	7.7275E-01	1.0286	0.9812	0.9752	0.9721	1.0129	1.0027	1.0000	1.0000	0.9883	0.1459	0.9883	0.9917
35.0	7.7503E-01	1.0330	0.9794	0.9731	0.9698	1.0150	1.0031	1.0000	1.0000	0.9874	0.0531	0.9874	0.9921
40.0	7.7743E-01	1.0374	0.9776	0.9708	0.9673	1.0171	1.0036	1.0000	1.0000	0.9865	-0.0388	0.9865	0.9924
45.0	7.7992E-01	1.0418	0.9757	0.9685	0.9648	1.0191	1.0040	1.0000	1.0000	0.9856	-0.1284	0.9856	0.9927
50.0	7.8249E-01	1.0462	0.9738	0.9661	0.9622	1.0211	1.0045	1.0000	1.0000	0.9846	-0.2171	0.9846	0.9930

Instr. see p. 22

POSITRON Z= -14 A= 29

p	F_0L_0	L_0	λ_2	λ_3	λ_4	μ_1	μ_2	Λ_1	Λ_2	ν_{12}	$\hat{\nu}_{12}$	η_{12}	$\hat{\eta}_{12}$
0.1	1.1133E-02	1.0037	1.9759	2.4602	2.7272	1.0000	1.0000	0.9995	0.9999	1.2391	0.7354	0.9830	0.9778
0.2	1.4027E-01	1.0037	1.2227	1.2876	1.3168	1.0000	1.0000	0.9995	0.9999	1.0472	0.7332	0.9835	0.9784
0.3	2.9078E-01	1.0037	1.0843	1.1037	1.1115	1.0000	1.0000	0.9995	0.9999	1.0121	0.7304	0.9840	0.9789
0.4	4.0395E-01	1.0038	1.0365	1.0426	1.0444	1.0001	1.0000	0.9995	0.9999	1.0001	0.7273	0.9845	0.9794
0.5	4.8437E-01	1.0038	1.0148	1.0152	1.0145	1.0001	1.0000	0.9996	0.9999	0.9948	0.7240	0.9849	0.9798
0.6	5.4205E-01	1.0038	1.0033	1.0009	0.9989	1.0002	1.0000	0.9996	0.9999	0.9920	0.7208	0.9852	0.9802
0.7	5.8427E-01	1.0039	0.9966	0.9925	0.9899	1.0002	1.0000	0.9996	0.9999	0.9905	0.7176	0.9855	0.9805
0.8	6.1587E-01	1.0039	0.9924	0.9873	0.9843	1.0003	1.0000	0.9996	0.9999	0.9895	0.7145	0.9858	0.9809
0.9	6.3997E-01	1.0039	0.9896	0.9840	0.9806	1.0003	1.0001	0.9996	0.9999	0.9890	0.7115	0.9861	0.9811
1.0	6.5869E-01	1.0040	0.9878	0.9817	0.9782	1.0004	1.0001	0.9997	0.9999	0.9886	0.7087	0.9863	0.9814
1.2	6.8525E-01	1.0041	0.9855	0.9790	0.9753	1.0005	1.0001	0.9997	0.9999	0.9882	0.7032	0.9867	0.9819
1.4	7.0262E-01	1.0042	0.9844	0.9776	0.9738	1.0006	1.0001	0.9997	0.9999	0.9881	0.6981	0.9870	0.9823
1.6	7.1447E-01	1.0044	0.9837	0.9769	0.9731	1.0007	1.0001	0.9997	0.9999	0.9881	0.6932	0.9873	0.9826
1.8	7.2284E-01	1.0045	0.9834	0.9765	0.9727	1.0008	1.0002	0.9998	1.0000	0.9882	0.6887	0.9875	0.9829
2.0	7.2892E-01	1.0047	0.9833	0.9764	0.9726	1.0009	1.0002	0.9998	1.0000	0.9882	0.6842	0.9877	0.9832
2.2	7.3345E-01	1.0048	0.9832	0.9764	0.9726	1.0010	1.0002	0.9998	1.0000	0.9883	0.6799	0.9879	0.9834
2.4	7.3689E-01	1.0050	0.9833	0.9764	0.9727	1.0011	1.0002	0.9998	1.0000	0.9884	0.6757	0.9881	0.9837
2.6	7.3954E-01	1.0052	0.9833	0.9765	0.9728	1.0012	1.0002	0.9998	1.0000	0.9885	0.6716	0.9883	0.9839
2.8	7.4163E-01	1.0053	0.9834	0.9767	0.9730	1.0013	1.0003	0.9998	1.0000	0.9886	0.6676	0.9884	0.9841
3.0	7.4328E-01	1.0055	0.9835	0.9768	0.9731	1.0014	1.0003	0.9998	1.0000	0.9887	0.6636	0.9885	0.9843
3.2	7.4461E-01	1.0057	0.9836	0.9769	0.9733	1.0015	1.0003	0.9999	1.0000	0.9888	0.6597	0.9886	0.9844
3.4	7.4569E-01	1.0059	0.9836	0.9771	0.9735	1.0016	1.0003	0.9999	1.0000	0.9889	0.6557	0.9887	0.9846
3.6	7.4657E-01	1.0061	0.9837	0.9772	0.9736	1.0017	1.0003	0.9999	1.0000	0.9889	0.6519	0.9888	0.9848
3.8	7.4730E-01	1.0062	0.9838	0.9773	0.9738	1.0018	1.0004	0.9999	1.0000	0.9890	0.6481	0.9889	0.9849
4.0	7.4791E-01	1.0064	0.9839	0.9775	0.9739	1.0019	1.0004	0.9999	1.0000	0.9891	0.6442	0.9890	0.9850
4.5	7.4904E-01	1.0069	0.9841	0.9778	0.9743	1.0022	1.0004	0.9999	1.0000	0.9892	0.6346	0.9892	0.9854
5.0	7.4978E-01	1.0074	0.9843	0.9780	0.9746	1.0024	1.0005	0.9999	1.0000	0.9893	0.6253	0.9893	0.9856
5.5	7.5030E-01	1.0078	0.9844	0.9783	0.9749	1.0027	1.0005	0.9999	1.0000	0.9894	0.6159	0.9894	0.9859
6.0	7.5066E-01	1.0083	0.9846	0.9785	0.9751	1.0029	1.0006	0.9999	1.0000	0.9895	0.6066	0.9895	0.9861
6.5	7.5093E-01	1.0088	0.9846	0.9786	0.9753	1.0031	1.0006	0.9999	1.0000	0.9896	0.5973	0.9896	0.9863
7.0	7.5114E-01	1.0093	0.9847	0.9787	0.9754	1.0034	1.0007	0.9999	1.0000	0.9896	0.5880	0.9896	0.9865
7.5	7.5130E-01	1.0098	0.9848	0.9788	0.9755	1.0036	1.0007	0.9999	1.0000	0.9897	0.5788	0.9896	0.9867
8.0	7.5144E-01	1.0102	0.9848	0.9789	0.9756	1.0039	1.0008	0.9999	1.0000	0.9897	0.5696	0.9897	0.9869
9.0	7.5167E-01	1.0112	0.9848	0.9790	0.9757	1.0044	1.0009	0.9999	1.0000	0.9897	0.5511	0.9897	0.9872
10.0	7.5187E-01	1.0122	0.9848	0.9790	0.9758	1.0048	1.0010	1.0000	1.0000	0.9897	0.5329	0.9897	0.9875
11.0	7.5207E-01	1.0131	0.9847	0.9789	0.9757	1.0053	1.0011	1.0000	1.0000	0.9896	0.5145	0.9897	0.9877
12.0	7.5228E-01	1.0141	0.9846	0.9788	0.9756	1.0058	1.0012	1.0000	1.0000	0.9896	0.4961	0.9896	0.9880
13.0	7.5251E-01	1.0151	0.9844	0.9786	0.9754	1.0063	1.0013	1.0000	1.0000	0.9895	0.4778	0.9895	0.9882
14.0	7.5276E-01	1.0161	0.9842	0.9784	0.9752	1.0068	1.0014	1.0000	1.0000	0.9894	0.4598	0.9894	0.9884
16.0	7.5331E-01	1.0180	0.9838	0.9779	0.9747	1.0077	1.0016	1.0000	1.0000	0.9892	0.4234	0.9892	0.9887
18.0	7.5393E-01	1.0200	0.9833	0.9773	0.9741	1.0087	1.0018	1.0000	1.0000	0.9890	0.3872	0.9890	0.9890
20.0	7.5462E-01	1.0219	0.9827	0.9767	0.9734	1.0096	1.0020	1.0000	1.0000	0.9887	0.3509	0.9887	0.9893
25.0	7.5657E-01	1.0268	0.9811	0.9748	0.9714	1.0119	1.0024	1.0000	1.0000	0.9879	0.2616	0.9879	0.9899
30.0	7.5878E-01	1.0317	0.9793	0.9726	0.9691	1.0143	1.0029	1.0000	1.0000	0.9870	0.1726	0.9870	0.9904
35.0	7.6118E-01	1.0365	0.9774	0.9703	0.9666	1.0166	1.0034	1.0000	1.0000	0.9860	0.0843	0.9861	0.9909
40.0	7.6372E-01	1.0414	0.9754	0.9678	0.9640	1.0188	1.0039	1.0000	1.0000	0.9851	-0.0022	0.9851	0.9913
45.0	7.6637E-01	1.0463	0.9733	0.9653	0.9612	1.0211	1.0044	1.0000	1.0000	0.9841	-0.0879	0.9841	0.9916
50.0	7.6911E-01	1.0512	0.9713	0.9628	0.9585	1.0233	1.0049	1.0000	1.0000	0.9830	-0.1724	0.9830	0.9920

POSITRON Z= -15 A= 31

p	F_0L_0	L_0	λ_2	λ_3	λ_4	μ_1	μ_2	Λ_1	Λ_2	ν_{12}	$\hat{\nu}_{12}$	η_{12}	$\hat{\eta}_{12}$
0.1	7.6099E-03	1.0042	2.1152	2.7094	3.0468	1.0000	1.0000	0.9995	0.9999	1.2742	0.7336	0.9807	0.9749
0.2	1.1940E-01	1.0042	1.2547	1.3310	1.3657	1.0000	1.0000	0.9995	0.9999	1.0542	0.7314	0.9813	0.9754
0.3	2.6402E-01	1.0042	1.0965	1.1190	1.1281	1.0001	1.0000	0.9995	0.9999	1.0140	0.7285	0.9818	0.9760
0.4	3.7740E-01	1.0042	1.0419	1.0489	1.0510	1.0001	1.0000	0.9995	0.9999	1.0003	0.7254	0.9823	0.9765
0.5	4.5959E-01	1.0043	1.0170	1.0176	1.0168	1.0001	1.0000	0.9995	0.9999	0.9941	0.7221	0.9828	0.9770
0.6	5.1916E-01	1.0043	1.0039	1.0012	0.9989	1.0002	1.0000	0.9995	0.9999	0.9910	0.7189	0.9832	0.9774
0.7	5.6307E-01	1.0043	0.9962	0.9916	0.9886	1.0002	1.0000	0.9996	0.9999	0.9892	0.7157	0.9835	0.9778
0.8	5.9605E-01	1.0044	0.9914	0.9857	0.9822	1.0003	1.0001	0.9996	0.9999	0.9881	0.7126	0.9838	0.9782
0.9	6.2128E-01	1.0044	0.9883	0.9818	0.9780	1.0003	1.0001	0.9996	0.9999	0.9875	0.7097	0.9841	0.9785
1.0	6.4091E-01	1.0045	0.9861	0.9792	0.9752	1.0004	1.0001	0.9996	0.9999	0.9870	0.7068	0.9844	0.9788
1.2	6.6880E-01	1.0046	0.9836	0.9761	0.9719	1.0005	1.0001	0.9997	0.9999	0.9866	0.7015	0.9848	0.9793
1.4	6.8706E-01	1.0047	0.9822	0.9745	0.9702	1.0006	1.0001	0.9997	0.9999	0.9865	0.6965	0.9852	0.9798
1.6	6.9952E-01	1.0049	0.9815	0.9737	0.9694	1.0007	1.0001	0.9997	0.9999	0.9865	0.6918	0.9855	0.9802
1.8	7.0832E-01	1.0051	0.9812	0.9733	0.9690	1.0009	1.0002	0.9997	0.9999	0.9866	0.6873	0.9858	0.9805
2.0	7.1470E-01	1.0052	0.9810	0.9732	0.9689	1.0010	1.0002	0.9998	1.0000	0.9866	0.6831	0.9861	0.9808
2.2	7.1945E-01	1.0054	0.9810	0.9732	0.9689	1.0011	1.0002	0.9998	1.0000	0.9868	0.6790	0.9863	0.9811
2.4	7.2305E-01	1.0056	0.9810	0.9733	0.9690	1.0012	1.0002	0.9998	1.0000	0.9869	0.6750	0.9865	0.9814
2.6	7.2582E-01	1.0058	0.9811	0.9734	0.9692	1.0013	1.0003	0.9998	1.0000	0.9870	0.6710	0.9867	0.9816
2.8	7.2799E-01	1.0060	0.9812	0.9735	0.9693	1.0014	1.0003	0.9998	1.0000	0.9871	0.6672	0.9868	0.9819
3.0	7.2971E-01	1.0062	0.9813	0.9737	0.9695	1.0015	1.0003	0.9998	1.0000	0.9872	0.6633	0.9870	0.9821
3.2	7.3108E-01	1.0064	0.9814	0.9739	0.9697	1.0016	1.0003	0.9998	1.0000	0.9873	0.6596	0.9871	0.9823
3.4	7.3220E-01	1.0066	0.9815	0.9740	0.9699	1.0018	1.0003	0.9998	1.0000	0.9874	0.6559	0.9872	0.9824
3.6	7.3311E-01	1.0068	0.9816	0.9742	0.9701	1.0019	1.0004	0.9999	1.0000	0.9875	0.6522	0.9873	0.9826
3.8	7.3385E-01	1.0070	0.9817	0.9744	0.9703	1.0020	1.0004	0.9999	1.0000	0.9876	0.6485	0.9874	0.9828
4.0	7.3447E-01	1.0072	0.9818	0.9745	0.9705	1.0021	1.0004	0.9999	1.0000	0.9876	0.6448	0.9875	0.9829
4.5	7.3561E-01	1.0077	0.9821	0.9749	0.9709	1.0024	1.0005	0.9999	1.0000	0.9878	0.6358	0.9877	0.9833
5.0	7.3635E-01	1.0082	0.9823	0.9752	0.9713	1.0026	1.0005	0.9999	1.0000	0.9879	0.6268	0.9879	0.9836
5.5	7.3685E-01	1.0087	0.9825	0.9755	0.9716	1.0029	1.0006	0.9999	1.0000	0.9880	0.6179	0.9880	0.9839
6.0	7.3720E-01	1.0092	0.9826	0.9757	0.9719	1.0032	1.0006	0.9999	1.0000	0.9881	0.6090	0.9881	0.9842
6.5	7.3745E-01	1.0098	0.9827	0.9759	0.9721	1.0034	1.0007	0.9999	1.0000	0.9882	0.6002	0.9882	0.9844
7.0	7.3763E-01	1.0103	0.9828	0.9760	0.9723	1.0037	1.0007	0.9999	1.0000	0.9883	0.5914	0.9883	0.9846
7.5	7.3777E-01	1.0108	0.9829	0.9762	0.9725	1.0040	1.0008	0.9999	1.0000	0.9883	0.5825	0.9883	0.9849
8.0	7.3789E-01	1.0113	0.9829	0.9763	0.9726	1.0042	1.0009	0.9999	1.0000	0.9884	0.5737	0.9884	0.9851
9.0	7.3808E-01	1.0124	0.9830	0.9764	0.9727	1.0048	1.0010	0.9999	1.0000	0.9884	0.5563	0.9884	0.9854
10.0	7.3825E-01	1.0135	0.9830	0.9764	0.9728	1.0053	1.0011	0.9999	1.0000	0.9884	0.5387	0.9884	0.9857
11.0	7.3842E-01	1.0145	0.9829	0.9764	0.9728	1.0058	1.0012	1.0000	1.0000	0.9884	0.5212	0.9884	0.9860
12.0	7.3861E-01	1.0156	0.9828	0.9763	0.9727	1.0064	1.0013	1.0000	1.0000	0.9883	0.5038	0.9883	0.9863
13.0	7.3882E-01	1.0167	0.9826	0.9761	0.9726	1.0069	1.0014	1.0000	1.0000	0.9883	0.4865	0.9883	0.9865
14.0	7.3906E-01	1.0177	0.9824	0.9759	0.9724	1.0074	1.0015	1.0000	1.0000	0.9882	0.4690	0.9882	0.9868
16.0	7.3959E-01	1.0199	0.9820	0.9755	0.9719	1.0085	1.0017	1.0000	1.0000	0.9880	0.4345	0.9880	0.9872
18.0	7.4021E-01	1.0220	0.9815	0.9749	0.9713	1.0095	1.0019	1.0000	1.0000	0.9877	0.4000	0.9877	0.9875
20.0	7.4090E-01	1.0241	0.9809	0.9742	0.9706	1.0105	1.0021	1.0000	1.0000	0.9874	0.3656	0.9874	0.9878
25.0	7.4291E-01	1.0295	0.9792	0.9722	0.9684	1.0131	1.0027	1.0000	1.0000	0.9866	0.2802	0.9866	0.9885
30.0	7.4520E-01	1.0348	0.9772	0.9699	0.9660	1.0156	1.0032	1.0000	1.0000	0.9856	0.1957	0.9856	0.9891
35.0	7.4772E-01	1.0402	0.9752	0.9674	0.9633	1.0181	1.0038	1.0000	1.0000	0.9846	0.1118	0.9846	0.9896
40.0	7.5039E-01	1.0456	0.9731	0.9648	0.9605	1.0206	1.0043	1.0000	1.0000	0.9836	0.0293	0.9836	0.9901
45.0	7.5318E-01	1.0509	0.9709	0.9621	0.9576	1.0231	1.0049	1.0000	1.0000	0.9825	-0.0523	0.9825	0.9905
50.0	7.5607E-01	1.0563	0.9687	0.9594	0.9546	1.0255	1.0054	1.0000	1.0000	0.9814	-0.1340	0.9814	0.9909

Instr. see p. 22

POSITRON Z= -16 A= 33

p	F_0L_0	L_0	λ_2	λ_3	λ_4	μ_1	μ_2	Λ_1	Λ_2	ν_{12}	$\hat{\nu}_{12}$	η_{12}	$\hat{\eta}_{12}$
0.1	5.1833E-03	1.0047	2.2628	2.9848	3.4073	1.0000	1.0000	0.9994	0.9999	1.3115	0.7318	0.9784	0.9717
0.2	1.0142E-01	1.0047	1.2886	1.3777	1.4185	1.0000	1.0000	0.9994	0.9999	1.0618	0.7295	0.9789	0.9723
0.3	2.3950E-01	1.0047	1.1094	1.1354	1.1459	1.0001	1.0000	0.9994	0.9999	1.0161	0.7266	0.9796	0.9729
0.4	3.5248E-01	1.0047	1.0475	1.0556	1.0580	1.0001	1.0000	0.9995	0.9999	1.0005	0.7234	0.9801	0.9735
0.5	4.3604E-01	1.0048	1.0194	1.0201	1.0192	1.0001	1.0000	0.9995	0.9999	0.9935	0.7202	0.9806	0.9740
0.6	4.9729E-01	1.0048	1.0045	1.0015	0.9990	1.0002	1.0000	0.9995	0.9999	0.9899	0.7169	0.9810	0.9745
0.7	5.4272E-01	1.0048	0.9958	0.9907	0.9873	1.0002	1.0000	0.9995	0.9999	0.9878	0.7137	0.9814	0.9749
0.8	5.7700E-01	1.0049	0.9904	0.9839	0.9800	1.0003	1.0001	0.9995	0.9999	0.9866	0.7106	0.9818	0.9753
0.9	6.0330E-01	1.0049	0.9868	0.9796	0.9753	1.0004	1.0001	0.9996	0.9999	0.9859	0.7077	0.9821	0.9757
1.0	6.2379E-01	1.0050	0.9844	0.9766	0.9721	1.0004	1.0001	0.9996	0.9999	0.9854	0.7049	0.9824	0.9760
1.2	6.5297E-01	1.0051	0.9815	0.9731	0.9684	1.0006	1.0001	0.9996	0.9999	0.9849	0.6996	0.9829	0.9766
1.4	6.7209E-01	1.0053	0.9800	0.9713	0.9665	1.0007	1.0001	0.9997	0.9999	0.9848	0.6947	0.9833	0.9771
1.6	6.8514E-01	1.0055	0.9792	0.9704	0.9656	1.0008	1.0002	0.9997	0.9999	0.9848	0.6902	0.9837	0.9776
1.8	6.9434E-01	1.0056	0.9788	0.9700	0.9651	1.0009	1.0002	0.9997	0.9999	0.9848	0.6859	0.9840	0.9780
2.0	7.0102E-01	1.0058	0.9786	0.9698	0.9650	1.0011	1.0002	0.9997	0.9999	0.9850	0.6818	0.9843	0.9783
2.2	7.0597E-01	1.0060	0.9786	0.9698	0.9650	1.0012	1.0002	0.9998	1.0000	0.9851	0.6778	0.9846	0.9786
2.4	7.0972E-01	1.0062	0.9786	0.9699	0.9651	1.0013	1.0003	0.9998	1.0000	0.9852	0.6740	0.9848	0.9789
2.6	7.1260E-01	1.0064	0.9787	0.9700	0.9653	1.0014	1.0003	0.9998	1.0000	0.9853	0.6702	0.9850	0.9792
2.8	7.1485E-01	1.0066	0.9788	0.9702	0.9655	1.0016	1.0003	0.9998	1.0000	0.9855	0.6665	0.9852	0.9795
3.0	7.1663E-01	1.0068	0.9789	0.9704	0.9657	1.0017	1.0003	0.9998	1.0000	0.9856	0.6628	0.9853	0.9797
3.2	7.1805E-01	1.0071	0.9791	0.9706	0.9660	1.0018	1.0004	0.9998	1.0000	0.9857	0.6592	0.9855	0.9799
3.4	7.1919E-01	1.0073	0.9792	0.9708	0.9662	1.0019	1.0004	0.9998	1.0000	0.9858	0.6557	0.9856	0.9802
3.6	7.2012E-01	1.0075	0.9793	0.9710	0.9664	1.0020	1.0004	0.9998	1.0000	0.9859	0.6521	0.9857	0.9804
3.8	7.2088E-01	1.0077	0.9795	0.9712	0.9667	1.0022	1.0004	0.9999	1.0000	0.9860	0.6487	0.9859	0.9805
4.0	7.2151E-01	1.0079	0.9796	0.9714	0.9669	1.0023	1.0005	0.9999	1.0000	0.9861	0.6452	0.9860	0.9807
4.5	7.2265E-01	1.0085	0.9799	0.9718	0.9674	1.0026	1.0005	0.9999	1.0000	0.9863	0.6365	0.9862	0.9811
5.0	7.2338E-01	1.0090	0.9801	0.9722	0.9678	1.0029	1.0006	0.9999	1.0000	0.9865	0.6279	0.9864	0.9815
5.5	7.2386E-01	1.0096	0.9804	0.9725	0.9682	1.0032	1.0006	0.9999	1.0000	0.9866	0.6194	0.9865	0.9818
6.0	7.2418E-01	1.0102	0.9805	0.9728	0.9685	1.0035	1.0007	0.9999	1.0000	0.9867	0.6109	0.9867	0.9821
6.5	7.2441E-01	1.0108	0.9807	0.9730	0.9688	1.0038	1.0008	0.9999	1.0000	0.9868	0.6025	0.9868	0.9824
7.0	7.2457E-01	1.0113	0.9808	0.9732	0.9690	1.0040	1.0008	0.9999	1.0000	0.9869	0.5941	0.9869	0.9826
7.5	7.2468E-01	1.0119	0.9809	0.9734	0.9692	1.0043	1.0009	0.9999	1.0000	0.9869	0.5856	0.9869	0.9829
8.0	7.2478E-01	1.0125	0.9810	0.9735	0.9694	1.0046	1.0009	0.9999	1.0000	0.9870	0.5773	0.9870	0.9831
9.0	7.2492E-01	1.0136	0.9811	0.9737	0.9696	1.0052	1.0010	0.9999	1.0000	0.9870	0.5606	0.9870	0.9835
10.0	7.2506E-01	1.0148	0.9811	0.9738	0.9697	1.0058	1.0012	0.9999	1.0000	0.9871	0.5438	0.9871	0.9839
11.0	7.2520E-01	1.0160	0.9810	0.9738	0.9698	1.0064	1.0013	0.9999	1.0000	0.9870	0.5271	0.9870	0.9842
12.0	7.2536E-01	1.0171	0.9809	0.9737	0.9697	1.0069	1.0014	1.0000	1.0000	0.9870	0.5104	0.9870	0.9845
13.0	7.2555E-01	1.0183	0.9808	0.9736	0.9696	1.0075	1.0015	1.0000	1.0000	0.9869	0.4937	0.9869	0.9848
14.0	7.2576E-01	1.0195	0.9806	0.9734	0.9694	1.0081	1.0016	1.0000	1.0000	0.9869	0.4772	0.9869	0.9850
16.0	7.2627E-01	1.0218	0.9802	0.9729	0.9689	1.0092	1.0019	1.0000	1.0000	0.9866	0.4443	0.9866	0.9855
18.0	7.2688E-01	1.0241	0.9796	0.9723	0.9683	1.0103	1.0021	1.0000	1.0000	0.9864	0.4113	0.9864	0.9859
20.0	7.2758E-01	1.0264	0.9790	0.9716	0.9676	1.0115	1.0023	1.0000	1.0000	0.9861	0.3785	0.9861	0.9863
25.0	7.2962E-01	1.0323	0.9772	0.9695	0.9654	1.0142	1.0029	1.0000	1.0000	0.9852	0.2968	0.9852	0.9871
30.0	7.3199E-01	1.0381	0.9752	0.9671	0.9628	1.0170	1.0035	1.0000	1.0000	0.9842	0.2161	0.9842	0.9877
35.0	7.3461E-01	1.0440	0.9730	0.9645	0.9600	1.0197	1.0041	1.0000	1.0000	0.9831	0.1361	0.9831	0.9883
40.0	7.3740E-01	1.0498	0.9708	0.9617	0.9570	1.0224	1.0047	1.0000	1.0000	0.9820	0.0571	0.9820	0.9888
45.0	7.4033E-01	1.0557	0.9684	0.9588	0.9539	1.0251	1.0053	1.0000	1.0000	0.9809	-0.0215	0.9809	0.9893
50.0	7.4337E-01	1.0615	0.9661	0.9559	0.9507	1.0277	1.0059	1.0000	1.0000	0.9797	-0.0984	0.9797	0.9897

Instr. see p. 22

POSITRON Z= -17 A= 35

p	F_0L_0	L_0	λ_2	λ_3	λ_4	μ_1	μ_2	Λ_1	Λ_2	ν_{12}	$\hat{\nu}_{12}$	η_{12}	$\hat{\eta}_{12}$
0.1	3.5195E-03	1.0052	2.4184	3.2879	3.8128	1.0000	1.0000	0.9994	0.9999	1.3512	0.7298	0.9759	0.9684
0.2	8.5992E-02	1.0052	1.3243	1.4275	1.4754	1.0000	1.0000	0.9994	0.9999	1.0698	0.7276	0.9765	0.9690
0.3	2.1706E-01	1.0052	1.1231	1.1527	1.1648	1.0001	1.0000	0.9994	0.9999	1.0183	0.7246	0.9771	0.9697
0.4	3.2907E-01	1.0053	1.0535	1.0628	1.0655	1.0001	1.0000	0.9994	0.9999	1.0007	0.7214	0.9777	0.9703
0.5	4.1366E-01	1.0053	1.0220	1.0228	1.0218	1.0002	1.0000	0.9994	0.9999	0.9928	0.7181	0.9783	0.9709
0.6	4.7637E-01	1.0053	1.0052	1.0018	0.9990	1.0002	1.0000	0.9995	0.9999	0.9887	0.7148	0.9788	0.9714
0.7	5.2321E-01	1.0054	0.9954	0.9897	0.9859	1.0003	1.0001	0.9995	0.9999	0.9864	0.7116	0.9792	0.9719
0.8	5.5870E-01	1.0054	0.9893	0.9821	0.9777	1.0003	1.0001	0.9995	0.9999	0.9851	0.7085	0.9796	0.9723
0.9	5.8601E-01	1.0055	0.9853	0.9772	0.9724	1.0004	1.0001	0.9995	0.9999	0.9842	0.7056	0.9800	0.9727
1.0	6.0733E-01	1.0056	0.9826	0.9738	0.9688	1.0005	1.0001	0.9996	0.9999	0.9837	0.7028	0.9803	0.9731
1.2	6.3774E-01	1.0057	0.9793	0.9699	0.9646	1.0006	1.0001	0.9996	0.9999	0.9832	0.6976	0.9808	0.9738
1.4	6.5768E-01	1.0059	0.9776	0.9679	0.9625	1.0007	1.0001	0.9996	0.9999	0.9830	0.6928	0.9813	0.9743
1.6	6.7130E-01	1.0061	0.9768	0.9669	0.9615	1.0009	1.0002	0.9997	0.9999	0.9830	0.6884	0.9817	0.9748
1.8	6.8089E-01	1.0062	0.9763	0.9665	0.9610	1.0010	1.0002	0.9997	0.9999	0.9831	0.6842	0.9821	0.9753
2.0	6.8784E-01	1.0064	0.9761	0.9663	0.9609	1.0012	1.0002	0.9997	0.9999	0.9832	0.6803	0.9824	0.9757
2.2	6.9299E-01	1.0067	0.9761	0.9663	0.9609	1.0013	1.0003	0.9997	0.9999	0.9833	0.6764	0.9827	0.9760
2.4	6.9689E-01	1.0069	0.9761	0.9664	0.9610	1.0014	1.0003	0.9998	1.0000	0.9835	0.6727	0.9830	0.9764
2.6	6.9987E-01	1.0071	0.9762	0.9665	0.9612	1.0016	1.0003	0.9998	1.0000	0.9836	0.6691	0.9832	0.9767
2.8	7.0220E-01	1.0073	0.9763	0.9667	0.9615	1.0017	1.0003	0.9998	1.0000	0.9838	0.6655	0.9834	0.9770
3.0	7.0403E-01	1.0075	0.9765	0.9670	0.9617	1.0018	1.0004	0.9998	1.0000	0.9839	0.6620	0.9836	0.9772
3.2	7.0549E-01	1.0078	0.9766	0.9672	0.9620	1.0020	1.0004	0.9998	1.0000	0.9840	0.6586	0.9838	0.9775
3.4	7.0666E-01	1.0080	0.9768	0.9674	0.9623	1.0021	1.0004	0.9998	1.0000	0.9841	0.6552	0.9839	0.9777
3.6	7.0761E-01	1.0083	0.9770	0.9677	0.9625	1.0022	1.0004	0.9998	1.0000	0.9843	0.6518	0.9841	0.9780
3.8	7.0838E-01	1.0085	0.9771	0.9679	0.9628	1.0023	1.0005	0.9998	1.0000	0.9844	0.6484	0.9842	0.9782
4.0	7.0901E-01	1.0087	0.9773	0.9681	0.9631	1.0025	1.0005	0.9998	1.0000	0.9845	0.6451	0.9843	0.9784
4.5	7.1014E-01	1.0093	0.9776	0.9686	0.9637	1.0028	1.0006	0.9999	1.0000	0.9847	0.6368	0.9846	0.9788
5.0	7.1086E-01	1.0099	0.9779	0.9690	0.9642	1.0031	1.0006	0.9999	1.0000	0.9849	0.6286	0.9848	0.9792
5.5	7.1131E-01	1.0106	0.9782	0.9694	0.9646	1.0034	1.0007	0.9999	1.0000	0.9851	0.6205	0.9850	0.9796
6.0	7.1161E-01	1.0112	0.9784	0.9698	0.9650	1.0038	1.0007	0.9999	1.0000	0.9852	0.6123	0.9852	0.9799
6.5	7.1180E-01	1.0118	0.9786	0.9700	0.9653	1.0041	1.0008	0.9999	1.0000	0.9853	0.6042	0.9853	0.9802
7.0	7.1193E-01	1.0124	0.9787	0.9703	0.9656	1.0044	1.0009	0.9999	1.0000	0.9854	0.5962	0.9854	0.9805
7.5	7.1202E-01	1.0130	0.9789	0.9705	0.9659	1.0047	1.0009	0.9999	1.0000	0.9855	0.5881	0.9855	0.9808
8.0	7.1209E-01	1.0137	0.9789	0.9706	0.9661	1.0050	1.0010	0.9999	1.0000	0.9855	0.5801	0.9855	0.9811
9.0	7.1218E-01	1.0149	0.9791	0.9709	0.9664	1.0057	1.0011	0.9999	1.0000	0.9856	0.5640	0.9856	0.9815
10.0	7.1227E-01	1.0162	0.9791	0.9710	0.9665	1.0063	1.0013	0.9999	1.0000	0.9857	0.5481	0.9857	0.9819
11.0	7.1238E-01	1.0174	0.9791	0.9710	0.9666	1.0069	1.0014	0.9999	1.0000	0.9857	0.5320	0.9857	0.9823
12.0	7.1251E-01	1.0187	0.9790	0.9710	0.9666	1.0075	1.0015	0.9999	1.0000	0.9856	0.5162	0.9856	0.9826
13.0	7.1267E-01	1.0200	0.9789	0.9709	0.9665	1.0081	1.0016	1.0000	1.0000	0.9856	0.5003	0.9856	0.9829
14.0	7.1287E-01	1.0212	0.9787	0.9707	0.9663	1.0088	1.0018	1.0000	1.0000	0.9855	0.4843	0.9855	0.9832
16.0	7.1335E-01	1.0238	0.9783	0.9703	0.9659	1.0100	1.0020	1.0000	1.0000	0.9853	0.4526	0.9853	0.9838
18.0	7.1394E-01	1.0263	0.9777	0.9697	0.9653	1.0112	1.0023	1.0000	1.0000	0.9850	0.4208	0.9850	0.9842
20.0	7.1463E-01	1.0288	0.9771	0.9689	0.9645	1.0124	1.0025	1.0000	1.0000	0.9847	0.3895	0.9847	0.9846
25.0	7.1670E-01	1.0352	0.9752	0.9668	0.9622	1.0154	1.0032	1.0000	1.0000	0.9838	0.3115	0.9838	0.9855
30.0	7.1914E-01	1.0415	0.9731	0.9643	0.9595	1.0184	1.0038	1.0000	1.0000	0.9827	0.2338	0.9828	0.9863
35.0	7.2184E-01	1.0479	0.9708	0.9615	0.9566	1.0214	1.0044	1.0000	1.0000	0.9816	0.1572	0.9816	0.9869
40.0	7.2475E-01	1.0542	0.9684	0.9586	0.9534	1.0243	1.0051	1.0000	1.0000	0.9804	0.0818	0.9804	0.9875
45.0	7.2781E-01	1.0605	0.9660	0.9555	0.9501	1.0272	1.0057	1.0000	1.0000	0.9792	0.0071	0.9792	0.9881
50.0	7.3099E-01	1.0669	0.9634	0.9524	0.9468	1.0300	1.0064	1.0000	1.0000	0.9780	-0.0675	0.9780	0.9886

Instr. see p. 22

POSITRON Z= -18 A= 38

p	F_0L_0	L_0	λ_2	λ_3	λ_4	μ_1	μ_2	Λ_1	Λ_2	ν_{12}	$\hat{\nu}_{12}$	η_{12}	$\hat{\eta}_{12}$
0.1	2.3829E-03	1.0058	2.5818	3.6206	4.2680	1.0000	1.0000	0.9993	0.9999	1.3931	0.7278	0.9733	0.9650
0.2	7.2775E-02	1.0058	1.3619	1.4808	1.5364	1.0000	1.0000	0.9993	0.9999	1.0783	0.7255	0.9740	0.9656
0.3	1.9654E-01	1.0058	1.1375	1.1712	1.1850	1.0001	1.0000	0.9993	0.9999	1.0207	0.7226	0.9747	0.9663
0.4	3.0708E-01	1.0058	1.0599	1.0704	1.0735	1.0001	1.0000	0.9994	0.9999	1.0010	0.7193	0.9753	0.9670
0.5	3.9235E-01	1.0059	1.0247	1.0256	1.0246	1.0002	1.0000	0.9994	0.9999	0.9921	0.7159	0.9759	0.9676
0.6	4.5631E-01	1.0059	1.0060	1.0022	0.9992	1.0002	1.0000	0.9994	0.9999	0.9876	0.7126	0.9764	0.9682
0.7	5.0442E-01	1.0060	0.9951	0.9887	0.9845	1.0003	1.0001	0.9994	0.9999	0.9850	0.7094	0.9769	0.9687
0.8	5.4104E-01	1.0060	0.9883	0.9803	0.9754	1.0004	1.0001	0.9995	0.9999	0.9835	0.7063	0.9773	0.9692
0.9	5.6930E-01	1.0061	0.9838	0.9748	0.9695	1.0004	1.0001	0.9995	0.9999	0.9825	0.7034	0.9777	0.9697
1.0	5.9140E-01	1.0062	0.9807	0.9711	0.9655	1.0005	1.0001	0.9995	0.9999	0.9819	0.7006	0.9781	0.9701
1.2	6.2299E-01	1.0063	0.9771	0.9667	0.9608	1.0007	1.0001	0.9996	0.9999	0.9813	0.6954	0.9787	0.9708
1.4	6.4373E-01	1.0065	0.9752	0.9645	0.9585	1.0008	1.0002	0.9996	0.9999	0.9811	0.6907	0.9793	0.9714
1.6	6.5788E-01	1.0067	0.9743	0.9633	0.9573	1.0010	1.0002	0.9996	0.9999	0.9811	0.6863	0.9797	0.9720
1.8	6.6786E-01	1.0069	0.9738	0.9628	0.9568	1.0011	1.0002	0.9997	0.9999	0.9812	0.6823	0.9801	0.9725
2.0	6.7508E-01	1.0071	0.9736	0.9626	0.9566	1.0013	1.0002	0.9997	0.9999	0.9813	0.6784	0.9805	0.9729
2.2	6.8042E-01	1.0074	0.9735	0.9626	0.9567	1.0014	1.0003	0.9997	0.9999	0.9815	0.6747	0.9808	0.9733
2.4	6.8445E-01	1.0076	0.9736	0.9628	0.9568	1.0016	1.0003	0.9997	0.9999	0.9817	0.6711	0.9811	0.9737
2.6	6.8754E-01	1.0078	0.9737	0.9629	0.9571	1.0017	1.0003	0.9998	1.0000	0.9818	0.6675	0.9814	0.9740
2.8	6.8994E-01	1.0081	0.9738	0.9632	0.9573	1.0018	1.0004	0.9998	1.0000	0.9820	0.6641	0.9816	0.9744
3.0	6.9182E-01	1.0083	0.9740	0.9634	0.9576	1.0020	1.0004	0.9998	1.0000	0.9821	0.6607	0.9818	0.9747
3.2	6.9331E-01	1.0086	0.9742	0.9637	0.9579	1.0021	1.0004	0.9998	1.0000	0.9823	0.6574	0.9820	0.9750
3.4	6.9450E-01	1.0088	0.9743	0.9640	0.9582	1.0023	1.0004	0.9998	1.0000	0.9824	0.6540	0.9822	0.9752
3.6	6.9547E-01	1.0091	0.9745	0.9642	0.9586	1.0024	1.0005	0.9998	1.0000	0.9826	0.6508	0.9824	0.9755
3.8	6.9625E-01	1.0094	0.9747	0.9645	0.9589	1.0026	1.0005	0.9998	1.0000	0.9827	0.6475	0.9825	0.9757
4.0	6.9688E-01	1.0096	0.9749	0.9647	0.9591	1.0027	1.0005	0.9998	1.0000	0.9828	0.6443	0.9827	0.9759
4.5	6.9801E-01	1.0103	0.9753	0.9653	0.9598	1.0031	1.0006	0.9999	1.0000	0.9831	0.6362	0.9830	0.9764
5.0	6.9870E-01	1.0109	0.9756	0.9658	0.9604	1.0034	1.0007	0.9999	1.0000	0.9833	0.6283	0.9832	0.9769
5.5	6.9913E-01	1.0116	0.9759	0.9662	0.9609	1.0038	1.0007	0.9999	1.0000	0.9835	0.6204	0.9834	0.9773
6.0	6.9940E-01	1.0123	0.9762	0.9666	0.9614	1.0041	1.0008	0.9999	1.0000	0.9836	0.6126	0.9836	0.9777
6.5	6.9957E-01	1.0130	0.9764	0.9670	0.9618	1.0044	1.0009	0.9999	1.0000	0.9838	0.6047	0.9837	0.9780
7.0	6.9967E-01	1.0136	0.9766	0.9672	0.9621	1.0048	1.0010	0.9999	1.0000	0.9839	0.5970	0.9839	0.9784
7.5	6.9973E-01	1.0143	0.9767	0.9675	0.9624	1.0051	1.0010	0.9999	1.0000	0.9840	0.5891	0.9840	0.9787
8.0	6.9977E-01	1.0150	0.9768	0.9677	0.9626	1.0055	1.0011	0.9999	1.0000	0.9840	0.5813	0.9840	0.9789
9.0	6.9982E-01	1.0164	0.9770	0.9680	0.9630	1.0062	1.0012	0.9999	1.0000	0.9841	0.5657	0.9841	0.9794
10.0	6.9987E-01	1.0177	0.9771	0.9681	0.9632	1.0068	1.0014	0.9999	1.0000	0.9842	0.5504	0.9842	0.9799
11.0	6.9994E-01	1.0191	0.9771	0.9682	0.9633	1.0075	1.0015	0.9999	1.0000	0.9842	0.5347	0.9842	0.9803
12.0	7.0005E-01	1.0205	0.9770	0.9682	0.9633	1.0082	1.0016	0.9999	1.0000	0.9842	0.5194	0.9842	0.9807
13.0	7.0019E-01	1.0219	0.9769	0.9681	0.9632	1.0089	1.0018	0.9999	1.0000	0.9841	0.5039	0.9841	0.9810
14.0	7.0037E-01	1.0233	0.9767	0.9679	0.9631	1.0095	1.0019	1.0000	1.0000	0.9841	0.4886	0.9841	0.9814
16.0	7.0083E-01	1.0260	0.9763	0.9675	0.9627	1.0109	1.0022	1.0000	1.0000	0.9838	0.4578	0.9839	0.9820
18.0	7.0141E-01	1.0288	0.9757	0.9669	0.9620	1.0122	1.0025	1.0000	1.0000	0.9836	0.4273	0.9836	0.9825
20.0	7.0211E-01	1.0315	0.9750	0.9661	0.9613	1.0135	1.0028	1.0000	1.0000	0.9833	0.3968	0.9833	0.9829
25.0	7.0423E-01	1.0384	0.9731	0.9639	0.9589	1.0168	1.0034	1.0000	1.0000	0.9823	0.3213	0.9823	0.9839
30.0	7.0677E-01	1.0453	0.9709	0.9612	0.9560	1.0200	1.0041	1.0000	1.0000	0.9812	0.2461	0.9812	0.9848
35.0	7.0960E-01	1.0523	0.9684	0.9583	0.9529	1.0232	1.0048	1.0000	1.0000	0.9800	0.1718	0.9800	0.9855
40.0	7.1265E-01	1.0592	0.9659	0.9551	0.9495	1.0264	1.0055	1.0000	1.0000	0.9787	0.0982	0.9787	0.9862
45.0	7.1587E-01	1.0661	0.9632	0.9519	0.9461	1.0295	1.0062	1.0000	1.0000	0.9774	0.0260	0.9774	0.9868
50.0	7.1922E-01	1.0731	0.9605	0.9486	0.9425	1.0326	1.0070	1.0000	1.0000	0.9761	-0.0446	0.9761	0.9873

Instr. see p. 22

POSITRON Z= -17 A= 35

p	F_0L_0	L_0	λ_2	λ_3	λ_4	μ_1	μ_2	Λ_1	Λ_2	ν_{12}	$\hat{\nu}_{12}$	η_{12}	$\hat{\eta}_{12}$
0.1	3.5195E-03	1.0052	2.4184	3.2879	3.8128	1.0000	1.0000	0.9994	0.9999	1.3512	0.7298	0.9759	0.9684
0.2	8.5992E-02	1.0052	1.3243	1.4275	1.4754	1.0000	1.0000	0.9994	0.9999	1.0698	0.7276	0.9765	0.9690
0.3	2.1706E-01	1.0052	1.1231	1.1527	1.1648	1.0001	1.0000	0.9994	0.9999	1.0183	0.7246	0.9771	0.9697
0.4	3.2907E-01	1.0053	1.0535	1.0628	1.0655	1.0001	1.0000	0.9994	0.9999	1.0007	0.7214	0.9777	0.9703
0.5	4.1366E-01	1.0053	1.0220	1.0228	1.0218	1.0002	1.0000	0.9994	0.9999	0.9928	0.7181	0.9783	0.9709
0.6	4.7637E-01	1.0053	1.0052	1.0018	0.9990	1.0002	1.0000	0.9995	0.9999	0.9887	0.7148	0.9788	0.9714
0.7	5.2321E-01	1.0054	0.9954	0.9897	0.9859	1.0003	1.0001	0.9995	0.9999	0.9864	0.7116	0.9792	0.9719
0.8	5.5870E-01	1.0054	0.9893	0.9821	0.9777	1.0003	1.0001	0.9995	0.9999	0.9851	0.7085	0.9796	0.9723
0.9	5.8601E-01	1.0055	0.9853	0.9772	0.9724	1.0004	1.0001	0.9995	0.9999	0.9842	0.7056	0.9800	0.9727
1.0	6.0733E-01	1.0056	0.9826	0.9738	0.9688	1.0005	1.0001	0.9996	0.9999	0.9837	0.7028	0.9803	0.9731
1.2	6.3774E-01	1.0057	0.9793	0.9699	0.9646	1.0006	1.0001	0.9996	0.9999	0.9832	0.6976	0.9808	0.9738
1.4	6.5768E-01	1.0059	0.9776	0.9679	0.9625	1.0007	1.0001	0.9996	0.9999	0.9830	0.6928	0.9813	0.9743
1.6	6.7130E-01	1.0061	0.9768	0.9669	0.9615	1.0009	1.0002	0.9997	0.9999	0.9830	0.6884	0.9817	0.9748
1.8	6.8089E-01	1.0062	0.9763	0.9665	0.9610	1.0010	1.0002	0.9997	0.9999	0.9831	0.6842	0.9821	0.9753
2.0	6.8784E-01	1.0064	0.9761	0.9663	0.9609	1.0012	1.0002	0.9997	0.9999	0.9832	0.6803	0.9824	0.9757
2.2	6.9299E-01	1.0067	0.9761	0.9663	0.9609	1.0013	1.0003	0.9997	0.9999	0.9833	0.6764	0.9827	0.9760
2.4	6.9689E-01	1.0069	0.9761	0.9664	0.9610	1.0014	1.0003	0.9998	1.0000	0.9835	0.6727	0.9830	0.9764
2.6	6.9987E-01	1.0071	0.9762	0.9665	0.9612	1.0016	1.0003	0.9998	1.0000	0.9836	0.6691	0.9832	0.9767
2.8	7.0220E-01	1.0073	0.9763	0.9667	0.9615	1.0017	1.0003	0.9998	1.0000	0.9838	0.6655	0.9834	0.9770
3.0	7.0403E-01	1.0075	0.9765	0.9670	0.9617	1.0018	1.0004	0.9998	1.0000	0.9839	0.6620	0.9836	0.9772
3.2	7.0549E-01	1.0078	0.9766	0.9672	0.9620	1.0020	1.0004	0.9998	1.0000	0.9840	0.6586	0.9838	0.9775
3.4	7.0666E-01	1.0080	0.9768	0.9674	0.9623	1.0021	1.0004	0.9998	1.0000	0.9841	0.6552	0.9839	0.9777
3.6	7.0761E-01	1.0083	0.9770	0.9677	0.9625	1.0022	1.0004	0.9998	1.0000	0.9843	0.6518	0.9841	0.9780
3.8	7.0838E-01	1.0085	0.9771	0.9679	0.9628	1.0023	1.0005	0.9998	1.0000	0.9844	0.6484	0.9842	0.9782
4.0	7.0901E-01	1.0087	0.9773	0.9681	0.9631	1.0025	1.0005	0.9998	1.0000	0.9845	0.6451	0.9843	0.9784
4.5	7.1014E-01	1.0093	0.9776	0.9686	0.9637	1.0028	1.0006	0.9999	1.0000	0.9847	0.6368	0.9846	0.9788
5.0	7.1086E-01	1.0099	0.9779	0.9690	0.9642	1.0031	1.0006	0.9999	1.0000	0.9849	0.6286	0.9848	0.9792
5.5	7.1131E-01	1.0106	0.9782	0.9694	0.9646	1.0034	1.0007	0.9999	1.0000	0.9851	0.6205	0.9850	0.9796
6.0	7.1161E-01	1.0112	0.9784	0.9698	0.9650	1.0038	1.0007	0.9999	1.0000	0.9852	0.6123	0.9852	0.9799
6.5	7.1180E-01	1.0118	0.9786	0.9700	0.9653	1.0041	1.0008	0.9999	1.0000	0.9853	0.6042	0.9853	0.9802
7.0	7.1193E-01	1.0124	0.9787	0.9703	0.9656	1.0044	1.0009	0.9999	1.0000	0.9854	0.5962	0.9854	0.9805
7.5	7.1202E-01	1.0130	0.9789	0.9705	0.9659	1.0047	1.0009	0.9999	1.0000	0.9855	0.5881	0.9855	0.9808
8.0	7.1209E-01	1.0137	0.9789	0.9706	0.9661	1.0050	1.0010	0.9999	1.0000	0.9855	0.5801	0.9855	0.9811
9.0	7.1218E-01	1.0149	0.9791	0.9709	0.9664	1.0057	1.0011	0.9999	1.0000	0.9856	0.5640	0.9856	0.9815
10.0	7.1227E-01	1.0162	0.9791	0.9710	0.9665	1.0063	1.0013	0.9999	1.0000	0.9857	0.5481	0.9857	0.9819
11.0	7.1238E-01	1.0174	0.9791	0.9710	0.9666	1.0069	1.0014	0.9999	1.0000	0.9857	0.5320	0.9857	0.9823
12.0	7.1251E-01	1.0187	0.9790	0.9710	0.9666	1.0075	1.0015	0.9999	1.0000	0.9856	0.5162	0.9856	0.9826
13.0	7.1267E-01	1.0200	0.9789	0.9709	0.9665	1.0081	1.0016	1.0000	1.0000	0.9856	0.5003	0.9856	0.9829
14.0	7.1287E-01	1.0212	0.9787	0.9707	0.9663	1.0088	1.0018	1.0000	1.0000	0.9855	0.4843	0.9855	0.9832
16.0	7.1335E-01	1.0238	0.9783	0.9703	0.9659	1.0100	1.0020	1.0000	1.0000	0.9853	0.4526	0.9853	0.9838
18.0	7.1394E-01	1.0263	0.9777	0.9697	0.9653	1.0112	1.0023	1.0000	1.0000	0.9850	0.4208	0.9850	0.9842
20.0	7.1463E-01	1.0288	0.9771	0.9689	0.9645	1.0124	1.0025	1.0000	1.0000	0.9847	0.3895	0.9847	0.9846
25.0	7.1670E-01	1.0352	0.9752	0.9668	0.9622	1.0154	1.0032	1.0000	1.0000	0.9838	0.3115	0.9838	0.9855
30.0	7.1914E-01	1.0415	0.9731	0.9643	0.9595	1.0184	1.0038	1.0000	1.0000	0.9827	0.2338	0.9828	0.9863
35.0	7.2184E-01	1.0479	0.9708	0.9615	0.9566	1.0214	1.0044	1.0000	1.0000	0.9816	0.1572	0.9816	0.9869
40.0	7.2475E-01	1.0542	0.9684	0.9586	0.9534	1.0243	1.0051	1.0000	1.0000	0.9804	0.0818	0.9804	0.9875
45.0	7.2781E-01	1.0605	0.9660	0.9555	0.9501	1.0272	1.0057	1.0000	1.0000	0.9792	0.0071	0.9792	0.9881
50.0	7.3099E-01	1.0669	0.9634	0.9524	0.9468	1.0300	1.0064	1.0000	1.0000	0.9780	-0.0675	0.9780	0.9886

Instr. see p. 22

POSITRON Z= -20 A= 44

p	F_0L_0	L_0	λ_2	λ_3	λ_4	μ_1	μ_2	Λ_1	Λ_2	ν_{12}	$\hat{\nu}_{12}$	η_{12}	$\hat{\eta}_{12}$
0.1	1.0848E-03	1.0070	2.9308	4.3806	5.3459	1.0000	1.0000	0.9992	0.9998	1.4836	0.7235	0.9679	0.9576
0.2	5.1881E-02	1.0070	1.4423	1.5973	1.6714	1.0000	1.0000	0.9992	0.9998	1.0968	0.7213	0.9686	0.9583
0.3	1.6075E-01	1.0070	1.1683	1.2111	1.2288	1.0001	1.0000	0.9992	0.9998	1.0259	0.7182	0.9694	0.9591
0.4	2.6713E-01	1.0071	1.0736	1.0867	1.0908	1.0001	1.0000	0.9993	0.9999	1.0016	0.7148	0.9701	0.9599
0.5	3.5289E-01	1.0071	1.0305	1.0319	1.0307	1.0002	1.0000	0.9993	0.9999	0.9908	0.7114	0.9708	0.9606
0.6	4.1878E-01	1.0071	1.0077	1.0032	0.9995	1.0003	1.0001	0.9993	0.9999	0.9851	0.7080	0.9714	0.9613
0.7	4.6908E-01	1.0072	0.9943	0.9866	0.9815	1.0003	1.0001	0.9993	0.9999	0.9820	0.7047	0.9720	0.9620
0.8	5.0772E-01	1.0073	0.9860	0.9763	0.9704	1.0004	1.0001	0.9994	0.9999	0.9800	0.7016	0.9725	0.9625
0.9	5.3772E-01	1.0074	0.9805	0.9696	0.9632	1.0005	1.0001	0.9994	0.9999	0.9789	0.6987	0.9730	0.9630
1.0	5.6129E-01	1.0074	0.9768	0.9651	0.9583	1.0006	1.0001	0.9994	0.9999	0.9781	0.6959	0.9734	0.9635
1.2	5.9508E-01	1.0076	0.9724	0.9597	0.9526	1.0008	1.0001	0.9995	0.9999	0.9774	0.6908	0.9742	0.9644
1.4	6.1733E-01	1.0078	0.9701	0.9571	0.9498	1.0009	1.0002	0.9995	0.9999	0.9771	0.6862	0.9748	0.9652
1.6	6.3251E-01	1.0081	0.9689	0.9557	0.9484	1.0011	1.0002	0.9996	0.9999	0.9771	0.6820	0.9754	0.9659
1.8	6.4321E-01	1.0083	0.9683	0.9551	0.9478	1.0013	1.0003	0.9996	0.9999	0.9772	0.6781	0.9759	0.9665
2.0	6.5093E-01	1.0086	0.9680	0.9548	0.9476	1.0015	1.0003	0.9996	0.9999	0.9774	0.6744	0.9764	0.9670
2.2	6.5663E-01	1.0088	0.9680	0.9549	0.9476	1.0016	1.0003	0.9997	0.9999	0.9776	0.6708	0.9768	0.9675
2.4	6.6091E-01	1.0091	0.9681	0.9550	0.9479	1.0018	1.0004	0.9997	0.9999	0.9778	0.6674	0.9771	0.9680
2.6	6.6418E-01	1.0094	0.9682	0.9553	0.9482	1.0020	1.0004	0.9997	0.9999	0.9780	0.6640	0.9774	0.9684
2.8	6.6670E-01	1.0097	0.9684	0.9556	0.9485	1.0022	1.0004	0.9997	0.9999	0.9782	0.6608	0.9777	0.9688
3.0	6.6867E-01	1.0100	0.9686	0.9559	0.9489	1.0023	1.0005	0.9997	0.9999	0.9784	0.6576	0.9780	0.9692
3.2	6.7022E-01	1.0103	0.9688	0.9562	0.9493	1.0025	1.0005	0.9998	1.0000	0.9786	0.6545	0.9782	0.9695
3.4	6.7146E-01	1.0106	0.9691	0.9565	0.9497	1.0027	1.0005	0.9998	1.0000	0.9788	0.6513	0.9785	0.9698
3.6	6.7244E-01	1.0109	0.9693	0.9569	0.9501	1.0028	1.0006	0.9998	1.0000	0.9789	0.6483	0.9787	0.9701
3.8	6.7323E-01	1.0112	0.9695	0.9572	0.9504	1.0030	1.0006	0.9998	1.0000	0.9791	0.6452	0.9789	0.9704
4.0	6.7386E-01	1.0115	0.9697	0.9575	0.9508	1.0032	1.0006	0.9998	1.0000	0.9792	0.6422	0.9791	0.9707
4.5	6.7496E-01	1.0123	0.9703	0.9583	0.9517	1.0036	1.0007	0.9998	1.0000	0.9796	0.6347	0.9794	0.9713
5.0	6.7560E-01	1.0130	0.9707	0.9589	0.9524	1.0040	1.0008	0.9998	1.0000	0.9799	0.6272	0.9798	0.9719
5.5	6.7596E-01	1.0138	0.9711	0.9595	0.9531	1.0044	1.0009	0.9999	1.0000	0.9801	0.6197	0.9800	0.9724
6.0	6.7616E-01	1.0146	0.9715	0.9600	0.9537	1.0048	1.0010	0.9999	1.0000	0.9803	0.6124	0.9803	0.9728
6.5	6.7626E-01	1.0154	0.9718	0.9604	0.9542	1.0052	1.0010	0.9999	1.0000	0.9805	0.6050	0.9805	0.9733
7.0	6.7629E-01	1.0162	0.9720	0.9608	0.9546	1.0056	1.0011	0.9999	1.0000	0.9806	0.5977	0.9806	0.9737
7.5	6.7629E-01	1.0170	0.9722	0.9611	0.9550	1.0060	1.0012	0.9999	1.0000	0.9808	0.5904	0.9807	0.9740
8.0	6.7627E-01	1.0178	0.9724	0.9614	0.9553	1.0064	1.0013	0.9999	1.0000	0.9809	0.5830	0.9809	0.9744
9.0	6.7621E-01	1.0194	0.9727	0.9618	0.9559	1.0072	1.0014	0.9999	1.0000	0.9810	0.5684	0.9810	0.9750
10.0	6.7617E-01	1.0210	0.9728	0.9621	0.9562	1.0080	1.0016	0.9999	1.0000	0.9811	0.5538	0.9811	0.9756
11.0	6.7616E-01	1.0226	0.9729	0.9623	0.9564	1.0088	1.0018	0.9999	1.0000	0.9812	0.5394	0.9812	0.9761
12.0	6.7620E-01	1.0242	0.9728	0.9623	0.9565	1.0096	1.0019	0.9999	1.0000	0.9812	0.5248	0.9812	0.9766
13.0	6.7629E-01	1.0258	0.9727	0.9623	0.9565	1.0104	1.0021	0.9999	1.0000	0.9811	0.5104	0.9811	0.9770
14.0	6.7642E-01	1.0275	0.9726	0.9621	0.9564	1.0111	1.0022	0.9999	1.0000	0.9811	0.4959	0.9811	0.9774
16.0	6.7683E-01	1.0307	0.9721	0.9617	0.9560	1.0127	1.0026	1.0000	1.0000	0.9808	0.4670	0.9809	0.9781
18.0	6.7738E-01	1.0339	0.9716	0.9611	0.9554	1.0143	1.0029	1.0000	1.0000	0.9806	0.4381	0.9806	0.9787
20.0	6.7808E-01	1.0372	0.9709	0.9603	0.9546	1.0158	1.0032	1.0000	1.0000	0.9802	0.4094	0.9802	0.9793
25.0	6.8027E-01	1.0452	0.9687	0.9579	0.9520	1.0196	1.0040	1.0000	1.0000	0.9792	0.3383	0.9792	0.9806
30.0	6.8296E-01	1.0534	0.9662	0.9549	0.9489	1.0234	1.0048	1.0000	1.0000	0.9780	0.2677	0.9780	0.9816
35.0	6.8601E-01	1.0615	0.9635	0.9517	0.9454	1.0271	1.0056	1.0000	1.0000	0.9766	0.1978	0.9767	0.9825
40.0	6.8933E-01	1.0696	0.9607	0.9482	0.9417	1.0307	1.0065	1.0000	1.0000	0.9752	0.1288	0.9752	0.9833
45.0	6.9284E-01	1.0777	0.9577	0.9446	0.9377	1.0344	1.0073	1.0000	1.0000	0.9738	0.0607	0.9738	0.9841
50.0	6.9652E-01	1.0859	0.9547	0.9408	0.9337	1.0379	1.0081	1.0000	1.0000	0.9723	-0.0062	0.9723	0.9848

Instr. see p. 22

POSITRON Z= -21 A= 45

p	F_0L_0	L_0	λ_2	λ_3	λ_4	μ_1	μ_2	Λ_1	Λ_2	ν_{12}	$\hat{\nu}_{12}$	η_{12}	$\hat{\eta}_{12}$
0.1	7.3008E-04	1.0076	3.1154	4.8105	5.9776	1.0000	1.0000	0.9991	0.9998	1.5320	0.7212	0.9650	0.9537
0.2	4.3726E-02	1.0076	1.4848	1.6603	1.7451	1.0001	1.0000	0.9992	0.9998	1.1067	0.7190	0.9657	0.9544
0.3	1.4527E-01	1.0076	1.1845	1.2322	1.2521	1.0001	1.0000	0.9992	0.9998	1.0287	0.7159	0.9665	0.9552
0.4	2.4912E-01	1.0077	1.0807	1.0953	1.0998	1.0001	1.0000	0.9992	0.9998	1.0020	0.7125	0.9673	0.9561
0.5	3.3474E-01	1.0077	1.0335	1.0350	1.0337	1.0002	1.0000	0.9992	0.9998	0.9900	0.7090	0.9680	0.9569
0.6	4.0136E-01	1.0078	1.0084	1.0035	0.9995	1.0003	1.0001	0.9993	0.9999	0.9838	0.7056	0.9687	0.9576
0.7	4.5260E-01	1.0078	0.9938	0.9853	0.9798	1.0004	1.0001	0.9993	0.9999	0.9803	0.7024	0.9693	0.9583
0.8	4.9215E-01	1.0079	0.9846	0.9740	0.9676	1.0005	1.0001	0.9993	0.9999	0.9782	0.6993	0.9699	0.9589
0.9	5.2296E-01	1.0080	0.9786	0.9667	0.9597	1.0005	1.0001	0.9994	0.9999	0.9769	0.6963	0.9704	0.9595
1.0	5.4721E-01	1.0081	0.9746	0.9617	0.9543	1.0006	1.0001	0.9994	0.9999	0.9760	0.6936	0.9709	0.9600
1.2	5.8205E-01	1.0083	0.9697	0.9559	0.9481	1.0008	1.0002	0.9995	0.9999	0.9752	0.6886	0.9717	0.9609
1.4	6.0502E-01	1.0085	0.9672	0.9529	0.9450	1.0010	1.0002	0.9995	0.9999	0.9749	0.6841	0.9724	0.9618
1.6	6.2070E-01	1.0087	0.9659	0.9515	0.9435	1.0012	1.0002	0.9995	0.9999	0.9749	0.6800	0.9730	0.9625
1.8	6.3173E-01	1.0090	0.9653	0.9508	0.9428	1.0014	1.0003	0.9996	0.9999	0.9751	0.6762	0.9736	0.9632
2.0	6.3969E-01	1.0093	0.9650	0.9505	0.9426	1.0016	1.0003	0.9996	0.9999	0.9752	0.6726	0.9741	0.9638
2.2	6.4556E-01	1.0096	0.9649	0.9506	0.9427	1.0017	1.0003	0.9996	0.9999	0.9755	0.6692	0.9745	0.9643
2.4	6.4996E-01	1.0098	0.9650	0.9507	0.9429	1.0019	1.0004	0.9997	0.9999	0.9757	0.6659	0.9749	0.9648
2.6	6.5330E-01	1.0101	0.9652	0.9510	0.9433	1.0021	1.0004	0.9997	0.9999	0.9759	0.6627	0.9753	0.9653
2.8	6.5588E-01	1.0105	0.9654	0.9514	0.9437	1.0023	1.0005	0.9997	0.9999	0.9762	0.6596	0.9756	0.9657
3.0	6.5789E-01	1.0108	0.9657	0.9517	0.9441	1.0025	1.0005	0.9997	0.9999	0.9764	0.6565	0.9759	0.9661
3.2	6.5946E-01	1.0111	0.9659	0.9521	0.9445	1.0027	1.0005	0.9997	0.9999	0.9766	0.6535	0.9762	0.9665
3.4	6.6070E-01	1.0114	0.9662	0.9525	0.9450	1.0028	1.0006	0.9998	1.0000	0.9768	0.6505	0.9764	0.9669
3.6	6.6169E-01	1.0117	0.9664	0.9529	0.9454	1.0030	1.0006	0.9998	1.0000	0.9770	0.6476	0.9767	0.9672
3.8	6.6247E-01	1.0120	0.9667	0.9532	0.9458	1.0032	1.0006	0.9998	1.0000	0.9771	0.6447	0.9769	0.9675
4.0	6.6310E-01	1.0124	0.9669	0.9536	0.9462	1.0034	1.0007	0.9998	1.0000	0.9773	0.6417	0.9771	0.9678
4.5	6.6416E-01	1.0132	0.9675	0.9544	0.9472	1.0038	1.0008	0.9998	1.0000	0.9777	0.6345	0.9775	0.9685
5.0	6.6476E-01	1.0140	0.9680	0.9552	0.9481	1.0042	1.0008	0.9998	1.0000	0.9780	0.6274	0.9779	0.9691
5.5	6.6508E-01	1.0148	0.9685	0.9558	0.9488	1.0047	1.0009	0.9998	1.0000	0.9783	0.6203	0.9782	0.9697
6.0	6.6523E-01	1.0157	0.9689	0.9564	0.9495	1.0051	1.0010	0.9999	1.0000	0.9785	0.6132	0.9785	0.9702
6.5	6.6528E-01	1.0165	0.9693	0.9569	0.9501	1.0055	1.0011	0.9999	1.0000	0.9787	0.6062	0.9787	0.9707
7.0	6.6527E-01	1.0174	0.9696	0.9573	0.9506	1.0059	1.0012	0.9999	1.0000	0.9789	0.5992	0.9789	0.9711
7.5	6.6522E-01	1.0182	0.9698	0.9577	0.9510	1.0064	1.0013	0.9999	1.0000	0.9791	0.5922	0.9790	0.9715
8.0	6.6516E-01	1.0191	0.9700	0.9581	0.9514	1.0068	1.0014	0.9999	1.0000	0.9792	0.5851	0.9792	0.9719
9.0	6.6502E-01	1.0208	0.9703	0.9586	0.9520	1.0076	1.0015	0.9999	1.0000	0.9794	0.5712	0.9794	0.9726
10.0	6.6491E-01	1.0225	0.9705	0.9589	0.9525	1.0085	1.0017	0.9999	1.0000	0.9795	0.5572	0.9795	0.9732
11.0	6.6484E-01	1.0242	0.9706	0.9591	0.9527	1.0093	1.0019	0.9999	1.0000	0.9795	0.5433	0.9795	0.9737
12.0	6.6483E-01	1.0259	0.9706	0.9592	0.9529	1.0101	1.0020	0.9999	1.0000	0.9796	0.5294	0.9796	0.9743
13.0	6.6487E-01	1.0276	0.9706	0.9592	0.9529	1.0110	1.0022	0.9999	1.0000	0.9795	0.5156	0.9796	0.9747
14.0	6.6496E-01	1.0293	0.9705	0.9591	0.9529	1.0118	1.0024	0.9999	1.0000	0.9795	0.5017	0.9795	0.9752
16.0	6.6529E-01	1.0327	0.9701	0.9588	0.9526	1.0135	1.0027	0.9999	1.0000	0.9793	0.4739	0.9793	0.9760
18.0	6.6580E-01	1.0361	0.9695	0.9582	0.9520	1.0151	1.0031	1.0000	1.0000	0.9790	0.4464	0.9791	0.9767
20.0	6.6645E-01	1.0396	0.9688	0.9574	0.9512	1.0167	1.0034	1.0000	1.0000	0.9787	0.4190	0.9787	0.9773
25.0	6.6859E-01	1.0482	0.9667	0.9550	0.9487	1.0208	1.0043	1.0000	1.0000	0.9777	0.3507	0.9777	0.9787
30.0	6.7127E-01	1.0568	0.9641	0.9520	0.9455	1.0247	1.0051	1.0000	1.0000	0.9764	0.2831	0.9764	0.9798
35.0	6.7433E-01	1.0654	0.9614	0.9487	0.9419	1.0287	1.0060	1.0000	1.0000	0.9751	0.2160	0.9751	0.9808
40.0	6.7769E-01	1.0740	0.9584	0.9451	0.9381	1.0325	1.0068	1.0000	1.0000	0.9736	0.1499	0.9736	0.9817
45.0	6.8126E-01	1.0826	0.9553	0.9414	0.9341	1.0364	1.0077	1.0000	1.0000	0.9721	0.0848	0.9721	0.9825
50.0	6.8501E-01	1.0913	0.9522	0.9375	0.9299	1.0401	1.0086	1.0000	1.0000	0.9706	0.0205	0.9706	0.9833

Instr. see p. 22

POSITRON Z= -22 A= 48

p	F_0L_0	L_0	λ_2	λ_3	λ_4	μ_1	μ_2	Λ_1	Λ_2	ν_{12}	$\hat{\nu}_{12}$	η_{12}	$\hat{\eta}_{12}$
0.1	4.9038E-04	1.0083	3.3070	5.2771	6.6807	1.0000	1.0000	0.9991	0.9998	1.5827	0.7189	0.9621	0.9497
0.2	3.6797E-02	1.0083	1.5289	1.7270	1.8238	1.0001	1.0000	0.9991	0.9998	1.1171	0.7166	0.9627	0.9504
0.3	1.3115E-01	1.0083	1.2015	1.2546	1.2767	1.0001	1.0000	0.9991	0.9998	1.0317	0.7135	0.9636	0.9512
0.4	2.3218E-01	1.0083	1.0882	1.1044	1.1094	1.0002	1.0000	0.9992	0.9998	1.0024	0.7101	0.9644	0.9521
0.5	3.1741E-01	1.0084	1.0367	1.0384	1.0370	1.0002	1.0000	0.9992	0.9998	0.9893	0.7066	0.9652	0.9530
0.6	3.8458E-01	1.0084	1.0093	1.0040	0.9996	1.0003	1.0001	0.9992	0.9998	0.9825	0.7032	0.9659	0.9538
0.7	4.3664E-01	1.0085	0.9934	0.9842	0.9781	1.0004	1.0001	0.9993	0.9999	0.9786	0.6999	0.9666	0.9545
0.8	4.7703E-01	1.0086	0.9834	0.9719	0.9648	1.0005	1.0001	0.9993	0.9999	0.9763	0.6967	0.9672	0.9552
0.9	5.0859E-01	1.0087	0.9768	0.9638	0.9562	1.0006	1.0001	0.9993	0.9999	0.9749	0.6938	0.9678	0.9558
1.0	5.3350E-01	1.0088	0.9724	0.9584	0.9504	1.0007	1.0001	0.9994	0.9999	0.9740	0.6911	0.9683	0.9564
1.2	5.6934E-01	1.0090	0.9671	0.9520	0.9436	1.0009	1.0002	0.9994	0.9999	0.9730	0.6861	0.9692	0.9574
1.4	5.9299E-01	1.0092	0.9643	0.9488	0.9402	1.0011	1.0002	0.9995	0.9999	0.9727	0.6816	0.9700	0.9583
1.6	6.0915E-01	1.0095	0.9629	0.9472	0.9386	1.0013	1.0002	0.9995	0.9999	0.9727	0.6776	0.9707	0.9591
1.8	6.2051E-01	1.0098	0.9622	0.9465	0.9378	1.0015	1.0003	0.9996	0.9999	0.9729	0.6738	0.9713	0.9599
2.0	6.2870E-01	1.0101	0.9620	0.9462	0.9376	1.0017	1.0003	0.9996	0.9999	0.9731	0.6703	0.9718	0.9605
2.2	6.3472E-01	1.0104	0.9619	0.9463	0.9377	1.0019	1.0004	0.9996	0.9999	0.9733	0.6670	0.9723	0.9611
2.4	6.3923E-01	1.0107	0.9620	0.9465	0.9380	1.0021	1.0004	0.9997	0.9999	0.9736	0.6638	0.9727	0.9617
2.6	6.4266E-01	1.0110	0.9622	0.9468	0.9384	1.0023	1.0004	0.9997	0.9999	0.9738	0.6607	0.9731	0.9622
2.8	6.4529E-01	1.0113	0.9624	0.9472	0.9388	1.0025	1.0005	0.9997	0.9999	0.9741	0.6577	0.9735	0.9626
3.0	6.4732E-01	1.0117	0.9627	0.9476	0.9393	1.0027	1.0005	0.9997	0.9999	0.9743	0.6547	0.9738	0.9631
3.2	6.4892E-01	1.0120	0.9630	0.9480	0.9398	1.0028	1.0006	0.9997	0.9999	0.9745	0.6517	0.9741	0.9635
3.4	6.5017E-01	1.0123	0.9633	0.9484	0.9403	1.0030	1.0006	0.9997	0.9999	0.9748	0.6488	0.9744	0.9639
3.6	6.5116E-01	1.0127	0.9636	0.9488	0.9407	1.0032	1.0006	0.9998	1.0000	0.9750	0.6460	0.9746	0.9642
3.8	6.5194E-01	1.0130	0.9639	0.9492	0.9412	1.0034	1.0007	0.9998	1.0000	0.9752	0.6431	0.9749	0.9646
4.0	6.5256E-01	1.0134	0.9641	0.9496	0.9417	1.0036	1.0007	0.9998	1.0000	0.9754	0.6403	0.9751	0.9649
4.5	6.5360E-01	1.0143	0.9648	0.9506	0.9427	1.0041	1.0008	0.9998	1.0000	0.9758	0.6333	0.9756	0.9657
5.0	6.5416E-01	1.0151	0.9654	0.9514	0.9437	1.0045	1.0009	0.9998	1.0000	0.9761	0.6264	0.9760	0.9663
5.5	6.5443E-01	1.0160	0.9659	0.9521	0.9445	1.0050	1.0010	0.9998	1.0000	0.9764	0.6195	0.9763	0.9670
6.0	6.5454E-01	1.0169	0.9663	0.9528	0.9453	1.0055	1.0011	0.9999	1.0000	0.9767	0.6126	0.9766	0.9675
6.5	6.5455E-01	1.0178	0.9667	0.9533	0.9460	1.0059	1.0012	0.9999	1.0000	0.9769	0.6058	0.9769	0.9680
7.0	6.5449E-01	1.0187	0.9671	0.9538	0.9465	1.0064	1.0013	0.9999	1.0000	0.9771	0.5990	0.9771	0.9685
7.5	6.5441E-01	1.0197	0.9674	0.9543	0.9470	1.0068	1.0014	0.9999	1.0000	0.9773	0.5922	0.9773	0.9689
8.0	6.5431E-01	1.0206	0.9676	0.9546	0.9475	1.0073	1.0014	0.9999	1.0000	0.9774	0.5854	0.9774	0.9694
9.0	6.5411E-01	1.0224	0.9680	0.9552	0.9482	1.0082	1.0016	0.9999	1.0000	0.9777	0.5718	0.9776	0.9701
10.0	6.5393E-01	1.0242	0.9682	0.9556	0.9487	1.0091	1.0018	0.9999	1.0000	0.9778	0.5582	0.9778	0.9708
11.0	6.5382E-01	1.0260	0.9684	0.9559	0.9490	1.0100	1.0020	0.9999	1.0000	0.9779	0.5447	0.9779	0.9714
12.0	6.5376E-01	1.0279	0.9684	0.9561	0.9492	1.0109	1.0022	0.9999	1.0000	0.9779	0.5312	0.9779	0.9720
13.0	6.5377E-01	1.0297	0.9683	0.9561	0.9493	1.0118	1.0024	0.9999	1.0000	0.9779	0.5176	0.9779	0.9725
14.0	6.5383E-01	1.0315	0.9682	0.9560	0.9493	1.0127	1.0025	0.9999	1.0000	0.9779	0.5042	0.9779	0.9730
16.0	6.5412E-01	1.0352	0.9679	0.9557	0.9490	1.0144	1.0029	0.9999	1.0000	0.9777	0.4772	0.9777	0.9738
18.0	6.5460E-01	1.0389	0.9673	0.9551	0.9484	1.0162	1.0033	1.0000	1.0000	0.9774	0.4503	0.9774	0.9746
20.0	6.5523E-01	1.0426	0.9666	0.9544	0.9477	1.0179	1.0036	1.0000	1.0000	0.9771	0.4236	0.9771	0.9753
25.0	6.5738E-01	1.0518	0.9644	0.9519	0.9451	1.0222	1.0046	1.0000	1.0000	0.9760	0.3572	0.9760	0.9768
30.0	6.6012E-01	1.0610	0.9618	0.9488	0.9418	1.0265	1.0055	1.0000	1.0000	0.9747	0.2914	0.9748	0.9781
35.0	6.6327E-01	1.0703	0.9588	0.9453	0.9381	1.0307	1.0064	1.0000	1.0000	0.9733	0.2261	0.9733	0.9792
40.0	6.6673E-01	1.0795	0.9558	0.9415	0.9341	1.0348	1.0073	1.0000	1.0000	0.9718	0.1619	0.9718	0.9801
45.0	6.7044E-01	1.0888	0.9525	0.9376	0.9298	1.0389	1.0083	1.0000	1.0000	0.9702	0.0982	0.9702	0.9810
50.0	6.7433E-01	1.0981	0.9492	0.9336	0.9254	1.0429	1.0092	1.0000	1.0000	0.9686	0.0357	0.9686	0.9819

Instr. see p. 22

POSITRON Z= -23 A= 51

p	F_0L_0	L_0	λ_2	λ_3	λ_4	μ_1	μ_2	Λ_1	Λ_2	ν_{12}	$\hat{\nu}_{12}$	η_{12}	$\hat{\eta}_{12}$
0.1	3.2890E-04	1.0090	3.5050	5.7816	7.4602	1.0000	1.0000	0.9990	0.9998	1.6354	0.7165	0.9590	0.9455
0.2	3.0932E-02	1.0090	1.5746	1.7971	1.9071	1.0001	1.0000	0.9990	0.9998	1.1279	0.7142	0.9597	0.9462
0.3	1.1832E-01	1.0090	1.2190	1.2778	1.3025	1.0001	1.0000	0.9991	0.9998	1.0348	0.7111	0.9606	0.9471
0.4	2.1633E-01	1.0090	1.0959	1.1138	1.1194	1.0002	1.0000	0.9991	0.9998	1.0028	0.7076	0.9615	0.9481
0.5	3.0096E-01	1.0091	1.0400	1.0420	1.0405	1.0002	1.0000	0.9991	0.9998	0.9885	0.7041	0.9623	0.9490
0.6	3.6854E-01	1.0091	1.0103	1.0046	0.9998	1.0003	1.0001	0.9992	0.9998	0.9811	0.7006	0.9631	0.9498
0.7	4.2132E-01	1.0092	0.9929	0.9830	0.9764	1.0004	1.0001	0.9992	0.9998	0.9769	0.6973	0.9638	0.9506
0.8	4.6248E-01	1.0093	0.9821	0.9696	0.9620	1.0005	1.0001	0.9992	0.9998	0.9743	0.6942	0.9644	0.9513
0.9	4.9475E-01	1.0094	0.9749	0.9609	0.9526	1.0006	1.0001	0.9993	0.9999	0.9728	0.6913	0.9650	0.9520
1.0	5.2027E-01	1.0095	0.9701	0.9550	0.9463	1.0007	1.0001	0.9993	0.9999	0.9718	0.6885	0.9656	0.9526
1.2	5.5707E-01	1.0097	0.9643	0.9481	0.9389	1.0009	1.0002	0.9994	0.9999	0.9708	0.6835	0.9666	0.9537
1.4	5.8139E-01	1.0100	0.9614	0.9446	0.9353	1.0012	1.0002	0.9994	0.9999	0.9705	0.6791	0.9674	0.9547
1.6	5.9801E-01	1.0103	0.9599	0.9429	0.9335	1.0014	1.0003	0.9995	0.9999	0.9704	0.6751	0.9682	0.9556
1.8	6.0969E-01	1.0106	0.9591	0.9421	0.9327	1.0016	1.0003	0.9995	0.9999	0.9706	0.6715	0.9689	0.9564
2.0	6.1810E-01	1.0109	0.9588	0.9418	0.9325	1.0018	1.0003	0.9996	0.9999	0.9708	0.6680	0.9694	0.9571
2.2	6.2428E-01	1.0112	0.9588	0.9418	0.9326	1.0020	1.0004	0.9996	0.9999	0.9711	0.6648	0.9700	0.9578
2.4	6.2889E-01	1.0115	0.9589	0.9421	0.9329	1.0022	1.0004	0.9996	0.9999	0.9714	0.6616	0.9704	0.9583
2.6	6.3239E-01	1.0119	0.9591	0.9424	0.9333	1.0024	1.0005	0.9997	0.9999	0.9716	0.6586	0.9709	0.9589
2.8	6.3507E-01	1.0122	0.9593	0.9428	0.9338	1.0026	1.0005	0.9997	0.9999	0.9719	0.6556	0.9713	0.9594
3.0	6.3714E-01	1.0126	0.9596	0.9433	0.9343	1.0028	1.0006	0.9997	0.9999	0.9722	0.6528	0.9716	0.9599
3.2	6.3875E-01	1.0130	0.9600	0.9437	0.9349	1.0030	1.0006	0.9997	0.9999	0.9724	0.6499	0.9720	0.9603
3.4	6.4001E-01	1.0133	0.9603	0.9442	0.9354	1.0033	1.0006	0.9997	0.9999	0.9727	0.6471	0.9723	0.9608
3.6	6.4100E-01	1.0137	0.9606	0.9447	0.9359	1.0035	1.0007	0.9997	0.9999	0.9729	0.6443	0.9726	0.9612
3.8	6.4177E-01	1.0141	0.9609	0.9451	0.9364	1.0037	1.0007	0.9998	1.0000	0.9731	0.6416	0.9728	0.9615
4.0	6.4238E-01	1.0144	0.9612	0.9455	0.9369	1.0039	1.0008	0.9998	1.0000	0.9733	0.6388	0.9731	0.9619
4.5	6.4338E-01	1.0154	0.9620	0.9466	0.9381	1.0044	1.0009	0.9998	1.0000	0.9738	0.6320	0.9736	0.9627
5.0	6.4390E-01	1.0163	0.9626	0.9475	0.9392	1.0049	1.0010	0.9998	1.0000	0.9742	0.6253	0.9740	0.9635
5.5	6.4413E-01	1.0173	0.9632	0.9483	0.9401	1.0053	1.0011	0.9998	1.0000	0.9745	0.6186	0.9744	0.9641
6.0	6.4419E-01	1.0182	0.9637	0.9490	0.9410	1.0058	1.0012	0.9998	1.0000	0.9748	0.6119	0.9747	0.9647
6.5	6.4415E-01	1.0192	0.9641	0.9497	0.9417	1.0063	1.0013	0.9999	1.0000	0.9751	0.6053	0.9750	0.9653
7.0	6.4405E-01	1.0202	0.9645	0.9502	0.9423	1.0068	1.0014	0.9999	1.0000	0.9753	0.5986	0.9753	0.9658
7.5	6.4392E-01	1.0211	0.9649	0.9507	0.9429	1.0073	1.0014	0.9999	1.0000	0.9755	0.5920	0.9755	0.9663
8.0	6.4379E-01	1.0221	0.9651	0.9511	0.9434	1.0078	1.0015	0.9999	1.0000	0.9757	0.5854	0.9756	0.9667
9.0	6.4351E-01	1.0241	0.9656	0.9518	0.9442	1.0088	1.0017	0.9999	1.0000	0.9759	0.5722	0.9759	0.9676
10.0	6.4328E-01	1.0260	0.9659	0.9523	0.9448	1.0097	1.0019	0.9999	1.0000	0.9761	0.5589	0.9761	0.9683
11.0	6.4311E-01	1.0280	0.9660	0.9526	0.9452	1.0107	1.0021	0.9999	1.0000	0.9762	0.5458	0.9762	0.9690
12.0	6.4301E-01	1.0299	0.9661	0.9528	0.9454	1.0116	1.0023	0.9999	1.0000	0.9762	0.5327	0.9762	0.9696
13.0	6.4298E-01	1.0319	0.9661	0.9529	0.9456	1.0126	1.0025	0.9999	1.0000	0.9762	0.5194	0.9762	0.9701
14.0	6.4301E-01	1.0339	0.9660	0.9529	0.9456	1.0135	1.0027	0.9999	1.0000	0.9762	0.5064	0.9762	0.9707
16.0	6.4325E-01	1.0378	0.9656	0.9526	0.9453	1.0154	1.0031	0.9999	1.0000	0.9761	0.4802	0.9761	0.9716
18.0	6.4370E-01	1.0417	0.9651	0.9520	0.9448	1.0173	1.0035	0.9999	1.0000	0.9758	0.4540	0.9758	0.9725
20.0	6.4432E-01	1.0456	0.9644	0.9513	0.9440	1.0191	1.0039	1.0000	1.0000	0.9755	0.4279	0.9755	0.9732
25.0	6.4646E-01	1.0555	0.9621	0.9487	0.9414	1.0237	1.0049	1.0000	1.0000	0.9744	0.3633	0.9744	0.9749
30.0	6.4924E-01	1.0654	0.9594	0.9455	0.9380	1.0282	1.0058	1.0000	1.0000	0.9730	0.2990	0.9730	0.9762
35.0	6.5247E-01	1.0753	0.9563	0.9419	0.9342	1.0327	1.0068	1.0000	1.0000	0.9715	0.2354	0.9715	0.9774
40.0	6.5604E-01	1.0852	0.9531	0.9380	0.9300	1.0371	1.0078	1.0000	1.0000	0.9700	0.1729	0.9700	0.9785
45.0	6.5986E-01	1.0951	0.9497	0.9338	0.9255	1.0415	1.0088	1.0000	1.0000	0.9683	0.1111	0.9683	0.9795
50.0	6.6389E-01	1.1051	0.9463	0.9296	0.9210	1.0457	1.0098	1.0000	1.0000	0.9666	0.0502	0.9666	0.9804

Instr. see p. 22

POSITRON Z= -24 A= 52

p	F_0L_0	L_0	λ_2	λ_3	λ_4	μ_1	μ_2	Λ_1	Λ_2	ν_{12}	$\hat{\nu}_{12}$	η_{12}	$\hat{\eta}_{12}$
0.1	2.2039E-04	1.0096	3.7083	6.3239	8.3203	1.0000	1.0000	0.9990	0.9998	1.6901	0.7140	0.9558	0.9412
0.2	2.5983E-02	1.0096	1.6214	1.8703	1.9947	1.0001	1.0000	0.9990	0.9998	1.1391	0.7117	0.9565	0.9419
0.3	1.0673E-01	1.0097	1.2368	1.3017	1.3291	1.0001	1.0000	0.9990	0.9998	1.0380	0.7086	0.9574	0.9428
0.4	2.0159E-01	1.0097	1.1037	1.1233	1.1294	1.0002	1.0000	0.9990	0.9998	1.0032	0.7051	0.9583	0.9438
0.5	2.8546E-01	1.0098	1.0432	1.0454	1.0438	1.0003	1.0000	0.9991	0.9998	0.9877	0.7015	0.9592	0.9448
0.6	3.5331E-01	1.0098	1.0111	1.0049	0.9997	1.0004	1.0001	0.9991	0.9998	0.9796	0.6981	0.9600	0.9457
0.7	4.0675E-01	1.0099	0.9923	0.9815	0.9744	1.0005	1.0001	0.9992	0.9998	0.9750	0.6947	0.9608	0.9465
0.8	4.4863E-01	1.0100	0.9805	0.9670	0.9588	1.0006	1.0001	0.9992	0.9998	0.9723	0.6916	0.9615	0.9473
0.9	4.8158E-01	1.0101	0.9728	0.9576	0.9486	1.0007	1.0001	0.9992	0.9998	0.9706	0.6887	0.9622	0.9480
1.0	5.0770E-01	1.0102	0.9676	0.9512	0.9418	1.0008	1.0001	0.9993	0.9999	0.9695	0.6860	0.9627	0.9486
1.2	5.4544E-01	1.0104	0.9614	0.9438	0.9339	1.0010	1.0002	0.9993	0.9999	0.9684	0.6811	0.9638	0.9499
1.4	5.7041E-01	1.0107	0.9582	0.9400	0.9299	1.0012	1.0002	0.9994	0.9999	0.9680	0.6767	0.9647	0.9509
1.6	5.8747E-01	1.0110	0.9565	0.9381	0.9280	1.0014	1.0003	0.9995	0.9999	0.9680	0.6728	0.9656	0.9519
1.8	5.9946E-01	1.0113	0.9557	0.9373	0.9272	1.0017	1.0003	0.9995	0.9999	0.9682	0.6692	0.9663	0.9527
2.0	6.0809E-01	1.0117	0.9554	0.9370	0.9269	1.0019	1.0004	0.9995	0.9999	0.9684	0.6659	0.9669	0.9535
2.2	6.1441E-01	1.0120	0.9553	0.9371	0.9271	1.0021	1.0004	0.9996	0.9999	0.9687	0.6628	0.9675	0.9542
2.4	6.1913E-01	1.0124	0.9555	0.9373	0.9274	1.0023	1.0005	0.9996	0.9999	0.9690	0.6597	0.9680	0.9548
2.6	6.2269E-01	1.0127	0.9557	0.9377	0.9279	1.0026	1.0005	0.9996	0.9999	0.9693	0.6568	0.9685	0.9554
2.8	6.2541E-01	1.0131	0.9560	0.9382	0.9284	1.0028	1.0005	0.9997	0.9999	0.9696	0.6540	0.9689	0.9560
3.0	6.2750E-01	1.0135	0.9563	0.9387	0.9290	1.0030	1.0006	0.9997	0.9999	0.9699	0.6512	0.9693	0.9565
3.2	6.2913E-01	1.0138	0.9567	0.9392	0.9296	1.0032	1.0006	0.9997	0.9999	0.9702	0.6485	0.9697	0.9570
3.4	6.3039E-01	1.0142	0.9570	0.9397	0.9302	1.0034	1.0007	0.9997	0.9999	0.9704	0.6458	0.9700	0.9574
3.6	6.3137E-01	1.0146	0.9574	0.9402	0.9307	1.0036	1.0007	0.9997	0.9999	0.9707	0.6431	0.9703	0.9579
3.8	6.3214E-01	1.0150	0.9578	0.9407	0.9313	1.0039	1.0008	0.9997	0.9999	0.9709	0.6404	0.9706	0.9583
4.0	6.3273E-01	1.0154	0.9581	0.9412	0.9319	1.0041	1.0008	0.9998	1.0000	0.9712	0.6378	0.9709	0.9587
4.5	6.3369E-01	1.0164	0.9589	0.9423	0.9332	1.0046	1.0009	0.9998	1.0000	0.9717	0.6313	0.9715	0.9596
5.0	6.3415E-01	1.0174	0.9596	0.9433	0.9344	1.0051	1.0010	0.9998	1.0000	0.9721	0.6248	0.9720	0.9604
5.5	6.3432E-01	1.0184	0.9603	0.9442	0.9354	1.0056	1.0011	0.9998	1.0000	0.9725	0.6184	0.9724	0.9611
6.0	6.3432E-01	1.0194	0.9609	0.9451	0.9363	1.0062	1.0012	0.9998	1.0000	0.9728	0.6120	0.9727	0.9617
6.5	6.3422E-01	1.0204	0.9614	0.9458	0.9372	1.0067	1.0013	0.9998	1.0000	0.9731	0.6056	0.9731	0.9624
7.0	6.3407E-01	1.0214	0.9618	0.9464	0.9379	1.0072	1.0014	0.9999	1.0000	0.9734	0.5992	0.9733	0.9629
7.5	6.3389E-01	1.0225	0.9622	0.9470	0.9385	1.0077	1.0015	0.9999	1.0000	0.9736	0.5929	0.9736	0.9634
8.0	6.3370E-01	1.0235	0.9625	0.9474	0.9391	1.0082	1.0016	0.9999	1.0000	0.9738	0.5865	0.9738	0.9639
9.0	6.3334E-01	1.0255	0.9630	0.9482	0.9400	1.0092	1.0018	0.9999	1.0000	0.9741	0.5738	0.9741	0.9648
10.0	6.3303E-01	1.0276	0.9634	0.9488	0.9407	1.0102	1.0020	0.9999	1.0000	0.9743	0.5611	0.9743	0.9656
11.0	6.3278E-01	1.0297	0.9636	0.9492	0.9412	1.0112	1.0022	0.9999	1.0000	0.9744	0.5484	0.9744	0.9664
12.0	6.3262E-01	1.0317	0.9637	0.9494	0.9415	1.0122	1.0024	0.9999	1.0000	0.9745	0.5358	0.9745	0.9670
13.0	6.3253E-01	1.0338	0.9637	0.9496	0.9417	1.0132	1.0027	0.9999	1.0000	0.9745	0.5231	0.9745	0.9676
14.0	6.3251E-01	1.0359	0.9637	0.9496	0.9418	1.0142	1.0029	0.9999	1.0000	0.9745	0.5105	0.9745	0.9682
16.0	6.3267E-01	1.0400	0.9634	0.9494	0.9417	1.0162	1.0033	0.9999	1.0000	0.9744	0.4852	0.9744	0.9692
18.0	6.3305E-01	1.0441	0.9629	0.9489	0.9412	1.0182	1.0037	0.9999	1.0000	0.9741	0.4600	0.9741	0.9702
20.0	6.3362E-01	1.0483	0.9622	0.9482	0.9405	1.0201	1.0041	1.0000	1.0000	0.9738	0.4349	0.9738	0.9710
25.0	6.3567E-01	1.0586	0.9599	0.9457	0.9379	1.0249	1.0051	1.0000	1.0000	0.9727	0.3726	0.9727	0.9728
30.0	6.3841E-01	1.0690	0.9572	0.9424	0.9345	1.0297	1.0061	1.0000	1.0000	0.9714	0.3109	0.9714	0.9742
35.0	6.4163E-01	1.0795	0.9541	0.9387	0.9306	1.0344	1.0072	1.0000	1.0000	0.9699	0.2496	0.9699	0.9756
40.0	6.4522E-01	1.0899	0.9508	0.9347	0.9263	1.0390	1.0082	1.0000	1.0000	0.9683	0.1892	0.9683	0.9767
45.0	6.4908E-01	1.1004	0.9473	0.9305	0.9218	1.0436	1.0093	1.0000	1.0000	0.9666	0.1298	0.9666	0.9778
50.0	6.5316E-01	1.1109	0.9437	0.9262	0.9171	1.0480	1.0103	1.0000	1.0000	0.9648	0.0712	0.9648	0.9788

Instr. see p. 22

POSITRON Z= -25 A= 55

p	F_0L_0	L_0	λ_2	λ_3	λ_4	μ_1	μ_2	Λ_1	Λ_2	ν_{12}	$\hat{\nu}_{12}$	η_{12}	$\hat{\eta}_{12}$
0.1	1.4747E-04	1.0104	3.9180	6.9085	9.2718	1.0000	1.0000	0.9989	0.9998	1.7470	0.7114	0.9526	0.9367
0.2	2.1799E-02	1.0104	1.6698	1.9473	2.0878	1.0001	1.0000	0.9989	0.9998	1.1509	0.7091	0.9533	0.9374
0.3	9.6191E-02	1.0104	1.2554	1.3268	1.3571	1.0001	1.0000	0.9990	0.9998	1.0414	0.7060	0.9542	0.9384
0.4	1.8774E-01	1.0105	1.1119	1.1334	1.1401	1.0002	1.0000	0.9990	0.9998	1.0038	0.7025	0.9552	0.9395
0.5	2.7065E-01	1.0105	1.0467	1.0491	1.0474	1.0003	1.0001	0.9990	0.9998	0.9869	0.6989	0.9561	0.9405
0.6	3.3864E-01	1.0106	1.0120	1.0054	0.9998	1.0004	1.0001	0.9991	0.9998	0.9781	0.6954	0.9570	0.9414
0.7	3.9262E-01	1.0107	0.9918	0.9802	0.9725	1.0005	1.0001	0.9991	0.9998	0.9732	0.6921	0.9578	0.9423
0.8	4.3515E-01	1.0108	0.9791	0.9646	0.9557	1.0006	1.0001	0.9991	0.9998	0.9702	0.6889	0.9585	0.9431
0.9	4.6873E-01	1.0109	0.9708	0.9544	0.9448	1.0007	1.0001	0.9992	0.9998	0.9683	0.6860	0.9592	0.9439
1.0	4.9541E-01	1.0110	0.9651	0.9476	0.9375	1.0008	1.0002	0.9992	0.9998	0.9672	0.6833	0.9599	0.9446
1.2	5.3404E-01	1.0112	0.9584	0.9395	0.9289	1.0011	1.0002	0.9993	0.9999	0.9660	0.6784	0.9610	0.9459
1.4	5.5964E-01	1.0115	0.9550	0.9355	0.9247	1.0013	1.0002	0.9994	0.9999	0.9656	0.6741	0.9620	0.9471
1.6	5.7713E-01	1.0118	0.9532	0.9335	0.9226	1.0015	1.0003	0.9994	0.9999	0.9656	0.6703	0.9629	0.9481
1.8	5.8942E-01	1.0122	0.9524	0.9326	0.9217	1.0018	1.0003	0.9995	0.9999	0.9657	0.6667	0.9637	0.9490
2.0	5.9825E-01	1.0125	0.9520	0.9323	0.9214	1.0020	1.0004	0.9995	0.9999	0.9660	0.6635	0.9644	0.9498
2.2	6.0472E-01	1.0129	0.9520	0.9323	0.9216	1.0023	1.0004	0.9996	0.9999	0.9663	0.6604	0.9650	0.9506
2.4	6.0953E-01	1.0133	0.9521	0.9326	0.9220	1.0025	1.0005	0.9996	0.9999	0.9666	0.6574	0.9656	0.9513
2.6	6.1316E-01	1.0137	0.9524	0.9330	0.9225	1.0027	1.0005	0.9996	0.9999	0.9670	0.6546	0.9661	0.9519
2.8	6.1592E-01	1.0140	0.9527	0.9335	0.9231	1.0030	1.0006	0.9996	0.9999	0.9673	0.6518	0.9665	0.9525
3.0	6.1803E-01	1.0144	0.9531	0.9341	0.9237	1.0032	1.0006	0.9997	0.9999	0.9676	0.6491	0.9670	0.9531
3.2	6.1967E-01	1.0149	0.9535	0.9346	0.9243	1.0034	1.0007	0.9997	0.9999	0.9679	0.6465	0.9674	0.9536
3.4	6.2093E-01	1.0153	0.9538	0.9352	0.9250	1.0037	1.0007	0.9997	0.9999	0.9682	0.6438	0.9677	0.9541
3.6	6.2191E-01	1.0157	0.9542	0.9357	0.9256	1.0039	1.0008	0.9997	0.9999	0.9685	0.6412	0.9681	0.9546
3.8	6.2266E-01	1.0161	0.9546	0.9363	0.9262	1.0041	1.0008	0.9997	0.9999	0.9688	0.6386	0.9684	0.9550
4.0	6.2324E-01	1.0165	0.9550	0.9368	0.9268	1.0043	1.0008	0.9997	0.9999	0.9690	0.6361	0.9687	0.9554
4.5	6.2416E-01	1.0176	0.9559	0.9381	0.9283	1.0049	1.0010	0.9998	1.0000	0.9696	0.6297	0.9693	0.9564
5.0	6.2456E-01	1.0186	0.9567	0.9392	0.9296	1.0054	1.0011	0.9998	1.0000	0.9701	0.6234	0.9699	0.9573
5.5	6.2468E-01	1.0197	0.9574	0.9402	0.9307	1.0060	1.0012	0.9998	1.0000	0.9705	0.6172	0.9703	0.9580
6.0	6.2463E-01	1.0208	0.9580	0.9411	0.9317	1.0065	1.0013	0.9998	1.0000	0.9708	0.6110	0.9707	0.9588
6.5	6.2448E-01	1.0219	0.9586	0.9419	0.9327	1.0071	1.0014	0.9998	1.0000	0.9712	0.6048	0.9711	0.9594
7.0	6.2428E-01	1.0229	0.9591	0.9426	0.9335	1.0076	1.0015	0.9998	1.0000	0.9714	0.5986	0.9714	0.9600
7.5	6.2405E-01	1.0240	0.9595	0.9432	0.9342	1.0082	1.0016	0.9999	1.0000	0.9717	0.5923	0.9716	0.9606
8.0	6.2382E-01	1.0251	0.9599	0.9437	0.9348	1.0087	1.0017	0.9999	1.0000	0.9719	0.5861	0.9719	0.9611
9.0	6.2338E-01	1.0273	0.9605	0.9446	0.9358	1.0098	1.0019	0.9999	1.0000	0.9722	0.5738	0.9722	0.9621
10.0	6.2300E-01	1.0295	0.9609	0.9453	0.9366	1.0109	1.0022	0.9999	1.0000	0.9725	0.5613	0.9724	0.9630
11.0	6.2270E-01	1.0317	0.9611	0.9457	0.9372	1.0119	1.0024	0.9999	1.0000	0.9726	0.5490	0.9726	0.9637
12.0	6.2248E-01	1.0339	0.9613	0.9460	0.9376	1.0130	1.0026	0.9999	1.0000	0.9727	0.5366	0.9727	0.9645
13.0	6.2235E-01	1.0361	0.9614	0.9462	0.9378	1.0141	1.0028	0.9999	1.0000	0.9728	0.5243	0.9728	0.9651
14.0	6.2229E-01	1.0383	0.9613	0.9463	0.9379	1.0151	1.0030	0.9999	1.0000	0.9728	0.5119	0.9728	0.9657
16.0	6.2239E-01	1.0427	0.9610	0.9461	0.9378	1.0172	1.0035	0.9999	1.0000	0.9726	0.4874	0.9727	0.9669
18.0	6.2273E-01	1.0471	0.9605	0.9456	0.9374	1.0193	1.0039	0.9999	1.0000	0.9724	0.4628	0.9724	0.9679
20.0	6.2327E-01	1.0515	0.9598	0.9449	0.9367	1.0214	1.0043	0.9999	1.0000	0.9721	0.4382	0.9721	0.9688
25.0	6.2531E-01	1.0625	0.9575	0.9424	0.9341	1.0265	1.0054	1.0000	1.0000	0.9710	0.3773	0.9710	0.9707
30.0	6.2807E-01	1.0736	0.9547	0.9391	0.9306	1.0315	1.0065	1.0000	1.0000	0.9696	0.3172	0.9696	0.9723
35.0	6.3136E-01	1.0847	0.9515	0.9353	0.9266	1.0365	1.0076	1.0000	1.0000	0.9680	0.2574	0.9680	0.9737
40.0	6.3503E-01	1.0958	0.9481	0.9311	0.9221	1.0414	1.0087	1.0000	1.0000	0.9664	0.1985	0.9664	0.9750
45.0	6.3899E-01	1.1070	0.9445	0.9267	0.9174	1.0462	1.0098	1.0000	1.0000	0.9646	0.1404	0.9646	0.9761
50.0	6.4320E-01	1.1182	0.9408	0.9222	0.9125	1.0510	1.0110	1.0000	1.0000	0.9628	0.0833	0.9628	0.9772

Instr. see p. 22

POSITRON Z= -26 A= 56

p	F_0L_0	L_0	λ_2	λ_3	λ_4	μ_1	μ_2	Λ_1	Λ_2	ν_{12}	$\hat{\nu}_{12}$	η_{12}	$\hat{\eta}_{12}$
0.1	9.8610E-05	1.0111	4.1323	7.5337	10.3168	1.0001	1.0000	0.9989	0.9998	1.8058	0.7087	0.9492	0.9321
0.2	1.8280E-02	1.0111	1.7191	2.0273	2.1853	1.0001	1.0000	0.9989	0.9998	1.1629	0.7065	0.9498	0.9328
0.3	8.6681E-02	1.0111	1.2742	1.3523	1.3857	1.0001	1.0000	0.9989	0.9998	1.0448	0.7033	0.9508	0.9338
0.4	1.7488E-01	1.0112	1.1202	1.1434	1.1508	1.0002	1.0000	0.9989	0.9998	1.0043	0.6998	0.9518	0.9349
0.5	2.5670E-01	1.0112	1.0500	1.0527	1.0509	1.0003	1.0001	0.9990	0.9998	0.9860	0.6962	0.9528	0.9360
0.6	3.2473E-01	1.0113	1.0128	1.0057	0.9997	1.0004	1.0001	0.9990	0.9998	0.9766	0.6927	0.9537	0.9370
0.7	3.7920E-01	1.0114	0.9911	0.9786	0.9703	1.0005	1.0001	0.9991	0.9998	0.9712	0.6894	0.9546	0.9379
0.8	4.2234E-01	1.0115	0.9775	0.9618	0.9523	1.0006	1.0001	0.9991	0.9998	0.9680	0.6862	0.9554	0.9388
0.9	4.5652E-01	1.0116	0.9685	0.9509	0.9405	1.0007	1.0001	0.9992	0.9998	0.9660	0.6833	0.9561	0.9396
1.0	4.8374E-01	1.0117	0.9624	0.9435	0.9327	1.0009	1.0002	0.9992	0.9998	0.9647	0.6806	0.9568	0.9404
1.2	5.2324E-01	1.0120	0.9552	0.9349	0.9235	1.0011	1.0002	0.9993	0.9999	0.9634	0.6758	0.9581	0.9418
1.4	5.4946E-01	1.0123	0.9515	0.9306	0.9190	1.0014	1.0003	0.9993	0.9999	0.9630	0.6715	0.9591	0.9430
1.6	5.6738E-01	1.0126	0.9496	0.9284	0.9168	1.0016	1.0003	0.9994	0.9999	0.9630	0.6678	0.9601	0.9441
1.8	5.7996E-01	1.0130	0.9487	0.9275	0.9158	1.0019	1.0004	0.9994	0.9999	0.9632	0.6644	0.9609	0.9451
2.0	5.8899E-01	1.0134	0.9484	0.9271	0.9155	1.0021	1.0004	0.9995	0.9999	0.9634	0.6612	0.9617	0.9460
2.2	5.9559E-01	1.0137	0.9483	0.9272	0.9157	1.0024	1.0005	0.9995	0.9999	0.9638	0.6582	0.9623	0.9468
2.4	6.0049E-01	1.0141	0.9485	0.9276	0.9161	1.0026	1.0005	0.9996	0.9999	0.9641	0.6553	0.9630	0.9475
2.6	6.0418E-01	1.0145	0.9488	0.9280	0.9167	1.0029	1.0006	0.9996	0.9999	0.9645	0.6526	0.9635	0.9482
2.8	6.0697E-01	1.0150	0.9491	0.9286	0.9173	1.0031	1.0006	0.9996	0.9999	0.9649	0.6499	0.9640	0.9489
3.0	6.0910E-01	1.0154	0.9495	0.9292	0.9180	1.0034	1.0006	0.9996	0.9999	0.9652	0.6473	0.9645	0.9495
3.2	6.1074E-01	1.0158	0.9500	0.9298	0.9187	1.0036	1.0007	0.9997	0.9999	0.9655	0.6448	0.9649	0.9500
3.4	6.1201E-01	1.0162	0.9504	0.9304	0.9194	1.0038	1.0007	0.9997	0.9999	0.9658	0.6422	0.9653	0.9506
3.6	6.1297E-01	1.0167	0.9508	0.9310	0.9201	1.0041	1.0008	0.9997	0.9999	0.9661	0.6397	0.9657	0.9511
3.8	6.1371E-01	1.0171	0.9513	0.9316	0.9208	1.0043	1.0008	0.9997	0.9999	0.9664	0.6373	0.9660	0.9515
4.0	6.1428E-01	1.0175	0.9517	0.9322	0.9215	1.0045	1.0009	0.9997	0.9999	0.9667	0.6348	0.9663	0.9520
4.5	6.1513E-01	1.0186	0.9527	0.9335	0.9230	1.0051	1.0010	0.9998	1.0000	0.9673	0.6287	0.9671	0.9530
5.0	6.1548E-01	1.0198	0.9535	0.9348	0.9244	1.0057	1.0011	0.9998	1.0000	0.9679	0.6226	0.9677	0.9540
5.5	6.1553E-01	1.0209	0.9543	0.9359	0.9257	1.0063	1.0012	0.9998	1.0000	0.9683	0.6166	0.9682	0.9548
6.0	6.1541E-01	1.0220	0.9550	0.9369	0.9268	1.0069	1.0013	0.9998	1.0000	0.9687	0.6106	0.9686	0.9556
6.5	6.1520E-01	1.0231	0.9557	0.9377	0.9279	1.0075	1.0015	0.9998	1.0000	0.9691	0.6046	0.9690	0.9563
7.0	6.1494E-01	1.0243	0.9562	0.9385	0.9287	1.0080	1.0016	0.9998	1.0000	0.9694	0.5987	0.9693	0.9569
7.5	6.1466E-01	1.0254	0.9567	0.9392	0.9295	1.0086	1.0017	0.9999	1.0000	0.9697	0.5927	0.9696	0.9575
8.0	6.1437E-01	1.0266	0.9571	0.9398	0.9302	1.0092	1.0018	0.9999	1.0000	0.9699	0.5867	0.9699	0.9581
9.0	6.1383E-01	1.0289	0.9578	0.9408	0.9314	1.0103	1.0020	0.9999	1.0000	0.9703	0.5748	0.9703	0.9592
10.0	6.1336E-01	1.0312	0.9583	0.9416	0.9323	1.0114	1.0023	0.9999	1.0000	0.9706	0.5629	0.9705	0.9601
11.0	6.1298E-01	1.0335	0.9586	0.9421	0.9330	1.0125	1.0025	0.9999	1.0000	0.9708	0.5510	0.9707	0.9609
12.0	6.1270E-01	1.0358	0.9588	0.9425	0.9335	1.0136	1.0027	0.9999	1.0000	0.9709	0.5391	0.9709	0.9617
13.0	6.1250E-01	1.0381	0.9589	0.9427	0.9338	1.0148	1.0029	0.9999	1.0000	0.9710	0.5271	0.9710	0.9624
14.0	6.1239E-01	1.0404	0.9589	0.9429	0.9340	1.0159	1.0032	0.9999	1.0000	0.9710	0.5152	0.9710	0.9631
16.0	6.1239E-01	1.0450	0.9587	0.9428	0.9340	1.0180	1.0036	0.9999	1.0000	0.9709	0.4915	0.9709	0.9643
18.0	6.1265E-01	1.0496	0.9582	0.9424	0.9336	1.0202	1.0041	0.9999	1.0000	0.9707	0.4677	0.9707	0.9654
20.0	6.1313E-01	1.0542	0.9576	0.9417	0.9330	1.0224	1.0045	0.9999	1.0000	0.9704	0.4441	0.9704	0.9664
25.0	6.1507E-01	1.0658	0.9553	0.9392	0.9304	1.0277	1.0057	1.0000	1.0000	0.9693	0.3854	0.9693	0.9684
30.0	6.1778E-01	1.0775	0.9524	0.9359	0.9270	1.0330	1.0068	1.0000	1.0000	0.9679	0.3271	0.9679	0.9702
35.0	6.2104E-01	1.0891	0.9492	0.9321	0.9229	1.0382	1.0080	1.0000	1.0000	0.9663	0.2694	0.9663	0.9717
40.0	6.2471E-01	1.1008	0.9457	0.9278	0.9184	1.0433	1.0091	1.0000	1.0000	0.9646	0.2125	0.9646	0.9731
45.0	6.2870E-01	1.1125	0.9420	0.9234	0.9136	1.0484	1.0103	1.0000	1.0000	0.9628	0.1561	0.9628	0.9743
50.0	6.3294E-01	1.1243	0.9382	0.9187	0.9086	1.0534	1.0115	1.0000	1.0000	0.9610	0.1008	0.9610	0.9755

Instr. see p. 22

POSITRON Z= -27 A= 57

p	F_0L_0	L_0	λ_2	λ_3	λ_4	μ_1	μ_2	Λ_1	Λ_2	ν_{12}	$\hat{\nu}_{12}$	η_{12}	$\hat{\eta}_{12}$
0.1	6.5884E-05	1.0118	4.3515	8.2025	11.4649	1.0001	1.0000	0.9988	0.9998	1.8664	0.7060	0.9457	0.9273
0.2	1.5319E-02	1.0119	1.7695	2.1106	2.2878	1.0001	1.0000	0.9988	0.9998	1.1754	0.7037	0.9463	0.9280
0.3	7.8079E-02	1.0119	1.2934	1.3787	1.4154	1.0002	1.0000	0.9989	0.9998	1.0484	0.7006	0.9473	0.9291
0.4	1.6287E-01	1.0119	1.1285	1.1538	1.1618	1.0002	1.0000	0.9989	0.9998	1.0048	0.6971	0.9484	0.9302
0.5	2.4348E-01	1.0120	1.0535	1.0563	1.0544	1.0003	1.0001	0.9989	0.9998	0.9851	0.6934	0.9494	0.9313
0.6	3.1144E-01	1.0121	1.0136	1.0059	0.9995	1.0004	1.0001	0.9990	0.9998	0.9750	0.6899	0.9504	0.9324
0.7	3.6631E-01	1.0122	0.9903	0.9769	0.9680	1.0005	1.0001	0.9990	0.9998	0.9692	0.6866	0.9513	0.9334
0.8	4.1001E-01	1.0123	0.9757	0.9589	0.9487	1.0007	1.0001	0.9991	0.9998	0.9657	0.6834	0.9522	0.9343
0.9	4.4476E-01	1.0124	0.9661	0.9473	0.9362	1.0008	1.0001	0.9991	0.9998	0.9635	0.6805	0.9530	0.9352
1.0	4.7251E-01	1.0125	0.9597	0.9394	0.9278	1.0009	1.0002	0.9992	0.9998	0.9622	0.6779	0.9537	0.9360
1.2	5.1285E-01	1.0128	0.9519	0.9301	0.9179	1.0012	1.0002	0.9992	0.9998	0.9608	0.6731	0.9550	0.9375
1.4	5.3966E-01	1.0131	0.9480	0.9255	0.9131	1.0014	1.0003	0.9993	0.9999	0.9603	0.6689	0.9562	0.9388
1.6	5.5800E-01	1.0134	0.9459	0.9232	0.9107	1.0017	1.0003	0.9994	0.9999	0.9603	0.6652	0.9572	0.9399
1.8	5.7086E-01	1.0138	0.9450	0.9222	0.9097	1.0020	1.0004	0.9994	0.9999	0.9605	0.6619	0.9581	0.9410
2.0	5.8008E-01	1.0142	0.9446	0.9219	0.9095	1.0022	1.0004	0.9995	0.9999	0.9608	0.6588	0.9589	0.9420
2.2	5.8682E-01	1.0146	0.9446	0.9220	0.9097	1.0025	1.0005	0.9995	0.9999	0.9612	0.6559	0.9596	0.9428
2.4	5.9181E-01	1.0150	0.9447	0.9223	0.9101	1.0028	1.0005	0.9995	0.9999	0.9615	0.6532	0.9603	0.9436
2.6	5.9554E-01	1.0154	0.9451	0.9228	0.9107	1.0030	1.0006	0.9996	0.9999	0.9619	0.6505	0.9609	0.9444
2.8	5.9837E-01	1.0159	0.9455	0.9234	0.9114	1.0033	1.0006	0.9996	0.9999	0.9623	0.6479	0.9614	0.9451
3.0	6.0052E-01	1.0163	0.9459	0.9241	0.9122	1.0035	1.0007	0.9996	0.9999	0.9627	0.6454	0.9619	0.9457
3.2	6.0216E-01	1.0168	0.9464	0.9247	0.9129	1.0038	1.0007	0.9996	0.9999	0.9631	0.6430	0.9624	0.9463
3.4	6.0342E-01	1.0172	0.9468	0.9254	0.9137	1.0040	1.0008	0.9997	0.9999	0.9634	0.6405	0.9628	0.9469
3.6	6.0437E-01	1.0177	0.9473	0.9261	0.9145	1.0043	1.0008	0.9997	0.9999	0.9637	0.6381	0.9632	0.9474
3.8	6.0509E-01	1.0181	0.9478	0.9267	0.9152	1.0045	1.0009	0.9997	0.9999	0.9640	0.6357	0.9636	0.9479
4.0	6.0564E-01	1.0186	0.9482	0.9274	0.9159	1.0048	1.0009	0.9997	0.9999	0.9643	0.6334	0.9639	0.9484
4.5	6.0643E-01	1.0197	0.9493	0.9288	0.9176	1.0054	1.0010	0.9997	0.9999	0.9650	0.6275	0.9647	0.9495
5.0	6.0671E-01	1.0209	0.9503	0.9302	0.9192	1.0060	1.0012	0.9998	1.0000	0.9656	0.6217	0.9654	0.9505
5.5	6.0669E-01	1.0221	0.9512	0.9314	0.9206	1.0066	1.0013	0.9998	1.0000	0.9661	0.6159	0.9659	0.9514
6.0	6.0651E-01	1.0233	0.9519	0.9325	0.9218	1.0072	1.0014	0.9998	1.0000	0.9666	0.6101	0.9664	0.9523
6.5	6.0623E-01	1.0245	0.9526	0.9335	0.9229	1.0078	1.0015	0.9998	1.0000	0.9669	0.6044	0.9668	0.9530
7.0	6.0591E-01	1.0257	0.9532	0.9343	0.9239	1.0084	1.0016	0.9998	1.0000	0.9673	0.5986	0.9672	0.9537
7.5	6.0556E-01	1.0269	0.9538	0.9351	0.9248	1.0090	1.0018	0.9998	1.0000	0.9676	0.5928	0.9675	0.9544
8.0	6.0522E-01	1.0281	0.9542	0.9358	0.9256	1.0096	1.0019	0.9999	1.0000	0.9679	0.5870	0.9678	0.9550
9.0	6.0458E-01	1.0305	0.9550	0.9369	0.9269	1.0108	1.0021	0.9999	1.0000	0.9683	0.5755	0.9683	0.9561
10.0	6.0402E-01	1.0329	0.9556	0.9378	0.9279	1.0120	1.0024	0.9999	1.0000	0.9686	0.5641	0.9686	0.9571
11.0	6.0356E-01	1.0353	0.9560	0.9384	0.9287	1.0131	1.0026	0.9999	1.0000	0.9688	0.5526	0.9688	0.9581
12.0	6.0320E-01	1.0377	0.9562	0.9389	0.9292	1.0143	1.0028	0.9999	1.0000	0.9690	0.5411	0.9690	0.9589
13.0	6.0294E-01	1.0401	0.9564	0.9392	0.9296	1.0154	1.0031	0.9999	1.0000	0.9691	0.5296	0.9691	0.9597
14.0	6.0277E-01	1.0425	0.9564	0.9394	0.9299	1.0166	1.0033	0.9999	1.0000	0.9691	0.5180	0.9691	0.9604
16.0	6.0268E-01	1.0474	0.9563	0.9394	0.9300	1.0189	1.0038	0.9999	1.0000	0.9691	0.4951	0.9691	0.9617
18.0	6.0286E-01	1.0522	0.9559	0.9391	0.9297	1.0212	1.0043	0.9999	1.0000	0.9689	0.4721	0.9689	0.9628
20.0	6.0327E-01	1.0571	0.9553	0.9385	0.9292	1.0234	1.0047	0.9999	1.0000	0.9686	0.4493	0.9686	0.9639
25.0	6.0509E-01	1.0692	0.9530	0.9360	0.9267	1.0290	1.0059	1.0000	1.0000	0.9675	0.3924	0.9676	0.9661
30.0	6.0774E-01	1.0814	0.9502	0.9327	0.9233	1.0345	1.0071	1.0000	1.0000	0.9662	0.3361	0.9662	0.9680
35.0	6.1097E-01	1.0936	0.9469	0.9288	0.9191	1.0400	1.0083	1.0000	1.0000	0.9646	0.2803	0.9646	0.9696
40.0	6.1464E-01	1.1059	0.9433	0.9246	0.9146	1.0453	1.0095	1.0000	1.0000	0.9628	0.2252	0.9628	0.9711
45.0	6.1865E-01	1.1182	0.9395	0.9200	0.9097	1.0506	1.0108	1.0000	1.0000	0.9610	0.1705	0.9610	0.9725
50.0	6.2292E-01	1.1305	0.9357	0.9152	0.9046	1.0558	1.0120	1.0000	1.0000	0.9591	0.1170	0.9591	0.9737

Instr. see p. 22

POSITRON Z= -28 A= 60

p	F_0L_0	L_0	λ_2	λ_3	λ_4	μ_1	μ_2	Λ_1	Λ_2	ν_{12}	$\hat{\nu}_{12}$	η_{12}	$\hat{\eta}_{12}$
0.1	4.3967E-05	1.0126	4.5762	8.9187	12.7277	1.0001	1.0000	0.9987	0.9997	1.9292	0.7032	0.9422	0.9225
0.2	1.2823E-02	1.0127	1.8214	2.1980	2.3962	1.0001	1.0000	0.9988	0.9998	1.1884	0.7010	0.9428	0.9232
0.3	7.0272E-02	1.0127	1.3133	1.4063	1.4465	1.0002	1.0000	0.9988	0.9998	1.0522	0.6978	0.9439	0.9243
0.4	1.5160E-01	1.0127	1.1373	1.1647	1.1734	1.0002	1.0000	0.9988	0.9998	1.0054	0.6943	0.9450	0.9255
0.5	2.3085E-01	1.0128	1.0572	1.0603	1.0583	1.0003	1.0001	0.9989	0.9998	0.9843	0.6907	0.9460	0.9266
0.6	2.9861E-01	1.0129	1.0146	1.0064	0.9995	1.0005	1.0001	0.9989	0.9998	0.9734	0.6871	0.9471	0.9277
0.7	3.5379E-01	1.0130	0.9897	0.9754	0.9659	1.0006	1.0001	0.9990	0.9998	0.9672	0.6837	0.9480	0.9288
0.8	3.9798E-01	1.0131	0.9741	0.9562	0.9453	1.0007	1.0001	0.9990	0.9998	0.9634	0.6806	0.9489	0.9298
0.9	4.3325E-01	1.0132	0.9639	0.9437	0.9319	1.0008	1.0002	0.9991	0.9998	0.9611	0.6777	0.9498	0.9307
1.0	4.6149E-01	1.0133	0.9569	0.9353	0.9230	1.0010	1.0002	0.9991	0.9998	0.9596	0.6750	0.9505	0.9316
1.2	5.0263E-01	1.0136	0.9487	0.9255	0.9125	1.0012	1.0002	0.9992	0.9998	0.9581	0.6702	0.9520	0.9331
1.4	5.3001E-01	1.0140	0.9445	0.9206	0.9073	1.0015	1.0003	0.9993	0.9999	0.9576	0.6661	0.9532	0.9345
1.6	5.4875E-01	1.0143	0.9423	0.9181	0.9048	1.0018	1.0003	0.9993	0.9999	0.9576	0.6625	0.9543	0.9358
1.8	5.6189E-01	1.0147	0.9413	0.9170	0.9038	1.0021	1.0004	0.9994	0.9999	0.9578	0.6592	0.9552	0.9369
2.0	5.7130E-01	1.0152	0.9409	0.9167	0.9035	1.0024	1.0005	0.9994	0.9999	0.9582	0.6562	0.9561	0.9379
2.2	5.7816E-01	1.0156	0.9409	0.9168	0.9037	1.0026	1.0005	0.9995	0.9999	0.9585	0.6533	0.9569	0.9389
2.4	5.8323E-01	1.0160	0.9411	0.9172	0.9042	1.0029	1.0006	0.9995	0.9999	0.9590	0.6506	0.9576	0.9397
2.6	5.8702E-01	1.0165	0.9414	0.9178	0.9049	1.0032	1.0006	0.9996	0.9999	0.9594	0.6481	0.9582	0.9405
2.8	5.8987E-01	1.0169	0.9418	0.9184	0.9056	1.0035	1.0007	0.9996	0.9999	0.9598	0.6455	0.9588	0.9413
3.0	5.9204E-01	1.0174	0.9423	0.9191	0.9064	1.0037	1.0007	0.9996	0.9999	0.9602	0.6431	0.9594	0.9420
3.2	5.9368E-01	1.0179	0.9428	0.9198	0.9072	1.0040	1.0008	0.9996	0.9999	0.9606	0.6407	0.9599	0.9426
3.4	5.9493E-01	1.0183	0.9433	0.9205	0.9081	1.0043	1.0008	0.9996	0.9999	0.9610	0.6383	0.9603	0.9432
3.6	5.9587E-01	1.0188	0.9438	0.9212	0.9089	1.0045	1.0009	0.9997	0.9999	0.9613	0.6360	0.9608	0.9438
3.8	5.9658E-01	1.0193	0.9444	0.9219	0.9097	1.0048	1.0009	0.9997	0.9999	0.9616	0.6337	0.9612	0.9443
4.0	5.9710E-01	1.0198	0.9448	0.9226	0.9105	1.0050	1.0010	0.9997	0.9999	0.9620	0.6314	0.9615	0.9449
4.5	5.9784E-01	1.0210	0.9460	0.9242	0.9123	1.0057	1.0011	0.9997	0.9999	0.9627	0.6256	0.9624	0.9461
5.0	5.9806E-01	1.0223	0.9471	0.9257	0.9140	1.0063	1.0012	0.9998	1.0000	0.9633	0.6200	0.9631	0.9471
5.5	5.9798E-01	1.0235	0.9480	0.9270	0.9155	1.0070	1.0014	0.9998	1.0000	0.9639	0.6143	0.9637	0.9481
6.0	5.9773E-01	1.0248	0.9489	0.9282	0.9168	1.0076	1.0015	0.9998	1.0000	0.9644	0.6087	0.9642	0.9490
6.5	5.9739E-01	1.0260	0.9496	0.9293	0.9180	1.0083	1.0016	0.9998	1.0000	0.9648	0.6031	0.9647	0.9498
7.0	5.9701E-01	1.0273	0.9503	0.9302	0.9191	1.0089	1.0017	0.9998	1.0000	0.9652	0.5974	0.9651	0.9506
7.5	5.9662E-01	1.0285	0.9509	0.9310	0.9201	1.0095	1.0019	0.9998	1.0000	0.9655	0.5918	0.9654	0.9513
8.0	5.9622E-01	1.0298	0.9514	0.9318	0.9209	1.0102	1.0020	0.9998	1.0000	0.9658	0.5862	0.9657	0.9519
9.0	5.9549E-01	1.0324	0.9522	0.9330	0.9224	1.0114	1.0022	0.9999	1.0000	0.9663	0.5750	0.9662	0.9531
10.0	5.9485E-01	1.0349	0.9529	0.9340	0.9235	1.0126	1.0025	0.9999	1.0000	0.9666	0.5637	0.9666	0.9542
11.0	5.9432E-01	1.0375	0.9533	0.9347	0.9243	1.0139	1.0028	0.9999	1.0000	0.9669	0.5525	0.9669	0.9552
12.0	5.9391E-01	1.0400	0.9537	0.9352	0.9250	1.0151	1.0030	0.9999	1.0000	0.9671	0.5412	0.9671	0.9561
13.0	5.9360E-01	1.0426	0.9538	0.9356	0.9255	1.0163	1.0033	0.9999	1.0000	0.9672	0.5299	0.9672	0.9569
14.0	5.9338E-01	1.0451	0.9539	0.9358	0.9258	1.0176	1.0035	0.9999	1.0000	0.9673	0.5187	0.9673	0.9577
16.0	5.9322E-01	1.0502	0.9538	0.9359	0.9260	1.0200	1.0040	0.9999	1.0000	0.9672	0.4963	0.9672	0.9591
18.0	5.9334E-01	1.0554	0.9534	0.9356	0.9258	1.0224	1.0045	0.9999	1.0000	0.9671	0.4739	0.9671	0.9603
20.0	5.9372E-01	1.0605	0.9528	0.9350	0.9252	1.0248	1.0050	0.9999	1.0000	0.9668	0.4515	0.9668	0.9614
25.0	5.9550E-01	1.0734	0.9506	0.9326	0.9228	1.0307	1.0063	1.0000	1.0000	0.9657	0.3959	0.9657	0.9638
30.0	5.9815E-01	1.0862	0.9476	0.9293	0.9192	1.0365	1.0075	1.0000	1.0000	0.9643	0.3407	0.9643	0.9658
35.0	6.0142E-01	1.0992	0.9442	0.9253	0.9150	1.0422	1.0088	1.0000	1.0000	0.9626	0.2861	0.9627	0.9676
40.0	6.0515E-01	1.1122	0.9406	0.9208	0.9103	1.0478	1.0101	1.0000	1.0000	0.9609	0.2320	0.9609	0.9692
45.0	6.0925E-01	1.1252	0.9367	0.9161	0.9053	1.0534	1.0114	1.0000	1.0000	0.9590	0.1786	0.9590	0.9707
50.0	6.1362E-01	1.1383	0.9326	0.9112	0.9000	1.0588	1.0127	1.0000	1.0000	0.9570	0.1261	0.9570	0.9720

POSITRON Z= -29 A= 65

p	F_0L_0	L_0	λ_2	λ_3	λ_4	μ_1	μ_2	Λ_1	Λ_2	ν_{12}	$\hat{\nu}_{12}$	η_{12}	$\hat{\eta}_{12}$
0.1	2.9309E-05	1.0135	4.8064	9.6844	14.1147	1.0001	1.0000	0.9987	0.9997	1.9940	0.7004	0.9387	0.9177
0.2	1.0723E-02	1.0135	1.8747	2.2896	2.5109	1.0001	1.0000	0.9987	0.9997	1.2019	0.6982	0.9394	0.9184
0.3	6.3196E-02	1.0136	1.3339	1.4351	1.4792	1.0002	1.0000	0.9987	0.9997	1.0563	0.6951	0.9404	0.9195
0.4	1.4102E-01	1.0136	1.1465	1.1761	1.1857	1.0003	1.0000	0.9988	0.9998	1.0061	0.6915	0.9415	0.9207
0.5	2.1878E-01	1.0137	1.0611	1.0646	1.0625	1.0004	1.0001	0.9988	0.9998	0.9836	0.6878	0.9427	0.9219
0.6	2.8621E-01	1.0138	1.0158	1.0072	0.9998	1.0005	1.0001	0.9988	0.9998	0.9719	0.6843	0.9437	0.9231
0.7	3.4162E-01	1.0139	0.9893	0.9741	0.9641	1.0006	1.0001	0.9989	0.9998	0.9652	0.6809	0.9448	0.9242
0.8	3.8623E-01	1.0140	0.9727	0.9537	0.9421	1.0008	1.0001	0.9990	0.9998	0.9612	0.6777	0.9457	0.9252
0.9	4.2198E-01	1.0141	0.9618	0.9404	0.9278	1.0009	1.0002	0.9990	0.9998	0.9587	0.6748	0.9466	0.9262
1.0	4.5066E-01	1.0143	0.9544	0.9314	0.9183	1.0010	1.0002	0.9991	0.9998	0.9571	0.6721	0.9474	0.9271
1.2	4.9256E-01	1.0146	0.9456	0.9209	0.9072	1.0013	1.0002	0.9991	0.9998	0.9555	0.6673	0.9489	0.9288
1.4	5.2049E-01	1.0149	0.9411	0.9157	0.9017	1.0016	1.0003	0.9992	0.9998	0.9550	0.6632	0.9502	0.9303
1.6	5.3961E-01	1.0153	0.9388	0.9131	0.8990	1.0019	1.0004	0.9993	0.9999	0.9549	0.6596	0.9514	0.9316
1.8	5.5302E-01	1.0158	0.9377	0.9120	0.8979	1.0022	1.0004	0.9994	0.9999	0.9552	0.6563	0.9524	0.9328
2.0	5.6261E-01	1.0162	0.9372	0.9116	0.8976	1.0025	1.0005	0.9994	0.9999	0.9555	0.6533	0.9533	0.9339
2.2	5.6959E-01	1.0167	0.9372	0.9117	0.8978	1.0028	1.0005	0.9994	0.9999	0.9559	0.6505	0.9542	0.9349
2.4	5.7474E-01	1.0171	0.9375	0.9122	0.8984	1.0031	1.0006	0.9995	0.9999	0.9564	0.6479	0.9549	0.9358
2.6	5.7858E-01	1.0176	0.9378	0.9128	0.8991	1.0034	1.0007	0.9995	0.9999	0.9568	0.6453	0.9556	0.9367
2.8	5.8146E-01	1.0181	0.9383	0.9134	0.8999	1.0037	1.0007	0.9996	0.9999	0.9573	0.6429	0.9563	0.9375
3.0	5.8364E-01	1.0186	0.9388	0.9142	0.9008	1.0040	1.0008	0.9996	0.9999	0.9577	0.6404	0.9568	0.9382
3.2	5.8529E-01	1.0191	0.9393	0.9150	0.9016	1.0043	1.0008	0.9996	0.9999	0.9581	0.6381	0.9574	0.9389
3.4	5.8653E-01	1.0196	0.9399	0.9157	0.9025	1.0045	1.0009	0.9996	0.9999	0.9585	0.6357	0.9579	0.9395
3.6	5.8746E-01	1.0201	0.9405	0.9165	0.9034	1.0048	1.0009	0.9996	0.9999	0.9589	0.6334	0.9583	0.9402
3.8	5.8816E-01	1.0206	0.9410	0.9172	0.9043	1.0051	1.0010	0.9997	0.9999	0.9593	0.6312	0.9588	0.9407
4.0	5.8866E-01	1.0211	0.9415	0.9180	0.9051	1.0054	1.0010	0.9997	0.9999	0.9596	0.6289	0.9592	0.9413
4.5	5.8934E-01	1.0225	0.9428	0.9197	0.9071	1.0061	1.0012	0.9997	0.9999	0.9604	0.6233	0.9601	0.9426
5.0	5.8950E-01	1.0238	0.9439	0.9213	0.9089	1.0068	1.0013	0.9997	0.9999	0.9611	0.6177	0.9608	0.9437
5.5	5.8936E-01	1.0251	0.9449	0.9227	0.9105	1.0075	1.0015	0.9998	1.0000	0.9617	0.6121	0.9615	0.9448
6.0	5.8905E-01	1.0265	0.9459	0.9240	0.9119	1.0081	1.0016	0.9998	1.0000	0.9622	0.6065	0.9621	0.9457
6.5	5.8866E-01	1.0278	0.9467	0.9251	0.9132	1.0088	1.0017	0.9998	1.0000	0.9627	0.6010	0.9625	0.9466
7.0	5.8823E-01	1.0291	0.9474	0.9261	0.9144	1.0095	1.0019	0.9998	1.0000	0.9631	0.5954	0.9630	0.9474
7.5	5.8779E-01	1.0305	0.9480	0.9270	0.9154	1.0102	1.0020	0.9998	1.0000	0.9634	0.5899	0.9634	0.9481
8.0	5.8735E-01	1.0318	0.9486	0.9278	0.9163	1.0108	1.0021	0.9998	1.0000	0.9637	0.5843	0.9637	0.9489
9.0	5.8654E-01	1.0346	0.9495	0.9292	0.9179	1.0122	1.0024	0.9999	1.0000	0.9643	0.5732	0.9642	0.9501
10.0	5.8583E-01	1.0373	0.9502	0.9302	0.9191	1.0135	1.0027	0.9999	1.0000	0.9647	0.5621	0.9646	0.9513
11.0	5.8525E-01	1.0400	0.9507	0.9310	0.9200	1.0148	1.0029	0.9999	1.0000	0.9649	0.5510	0.9649	0.9524
12.0	5.8479E-01	1.0427	0.9511	0.9316	0.9207	1.0161	1.0032	0.9999	1.0000	0.9651	0.5399	0.9651	0.9533
13.0	5.8444E-01	1.0454	0.9513	0.9320	0.9213	1.0174	1.0035	0.9999	1.0000	0.9653	0.5288	0.9653	0.9542
14.0	5.8420E-01	1.0482	0.9514	0.9322	0.9216	1.0187	1.0037	0.9999	1.0000	0.9653	0.5176	0.9653	0.9550
16.0	5.8398E-01	1.0536	0.9513	0.9324	0.9219	1.0213	1.0043	0.9999	1.0000	0.9653	0.4955	0.9653	0.9565
18.0	5.8409E-01	1.0591	0.9509	0.9321	0.9217	1.0238	1.0048	0.9999	1.0000	0.9652	0.4733	0.9652	0.9578
20.0	5.8445E-01	1.0646	0.9503	0.9315	0.9211	1.0264	1.0053	0.9999	1.0000	0.9649	0.4511	0.9649	0.9590
25.0	5.8625E-01	1.0783	0.9479	0.9290	0.9186	1.0326	1.0067	1.0000	1.0000	0.9638	0.3962	0.9638	0.9616
30.0	5.8897E-01	1.0921	0.9449	0.9255	0.9149	1.0388	1.0080	1.0000	1.0000	0.9623	0.3415	0.9623	0.9638
35.0	5.9235E-01	1.1059	0.9413	0.9213	0.9105	1.0449	1.0094	1.0000	1.0000	0.9606	0.2874	0.9606	0.9657
40.0	5.9621E-01	1.1198	0.9375	0.9167	0.9056	1.0509	1.0107	1.0000	1.0000	0.9587	0.2342	0.9587	0.9674
45.0	6.0046E-01	1.1337	0.9334	0.9118	0.9003	1.0568	1.0121	1.0000	1.0000	0.9567	0.1813	0.9568	0.9690
50.0	6.0500E-01	1.1477	0.9292	0.9066	0.8948	1.0625	1.0135	1.0000	1.0000	0.9547	0.1292	0.9547	0.9704

Instr. see p. 22

POSITRON Z= -30 A= 67

p	F_0L_0	L_0	λ_2	λ_3	λ_4	μ_1	μ_2	Λ_1	Λ_2	ν_{12}	$\hat{\nu}_{12}$	η_{12}	$\hat{\eta}_{12}$
0.1	1.9538E-05	1.0143	5.0392	10.4943	15.6230	1.0001	1.0000	0.9986	0.9997	2.0605	0.6975	0.9350	0.9125
0.2	8.9671E-03	1.0144	1.9283	2.3836	2.6299	1.0001	1.0000	0.9986	0.9997	1.2157	0.6953	0.9357	0.9133
0.3	5.6848E-02	1.0144	1.3544	1.4640	1.5121	1.0002	1.0000	0.9987	0.9997	1.0603	0.6922	0.9367	0.9144
0.4	1.3125E-01	1.0145	1.1554	1.1873	1.1977	1.0003	1.0000	0.9987	0.9997	1.0068	0.6886	0.9379	0.9156
0.5	2.0747E-01	1.0145	1.0648	1.0686	1.0664	1.0004	1.0001	0.9987	0.9998	0.9827	0.6849	0.9390	0.9169
0.6	2.7454E-01	1.0146	1.0167	1.0075	0.9997	1.0005	1.0001	0.9988	0.9998	0.9702	0.6813	0.9402	0.9181
0.7	3.3013E-01	1.0147	0.9885	0.9723	0.9617	1.0007	1.0001	0.9988	0.9998	0.9631	0.6779	0.9412	0.9193
0.8	3.7516E-01	1.0148	0.9708	0.9506	0.9383	1.0008	1.0001	0.9989	0.9998	0.9588	0.6748	0.9422	0.9204
0.9	4.1137E-01	1.0150	0.9593	0.9365	0.9232	1.0009	1.0002	0.9990	0.9998	0.9561	0.6719	0.9432	0.9214
1.0	4.4050E-01	1.0151	0.9514	0.9270	0.9131	1.0011	1.0002	0.9990	0.9998	0.9544	0.6692	0.9441	0.9224
1.2	4.8315E-01	1.0155	0.9421	0.9159	0.9013	1.0014	1.0003	0.9991	0.9998	0.9527	0.6644	0.9456	0.9242
1.4	5.1163E-01	1.0158	0.9373	0.9104	0.8955	1.0017	1.0003	0.9992	0.9998	0.9521	0.6604	0.9470	0.9258
1.6	5.3113E-01	1.0163	0.9349	0.9076	0.8927	1.0020	1.0004	0.9993	0.9999	0.9521	0.6568	0.9483	0.9272
1.8	5.4480E-01	1.0167	0.9337	0.9064	0.8915	1.0023	1.0004	0.9993	0.9999	0.9523	0.6536	0.9494	0.9284
2.0	5.5456E-01	1.0172	0.9332	0.9060	0.8912	1.0027	1.0005	0.9994	0.9999	0.9527	0.6507	0.9504	0.9296
2.2	5.6166E-01	1.0176	0.9332	0.9062	0.8915	1.0030	1.0006	0.9994	0.9999	0.9531	0.6480	0.9512	0.9307
2.4	5.6689E-01	1.0181	0.9335	0.9066	0.8920	1.0033	1.0006	0.9995	0.9999	0.9536	0.6454	0.9521	0.9316
2.6	5.7077E-01	1.0186	0.9339	0.9073	0.8928	1.0036	1.0007	0.9995	0.9999	0.9541	0.6429	0.9528	0.9325
2.8	5.7368E-01	1.0191	0.9344	0.9080	0.8937	1.0039	1.0007	0.9995	0.9999	0.9546	0.6405	0.9535	0.9334
3.0	5.7587E-01	1.0197	0.9350	0.9088	0.8946	1.0042	1.0008	0.9996	0.9999	0.9550	0.6382	0.9541	0.9342
3.2	5.7751E-01	1.0202	0.9355	0.9097	0.8955	1.0045	1.0009	0.9996	0.9999	0.9555	0.6359	0.9547	0.9349
3.4	5.7874E-01	1.0207	0.9361	0.9105	0.8965	1.0048	1.0009	0.9996	0.9999	0.9559	0.6336	0.9552	0.9356
3.6	5.7966E-01	1.0213	0.9367	0.9113	0.8974	1.0051	1.0010	0.9996	0.9999	0.9563	0.6314	0.9557	0.9363
3.8	5.8033E-01	1.0218	0.9373	0.9121	0.8984	1.0054	1.0010	0.9996	0.9999	0.9567	0.6292	0.9562	0.9369
4.0	5.8081E-01	1.0224	0.9379	0.9129	0.8993	1.0057	1.0011	0.9997	0.9999	0.9571	0.6270	0.9566	0.9375
4.5	5.8142E-01	1.0237	0.9392	0.9148	0.9014	1.0064	1.0012	0.9997	0.9999	0.9579	0.6215	0.9576	0.9388
5.0	5.8150E-01	1.0251	0.9405	0.9165	0.9033	1.0071	1.0014	0.9997	0.9999	0.9587	0.6161	0.9584	0.9401
5.5	5.8129E-01	1.0265	0.9416	0.9180	0.9051	1.0078	1.0015	0.9998	1.0000	0.9593	0.6107	0.9591	0.9412
6.0	5.8091E-01	1.0279	0.9426	0.9194	0.9066	1.0085	1.0017	0.9998	1.0000	0.9599	0.6053	0.9597	0.9422
6.5	5.8045E-01	1.0293	0.9435	0.9206	0.9080	1.0092	1.0018	0.9998	1.0000	0.9604	0.6000	0.9603	0.9431
7.0	5.7996E-01	1.0307	0.9442	0.9217	0.9093	1.0100	1.0019	0.9998	1.0000	0.9608	0.5945	0.9607	0.9440
7.5	5.7945E-01	1.0322	0.9449	0.9227	0.9104	1.0107	1.0021	0.9998	1.0000	0.9612	0.5892	0.9611	0.9448
8.0	5.7896E-01	1.0336	0.9456	0.9236	0.9114	1.0114	1.0022	0.9998	1.0000	0.9616	0.5838	0.9615	0.9455
9.0	5.7804E-01	1.0364	0.9466	0.9250	0.9131	1.0127	1.0025	0.9999	1.0000	0.9621	0.5730	0.9621	0.9469
10.0	5.7724E-01	1.0393	0.9474	0.9262	0.9144	1.0141	1.0028	0.9999	1.0000	0.9626	0.5622	0.9625	0.9481
11.0	5.7658E-01	1.0421	0.9479	0.9271	0.9155	1.0155	1.0031	0.9999	1.0000	0.9629	0.5513	0.9629	0.9493
12.0	5.7604E-01	1.0450	0.9484	0.9278	0.9163	1.0169	1.0033	0.9999	1.0000	0.9631	0.5406	0.9631	0.9503
13.0	5.7563E-01	1.0478	0.9486	0.9282	0.9169	1.0182	1.0036	0.9999	1.0000	0.9633	0.5297	0.9633	0.9512
14.0	5.7533E-01	1.0507	0.9488	0.9285	0.9173	1.0196	1.0039	0.9999	1.0000	0.9634	0.5189	0.9634	0.9521
16.0	5.7502E-01	1.0564	0.9488	0.9288	0.9177	1.0223	1.0045	0.9999	1.0000	0.9634	0.4973	0.9634	0.9537
18.0	5.7505E-01	1.0622	0.9484	0.9286	0.9176	1.0250	1.0050	0.9999	1.0000	0.9633	0.4757	0.9633	0.9551
20.0	5.7535E-01	1.0679	0.9478	0.9280	0.9171	1.0276	1.0056	0.9999	1.0000	0.9630	0.4541	0.9630	0.9564
25.0	5.7705E-01	1.0823	0.9455	0.9256	0.9146	1.0342	1.0070	0.9999	1.0000	0.9619	0.4006	0.9619	0.9592
30.0	5.7972E-01	1.0967	0.9424	0.9221	0.9110	1.0406	1.0084	1.0000	1.0000	0.9604	0.3473	0.9604	0.9615
35.0	5.8309E-01	1.1113	0.9388	0.9179	0.9065	1.0470	1.0098	1.0000	1.0000	0.9587	0.2947	0.9587	0.9635
40.0	5.8697E-01	1.1258	0.9349	0.9131	0.9015	1.0532	1.0112	1.0000	1.0000	0.9568	0.2425	0.9568	0.9653
45.0	5.9125E-01	1.1405	0.9307	0.9081	0.8961	1.0593	1.0126	1.0000	1.0000	0.9548	0.1912	0.9548	0.9670
50.0	5.9585E-01	1.1552	0.9264	0.9028	0.8905	1.0654	1.0141	1.0000	1.0000	0.9527	0.1403	0.9527	0.9686

POSITRON Z= -31 A= 69

p	F_0L_0	L_0	λ_2	λ_3	λ_4	μ_1	μ_2	Λ_1	Λ_2	ν_{12}	$\hat{\nu}_{12}$	η_{12}	$\hat{\eta}_{12}$
0.1	1.3018E-05	1.0152	5.2757	11.3533	17.2673	1.0001	1.0000	0.9985	0.9997	2.1287	0.6945	0.9312	0.9073
0.2	7.4956E-03	1.0152	1.9828	2.4812	2.7544	1.0001	1.0000	0.9986	0.9997	1.2298	0.6923	0.9319	0.9081
0.3	5.1123E-02	1.0153	1.3752	1.4937	1.5461	1.0002	1.0000	0.9986	0.9997	1.0644	0.6892	0.9329	0.9092
0.4	1.2214E-01	1.0153	1.1645	1.1988	1.2100	1.0003	1.0000	0.9986	0.9997	1.0075	0.6856	0.9341	0.9105
0.5	1.9676E-01	1.0154	1.0685	1.0726	1.0702	1.0004	1.0001	0.9987	0.9997	0.9818	0.6819	0.9354	0.9118
0.6	2.6338E-01	1.0155	1.0175	1.0077	0.9994	1.0005	1.0001	0.9987	0.9998	0.9685	0.6783	0.9365	0.9131
0.7	3.1910E-01	1.0156	0.9876	0.9705	0.9592	1.0007	1.0001	0.9988	0.9998	0.9609	0.6749	0.9377	0.9143
0.8	3.6449E-01	1.0157	0.9689	0.9475	0.9344	1.0008	1.0001	0.9989	0.9998	0.9563	0.6718	0.9387	0.9155
0.9	4.0114E-01	1.0159	0.9567	0.9326	0.9185	1.0010	1.0002	0.9989	0.9998	0.9535	0.6689	0.9397	0.9166
1.0	4.3070E-01	1.0160	0.9483	0.9225	0.9078	1.0011	1.0002	0.9990	0.9998	0.9517	0.6662	0.9406	0.9176
1.2	4.7408E-01	1.0164	0.9385	0.9108	0.8953	1.0015	1.0003	0.9991	0.9998	0.9498	0.6615	0.9423	0.9195
1.4	5.0310E-01	1.0168	0.9334	0.9049	0.8892	1.0018	1.0003	0.9991	0.9998	0.9492	0.6575	0.9438	0.9211
1.6	5.2297E-01	1.0172	0.9308	0.9020	0.8862	1.0021	1.0004	0.9992	0.9998	0.9491	0.6540	0.9451	0.9226
1.8	5.3689E-01	1.0177	0.9296	0.9007	0.8849	1.0025	1.0005	0.9993	0.9999	0.9494	0.6509	0.9462	0.9240
2.0	5.4683E-01	1.0181	0.9291	0.9003	0.8847	1.0028	1.0005	0.9993	0.9999	0.9498	0.6480	0.9473	0.9252
2.2	5.5404E-01	1.0186	0.9291	0.9005	0.8849	1.0031	1.0006	0.9994	0.9999	0.9503	0.6454	0.9482	0.9263
2.4	5.5934E-01	1.0192	0.9294	0.9010	0.8856	1.0034	1.0007	0.9994	0.9999	0.9508	0.6428	0.9491	0.9274
2.6	5.6327E-01	1.0197	0.9299	0.9017	0.8864	1.0038	1.0007	0.9995	0.9999	0.9513	0.6404	0.9499	0.9283
2.8	5.6620E-01	1.0202	0.9304	0.9025	0.8873	1.0041	1.0008	0.9995	0.9999	0.9518	0.6381	0.9506	0.9292
3.0	5.6839E-01	1.0208	0.9310	0.9033	0.8883	1.0044	1.0008	0.9995	0.9999	0.9523	0.6359	0.9513	0.9300
3.2	5.7003E-01	1.0213	0.9316	0.9042	0.8893	1.0047	1.0009	0.9996	0.9999	0.9528	0.6336	0.9519	0.9308
3.4	5.7124E-01	1.0219	0.9323	0.9051	0.8903	1.0050	1.0010	0.9996	0.9999	0.9532	0.6314	0.9525	0.9316
3.6	5.7214E-01	1.0225	0.9329	0.9060	0.8913	1.0053	1.0010	0.9996	0.9999	0.9537	0.6293	0.9530	0.9323
3.8	5.7279E-01	1.0230	0.9335	0.9069	0.8923	1.0056	1.0011	0.9996	0.9999	0.9541	0.6271	0.9535	0.9329
4.0	5.7324E-01	1.0236	0.9342	0.9077	0.8933	1.0059	1.0011	0.9996	0.9999	0.9545	0.6250	0.9540	0.9336
4.5	5.7378E-01	1.0250	0.9356	0.9097	0.8956	1.0067	1.0013	0.9997	0.9999	0.9554	0.6197	0.9550	0.9350
5.0	5.7379E-01	1.0265	0.9369	0.9116	0.8976	1.0074	1.0014	0.9997	0.9999	0.9562	0.6145	0.9559	0.9363
5.5	5.7349E-01	1.0279	0.9381	0.9132	0.8995	1.0082	1.0016	0.9997	0.9999	0.9569	0.6092	0.9567	0.9375
6.0	5.7304E-01	1.0294	0.9392	0.9147	0.9012	1.0089	1.0017	0.9998	1.0000	0.9575	0.6040	0.9573	0.9386
6.5	5.7251E-01	1.0309	0.9402	0.9160	0.9027	1.0097	1.0019	0.9998	1.0000	0.9581	0.5988	0.9579	0.9395
7.0	5.7195E-01	1.0324	0.9410	0.9172	0.9041	1.0104	1.0020	0.9998	1.0000	0.9585	0.5936	0.9584	0.9405
7.5	5.7138E-01	1.0339	0.9418	0.9183	0.9053	1.0112	1.0022	0.9998	1.0000	0.9590	0.5883	0.9589	0.9413
8.0	5.7082E-01	1.0353	0.9425	0.9192	0.9064	1.0119	1.0023	0.9998	1.0000	0.9593	0.5831	0.9593	0.9421
9.0	5.6980E-01	1.0383	0.9436	0.9208	0.9082	1.0133	1.0026	0.9998	1.0000	0.9600	0.5726	0.9599	0.9436
10.0	5.6891E-01	1.0413	0.9445	0.9221	0.9097	1.0148	1.0029	0.9999	1.0000	0.9605	0.5621	0.9604	0.9449
11.0	5.6816E-01	1.0443	0.9451	0.9231	0.9109	1.0162	1.0032	0.9999	1.0000	0.9608	0.5515	0.9608	0.9461
12.0	5.6755E-01	1.0473	0.9456	0.9239	0.9118	1.0177	1.0035	0.9999	1.0000	0.9611	0.5410	0.9611	0.9472
13.0	5.6707E-01	1.0503	0.9459	0.9244	0.9124	1.0191	1.0038	0.9999	1.0000	0.9613	0.5305	0.9613	0.9482
14.0	5.6671E-01	1.0533	0.9461	0.9248	0.9129	1.0205	1.0041	0.9999	1.0000	0.9614	0.5199	0.9614	0.9491
16.0	5.6630E-01	1.0593	0.9462	0.9251	0.9134	1.0233	1.0047	0.9999	1.0000	0.9615	0.4989	0.9615	0.9508
18.0	5.6625E-01	1.0653	0.9459	0.9250	0.9134	1.0261	1.0052	0.9999	1.0000	0.9614	0.4778	0.9614	0.9523
20.0	5.6649E-01	1.0713	0.9453	0.9245	0.9130	1.0289	1.0058	0.9999	1.0000	0.9611	0.4568	0.9611	0.9537
25.0	5.6808E-01	1.0864	0.9431	0.9222	0.9106	1.0357	1.0073	0.9999	1.0000	0.9600	0.4045	0.9601	0.9566
30.0	5.7070E-01	1.1015	0.9400	0.9186	0.9070	1.0424	1.0087	1.0000	1.0000	0.9585	0.3526	0.9586	0.9591
35.0	5.7405E-01	1.1167	0.9363	0.9144	0.9025	1.0491	1.0102	1.0000	1.0000	0.9568	0.3011	0.9568	0.9613
40.0	5.7794E-01	1.1320	0.9323	0.9096	0.8974	1.0556	1.0117	1.0000	1.0000	0.9549	0.2503	0.9549	0.9632
45.0	5.8225E-01	1.1474	0.9280	0.9044	0.8919	1.0620	1.0132	1.0000	1.0000	0.9528	0.2000	0.9528	0.9650
50.0	5.8689E-01	1.1628	0.9236	0.8990	0.8862	1.0683	1.0147	1.0000	1.0000	0.9507	0.1505	0.9507	0.9667

 Instr. see p. 22

POSITRON Z= -32 A= 72

p	F_0L_0	L_0	λ_2	λ_3	λ_4	μ_1	μ_2	Λ_1	Λ_2	ν_{12}	$\hat{\nu}_{12}$	η_{12}	$\hat{\eta}_{12}$
0.1	8.6677E-06	1.0161	5.5161	12.2643	19.0595	1.0001	1.0000	0.9985	0.9997	2.1989	0.6915	0.9273	0.9020
0.2	6.2615E-03	1.0161	2.0382	2.5825	2.8851	1.0002	1.0000	0.9985	0.9997	1.2444	0.6893	0.9280	0.9028
0.3	4.5952E-02	1.0162	1.3964	1.5244	1.5813	1.0002	1.0000	0.9985	0.9997	1.0687	0.6862	0.9291	0.9039
0.4	1.1363E-01	1.0162	1.1738	1.2106	1.2226	1.0003	1.0000	0.9986	0.9997	1.0082	0.6826	0.9304	0.9053
0.5	1.8657E-01	1.0163	1.0723	1.0768	1.0743	1.0004	1.0001	0.9986	0.9997	0.9810	0.6789	0.9316	0.9066
0.6	2.5265E-01	1.0164	1.0184	1.0081	0.9993	1.0006	1.0001	0.9987	0.9997	0.9668	0.6753	0.9328	0.9080
0.7	3.0843E-01	1.0165	0.9868	0.9687	0.9567	1.0007	1.0001	0.9987	0.9998	0.9587	0.6719	0.9340	0.9092
0.8	3.5414E-01	1.0166	0.9671	0.9444	0.9306	1.0009	1.0002	0.9988	0.9998	0.9538	0.6688	0.9351	0.9105
0.9	3.9119E-01	1.0168	0.9541	0.9286	0.9137	1.0010	1.0002	0.9989	0.9998	0.9508	0.6658	0.9362	0.9116
1.0	4.2115E-01	1.0170	0.9453	0.9180	0.9024	1.0012	1.0002	0.9989	0.9998	0.9489	0.6632	0.9371	0.9127
1.2	4.6523E-01	1.0173	0.9348	0.9056	0.8893	1.0016	1.0003	0.9990	0.9998	0.9469	0.6585	0.9389	0.9147
1.4	4.9476E-01	1.0177	0.9295	0.8994	0.8828	1.0019	1.0004	0.9991	0.9998	0.9462	0.6545	0.9405	0.9164
1.6	5.1500E-01	1.0182	0.9268	0.8963	0.8797	1.0022	1.0004	0.9992	0.9998	0.9462	0.6511	0.9418	0.9180
1.8	5.2917E-01	1.0187	0.9255	0.8950	0.8784	1.0026	1.0005	0.9993	0.9999	0.9464	0.6480	0.9431	0.9194
2.0	5.3928E-01	1.0192	0.9250	0.8946	0.8781	1.0029	1.0006	0.9993	0.9999	0.9469	0.6452	0.9442	0.9207
2.2	5.4660E-01	1.0197	0.9250	0.8948	0.8784	1.0033	1.0006	0.9994	0.9999	0.9474	0.6426	0.9452	0.9219
2.4	5.5196E-01	1.0203	0.9253	0.8953	0.8791	1.0036	1.0007	0.9994	0.9999	0.9479	0.6402	0.9461	0.9230
2.6	5.5593E-01	1.0208	0.9258	0.8961	0.8799	1.0039	1.0007	0.9995	0.9999	0.9485	0.6378	0.9470	0.9240
2.8	5.5888E-01	1.0214	0.9264	0.8969	0.8809	1.0043	1.0008	0.9995	0.9999	0.9490	0.6355	0.9478	0.9250
3.0	5.6108E-01	1.0220	0.9270	0.8978	0.8820	1.0046	1.0009	0.9995	0.9999	0.9495	0.6333	0.9485	0.9259
3.2	5.6271E-01	1.0225	0.9277	0.8988	0.8831	1.0049	1.0009	0.9995	0.9999	0.9501	0.6312	0.9491	0.9267
3.4	5.6391E-01	1.0231	0.9284	0.8997	0.8842	1.0053	1.0010	0.9996	0.9999	0.9505	0.6291	0.9497	0.9275
3.6	5.6479E-01	1.0237	0.9291	0.9007	0.8852	1.0056	1.0011	0.9996	0.9999	0.9510	0.6270	0.9503	0.9282
3.8	5.6541E-01	1.0243	0.9297	0.9016	0.8863	1.0059	1.0011	0.9996	0.9999	0.9515	0.6249	0.9508	0.9289
4.0	5.6584E-01	1.0249	0.9304	0.9025	0.8873	1.0062	1.0012	0.9996	0.9999	0.9519	0.6228	0.9513	0.9296
4.5	5.6630E-01	1.0264	0.9320	0.9047	0.8897	1.0070	1.0013	0.9997	0.9999	0.9529	0.6177	0.9524	0.9311
5.0	5.6623E-01	1.0280	0.9334	0.9066	0.8920	1.0078	1.0015	0.9997	0.9999	0.9537	0.6125	0.9534	0.9325
5.5	5.6586E-01	1.0295	0.9347	0.9084	0.8940	1.0086	1.0017	0.9997	0.9999	0.9545	0.6074	0.9542	0.9338
6.0	5.6534E-01	1.0310	0.9359	0.9100	0.8958	1.0094	1.0018	0.9998	1.0000	0.9551	0.6023	0.9549	0.9349
6.5	5.6474E-01	1.0326	0.9369	0.9114	0.8974	1.0102	1.0020	0.9998	1.0000	0.9557	0.5972	0.9556	0.9359
7.0	5.6410E-01	1.0341	0.9378	0.9127	0.8988	1.0110	1.0021	0.9998	1.0000	0.9562	0.5921	0.9561	0.9369
7.5	5.6348E-01	1.0357	0.9386	0.9138	0.9001	1.0117	1.0023	0.9998	1.0000	0.9567	0.5870	0.9566	0.9378
8.0	5.6287E-01	1.0373	0.9394	0.9149	0.9013	1.0125	1.0024	0.9998	1.0000	0.9571	0.5819	0.9570	0.9387
9.0	5.6174E-01	1.0404	0.9406	0.9166	0.9033	1.0140	1.0027	0.9998	1.0000	0.9578	0.5717	0.9577	0.9402
10.0	5.6075E-01	1.0435	0.9415	0.9180	0.9049	1.0155	1.0030	0.9999	1.0000	0.9583	0.5613	0.9583	0.9416
11.0	5.5993E-01	1.0467	0.9423	0.9191	0.9062	1.0171	1.0034	0.9999	1.0000	0.9587	0.5510	0.9587	0.9429
12.0	5.5925E-01	1.0498	0.9428	0.9199	0.9072	1.0186	1.0037	0.9999	1.0000	0.9590	0.5407	0.9590	0.9441
13.0	5.5871E-01	1.0530	0.9432	0.9205	0.9080	1.0200	1.0040	0.9999	1.0000	0.9592	0.5304	0.9592	0.9452
14.0	5.5829E-01	1.0561	0.9434	0.9210	0.9085	1.0215	1.0043	0.9999	1.0000	0.9594	0.5201	0.9594	0.9461
16.0	5.5780E-01	1.0624	0.9435	0.9214	0.9091	1.0245	1.0049	0.9999	1.0000	0.9595	0.4995	0.9595	0.9480
18.0	5.5768E-01	1.0687	0.9433	0.9214	0.9092	1.0274	1.0055	0.9999	1.0000	0.9594	0.4789	0.9594	0.9496
20.0	5.5787E-01	1.0750	0.9428	0.9209	0.9088	1.0303	1.0061	0.9999	1.0000	0.9592	0.4581	0.9592	0.9510
25.0	5.5938E-01	1.0909	0.9405	0.9186	0.9065	1.0375	1.0076	0.9999	1.0000	0.9581	0.4070	0.9581	0.9541
30.0	5.6198E-01	1.1068	0.9374	0.9150	0.9029	1.0445	1.0092	1.0000	1.0000	0.9566	0.3560	0.9566	0.9568
35.0	5.6534E-01	1.1228	0.9336	0.9107	0.8983	1.0514	1.0107	1.0000	1.0000	0.9548	0.3056	0.9548	0.9591
40.0	5.6927E-01	1.1389	0.9295	0.9058	0.8931	1.0583	1.0123	1.0000	1.0000	0.9528	0.2557	0.9528	0.9611
45.0	5.7364E-01	1.1551	0.9251	0.9005	0.8875	1.0650	1.0138	1.0000	1.0000	0.9507	0.2064	0.9507	0.9630
50.0	5.7835E-01	1.1713	0.9206	0.8950	0.8815	1.0715	1.0154	1.0000	1.0000	0.9485	0.1578	0.9485	0.9648

Instr. see p. 22

POSITRON Z= -33 A= 73

p	F_0L_0	L_0	λ_2	λ_3	λ_4	μ_1	μ_2	Λ_1	Λ_2	ν_{12}	$\hat{\nu}_{12}$	η_{12}	$\hat{\eta}_{12}$
0.1	5.7717E-06	1.0170	5.7583	13.2226	20.9984	1.0001	1.0000	0.9984	0.9997	2.2705	0.6883	0.9233	0.8965
0.2	5.2312E-03	1.0170	2.0939	2.6864	3.0206	1.0002	1.0000	0.9984	0.9997	1.2593	0.6862	0.9240	0.8973
0.3	4.1314E-02	1.0170	1.4175	1.5552	1.6168	1.0002	1.0000	0.9985	0.9997	1.0731	0.6831	0.9251	0.8984
0.4	1.0576E-01	1.0171	1.1829	1.2222	1.2351	1.0003	1.0001	0.9985	0.9997	1.0089	0.6795	0.9264	0.8998
0.5	1.7699E-01	1.0172	1.0759	1.0806	1.0780	1.0005	1.0001	0.9986	0.9997	0.9800	0.6758	0.9277	0.9012
0.6	2.4251E-01	1.0173	1.0191	1.0081	0.9988	1.0006	1.0001	0.9986	0.9997	0.9649	0.6722	0.9290	0.9026
0.7	2.9832E-01	1.0174	0.9858	0.9666	0.9539	1.0008	1.0001	0.9987	0.9997	0.9564	0.6688	0.9302	0.9039
0.8	3.4433E-01	1.0175	0.9649	0.9409	0.9263	1.0009	1.0002	0.9988	0.9998	0.9512	0.6656	0.9314	0.9052
0.9	3.8177E-01	1.0177	0.9513	0.9243	0.9086	1.0011	1.0002	0.9988	0.9998	0.9479	0.6628	0.9324	0.9064
1.0	4.1214E-01	1.0179	0.9420	0.9131	0.8967	1.0013	1.0002	0.9989	0.9998	0.9459	0.6601	0.9335	0.9075
1.2	4.5691E-01	1.0183	0.9309	0.9000	0.8828	1.0016	1.0003	0.9990	0.9998	0.9438	0.6555	0.9353	0.9096
1.4	4.8696E-01	1.0187	0.9253	0.8935	0.8760	1.0020	1.0004	0.9991	0.9998	0.9431	0.6516	0.9370	0.9115
1.6	5.0756E-01	1.0192	0.9225	0.8903	0.8727	1.0023	1.0004	0.9992	0.9998	0.9430	0.6482	0.9384	0.9131
1.8	5.2198E-01	1.0197	0.9211	0.8889	0.8714	1.0027	1.0005	0.9992	0.9998	0.9433	0.6452	0.9398	0.9146
2.0	5.3225E-01	1.0202	0.9206	0.8885	0.8711	1.0031	1.0006	0.9993	0.9999	0.9438	0.6425	0.9409	0.9160
2.2	5.3968E-01	1.0207	0.9206	0.8887	0.8714	1.0034	1.0006	0.9993	0.9999	0.9443	0.6400	0.9420	0.9173
2.4	5.4511E-01	1.0213	0.9209	0.8893	0.8721	1.0038	1.0007	0.9994	0.9999	0.9449	0.6376	0.9430	0.9184
2.6	5.4911E-01	1.0219	0.9215	0.8901	0.8731	1.0041	1.0008	0.9994	0.9999	0.9455	0.6353	0.9439	0.9195
2.8	5.5207E-01	1.0225	0.9221	0.8910	0.8741	1.0045	1.0008	0.9995	0.9999	0.9461	0.6331	0.9447	0.9205
3.0	5.5427E-01	1.0231	0.9228	0.8920	0.8753	1.0048	1.0009	0.9995	0.9999	0.9466	0.6310	0.9455	0.9215
3.2	5.5589E-01	1.0237	0.9235	0.8930	0.8764	1.0051	1.0010	0.9995	0.9999	0.9472	0.6289	0.9462	0.9223
3.4	5.5707E-01	1.0243	0.9242	0.8940	0.8776	1.0055	1.0010	0.9996	0.9999	0.9477	0.6269	0.9468	0.9232
3.6	5.5793E-01	1.0249	0.9250	0.8950	0.8787	1.0058	1.0011	0.9996	0.9999	0.9482	0.6248	0.9474	0.9240
3.8	5.5852E-01	1.0255	0.9257	0.8960	0.8799	1.0062	1.0012	0.9996	0.9999	0.9487	0.6228	0.9480	0.9247
4.0	5.5892E-01	1.0261	0.9264	0.8970	0.8809	1.0065	1.0012	0.9996	0.9999	0.9491	0.6208	0.9485	0.9254
4.5	5.5930E-01	1.0277	0.9281	0.8993	0.8835	1.0073	1.0014	0.9997	0.9999	0.9502	0.6159	0.9497	0.9270
5.0	5.5914E-01	1.0293	0.9296	0.9014	0.8859	1.0081	1.0016	0.9997	0.9999	0.9511	0.6109	0.9507	0.9285
5.5	5.5868E-01	1.0309	0.9310	0.9033	0.8881	1.0090	1.0017	0.9997	0.9999	0.9519	0.6060	0.9516	0.9298
6.0	5.5808E-01	1.0325	0.9323	0.9050	0.8900	1.0098	1.0019	0.9997	0.9999	0.9526	0.6011	0.9524	0.9310
6.5	5.5739E-01	1.0341	0.9334	0.9065	0.8917	1.0106	1.0020	0.9998	1.0000	0.9532	0.5962	0.9531	0.9321
7.0	5.5669E-01	1.0357	0.9344	0.9079	0.8933	1.0114	1.0022	0.9998	1.0000	0.9538	0.5912	0.9537	0.9332
7.5	5.5599E-01	1.0374	0.9353	0.9091	0.8947	1.0122	1.0024	0.9998	1.0000	0.9543	0.5863	0.9542	0.9341
8.0	5.5531E-01	1.0390	0.9361	0.9103	0.8960	1.0130	1.0025	0.9998	1.0000	0.9547	0.5813	0.9547	0.9350
9.0	5.5406E-01	1.0422	0.9374	0.9122	0.8982	1.0146	1.0028	0.9998	1.0000	0.9555	0.5714	0.9554	0.9367
10.0	5.5297E-01	1.0455	0.9385	0.9137	0.8999	1.0162	1.0032	0.9998	1.0000	0.9561	0.5614	0.9560	0.9382
11.0	5.5204E-01	1.0488	0.9393	0.9149	0.9013	1.0177	1.0035	0.9999	1.0000	0.9565	0.5514	0.9565	0.9395
12.0	5.5128E-01	1.0520	0.9399	0.9158	0.9025	1.0193	1.0038	0.9999	1.0000	0.9569	0.5414	0.9568	0.9408
13.0	5.5066E-01	1.0553	0.9404	0.9166	0.9033	1.0208	1.0041	0.9999	1.0000	0.9571	0.5314	0.9571	0.9419
14.0	5.5018E-01	1.0586	0.9407	0.9171	0.9040	1.0224	1.0044	0.9999	1.0000	0.9573	0.5214	0.9573	0.9430
16.0	5.4956E-01	1.0651	0.9409	0.9176	0.9047	1.0254	1.0051	0.9999	1.0000	0.9574	0.5013	0.9575	0.9449
18.0	5.4934E-01	1.0717	0.9407	0.9177	0.9049	1.0285	1.0057	0.9999	1.0000	0.9574	0.4813	0.9574	0.9466
20.0	5.4945E-01	1.0783	0.9402	0.9173	0.9046	1.0315	1.0063	0.9999	1.0000	0.9572	0.4613	0.9572	0.9481
25.0	5.5081E-01	1.0948	0.9381	0.9151	0.9025	1.0389	1.0079	0.9999	1.0000	0.9562	0.4114	0.9562	0.9514
30.0	5.5330E-01	1.1113	0.9349	0.9116	0.8989	1.0462	1.0095	1.0000	1.0000	0.9547	0.3619	0.9547	0.9542
35.0	5.5659E-01	1.1280	0.9312	0.9073	0.8943	1.0534	1.0111	1.0000	1.0000	0.9529	0.3128	0.9529	0.9566
40.0	5.6048E-01	1.1447	0.9270	0.9024	0.8891	1.0604	1.0127	1.0000	1.0000	0.9509	0.2643	0.9509	0.9588
45.0	5.6483E-01	1.1615	0.9226	0.8970	0.8835	1.0674	1.0143	1.0000	1.0000	0.9488	0.2163	0.9488	0.9608
50.0	5.6955E-01	1.1785	0.9180	0.8914	0.8775	1.0742	1.0160	1.0000	1.0000	0.9466	0.1689	0.9466	0.9627

Instr. see p. 22

POSITRON Z= -34 A= 80

p	F_0L_0	L_0	λ_2	λ_3	λ_4	μ_1	μ_2	Λ_1	Λ_2	ν_{12}	$\hat{\nu}_{12}$	η_{12}	$\hat{\eta}_{12}$
0.1	3.8364E-06	1.0180	6.0074	14.2462	23.1245	1.0001	1.0000	0.9983	0.9997	2.3445	0.6853	0.9196	0.8912
0.2	4.3627E-03	1.0180	2.1517	2.7963	3.1653	1.0002	1.0000	0.9983	0.9997	1.2749	0.6832	0.9203	0.8920
0.3	3.7083E-02	1.0181	1.4399	1.5882	1.6551	1.0003	1.0000	0.9984	0.9997	1.0779	0.6801	0.9214	0.8932
0.4	9.8279E-02	1.0181	1.1929	1.2351	1.2491	1.0004	1.0001	0.9984	0.9997	1.0100	0.6765	0.9227	0.8946
0.5	1.6768E-01	1.0182	1.0803	1.0855	1.0829	1.0005	1.0001	0.9985	0.9997	0.9793	0.6728	0.9240	0.8960
0.6	2.3247E-01	1.0183	1.0204	1.0090	0.9993	1.0007	1.0001	0.9985	0.9997	0.9634	0.6692	0.9253	0.8975
0.7	2.8818E-01	1.0185	0.9854	0.9652	0.9519	1.0008	1.0001	0.9986	0.9997	0.9543	0.6658	0.9266	0.8989
0.8	3.3439E-01	1.0186	0.9634	0.9382	0.9229	1.0010	1.0002	0.9987	0.9997	0.9488	0.6626	0.9278	0.9002
0.9	3.7215E-01	1.0188	0.9490	0.9207	0.9042	1.0012	1.0002	0.9987	0.9998	0.9454	0.6597	0.9290	0.9014
1.0	4.0286E-01	1.0190	0.9392	0.9090	0.8917	1.0014	1.0002	0.9988	0.9998	0.9432	0.6570	0.9300	0.9026
1.2	4.4825E-01	1.0194	0.9276	0.8952	0.8772	1.0017	1.0003	0.9989	0.9998	0.9409	0.6524	0.9320	0.9048
1.4	4.7876E-01	1.0198	0.9217	0.8883	0.8700	1.0021	1.0004	0.9990	0.9998	0.9402	0.6485	0.9337	0.9068
1.6	4.9970E-01	1.0203	0.9186	0.8850	0.8666	1.0025	1.0005	0.9991	0.9998	0.9401	0.6451	0.9353	0.9085
1.8	5.1435E-01	1.0209	0.9172	0.8835	0.8651	1.0029	1.0005	0.9992	0.9998	0.9404	0.6421	0.9367	0.9101
2.0	5.2477E-01	1.0214	0.9167	0.8831	0.8648	1.0033	1.0006	0.9992	0.9999	0.9409	0.6394	0.9379	0.9116
2.2	5.3230E-01	1.0220	0.9167	0.8833	0.8652	1.0036	1.0007	0.9993	0.9999	0.9415	0.6369	0.9391	0.9129
2.4	5.3780E-01	1.0226	0.9171	0.8839	0.8659	1.0040	1.0008	0.9994	0.9999	0.9421	0.6346	0.9401	0.9141
2.6	5.4183E-01	1.0232	0.9176	0.8847	0.8669	1.0044	1.0008	0.9994	0.9999	0.9427	0.6323	0.9410	0.9153
2.8	5.4481E-01	1.0239	0.9183	0.8857	0.8681	1.0048	1.0009	0.9994	0.9999	0.9433	0.6301	0.9419	0.9163
3.0	5.4701E-01	1.0245	0.9190	0.8868	0.8692	1.0051	1.0010	0.9995	0.9999	0.9439	0.6280	0.9427	0.9173
3.2	5.4863E-01	1.0251	0.9198	0.8878	0.8705	1.0055	1.0010	0.9995	0.9999	0.9445	0.6260	0.9435	0.9182
3.4	5.4980E-01	1.0258	0.9206	0.8889	0.8717	1.0059	1.0011	0.9995	0.9999	0.9451	0.6239	0.9441	0.9191
3.6	5.5063E-01	1.0265	0.9213	0.8900	0.8729	1.0062	1.0012	0.9996	0.9999	0.9456	0.6219	0.9448	0.9199
3.8	5.5120E-01	1.0271	0.9221	0.8910	0.8741	1.0066	1.0012	0.9996	0.9999	0.9461	0.6199	0.9454	0.9207
4.0	5.5158E-01	1.0278	0.9229	0.8920	0.8753	1.0069	1.0013	0.9996	0.9999	0.9466	0.6180	0.9459	0.9215
4.5	5.5189E-01	1.0295	0.9246	0.8945	0.8780	1.0078	1.0015	0.9996	0.9999	0.9477	0.6130	0.9472	0.9232
5.0	5.5166E-01	1.0312	0.9263	0.8967	0.8805	1.0087	1.0017	0.9997	0.9999	0.9486	0.6081	0.9483	0.9247
5.5	5.5114E-01	1.0329	0.9277	0.8987	0.8828	1.0096	1.0018	0.9997	0.9999	0.9495	0.6032	0.9492	0.9262
6.0	5.5047E-01	1.0346	0.9291	0.9005	0.8848	1.0104	1.0020	0.9997	0.9999	0.9503	0.5984	0.9500	0.9274
6.5	5.4973E-01	1.0363	0.9302	0.9021	0.8866	1.0113	1.0022	0.9997	0.9999	0.9509	0.5934	0.9508	0.9286
7.0	5.4897E-01	1.0380	0.9313	0.9036	0.8883	1.0122	1.0023	0.9998	1.0000	0.9515	0.5885	0.9514	0.9297
7.5	5.4822E-01	1.0397	0.9323	0.9049	0.8898	1.0130	1.0025	0.9998	1.0000	0.9521	0.5836	0.9519	0.9307
8.0	5.4750E-01	1.0415	0.9331	0.9061	0.8911	1.0139	1.0027	0.9998	1.0000	0.9525	0.5787	0.9524	0.9317
9.0	5.4616E-01	1.0449	0.9345	0.9081	0.8934	1.0155	1.0030	0.9998	1.0000	0.9533	0.5687	0.9532	0.9334
10.0	5.4501E-01	1.0484	0.9356	0.9097	0.8953	1.0172	1.0034	0.9998	1.0000	0.9539	0.5588	0.9539	0.9350
11.0	5.4403E-01	1.0519	0.9365	0.9110	0.8968	1.0189	1.0037	0.9999	1.0000	0.9544	0.5489	0.9544	0.9364
12.0	5.4322E-01	1.0554	0.9372	0.9120	0.8980	1.0205	1.0040	0.9999	1.0000	0.9548	0.5389	0.9548	0.9377
13.0	5.4257E-01	1.0589	0.9376	0.9127	0.8989	1.0222	1.0044	0.9999	1.0000	0.9550	0.5289	0.9550	0.9390
14.0	5.4206E-01	1.0624	0.9379	0.9133	0.8996	1.0238	1.0047	0.9999	1.0000	0.9552	0.5189	0.9552	0.9401
16.0	5.4141E-01	1.0694	0.9382	0.9139	0.9004	1.0271	1.0054	0.9999	1.0000	0.9554	0.4989	0.9554	0.9421
18.0	5.4117E-01	1.0764	0.9380	0.9139	0.9006	1.0303	1.0061	0.9999	1.0000	0.9553	0.4789	0.9554	0.9439
20.0	5.4128E-01	1.0834	0.9375	0.9136	0.9003	1.0335	1.0067	0.9999	1.0000	0.9551	0.4589	0.9551	0.9455
25.0	5.4268E-01	1.1010	0.9352	0.9113	0.8980	1.0414	1.0084	0.9999	1.0000	0.9541	0.4093	0.9541	0.9491
30.0	5.4528E-01	1.1187	0.9319	0.9076	0.8942	1.0491	1.0101	0.9999	1.0000	0.9525	0.3598	0.9525	0.9520
35.0	5.4871E-01	1.1364	0.9280	0.9030	0.8895	1.0567	1.0118	1.0000	1.0000	0.9506	0.3108	0.9506	0.9546
40.0	5.5277E-01	1.1543	0.9236	0.8978	0.8840	1.0642	1.0135	1.0000	1.0000	0.9485	0.2623	0.9486	0.9569
45.0	5.5730E-01	1.1723	0.9190	0.8922	0.8780	1.0716	1.0152	1.0000	1.0000	0.9463	0.2145	0.9463	0.9591
50.0	5.6222E-01	1.1904	0.9142	0.8863	0.8717	1.0788	1.0170	1.0000	1.0000	0.9440	0.1675	0.9440	0.9611

POSITRON Z= -35 A= 79

p	F_0L_0	L_0	λ_2	λ_3	λ_4	μ_1	μ_2	Λ_1	Λ_2	ν_{12}	$\hat{\nu}_{12}$	η_{12}	$\hat{\eta}_{12}$
0.1	2.5543E-06	1.0189	6.2530	15.3046	25.3899	1.0001	1.0000	0.9983	0.9997	2.4193	0.6819	0.9153	0.8854
0.2	3.6445E-03	1.0189	2.2079	2.9060	3.3114	1.0002	1.0000	0.9983	0.9997	1.2903	0.6798	0.9160	0.8861
0.3	3.3344E-02	1.0189	1.4611	1.6197	1.6919	1.0003	1.0000	0.9983	0.9997	1.0823	0.6768	0.9171	0.8873
0.4	9.1503E-02	1.0190	1.2019	1.2466	1.2615	1.0004	1.0001	0.9984	0.9997	1.0106	0.6732	0.9184	0.8888
0.5	1.5918E-01	1.0191	1.0836	1.0890	1.0862	1.0005	1.0001	0.9984	0.9997	0.9782	0.6695	0.9198	0.8903
0.6	2.2332E-01	1.0192	1.0208	1.0086	0.9982	1.0007	1.0001	0.9985	0.9997	0.9614	0.6659	0.9212	0.8918
0.7	2.7899E-01	1.0193	0.9839	0.9626	0.9485	1.0009	1.0001	0.9986	0.9997	0.9517	0.6625	0.9225	0.8932
0.8	3.2545E-01	1.0195	0.9609	0.9343	0.9181	1.0010	1.0002	0.9986	0.9997	0.9459	0.6594	0.9238	0.8946
0.9	3.6357E-01	1.0197	0.9457	0.9159	0.8985	1.0012	1.0002	0.9987	0.9997	0.9423	0.6565	0.9250	0.8959
1.0	3.9466E-01	1.0198	0.9355	0.9035	0.8854	1.0014	1.0003	0.9988	0.9998	0.9400	0.6538	0.9261	0.8972
1.2	4.4073E-01	1.0203	0.9233	0.8891	0.8701	1.0018	1.0003	0.9989	0.9998	0.9376	0.6493	0.9282	0.8995
1.4	4.7176E-01	1.0208	0.9170	0.8819	0.8627	1.0022	1.0004	0.9990	0.9998	0.9368	0.6454	0.9300	0.9015
1.6	4.9305E-01	1.0213	0.9139	0.8784	0.8591	1.0026	1.0005	0.9991	0.9998	0.9367	0.6422	0.9316	0.9033
1.8	5.0794E-01	1.0218	0.9124	0.8769	0.8576	1.0030	1.0005	0.9992	0.9998	0.9371	0.6393	0.9331	0.9050
2.0	5.1853E-01	1.0224	0.9119	0.8765	0.8573	1.0034	1.0006	0.9992	0.9998	0.9376	0.6367	0.9344	0.9065
2.2	5.2616E-01	1.0230	0.9119	0.8767	0.8577	1.0038	1.0007	0.9993	0.9999	0.9382	0.6343	0.9356	0.9079
2.4	5.3171E-01	1.0236	0.9123	0.8774	0.8585	1.0041	1.0008	0.9993	0.9999	0.9388	0.6320	0.9367	0.9092
2.6	5.3578E-01	1.0243	0.9129	0.8783	0.8595	1.0045	1.0008	0.9994	0.9999	0.9395	0.6298	0.9377	0.9104
2.8	5.3877E-01	1.0249	0.9136	0.8793	0.8607	1.0049	1.0009	0.9994	0.9999	0.9401	0.6278	0.9386	0.9115
3.0	5.4096E-01	1.0256	0.9144	0.8804	0.8620	1.0053	1.0010	0.9995	0.9999	0.9408	0.6258	0.9395	0.9126
3.2	5.4256E-01	1.0262	0.9152	0.8815	0.8633	1.0057	1.0011	0.9995	0.9999	0.9414	0.6238	0.9403	0.9136
3.4	5.4371E-01	1.0269	0.9160	0.8827	0.8646	1.0060	1.0011	0.9995	0.9999	0.9420	0.6219	0.9410	0.9145
3.6	5.4451E-01	1.0276	0.9169	0.8838	0.8659	1.0064	1.0012	0.9995	0.9999	0.9425	0.6199	0.9417	0.9154
3.8	5.4505E-01	1.0283	0.9177	0.8850	0.8672	1.0068	1.0013	0.9996	0.9999	0.9431	0.6180	0.9423	0.9162
4.0	5.4538E-01	1.0289	0.9185	0.8860	0.8684	1.0071	1.0013	0.9996	0.9999	0.9436	0.6161	0.9429	0.9170
4.5	5.4560E-01	1.0307	0.9204	0.8886	0.8713	1.0081	1.0015	0.9996	0.9999	0.9448	0.6115	0.9443	0.9188
5.0	5.4527E-01	1.0324	0.9222	0.8910	0.8740	1.0090	1.0017	0.9997	0.9999	0.9458	0.6068	0.9454	0.9204
5.5	5.4465E-01	1.0342	0.9238	0.8932	0.8764	1.0099	1.0019	0.9997	0.9999	0.9467	0.6021	0.9464	0.9219
6.0	5.4388E-01	1.0359	0.9252	0.8951	0.8786	1.0107	1.0021	0.9997	0.9999	0.9476	0.5974	0.9473	0.9233
6.5	5.4305E-01	1.0377	0.9265	0.8968	0.8806	1.0116	1.0022	0.9997	0.9999	0.9483	0.5927	0.9481	0.9245
7.0	5.4220E-01	1.0395	0.9276	0.8984	0.8824	1.0125	1.0024	0.9998	1.0000	0.9489	0.5880	0.9488	0.9257
7.5	5.4137E-01	1.0413	0.9286	0.8998	0.8840	1.0134	1.0026	0.9998	1.0000	0.9495	0.5833	0.9494	0.9267
8.0	5.4057E-01	1.0430	0.9296	0.9011	0.8854	1.0143	1.0028	0.9998	1.0000	0.9500	0.5785	0.9499	0.9278
9.0	5.3909E-01	1.0466	0.9311	0.9033	0.8879	1.0160	1.0031	0.9998	1.0000	0.9509	0.5690	0.9508	0.9296
10.0	5.3781E-01	1.0502	0.9324	0.9051	0.8900	1.0177	1.0034	0.9998	1.0000	0.9516	0.5595	0.9515	0.9313
11.0	5.3671E-01	1.0538	0.9334	0.9065	0.8916	1.0194	1.0038	0.9998	1.0000	0.9521	0.5499	0.9521	0.9326
12.0	5.3580E-01	1.0574	0.9341	0.9076	0.8929	1.0211	1.0041	0.9999	1.0000	0.9525	0.5403	0.9525	0.9342
13.0	5.3505E-01	1.0610	0.9347	0.9085	0.8940	1.0228	1.0045	0.9999	1.0000	0.9528	0.5307	0.9528	0.9354
14.0	5.3445E-01	1.0645	0.9351	0.9092	0.8948	1.0245	1.0048	0.9999	1.0000	0.9531	0.5210	0.9531	0.9366
16.0	5.3365E-01	1.0717	0.9354	0.9100	0.8958	1.0279	1.0055	0.9999	1.0000	0.9533	0.5018	0.9533	0.9388
18.0	5.3328E-01	1.0790	0.9354	0.9102	0.8962	1.0312	1.0062	0.9999	1.0000	0.9533	0.4825	0.9533	0.9407
20.0	5.3326E-01	1.0862	0.9350	0.9099	0.8960	1.0345	1.0069	0.9999	1.0000	0.9531	0.4632	0.9532	0.9424
25.0	5.3443E-01	1.1043	0.9328	0.9079	0.8940	1.0425	1.0086	0.9999	1.0000	0.9522	0.4152	0.9522	0.9461
30.0	5.3685E-01	1.1225	0.9297	0.9043	0.8904	1.0505	1.0104	0.9999	1.0000	0.9507	0.3674	0.9507	0.9492
35.0	5.4013E-01	1.1408	0.9258	0.8999	0.8858	1.0583	1.0121	1.0000	1.0000	0.9488	0.3202	0.9488	0.9519
40.0	5.4407E-01	1.1592	0.9215	0.8947	0.8804	1.0660	1.0138	1.0000	1.0000	0.9468	0.2732	0.9468	0.9544
45.0	5.4851E-01	1.1778	0.9168	0.8892	0.8745	1.0735	1.0156	1.0000	1.0000	0.9445	0.2270	0.9445	0.9567
50.0	5.5335E-01	1.1964	0.9120	0.8833	0.8682	1.0810	1.0174	1.0000	1.0000	0.9422	0.1812	0.9422	0.9588

 Instr. see p. 22

POSITRON Z= -36 A= 82

p	F_0L_0	L_0	λ_2	λ_3	λ_4	μ_1	μ_2	Λ_1	Λ_2	ν_{12}	$\hat{\nu}_{12}$	η_{12}	$\hat{\eta}_{12}$
0.1	1.6986E-06	1.0198	6.5037	16.4258	27.8547	1.0002	1.0000	0.9982	0.9996	2.4961	0.6787	0.9111	0.8796
0.2	3.0409E-03	1.0199	2.2657	3.0209	3.4660	1.0002	1.0000	0.9982	0.9997	1.3063	0.6766	0.9118	0.8804
0.3	2.9949E-02	1.0199	1.4831	1.6530	1.7307	1.0003	1.0000	0.9983	0.9997	1.0871	0.6735	0.9130	0.8816
0.4	8.5108E-02	1.0200	1.2114	1.2590	1.2749	1.0004	1.0001	0.9983	0.9997	1.0115	0.6700	0.9143	0.8831
0.5	1.5097E-01	1.0201	1.0875	1.0932	1.0902	1.0006	1.0001	0.9984	0.9997	0.9773	0.6663	0.9158	0.8847
0.6	2.1435E-01	1.0202	1.0216	1.0088	0.9979	1.0007	1.0001	0.9984	0.9997	0.9595	0.6627	0.9172	0.8862
0.7	2.6989E-01	1.0203	0.9829	0.9605	0.9457	1.0009	1.0002	0.9985	0.9997	0.9494	0.6593	0.9186	0.8877
0.8	3.1653E-01	1.0205	0.9587	0.9308	0.9138	1.0011	1.0002	0.9986	0.9997	0.9432	0.6561	0.9199	0.8891
0.9	3.5495E-01	1.0207	0.9429	0.9115	0.8933	1.0013	1.0002	0.9986	0.9997	0.9394	0.6533	0.9211	0.8905
1.0	3.8638E-01	1.0209	0.9321	0.8986	0.8795	1.0015	1.0003	0.9987	0.9997	0.9369	0.6506	0.9223	0.8918
1.2	4.3307E-01	1.0213	0.9193	0.8834	0.8636	1.0019	1.0003	0.9988	0.9998	0.9344	0.6461	0.9245	0.8942
1.4	4.6457E-01	1.0218	0.9128	0.8759	0.8558	1.0023	1.0004	0.9989	0.9998	0.9336	0.6423	0.9264	0.8963
1.6	4.8620E-01	1.0224	0.9095	0.8723	0.8520	1.0027	1.0005	0.9990	0.9998	0.9335	0.6391	0.9281	0.8983
1.8	5.0132E-01	1.0229	0.9079	0.8707	0.8505	1.0031	1.0006	0.9991	0.9998	0.9338	0.6362	0.9296	0.9000
2.0	5.1206E-01	1.0235	0.9074	0.8702	0.8502	1.0035	1.0007	0.9992	0.9998	0.9344	0.6337	0.9310	0.9016
2.2	5.1979E-01	1.0242	0.9074	0.8705	0.8506	1.0039	1.0007	0.9993	0.9999	0.9350	0.6314	0.9323	0.9031
2.4	5.2540E-01	1.0248	0.9078	0.8712	0.8515	1.0043	1.0008	0.9993	0.9999	0.9357	0.6292	0.9335	0.9045
2.6	5.2950E-01	1.0255	0.9085	0.8722	0.8526	1.0047	1.0009	0.9994	0.9999	0.9364	0.6271	0.9345	0.9057
2.8	5.3250E-01	1.0262	0.9092	0.8733	0.8538	1.0051	1.0010	0.9994	0.9999	0.9371	0.6251	0.9355	0.9069
3.0	5.3469E-01	1.0269	0.9101	0.8744	0.8552	1.0055	1.0010	0.9994	0.9999	0.9378	0.6231	0.9364	0.9080
3.2	5.3627E-01	1.0275	0.9109	0.8757	0.8566	1.0059	1.0011	0.9995	0.9999	0.9384	0.6212	0.9372	0.9090
3.4	5.3739E-01	1.0283	0.9118	0.8769	0.8579	1.0063	1.0012	0.9995	0.9999	0.9390	0.6193	0.9380	0.9100
3.6	5.3817E-01	1.0290	0.9127	0.8781	0.8593	1.0067	1.0013	0.9995	0.9999	0.9396	0.6174	0.9387	0.9109
3.8	5.3868E-01	1.0297	0.9136	0.8793	0.8606	1.0071	1.0013	0.9995	0.9999	0.9402	0.6156	0.9394	0.9118
4.0	5.3898E-01	1.0304	0.9144	0.8804	0.8619	1.0075	1.0014	0.9996	0.9999	0.9408	0.6137	0.9400	0.9126
4.5	5.3911E-01	1.0322	0.9165	0.8832	0.8650	1.0084	1.0016	0.9996	0.9999	0.9420	0.6092	0.9415	0.9145
5.0	5.3869E-01	1.0340	0.9183	0.8857	0.8679	1.0094	1.0018	0.9997	0.9999	0.9431	0.6046	0.9427	0.9163
5.5	5.3798E-01	1.0359	0.9200	0.8880	0.8704	1.0103	1.0020	0.9997	0.9999	0.9441	0.6001	0.9438	0.9178
6.0	5.3713E-01	1.0377	0.9215	0.8900	0.8727	1.0112	1.0021	0.9997	0.9999	0.9450	0.5955	0.9447	0.9193
6.5	5.3622E-01	1.0396	0.9229	0.8919	0.8748	1.0122	1.0023	0.9997	0.9999	0.9457	0.5909	0.9455	0.9206
7.0	5.3530E-01	1.0414	0.9241	0.8936	0.8767	1.0131	1.0025	0.9998	0.9999	0.9464	0.5863	0.9463	0.9218
7.5	5.3440E-01	1.0433	0.9252	0.8951	0.8785	1.0140	1.0027	0.9998	1.0000	0.9470	0.5817	0.9469	0.9229
8.0	5.3353E-01	1.0451	0.9262	0.8964	0.8800	1.0149	1.0029	0.9998	1.0000	0.9476	0.5771	0.9475	0.9240
9.0	5.3194E-01	1.0489	0.9279	0.8988	0.8827	1.0167	1.0032	0.9998	1.0000	0.9485	0.5677	0.9484	0.9259
10.0	5.3055E-01	1.0526	0.9292	0.9007	0.8849	1.0185	1.0036	0.9998	1.0000	0.9492	0.5584	0.9492	0.9277
11.0	5.2937E-01	1.0564	0.9303	0.9022	0.8866	1.0203	1.0040	0.9998	1.0000	0.9498	0.5490	0.9498	0.9293
12.0	5.2838E-01	1.0601	0.9311	0.9035	0.8881	1.0221	1.0043	0.9999	1.0000	0.9503	0.5396	0.9503	0.9307
13.0	5.2756E-01	1.0639	0.9317	0.9044	0.8892	1.0239	1.0047	0.9999	1.0000	0.9506	0.5302	0.9506	0.9321
14.0	5.2690E-01	1.0676	0.9322	0.9051	0.8901	1.0256	1.0050	0.9999	1.0000	0.9509	0.5207	0.9509	0.9333
16.0	5.2599E-01	1.0752	0.9327	0.9061	0.8912	1.0291	1.0058	0.9999	1.0000	0.9512	0.5018	0.9512	0.9356
18.0	5.2554E-01	1.0827	0.9326	0.9064	0.8917	1.0326	1.0065	0.9999	1.0000	0.9512	0.4829	0.9512	0.9376
20.0	5.2547E-01	1.0902	0.9323	0.9062	0.8916	1.0360	1.0072	0.9999	1.0000	0.9511	0.4640	0.9511	0.9394
25.0	5.2654E-01	1.1092	0.9302	0.9042	0.8898	1.0444	1.0090	0.9999	1.0000	0.9501	0.4168	0.9501	0.9433
30.0	5.2890E-01	1.1282	0.9270	0.9006	0.8862	1.0527	1.0108	0.9999	1.0000	0.9486	0.3695	0.9486	0.9467
35.0	5.3218E-01	1.1474	0.9231	0.8961	0.8815	1.0608	1.0126	1.0000	1.0000	0.9468	0.3234	0.9468	0.9495
40.0	5.3613E-01	1.1667	0.9187	0.8909	0.8760	1.0688	1.0144	1.0000	1.0000	0.9446	0.2773	0.9447	0.9521
45.0	5.4061E-01	1.1861	0.9139	0.8852	0.8699	1.0767	1.0163	1.0000	1.0000	0.9424	0.2318	0.9424	0.9545
50.0	5.4550E-01	1.2057	0.9090	0.8792	0.8635	1.0844	1.0181	1.0000	1.0000	0.9400	0.1868	0.9400	0.9568

Instr. see p. 22

POSITRON Z= -37 A= 85

p	F_0L_0	L_0	λ_2	λ_3	λ_4	μ_1	μ_2	Λ_1	Λ_2	ν_{12}	$\hat{\nu}_{12}$	η_{12}	$\hat{\eta}_{12}$
0.1	1.1293E-06	1.0208	6.7560	17.6012	30.5109	1.0002	1.0000	0.9981	0.9996	2.5746	0.6753	0.9069	0.8738
0.2	2.5368E-03	1.0208	2.3237	3.1390	3.6267	1.0002	1.0000	0.9981	0.9996	1.3227	0.6733	0.9076	0.8746
0.3	2.6896E-02	1.0209	1.5052	1.6867	1.7704	1.0003	1.0000	0.9982	0.9996	1.0920	0.6703	0.9088	0.8758
0.4	7.9160E-02	1.0210	1.2210	1.2715	1.2885	1.0004	1.0001	0.9982	0.9997	1.0124	0.6667	0.9102	0.8773
0.5	1.4319E-01	1.0211	1.0913	1.0974	1.0943	1.0006	1.0001	0.9983	0.9997	0.9764	0.6630	0.9116	0.8789
0.6	2.0577E-01	1.0212	1.0223	1.0088	0.9974	1.0008	1.0001	0.9984	0.9997	0.9577	0.6594	0.9131	0.8805
0.7	2.6114E-01	1.0214	0.9818	0.9583	0.9427	1.0009	1.0002	0.9984	0.9997	0.9469	0.6560	0.9145	0.8821
0.8	3.0792E-01	1.0215	0.9565	0.9272	0.9094	1.0011	1.0002	0.9985	0.9997	0.9405	0.6529	0.9159	0.8836
0.9	3.4663E-01	1.0217	0.9399	0.9070	0.8879	1.0013	1.0002	0.9986	0.9997	0.9364	0.6500	0.9172	0.8850
1.0	3.7839E-01	1.0219	0.9286	0.8935	0.8736	1.0016	1.0003	0.9987	0.9997	0.9338	0.6474	0.9184	0.8863
1.2	4.2568E-01	1.0224	0.9152	0.8777	0.8569	1.0020	1.0004	0.9988	0.9998	0.9312	0.6429	0.9207	0.8888
1.4	4.5765E-01	1.0229	0.9084	0.8698	0.8488	1.0024	1.0004	0.9989	0.9998	0.9303	0.6392	0.9227	0.8911
1.6	4.7961E-01	1.0235	0.9049	0.8660	0.8449	1.0028	1.0005	0.9990	0.9998	0.9302	0.6360	0.9245	0.8931
1.8	4.9496E-01	1.0241	0.9033	0.8643	0.8433	1.0033	1.0006	0.9991	0.9998	0.9305	0.6332	0.9261	0.8950
2.0	5.0585E-01	1.0247	0.9028	0.8639	0.8430	1.0037	1.0007	0.9992	0.9998	0.9311	0.6307	0.9276	0.8966
2.2	5.1367E-01	1.0254	0.9028	0.8642	0.8434	1.0041	1.0008	0.9992	0.9998	0.9318	0.6284	0.9289	0.8982
2.4	5.1934E-01	1.0261	0.9033	0.8650	0.8443	1.0045	1.0008	0.9993	0.9999	0.9325	0.6263	0.9301	0.8996
2.6	5.2346E-01	1.0267	0.9039	0.8660	0.8455	1.0050	1.0009	0.9993	0.9999	0.9332	0.6242	0.9313	0.9009
2.8	5.2647E-01	1.0275	0.9047	0.8671	0.8468	1.0054	1.0010	0.9994	0.9999	0.9340	0.6223	0.9323	0.9022
3.0	5.2865E-01	1.0282	0.9056	0.8684	0.8483	1.0058	1.0011	0.9994	0.9999	0.9347	0.6204	0.9332	0.9033
3.2	5.3022E-01	1.0289	0.9066	0.8697	0.8497	1.0062	1.0012	0.9994	0.9999	0.9354	0.6185	0.9341	0.9044
3.4	5.3132E-01	1.0296	0.9075	0.8709	0.8512	1.0066	1.0012	0.9995	0.9999	0.9360	0.6167	0.9349	0.9054
3.6	5.3206E-01	1.0304	0.9084	0.8722	0.8526	1.0070	1.0013	0.9995	0.9999	0.9367	0.6149	0.9357	0.9064
3.8	5.3254E-01	1.0311	0.9094	0.8735	0.8540	1.0074	1.0014	0.9995	0.9999	0.9373	0.6131	0.9364	0.9073
4.0	5.3281E-01	1.0319	0.9103	0.8747	0.8554	1.0078	1.0015	0.9995	0.9999	0.9378	0.6113	0.9371	0.9082
4.5	5.3285E-01	1.0338	0.9124	0.8776	0.8586	1.0088	1.0017	0.9996	0.9999	0.9392	0.6069	0.9386	0.9102
5.0	5.3233E-01	1.0357	0.9144	0.8803	0.8616	1.0098	1.0018	0.9996	0.9999	0.9403	0.6024	0.9399	0.9120
5.5	5.3154E-01	1.0376	0.9162	0.8827	0.8643	1.0108	1.0020	0.9997	0.9999	0.9414	0.5980	0.9410	0.9137
6.0	5.3060E-01	1.0395	0.9178	0.8848	0.8668	1.0118	1.0022	0.9997	0.9999	0.9423	0.5935	0.9420	0.9152
6.5	5.2961E-01	1.0414	0.9193	0.8868	0.8690	1.0127	1.0024	0.9997	0.9999	0.9431	0.5891	0.9429	0.9165
7.0	5.2861E-01	1.0434	0.9206	0.8886	0.8710	1.0137	1.0026	0.9997	0.9999	0.9439	0.5846	0.9437	0.9178
7.5	5.2764E-01	1.0453	0.9218	0.8902	0.8729	1.0146	1.0028	0.9998	1.0000	0.9445	0.5801	0.9444	0.9190
8.0	5.2671E-01	1.0473	0.9228	0.8917	0.8745	1.0156	1.0030	0.9998	1.0000	0.9451	0.5755	0.9450	0.9201
9.0	5.2499E-01	1.0512	0.9246	0.8942	0.8774	1.0175	1.0034	0.9996	1.0000	0.9461	0.5664	0.9460	0.9222
10.0	5.2351E-01	1.0551	0.9261	0.8962	0.8797	1.0194	1.0037	0.9998	1.0000	0.9469	0.5573	0.9468	0.9240
11.0	5.2223E-01	1.0590	0.9272	0.8979	0.8816	1.0212	1.0041	0.9998	1.0000	0.9475	0.5480	0.9475	0.9257
12.0	5.2116E-01	1.0629	0.9281	0.8992	0.8831	1.0231	1.0045	0.9999	1.0000	0.9480	0.5389	0.9480	0.9273
13.0	5.2028E-01	1.0668	0.9288	0.9003	0.8844	1.0249	1.0049	0.9999	1.0000	0.9484	0.5296	0.9484	0.9287
14.0	5.1955E-01	1.0708	0.9293	0.9011	0.8854	1.0267	1.0052	0.9999	1.0000	0.9487	0.5203	0.9487	0.9300
16.0	5.1854E-01	1.0786	0.9298	0.9021	0.8866	1.0304	1.0060	0.9999	1.0000	0.9490	0.5018	0.9490	0.9324
18.0	5.1801E-01	1.0865	0.9299	0.9025	0.8872	1.0340	1.0067	0.9999	1.0000	0.9491	0.4832	0.9491	0.9345
20.0	5.1787E-01	1.0944	0.9296	0.9024	0.8872	1.0375	1.0075	0.9999	1.0000	0.9490	0.4646	0.9490	0.9364
25.0	5.1883E-01	1.1142	0.9275	0.9004	0.8854	1.0463	1.0094	0.9999	1.0000	0.9481	0.4183	0.9481	0.9405
30.0	5.2114E-01	1.1341	0.9243	0.8969	0.8819	1.0549	1.0112	0.9999	1.0000	0.9466	0.3721	0.9466	0.9440
35.0	5.2439E-01	1.1542	0.9203	0.8924	0.8771	1.0634	1.0131	1.0000	1.0000	0.9447	0.3265	0.9447	0.9471
40.0	5.2836E-01	1.1743	0.9159	0.8871	0.8716	1.0717	1.0150	1.0000	1.0000	0.9425	0.2811	0.9425	0.9498
45.0	5.3286E-01	1.1947	0.9110	0.8813	0.8654	1.0799	1.0169	1.0000	1.0000	0.9402	0.2364	0.9402	0.9523
50.0	5.3780E-01	1.2152	0.9060	0.8751	0.8589	1.0879	1.0189	1.0000	1.0000	0.9378	0.1922	0.9378	0.9547

Instr. see p. 22

POSITRON Z= −38 A= 86

p	F_0L_0	L_0	λ_2	λ_3	λ_4	μ_1	μ_2	Λ_1	Λ_2	ν_{12}	$\hat{\nu}_{12}$	η_{12}	$\hat{\eta}_{12}$
0.1	7.5113E-07	1.0218	7.0079	18.8246	33.3552	1.0002	1.0000	0.9980	0.9996	2.6542	0.6719	0.9025	0.8678
0.2	2.1171E-03	1.0218	2.3815	3.2591	3.7921	1.0002	1.0000	0.9981	0.9996	1.3392	0.6698	0.9032	0.8685
0.3	2.4165E-02	1.0219	1.5270	1.7204	1.8103	1.0003	1.0000	0.9981	0.9996	1.0969	0.6668	0.9044	0.8698
0.4	7.3668E-02	1.0220	1.2302	1.2836	1.3017	1.0005	1.0001	0.9982	0.9996	1.0132	0.6633	0.9058	0.8713
0.5	1.3591E-01	1.0221	1.0947	1.1011	1.0978	1.0006	1.0001	0.9982	0.9997	0.9754	0.6597	0.9073	0.8729
0.6	1.9768E-01	1.0222	1.0227	1.0085	0.9964	1.0008	1.0001	0.9983	0.9997	0.9556	0.6561	0.9088	0.8746
0.7	2.5287E-01	1.0223	0.9804	0.9556	0.9393	1.0010	1.0002	0.9984	0.9997	0.9444	0.6527	0.9103	0.8762
0.8	2.9980E-01	1.0225	0.9540	0.9232	0.9045	1.0012	1.0002	0.9985	0.9997	0.9375	0.6495	0.9117	0.8778
0.9	3.3879E-01	1.0227	0.9366	0.9021	0.8821	1.0014	1.0002	0.9985	0.9997	0.9333	0.6467	0.9131	0.8792
1.0	3.7087E-01	1.0229	0.9248	0.8880	0.8671	1.0016	1.0003	0.9986	0.9997	0.9306	0.6441	0.9143	0.8806
1.2	4.1878E-01	1.0234	0.9108	0.8715	0.8498	1.0021	1.0004	0.9987	0.9998	0.9278	0.6397	0.9167	0.8833
1.4	4.5122E-01	1.0239	0.9037	0.8633	0.8413	1.0025	1.0004	0.9989	0.9998	0.9268	0.6360	0.9188	0.8856
1.6	4.7352E-01	1.0245	0.9001	0.8593	0.8372	1.0030	1.0005	0.9990	0.9998	0.9267	0.6329	0.9207	0.8877
1.8	4.8910E-01	1.0252	0.8984	0.8576	0.8356	1.0034	1.0006	0.9991	0.9998	0.9271	0.6302	0.9224	0.8896
2.0	5.0014E-01	1.0258	0.8978	0.8572	0.8353	1.0038	1.0007	0.9991	0.9998	0.9277	0.6277	0.9240	0.8914
2.2	5.0806E-01	1.0265	0.8979	0.8575	0.8358	1.0043	1.0008	0.9992	0.9998	0.9284	0.6255	0.9254	0.8930
2.4	5.1378E-01	1.0272	0.8984	0.8583	0.8368	1.0047	1.0009	0.9993	0.9999	0.9291	0.6235	0.9266	0.8945
2.6	5.1793E-01	1.0279	0.8991	0.8594	0.8380	1.0052	1.0009	0.9993	0.9999	0.9299	0.6215	0.9278	0.8959
2.8	5.2094E-01	1.0287	0.9000	0.8606	0.8394	1.0056	1.0010	0.9993	0.9999	0.9307	0.6196	0.9289	0.8972
3.0	5.2311E-01	1.0294	0.9009	0.8619	0.8409	1.0060	1.0011	0.9994	0.9999	0.9314	0.6178	0.9299	0.8984
3.2	5.2465E-01	1.0302	0.9019	0.8633	0.8424	1.0064	1.0012	0.9994	0.9999	0.9322	0.6160	0.9308	0.8995
3.4	5.2573E-01	1.0309	0.9029	0.8646	0.8440	1.0069	1.0013	0.9995	0.9999	0.9329	0.6142	0.9317	0.9006
3.6	5.2644E-01	1.0317	0.9039	0.8660	0.8455	1.0073	1.0014	0.9995	0.9999	0.9335	0.6125	0.9325	0.9016
3.8	5.2688E-01	1.0325	0.9049	0.8673	0.8470	1.0077	1.0014	0.9995	0.9999	0.9342	0.6108	0.9333	0.9026
4.0	5.2711E-01	1.0332	0.9058	0.8686	0.8484	1.0081	1.0015	0.9995	0.9999	0.9348	0.6091	0.9340	0.9035
4.5	5.2705E-01	1.0352	0.9081	0.8717	0.8519	1.0091	1.0017	0.9996	0.9999	0.9362	0.6048	0.9356	0.9056
5.0	5.2643E-01	1.0372	0.9102	0.8745	0.8550	1.0102	1.0019	0.9996	0.9999	0.9374	0.6005	0.9370	0.9075
5.5	5.2553E-01	1.0391	0.9121	0.8770	0.8579	1.0112	1.0021	0.9997	0.9999	0.9385	0.5962	0.9382	0.9093
6.0	5.2450E-01	1.0411	0.9138	0.8794	0.8605	1.0122	1.0023	0.9997	0.9999	0.9395	0.5919	0.9392	0.9108
6.5	5.2342E-01	1.0432	0.9154	0.8815	0.8629	1.0132	1.0025	0.9997	0.9999	0.9404	0.5876	0.9402	0.9123
7.0	5.2233E-01	1.0452	0.9168	0.8834	0.8650	1.0142	1.0027	0.9997	0.9999	0.9412	0.5832	0.9410	0.9136
7.5	5.2128E-01	1.0472	0.9181	0.8851	0.8669	1.0152	1.0029	0.9998	0.9999	0.9419	0.5789	0.9417	0.9149
8.0	5.2027E-01	1.0492	0.9192	0.8866	0.8687	1.0162	1.0031	0.9998	1.0000	0.9425	0.5745	0.9424	0.9161
9.0	5.1843E-01	1.0532	0.9211	0.8893	0.8717	1.0181	1.0035	0.9998	1.0000	0.9436	0.5656	0.9435	0.9182
10.0	5.1681E-01	1.0573	0.9227	0.8915	0.8743	1.0201	1.0039	0.9998	1.0000	0.9444	0.5568	0.9444	0.9202
11.0	5.1543E-01	1.0613	0.9240	0.8933	0.8763	1.0220	1.0042	0.9998	1.0000	0.9451	0.5478	0.9451	0.9219
12.0	5.1427E-01	1.0654	0.9250	0.8948	0.8780	1.0239	1.0046	0.9998	1.0000	0.9457	0.5388	0.9456	0.9236
13.0	5.1329E-01	1.0695	0.9257	0.8959	0.8794	1.0258	1.0050	0.9999	1.0000	0.9461	0.5298	0.9461	0.9250
14.0	5.1249E-01	1.0735	0.9263	0.8968	0.8804	1.0277	1.0054	0.9999	1.0000	0.9464	0.5208	0.9464	0.9264
16.0	5.1135E-01	1.0817	0.9270	0.8980	0.8819	1.0314	1.0062	0.9999	1.0000	0.9468	0.5027	0.9468	0.9289
18.0	5.1070E-01	1.0898	0.9271	0.8985	0.8826	1.0351	1.0070	0.9999	1.0000	0.9469	0.4846	0.9470	0.9312
20.0	5.1046E-01	1.0980	0.9269	0.8985	0.8828	1.0388	1.0077	0.9999	1.0000	0.9469	0.4665	0.9469	0.9332
25.0	5.1124E-01	1.1185	0.9250	0.8968	0.8812	1.0479	1.0097	0.9999	1.0000	0.9460	0.4214	0.9460	0.9375
30.0	5.1342E-01	1.1391	0.9218	0.8934	0.8778	1.0568	1.0116	0.9999	1.0000	0.9446	0.3760	0.9446	0.9412
35.0	5.1659E-01	1.1599	0.9179	0.8889	0.8731	1.0655	1.0135	1.0000	1.0000	0.9427	0.3317	0.9427	0.9444
40.0	5.2048E-01	1.1809	0.9133	0.8836	0.8675	1.0741	1.0155	1.0000	1.0000	0.9405	0.2874	0.9406	0.9473
45.0	5.2495E-01	1.2019	0.9085	0.8778	0.8614	1.0825	1.0175	1.0000	1.0000	0.9382	0.2437	0.9382	0.9499
50.0	5.2986E-01	1.2232	0.9034	0.8716	0.8548	1.0908	1.0195	1.0000	1.0000	0.9358	0.2006	0.9358	0.9524

Instr. see p. 22

POSITRON Z= -39 A= 89

p	F_0L_0	L_0	λ_2	λ_3	λ_4	μ_1	μ_2	Λ_1	Λ_2	ν_{12}	$\hat{\nu}_{12}$	η_{12}	$\hat{\eta}_{12}$
0.1	4.9925E-07	1.0228	7.2625	20.1090	36.4233	1.0002	1.0000	0.9980	0.9996	2.7356	0.6684	0.8982	0.8617
0.2	1.7656E-03	1.0228	2.4400	3.3835	3.9653	1.0003	1.0000	0.9980	0.9996	1.3562	0.6664	0.8989	0.8625
0.3	2.1698E-02	1.0229	1.5492	1.7551	1.8516	1.0004	1.0000	0.9980	0.9996	1.1020	0.6635	0.9000	0.8638
0.4	6.8522E-02	1.0230	1.2397	1.2962	1.3155	1.0005	1.0001	0.9981	0.9996	1.0141	0.6600	0.9015	0.8653
0.5	1.2894E-01	1.0231	1.0985	1.1052	1.1018	1.0007	1.0001	0.9982	0.9996	0.9745	0.6563	0.9030	0.8670
0.6	1.8984E-01	1.0232	1.0233	1.0084	0.9958	1.0008	1.0001	0.9982	0.9997	0.9537	0.6527	0.9046	0.8687
0.7	2.4478E-01	1.0234	0.9792	0.9532	0.9362	1.0010	1.0002	0.9983	0.9997	0.9419	0.6493	0.9061	0.8704
0.8	2.9180E-01	1.0236	0.9516	0.9194	0.8999	1.0013	1.0002	0.9984	0.9997	0.9347	0.6462	0.9076	0.8720
0.9	3.3105E-01	1.0238	0.9334	0.8974	0.8765	1.0015	1.0003	0.9985	0.9997	0.9302	0.6434	0.9090	0.8735
1.0	3.6343E-01	1.0240	0.9211	0.8827	0.8609	1.0017	1.0003	0.9986	0.9997	0.9273	0.6408	0.9103	0.8750
1.2	4.1191E-01	1.0245	0.9065	0.8655	0.8429	1.0022	1.0004	0.9987	0.9997	0.9244	0.6364	0.9128	0.8777
1.4	4.4481E-01	1.0251	0.8991	0.8570	0.8341	1.0026	1.0005	0.9988	0.9998	0.9234	0.6328	0.9150	0.8801
1.6	4.6743E-01	1.0257	0.8954	0.8529	0.8298	1.0031	1.0006	0.9989	0.9998	0.9233	0.6297	0.9170	0.8824
1.8	4.8323E-01	1.0263	0.8936	0.8511	0.8281	1.0036	1.0006	0.9990	0.9998	0.9236	0.6270	0.9187	0.8844
2.0	4.9441E-01	1.0270	0.8930	0.8506	0.8278	1.0040	1.0007	0.9991	0.9998	0.9243	0.6247	0.9204	0.8862
2.2	5.0242E-01	1.0277	0.8931	0.8510	0.8284	1.0045	1.0008	0.9992	0.9998	0.9250	0.6225	0.9218	0.8879
2.4	5.0819E-01	1.0285	0.8937	0.8519	0.8294	1.0049	1.0009	0.9992	0.9998	0.9258	0.6205	0.9232	0.8895
2.6	5.1236E-01	1.0292	0.8944	0.8530	0.8307	1.0054	1.0010	0.9993	0.9999	0.9266	0.6186	0.9244	0.8909
2.8	5.1538E-01	1.0300	0.8953	0.8543	0.8322	1.0058	1.0011	0.9993	0.9999	0.9274	0.6168	0.9256	0.8923
3.0	5.1754E-01	1.0308	0.8963	0.8556	0.8338	1.0063	1.0012	0.9994	0.9999	0.9282	0.6150	0.9266	0.8935
3.2	5.1906E-01	1.0316	0.8974	0.8571	0.8354	1.0067	1.0012	0.9994	0.9999	0.9290	0.6133	0.9276	0.8947
3.4	5.2010E-01	1.0323	0.8984	0.8585	0.8370	1.0072	1.0013	0.9994	0.9999	0.9297	0.6116	0.9285	0.8959
3.6	5.2079E-01	1.0331	0.8995	0.8599	0.8386	1.0076	1.0014	0.9995	0.9999	0.9304	0.6099	0.9294	0.8969
3.8	5.2119E-01	1.0339	0.9005	0.8613	0.8401	1.0080	1.0015	0.9995	0.9999	0.9311	0.6082	0.9302	0.8979
4.0	5.2138E-01	1.0348	0.9015	0.8627	0.8417	1.0085	1.0016	0.9995	0.9999	0.9318	0.6065	0.9309	0.8989
4.5	5.2122E-01	1.0368	0.9039	0.8659	0.8453	1.0095	1.0018	0.9996	0.9999	0.9332	0.6024	0.9326	0.9011
5.0	5.2051E-01	1.0389	0.9061	0.8689	0.8486	1.0106	1.0020	0.9996	0.9999	0.9346	0.5982	0.9341	0.9031
5.5	5.1951E-01	1.0409	0.9082	0.8716	0.8516	1.0117	1.0022	0.9996	0.9999	0.9357	0.5940	0.9354	0.9049
6.0	5.1839E-01	1.0430	0.9100	0.8740	0.8544	1.0127	1.0024	0.9997	0.9999	0.9368	0.5898	0.9365	0.9066
6.5	5.1722E-01	1.0451	0.9116	0.8762	0.8569	1.0138	1.0026	0.9997	0.9999	0.9377	0.5856	0.9375	0.9081
7.0	5.1606E-01	1.0472	0.9131	0.8782	0.8591	1.0148	1.0028	0.9997	0.9999	0.9385	0.5814	0.9383	0.9095
7.5	5.1494E-01	1.0493	0.9145	0.8801	0.8612	1.0158	1.0030	0.9997	0.9999	0.9393	0.5771	0.9391	0.9108
8.0	5.1386E-01	1.0514	0.9157	0.8817	0.8630	1.0169	1.0032	0.9998	1.0000	0.9399	0.5728	0.9398	0.9121
9.0	5.1189E-01	1.0556	0.9177	0.8846	0.8663	1.0189	1.0036	0.9998	1.0000	0.9411	0.5641	0.9410	0.9144
10.0	5.1017E-01	1.0598	0.9194	0.8869	0.8689	1.0209	1.0040	0.9998	1.0000	0.9420	0.5555	0.9419	0.9164
11.0	5.0869E-01	1.0641	0.9208	0.8888	0.8711	1.0229	1.0044	0.9998	1.0000	0.9427	0.5467	0.9427	0.9182
12.0	5.0744E-01	1.0683	0.9218	0.8904	0.8729	1.0249	1.0048	0.9998	1.0000	0.9433	0.5379	0.9433	0.9199
13.0	5.0640E-01	1.0725	0.9227	0.8916	0.8744	1.0269	1.0052	0.9999	1.0000	0.9438	0.5291	0.9438	0.9215
14.0	5.0553E-01	1.0768	0.9233	0.8926	0.8756	1.0289	1.0056	0.9999	1.0000	0.9442	0.5202	0.9441	0.9229
16.0	5.0427E-01	1.0853	0.9241	0.8940	0.8772	1.0328	1.0064	0.9999	1.0000	0.9446	0.5024	0.9446	0.9256
18.0	5.0354E-01	1.0938	0.9243	0.8946	0.8780	1.0366	1.0072	0.9999	1.0000	0.9448	0.4847	0.9448	0.9279
20.0	5.0323E-01	1.1023	0.9241	0.8946	0.8783	1.0404	1.0080	0.9999	1.0000	0.9447	0.4669	0.9447	0.9300
25.0	5.0388E-01	1.1237	0.9223	0.8930	0.8768	1.0499	1.0100	0.9999	1.0000	0.9439	0.4225	0.9439	0.9346
30.0	5.0600E-01	1.1453	0.9191	0.8896	0.8734	1.0591	1.0120	0.9999	1.0000	0.9425	0.3781	0.9425	0.9385
35.0	5.0914E-01	1.1669	0.9151	0.8851	0.8687	1.0682	1.0141	0.9999	1.0000	0.9406	0.3343	0.9406	0.9418
40.0	5.1304E-01	1.1888	0.9105	0.8797	0.8631	1.0771	1.0161	1.0000	1.0000	0.9384	0.2908	0.9384	0.9449
45.0	5.1753E-01	1.2108	0.9056	0.8738	0.8568	1.0858	1.0181	1.0000	1.0000	0.9360	0.2477	0.9361	0.9476
50.0	5.2248E-01	1.2331	0.9004	0.8675	0.8501	1.0944	1.0202	1.0000	1.0000	0.9336	0.2052	0.9336	0.9503

 Instr. see p. 22

POSITRON Z= -40 A= 90

p	F_0L_0	L_0	λ_2	λ_3	λ_4	μ_1	μ_2	Λ_1	Λ_2	ν_{12}	$\hat{\nu}_{12}$	η_{12}	$\hat{\eta}_{12}$
0.1	3.3201E-07	1.0238	7.5157	21.4401	39.6945	1.0002	1.0000	0.9979	0.9996	2.8182	0.6648	0.8936	0.8554
0.2	1.4732E-03	1.0238	2.4980	3.5096	4.1430	1.0003	1.0000	0.9979	0.9996	1.3734	0.6629	0.8943	0.8562
0.3	1.9494E-02	1.0239	1.5710	1.7896	1.8929	1.0004	1.0000	0.9980	0.9996	1.1070	0.6599	0.8955	0.8575
0.4	6.3774E-02	1.0240	1.2489	1.3084	1.3289	1.0005	1.0001	0.9980	0.9996	1.0150	0.6565	0.8970	0.8591
0.5	1.2241E-01	1.0241	1.1018	1.1088	1.1052	1.0007	1.0001	0.9981	0.9996	0.9734	0.6528	0.8985	0.8608
0.6	1.8245E-01	1.0242	1.0235	1.0078	0.9946	1.0009	1.0001	0.9982	0.9996	0.9516	0.6492	0.9002	0.8626
0.7	2.3715E-01	1.0244	0.9776	0.9504	0.9325	1.0011	1.0002	0.9983	0.9997	0.9392	0.6459	0.9017	0.8643
0.8	2.8426E-01	1.0246	0.9488	0.9151	0.8947	1.0013	1.0002	0.9984	0.9997	0.9316	0.6428	0.9033	0.8659
0.9	3.2376E-01	1.0248	0.9300	0.8923	0.8704	1.0015	1.0003	0.9984	0.9997	0.9269	0.6400	0.9047	0.8675
1.0	3.5644E-01	1.0250	0.9171	0.8770	0.8543	1.0018	1.0003	0.9985	0.9997	0.9239	0.6374	0.9061	0.8691
1.2	4.0552E-01	1.0256	0.9019	0.8591	0.8355	1.0022	1.0004	0.9987	0.9997	0.9208	0.6331	0.9087	0.8719
1.4	4.3888E-01	1.0261	0.8942	0.8502	0.8263	1.0027	1.0005	0.9988	0.9998	0.9197	0.6295	0.9110	0.8744
1.6	4.6183E-01	1.0268	0.8903	0.8459	0.8220	1.0032	1.0006	0.9989	0.9998	0.9196	0.6265	0.9130	0.8767
1.8	4.7786E-01	1.0275	0.8885	0.8441	0.8202	1.0037	1.0007	0.9990	0.9998	0.9200	0.6239	0.9149	0.8788
2.0	4.8919E-01	1.0282	0.8879	0.8437	0.8199	1.0042	1.0008	0.9991	0.9998	0.9207	0.6216	0.9166	0.8808
2.2	4.9728E-01	1.0289	0.8880	0.8441	0.8205	1.0046	1.0008	0.9991	0.9998	0.9215	0.6195	0.9181	0.8825
2.4	5.0310E-01	1.0297	0.8886	0.8450	0.8216	1.0051	1.0009	0.9992	0.9998	0.9223	0.6176	0.9196	0.8842
2.6	5.0729E-01	1.0305	0.8894	0.8462	0.8230	1.0056	1.0010	0.9993	0.9999	0.9232	0.6158	0.9209	0.8857
2.8	5.1031E-01	1.0312	0.8903	0.8475	0.8245	1.0060	1.0011	0.9993	0.9999	0.9240	0.6140	0.9221	0.8871
3.0	5.1246E-01	1.0320	0.8914	0.8490	0.8262	1.0065	1.0012	0.9993	0.9999	0.9249	0.6123	0.9232	0.8884
3.2	5.1396E-01	1.0329	0.8925	0.8505	0.8279	1.0070	1.0013	0.9994	0.9999	0.9257	0.6106	0.9242	0.8897
3.4	5.1497E-01	1.0337	0.8936	0.8520	0.8295	1.0074	1.0014	0.9994	0.9999	0.9264	0.6090	0.9252	0.8909
3.6	5.1562E-01	1.0345	0.8947	0.8535	0.8312	1.0079	1.0014	0.9994	0.9999	0.9272	0.6074	0.9261	0.8920
3.8	5.1598E-01	1.0353	0.8958	0.8549	0.8329	1.0083	1.0015	0.9995	0.9999	0.9279	0.6058	0.9269	0.8930
4.0	5.1613E-01	1.0362	0.8969	0.8564	0.8345	1.0088	1.0016	0.9995	0.9999	0.9286	0.6042	0.9277	0.8941
4.5	5.1586E-01	1.0383	0.8995	0.8598	0.8383	1.0099	1.0018	0.9996	0.9999	0.9301	0.6002	0.9295	0.8964
5.0	5.1504E-01	1.0404	0.9018	0.8629	0.8418	1.0110	1.0020	0.9996	0.9999	0.9315	0.5962	0.9310	0.8985
5.5	5.1394E-01	1.0426	0.9039	0.8658	0.8450	1.0121	1.0023	0.9996	0.9999	0.9328	0.5921	0.9324	0.9004
6.0	5.1271E-01	1.0447	0.9059	0.8684	0.8479	1.0132	1.0025	0.9997	0.9999	0.9339	0.5881	0.9336	0.9021
6.5	5.1145E-01	1.0469	0.9076	0.8707	0.8505	1.0142	1.0027	0.9997	0.9999	0.9349	0.5840	0.9346	0.9037
7.0	5.1020E-01	1.0491	0.9092	0.8728	0.8529	1.0153	1.0029	0.9997	0.9999	0.9358	0.5799	0.9355	0.9052
7.5	5.0899E-01	1.0512	0.9106	0.8748	0.8551	1.0164	1.0031	0.9997	0.9999	0.9365	0.5757	0.9364	0.9066
8.0	5.0784E-01	1.0534	0.9119	0.8765	0.8571	1.0174	1.0033	0.9998	0.9999	0.9373	0.5716	0.9371	0.9079
9.0	5.0572E-01	1.0578	0.9142	0.8796	0.8605	1.0195	1.0037	0.9998	1.0000	0.9385	0.5632	0.9384	0.9102
10.0	5.0387E-01	1.0621	0.9160	0.8821	0.8633	1.0216	1.0041	0.9998	1.0000	0.9395	0.5547	0.9394	0.9124
11.0	5.0229E-01	1.0665	0.9174	0.8841	0.8657	1.0237	1.0045	0.9998	1.0000	0.9403	0.5462	0.9402	0.9143
12.0	5.0094E-01	1.0709	0.9186	0.8858	0.8676	1.0258	1.0050	0.9998	1.0000	0.9409	0.5376	0.9409	0.9161
13.0	4.9980E-01	1.0753	0.9195	0.8872	0.8692	1.0278	1.0054	0.9998	1.0000	0.9414	0.5291	0.9414	0.9177
14.0	4.9886E-01	1.0797	0.9202	0.8883	0.8705	1.0298	1.0058	0.9999	1.0000	0.9418	0.5204	0.9418	0.9193
16.0	4.9746E-01	1.0884	0.9211	0.8898	0.8723	1.0339	1.0066	0.9999	1.0000	0.9423	0.5031	0.9423	0.9220
18.0	4.9660E-01	1.0972	0.9215	0.8906	0.8733	1.0378	1.0074	0.9999	1.0000	0.9426	0.4858	0.9426	0.9245
20.0	4.9620E-01	1.1061	0.9213	0.8907	0.8737	1.0418	1.0083	0.9999	1.0000	0.9426	0.4684	0.9426	0.9267
25.0	4.9665E-01	1.1282	0.9196	0.8893	0.8725	1.0515	1.0103	0.9999	1.0000	0.9418	0.4250	0.9418	0.9315
30.0	4.9863E-01	1.1505	0.9166	0.8860	0.8692	1.0610	1.0124	0.9999	1.0000	0.9404	0.3818	0.9404	0.9355
35.0	5.0167E-01	1.1729	0.9126	0.8815	0.8646	1.0704	1.0145	0.9999	1.0000	0.9385	0.3390	0.9386	0.9390
40.0	5.0550E-01	1.1956	0.9080	0.8762	0.8590	1.0796	1.0166	1.0000	1.0000	0.9364	0.2963	0.9364	0.9422
45.0	5.0994E-01	1.2184	0.9030	0.8703	0.8527	1.0886	1.0187	1.0000	1.0000	0.9340	0.2542	0.9340	0.9451
50.0	5.1486E-01	1.2414	0.8978	0.8639	0.8460	1.0974	1.0208	1.0000	1.0000	0.9315	0.2127	0.9315	0.9479

POSITRON Z= -41 A= 91

p	F_0L_0	L_0	λ_2	λ_3	λ_4	μ_1	μ_2	Λ_1	Λ_2	ν_{12}	$\hat{\nu}_{12}$	η_{12}	$\hat{\eta}_{12}$
0.1	2.2078E-07	1.0248	7.7690	22.8241	43.1920	1.0002	1.0000	0.9978	0.9996	2.9020	0.6612	0.8890	0.8491
0.2	1.2293E-03	1.0249	2.5558	3.6386	4.3268	1.0003	1.0000	0.9979	0.9996	1.3908	0.6593	0.8897	0.8499
0.3	1.7514E-02	1.0249	1.5927	1.8245	1.9348	1.0004	1.0000	0.9979	0.9996	1.1121	0.6563	0.8909	0.8512
0.4	5.9361E-02	1.0250	1.2580	1.3206	1.3422	1.0006	1.0001	0.9980	0.9996	1.0158	0.6529	0.8924	0.8528
0.5	1.1623E-01	1.0251	1.1051	1.1123	1.1085	1.0007	1.0001	0.9981	0.9996	0.9723	0.6493	0.8940	0.8545
0.6	1.7538E-01	1.0253	1.0237	1.0072	0.9933	1.0009	1.0001	0.9981	0.9996	0.9495	0.6457	0.8956	0.8563
0.7	2.2981E-01	1.0254	0.9759	0.9474	0.9287	1.0011	1.0002	0.9982	0.9997	0.9364	0.6424	0.8973	0.8581
0.8	2.7700E-01	1.0256	0.9460	0.9107	0.8894	1.0014	1.0002	0.9983	0.9997	0.9286	0.6393	0.8988	0.8598
0.9	3.1673E-01	1.0259	0.9264	0.8870	0.8642	1.0016	1.0003	0.9984	0.9997	0.9236	0.6365	0.9003	0.8615
1.0	3.4972E-01	1.0261	0.9130	0.8711	0.8474	1.0018	1.0003	0.9985	0.9997	0.9204	0.6340	0.9018	0.8631
1.2	3.9937E-01	1.0266	0.8972	0.8525	0.8280	1.0023	1.0004	0.9986	0.9997	0.9172	0.6297	0.9045	0.8660
1.4	4.3319E-01	1.0272	0.8892	0.8434	0.8185	1.0028	1.0005	0.9987	0.9998	0.9160	0.6262	0.9069	0.8686
1.6	4.5648E-01	1.0279	0.8852	0.8389	0.8140	1.0033	1.0006	0.9989	0.9998	0.9159	0.6233	0.9090	0.8710
1.8	4.7272E-01	1.0286	0.8833	0.8370	0.8122	1.0038	1.0007	0.9990	0.9998	0.9164	0.6207	0.9110	0.8732
2.0	4.8420E-01	1.0293	0.8827	0.8366	0.8119	1.0043	1.0008	0.9990	0.9998	0.9170	0.6185	0.9127	0.8752
2.2	4.9238E-01	1.0301	0.8828	0.8370	0.8125	1.0048	1.0009	0.9991	0.9998	0.9179	0.6165	0.9144	0.8771
2.4	4.9825E-01	1.0309	0.8834	0.8380	0.8136	1.0053	1.0010	0.9992	0.9998	0.9187	0.6146	0.9159	0.8788
2.6	5.0246E-01	1.0317	0.8843	0.8392	0.8151	1.0058	1.0010	0.9992	0.9998	0.9196	0.6129	0.9172	0.8804
2.8	5.0547E-01	1.0325	0.8853	0.8407	0.8167	1.0063	1.0011	0.9993	0.9999	0.9205	0.6112	0.9185	0.8819
3.0	5.0761E-01	1.0333	0.8864	0.8422	0.8185	1.0067	1.0012	0.9993	0.9999	0.9214	0.6096	0.9196	0.8832
3.2	5.0908E-01	1.0342	0.8876	0.8438	0.8202	1.0072	1.0013	0.9994	0.9999	0.9223	0.6080	0.9207	0.8845
3.4	5.1006E-01	1.0350	0.8887	0.8453	0.8220	1.0077	1.0014	0.9994	0.9999	0.9231	0.6064	0.9217	0.8858
3.6	5.1067E-01	1.0359	0.8899	0.8469	0.8238	1.0082	1.0015	0.9994	0.9999	0.9239	0.6048	0.9227	0.8869
3.8	5.1099E-01	1.0368	0.8911	0.8485	0.8255	1.0086	1.0016	0.9995	0.9999	0.9246	0.6033	0.9236	0.8881
4.0	5.1110E-01	1.0376	0.8922	0.8500	0.8272	1.0091	1.0017	0.9995	0.9999	0.9253	0.6017	0.9244	0.8891
4.5	5.1072E-01	1.0398	0.8949	0.8536	0.8312	1.0102	1.0019	0.9995	0.9999	0.9270	0.5979	0.9263	0.8915
5.0	5.0978E-01	1.0420	0.8974	0.8569	0.8349	1.0114	1.0021	0.9996	0.9999	0.9285	0.5940	0.9279	0.8937
5.5	5.0857E-01	1.0442	0.8996	0.8599	0.8382	1.0125	1.0023	0.9996	0.9999	0.9298	0.5902	0.9293	0.8957
6.0	5.0725E-01	1.0465	0.9017	0.8626	0.8413	1.0136	1.0025	0.9997	0.9999	0.9309	0.5862	0.9306	0.8975
6.5	5.0589E-01	1.0487	0.9035	0.8651	0.8441	1.0147	1.0028	0.9997	0.9999	0.9320	0.5823	0.9317	0.8992
7.0	5.0454E-01	1.0509	0.9052	0.8673	0.8466	1.0158	1.0030	0.9997	0.9999	0.9329	0.5783	0.9327	0.9008
7.5	5.0325E-01	1.0532	0.9067	0.8694	0.8489	1.0169	1.0032	0.9997	0.9999	0.9338	0.5743	0.9336	0.9022
8.0	5.0201E-01	1.0554	0.9081	0.8712	0.8510	1.0180	1.0034	0.9997	0.9999	0.9345	0.5703	0.9344	0.9036
9.0	4.9975E-01	1.0599	0.9105	0.8745	0.8546	1.0202	1.0038	0.9998	1.0000	0.9358	0.5621	0.9357	0.9061
10.0	4.9778E-01	1.0645	0.9124	0.8772	0.8577	1.0224	1.0043	0.9998	1.0000	0.9369	0.5539	0.9368	0.9083
11.0	4.9608E-01	1.0690	0.9140	0.8794	0.8602	1.0245	1.0047	0.9998	1.0000	0.9377	0.5456	0.9377	0.9103
12.0	4.9463E-01	1.0735	0.9153	0.8812	0.8623	1.0266	1.0051	0.9998	1.0000	0.9384	0.5373	0.9384	0.9122
13.0	4.9340E-01	1.0780	0.9163	0.8827	0.8640	1.0287	1.0055	0.9998	1.0000	0.9390	0.5289	0.9390	0.9139
14.0	4.9237E-01	1.0826	0.9171	0.8839	0.8654	1.0308	1.0060	0.9999	1.0000	0.9394	0.5205	0.9394	0.9155
16.0	4.9083E-01	1.0917	0.9181	0.8856	0.8674	1.0350	1.0068	0.9999	1.0000	0.9400	0.5036	0.9400	0.9184
18.0	4.8985E-01	1.1008	0.9186	0.8865	0.8686	1.0391	1.0077	0.9999	1.0000	0.9403	0.4867	0.9403	0.9210
20.0	4.8934E-01	1.1099	0.9185	0.8868	0.8691	1.0432	1.0085	0.9999	1.0000	0.9404	0.4698	0.9404	0.9233
25.0	4.8960E-01	1.1328	0.9170	0.8856	0.8681	1.0532	1.0106	0.9999	1.0000	0.9397	0.4274	0.9397	0.9283
30.0	4.9144E-01	1.1558	0.9140	0.8824	0.8650	1.0630	1.0128	0.9999	1.0000	0.9383	0.3852	0.9384	0.9325
35.0	4.9438E-01	1.1791	0.9101	0.8780	0.8604	1.0726	1.0149	0.9999	1.0000	0.9365	0.3432	0.9365	0.9362
40.0	4.9813E-01	1.2025	0.9055	0.8727	0.8549	1.0821	1.0171	1.0000	1.0000	0.9343	0.3016	0.9344	0.9395
45.0	5.0252E-01	1.2261	0.9005	0.8667	0.8486	1.0914	1.0192	1.0000	1.0000	0.9320	0.2604	0.9320	0.9426
50.0	5.0740E-01	1.2500	0.8952	0.8603	0.8419	1.1005	1.0214	1.0000	1.0000	0.9295	0.2197	0.9295	0.9455

Instr. see p. 22

POSITRON Z= -42 A= 94

p	F_0L_0	L_0	λ_2	λ_3	λ_4	μ_1	μ_2	Λ_1	Λ_2	ν_{12}	$\hat{\nu}_{12}$	η_{12}	$\hat{\eta}_{12}$
0.1	1.4672E-07	1.0259	8.0245	24.2711	46.9465	1.0003	1.0000	0.9977	0.9996	2.9876	0.6576	0.8844	0.8428
0.2	1.0251E-03	1.0259	2.6144	3.7718	4.5189	1.0003	1.0000	0.9978	0.9996	1.4087	0.6557	0.8851	0.8435
0.3	1.5726E-02	1.0260	1.6148	1.8604	1.9783	1.0004	1.0000	0.9978	0.9996	1.1175	0.6528	0.8863	0.8448
0.4	5.5224E-02	1.0261	1.2673	1.3333	1.3562	1.0006	1.0001	0.9979	0.9996	1.0168	0.6494	0.8878	0.8465
0.5	1.1032E-01	1.0262	1.1086	1.1162	1.1122	1.0008	1.0001	0.9980	0.9996	0.9713	0.6458	0.8895	0.8483
0.6	1.6852E-01	1.0264	1.0241	1.0068	0.9923	1.0010	1.0001	0.9981	0.9996	0.9475	0.6422	0.8912	0.8501
0.7	2.2262E-01	1.0265	0.9745	0.9447	0.9252	1.0012	1.0002	0.9982	0.9996	0.9338	0.6389	0.8929	0.8520
0.8	2.6984E-01	1.0268	0.9434	0.9066	0.8845	1.0014	1.0002	0.9982	0.9997	0.9255	0.6358	0.8945	0.8537
0.9	3.0977E-01	1.0270	0.9230	0.8820	0.8583	1.0017	1.0003	0.9983	0.9997	0.9204	0.6331	0.8960	0.8554
1.0	3.4302E-01	1.0272	0.9091	0.8655	0.8409	1.0019	1.0003	0.9984	0.9997	0.9171	0.6306	0.8975	0.8571
1.2	3.9321E-01	1.0278	0.8927	0.8462	0.8207	1.0024	1.0004	0.9986	0.9997	0.9136	0.6263	0.9003	0.8601
1.4	4.2747E-01	1.0284	0.8844	0.8367	0.8109	1.0030	1.0005	0.9987	0.9997	0.9124	0.6229	0.9028	0.8629
1.6	4.5107E-01	1.0291	0.8802	0.8321	0.8063	1.0035	1.0006	0.9988	0.9998	0.9123	0.6200	0.9051	0.8654
1.8	4.6754E-01	1.0298	0.8782	0.8302	0.8044	1.0040	1.0007	0.9989	0.9998	0.9127	0.6175	0.9071	0.8677
2.0	4.7915E-01	1.0306	0.8776	0.8297	0.8041	1.0045	1.0008	0.9990	0.9998	0.9135	0.6154	0.9090	0.8697
2.2	4.8742E-01	1.0314	0.8778	0.8302	0.8048	1.0050	1.0009	0.9991	0.9998	0.9143	0.6134	0.9107	0.8717
2.4	4.9333E-01	1.0322	0.8784	0.8312	0.8060	1.0055	1.0010	0.9991	0.9998	0.9152	0.6116	0.9122	0.8734
2.6	4.9756E-01	1.0331	0.8793	0.8325	0.8075	1.0060	1.0011	0.9992	0.9998	0.9162	0.6099	0.9136	0.8751
2.8	5.0057E-01	1.0339	0.8804	0.8340	0.8092	1.0065	1.0012	0.9993	0.9999	0.9171	0.6083	0.9150	0.8766
3.0	5.0269E-01	1.0348	0.8815	0.8356	0.8110	1.0070	1.0013	0.9993	0.9999	0.9180	0.6067	0.9162	0.8781
3.2	5.0414E-01	1.0357	0.8828	0.8373	0.8129	1.0075	1.0014	0.9993	0.9999	0.9189	0.6051	0.9173	0.8795
3.4	5.0508E-01	1.0365	0.8840	0.8389	0.8147	1.0080	1.0015	0.9994	0.9999	0.9198	0.6036	0.9184	0.8807
3.6	5.0566E-01	1.0374	0.8852	0.8406	0.8166	1.0085	1.0015	0.9994	0.9999	0.9206	0.6021	0.9194	0.8820
3.8	5.0594E-01	1.0383	0.8865	0.8422	0.8184	1.0090	1.0016	0.9994	0.9999	0.9214	0.6006	0.9203	0.8831
4.0	5.0600E-01	1.0392	0.8876	0.8438	0.8201	1.0095	1.0017	0.9995	0.9999	0.9222	0.5991	0.9212	0.8842
4.5	5.0551E-01	1.0415	0.8905	0.8475	0.8243	1.0107	1.0020	0.9995	0.9999	0.9239	0.5954	0.9232	0.8868
5.0	5.0447E-01	1.0438	0.8931	0.8510	0.8282	1.0118	1.0022	0.9996	0.9999	0.9254	0.5916	0.9249	0.8890
5.5	5.0316E-01	1.0461	0.8955	0.8541	0.8317	1.0130	1.0024	0.9996	0.9999	0.9268	0.5878	0.9264	0.8911
6.0	5.0173E-01	1.0484	0.8976	0.8570	0.8349	1.0142	1.0026	0.9996	0.9999	0.9280	0.5840	0.9277	0.8930
6.5	5.0028E-01	1.0508	0.8996	0.8596	0.8378	1.0153	1.0029	0.9997	0.9999	0.9292	0.5802	0.9289	0.8948
7.0	4.9886E-01	1.0531	0.9014	0.8620	0.8405	1.0165	1.0031	0.9997	0.9999	0.9301	0.5763	0.9299	0.8964
7.5	4.9748E-01	1.0554	0.9030	0.8641	0.8429	1.0176	1.0033	0.9997	0.9999	0.9310	0.5724	0.9308	0.8979
8.0	4.9617E-01	1.0578	0.9044	0.8661	0.8451	1.0188	1.0035	0.9997	0.9999	0.9318	0.5684	0.9317	0.8993
9.0	4.9378E-01	1.0625	0.9069	0.8695	0.8489	1.0210	1.0040	0.9998	1.0000	0.9332	0.5605	0.9331	0.9019
10.0	4.9169E-01	1.0672	0.9090	0.8723	0.8521	1.0233	1.0044	0.9998	1.0000	0.9343	0.5524	0.9343	0.9043
11.0	4.8989E-01	1.0719	0.9107	0.8747	0.8548	1.0255	1.0049	0.9998	1.0000	0.9353	0.5442	0.9352	0.9064
12.0	4.8835E-01	1.0766	0.9120	0.8766	0.8570	1.0277	1.0053	0.9998	1.0000	0.9360	0.5361	0.9360	0.9083
13.0	4.8704E-01	1.0813	0.9131	0.8782	0.8589	1.0299	1.0057	0.9998	1.0000	0.9366	0.5279	0.9366	0.9101
14.0	4.8594E-01	1.0860	0.9140	0.8795	0.8604	1.0321	1.0062	0.9999	1.0000	0.9371	0.5196	0.9371	0.9118
16.0	4.8428E-01	1.0955	0.9151	0.8814	0.8625	1.0364	1.0071	0.9999	1.0000	0.9377	0.5030	0.9377	0.9148
18.0	4.8322E-01	1.1049	0.9157	0.8824	0.8638	1.0406	1.0080	0.9999	1.0000	0.9381	0.4864	0.9381	0.9175
20.0	4.8263E-01	1.1144	0.9157	0.8828	0.8644	1.0449	1.0088	0.9999	1.0000	0.9381	0.4697	0.9382	0.9200
25.0	4.8276E-01	1.1383	0.9142	0.8817	0.8636	1.0552	1.0110	0.9999	1.0000	0.9375	0.4281	0.9376	0.9252
30.0	4.8451E-01	1.1623	0.9113	0.8786	0.8606	1.0654	1.0132	0.9999	1.0000	0.9362	0.3865	0.9362	0.9296
35.0	4.8741E-01	1.1865	0.9073	0.8741	0.8560	1.0754	1.0155	0.9999	1.0000	0.9343	0.3451	0.9343	0.9335
40.0	4.9116E-01	1.2109	0.9027	0.8688	0.8504	1.0852	1.0177	1.0000	1.0000	0.9322	0.3041	0.9322	0.9370
45.0	4.9555E-01	1.2356	0.8976	0.8627	0.8441	1.0948	1.0199	1.0000	1.0000	0.9298	0.2636	0.9298	0.9402
50.0	5.0046E-01	1.2605	0.8922	0.8563	0.8372	1.1042	1.0222	1.0000	1.0000	0.9272	0.2234	0.9272	0.9432

Instr. see p. 22

POSITRON Z= -43 A= 97

p	F_0L_0	L_0	λ_2	λ_3	λ_4	μ_1	μ_2	Λ_1	Λ_2	ν_{12}	$\hat{\nu}_{12}$	η_{12}	$\hat{\eta}_{12}$
0.1	9.7507E-08	1.0270	8.2797	25.7724	50.9496	1.0003	1.0000	0.9977	0.9996	3.0745	0.6540	0.8798	0.8363
0.2	8.5476E-04	1.0270	2.6729	3.9078	4.7174	1.0004	1.0000	0.9977	0.9996	1.4269	0.6521	0.8805	0.8371
0.3	1.4120E-02	1.0271	1.6368	1.8967	2.0224	1.0005	1.0001	0.9978	0.9996	1.1230	0.6492	0.8817	0.8384
0.4	5.1379E-02	1.0272	1.2766	1.3460	1.3702	1.0006	1.0001	0.9978	0.9996	1.0179	0.6458	0.8832	0.8401
0.5	1.0471E-01	1.0273	1.1120	1.1200	1.1158	1.0008	1.0001	0.9979	0.9996	0.9703	0.6422	0.8849	0.8419
0.6	1.6196E-01	1.0275	1.0243	1.0063	0.9912	1.0010	1.0002	0.9980	0.9996	0.9454	0.6387	0.8867	0.8438
0.7	2.1571E-01	1.0277	0.9729	0.9419	0.9216	1.0013	1.0002	0.9981	0.9996	0.9311	0.6354	0.8884	0.8457
0.8	2.6293E-01	1.0279	0.9407	0.9023	0.8793	1.0015	1.0002	0.9982	0.9996	0.9225	0.6323	0.8901	0.8475
0.9	3.0305E-01	1.0281	0.9195	0.8768	0.8523	1.0017	1.0003	0.9983	0.9997	0.9171	0.6296	0.8917	0.8493
1.0	3.3655E-01	1.0284	0.9051	0.8597	0.8342	1.0020	1.0003	0.9983	0.9997	0.9136	0.6271	0.8932	0.8510
1.2	3.8728E-01	1.0290	0.8880	0.8398	0.8134	1.0025	1.0004	0.9985	0.9997	0.9100	0.6229	0.8961	0.8542
1.4	4.2197E-01	1.0296	0.8794	0.8300	0.8032	1.0031	1.0005	0.9986	0.9997	0.9087	0.6195	0.8987	0.8570
1.6	4.4589E-01	1.0303	0.8751	0.8252	0.7984	1.0036	1.0006	0.9988	0.9998	0.9086	0.6167	0.9011	0.8596
1.8	4.6256E-01	1.0311	0.8731	0.8232	0.7965	1.0042	1.0007	0.9989	0.9998	0.9091	0.6143	0.9032	0.8620
2.0	4.7431E-01	1.0319	0.8725	0.8228	0.7962	1.0047	1.0008	0.9990	0.9998	0.9098	0.6122	0.9051	0.8642
2.2	4.8266E-01	1.0327	0.8726	0.8233	0.7969	1.0052	1.0009	0.9990	0.9998	0.9107	0.6103	0.9069	0.8662
2.4	4.8862E-01	1.0336	0.8733	0.8243	0.7982	1.0058	1.0010	0.9991	0.9998	0.9117	0.6085	0.9085	0.8680
2.6	4.9287E-01	1.0345	0.8743	0.8257	0.7998	1.0063	1.0011	0.9992	0.9998	0.9127	0.6069	0.9100	0.8697
2.8	4.9588E-01	1.0353	0.8754	0.8273	0.8016	1.0068	1.0012	0.9992	0.9998	0.9136	0.6053	0.9114	0.8714
3.0	4.9797E-01	1.0362	0.8766	0.8289	0.8034	1.0073	1.0013	0.9993	0.9999	0.9146	0.6038	0.9127	0.8729
3.2	4.9939E-01	1.0372	0.8779	0.8307	0.8054	1.0078	1.0014	0.9993	0.9999	0.9155	0.6023	0.9139	0.8743
3.4	5.0031E-01	1.0381	0.8792	0.8324	0.8073	1.0083	1.0015	0.9994	0.9999	0.9164	0.6008	0.9150	0.8756
3.6	5.0084E-01	1.0390	0.8805	0.8341	0.8092	1.0089	1.0016	0.9994	0.9999	0.9173	0.5994	0.9160	0.8769
3.8	5.0108E-01	1.0399	0.8818	0.8358	0.8111	1.0094	1.0017	0.9994	0.9999	0.9181	0.5979	0.9170	0.8781
4.0	5.0111E-01	1.0409	0.8830	0.8375	0.8130	1.0099	1.0018	0.9994	0.9999	0.9189	0.5964	0.9179	0.8793
4.5	5.0050E-01	1.0432	0.8860	0.8414	0.8174	1.0111	1.0020	0.9995	0.9999	0.9207	0.5928	0.9200	0.8819
5.0	4.9935E-01	1.0456	0.8887	0.8450	0.8214	1.0123	1.0023	0.9996	0.9999	0.9224	0.5892	0.9218	0.8843
5.5	4.9793E-01	1.0480	0.8912	0.8483	0.8251	1.0136	1.0025	0.9996	0.9999	0.9238	0.5855	0.9233	0.8865
6.0	4.9641E-01	1.0504	0.8935	0.8513	0.8284	1.0148	1.0027	0.9996	0.9999	0.9251	0.5818	0.9247	0.8884
6.5	4.9487E-01	1.0529	0.8956	0.8541	0.8315	1.0160	1.0030	0.9997	0.9999	0.9263	0.5780	0.9260	0.8903
7.0	4.9335E-01	1.0553	0.8974	0.8566	0.8342	1.0172	1.0032	0.9997	0.9999	0.9273	0.5742	0.9271	0.8920
7.5	4.9190E-01	1.0577	0.8991	0.8588	0.8368	1.0184	1.0034	0.9997	0.9999	0.9283	0.5704	0.9280	0.8936
8.0	4.9051E-01	1.0602	0.9007	0.8609	0.8391	1.0195	1.0037	0.9997	0.9999	0.9291	0.5666	0.9289	0.8950
9.0	4.8799E-01	1.0650	0.9033	0.8645	0.8431	1.0219	1.0041	0.9998	1.0000	0.9306	0.5587	0.9304	0.8978
10.0	4.8578E-01	1.0699	0.9055	0.8675	0.8465	1.0242	1.0046	0.9998	1.0000	0.9318	0.5508	0.9317	0.9002
11.0	4.8388E-01	1.0748	0.9073	0.8700	0.8493	1.0265	1.0050	0.9998	1.0000	0.9327	0.5429	0.9327	0.9024
12.0	4.8225E-01	1.0797	0.9088	0.8720	0.8517	1.0288	1.0055	0.9998	1.0000	0.9335	0.5349	0.9335	0.9044
13.0	4.8086E-01	1.0846	0.9099	0.8737	0.8536	1.0311	1.0059	0.9998	1.0000	0.9342	0.5268	0.9341	0.9063
14.0	4.7969E-01	1.0895	0.9109	0.8751	0.8553	1.0333	1.0064	0.9998	1.0000	0.9347	0.5187	0.9347	0.9081
16.0	4.7791E-01	1.0994	0.9121	0.8771	0.8576	1.0378	1.0073	0.9999	1.0000	0.9354	0.5024	0.9354	0.9112
18.0	4.7674E-01	1.1092	0.9127	0.8783	0.8591	1.0422	1.0082	0.9999	1.0000	0.9358	0.4860	0.9358	0.9141
20.0	4.7608E-01	1.1191	0.9128	0.8787	0.8597	1.0466	1.0091	0.9999	1.0000	0.9359	0.4696	0.9359	0.9166
25.0	4.7606E-01	1.1439	0.9115	0.8778	0.8591	1.0573	1.0114	0.9999	1.0000	0.9354	0.4286	0.9354	0.9221
30.0	4.7774E-01	1.1689	0.9085	0.8748	0.8561	1.0679	1.0137	0.9999	1.0000	0.9340	0.3877	0.9340	0.9267
35.0	4.8059E-01	1.1941	0.9045	0.8703	0.8516	1.0782	1.0160	0.9999	1.0000	0.9322	0.3469	0.9322	0.9308
40.0	4.8432E-01	1.2195	0.8998	0.8649	0.8459	1.0883	1.0183	0.9999	1.0000	0.9300	0.3065	0.9300	0.9344
45.0	4.8872E-01	1.2452	0.8947	0.8588	0.8395	1.0983	1.0206	1.0000	1.0000	0.9275	0.2665	0.9276	0.9378
50.0	4.9365E-01	1.2712	0.8892	0.8522	0.8326	1.1081	1.0230	1.0000	1.0000	0.9250	0.2270	0.9250	0.9409

Instr. see p. 22

POSITRON Z= -44 A= 98

p	F_0L_0	L_0	λ_2	λ_3	λ_4	μ_1	μ_2	Λ_1	Λ_2	ν_{12}	$\hat{\nu}_{12}$	η_{12}	$\hat{\eta}_{12}$
0.1	6.4844E-08	1.0281	8.5315	27.3149	55.1833	1.0003	1.0000	0.9976	0.9995	3.1623	0.6501	0.8749	0.8297
0.2	7.1324E-04	1.0281	2.7302	4.0447	4.9199	1.0004	1.0000	0.9976	0.9995	1.4451	0.6483	0.8757	0.8305
0.3	1.2688E-02	1.0282	1.6582	1.9324	2.0662	1.0005	1.0001	0.9977	0.9996	1.1284	0.6455	0.8769	0.8318
0.4	4.7838E-02	1.0283	1.2854	1.3580	1.3835	1.0007	1.0001	0.9978	0.9996	1.0188	0.6421	0.8784	0.8335
0.5	9.9478E-02	1.0284	1.1150	1.1232	1.1188	1.0008	1.0001	0.9978	0.9996	0.9692	0.6385	0.8801	0.8354
0.6	1.5580E-01	1.0286	1.0242	1.0053	0.9895	1.0011	1.0002	0.9979	0.9996	0.9432	0.6350	0.8819	0.8373
0.7	2.0920E-01	1.0288	0.9709	0.9385	0.9174	1.0013	1.0002	0.9980	0.9996	0.9283	0.6318	0.8837	0.8392
0.8	2.5644E-01	1.0290	0.9375	0.8976	0.8737	1.0016	1.0002	0.9981	0.9996	0.9192	0.6287	0.8854	0.8411
0.9	2.9676E-01	1.0292	0.9156	0.8712	0.8457	1.0018	1.0003	0.9982	0.9997	0.9136	0.6260	0.8871	0.8429
1.0	3.3053E-01	1.0295	0.9007	0.8535	0.8270	1.0021	1.0003	0.9983	0.9997	0.9099	0.6236	0.8887	0.8447
1.2	3.8181E-01	1.0301	0.8830	0.8329	0.8055	1.0026	1.0004	0.9985	0.9997	0.9062	0.6194	0.8917	0.8480
1.4	4.1696E-01	1.0308	0.8741	0.8227	0.7950	1.0032	1.0005	0.9986	0.9997	0.9049	0.6161	0.8944	0.8509
1.6	4.4120E-01	1.0315	0.8696	0.8178	0.7901	1.0038	1.0007	0.9987	0.9998	0.9047	0.6133	0.8968	0.8536
1.8	4.5809E-01	1.0323	0.8676	0.8158	0.7881	1.0043	1.0008	0.9988	0.9998	0.9052	0.6110	0.8990	0.8561
2.0	4.6999E-01	1.0331	0.8670	0.8154	0.7879	1.0049	1.0009	0.9989	0.9998	0.9060	0.6090	0.9011	0.8583
2.2	4.7842E-01	1.0340	0.8672	0.8159	0.7886	1.0054	1.0010	0.9990	0.9998	0.9069	0.6071	0.9029	0.8604
2.4	4.8442E-01	1.0349	0.8679	0.8170	0.7899	1.0060	1.0011	0.9991	0.9998	0.9079	0.6055	0.9046	0.8624
2.6	4.8868E-01	1.0358	0.8689	0.8184	0.7916	1.0065	1.0012	0.9991	0.9998	0.9090	0.6039	0.9062	0.8641
2.8	4.9168E-01	1.0367	0.8701	0.8201	0.7935	1.0070	1.0013	0.9992	0.9998	0.9100	0.6024	0.9076	0.8658
3.0	4.9376E-01	1.0376	0.8713	0.8219	0.7954	1.0076	1.0014	0.9993	0.9999	0.9110	0.6009	0.9090	0.8674
3.2	4.9515E-01	1.0385	0.8727	0.8237	0.7975	1.0081	1.0015	0.9993	0.9999	0.9120	0.5995	0.9102	0.8689
3.4	4.9603E-01	1.0395	0.8740	0.8255	0.7995	1.0086	1.0016	0.9993	0.9999	0.9129	0.5981	0.9114	0.8703
3.6	4.9652E-01	1.0405	0.8754	0.8273	0.8015	1.0092	1.0017	0.9994	0.9999	0.9138	0.5967	0.9125	0.8716
3.8	4.9671E-01	1.0414	0.8768	0.8290	0.8035	1.0097	1.0017	0.9994	0.9999	0.9147	0.5953	0.9135	0.8729
4.0	4.9669E-01	1.0424	0.8781	0.8308	0.8054	1.0102	1.0018	0.9994	0.9999	0.9155	0.5939	0.9144	0.8741
4.5	4.9596E-01	1.0448	0.8812	0.8349	0.8100	1.0115	1.0021	0.9995	0.9999	0.9174	0.5904	0.9166	0.8768
5.0	4.9469E-01	1.0473	0.8841	0.8387	0.8142	1.0128	1.0023	0.9995	0.9999	0.9191	0.5869	0.9185	0.8793
5.5	4.9315E-01	1.0498	0.8867	0.8422	0.8180	1.0140	1.0026	0.9996	0.9999	0.9207	0.5834	0.9202	0.8816
6.0	4.9152E-01	1.0523	0.8891	0.8453	0.8215	1.0153	1.0028	0.9996	0.9999	0.9220	0.5798	0.9216	0.8836
6.5	4.8987E-01	1.0548	0.8913	0.8482	0.8247	1.0165	1.0030	0.9996	0.9999	0.9232	0.5762	0.9229	0.8855
7.0	4.8826E-01	1.0573	0.8933	0.8508	0.8277	1.0177	1.0033	0.9997	0.9999	0.9243	0.5725	0.9241	0.8873
7.5	4.8672E-01	1.0598	0.8950	0.8532	0.8303	1.0190	1.0035	0.9997	0.9999	0.9253	0.5688	0.9251	0.8890
8.0	4.8525E-01	1.0623	0.8967	0.8554	0.8328	1.0202	1.0038	0.9997	0.9999	0.9262	0.5651	0.9261	0.8905
9.0	4.8257E-01	1.0673	0.8995	0.8592	0.8370	1.0226	1.0042	0.9997	0.9999	0.9278	0.5575	0.9277	0.8933
10.0	4.8023E-01	1.0724	0.9018	0.8624	0.8406	1.0250	1.0047	0.9998	1.0000	0.9291	0.5498	0.9290	0.8959
11.0	4.7820E-01	1.0774	0.9037	0.8650	0.8436	1.0274	1.0052	0.9998	1.0000	0.9301	0.5420	0.9300	0.8982
12.0	4.7646E-01	1.0825	0.9053	0.8672	0.8461	1.0297	1.0056	0.9998	1.0000	0.9310	0.5342	0.9309	0.9003
13.0	4.7498E-01	1.0876	0.9066	0.8690	0.8482	1.0321	1.0061	0.9998	1.0000	0.9317	0.5263	0.9316	0.9023
14.0	4.7372E-01	1.0926	0.9076	0.8705	0.8500	1.0344	1.0066	0.9998	1.0000	0.9322	0.5184	0.9322	0.9041
16.0	4.7179E-01	1.1028	0.9090	0.8727	0.8525	1.0390	1.0075	0.9999	1.0000	0.9330	0.5025	0.9330	0.9074
18.0	4.7050E-01	1.1129	0.9097	0.8740	0.8541	1.0435	1.0085	0.9999	1.0000	0.9335	0.4865	0.9335	0.9104
20.0	4.6973E-01	1.1231	0.9099	0.8746	0.8550	1.0480	1.0094	0.9999	1.0000	0.9336	0.4705	0.9336	0.9130
25.0	4.6950E-01	1.1487	0.9087	0.8740	0.8546	1.0591	1.0117	0.9999	1.0000	0.9332	0.4304	0.9332	0.9187
30.0	4.7102E-01	1.1745	0.9059	0.8710	0.8518	1.0699	1.0141	0.9999	1.0000	0.9319	0.3903	0.9319	0.9236
35.0	4.7377E-01	1.2006	0.9019	0.8667	0.8474	1.0806	1.0164	0.9999	1.0000	0.9301	0.3505	0.9301	0.9278
40.0	4.7740E-01	1.2269	0.8973	0.8613	0.8418	1.0910	1.0188	0.9999	1.0000	0.9279	0.3110	0.9279	0.9316
45.0	4.8174E-01	1.2534	0.8921	0.8552	0.8354	1.1012	1.0212	1.0000	1.0000	0.9255	0.2718	0.9255	0.9351
50.0	4.8662E-01	1.2803	0.8866	0.8486	0.8284	1.1113	1.0236	1.0000	1.0000	0.9229	0.2329	0.9229	0.9384

Instr. see p. 22

POSITRON Z= -45 A= 103

p	F_0L_0	L_0	λ_2	λ_3	λ_4	μ_1	μ_2	Λ_1	Λ_2	ν_{12}	$\hat{\nu}_{12}$	η_{12}	$\hat{\eta}_{12}$
0.1	4.3068E-08	1.0292	8.7879	28.9345	59.7382	1.0003	1.0000	0.9975	0.9995	3.2521	0.6465	0.8703	0.8232
0.2	5.9439E-04	1.0293	2.7891	4.1878	5.1337	1.0004	1.0000	0.9975	0.9995	1.4641	0.6447	0.8711	0.8240
0.3	1.1386E-02	1.0293	1.6804	1.9701	2.1126	1.0005	1.0001	0.9976	0.9995	1.1341	0.6419	0.8723	0.8254
0.4	4.4487E-02	1.0295	1.2948	1.3712	1.3982	1.0007	1.0001	0.9977	0.9996	1.0200	0.6385	0.8738	0.8271
0.5	9.4394E-02	1.0296	1.1186	1.1272	1.1228	1.0009	1.0001	0.9978	0.9996	0.9683	0.6350	0.8756	0.8290
0.6	1.4969E-01	1.0298	1.0246	1.0050	0.9886	1.0011	1.0002	0.9978	0.9996	0.9412	0.6315	0.8774	0.8310
0.7	2.0268E-01	1.0300	0.9695	0.9359	0.9139	1.0014	1.0002	0.9979	0.9996	0.9256	0.6282	0.8792	0.8330
0.8	2.4985E-01	1.0302	0.9349	0.8935	0.8687	1.0016	1.0003	0.9980	0.9996	0.9162	0.6252	0.8810	0.8349
0.9	2.9030E-01	1.0304	0.9122	0.8662	0.8398	1.0019	1.0003	0.9981	0.9996	0.9103	0.6225	0.8827	0.8368
1.0	3.2430E-01	1.0307	0.8968	0.8479	0.8206	1.0022	1.0004	0.9982	0.9997	0.9065	0.6201	0.8844	0.8386
1.2	3.7606E-01	1.0314	0.8785	0.8266	0.7983	1.0028	1.0005	0.9984	0.9997	0.9026	0.6160	0.8875	0.8420
1.4	4.1161E-01	1.0321	0.8693	0.8161	0.7875	1.0034	1.0006	0.9985	0.9997	0.9012	0.6127	0.8903	0.8450
1.6	4.3615E-01	1.0329	0.8647	0.8111	0.7824	1.0039	1.0007	0.9987	0.9997	0.9011	0.6100	0.8928	0.8478
1.8	4.5325E-01	1.0337	0.8626	0.8090	0.7804	1.0045	1.0008	0.9988	0.9998	0.9015	0.6077	0.8951	0.8504
2.0	4.6527E-01	1.0345	0.8619	0.8086	0.7802	1.0051	1.0009	0.9989	0.9998	0.9023	0.6057	0.8972	0.8527
2.2	4.7378E-01	1.0354	0.8621	0.8091	0.7809	1.0057	1.0010	0.9990	0.9998	0.9033	0.6039	0.8992	0.8549
2.4	4.7982E-01	1.0364	0.8629	0.8103	0.7823	1.0063	1.0011	0.9990	0.9998	0.9044	0.6023	0.9009	0.8569
2.6	4.8409E-01	1.0373	0.8639	0.8118	0.7841	1.0068	1.0012	0.9991	0.9998	0.9054	0.6008	0.9026	0.8588
2.8	4.8709E-01	1.0383	0.8652	0.8135	0.7860	1.0074	1.0013	0.9992	0.9998	0.9065	0.5993	0.9041	0.8605
3.0	4.8914E-01	1.0392	0.8665	0.8153	0.7881	1.0079	1.0014	0.9992	0.9998	0.9076	0.5979	0.9055	0.8622
3.2	4.9051E-01	1.0402	0.8679	0.8172	0.7902	1.0085	1.0015	0.9993	0.9999	0.9086	0.5965	0.9068	0.8637
3.4	4.9135E-01	1.0412	0.8693	0.8191	0.7923	1.0090	1.0016	0.9993	0.9999	0.9096	0.5951	0.9080	0.8651
3.6	4.9180E-01	1.0422	0.8708	0.8210	0.7944	1.0096	1.0017	0.9993	0.9999	0.9105	0.5938	0.9091	0.8665
3.8	4.9196E-01	1.0432	0.8722	0.8228	0.7964	1.0101	1.0018	0.9994	0.9999	0.9114	0.5924	0.9102	0.8678
4.0	4.9189E-01	1.0442	0.8735	0.8246	0.7984	1.0107	1.0019	0.9994	0.9999	0.9123	0.5911	0.9112	0.8691
4.5	4.9105E-01	1.0468	0.8768	0.8289	0.8032	1.0120	1.0022	0.9995	0.9999	0.9143	0.5877	0.9135	0.8720
5.0	4.8967E-01	1.0494	0.8798	0.8329	0.8076	1.0133	1.0024	0.9995	0.9999	0.9161	0.5842	0.9154	0.8745
5.5	4.8804E-01	1.0520	0.8826	0.8365	0.8116	1.0147	1.0027	0.9996	0.9999	0.9177	0.5808	0.9172	0.8769
6.0	4.8630E-01	1.0546	0.8851	0.8398	0.8152	1.0160	1.0029	0.9996	0.9999	0.9191	0.5772	0.9187	0.8790
6.5	4.8457E-01	1.0572	0.8873	0.8428	0.8186	1.0173	1.0032	0.9996	0.9999	0.9204	0.5737	0.9200	0.8810
7.0	4.8288E-01	1.0598	0.8894	0.8455	0.8216	1.0186	1.0034	0.9997	0.9999	0.9215	0.5701	0.9213	0.8829
7.5	4.8125E-01	1.0624	0.8913	0.8480	0.8244	1.0198	1.0037	0.9997	0.9999	0.9226	0.5664	0.9223	0.8846
8.0	4.7972E-01	1.0650	0.8930	0.8503	0.8269	1.0211	1.0039	0.9997	0.9999	0.9235	0.5627	0.9233	0.8862
9.0	4.7691E-01	1.0703	0.8959	0.8543	0.8314	1.0236	1.0044	0.9997	0.9999	0.9251	0.5552	0.9250	0.8892
10.0	4.7446E-01	1.0756	0.8984	0.8576	0.8351	1.0261	1.0049	0.9998	1.0000	0.9265	0.5477	0.9264	0.8918
11.0	4.7235E-01	1.0809	0.9004	0.8604	0.8383	1.0286	1.0054	0.9998	1.0000	0.9276	0.5400	0.9275	0.8943
12.0	4.7053E-01	1.0862	0.9020	0.8627	0.8409	1.0311	1.0059	0.9998	1.0000	0.9285	0.5323	0.9284	0.8965
13.0	4.6898E-01	1.0914	0.9034	0.8646	0.8431	1.0335	1.0064	0.9998	1.0000	0.9292	0.5245	0.9292	0.8985
14.0	4.6766E-01	1.0967	0.9045	0.8662	0.8449	1.0359	1.0069	0.9998	1.0000	0.9298	0.5167	0.9298	0.9004
16.0	4.6563E-01	1.1074	0.9060	0.8685	0.8476	1.0407	1.0078	0.9999	1.0000	0.9307	0.5009	0.9306	0.9039
18.0	4.6427E-01	1.1180	0.9068	0.8699	0.8494	1.0455	1.0088	0.9999	1.0000	0.9311	0.4851	0.9311	0.9069
20.0	4.6344E-01	1.1287	0.9070	0.8706	0.8503	1.0501	1.0098	0.9999	1.0000	0.9313	0.4692	0.9313	0.9097
25.0	4.6313E-01	1.1554	0.9059	0.8700	0.8500	1.0617	1.0122	0.9999	1.0000	0.9309	0.4294	0.9309	0.9157
30.0	4.6462E-01	1.1825	0.9030	0.8670	0.8472	1.0729	1.0147	0.9999	1.0000	0.9296	0.3897	0.9296	0.9207
35.0	4.6738E-01	1.2097	0.8990	0.8626	0.8427	1.0840	1.0171	0.9999	1.0000	0.9278	0.3502	0.9278	0.9251
40.0	4.7106E-01	1.2373	0.8942	0.8571	0.8370	1.0948	1.0196	0.9999	1.0000	0.9255	0.3109	0.9256	0.9291
45.0	4.7545E-01	1.2651	0.8889	0.8509	0.8305	1.1055	1.0221	1.0000	1.0000	0.9231	0.2721	0.9231	0.9328
50.0	4.8042E-01	1.2933	0.8832	0.8441	0.8233	1.1159	1.0246	1.0000	1.0000	0.9204	0.2336	0.9204	0.9362

Instr. see p. 22

POSITRON Z= -46 A= 104

p	F_0L_0	L_0	λ_2	λ_3	λ_4	μ_1	μ_2	Λ_1	Λ_2	ν_{12}	$\hat{\nu}_{12}$	η_{12}	$\hat{\eta}_{12}$
0.1	2.8645E-08	1.0303	9.0371	30.5797	64.5071	1.0004	1.0000	0.9974	0.9995	3.3422	0.6426	0.8653	0.8164
0.2	4.9605E-04	1.0304	2.8457	4.3295	5.3487	1.0004	1.0000	0.9975	0.9995	1.4828	0.6408	0.8661	0.8172
0.3	1.0233E-02	1.0304	1.7014	2.0061	2.1574	1.0006	1.0001	0.9975	0.9995	1.1397	0.6380	0.8673	0.8186
0.4	4.1431E-02	1.0305	1.3033	1.3831	1.4114	1.0007	1.0001	0.9976	0.9995	1.0209	0.6347	0.8689	0.8203
0.5	8.9706E-02	1.0307	1.1213	1.1302	1.1255	1.0009	1.0001	0.9977	0.9996	0.9671	0.6312	0.8707	0.8222
0.6	1.4406E-01	1.0309	1.0243	1.0037	0.9866	1.0012	1.0002	0.9978	0.9996	0.9389	0.6278	0.8725	0.8243
0.7	1.9667E-01	1.0311	0.9673	0.9323	0.9095	1.0014	1.0002	0.9979	0.9996	0.9227	0.6245	0.8744	0.8263
0.8	2.4383E-01	1.0313	0.9316	0.8886	0.8628	1.0017	1.0003	0.9980	0.9996	0.9129	0.6216	0.8762	0.8283
0.9	2.8445E-01	1.0316	0.9081	0.8604	0.8330	1.0020	1.0003	0.9981	0.9996	0.9067	0.6189	0.8780	0.8302
1.0	3.1871E-01	1.0319	0.8922	0.8415	0.8132	1.0023	1.0004	0.9982	0.9996	0.9028	0.6165	0.8797	0.8321
1.2	3.7101E-01	1.0325	0.8733	0.8195	0.7902	1.0029	1.0005	0.9984	0.9997	0.8986	0.6124	0.8829	0.8356
1.4	4.0701E-01	1.0333	0.8638	0.8087	0.7791	1.0035	1.0006	0.9985	0.9997	0.8972	0.6092	0.8858	0.8388
1.6	4.3187E-01	1.0341	0.8590	0.8035	0.7739	1.0041	1.0007	0.9986	0.9997	0.8970	0.6066	0.8885	0.8417
1.8	4.4919E-01	1.0349	0.8569	0.8014	0.7719	1.0047	1.0008	0.9988	0.9998	0.8975	0.6043	0.8909	0.8443
2.0	4.6135E-01	1.0358	0.8562	0.8010	0.7716	1.0053	1.0009	0.9989	0.9998	0.8984	0.6024	0.8931	0.8467
2.2	4.6994E-01	1.0367	0.8565	0.8016	0.7725	1.0059	1.0010	0.9989	0.9998	0.8994	0.6007	0.8951	0.8490
2.4	4.7602E-01	1.0377	0.8572	0.8028	0.7739	1.0065	1.0011	0.9990	0.9998	0.9005	0.5992	0.8969	0.8511
2.6	4.8031E-01	1.0386	0.8584	0.8044	0.7757	1.0071	1.0012	0.9991	0.9998	0.9016	0.5977	0.8986	0.8530
2.8	4.8329E-01	1.0396	0.8597	0.8062	0.7777	1.0076	1.0013	0.9991	0.9998	0.9027	0.5963	0.9002	0.8548
3.0	4.8533E-01	1.0406	0.8611	0.8081	0.7799	1.0082	1.0015	0.9992	0.9998	0.9038	0.5949	0.9016	0.8565
3.2	4.8665E-01	1.0417	0.8625	0.8100	0.7821	1.0088	1.0016	0.9992	0.9999	0.9049	0.5936	0.9030	0.8581
3.4	4.8746E-01	1.0427	0.8640	0.8120	0.7843	1.0093	1.0017	0.9993	0.9999	0.9059	0.5923	0.9043	0.8596
3.6	4.8786E-01	1.0437	0.8655	0.8140	0.7864	1.0099	1.0018	0.9993	0.9999	0.9069	0.5910	0.9055	0.8611
3.8	4.8797E-01	1.0447	0.8670	0.8159	0.7886	1.0105	1.0019	0.9994	0.9999	0.9079	0.5897	0.9066	0.8624
4.0	4.8785E-01	1.0458	0.8684	0.8178	0.7907	1.0110	1.0020	0.9994	0.9999	0.9088	0.5884	0.9076	0.8637
4.5	4.8688E-01	1.0484	0.8719	0.8223	0.7956	1.0124	1.0022	0.9995	0.9999	0.9109	0.5852	0.9100	0.8667
5.0	4.8537E-01	1.0511	0.8750	0.8264	0.8002	1.0138	1.0025	0.9995	0.9999	0.9127	0.5819	0.9121	0.8694
5.5	4.8362E-01	1.0538	0.8779	0.8302	0.8044	1.0151	1.0027	0.9996	0.9999	0.9144	0.5786	0.9139	0.8718
6.0	4.8177E-01	1.0565	0.8805	0.8336	0.8082	1.0165	1.0030	0.9996	0.9999	0.9159	0.5752	0.9155	0.8741
6.5	4.7993E-01	1.0591	0.8829	0.8368	0.8117	1.0178	1.0033	0.9996	0.9999	0.9173	0.5717	0.9169	0.8761
7.0	4.7813E-01	1.0619	0.8851	0.8396	0.8149	1.0192	1.0035	0.9997	0.9999	0.9185	0.5682	0.9182	0.8781
7.5	4.7642E-01	1.0646	0.8871	0.8422	0.8178	1.0205	1.0038	0.9997	0.9999	0.9196	0.5647	0.9193	0.8799
8.0	4.7479E-01	1.0673	0.8889	0.8446	0.8205	1.0218	1.0040	0.9997	0.9999	0.9206	0.5611	0.9204	0.8815
9.0	4.7183E-01	1.0727	0.8920	0.8488	0.8251	1.0244	1.0045	0.9997	0.9999	0.9223	0.5539	0.9221	0.8846
10.0	4.6924E-01	1.0781	0.8946	0.8523	0.8291	1.0270	1.0050	0.9998	1.0000	0.9237	0.5465	0.9236	0.8874
11.0	4.6700E-01	1.0836	0.8967	0.8553	0.8324	1.0295	1.0055	0.9998	1.0000	0.9249	0.5390	0.9248	0.8899
12.0	4.6507E-01	1.0890	0.8985	0.8577	0.8352	1.0320	1.0060	0.9998	1.0000	0.9258	0.5315	0.9258	0.8922
13.0	4.6342E-01	1.0945	0.9000	0.8598	0.8375	1.0345	1.0065	0.9998	1.0000	0.9266	0.5239	0.9266	0.8943
14.0	4.6202E-01	1.1000	0.9011	0.8615	0.8395	1.0370	1.0070	0.9998	1.0000	0.9273	0.5163	0.9273	0.8963
16.0	4.5983E-01	1.1109	0.9028	0.8640	0.8425	1.0420	1.0080	0.9999	1.0000	0.9282	0.5009	0.9282	0.8999
18.0	4.5834E-01	1.1219	0.9037	0.8656	0.8444	1.0468	1.0090	0.9999	1.0000	0.9288	0.4854	0.9288	0.9031
20.0	4.5740E-01	1.1329	0.9041	0.8664	0.8454	1.0517	1.0100	0.9999	1.0000	0.9290	0.4699	0.9290	0.9060
25.0	4.5687E-01	1.1605	0.9031	0.8660	0.8455	1.0635	1.0126	0.9999	1.0000	0.9287	0.4310	0.9287	0.9122
30.0	4.5821E-01	1.1884	0.9003	0.8633	0.8428	1.0751	1.0151	0.9999	1.0000	0.9275	0.3921	0.9275	0.9175
35.0	4.6084E-01	1.2165	0.8964	0.8589	0.8384	1.0863	1.0176	0.9999	1.0000	0.9256	0.3539	0.9257	0.9220
40.0	4.6443E-01	1.2450	0.8916	0.8535	0.8328	1.0976	1.0201	0.9999	1.0000	0.9234	0.3149	0.9234	0.9262
45.0	4.6875E-01	1.2737	0.8863	0.8473	0.8263	1.1085	1.0227	1.0000	1.0000	0.9210	0.2767	0.9210	0.9300
50.0	4.7366E-01	1.3028	0.8807	0.8405	0.8192	1.1192	1.0252	1.0000	1.0000	0.9183	0.2390	0.9183	0.9336

Instr. see p. 22

POSITRON Z= -47 A= 107

p	F_0L_0	L_0	λ_2	λ_3	λ_4	μ_1	μ_2	Λ_1	Λ_2	ν_{12}	$\hat{\nu}_{12}$	η_{12}	$\hat{\eta}_{12}$
0.1	1.9041E-08	1.0315	9.2876	32.2885	69.5890	1.0004	1.0000	0.9973	0.9995	3.4338	0.6387	0.8605	0.8096
0.2	4.1373E-04	1.0315	2.9028	4.4754	5.5727	1.0005	1.0000	0.9974	0.9995	1.5019	0.6370	0.8612	0.8105
0.3	9.1907E-03	1.0316	1.7226	2.0433	2.2037	1.0006	1.0001	0.9974	0.9995	1.1454	0.6342	0.8624	0.8118
0.4	3.8565E-02	1.0317	1.3120	1.3954	1.4253	1.0008	1.0001	0.9975	0.9995	1.0220	0.6310	0.8640	0.8136
0.5	8.5209E-02	1.0318	1.1242	1.1335	1.1286	1.0010	1.0001	0.9976	0.9995	0.9660	0.6275	0.8658	0.8155
0.6	1.3857E-01	1.0320	1.0241	1.0027	0.9850	1.0012	1.0002	0.9977	0.9996	0.9367	0.6241	0.8678	0.8176
0.7	1.9075E-01	1.0322	0.9654	0.9289	0.9053	1.0015	1.0002	0.9978	0.9996	0.9198	0.6209	0.8697	0.8197
0.8	2.3786E-01	1.0325	0.9285	0.8839	0.8572	1.0018	1.0003	0.9979	0.9996	0.9096	0.6179	0.8716	0.8218
0.9	2.7862E-01	1.0328	0.9043	0.8548	0.8265	1.0021	1.0003	0.9980	0.9996	0.9032	0.6153	0.8734	0.8237
1.0	3.1310E-01	1.0331	0.8878	0.8353	0.8061	1.0024	1.0004	0.9981	0.9996	0.8991	0.6129	0.8751	0.8257
1.2	3.6591E-01	1.0338	0.8683	0.8127	0.7825	1.0030	1.0005	0.9983	0.9997	0.8948	0.6089	0.8785	0.8293
1.4	4.0233E-01	1.0345	0.8585	0.8016	0.7711	1.0036	1.0006	0.9985	0.9997	0.8933	0.6057	0.8815	0.8325
1.6	4.2750E-01	1.0353	0.8536	0.7962	0.7657	1.0043	1.0007	0.9986	0.9997	0.8931	0.6031	0.8842	0.8355
1.8	4.4503E-01	1.0362	0.8514	0.7940	0.7636	1.0049	1.0008	0.9987	0.9998	0.8936	0.6010	0.8867	0.8383
2.0	4.5733E-01	1.0372	0.8507	0.7936	0.7634	1.0055	1.0009	0.9988	0.9998	0.8945	0.5991	0.8890	0.8408
2.2	4.6599E-01	1.0381	0.8510	0.7943	0.7643	1.0061	1.0011	0.9989	0.9998	0.8956	0.5975	0.8910	0.8431
2.4	4.7211E-01	1.0391	0.8518	0.7955	0.7658	1.0067	1.0012	0.9990	0.9998	0.8967	0.5960	0.8930	0.8453
2.6	4.7641E-01	1.0401	0.8530	0.7972	0.7676	1.0073	1.0013	0.9991	0.9998	0.8979	0.5946	0.8947	0.8473
2.8	4.7938E-01	1.0411	0.8543	0.7991	0.7698	1.0079	1.0014	0.9991	0.9998	0.8990	0.5932	0.8964	0.8492
3.0	4.8139E-01	1.0422	0.8558	0.8010	0.7720	1.0085	1.0015	0.9992	0.9998	0.9002	0.5919	0.8979	0.8509
3.2	4.8269E-01	1.0432	0.8574	0.8031	0.7743	1.0091	1.0016	0.9992	0.9998	0.9013	0.5907	0.8993	0.8526
3.4	4.8345E-01	1.0443	0.8589	0.8051	0.7765	1.0097	1.0017	0.9993	0.9999	0.9024	0.5894	0.9006	0.8542
3.6	4.8381E-01	1.0454	0.8605	0.8072	0.7788	1.0103	1.0018	0.9993	0.9999	0.9034	0.5882	0.9019	0.8557
3.8	4.8387E-01	1.0465	0.8620	0.8092	0.7810	1.0109	1.0019	0.9993	0.9999	0.9044	0.5869	0.9030	0.8571
4.0	4.8370E-01	1.0475	0.8635	0.8112	0.7832	1.0115	1.0020	0.9994	0.9999	0.9053	0.5857	0.9041	0.8584
4.5	4.8261E-01	1.0503	0.8671	0.8158	0.7883	1.0129	1.0023	0.9994	0.9999	0.9075	0.5825	0.9066	0.8615
5.0	4.8098E-01	1.0530	0.8704	0.8201	0.7931	1.0143	1.0026	0.9995	0.9999	0.9095	0.5793	0.9088	0.8643
5.5	4.7910E-01	1.0558	0.8734	0.8241	0.7975	1.0157	1.0028	0.9995	0.9999	0.9112	0.5761	0.9107	0.8669
6.0	4.7715E-01	1.0586	0.8762	0.8277	0.8014	1.0171	1.0031	0.9996	0.9999	0.9128	0.5728	0.9123	0.8692
6.5	4.7521E-01	1.0614	0.8787	0.8309	0.8051	1.0185	1.0034	0.9996	0.9999	0.9142	0.5694	0.9138	0.8714
7.0	4.7332E-01	1.0642	0.8810	0.8339	0.8084	1.0199	1.0036	0.9996	0.9999	0.9155	0.5661	0.9152	0.8734
7.5	4.7152E-01	1.0670	0.8830	0.8367	0.8114	1.0212	1.0039	0.9997	0.9999	0.9166	0.5626	0.9164	0.8752
8.0	4.6981E-01	1.0698	0.8849	0.8392	0.8142	1.0226	1.0042	0.9997	0.9999	0.9177	0.5591	0.9175	0.8770
9.0	4.6670E-01	1.0755	0.8882	0.8436	0.8191	1.0253	1.0047	0.9997	0.9999	0.9195	0.5520	0.9193	0.8802
10.0	4.6399E-01	1.0811	0.8909	0.8472	0.8232	1.0280	1.0052	0.9998	0.9999	0.9210	0.5448	0.9209	0.8830
11.0	4.6163E-01	1.0867	0.8932	0.8503	0.8267	1.0306	1.0057	0.9998	1.0000	0.9222	0.5375	0.9221	0.8857
12.0	4.5961E-01	1.0924	0.8951	0.8529	0.8297	1.0332	1.0062	0.9998	1.0000	0.9232	0.5301	0.9232	0.8881
13.0	4.5787E-01	1.0981	0.8966	0.8551	0.8321	1.0358	1.0068	0.9998	1.0000	0.9241	0.5226	0.9240	0.8903
14.0	4.5639E-01	1.1037	0.8979	0.8569	0.8342	1.0384	1.0073	0.9998	1.0000	0.9248	0.5151	0.9247	0.8923
16.0	4.5407E-01	1.1151	0.8997	0.8596	0.8374	1.0435	1.0083	0.9998	1.0000	0.9258	0.5000	0.9258	0.8961
18.0	4.5247E-01	1.1265	0.9007	0.8613	0.8394	1.0485	1.0093	0.9999	1.0000	0.9264	0.4848	0.9264	0.8994
20.0	4.5145E-01	1.1379	0.9011	0.8623	0.8406	1.0535	1.0104	0.9999	1.0000	0.9267	0.4695	0.9267	0.9024
25.0	4.5076E-01	1.1666	0.9003	0.8621	0.8409	1.0658	1.0130	0.9999	1.0000	0.9264	0.4312	0.9264	0.9089
30.0	4.5200E-01	1.1955	0.8975	0.8594	0.8383	1.0777	1.0155	0.9999	1.0000	0.9252	0.3928	0.9252	0.9144
35.0	4.5457E-01	1.2247	0.8936	0.8551	0.8340	1.0895	1.0181	0.9999	1.0000	0.9234	0.3545	0.9234	0.9191
40.0	4.5813E-01	1.2543	0.8888	0.8496	0.8283	1.1010	1.0208	0.9999	1.0000	0.9212	0.3167	0.9212	0.9235
45.0	4.6244E-01	1.2842	0.8834	0.8433	0.8218	1.1122	1.0234	1.0000	1.0000	0.9187	0.2791	0.9187	0.9274
50.0	4.6736E-01	1.3145	0.8777	0.8365	0.8146	1.1233	1.0261	1.0000	1.0000	0.9160	0.2419	0.9161	0.9312

Instr. see p. 22

POSITRON Z= -48 A= 109

p	F_0L_0	L_0	λ_2	λ_3	λ_4	μ_1	μ_2	Λ_1	Λ_2	ν_{12}	$\hat{\nu}_{12}$	η_{12}	$\hat{\eta}_{12}$
0.1	1.2663E-08	1.0326	9.5344	34.0389	74.9408	1.0004	1.0000	0.9973	0.9995	3.5263	0.6348	0.8554	0.8027
0.2	3.4524E-04	1.0326	2.9588	4.6223	5.8013	1.0005	1.0000	0.9973	0.9995	1.5212	0.6330	0.8562	0.8036
0.3	8.2588E-03	1.0327	1.7433	2.0800	2.2499	1.0006	1.0001	0.9974	0.9995	1.1512	0.6304	0.8574	0.8049
0.4	3.5915E-02	1.0328	1.3204	1.4074	1.4386	1.0008	1.0001	0.9974	0.9995	1.0230	0.6271	0.8590	0.8067
0.5	8.0983E-02	1.0330	1.1269	1.1364	1.1312	1.0010	1.0001	0.9975	0.9995	0.9649	0.6237	0.8609	0.8087
0.6	1.3337E-01	1.0332	1.0237	1.0014	0.9829	1.0013	1.0002	0.9976	0.9995	0.9344	0.6203	0.8628	0.8108
0.7	1.8514E-01	1.0334	0.9631	0.9253	0.9008	1.0016	1.0002	0.9977	0.9996	0.9169	0.6171	0.8648	0.8129
0.8	2.3219E-01	1.0337	0.9251	0.8788	0.8513	1.0019	1.0003	0.9979	0.9996	0.9062	0.6142	0.8667	0.8150
0.9	2.7309E-01	1.0339	0.9001	0.8489	0.8197	1.0022	1.0003	0.9980	0.9996	0.8996	0.6116	0.8686	0.8171
1.0	3.0781E-01	1.0343	0.8832	0.8289	0.7986	1.0025	1.0004	0.9981	0.9996	0.8953	0.6092	0.8704	0.8191
1.2	3.6113E-01	1.0350	0.8631	0.8056	0.7744	1.0031	1.0005	0.9982	0.9997	0.8908	0.6053	0.8738	0.8228
1.4	3.9799E-01	1.0358	0.8530	0.7942	0.7627	1.0038	1.0006	0.9984	0.9997	0.8892	0.6022	0.8769	0.8262
1.6	4.2348E-01	1.0366	0.8479	0.7887	0.7572	1.0044	1.0007	0.9986	0.9997	0.8890	0.5997	0.8798	0.8293
1.8	4.4122E-01	1.0375	0.8457	0.7864	0.7551	1.0051	1.0009	0.9987	0.9997	0.8896	0.5976	0.8824	0.8321
2.0	4.5365E-01	1.0385	0.8450	0.7860	0.7549	1.0057	1.0010	0.9988	0.9998	0.8905	0.5958	0.8847	0.8347
2.2	4.6240E-01	1.0395	0.8453	0.7867	0.7558	1.0063	1.0011	0.9989	0.9998	0.8916	0.5942	0.8869	0.8371
2.4	4.6855E-01	1.0405	0.8462	0.7880	0.7573	1.0070	1.0012	0.9990	0.9998	0.8928	0.5928	0.8889	0.8394
2.6	4.7286E-01	1.0416	0.8474	0.7898	0.7593	1.0076	1.0013	0.9990	0.9998	0.8940	0.5914	0.8907	0.8414
2.8	4.7582E-01	1.0426	0.8488	0.7917	0.7615	1.0082	1.0014	0.9991	0.9998	0.8952	0.5902	0.8924	0.8434
3.0	4.7780E-01	1.0437	0.8504	0.7938	0.7638	1.0088	1.0015	0.9991	0.9998	0.8964	0.5889	0.8940	0.8452
3.2	4.7906E-01	1.0448	0.8520	0.7959	0.7662	1.0095	1.0017	0.9992	0.9998	0.8976	0.5877	0.8955	0.8469
3.4	4.7978E-01	1.0459	0.8536	0.7980	0.7685	1.0101	1.0018	0.9992	0.9999	0.8987	0.5865	0.8969	0.8486
3.6	4.8009E-01	1.0470	0.8552	0.8002	0.7709	1.0107	1.0019	0.9993	0.9999	0.8998	0.5853	0.8982	0.8501
3.8	4.8010E-01	1.0481	0.8568	0.8023	0.7732	1.0113	1.0020	0.9993	0.9999	0.9008	0.5841	0.8994	0.8516
4.0	4.7988E-01	1.0492	0.8584	0.8043	0.7754	1.0119	1.0021	0.9994	0.9999	0.9018	0.5829	0.9005	0.8530
4.5	4.7866E-01	1.0521	0.8621	0.8092	0.7808	1.0134	1.0024	0.9994	0.9999	0.9041	0.5799	0.9031	0.8562
5.0	4.7690E-01	1.0549	0.8656	0.8136	0.7858	1.0148	1.0026	0.9995	0.9999	0.9061	0.5769	0.9054	0.8591
5.5	4.7490E-01	1.0578	0.8687	0.8177	0.7903	1.0163	1.0029	0.9995	0.9999	0.9079	0.5737	0.9073	0.8617
6.0	4.7283E-01	1.0607	0.8716	0.8215	0.7944	1.0177	1.0032	0.9996	0.9999	0.9096	0.5705	0.9091	0.8642
6.5	4.7078E-01	1.0636	0.8742	0.8249	0.7982	1.0192	1.0035	0.9996	0.9999	0.9111	0.5673	0.9107	0.8664
7.0	4.6879E-01	1.0665	0.8766	0.8280	0.8017	1.0206	1.0037	0.9996	0.9999	0.9124	0.5640	0.9121	0.8685
7.5	4.6689E-01	1.0694	0.8788	0.8309	0.8049	1.0220	1.0040	0.9997	0.9999	0.9136	0.5606	0.9133	0.8704
8.0	4.6510E-01	1.0723	0.8808	0.8335	0.8078	1.0234	1.0043	0.9997	0.9999	0.9147	0.5572	0.9145	0.8722
9.0	4.6184E-01	1.0781	0.8842	0.8381	0.8129	1.0262	1.0048	0.9997	0.9999	0.9166	0.5503	0.9164	0.8756
10.0	4.5899E-01	1.0839	0.8871	0.8420	0.8172	1.0289	1.0053	0.9997	0.9999	0.9182	0.5433	0.9181	0.8786
11.0	4.5651E-01	1.0898	0.8895	0.8452	0.8209	1.0316	1.0059	0.9998	1.0000	0.9195	0.5361	0.9194	0.8813
12.0	4.5438E-01	1.0956	0.8915	0.8480	0.8240	1.0343	1.0064	0.9998	1.0000	0.9206	0.5289	0.9205	0.8838
13.0	4.5255E-01	1.1015	0.8931	0.8503	0.8266	1.0370	1.0070	0.9998	1.0000	0.9215	0.5216	0.9214	0.8861
14.0	4.5098E-01	1.1073	0.8945	0.8522	0.8288	1.0397	1.0075	0.9998	1.0000	0.9222	0.5142	0.9222	0.8882
16.0	4.4852E-01	1.1191	0.8964	0.8551	0.8321	1.0449	1.0085	0.9998	1.0000	0.9233	0.4994	0.9233	0.8921
18.0	4.4680E-01	1.1308	0.8976	0.8570	0.8344	1.0501	1.0096	0.9999	1.0000	0.9240	0.4845	0.9240	0.8956
20.0	4.4567E-01	1.1426	0.8981	0.8580	0.8357	1.0553	1.0107	0.9999	1.0000	0.9243	0.4695	0.9243	0.8987
25.0	4.4479E-01	1.1723	0.8974	0.8581	0.8362	1.0679	1.0133	0.9999	1.0000	0.9241	0.4319	0.9242	0.9054
30.0	4.4590E-01	1.2022	0.8948	0.8555	0.8339	1.0802	1.0160	0.9999	1.0000	0.9230	0.3942	0.9230	0.9111
35.0	4.4837E-01	1.2325	0.8909	0.8513	0.8296	1.0923	1.0187	0.9999	1.0000	0.9212	0.3566	0.9212	0.9161
40.0	4.5186E-01	1.2631	0.8861	0.8459	0.8240	1.1041	1.0213	0.9999	1.0000	0.9190	0.3194	0.9190	0.9206
45.0	4.5613E-01	1.2940	0.8807	0.8396	0.8174	1.1157	1.0241	0.9999	1.0000	0.9165	0.2823	0.9165	0.9247
50.0	4.6102E-01	1.3254	0.8749	0.8327	0.8102	1.1271	1.0268	1.0000	1.0000	0.9138	0.2456	0.9139	0.9286

POSITRON Z= -49 A= 113

p	F_0L_0	L_0	λ_2	λ_3	λ_4	μ_1	μ_2	Λ_1	Λ_2	ν_{12}	$\hat{\nu}_{12}$	η_{12}	$\hat{\eta}_{12}$
0.1	8.4161E-09	1.0338	9.7824	35.8535	80.6298	1.0004	1.0000	0.9972	0.9995	3.6203	0.6309	0.8505	0.7959
0.2	2.8791E-04	1.0338	3.0153	4.7735	6.0393	1.0005	1.0000	0.9972	0.9995	1.5410	0.6292	0.8512	0.7967
0.3	7.4171E-03	1.0339	1.7643	2.1177	2.2977	1.0007	1.0001	0.9973	0.9995	1.1572	0.6265	0.8525	0.7981
0.4	3.3429E-02	1.0341	1.3290	1.4198	1.4527	1.0009	1.0001	0.9973	0.9995	1.0242	0.6233	0.8541	0.7999
0.5	7.6926E-02	1.0342	1.1298	1.1396	1.1343	1.0011	1.0001	0.9974	0.9995	0.9639	0.6199	0.8560	0.8019
0.6	1.2831E-01	1.0344	1.0235	1.0004	0.9812	1.0014	1.0002	0.9976	0.9995	0.9322	0.6166	0.8580	0.8041
0.7	1.7960E-01	1.0346	0.9611	0.9220	0.8966	1.0017	1.0002	0.9977	0.9996	0.9140	0.6134	0.8600	0.8063
0.8	2.2655E-01	1.0349	0.9220	0.8741	0.8457	1.0020	1.0003	0.9978	0.9996	0.9030	0.6105	0.8620	0.8084
0.9	2.6756E-01	1.0352	0.8962	0.8433	0.8131	1.0023	1.0004	0.9979	0.9996	0.8960	0.6079	0.8639	0.8105
1.0	3.0248E-01	1.0355	0.8788	0.8227	0.7916	1.0026	1.0004	0.9980	0.9996	0.8916	0.6056	0.8658	0.8125
1.2	3.5628E-01	1.0363	0.8581	0.7987	0.7667	1.0033	1.0005	0.9982	0.9997	0.8869	0.6017	0.8693	0.8164
1.4	3.9355E-01	1.0371	0.8477	0.7870	0.7546	1.0039	1.0006	0.9984	0.9997	0.8853	0.5987	0.8725	0.8199
1.6	4.1934E-01	1.0380	0.8425	0.7814	0.7490	1.0046	1.0008	0.9985	0.9997	0.8851	0.5962	0.8755	0.8231
1.8	4.3729E-01	1.0389	0.8402	0.7791	0.7469	1.0053	1.0009	0.9986	0.9997	0.8856	0.5942	0.8781	0.8260
2.0	4.4985E-01	1.0399	0.8395	0.7787	0.7467	1.0060	1.0010	0.9987	0.9998	0.8866	0.5925	0.8806	0.8287
2.2	4.5867E-01	1.0410	0.8398	0.7794	0.7476	1.0066	1.0011	0.9988	0.9998	0.8877	0.5909	0.8828	0.8312
2.4	4.6486E-01	1.0420	0.8407	0.7808	0.7492	1.0073	1.0012	0.9989	0.9998	0.8889	0.5895	0.8849	0.8335
2.6	4.6917E-01	1.0431	0.8420	0.7826	0.7513	1.0079	1.0014	0.9990	0.9998	0.8902	0.5883	0.8868	0.8357
2.8	4.7212E-01	1.0442	0.8435	0.7846	0.7536	1.0086	1.0015	0.9991	0.9998	0.8915	0.5870	0.8886	0.8377
3.0	4.7408E-01	1.0453	0.8451	0.7868	0.7559	1.0092	1.0016	0.9991	0.9998	0.8927	0.5858	0.8902	0.8396
3.2	4.7530E-01	1.0465	0.8468	0.7890	0.7584	1.0099	1.0017	0.9992	0.9998	0.8939	0.5847	0.8918	0.8414
3.4	4.7598E-01	1.0476	0.8485	0.7912	0.7608	1.0105	1.0018	0.9992	0.9998	0.8951	0.5835	0.8932	0.8431
3.6	4.7625E-01	1.0488	0.8502	0.7934	0.7633	1.0111	1.0019	0.9993	0.9999	0.8962	0.5824	0.8945	0.8447
3.8	4.7620E-01	1.0499	0.8518	0.7956	0.7657	1.0117	1.0021	0.9993	0.9999	0.8973	0.5812	0.8958	0.8462
4.0	4.7593E-01	1.0511	0.8535	0.7977	0.7680	1.0124	1.0022	0.9993	0.9999	0.8983	0.5801	0.8970	0.8476
4.5	4.7459E-01	1.0540	0.8574	0.8027	0.7736	1.0139	1.0025	0.9994	0.9999	0.9007	0.5772	0.8997	0.8510
5.0	4.7270E-01	1.0570	0.8610	0.8074	0.7787	1.0154	1.0027	0.9995	0.9999	0.9028	0.5742	0.9020	0.8540
5.5	4.7059E-01	1.0600	0.8643	0.8117	0.7834	1.0169	1.0030	0.9995	0.9999	0.9047	0.5711	0.9041	0.8567
6.0	4.6841E-01	1.0630	0.8673	0.8155	0.7877	1.0184	1.0033	0.9996	0.9999	0.9064	0.5680	0.9059	0.8592
6.5	4.6625E-01	1.0660	0.8700	0.8191	0.7916	1.0199	1.0036	0.9996	0.9999	0.9080	0.5649	0.9076	0.8616
7.0	4.6417E-01	1.0690	0.8725	0.8224	0.7952	1.0214	1.0039	0.9996	0.9999	0.9094	0.5616	0.9090	0.8637
7.5	4.6219E-01	1.0720	0.8747	0.8253	0.7985	1.0229	1.0041	0.9996	0.9999	0.9106	0.5583	0.9104	0.8657
8.0	4.6031E-01	1.0750	0.8768	0.8281	0.8015	1.0243	1.0044	0.9997	0.9999	0.9118	0.5550	0.9116	0.8676
9.0	4.5691E-01	1.0811	0.8804	0.8329	0.8069	1.0272	1.0050	0.9997	0.9999	0.9138	0.5482	0.9136	0.8711
10.0	4.5393E-01	1.0872	0.8834	0.8369	0.8114	1.0301	1.0055	0.9997	0.9999	0.9154	0.5413	0.9153	0.8742
11.0	4.5136E-01	1.0932	0.8859	0.8403	0.8152	1.0329	1.0061	0.9998	1.0000	0.9168	0.5343	0.9167	0.8770
12.0	4.4913E-01	1.0993	0.8880	0.8432	0.8184	1.0357	1.0066	0.9998	1.0000	0.9179	0.5271	0.9179	0.8796
13.0	4.4721E-01	1.1054	0.8898	0.8456	0.8212	1.0385	1.0072	0.9998	1.0000	0.9189	0.5200	0.9188	0.8820
14.0	4.4558E-01	1.1115	0.8912	0.8476	0.8235	1.0412	1.0077	0.9998	1.0000	0.9197	0.5127	0.9196	0.8842
16.0	4.4299E-01	1.1237	0.8933	0.8507	0.8270	1.0467	1.0088	0.9998	1.0000	0.9208	0.4981	0.9208	0.8882
18.0	4.4118E-01	1.1359	0.8945	0.8527	0.8294	1.0521	1.0099	0.9999	1.0000	0.9215	0.4834	0.9215	0.8918
20.0	4.3997E-01	1.1482	0.8951	0.8538	0.8309	1.0574	1.0110	0.9999	1.0000	0.9219	0.4686	0.9219	0.8951
25.0	4.3896E-01	1.1790	0.8945	0.8540	0.8316	1.0704	1.0138	0.9999	1.0000	0.9218	0.4314	0.9218	0.9021
30.0	4.3999E-01	1.2102	0.8919	0.8516	0.8293	1.0832	1.0165	0.9999	1.0000	0.9207	0.3941	0.9207	0.9080
35.0	4.4243E-01	1.2417	0.8880	0.8473	0.8250	1.0957	1.0193	0.9999	1.0000	0.9189	0.3569	0.9189	0.9132
40.0	4.4591E-01	1.2735	0.8831	0.8418	0.8193	1.1079	1.0221	0.9999	1.0000	0.9167	0.3200	0.9167	0.9179
45.0	4.5019E-01	1.3058	0.8777	0.8354	0.8127	1.1199	1.0249	0.9999	1.0000	0.9142	0.2833	0.9142	0.9222
50.0	4.5511E-01	1.3385	0.8718	0.8285	0.8054	1.1316	1.0277	1.0000	1.0000	0.9115	0.2471	0.9115	0.9262

POSITRON Z= -50 A= 116

p	F_0L_0	L_0	λ_2	λ_3	λ_4	μ_1	μ_2	Λ_1	Λ_2	ν_{12}	$\hat{\nu}_{12}$	η_{12}	$\hat{\eta}_{12}$
0.1	5.5965E-09	1.0350	10.0262	37.7062	86.6026	1.0005	1.0000	0.9971	0.9994	3.7150	0.6269	0.8454	0.7889
0.2	2.4023E-04	1.0350	3.0706	4.9255	6.2818	1.0006	1.0000	0.9971	0.9995	1.5608	0.6252	0.8462	0.7897
0.3	6.6645E-03	1.0351	1.7847	2.1550	2.3452	1.0007	1.0001	0.9972	0.9995	1.1631	0.6226	0.8474	0.7911
0.4	3.1131E-02	1.0352	1.3371	1.4317	1.4662	1.0009	1.0001	0.9973	0.9995	1.0253	0.6195	0.8491	0.7929
0.5	7.3113E-02	1.0354	1.1323	1.1424	1.1370	1.0012	1.0001	0.9974	0.9995	0.9628	0.6161	0.8510	0.7950
0.6	1.2351E-01	1.0356	1.0230	0.9990	0.9791	1.0014	1.0002	0.9975	0.9995	0.9299	0.6128	0.8530	0.7972
0.7	1.7435E-01	1.0359	0.9588	0.9183	0.8921	1.0017	1.0002	0.9976	0.9995	0.9111	0.6096	0.8551	0.7994
0.8	2.2120E-01	1.0361	0.9186	0.8691	0.8397	1.0020	1.0003	0.9977	0.9996	0.8996	0.6068	0.8571	0.8016
0.9	2.6232E-01	1.0364	0.8921	0.8374	0.8063	1.0024	1.0004	0.9978	0.9996	0.8924	0.6042	0.8591	0.8038
1.0	2.9745E-01	1.0368	0.8741	0.8162	0.7841	1.0027	1.0004	0.9979	0.9996	0.8877	0.6019	0.8610	0.8059
1.2	3.5174E-01	1.0375	0.8528	0.7916	0.7586	1.0034	1.0005	0.9981	0.9996	0.8829	0.5981	0.8647	0.8098
1.4	3.8944E-01	1.0384	0.8421	0.7796	0.7463	1.0041	1.0007	0.9983	0.9997	0.8812	0.5951	0.8680	0.8134
1.6	4.1554E-01	1.0393	0.8368	0.7739	0.7406	1.0048	1.0008	0.9985	0.9997	0.8810	0.5927	0.8710	0.8167
1.8	4.3369E-01	1.0403	0.8344	0.7715	0.7384	1.0055	1.0009	0.9986	0.9997	0.8815	0.5908	0.8738	0.8197
2.0	4.4639E-01	1.0413	0.8338	0.7711	0.7382	1.0062	1.0010	0.9987	0.9998	0.8825	0.5891	0.8763	0.8225
2.2	4.5529E-01	1.0424	0.8341	0.7719	0.7392	1.0069	1.0012	0.9988	0.9998	0.8837	0.5876	0.8786	0.8251
2.4	4.6151E-01	1.0435	0.8351	0.7733	0.7409	1.0076	1.0013	0.9989	0.9998	0.8850	0.5863	0.8808	0.8275
2.6	4.6583E-01	1.0446	0.8364	0.7752	0.7430	1.0082	1.0014	0.9990	0.9998	0.8863	0.5851	0.8828	0.8297
2.8	4.6876E-01	1.0458	0.8380	0.7773	0.7453	1.0089	1.0015	0.9990	0.9998	0.8876	0.5839	0.8846	0.8318
3.0	4.7069E-01	1.0469	0.8396	0.7795	0.7478	1.0096	1.0016	0.9991	0.9998	0.8889	0.5827	0.8863	0.8338
3.2	4.7187E-01	1.0481	0.8414	0.7818	0.7504	1.0102	1.0018	0.9991	0.9998	0.8902	0.5816	0.8879	0.8356
3.4	4.7250E-01	1.0493	0.8432	0.7841	0.7529	1.0109	1.0019	0.9992	0.9998	0.8914	0.5805	0.8894	0.8374
3.6	4.7272E-01	1.0505	0.8449	0.7864	0.7554	1.0115	1.0020	0.9992	0.9999	0.8925	0.5794	0.8908	0.8390
3.8	4.7263E-01	1.0517	0.8466	0.7887	0.7579	1.0122	1.0021	0.9993	0.9999	0.8937	0.5783	0.8921	0.8406
4.0	4.7230E-01	1.0529	0.8484	0.7909	0.7603	1.0128	1.0022	0.9993	0.9999	0.8947	0.5772	0.8934	0.8421
4.5	4.7082E-01	1.0560	0.8524	0.7961	0.7661	1.0144	1.0025	0.9994	0.9999	0.8972	0.5744	0.8962	0.8456
5.0	4.6881E-01	1.0590	0.8562	0.8009	0.7714	1.0160	1.0028	0.9994	0.9999	0.8994	0.5716	0.8986	0.8487
5.5	4.6657E-01	1.0621	0.8596	0.8053	0.7762	1.0176	1.0031	0.9995	0.9999	0.9014	0.5686	0.9008	0.8515
6.0	4.6427E-01	1.0652	0.8627	0.8094	0.7807	1.0191	1.0034	0.9995	0.9999	0.9032	0.5656	0.9027	0.8541
6.5	4.6201E-01	1.0683	0.8656	0.8131	0.7848	1.0207	1.0037	0.9996	0.9999	0.9048	0.5625	0.9044	0.8565
7.0	4.5983E-01	1.0715	0.8682	0.8165	0.7885	1.0222	1.0040	0.9996	0.9999	0.9063	0.5594	0.9059	0.8588
7.5	4.5775E-01	1.0746	0.8705	0.8196	0.7920	1.0237	1.0043	0.9996	0.9999	0.9076	0.5562	0.9073	0.8609
8.0	4.5579E-01	1.0777	0.8727	0.8224	0.7951	1.0252	1.0046	0.9997	0.9999	0.9088	0.5529	0.9085	0.8628
9.0	4.5223E-01	1.0840	0.8765	0.8274	0.8007	1.0282	1.0051	0.9997	0.9999	0.9109	0.5463	0.9107	0.8664
10.0	4.4912E-01	1.0903	0.8796	0.8317	0.8054	1.0311	1.0057	0.9997	0.9999	0.9126	0.5395	0.9125	0.8696
11.0	4.4643E-01	1.0966	0.8823	0.8352	0.8094	1.0341	1.0063	0.9998	0.9999	0.9140	0.5326	0.9139	0.8726
12.0	4.4410E-01	1.1029	0.8845	0.8382	0.8127	1.0370	1.0068	0.9998	1.0000	0.9152	0.5256	0.9152	0.8753
13.0	4.4210E-01	1.1092	0.8863	0.8408	0.8156	1.0398	1.0074	0.9998	1.0000	0.9162	0.5186	0.9162	0.8777
14.0	4.4038E-01	1.1155	0.8878	0.8429	0.8181	1.0427	1.0080	0.9998	1.0000	0.9171	0.5115	0.9170	0.8801
16.0	4.3766E-01	1.1281	0.8900	0.8462	0.8218	1.0483	1.0091	0.9998	1.0000	0.9183	0.4971	0.9183	0.8842
18.0	4.3573E-01	1.1408	0.8914	0.8483	0.8244	1.0539	1.0102	0.9999	1.0000	0.9191	0.4826	0.9191	0.8880
20.0	4.3443E-01	1.1535	0.8921	0.8496	0.8260	1.0593	1.0114	0.9999	1.0000	0.9195	0.4680	0.9195	0.8913
25.0	4.3325E-01	1.1855	0.8916	0.8500	0.8269	1.0728	1.0142	0.9999	1.0000	0.9195	0.4314	0.9195	0.8986
30.0	4.3417E-01	1.2177	0.8891	0.8476	0.8247	1.0860	1.0170	0.9999	1.0000	0.9184	0.3946	0.9184	0.9048
35.0	4.3654E-01	1.2504	0.8852	0.8434	0.8205	1.0988	1.0199	0.9999	1.0000	0.9167	0.3579	0.9167	0.9102
40.0	4.3998E-01	1.2835	0.8803	0.8379	0.8148	1.1114	1.0227	0.9999	1.0000	0.9144	0.3215	0.9145	0.9150
45.0	4.4424E-01	1.3170	0.8748	0.8315	0.8082	1.1238	1.0256	0.9999	1.0000	0.9119	0.2853	0.9119	0.9195
50.0	4.4915E-01	1.3509	0.8689	0.8245	0.8008	1.1359	1.0285	1.0000	1.0000	0.9092	0.2495	0.9092	0.9237

POSITRON Z= -51 A= 121

p	F_0L_0	L_0	λ_2	λ_3	λ_4	μ_1	μ_2	Λ_1	Λ_2	ν_{12}	$\hat{\nu}_{12}$	η_{12}	$\hat{\eta}_{12}$
0.1	3.7194E-09	1.0362	10.2709	39.6232	92.9369	1.0005	1.0000	0.9969	0.9994	3.8112	0.6230	0.8404	0.7819
0.2	2.0032E-04	1.0363	3.1263	5.0816	6.5340	1.0006	1.0000	0.9970	0.9994	1.5810	0.6213	0.8412	0.7828
0.3	5.9849E-03	1.0364	1.8054	2.1933	2.3944	1.0008	1.0001	0.9971	0.9994	1.1693	0.6187	0.8425	0.7842
0.4	2.8975E-02	1.0365	1.3456	1.4442	1.4804	1.0010	1.0001	0.9972	0.9995	1.0266	0.6156	0.8441	0.7860
0.5	6.9454E-02	1.0367	1.1351	1.1457	1.1401	1.0012	1.0002	0.9973	0.9995	0.9618	0.6123	0.8461	0.7881
0.6	1.1883E-01	1.0369	1.0228	0.9979	0.9774	1.0015	1.0002	0.9974	0.9995	0.9278	0.6090	0.8481	0.7904
0.7	1.6917E-01	1.0371	0.9568	0.9149	0.8879	1.0018	1.0003	0.9975	0.9995	0.9082	0.6059	0.8502	0.7926
0.8	2.1588E-01	1.0374	0.9154	0.8643	0.8341	1.0022	1.0003	0.9976	0.9995	0.8963	0.6031	0.8523	0.7949
0.9	2.5708E-01	1.0377	0.8882	0.8318	0.7998	1.0025	1.0004	0.9977	0.9996	0.8888	0.6005	0.8544	0.7971
1.0	2.9239E-01	1.0381	0.8697	0.8101	0.7771	1.0028	1.0004	0.9978	0.9996	0.8840	0.5983	0.8563	0.7993
1.2	3.4713E-01	1.0389	0.8478	0.7848	0.7509	1.0036	1.0006	0.9981	0.9996	0.8790	0.5945	0.8601	0.8033
1.4	3.8522E-01	1.0398	0.8368	0.7725	0.7383	1.0043	1.0007	0.9982	0.9997	0.8772	0.5916	0.8635	0.8070
1.6	4.1162E-01	1.0408	0.8313	0.7666	0.7324	1.0050	1.0008	0.9984	0.9997	0.8770	0.5892	0.8666	0.8104
1.8	4.2997E-01	1.0418	0.8289	0.7642	0.7302	1.0057	1.0010	0.9985	0.9997	0.8776	0.5873	0.8695	0.8136
2.0	4.4280E-01	1.0429	0.8282	0.7638	0.7300	1.0065	1.0011	0.9987	0.9997	0.8786	0.5857	0.8721	0.8164
2.2	4.5176E-01	1.0440	0.8286	0.7646	0.7311	1.0072	1.0012	0.9988	0.9998	0.8798	0.5843	0.8746	0.8191
2.4	4.5802E-01	1.0451	0.8296	0.7662	0.7328	1.0079	1.0013	0.9988	0.9998	0.8811	0.5830	0.8766	0.8216
2.6	4.6234E-01	1.0463	0.8310	0.7681	0.7350	1.0086	1.0015	0.9989	0.9998	0.8825	0.5818	0.8788	0.8239
2.8	4.6526E-01	1.0475	0.8326	0.7703	0.7375	1.0093	1.0016	0.9990	0.9998	0.8839	0.5807	0.8807	0.8260
3.0	4.6716E-01	1.0487	0.8344	0.7726	0.7400	1.0100	1.0017	0.9991	0.9998	0.8852	0.5796	0.8825	0.8281
3.2	4.6830E-01	1.0499	0.8362	0.7749	0.7426	1.0107	1.0018	0.9991	0.9998	0.8865	0.5785	0.8842	0.8300
3.4	4.6889E-01	1.0512	0.8380	0.7773	0.7453	1.0114	1.0020	0.9992	0.9998	0.8878	0.5775	0.8857	0.8318
3.6	4.6906E-01	1.0524	0.8399	0.7797	0.7479	1.0120	1.0021	0.9992	0.9998	0.8890	0.5764	0.8872	0.8335
3.8	4.6892E-01	1.0537	0.8417	0.7820	0.7504	1.0127	1.0022	0.9992	0.9999	0.8901	0.5754	0.8885	0.8351
4.0	4.6854E-01	1.0549	0.8434	0.7843	0.7529	1.0134	1.0023	0.9993	0.9999	0.8912	0.5743	0.8898	0.8367
4.5	4.6692E-01	1.0581	0.8476	0.7897	0.7589	1.0150	1.0026	0.9994	0.9999	0.8938	0.5716	0.8927	0.8403
5.0	4.6478E-01	1.0613	0.8515	0.7947	0.7644	1.0167	1.0029	0.9994	0.9999	0.8961	0.5688	0.8953	0.8435
5.5	4.6243E-01	1.0645	0.8551	0.7993	0.7694	1.0183	1.0032	0.9995	0.9999	0.8982	0.5659	0.8975	0.8465
6.0	4.6002E-01	1.0677	0.8583	0.8035	0.7740	1.0199	1.0035	0.9995	0.9999	0.9001	0.5630	0.8995	0.8492
6.5	4.5766E-01	1.0710	0.8613	0.8073	0.7782	1.0215	1.0038	0.9996	0.9999	0.9017	0.5600	0.9013	0.8516
7.0	4.5538E-01	1.0742	0.8640	0.8108	0.7821	1.0231	1.0041	0.9996	0.9999	0.9032	0.5569	0.9029	0.8540
7.5	4.5322E-01	1.0775	0.8665	0.8141	0.7857	1.0247	1.0044	0.9996	0.9999	0.9046	0.5537	0.9043	0.8561
8.0	4.5118E-01	1.0807	0.8687	0.8170	0.7889	1.0263	1.0047	0.9996	0.9999	0.9059	0.5505	0.9056	0.8582
9.0	4.4748E-01	1.0873	0.8727	0.8222	0.7947	1.0293	1.0053	0.9997	0.9999	0.9080	0.5440	0.9078	0.8619
10.0	4.4425E-01	1.0938	0.8760	0.8266	0.7996	1.0324	1.0059	0.9997	0.9999	0.9098	0.5373	0.9097	0.8652
11.0	4.4146E-01	1.1004	0.8787	0.8303	0.8037	1.0354	1.0065	0.9997	0.9999	0.9113	0.5305	0.9112	0.8683
12.0	4.3904E-01	1.1069	0.8810	0.8334	0.8072	1.0385	1.0071	0.9998	1.0000	0.9126	0.5236	0.9125	0.8711
13.0	4.3696E-01	1.1135	0.8829	0.8361	0.8102	1.0414	1.0077	0.9998	1.0000	0.9136	0.5167	0.9136	0.8736
14.0	4.3517E-01	1.1200	0.8845	0.8383	0.8128	1.0444	1.0083	0.9998	1.0000	0.9145	0.5096	0.9145	0.8760
16.0	4.3234E-01	1.1332	0.8869	0.8417	0.8167	1.0502	1.0094	0.9998	1.0000	0.9158	0.4954	0.9158	0.8804
18.0	4.3032E-01	1.1464	0.8883	0.8440	0.8194	1.0560	1.0106	0.9998	1.0000	0.9166	0.4811	0.9166	0.8842
20.0	4.2895E-01	1.1596	0.8891	0.8454	0.8211	1.0617	1.0118	0.9999	1.0000	0.9171	0.4666	0.9171	0.8877
25.0	4.2766E-01	1.1929	0.8887	0.8459	0.8222	1.0756	1.0147	0.9999	1.0000	0.9172	0.4303	0.9172	0.8953
30.0	4.2853E-01	1.2266	0.8861	0.8436	0.8200	1.0892	1.0176	0.9999	1.0000	0.9161	0.3938	0.9161	0.9017
35.0	4.3089E-01	1.2606	0.8822	0.8393	0.8158	1.1026	1.0206	0.9999	1.0000	0.9143	0.3575	0.9143	0.9073
40.0	4.3434E-01	1.2951	0.8772	0.8337	0.8101	1.1156	1.0236	0.9999	1.0000	0.9121	0.3213	0.9121	0.9123
45.0	4.3863E-01	1.3301	0.8716	0.8272	0.8033	1.1284	1.0265	0.9999	1.0000	0.9095	0.2854	0.9095	0.9170
50.0	4.4360E-01	1.3656	0.8656	0.8201	0.7958	1.1410	1.0295	1.0000	1.0000	0.9067	0.2498	0.9067	0.9213

 Instr. see p. 22

POSITRON Z= -52 A= 121

p	F_0L_0	L_0	λ_2	λ_3	λ_4	μ_1	μ_2	Λ_1	Λ_2	ν_{12}	$\hat{\nu}_{12}$	η_{12}	$\hat{\eta}_{12}$
0.1	2.4769E-09	1.0373	10.5027	41.5336	99.4542	1.0006	1.0000	0.9969	0.9994	3.9068	0.6188	0.8350	0.7746
0.2	1.6739E-04	1.0374	3.1782	5.2328	6.7829	1.0007	1.0000	0.9969	0.9994	1.6008	0.6171	0.8358	0.7754
0.3	5.3855E-03	1.0375	1.8240	2.2288	2.4405	1.0008	1.0001	0.9970	0.9994	1.1750	0.6146	0.8371	0.7768
0.4	2.7024E-02	1.0376	1.3525	1.4547	1.4923	1.0010	1.0001	0.9971	0.9995	1.0274	0.6115	0.8388	0.7787
0.5	6.6115E-02	1.0378	1.1366	1.1473	1.1413	1.0013	1.0002	0.9972	0.9995	0.9604	0.6082	0.8407	0.7808
0.6	1.1457E-01	1.0380	1.0215	0.9954	0.9742	1.0016	1.0002	0.9973	0.9995	0.9252	0.6050	0.8428	0.7831
0.7	1.6449E-01	1.0383	0.9537	0.9103	0.8823	1.0019	1.0003	0.9974	0.9995	0.9049	0.6019	0.8449	0.7854
0.8	2.1115E-01	1.0386	0.9112	0.8584	0.8272	1.0022	1.0003	0.9976	0.9995	0.8926	0.5991	0.8471	0.7877
0.9	2.5249E-01	1.0389	0.8833	0.8251	0.7921	1.0026	1.0004	0.9977	0.9996	0.8848	0.5966	0.8492	0.7900
1.0	2.8804E-01	1.0393	0.8644	0.8028	0.7688	1.0029	1.0004	0.9978	0.9996	0.8799	0.5944	0.8512	0.7922
1.2	3.4333E-01	1.0401	0.8419	0.7770	0.7421	1.0037	1.0006	0.9980	0.9996	0.8746	0.5907	0.8551	0.7963
1.4	3.8188E-01	1.0410	0.8306	0.7643	0.7292	1.0044	1.0007	0.9982	0.9997	0.8728	0.5879	0.8586	0.8002
1.6	4.0862E-01	1.0420	0.8250	0.7583	0.7232	1.0052	1.0008	0.9984	0.9997	0.8725	0.5856	0.8618	0.8037
1.8	4.2721E-01	1.0431	0.8226	0.7559	0.7210	1.0059	1.0010	0.9985	0.9997	0.8731	0.5838	0.8648	0.8069
2.0	4.4017E-01	1.0442	0.8219	0.7556	0.7208	1.0067	1.0011	0.9986	0.9997	0.8742	0.5823	0.8675	0.8099
2.2	4.4922E-01	1.0453	0.8223	0.7564	0.7219	1.0074	1.0012	0.9987	0.9998	0.8755	0.5809	0.8700	0.8126
2.4	4.5551E-01	1.0465	0.8234	0.7580	0.7238	1.0081	1.0014	0.9988	0.9998	0.8768	0.5797	0.8723	0.8152
2.6	4.5983E-01	1.0477	0.8248	0.7600	0.7260	1.0089	1.0015	0.9989	0.9998	0.8783	0.5786	0.8745	0.8175
2.8	4.6274E-01	1.0489	0.8265	0.7623	0.7285	1.0096	1.0016	0.9990	0.9998	0.8797	0.5776	0.8764	0.8198
3.0	4.6460E-01	1.0502	0.8284	0.7647	0.7312	1.0103	1.0017	0.9990	0.9998	0.8811	0.5765	0.8783	0.8219
3.2	4.6570E-01	1.0514	0.8302	0.7671	0.7339	1.0110	1.0019	0.9991	0.9998	0.8824	0.5755	0.8800	0.8239
3.4	4.6623E-01	1.0527	0.8321	0.7696	0.7366	1.0117	1.0020	0.9991	0.9998	0.8837	0.5745	0.8816	0.8257
3.6	4.6634E-01	1.0540	0.8341	0.7720	0.7393	1.0124	1.0021	0.9992	0.9998	0.8850	0.5736	0.8831	0.8275
3.8	4.6614E-01	1.0552	0.8359	0.7744	0.7420	1.0131	1.0022	0.9992	0.9999	0.8862	0.5726	0.8845	0.8292
4.0	4.6570E-01	1.0565	0.8378	0.7768	0.7445	1.0138	1.0024	0.9993	0.9999	0.8873	0.5715	0.8859	0.8308
4.5	4.6393E-01	1.0598	0.8422	0.7824	0.7507	1.0155	1.0027	0.9993	0.9999	0.8900	0.5690	0.8889	0.8345
5.0	4.6164E-01	1.0631	0.8462	0.7876	0.7564	1.0172	1.0030	0.9994	0.9999	0.8924	0.5663	0.8916	0.8378
5.5	4.5914E-01	1.0664	0.8499	0.7924	0.7616	1.0188	1.0033	0.9995	0.9999	0.8946	0.5636	0.8939	0.8409
6.0	4.5660E-01	1.0697	0.8533	0.7968	0.7664	1.0205	1.0036	0.9995	0.9999	0.8965	0.5608	0.8960	0.8437
6.5	4.5411E-01	1.0730	0.8564	0.8008	0.7708	1.0221	1.0039	0.9996	0.9999	0.8983	0.5579	0.8978	0.8463
7.0	4.5172E-01	1.0764	0.8593	0.8044	0.7748	1.0238	1.0042	0.9996	0.9999	0.8999	0.5549	0.8995	0.8487
7.5	4.4945E-01	1.0797	0.8618	0.8078	0.7785	1.0254	1.0045	0.9996	0.9999	0.9013	0.5519	0.9010	0.8509
8.0	4.4731E-01	1.0831	0.8642	0.8109	0.7819	1.0270	1.0048	0.9996	0.9999	0.9026	0.5488	0.9024	0.8530
9.0	4.4342E-01	1.0898	0.8683	0.8163	0.7880	1.0302	1.0054	0.9997	0.9999	0.9049	0.5425	0.9047	0.8568
10.0	4.4004E-01	1.0965	0.8718	0.8209	0.7931	1.0333	1.0060	0.9997	0.9999	0.9068	0.5360	0.9067	0.8603
11.0	4.3710E-01	1.1032	0.8747	0.8248	0.7974	1.0364	1.0066	0.9997	0.9999	0.9084	0.5294	0.9083	0.8635
12.0	4.3455E-01	1.1099	0.8772	0.8281	0.8011	1.0395	1.0072	0.9998	1.0000	0.9097	0.5227	0.9096	0.8663
13.0	4.3236E-01	1.1166	0.8792	0.8309	0.8043	1.0425	1.0078	0.9998	1.0000	0.9108	0.5160	0.9108	0.8690
14.0	4.3047E-01	1.1234	0.8809	0.8333	0.8070	1.0456	1.0084	0.9998	1.0000	0.9118	0.5091	0.9117	0.8715
16.0	4.2745E-01	1.1369	0.8834	0.8370	0.8112	1.0515	1.0096	0.9998	1.0000	0.9132	0.4952	0.9132	0.8760
18.0	4.2528E-01	1.1504	0.8851	0.8394	0.8141	1.0574	1.0108	0.9998	1.0000	0.9141	0.4812	0.9141	0.8800
20.0	4.2378E-01	1.1640	0.8859	0.8410	0.8160	1.0633	1.0120	0.9999	1.0000	0.9146	0.4672	0.9147	0.8836
25.0	4.2222E-01	1.1982	0.8858	0.8419	0.8175	1.0775	1.0150	0.9999	1.0000	0.9148	0.4316	0.9148	0.8915
30.0	4.2288E-01	1.2328	0.8835	0.8398	0.8156	1.0915	1.0180	0.9999	1.0000	0.9139	0.3960	0.9139	0.8981
35.0	4.2507E-01	1.2677	0.8796	0.8357	0.8116	1.1051	1.0210	0.9999	1.0000	0.9121	0.3605	0.9122	0.9039
40.0	4.2838E-01	1.3032	0.8748	0.8302	0.8060	1.1184	1.0240	0.9999	1.0000	0.9099	0.3250	0.9100	0.9091
45.0	4.3256E-01	1.3391	0.8692	0.8238	0.7993	1.1315	1.0271	0.9999	1.0000	0.9074	0.2898	0.9074	0.9139
50.0	4.3743E-01	1.3756	0.8632	0.8167	0.7919	1.1444	1.0301	0.9999	1.0000	0.9046	0.2549	0.9047	0.9184

POSITRON Z= -53 A= 127

p	F_0L_0	L_0	λ_2	λ_3	λ_4	μ_1	μ_2	Λ_1	Λ_2	ν_{12}	$\hat{\nu}_{12}$	η_{12}	$\hat{\eta}_{12}$
0.1	1.6460E-09	1.0386	10.7432	43.5479	106.4610	1.0006	1.0000	0.9968	0.9994	4.0052	0.6148	0.8300	0.7676
0.2	1.3958E-04	1.0387	3.2329	5.3934	7.0490	1.0007	1.0000	0.9968	0.9994	1.6216	0.6132	0.8308	0.7684
0.3	4.8361E-03	1.0388	1.8443	2.2675	2.4909	1.0009	1.0001	0.9969	0.9994	1.1815	0.6107	0.8321	0.7699
0.4	2.5153E-02	1.0389	1.3607	1.4671	1.5066	1.0011	1.0001	0.9970	0.9994	1.0288	0.6077	0.8338	0.7717
0.5	6.2809E-02	1.0391	1.1393	1.1504	1.1444	1.0014	1.0002	0.9971	0.9995	0.9595	0.6044	0.8356	0.7739
0.6	1.1025E-01	1.0393	1.0211	0.9943	0.9724	1.0017	1.0002	0.9972	0.9995	0.9231	0.6012	0.8379	0.7762
0.7	1.5963E-01	1.0396	0.9517	0.9069	0.8781	1.0020	1.0003	0.9973	0.9995	0.9021	0.5982	0.8401	0.7786
0.8	2.0611E-01	1.0399	0.9080	0.8537	0.8216	1.0024	1.0003	0.9975	0.9995	0.8893	0.5954	0.8423	0.7809
0.9	2.4750E-01	1.0403	0.8794	0.8195	0.7856	1.0027	1.0004	0.9976	0.9995	0.8813	0.5929	0.8444	0.7833
1.0	2.8322E-01	1.0406	0.8599	0.7967	0.7618	1.0031	1.0005	0.9977	0.9996	0.8761	0.5907	0.8465	0.7855
1.2	3.3893E-01	1.0415	0.8369	0.7702	0.7344	1.0039	1.0006	0.9979	0.9996	0.8707	0.5871	0.8505	0.7898
1.4	3.7787E-01	1.0425	0.8253	0.7572	0.7213	1.0046	1.0007	0.9981	0.9996	0.8688	0.5843	0.8541	0.7937
1.6	4.0489E-01	1.0435	0.8196	0.7511	0.7152	1.0054	1.0009	0.9983	0.9997	0.8685	0.5821	0.8575	0.7973
1.8	4.2367E-01	1.0446	0.8171	0.7487	0.7129	1.0062	1.0010	0.9984	0.9997	0.8691	0.5804	0.8605	0.8006
2.0	4.3676E-01	1.0457	0.8164	0.7483	0.7127	1.0070	1.0011	0.9986	0.9997	0.8702	0.5789	0.8633	0.8037
2.2	4.4588E-01	1.0469	0.8168	0.7492	0.7139	1.0077	1.0013	0.9987	0.9997	0.8715	0.5776	0.8659	0.8065
2.4	4.5220E-01	1.0482	0.8179	0.7508	0.7158	1.0085	1.0014	0.9988	0.9998	0.8730	0.5764	0.8683	0.8092
2.6	4.5652E-01	1.0494	0.8195	0.7529	0.7181	1.0093	1.0015	0.9989	0.9998	0.8744	0.5754	0.8705	0.8116
2.8	4.5940E-01	1.0507	0.8212	0.7553	0.7207	1.0100	1.0017	0.9989	0.9998	0.8759	0.5744	0.8725	0.8139
3.0	4.6124E-01	1.0520	0.8231	0.7577	0.7235	1.0107	1.0018	0.9990	0.9998	0.8773	0.5734	0.8744	0.8161
3.2	4.6229E-01	1.0533	0.8251	0.7603	0.7263	1.0115	1.0019	0.9991	0.9998	0.8787	0.5724	0.8762	0.8182
3.4	4.6278E-01	1.0546	0.8270	0.7628	0.7291	1.0122	1.0021	0.9991	0.9998	0.8801	0.5715	0.8779	0.8201
3.6	4.6284E-01	1.0560	0.8290	0.7654	0.7318	1.0129	1.0022	0.9992	0.9998	0.8814	0.5705	0.8795	0.8219
3.8	4.6259E-01	1.0573	0.8310	0.7678	0.7346	1.0137	1.0023	0.9992	0.9998	0.8826	0.5695	0.8809	0.8237
4.0	4.6210E-01	1.0587	0.8329	0.7703	0.7372	1.0144	1.0025	0.9992	0.9999	0.8838	0.5686	0.8823	0.8253
4.5	4.6020E-01	1.0621	0.8374	0.7761	0.7436	1.0162	1.0028	0.9993	0.9999	0.8866	0.5661	0.8855	0.8292
5.0	4.5778E-01	1.0655	0.8416	0.7815	0.7494	1.0179	1.0031	0.9994	0.9999	0.8891	0.5635	0.8882	0.8326
5.5	4.5516E-01	1.0689	0.8455	0.7864	0.7548	1.0197	1.0034	0.9994	0.9999	0.8914	0.5608	0.8906	0.8358
6.0	4.5251E-01	1.0724	0.8490	0.7909	0.7598	1.0214	1.0037	0.9995	0.9999	0.8934	0.5581	0.8928	0.8387
6.5	4.4992E-01	1.0759	0.8522	0.7950	0.7643	1.0231	1.0041	0.9995	0.9999	0.8952	0.5552	0.8947	0.8413
7.0	4.4743E-01	1.0793	0.8551	0.7988	0.7685	1.0248	1.0044	0.9996	0.9999	0.8968	0.5523	0.8964	0.8438
7.5	4.4508E-01	1.0828	0.8578	0.8023	0.7723	1.0265	1.0047	0.9996	0.9999	0.8983	0.5493	0.8980	0.8461
8.0	4.4286E-01	1.0863	0.8603	0.8055	0.7758	1.0281	1.0050	0.9996	0.9999	0.8997	0.5463	0.8994	0.8483
9.0	4.3883E-01	1.0933	0.8646	0.8111	0.7820	1.0314	1.0056	0.9997	0.9999	0.9021	0.5401	0.9019	0.8523
10.0	4.3533E-01	1.1003	0.8682	0.8159	0.7873	1.0347	1.0063	0.9997	0.9999	0.9040	0.5337	0.9039	0.8559
11.0	4.3229E-01	1.1073	0.8712	0.8199	0.7918	1.0379	1.0069	0.9997	0.9999	0.9057	0.5271	0.9056	0.8591
12.0	4.2966E-01	1.1143	0.8737	0.8234	0.7957	1.0411	1.0075	0.9998	0.9999	0.9071	0.5205	0.9070	0.8621
13.0	4.2739E-01	1.1213	0.8758	0.8263	0.7989	1.0443	1.0081	0.9998	1.0000	0.9082	0.5138	0.9082	0.8649
14.0	4.2544E-01	1.1284	0.8776	0.8287	0.8017	1.0475	1.0088	0.9998	1.0000	0.9092	0.5070	0.9092	0.8674
16.0	4.2232E-01	1.1425	0.8803	0.8325	0.8061	1.0537	1.0100	0.9998	1.0000	0.9107	0.4932	0.9107	0.8721
18.0	4.2007E-01	1.1566	0.8820	0.8351	0.8091	1.0598	1.0112	0.9998	1.0000	0.9116	0.4793	0.9116	0.8763
20.0	4.1851E-01	1.1708	0.8829	0.8368	0.8111	1.0658	1.0125	0.9999	1.0000	0.9122	0.4653	0.9122	0.8800
25.0	4.1686E-01	1.2064	0.8829	0.8378	0.8127	1.0807	1.0156	0.9999	1.0000	0.9124	0.4300	0.9125	0.8862
30.0	4.1749E-01	1.2425	0.8805	0.8357	0.8109	1.0951	1.0187	0.9999	1.0000	0.9115	0.3946	0.9115	0.8950
35.0	4.1969E-01	1.2790	0.8766	0.8315	0.8068	1.1092	1.0218	0.9999	1.0000	0.9097	0.3592	0.9098	0.9010
40.0	4.2304E-01	1.3161	0.8716	0.8259	0.8011	1.1231	1.0249	0.9999	1.0000	0.9075	0.3240	0.9075	0.9064
45.0	4.2727E-01	1.3537	0.8659	0.8194	0.7943	1.1366	1.0281	0.9999	1.0000	0.9049	0.2889	0.9049	0.9114
50.0	4.3221E-01	1.3919	0.8598	0.8121	0.7867	1.1499	1.0312	0.9999	1.0000	0.9021	0.2542	0.9021	0.9161

Instr. see p. 22

POSITRON Z= -54 A= 128

p	F_0L_0	L_0	λ_2	λ_3	λ_4	μ_1	μ_2	Λ_1	Λ_2	ν_{12}	$\hat{\nu}_{12}$	η_{12}	$\hat{\eta}_{12}$
0.1	1.0962E-09	1.0398	10.9695	45.5449	113.6398	1.0006	1.0000	0.9967	0.9994	4.1026	0.6105	0.8246	0.7602
0.2	1.1663E-04	1.0398	3.2835	5.5481	7.3107	1.0008	1.0001	0.9967	0.9994	1.6418	0.6089	0.8253	0.7610
0.3	4.3520E-03	1.0399	1.8623	2.3031	2.5378	1.0009	1.0001	0.9968	0.9994	1.1874	0.6065	0.8266	0.7624
0.4	2.3461E-02	1.0401	1.3673	1.4774	1.5185	1.0012	1.0001	0.9969	0.9994	1.0298	0.6035	0.8284	0.7643
0.5	5.9798E-02	1.0403	1.1407	1.1518	1.1455	1.0014	1.0002	0.9970	0.9994	0.9582	0.6003	0.8304	0.7665
0.6	1.0632E-01	1.0405	1.0197	0.9917	0.9691	1.0017	1.0002	0.9971	0.9995	0.9205	0.5972	0.8325	0.7689
0.7	1.5525E-01	1.0408	0.9485	0.9022	0.8725	1.0021	1.0003	0.9973	0.9995	0.8988	0.5942	0.8348	0.7713
0.8	2.0164E-01	1.0411	0.9038	0.8477	0.8147	1.0024	1.0003	0.9974	0.9995	0.8856	0.5915	0.8370	0.7737
0.9	2.4315E-01	1.0415	0.8745	0.8127	0.7779	1.0028	1.0004	0.9975	0.9995	0.8773	0.5890	0.8392	0.7761
1.0	2.7910E-01	1.0419	0.8545	0.7894	0.7536	1.0032	1.0005	0.9977	0.9996	0.8719	0.5869	0.8413	0.7784
1.2	3.3533E-01	1.0427	0.8309	0.7623	0.7257	1.0040	1.0006	0.9979	0.9996	0.8663	0.5833	0.8454	0.7828
1.4	3.7473E-01	1.0437	0.8191	0.7491	0.7122	1.0048	1.0008	0.9981	0.9996	0.8643	0.5806	0.8492	0.7868
1.6	4.0209E-01	1.0448	0.8133	0.7429	0.7060	1.0056	1.0009	0.9983	0.9997	0.8640	0.5785	0.8526	0.7905
1.8	4.2110E-01	1.0459	0.8107	0.7404	0.7037	1.0064	1.0010	0.9984	0.9997	0.8647	0.5768	0.8558	0.7939
2.0	4.3433E-01	1.0471	0.8101	0.7401	0.7036	1.0072	1.0012	0.9985	0.9997	0.8658	0.5754	0.8587	0.7971
2.2	4.4352E-01	1.0483	0.8105	0.7410	0.7048	1.0080	1.0013	0.9986	0.9997	0.8672	0.5742	0.8613	0.8000
2.4	4.4987E-01	1.0496	0.8117	0.7427	0.7068	1.0088	1.0014	0.9987	0.9998	0.8686	0.5731	0.8638	0.8027
2.6	4.5420E-01	1.0509	0.8133	0.7449	0.7092	1.0096	1.0016	0.9988	0.9998	0.8702	0.5721	0.8661	0.8052
2.8	4.5706E-01	1.0522	0.8151	0.7473	0.7119	1.0103	1.0017	0.9989	0.9998	0.8717	0.5712	0.8682	0.8076
3.0	4.5885E-01	1.0536	0.8171	0.7499	0.7147	1.0111	1.0019	0.9990	0.9998	0.8732	0.5702	0.8702	0.8099
3.2	4.5987E-01	1.0549	0.8191	0.7525	0.7176	1.0118	1.0020	0.9990	0.9998	0.8746	0.5693	0.8720	0.8120
3.4	4.6030E-01	1.0563	0.8212	0.7551	0.7205	1.0126	1.0021	0.9991	0.9998	0.8760	0.5684	0.8738	0.8140
3.6	4.6030E-01	1.0576	0.8232	0.7577	0.7234	1.0133	1.0023	0.9991	0.9998	0.8774	0.5675	0.8754	0.8158
3.8	4.5999E-01	1.0590	0.8252	0.7603	0.7262	1.0141	1.0024	0.9992	0.9998	0.8787	0.5666	0.8769	0.8176
4.0	4.5943E-01	1.0604	0.8272	0.7628	0.7289	1.0148	1.0025	0.9992	0.9998	0.8799	0.5657	0.8784	0.8194
4.5	4.5737E-01	1.0639	0.8319	0.7689	0.7355	1.0167	1.0028	0.9993	0.9999	0.8828	0.5634	0.8816	0.8233
5.0	4.5480E-01	1.0674	0.8363	0.7744	0.7416	1.0185	1.0032	0.9994	0.9999	0.8854	0.5609	0.8845	0.8269
5.5	4.5203E-01	1.0710	0.8403	0.7795	0.7471	1.0203	1.0035	0.9994	0.9999	0.8877	0.5584	0.8870	0.8301
6.0	4.4925E-01	1.0745	0.8440	0.7842	0.7523	1.0220	1.0038	0.9995	0.9999	0.8898	0.5557	0.8892	0.8331
6.5	4.4653E-01	1.0781	0.8473	0.7885	0.7570	1.0238	1.0042	0.9995	0.9999	0.8917	0.5530	0.8912	0.8359
7.0	4.4393E-01	1.0817	0.8504	0.7924	0.7613	1.0255	1.0045	0.9996	0.9999	0.8935	0.5502	0.8930	0.8384
7.5	4.4147E-01	1.0853	0.8532	0.7960	0.7652	1.0273	1.0048	0.9996	0.9999	0.8950	0.5473	0.8947	0.8406
8.0	4.3915E-01	1.0889	0.8557	0.7994	0.7689	1.0290	1.0051	0.9996	0.9999	0.8964	0.5444	0.8961	0.8431
9.0	4.3494E-01	1.0960	0.8602	0.8052	0.7753	1.0324	1.0058	0.9997	0.9999	0.8989	0.5383	0.8987	0.8472
10.0	4.3128E-01	1.1032	0.8640	0.8102	0.7808	1.0357	1.0064	0.9997	0.9999	0.9010	0.5321	0.9008	0.8509
11.0	4.2810E-01	1.1105	0.8672	0.8144	0.7855	1.0390	1.0071	0.9997	0.9999	0.9027	0.5257	0.9026	0.8543
12.0	4.2535E-01	1.1177	0.8699	0.8180	0.7896	1.0423	1.0077	0.9997	0.9999	0.9042	0.5193	0.9041	0.8574
13.0	4.2297E-01	1.1249	0.8721	0.8211	0.7930	1.0456	1.0083	0.9998	1.0000	0.9054	0.5127	0.9053	0.8602
14.0	4.2092E-01	1.1321	0.8740	0.8237	0.7960	1.0488	1.0090	0.9998	1.0000	0.9064	0.5061	0.9064	0.8629
16.0	4.1763E-01	1.1466	0.8768	0.8277	0.8006	1.0552	1.0102	0.9998	1.0000	0.9080	0.4927	0.9080	0.8677
18.0	4.1523E-01	1.1611	0.8787	0.8305	0.8038	1.0615	1.0115	0.9998	1.0000	0.9091	0.4791	0.9091	0.8720
20.0	4.1355E-01	1.1757	0.8798	0.8324	0.8060	1.0677	1.0128	0.9999	1.0000	0.9097	0.4653	0.9097	0.8759
25.0	4.1166E-01	1.2124	0.8800	0.8337	0.8080	1.0829	1.0159	0.9999	1.0000	0.9101	0.4307	0.9101	0.8843
30.0	4.1211E-01	1.2495	0.8777	0.8318	0.8064	1.0977	1.0191	0.9999	1.0000	0.9092	0.3960	0.9092	0.8914
35.0	4.1416E-01	1.2872	0.8739	0.8278	0.8024	1.1121	1.0223	0.9999	1.0000	0.9075	0.3612	0.9075	0.8976
40.0	4.1740E-01	1.3253	0.8690	0.8223	0.7969	1.1263	1.0255	0.9999	1.0000	0.9053	0.3266	0.9053	0.9032
45.0	4.2154E-01	1.3641	0.8634	0.8158	0.7901	1.1402	1.0287	0.9999	1.0000	0.9027	0.2922	0.9028	0.9084
50.0	4.2640E-01	1.4034	0.8572	0.8086	0.7826	1.1538	1.0320	0.9999	1.0000	0.8999	0.2580	0.9000	0.9132

Instr. see p. 22

POSITRON Z= −55 A= 133

p	F_0L_0	L_0	λ_2	λ_3	λ_4	μ_1	μ_2	Λ_1	Λ_2	ν_{12}	$\hat{\nu}_{12}$	η_{12}	$\hat{\eta}_{12}$
0.1	7.2908E-10	1.0410	11.2004	47.6257	121.2774	1.0007	1.0000	0.9966	0.9994	4.2022	0.6065	0.8194	0.7530
0.2	9.7331E-05	1.0411	3.3356	5.7096	7.5864	1.0008	1.0001	0.9966	0.9994	1.6627	0.6049	0.8202	0.7538
0.3	3.9111E-03	1.0412	1.8813	2.3409	2.5878	1.0010	1.0001	0.9967	0.9994	1.1938	0.6025	0.8215	0.7553
0.4	2.1854E-02	1.0414	1.3748	1.4889	1.5319	1.0012	1.0001	0.9968	0.9994	1.0311	0.5996	0.8232	0.7572
0.5	5.6856E-02	1.0416	1.1427	1.1543	1.1478	1.0015	1.0002	0.9969	0.9994	0.9572	0.5964	0.8252	0.7594
0.6	1.0240E-01	1.0418	1.0189	0.9900	0.9667	1.0018	1.0002	0.9970	0.9994	0.9182	0.5933	0.8274	0.7618
0.7	1.5081E-01	1.0421	0.9460	0.8982	0.8678	1.0022	1.0003	0.9972	0.9995	0.8958	0.5903	0.8297	0.7642
0.8	1.9703E-01	1.0424	0.9002	0.8425	0.8086	1.0026	1.0004	0.9973	0.9995	0.8822	0.5876	0.8320	0.7667
0.9	2.3859E-01	1.0428	0.8702	0.8067	0.7710	1.0030	1.0004	0.9974	0.9995	0.8736	0.5852	0.8342	0.7691
1.0	2.7470E-01	1.0432	0.8497	0.7828	0.7462	1.0034	1.0005	0.9976	0.9995	0.8680	0.5831	0.8364	0.7715
1.2	3.3138E-01	1.0441	0.8256	0.7552	0.7176	1.0042	1.0006	0.9978	0.9996	0.8622	0.5796	0.8406	0.7760
1.4	3.7118E-01	1.0452	0.8135	0.7417	0.7040	1.0050	1.0008	0.9980	0.9996	0.8601	0.5770	0.8445	0.7801
1.6	3.9884E-01	1.0463	0.8075	0.7353	0.6977	1.0059	1.0009	0.9982	0.9997	0.8598	0.5749	0.8480	0.7839
1.8	4.1806E-01	1.0474	0.8049	0.7328	0.6953	1.0067	1.0011	0.9983	0.9997	0.8605	0.5733	0.8513	0.7874
2.0	4.3141E-01	1.0487	0.8043	0.7325	0.6952	1.0075	1.0012	0.9985	0.9997	0.8616	0.5720	0.8543	0.7907
2.2	4.4068E-01	1.0500	0.8048	0.7335	0.6965	1.0083	1.0014	0.9986	0.9997	0.8630	0.5708	0.8570	0.7937
2.4	4.4706E-01	1.0513	0.8060	0.7353	0.6985	1.0091	1.0015	0.9987	0.9998	0.8646	0.5698	0.8596	0.7965
2.6	4.5138E-01	1.0526	0.8076	0.7375	0.7010	1.0100	1.0016	0.9988	0.9998	0.8661	0.5688	0.8619	0.7991
2.8	4.5422E-01	1.0540	0.8095	0.7400	0.7038	1.0108	1.0018	0.9989	0.9998	0.8677	0.5679	0.8641	0.8016
3.0	4.5598E-01	1.0554	0.8116	0.7426	0.7067	1.0115	1.0019	0.9989	0.9998	0.8693	0.5671	0.8662	0.8039
3.2	4.5695E-01	1.0568	0.8137	0.7454	0.7097	1.0123	1.0021	0.9990	0.9998	0.8708	0.5662	0.8681	0.8060
3.4	4.5733E-01	1.0582	0.8158	0.7481	0.7126	1.0131	1.0022	0.9991	0.9998	0.8722	0.5653	0.8699	0.8081
3.6	4.5728E-01	1.0596	0.8179	0.7508	0.7156	1.0139	1.0023	0.9991	0.9998	0.8736	0.5645	0.8716	0.8100
3.8	4.5690E-01	1.0611	0.8200	0.7534	0.7185	1.0147	1.0025	0.9992	0.9998	0.8750	0.5636	0.8731	0.8119
4.0	4.5629E-01	1.0625	0.8221	0.7560	0.7213	1.0154	1.0026	0.9992	0.9998	0.8763	0.5627	0.8746	0.8137
4.5	4.5408E-01	1.0661	0.8269	0.7622	0.7281	1.0173	1.0029	0.9993	0.9999	0.8793	0.5605	0.8780	0.8178
5.0	4.5137E-01	1.0698	0.8315	0.7680	0.7343	1.0192	1.0033	0.9994	0.9999	0.8819	0.5581	0.8810	0.8214
5.5	4.4848E-01	1.0735	0.8356	0.7733	0.7401	1.0211	1.0036	0.9994	0.9999	0.8844	0.5556	0.8836	0.8248
6.0	4.4557E-01	1.0772	0.8394	0.7781	0.7454	1.0229	1.0040	0.9995	0.9999	0.8865	0.5530	0.8859	0.8279
6.5	4.4275E-01	1.0809	0.8429	0.7825	0.7502	1.0247	1.0043	0.9995	0.9999	0.8885	0.5504	0.8880	0.8307
7.0	4.4004E-01	1.0846	0.8460	0.7866	0.7546	1.0265	1.0046	0.9995	0.9999	0.8903	0.5476	0.8899	0.8334
7.5	4.3749E-01	1.0883	0.8489	0.7903	0.7587	1.0283	1.0050	0.9996	0.9999	0.8919	0.5448	0.8915	0.8359
8.0	4.3508E-01	1.0921	0.8516	0.7938	0.7625	1.0301	1.0053	0.9996	0.9999	0.8934	0.5419	0.8931	0.8382
9.0	4.3073E-01	1.0995	0.8563	0.7998	0.7692	1.0336	1.0060	0.9996	0.9999	0.8959	0.5360	0.8957	0.8424
10.0	4.2694E-01	1.1070	0.8602	0.8050	0.7749	1.0371	1.0066	0.9997	0.9999	0.8981	0.5299	0.8979	0.8463
11.0	4.2365E-01	1.1145	0.8635	0.8094	0.7798	1.0406	1.0073	0.9997	0.9999	0.8999	0.5236	0.8998	0.8498
12.0	4.2080E-01	1.1220	0.8663	0.8131	0.7839	1.0440	1.0079	0.9997	0.9999	0.9014	0.5172	0.9013	0.8530
13.0	4.1834E-01	1.1295	0.8686	0.8163	0.7875	1.0473	1.0086	0.9998	1.0000	0.9027	0.5108	0.9026	0.8559
14.0	4.1621E-01	1.1370	0.8706	0.8190	0.7906	1.0507	1.0093	0.9998	1.0000	0.9038	0.5042	0.9037	0.8587
16.0	4.1280E-01	1.1521	0.8736	0.8232	0.7954	1.0573	1.0106	0.9998	1.0000	0.9054	0.4909	0.9054	0.8637
18.0	4.1031E-01	1.1672	0.8755	0.8261	0.7988	1.0638	1.0119	0.9998	1.0000	0.9066	0.4775	0.9065	0.8681
20.0	4.0855E-01	1.1823	0.8767	0.8281	0.8011	1.0702	1.0132	0.9999	1.0000	0.9073	0.4639	0.9073	0.8721
25.0	4.0653E-01	1.2205	0.8770	0.8295	0.8032	1.0859	1.0165	0.9999	1.0000	0.9077	0.4296	0.9077	0.8808
30.0	4.0691E-01	1.2591	0.8748	0.8277	0.8017	1.1012	1.0197	0.9999	1.0000	0.9068	0.3951	0.9068	0.8882
35.0	4.0894E-01	1.2982	0.8709	0.8237	0.7977	1.1162	1.0230	0.9999	1.0000	0.9051	0.3606	0.9051	0.8946
40.0	4.1217E-01	1.3380	0.8660	0.8182	0.7921	1.1308	1.0263	0.9999	1.0000	0.9029	0.3262	0.9029	0.9004
45.0	4.1633E-01	1.3783	0.8602	0.8116	0.7853	1.1451	1.0297	0.9999	1.0000	0.9003	0.2921	0.9003	0.9058
50.0	4.2123E-01	1.4194	0.8540	0.8042	0.7776	1.1592	1.0330	0.9999	1.0000	0.8975	0.2582	0.8975	0.9108

Instr. see p. 22

POSITRON Z= -56 A= 134

p	F_0L_0	L_0	λ_2	λ_3	λ_4	μ_1	μ_2	Λ_1	Λ_2	ν_{12}	$\hat{\nu}_{12}$	η_{12}	$\hat{\eta}_{12}$
0.1	4.8578E-10	1.0422	11.4180	49.6897	129.0996	1.0007	1.0000	0.9965	0.9993	4.3010	0.6021	0.8138	0.7454
0.2	8.1370E-05	1.0423	3.3839	5.8657	7.8581	1.0009	1.0001	0.9965	0.9994	1.6832	0.6006	0.8146	0.7463
0.3	3.5213E-03	1.0424	1.8983	2.3759	2.6347	1.0010	1.0001	0.9966	0.9994	1.1997	0.5982	0.8159	0.7477
0.4	2.0394E-02	1.0425	1.3808	1.4986	1.5432	1.0013	1.0001	0.9967	0.9994	1.0320	0.5953	0.8176	0.7496
0.5	5.4159E-02	1.0427	1.1437	1.1552	1.1485	1.0016	1.0002	0.9968	0.9994	0.9558	0.5922	0.8197	0.7519
0.6	9.8802E-02	1.0430	1.0171	0.9871	0.9629	1.0019	1.0002	0.9970	0.9994	0.9156	0.5892	0.8219	0.7543
0.7	1.4677E-01	1.0433	0.9426	0.8932	0.8618	1.0023	1.0003	0.9971	0.9995	0.8925	0.5863	0.8243	0.7568
0.8	1.9289E-01	1.0436	0.8958	0.8363	0.8014	1.0027	1.0004	0.9972	0.9995	0.8784	0.5836	0.8266	0.7593
0.9	2.3456E-01	1.0440	0.8651	0.7997	0.7631	1.0031	1.0004	0.9974	0.9995	0.8695	0.5813	0.8289	0.7618
1.0	2.7090E-01	1.0444	0.8442	0.7754	0.7378	1.0035	1.0005	0.9975	0.9995	0.8637	0.5792	0.8311	0.7642
1.2	3.2811E-01	1.0454	0.8195	0.7472	0.7087	1.0043	1.0007	0.9977	0.9996	0.8577	0.5758	0.8354	0.7688
1.4	3.6837E-01	1.0464	0.8071	0.7335	0.6948	1.0052	1.0008	0.9980	0.9996	0.8556	0.5732	0.8394	0.7730
1.6	3.9638E-01	1.0476	0.8010	0.7270	0.6884	1.0061	1.0009	0.9981	0.9997	0.8553	0.5713	0.8431	0.7770
1.8	4.1582E-01	1.0488	0.7984	0.7245	0.6861	1.0069	1.0011	0.9983	0.9997	0.8559	0.5697	0.8464	0.7806
2.0	4.2932E-01	1.0501	0.7978	0.7242	0.6860	1.0078	1.0012	0.9984	0.9997	0.8571	0.5684	0.8495	0.7839
2.2	4.3866E-01	1.0514	0.7983	0.7252	0.6873	1.0086	1.0014	0.9986	0.9997	0.8586	0.5673	0.8524	0.7870
2.4	4.4507E-01	1.0527	0.7996	0.7270	0.6894	1.0094	1.0015	0.9987	0.9997	0.8602	0.5664	0.8550	0.7899
2.6	4.4940E-01	1.0541	0.8013	0.7293	0.6920	1.0103	1.0017	0.9988	0.9998	0.8618	0.5655	0.8574	0.7926
2.8	4.5221E-01	1.0555	0.8033	0.7319	0.6948	1.0111	1.0018	0.9988	0.9998	0.8634	0.5647	0.8597	0.7951
3.0	4.5393E-01	1.0570	0.8054	0.7347	0.6978	1.0119	1.0020	0.9989	0.9998	0.8650	0.5639	0.8618	0.7975
3.2	4.5485E-01	1.0584	0.8076	0.7374	0.7009	1.0127	1.0021	0.9990	0.9998	0.8666	0.5631	0.8638	0.7997
3.4	4.5517E-01	1.0599	0.8098	0.7403	0.7040	1.0135	1.0022	0.9990	0.9998	0.8681	0.5623	0.8656	0.8018
3.6	4.5506E-01	1.0613	0.8120	0.7430	0.7070	1.0143	1.0024	0.9991	0.9998	0.8695	0.5615	0.8674	0.8038
3.8	4.5462E-01	1.0628	0.8141	0.7458	0.7100	1.0151	1.0025	0.9991	0.9998	0.8709	0.5607	0.8690	0.8057
4.0	4.5394E-01	1.0643	0.8163	0.7485	0.7129	1.0159	1.0027	0.9992	0.9998	0.8723	0.5599	0.8706	0.8076
4.5	4.5156E-01	1.0680	0.8213	0.7549	0.7199	1.0179	1.0030	0.9993	0.9999	0.8754	0.5577	0.8741	0.8118
5.0	4.4869E-01	1.0718	0.8260	0.7608	0.7263	1.0198	1.0034	0.9993	0.9999	0.8782	0.5555	0.8771	0.8156
5.5	4.4565E-01	1.0756	0.8303	0.7663	0.7323	1.0217	1.0037	0.9994	0.9999	0.8807	0.5531	0.8799	0.8191
6.0	4.4260E-01	1.0794	0.8343	0.7713	0.7377	1.0236	1.0040	0.9995	0.9999	0.8829	0.5507	0.8823	0.8222
6.5	4.3964E-01	1.0832	0.8379	0.7759	0.7427	1.0255	1.0044	0.9995	0.9999	0.8850	0.5481	0.8844	0.8252
7.0	4.3682E-01	1.0870	0.8412	0.7801	0.7473	1.0273	1.0047	0.9995	0.9999	0.8868	0.5455	0.8864	0.8279
7.5	4.3415E-01	1.0909	0.8442	0.7840	0.7516	1.0292	1.0051	0.9996	0.9999	0.8885	0.5427	0.8881	0.8305
8.0	4.3164E-01	1.0947	0.8470	0.7875	0.7555	1.0310	1.0054	0.9996	0.9999	0.8901	0.5399	0.8897	0.8329
9.0	4.2710E-01	1.1024	0.8518	0.7938	0.7624	1.0346	1.0061	0.9996	0.9999	0.8927	0.5342	0.8925	0.8373
10.0	4.2315E-01	1.1101	0.8559	0.7992	0.7683	1.0382	1.0068	0.9997	0.9999	0.8950	0.5283	0.8948	0.8412
11.0	4.1972E-01	1.1177	0.8594	0.8038	0.7734	1.0417	1.0074	0.9997	0.9999	0.8969	0.5222	0.8968	0.8448
12.0	4.1674E-01	1.1255	0.8623	0.8077	0.7777	1.0452	1.0081	0.9997	0.9999	0.8985	0.5160	0.8984	0.8481
13.0	4.1417E-01	1.1332	0.8648	0.8110	0.7815	1.0467	1.0088	0.9998	0.9999	0.8998	0.5096	0.8998	0.8512
14.0	4.1193E-01	1.1409	0.8669	0.8139	0.7847	1.0521	1.0095	0.9998	1.0000	0.9010	0.5032	0.9009	0.8540
16.0	4.0835E-01	1.1564	0.8701	0.8184	0.7898	1.0569	1.0108	0.9998	1.0000	0.9027	0.4902	0.9027	0.8592
18.0	4.0571E-01	1.1719	0.8722	0.8215	0.7934	1.0655	1.0122	0.9998	1.0000	0.9039	0.4771	0.9039	0.8638
20.0	4.0383E-01	1.1875	0.8735	0.8236	0.7959	1.0721	1.0135	0.9998	1.0000	0.9047	0.4637	0.9047	0.8679
25.0	4.0156E-01	1.2267	0.8740	0.8254	0.7984	1.0882	1.0168	0.9999	1.0000	0.9053	0.4301	0.9053	0.8769
30.0	4.0175E-01	1.2664	0.8720	0.8238	0.7971	1.1039	1.0202	0.9999	1.0000	0.9045	0.3963	0.9045	0.8845
35.0	4.0364E-01	1.3067	0.8683	0.8199	0.7934	1.1192	1.0235	0.9999	1.0000	0.9029	0.3624	0.9029	0.8911
40.0	4.0675E-01	1.3476	0.8634	0.8145	0.7879	1.1342	1.0269	0.9999	1.0000	0.9007	0.3286	0.9007	0.8971
45.0	4.1082E-01	1.3892	0.8577	0.8080	0.7812	1.1488	1.0303	0.9999	1.0000	0.8981	0.2950	0.8982	0.9026
50.0	4.1564E-01	1.4315	0.8514	0.8007	0.7735	1.1632	1.0337	0.9999	1.0000	0.8953	0.2617	0.8953	0.9078

POSITRON Z= -57 A= 131

p	F_0L_0	L_0	λ_2	λ_3	λ_4	μ_1	μ_2	Λ_1	Λ_2	ν_{12}	$\hat{\nu}_{12}$	η_{12}	$\hat{\eta}_{12}$
0.1	3.2431E-10	1.0432	11.6207	51.7216	137.0611	1.0008	1.0000	0.9964	0.9993	4.3987	0.5975	0.8078	0.7374
0.2	6.8162E-05	1.0433	3.4281	6.0152	8.1239	1.0009	1.0001	0.9965	0.9993	1.7031	0.5960	0.8086	0.7383
0.3	3.1766E-03	1.0434	1.9130	2.4076	2.6780	1.0011	1.0001	0.9966	0.9994	1.2052	0.5937	0.8099	0.7398
0.4	1.9070E-02	1.0436	1.3853	1.5062	1.5521	1.0013	1.0001	0.9967	0.9994	1.0325	0.5908	0.8117	0.7417
0.5	5.1693E-02	1.0438	1.1434	1.1546	1.1474	1.0016	1.0002	0.9968	0.9994	0.9539	0.5878	0.8138	0.7440
0.6	9.5528E-02	1.0440	1.0142	0.9828	0.9578	1.0020	1.0002	0.9969	0.9994	0.9126	0.5848	0.8161	0.7464
0.7	1.4312E-01	1.0444	0.9382	0.8871	0.8547	1.0023	1.0003	0.9971	0.9995	0.8887	0.5819	0.8184	0.7490
0.8	1.8922E-01	1.0447	0.8905	0.8290	0.7931	1.0027	1.0004	0.9972	0.9995	0.8742	0.5794	0.8208	0.7515
0.9	2.3108E-01	1.0451	0.8591	0.7918	0.7541	1.0032	1.0004	0.9973	0.9995	0.8650	0.5770	0.8231	0.7540
1.0	2.6770E-01	1.0455	0.8378	0.7670	0.7284	1.0036	1.0005	0.9975	0.9995	0.8591	0.5750	0.8255	0.7565
1.2	3.2554E-01	1.0465	0.8126	0.7382	0.6989	1.0044	1.0007	0.9977	0.9996	0.8528	0.5717	0.8299	0.7612
1.4	3.6634E-01	1.0476	0.8000	0.7243	0.6847	1.0053	1.0008	0.9979	0.9996	0.8506	0.5693	0.8339	0.7656
1.6	3.9473E-01	1.0487	0.7938	0.7178	0.6783	1.0062	1.0010	0.9981	0.9996	0.8503	0.5674	0.8377	0.7696
1.8	4.1444E-01	1.0500	0.7912	0.7152	0.6759	1.0071	1.0011	0.9983	0.9997	0.8510	0.5660	0.8412	0.7733
2.0	4.2810E-01	1.0513	0.7906	0.7150	0.6759	1.0079	1.0013	0.9984	0.9997	0.8522	0.5648	0.8443	0.7767
2.2	4.3753E-01	1.0526	0.7912	0.7161	0.6773	1.0088	1.0014	0.9985	0.9997	0.8537	0.5638	0.8473	0.7799
2.4	4.4398E-01	1.0540	0.7925	0.7180	0.6794	1.0097	1.0016	0.9987	0.9997	0.8553	0.5629	0.8500	0.7828
2.6	4.4830E-01	1.0554	0.7943	0.7203	0.6821	1.0105	1.0017	0.9987	0.9998	0.8570	0.5621	0.8525	0.7856
2.8	4.5109E-01	1.0568	0.7963	0.7230	0.6850	1.0113	1.0018	0.9988	0.9998	0.8587	0.5614	0.8549	0.7882
3.0	4.5276E-01	1.0583	0.7985	0.7258	0.6881	1.0122	1.0020	0.9989	0.9998	0.8604	0.5607	0.8571	0.7906
3.2	4.5362E-01	1.0598	0.8008	0.7287	0.6913	1.0130	1.0021	0.9990	0.9998	0.8620	0.5600	0.8591	0.7929
3.4	4.5388E-01	1.0612	0.8031	0.7316	0.6944	1.0138	1.0023	0.9990	0.9998	0.8635	0.5592	0.8610	0.7951
3.6	4.5369E-01	1.0627	0.8054	0.7345	0.6976	1.0146	1.0024	0.9991	0.9998	0.8650	0.5585	0.8628	0.7972
3.8	4.5318E-01	1.0642	0.8076	0.7374	0.7007	1.0154	1.0026	0.9991	0.9998	0.8665	0.5578	0.8645	0.7992
4.0	4.5242E-01	1.0658	0.8098	0.7401	0.7037	1.0162	1.0027	0.9992	0.9998	0.8679	0.5570	0.8661	0.8010
4.5	4.4986E-01	1.0696	0.8151	0.7468	0.7109	1.0182	1.0030	0.9993	0.9999	0.8711	0.5551	0.8698	0.8054
5.0	4.4680E-01	1.0734	0.8200	0.7529	0.7175	1.0202	1.0034	0.9993	0.9999	0.8740	0.5530	0.8730	0.8093
5.5	4.4358E-01	1.0773	0.8245	0.7585	0.7236	1.0221	1.0037	0.9994	0.9999	0.8766	0.5508	0.8758	0.8129
6.0	4.4037E-01	1.0811	0.8286	0.7637	0.7293	1.0241	1.0041	0.9994	0.9999	0.8790	0.5485	0.8783	0.8161
6.5	4.3726E-01	1.0850	0.8323	0.7685	0.7345	1.0260	1.0044	0.9995	0.9999	0.8811	0.5461	0.8806	0.8192
7.0	4.3430E-01	1.0889	0.8358	0.7729	0.7393	1.0279	1.0048	0.9995	0.9999	0.8831	0.5436	0.8826	0.8220
7.5	4.3150E-01	1.0928	0.8389	0.7769	0.7437	1.0297	1.0051	0.9996	0.9999	0.8848	0.5410	0.8844	0.8246
8.0	4.2886E-01	1.0967	0.8418	0.7806	0.7477	1.0316	1.0055	0.9996	0.9999	0.8864	0.5383	0.8861	0.8271
9.0	4.2410E-01	1.1045	0.8469	0.7872	0.7549	1.0353	1.0062	0.9996	0.9999	0.8893	0.5328	0.8890	0.8316
10.0	4.1995E-01	1.1123	0.8512	0.7928	0.7611	1.0389	1.0069	0.9997	0.9999	0.8916	0.5272	0.8914	0.8357
11.0	4.1634E-01	1.1202	0.8549	0.7976	0.7664	1.0425	1.0075	0.9997	0.9999	0.8936	0.5213	0.8935	0.8394
12.0	4.1321E-01	1.1280	0.8580	0.8017	0.7710	1.0461	1.0082	0.9997	0.9999	0.8953	0.5153	0.8952	0.8428
13.0	4.1049E-01	1.1359	0.8606	0.8053	0.7749	1.0496	1.0089	0.9998	0.9999	0.8968	0.5092	0.8967	0.8460
14.0	4.0813E-01	1.1437	0.8628	0.8083	0.7783	1.0531	1.0096	0.9998	1.0000	0.8980	0.5030	0.8979	0.8489
16.0	4.0431E-01	1.1595	0.8663	0.8131	0.7838	1.0600	1.0109	0.9998	1.0000	0.8999	0.4905	0.8999	0.8542
18.0	4.0147E-01	1.1753	0.8686	0.8166	0.7877	1.0667	1.0123	0.9998	1.0000	0.9012	0.4777	0.9012	0.8590
20.0	3.9941E-01	1.1911	0.8701	0.8189	0.7905	1.0734	1.0136	0.9998	1.0000	0.9021	0.4648	0.9021	0.8632
25.0	3.9677E-01	1.2310	0.8711	0.8212	0.7935	1.0897	1.0170	0.9999	1.0000	0.9029	0.4322	0.9029	0.8725
30.0	3.9666E-01	1.2714	0.8694	0.8200	0.7927	1.1057	1.0204	0.9999	1.0000	0.9022	0.3993	0.9023	0.8803
35.0	3.9828E-01	1.3124	0.8659	0.8165	0.7893	1.1212	1.0238	0.9999	1.0000	0.9008	0.3664	0.9008	0.8872
40.0	4.0117E-01	1.3541	0.8612	0.8113	0.7841	1.1360	1.0271	0.9999	1.0000	0.8987	0.3342	0.8987	0.8931
45.0	4.0502E-01	1.3964	0.8556	0.8050	0.7776	1.1513	1.0306	0.9999	1.0000	0.8962	0.3007	0.8962	0.8990
50.0	4.0965E-01	1.4394	0.8495	0.7979	0.7702	1.1659	1.0341	0.9999	1.0000	0.8934	0.2683	0.8934	0.9043

Instr. see p. 22

POSITRON Z= -58 A= 136

p	F_0L_0	L_0	λ_2	λ_3	λ_4	μ_1	μ_2	Λ_1	Λ_2	ν_{12}	$\hat{\nu}_{12}$	η_{12}	$\hat{\eta}_{12}$
0.1	2.1584E-10	1.0445	11.8381	53.8958	145.6889	1.0009	1.0000	0.9963	0.9993	4.5001	0.5933	0.8025	0.7301
0.2	5.6921E-05	1.0446	3.4767	6.1784	8.4146	1.0010	1.0001	0.9964	0.9993	1.7244	0.5918	0.8033	0.7309
0.3	2.8567E-03	1.0447	1.9305	2.4443	2.7279	1.0012	1.0001	0.9964	0.9993	1.2117	0.5895	0.8046	0.7324
0.4	1.7776E-02	1.0449	1.3917	1.5168	1.5647	1.0014	1.0001	0.9966	0.9994	1.0338	0.5868	0.8064	0.7344
0.5	4.9188E-02	1.0451	1.1448	1.1563	1.1489	1.0017	1.0002	0.9967	0.9994	0.9528	0.5838	0.8085	0.7366
0.6	9.2080E-02	1.0454	1.0129	0.9806	0.9548	1.0021	1.0002	0.9968	0.9994	0.9102	0.5808	0.8108	0.7391
0.7	1.3915E-01	1.0457	0.9353	0.8827	0.8495	1.0025	1.0003	0.9970	0.9994	0.8856	0.5780	0.8132	0.7417
0.8	1.8506E-01	1.0460	0.8866	0.8234	0.7867	1.0029	1.0004	0.9971	0.9995	0.8706	0.5754	0.8156	0.7443
0.9	2.2696E-01	1.0464	0.8545	0.7855	0.7470	1.0033	1.0005	0.9973	0.9995	0.8611	0.5732	0.8180	0.7469
1.0	2.6375E-01	1.0469	0.8327	0.7602	0.7207	1.0037	1.0005	0.9974	0.9995	0.8550	0.5712	0.8204	0.7494
1.2	3.2204E-01	1.0479	0.8070	0.7309	0.6907	1.0046	1.0007	0.9976	0.9996	0.8486	0.5680	0.8249	0.7542
1.4	3.6325E-01	1.0490	0.7941	0.7167	0.6763	1.0056	1.0008	0.9979	0.9996	0.8463	0.5656	0.8291	0.7587
1.6	3.9196E-01	1.0502	0.7878	0.7100	0.6697	1.0065	1.0010	0.9981	0.9996	0.8459	0.5638	0.8330	0.7628
1.8	4.1188E-01	1.0515	0.7852	0.7075	0.6674	1.0074	1.0011	0.9982	0.9997	0.8467	0.5624	0.8365	0.7666
2.0	4.2567E-01	1.0529	0.7846	0.7072	0.6674	1.0083	1.0013	0.9984	0.9997	0.8479	0.5613	0.8398	0.7701
2.2	4.3517E-01	1.0543	0.7852	0.7084	0.6688	1.0092	1.0015	0.9985	0.9997	0.8495	0.5604	0.8428	0.7734
2.4	4.4164E-01	1.0557	0.7866	0.7104	0.6710	1.0101	1.0016	0.9986	0.9997	0.8511	0.5595	0.8456	0.7764
2.6	4.4595E-01	1.0572	0.7884	0.7128	0.6738	1.0109	1.0018	0.9987	0.9998	0.8529	0.5588	0.8482	0.7793
2.8	4.4872E-01	1.0586	0.7906	0.7156	0.6768	1.0118	1.0019	0.9988	0.9998	0.8546	0.5581	0.8507	0.7819
3.0	4.5035E-01	1.0602	0.7928	0.7185	0.6800	1.0127	1.0020	0.9989	0.9998	0.8563	0.5574	0.8529	0.7845
3.2	4.5116E-01	1.0617	0.7952	0.7214	0.6832	1.0135	1.0022	0.9989	0.9998	0.8580	0.5568	0.8550	0.7868
3.4	4.5136E-01	1.0632	0.7975	0.7244	0.6865	1.0144	1.0023	0.9990	0.9998	0.8596	0.5561	0.8570	0.7891
3.6	4.5111E-01	1.0648	0.7999	0.7274	0.6897	1.0152	1.0025	0.9990	0.9998	0.8612	0.5554	0.8589	0.7912
3.8	4.5054E-01	1.0663	0.8022	0.7303	0.6928	1.0161	1.0026	0.9991	0.9998	0.8627	0.5547	0.8606	0.7932
4.0	4.4972E-01	1.0679	0.8045	0.7332	0.6959	1.0169	1.0028	0.9991	0.9998	0.8641	0.5540	0.8623	0.7952
4.5	4.4700E-01	1.0719	0.8100	0.7400	0.7033	1.0190	1.0031	0.9992	0.9999	0.8674	0.5521	0.8661	0.7996
5.0	4.4379E-01	1.0758	0.8150	0.7463	0.7102	1.0210	1.0035	0.9993	0.9999	0.8704	0.5501	0.8694	0.8037
5.5	4.4043E-01	1.0799	0.8196	0.7521	0.7165	1.0230	1.0039	0.9994	0.9999	0.8731	0.5480	0.8723	0.8074
6.0	4.3709E-01	1.0839	0.8239	0.7575	0.7223	1.0250	1.0042	0.9994	0.9999	0.8756	0.5458	0.8749	0.8107
6.5	4.3387E-01	1.0879	0.8277	0.7624	0.7276	1.0270	1.0046	0.9995	0.9999	0.8778	0.5434	0.8772	0.8138
7.0	4.3079E-01	1.0919	0.8313	0.7669	0.7325	1.0290	1.0049	0.9995	0.9999	0.8798	0.5410	0.8793	0.8168
7.5	4.2789E-01	1.0960	0.8346	0.7711	0.7371	1.0309	1.0053	0.9995	0.9999	0.8816	0.5385	0.8812	0.8195
8.0	4.2517E-01	1.1000	0.8376	0.7749	0.7412	1.0328	1.0057	0.9996	0.9999	0.8833	0.5359	0.8829	0.8220
9.0	4.2025E-01	1.1081	0.8428	0.7817	0.7487	1.0367	1.0064	0.9996	0.9999	0.8862	0.5305	0.8860	0.8267
10.0	4.1596E-01	1.1163	0.8473	0.7875	0.7550	1.0404	1.0071	0.9997	0.9999	0.8887	0.5249	0.8885	0.8309
11.0	4.1224E-01	1.1244	0.8511	0.7925	0.7605	1.0441	1.0078	0.9997	0.9999	0.8907	0.5191	0.8906	0.8347
12.0	4.0900E-01	1.1325	0.8543	0.7967	0.7653	1.0478	1.0085	0.9997	0.9999	0.8925	0.5132	0.8924	0.8382
13.0	4.0620E-01	1.1407	0.8570	0.8004	0.7693	1.0515	1.0092	0.9997	0.9999	0.8940	0.5072	0.8939	0.8415
14.0	4.0376E-01	1.1488	0.8594	0.8035	0.7729	1.0551	1.0099	0.9998	1.0000	0.8953	0.5011	0.8952	0.8445
16.0	3.9981E-01	1.1652	0.8630	0.8085	0.7785	1.0622	1.0113	0.9998	1.0000	0.8972	0.4887	0.8972	0.8500
18.0	3.9688E-01	1.1816	0.8654	0.8121	0.7826	1.0692	1.0127	0.9998	1.0000	0.8986	0.4760	0.8986	0.8549
20.0	3.9474E-01	1.1981	0.8670	0.8145	0.7855	1.0761	1.0141	0.9998	1.0000	0.8996	0.4632	0.8996	0.8593
25.0	3.9197E-01	1.2395	0.8681	0.8170	0.7887	1.0930	1.0176	0.9999	1.0000	0.9004	0.4309	0.9004	0.8689
30.0	3.9178E-01	1.2815	0.8664	0.8159	0.7880	1.1094	1.0211	0.9999	1.0000	0.8998	0.3982	0.8998	0.8770
35.0	3.9337E-01	1.3242	0.8629	0.8124	0.7846	1.1254	1.0245	0.9999	1.0000	0.8983	0.3656	0.8984	0.8840
40.0	3.9624E-01	1.3675	0.8581	0.8072	0.7794	1.1411	1.0281	0.9999	1.0000	0.8962	0.3330	0.8963	0.8904
45.0	4.0011E-01	1.4116	0.8525	0.8008	0.7728	1.1565	1.0316	0.9999	1.0000	0.8937	0.3004	0.8938	0.8963
50.0	4.0476E-01	1.4564	0.8463	0.7936	0.7653	1.1715	1.0351	0.9999	1.0000	0.8909	0.2681	0.8910	0.9018

Instr. see p. 22

POSITRON Z= -59 A= 138

p	F_0L_0	L_0	λ_2	λ_3	λ_4	μ_1	μ_2	Λ_1	Λ_2	ν_{12}	$\hat{\nu}_{12}$	η_{12}	$\hat{\eta}_{12}$
0.1	1.4388E-10	1.0457	12.0426	56.0488	154.5042	1.0009	1.0000	0.9962	0.9993	4.6008	0.5889	0.7968	0.7224
0.2	4.7608E-05	1.0458	3.5218	6.3363	8.7017	1.0010	1.0001	0.9963	0.9993	1.7453	0.5874	0.7976	0.7233
0.3	2.5731E-03	1.0459	1.9460	2.4785	2.7750	1.0012	1.0001	0.9963	0.9993	1.2178	0.5852	0.7990	0.7247
0.4	1.6596E-02	1.0461	1.3970	1.5257	1.5753	1.0015	1.0001	0.9965	0.9993	1.0347	0.5825	0.8008	0.7267
0.5	4.6879E-02	1.0463	1.1453	1.1568	1.1491	1.0018	1.0002	0.9966	0.9994	0.9514	0.5795	0.8029	0.7290
0.6	8.8899E-02	1.0466	1.0108	0.9773	0.9508	1.0022	1.0003	0.9967	0.9994	0.9075	0.5766	0.8052	0.7316
0.7	1.3550E-01	1.0469	0.9317	0.8775	0.8433	1.0026	1.0003	0.9969	0.9994	0.8822	0.5739	0.8077	0.7342
0.8	1.8129E-01	1.0473	0.8819	0.8171	0.7794	1.0030	1.0004	0.9970	0.9995	0.8667	0.5714	0.8101	0.7368
0.9	2.2329E-01	1.0477	0.8493	0.7784	0.7390	1.0034	1.0005	0.9972	0.9995	0.8569	0.5692	0.8126	0.7395
1.0	2.6028E-01	1.0482	0.8271	0.7526	0.7123	1.0039	1.0005	0.9973	0.9995	0.8507	0.5672	0.8150	0.7420
1.2	3.1910E-01	1.0492	0.8008	0.7228	0.6818	1.0048	1.0007	0.9976	0.9996	0.8440	0.5641	0.8196	0.7469
1.4	3.6078E-01	1.0503	0.7877	0.7084	0.6672	1.0058	1.0009	0.9978	0.9996	0.8416	0.5618	0.8239	0.7515
1.6	3.8984E-01	1.0516	0.7813	0.7017	0.6606	1.0067	1.0010	0.9980	0.9996	0.8413	0.5601	0.8279	0.7557
1.8	4.0999E-01	1.0529	0.7786	0.6991	0.6582	1.0076	1.0012	0.9982	0.9997	0.8420	0.5587	0.8316	0.7596
2.0	4.2392E-01	1.0543	0.7780	0.6989	0.6582	1.0086	1.0013	0.9983	0.9997	0.8433	0.5577	0.8350	0.7632
2.2	4.3350E-01	1.0558	0.7787	0.7001	0.6597	1.0095	1.0015	0.9985	0.9997	0.8449	0.5568	0.8381	0.7666
2.4	4.4000E-01	1.0572	0.7801	0.7021	0.6620	1.0104	1.0016	0.9986	0.9997	0.8466	0.5561	0.8410	0.7697
2.6	4.4431E-01	1.0587	0.7821	0.7047	0.6648	1.0113	1.0018	0.9987	0.9997	0.8484	0.5554	0.8437	0.7726
2.8	4.4704E-01	1.0603	0.7843	0.7075	0.6679	1.0122	1.0019	0.9988	0.9998	0.8502	0.5548	0.8462	0.7754
3.0	4.4863E-01	1.0618	0.7866	0.7105	0.6712	1.0131	1.0021	0.9988	0.9998	0.8520	0.5542	0.8485	0.7779
3.2	4.4938E-01	1.0634	0.7890	0.7135	0.6745	1.0140	1.0023	0.9989	0.9998	0.8537	0.5536	0.8507	0.7804
3.4	4.4952E-01	1.0650	0.7915	0.7166	0.6778	1.0149	1.0024	0.9990	0.9998	0.8554	0.5530	0.8527	0.7827
3.6	4.4921E-01	1.0666	0.7939	0.7197	0.6811	1.0157	1.0026	0.9990	0.9998	0.8570	0.5523	0.8546	0.7849
3.8	4.4856E-01	1.0682	0.7963	0.7227	0.6844	1.0166	1.0027	0.9991	0.9998	0.8585	0.5517	0.8565	0.7870
4.0	4.4767E-01	1.0698	0.7987	0.7256	0.6876	1.0175	1.0029	0.9991	0.9998	0.8600	0.5510	0.8582	0.7890
4.5	4.4478E-01	1.0739	0.8043	0.7326	0.6952	1.0196	1.0032	0.9992	0.9998	0.8635	0.5493	0.8621	0.7936
5.0	4.4140E-01	1.0780	0.8095	0.7392	0.7022	1.0217	1.0036	0.9993	0.9999	0.8666	0.5474	0.8655	0.7977
5.5	4.3789E-01	1.0821	0.8143	0.7452	0.7087	1.0238	1.0040	0.9994	0.9999	0.8694	0.5454	0.8685	0.8015
6.0	4.3440E-01	1.0863	0.8187	0.7507	0.7147	1.0258	1.0043	0.9994	0.9999	0.8719	0.5432	0.8712	0.8050
6.5	4.3104E-01	1.0904	0.8227	0.7557	0.7202	1.0279	1.0047	0.9995	0.9999	0.8742	0.5410	0.8736	0.8082
7.0	4.2784E-01	1.0946	0.8264	0.7604	0.7252	1.0299	1.0051	0.9995	0.9999	0.8763	0.5386	0.8758	0.8112
7.5	4.2483E-01	1.0988	0.8298	0.7647	0.7299	1.0319	1.0054	0.9995	0.9999	0.8782	0.5362	0.8778	0.8140
8.0	4.2200E-01	1.1029	0.8329	0.7687	0.7342	1.0339	1.0058	0.9996	0.9999	0.8799	0.5337	0.8796	0.8166
9.0	4.1689E-01	1.1113	0.8384	0.7757	0.7419	1.0378	1.0065	0.9996	0.9999	0.8829	0.5284	0.8827	0.8214
10.0	4.1244E-01	1.1196	0.8430	0.7817	0.7485	1.0417	1.0072	0.9997	0.9999	0.8855	0.5230	0.8853	0.8258
11.0	4.0858E-01	1.1280	0.8470	0.7869	0.7542	1.0455	1.0080	0.9997	0.9999	0.8877	0.5174	0.8875	0.8297
12.0	4.0522E-01	1.1364	0.8503	0.7913	0.7591	1.0493	1.0087	0.9997	0.9999	0.8895	0.5116	0.8894	0.8333
13.0	4.0230E-01	1.1448	0.8532	0.7951	0.7633	1.0530	1.0094	0.9997	0.9999	0.8911	0.5057	0.8910	0.8367
14.0	3.9976E-01	1.1532	0.8556	0.7984	0.7670	1.0567	1.0101	0.9998	1.0000	0.8924	0.4997	0.8923	0.8398
16.0	3.9565E-01	1.1700	0.8594	0.8036	0.7729	1.0640	1.0116	0.9998	1.0000	0.8945	0.4876	0.8945	0.8455
18.0	3.9257E-01	1.1869	0.8620	0.8074	0.7772	1.0712	1.0130	0.9998	1.0000	0.8960	0.4751	0.8960	0.8505
20.0	3.9032E-01	1.2039	0.8638	0.8100	0.7803	1.0783	1.0144	0.9998	1.0000	0.8970	0.4626	0.8970	0.8551
25.0	3.8732E-01	1.2466	0.8651	0.8128	0.7839	1.0956	1.0180	0.9999	1.0000	0.8979	0.4308	0.8980	0.8649
30.0	3.8697E-01	1.2899	0.8636	0.8119	0.7834	1.1125	1.0215	0.9999	1.0000	0.8975	0.3986	0.8975	0.8732
35.0	3.8843E-01	1.3339	0.8601	0.8085	0.7802	1.1289	1.0251	0.9999	1.0000	0.8960	0.3664	0.8960	0.8805
40.0	3.9121E-01	1.3786	0.8554	0.8034	0.7750	1.1450	1.0287	0.9999	1.0000	0.8940	0.3342	0.8940	0.8871
45.0	3.9500E-01	1.4241	0.8498	0.7971	0.7685	1.1608	1.0323	0.9999	1.0000	0.8915	0.3021	0.8915	0.8932
50.0	3.9960E-01	1.4704	0.8436	0.7898	0.7610	1.1762	1.0360	0.9999	1.0000	0.8887	0.2704	0.8887	0.8988

Instr. see p. 22

POSITRON Z= -60 A= 141

p	F_0L_0	L_0	λ_2	λ_3	λ_4	μ_1	μ_2	Λ_1	Λ_2	ν_{12}	$\hat{\nu}_{12}$	η_{12}	$\hat{\eta}_{12}$
0.1	9.5893E-11	1.0469	12.2444	58.2395	163.6930	1.0010	1.0000	0.9961	0.9993	4.7024	0.5845	0.7912	0.7147
0.2	3.9812E-05	1.0470	3.5663	6.4958	8.9956	1.0011	1.0001	0.9962	0.9993	1.7664	0.5831	0.7920	0.7156
0.3	2.3173E-03	1.0471	1.9614	2.5128	2.8227	1.0013	1.0001	0.9962	0.9993	1.2241	0.5809	0.7934	0.7171
0.4	1.5493E-02	1.0473	1.4021	1.5347	1.5861	1.0016	1.0001	0.9964	0.9993	1.0358	0.5782	0.7952	0.7191
0.5	4.4671E-02	1.0475	1.1457	1.1573	1.1493	1.0019	1.0002	0.9965	0.9994	0.9501	0.5753	0.7973	0.7214
0.6	8.5817E-02	1.0478	1.0087	0.9741	0.9468	1.0023	1.0003	0.9966	0.9994	0.9049	0.5725	0.7997	0.7240
0.7	1.3194E-01	1.0482	0.9281	0.8723	0.8373	1.0027	1.0003	0.9968	0.9994	0.8788	0.5698	0.8022	0.7267
0.8	1.7758E-01	1.0485	0.8774	0.8108	0.7722	1.0031	1.0004	0.9969	0.9994	0.8629	0.5673	0.8047	0.7293
0.9	2.1965E-01	1.0490	0.8441	0.7714	0.7311	1.0036	1.0005	0.9971	0.9995	0.8528	0.5651	0.8072	0.7320
1.0	2.5685E-01	1.0495	0.8214	0.7452	0.7040	1.0041	1.0006	0.9972	0.9995	0.8463	0.5632	0.8097	0.7346
1.2	3.1616E-01	1.0505	0.7947	0.7149	0.6730	1.0050	1.0007	0.9975	0.9995	0.8395	0.5602	0.8144	0.7397
1.4	3.5831E-01	1.0517	0.7813	0.7003	0.6582	1.0060	1.0009	0.9977	0.9996	0.8370	0.5580	0.8188	0.7443
1.6	3.8771E-01	1.0530	0.7748	0.6934	0.6515	1.0070	1.0010	0.9980	0.9996	0.8367	0.5563	0.8229	0.7486
1.8	4.0809E-01	1.0544	0.7721	0.6908	0.6491	1.0079	1.0012	0.9981	0.9997	0.8374	0.5551	0.8267	0.7526
2.0	4.2216E-01	1.0558	0.7715	0.6907	0.6492	1.0089	1.0014	0.9983	0.9997	0.8387	0.5541	0.8302	0.7563
2.2	4.3181E-01	1.0573	0.7723	0.6919	0.6507	1.0098	1.0015	0.9984	0.9997	0.8404	0.5533	0.8334	0.7598
2.4	4.3834E-01	1.0588	0.7738	0.6940	0.6531	1.0108	1.0017	0.9985	0.9997	0.8422	0.5527	0.8363	0.7630
2.6	4.4264E-01	1.0604	0.7758	0.6966	0.6560	1.0117	1.0018	0.9986	0.9997	0.8440	0.5520	0.8391	0.7660
2.8	4.4535E-01	1.0620	0.7780	0.6996	0.6592	1.0126	1.0020	0.9987	0.9998	0.8459	0.5515	0.8417	0.7688
3.0	4.4689E-01	1.0636	0.7805	0.7026	0.6626	1.0136	1.0022	0.9988	0.9998	0.8477	0.5509	0.8441	0.7715
3.2	4.4759E-01	1.0652	0.7830	0.7058	0.6660	1.0145	1.0023	0.9989	0.9998	0.8495	0.5504	0.8463	0.7740
3.4	4.4766E-01	1.0669	0.7855	0.7089	0.6694	1.0154	1.0025	0.9989	0.9998	0.8512	0.5498	0.8484	0.7763
3.6	4.4728E-01	1.0685	0.7880	0.7121	0.6728	1.0163	1.0026	0.9990	0.9998	0.8529	0.5492	0.8504	0.7786
3.8	4.4656E-01	1.0702	0.7905	0.7152	0.6761	1.0172	1.0028	0.9990	0.9998	0.8545	0.5486	0.8523	0.7807
4.0	4.4560E-01	1.0718	0.7929	0.7182	0.6794	1.0181	1.0029	0.9991	0.9998	0.8560	0.5480	0.8541	0.7828
4.5	4.4254E-01	1.0761	0.7987	0.7254	0.6872	1.0203	1.0033	0.9992	0.9998	0.8596	0.5464	0.8581	0.7875
5.0	4.3899E-01	1.0803	0.8041	0.7321	0.6944	1.0224	1.0037	0.9993	0.9999	0.8628	0.5446	0.8616	0.7918
5.5	4.3532E-01	1.0846	0.8091	0.7383	0.7011	1.0246	1.0041	0.9993	0.9999	0.8657	0.5427	0.8647	0.7957
6.0	4.3169E-01	1.0888	0.8136	0.7440	0.7072	1.0267	1.0044	0.9994	0.9999	0.8683	0.5406	0.8675	0.7993
6.5	4.2820E-01	1.0931	0.8178	0.7492	0.7129	1.0288	1.0048	0.9994	0.9999	0.8707	0.5384	0.8700	0.8026
7.0	4.2488E-01	1.0974	0.8216	0.7540	0.7181	1.0309	1.0052	0.9995	0.9999	0.8728	0.5362	0.8723	0.8056
7.5	4.2175E-01	1.1017	0.8251	0.7584	0.7229	1.0330	1.0056	0.9995	0.9999	0.8748	0.5338	0.8743	0.8085
8.0	4.1882E-01	1.1060	0.8283	0.7625	0.7273	1.0350	1.0059	0.9996	0.9999	0.8766	0.5314	0.8762	0.8112
9.0	4.1353E-01	1.1146	0.8340	0.7698	0.7352	1.0391	1.0067	0.9996	0.9999	0.8797	0.5263	0.8794	0.8162
10.0	4.0892E-01	1.1233	0.8388	0.7760	0.7420	1.0430	1.0074	0.9996	0.9999	0.8824	0.5210	0.8822	0.8207
11.0	4.0492E-01	1.1319	0.8429	0.7813	0.7479	1.0470	1.0082	0.9997	0.9999	0.8846	0.5155	0.8845	0.8247
12.0	4.0144E-01	1.1406	0.8464	0.7859	0.7530	1.0509	1.0089	0.9997	0.9999	0.8865	0.5098	0.8864	0.8285
13.0	3.9842E-01	1.1492	0.8494	0.7899	0.7574	1.0547	1.0097	0.9997	0.9999	0.8882	0.5040	0.8881	0.8319
14.0	3.9579E-01	1.1579	0.8519	0.7933	0.7612	1.0585	1.0104	0.9998	0.9999	0.8896	0.4982	0.8895	0.8351
16.0	3.9152E-01	1.1753	0.8559	0.7988	0.7673	1.0661	1.0119	0.9998	1.0000	0.8918	0.4862	0.8917	0.8410
18.0	3.8831E-01	1.1927	0.8587	0.8027	0.7719	1.0735	1.0133	0.9998	1.0000	0.8933	0.4740	0.8933	0.8462
20.0	3.8595E-01	1.2103	0.8605	0.8055	0.7751	1.0808	1.0148	0.9998	1.0000	0.8944	0.4616	0.8944	0.8509
25.0	3.8275E-01	1.2544	0.8621	0.8086	0.7790	1.0986	1.0184	0.9999	1.0000	0.8955	0.4302	0.8955	0.8610
30.0	3.8227E-01	1.2991	0.8607	0.8079	0.7787	1.1159	1.0221	0.9999	1.0000	0.8951	0.3985	0.8951	0.8696
35.0	3.8363E-01	1.3446	0.8573	0.8046	0.7756	1.1328	1.0258	0.9999	1.0000	0.8937	0.3666	0.8937	0.8771
40.0	3.8635E-01	1.3908	0.8526	0.7995	0.7705	1.1497	1.0295	0.9999	1.0000	0.8916	0.3338	0.8916	0.8841
45.0	3.9009E-01	1.4379	0.8470	0.7932	0.7641	1.1655	1.0331	0.9999	1.0000	0.8891	0.3031	0.8892	0.8902
50.0	3.9465E-01	1.4859	0.8407	0.7859	0.7565	1.1813	1.0369	0.9999	1.0000	0.8864	0.2716	0.8864	0.8960

Instr. see p. 22

POSITRON Z= -61 A= 143

p	F_0L_0	L_0	λ_2	λ_3	λ_4	μ_1	μ_2	Λ_1	Λ_2	ν_{12}	$\hat{\nu}_{12}$	η_{12}	$\hat{\eta}_{12}$
0.1	6.3959E-11	1.0481	12.4376	60.4296	173.1444	1.0010	1.0000	0.9960	0.9993	4.8039	0.5800	0.7855	0.7069
0.2	3.3318E-05	1.0482	3.6086	6.6529	9.2903	1.0012	1.0001	0.9961	0.9993	1.7874	0.5786	0.7863	0.7079
0.3	2.0884E-03	1.0483	1.9757	2.5457	2.8690	1.0014	1.0001	0.9961	0.9993	1.2302	0.5765	0.7876	0.7093
0.4	1.4473E-02	1.0485	1.4066	1.5428	1.5960	1.0017	1.0002	0.9963	0.9993	1.0367	0.5739	0.7894	0.7113
0.5	4.2600E-02	1.0487	1.1457	1.1572	1.1489	1.0020	1.0002	0.9964	0.9993	0.9486	0.5710	0.7916	0.7137
0.6	8.2906E-02	1.0490	1.0062	0.9704	0.9423	1.0024	1.0003	0.9966	0.9994	0.9022	0.5682	0.7940	0.7163
0.7	1.2857E-01	1.0494	0.9241	0.8667	0.8308	1.0028	1.0003	0.9967	0.9994	0.8753	0.5656	0.7965	0.7190
0.8	1.7409E-01	1.0498	0.8725	0.8041	0.7647	1.0033	1.0004	0.9969	0.9994	0.8589	0.5632	0.7991	0.7217
0.9	2.1625E-01	1.0502	0.8386	0.7641	0.7229	1.0037	1.0005	0.9970	0.9995	0.8486	0.5611	0.8017	0.7244
1.0	2.5366E-01	1.0507	0.8155	0.7375	0.6954	1.0042	1.0006	0.9972	0.9995	0.8419	0.5592	0.8042	0.7271
1.2	3.1352E-01	1.0518	0.7883	0.7067	0.6640	1.0052	1.0007	0.9974	0.9995	0.8348	0.5563	0.8090	0.7322
1.4	3.5614E-01	1.0530	0.7747	0.6919	0.6490	1.0062	1.0009	0.9977	0.9996	0.8323	0.5541	0.8136	0.7370
1.6	3.8591E-01	1.0544	0.7681	0.6850	0.6422	1.0072	1.0011	0.9979	0.9996	0.8319	0.5526	0.8178	0.7414
1.8	4.0653E-01	1.0558	0.7654	0.6823	0.6399	1.0082	1.0012	0.9981	0.9996	0.8327	0.5514	0.8216	0.7455
2.0	4.2075E-01	1.0573	0.7648	0.6822	0.6400	1.0092	1.0014	0.9982	0.9997	0.8341	0.5505	0.8252	0.7493
2.2	4.3047E-01	1.0588	0.7656	0.6835	0.6416	1.0102	1.0016	0.9984	0.9997	0.8357	0.5498	0.8285	0.7528
2.4	4.3703E-01	1.0604	0.7672	0.6857	0.6440	1.0111	1.0017	0.9985	0.9997	0.8376	0.5492	0.8316	0.7561
2.6	4.4133E-01	1.0620	0.7693	0.6884	0.6470	1.0121	1.0019	0.9986	0.9997	0.8395	0.5486	0.8344	0.7592
2.8	4.4399E-01	1.0636	0.7716	0.6914	0.6503	1.0131	1.0020	0.9987	0.9998	0.8414	0.5481	0.8371	0.7621
3.0	4.4549E-01	1.0653	0.7741	0.6945	0.6537	1.0140	1.0022	0.9988	0.9998	0.8433	0.5476	0.8395	0.7648
3.2	4.4613E-01	1.0670	0.7767	0.6978	0.6572	1.0150	1.0024	0.9988	0.9998	0.8451	0.5471	0.8419	0.7674
3.4	4.4613E-01	1.0687	0.7793	0.7010	0.6607	1.0159	1.0025	0.9989	0.9998	0.8469	0.5466	0.8440	0.7698
3.6	4.4568E-01	1.0704	0.7819	0.7043	0.6642	1.0168	1.0027	0.9990	0.9998	0.8486	0.5461	0.8461	0.7721
3.8	4.4489E-01	1.0721	0.7845	0.7074	0.6676	1.0177	1.0028	0.9990	0.9998	0.8502	0.5456	0.8480	0.7743
4.0	4.4386E-01	1.0738	0.7870	0.7105	0.6709	1.0187	1.0030	0.9991	0.9998	0.8518	0.5450	0.8499	0.7764
4.5	4.4060E-01	1.0781	0.7930	0.7180	0.6789	1.0209	1.0034	0.9992	0.9998	0.8555	0.5435	0.8540	0.7813
5.0	4.3688E-01	1.0825	0.7985	0.7249	0.6864	1.0232	1.0038	0.9993	0.9999	0.8588	0.5418	0.8576	0.7857
5.5	4.3305E-01	1.0869	0.8037	0.7312	0.6932	1.0254	1.0042	0.9993	0.9999	0.8618	0.5400	0.8609	0.7897
6.0	4.2926E-01	1.0913	0.8083	0.7371	0.6995	1.0276	1.0046	0.9994	0.9999	0.8645	0.5380	0.8638	0.7934
6.5	4.2563E-01	1.0957	0.8126	0.7425	0.7053	1.0297	1.0049	0.9994	0.9999	0.8670	0.5359	0.8664	0.7968
7.0	4.2218E-01	1.1001	0.8166	0.7474	0.7107	1.0319	1.0053	0.9995	0.9999	0.8692	0.5338	0.8687	0.8000
7.5	4.1894E-01	1.1046	0.8202	0.7520	0.7156	1.0340	1.0057	0.9995	0.9999	0.8713	0.5315	0.8708	0.8029
8.0	4.1590E-01	1.1090	0.8235	0.7562	0.7202	1.0361	1.0061	0.9995	0.9999	0.8731	0.5291	0.8727	0.8057
9.0	4.1041E-01	1.1179	0.8294	0.7637	0.7284	1.0403	1.0069	0.9996	0.9999	0.8764	0.5242	0.8761	0.8108
10.0	4.0564E-01	1.1268	0.8344	0.7701	0.7354	1.0444	1.0076	0.9996	0.9999	0.8792	0.5190	0.8789	0.8154
11.0	4.0149E-01	1.1357	0.8387	0.7757	0.7415	1.0484	1.0084	0.9997	0.9999	0.8815	0.5136	0.8813	0.8196
12.0	3.9788E-01	1.1446	0.8424	0.7804	0.7467	1.0524	1.0091	0.9997	0.9999	0.8835	0.5081	0.8834	0.8234
13.0	3.9475E-01	1.1535	0.8455	0.7846	0.7513	1.0564	1.0099	0.9997	0.9999	0.8852	0.5025	0.8851	0.8270
14.0	3.9202E-01	1.1624	0.8481	0.7881	0.7553	1.0603	1.0106	0.9998	0.9999	0.8867	0.4967	0.8866	0.8303
16.0	3.8757E-01	1.1803	0.8523	0.7938	0.7617	1.0680	1.0121	0.9998	1.0000	0.8890	0.4850	0.8889	0.8363
18.0	3.8422E-01	1.1983	0.8552	0.7980	0.7664	1.0756	1.0136	0.9998	1.0000	0.8906	0.4730	0.8906	0.8417
20.0	3.8174E-01	1.2163	0.8572	0.8010	0.7698	1.0831	1.0151	0.9998	1.0000	0.8917	0.4608	0.8917	0.8465
25.0	3.7831E-01	1.2618	0.8590	0.8044	0.7741	1.1013	1.0189	0.9999	1.0000	0.8930	0.4300	0.8930	0.8570
30.0	3.7765E-01	1.3079	0.8578	0.8039	0.7741	1.1191	1.0226	0.9999	1.0000	0.8927	0.3987	0.8927	0.8658
35.0	3.7889E-01	1.3548	0.8545	0.8007	0.7711	1.1364	1.0264	0.9999	1.0000	0.8913	0.3673	0.8914	0.8736
40.0	3.8151E-01	1.4024	0.8499	0.7957	0.7662	1.1533	1.0301	0.9999	1.0000	0.8893	0.3360	0.8893	0.8806
45.0	3.8517E-01	1.4510	0.8443	0.7894	0.7597	1.1699	1.0339	0.9999	1.0000	0.8869	0.3047	0.8869	0.8870
50.0	3.8968E-01	1.5005	0.8380	0.7822	0.7523	1.1861	1.0377	0.9999	1.0000	0.8841	0.2736	0.8841	0.8930

Instr. see p. 22

POSITRON Z= -62 A= 144

p	F_0L_0	L_0	λ_2	λ_3	λ_4	μ_1	μ_2	Λ_1	Λ_2	ν_{12}	$\hat{\nu}_{12}$	η_{12}	$\hat{\eta}_{12}$
0.1	4.2693E-11	1.0492	12.6219	62.6137	182.8407	1.0011	1.0000	0.9959	0.9993	4.9053	0.5754	0.7796	0.6990
0.2	2.7905E-05	1.0493	3.6484	6.8070	9.5846	1.0012	1.0001	0.9960	0.9993	1.8083	0.5740	0.7804	0.6999
0.3	1.8837E-03	1.0494	1.9888	2.5771	2.9139	1.0015	1.0001	0.9961	0.9993	1.2362	0.5720	0.7817	0.7014
0.4	1.3531E-02	1.0496	1.4103	1.5500	1.6048	1.0017	1.0002	0.9962	0.9993	1.0375	0.5694	0.7836	0.7034
0.5	4.0657E-02	1.0499	1.1451	1.1564	1.1478	1.0021	1.0002	0.9963	0.9993	0.9470	0.5666	0.7858	0.7058
0.6	8.0158E-02	1.0502	1.0033	0.9661	0.9372	1.0025	1.0003	0.9965	0.9994	0.8993	0.5639	0.7882	0.7084
0.7	1.2539E-01	1.0505	0.9198	0.8607	0.8239	1.0029	1.0004	0.9966	0.9994	0.8717	0.5613	0.7908	0.7112
0.8	1.7080E-01	1.0510	0.8673	0.7971	0.7568	1.0034	1.0004	0.9968	0.9994	0.8548	0.5590	0.7934	0.7140
0.9	2.1308E-01	1.0514	0.8328	0.7565	0.7144	1.0039	1.0005	0.9970	0.9994	0.8442	0.5569	0.7960	0.7167
1.0	2.5073E-01	1.0519	0.8094	0.7294	0.6865	1.0044	1.0006	0.9971	0.9995	0.8373	0.5551	0.7986	0.7195
1.2	3.1115E-01	1.0531	0.7817	0.6982	0.6547	1.0054	1.0008	0.9974	0.9995	0.8300	0.5522	0.8035	0.7247
1.4	3.5429E-01	1.0543	0.7679	0.6832	0.6396	1.0064	1.0009	0.9976	0.9996	0.8274	0.5502	0.8082	0.7295
1.6	3.8443E-01	1.0557	0.7612	0.6762	0.6327	1.0074	1.0011	0.9979	0.9996	0.8270	0.5487	0.8125	0.7341
1.8	4.0531E-01	1.0572	0.7584	0.6736	0.6303	1.0085	1.0013	0.9980	0.9996	0.8278	0.5476	0.8164	0.7382
2.0	4.1968E-01	1.0587	0.7579	0.6735	0.6305	1.0095	1.0014	0.9982	0.9997	0.8292	0.5468	0.8201	0.7421
2.2	4.2949E-01	1.0603	0.7587	0.6749	0.6321	1.0105	1.0016	0.9983	0.9997	0.8309	0.5462	0.8235	0.7457
2.4	4.3607E-01	1.0619	0.7604	0.6771	0.6347	1.0115	1.0018	0.9985	0.9997	0.8328	0.5456	0.8267	0.7491
2.6	4.4036E-01	1.0635	0.7625	0.6799	0.6377	1.0125	1.0019	0.9986	0.9997	0.8348	0.5452	0.8296	0.7523
2.8	4.4300E-01	1.0652	0.7650	0.6830	0.6411	1.0135	1.0021	0.9987	0.9997	0.8368	0.5447	0.8323	0.7552
3.0	4.4444E-01	1.0669	0.7675	0.6862	0.6446	1.0145	1.0023	0.9987	0.9998	0.8387	0.5443	0.8349	0.7580
3.2	4.4501E-01	1.0686	0.7702	0.6895	0.6482	1.0154	1.0024	0.9988	0.9998	0.8406	0.5439	0.8373	0.7607
3.4	4.4495E-01	1.0704	0.7729	0.6929	0.6518	1.0164	1.0026	0.9989	0.9998	0.8424	0.5434	0.8395	0.7632
3.6	4.4442E-01	1.0721	0.7756	0.6962	0.6553	1.0173	1.0027	0.9989	0.9998	0.8442	0.5430	0.8416	0.7655
3.8	4.4355E-01	1.0739	0.7782	0.6995	0.6588	1.0183	1.0029	0.9990	0.9998	0.8459	0.5425	0.8436	0.7678
4.0	4.4244E-01	1.0757	0.7808	0.7027	0.6623	1.0192	1.0031	0.9991	0.9998	0.8475	0.5420	0.8455	0.7700
4.5	4.3899E-01	1.0801	0.7870	0.7103	0.6705	1.0216	1.0035	0.9992	0.9998	0.8514	0.5406	0.8498	0.7749
5.0	4.3508E-01	1.0846	0.7928	0.7174	0.6781	1.0239	1.0039	0.9992	0.9999	0.8548	0.5390	0.8536	0.7795
5.5	4.3107E-01	1.0891	0.7980	0.7239	0.6851	1.0261	1.0043	0.9993	0.9999	0.8579	0.5373	0.8569	0.7836
6.0	4.2713E-01	1.0936	0.8029	0.7299	0.6916	1.0284	1.0046	0.9994	0.9999	0.8607	0.5355	0.8599	0.7874
6.5	4.2335E-01	1.0982	0.8073	0.7355	0.6976	1.0306	1.0050	0.9994	0.9999	0.8632	0.5335	0.8626	0.7909
7.0	4.1976E-01	1.1027	0.8114	0.7406	0.7031	1.0328	1.0054	0.9995	0.9999	0.8655	0.5314	0.8650	0.7941
7.5	4.1639E-01	1.1C73	0.8152	0.7453	0.7082	1.0350	1.0058	0.9995	0.9999	0.8677	0.5292	0.8672	0.7972
8.0	4.1324E-01	1.1118	0.8186	0.7497	0.7129	1.0371	1.0062	0.9995	0.9999	0.8696	0.5270	0.8692	0.8000
9.0	4.0754E-01	1.1209	0.8247	0.7574	0.7213	1.0414	1.0070	0.9996	0.9999	0.8730	0.5222	0.8727	0.8053
10.0	4.0259E-01	1.1301	0.8299	0.7641	0.7286	1.0456	1.0078	0.9996	0.9999	0.8758	0.5172	0.8756	0.8100
11.0	3.9828E-01	1.1392	0.8344	0.7698	0.7349	1.0498	1.0085	0.9997	0.9999	0.8783	0.5120	0.8781	0.8143
12.0	3.9454E-01	1.1484	0.8382	0.7748	0.7403	1.0539	1.0093	0.9997	0.9999	0.8804	0.5066	0.8802	0.8183
13.0	3.9128E-01	1.1575	0.8414	0.7791	0.7451	1.0579	1.0101	0.9997	0.9999	0.8822	0.5011	0.8821	0.8219
14.0	3.8844E-01	1.1667	0.8442	0.7828	0.7492	1.0620	1.0109	0.9997	0.9999	0.8837	0.4955	0.8836	0.8253
16.0	3.8381E-01	1.1851	0.8486	0.7887	0.7559	1.0699	1.0124	0.9998	1.0000	0.8861	0.4840	0.8861	0.8315
18.0	3.8030E-01	1.2035	0.8517	0.7931	0.7608	1.0776	1.0139	0.9998	1.0000	0.8879	0.4722	0.8878	0.8370
20.0	3.7768E-01	1.2221	0.8539	0.7963	0.7645	1.0853	1.0154	0.9998	1.0000	0.8891	0.4603	0.8891	0.8420
25.0	3.7399E-01	1.2688	0.8560	0.8001	0.7692	1.1040	1.0193	0.9999	1.0000	0.8905	0.4300	0.8905	0.8528
30.0	3.7314E-01	1.3162	0.8549	0.7999	0.7694	1.1221	1.0231	0.9999	1.0000	0.8903	0.3993	0.8903	0.8619
35.0	3.7422E-01	1.3643	0.8518	0.7969	0.7667	1.1399	1.0269	0.9999	1.0000	0.8890	0.3685	0.8890	0.8698
40.0	3.7671E-01	1.4134	0.8472	0.7920	0.7619	1.1572	1.0307	0.9999	1.0000	0.8871	0.3376	0.8871	0.8770
45.0	3.8027E-01	1.4634	0.8417	0.7858	0.7556	1.1741	1.0346	0.9999	1.0000	0.8846	0.3068	0.8847	0.8836
50.0	3.8468E-01	1.5143	0.8355	0.7787	0.7482	1.1907	1.0385	0.9999	1.0000	0.8819	0.2762	0.8819	0.8898

Instr. see p. 22

POSITRON Z= -63 A= 147

p	F_0L_0	L_0	λ_2	λ_3	λ_4	μ_1	μ_2	Λ_1	Λ_2	ν_{12}	$\hat{\nu}_{12}$	η_{12}	$\hat{\eta}_{12}$
0.1	2.8481E-11	1.0504	12.8060	64.8470	192.9711	1.0012	1.0001	0.9958	0.9992	5.0079	0.5709	0.7738	0.6912
0.2	2.3357E-05	1.0505	3.6884	6.9643	9.8887	1.0013	1.0001	0.9958	0.9992	1.8296	0.5696	0.7746	0.6921
0.3	1.6979E-03	1.0506	2.0021	2.6094	2.9603	1.0015	1.0001	0.9959	0.9993	1.2424	0.5676	0.7760	0.6936
0.4	1.2643E-02	1.0508	1.4143	1.5577	1.6143	1.0018	1.0002	0.9961	0.9993	1.0384	0.5651	0.7778	0.6957
0.5	3.8779E-02	1.0511	1.1447	1.1561	1.1471	1.0022	1.0002	0.9962	0.9993	0.9456	0.5623	0.7801	0.6981
0.6	7.7455E-02	1.0514	1.0006	0.9623	0.9326	1.0026	1.0003	0.9964	0.9993	0.8966	0.5596	0.7825	0.7007
0.7	1.2222E-01	1.0518	0.9157	0.8551	0.8174	1.0031	1.0004	0.9965	0.9994	0.8682	0.5571	0.7851	0.7035
0.8	1.6748E-01	1.0522	0.8624	0.7905	0.7493	1.0035	1.0004	0.9967	0.9994	0.8509	0.5548	0.7878	0.7063
0.9	2.0984E-01	1.0527	0.8273	0.7492	0.7063	1.0040	1.0005	0.9969	0.9994	0.8399	0.5528	0.7904	0.7091
1.0	2.4769E-01	1.0532	0.8035	0.7218	0.6780	1.0046	1.0006	0.9970	0.9995	0.8329	0.5510	0.7931	0.7119
1.2	3.0864E-01	1.0544	0.7753	0.6901	0.6458	1.0056	1.0008	0.9973	0.9995	0.8254	0.5483	0.7981	0.7172
1.4	3.5226E-01	1.0557	0.7613	0.6749	0.6305	1.0067	1.0009	0.9976	0.9996	0.8226	0.5463	0.8029	0.7222
1.6	3.8276E-01	1.0571	0.7545	0.6678	0.6236	1.0077	1.0011	0.9978	0.9996	0.8222	0.5449	0.8073	0.7268
1.8	4.0388E-01	1.0586	0.7517	0.6652	0.6212	1.0088	1.0013	0.9980	0.9996	0.8231	0.5439	0.8114	0.7311
2.0	4.1840E-01	1.0602	0.7513	0.6651	0.6214	1.0098	1.0015	0.9982	0.9997	0.8245	0.5432	0.8151	0.7351
2.2	4.2829E-01	1.0618	0.7521	0.6666	0.6231	1.0109	1.0016	0.9983	0.9997	0.8263	0.5426	0.8186	0.7388
2.4	4.3489E-01	1.0635	0.7538	0.6689	0.6257	1.0119	1.0018	0.9984	0.9997	0.8282	0.5422	0.8219	0.7422
2.6	4.3917E-01	1.0652	0.7561	0.6717	0.6288	1.0129	1.0020	0.9985	0.9997	0.8303	0.5417	0.8249	0.7455
2.8	4.4177E-01	1.0669	0.7586	0.6749	0.6323	1.0140	1.0021	0.9986	0.9997	0.8323	0.5414	0.8277	0.7485
3.0	4.4316E-01	1.0687	0.7612	0.6782	0.6359	1.0150	1.0023	0.9987	0.9998	0.8343	0.5410	0.8303	0.7514
3.2	4.4367E-01	1.0705	0.7640	0.6817	0.6395	1.0160	1.0025	0.9988	0.9998	0.8362	0.5406	0.8328	0.7541
3.4	4.4354E-01	1.0723	0.7668	0.6851	0.6432	1.0170	1.0026	0.9989	0.9998	0.8381	0.5402	0.8351	0.7566
3.6	4.4294E-01	1.0741	0.7695	0.6885	0.6469	1.0180	1.0028	0.9989	0.9998	0.8399	0.5398	0.8373	0.7591
3.8	4.4199E-01	1.0759	0.7723	0.6918	0.6505	1.0189	1.0030	0.9990	0.9998	0.8417	0.5394	0.8393	0.7614
4.0	4.4080E-01	1.0777	0.7749	0.6951	0.6540	1.0199	1.0031	0.9990	0.9998	0.8434	0.5389	0.8413	0.7636
4.5	4.3716E-01	1.0823	0.7813	0.7030	0.6624	1.0223	1.0035	0.9991	0.9998	0.8473	0.5376	0.8457	0.7687
5.0	4.3308E-01	1.0870	0.7872	0.7102	0.6702	1.0247	1.0040	0.9992	0.9999	0.8509	0.5362	0.8496	0.7734
5.5	4.2890E-01	1.0916	0.7927	0.7169	0.6774	1.0270	1.0044	0.9993	0.9999	0.8541	0.5346	0.8530	0.7776
6.0	4.2480E-01	1.0963	0.7977	0.7231	0.6840	1.0293	1.0048	0.9994	0.9999	0.8570	0.5328	0.8561	0.7815
6.5	4.2088E-01	1.1010	0.8022	0.7288	0.6902	1.0316	1.0052	0.9994	0.9999	0.8596	0.5309	0.8589	0.7851
7.0	4.1716E-01	1.1056	0.8065	0.7341	0.6958	1.0339	1.0056	0.9995	0.9999	0.8620	0.5289	0.8614	0.7884
7.5	4.1368E-01	1.1103	0.8103	0.7389	0.7011	1.0361	1.0060	0.9995	0.9999	0.8641	0.5268	0.8637	0.7916
8.0	4.1041E-01	1.1150	0.8139	0.7434	0.7059	1.0384	1.0064	0.9995	0.9999	0.8661	0.5246	0.8657	0.7945
9.0	4.0452E-01	1.1245	0.8202	0.7514	0.7146	1.0428	1.0072	0.9996	0.9999	0.8697	0.5200	0.8694	0.7999
10.0	3.9940E-01	1.1339	0.8256	0.7583	0.7220	1.0471	1.0080	0.9996	0.9999	0.8726	0.5151	0.8724	0.8048
11.0	3.9496E-01	1.1433	0.8302	0.7642	0.7285	1.0514	1.0088	0.9997	0.9999	0.8752	0.5100	0.8750	0.8092
12.0	3.9109E-01	1.1528	0.8342	0.7693	0.7341	1.0556	1.0095	0.9997	0.9999	0.8773	0.5047	0.8772	0.8132
13.0	3.8772E-01	1.1622	0.8375	0.7738	0.7390	1.0598	1.0103	0.9997	0.9999	0.8792	0.4993	0.8791	0.8170
14.0	3.8478E-01	1.1717	0.8404	0.7776	0.7433	1.0639	1.0111	0.9997	0.9999	0.8808	0.4938	0.8807	0.8205
16.0	3.7999E-01	1.1906	0.8450	0.7838	0.7502	1.0721	1.0127	0.9998	1.0000	0.8833	0.4825	0.8833	0.8269
18.0	3.7635E-01	1.2097	0.8483	0.7884	0.7554	1.0801	1.0143	0.9998	1.0000	0.8851	0.4710	0.8851	0.8325
20.0	3.7362E-01	1.2288	0.8506	0.7917	0.7593	1.0879	1.0158	0.9998	1.0000	0.8864	0.4592	0.8864	0.8377
25.0	3.6971E-01	1.2770	0.8529	0.7959	0.7643	1.1071	1.0197	0.9999	1.0000	0.8880	0.4293	0.8880	0.8488
30.0	3.6872E-01	1.3259	0.8520	0.7958	0.7647	1.1258	1.0236	0.9999	1.0000	0.8878	0.3990	0.8879	0.8581
35.0	3.6970E-01	1.3757	0.8489	0.7929	0.7621	1.1439	1.0275	0.9999	1.0000	0.8866	0.3685	0.8867	0.8663
40.0	3.7211E-01	1.4264	0.8444	0.7881	0.7574	1.1617	1.0315	0.9999	1.0000	0.8847	0.3380	0.8847	0.8737
45.0	3.7562E-01	1.4781	0.8389	0.7820	0.7511	1.1791	1.0354	0.9999	1.0000	0.8823	0.3075	0.8823	0.8805
50.0	3.8000E-01	1.5309	0.8327	0.7748	0.7437	1.1961	1.0394	0.9999	1.0000	0.8795	0.2772	0.8796	0.8869

Instr. see p. 22

POSITRON Z= -64 A= 154

p	F_0L_0	L_0	λ_2	λ_3	λ_4	μ_1	μ_2	Λ_1	Λ_2	ν_{12}	$\hat{\nu}_{12}$	η_{12}	$\hat{\eta}_{12}$
0.1	1.8970E-11	1.0517	12.9962	67.1743	203.6990	1.0013	1.0001	0.9956	0.9992	5.1129	0.5666	0.7684	0.6837
0.2	1.9520E-05	1.0518	3.7305	7.1296	10.2102	1.0014	1.0001	0.9957	0.9992	1.8517	0.5653	0.7692	0.6846
0.3	1.5281E-03	1.0520	2.0167	2.6444	3.0105	1.0017	1.0001	0.9958	0.9992	1.2492	0.5634	0.7706	0.6861
0.4	1.1794E-02	1.0522	1.4193	1.5669	1.6257	1.0020	1.0002	0.9959	0.9993	1.0399	0.5609	0.7724	0.6882
0.5	3.6930E-02	1.0524	1.1452	1.1568	1.1478	1.0023	1.0002	0.9961	0.9993	0.9445	0.5582	0.7747	0.6906
0.6	7.4728E-02	1.0528	0.9987	0.9594	0.9291	1.0028	1.0003	0.9962	0.9993	0.8942	0.5556	0.7772	0.6933
0.7	1.1894E-01	1.0532	0.9124	0.8504	0.8120	1.0032	1.0004	0.9964	0.9994	0.8651	0.5531	0.7798	0.6961
0.8	1.6398E-01	1.0536	0.8581	0.7847	0.7427	1.0037	1.0005	0.9966	0.9994	0.8473	0.5509	0.7825	0.6990
0.9	2.0634E-01	1.0541	0.8224	0.7428	0.6991	1.0042	1.0005	0.9968	0.9994	0.8360	0.5489	0.7852	0.7018
1.0	2.4432E-01	1.0547	0.7982	0.7149	0.6704	1.0048	1.0006	0.9969	0.9994	0.8287	0.5472	0.7879	0.7047
1.2	3.0570E-01	1.0559	0.7696	0.6828	0.6377	1.0059	1.0008	0.9972	0.9995	0.8211	0.5445	0.7931	0.7101
1.4	3.4973E-01	1.0572	0.7554	0.6674	0.6223	1.0070	1.0010	0.9975	0.9995	0.8182	0.5426	0.7979	0.7152
1.6	3.8054E-01	1.0587	0.7485	0.6602	0.6153	1.0081	1.0012	0.9977	0.9996	0.8178	0.5413	0.8025	0.7199
1.8	4.0187E-01	1.0603	0.7457	0.6576	0.6129	1.0092	1.0013	0.9979	0.9996	0.8186	0.5404	0.8066	0.7243
2.0	4.1652E-01	1.0619	0.7452	0.6576	0.6131	1.0103	1.0015	0.9981	0.9996	0.8201	0.5397	0.8105	0.7283
2.2	4.2647E-01	1.0636	0.7462	0.6590	0.6149	1.0114	1.0017	0.9982	0.9997	0.8220	0.5392	0.8141	0.7321
2.4	4.3310E-01	1.0654	0.7479	0.6614	0.6176	1.0125	1.0019	0.9984	0.9997	0.8240	0.5387	0.8174	0.7357
2.6	4.3736E-01	1.0671	0.7502	0.6644	0.6208	1.0135	1.0020	0.9985	0.9997	0.8260	0.5384	0.8205	0.7390
2.8	4.3992E-01	1.0690	0.7528	0.6676	0.6243	1.0146	1.0022	0.9986	0.9997	0.8281	0.5380	0.8234	0.7421
3.0	4.4126E-01	1.0708	0.7555	0.6710	0.6280	1.0156	1.0024	0.9987	0.9997	0.8302	0.5377	0.8261	0.7450
3.2	4.4171E-01	1.0726	0.7584	0.6745	0.6317	1.0167	1.0026	0.9987	0.9998	0.8322	0.5374	0.8287	0.7478
3.4	4.4152E-01	1.0745	0.7612	0.6780	0.6355	1.0177	1.0027	0.9988	0.9998	0.8341	0.5370	0.8310	0.7504
3.6	4.4085E-01	1.0764	0.7641	0.6815	0.6392	1.0187	1.0029	0.9989	0.9998	0.8360	0.5366	0.8333	0.7529
3.8	4.3983E-01	1.0783	0.7669	0.6849	0.6429	1.0197	1.0031	0.9989	0.9998	0.8378	0.5362	0.8354	0.7553
4.0	4.3857E-01	1.0802	0.7696	0.6883	0.6465	1.0208	1.0032	0.9990	0.9998	0.8395	0.5358	0.8374	0.7576
4.5	4.3476E-01	1.0850	0.7762	0.6963	0.6550	1.0233	1.0037	0.9991	0.9998	0.8436	0.5346	0.8419	0.7628
5.0	4.3051E-01	1.0898	0.7822	0.7038	0.6630	1.0257	1.0041	0.9992	0.9998	0.8472	0.5332	0.8459	0.7676
5.5	4.2618E-01	1.0946	0.7878	0.7107	0.6704	1.0282	1.0045	0.9993	0.9999	0.8505	0.5317	0.8495	0.7720
6.0	4.2195E-01	1.0995	0.7930	0.7170	0.6772	1.0306	1.0049	0.9993	0.9999	0.8535	0.5300	0.8526	0.7759
6.5	4.1790E-01	1.1044	0.7977	0.7228	0.6835	1.0329	1.0053	0.9994	0.9999	0.8562	0.5281	0.8555	0.7796
7.0	4.1407E-01	1.1092	0.8020	0.7282	0.6893	1.0353	1.0058	0.9994	0.9999	0.8587	0.5262	0.8581	0.7831
7.5	4.1048E-01	1.1141	0.8060	0.7332	0.6946	1.0376	1.0062	0.9995	0.9999	0.8609	0.5241	0.8604	0.7863
8.0	4.0713E-01	1.1190	0.8097	0.7378	0.6996	1.0399	1.0066	0.9995	0.9999	0.8630	0.5219	0.8625	0.7893
9.0	4.0108E-01	1.1288	0.8162	0.7460	0.7085	1.0445	1.0074	0.9996	0.9999	0.8666	0.5174	0.8663	0.7949
10.0	3.9583E-01	1.1386	0.8217	0.7530	0.7161	1.0490	1.0082	0.9996	0.9999	0.8696	0.5125	0.8694	0.7999
11.0	3.9127E-01	1.1484	0.8265	0.7591	0.7228	1.0535	1.0091	0.9997	0.9999	0.8722	0.5075	0.8721	0.8044
12.0	3.8731E-01	1.1582	0.8305	0.7644	0.7285	1.0578	1.0099	0.9997	0.9999	0.8745	0.5023	0.8743	0.8086
13.0	3.8386E-01	1.1681	0.8340	0.7689	0.7336	1.0622	1.0107	0.9997	0.9999	0.8764	0.4969	0.8763	0.8125
14.0	3.8085E-01	1.1779	0.8370	0.7729	0.7379	1.0665	1.0115	0.9997	0.9999	0.8780	0.4915	0.8780	0.8161
16.0	3.7594E-01	1.1977	0.8417	0.7793	0.7450	1.0749	1.0131	0.9998	1.0000	0.8806	0.4802	0.8806	0.8226
18.0	3.7221E-01	1.2175	0.8451	0.7840	0.7504	1.0832	1.0147	0.9998	1.0000	0.8825	0.4687	0.8825	0.8285
20.0	3.6942E-01	1.2374	0.8474	0.7874	0.7543	1.0913	1.0164	0.9998	1.0000	0.8839	0.4570	0.8839	0.8337
25.0	3.6542E-01	1.2876	0.8499	0.7917	0.7595	1.1112	1.0204	0.9999	1.0000	0.8855	0.4273	0.8855	0.8452
30.0	3.6439E-01	1.3386	0.8490	0.7916	0.7600	1.1305	1.0244	0.9999	1.0000	0.8854	0.3971	0.8854	0.8548
35.0	3.6537E-01	1.3905	0.8458	0.7887	0.7574	1.1493	1.0284	0.9999	1.0000	0.8841	0.3666	0.8842	0.8633
40.0	3.6781E-01	1.4435	0.8412	0.7839	0.7525	1.1676	1.0325	0.9999	1.0000	0.8822	0.3362	0.8822	0.8709
45.0	3.7137E-01	1.4975	0.8356	0.7776	0.7462	1.1856	1.0366	0.9999	1.0000	0.8797	0.3058	0.8798	0.8779
50.0	3.7581E-01	1.5527	0.8293	0.7702	0.7386	1.2031	1.0407	0.9999	1.0000	0.8769	0.2756	0.8770	0.8845

POSITRON Z= -65 A= 155

p	F_0L_0	L_0	λ_2	λ_3	λ_4	μ_1	μ_2	Λ_1	Λ_2	ν_{12}	$\hat{\nu}_{12}$	η_{12}	$\hat{\eta}_{12}$
0.1	1.2675E-11	1.0528	13.1613	69.3792	214.2976	1.0014	1.0001	0.9955	0.9992	5.2148	0.5619	0.7623	0.6756
0.2	1.6364E-05	1.0529	3.7655	7.2801	10.5126	1.0015	1.0001	0.9956	0.9992	1.8726	0.5607	0.7632	0.6766
0.3	1.3796E-03	1.0531	2.0276	2.6735	3.0537	1.0017	1.0001	0.9957	0.9992	1.2551	0.5587	0.7645	0.6781
0.4	1.1037E-02	1.0533	1.4218	1.5726	1.6332	1.0021	1.0002	0.9958	0.9993	1.0406	0.5564	0.7664	0.6801
0.5	3.5281E-02	1.0536	1.1438	1.1551	1.1457	1.0024	1.0002	0.9960	0.9993	0.9428	0.5538	0.7687	0.6826
0.6	7.2324E-02	1.0539	0.9952	0.9546	0.9234	1.0029	1.0003	0.9962	0.9993	0.8912	0.5512	0.7712	0.6853
0.7	1.1612E-01	1.0543	0.9076	0.8439	0.8046	1.0034	1.0004	0.9963	0.9993	0.8613	0.5487	0.7739	0.6881
0.8	1.6106E-01	1.0548	0.8525	0.7773	0.7345	1.0039	1.0005	0.9965	0.9994	0.8431	0.5466	0.7766	0.6910
0.9	2.0354E-01	1.0553	0.8163	0.7348	0.6903	1.0044	1.0006	0.9967	0.9994	0.8315	0.5446	0.7794	0.6940
1.0	2.4177E-01	1.0558	0.7917	0.7066	0.6613	1.0050	1.0006	0.9969	0.9994	0.8240	0.5430	0.7821	0.6968
1.2	3.0374E-01	1.0571	0.7627	0.6742	0.6283	1.0061	1.0008	0.9972	0.9995	0.8161	0.5404	0.7874	0.7023
1.4	3.4831E-01	1.0585	0.7483	0.6585	0.6127	1.0072	1.0010	0.9974	0.9995	0.8132	0.5386	0.7924	0.7075
1.6	3.7952E-01	1.0600	0.7414	0.6514	0.6057	1.0084	1.0012	0.9977	0.9996	0.8128	0.5374	0.7970	0.7123
1.8	4.0112E-01	1.0616	0.7385	0.6487	0.6033	1.0095	1.0014	0.9979	0.9996	0.8136	0.5365	0.8013	0.7168
2.0	4.1593E-01	1.0633	0.7381	0.6487	0.6036	1.0106	1.0015	0.9980	0.9996	0.8151	0.5359	0.8053	0.7210
2.2	4.2596E-01	1.0651	0.7391	0.6503	0.6054	1.0117	1.0017	0.9982	0.9997	0.8170	0.5355	0.8090	0.7249
2.4	4.3261E-01	1.0669	0.7409	0.6527	0.6082	1.0129	1.0019	0.9983	0.9997	0.8191	0.5352	0.8124	0.7285
2.6	4.3686E-01	1.0687	0.7433	0.6557	0.6114	1.0140	1.0021	0.9984	0.9997	0.8212	0.5349	0.8156	0.7319
2.8	4.3939E-01	1.0706	0.7460	0.6591	0.6150	1.0150	1.0023	0.9985	0.9997	0.8234	0.5346	0.8185	0.7351
3.0	4.4067E-01	1.0724	0.7488	0.6626	0.6188	1.0161	1.0024	0.9986	0.9997	0.8255	0.5343	0.8213	0.7381
3.2	4.4105E-01	1.0743	0.7517	0.6662	0.6226	1.0172	1.0026	0.9987	0.9998	0.8275	0.5341	0.8239	0.7409
3.4	4.4078E-01	1.0763	0.7547	0.6698	0.6265	1.0183	1.0028	0.9988	0.9998	0.8295	0.5338	0.8264	0.7436
3.6	4.4002E-01	1.0782	0.7576	0.6733	0.6303	1.0193	1.0030	0.9989	0.9998	0.8315	0.5335	0.8287	0.7462
3.8	4.3892E-01	1.0801	0.7605	0.6769	0.6341	1.0204	1.0031	0.9989	0.9998	0.8333	0.5331	0.8308	0.7486
4.0	4.3758E-01	1.0821	0.7633	0.6803	0.6377	1.0214	1.0033	0.9990	0.9998	0.8351	0.5327	0.8329	0.7509
4.5	4.3355E-01	1.0870	0.7700	0.6885	0.6465	1.0240	1.0037	0.9991	0.9998	0.8393	0.5317	0.8376	0.7563
5.0	4.2910E-01	1.0920	0.7763	0.6962	0.6547	1.0265	1.0042	0.9992	0.9998	0.8431	0.5304	0.8417	0.7612
5.5	4.2459E-01	1.0969	0.7821	0.7032	0.6622	1.0290	1.0046	0.9993	0.9999	0.8465	0.5290	0.8454	0.7657
6.0	4.2018E-01	1.1019	0.7874	0.7098	0.6692	1.0315	1.0050	0.9993	0.9999	0.8495	0.5274	0.8487	0.7698
6.5	4.1598E-01	1.1069	0.7922	0.7158	0.6756	1.0339	1.0055	0.9994	0.9999	0.8523	0.5256	0.8516	0.7735
7.0	4.1201E-01	1.1119	0.7967	0.7213	0.6816	1.0363	1.0059	0.9994	0.9999	0.8549	0.5238	0.8543	0.7771
7.5	4.0829E-01	1.1169	0.8008	0.7264	0.6871	1.0387	1.0063	0.9995	0.9999	0.8572	0.5218	0.8567	0.7804
8.0	4.0481E-01	1.1220	0.8046	0.7312	0.6922	1.0411	1.0067	0.9995	0.9999	0.8593	0.5197	0.8589	0.7835
9.0	3.9854E-01	1.1320	0.8113	0.7396	0.7013	1.0458	1.0076	0.9996	0.9999	0.8631	0.5153	0.8628	0.7892
10.0	3.9310E-01	1.1421	0.8171	0.7469	0.7092	1.0504	1.0084	0.9996	0.9999	0.8662	0.5106	0.8660	0.7943
11.0	3.8838E-01	1.1522	0.8220	0.7532	0.7161	1.0549	1.0092	0.9996	0.9999	0.8690	0.5057	0.8688	0.7990
12.0	3.8427E-01	1.1622	0.8262	0.7586	0.7220	1.0594	1.0101	0.9997	0.9999	0.8713	0.5007	0.8711	0.8033
13.0	3.8069E-01	1.1723	0.8299	0.7633	0.7272	1.0639	1.0109	0.9997	0.9999	0.8733	0.4955	0.8732	0.8073
14.0	3.7757E-01	1.1825	0.8330	0.7675	0.7318	1.0683	1.0117	0.9997	0.9999	0.8750	0.4901	0.8749	0.8110
16.0	3.7247E-01	1.2027	0.8379	0.7741	0.7392	1.0769	1.0134	0.9998	1.0000	0.8777	0.4791	0.8777	0.8177
18.0	3.6858E-01	1.2231	0.8415	0.7791	0.7448	1.0854	1.0150	0.9998	1.0000	0.8797	0.4679	0.8797	0.8237
20.0	3.6565E-01	1.2435	0.8440	0.7827	0.7489	1.0937	1.0167	0.9998	1.0000	0.8812	0.4564	0.8812	0.8291
25.0	3.6138E-01	1.2951	0.8468	0.7874	0.7545	1.1140	1.0208	0.9999	1.0000	0.8829	0.4272	0.8829	0.8408
30.0	3.6015E-01	1.3475	0.8461	0.7876	0.7553	1.1338	1.0249	0.9999	1.0000	0.8829	0.3974	0.8829	0.8508
35.0	3.6097E-01	1.4009	0.8431	0.7849	0.7529	1.1530	1.0290	0.9999	1.0000	0.8818	0.3675	0.8818	0.8594
40.0	3.6328E-01	1.4553	0.8386	0.7802	0.7483	1.1717	1.0331	0.9999	1.0000	0.8799	0.3376	0.8799	0.8673
45.0	3.6673E-01	1.5108	0.8330	0.7740	0.7420	1.1901	1.0373	0.9999	1.0000	0.8775	0.3076	0.8775	0.8745
50.0	3.7107E-01	1.5676	0.8267	0.7667	0.7346	1.2080	1.0415	0.9999	1.0000	0.8747	0.2779	0.8747	0.8812

Instr. see p. 22

POSITRON   Z= -66   A= 160

p	F_0L_0	L_0	λ_2	λ_3	λ_4	μ_1	μ_2	Λ_1	Λ_2	ν_{12}	$\hat{\nu}_{12}$	η_{12}	$\hat{\eta}_{12}$
0.1	8.4554E-12	1.0541	13.3326	71.6750	225.4959	1.0014	1.0001	0.9954	0.9992	5.3192	0.5575	0.7566	0.6678
0.2	1.3697E-05	1.0541	3.8025	7.4383	10.8321	1.0016	1.0001	0.9955	0.9992	1.8944	0.5562	0.7574	0.6688
0.3	1.2436E-03	1.0543	2.0397	2.7052	3.1006	1.0019	1.0001	0.9956	0.9992	1.2616	0.5544	0.7588	0.6703
0.4	1.0313E-02	1.0545	1.4252	1.5799	1.6424	1.0022	1.0002	0.9957	0.9992	1.0417	0.5520	0.7607	0.6724
0.5	3.3652E-02	1.0548	1.1432	1.1546	1.1449	1.0026	1.0002	0.9959	0.9993	0.9415	0.5495	0.7630	0.6749
0.6	6.9889E-02	1.0552	0.9924	0.9507	0.9189	1.0030	1.0003	0.9960	0.9993	0.8886	0.5470	0.7656	0.6776
0.7	1.1319E-01	1.0556	0.9035	0.8384	0.7982	1.0035	1.0004	0.9962	0.9993	0.8580	0.5446	0.7683	0.6805
0.8	1.5794E-01	1.0561	0.8476	0.7708	0.7271	1.0041	1.0005	0.9964	0.9994	0.8392	0.5425	0.7711	0.6834
0.9	2.0047E-01	1.0566	0.8109	0.7278	0.6824	1.0046	1.0006	0.9966	0.9994	0.8273	0.5406	0.7739	0.6864
1.0	2.3887E-01	1.0572	0.7859	0.6992	0.6531	1.0052	1.0007	0.9968	0.9994	0.8196	0.5390	0.7767	0.6893
1.2	3.0135E-01	1.0585	0.7565	0.6663	0.6197	1.0064	1.0008	0.9971	0.9995	0.8115	0.5365	0.7821	0.6949
1.4	3.4638E-01	1.0600	0.7419	0.6505	0.6040	1.0075	1.0010	0.9974	0.9995	0.8085	0.5348	0.7872	0.7002
1.6	3.7795E-01	1.0615	0.7349	0.6433	0.5969	1.0087	1.0012	0.9976	0.9996	0.8081	0.5336	0.7919	0.7051
1.8	3.9978E-01	1.0632	0.7320	0.6406	0.5946	1.0099	1.0014	0.9978	0.9996	0.8089	0.5329	0.7963	0.7097
2.0	4.1473E-01	1.0650	0.7316	0.6407	0.5949	1.0111	1.0016	0.9980	0.9996	0.8105	0.5323	0.8004	0.7140
2.2	4.2483E-01	1.0668	0.7327	0.6423	0.5968	1.0122	1.0018	0.9981	0.9997	0.8124	0.5320	0.8041	0.7179
2.4	4.3150E-01	1.0686	0.7346	0.6448	0.5996	1.0134	1.0020	0.9983	0.9997	0.8145	0.5317	0.8077	0.7216
2.6	4.3574E-01	1.0705	0.7370	0.6479	0.6029	1.0145	1.0021	0.9984	0.9997	0.8167	0.5315	0.8109	0.7251
2.8	4.3822E-01	1.0724	0.7397	0.6513	0.6066	1.0156	1.0023	0.9985	0.9997	0.8189	0.5312	0.8140	0.7284
3.0	4.3945E-01	1.0744	0.7427	0.6549	0.6104	1.0168	1.0025	0.9986	0.9997	0.8211	0.5310	0.8168	0.7314
3.2	4.3976E-01	1.0764	0.7457	0.6586	0.6144	1.0179	1.0027	0.9987	0.9998	0.8232	0.5308	0.8195	0.7343
3.4	4.3941E-01	1.0783	0.7487	0.6622	0.6183	1.0190	1.0029	0.9988	0.9998	0.8253	0.5306	0.8220	0.7371
3.6	4.3858E-01	1.0803	0.7517	0.6659	0.6222	1.0201	1.0030	0.9988	0.9998	0.8273	0.5303	0.8244	0.7397
3.8	4.3740E-01	1.0824	0.7547	0.6695	0.6260	1.0211	1.0032	0.9989	0.9998	0.8292	0.5300	0.8266	0.7422
4.0	4.3598E-01	1.0844	0.7576	0.6730	0.6298	1.0222	1.0034	0.9989	0.9998	0.8310	0.5296	0.8287	0.7446
4.5	4.3176E-01	1.0895	0.7645	0.6815	0.6387	1.0249	1.0039	0.9991	0.9998	0.8353	0.5287	0.8336	0.7501
5.0	4.2713E-01	1.0946	0.7709	0.6893	0.6471	1.0275	1.0043	0.9992	0.9998	0.8392	0.5275	0.8378	0.7552
5.5	4.2244E-01	1.0997	0.7769	0.6965	0.6548	1.0301	1.0047	0.9992	0.9999	0.8427	0.5261	0.8416	0.7597
6.0	4.1789E-01	1.1049	0.7823	0.7032	0.6619	1.0326	1.0052	0.9993	0.9999	0.8459	0.5246	0.8450	0.7639
6.5	4.1354E-01	1.1101	0.7873	0.7094	0.6685	1.0352	1.0056	0.9994	0.9999	0.8487	0.5229	0.8480	0.7678
7.0	4.0944E-01	1.1153	0.7919	0.7150	0.6746	1.0377	1.0060	0.9994	0.9999	0.8514	0.5211	0.8507	0.7714
7.5	4.0561E-01	1.1205	0.7962	0.7203	0.6803	1.0401	1.0065	0.9994	0.9999	0.8537	0.5192	0.8532	0.7748
8.0	4.0202E-01	1.1257	0.8001	0.7251	0.6855	1.0426	1.0069	0.9995	0.9999	0.8559	0.5172	0.8555	0.7780
9.0	3.9557E-01	1.1361	0.8070	0.7338	0.6948	1.0474	1.0078	0.9995	0.9999	0.8598	0.5129	0.8595	0.7839
10.0	3.8997E-01	1.1465	0.8129	0.7413	0.7029	1.0522	1.0086	0.9996	0.9999	0.8631	0.5083	0.8628	0.7891
11.0	3.8512E-01	1.1569	0.8180	0.7478	0.7100	1.0569	1.0095	0.9996	0.9999	0.8659	0.5035	0.8657	0.7939
12.0	3.8090E-01	1.1674	0.8223	0.7534	0.7161	1.0615	1.0104	0.9997	0.9999	0.8683	0.4985	0.8681	0.7983
13.0	3.7722E-01	1.1778	0.8261	0.7582	0.7214	1.0661	1.0112	0.9997	0.9999	0.8703	0.4934	0.8702	0.8024
14.0	3.7401E-01	1.1883	0.8293	0.7625	0.7261	1.0707	1.0121	0.9997	0.9999	0.8721	0.4881	0.8720	0.8062
16.0	3.6876E-01	1.2093	0.8344	0.7693	0.7337	1.0796	1.0138	0.9998	0.9999	0.8749	0.4773	0.8749	0.8131
18.0	3.6476E-01	1.2304	0.8382	0.7745	0.7395	1.0883	1.0155	0.9998	1.0000	0.8770	0.4661	0.8770	0.8193
20.0	3.6174E-01	1.2515	0.8408	0.7782	0.7438	1.0969	1.0171	0.9998	1.0000	0.8785	0.4548	0.8785	0.8249
25.0	3.5731E-01	1.3050	0.8437	0.7831	0.7496	1.1179	1.0214	0.9999	1.0000	0.8804	0.4258	0.8804	0.8369
30.0	3.5598E-01	1.3593	0.8431	0.7835	0.7506	1.1382	1.0256	0.9999	1.0000	0.8805	0.3963	0.8805	0.8471
35.0	3.5675E-01	1.4147	0.8402	0.7809	0.7483	1.1579	1.0298	0.9999	1.0000	0.8794	0.3666	0.8794	0.8561
40.0	3.5903E-01	1.4712	0.8356	0.7761	0.7436	1.1772	1.0341	0.9999	1.0000	0.8775	0.3369	0.8775	0.8642
45.0	3.6246E-01	1.5289	0.8300	0.7699	0.7373	1.1961	1.0383	0.9999	1.0000	0.8750	0.3071	0.8751	0.8716
50.0	3.6682E-01	1.5879	0.8236	0.7626	0.7298	1.2145	1.0426	0.9999	1.0000	0.8723	0.2776	0.8723	0.8785

POSITRON Z= -67 A= 161

p	F_0L_0	L_0	λ_2	λ_3	λ_4	μ_1	μ_2	Λ_1	Λ_2	ν_{12}	$\hat{\nu}_{12}$	η_{12}	$\hat{\eta}_{12}$
0.1	5.6537E-12	1.0551	13.4838	73.8734	236.6437	1.0015	1.0001	0.9953	0.9992	5.4213	0.5527	0.7504	0.6597
0.2	1.1491E-05	1.0552	3.8340	7.5848	11.1375	1.0017	1.0001	0.9954	0.9992	1.9153	0.5515	0.7513	0.6606
0.3	1.1235E-03	1.0554	2.0491	2.7324	3.1421	1.0020	1.0001	0.9955	0.9992	1.2675	0.5497	0.7527	0.6622
0.4	9.6577E-03	1.0556	1.4268	1.5846	1.6488	1.0023	1.0002	0.9956	0.9992	1.0423	0.5474	0.7546	0.6643
0.5	3.2172E-02	1.0559	1.1412	1.1522	1.1421	1.0027	1.0002	0.9958	0.9992	0.9397	0.5449	0.7569	0.6667
0.6	6.7691E-02	1.0563	0.9884	0.9454	0.9127	1.0032	1.0003	0.9960	0.9993	0.8855	0.5425	0.7595	0.6695
0.7	1.1059E-01	1.0567	0.8984	0.8316	0.7906	1.0037	1.0004	0.9961	0.9993	0.8541	0.5402	0.7623	0.6724
0.8	1.5525E-01	1.0572	0.8418	0.7632	0.7187	1.0042	1.0005	0.9963	0.9993	0.8349	0.5381	0.7651	0.6754
0.9	1.9790E-01	1.0578	0.8045	0.7196	0.6735	1.0048	1.0006	0.9965	0.9994	0.8227	0.5363	0.7680	0.6784
1.0	2.3656E-01	1.0584	0.7793	0.6907	0.6439	1.0054	1.0007	0.9967	0.9994	0.8148	0.5347	0.7708	0.6814
1.2	2.9966E-01	1.0597	0.7494	0.6575	0.6102	1.0066	1.0009	0.9970	0.9995	0.8065	0.5324	0.7763	0.6871
1.4	3.4525E-01	1.0612	0.7347	0.6416	0.5944	1.0078	1.0011	0.9973	0.9995	0.8034	0.5308	0.7815	0.6925
1.6	3.7723E-01	1.0628	0.7276	0.6343	0.5873	1.0090	1.0012	0.9975	0.9996	0.8029	0.5297	0.7864	0.6975
1.8	3.9934E-01	1.0645	0.7248	0.6317	0.5850	1.0102	1.0014	0.9978	0.9996	0.8038	0.5290	0.7909	0.7022
2.0	4.1446E-01	1.0663	0.7244	0.6318	0.5853	1.0114	1.0016	0.9979	0.9996	0.8054	0.5286	0.7950	0.7065
2.2	4.2464E-01	1.0682	0.7255	0.6334	0.5873	1.0126	1.0018	0.9981	0.9997	0.8074	0.5283	0.7989	0.7105
2.4	4.3134E-01	1.0701	0.7275	0.6361	0.5901	1.0138	1.0020	0.9982	0.9997	0.8096	0.5281	0.8025	0.7143
2.6	4.3556E-01	1.0721	0.7300	0.6392	0.5936	1.0150	1.0022	0.9984	0.9997	0.8118	0.5279	0.8059	0.7179
2.8	4.3800E-01	1.0740	0.7328	0.6427	0.5973	1.0161	1.0024	0.9985	0.9997	0.8141	0.5278	0.8090	0.7212
3.0	4.3917E-01	1.0760	0.7358	0.6464	0.6013	1.0173	1.0026	0.9986	0.9997	0.8163	0.5276	0.8119	0.7244
3.2	4.3941E-01	1.0781	0.7389	0.6502	0.6053	1.0184	1.0027	0.9987	0.9997	0.8185	0.5275	0.8147	0.7273
3.4	4.3897E-01	1.0801	0.7420	0.6539	0.6093	1.0195	1.0029	0.9987	0.9998	0.8206	0.5273	0.8173	0.7302
3.6	4.3806E-01	1.0822	0.7451	0.6577	0.6132	1.0207	1.0031	0.9988	0.9998	0.8227	0.5271	0.8197	0.7329
3.8	4.3679E-01	1.0842	0.7481	0.6614	0.6172	1.0218	1.0033	0.9989	0.9998	0.8246	0.5268	0.8220	0.7354
4.0	4.3527E-01	1.0863	0.7511	0.6650	0.6210	1.0229	1.0035	0.9989	0.9998	0.8265	0.5265	0.8242	0.7379
4.5	4.3083E-01	1.0915	0.7583	0.6736	0.6302	1.0256	1.0039	0.9990	0.9998	0.8310	0.5257	0.8292	0.7435
5.0	4.2599E-01	1.0968	0.7649	0.6816	0.6387	1.0283	1.0044	0.9991	0.9998	0.8349	0.5246	0.8335	0.7486
5.5	4.2111E-01	1.1021	0.7710	0.6890	0.6466	1.0310	1.0048	0.9992	0.9999	0.8386	0.5234	0.8374	0.7533
6.0	4.1637E-01	1.1074	0.7766	0.6959	0.6539	1.0336	1.0053	0.9993	0.9999	0.8418	0.5220	0.8409	0.7576
6.5	4.1186E-01	1.1127	0.7818	0.7022	0.6606	1.0362	1.0057	0.9993	0.9999	0.8448	0.5204	0.8440	0.7616
7.0	4.0761E-01	1.1180	0.7865	0.7080	0.6669	1.0388	1.0062	0.9994	0.9999	0.8475	0.5187	0.8469	0.7653
7.5	4.0363E-01	1.1234	0.7909	0.7134	0.6727	1.0413	1.0066	0.9994	0.9999	0.8500	0.5169	0.8494	0.7688
8.0	3.9992E-01	1.1287	0.7949	0.7184	0.6780	1.0438	1.0070	0.9995	0.9999	0.8522	0.5149	0.8518	0.7721
9.0	3.9324E-01	1.1394	0.8021	0.7274	0.6876	1.0488	1.0079	0.9995	0.9999	0.8562	0.5108	0.8559	0.7781
10.0	3.8746E-01	1.1501	0.8082	0.7351	0.6960	1.0537	1.0088	0.9996	0.9999	0.8596	0.5064	0.8594	0.7835
11.0	3.8243E-01	1.1608	0.8135	0.7417	0.7032	1.0585	1.0097	0.9996	0.9999	0.8625	0.5017	0.8623	0.7884
12.0	3.7807E-01	1.1715	0.8180	0.7475	0.7095	1.0632	1.0106	0.9997	0.9999	0.8650	0.4969	0.8649	0.7929
13.0	3.7426E-01	1.1822	0.8219	0.7526	0.7151	1.0679	1.0114	0.9997	0.9999	0.8672	0.4918	0.8671	0.7971
14.0	3.7093E-01	1.1930	0.8252	0.7570	0.7199	1.0726	1.0123	0.9997	0.9999	0.8690	0.4867	0.8689	0.8010
16.0	3.6548E-01	1.2145	0.8306	0.7641	0.7278	1.0817	1.0140	0.9998	0.9999	0.8720	0.4761	0.8719	0.8081
18.0	3.6132E-01	1.2362	0.8345	0.7695	0.7338	1.0907	1.0158	0.9998	1.0000	0.8742	0.4652	0.8742	0.8144
20.0	3.5816E-01	1.2579	0.8373	0.7735	0.7383	1.0995	1.0175	0.9998	1.0000	0.8758	0.4541	0.8758	0.8202
25.0	3.5345E-01	1.3128	0.8406	0.7788	0.7446	1.1209	1.0218	0.9998	1.0000	0.8778	0.4256	0.8778	0.8326
30.0	3.5192E-01	1.3686	0.8402	0.7795	0.7459	1.1416	1.0261	0.9999	1.0000	0.8780	0.3966	0.8780	0.8430
35.0	3.5252E-01	1.4255	0.8374	0.7770	0.7438	1.1618	1.0304	0.9999	1.0000	0.8770	0.3674	0.8770	0.8522
40.0	3.5467E-01	1.4836	0.8330	0.7724	0.7394	1.1815	1.0347	0.9999	1.0000	0.8752	0.3381	0.8752	0.8605
45.0	3.5799E-01	1.5429	0.8275	0.7663	0.7332	1.2008	1.0390	0.9999	1.0000	0.8728	0.3088	0.8728	0.8681
50.0	3.6225E-01	1.6037	0.8211	0.7591	0.7258	1.2197	1.0434	0.9999	1.0000	0.8700	0.2797	0.8701	0.8752

 Instr. see p. 22

POSITRON Z= -68 A= 166

p	F_0L_0	L_0	λ_2	λ_3	λ_4	μ_1	μ_2	Λ_1	Λ_2	ν_{12}	$\hat{\nu}_{12}$	η_{12}	$\hat{\eta}_{12}$
0.1	3.7742E-12	1.0563	13.6414	76.1639	248.4127	1.0016	1.0001	0.9951	0.9991	5.5259	0.5482	0.7446	0.6518
0.2	9.6253E-06	1.0564	3.8676	7.7393	11.4606	1.0018	1.0001	0.9952	0.9992	1.9371	0.5470	0.7455	0.6528
0.3	1.0134E-03	1.0566	2.0597	2.7623	3.1876	1.0021	1.0001	0.9953	0.9992	1.2739	0.5452	0.7469	0.6543
0.4	9.0297E-03	1.0568	1.4293	1.5908	1.6570	1.0024	1.0002	0.9955	0.9992	1.0434	0.5430	0.7489	0.6564
0.5	3.0707E-02	1.0571	1.1400	1.1510	1.1406	1.0029	1.0003	0.9956	0.9992	0.9384	0.5406	0.7512	0.6589
0.6	6.5457E-02	1.0575	0.9852	0.9410	0.9077	1.0033	1.0003	0.9958	0.9993	0.8828	0.5382	0.7538	0.6617
0.7	1.0787E-01	1.0580	0.8940	0.8257	0.7839	1.0039	1.0004	0.9960	0.9993	0.8506	0.5360	0.7566	0.6647
0.8	1.5236E-01	1.0585	0.8366	0.7565	0.7111	1.0044	1.0005	0.9962	0.9993	0.8309	0.5340	0.7595	0.6677
0.9	1.9506E-01	1.0590	0.7989	0.7124	0.6655	1.0050	1.0006	0.9964	0.9994	0.8184	0.5322	0.7624	0.6708
1.0	2.3391E-01	1.0597	0.7733	0.6831	0.6356	1.0056	1.0007	0.9966	0.9994	0.8104	0.5307	0.7653	0.6738
1.2	2.9752E-01	1.0611	0.7431	0.6496	0.6016	1.0069	1.0009	0.9969	0.9994	0.8018	0.5284	0.7709	0.6796
1.4	3.4360E-01	1.0626	0.7281	0.6335	0.5856	1.0081	1.0011	0.9972	0.9995	0.7986	0.5269	0.7762	0.6851
1.6	3.7594E-01	1.0643	0.7210	0.6261	0.5785	1.0094	1.0013	0.9975	0.9995	0.7981	0.5259	0.7812	0.6902
1.8	3.9830E-01	1.0661	0.7181	0.6236	0.5762	1.0106	1.0015	0.9977	0.9996	0.7990	0.5253	0.7858	0.6949
2.0	4.1356E-01	1.0679	0.7178	0.6237	0.5766	1.0119	1.0017	0.9979	0.9996	0.8007	0.5249	0.7900	0.6994
2.2	4.2382E-01	1.0699	0.7190	0.6254	0.5786	1.0131	1.0019	0.9980	0.9996	0.8027	0.5247	0.7940	0.7035
2.4	4.3053E-01	1.0718	0.7210	0.6281	0.5815	1.0144	1.0021	0.9982	0.9997	0.8050	0.5246	0.7977	0.7074
2.6	4.3473E-01	1.0739	0.7236	0.6313	0.5850	1.0156	1.0023	0.9983	0.9997	0.8073	0.5245	0.8011	0.7110
2.8	4.3713E-01	1.0759	0.7265	0.6349	0.5889	1.0168	1.0024	0.9984	0.9997	0.8096	0.5244	0.8044	0.7144
3.0	4.3824E-01	1.0780	0.7296	0.6387	0.5929	1.0180	1.0026	0.9985	0.9997	0.8119	0.5243	0.8074	0.7176
3.2	4.3840E-01	1.0801	0.7328	0.6425	0.5970	1.0191	1.0028	0.9986	0.9997	0.8141	0.5242	0.8102	0.7207
3.4	4.3789E-01	1.0822	0.7359	0.6464	0.6011	1.0203	1.0030	0.9987	0.9998	0.8163	0.5240	0.8129	0.7236
3.6	4.3689E-01	1.0843	0.7391	0.6502	0.6051	1.0215	1.0032	0.9988	0.9998	0.8184	0.5239	0.8154	0.7263
3.8	4.3554E-01	1.0865	0.7422	0.6539	0.6091	1.0226	1.0034	0.9988	0.9998	0.8204	0.5237	0.8177	0.7289
4.0	4.3394E-01	1.0886	0.7453	0.6576	0.6130	1.0238	1.0036	0.9989	0.9998	0.8224	0.5234	0.8200	0.7314
4.5	4.2929E-01	1.0940	0.7526	0.6665	0.6224	1.0266	1.0040	0.9990	0.9998	0.8269	0.5227	0.8251	0.7372
5.0	4.2426E-01	1.0995	0.7594	0.6747	0.6311	1.0294	1.0045	0.9991	0.9998	0.8310	0.5217	0.8296	0.7425
5.5	4.1920E-01	1.1050	0.7657	0.6823	0.6391	1.0321	1.0050	0.9992	0.9998	0.8347	0.5205	0.8335	0.7473
6.0	4.1430E-01	1.1104	0.7715	0.6893	0.6466	1.0348	1.0054	0.9993	0.9999	0.8381	0.5192	0.8371	0.7517
6.5	4.0965E-01	1.1159	0.7768	0.6957	0.6535	1.0375	1.0059	0.9993	0.9999	0.8411	0.5177	0.8404	0.7558
7.0	4.0527E-01	1.1215	0.7817	0.7017	0.6599	1.0402	1.0063	0.9994	0.9999	0.8439	0.5160	0.8433	0.7596
7.5	4.0117E-01	1.1270	0.7861	0.7072	0.6658	1.0428	1.0068	0.9994	0.9999	0.8465	0.5143	0.8459	0.7632
8.0	3.9735E-01	1.1325	0.7903	0.7124	0.6713	1.0454	1.0072	0.9995	0.9999	0.8488	0.5124	0.8483	0.7665
9.0	3.9047E-01	1.1436	0.7977	0.7215	0.6811	1.0505	1.0081	0.9995	0.9999	0.8529	0.5084	0.8526	0.7727
10.0	3.8453E-01	1.1546	0.8040	0.7294	0.6896	1.0556	1.0090	0.9996	0.9999	0.8564	0.5040	0.8562	0.7782
11.0	3.7936E-01	1.1657	0.8094	0.7363	0.6971	1.0606	1.0099	0.9996	0.9999	0.8594	0.4995	0.8592	0.7833
12.0	3.7488E-01	1.1768	0.8140	0.7422	0.7035	1.0655	1.0108	0.9997	0.9999	0.8620	0.4947	0.8618	0.7879
13.0	3.7097E-01	1.1879	0.8181	0.7474	0.7092	1.0703	1.0117	0.9997	0.9999	0.8642	0.4898	0.8641	0.7922
14.0	3.6755E-01	1.1990	0.8215	0.7520	0.7142	1.0751	1.0126	0.9997	0.9999	0.8661	0.4847	0.8660	0.7962
16.0	3.6196E-01	1.2213	0.8271	0.7593	0.7223	1.0845	1.0144	0.9997	0.9999	0.8692	0.4742	0.8691	0.8035
18.0	3.5768E-01	1.2437	0.8311	0.7649	0.7286	1.0937	1.0162	0.9998	1.0000	0.8715	0.4634	0.8714	0.8100
20.0	3.5442E-01	1.2662	0.8340	0.7690	0.7332	1.1028	1.0180	0.9998	1.0000	0.8731	0.4524	0.8731	0.8159
25.0	3.4955E-01	1.3231	0.8375	0.7745	0.7398	1.1248	1.0224	0.9998	1.0000	0.8753	0.4242	0.8753	0.8286
30.0	3.4792E-01	1.3809	0.8372	0.7753	0.7412	1.1462	1.0268	0.9999	1.0000	0.8755	0.3955	0.8755	0.8393
35.0	3.4846E-01	1.4399	0.8345	0.7730	0.7392	1.1670	1.0312	0.9999	1.0000	0.8745	0.3665	0.8746	0.8488
40.0	3.5057E-01	1.5002	0.8300	0.7684	0.7348	1.1873	1.0356	0.9999	1.0000	0.8727	0.3374	0.8727	0.8573
45.0	3.5387E-01	1.5618	0.8245	0.7622	0.7286	1.2071	1.0401	0.9999	1.0000	0.8703	0.3083	0.8704	0.8651
50.0	3.5813E-01	1.6249	0.8181	0.7549	0.7211	1.2265	1.0446	0.9999	1.0000	0.8676	0.2793	0.8676	0.8725

Instr. see p. 22

POSITRON Z= -69 A= 167

p	F_0L_0	L_0	λ_2	λ_3	λ_4	μ_1	μ_2	Λ_1	Λ_2	ν_{12}	$\hat{\nu}_{12}$	η_{12}	$\hat{\eta}_{12}$
0.1	2.5255E-12	1.0574	13.7778	78.3400	260.0602	1.0017	1.0001	0.9950	0.9991	5.6279	0.5434	0.7384	0.6436
0.2	8.0813E-06	1.0575	3.8954	7.8809	11.7670	1.0019	1.0001	0.9951	0.9991	1.9579	0.5422	0.7393	0.6446
0.3	9.1627E-04	1.0576	2.0674	2.7874	3.2272	1.0022	1.0001	0.9952	0.9992	1.2797	0.5405	0.7407	0.6461
0.4	8.4626E-03	1.0579	1.4299	1.5944	1.6622	1.0026	1.0002	0.9954	0.9992	1.0440	0.5383	0.7427	0.6482
0.5	2.9379E-02	1.0582	1.1374	1.1479	1.1371	1.0030	1.0003	0.9956	0.9992	0.9365	0.5360	0.7450	0.6507
0.6	6.3447E-02	1.0586	0.9808	0.9352	0.9011	1.0035	1.0003	0.9957	0.9992	0.8796	0.5337	0.7477	0.6536
0.7	1.0548E-01	1.0590	0.8885	0.8186	0.7759	1.0040	1.0004	0.9959	0.9993	0.8467	0.5315	0.7505	0.6565
0.8	1.4988E-01	1.0596	0.8305	0.7486	0.7024	1.0046	1.0005	0.9961	0.9993	0.8265	0.5295	0.7534	0.6596
0.9	1.9272E-01	1.0602	0.7923	0.7040	0.6564	1.0052	1.0006	0.9963	0.9993	0.8137	0.5279	0.7564	0.6627
1.0	2.3183E-01	1.0608	0.7664	0.6745	0.6262	1.0058	1.0007	0.9965	0.9994	0.8055	0.5264	0.7593	0.6657
1.2	2.9609E-01	1.0622	0.7359	0.6407	0.5921	1.0071	1.0009	0.9969	0.9994	0.7967	0.5243	0.7650	0.6717
1.4	3.4275E-01	1.0638	0.7207	0.6245	0.5760	1.0084	1.0011	0.9972	0.9995	0.7934	0.5228	0.7705	0.6772
1.6	3.7554E-01	1.0656	0.7136	0.6171	0.5689	1.0097	1.0013	0.9974	0.9995	0.7929	0.5220	0.7755	0.6824
1.8	3.9818E-01	1.0674	0.7107	0.6146	0.5666	1.0110	1.0015	0.9976	0.9996	0.7938	0.5214	0.7802	0.6873
2.0	4.1361E-01	1.0693	0.7104	0.6147	0.5670	1.0123	1.0017	0.9978	0.9996	0.7955	0.5211	0.7846	0.6918
2.2	4.2396E-01	1.0713	0.7117	0.6165	0.5691	1.0136	1.0019	0.9980	0.9996	0.7976	0.5210	0.7887	0.6960
2.4	4.3070E-01	1.0733	0.7138	0.6193	0.5721	1.0148	1.0021	0.9982	0.9997	0.7999	0.5209	0.7925	0.7000
2.6	4.3488E-01	1.0754	0.7165	0.6226	0.5757	1.0161	1.0023	0.9983	0.9997	0.8023	0.5209	0.7960	0.7037
2.8	4.3723E-01	1.0775	0.7195	0.6263	0.5796	1.0173	1.0025	0.9984	0.9997	0.8047	0.5209	0.7993	0.7072
3.0	4.3827E-01	1.0796	0.7226	0.6301	0.5837	1.0185	1.0027	0.9985	0.9997	0.8070	0.5209	0.8024	0.7105
3.2	4.3835E-01	1.0818	0.7259	0.6340	0.5878	1.0197	1.0029	0.9986	0.9997	0.8093	0.5208	0.8053	0.7136
3.4	4.3775E-01	1.0840	0.7292	0.6380	0.5920	1.0209	1.0031	0.9987	0.9998	0.8116	0.5207	0.8080	0.7165
3.6	4.3666E-01	1.0861	0.7324	0.6419	0.5961	1.0221	1.0033	0.9987	0.9998	0.8137	0.5206	0.8106	0.7193
3.8	4.3521E-01	1.0883	0.7356	0.6457	0.6002	1.0233	1.0035	0.9988	0.9998	0.8158	0.5205	0.8130	0.7220
4.0	4.3352E-01	1.0906	0.7388	0.6495	0.6042	1.0245	1.0036	0.9989	0.9998	0.8178	0.5203	0.8153	0.7246
4.5	4.2863E-01	1.0961	0.7463	0.6585	0.6137	1.0274	1.0041	0.9990	0.9998	0.8225	0.5197	0.8206	0.7305
5.0	4.2338E-01	1.1017	0.7533	0.6669	0.6226	1.0303	1.0046	0.9991	0.9998	0.8267	0.5188	0.8252	0.7359
5.5	4.1811E-01	1.1073	0.7598	0.6747	0.6309	1.0331	1.0051	0.9992	0.9998	0.8305	0.5177	0.8293	0.7408
6.0	4.1303E-01	1.1130	0.7657	0.6819	0.6385	1.0359	1.0055	0.9993	0.9999	0.8340	0.5165	0.8330	0.7453
6.5	4.0820E-01	1.1186	0.7711	0.6885	0.6456	1.0386	1.0060	0.9993	0.9999	0.8371	0.5151	0.8363	0.7495
7.0	4.0366E-01	1.1243	0.7762	0.6947	0.6521	1.0413	1.0065	0.9994	0.9999	0.8400	0.5136	0.8393	0.7534
7.5	3.9942E-01	1.1299	0.7808	0.7003	0.6581	1.0440	1.0069	0.9994	0.9999	0.8426	0.5119	0.8421	0.7571
8.0	3.9547E-01	1.1356	0.7851	0.7056	0.6638	1.0467	1.0074	0.9995	0.9999	0.8450	0.5101	0.8446	0.7605
9.0	3.8836E-01	1.1469	0.7927	0.7150	0.6739	1.0520	1.0083	0.9995	0.9999	0.8493	0.5062	0.8489	0.7668
10.0	3.8221E-01	1.1583	0.7992	0.7231	0.6826	1.0572	1.0092	0.9996	0.9999	0.8529	0.5021	0.8527	0.7725
11.0	3.7688E-01	1.1697	0.8048	0.7302	0.6903	1.0622	1.0101	0.9996	0.9999	0.8560	0.4976	0.8558	0.7776
12.0	3.7224E-01	1.1811	0.8096	0.7363	0.6969	1.0673	1.0110	0.9996	0.9999	0.8587	0.4930	0.8585	0.7824
13.0	3.6819E-01	1.1925	0.8138	0.7417	0.7028	1.0722	1.0120	0.9997	0.9999	0.8610	0.4882	0.8609	0.7868
14.0	3.6466E-01	1.2039	0.8174	0.7464	0.7079	1.0771	1.0129	0.9997	0.9999	0.8630	0.4832	0.8629	0.7909
16.0	3.5886E-01	1.2268	0.8232	0.7541	0.7164	1.0868	1.0147	0.9997	0.9999	0.8662	0.4730	0.8661	0.7964
18.0	3.5441E-01	1.2497	0.8275	0.7599	0.7228	1.0962	1.0165	0.9998	1.0000	0.8686	0.4624	0.8685	0.8050
20.0	3.5101E-01	1.2728	0.8305	0.7642	0.7277	1.1055	1.0183	0.9998	1.0000	0.8703	0.4516	0.8703	0.8111
25.0	3.4586E-01	1.3312	0.8343	0.7702	0.7348	1.1280	1.0228	0.9998	1.0000	0.8727	0.4240	0.8727	0.8241
30.0	3.4402E-01	1.3906	0.8343	0.7713	0.7365	1.1499	1.0273	0.9999	1.0000	0.8730	0.3957	0.8731	0.8352
35.0	3.4439E-01	1.4512	0.8317	0.7691	0.7348	1.1711	1.0318	0.9999	1.0000	0.8722	0.3672	0.8722	0.8448
40.0	3.4637E-01	1.5132	0.8274	0.7647	0.7305	1.1918	1.0363	0.9999	1.0000	0.8704	0.3385	0.8704	0.8535
45.0	3.4956E-01	1.5766	0.8219	0.7587	0.7244	1.2121	1.0408	0.9999	1.0000	0.8681	0.3098	0.8681	0.8616
50.0	3.5372E-01	1.6415	0.8155	0.7515	0.7171	1.2318	1.0454	0.9999	1.0000	0.8654	0.2812	0.8654	0.8691

 Instr. see p. 22

POSITRON Z= −70 A= 172

p	F_0L_0	L_0	λ_2	λ_3	λ_4	μ_1	μ_2	Λ_1	Λ_2	ν_{12}	$\hat{\nu}_{12}$	η_{12}	$\hat{\eta}_{12}$
0.1	1.6871E−12	1.0585	13.9211	80.6098	272.3508	1.0019	1.0001	0.9949	0.9991	5.7324	0.5388	0.7325	0.6357
0.2	6.7737E−06	1.0586	3.9254	8.0307	12.0919	1.0020	1.0001	0.9950	0.9991	1.9796	0.5377	0.7334	0.6366
0.3	8.2707E−04	1.0588	2.0764	2.8153	3.2709	1.0023	1.0001	0.9951	0.9991	1.2861	0.5360	0.7349	0.6382
0.4	7.9179E−03	1.0590	1.4316	1.5995	1.6693	1.0027	1.0002	0.9952	0.9992	1.0450	0.5339	0.7368	0.6403
0.5	2.8061E−02	1.0594	1.1356	1.1460	1.1350	1.0032	1.0003	0.9954	0.9992	0.9350	0.5316	0.7392	0.6429
0.6	6.1397E−02	1.0598	0.9772	0.9305	0.8956	1.0037	1.0004	0.9956	0.9992	0.8769	0.5294	0.7419	0.6457
0.7	1.0296E−01	1.0603	0.8838	0.8124	0.7689	1.0042	1.0004	0.9958	0.9993	0.8432	0.5272	0.7447	0.6487
0.8	1.4720E−01	1.0608	0.8251	0.7416	0.6947	1.0048	1.0005	0.9960	0.9993	0.8225	0.5254	0.7477	0.6518
0.9	1.9010E−01	1.0614	0.7865	0.6966	0.6482	1.0055	1.0006	0.9962	0.9993	0.8094	0.5237	0.7507	0.6549
1.0	2.2940E−01	1.0621	0.7602	0.6668	0.6178	1.0061	1.0007	0.9964	0.9994	0.8009	0.5223	0.7537	0.6580
1.2	2.9420E−01	1.0636	0.7293	0.6326	0.5834	1.0074	1.0009	0.9968	0.9994	0.7919	0.5203	0.7595	0.6641
1.4	3.4137E−01	1.0652	0.7141	0.6163	0.5672	1.0088	1.0011	0.9971	0.9995	0.7886	0.5189	0.7651	0.6697
1.6	3.7453E−01	1.0670	0.7068	0.6089	0.5601	1.0101	1.0013	0.9973	0.9995	0.7881	0.5181	0.7702	0.6750
1.8	3.9743E−01	1.0689	0.7040	0.6064	0.5578	1.0115	1.0015	0.9976	0.9996	0.7890	0.5177	0.7751	0.6800
2.0	4.1302E−01	1.0709	0.7037	0.6066	0.5583	1.0128	1.0018	0.9978	0.9996	0.7907	0.5175	0.7795	0.6846
2.2	4.2343E−01	1.0730	0.7050	0.6084	0.5604	1.0141	1.0020	0.9979	0.9996	0.7929	0.5174	0.7837	0.6889
2.4	4.3019E−01	1.0751	0.7072	0.6112	0.5635	1.0154	1.0022	0.9981	0.9997	0.7952	0.5174	0.7876	0.6929
2.6	4.3435E−01	1.0772	0.7100	0.6146	0.5671	1.0167	1.0024	0.9982	0.9997	0.7976	0.5174	0.7912	0.6967
2.8	4.3666E−01	1.0794	0.7130	0.6184	0.5711	1.0180	1.0026	0.9984	0.9997	0.8001	0.5175	0.7946	0.7003
3.0	4.3763E−01	1.0816	0.7163	0.6223	0.5753	1.0192	1.0028	0.9985	0.9997	0.8025	0.5175	0.7978	0.7036
3.2	4.3764E−01	1.0838	0.7197	0.6263	0.5795	1.0205	1.0030	0.9985	0.9997	0.8049	0.5175	0.8007	0.7068
3.4	4.3695E−01	1.0861	0.7230	0.6304	0.5838	1.0217	1.0032	0.9986	0.9997	0.8072	0.5175	0.8035	0.7098
3.6	4.3577E−01	1.0883	0.7264	0.6343	0.5880	1.0230	1.0033	0.9987	0.9998	0.8094	0.5174	0.8062	0.7127
3.8	4.3423E−01	1.0906	0.7297	0.6383	0.5921	1.0242	1.0035	0.9988	0.9998	0.8115	0.5173	0.8087	0.7155
4.0	4.3245E−01	1.0929	0.7329	0.6421	0.5962	1.0254	1.0037	0.9988	0.9998	0.8136	0.5172	0.8110	0.7181
4.5	4.2735E−01	1.0986	0.7406	0.6514	0.6059	1.0284	1.0042	0.9990	0.9998	0.8184	0.5166	0.8164	0.7241
5.0	4.2189E−01	1.1044	0.7478	0.6599	0.6150	1.0314	1.0047	0.9991	0.9998	0.8227	0.5159	0.8212	0.7296
5.5	4.1644E−01	1.1102	0.7544	0.6679	0.6234	1.0343	1.0052	0.9992	0.9998	0.8266	0.5149	0.8254	0.7347
6.0	4.1118E−01	1.1161	0.7605	0.6752	0.6312	1.0372	1.0057	0.9992	0.9999	0.8302	0.5137	0.8292	0.7393
6.5	4.0621E−01	1.1219	0.7661	0.6820	0.6384	1.0400	1.0062	0.9993	0.9999	0.8334	0.5124	0.8326	0.7436
7.0	4.0153E−01	1.1278	0.7712	0.6883	0.6450	1.0429	1.0066	0.9993	0.9999	0.8364	0.5109	0.8357	0.7476
7.5	3.9716E−01	1.1336	0.7760	0.6941	0.6512	1.0456	1.0071	0.9994	0.9999	0.8391	0.5093	0.8385	0.7513
8.0	3.9309E−01	1.1395	0.7804	0.6995	0.6570	1.0484	1.0076	0.9994	0.9999	0.8416	0.5076	0.8411	0.7548
9.0	3.8579E−01	1.1512	0.7882	0.7091	0.6673	1.0538	1.0085	0.9995	0.9999	0.8459	0.5038	0.8456	0.7613
10.0	3.7947E−01	1.1630	0.7949	0.7174	0.6763	1.0592	1.0095	0.9996	0.9999	0.8497	0.4997	0.8494	0.7671
11.0	3.7399E−01	1.1747	0.8006	0.7247	0.6841	1.0644	1.0104	0.9996	0.9999	0.8529	0.4954	0.8526	0.7724
12.0	3.6923E−01	1.1865	0.8056	0.7310	0.6909	1.0696	1.0113	0.9996	0.9999	0.8556	0.4908	0.8554	0.7773
13.0	3.6508E−01	1.1983	0.8099	0.7365	0.6969	1.0747	1.0123	0.9997	0.9999	0.8580	0.4861	0.8579	0.7818
14.0	3.6146E−01	1.2101	0.8136	0.7413	0.7022	1.0798	1.0132	0.9997	0.9999	0.8600	0.4812	0.8599	0.7860
16.0	3.5550E−01	1.2338	0.8196	0.7492	0.7109	1.0897	1.0151	0.9997	0.9999	0.8633	0.4711	0.8633	0.7937
18.0	3.5093E−01	1.2576	0.8241	0.7552	0.7175	1.0994	1.0169	0.9998	1.0000	0.8658	0.4607	0.8658	0.8005
20.0	3.4744E−01	1.2815	0.8272	0.7597	0.7226	1.1090	1.0188	0.9998	1.0000	0.8676	0.4500	0.8676	0.8067
25.0	3.4212E−01	1.3419	0.8312	0.7659	0.7299	1.1322	1.0234	0.9998	1.0000	0.8701	0.4225	0.8701	0.8201
30.0	3.4017E−01	1.4034	0.8313	0.7672	0.7318	1.1547	1.0280	0.9999	1.0000	0.8706	0.3945	0.8706	0.8314
35.0	3.4048E−01	1.4663	0.8287	0.7651	0.7301	1.1765	1.0326	0.9999	1.0000	0.8697	0.3662	0.8697	0.8413
40.0	3.4241E−01	1.5305	0.8244	0.7607	0.7259	1.1978	1.0372	0.9999	1.0000	0.8680	0.3377	0.8680	0.8503
45.0	3.4559E−01	1.5964	0.8189	0.7546	0.7198	1.2186	1.0421	0.9999	1.0000	0.8656	0.3095	0.8657	0.8587
50.0	3.4974E−01	1.6638	0.8125	0.7474	0.7124	1.2389	1.0465	0.9999	1.0000	0.8629	0.2808	0.8629	0.8663

POSITRON Z= -71 A= 170

p	F_0L_0	L_0	λ_2	λ_3	λ_4	μ_1	μ_2	Λ_1	Λ_2	ν_{12}	$\hat{\nu}_{12}$	η_{12}	$\hat{\eta}_{12}$
0.1	1.1316E-12	1.0594	14.0306	82.6592	284.1011	1.0020	1.0001	0.9948	0.9991	5.8320	0.5337	0.7259	0.6271
0.2	5.7006E-06	1.0595	3.9461	8.1576	12.3828	1.0021	1.0001	0.9949	0.9991	1.9995	0.5326	0.7268	0.6281
0.3	7.4956E-04	1.0597	2.0806	2.8350	3.3044	1.0024	1.0001	0.9950	0.9991	1.2913	0.5310	0.7283	0.6297
0.4	7.4380E-03	1.0599	1.4301	1.6001	1.6712	1.0028	1.0002	0.9952	0.9992	1.0450	0.5290	0.7302	0.6318
0.5	2.6910E-02	1.0603	1.1314	1.1410	1.1294	1.0033	1.0003	0.9953	0.9992	0.9327	0.5268	0.7326	0.6343
0.6	5.9652E-02	1.0607	0.9716	0.9232	0.8875	1.0038	1.0004	0.9955	0.9992	0.8733	0.5246	0.7353	0.6372
0.7	1.0091E-01	1.0612	0.8773	0.8040	0.7597	1.0044	1.0004	0.9957	0.9993	0.8388	0.5225	0.7382	0.6402
0.8	1.4515E-01	1.0617	0.8180	0.7327	0.6850	1.0050	1.0005	0.9960	0.9993	0.8177	0.5207	0.7412	0.6434
0.9	1.8826E-01	1.0624	0.7790	0.6874	0.6383	1.0056	1.0006	0.9962	0.9993	0.8043	0.5191	0.7443	0.6465
1.0	2.2791E-01	1.0630	0.7526	0.6574	0.6077	1.0063	1.0007	0.9963	0.9994	0.7956	0.5178	0.7473	0.6497
1.2	2.9348E-01	1.0646	0.7214	0.6230	0.5731	1.0077	1.0009	0.9967	0.9994	0.7864	0.5159	0.7533	0.6558
1.4	3.4134E-01	1.0663	0.7060	0.6066	0.5569	1.0090	1.0012	0.9970	0.9995	0.7830	0.5147	0.7589	0.6615
1.6	3.7501E-01	1.0681	0.6987	0.5992	0.5498	1.0104	1.0014	0.9973	0.9995	0.7824	0.5140	0.7642	0.6669
1.8	3.9824E-01	1.0700	0.6960	0.5967	0.5475	1.0118	1.0016	0.9975	0.9996	0.7834	0.5137	0.7691	0.6720
2.0	4.1402E-01	1.0721	0.6958	0.5970	0.5481	1.0131	1.0018	0.9977	0.9996	0.7852	0.5135	0.7737	0.6767
2.2	4.2454E-01	1.0742	0.6971	0.5989	0.5503	1.0145	1.0020	0.9979	0.9996	0.7874	0.5135	0.7780	0.6810
2.4	4.3133E-01	1.0763	0.6994	0.6018	0.5534	1.0158	1.0022	0.9981	0.9996	0.7898	0.5136	0.7820	0.6852
2.6	4.3547E-01	1.0785	0.7022	0.6053	0.5571	1.0171	1.0024	0.9982	0.9997	0.7923	0.5137	0.7857	0.6890
2.8	4.3772E-01	1.0807	0.7054	0.6091	0.5612	1.0184	1.0026	0.9983	0.9997	0.7948	0.5139	0.7892	0.6926
3.0	4.3861E-01	1.0830	0.7087	0.6131	0.5655	1.0197	1.0028	0.9984	0.9997	0.7973	0.5140	0.7924	0.6961
3.2	4.3853E-01	1.0853	0.7122	0.6172	0.5698	1.0210	1.0030	0.9985	0.9997	0.7997	0.5140	0.7955	0.6993
3.4	4.3775E-01	1.0876	0.7156	0.6213	0.5741	1.0223	1.0032	0.9986	0.9997	0.8021	0.5141	0.7983	0.7024
3.6	4.3646E-01	1.0899	0.7191	0.6254	0.5784	1.0236	1.0034	0.9987	0.9998	0.8043	0.5141	0.8010	0.7053
3.8	4.3481E-01	1.0922	0.7224	0.6294	0.5826	1.0248	1.0036	0.9988	0.9998	0.8065	0.5140	0.8036	0.7081
4.0	4.3292E-01	1.0945	0.7258	0.6334	0.5868	1.0261	1.0038	0.9988	0.9998	0.8086	0.5140	0.8060	0.7108
4.5	4.2755E-01	1.1004	0.7337	0.6428	0.5967	1.0291	1.0043	0.9990	0.9998	0.8136	0.5136	0.8116	0.7170
5.0	4.2183E-01	1.1063	0.7411	0.6516	0.6059	1.0322	1.0048	0.9991	0.9998	0.8180	0.5130	0.8165	0.7226
5.5	4.1615E-01	1.1122	0.7479	0.6597	0.6145	1.0351	1.0053	0.9991	0.9998	0.8221	0.5121	0.8208	0.7278
6.0	4.1068E-01	1.1182	0.7541	0.6672	0.6225	1.0381	1.0057	0.9992	0.9999	0.8258	0.5111	0.8247	0.7325
6.5	4.0550E-01	1.1242	0.7599	0.6742	0.6298	1.0410	1.0062	0.9993	0.9999	0.8291	0.5099	0.8282	0.7369
7.0	4.0064E-01	1.1301	0.7652	0.6806	0.6366	1.0439	1.0067	0.9993	0.9999	0.8321	0.5085	0.8314	0.7410
7.5	3.9610E-01	1.1361	0.7701	0.6866	0.6430	1.0467	1.0072	0.9994	0.9999	0.8349	0.5070	0.8343	0.7448
8.0	3.9188E-01	1.1421	0.7746	0.6921	0.6489	1.0495	1.0077	0.9994	0.9999	0.8375	0.5054	0.8370	0.7484
9.0	3.8431E-01	1.1541	0.7827	0.7020	0.6595	1.0551	1.0086	0.9995	0.9999	0.8420	0.5018	0.8417	0.7550
10.0	3.7775E-01	1.1661	0.7896	0.7106	0.6687	1.0605	1.0096	0.9996	0.9999	0.8459	0.4979	0.8456	0.7610
11.0	3.7207E-01	1.1781	0.7956	0.7181	0.6767	1.0659	1.0105	0.9996	0.9999	0.8492	0.4938	0.8490	0.7664
12.0	3.6713E-01	1.1901	0.8008	0.7246	0.6838	1.0712	1.0115	0.9996	0.9999	0.8521	0.4894	0.8519	0.7714
13.0	3.6282E-01	1.2021	0.8053	0.7303	0.6900	1.0764	1.0124	0.9997	0.9999	0.8545	0.4848	0.8544	0.7760
14.0	3.5905E-01	1.2141	0.8092	0.7353	0.6955	1.0815	1.0134	0.9997	0.9999	0.8567	0.4801	0.8566	0.7803
16.0	3.5285E-01	1.2383	0.8154	0.7436	0.7045	1.0916	1.0153	0.9997	0.9999	0.8602	0.4703	0.8601	0.7882
18.0	3.4807E-01	1.2626	0.8201	0.7499	0.7115	1.1016	1.0171	0.9998	1.0000	0.8628	0.4601	0.8628	0.7952
20.0	3.4439E-01	1.2869	0.8235	0.7546	0.7168	1.1113	1.0190	0.9998	1.0000	0.8647	0.4498	0.8647	0.8015
25.0	3.3871E-01	1.3486	0.8280	0.7614	0.7247	1.1349	1.0237	0.9998	1.0000	0.8674	0.4230	0.8674	0.8152
30.0	3.3647E-01	1.4114	0.8284	0.7631	0.7271	1.1577	1.0283	0.9999	1.0000	0.8681	0.3956	0.8681	0.8268
35.0	3.3653E-01	1.4755	0.8261	0.7614	0.7258	1.1799	1.0330	0.9999	1.0000	0.8674	0.3679	0.8674	0.8369
40.0	3.3826E-01	1.5411	0.8220	0.7572	0.7219	1.2016	1.0377	0.9999	1.0000	0.8657	0.3400	0.8658	0.8461
45.0	3.4124E-01	1.6083	0.8167	0.7514	0.7161	1.2227	1.0423	0.9999	1.0000	0.8635	0.3120	0.8635	0.8545
50.0	3.4522E-01	1.6773	0.8104	0.7443	0.7089	1.2434	1.0471	0.9999	1.0000	0.8608	0.2842	0.8608	0.8624

Instr. see p. 22

POSITRON Z= −72 A= 178

p	F_0L_0	L_0	λ_2	λ_3	λ_4	μ_1	μ_2	Λ_1	Λ_2	ν_{12}	$\hat{\nu}_{12}$	η_{12}	$\hat{\eta}_{12}$
0.1	7.5536E-13	1.0606	14.1705	84.9809	297.1964	1.0021	1.0001	0.9946	0.9991	5.9384	0.5293	0.7202	0.6194
0.2	4.7742E-06	1.0607	3.9755	8.3104	12.7224	1.0023	1.0001	0.9947	0.9991	2.0218	0.5282	0.7211	0.6204
0.3	6.7601E-04	1.0609	2.0896	2.8637	3.3500	1.0026	1.0002	0.9948	0.9991	1.2980	0.5266	0.7226	0.6219
0.4	6.9534E-03	1.0612	1.4319	1.6058	1.6791	1.0030	1.0002	0.9950	0.9991	1.0463	0.5247	0.7246	0.6241
0.5	2.5682E-02	1.0615	1.1300	1.1396	1.1279	1.0035	1.0003	0.9952	0.9992	0.9315	0.5225	0.7270	0.6267
0.6	5.7676E-02	1.0620	0.9683	0.9190	0.8826	1.0040	1.0004	0.9954	0.9992	0.8708	0.5204	0.7297	0.6295
0.7	9.8426E-02	1.0625	0.8730	0.7984	0.7534	1.0046	1.0005	0.9956	0.9992	0.8355	0.5184	0.7327	0.6326
0.8	1.4243E-01	1.0630	0.8130	0.7263	0.6779	1.0053	1.0006	0.9958	0.9993	0.8139	0.5167	0.7357	0.6358
0.9	1.8555E-01	1.0637	0.7736	0.6805	0.6307	1.0059	1.0007	0.9960	0.9993	0.8002	0.5151	0.7388	0.6390
1.0	2.2534E-01	1.0644	0.7468	0.6502	0.5999	1.0066	1.0008	0.9962	0.9993	0.7913	0.5139	0.7419	0.6422
1.2	2.9138E-01	1.0660	0.7153	0.6155	0.5650	1.0080	1.0010	0.9966	0.9994	0.7819	0.5120	0.7480	0.6484
1.4	3.3970E-01	1.0678	0.6998	0.5990	0.5487	1.0095	1.0012	0.9969	0.9995	0.7784	0.5109	0.7537	0.6542
1.6	3.7372E-01	1.0697	0.6924	0.5916	0.5416	1.0109	1.0014	0.9972	0.9995	0.7778	0.5103	0.7591	0.6597
1.8	3.9718E-01	1.0717	0.6897	0.5891	0.5394	1.0123	1.0016	0.9974	0.9995	0.7788	0.5100	0.7642	0.6649
2.0	4.1310E-01	1.0738	0.6895	0.5894	0.5400	1.0137	1.0018	0.9977	0.9996	0.7806	0.5100	0.7689	0.6696
2.2	4.2368E-01	1.0760	0.6909	0.5914	0.5422	1.0151	1.0020	0.9978	0.9996	0.7828	0.5101	0.7732	0.6741
2.4	4.3048E-01	1.0782	0.6932	0.5943	0.5454	1.0165	1.0023	0.9980	0.9996	0.7853	0.5102	0.7773	0.6783
2.6	4.3460E-01	1.0805	0.6962	0.5979	0.5492	1.0179	1.0025	0.9981	0.9997	0.7878	0.5104	0.7811	0.6822
2.8	4.3680E-01	1.0828	0.6994	0.6018	0.5534	1.0193	1.0027	0.9983	0.9997	0.7904	0.5105	0.7847	0.6860
3.0	4.3762E-01	1.0852	0.7028	0.6059	0.5577	1.0206	1.0029	0.9984	0.9997	0.7930	0.5107	0.7880	0.6895
3.2	4.3746E-01	1.0875	0.7064	0.6101	0.5621	1.0220	1.0031	0.9985	0.9997	0.7954	0.5108	0.7911	0.6928
3.4	4.3659E-01	1.0899	0.7099	0.6143	0.5665	1.0233	1.0033	0.9986	0.9997	0.7979	0.5109	0.7941	0.6959
3.6	4.3522E-01	1.0923	0.7134	0.6184	0.5709	1.0246	1.0035	0.9986	0.9997	0.8002	0.5109	0.7968	0.6989
3.8	4.3349E-01	1.0947	0.7169	0.6225	0.5752	1.0259	1.0037	0.9987	0.9998	0.8024	0.5109	0.7995	0.7018
4.0	4.3150E-01	1.0972	0.7203	0.6265	0.5794	1.0272	1.0039	0.9988	0.9998	0.8046	0.5109	0.8019	0.7045
4.5	4.2593E-01	1.1033	0.7284	0.6362	0.5895	1.0304	1.0044	0.9989	0.9998	0.8097	0.5106	0.8076	0.7109
5.0	4.2002E-01	1.1094	0.7359	0.6451	0.5989	1.0335	1.0049	0.9990	0.9998	0.8142	0.5100	0.8126	0.7166
5.5	4.1416E-01	1.1156	0.7429	0.6534	0.6076	1.0366	1.0054	0.9991	0.9998	0.8184	0.5092	0.8171	0.7219
6.0	4.0853E-01	1.1218	0.7493	0.6611	0.6157	1.0397	1.0059	0.9992	0.9998	0.8222	0.5082	0.8211	0.7267
6.5	4.0320E-01	1.1280	0.7552	0.6682	0.6232	1.0427	1.0064	0.9993	0.9999	0.8256	0.5071	0.8247	0.7312
7.0	3.9822E-01	1.1342	0.7607	0.6747	0.6302	1.0457	1.0069	0.9993	0.9999	0.8287	0.5057	0.8280	0.7354
7.5	3.9357E-01	1.1404	0.7657	0.6808	0.6366	1.0487	1.0074	0.9994	0.9999	0.8316	0.5043	0.8310	0.7393
8.0	3.8924E-01	1.1466	0.7703	0.6865	0.6426	1.0516	1.0079	0.9994	0.9999	0.8342	0.5027	0.8337	0.7430
9.0	3.8149E-01	1.1591	0.7786	0.6966	0.6534	1.0573	1.0089	0.9995	0.9999	0.8388	0.4992	0.8385	0.7497
10.0	3.7479E-01	1.1716	0.7857	0.7054	0.6628	1.0630	1.0099	0.9995	0.9999	0.8428	0.4954	0.8425	0.7558
11.0	3.6898E-01	1.1840	0.7918	0.7130	0.6710	1.0685	1.0109	0.9996	0.9999	0.8462	0.4912	0.8460	0.7614
12.0	3.6394E-01	1.1965	0.7971	0.7197	0.6782	1.0740	1.0119	0.9996	0.9999	0.8491	0.4869	0.8490	0.7665
13.0	3.5955E-01	1.2090	0.8017	0.7255	0.6846	1.0794	1.0128	0.9997	0.9999	0.8517	0.4823	0.8515	0.7713
14.0	3.5570E-01	1.2216	0.8057	0.7306	0.6901	1.0847	1.0138	0.9997	0.9999	0.8539	0.4777	0.8538	0.7757
16.0	3.4938E-01	1.2467	0.8121	0.7390	0.6994	1.0952	1.0158	0.9997	0.9999	0.8574	0.4679	0.8574	0.7837
18.0	3.4451E-01	1.2720	0.8169	0.7455	0.7065	1.1055	1.0177	0.9998	0.9999	0.8601	0.4578	0.8601	0.7909
20.0	3.4078E-01	1.2973	0.8204	0.7503	0.7119	1.1155	1.0196	0.9998	1.0000	0.8621	0.4475	0.8621	0.7974
25.0	3.3500E-01	1.3615	0.8249	0.7572	0.7200	1.1399	1.0244	0.9998	1.0000	0.8649	0.4208	0.8649	0.8115
30.0	3.3273E-01	1.4270	0.8253	0.7590	0.7224	1.1635	1.0292	0.9999	1.0000	0.8655	0.3934	0.8656	0.8234
35.0	3.3280E-01	1.4939	0.8230	0.7572	0.7211	1.1865	1.0340	0.9999	1.0000	0.8648	0.3657	0.8649	0.8338
40.0	3.3455E-01	1.5624	0.8189	0.7530	0.7171	1.2088	1.0388	0.9999	1.0000	0.8632	0.3379	0.8632	0.8432
45.0	3.3759E-01	1.6327	0.8134	0.7471	0.7112	1.2306	1.0437	0.9999	1.0000	0.8609	0.3100	0.8609	0.8519
50.0	3.4163E-01	1.7048	0.8070	0.7398	0.7038	1.2519	1.0486	0.9999	1.0000	0.8582	0.2822	0.8582	0.8600

Instr. see p. 22

POSITRON Z= -73 A= 179

p	F_0L_0	L_0	λ_2	λ_3	λ_4	μ_1	μ_2	Λ_1	Λ_2	ν_{12}	$\hat{\nu}_{12}$	η_{12}	$\hat{\eta}_{12}$
0.1	5.0626E-13	1.0615	14.2755	87.0654	309.6663	1.0022	1.0001	0.9945	0.9990	6.0396	0.5243	0.7138	0.6110
0.2	4.0148E-06	1.0616	3.9954	8.4392	13.0254	1.0024	1.0001	0.9946	0.9991	2.0423	0.5233	0.7147	0.6120
0.3	6.1219E-04	1.0618	2.0938	2.8840	3.3850	1.0027	1.0002	0.9947	0.9991	1.3036	0.5218	0.7162	0.6136
0.4	6.5270E-03	1.0621	1.4306	1.6069	1.6817	1.0032	1.0002	0.9949	0.9991	1.0467	0.5199	0.7182	0.6157
0.5	2.4610E-02	1.0625	1.1261	1.1351	1.1229	1.0037	1.0003	0.9951	0.9991	0.9295	0.5178	0.7207	0.6183
0.6	5.5994E-02	1.0629	0.9630	0.9122	0.8751	1.0042	1.0004	0.9953	0.9992	0.8674	0.5158	0.7234	0.6212
0.7	9.6393E-02	1.0634	0.8669	0.7906	0.7448	1.0048	1.0005	0.9955	0.9992	0.8314	0.5139	0.7264	0.6243
0.8	1.4034E-01	1.0640	0.8064	0.7179	0.6688	1.0055	1.0006	0.9957	0.9993	0.8094	0.5122	0.7295	0.6275
0.9	1.8362E-01	1.0647	0.7666	0.6718	0.6214	1.0062	1.0007	0.9959	0.9993	0.7953	0.5107	0.7326	0.6308
1.0	2.2371E-01	1.0654	0.7396	0.6414	0.5904	1.0069	1.0008	0.9961	0.9993	0.7862	0.5095	0.7358	0.6340
1.2	2.9046E-01	1.0671	0.7078	0.6065	0.5554	1.0083	1.0010	0.9965	0.9994	0.7766	0.5078	0.7420	0.6403
1.4	3.3942E-01	1.0689	0.6922	0.5899	0.5391	1.0098	1.0012	0.9969	0.9994	0.7730	0.5068	0.7478	0.6462
1.6	3.7392E-01	1.0709	0.6848	0.5825	0.5320	1.0113	1.0014	0.9971	0.9995	0.7724	0.5063	0.7533	0.6518
1.8	3.9770E-01	1.0729	0.6821	0.5800	0.5298	1.0127	1.0017	0.9974	0.9995	0.7734	0.5061	0.7585	0.6570
2.0	4.1381E-01	1.0751	0.6819	0.5804	0.5304	1.0142	1.0019	0.9976	0.9996	0.7753	0.5061	0.7633	0.6619
2.2	4.2448E-01	1.0773	0.6834	0.5824	0.5327	1.0156	1.0021	0.9978	0.9996	0.7776	0.5063	0.7677	0.6665
2.4	4.3130E-01	1.0796	0.6858	0.5855	0.5360	1.0171	1.0023	0.9980	0.9996	0.7801	0.5065	0.7719	0.6707
2.6	4.3539E-01	1.0820	0.6888	0.5891	0.5399	1.0185	1.0025	0.9981	0.9997	0.7827	0.5067	0.7758	0.6748
2.8	4.3753E-01	1.0844	0.6922	0.5931	0.5441	1.0199	1.0027	0.9982	0.9997	0.7853	0.5070	0.7794	0.6785
3.0	4.3828E-01	1.0868	0.6957	0.5973	0.5485	1.0213	1.0029	0.9983	0.9997	0.7879	0.5072	0.7828	0.6821
3.2	4.3803E-01	1.0892	0.6993	0.6015	0.5530	1.0227	1.0031	0.9984	0.9997	0.7905	0.5074	0.7860	0.6855
3.4	4.3706E-01	1.0917	0.7029	0.6058	0.5575	1.0240	1.0034	0.9985	0.9997	0.7930	0.5075	0.7891	0.6887
3.6	4.3558E-01	1.0941	0.7065	0.6101	0.5619	1.0254	1.0036	0.9986	0.9997	0.7954	0.5076	0.7919	0.6918
3.8	4.3373E-01	1.0966	0.7101	0.6142	0.5663	1.0267	1.0038	0.9987	0.9998	0.7977	0.5077	0.7946	0.6947
4.0	4.3164E-01	1.0991	0.7136	0.6183	0.5706	1.0280	1.0040	0.9988	0.9998	0.7999	0.5077	0.7972	0.6975
4.5	4.2580E-01	1.1054	0.7219	0.6281	0.5808	1.0313	1.0045	0.9989	0.9998	0.8051	0.5075	0.8030	0.7040
5.0	4.1965E-01	1.1117	0.7296	0.6373	0.5904	1.0346	1.0050	0.9990	0.9998	0.8098	0.5071	0.8081	0.7098
5.5	4.1356E-01	1.1180	0.7368	0.6458	0.5993	1.0377	1.0055	0.9991	0.9998	0.8141	0.5064	0.8127	0.7152
6.0	4.0772E-01	1.1244	0.7434	0.6536	0.6076	1.0409	1.0060	0.9992	0.9998	0.8179	0.5055	0.8168	0.7201
6.5	4.0221E-01	1.1308	0.7494	0.6609	0.6152	1.0440	1.0066	0.9992	0.9999	0.8214	0.5044	0.8205	0.7247
7.0	3.9705E-01	1.1371	0.7550	0.6676	0.6223	1.0470	1.0071	0.9993	0.9999	0.8247	0.5032	0.8239	0.7290
7.5	3.9225E-01	1.1435	0.7602	0.6738	0.6289	1.0501	1.0076	0.9994	0.9999	0.8276	0.5018	0.8270	0.7330
8.0	3.8778E-01	1.1499	0.7649	0.6796	0.6351	1.0531	1.0081	0.9994	0.9999	0.8303	0.5003	0.8298	0.7368
9.0	3.7977E-01	1.1627	0.7734	0.6900	0.6461	1.0590	1.0091	0.9995	0.9999	0.8351	0.4970	0.8347	0.7437
10.0	3.7286E-01	1.1755	0.7807	0.6990	0.6557	1.0648	1.0101	0.9995	0.9999	0.8392	0.4933	0.8389	0.7499
11.0	3.6687E-01	1.1883	0.7871	0.7068	0.6642	1.0705	1.0111	0.9996	0.9999	0.8427	0.4893	0.8425	0.7556
12.0	3.6166E-01	1.2011	0.7925	0.7137	0.6716	1.0761	1.0121	0.9996	0.9999	0.8457	0.4851	0.8456	0.7609
13.0	3.5713E-01	1.2139	0.7973	0.7197	0.6781	1.0816	1.0131	0.9996	0.9999	0.8484	0.4807	0.8482	0.7657
14.0	3.5316E-01	1.2268	0.8014	0.7250	0.6838	1.0871	1.0141	0.9997	0.9999	0.8507	0.4761	0.8506	0.7702
16.0	3.4662E-01	1.2526	0.8081	0.7337	0.6933	1.0978	1.0160	0.9997	0.9999	0.8544	0.4666	0.8543	0.7785
18.0	3.4158E-01	1.2785	0.8131	0.7404	0.7007	1.1083	1.0180	0.9998	0.9999	0.8572	0.4567	0.8571	0.7858
20.0	3.3770E-01	1.3045	0.8168	0.7454	0.7064	1.1185	1.0200	0.9998	1.0000	0.8593	0.4466	0.8593	0.7925
25.0	3.3163E-01	1.3704	0.8217	0.7528	0.7149	1.1435	1.0249	0.9998	1.0000	0.8622	0.4203	0.8623	0.8069
30.0	3.2914E-01	1.4376	0.8224	0.7549	0.7177	1.1676	1.0297	0.9999	1.0000	0.8630	0.3934	0.8631	0.8190
35.0	3.2905E-01	1.5064	0.8203	0.7533	0.7166	1.1910	1.0346	0.9999	1.0000	0.8624	0.3662	0.8624	0.8297
40.0	3.3066E-01	1.5768	0.8162	0.7493	0.7128	1.2138	1.0395	0.9999	1.0000	0.8609	0.3387	0.8609	0.8393
45.0	3.3358E-01	1.6491	0.8108	0.7435	0.7071	1.2361	1.0447	0.9999	1.0000	0.8586	0.3116	0.8587	0.8483
50.0	3.3752E-01	1.7233	0.8045	0.7364	0.6999	1.2579	1.0494	0.9999	1.0000	0.8560	0.2838	0.8560	0.8565

POSITRON Z= -74 A= 180

p	F_0L_0	L_0	λ_2	λ_3	λ_4	μ_1	μ_2	Λ_1	Λ_2	ν_{12}	$\hat{\nu}_{12}$	η_{12}	$\hat{\eta}_{12}$
0.1	3.3947E-13	1.0624	14.3719	89.1126	322.2772	1.0024	1.0001	0.9944	0.9990	6.1403	0.5194	0.7074	0.6026
0.2	3.3777E-06	1.0625	4.0132	8.5637	13.3257	1.0026	1.0001	0.9945	0.9990	2.0627	0.5184	0.7083	0.6036
0.3	5.5463E-04	1.0627	2.0970	2.9029	3.4186	1.0029	1.0002	0.9946	0.9991	1.3091	0.5169	0.7098	0.6052
0.4	6.1294E-03	1.0630	1.4287	1.6073	1.6836	1.0033	1.0002	0.9948	0.9991	1.0470	0.5151	0.7118	0.6074
0.5	2.3593E-02	1.0634	1.1218	1.1301	1.1175	1.0038	1.0003	0.9950	0.9991	0.9274	0.5131	0.7143	0.6100
0.6	5.4385E-02	1.0638	0.9575	0.9052	0.8673	1.0044	1.0004	0.9952	0.9992	0.8640	0.5111	0.7171	0.6129
0.7	9.4445E-02	1.0644	0.8605	0.7826	0.7361	1.0050	1.0005	0.9954	0.9992	0.8273	0.5093	0.7201	0.6160
0.8	1.3834E-01	1.0650	0.7996	0.7095	0.6596	1.0057	1.0006	0.9956	0.9992	0.8047	0.5076	0.7232	0.6193
0.9	1.8180E-01	1.0657	0.7594	0.6631	0.6120	1.0064	1.0007	0.9958	0.9993	0.7904	0.5062	0.7264	0.6225
1.0	2.2219E-01	1.0664	0.7322	0.6324	0.5809	1.0072	1.0008	0.9961	0.9993	0.7811	0.5051	0.7296	0.6258
1.2	2.8968E-01	1.0681	0.7002	0.5974	0.5458	1.0086	1.0010	0.9964	0.9994	0.7712	0.5035	0.7359	0.6322
1.4	3.3930E-01	1.0700	0.6845	0.5808	0.5294	1.0102	1.0012	0.9968	0.9994	0.7676	0.5026	0.7419	0.6382
1.6	3.7429E-01	1.0720	0.6771	0.5733	0.5223	1.0117	1.0015	0.9971	0.9995	0.7670	0.5022	0.7475	0.6439
1.8	3.9840E-01	1.0742	0.6744	0.5709	0.5202	1.0132	1.0017	0.9973	0.9995	0.7680	0.5021	0.7527	0.6492
2.0	4.1470E-01	1.0764	0.6743	0.5713	0.5209	1.0147	1.0019	0.9976	0.9996	0.7699	0.5022	0.7576	0.6541
2.2	4.2546E-01	1.0787	0.6759	0.5735	0.5233	1.0162	1.0021	0.9978	0.9996	0.7722	0.5025	0.7622	0.6588
2.4	4.3230E-01	1.0810	0.6784	0.5766	0.5266	1.0176	1.0023	0.9979	0.9996	0.7748	0.5028	0.7665	0.6631
2.6	4.3637E-01	1.0835	0.6815	0.5803	0.5306	1.0191	1.0026	0.9981	0.9997	0.7775	0.5031	0.7704	0.6672
2.8	4.3845E-01	1.0859	0.6849	0.5844	0.5349	1.0205	1.0028	0.9982	0.9997	0.7802	0.5034	0.7742	0.6711
3.0	4.3911E-01	1.0884	0.6885	0.5886	0.5393	1.0220	1.0030	0.9983	0.9997	0.7829	0.5037	0.7777	0.6747
3.2	4.3876E-01	1.0909	0.6922	0.5930	0.5439	1.0234	1.0032	0.9984	0.9997	0.7855	0.5039	0.7809	0.6782
3.4	4.3769E-01	1.0934	0.6959	0.5973	0.5484	1.0248	1.0034	0.9985	0.9997	0.7880	0.5041	0.7840	0.6814
3.6	4.3609E-01	1.0959	0.6996	0.6017	0.5529	1.0262	1.0036	0.9986	0.9997	0.7905	0.5043	0.7870	0.6846
3.8	4.3414E-01	1.0985	0.7033	0.6059	0.5574	1.0275	1.0038	0.9987	0.9998	0.7929	0.5044	0.7897	0.6875
4.0	4.3194E-01	1.1010	0.7068	0.6101	0.5617	1.0289	1.0041	0.9987	0.9998	0.7951	0.5045	0.7923	0.6904
4.5	4.2582E-01	1.1075	0.7154	0.6201	0.5722	1.0323	1.0046	0.9989	0.9998	0.8005	0.5044	0.7983	0.6970
5.0	4.1942E-01	1.1140	0.7233	0.6294	0.5819	1.0356	1.0051	0.9990	0.9998	0.8053	0.5041	0.8036	0.7030
5.5	4.1310E-01	1.1205	0.7306	0.6381	0.5910	1.0389	1.0056	0.9991	0.9998	0.8097	0.5035	0.8083	0.7085
6.0	4.0704E-01	1.1270	0.7374	0.6461	0.5994	1.0421	1.0062	0.9992	0.9998	0.8137	0.5028	0.8125	0.7135
6.5	4.0134E-01	1.1335	0.7436	0.6535	0.6072	1.0453	1.0067	0.9992	0.9999	0.8173	0.5018	0.8163	0.7182
7.0	3.9600E-01	1.1401	0.7493	0.6604	0.6145	1.0484	1.0072	0.9993	0.9999	0.8206	0.5006	0.8198	0.7226
7.5	3.9104E-01	1.1466	0.7546	0.6667	0.6212	1.0515	1.0077	0.9993	0.9999	0.8236	0.4994	0.8230	0.7267
8.0	3.8642E-01	1.1532	0.7595	0.6727	0.6275	1.0546	1.0082	0.9994	0.9999	0.8264	0.4979	0.8259	0.7305
9.0	3.7816E-01	1.1663	0.7683	0.6833	0.6388	1.0607	1.0092	0.9995	0.9999	0.8313	0.4948	0.8309	0.7376
10.0	3.7102E-01	1.1794	0.7758	0.6925	0.6486	1.0666	1.0103	0.9995	0.9999	0.8356	0.4912	0.8353	0.7440
11.0	3.6485E-01	1.1925	0.7823	0.7006	0.6573	1.0724	1.0113	0.9996	0.9999	0.8392	0.4874	0.8390	0.7498
12.0	3.5948E-01	1.2057	0.7880	0.7077	0.6649	1.0782	1.0123	0.9996	0.9999	0.8423	0.4833	0.8421	0.7552
13.0	3.5480E-01	1.2189	0.7929	0.7139	0.6715	1.0838	1.0133	0.9996	0.9999	0.8451	0.4790	0.8449	0.7601
14.0	3.5070E-01	1.2321	0.7972	0.7193	0.6775	1.0894	1.0143	0.9997	0.9999	0.8474	0.4745	0.8473	0.7647
16.0	3.4395E-01	1.2585	0.8041	0.7283	0.6873	1.1004	1.0163	0.9997	0.9999	0.8513	0.4653	0.8512	0.7732
18.0	3.3873E-01	1.2851	0.8094	0.7353	0.6949	1.1111	1.0183	0.9997	0.9999	0.8542	0.4556	0.8542	0.7807
20.0	3.3470E-01	1.3119	0.8132	0.7405	0.7008	1.1217	1.0203	0.9998	1.0000	0.8564	0.4456	0.8564	0.7875
25.0	3.2834E-01	1.3795	0.8185	0.7484	0.7098	1.1472	1.0253	0.9998	1.0000	0.8596	0.4199	0.8596	0.8022
30.0	3.2564E-01	1.4485	0.8194	0.7507	0.7130	1.1718	1.0303	0.9999	1.0000	0.8605	0.3934	0.8605	0.8147
35.0	3.2537E-01	1.5191	0.8175	0.7495	0.7122	1.1957	1.0352	0.9999	1.0000	0.8600	0.3666	0.8600	0.8256
40.0	3.2685E-01	1.5915	0.8136	0.7456	0.7086	1.2190	1.0402	0.9999	1.0000	0.8585	0.3395	0.8585	0.8355
45.0	3.2964E-01	1.6658	0.8083	0.7400	0.7030	1.2418	1.0452	0.9999	1.0000	0.8563	0.3124	0.8564	0.8445
50.0	3.3348E-01	1.7422	0.8020	0.7329	0.6959	1.2640	1.0502	0.9999	1.0000	0.8537	0.2853	0.8537	0.8530

Instr. see p. 22

POSITRON Z= -75 A= 183

p	F_0L_0	L_0	λ_2	λ_3	λ_4	μ_1	μ_2	Λ_1	Λ_2	ν_{12}	$\hat{\nu}_{12}$	η_{12}	$\hat{\eta}_{12}$
0.1	2.2748E-13	1.0633	14.4683	91.1891	335.2964	1.0025	1.0001	0.9942	0.9990	6.2422	0.5145	0.7011	0.5944
0.2	2.8399E-06	1.0634	4.0312	8.6905	13.6343	1.0027	1.0001	0.9943	0.9990	2.0834	0.5136	0.7020	0.5954
0.3	5.0216E-04	1.0636	2.1005	2.9226	3.4535	1.0031	1.0002	0.9945	0.9991	1.3148	0.5121	0.7035	0.5970
0.4	5.7522E-03	1.0639	1.4271	1.6082	1.6861	1.0035	1.0002	0.9946	0.9991	1.0475	0.5104	0.7056	0.5991
0.5	2.2603E-02	1.0643	1.1178	1.1256	1.1126	1.0040	1.0003	0.9948	0.9991	0.9255	0.5084	0.7080	0.6018
0.6	5.2787E-02	1.0648	0.9522	0.8986	0.8600	1.0046	1.0004	0.9951	0.9992	0.8608	0.5065	0.7109	0.6047
0.7	9.2474E-02	1.0654	0.8545	0.7751	0.7278	1.0053	1.0005	0.9953	0.9992	0.8233	0.5048	0.7139	0.6079
0.8	1.3629E-01	1.0660	0.7931	0.7014	0.6509	1.0060	1.0006	0.9955	0.9992	0.8003	0.5032	0.7171	0.6111
0.9	1.7987E-01	1.0667	0.7526	0.6547	0.6030	1.0067	1.0007	0.9957	0.9993	0.7857	0.5019	0.7203	0.6144
1.0	2.2053E-01	1.0675	0.7252	0.6239	0.5718	1.0075	1.0008	0.9960	0.9993	0.7762	0.5008	0.7236	0.6178
1.2	2.8871E-01	1.0692	0.6929	0.5887	0.5366	1.0090	1.0010	0.9964	0.9994	0.7661	0.4993	0.7300	0.6242
1.4	3.3896E-01	1.0712	0.6771	0.5720	0.5202	1.0106	1.0013	0.9967	0.9994	0.7623	0.4985	0.7360	0.6303
1.6	3.7442E-01	1.0732	0.6697	0.5646	0.5131	1.0121	1.0015	0.9970	0.9995	0.7617	0.4982	0.7418	0.6361
1.8	3.9884E-01	1.0755	0.6670	0.5622	0.5111	1.0137	1.0017	0.9973	0.9995	0.7627	0.4982	0.7471	0.6415
2.0	4.1532E-01	1.0778	0.6670	0.5627	0.5118	1.0152	1.0019	0.9975	0.9996	0.7647	0.4984	0.7521	0.6465
2.2	4.2616E-01	1.0801	0.6686	0.5649	0.5142	1.0168	1.0022	0.9977	0.9996	0.7671	0.4987	0.7568	0.6512
2.4	4.3302E-01	1.0826	0.6712	0.5681	0.5176	1.0183	1.0024	0.9979	0.9996	0.7697	0.4991	0.7612	0.6557
2.6	4.3706E-01	1.0850	0.6744	0.5719	0.5217	1.0198	1.0026	0.9980	0.9996	0.7725	0.4995	0.7652	0.6598
2.8	4.3907E-01	1.0876	0.6779	0.5760	0.5260	1.0213	1.0028	0.9982	0.9997	0.7752	0.4999	0.7691	0.6638
3.0	4.3966E-01	1.0901	0.6816	0.5804	0.5306	1.0227	1.0031	0.9983	0.9997	0.7780	0.5002	0.7726	0.6675
3.2	4.3921E-01	1.0927	0.6854	0.5848	0.5352	1.0242	1.0033	0.9984	0.9997	0.7806	0.5005	0.7760	0.6710
3.4	4.3803E-01	1.0953	0.6892	0.5892	0.5398	1.0256	1.0035	0.9985	0.9997	0.7832	0.5008	0.7792	0.6743
3.6	4.3633E-01	1.0979	0.6930	0.5937	0.5444	1.0271	1.0037	0.9986	0.9997	0.7858	0.5010	0.7822	0.6775
3.8	4.3427E-01	1.1005	0.6967	0.5980	0.5489	1.0285	1.0039	0.9986	0.9997	0.7882	0.5012	0.7850	0.6806
4.0	4.3196E-01	1.1032	0.7004	0.6022	0.5533	1.0299	1.0041	0.9987	0.9998	0.7905	0.5013	0.7877	0.6835
4.5	4.2558E-01	1.1098	0.7091	0.6124	0.5640	1.0334	1.0047	0.9988	0.9998	0.7960	0.5013	0.7938	0.6902
5.0	4.1892E-01	1.1165	0.7172	0.6219	0.5739	1.0368	1.0052	0.9990	0.9998	0.8010	0.5011	0.7992	0.6963
5.5	4.1237E-01	1.1232	0.7247	0.6308	0.5831	1.0402	1.0058	0.9991	0.9998	0.8054	0.5007	0.8040	0.7019
6.0	4.0612E-01	1.1299	0.7317	0.6389	0.5917	1.0435	1.0063	0.9991	0.9998	0.8095	0.5000	0.8084	0.7071
6.5	4.0023E-01	1.1366	0.7380	0.6465	0.5996	1.0468	1.0068	0.9992	0.9999	0.8133	0.4991	0.8123	0.7119
7.0	3.9472E-01	1.1434	0.7439	0.6535	0.6070	1.0500	1.0073	0.9993	0.9999	0.8167	0.4980	0.8159	0.7163
7.5	3.8960E-01	1.1501	0.7493	0.6600	0.6139	1.0532	1.0079	0.9993	0.9999	0.8198	0.4968	0.8191	0.7205
8.0	3.8485E-01	1.1569	0.7544	0.6661	0.6203	1.0564	1.0084	0.9994	0.9999	0.8226	0.4954	0.8221	0.7244
9.0	3.7634E-01	1.1704	0.7633	0.6770	0.6318	1.0626	1.0094	0.9995	0.9999	0.8277	0.4924	0.8273	0.7317
10.0	3.6901E-01	1.1839	0.7711	0.6864	0.6418	1.0687	1.0105	0.9995	0.9999	0.8321	0.4890	0.8317	0.7382
11.0	3.6266E-01	1.1974	0.7778	0.6947	0.6507	1.0747	1.0115	0.9996	0.9999	0.8358	0.4852	0.8355	0.7442
12.0	3.5714E-01	1.2110	0.7836	0.7019	0.6585	1.0806	1.0126	0.9996	0.9999	0.8390	0.4813	0.8388	0.7496
13.0	3.5233E-01	1.2245	0.7887	0.7083	0.6653	1.0864	1.0136	0.9996	0.9999	0.8418	0.4771	0.8417	0.7547
14.0	3.4812E-01	1.2381	0.7931	0.7139	0.6714	1.0922	1.0146	0.9997	0.9999	0.8443	0.4727	0.8442	0.7594
16.0	3.4118E-01	1.2654	0.8003	0.7232	0.6815	1.1035	1.0167	0.9997	0.9999	0.8483	0.4636	0.8482	0.7681
18.0	3.3580E-01	1.2928	0.8057	0.7303	0.6894	1.1145	1.0187	0.9997	0.9999	0.8513	0.4541	0.8513	0.7757
20.0	3.3164E-01	1.3203	0.8097	0.7358	0.6954	1.1253	1.0208	0.9998	1.0000	0.8536	0.4443	0.8536	0.7827
25.0	3.2505E-01	1.3901	0.8153	0.7440	0.7048	1.1514	1.0258	0.9998	1.0000	0.8569	0.4189	0.8569	0.7978
30.0	3.2218E-01	1.4612	0.8164	0.7466	0.7083	1.1767	1.0309	0.9999	1.0000	0.8580	0.3927	0.8580	0.8105
35.0	3.2179E-01	1.5341	0.8146	0.7455	0.7077	1.2012	1.0360	0.9999	1.0000	0.8575	0.3662	0.8576	0.8217
40.0	3.2317E-01	1.6088	0.8108	0.7418	0.7042	1.2250	1.0410	0.9999	1.0000	0.8561	0.3394	0.8561	0.8318
45.0	3.2589E-01	1.6856	0.8055	0.7362	0.6987	1.2483	1.0461	0.9999	1.0000	0.8540	0.3126	0.8540	0.8411
50.0	3.2967E-01	1.7645	0.7993	0.7292	0.6916	1.2711	1.0513	0.9999	1.0000	0.8514	0.2858	0.8514	0.8498

Instr. see p. 22

POSITRON Z= -76 A= 186

p	F_0L_0	L_0	λ_2	λ_3	λ_4	μ_1	μ_2	Λ_1	Λ_2	ν_{12}	$\hat{\nu}_{12}$	η_{12}	$\hat{\eta}_{12}$
0.1	1.5250E-13	1.0642	14.5564	93.2263	348.4461	1.0027	1.0001	0.9941	0.9990	6.3436	0.5096	0.6948	0.5861
0.2	2.3887E-06	1.0643	4.0472	8.8132	13.9401	1.0029	1.0001	0.9942	0.9990	2.1040	0.5087	0.6957	0.5871
0.3	4.5484E-04	1.0645	2.1031	2.9409	3.4870	1.0032	1.0002	0.9943	0.9990	1.3205	0.5074	0.6972	0.5887
0.4	5.4005E-03	1.0648	1.4250	1.6085	1.6880	1.0037	1.0002	0.9945	0.9991	1.0479	0.5056	0.6993	0.5909
0.5	2.1663E-02	1.0652	1.1135	1.1207	1.1073	1.0042	1.0003	0.9947	0.9991	0.9235	0.5038	0.7018	0.5935
0.6	5.1257E-02	1.0657	0.9467	0.8918	0.8524	1.0049	1.0004	0.9949	0.9991	0.8576	0.5020	0.7046	0.5965
0.7	9.0583E-02	1.0663	0.8483	0.7674	0.7194	1.0055	1.0005	0.9952	0.9992	0.8193	0.5003	0.7077	0.5997
0.8	1.3431E-01	1.0670	0.7864	0.6933	0.6421	1.0063	1.0006	0.9954	0.9992	0.7958	0.4988	0.7109	0.6030
0.9	1.7804E-01	1.0677	0.7457	0.6463	0.5940	1.0070	1.0007	0.9956	0.9992	0.7809	0.4975	0.7142	0.6063
1.0	2.1899E-01	1.0685	0.7181	0.6153	0.5626	1.0078	1.0008	0.9959	0.9993	0.7712	0.4965	0.7175	0.6097
1.2	2.8787E-01	1.0703	0.6856	0.5800	0.5274	1.0094	1.0011	0.9963	0.9994	0.7609	0.4951	0.7240	0.6162
1.4	3.3877E-01	1.0723	0.6697	0.5633	0.5110	1.0110	1.0013	0.9966	0.9994	0.7570	0.4944	0.7302	0.6224
1.6	3.7470E-01	1.0744	0.6623	0.5559	0.5040	1.0126	1.0015	0.9969	0.9995	0.7564	0.4942	0.7360	0.6283
1.8	3.9944E-01	1.0767	0.6596	0.5535	0.5019	1.0142	1.0018	0.9972	0.9995	0.7575	0.4943	0.7415	0.6337
2.0	4.1610E-01	1.0791	0.6597	0.5541	0.5027	1.0158	1.0020	0.9974	0.9995	0.7594	0.4946	0.7466	0.6389
2.2	4.2703E-01	1.0816	0.6614	0.5563	0.5052	1.0174	1.0022	0.9976	0.9996	0.7619	0.4950	0.7514	0.6437
2.4	4.3391E-01	1.0841	0.6641	0.5596	0.5087	1.0190	1.0024	0.9978	0.9996	0.7646	0.4955	0.7558	0.6482
2.6	4.3791E-01	1.0866	0.6673	0.5635	0.5128	1.0205	1.0027	0.9980	0.9996	0.7674	0.4959	0.7600	0.6524
2.8	4.3987E-01	1.0892	0.6709	0.5677	0.5172	1.0220	1.0029	0.9981	0.9997	0.7702	0.4963	0.7639	0.6564
3.0	4.4036E-01	1.0919	0.6747	0.5721	0.5218	1.0236	1.0031	0.9982	0.9997	0.7730	0.4968	0.7676	0.6602
3.2	4.3982E-01	1.0945	0.6786	0.5766	0.5265	1.0251	1.0034	0.9983	0.9997	0.7758	0.4971	0.7710	0.6638
3.4	4.3853E-01	1.0972	0.6825	0.5811	0.5312	1.0266	1.0036	0.9984	0.9997	0.7784	0.4974	0.7743	0.6672
3.6	4.3672E-01	1.0999	0.6863	0.5856	0.5359	1.0280	1.0038	0.9985	0.9997	0.7810	0.4977	0.7773	0.6704
3.8	4.3454E-01	1.1026	0.6902	0.5900	0.5404	1.0295	1.0040	0.9986	0.9997	0.7835	0.4979	0.7802	0.6735
4.0	4.3212E-01	1.1053	0.6939	0.5944	0.5449	1.0310	1.0042	0.9987	0.9998	0.7859	0.4981	0.7830	0.6765
4.5	4.2546E-01	1.1121	0.7028	0.6048	0.5557	1.0345	1.0048	0.9988	0.9998	0.7915	0.4983	0.7892	0.6834
5.0	4.1855E-01	1.1190	0.7111	0.6144	0.5658	1.0381	1.0053	0.9989	0.9998	0.7966	0.4982	0.7948	0.6896
5.5	4.1177E-01	1.1259	0.7188	0.6234	0.5752	1.0415	1.0059	0.9990	0.9998	0.8012	0.4978	0.7997	0.6953
6.0	4.0530E-01	1.1329	0.7259	0.6318	0.5839	1.0450	1.0064	0.9991	0.9998	0.8054	0.4972	0.8042	0.7006
6.5	3.9922E-01	1.1398	0.7324	0.6395	0.5920	1.0483	1.0070	0.9992	0.9998	0.8092	0.4964	0.8082	0.7055
7.0	3.9355E-01	1.1467	0.7385	0.6466	0.5995	1.0517	1.0075	0.9993	0.9999	0.8127	0.4954	0.8119	0.7100
7.5	3.8827E-01	1.1537	0.7440	0.6533	0.6065	1.0550	1.0080	0.9993	0.9999	0.8159	0.4942	0.8152	0.7143
8.0	3.8338E-01	1.1606	0.7492	0.6595	0.6130	1.0582	1.0086	0.9994	0.9999	0.8188	0.4930	0.8182	0.7183
9.0	3.7462E-01	1.1745	0.7584	0.6706	0.6248	1.0646	1.0096	0.9994	0.9999	0.8241	0.4900	0.8236	0.7257
10.0	3.6708E-01	1.1885	0.7663	0.6803	0.6351	1.0709	1.0107	0.9995	0.9999	0.8285	0.4867	0.8282	0.7324
11.0	3.6055E-01	1.2024	0.7732	0.6887	0.6441	1.0771	1.0118	0.9995	0.9999	0.8324	0.4831	0.8321	0.7385
12.0	3.5488E-01	1.2163	0.7792	0.6962	0.6521	1.0831	1.0128	0.9996	0.9999	0.8357	0.4792	0.8355	0.7441
13.0	3.4994E-01	1.2303	0.7844	0.7027	0.6591	1.0891	1.0139	0.9996	0.9999	0.8386	0.4751	0.8384	0.7493
14.0	3.4561E-01	1.2443	0.7890	0.7085	0.6653	1.0950	1.0149	0.9997	0.9999	0.8411	0.4709	0.8410	0.7541
16.0	3.3848E-01	1.2724	0.7964	0.7180	0.6757	1.1066	1.0170	0.9997	0.9999	0.8452	0.4620	0.8452	0.7629
18.0	3.3294E-01	1.3006	0.8020	0.7254	0.6838	1.1179	1.0191	0.9997	0.9999	0.8484	0.4526	0.8483	0.7708
20.0	3.2865E-01	1.3290	0.8062	0.7311	0.6901	1.1290	1.0212	0.9998	1.0000	0.8508	0.4429	0.8507	0.7779
25.0	3.2182E-01	1.4008	0.8121	0.7396	0.6998	1.1558	1.0264	0.9998	1.0000	0.8543	0.4179	0.8543	0.7933
30.0	3.1879E-01	1.4742	0.8134	0.7425	0.7035	1.1817	1.0315	0.9999	1.0000	0.8555	0.3920	0.8555	0.8064
35.0	3.1827E-01	1.5494	0.8117	0.7416	0.7031	1.2068	1.0367	0.9999	1.0000	0.8551	0.3658	0.8551	0.8178
40.0	3.1956E-01	1.6265	0.8080	0.7380	0.6998	1.2312	1.0419	0.9999	1.0000	0.8537	0.3393	0.8537	0.8281
45.0	3.2220E-01	1.7058	0.8028	0.7324	0.6944	1.2550	1.0471	0.9999	1.0000	0.8516	0.3128	0.8516	0.8376
50.0	3.2592E-01	1.7874	0.7965	0.7255	0.6874	1.2783	1.0523	0.9999	1.0000	0.8490	0.2862	0.8491	0.8465

Instr. see p. 22

POSITRON Z= -77 A= 191

p	F_0L_0	L_0	λ_2	λ_3	λ_4	μ_1	μ_2	Λ_1	Λ_2	ν_{12}	$\hat{\nu}_{12}$	η_{12}	$\hat{\eta}_{12}$
0.1	1.0216E-13	1.0651	14.6447	95.2945	362.0224	1.0029	1.0001	0.9939	0.9990	6.4462	0.5049	0.6886	0.5780
0.2	2.0077E-06	1.0652	4.0634	8.9383	14.2548	1.0031	1.0001	0.9940	0.9990	2.1251	0.5040	0.6895	0.5790
0.3	4.1169E-04	1.0654	2.1059	2.9601	3.5221	1.0034	1.0002	0.9942	0.9990	1.3264	0.5027	0.6911	0.5806
0.4	5.0667E-03	1.0658	1.4232	1.6093	1.6906	1.0039	1.0002	0.9943	0.9990	1.0486	0.5010	0.6932	0.5828
0.5	2.0748E-02	1.0662	1.1096	1.1163	1.1026	1.0045	1.0003	0.9946	0.9991	0.9218	0.4992	0.6957	0.5854
0.6	4.9736E-02	1.0667	0.9416	0.8854	0.8454	1.0051	1.0004	0.9948	0.9991	0.8545	0.4975	0.6986	0.5884
0.7	8.8667E-02	1.0673	0.8425	0.7601	0.7114	1.0058	1.0005	0.9950	0.9992	0.8155	0.4958	0.7017	0.5916
0.8	1.3227E-01	1.0680	0.7801	0.6855	0.6337	1.0066	1.0006	0.9953	0.9992	0.7915	0.4944	0.7049	0.5950
0.9	1.7611E-01	1.0687	0.7391	0.6383	0.5854	1.0073	1.0007	0.9955	0.9992	0.7763	0.4932	0.7083	0.5984
1.0	2.1730E-01	1.0696	0.7113	0.6072	0.5540	1.0081	1.0009	0.9957	0.9993	0.7664	0.4923	0.7116	0.6017
1.2	2.8683E-01	1.0714	0.6786	0.5717	0.5186	1.0098	1.0011	0.9962	0.9993	0.7558	0.4910	0.7182	0.6084
1.4	3.3833E-01	1.0735	0.6626	0.5549	0.5022	1.0115	1.0013	0.9965	0.9994	0.7519	0.4904	0.7245	0.6147
1.6	3.7473E-01	1.0757	0.6552	0.5476	0.4952	1.0131	1.0016	0.9969	0.9995	0.7512	0.4903	0.7305	0.6206
1.8	3.9976E-01	1.0781	0.6526	0.5452	0.4932	1.0148	1.0018	0.9971	0.9995	0.7523	0.4905	0.7361	0.6262
2.0	4.1660E-01	1.0805	0.6527	0.5458	0.4941	1.0165	1.0020	0.9974	0.9995	0.7543	0.4909	0.7413	0.6314
2.2	4.2761E-01	1.0831	0.6544	0.5482	0.4966	1.0181	1.0023	0.9976	0.9996	0.7569	0.4913	0.7461	0.6363
2.4	4.3450E-01	1.0857	0.6572	0.5515	0.5002	1.0197	1.0025	0.9978	0.9996	0.7596	0.4919	0.7507	0.6408
2.6	4.3846E-01	1.0883	0.6605	0.5554	0.5043	1.0213	1.0027	0.9979	0.9996	0.7625	0.4924	0.7549	0.6452
2.8	4.4035E-01	1.0910	0.6642	0.5598	0.5088	1.0229	1.0030	0.9981	0.9997	0.7654	0.4929	0.7589	0.6492
3.0	4.4076E-01	1.0937	0.6681	0.5643	0.5135	1.0245	1.0032	0.9982	0.9997	0.7683	0.4933	0.7627	0.6531
3.2	4.4011E-01	1.0965	0.6721	0.5688	0.5183	1.0260	1.0034	0.9983	0.9997	0.7711	0.4938	0.7662	0.6567
3.4	4.3872E-01	1.0993	0.6761	0.5734	0.5230	1.0276	1.0037	0.9984	0.9997	0.7738	0.4941	0.7695	0.6602
3.6	4.3680E-01	1.1020	0.6800	0.5780	0.5277	1.0291	1.0039	0.9985	0.9997	0.7764	0.4944	0.7726	0.6635
3.8	4.3451E-01	1.1048	0.6839	0.5825	0.5324	1.0306	1.0041	0.9986	0.9997	0.7790	0.4947	0.7756	0.6667
4.0	4.3197E-01	1.1077	0.6877	0.5869	0.5370	1.0321	1.0043	0.9986	0.9998	0.7814	0.4949	0.7784	0.6697
4.5	4.2505E-01	1.1147	0.6969	0.5975	0.5479	1.0358	1.0049	0.9988	0.9998	0.7872	0.4952	0.7848	0.6767
5.0	4.1789E-01	1.1218	0.7053	0.6073	0.5582	1.0395	1.0055	0.9989	0.9998	0.7924	0.4952	0.7905	0.6831
5.5	4.1089E-01	1.1290	0.7132	0.6165	0.5677	1.0431	1.0060	0.9990	0.9998	0.7971	0.4949	0.7956	0.6889
6.0	4.0422E-01	1.1361	0.7204	0.6250	0.5765	1.0466	1.0066	0.9991	0.9998	0.8014	0.4944	0.8002	0.6943
6.5	3.9796E-01	1.1433	0.7271	0.6328	0.5848	1.0501	1.0071	0.9992	0.9998	0.8053	0.4936	0.8043	0.6993
7.0	3.9212E-01	1.1505	0.7333	0.6401	0.5924	1.0535	1.0077	0.9992	0.9999	0.8089	0.4927	0.8080	0.7039
7.5	3.8670E-01	1.1576	0.7390	0.6469	0.5995	1.0569	1.0082	0.9993	0.9999	0.8122	0.4916	0.8114	0.7083
8.0	3.8167E-01	1.1648	0.7443	0.6532	0.6062	1.0603	1.0088	0.9993	0.9999	0.8152	0.4904	0.8146	0.7124
9.0	3.7269E-01	1.1792	0.7537	0.6646	0.6182	1.0669	1.0099	0.9994	0.9999	0.8205	0.4876	0.8201	0.7199
10.0	3.6495E-01	1.1936	0.7618	0.6745	0.6286	1.0734	1.0110	0.9995	0.9999	0.8251	0.4844	0.8248	0.7268
11.0	3.5826E-01	1.2080	0.7689	0.6831	0.6379	1.0797	1.0121	0.9995	0.9999	0.8291	0.4808	0.8288	0.7330
12.0	3.5245E-01	1.2224	0.7750	0.6907	0.6460	1.0860	1.0131	0.9996	0.9999	0.8325	0.4770	0.8323	0.7387
13.0	3.4739E-01	1.2368	0.7804	0.6974	0.6531	1.0922	1.0142	0.9996	0.9999	0.8355	0.4730	0.8353	0.7440
14.0	3.4296E-01	1.2513	0.7851	0.7033	0.6595	1.0982	1.0153	0.9996	0.9999	0.8380	0.4688	0.8379	0.7490
16.0	3.3566E-01	1.2803	0.7927	0.7131	0.6701	1.1101	1.0174	0.9997	0.9999	0.8423	0.4600	0.8422	0.7560
18.0	3.2999E-01	1.3095	0.7985	0.7206	0.6784	1.1218	1.0196	0.9997	0.9999	0.8455	0.4508	0.8455	0.7660
20.0	3.2560E-01	1.3388	0.8028	0.7265	0.6849	1.1332	1.0217	0.9998	0.9999	0.8480	0.4412	0.8480	0.7733
25.0	3.1858E-01	1.4132	0.8090	0.7353	0.6950	1.1608	1.0270	0.9998	1.0000	0.8517	0.4164	0.8517	0.7891
30.0	3.1542E-01	1.4891	0.8104	0.7384	0.6989	1.1874	1.0323	0.9999	1.0000	0.8529	0.3908	0.8529	0.8024
35.0	3.1483E-01	1.5670	0.8088	0.7376	0.6986	1.2131	1.0376	0.9999	1.0000	0.8526	0.3648	0.8526	0.8141
40.0	3.1606E-01	1.6470	0.8051	0.7340	0.6953	1.2382	1.0429	0.9999	1.0000	0.8512	0.3385	0.8513	0.8247
45.0	3.1868E-01	1.7293	0.7999	0.7285	0.6899	1.2627	1.0482	0.9999	1.0000	0.8492	0.3121	0.8492	0.8345
50.0	3.2238E-01	1.8140	0.7936	0.7215	0.6829	1.2866	1.0536	0.9999	1.0000	0.8466	0.2857	0.8466	0.8436

Instr. see p. 22

POSITRON Z= -78 A= 194

p	F_0L_0	L_0	λ_2	λ_3	λ_4	μ_1	μ_2	Λ_1	Λ_2	ν_{12}	$\hat{\nu}_{12}$	η_{12}	$\hat{\eta}_{12}$
0.1	6.8545E-14	1.0658	14.7162	97.2443	375.3836	1.0031	1.0001	0.9938	0.9990	6.5466	0.4999	0.6822	0.5696
0.2	1.6902E-06	1.0660	4.0752	9.0521	14.5534	1.0033	1.0001	0.9939	0.9990	2.1454	0.4991	0.6832	0.5707
0.3	3.7321E-04	1.0662	2.1067	2.9756	3.5526	1.0036	1.0002	0.9940	0.9990	1.3319	0.4978	0.6847	0.5723
0.4	4.7609E-03	1.0666	1.4200	1.6082	1.6910	1.0041	1.0003	0.9942	0.9990	1.0489	0.4963	0.6868	0.5745
0.5	1.9902E-02	1.0670	1.1046	1.1107	1.0966	1.0047	1.0003	0.9944	0.9991	0.9197	0.4945	0.6893	0.5772
0.6	4.8335E-02	1.0675	0.9357	0.8781	0.8375	1.0054	1.0004	0.9947	0.9991	0.8512	0.4928	0.6923	0.5802
0.7	8.6926E-02	1.0682	0.8360	0.7521	0.7028	1.0061	1.0005	0.9949	0.9991	0.8114	0.4913	0.6954	0.5834
0.8	1.3047E-01	1.0689	0.7732	0.6771	0.6247	1.0069	1.0006	0.9952	0.9992	0.7870	0.4899	0.6987	0.5868
0.9	1.7446E-01	1.0696	0.7320	0.6297	0.5763	1.0077	1.0008	0.9954	0.9992	0.7714	0.4888	0.7021	0.5902
1.0	2.1595E-01	1.0705	0.7040	0.5985	0.5448	1.0085	1.0009	0.9956	0.9993	0.7613	0.4879	0.7055	0.5936
1.2	2.8623E-01	1.0724	0.6711	0.5629	0.5094	1.0102	1.0011	0.9961	0.9993	0.7505	0.4868	0.7122	0.6003
1.4	3.3841E-01	1.0746	0.6551	0.5462	0.4931	1.0119	1.0014	0.9965	0.9994	0.7465	0.4863	0.7186	0.6067
1.6	3.7531E-01	1.0769	0.6477	0.5388	0.4861	1.0137	1.0016	0.9968	0.9994	0.7459	0.4863	0.7247	0.6127
1.8	4.0068E-01	1.0793	0.6451	0.5365	0.4841	1.0154	1.0018	0.9971	0.9995	0.7470	0.4866	0.7304	0.6184
2.0	4.1771E-01	1.0818	0.6453	0.5372	0.4850	1.0171	1.0021	0.9973	0.9995	0.7490	0.4870	0.7357	0.6237
2.2	4.2881E-01	1.0844	0.6471	0.5396	0.4876	1.0188	1.0023	0.9975	0.9996	0.7516	0.4876	0.7407	0.6286
2.4	4.3572E-01	1.0871	0.6499	0.5430	0.4913	1.0204	1.0026	0.9977	0.9996	0.7544	0.4882	0.7453	0.6333
2.6	4.3964E-01	1.0899	0.6534	0.5470	0.4955	1.0221	1.0028	0.9979	0.9996	0.7574	0.4888	0.7496	0.6377
2.8	4.4146E-01	1.0926	0.6571	0.5514	0.5000	1.0237	1.0030	0.9980	0.9996	0.7603	0.4893	0.7537	0.6418
3.0	4.4178E-01	1.0954	0.6611	0.5560	0.5048	1.0254	1.0033	0.9981	0.9997	0.7632	0.4899	0.7575	0.6457
3.2	4.4102E-01	1.0983	0.6652	0.5607	0.5096	1.0270	1.0035	0.9983	0.9997	0.7661	0.4903	0.7611	0.6495
3.4	4.3951E-01	1.1011	0.6692	0.5653	0.5144	1.0286	1.0037	0.9984	0.9997	0.7689	0.4908	0.7645	0.6530
3.6	4.3747E-01	1.1040	0.6733	0.5700	0.5192	1.0301	1.0040	0.9985	0.9997	0.7716	0.4911	0.7677	0.6564
3.8	4.3506E-01	1.1069	0.6773	0.5745	0.5240	1.0317	1.0042	0.9985	0.9997	0.7742	0.4914	0.7708	0.6596
4.0	4.3240E-01	1.1098	0.6812	0.5790	0.5286	1.0333	1.0044	0.9986	0.9997	0.7767	0.4917	0.7737	0.6627
4.5	4.2519E-01	1.1171	0.6905	0.5898	0.5397	1.0371	1.0050	0.9988	0.9998	0.7826	0.4921	0.7802	0.6698
5.0	4.1776E-01	1.1244	0.6992	0.5998	0.5501	1.0408	1.0056	0.9989	0.9998	0.7879	0.4922	0.7861	0.6763
5.5	4.1051E-01	1.1317	0.7072	0.6091	0.5598	1.0445	1.0062	0.9990	0.9998	0.7928	0.4920	0.7913	0.6822
6.0	4.0362E-01	1.1391	0.7146	0.6178	0.5688	1.0482	1.0067	0.9991	0.9998	0.7972	0.4916	0.7959	0.6877
6.5	3.9716E-01	1.1465	0.7215	0.6258	0.5772	1.0518	1.0073	0.9992	0.9998	0.8012	0.4909	0.8002	0.6928
7.0	3.9114E-01	1.1539	0.7278	0.6332	0.5849	1.0553	1.0079	0.9992	0.9999	0.8049	0.4901	0.8040	0.6976
7.5	3.8556E-01	1.1613	0.7336	0.6402	0.5922	1.0588	1.0084	0.9993	0.9999	0.8082	0.4891	0.8075	0.7020
8.0	3.8038E-01	1.1686	0.7390	0.6466	0.5989	1.0623	1.0090	0.9993	0.9999	0.8113	0.4879	0.8107	0.7062
9.0	3.7114E-01	1.1834	0.7487	0.6582	0.6112	1.0691	1.0101	0.9994	0.9999	0.8168	0.4852	0.8164	0.7139
10.0	3.6319E-01	1.1983	0.7571	0.6683	0.6218	1.0757	1.0112	0.9995	0.9999	0.8215	0.4821	0.8212	0.7209
11.0	3.5632E-01	1.2131	0.7643	0.6771	0.6313	1.0823	1.0123	0.9995	0.9999	0.8256	0.4787	0.8253	0.7272
12.0	3.5035E-01	1.2279	0.7706	0.6849	0.6395	1.0887	1.0134	0.9996	0.9999	0.8291	0.4750	0.8289	0.7331
13.0	3.4516E-01	1.2428	0.7761	0.6918	0.6469	1.0950	1.0145	0.9996	0.9999	0.8322	0.4711	0.8320	0.7385
14.0	3.4061E-01	1.2577	0.7809	0.6978	0.6534	1.1013	1.0156	0.9996	0.9999	0.8348	0.4670	0.8347	0.7435
16.0	3.3310E-01	1.2876	0.7888	0.7079	0.6643	1.1135	1.0178	0.9997	0.9999	0.8392	0.4583	0.8392	0.7527
18.0	3.2728E-01	1.3176	0.7948	0.7157	0.6728	1.1254	1.0200	0.9997	0.9999	0.8426	0.4493	0.8425	0.7610
20.0	3.2275E-01	1.3478	0.7993	0.7217	0.6795	1.1371	1.0222	0.9998	0.9999	0.8451	0.4399	0.8451	0.7684
25.0	3.1548E-01	1.4244	0.8058	0.7310	0.6900	1.1654	1.0276	0.9998	1.0000	0.8490	0.4154	0.8490	0.7846
30.0	3.1216E-01	1.5028	0.8074	0.7343	0.6942	1.1926	1.0329	0.9998	1.0000	0.8504	0.3901	0.8504	0.7982
35.0	3.1144E-01	1.5831	0.8059	0.7336	0.6941	1.2190	1.0383	0.9999	1.0000	0.8501	0.3643	0.8502	0.8102
40.0	3.1258E-01	1.6657	0.8023	0.7302	0.6909	1.2447	1.0437	0.9999	1.0000	0.8488	0.3383	0.8489	0.8210
45.0	3.1511E-01	1.7507	0.7971	0.7248	0.6856	1.2698	1.0492	0.9999	1.0000	0.8468	0.3122	0.8468	0.8310
50.0	3.1875E-01	1.8383	0.7909	0.7178	0.6787	1.2942	1.0546	0.9999	1.0000	0.8442	0.2861	0.8443	0.8403

POSITRON Z= -79 A= 193

p	F_0L_0	L_0	λ_2	λ_3	λ_4	μ_1	μ_2	Λ_1	Λ_2	ν_{12}	$\hat{\nu}_{12}$	η_{12}	$\hat{\eta}_{12}$
0.1	4.6120E-14	1.0664	14.7608	98.9831	388.1094	1.0032	1.0001	0.9937	0.9989	6.6429	0.4947	0.6753	0.5610
0.2	1.4269E-06	1.0666	4.0798	9.1461	14.8209	1.0035	1.0001	0.9938	0.9990	2.1643	0.4939	0.6763	0.5620
0.3	3.3928E-04	1.0668	2.1038	2.9848	3.5750	1.0038	1.0002	0.9939	0.9990	1.3365	0.4927	0.6779	0.5636
0.4	4.4861E-03	1.0671	1.4146	1.6039	1.6877	1.0043	1.0003	0.9941	0.9990	1.0486	0.4912	0.6800	0.5658
0.5	1.9144E-02	1.0676	1.0980	1.1029	1.0882	1.0049	1.0003	0.9943	0.9990	0.9170	0.4895	0.6825	0.5685
0.6	4.7105E-02	1.0681	0.9284	0.8692	0.8278	1.0056	1.0004	0.9946	0.9991	0.8473	0.4879	0.6855	0.5715
0.7	8.5456E-02	1.0688	0.8283	0.7427	0.6927	1.0063	1.0005	0.9948	0.9991	0.8068	0.4864	0.6887	0.5748
0.8	1.2905E-01	1.0695	0.7653	0.6676	0.6146	1.0071	1.0007	0.9951	0.9992	0.7819	0.4852	0.6920	0.5782
0.9	1.7331E-01	1.0703	0.7238	0.6201	0.5661	1.0080	1.0008	0.9953	0.9992	0.7660	0.4841	0.6954	0.5816
1.0	2.1521E-01	1.0712	0.6958	0.5889	0.5347	1.0088	1.0009	0.9956	0.9992	0.7558	0.4833	0.6989	0.5851
1.2	2.8642E-01	1.0732	0.6628	0.5533	0.4993	1.0106	1.0011	0.9960	0.9993	0.7448	0.4823	0.7057	0.5918
1.4	3.3942E-01	1.0754	0.6467	0.5365	0.4831	1.0123	1.0014	0.9964	0.9994	0.7407	0.4819	0.7122	0.5983
1.6	3.7692E-01	1.0777	0.6394	0.5292	0.4761	1.0141	1.0016	0.9967	0.9994	0.7400	0.4820	0.7184	0.6044
1.8	4.0268E-01	1.0802	0.6368	0.5270	0.4742	1.0159	1.0019	0.9970	0.9995	0.7411	0.4824	0.7242	0.6101
2.0	4.1994E-01	1.0828	0.6371	0.5278	0.4752	1.0176	1.0021	0.9973	0.9995	0.7432	0.4830	0.7296	0.6155
2.2	4.3115E-01	1.0855	0.6390	0.5302	0.4779	1.0194	1.0024	0.9975	0.9996	0.7459	0.4836	0.7347	0.6206
2.4	4.3808E-01	1.0882	0.6419	0.5337	0.4816	1.0211	1.0026	0.9977	0.9996	0.7488	0.4843	0.7394	0.6253
2.6	4.4197E-01	1.0910	0.6454	0.5378	0.4858	1.0228	1.0028	0.9978	0.9996	0.7518	0.4850	0.7439	0.6297
2.8	4.4371E-01	1.0939	0.6493	0.5423	0.4905	1.0245	1.0031	0.9980	0.9996	0.7548	0.4856	0.7480	0.6340
3.0	4.4392E-01	1.0967	0.6534	0.5469	0.4953	1.0261	1.0033	0.9981	0.9997	0.7578	0.4862	0.7520	0.6379
3.2	4.4304E-01	1.0996	0.6575	0.5517	0.5002	1.0278	1.0036	0.9982	0.9997	0.7607	0.4868	0.7556	0.6417
3.4	4.4140E-01	1.1026	0.6617	0.5564	0.5051	1.0294	1.0038	0.9983	0.9997	0.7636	0.4873	0.7591	0.6453
3.6	4.3923E-01	1.1055	0.6658	0.5611	0.5099	1.0310	1.0040	0.9984	0.9997	0.7663	0.4877	0.7624	0.6488
3.8	4.3668E-01	1.1085	0.6699	0.5658	0.5147	1.0326	1.0043	0.9985	0.9997	0.7690	0.4881	0.7655	0.6520
4.0	4.3388E-01	1.1114	0.6739	0.5703	0.5194	1.0342	1.0045	0.9986	0.9997	0.7716	0.4884	0.7684	0.6552
4.5	4.2634E-01	1.1189	0.6834	0.5813	0.5307	1.0381	1.0051	0.9987	0.9998	0.7776	0.4889	0.7752	0.6624
5.0	4.1860E-01	1.1264	0.6923	0.5915	0.5412	1.0420	1.0057	0.9989	0.9998	0.7831	0.4891	0.7812	0.6690
5.5	4.1106E-01	1.1339	0.7005	0.6010	0.5511	1.0458	1.0062	0.9990	0.9998	0.7880	0.4891	0.7865	0.6751
6.0	4.0392E-01	1.1414	0.7081	0.6098	0.5602	1.0495	1.0068	0.9991	0.9998	0.7926	0.4888	0.7913	0.6807
6.5	3.9722E-01	1.1490	0.7151	0.6180	0.5687	1.0532	1.0074	0.9992	0.9998	0.7967	0.4882	0.7956	0.6859
7.0	3.9099E-01	1.1565	0.7216	0.6256	0.5767	1.0568	1.0080	0.9992	0.9999	0.8005	0.4875	0.7996	0.6907
7.5	3.8521E-01	1.1641	0.7276	0.6326	0.5840	1.0604	1.0085	0.9993	0.9999	0.8039	0.4866	0.8032	0.6952
8.0	3.7986E-01	1.1717	0.7331	0.6392	0.5909	1.0639	1.0091	0.9993	0.9999	0.8071	0.4855	0.8065	0.6995
9.0	3.7031E-01	1.1868	0.7431	0.6511	0.6034	1.0709	1.0102	0.9994	0.9999	0.8128	0.4830	0.8123	0.7074
10.0	3.6210E-01	1.2019	0.7517	0.6614	0.6143	1.0777	1.0114	0.9995	0.9999	0.8176	0.4801	0.8173	0.7145
11.0	3.5500E-01	1.2171	0.7591	0.6705	0.6239	1.0844	1.0125	0.9995	0.9999	0.8218	0.4768	0.8215	0.7210
12.0	3.4885E-01	1.2323	0.7656	0.6785	0.6324	1.0909	1.0136	0.9996	0.9999	0.8254	0.4733	0.8252	0.7269
13.0	3.4348E-01	1.2475	0.7713	0.6855	0.6400	1.0974	1.0147	0.9996	0.9999	0.8286	0.4695	0.8284	0.7325
14.0	3.3878E-01	1.2627	0.7763	0.6918	0.6467	1.1038	1.0158	0.9996	0.9999	0.8314	0.4655	0.8313	0.7376
16.0	3.3101E-01	1.2933	0.7845	0.7021	0.6579	1.1163	1.0181	0.9997	0.9999	0.8359	0.4571	0.8359	0.7470
18.0	3.2498E-01	1.3240	0.7907	0.7103	0.6667	1.1284	1.0203	0.9997	0.9999	0.8394	0.4483	0.8394	0.7554
20.0	3.2027E-01	1.3549	0.7955	0.7166	0.6737	1.1404	1.0225	0.9998	0.9999	0.8421	0.4391	0.8421	0.7630
25.0	3.1265E-01	1.4332	0.8024	0.7263	0.6847	1.1692	1.0279	0.9998	1.0000	0.8462	0.4152	0.8462	0.7795
30.0	3.0905E-01	1.5134	0.8044	0.7301	0.6894	1.1970	1.0334	0.9998	1.0000	0.8478	0.3904	0.8478	0.7934
35.0	3.0811E-01	1.5956	0.8032	0.7298	0.6896	1.2238	1.0388	0.9999	1.0000	0.8477	0.3651	0.8477	0.8057
40.0	3.0906E-01	1.6801	0.7998	0.7266	0.6868	1.2500	1.0443	0.9999	1.0000	0.8465	0.3396	0.8465	0.8167
45.0	3.1142E-01	1.7672	0.7947	0.7214	0.6817	1.2755	1.0498	0.9999	1.0000	0.8446	0.3139	0.8446	0.8268
50.0	3.1491E-01	1.8569	0.7886	0.7147	0.6750	1.3004	1.0553	0.9999	1.0000	0.8421	0.2882	0.8421	0.8363

Instr. see p. 22

POSITRON Z= -80 A= 200

p	F_0L_0	L_0	λ_2	λ_3	λ_4	μ_1	μ_2	Λ_1	Λ_2	ν_{12}	$\hat{\nu}_{12}$	η_{12}	$\hat{\eta}_{12}$
0.1	3.0901E-14	1.0672	14.8331	100.9865	402.2705	1.0035	1.0001	0.9935	0.9989	6.7456	0.4900	0.6693	0.5530
0.2	1.1995E-06	1.0674	4.0923	9.2645	15.1365	1.0037	1.0001	0.9936	0.9989	2.1855	0.4893	0.6703	0.5540
0.3	3.0713E-04	1.0676	2.1052	3.0020	3.6085	1.0041	1.0002	0.9937	0.9990	1.3426	0.4881	0.6718	0.5556
0.4	4.2093E-03	1.0680	1.4121	1.6040	1.6896	1.0046	1.0003	0.9939	0.9990	1.0493	0.4866	0.6740	0.5579
0.5	1.8337E-02	1.0685	1.0937	1.0983	1.0834	1.0052	1.0004	0.9941	0.9990	0.9154	0.4851	0.6766	0.5605
0.6	4.5711E-02	1.0690	0.9231	0.8628	0.8208	1.0059	1.0005	0.9944	0.9991	0.8444	0.4835	0.6795	0.5636
0.7	8.3656E-02	1.0697	0.8224	0.7356	0.6850	1.0067	1.0006	0.9947	0.9991	0.8031	0.4821	0.6827	0.5669
0.8	1.2710E-01	1.0705	0.7591	0.6601	0.6065	1.0075	1.0007	0.9949	0.9991	0.7777	0.4809	0.6861	0.5703
0.9	1.7144E-01	1.0713	0.7174	0.6124	0.5579	1.0084	1.0008	0.9952	0.9992	0.7615	0.4799	0.6896	0.5738
1.0	2.1357E-01	1.0722	0.6891	0.5810	0.5264	1.0093	1.0009	0.9954	0.9992	0.7510	0.4792	0.6931	0.5773
1.2	2.8541E-01	1.0743	0.6560	0.5453	0.4910	1.0111	1.0012	0.9959	0.9993	0.7398	0.4783	0.7000	0.5841
1.4	3.3901E-01	1.0765	0.6399	0.5286	0.4748	1.0129	1.0014	0.9963	0.9994	0.7357	0.4780	0.7067	0.5907
1.6	3.7697E-01	1.0790	0.6325	0.5213	0.4679	1.0148	1.0017	0.9966	0.9994	0.7350	0.4782	0.7129	0.5969
1.8	4.0303E-01	1.0816	0.6300	0.5191	0.4660	1.0166	1.0019	0.9969	0.9995	0.7361	0.4787	0.7189	0.6027
2.0	4.2046E-01	1.0843	0.6303	0.5199	0.4671	1.0184	1.0022	0.9972	0.9995	0.7383	0.4793	0.7244	0.6081
2.2	4.3174E-01	1.0871	0.6323	0.5224	0.4698	1.0202	1.0024	0.9974	0.9995	0.7409	0.4800	0.7296	0.6132
2.4	4.3867E-01	1.0899	0.6353	0.5260	0.4735	1.0220	1.0027	0.9976	0.9996	0.7439	0.4808	0.7344	0.6181
2.6	4.4252E-01	1.0928	0.6389	0.5302	0.4779	1.0238	1.0029	0.9978	0.9996	0.7470	0.4815	0.7389	0.6226
2.8	4.4418E-01	1.0958	0.6429	0.5347	0.4826	1.0255	1.0032	0.9979	0.9996	0.7501	0.4822	0.7432	0.6269
3.0	4.4430E-01	1.0987	0.6470	0.5395	0.4874	1.0272	1.0034	0.9981	0.9997	0.7531	0.4829	0.7472	0.6309
3.2	4.4332E-01	1.1017	0.6512	0.5443	0.4924	1.0289	1.0037	0.9982	0.9997	0.7561	0.4835	0.7509	0.6348
3.4	4.4156E-01	1.1048	0.6555	0.5491	0.4974	1.0306	1.0039	0.9983	0.9997	0.7590	0.4840	0.7545	0.6385
3.6	4.3927E-01	1.1078	0.6597	0.5539	0.5023	1.0323	1.0041	0.9984	0.9997	0.7619	0.4845	0.7578	0.6419
3.8	4.3661E-01	1.1109	0.6639	0.5586	0.5071	1.0340	1.0044	0.9985	0.9997	0.7646	0.4849	0.7610	0.6453
4.0	4.3369E-01	1.1140	0.6679	0.5633	0.5119	1.0356	1.0046	0.9985	0.9997	0.7672	0.4852	0.7640	0.6485
4.5	4.2587E-01	1.1217	0.6777	0.5744	0.5233	1.0397	1.0052	0.9987	0.9998	0.7734	0.4859	0.7709	0.6559
5.0	4.1788E-01	1.1295	0.6868	0.5847	0.5340	1.0437	1.0058	0.9988	0.9998	0.7790	0.4862	0.7770	0.6626
5.5	4.1012E-01	1.1373	0.6951	0.5944	0.5440	1.0476	1.0064	0.9990	0.9998	0.7840	0.4862	0.7825	0.6688
6.0	4.0277E-01	1.1451	0.7029	0.6034	0.5533	1.0515	1.0070	0.9991	0.9998	0.7887	0.4859	0.7874	0.6745
6.5	3.9589E-01	1.1529	0.7100	0.6117	0.5619	1.0553	1.0076	0.9991	0.9998	0.7929	0.4854	0.7918	0.6798
7.0	3.8950E-01	1.1607	0.7167	0.6194	0.5700	1.0591	1.0082	0.9992	0.9998	0.7967	0.4848	0.7958	0.6847
7.5	3.8357E-01	1.1686	0.7228	0.6266	0.5775	1.0628	1.0088	0.9993	0.9999	0.8003	0.4839	0.7995	0.6893
8.0	3.7809E-01	1.1764	0.7284	0.6333	0.5845	1.0664	1.0094	0.9993	0.9999	0.8036	0.4829	0.8029	0.6937
9.0	3.6832E-01	1.1921	0.7386	0.6454	0.5971	1.0736	1.0105	0.9994	0.9999	0.8093	0.4805	0.8089	0.7017
10.0	3.5992E-01	1.2078	0.7474	0.6559	0.6082	1.0807	1.0117	0.9995	0.9999	0.8143	0.4776	0.8139	0.7090
11.0	3.5268E-01	1.2235	0.7550	0.6651	0.6180	1.0876	1.0128	0.9995	0.9999	0.8186	0.4744	0.8183	0.7156
12.0	3.4639E-01	1.2392	0.7617	0.6733	0.6267	1.0944	1.0140	0.9996	0.9999	0.8223	0.4709	0.8221	0.7217
13.0	3.4091E-01	1.2550	0.7675	0.6805	0.6344	1.1010	1.0151	0.9996	0.9999	0.8256	0.4672	0.8254	0.7274
14.0	3.3612E-01	1.2708	0.7726	0.6868	0.6412	1.1076	1.0163	0.9996	0.9999	0.8284	0.4633	0.8283	0.7326
16.0	3.2821E-01	1.3025	0.7809	0.6974	0.6526	1.1205	1.0185	0.9997	0.9999	0.8331	0.4550	0.8330	0.7422
18.0	3.2206E-01	1.3343	0.7874	0.7057	0.6616	1.1331	1.0208	0.9997	0.9999	0.8367	0.4462	0.8366	0.7508
20.0	3.1726E-01	1.3664	0.7922	0.7121	0.6687	1.1454	1.0231	0.9998	0.9999	0.8394	0.4371	0.8394	0.7586
25.0	3.0950E-01	1.4478	0.7993	0.7222	0.6800	1.1751	1.0287	0.9998	1.0000	0.8436	0.4133	0.8436	0.7754
30.0	3.0583E-01	1.5310	0.8014	0.7260	0.6848	1.2037	1.0343	0.9998	1.0000	0.8452	0.3886	0.8453	0.7897
35.0	3.0485E-01	1.6165	0.8002	0.7258	0.6851	1.2314	1.0399	0.9999	1.0000	0.8452	0.3634	0.8452	0.8022
40.0	3.0579E-01	1.7045	0.7967	0.7226	0.6822	1.2583	1.0455	0.9999	1.0000	0.8440	0.3379	0.8440	0.8135
45.0	3.0816E-01	1.7953	0.7917	0.7173	0.6771	1.2845	1.0511	0.9999	1.0000	0.8420	0.3123	0.8421	0.8239
50.0	3.1167E-01	1.8889	0.7855	0.7105	0.6703	1.3101	1.0568	0.9999	1.0000	0.8395	0.2867	0.8396	0.8336

Instr. see p. 22

POSITRON Z= -81 A= 201

p	F_0L_0	L_0	λ_2	λ_3	λ_4	μ_1	μ_2	Λ_1	Λ_2	ν_{12}	$\hat{\nu}_{12}$	η_{12}	$\hat{\eta}_{12}$
0.1	2.0787E-14	1.0678	14.8693	102.6867	415.3597	1.0037	1.0001	0.9933	0.9989	6.8422	0.4849	0.6626	0.5444
0.2	1.0124E-06	1.0679	4.0950	9.3552	15.4054	1.0039	1.0002	0.9934	0.9989	2.2046	0.4841	0.6635	0.5454
0.3	2.7914E-04	1.0682	2.1018	3.0104	3.6305	1.0043	1.0002	0.9936	0.9989	1.3474	0.4830	0.6651	0.5471
0.4	3.9654E-03	1.0686	1.4065	1.5995	1.6863	1.0048	1.0003	0.9938	0.9990	1.0491	0.4816	0.6673	0.5493
0.5	1.7634E-02	1.0690	1.0870	1.0906	1.0752	1.0055	1.0004	0.9940	0.9990	0.9129	0.4802	0.6699	0.5520
0.6	4.4536E-02	1.0696	0.9159	0.8542	0.8115	1.0062	1.0005	0.9943	0.9990	0.8406	0.4787	0.6729	0.5551
0.7	8.2221E-02	1.0703	0.8149	0.7266	0.6753	1.0070	1.0006	0.9945	0.9991	0.7986	0.4774	0.6761	0.5584
0.8	1.2568E-01	1.0711	0.7513	0.6509	0.5968	1.0078	1.0007	0.9948	0.9991	0.7728	0.4762	0.6796	0.5618
0.9	1.7027E-01	1.0720	0.7095	0.6031	0.5482	1.0087	1.0008	0.9951	0.9992	0.7563	0.4753	0.6831	0.5653
1.0	2.1278E-01	1.0729	0.6812	0.5717	0.5167	1.0096	1.0009	0.9953	0.9992	0.7456	0.4747	0.6866	0.5689
1.2	2.8553E-01	1.0750	0.6479	0.5361	0.4814	1.0115	1.0012	0.9958	0.9993	0.7342	0.4739	0.6937	0.5758
1.4	3.3993E-01	1.0773	0.6318	0.5194	0.4652	1.0134	1.0014	0.9962	0.9994	0.7300	0.4737	0.7004	0.5824
1.6	3.7848E-01	1.0799	0.6245	0.5121	0.4584	1.0153	1.0017	0.9966	0.9994	0.7293	0.4740	0.7068	0.5887
1.8	4.0493E-01	1.0825	0.6220	0.5101	0.4566	1.0172	1.0020	0.9969	0.9995	0.7304	0.4746	0.7128	0.5946
2.0	4.2258E-01	1.0853	0.6224	0.5109	0.4577	1.0191	1.0022	0.9971	0.9995	0.7326	0.4753	0.7185	0.6001
2.2	4.3396E-01	1.0882	0.6245	0.5135	0.4605	1.0209	1.0025	0.9974	0.9995	0.7354	0.4761	0.7237	0.6053
2.4	4.4091E-01	1.0911	0.6276	0.5171	0.4643	1.0228	1.0027	0.9976	0.9996	0.7384	0.4770	0.7287	0.6102
2.6	4.4471E-01	1.0941	0.6313	0.5214	0.4687	1.0246	1.0030	0.9977	0.9996	0.7415	0.4778	0.7333	0.6148
2.8	4.4629E-01	1.0971	0.6353	0.5260	0.4735	1.0264	1.0032	0.9979	0.9996	0.7447	0.4785	0.7376	0.6192
3.0	4.4630E-01	1.1002	0.6396	0.5308	0.4784	1.0282	1.0035	0.9980	0.9997	0.7478	0.4793	0.7417	0.6233
3.2	4.4519E-01	1.1033	0.6439	0.5357	0.4834	1.0299	1.0037	0.9981	0.9997	0.7509	0.4799	0.7456	0.6272
3.4	4.4331E-01	1.1064	0.6482	0.5406	0.4885	1.0317	1.0040	0.9983	0.9997	0.7538	0.4805	0.7492	0.6309
3.6	4.4087E-01	1.1095	0.6525	0.5455	0.4934	1.0334	1.0042	0.9983	0.9997	0.7567	0.4811	0.7526	0.6345
3.8	4.3807E-01	1.1126	0.6568	0.5503	0.4984	1.0351	1.0045	0.9984	0.9997	0.7595	0.4815	0.7559	0.6379
4.0	4.3502E-01	1.1158	0.6609	0.5550	0.5032	1.0368	1.0047	0.9985	0.9997	0.7622	0.4820	0.7589	0.6411
4.5	4.2686E-01	1.1237	0.6709	0.5662	0.5148	1.0410	1.0053	0.9987	0.9998	0.7685	0.4827	0.7660	0.6487
5.0	4.1855E-01	1.1317	0.6801	0.5768	0.5256	1.0451	1.0059	0.9988	0.9998	0.7742	0.4831	0.7722	0.6555
5.5	4.1051E-01	1.1397	0.6887	0.5866	0.5357	1.0491	1.0065	0.9989	0.9998	0.7794	0.4833	0.7778	0.6618
6.0	4.0290E-01	1.1477	0.6966	0.5957	0.5452	1.0531	1.0071	0.9990	0.9998	0.7842	0.4831	0.7829	0.6676
6.5	3.9579E-01	1.1558	0.7040	0.6042	0.5539	1.0570	1.0077	0.9991	0.9998	0.7885	0.4827	0.7874	0.6730
7.0	3.8919E-01	1.1638	0.7107	0.6121	0.5621	1.0609	1.0083	0.9992	0.9998	0.7925	0.4821	0.7915	0.6780
7.5	3.8308E-01	1.1719	0.7170	0.6194	0.5697	1.0647	1.0089	0.9992	0.9999	0.7961	0.4813	0.7953	0.6827
8.0	3.7743E-01	1.1799	0.7228	0.6262	0.5769	1.0684	1.0095	0.9993	0.9999	0.7995	0.4804	0.7988	0.6872
9.0	3.6736E-01	1.1960	0.7332	0.6386	0.5898	1.0758	1.0107	0.9994	0.9999	0.8054	0.4781	0.8049	0.6954
10.0	3.5871E-01	1.2121	0.7422	0.6493	0.6011	1.0830	1.0119	0.9994	0.9999	0.8105	0.4754	0.8101	0.7028
11.0	3.5125E-01	1.2282	0.7500	0.6588	0.6111	1.0901	1.0130	0.9995	0.9999	0.8149	0.4723	0.8146	0.7095
12.0	3.4478E-01	1.2444	0.7569	0.6671	0.6199	1.0971	1.0142	0.9995	0.9999	0.8187	0.4690	0.8185	0.7157
13.0	3.3914E-01	1.2606	0.7629	0.6745	0.6278	1.1039	1.0154	0.9996	0.9999	0.8221	0.4654	0.8219	0.7215
14.0	3.3421E-01	1.2768	0.7681	0.6810	0.6347	1.1106	1.0165	0.9996	0.9999	0.8250	0.4616	0.8249	0.7269
16.0	3.2606E-01	1.3093	0.7768	0.6919	0.6465	1.1238	1.0189	0.9997	0.9999	0.8298	0.4535	0.8298	0.7367
18.0	3.1973E-01	1.3420	0.7834	0.7005	0.6558	1.1367	1.0212	0.9997	0.9999	0.8336	0.4449	0.8335	0.7454
20.0	3.1477E-01	1.3749	0.7885	0.7072	0.6631	1.1493	1.0234	0.9997	0.9999	0.8364	0.4360	0.8364	0.7534
25.0	3.0670E-01	1.4584	0.7960	0.7177	0.6749	1.1797	1.0292	0.9998	1.0000	0.8409	0.4126	0.8409	0.7706
30.0	3.0281E-01	1.5440	0.7984	0.7219	0.6800	1.2089	1.0348	0.9998	1.0000	0.8427	0.3883	0.8427	0.7851
35.0	3.0165E-01	1.6318	0.7974	0.7219	0.6806	1.2371	1.0405	0.9999	1.0000	0.8427	0.3635	0.8427	0.7979
40.0	3.0244E-01	1.7223	0.7941	0.7189	0.6780	1.2646	1.0462	0.9999	1.0000	0.8416	0.3384	0.8417	0.8094
45.0	3.0469E-01	1.8157	0.7891	0.7138	0.6731	1.2914	1.0519	0.9999	1.0000	0.8397	0.3131	0.8398	0.8200
50.0	3.0809E-01	1.9120	0.7830	0.7071	0.6664	1.3175	1.0577	0.9999	1.0000	0.8373	0.2878	0.8373	0.8299

Instr. see p. 22

POSITRON Z= -82 A= 203

p	F_0L_0	L_0	λ_2	λ_3	λ_4	μ_1	μ_2	Λ_1	Λ_2	ν_{12}	$\hat{\nu}_{12}$	η_{12}	$\hat{\eta}_{12}$
0.1	1.3982E-14	1.0683	14.9011	104.3623	428.5953	1.0039	1.0001	0.9932	0.9989	6.9389	0.4798	0.6559	0.5359
0.2	8.5439E-07	1.0684	4.0968	9.4438	15.6743	1.0041	1.0002	0.9933	0.9989	2.2238	0.4791	0.6569	0.5370
0.3	2.5368E-04	1.0687	2.0980	3.0183	3.6521	1.0046	1.0002	0.9934	0.9989	1.3523	0.4780	0.6585	0.5386
0.4	3.7352E-03	1.0691	1.4007	1.5949	1.6829	1.0051	1.0003	0.9936	0.9990	1.0490	0.4767	0.6607	0.5409
0.5	1.6957E-02	1.0696	1.0804	1.0830	1.0671	1.0058	1.0004	0.9939	0.9990	0.9105	0.4753	0.6633	0.5436
0.6	4.3387E-02	1.0702	0.9087	0.8456	0.8022	1.0065	1.0005	0.9941	0.9990	0.8370	0.4739	0.6663	0.5467
0.7	8.0801E-02	1.0709	0.8074	0.7176	0.6658	1.0073	1.0006	0.9944	0.9991	0.7942	0.4727	0.6696	0.5500
0.8	1.2427E-01	1.0717	0.7436	0.6419	0.5873	1.0082	1.0007	0.9947	0.9991	0.7680	0.4716	0.6731	0.5535
0.9	1.6908E-01	1.0726	0.7017	0.5941	0.5387	1.0091	1.0008	0.9950	0.9992	0.7512	0.4708	0.6767	0.5570
1.0	2.1197E-01	1.0736	0.6733	0.5627	0.5072	1.0100	1.0010	0.9952	0.9992	0.7403	0.4702	0.6802	0.5606
1.2	2.8561E-01	1.0757	0.6400	0.5270	0.4721	1.0120	1.0012	0.9957	0.9993	0.7287	0.4695	0.6874	0.5675
1.4	3.4081E-01	1.0781	0.6239	0.5104	0.4559	1.0139	1.0015	0.9961	0.9993	0.7243	0.4695	0.6942	0.5743
1.6	3.7995E-01	1.0807	0.6166	0.5032	0.4492	1.0159	1.0017	0.9965	0.9994	0.7236	0.4699	0.7007	0.5806
1.8	4.0679E-01	1.0835	0.6142	0.5012	0.4475	1.0178	1.0020	0.9968	0.9995	0.7248	0.4706	0.7069	0.5866
2.0	4.2466E-01	1.0863	0.6146	0.5021	0.4486	1.0198	1.0023	0.9971	0.9995	0.7270	0.4714	0.7126	0.5922
2.2	4.3614E-01	1.0893	0.6168	0.5047	0.4515	1.0217	1.0025	0.9973	0.9995	0.7298	0.4723	0.7180	0.5975
2.4	4.4310E-01	1.0923	0.6200	0.5085	0.4553	1.0236	1.0028	0.9975	0.9996	0.7329	0.4732	0.7230	0.6024
2.6	4.4686E-01	1.0954	0.6238	0.5128	0.4598	1.0255	1.0030	0.9977	0.9996	0.7361	0.4741	0.7277	0.6071
2.8	4.4835E-01	1.0985	0.6279	0.5175	0.4646	1.0273	1.0033	0.9978	0.9996	0.7393	0.4749	0.7322	0.6115
3.0	4.4824E-01	1.1016	0.6322	0.5224	0.4696	1.0291	1.0035	0.9980	0.9996	0.7425	0.4757	0.7363	0.6157
3.2	4.4701E-01	1.1048	0.6367	0.5273	0.4747	1.0310	1.0038	0.9981	0.9997	0.7457	0.4764	0.7403	0.6197
3.4	4.4498E-01	1.1080	0.6411	0.5323	0.4798	1.0328	1.0040	0.9982	0.9997	0.7487	0.4771	0.7440	0.6235
3.6	4.4241E-01	1.1112	0.6455	0.5372	0.4848	1.0345	1.0043	0.9983	0.9997	0.7517	0.4777	0.7475	0.6271
3.8	4.3946E-01	1.1145	0.6498	0.5421	0.4898	1.0363	1.0045	0.9984	0.9997	0.7545	0.4782	0.7508	0.6305
4.0	4.3627E-01	1.1177	0.6541	0.5469	0.4947	1.0381	1.0048	0.9985	0.9997	0.7573	0.4787	0.7539	0.6338
4.5	4.2777E-01	1.1259	0.6642	0.5583	0.5064	1.0424	1.0054	0.9987	0.9998	0.7637	0.4796	0.7611	0.6415
5.0	4.1915E-01	1.1341	0.6737	0.5691	0.5174	1.0466	1.0060	0.9988	0.9998	0.7696	0.4801	0.7675	0.6485
5.5	4.1082E-01	1.1423	0.6824	0.5790	0.5277	1.0507	1.0067	0.9989	0.9998	0.7749	0.4803	0.7733	0.6549
6.0	4.0296E-01	1.1506	0.6905	0.5883	0.5372	1.0548	1.0073	0.9990	0.9998	0.7798	0.4803	0.7784	0.6608
6.5	3.9562E-01	1.1588	0.6980	0.5969	0.5461	1.0588	1.0079	0.9991	0.9998	0.7842	0.4800	0.7831	0.6663
7.0	3.8881E-01	1.1671	0.7049	0.6049	0.5544	1.0628	1.0085	0.9992	0.9998	0.7883	0.4795	0.7873	0.6714
7.5	3.8252E-01	1.1754	0.7113	0.6124	0.5622	1.0667	1.0091	0.9992	0.9999	0.7920	0.4788	0.7912	0.6762
8.0	3.7670E-01	1.1836	0.7173	0.6194	0.5694	1.0706	1.0097	0.9993	0.9999	0.7954	0.4779	0.7947	0.6807
9.0	3.6633E-01	1.2002	0.7279	0.6319	0.5825	1.0782	1.0109	0.9994	0.9999	0.8015	0.4758	0.8010	0.6891
10.0	3.5744E-01	1.2168	0.7371	0.6429	0.5941	1.0856	1.0121	0.9994	0.9999	0.8068	0.4732	0.8064	0.6966
11.0	3.4977E-01	1.2333	0.7452	0.6526	0.6043	1.0929	1.0133	0.9995	0.9999	0.8113	0.4702	0.8110	0.7035
12.0	3.4313E-01	1.2499	0.7522	0.6611	0.6133	1.1000	1.0145	0.9995	0.9999	0.8152	0.4670	0.8150	0.7098
13.0	3.3734E-01	1.2666	0.7584	0.6687	0.6213	1.1070	1.0157	0.9996	0.9999	0.8187	0.4635	0.8185	0.7157
14.0	3.3227E-01	1.2832	0.7638	0.6753	0.6284	1.1139	1.0168	0.9996	0.9999	0.8217	0.4598	0.8215	0.7212
16.0	3.2389E-01	1.3166	0.7727	0.6865	0.6405	1.1275	1.0192	0.9997	0.9999	0.8267	0.4519	0.8266	0.7312
18.0	3.1737E-01	1.3503	0.7796	0.6954	0.6500	1.1406	1.0215	0.9997	0.9999	0.8305	0.4435	0.8305	0.7401
20.0	3.1227E-01	1.3841	0.7848	0.7022	0.6576	1.1535	1.0239	0.9997	0.9999	0.8335	0.4347	0.8335	0.7482
25.0	3.0392E-01	1.4700	0.7927	0.7132	0.6698	1.1846	1.0297	0.9998	1.0000	0.8381	0.4117	0.8381	0.7658
30.0	2.9982E-01	1.5581	0.7953	0.7177	0.6753	1.2145	1.0355	0.9998	1.0000	0.8401	0.3878	0.8401	0.7806
35.0	2.9852E-01	1.6486	0.7945	0.7180	0.6761	1.2434	1.0412	0.9999	1.0000	0.8402	0.3633	0.8403	0.7937
40.0	2.9918E-01	1.7418	0.7914	0.7152	0.6737	1.2715	1.0470	0.9999	1.0000	0.8392	0.3385	0.8393	0.8054
45.0	3.0133E-01	1.8380	0.7865	0.7102	0.6689	1.2989	1.0528	0.9999	1.0000	0.8374	0.3135	0.8374	0.8163
50.0	3.0464E-01	1.9374	0.7805	0.7036	0.6624	1.3256	1.0588	0.9999	1.0000	0.8350	0.2885	0.8350	0.8264

Instr. see p. 22

POSITRON Z= -83 A= 207

p	F_0L_0	L_0	λ_2	λ_3	λ_4	μ_1	μ_2	Λ_1	Λ_2	ν_{12}	$\hat{\nu}_{12}$	η_{12}	$\hat{\eta}_{12}$
0.1	9.3980E-15	1.0688	14.9337	106.0601	442.1909	1.0041	1.0001	0.9930	0.9989	7.0366	0.4748	0.6494	0.5276
0.2	7.2052E-07	1.0689	4.0990	9.5345	15.9509	1.0044	1.0002	0.9931	0.9989	2.2434	0.4742	0.6504	0.5287
0.3	2.3037E-04	1.0692	2.0946	3.0272	3.6753	1.0048	1.0002	0.9933	0.9989	1.3574	0.4731	0.6521	0.5304
0.4	3.5158E-03	1.0696	1.3953	1.5910	1.6802	1.0054	1.0003	0.9935	0.9989	1.0491	0.4719	0.6542	0.5326
0.5	1.6293E-02	1.0701	1.0741	1.0759	1.0596	1.0061	1.0004	0.9937	0.9990	0.9083	0.4706	0.6569	0.5353
0.6	4.2235E-02	1.0707	0.9019	0.8375	0.7936	1.0068	1.0005	0.9940	0.9990	0.8335	0.4693	0.6600	0.5384
0.7	7.9346E-02	1.0715	0.8003	0.7092	0.6568	1.0077	1.0006	0.9943	0.9991	0.7901	0.4681	0.6633	0.5418
0.8	1.2277E-01	1.0723	0.7363	0.6333	0.5782	1.0086	1.0007	0.9946	0.9991	0.7633	0.4671	0.6668	0.5453
0.9	1.6778E-01	1.0732	0.6943	0.5854	0.5297	1.0095	1.0008	0.9948	0.9991	0.7462	0.4663	0.6704	0.5488
1.0	2.1101E-01	1.0743	0.6658	0.5540	0.4982	1.0105	1.0010	0.9951	0.9992	0.7352	0.4658	0.6740	0.5524
1.2	2.8548E-01	1.0765	0.6324	0.5184	0.4631	1.0125	1.0012	0.9956	0.9993	0.7233	0.4653	0.6813	0.5595
1.4	3.4144E-01	1.0790	0.6163	0.5018	0.4471	1.0145	1.0015	0.9960	0.9993	0.7189	0.4654	0.6882	0.5663
1.6	3.8114E-01	1.0817	0.6090	0.4947	0.4404	1.0166	1.0018	0.9964	0.9994	0.7182	0.4659	0.6949	0.5727
1.8	4.0834E-01	1.0845	0.6067	0.4927	0.4388	1.0186	1.0020	0.9967	0.9994	0.7194	0.4666	0.7011	0.5787
2.0	4.2642E-01	1.0875	0.6072	0.4937	0.4400	1.0206	1.0023	0.9970	0.9995	0.7217	0.4675	0.7069	0.5844
2.2	4.3799E-01	1.0905	0.6095	0.4964	0.4429	1.0226	1.0026	0.9972	0.9995	0.7245	0.4685	0.7124	0.5898
2.4	4.4496E-01	1.0936	0.6127	0.5002	0.4468	1.0245	1.0028	0.9974	0.9996	0.7276	0.4695	0.7176	0.5948
2.6	4.4866E-01	1.0968	0.6166	0.5046	0.4513	1.0265	1.0031	0.9976	0.9996	0.7309	0.4704	0.7224	0.5996
2.8	4.5007E-01	1.1000	0.6208	0.5094	0.4562	1.0284	1.0034	0.9978	0.9996	0.7342	0.4714	0.7269	0.6041
3.0	4.4985E-01	1.1033	0.6253	0.5143	0.4613	1.0303	1.0036	0.9979	0.9996	0.7375	0.4722	0.7311	0.6083
3.2	4.4849E-01	1.1066	0.6298	0.5194	0.4664	1.0321	1.0039	0.9981	0.9997	0.7407	0.4730	0.7351	0.6124
3.4	4.4633E-01	1.1099	0.6343	0.5244	0.4715	1.0340	1.0041	0.9982	0.9997	0.7438	0.4737	0.7389	0.6162
3.6	4.4361E-01	1.1132	0.6388	0.5294	0.4766	1.0358	1.0044	0.9983	0.9997	0.7468	0.4744	0.7425	0.6199
3.8	4.4052E-01	1.1165	0.6432	0.5344	0.4817	1.0377	1.0046	0.9984	0.9997	0.7497	0.4749	0.7459	0.6234
4.0	4.3719E-01	1.1199	0.6475	0.5392	0.4866	1.0395	1.0049	0.9985	0.9997	0.7525	0.4755	0.7491	0.6267
4.5	4.2836E-01	1.1283	0.6579	0.5508	0.4985	1.0439	1.0055	0.9986	0.9998	0.7591	0.4765	0.7565	0.6345
5.0	4.1943E-01	1.1368	0.6675	0.5617	0.5096	1.0483	1.0062	0.9988	0.9998	0.7651	0.4771	0.7630	0.6416
5.5	4.1083E-01	1.1453	0.6764	0.5718	0.5200	1.0526	1.0068	0.9989	C.9998	0.7706	0.4774	0.7689	0.6482
6.0	4.0272E-01	1.1538	0.6847	0.5813	0.5297	1.0568	1.0074	0.9990	0.9998	0.7755	0.4774	0.7741	0.6542
6.5	3.9516E-01	1.1623	0.6923	0.5900	0.5387	1.0609	1.0080	0.9991	0.9998	0.7800	0.4772	0.7789	0.6598
7.0	3.8816E-01	1.1708	0.6994	0.5982	0.5472	1.0650	1.0087	0.9991	0.9998	0.7842	0.4768	0.7832	0.6650
7.5	3.8169E-01	1.1793	0.7060	0.6058	0.5550	1.0691	1.0093	0.9992	0.9998	0.7880	0.4761	0.7872	0.6699
8.0	3.7571E-01	1.1879	0.7120	0.6129	0.5624	1.0730	1.0099	0.9993	0.9999	0.7915	0.4753	0.7908	0.6745
9.0	3.6507E-01	1.2049	0.7229	0.6256	0.5757	1.0809	1.0111	0.9994	0.9999	0.7978	0.4733	0.7973	0.6830
10.0	3.5595E-01	1.2220	0.7324	0.6369	0.5875	1.0885	1.0124	0.9994	0.9999	0.8031	0.4708	0.8028	0.6907
11.0	3.4809E-01	1.2391	0.7406	0.6467	0.5978	1.0960	1.0136	0.9995	0.9999	0.8078	0.4680	0.8075	0.6977
12.0	3.4128E-01	1.2562	0.7478	0.6554	0.6070	1.1033	1.0148	0.9995	0.9999	0.8118	0.4648	0.8116	0.7042
13.0	3.3535E-01	1.2734	0.7541	0.6631	0.6152	1.1106	1.0160	0.9996	0.9999	0.8154	0.4614	0.8152	0.7102
14.0	3.3016E-01	1.2906	0.7596	0.6700	0.6225	1.1177	1.0172	0.9996	0.9999	0.8185	0.4578	0.8183	0.7158
16.0	3.2159E-01	1.3250	0.7688	0.6814	0.6348	1.1316	1.0196	0.9997	0.9999	0.8236	0.4500	0.8235	0.7259
18.0	3.1491E-01	1.3597	0.7759	0.6904	0.6445	1.1451	1.0220	0.9997	0.9999	0.8276	0.4418	0.8275	0.7351
20.0	3.0968E-01	1.3947	0.7813	0.6975	0.6523	1.1563	1.0244	0.9997	0.9999	0.8306	0.4331	0.8306	0.7433
25.0	3.0110E-01	1.4834	0.7895	0.7088	0.6649	1.1903	1.0303	0.9998	1.0000	0.8355	0.4104	0.8355	0.7612
30.0	2.9685E-01	1.5743	0.7923	0.7136	0.6706	1.2209	1.0362	0.9998	1.0000	0.8375	0.3867	0.8375	0.7764
35.0	2.9543E-01	1.6679	0.7916	0.7140	0.6717	1.2506	1.0421	0.9999	1.0000	0.8378	0.3624	0.8378	0.7897
40.0	2.9602E-01	1.7644	0.7885	0.7113	0.6694	1.2794	1.0480	0.9999	1.0000	0.8368	0.3378	0.8368	0.8017
45.0	2.9810E-01	1.8640	0.7837	0.7064	0.6646	1.3074	1.0539	0.9999	1.0000	0.8350	0.3131	0.8350	0.8128
50.0	3.0138E-01	1.9671	0.7777	0.6998	0.6581	1.3328	1.0594	0.9999	1.0000	0.8326	0.2898	0.8326	C.8221

 Instr. see p. 22

POSITRON Z= -84 A= 208

p	F_0L_0	L_0	λ_2	λ_3	λ_4	μ_1	μ_2	Λ_1	Λ_2	ν_{12}	$\hat{\nu}_{12}$	η_{12}	$\hat{\eta}_{12}$
0.1	6.3318E-15	1.0690	14.9427	107.5498	455.0728	1.0044	1.0001	0.9928	0.9988	7.1304	0.4696	0.6426	0.5191
0.2	6.0907E-07	1.0692	4.0950	9.6070	16.1971	1.0047	1.0002	0.9929	0.9989	2.2618	0.4690	0.6436	0.5201
0.3	2.0970E-04	1.0695	2.0881	3.0304	3.6912	1.0051	1.0002	0.9931	0.9989	1.3619	0.4680	0.6452	0.5218
0.4	3.3170E-03	1.0699	1.3880	1.5842	1.6744	1.0057	1.0003	0.9933	0.9989	1.0487	0.4669	0.6474	0.5240
0.5	1.5692E-02	1.0705	1.0664	1.0670	1.0502	1.0064	1.0004	0.9936	0.9990	0.9056	0.4656	0.6501	0.5268
0.6	4.1210E-02	1.0711	0.8940	0.8281	0.7835	1.0072	1.0005	0.9939	0.9990	0.8296	0.4644	0.6532	0.5299
0.7	7.8100E-02	1.0719	0.7922	0.6997	0.6468	1.0081	1.0006	0.9942	0.9990	0.7854	0.4633	0.6566	0.5332
0.8	1.2158E-01	1.0727	0.7282	0.6238	0.5683	1.0090	1.0007	0.9944	0.9991	0.7583	0.4624	0.6601	0.5368
0.9	1.6688E-01	1.0737	0.6861	0.5760	0.5198	1.0100	1.0009	0.9947	0.9991	0.7409	0.4617	0.6638	0.5404
1.0	2.1053E-01	1.0747	0.6575	0.5446	0.4885	1.0110	1.0010	0.9950	0.9992	0.7296	0.4612	0.6675	0.5440
1.2	2.8600E-01	1.0770	0.6242	0.5091	0.4536	1.0130	1.0013	0.9955	0.9993	0.7176	0.4608	0.6748	0.5511
1.4	3.4285E-01	1.0796	0.6081	0.4926	0.4377	1.0151	1.0015	0.9959	0.9993	0.7131	0.4611	0.6819	0.5580
1.6	3.8320E-01	1.0824	0.6009	0.4855	0.4311	1.0172	1.0018	0.9963	0.9994	0.7123	0.4617	0.6886	0.5645
1.8	4.1082E-01	1.0853	0.5986	0.4837	0.4295	1.0193	1.0021	0.9966	0.9994	0.7136	0.4625	0.6950	0.5706
2.0	4.2914E-01	1.0883	0.5992	0.4847	0.4308	1.0213	1.0024	0.9969	0.9995	0.7159	0.4635	0.7009	0.5763
2.2	4.4082E-01	1.0915	0.6015	0.4875	0.4337	1.0234	1.0026	0.9972	0.9995	0.7188	0.4646	0.7065	0.5818
2.4	4.4780E-01	1.0947	0.6049	0.4914	0.4377	1.0254	1.0029	0.9974	0.9996	0.7220	0.4656	0.7117	0.5869
2.6	4.5145E-01	1.0979	0.6088	0.4958	0.4423	1.0274	1.0032	0.9976	0.9996	0.7253	0.4667	0.7166	0.5917
2.8	4.5276E-01	1.1012	0.6132	0.5007	0.4472	1.0294	1.0034	0.9977	0.9996	0.7287	0.4677	0.7212	0.5962
3.0	4.5241E-01	1.1046	0.6177	0.5057	0.4523	1.0313	1.0037	0.9979	0.9996	0.7320	0.4686	0.7256	0.6006
3.2	4.5091E-01	1.1079	0.6223	0.5108	0.4575	1.0332	1.0039	0.9980	0.9997	0.7353	0.4694	0.7297	0.6047
3.4	4.4860E-01	1.1113	0.6269	0.5159	0.4627	1.0352	1.0042	0.9981	0.9997	0.7385	0.4702	0.7335	0.6086
3.6	4.4573E-01	1.1148	0.6315	0.5210	0.4679	1.0371	1.0045	0.9982	0.9997	0.7416	0.4709	0.7372	0.6123
3.8	4.4248E-01	1.1182	0.6360	0.5260	0.4730	1.0389	1.0047	0.9983	0.9997	0.7445	0.4716	0.7406	0.6159
4.0	4.3899E-01	1.1216	0.6404	0.5309	0.4780	1.0408	1.0050	0.9984	0.9997	0.7474	0.4721	0.7439	0.6193
4.5	4.2979E-01	1.1303	0.6510	0.5427	0.4900	1.0454	1.0056	0.9986	0.9997	0.7542	0.4733	0.7515	0.6272
5.0	4.2052E-01	1.1390	0.6608	0.5538	0.5013	1.0499	1.0063	0.9988	0.9998	0.7603	0.4740	0.7581	0.6344
5.5	4.1161E-01	1.1477	0.6699	0.5641	0.5118	1.0543	1.0069	0.9989	0.9998	0.7659	0.4744	0.7641	0.6411
6.0	4.0322E-01	1.1564	0.6783	0.5736	0.5216	1.0586	1.0075	0.9990	0.9998	0.7709	0.4746	0.7695	0.6472
6.5	3.9541E-01	1.1652	0.6861	0.5825	0.5308	1.0629	1.0082	0.9991	0.9998	0.7756	0.4744	0.7744	0.6529
7.0	3.8818E-01	1.1739	0.6934	0.5908	0.5393	1.0671	1.0088	0.9991	0.9998	0.7798	0.4741	0.7788	0.6582
7.5	3.8151E-01	1.1827	0.7001	0.5986	0.5473	1.0712	1.0094	0.9992	0.9998	0.7837	0.4735	0.7829	0.6632
8.0	3.7534E-01	1.1914	0.7063	0.6058	0.5548	1.0753	1.0101	0.9993	0.9999	0.7873	0.4728	0.7866	0.6679
9.0	3.6439E-01	1.2090	0.7174	0.6188	0.5683	1.0833	1.0113	0.9993	0.9999	0.7938	0.4710	0.7932	0.6765
10.0	3.5500E-01	1.2265	0.7271	0.6302	0.5803	1.0911	1.0126	0.9994	0.9999	0.7993	0.4686	0.7989	0.6843
11.0	3.4692E-01	1.2440	0.7355	0.6403	0.5909	1.0988	1.0138	0.9995	0.9999	0.8040	0.4659	0.8037	0.6915
12.0	3.3991E-01	1.2616	0.7429	0.6492	0.6002	1.1064	1.0150	0.9995	0.9999	0.8082	0.4629	0.8080	0.6981
13.0	3.3381E-01	1.2792	0.7494	0.6571	0.6086	1.1138	1.0163	0.9996	0.9999	0.8118	0.4595	0.8116	0.7042
14.0	3.2848E-01	1.2968	0.7551	0.6641	0.6160	1.1210	1.0175	0.9996	0.9999	0.8150	0.4560	0.8149	0.7099
16.0	3.1966E-01	1.3322	0.7646	0.6759	0.6286	1.1353	1.0199	0.9997	0.9999	0.8203	0.4485	0.8202	0.7203
18.0	3.1278E-01	1.3678	0.7719	0.6852	0.6386	1.1491	1.0224	0.9997	0.9999	0.8244	0.4404	0.8244	0.7296
20.0	3.0739E-01	1.4037	0.7775	0.6925	0.6466	1.1627	1.0248	0.9997	0.9999	0.8276	0.4319	0.8276	0.7380
25.0	2.9850E-01	1.4948	0.7861	0.7043	0.6597	1.1954	1.0308	0.9998	1.0000	0.8327	0.4096	0.8327	0.7563
30.0	2.9402E-01	1.5883	0.7893	0.7095	0.6659	1.2267	1.0368	0.9998	1.0000	0.8349	0.3863	0.8349	0.7718
35.0	2.9242E-01	1.6844	0.7888	0.7101	0.6672	1.2569	1.0428	0.9999	1.0000	0.8353	0.3624	0.8353	0.7853
40.0	2.9286E-01	1.7837	0.7859	0.7077	0.6652	1.2863	1.0488	0.9999	1.0000	0.8344	0.3381	0.8344	0.7976
45.0	2.9482E-01	1.8862	0.7812	0.7029	0.6606	1.3150	1.0548	0.9999	1.0000	0.8327	0.3137	0.8327	0.8088
50.0	2.9797E-01	1.9923	0.7752	0.6964	0.6542	1.3429	1.0608	0.9999	1.0000	0.8303	0.2892	0.8304	0.8194

POSITRON Z= -85 A= 211

p	F_0L_0	L_0	λ_2	λ_3	λ_4	μ_1	μ_2	Λ_1	Λ_2	ν_{12}	$\hat{\nu}_{12}$	η_{12}	$\hat{\eta}_{12}$
0.1	4.2628E-15	1.0693	14.9526	109.0575	468.2688	1.0047	1.0001	0.9927	0.9988	7.2253	0.4645	0.6360	0.5107
0.2	5.1447E-07	1.0695	4.0914	9.6814	16.4501	1.0050	1.0002	0.9928	0.9988	2.2806	0.4640	0.6370	0.5117
0.3	1.9074E-04	1.0698	2.0821	3.0345	3.7086	1.0054	1.0002	0.9929	0.9989	1.3666	0.4630	0.6386	0.5134
0.4	3.1272E-03	1.0702	1.3811	1.5781	1.6694	1.0060	1.0003	0.9932	0.9989	1.0485	0.4619	0.6408	0.5157
0.5	1.5102E-02	1.0708	1.0590	1.0587	1.0414	1.0067	1.0004	0.9934	0.9989	0.9031	0.4607	0.6436	0.5184
0.6	4.0180E-02	1.0715	0.8864	0.8192	0.7741	1.0076	1.0005	0.9937	0.9990	0.8259	0.4596	0.6467	0.5215
0.7	7.6814E-02	1.0723	0.7845	0.6907	0.6372	1.0085	1.0006	0.9940	0.9990	0.7810	0.4586	0.6501	0.5249
0.8	1.2031E-01	1.0731	0.7204	0.6148	0.5588	1.0094	1.0008	0.9943	0.9991	0.7534	0.4578	0.6536	0.5284
0.9	1.6585E-01	1.0741	0.6782	0.5670	0.5105	1.0104	1.0009	0.9946	0.9991	0.7358	0.4571	0.6573	0.5321
1.0	2.0990E-01	1.0752	0.6497	0.5357	0.4793	1.0115	1.0010	0.9949	0.9992	0.7243	0.4567	0.6611	0.5357
1.2	2.8631E-01	1.0776	0.6163	0.5002	0.4445	1.0136	1.0013	0.9954	0.9992	0.7120	0.4565	0.6685	0.5429
1.4	3.4400E-01	1.0803	0.6002	0.4838	0.4287	1.0157	1.0016	0.9958	0.9993	0.7074	0.4568	0.6757	0.5498
1.6	3.8498E-01	1.0831	0.5930	0.4768	0.4222	1.0179	1.0018	0.9962	0.9994	0.7067	0.4576	0.6826	0.5564
1.8	4.1300E-01	1.0861	0.5909	0.4750	0.4207	1.0200	1.0021	0.9966	0.9994	0.7080	0.4585	0.6890	0.5626
2.0	4.3155E-01	1.0893	0.5915	0.4761	0.4220	1.0222	1.0024	0.9969	0.9995	0.7103	0.4596	0.6951	0.5684
2.2	4.4333E-01	1.0925	0.5939	0.4790	0.4250	1.0243	1.0027	0.9971	0.9995	0.7133	0.4607	0.7008	0.5739
2.4	4.5030E-01	1.0958	0.5974	0.4829	0.4291	1.0264	1.0029	0.9973	0.9995	0.7165	0.4619	0.7061	0.5791
2.6	4.5390E-01	1.0991	0.6014	0.4875	0.4337	1.0284	1.0032	0.9975	0.9996	0.7199	0.4630	0.7111	0.5840
2.8	4.5511E-01	1.1026	0.6059	0.4924	0.4387	1.0305	1.0035	0.9977	0.9996	0.7234	0.4640	0.7158	0.5886
3.0	4.5464E-01	1.1060	0.6105	0.4975	0.4439	1.0325	1.0038	0.9978	0.9996	0.7268	0.4650	0.7202	0.5930
3.2	4.5299E-01	1.1095	0.6152	0.5027	0.4491	1.0345	1.0040	0.9980	0.9996	0.7301	0.4659	0.7244	0.5972
3.4	4.5053E-01	1.1130	0.6199	0.5078	0.4544	1.0365	1.0043	0.9981	0.9997	0.7334	0.4668	0.7283	0.6011
3.6	4.4751E-01	1.1165	0.6245	0.5130	0.4596	1.0384	1.0046	0.9982	0.9997	0.7365	0.4676	0.7321	0.6049
3.8	4.4411E-01	1.1200	0.6291	0.5181	0.4647	1.0404	1.0048	0.9983	0.9997	0.7396	0.4682	0.7356	0.6085
4.0	4.4046E-01	1.1236	0.6336	0.5230	0.4698	1.0423	1.0051	0.9984	0.9997	0.7425	0.4689	0.7389	0.6120
4.5	4.3089E-01	1.1325	0.6444	0.5350	0.4819	1.0470	1.0057	0.9986	0.9997	0.7494	0.4701	0.7466	0.6201
5.0	4.2129E-01	1.1415	0.6544	0.5462	0.4933	1.0516	1.0064	0.9987	0.9998	0.7556	0.4710	0.7535	0.6274
5.5	4.1207E-01	1.1505	0.6637	0.5567	0.5040	1.0562	1.0070	0.9988	0.9998	0.7613	0.4715	0.7596	0.6342
6.0	4.0342E-01	1.1594	0.6723	0.5664	0.5139	1.0606	1.0077	0.9990	0.9998	0.7665	0.4717	0.7651	0.6404
6.5	3.9537E-01	1.1685	0.6803	0.5754	0.5232	1.0650	1.0083	0.9990	0.9998	0.7713	0.4717	0.7701	0.6462
7.0	3.8793E-01	1.1775	0.6876	0.5839	0.5319	1.0694	1.0090	0.9991	0.9998	0.7756	0.4714	0.7746	0.6516
7.5	3.8105E-01	1.1865	0.6945	0.5917	0.5400	1.0736	1.0096	0.9992	0.9998	0.7796	0.4709	0.7788	0.6567
8.0	3.7472E-01	1.1955	0.7008	0.5991	0.5476	1.0778	1.0103	0.9992	0.9999	0.7833	0.4703	0.7826	0.6614
9.0	3.6346E-01	1.2135	0.7122	0.6123	0.5613	1.0861	1.0115	0.9993	0.9999	0.7899	0.4685	0.7893	0.6702
10.0	3.5382E-01	1.2316	0.7221	0.6240	0.5735	1.0941	1.0128	0.9994	0.9999	0.7955	0.4663	0.7951	0.6782
11.0	3.4553E-01	1.2496	0.7307	0.6342	0.5842	1.1020	1.0141	0.9995	0.9999	0.8004	0.4637	0.8001	0.6855
12.0	3.3835E-01	1.2677	0.7383	0.6433	0.5938	1.1098	1.0153	0.9995	0.9999	0.8047	0.4607	0.8045	0.6922
13.0	3.3210E-01	1.2859	0.7450	0.6514	0.6023	1.1174	1.0166	0.9996	0.9999	0.8084	0.4575	0.8082	0.6984
14.0	3.2663E-01	1.3040	0.7508	0.6585	0.6099	1.1249	1.0178	0.9996	0.9999	0.8117	0.4541	0.8116	0.7043
16.0	3.1759E-01	1.3405	0.7605	0.6706	0.6227	1.1395	1.0203	0.9996	0.9999	0.8172	0.4467	0.8171	0.7149
18.0	3.1054E-01	1.3772	0.7681	0.6801	0.6330	1.1537	1.0228	0.9997	0.9999	0.8214	0.4388	0.8213	0.7243
20.0	3.0500E-01	1.4141	0.7739	0.6877	0.6412	1.1676	1.0253	0.9997	0.9999	0.8247	0.4305	0.8246	0.7329
25.0	2.9585E-01	1.5080	0.7828	0.6999	0.6547	1.2011	1.0314	0.9998	1.0000	0.8299	0.4085	0.8299	0.7516
30.0	2.9119E-01	1.6044	0.7862	0.7053	0.6612	1.2332	1.0375	0.9998	1.0000	0.8323	0.3854	0.8323	0.7673
35.0	2.8946E-01	1.7037	0.7859	0.7062	0.6627	1.2642	1.0436	0.9999	1.0000	0.8328	0.3618	0.8328	0.7812
40.0	2.8979E-01	1.8061	0.7831	0.7039	0.6608	1.2943	1.0497	0.9999	1.0000	0.8320	0.3378	0.8320	0.7937
45.0	2.9166E-01	1.9121	0.7785	0.6992	0.6564	1.3236	1.0558	0.9999	1.0000	0.8303	0.3136	0.8303	0.8052
50.0	2.9475E-01	2.0219	0.7726	0.6928	0.6501	1.3522	1.0620	0.9999	1.0000	0.8280	0.2894	0.8280	0.8160

Instr. see p. 22

POSITRON Z= -86 A= 212

p	F_0L_0	L_0	λ_2	λ_3	λ_4	μ_1	μ_2	Λ_1	Λ_2	ν_{12}	$\hat{\nu}_{12}$	η_{12}	$\hat{\eta}_{12}$
0.1	2.8751E-15	1.0695	14.9433	110.3892	480.8567	1.0050	1.0001	0.9925	0.9988	7.3169	0.4593	0.6291	0.5021
0.2	4.3536E-07	1.0696	4.0829	9.7408	16.6778	1.0053	1.0002	0.9926	0.9988	2.2984	0.4587	0.6301	0.5031
0.3	1.7381E-04	1.0700	2.0737	3.0342	3.7201	1.0057	1.0002	0.9928	0.9988	1.3707	0.4579	0.6317	0.5048
0.4	2.9535E-03	1.0704	1.3727	1.5699	1.6619	1.0063	1.0003	0.9930	0.9989	1.0478	0.4568	0.6340	0.5071
0.5	1.4560E-02	1.0710	1.0506	1.0490	1.0313	1.0071	1.0004	0.9933	0.9989	0.9003	0.4557	0.6367	0.5098
0.6	3.9245E-02	1.0717	0.8780	0.8093	0.7636	1.0079	1.0005	0.9936	0.9990	0.8219	0.4547	0.6399	0.5130
0.7	7.5684E-02	1.0725	0.7761	0.6809	0.6270	1.0089	1.0006	0.9939	0.9990	0.7763	0.4537	0.6433	0.5164
0.8	1.1926E-01	1.0734	0.7120	0.6051	0.5488	1.0098	1.0008	0.9942	0.9991	0.7483	0.4530	0.6469	0.5199
0.9	1.6512E-01	1.0744	0.6698	0.5575	0.5007	1.0109	1.0009	0.9945	0.9991	0.7303	0.4525	0.6506	0.5236
1.0	2.0964E-01	1.0755	0.6413	0.5262	0.4696	1.0120	1.0010	0.9948	0.9991	0.7187	0.4521	0.6544	0.5272
1.2	2.8711E-01	1.0780	0.6079	0.4909	0.4350	1.0141	1.0013	0.9953	0.9992	0.7062	0.4520	0.6620	0.5345
1.4	3.4575E-01	1.0807	0.5919	0.4746	0.4194	1.0164	1.0016	0.9958	0.9993	0.7015	0.4525	0.6693	0.5415
1.6	3.8742E-01	1.0836	0.5848	0.4678	0.4130	1.0186	1.0019	0.9962	0.9994	0.7008	0.4533	0.6762	0.5481
1.8	4.1590E-01	1.0867	0.5827	0.4660	0.4115	1.0208	1.0022	0.9965	0.9994	0.7021	0.4544	0.6828	0.5544
2.0	4.3470E-01	1.0900	0.5835	0.4672	0.4129	1.0230	1.0024	0.9968	0.9995	0.7045	0.4556	0.6890	0.5603
2.2	4.4658E-01	1.0933	0.5860	0.4701	0.4160	1.0252	1.0027	0.9971	0.9995	0.7075	0.4568	0.6948	0.5659
2.4	4.5357E-01	1.0967	0.5895	0.4741	0.4201	1.0273	1.0030	0.9973	0.9995	0.7108	0.4580	0.7002	0.5711
2.6	4.5710E-01	1.1001	0.5936	0.4787	0.4248	1.0295	1.0033	0.9975	0.9996	0.7143	0.4592	0.7053	0.5761
2.8	4.5821E-01	1.1037	0.5982	0.4837	0.4298	1.0316	1.0035	0.9976	0.9996	0.7178	0.4603	0.7101	0.5808
3.0	4.5760E-01	1.1072	0.6029	0.4889	0.4350	1.0336	1.0038	0.9978	0.9996	0.7213	0.4614	0.7146	0.5852
3.2	4.5580E-01	1.1108	0.6076	0.4941	0.4403	1.0357	1.0041	0.9979	0.9996	0.7247	0.4624	0.7188	0.5894
3.4	4.5317E-01	1.1144	0.6124	0.4994	0.4456	1.0377	1.0044	0.9981	0.9997	0.7280	0.4633	0.7229	0.5935
3.6	4.4998E-01	1.1180	0.6172	0.5046	0.4509	1.0397	1.0046	0.9982	0.9997	0.7312	0.4641	0.7267	0.5973
3.8	4.4641E-01	1.1216	0.6219	0.5097	0.4561	1.0417	1.0049	0.9983	0.9997	0.7343	0.4648	0.7303	0.6010
4.0	4.4260E-01	1.1253	0.6264	0.5148	0.4612	1.0437	1.0052	0.9984	0.9997	0.7373	0.4655	0.7337	0.6045
4.5	4.3263E-01	1.1344	0.6374	0.5269	0.4735	1.0486	1.0058	0.9985	0.9997	0.7444	0.4669	0.7415	0.6127
5.0	4.2266E-01	1.1436	0.6476	0.5383	0.4850	1.0533	1.0065	0.9987	0.9998	0.7508	0.4679	0.7485	0.6202
5.5	4.1312E-01	1.1529	0.6571	0.5489	0.4958	1.0580	1.0072	0.9988	0.9998	0.7566	0.4685	0.7548	0.6270
6.0	4.0416E-01	1.1621	0.6659	0.5587	0.5059	1.0626	1.0078	0.9989	0.9998	0.7619	0.4688	0.7604	0.6333
6.5	3.9585E-01	1.1713	0.6740	0.5679	0.5153	1.0671	1.0085	0.9990	0.9998	0.7668	0.4689	0.7655	0.6392
7.0	3.8817E-01	1.1806	0.6815	0.5765	0.5241	1.0716	1.0091	0.9991	0.9998	0.7712	0.4687	0.7702	0.6447
7.5	3.8108E-01	1.1898	0.6885	0.5845	0.5323	1.0759	1.0098	0.9992	0.9998	0.7753	0.4683	0.7744	0.6499
8.0	3.7455E-01	1.1991	0.6950	0.5920	0.5400	1.0803	1.0104	0.9992	0.9999	0.7791	0.4677	0.7783	0.6547
9.0	3.6296E-01	1.2176	0.7067	0.6055	0.5540	1.0887	1.0117	0.9993	0.9999	0.7858	0.4662	0.7853	0.6637
10.0	3.5305E-01	1.2361	0.7168	0.6173	0.5663	1.0970	1.0130	0.9994	0.9999	0.7916	0.4641	0.7912	0.6718
11.0	3.4452E-01	1.2547	0.7256	0.6278	0.5773	1.1051	1.0143	0.9995	0.9999	0.7966	0.4616	0.7963	0.6792
12.0	3.3714E-01	1.2733	0.7334	0.6371	0.5870	1.1131	1.0156	0.9995	0.9999	0.8010	0.4588	0.8008	0.6861
13.0	3.3071E-01	1.2919	0.7402	0.6453	0.5957	1.1208	1.0169	0.9995	0.9999	0.8049	0.4557	0.8047	0.6924
14.0	3.2509E-01	1.3105	0.7463	0.6527	0.6034	1.1285	1.0181	0.9996	0.9999	0.8082	0.4523	0.8081	0.6983
16.0	3.1580E-01	1.3479	0.7562	0.6650	0.6166	1.1435	1.0207	0.9996	0.9999	0.8138	0.4451	0.8137	0.7091
18.0	3.0854E-01	1.3856	0.7640	0.6748	0.6271	1.1581	1.0232	0.9997	0.9999	0.8182	0.4374	0.8182	0.7188
20.0	3.0284E-01	1.4236	0.7701	0.6826	0.6355	1.1723	1.0257	0.9997	0.9999	0.8216	0.4293	0.8216	0.7276
25.0	2.9336E-01	1.5200	0.7794	0.6953	0.6496	1.2065	1.0319	0.9998	1.0000	0.8271	0.4077	0.8271	0.7465
30.0	2.8847E-01	1.6191	0.7831	0.7011	0.6564	1.2393	1.0381	0.9998	1.0000	0.8297	0.3850	0.8297	0.7626
35.0	2.8656E-01	1.7211	0.7831	0.7023	0.6583	1.2710	1.0443	0.9999	1.0000	0.8303	0.3617	0.8303	0.7767
40.0	2.8675E-01	1.8266	0.7804	0.7002	0.6566	1.3017	1.0504	0.9999	1.0000	0.8296	0.3380	0.8296	0.7895
45.0	2.8849E-01	1.9357	0.7759	0.6957	0.6524	1.3317	1.0567	0.9999	1.0000	0.8280	0.3141	0.8280	0.8012
50.0	2.9148E-01	2.0488	0.7701	0.6894	0.6462	1.3588	1.0628	0.9999	1.0000	0.8257	0.2922	0.8257	0.8113

Instr. see p. 22

POSITRON Z= -87 A= 220

p	F_0L_0	L_0	λ_2	λ_3	λ_4	μ_1	μ_2	Λ_1	Λ_2	ν_{12}	$\hat{\nu}_{12}$	η_{12}	$\hat{\eta}_{12}$
0.1	1.9314E-15	1.0698	14.9614	111.9896	494.9831	1.0053	1.0002	0.9923	0.9988	7.4152	0.4546	0.6229	0.4941
0.2	3.6692E-07	1.0700	4.0819	9.8244	16.9549	1.0056	1.0002	0.9924	0.9988	2.3184	0.4540	0.6239	0.4952
0.3	1.5774E-04	1.0703	2.0693	3.0417	3.7426	1.0061	1.0003	0.9926	0.9988	1.3762	0.4532	0.6256	0.4969
0.4	2.7782E-03	1.0708	1.3671	1.5658	1.6594	1.0067	1.0003	0.9928	0.9989	1.0483	0.4522	0.6279	0.4991
0.5	1.3981E-02	1.0714	1.0444	1.0422	1.0243	1.0075	1.0004	0.9931	0.9989	0.8984	0.4512	0.6306	0.5019
0.6	3.8178E-02	1.0721	0.8715	0.8018	0.7557	1.0084	1.0005	0.9934	0.9989	0.8188	0.4502	0.6338	0.5050
0.7	7.4271E-02	1.0729	0.7694	0.6731	0.6188	1.0094	1.0007	0.9937	0.9990	0.7725	0.4494	0.6373	0.5085
0.8	1.1775E-01	1.0739	0.7051	0.5973	0.5406	1.0104	1.0008	0.9940	0.9990	0.7439	0.4487	0.6409	0.5120
0.9	1.6373E-01	1.0749	0.6629	0.5496	0.4925	1.0115	1.0009	0.9943	0.9991	0.7257	0.4482	0.6447	0.5157
1.0	2.0854E-01	1.0761	0.6343	0.5184	0.4615	1.0126	1.0011	0.9946	0.9991	0.7139	0.4480	0.6485	0.5194
1.2	2.8677E-01	1.0787	0.6009	0.4832	0.4271	1.0149	1.0014	0.9951	0.9992	0.7012	0.4480	0.6562	0.5267
1.4	3.4613E-01	1.0815	0.5849	0.4669	0.4115	1.0172	1.0016	0.9956	0.9993	0.6964	0.4486	0.6636	0.5338
1.6	3.8834E-01	1.0846	0.5779	0.4601	0.4052	1.0195	1.0019	0.9960	0.9993	0.6956	0.4495	0.6707	0.5405
1.8	4.1716E-01	1.0878	0.5758	0.4584	0.4038	1.0218	1.0022	0.9964	0.9994	0.6970	0.4506	0.6773	0.5469
2.0	4.3615E-01	1.0911	0.5767	0.4597	0.4052	1.0241	1.0025	0.9967	0.9994	0.6994	0.4519	0.6836	0.5529
2.2	4.4810E-01	1.0946	0.5793	0.4627	0.4084	1.0264	1.0028	0.9970	0.9995	0.7025	0.4532	0.6895	0.5565
2.4	4.5508E-01	1.0981	0.5829	0.4667	0.4125	1.0286	1.0031	0.9972	0.9995	0.7059	0.4545	0.6950	0.5638
2.6	4.5855E-01	1.1017	0.5871	0.4714	0.4173	1.0308	1.0034	0.9974	0.9996	0.7094	0.4557	0.7002	0.5689
2.8	4.5955E-01	1.1054	0.5917	0.4765	0.4224	1.0330	1.0036	0.9976	0.9996	0.7130	0.4569	0.7051	0.5736
3.0	4.5882E-01	1.1090	0.5965	0.4817	0.4276	1.0352	1.0039	0.9977	0.9996	0.7165	0.4580	0.7097	0.5781
3.2	4.5688E-01	1.1127	0.6014	0.4870	0.4330	1.0373	1.0042	0.9979	0.9996	0.7200	0.4591	0.7140	0.5824
3.4	4.5411E-01	1.1165	0.6063	0.4923	0.4383	1.0394	1.0045	0.9980	0.9997	0.7234	0.4600	0.7181	0.5865
3.6	4.5077E-01	1.1202	0.6111	0.4976	0.4437	1.0415	1.0048	0.9981	0.9997	0.7267	0.4609	0.7220	0.5904
3.8	4.4705E-01	1.1240	0.6159	0.5028	0.4489	1.0436	1.0050	0.9982	0.9997	0.7298	0.4617	0.7257	0.5941
4.0	4.4310E-01	1.1278	0.6205	0.5079	0.4541	1.0456	1.0053	0.9983	0.9997	0.7329	0.4624	0.7292	0.5977
4.5	4.3279E-01	1.1373	0.6317	0.5202	0.4665	1.0507	1.0060	0.9985	0.9997	0.7401	0.4639	0.7372	0.6061
5.0	4.2252E-01	1.1469	0.6421	0.5317	0.4781	1.0556	1.0067	0.9987	0.9998	0.7466	0.4649	0.7443	0.6136
5.5	4.1270E-01	1.1564	0.6518	0.5425	0.4890	1.0605	1.0074	0.9988	0.9998	0.7525	0.4656	0.7507	0.6206
6.0	4.0351E-01	1.1660	0.6607	0.5525	0.4992	1.0652	1.0080	0.9989	0.9998	0.7579	0.4660	0.7564	0.6270
6.5	3.9499E-01	1.1756	0.6690	0.5618	0.5087	1.0699	1.0087	0.9990	0.9998	0.7629	0.4661	0.7616	0.6330
7.0	3.8712E-01	1.1852	0.6766	0.5705	0.5176	1.0745	1.0094	0.9991	0.9998	0.7675	0.4660	0.7664	0.6386
7.5	3.7987E-01	1.1948	0.6837	0.5786	0.5260	1.0791	1.0101	0.9991	0.9998	0.7716	0.4656	0.7707	0.6439
8.0	3.7319E-01	1.2044	0.6903	0.5862	0.5338	1.0836	1.0107	0.9992	0.9998	0.7755	0.4651	0.7747	0.6488
9.0	3.6135E-01	1.2236	0.7022	0.5999	0.5479	1.0923	1.0121	0.9993	0.9999	0.7824	0.4636	0.7818	0.6580
10.0	3.5123E-01	1.2429	0.7125	0.6120	0.5605	1.1009	1.0134	0.9994	0.9999	0.7883	0.4616	0.7878	0.6662
11.0	3.4254E-01	1.2621	0.7215	0.6226	0.5716	1.1093	1.0147	0.9994	0.9999	0.7934	0.4592	0.7931	0.6738
12.0	3.3502E-01	1.2814	0.7295	0.6320	0.5814	1.1175	1.0160	0.9995	0.9999	0.7979	0.4564	0.7976	0.6808
13.0	3.2848E-01	1.3007	0.7364	0.6404	0.5902	1.1255	1.0173	0.9995	0.9999	0.8018	0.4533	0.8016	0.6872
14.0	3.2276E-01	1.3200	0.7425	0.6479	0.5981	1.1335	1.0186	0.9996	0.9999	0.8052	0.4500	0.8051	0.6933
16.0	3.1332E-01	1.3589	0.7527	0.6604	0.6115	1.1489	1.0212	0.9996	0.9999	0.8110	0.4429	0.8109	0.7043
18.0	3.0595E-01	1.3980	0.7607	0.6704	0.6222	1.1640	1.0238	0.9997	0.9999	0.8154	0.4352	0.8154	0.7141
20.0	3.0016E-01	1.4375	0.7668	0.6783	0.6307	1.1787	1.0264	0.9997	0.9999	0.8189	0.4271	0.8189	0.7231
25.0	2.9055E-01	1.5377	0.7764	0.6912	0.6450	1.2140	1.0327	0.9998	1.0000	0.8245	0.4056	0.8245	0.7424
30.0	2.8560E-01	1.6408	0.7802	0.6971	0.6519	1.2478	1.0391	0.9998	1.0000	0.8271	0.3830	0.8271	0.7589
35.0	2.8366E-01	1.7472	0.7801	0.6984	0.6538	1.2804	1.0454	0.9999	1.0000	0.8278	0.3598	0.8278	0.7733
40.0	2.8385E-01	1.8571	0.7774	0.6962	0.6522	1.3121	1.0518	0.9999	1.0000	0.8271	0.3362	0.8271	0.7863
45.0	2.8561E-01	1.9711	0.7729	0.6917	0.6479	1.3429	1.0581	0.9999	1.0000	0.8254	0.3123	0.8255	0.7983
50.0	2.8863E-01	2.0893	0.7670	0.6853	0.6416	1.3708	1.0640	0.9999	1.0000	0.8232	0.2900	0.8232	0.8084

POSITRON Z= -88 A= 224

p	F_0L_0	L_0	λ_2	λ_3	λ_4	μ_1	μ_2	Λ_1	Λ_2	ν_{12}	$\hat{\nu}_{12}$	η_{12}	$\hat{\eta}_{12}$
0.1	1.3014E-15	1.0698	14.9501	113.3126	507.9786	1.0057	1.0002	0.9921	0.9987	7.5081	0.4495	0.6162	0.4858
0.2	3.1021E-07	1.0700	4.0733	9.8841	17.1893	1.0060	1.0002	0.9922	0.9988	2.3367	0.4490	0.6173	0.4868
0.3	1.4360E-04	1.0704	2.0610	3.0420	3.7554	1.0065	1.0003	0.9924	0.9988	1.3807	0.4483	0.6190	0.4885
0.4	2.6215E-03	1.0709	1.3590	1.5582	1.6528	1.0071	1.0003	0.9926	0.9988	1.0480	0.4473	0.6213	0.4908
0.5	1.3466E-02	1.0715	1.0364	1.0332	1.0149	1.0079	1.0004	0.9929	0.9989	0.8958	0.4464	0.6241	0.4936
0.6	3.7254E-02	1.0722	0.8635	0.7926	0.7460	1.0089	1.0006	0.9932	0.9989	0.8151	0.4455	0.6273	0.4968
0.7	7.3108E-02	1.0731	0.7614	0.6640	0.6093	1.0099	1.0007	0.9935	0.9990	0.7680	0.4447	0.6308	0.5002
0.8	1.1661E-01	1.0741	0.6972	0.5882	0.5313	1.0109	1.0008	0.9938	0.9990	0.7391	0.4441	0.6345	0.5038
0.9	1.6285E-01	1.0752	0.6550	0.5407	0.4834	1.0121	1.0009	0.9942	0.9991	0.7206	0.4437	0.6383	0.5075
1.0	2.0807E-01	1.0764	0.6264	0.5096	0.4525	1.0132	1.0011	0.9945	0.9991	0.7085	0.4435	0.6422	0.5112
1.2	2.8730E-01	1.0790	0.5931	0.4745	0.4183	1.0156	1.0014	0.9950	0.9992	0.6956	0.4437	0.6499	0.5186
1.4	3.4754E-01	1.0820	0.5771	0.4584	0.4029	1.0180	1.0017	0.9955	0.9993	0.6908	0.4444	0.6574	0.5257
1.6	3.9041E-01	1.0851	0.5702	0.4516	0.3966	1.0204	1.0020	0.9959	0.9993	0.6900	0.4454	0.6646	0.5325
1.8	4.1965E-01	1.0885	0.5682	0.4500	0.3953	1.0228	1.0023	0.9963	0.9994	0.6914	0.4467	0.6714	0.5390
2.0	4.3887E-01	1.0919	0.5691	0.4514	0.3968	1.0251	1.0026	0.9966	0.9994	0.6939	0.4480	0.6778	0.5450
2.2	4.5093E-01	1.0955	0.5718	0.4545	0.4000	1.0275	1.0029	0.9969	0.9995	0.6970	0.4494	0.6838	0.5507
2.4	4.5790E-01	1.0991	0.5755	0.4586	0.4042	1.0298	1.0031	0.9971	0.9995	0.7004	0.4508	0.6894	0.5561
2.6	4.6130E-01	1.1029	0.5798	0.4633	0.4090	1.0321	1.0034	0.9973	0.9996	0.7040	0.4521	0.6947	0.5612
2.8	4.6219E-01	1.1066	0.5845	0.4684	0.4142	1.0343	1.0037	0.9975	0.9996	0.7077	0.4533	0.6997	0.5660
3.0	4.6132E-01	1.1104	0.5894	0.4737	0.4195	1.0366	1.0040	0.9977	0.9996	0.7113	0.4545	0.7044	0.5706
3.2	4.5922E-01	1.1143	0.5944	0.4791	0.4249	1.0388	1.0043	0.9978	0.9996	0.7148	0.4556	0.7088	0.5750
3.4	4.5628E-01	1.1181	0.5993	0.4845	0.4303	1.0410	1.0046	0.9980	0.9996	0.7183	0.4566	0.7129	0.5791
3.6	4.5277E-01	1.1220	0.6043	0.4898	0.4356	1.0431	1.0049	0.9981	0.9997	0.7216	0.4575	0.7169	0.5831
3.8	4.4889E-01	1.1259	0.6091	0.4951	0.4409	1.0453	1.0051	0.9982	0.9997	0.7249	0.4584	0.7207	0.5869
4.0	4.4477E-01	1.1298	0.6139	0.5003	0.4461	1.0474	1.0054	0.9983	0.9997	0.7280	0.4591	0.7242	0.5905
4.5	4.3407E-01	1.1396	0.6252	0.5127	0.4587	1.0526	1.0061	0.9985	0.9997	0.7353	0.4607	0.7324	0.5989
5.0	4.2344E-01	1.1495	0.6358	0.5244	0.4705	1.0577	1.0068	0.9986	0.9998	0.7420	0.4619	0.7396	0.6067
5.5	4.1331E-01	1.1593	0.6457	0.5353	0.4815	1.0627	1.0075	0.9988	0.9998	0.7480	0.4627	0.7462	0.6137
6.0	4.0383E-01	1.1692	0.6548	0.5454	0.4918	1.0677	1.0082	0.9989	0.9998	0.7536	0.4631	0.7520	0.6203
6.5	3.9505E-01	1.1791	0.6632	0.5549	0.5014	1.0725	1.0089	0.9990	0.9998	0.7586	0.4633	0.7574	0.6264
7.0	3.8696E-01	1.1890	0.6710	0.5637	0.5104	1.0772	1.0096	0.9991	0.9998	0.7633	0.4633	0.7622	0.6321
7.5	3.7951E-01	1.1989	0.6782	0.5720	0.5189	1.0819	1.0103	0.9991	0.9998	0.7676	0.4630	0.7666	0.6374
8.0	3.7265E-01	1.2088	0.6850	0.5797	0.5268	1.0865	1.0109	0.9992	0.9998	0.7715	0.4625	0.7707	0.6424
9.0	3.6050E-01	1.2286	0.6971	0.5936	0.5412	1.0956	1.0123	0.9993	0.9999	0.7785	0.4612	0.7779	0.6517
10.0	3.5013E-01	1.2485	0.7076	0.6059	0.5539	1.1044	1.0137	0.9994	0.9999	0.7846	0.4592	0.7841	0.6601
11.0	3.4122E-01	1.2683	0.7168	0.6167	0.5651	1.1130	1.0150	0.9994	0.9999	0.7898	0.4569	0.7895	0.6678
12.0	3.3353E-01	1.2882	0.7249	0.6263	0.5752	1.1215	1.0163	0.9995	0.9999	0.7944	0.4542	0.7941	0.6749
13.0	3.2683E-01	1.3081	0.7320	0.6348	0.5841	1.1298	1.0177	0.9995	0.9999	0.7984	0.4512	0.7982	0.6815
14.0	3.2098E-01	1.3280	0.7383	0.6424	0.5921	1.1379	1.0190	0.9996	0.9999	0.8019	0.4480	0.8018	0.6877
16.0	3.1133E-01	1.3681	0.7488	0.6553	0.6057	1.1538	1.0216	0.9996	0.9999	0.8078	0.4411	0.8077	0.6989
18.0	3.0379E-01	1.4085	0.7569	0.6655	0.6167	1.1693	1.0243	0.9997	0.9999	0.8124	0.4335	0.8123	0.7089
20.0	2.9787E-01	1.4492	0.7632	0.6736	0.6254	1.1844	1.0269	0.9997	0.9999	0.8160	0.4255	0.8160	0.7180
25.0	2.8802E-01	1.5528	0.7731	0.6868	0.6401	1.2206	1.0334	0.9998	1.0000	0.8218	0.4043	0.8218	0.7378
30.0	2.8290E-01	1.6593	0.7771	0.6930	0.6473	1.2553	1.0399	0.9998	1.0000	0.8245	0.3819	0.8245	0.7545
35.0	2.8084E-01	1.7693	0.7772	0.6944	0.6494	1.2888	1.0463	0.9998	1.0000	0.8252	0.3589	0.8253	0.7692
40.0	2.8094E-01	1.8831	0.7746	0.6924	0.6479	1.3212	1.0528	0.9999	1.0000	0.8246	0.3355	0.8246	0.7825
45.0	2.8264E-01	2.0011	0.7701	0.6880	0.6437	1.3528	1.0593	0.9999	1.0000	0.8230	0.3118	0.8231	0.7947
50.0	2.8560E-01	2.1237	0.7642	0.6817	0.6375	1.3836	1.0658	0.9999	1.0000	0.8208	0.2880	0.8208	0.8061

POSITRON Z= -89 A= 226

p	F_0L_0	L_0	λ_2	λ_3	λ_4	μ_1	μ_2	Λ_1	Λ_2	ν_{12}	$\hat{\nu}_{12}$	η_{12}	$\hat{\eta}_{12}$
0.1	8.7859E-16	1.0697	14.9194	114.4463	520.2382	1.0060	1.0002	0.9919	0.9987	7.5974	0.4443	0.6094	0.4773
0.2	2.6275E-07	1.0699	4.0595	9.9279	17.3958	1.0063	1.0002	0.9920	0.9987	2.3539	0.4438	0.6104	0.4783
0.3	1.3097E-04	1.0702	2.0504	3.0378	3.7622	1.0069	1.0003	0.9922	0.9988	1.3846	0.4431	0.6121	0.4800
0.4	2.4781E-03	1.0707	1.3495	1.5485	1.6437	1.0076	1.0004	0.9924	0.9988	1.0472	0.4423	0.6144	0.4823
0.5	1.2994E-02	1.0714	1.0274	1.0229	1.0041	1.0084	1.0005	0.9927	0.9989	0.8930	0.4414	0.6172	0.4851
0.6	3.6419E-02	1.0721	0.8547	0.7825	0.7354	1.0093	1.0006	0.9930	0.9989	0.8110	0.4406	0.6205	0.4883
0.7	7.2093E-02	1.0730	0.7528	0.6541	0.5990	1.0104	1.0007	0.9934	0.9990	0.7633	0.4399	0.6240	0.4917
0.8	1.1569E-01	1.0741	0.6887	0.5786	0.5214	1.0115	1.0008	0.9937	0.9990	0.7339	0.4394	0.6277	0.4954
0.9	1.6227E-01	1.0752	0.6465	0.5313	0.4738	1.0126	1.0010	0.9940	0.9991	0.7151	0.4391	0.6316	0.4991
1.0	2.0798E-01	1.0764	0.6180	0.5003	0.4431	1.0138	1.0011	0.9943	0.9991	0.7029	0.4389	0.6355	0.5028
1.2	2.8833E-01	1.0792	0.5847	0.4655	0.4092	1.0163	1.0014	0.9949	0.9992	0.6898	0.4392	0.6434	0.5103
1.4	3.4958E-01	1.0822	0.5689	0.4494	0.3939	1.0187	1.0017	0.9954	0.9993	0.6849	0.4400	0.6510	0.5175
1.6	3.9317E-01	1.0855	0.5620	0.4428	0.3877	1.0212	1.0020	0.9958	0.9993	0.6841	0.4412	0.6583	0.5243
1.8	4.2288E-01	1.0889	0.5601	0.4413	0.3865	1.0237	1.0023	0.9962	0.9994	0.6855	0.4425	0.6652	0.5308
2.0	4.4237E-01	1.0925	0.5612	0.4427	0.3881	1.0261	1.0026	0.9965	0.9994	0.6880	0.4440	0.6717	0.5370
2.2	4.5453E-01	1.0962	0.5639	0.4459	0.3913	1.0286	1.0029	0.9968	0.9995	0.6912	0.4455	0.6778	0.5427
2.4	4.6150E-01	1.0999	0.5677	0.4500	0.3956	1.0309	1.0032	0.9971	0.9995	0.6948	0.4469	0.6835	0.5482
2.6	4.6483E-01	1.1037	0.5722	0.4549	0.4004	1.0333	1.0035	0.9973	0.9995	0.6984	0.4483	0.6889	0.5533
2.8	4.6559E-01	1.1076	0.5769	0.4600	0.4056	1.0356	1.0038	0.9975	0.9996	0.7021	0.4496	0.6940	0.5582
3.0	4.6457E-01	1.1115	0.5819	0.4654	0.4110	1.0380	1.0041	0.9976	0.9996	0.7058	0.4509	0.6987	0.5629
3.2	4.6230E-01	1.1155	0.5870	0.4708	0.4164	1.0402	1.0044	0.9978	0.9996	0.7094	0.4520	0.7032	0.5673
3.4	4.5919E-01	1.1194	0.5920	0.4763	0.4219	1.0425	1.0047	0.9979	0.9996	0.7129	0.4531	0.7075	0.5715
3.6	4.5549E-01	1.1234	0.5970	0.4817	0.4273	1.0447	1.0050	0.9980	0.9997	0.7164	0.4541	0.7115	0.5755
3.8	4.5143E-01	1.1274	0.6020	0.4870	0.4326	1.0469	1.0052	0.9981	0.9997	0.7197	0.4550	0.7154	0.5793
4.0	4.4713E-01	1.1315	0.6068	0.4923	0.4379	1.0491	1.0055	0.9982	0.9997	0.7228	0.4558	0.7190	0.5830
4.5	4.3599E-01	1.1416	0.6184	0.5049	0.4505	1.0545	1.0062	0.9984	0.9997	0.7303	0.4575	0.7273	0.5916
5.0	4.2497E-01	1.1517	0.6292	0.5167	0.4624	1.0598	1.0069	0.9986	0.9998	0.7371	0.4588	0.7347	0.5994
5.5	4.1449E-01	1.1619	0.6392	0.5277	0.4736	1.0649	1.0077	0.9987	0.9998	0.7433	0.4597	0.7414	0.6066
6.0	4.0470E-01	1.1720	0.6484	0.5380	0.4840	1.0700	1.0084	0.9989	0.9998	0.7490	0.4602	0.7474	0.6133
6.5	3.9564E-01	1.1822	0.6570	0.5476	0.4937	1.0750	1.0091	0.9990	0.9998	0.7541	0.4605	0.7528	0.6194
7.0	3.8730E-01	1.1924	0.6650	0.5566	0.5029	1.0799	1.0098	0.9990	0.9998	0.7589	0.4606	0.7578	0.6252
7.5	3.7963E-01	1.2026	0.6724	0.5649	0.5114	1.0847	1.0104	0.9991	0.9998	0.7633	0.4604	0.7623	0.6307
8.0	3.7257E-01	1.2128	0.6793	0.5727	0.5194	1.0894	1.0111	0.9992	0.9998	0.7673	0.4600	0.7665	0.6358
9.0	3.6008E-01	1.2331	0.6916	0.5869	0.5340	1.0987	1.0125	0.9993	0.9999	0.7745	0.4587	0.7739	0.6452
10.0	3.4943E-01	1.2535	0.7024	0.5994	0.5469	1.1078	1.0139	0.9994	0.9999	0.7807	0.4570	0.7802	0.6538
11.0	3.4028E-01	1.2739	0.7118	0.6104	0.5583	1.1167	1.0153	0.9994	0.9999	0.7861	0.4547	0.7857	0.6616
12.0	3.3238E-01	1.2944	0.7201	0.6202	0.5686	1.1253	1.0166	0.9995	0.9999	0.7908	0.4521	0.7905	0.6688
13.0	3.2552E-01	1.3148	0.7274	0.6289	0.5777	1.1339	1.0180	0.9995	0.9999	0.7949	0.4493	0.7947	0.6755
14.0	3.1952E-01	1.3354	0.7338	0.6367	0.5858	1.1422	1.0193	0.9996	0.9999	0.7985	0.4462	0.7983	0.6818
16.0	3.0961E-01	1.3766	0.7446	0.6498	0.5997	1.1585	1.0220	0.9996	0.9999	0.8045	0.4394	0.8044	0.6932
18.0	3.0188E-01	1.4181	0.7529	0.6603	0.6109	1.1744	1.0247	0.9997	0.9999	0.8093	0.4320	0.8092	0.7034
20.0	2.9579E-01	1.4600	0.7595	0.6686	0.6199	1.1898	1.0273	0.9997	0.9999	0.8130	0.4242	0.8129	0.7127
25.0	2.8564E-01	1.5666	0.7698	0.6823	0.6350	1.2270	1.0340	0.9998	1.0000	0.8190	0.4033	0.8190	0.7328
30.0	2.8031E-01	1.6763	0.7741	0.6889	0.6426	1.2625	1.0405	0.9998	1.0000	0.8219	0.3812	0.8219	0.7499
35.0	2.7810E-01	1.7896	0.7744	0.6905	0.6450	1.2967	1.0471	0.9998	1.0000	0.8227	0.3585	0.8228	0.7648
40.0	2.7807E-01	1.9070	0.7719	0.6887	0.6436	1.3299	1.0537	0.9999	1.0000	0.8222	0.3354	0.8222	0.7783
45.0	2.7966E-01	2.0288	0.7675	0.6844	0.6396	1.3622	1.0603	0.9999	1.0000	0.8207	0.3120	0.8207	0.7908
50.0	2.8252E-01	2.1554	0.7617	0.6782	0.6336	1.3936	1.0669	0.9999	1.0000	0.8185	0.2885	0.8185	0.8024

Instr. see p. 22

POSITRON Z= -90 A= 228

p	F_0L_0	L_0	λ_2	λ_3	λ_4	μ_1	μ_2	Λ_1	Λ_2	ν_{12}	$\hat{\nu}_{12}$	η_{12}	$\hat{\eta}_{12}$
0.1	5.9346E-16	1.0694	14.8798	115.4892	532.2233	1.0064	1.0002	0.9917	0.9987	7.6855	0.4391	0.6024	0.4688
0.2	2.2267E-07	1.0696	4.0436	9.9646	17.5909	1.0067	1.0002	0.9918	0.9987	2.3708	0.4386	0.6035	0.4698
0.3	1.1952E-04	1.0700	2.0388	3.0317	3.7667	1.0073	1.0003	0.9920	0.9988	1.3883	0.4380	0.6052	0.4716
0.4	2.3439E-03	1.0705	1.3394	1.5380	1.6338	1.0080	1.0004	0.9923	0.9988	1.0464	0.4372	0.6076	0.4739
0.5	1.2546E-02	1.0712	1.0180	1.0123	0.9930	1.0088	1.0005	0.9926	0.9988	0.8900	0.4364	0.6104	0.4766
0.6	3.5621E-02	1.0720	0.8457	0.7722	0.7246	1.0098	1.0006	0.9929	0.9989	0.8069	0.4356	0.6137	0.4798
0.7	7.1130E-02	1.0729	0.7440	0.6442	0.5887	1.0109	1.0007	0.9932	0.9989	0.7585	0.4350	0.6172	0.4833
0.8	1.1483E-01	1.0739	0.6800	0.5690	0.5115	1.0120	1.0008	0.9936	0.9990	0.7287	0.4346	0.6210	0.4869
0.9	1.6177E-01	1.0751	0.6380	0.5219	0.4641	1.0132	1.0010	0.9939	0.9990	0.7097	0.4344	0.6249	0.4906
1.0	2.0799E-01	1.0764	0.6095	0.4911	0.4337	1.0145	1.0011	0.9942	0.9991	0.6973	0.4343	0.6289	0.4944
1.2	2.8952E-01	1.0792	0.5764	0.4564	0.4000	1.0170	1.0014	0.9948	0.9992	0.6840	0.4348	0.6368	0.5019
1.4	3.5179E-01	1.0824	0.5606	0.4405	0.3849	1.0195	1.0017	0.9953	0.9992	0.6790	0.4357	0.6446	0.5092
1.6	3.9614E-01	1.0857	0.5538	0.4340	0.3789	1.0221	1.0020	0.9958	0.9993	0.6782	0.4370	0.6520	0.5161
1.8	4.2634E-01	1.0893	0.5521	0.4326	0.3778	1.0247	1.0024	0.9961	0.9994	0.6796	0.4384	0.6590	0.5227
2.0	4.4609E-01	1.0930	0.5532	0.4341	0.3794	1.0272	1.0027	0.9965	0.9994	0.6822	0.4400	0.6656	0.5289
2.2	4.5836E-01	1.0967	0.5560	0.4373	0.3827	1.0297	1.0030	0.9968	0.9995	0.6854	0.4415	0.6718	0.5347
2.4	4.6533E-01	1.1006	0.5599	0.4416	0.3870	1.0322	1.0033	0.9970	0.9995	0.6890	0.4431	0.6776	0.5402
2.6	4.6858E-01	1.1045	0.5645	0.4464	0.3919	1.0346	1.0036	0.9972	0.9995	0.6928	0.4445	0.6831	0.5454
2.8	4.6922E-01	1.1085	0.5693	0.4517	0.3971	1.0370	1.0039	0.9974	0.9996	0.6965	0.4459	0.6882	0.5504
3.0	4.6803E-01	1.1126	0.5744	0.4571	0.4025	1.0394	1.0042	0.9976	0.9996	0.7003	0.4472	0.6931	0.5551
3.2	4.6559E-01	1.1166	0.5795	0.4626	0.4080	1.0418	1.0045	0.9977	0.9996	0.7040	0.4485	0.6977	0.5596
3.4	4.6228E-01	1.1207	0.5847	0.4681	0.4135	1.0441	1.0047	0.9979	0.9996	0.7076	0.4496	0.7020	0.5638
3.6	4.5840E-01	1.1248	0.5898	0.4736	0.4190	1.0464	1.0050	0.9980	0.9997	0.7110	0.4506	0.7061	0.5679
3.8	4.5414E-01	1.1290	0.5948	0.4790	0.4244	1.0487	1.0053	0.9981	0.9997	0.7144	0.4516	0.7100	0.5718
4.0	4.4966E-01	1.1331	0.5997	0.4843	0.4297	1.0509	1.0056	0.9982	0.9997	0.7177	0.4525	0.7137	0.5755
4.5	4.3807E-01	1.1435	0.6115	0.4970	0.4424	1.0565	1.0063	0.9984	0.9997	0.7253	0.4543	0.7222	0.5842
5.0	4.2665E-01	1.1539	0.6225	0.5090	0.4544	1.0619	1.0071	0.9986	0.9997	0.7322	0.4557	0.7298	0.5922
5.5	4.1580E-01	1.1644	0.6327	0.5202	0.4657	1.0672	1.0078	0.9987	0.9998	0.7386	0.4567	0.7366	0.5995
6.0	4.0569E-01	1.1748	0.6421	0.5306	0.4762	1.0724	1.0085	0.9988	0.9998	0.7443	0.4574	0.7427	0.6062
6.5	3.9635E-01	1.1853	0.6508	0.5403	0.4861	1.0776	1.0092	0.9989	0.9998	0.7496	0.4577	0.7483	0.6125
7.0	3.8775E-01	1.1958	0.6590	0.5494	0.4953	1.0826	1.0099	0.9990	0.9998	0.7545	0.4578	0.7533	0.6184
7.5	3.7985E-01	1.2062	0.6665	0.5579	0.5040	1.0876	1.0106	0.9991	0.9998	0.7589	0.4577	0.7580	0.6239
8.0	3.7259E-01	1.2167	0.6735	0.5658	0.5121	1.0924	1.0113	0.9992	0.9998	0.7631	0.4574	0.7622	0.6291
9.0	3.5974E-01	1.2377	0.6861	0.5802	0.5268	1.1020	1.0127	0.9993	0.9999	0.7704	0.4563	0.7698	0.6387
10.0	3.4880E-01	1.2586	0.6971	0.5929	0.5399	1.1113	1.0141	0.9993	0.9999	0.7768	0.4546	0.7763	0.6474
11.0	3.3942E-01	1.2796	0.7068	0.6042	0.5516	1.1204	1.0155	0.9994	0.9999	0.7823	0.4525	0.7819	0.6554
12.0	3.3131E-01	1.3006	0.7152	0.6141	0.5620	1.1294	1.0169	0.9995	0.9999	0.7871	0.4501	0.7868	0.6627
13.0	3.2427E-01	1.3217	0.7227	0.6230	0.5712	1.1381	1.0183	0.9995	0.9999	0.7913	0.4473	0.7911	0.6695
14.0	3.1812E-01	1.3428	0.7293	0.6309	0.5796	1.1467	1.0197	0.9996	0.9999	0.7950	0.4443	0.7949	0.6759
16.0	3.0796E-01	1.3852	0.7403	0.6443	0.5937	1.1634	1.0224	0.9996	0.9999	0.8012	0.4377	0.8011	0.6875
18.0	3.0002E-01	1.4279	0.7489	0.6551	0.6051	1.1797	1.0251	0.9997	0.9999	0.8061	0.4304	0.8060	0.6979
20.0	2.9378E-01	1.4710	0.7557	0.6636	0.6143	1.1955	1.0278	0.9997	0.9999	0.8099	0.4228	0.8099	0.7074
25.0	2.8332E-01	1.5807	0.7664	0.6778	0.6299	1.2336	1.0345	0.9998	0.9999	0.8162	0.4022	0.8162	0.7278
30.0	2.7778E-01	1.6937	0.7710	0.6847	0.6379	1.2699	1.0412	0.9998	1.0000	0.8193	0.3805	0.8193	0.7452
35.0	2.7541E-01	1.8106	0.7715	0.6866	0.6405	1.3049	1.0479	0.9998	1.0000	0.8202	0.3581	0.8202	0.7604
40.0	2.7525E-01	1.9317	0.7692	0.6850	0.6394	1.3388	1.0546	0.9999	1.0000	0.8198	0.3352	0.8198	0.7742
45.0	2.7673E-01	2.0574	0.7649	0.6809	0.6356	1.3718	1.0612	0.9999	1.0000	0.8183	0.3121	0.8184	0.7869
50.0	2.7950E-01	2.1881	0.7592	0.6748	0.6297	1.4040	1.0680	0.9999	1.0000	0.8162	0.2888	0.8162	0.7987

POSITRON Z= −91 A= 229

p	F_0L_0	L_0	λ_2	λ_3	λ_4	μ_1	μ_2	Λ_1	Λ_2	ν_{12}	$\hat{\nu}_{12}$	η_{12}	$\hat{\eta}_{12}$
0.1	4.0138E-16	1.0689	14.8257	116.3809	543.6119	1.0068	1.0002	0.9915	0.9987	7.7708	0.4337	0.5954	0.4602
0.2	1.8895E-07	1.0692	4.0239	9.9892	17.7645	1.0072	1.0002	0.9916	0.9987	2.3870	0.4333	0.5965	0.4613
0.3	1.0921E-04	1.0695	2.0255	3.0223	3.7668	1.0077	1.0003	0.9918	0.9987	1.3917	0.4327	0.5982	0.4630
0.4	2.2197E-03	1.0701	1.3283	1.5261	1.6221	1.0084	1.0004	0.9921	0.9988	1.0452	0.4320	0.6005	0.4653
0.5	1.2128E-02	1.0708	1.0079	1.0007	0.9809	1.0093	1.0005	0.9924	0.9988	0.8868	0.4313	0.6034	0.4681
0.6	3.4885E-02	1.0716	0.8362	0.7613	0.7132	1.0103	1.0006	0.9927	0.9989	0.8026	0.4306	0.6067	0.4713
0.7	7.0267E-02	1.0725	0.7348	0.6338	0.5780	1.0114	1.0007	0.9931	0.9989	0.7536	0.4301	0.6103	0.4747
0.8	1.1413E-01	1.0736	0.6710	0.5590	0.5013	1.0126	1.0009	0.9934	0.9990	0.7234	0.4298	0.6141	0.4784
0.9	1.6147E-01	1.0749	0.6291	0.5121	0.4543	1.0138	1.0010	0.9938	0.9990	0.7040	0.4296	0.6181	0.4821
1.0	2.0826E-01	1.0762	0.6008	0.4815	0.4241	1.0151	1.0012	0.9941	0.9991	0.6915	0.4296	0.6221	0.4859
1.2	2.9105E-01	1.0791	0.5678	0.4472	0.3908	1.0177	1.0015	0.9947	0.9992	0.6780	0.4302	0.6302	0.4935
1.4	3.5444E-01	1.0823	0.5521	0.4315	0.3759	1.0204	1.0018	0.9952	0.9992	0.6730	0.4313	0.6380	0.5008
1.6	3.9961E-01	1.0858	0.5455	0.4251	0.3700	1.0230	1.0021	0.9957	0.9993	0.6721	0.4327	0.6455	0.5078
1.8	4.3032E-01	1.0894	0.5438	0.4237	0.3689	1.0257	1.0024	0.9961	0.9994	0.6736	0.4342	0.6526	0.5144
2.0	4.5036E-01	1.0932	0.5450	0.4253	0.3706	1.0283	1.0027	0.9964	0.9994	0.6762	0.4359	0.6593	0.5207
2.2	4.6274E-01	1.0971	0.5479	0.4286	0.3740	1.0308	1.0030	0.9967	0.9995	0.6795	0.4375	0.6656	0.5266
2.4	4.6971E-01	1.1011	0.5519	0.4329	0.3783	1.0334	1.0033	0.9969	0.9995	0.6832	0.4391	0.6716	0.5322
2.6	4.7287E-01	1.1052	0.5565	0.4378	0.3832	1.0359	1.0036	0.9972	0.9995	0.6870	0.4407	0.6771	0.5374
2.8	4.7337E-01	1.1093	0.5615	0.4431	0.3885	1.0384	1.0039	0.9974	0.9996	0.6908	0.4422	0.6824	0.5424
3.0	4.7202E-01	1.1134	0.5667	0.4486	0.3940	1.0409	1.0042	0.9975	0.9996	0.6946	0.4436	0.6873	0.5472
3.2	4.6939E-01	1.1176	0.5719	0.4542	0.3995	1.0433	1.0045	0.9977	0.9996	0.6984	0.4448	0.6920	0.5517
3.4	4.6588E-01	1.1218	0.5771	0.4598	0.4050	1.0457	1.0048	0.9978	0.9996	0.7020	0.4460	0.6964	0.5560
3.6	4.6179E-01	1.1260	0.5823	0.4653	0.4105	1.0481	1.0051	0.9980	0.9997	0.7056	0.4471	0.7006	0.5602
3.8	4.5733E-01	1.1303	0.5874	0.4707	0.4160	1.0504	1.0054	0.9981	0.9997	0.7090	0.4482	0.7046	0.5641
4.0	4.5265E-01	1.1345	0.5924	0.4761	0.4213	1.0528	1.0057	0.9982	0.9997	0.7123	0.4491	0.7084	0.5679
4.5	4.4059E-01	1.1452	0.6044	0.4890	0.4342	1.0585	1.0065	0.9984	0.9997	0.7201	0.4510	0.7170	0.5767
5.0	4.2873E-01	1.1559	0.6156	0.5011	0.4463	1.0641	1.0072	0.9986	0.9997	0.7272	0.4526	0.7247	0.5848
5.5	4.1750E-01	1.1667	0.6259	0.5124	0.4576	1.0695	1.0079	0.9987	0.9998	0.7337	0.4537	0.7317	0.5922
6.0	4.0705E-01	1.1774	0.6355	0.5230	0.4683	1.0749	1.0086	0.9988	0.9998	0.7396	0.4544	0.7379	0.5990
6.5	3.9740E-01	1.1882	0.6445	0.5328	0.4782	1.0802	1.0094	0.9989	0.9998	0.7450	0.4549	0.7436	0.6054
7.0	3.8853E-01	1.1989	0.6527	0.5420	0.4876	1.0854	1.0101	0.9990	0.9998	0.7499	0.4551	0.7488	0.6114
7.5	3.8039E-01	1.2097	0.6604	0.5506	0.4963	1.0905	1.0108	0.9991	0.9998	0.7545	0.4551	0.7535	0.6170
8.0	3.7290E-01	1.2204	0.6676	0.5587	0.5045	1.0955	1.0115	0.9991	0.9998	0.7587	0.4549	0.7579	0.6223
9.0	3.5968E-01	1.2420	0.6805	0.5733	0.5195	1.1053	1.0129	0.9992	0.9999	0.7662	0.4539	0.7656	0.6320
10.0	3.4843E-01	1.2635	0.6917	0.5863	0.5328	1.1149	1.0144	0.9993	0.9999	0.7727	0.4524	0.7723	0.6409
11.0	3.3879E-01	1.2850	0.7015	0.5977	0.5446	1.1242	1.0158	0.9994	0.9999	0.7784	0.4504	0.7780	0.6490
12.0	3.3046E-01	1.3066	0.7102	0.6079	0.5552	1.1334	1.0172	0.9995	0.9999	0.7833	0.4480	0.7830	0.6564
13.0	3.2323E-01	1.3282	0.7179	0.6169	0.5646	1.1424	1.0186	0.9995	0.9999	0.7877	0.4454	0.7874	0.6634
14.0	3.1691E-01	1.3499	0.7246	0.6250	0.5731	1.1512	1.0200	0.9995	0.9999	0.7915	0.4425	0.7913	0.6698
16.0	3.0648E-01	1.3935	0.7359	0.6387	0.5876	1.1683	1.0227	0.9996	0.9999	0.7979	0.4360	0.7977	0.6817
18.0	2.9833E-01	1.4374	0.7448	0.6497	0.5993	1.1850	1.0255	0.9997	0.9999	0.8029	0.4290	0.8028	0.6922
20.0	2.9190E-01	1.4817	0.7518	0.6585	0.6087	1.2012	1.0282	0.9997	0.9999	0.8068	0.4215	0.8068	0.7018
25.0	2.8112E-01	1.5944	0.7630	0.6732	0.6248	1.2401	1.0351	0.9998	0.9999	0.8134	0.4013	0.8134	0.7227
30.0	2.7534E-01	1.7107	0.7679	0.6805	0.6331	1.2773	1.0418	0.9998	1.0000	0.8166	0.3800	0.8166	0.7404
35.0	2.7279E-01	1.8309	0.7686	0.6827	0.6361	1.3131	1.0486	0.9998	1.0000	0.8177	0.3578	0.8177	0.7559
40.0	2.7247E-01	1.9556	0.7665	0.6814	0.6352	1.3477	1.0554	0.9999	1.0000	0.8173	0.3353	0.8174	0.7699
45.0	2.7382E-01	2.0851	0.7624	0.6774	0.6316	1.3814	1.0621	0.9999	1.0000	0.8160	0.3124	0.8160	0.7828
50.0	2.7648E-01	2.2199	0.7568	0.6715	0.6259	1.4143	1.0690	0.9999	1.0000	0.8139	0.2895	0.8140	0.7948

POSITRON Z= -92 A= 231

p	F_0L_0	L_0	λ_2	λ_3	λ_4	μ_1	μ_2	Λ_1	Λ_2	ν_{12}	$\hat{\nu}_{12}$	η_{12}	$\hat{\eta}_{12}$
0.1	2.7144E-16	1.0684	14.7682	117.2307	554.9382	1.0072	1.0002	0.9913	0.9987	7.8560	0.4284	0.5884	0.4517
0.2	1.6032E-07	1.0686	4.0035	10.0110	17.9344	1.0076	1.0002	0.9914	0.9987	2.4031	0.4281	0.5895	0.4528
0.3	9.9777E-05	1.0690	2.0120	3.0124	3.7663	1.0082	1.0003	0.9916	0.9987	1.3950	0.4275	0.5913	0.4545
0.4	2.1019E-03	1.0696	1.3172	1.5141	1.6105	1.0089	1.0004	0.9919	0.9988	1.0441	0.4269	0.5936	0.4568
0.5	1.1723E-02	1.0703	0.9979	0.9893	0.9690	1.0098	1.0005	0.9922	0.9988	0.8837	0.4263	0.5965	0.4596
0.6	3.4159E-02	1.0711	0.8267	0.7505	0.7021	1.0109	1.0006	0.9926	0.9989	0.7984	0.4257	0.5998	0.4628
0.7	6.9406E-02	1.0721	0.7257	0.6236	0.5675	1.0120	1.0007	0.9929	0.9989	0.7487	0.4252	0.6035	0.4663
0.8	1.1341E-01	1.0733	0.6621	0.5492	0.4913	1.0132	1.0009	0.9933	0.9990	0.7181	0.4250	0.6073	0.4699
0.9	1.6115E-01	1.0745	0.6204	0.5026	0.4447	1.0145	1.0010	0.9936	0.9990	0.6985	0.4249	0.6113	0.4737
1.0	2.0850E-01	1.0759	0.5921	0.4722	0.4148	1.0158	1.0012	0.9939	0.9991	0.6857	0.4250	0.6154	0.4775
1.2	2.9256E-01	1.0789	0.5593	0.4381	0.3818	1.0185	1.0015	0.9946	0.9991	0.6721	0.4257	0.6235	0.4851
1.4	3.5706E-01	1.0822	0.5438	0.4226	0.3671	1.0213	1.0018	0.9951	0.9992	0.6670	0.4269	0.6315	0.4925
1.6	4.0303E-01	1.0858	0.5372	0.4164	0.3613	1.0240	1.0021	0.9956	0.9993	0.6661	0.4284	0.6391	0.4996
1.8	4.3426E-01	1.0896	0.5357	0.4151	0.3603	1.0267	1.0024	0.9960	0.9994	0.6676	0.4301	0.6463	0.5062
2.0	4.5459E-01	1.0935	0.5370	0.4168	0.3621	1.0294	1.0028	0.9963	0.9994	0.6703	0.4318	0.6531	0.5126
2.2	4.6708E-01	1.0975	0.5400	0.4201	0.3655	1.0321	1.0031	0.9966	0.9994	0.6737	0.4336	0.6596	0.5185
2.4	4.7404E-01	1.1016	0.5441	0.4245	0.3699	1.0347	1.0034	0.9969	0.9995	0.6774	0.4353	0.6656	0.5242
2.6	4.7711E-01	1.1058	0.5488	0.4295	0.3748	1.0373	1.0037	0.9971	0.9995	0.6813	0.4369	0.6712	0.5295
2.8	4.7747E-01	1.1100	0.5539	0.4348	0.3802	1.0399	1.0040	0.9973	0.9996	0.6852	0.4385	0.6766	0.5346
3.0	4.7594E-01	1.1143	0.5591	0.4404	0.3856	1.0425	1.0043	0.9975	0.9996	0.6891	0.4399	0.6816	0.5394
3.2	4.7311E-01	1.1186	0.5644	0.4460	0.3912	1.0450	1.0046	0.9976	0.9996	0.6929	0.4413	0.6864	0.5440
3.4	4.6940E-01	1.1229	0.5698	0.4516	0.3968	1.0475	1.0049	0.9978	0.9996	0.6966	0.4425	0.6909	0.5483
3.6	4.6510E-01	1.1273	0.5750	0.4572	0.4023	1.0499	1.0052	0.9979	0.9996	0.7002	0.4437	0.6952	0.5525
3.8	4.6044E-01	1.1316	0.5802	0.4627	0.4078	1.0523	1.0055	0.9980	0.9997	0.7037	0.4448	0.6992	0.5565
4.0	4.5555E-01	1.1360	0.5853	0.4681	0.4132	1.0547	1.0058	0.9981	0.9997	0.7071	0.4457	0.7030	0.5603
4.5	4.4301E-01	1.1470	0.5975	0.4812	0.4261	1.0606	1.0066	0.9984	0.9997	0.7150	0.4478	0.7119	0.5693
5.0	4.3072E-01	1.1580	0.6088	0.4934	0.4383	1.0664	1.0073	0.9985	0.9997	0.7223	0.4494	0.7198	0.5775
5.5	4.1910E-01	1.1691	0.6194	0.5049	0.4498	1.0721	1.0081	0.9987	0.9998	0.7289	0.4507	0.7268	0.5850
6.0	4.0830E-01	1.1801	0.6292	0.5156	0.4606	1.0776	1.0088	0.9988	0.9998	0.7349	0.4515	0.7332	0.5919
6.5	3.9835E-01	1.1912	0.6382	0.5255	0.4706	1.0830	1.0095	0.9989	0.9998	0.7404	0.4521	0.7390	0.5984
7.0	3.8921E-01	1.2023	0.6467	0.5349	0.4801	1.0884	1.0103	0.9990	0.9998	0.7455	0.4524	0.7443	0.6045
7.5	3.8082E-01	1.2133	0.6545	0.5436	0.4889	1.0936	1.0110	0.9991	0.9998	0.7501	0.4524	0.7491	0.6102
8.0	3.7312E-01	1.2244	0.6618	0.5518	0.4972	1.0988	1.0117	0.9991	0.9998	0.7544	0.4523	0.7536	0.6155
9.0	3.5953E-01	1.2466	0.6749	0.5666	0.5124	1.1089	1.0132	0.9992	0.9999	0.7621	0.4514	0.7615	0.6254
10.0	3.4798E-01	1.2687	0.6864	0.5798	0.5259	1.1187	1.0146	0.9993	0.9999	0.7688	0.4500	0.7683	0.6344
11.0	3.3808E-01	1.2909	0.6964	0.5914	0.5379	1.1283	1.0160	0.9994	0.9999	0.7746	0.4482	0.7742	0.6426
12.0	3.2954E-01	1.3131	0.7053	0.6017	0.5486	1.1378	1.0175	0.9995	0.9999	0.7796	0.4459	0.7793	0.6502
13.0	3.2212E-01	1.3353	0.7131	0.6110	0.5582	1.1470	1.0189	0.9995	0.9999	0.7841	0.4434	0.7838	0.6573
14.0	3.1564E-01	1.3576	0.7201	0.6192	0.5668	1.1560	1.0203	0.9995	0.9999	0.7880	0.4406	0.7878	0.6639
16.0	3.0495E-01	1.4025	0.7317	0.6332	0.5816	1.1736	1.0231	0.9996	0.9999	0.7945	0.4343	0.7944	0.6759
18.0	2.9660E-01	1.4476	0.7408	0.6445	0.5935	1.1907	1.0259	0.9997	0.9999	0.7997	0.4274	0.7996	0.6867
20.0	2.9000E-01	1.4932	0.7480	0.6535	0.6032	1.2073	1.0287	0.9997	0.9999	0.8038	0.4201	0.8037	0.6964
25.0	2.7891E-01	1.6093	0.7596	0.6687	0.6197	1.2472	1.0357	0.9998	0.9999	0.8105	0.4003	0.8105	0.7177
30.0	2.7292E-01	1.7291	0.7648	0.6763	0.6284	1.2853	1.0425	0.9998	1.0000	0.8139	0.3792	0.8140	0.7356
35.0	2.7020E-01	1.8530	0.7658	0.6788	0.6317	1.3219	1.0494	0.9998	1.0000	0.8152	0.3574	0.8152	0.7514
40.0	2.6976E-01	1.9817	0.7638	0.6777	0.6310	1.3574	1.0563	0.9999	1.0000	0.8149	0.3351	0.8149	0.7657
45.0	2.7100E-01	2.1155	0.7598	0.6738	0.6276	1.3918	1.0631	0.9999	1.0000	0.8136	0.3125	0.8137	0.7788
50.0	2.7356E-01	2.2548	0.7542	0.6680	0.6220	1.4254	1.0701	0.9999	1.0000	0.8116	0.2897	0.8117	0.7910

POSITRON Z= -93 A= 235

p	F_0L_0	L_0	λ_2	λ_3	λ_4	μ_1	μ_2	Λ_1	Λ_2	ν_{12}	$\hat{\nu}_{12}$	η_{12}	$\hat{\eta}_{12}$
0.1	1.8342E-16	1.0678	14.7130	118.0989	566.5217	1.0077	1.0002	0.9911	0.9986	7.9423	0.4233	0.5816	0.4434
0.2	1.3592E-07	1.0680	3.9839	10.0353	18.1113	1.0081	1.0002	0.9912	0.9987	2.4197	0.4230	0.5827	0.4445
0.3	9.1084E-05	1.0685	1.9990	3.0037	3.7675	1.0087	1.0003	0.9914	0.9987	1.3987	0.4225	0.5845	0.4462
0.4	1.9887E-03	1.0690	1.3065	1.5028	1.5999	1.0095	1.0004	0.9917	0.9987	1.0433	0.4219	0.5869	0.4485
0.5	1.1321E-02	1.0698	0.9883	0.9785	0.9578	1.0104	1.0005	0.9920	0.9988	0.8808	0.4213	0.5898	0.4513
0.6	3.3421E-02	1.0706	0.8177	0.7404	0.6915	1.0115	1.0006	0.9924	0.9988	0.7944	0.4209	0.5931	0.4545
0.7	6.8497E-02	1.0717	0.7170	0.6139	0.5576	1.0127	1.0008	0.9927	0.9989	0.7441	0.4205	0.5968	0.4580
0.8	1.1260E-01	1.0729	0.6536	0.5399	0.4818	1.0140	1.0009	0.9931	0.9989	0.7130	0.4203	0.6007	0.4617
0.9	1.6069E-01	1.0742	0.6120	0.4936	0.4356	1.0153	1.0011	0.9934	0.9990	0.6932	0.4203	0.6048	0.4655
1.0	2.0856E-01	1.0756	0.5839	0.4634	0.4059	1.0166	1.0012	0.9938	0.9990	0.6802	0.4205	0.6089	0.4693
1.2	2.9381E-01	1.0787	0.5512	0.4296	0.3732	1.0194	1.0015	0.9944	0.9991	0.6664	0.4214	0.6171	0.4770
1.4	3.5938E-01	1.0821	0.5358	0.4142	0.3587	1.0223	1.0018	0.9950	0.9992	0.6612	0.4227	0.6252	0.4844
1.6	4.0613E-01	1.0858	0.5294	0.4081	0.3530	1.0251	1.0022	0.9955	0.9993	0.6604	0.4243	0.6329	0.4915
1.8	4.3786E-01	1.0897	0.5279	0.4069	0.3522	1.0280	1.0025	0.9959	0.9993	0.6618	0.4261	0.6403	0.4983
2.0	4.5845E-01	1.0938	0.5293	0.4086	0.3540	1.0308	1.0028	0.9962	0.9994	0.6646	0.4279	0.6472	0.5046
2.2	4.7104E-01	1.0979	0.5325	0.4120	0.3575	1.0335	1.0031	0.9965	0.9994	0.6680	0.4297	0.6537	0.5107
2.4	4.7798E-01	1.1022	0.5366	0.4165	0.3619	1.0363	1.0034	0.9966	0.9995	0.6718	0.4315	0.6598	0.5164
2.6	4.8096E-01	1.1065	0.5414	0.4215	0.3669	1.0390	1.0038	0.9970	0.9995	0.6757	0.4332	0.6656	0.5218
2.8	4.8117E-01	1.1109	0.5466	0.4270	0.3722	1.0416	1.0041	0.9973	0.9995	0.6797	0.4348	0.6710	0.5269
3.0	4.7946E-01	1.1153	0.5519	0.4326	0.3778	1.0442	1.0044	0.9974	0.9996	0.6837	0.4363	0.6761	0.5318
3.2	4.7644E-01	1.1197	0.5573	0.4382	0.3834	1.0469	1.0047	0.9976	0.9996	0.6876	0.4378	0.6810	0.5364
3.4	4.7252E-01	1.1242	0.5627	0.4439	0.3890	1.0494	1.0050	0.9977	0.9996	0.6914	0.4391	0.6856	0.5408
3.6	4.6802E-01	1.1287	0.5681	0.4496	0.3946	1.0520	1.0053	0.9979	0.9996	0.6951	0.4403	0.6899	0.5451
3.8	4.6315E-01	1.1332	0.5734	0.4551	0.4001	1.0545	1.0056	0.9980	0.9997	0.6986	0.4414	0.6940	0.5491
4.0	4.5806E-01	1.1377	0.5785	0.4606	0.4055	1.0570	1.0059	0.9981	0.9997	0.7021	0.4424	0.6979	0.5530
4.5	4.4505E-01	1.1491	0.5909	0.4738	0.4185	1.0631	1.0067	0.9983	0.9997	0.7101	0.4447	0.7069	0.5621
5.0	4.3234E-01	1.1605	0.6024	0.4861	0.4309	1.0690	1.0075	0.9985	0.9997	0.7175	0.4464	0.7150	0.5704
5.5	4.2034E-01	1.1719	0.6132	0.4977	0.4424	1.0749	1.0082	0.9986	0.9998	0.7242	0.4477	0.7222	0.5780
6.0	4.0921E-01	1.1833	0.6231	0.5085	0.4533	1.0806	1.0090	0.9988	0.9998	0.7304	0.4486	0.7287	0.5850
6.5	3.9897E-01	1.1947	0.6323	0.5187	0.4634	1.0862	1.0097	0.9989	0.9998	0.7360	0.4493	0.7346	0.5916
7.0	3.8957E-01	1.2061	0.6409	0.5281	0.4730	1.0917	1.0105	0.9990	0.9998	0.7412	0.4496	0.7400	0.5978
7.5	3.8095E-01	1.2175	0.6489	0.5370	0.4819	1.0971	1.0112	0.9990	0.9998	0.7459	0.4498	0.7449	0.6035
8.0	3.7304E-01	1.2289	0.6563	0.5453	0.4903	1.1024	1.0120	0.9991	0.9998	0.7503	0.4497	0.7495	0.6090
9.0	3.5910E-01	1.2518	0.6697	0.5603	0.5057	1.1129	1.0134	0.9992	0.9999	0.7582	0.4490	0.7575	0.6191
10.0	3.4726E-01	1.2746	0.6814	0.5736	0.5193	1.1230	1.0149	0.9993	0.9999	0.7650	0.4477	0.7645	0.6282
11.0	3.3713E-01	1.2975	0.6916	0.5855	0.5315	1.1329	1.0164	0.9994	0.9999	0.7709	0.4459	0.7705	0.6366
12.0	3.2839E-01	1.3204	0.7007	0.5960	0.5424	1.1426	1.0178	0.9994	0.9999	0.7761	0.4437	0.7758	0.6443
13.0	3.2080E-01	1.3434	0.7087	0.6054	0.5521	1.1521	1.0193	0.9995	0.9999	0.7806	0.4413	0.7804	0.6515
14.0	3.1418E-01	1.3664	0.7158	0.6138	0.5609	1.1614	1.0207	0.9995	0.9999	0.7846	0.4385	0.7844	0.6582
16.0	3.0326E-01	1.4126	0.7276	0.6281	0.5759	1.1796	1.0236	0.9996	0.9999	0.7913	0.4324	0.7912	0.6704
18.0	2.9472E-01	1.4593	0.7370	0.6395	0.5880	1.1971	1.0264	0.9996	0.9999	0.7966	0.4257	0.7965	0.6813
20.0	2.8798E-01	1.5064	0.7443	0.6488	0.5979	1.2142	1.0293	0.9997	0.9999	0.8008	0.4184	0.8008	0.6913
25.0	2.7664E-01	1.6263	0.7563	0.6643	0.6148	1.2552	1.0364	0.9998	0.9999	0.8078	0.3989	0.8078	0.7129
30.0	2.7047E-01	1.7502	0.7617	0.6723	0.6238	1.2942	1.0434	0.9998	1.0000	0.8113	0.3780	0.8113	0.7312
35.0	2.6763E-01	1.8785	0.7629	0.6749	0.6273	1.3318	1.0503	0.9998	1.0000	0.8126	0.3564	0.8127	0.7473
40.0	2.6710E-01	2.0118	0.7610	0.6739	0.6268	1.3682	1.0573	0.9999	1.0000	0.8125	0.3343	0.8125	0.7618
45.0	2.6827E-01	2.1505	0.7570	0.6702	0.6234	1.4035	1.0643	0.9999	1.0000	0.8112	0.3119	0.8113	0.7751
50.0	2.7077E-01	2.2951	0.7516	0.6644	0.6179	1.4379	1.0714	0.9999	1.0000	0.8093	0.2893	0.8093	0.7876

Instr. see p. 22

POSITRON Z= -94 A= 238

p	F_0L_0	L_0	λ_2	λ_3	λ_4	μ_1	μ_2	Λ_1	Λ_2	ν_{12}	$\hat{\nu}_{12}$	η_{12}	$\hat{\eta}_{12}$
0.1	1.2410E-16	1.0670	14.6436	118.8105	577.4245	1.0082	1.0002	0.9908	0.9986	8.0257	0.4181	0.5747	0.4350
0.2	1.1538E-07	1.0673	3.9607	10.0473	18.2652	1.0086	1.0002	0.9910	0.9986	2.4355	0.4178	0.5758	0.4361
0.3	8.3255E-05	1.0677	1.9843	2.9917	3.7643	1.0092	1.0003	0.9912	0.9987	1.4020	0.4174	0.5776	0.4378
0.4	1.8839E-03	1.0683	1.2948	1.4902	1.5875	1.0100	1.0004	0.9915	0.9987	1.0422	0.4169	0.5800	0.4401
0.5	1.0948E-02	1.0690	0.9780	0.9668	0.9458	1.0110	1.0005	0.9918	0.9988	0.8776	0.4164	0.5829	0.4430
0.6	3.2739E-02	1.0700	0.8081	0.7296	0.6804	1.0122	1.0006	0.9922	0.9988	0.7902	0.4159	0.5863	0.4462
0.7	6.7684E-02	1.0710	0.7078	0.6039	0.5473	1.0134	1.0008	0.9925	0.9989	0.7392	0.4157	0.5900	0.4497
0.8	1.1194E-01	1.0722	0.6447	0.5303	0.4721	1.0147	1.0009	0.9929	0.9989	0.7078	0.4156	0.5940	0.4534
0.9	1.6043E-01	1.0736	0.6033	0.4843	0.4262	1.0161	1.0011	0.9933	0.9990	0.6877	0.4157	0.5981	0.4572
1.0	2.0887E-01	1.0751	0.5754	0.4543	0.3969	1.0175	1.0012	0.9936	0.9990	0.6746	0.4159	0.6022	0.4610
1.2	2.9543E-01	1.0783	0.5429	0.4208	0.3645	1.0204	1.0016	0.9943	0.9991	0.6605	0.4169	0.6106	0.4687
1.4	3.6215E-01	1.0819	0.5277	0.4056	0.3502	1.0233	1.0019	0.9949	0.9992	0.6553	0.4184	0.6188	0.4762
1.6	4.0973E-01	1.0857	0.5213	0.3996	0.3447	1.0263	1.0022	0.9954	0.9993	0.6544	0.4201	0.6266	0.4834
1.8	4.4199E-01	1.0897	0.5199	0.3986	0.3439	1.0292	1.0025	0.9958	0.9993	0.6559	0.4220	0.6340	0.4902
2.0	4.6286E-01	1.0939	0.5215	0.4003	0.3458	1.0321	1.0029	0.9961	0.9994	0.6587	0.4239	0.6410	0.4966
2.2	4.7556E-01	1.0982	0.5247	0.4038	0.3493	1.0350	1.0032	0.9965	0.9994	0.6622	0.4258	0.6477	0.5027
2.4	4.8248E-01	1.1026	0.5290	0.4083	0.3538	1.0378	1.0035	0.9967	0.9995	0.6661	0.4277	0.6539	0.5085
2.6	4.8536E-01	1.1070	0.5338	0.4135	0.3588	1.0406	1.0038	0.9970	0.9995	0.6701	0.4295	0.6597	0.5139
2.8	4.8541E-01	1.1115	0.5391	0.4189	0.3642	1.0434	1.0042	0.9972	0.9995	0.6741	0.4311	0.6653	0.5191
3.0	4.8352E-01	1.1161	0.5445	0.4246	0.3698	1.0461	1.0045	0.9974	0.9996	0.6782	0.4327	0.6705	0.5241
3.2	4.8029E-01	1.1207	0.5500	0.4303	0.3754	1.0488	1.0048	0.9975	0.9996	0.6821	0.4342	0.6754	0.5287
3.4	4.7616E-01	1.1253	0.5555	0.4361	0.3810	1.0514	1.0051	0.9977	0.9996	0.6860	0.4356	0.6801	0.5332
3.6	4.7143E-01	1.1299	0.5609	0.4417	0.3867	1.0541	1.0054	0.9978	0.9996	0.6898	0.4369	0.6845	0.5375
3.8	4.6634E-01	1.1346	0.5663	0.4474	0.3922	1.0567	1.0057	0.9979	0.9997	0.6934	0.4380	0.6887	0.5416
4.0	4.6104E-01	1.1392	0.5715	0.4529	0.3976	1.0592	1.0060	0.9981	0.9997	0.6969	0.4391	0.6927	0.5455
4.5	4.4753E-01	1.1509	0.5841	0.4662	0.4108	1.0655	1.0068	0.9983	0.9997	0.7051	0.4414	0.7019	0.5547
5.0	4.3436E-01	1.1627	0.5958	0.4787	0.4232	1.0717	1.0076	0.9985	0.9997	0.7126	0.4433	0.7100	0.5631
5.5	4.2196E-01	1.1744	0.6067	0.4904	0.4349	1.0777	1.0084	0.9986	0.9998	0.7195	0.4447	0.7174	0.5709
6.0	4.1048E-01	1.1862	0.6169	0.5014	0.4458	1.0836	1.0091	0.9987	0.9998	0.7258	0.4457	0.7240	0.5780
6.5	3.9992E-01	1.1979	0.6262	0.5116	0.4561	1.0894	1.0099	0.9989	0.9998	0.7315	0.4465	0.7300	0.5847
7.0	3.9024E-01	1.2097	0.6350	0.5212	0.4657	1.0951	1.0107	0.9989	0.9998	0.7368	0.4469	0.7356	0.5909
7.5	3.8138E-01	1.2215	0.6431	0.5301	0.4748	1.1007	1.0114	0.9990	0.9998	0.7416	0.4471	0.7406	0.5968
8.0	3.7325E-01	1.2332	0.6506	0.5385	0.4833	1.1062	1.0122	0.9991	0.9998	0.7461	0.4471	0.7452	0.6023
9.0	3.5894E-01	1.2568	0.6643	0.5538	0.4988	1.1169	1.0137	0.9992	0.9999	0.7541	0.4465	0.7535	0.6126
10.0	3.4679E-01	1.2803	0.6762	0.5673	0.5126	1.1274	1.0152	0.9993	0.9999	0.7611	0.4453	0.7606	0.6218
11.0	3.3640E-01	1.3039	0.6867	0.5794	0.5249	1.1376	1.0167	0.9994	0.9999	0.7671	0.4437	0.7667	0.6303
12.0	3.2745E-01	1.3275	0.6959	0.5900	0.5360	1.1476	1.0181	0.9994	0.9999	0.7724	0.4416	0.7721	0.6382
13.0	3.1969E-01	1.3511	0.7041	0.5996	0.5459	1.1573	1.0196	0.9995	0.9999	0.7771	0.4392	0.7768	0.6455
14.0	3.1291E-01	1.3748	0.7113	0.6082	0.5548	1.1669	1.0211	0.9995	0.9999	0.7812	0.4366	0.7810	0.6523
16.0	3.0173E-01	1.4225	0.7234	0.6227	0.5700	1.1855	1.0240	0.9996	0.9999	0.7880	0.4306	0.7879	0.6647
18.0	2.9300E-01	1.4706	0.7330	0.6344	0.5824	1.2036	1.0269	0.9996	0.9999	0.7935	0.4240	0.7934	0.6758
20.0	2.8610E-01	1.5191	0.7406	0.6439	0.5925	1.2211	1.0298	0.9997	0.9999	0.7978	0.4169	0.7977	0.6860
25.0	2.7447E-01	1.6429	0.7529	0.6599	0.6098	1.2632	1.0370	0.9998	0.9999	0.8050	0.3976	0.8050	0.7079
30.0	2.6811E-01	1.7708	0.7586	0.6681	0.6192	1.3032	1.0441	0.9998	1.0000	0.8087	0.3770	0.8087	0.7266
35.0	2.6513E-01	1.9034	0.7600	0.6710	0.6229	1.3417	1.0512	0.9998	1.0000	0.8101	0.3557	0.8101	0.7429
40.0	2.6448E-01	2.0413	0.7583	0.6702	0.6226	1.3789	1.0583	0.9999	1.0000	0.8100	0.3338	0.8100	0.7577
45.0	2.6556E-01	2.1849	0.7544	0.6666	0.6194	1.4151	1.0654	0.9999	1.0000	0.8088	0.3116	0.8089	0.7713
50.0	2.6799E-01	2.3347	0.7490	0.6609	0.6139	1.4504	1.0726	0.9999	1.0000	0.8069	0.2892	0.8069	0.7840

POSITRON Z= -95 A= 239

p	F_0L_0	L_0	λ_2	λ_3	λ_4	μ_1	μ_3	Λ_1	Λ_2	ν_{12}	$\hat{\nu}_{12}$	η_{12}	$\hat{\eta}_{12}$
0.1	8.4139E-17	1.0659	14.5542	119.3002	587.2541	1.0088	1.0002	0.9906	0.9986	8.1048	0.4127	0.5675	0.4265
0.2	9.8142E-08	1.0662	3.9323	10.0415	18.3848	1.0092	1.0002	0.9908	0.9986	2.4500	0.4124	0.5686	0.4275
0.3	7.6256E-05	1.0666	1.9672	2.9747	3.7543	1.0098	1.0003	0.9910	0.9987	1.4045	0.4121	0.5705	0.4293
0.4	1.7884E-03	1.0672	1.2816	1.4753	1.5725	1.0106	1.0004	0.9913	0.9987	1.0406	0.4116	0.5729	0.4316
0.5	1.0608E-02	1.0680	0.9666	0.9539	0.9323	1.0117	1.0005	0.9916	0.9988	0.8741	0.4112	0.5758	0.4344
0.6	3.2137E-02	1.0690	0.7977	0.7180	0.6684	1.0128	1.0007	0.9920	0.9988	0.7856	0.4109	0.5793	0.4376
0.7	6.7016E-02	1.0701	0.6980	0.5931	0.5363	1.0141	1.0008	0.9924	0.9989	0.7340	0.4107	0.5830	0.4411
0.8	1.1150E-01	1.0714	0.6353	0.5200	0.4618	1.0154	1.0009	0.9928	0.9989	0.7022	0.4107	0.5870	0.4448
0.9	1.6049E-01	1.0727	0.5941	0.4745	0.4164	1.0169	1.0011	0.9931	0.9990	0.6819	0.4108	0.5911	0.4487
1.0	2.0960E-01	1.0743	0.5663	0.4448	0.3874	1.0183	1.0013	0.9935	0.9990	0.6686	0.4112	0.5953	0.4525
1.2	2.9764E-01	1.0776	0.5341	0.4117	0.3555	1.0213	1.0016	0.9942	0.9991	0.6544	0.4123	0.6038	0.4603
1.4	3.6564E-01	1.0813	0.5190	0.3967	0.3414	1.0244	1.0019	0.9947	0.9992	0.6491	0.4139	0.6120	0.4678
1.6	4.1415E-01	1.0852	0.5128	0.3908	0.3360	1.0274	1.0022	0.9953	0.9993	0.6482	0.4157	0.6200	0.4750
1.8	4.4700E-01	1.0894	0.5116	0.3899	0.3353	1.0304	1.0026	0.9957	0.9993	0.6498	0.4177	0.6275	0.4819
2.0	4.6819E-01	1.0937	0.5132	0.3917	0.3373	1.0334	1.0029	0.9961	0.9994	0.6526	0.4198	0.6347	0.4884
2.2	4.8101E-01	1.0981	0.5165	0.3953	0.3408	1.0364	1.0032	0.9964	0.9994	0.6562	0.4218	0.6414	0.4945
2.4	4.8791E-01	1.1026	0.5209	0.3998	0.3453	1.0393	1.0036	0.9967	0.9995	0.6601	0.4237	0.6477	0.5004
2.6	4.9068E-01	1.1072	0.5259	0.4050	0.3504	1.0422	1.0039	0.9969	0.9995	0.6642	0.4256	0.6537	0.5059
2.8	4.9057E-01	1.1118	0.5312	0.4105	0.3558	1.0451	1.0042	0.9971	0.9995	0.6683	0.4273	0.6593	0.5111
3.0	4.8847E-01	1.1165	0.5367	0.4162	0.3614	1.0479	1.0045	0.9973	0.9996	0.6724	0.4290	0.6646	0.5161
3.2	4.8501E-01	1.1212	0.5423	0.4220	0.3671	1.0507	1.0049	0.9975	0.9996	0.6764	0.4306	0.6696	0.5208
3.4	4.8064E-01	1.1260	0.5479	0.4278	0.3728	1.0534	1.0052	0.9976	0.9996	0.6804	0.4320	0.6743	0.5254
3.6	4.7567E-01	1.1308	0.5534	0.4336	0.3784	1.0561	1.0055	0.9978	0.9996	0.6842	0.4333	0.6788	0.5297
3.8	4.7035E-01	1.1355	0.5588	0.4392	0.3840	1.0588	1.0058	0.9979	0.9996	0.6879	0.4346	0.6831	0.5339
4.0	4.6481E-01	1.1403	0.5642	0.4448	0.3895	1.0615	1.0061	0.9980	0.9997	0.6915	0.4357	0.6872	0.5378
4.5	4.5074E-01	1.1524	0.5769	0.4582	0.4027	1.0680	1.0069	0.9983	0.9997	0.6998	0.4382	0.6965	0.5472
5.0	4.3708E-01	1.1644	0.5888	0.4709	0.4152	1.0743	1.0077	0.9984	0.9997	0.7075	0.4401	0.7049	0.5557
5.5	4.2424E-01	1.1765	0.5999	0.4827	0.4270	1.0805	1.0085	0.9986	0.9998	0.7145	0.4416	0.7123	0.5635
6.0	4.1237E-01	1.1886	0.6102	0.4938	0.4380	1.0866	1.0093	0.9987	0.9998	0.7209	0.4428	0.7191	0.5707
6.5	4.0146E-01	1.2007	0.6198	0.5041	0.4484	1.0926	1.0101	0.9988	0.9998	0.7267	0.4436	0.7253	0.5775
7.0	3.9148E-01	1.2128	0.6286	0.5138	0.4581	1.0984	1.0108	0.9989	0.9998	0.7321	0.4441	0.7309	0.5838
7.5	3.8234E-01	1.2249	0.6369	0.5229	0.4672	1.1042	1.0116	0.9990	0.9998	0.7371	0.4444	0.7360	0.5898
8.0	3.7397E-01	1.2370	0.6446	0.5314	0.4758	1.1098	1.0124	0.9991	0.9998	0.7417	0.4445	0.7408	0.5954
9.0	3.5924E-01	1.2612	0.6585	0.5469	0.4915	1.1209	1.0139	0.9992	0.9998	0.7499	0.4440	0.7492	0.6058
10.0	3.4676E-01	1.2853	0.6707	0.5607	0.5055	1.1317	1.0154	0.9993	0.9999	0.7570	0.4430	0.7564	0.6152
11.0	3.3608E-01	1.3096	0.6813	0.5729	0.5180	1.1421	1.0169	0.9994	0.9999	0.7631	0.4415	0.7627	0.6238
12.0	3.2689E-01	1.3338	0.6908	0.5838	0.5292	1.1524	1.0184	0.9994	0.9999	0.7686	0.4395	0.7682	0.6318
13.0	3.1892E-01	1.3581	0.6991	0.5935	0.5393	1.1624	1.0199	0.9995	0.9999	0.7733	0.4373	0.7731	0.6392
14.0	3.1197E-01	1.3825	0.7065	0.6022	0.5483	1.1722	1.0214	0.9995	0.9999	0.7775	0.4347	0.7773	0.6461
16.0	3.0050E-01	1.4315	0.7190	0.6171	0.5639	1.1913	1.0244	0.9996	0.9999	0.7846	0.4289	0.7845	0.6587
18.0	2.9153E-01	1.4809	0.7288	0.6291	0.5765	1.2098	1.0273	0.9996	0.9999	0.7902	0.4225	0.7901	0.6701
20.0	2.8446E-01	1.5309	0.7366	0.6387	0.5868	1.2278	1.0303	0.9997	0.9999	0.7946	0.4155	0.7946	0.6803
25.0	2.7248E-01	1.6582	0.7494	0.6553	0.6047	1.2709	1.0375	0.9998	0.9999	0.8021	0.3967	0.8021	0.7027
30.0	2.6588E-01	1.7898	0.7555	0.6639	0.6144	1.3119	1.0448	0.9998	1.0000	0.8060	0.3764	0.8060	0.7216
35.0	2.6270E-01	1.9264	0.7571	0.6671	0.6185	1.3512	1.0520	0.9998	1.0000	0.8076	0.3553	0.8076	0.7383
40.0	2.6191E-01	2.0685	0.7556	0.6665	0.6184	1.3893	1.0592	0.9999	1.0000	0.8076	0.3337	0.8076	0.7533
45.0	2.6286E-01	2.2166	0.7519	0.6631	0.6154	1.4263	1.0664	0.9999	1.0000	0.8065	0.3118	0.8065	0.7671
50.0	2.6517E-01	2.3712	0.7465	0.6576	0.6102	1.4623	1.0736	0.9999	1.0000	0.8046	0.2897	0.8047	0.7800

Instr. see p. 22

POSITRON   Z= -96   A= 243

p	F_0L_0	L_0	λ_2	λ_3	λ_4	μ_1	μ_2	Λ_1	Λ_2	ν_{12}	$\hat{\nu}_{12}$	η_{12}	$\hat{\eta}_{12}$
0.1	5.6956E-17	1.0648	14.4733	119.8684	597.6283	1.0094	1.0002	0.9904	0.9986	8.1864	0.4075	0.5607	0.4182
0.2	8.3351E-08	1.0651	3.9064	10.0435	18.5212	1.0098	1.0003	0.9905	0.9986	2.4653	0.4073	0.5618	0.4193
0.3	6.9735E-05	1.0656	1.9515	2.9605	3.7482	1.0104	1.0003	0.9908	0.9986	1.4076	0.4070	0.5636	0.4210
0.4	1.6950E-03	1.0662	1.2694	1.4619	1.5595	1.0113	1.0004	0.9911	0.9987	1.0394	0.4066	0.5661	0.4233
0.5	1.0263E-02	1.0670	0.9560	0.9421	0.9202	1.0124	1.0005	0.9914	0.9987	0.8710	0.4063	0.5691	0.4262
0.6	3.1495E-02	1.0680	0.7880	0.7073	0.6574	1.0136	1.0007	0.9918	0.9988	0.7814	0.4060	0.5725	0.4294
0.7	6.6249E-02	1.0692	0.6888	0.5831	0.5262	1.0149	1.0008	0.9922	0.9988	0.7292	0.4059	0.5763	0.4329
0.8	1.1089E-01	1.0705	0.6265	0.5106	0.4523	1.0163	1.0010	0.9926	0.9989	0.6970	0.4060	0.5803	0.4366
0.9	1.6030E-01	1.0719	0.5855	0.4654	0.4074	1.0178	1.0011	0.9930	0.9989	0.6764	0.4062	0.5845	0.4405
1.0	2.1000E-01	1.0735	0.5579	0.4360	0.3787	1.0193	1.0013	0.9933	0.9990	0.6630	0.4066	0.5887	0.4443
1.2	2.9938E-01	1.0769	0.5259	0.4031	0.3471	1.0224	1.0016	0.9940	0.9991	0.6486	0.4079	0.5973	0.4521
1.4	3.6857E-01	1.0808	0.5110	0.3884	0.3333	1.0256	1.0020	0.9946	0.9992	0.6432	0.4096	0.6057	0.4597
1.6	4.1795E-01	1.0848	0.5049	0.3826	0.3280	1.0287	1.0023	0.9951	0.9992	0.6424	0.4116	0.6137	0.4670
1.8	4.5134E-01	1.0891	0.5038	0.3818	0.3274	1.0319	1.0026	0.9956	0.9993	0.6439	0.4137	0.6214	0.4739
2.0	4.7282E-01	1.0936	0.5055	0.3837	0.3294	1.0350	1.0030	0.9960	0.9994	0.6468	0.4158	0.6286	0.4805
2.2	4.8574E-01	1.0982	0.5089	0.3873	0.3330	1.0381	1.0033	0.9963	0.9994	0.6504	0.4179	0.6354	0.4867
2.4	4.9261E-01	1.1028	0.5134	0.3920	0.3375	1.0411	1.0036	0.9966	0.9995	0.6544	0.4199	0.6418	0.4925
2.6	4.9526E-01	1.1076	0.5185	0.3972	0.3427	1.0441	1.0040	0.9968	0.9995	0.6586	0.4219	0.6479	0.4981
2.8	4.9498E-01	1.1124	0.5239	0.4028	0.3481	1.0471	1.0043	0.9971	0.9995	0.6628	0.4237	0.6536	0.5034
3.0	4.9267E-01	1.1172	0.5295	0.4085	0.3537	1.0500	1.0046	0.9973	0.9996	0.6669	0.4254	0.6590	0.5085
3.2	4.8899E-01	1.1221	0.5352	0.4144	0.3594	1.0529	1.0050	0.9974	0.9996	0.6710	0.4270	0.6641	0.5133
3.4	4.8439E-01	1.1270	0.5408	0.4202	0.3651	1.0557	1.0053	0.9976	0.9996	0.6750	0.4285	0.6689	0.5178
3.6	4.7919E-01	1.1319	0.5464	0.4260	0.3708	1.0585	1.0056	0.9977	0.9996	0.6789	0.4299	0.6735	0.5222
3.8	4.7363E-01	1.1369	0.5519	0.4317	0.3764	1.0613	1.0059	0.9979	0.9996	0.6827	0.4312	0.6779	0.5264
4.0	4.6787E-01	1.1418	0.5574	0.4374	0.3819	1.0641	1.0063	0.9980	0.9997	0.6863	0.4324	0.6820	0.5305
4.5	4.5328E-01	1.1542	0.5703	0.4509	0.3953	1.0708	1.0071	0.9982	0.9997	0.6949	0.4350	0.6915	0.5399
5.0	4.3914E-01	1.1667	0.5824	0.4637	0.4079	1.0774	1.0079	0.9984	0.9997	0.7027	0.4370	0.7000	0.5485
5.5	4.2589E-01	1.1792	0.5937	0.4756	0.4197	1.0838	1.0087	0.9986	0.9997	0.7098	0.4387	0.7076	0.5564
6.0	4.1365E-01	1.1916	0.6041	0.4868	0.4308	1.0901	1.0095	0.9987	0.9998	0.7163	0.4399	0.7145	0.5638
6.5	4.0242E-01	1.2041	0.6138	0.4973	0.4413	1.0962	1.0103	0.9988	0.9998	0.7223	0.4408	0.7208	0.5706
7.0	3.9215E-01	1.2166	0.6228	0.5071	0.4511	1.1023	1.0110	0.9989	0.9998	0.7278	0.4414	0.7265	0.5771
7.5	3.8276E-01	1.2291	0.6312	0.5163	0.4603	1.1082	1.0118	0.9990	0.9998	0.7328	0.4417	0.7318	0.5831
8.0	3.7417E-01	1.2416	0.6391	0.5249	0.4690	1.1141	1.0126	0.9991	0.9998	0.7375	0.4419	0.7366	0.5888
9.0	3.5906E-01	1.2665	0.6532	0.5406	0.4849	1.1255	1.0142	0.9992	0.9998	0.7459	0.4416	0.7452	0.5993
10.0	3.4627E-01	1.2915	0.6656	0.5546	0.4990	1.1366	1.0157	0.9993	0.9999	0.7531	0.4406	0.7526	0.6089
11.0	3.3534E-01	1.3165	0.6765	0.5669	0.5117	1.1474	1.0172	0.9993	0.9999	0.7594	0.4392	0.7590	0.6177
12.0	3.2594E-01	1.3415	0.6861	0.5780	0.5230	1.1579	1.0188	0.9994	0.9999	0.7650	0.4373	0.7646	0.6258
13.0	3.1779E-01	1.3666	0.6946	0.5879	0.5332	1.1683	1.0203	0.9995	0.9999	0.7698	0.4351	0.7696	0.6333
14.0	3.1069E-01	1.3918	0.7022	0.5967	0.5424	1.1784	1.0218	0.9995	0.9999	0.7741	0.4327	0.7739	0.6403
16.0	2.9897E-01	1.4424	0.7149	0.6119	0.5582	1.1980	1.0249	0.9996	0.9999	0.7814	0.4270	0.7812	0.6531
18.0	2.8982E-01	1.4934	0.7250	0.6241	0.5711	1.2171	1.0279	0.9996	0.9999	0.7871	0.4207	0.7870	0.6647
20.0	2.8259E-01	1.5450	0.7330	0.6340	0.5816	1.2356	1.0309	0.9997	0.9999	0.7916	0.4139	0.7916	0.6751
25.0	2.7034E-01	1.6766	0.7461	0.6509	0.5998	1.2798	1.0383	0.9997	0.9999	0.7993	0.3953	0.7993	0.6978
30.0	2.6357E-01	1.8128	0.7524	0.6598	0.6098	1.3219	1.0456	0.9998	1.0000	0.8034	0.3752	0.8034	0.7171
35.0	2.6027E-01	1.9542	0.7542	0.6633	0.6141	1.3623	1.0529	0.9998	1.0000	0.8050	0.3543	0.8051	0.7340
40.0	2.5938E-01	2.1016	0.7528	0.6628	0.6142	1.4014	1.0603	0.9999	1.0000	0.8051	0.3330	0.8052	0.7493
45.0	2.6026E-01	2.2553	0.7491	0.6595	0.6113	1.4393	1.0676	0.9999	1.0000	0.8041	0.3112	0.8041	0.7634
50.0	2.6251E-01	2.4158	0.7439	0.6541	0.6062	1.4762	1.0750	0.9999	1.0000	0.8023	0.2893	0.8023	0.7765

POSITRON Z= -97 A= 245

p	F_0L_0	L_0	λ_2	λ_3	λ_4	μ_1	μ_2	Λ_1	Λ_2	ν_{12}	$\hat{\nu}_{12}$	η_{12}	$\hat{\eta}_{12}$
0.1	3.8637E-17	1.0635	14.3725	120.2089	606.8486	1.0100	1.0002	0.9901	0.9986	8.2635	0.4022	0.5536	0.4098
0.2	7.0937E-08	1.0637	3.8753	10.0274	18.6219	1.0104	1.0003	0.9903	0.9986	2.4794	0.4020	0.5547	0.4108
0.3	6.3905E-05	1.0642	1.9333	2.9414	3.7352	1.0111	1.0003	0.9905	0.9986	1.4100	0.4017	0.5565	0.4126
0.4	1.6098E-03	1.0649	1.2557	1.4463	1.5438	1.0120	1.0004	0.9908	0.9987	1.0377	0.4014	0.5590	0.4149
0.5	9.9489E-03	1.0657	0.9444	0.9289	0.9066	1.0131	1.0006	0.9912	0.9987	0.8674	0.4012	0.5620	0.4177
0.6	3.0930E-02	1.0667	0.7775	0.6956	0.6456	1.0143	1.0007	0.9916	0.9988	0.7769	0.4010	0.5655	0.4210
0.7	6.5624E-02	1.0679	0.6790	0.5725	0.5155	1.0157	1.0008	0.9920	0.9988	0.7241	0.4010	0.5693	0.4245
0.8	1.1051E-01	1.0693	0.6171	0.5006	0.4423	1.0172	1.0010	0.9924	0.9989	0.6915	0.4011	0.5734	0.4282
0.9	1.6042E-01	1.0708	0.5764	0.4558	0.3979	1.0187	1.0011	0.9928	0.9989	0.6706	0.4015	0.5776	0.4321
1.0	2.1082E-01	1.0724	0.5490	0.4267	0.3695	1.0203	1.0013	0.9932	0.9990	0.6571	0.4020	0.5819	0.4360
1.2	3.0173E-01	1.0760	0.5173	0.3943	0.3384	1.0235	1.0016	0.9939	0.9991	0.6425	0.4034	0.5905	0.4438
1.4	3.7226E-01	1.0799	0.5026	0.3797	0.3248	1.0268	1.0020	0.9945	0.9992	0.6371	0.4052	0.5990	0.4514
1.6	4.2260E-01	1.0841	0.4966	0.3741	0.3197	1.0301	1.0023	0.9950	0.9992	0.6362	0.4073	0.6072	0.4587
1.8	4.5660E-01	1.0886	0.4956	0.3734	0.3192	1.0333	1.0027	0.9955	0.9993	0.6378	0.4095	0.6149	0.4657
2.0	4.7840E-01	1.0932	0.4974	0.3754	0.3212	1.0366	1.0030	0.9959	0.9994	0.6407	0.4117	0.6223	0.4723
2.2	4.9143E-01	1.0979	0.5009	0.3791	0.3249	1.0397	1.0034	0.9962	0.9994	0.6444	0.4139	0.6292	0.4786
2.4	4.9828E-01	1.1027	0.5055	0.3838	0.3295	1.0429	1.0037	0.9965	0.9994	0.6485	0.4160	0.6357	0.4845
2.6	5.0080E-01	1.1076	0.5107	0.3890	0.3346	1.0460	1.0040	0.9968	0.9995	0.6527	0.4180	0.6419	0.4902
2.8	5.0033E-01	1.1125	0.5162	0.3947	0.3401	1.0491	1.0044	0.9970	0.9995	0.6570	0.4199	0.6477	0.4955
3.0	4.9780E-01	1.1175	0.5219	0.4005	0.3457	1.0521	1.0047	0.9972	0.9995	0.6612	0.4217	0.6532	0.5006
3.2	4.9388E-01	1.1226	0.5276	0.4064	0.3515	1.0551	1.0051	0.9974	0.9996	0.6654	0.4234	0.6583	0.5054
3.4	4.8902E-01	1.1276	0.5334	0.4122	0.3572	1.0580	1.0054	0.9975	0.9996	0.6695	0.4250	0.6633	0.5101
3.6	4.8357E-01	1.1327	0.5391	0.4181	0.3629	1.0609	1.0057	0.9977	0.9996	0.6734	0.4264	0.6679	0.5145
3.8	4.7776E-01	1.1378	0.5447	0.4239	0.3685	1.0638	1.0060	0.9978	0.9996	0.6773	0.4278	0.6723	0.5188
4.0	4.7175E-01	1.1429	0.5501	0.4295	0.3741	1.0666	1.0064	0.9979	0.9997	0.6810	0.4290	0.6765	0.5228
4.5	4.5657E-01	1.1557	0.5633	0.4432	0.3875	1.0736	1.0072	0.9982	0.9997	0.6897	0.4317	0.6862	0.5324
5.0	4.4192E-01	1.1685	0.5756	0.4561	0.4001	1.0804	1.0080	0.9984	0.9997	0.6976	0.4339	0.6949	0.5411
5.5	4.2820E-01	1.1814	0.5870	0.4682	0.4121	1.0870	1.0088	0.9985	0.9997	0.7049	0.4356	0.7026	0.5491
6.0	4.1556E-01	1.1942	0.5976	0.4795	0.4233	1.0935	1.0096	0.9987	0.9998	0.7115	0.4369	0.7097	0.5566
6.5	4.0398E-01	1.2071	0.6075	0.4901	0.4338	1.0999	1.0104	0.9988	0.9998	0.7176	0.4379	0.7161	0.5635
7.0	3.9339E-01	1.2199	0.6167	0.5000	0.4437	1.1061	1.0112	0.9989	0.9998	0.7232	0.4386	0.7219	0.5700
7.5	3.8373E-01	1.2328	0.6252	0.5093	0.4530	1.1122	1.0120	0.9990	0.9998	0.7284	0.4391	0.7273	0.5762
8.0	3.7488E-01	1.2456	0.6332	0.5180	0.4618	1.1183	1.0128	0.9990	0.9998	0.7331	0.4393	0.7322	0.5820
9.0	3.5935E-01	1.2713	0.6476	0.5339	0.4779	1.1300	1.0144	0.9992	0.9998	0.7417	0.4391	0.7410	0.5926
10.0	3.4622E-01	1.2970	0.6602	0.5481	0.4922	1.1414	1.0160	0.9993	0.9999	0.7491	0.4383	0.7485	0.6023
11.0	3.3501E-01	1.3227	0.6713	0.5606	0.5050	1.1526	1.0175	0.9993	0.9999	0.7555	0.4369	0.7551	0.6112
12.0	3.2537E-01	1.3485	0.6811	0.5719	0.5165	1.1634	1.0191	0.9994	0.9999	0.7612	0.4352	0.7608	0.6194
13.0	3.1701E-01	1.3743	0.6898	0.5819	0.5268	1.1740	1.0206	0.9995	0.9999	0.7661	0.4331	0.7659	0.6271
14.0	3.0973E-01	1.4002	0.6975	0.5909	0.5362	1.1844	1.0222	0.9995	0.9999	0.7706	0.4307	0.7703	0.6342
16.0	2.9773E-01	1.4523	0.7105	0.6064	0.5522	1.2046	1.0253	0.9996	0.9999	0.7780	0.4253	0.7778	0.6473
18.0	2.8836E-01	1.5049	0.7209	0.6188	0.5653	1.2242	1.0283	0.9996	0.9999	0.7838	0.4191	0.7837	0.6590
20.0	2.8095E-01	1.5580	0.7291	0.6290	0.5760	1.2432	1.0314	0.9997	0.9999	0.7885	0.4124	0.7885	0.6696
25.0	2.6839E-01	1.6937	0.7427	0.6464	0.5948	1.2885	1.0389	0.9997	0.9999	0.7964	0.3941	0.7964	0.6927
30.0	2.6139E-01	1.8342	0.7493	0.6557	0.6051	1.3317	1.0464	0.9998	1.0000	0.8007	0.3744	0.8007	0.7123
35.0	2.5793E-01	1.9802	0.7513	0.6594	0.6097	1.3731	1.0538	0.9998	1.0000	0.8025	0.3538	0.8025	0.7295
40.0	2.5690E-01	2.1324	0.7501	0.6591	0.6101	1.4130	1.0612	0.9999	1.0000	0.8027	0.3326	0.8027	0.7450
45.0	2.5767E-01	2.2913	0.7466	0.6560	0.6073	1.4519	1.0687	0.9999	1.0000	0.8017	0.3111	0.8018	0.7593
50.0	2.5982E-01	2.4575	0.7414	0.6507	0.6023	1.4896	1.0761	0.9999	1.0000	0.8000	0.2894	0.8000	0.7726

 Instr. see p. 22

POSITRON Z= −98 A= 248

p	F_0L_0	L_0	λ_2	λ_3	λ_4	μ_1	μ_2	Λ_1	Λ_2	ν_{12}	$\hat{\nu}_{12}$	η_{12}	$\hat{\eta}_{12}$
0.1	2.6206E-17	1.0619	14.2693	120.5070	615.9180	1.0107	1.0002	0.9899	0.9985	8.3402	0.3969	0.5465	0.4014
0.2	6.0366E-08	1.0622	3.8437	10.0089	18.7182	1.0111	1.0003	0.9900	0.9986	2.4934	0.3968	0.5477	0.4025
0.3	5.8556E-05	1.0627	1.9150	2.9220	3.7219	1.0118	1.0003	0.9903	0.9986	1.4125	0.3965	0.5495	0.4043
0.4	1.5288E-03	1.0634	1.2421	1.4308	1.5282	1.0128	1.0004	0.9906	0.9986	1.0361	0.3963	0.5520	0.4066
0.5	9.6437E-03	1.0643	0.9329	0.9160	0.8932	1.0139	1.0006	0.9910	0.9987	0.8640	0.3961	0.5551	0.4094
0.6	3.0371E-02	1.0653	0.7671	0.6842	0.6339	1.0152	1.0007	0.9914	0.9987	0.7724	0.3960	0.5586	0.4127
0.7	6.4996E-02	1.0666	0.6693	0.5620	0.5050	1.0166	1.0008	0.9918	0.9988	0.7190	0.3961	0.5624	0.4162
0.8	1.1011E-01	1.0680	0.6078	0.4908	0.4326	1.0181	1.0010	0.9922	0.9989	0.6861	0.3963	0.5665	0.4199
0.9	1.6053E-01	1.0695	0.5674	0.4465	0.3886	1.0197	1.0012	0.9926	0.9989	0.6650	0.3968	0.5708	0.4238
1.0	2.1160E-01	1.0712	0.5403	0.4177	0.3606	1.0214	1.0013	0.9930	0.9990	0.6512	0.3973	0.5751	0.4277
1.2	3.0404E-01	1.0749	0.5088	0.3856	0.3300	1.0247	1.0017	0.9937	0.9991	0.6365	0.3989	0.5839	0.4356
1.4	3.7591E-01	1.0790	0.4943	0.3713	0.3166	1.0281	1.0020	0.9944	0.9992	0.6310	0.4009	0.5925	0.4432
1.6	4.2722E-01	1.0834	0.4885	0.3659	0.3116	1.0315	1.0024	0.9949	0.9992	0.6302	0.4031	0.6007	0.4506
1.8	4.6182E-01	1.0879	0.4876	0.3652	0.3112	1.0349	1.0027	0.9954	0.9993	0.6318	0.4054	0.6086	0.4576
2.0	4.8393E-01	1.0927	0.4895	0.3673	0.3133	1.0383	1.0031	0.9958	0.9993	0.6348	0.4077	0.6160	0.4643
2.2	4.9707E-01	1.0976	0.4931	0.3711	0.3170	1.0416	1.0034	0.9961	0.9994	0.6385	0.4100	0.6231	0.4706
2.4	5.0388E-01	1.1025	0.4978	0.3758	0.3216	1.0448	1.0038	0.9964	0.9994	0.6426	0.4122	0.6297	0.4766
2.6	5.0628E-01	1.1076	0.5030	0.3811	0.3268	1.0480	1.0041	0.9967	0.9995	0.6469	0.4143	0.6359	0.4823
2.8	5.0561E-01	1.1127	0.5086	0.3868	0.3323	1.0512	1.0045	0.9969	0.9995	0.6513	0.4162	0.6418	0.4877
3.0	5.0285E-01	1.1179	0.5144	0.3926	0.3380	1.0544	1.0048	0.9971	0.9995	0.6556	0.4181	0.6474	0.4928
3.2	4.9868E-01	1.1230	0.5203	0.3986	0.3437	1.0574	1.0051	0.9973	0.9996	0.6598	0.4198	0.6527	0.4977
3.4	4.9356E-01	1.1283	0.5261	0.4045	0.3495	1.0605	1.0055	0.9975	0.9996	0.6640	0.4215	0.6577	0.5024
3.6	4.8785E-01	1.1335	0.5319	0.4104	0.3552	1.0635	1.0058	0.9976	0.9996	0.6680	0.4230	0.6624	0.5069
3.8	4.8178E-01	1.1387	0.5375	0.4162	0.3609	1.0665	1.0062	0.9978	0.9996	0.6719	0.4244	0.6669	0.5112
4.0	4.7552E-01	1.1440	0.5431	0.4219	0.3665	1.0694	1.0065	0.9979	0.9996	0.6757	0.4257	0.6712	0.5153
4.5	4.5976E-01	1.1572	0.5564	0.4357	0.3799	1.0766	1.0073	0.9981	0.9997	0.6845	0.4285	0.6810	0.5250
5.0	4.4458E-01	1.1705	0.5689	0.4487	0.3927	1.0837	1.0082	0.9983	0.9997	0.6926	0.4308	0.6898	0.5338
5.5	4.3040E-01	1.1837	0.5805	0.4609	0.4047	1.0905	1.0090	0.9985	0.9997	0.7000	0.4326	0.6978	0.5420
6.0	4.1735E-01	1.1969	0.5913	0.4723	0.4160	1.0972	1.0098	0.9986	0.9998	0.7068	0.4340	0.7049	0.5495
6.5	4.0541E-01	1.2102	0.6013	0.4830	0.4266	1.1038	1.0106	0.9988	0.9998	0.7130	0.4351	0.7114	0.5565
7.0	3.9452E-01	1.2234	0.6106	0.4931	0.4366	1.1102	1.0114	0.9989	0.9998	0.7187	0.4359	0.7174	0.5631
7.5	3.8457E-01	1.2367	0.6193	0.5025	0.4460	1.1166	1.0122	0.9990	0.9998	0.7240	0.4364	0.7228	0.5693
8.0	3.7549E-01	1.2499	0.6274	0.5113	0.4548	1.1228	1.0131	0.9990	0.9998	0.7288	0.4366	0.7279	0.5752
9.0	3.5954E-01	1.2764	0.6421	0.5275	0.4710	1.1349	1.0147	0.9991	0.9998	0.7375	0.4366	0.7368	0.5860
10.0	3.4606E-01	1.3029	0.6549	0.5418	0.4855	1.1467	1.0163	0.9992	0.9999	0.7451	0.4359	0.7445	0.5959
11.0	3.3458E-01	1.3294	0.6662	0.5545	0.4985	1.1581	1.0179	0.9993	0.9999	0.7517	0.4347	0.7512	0.6049
12.0	3.2470E-01	1.3560	0.6762	0.5659	0.5102	1.1693	1.0194	0.9994	0.9999	0.7574	0.4330	0.7571	0.6132
13.0	3.1616E-01	1.3826	0.6851	0.5761	0.5206	1.1803	1.0210	0.9994	0.9999	0.7625	0.4310	0.7623	0.6210
14.0	3.0871E-01	1.4093	0.6930	0.5853	0.5301	1.1910	1.0226	0.9995	0.9999	0.7670	0.4287	0.7668	0.6283
16.0	2.9644E-01	1.4631	0.7063	0.6010	0.5464	1.2118	1.0257	0.9996	0.9999	0.7746	0.4234	0.7745	0.6415
18.0	2.8685E-01	1.5173	0.7169	0.6137	0.5597	1.2319	1.0288	0.9996	0.9999	0.7806	0.4174	0.7806	0.6534
20.0	2.7928E-01	1.5722	0.7253	0.6241	0.5707	1.2514	1.0319	0.9997	0.9999	0.7854	0.4109	0.7854	0.6642
25.0	2.6641E-01	1.7122	0.7393	0.6419	0.5898	1.2980	1.0395	0.9997	0.9999	0.7936	0.3928	0.7936	0.6876
30.0	2.5921E-01	1.8574	0.7462	0.6515	0.6005	1.3422	1.0471	0.9998	1.0000	0.7980	0.3733	0.7980	0.7076
35.0	2.5560E-01	2.0085	0.7484	0.6555	0.6053	1.3847	1.0547	0.9998	1.0000	0.8000	0.3530	0.8000	0.7251
40.0	2.5446E-01	2.1661	0.7473	0.6554	0.6059	1.4257	1.0623	0.9999	1.0000	0.8002	0.3320	0.8003	0.7406
45.0	2.5514E-01	2.3308	0.7439	0.6524	0.6033	1.4655	1.0698	0.9999	1.0000	0.7993	0.3108	0.7994	0.7554
50.0	2.5722E-01	2.5032	0.7388	0.6472	0.5984	1.5042	1.0774	0.9999	1.0000	0.7976	0.2692	0.7977	0.7689

POSITRON Z= -99 A= 251

p	F_0L_0	L_0	λ_2	λ_3	λ_4	μ_1	μ_2	Λ_1	Λ_2	ν_{12}	$\hat{\nu}_{12}$	η_{12}	$\hat{\eta}_{12}$
0.1	1.7787E-17	1.0602	14.1577	120.6967	624.4578	1.0114	1.0002	0.9896	0.9985	8.4150	0.3916	0.5395	0.3931
0.2	5.1403E-08	1.0605	3.8101	9.9827	18.7993	1.0119	1.0003	0.9898	0.9985	2.5070	0.3915	0.5406	0.3942
0.3	5.3689E-05	1.0610	1.8958	2.9008	3.7060	1.0126	1.0004	0.9900	0.9986	1.4147	0.3913	0.5425	0.3960
0.4	1.4527E-03	1.0617	1.2279	1.4147	1.5119	1.0136	1.0005	0.9904	0.9986	1.0344	0.3912	0.5450	0.3983
0.5	9.3536E-03	1.0626	0.9210	0.9027	0.8796	1.0148	1.0006	0.9907	0.9987	0.8604	0.3911	0.5481	0.4011
0.6	2.9840E-02	1.0637	0.7565	0.6727	0.6222	1.0161	1.0007	0.9912	0.9987	0.7679	0.3911	0.5516	0.4044
0.7	6.4413E-02	1.0650	0.6594	0.5515	0.4944	1.0176	1.0009	0.9916	0.9988	0.7139	0.3912	0.5555	0.4079
0.8	1.0977E-01	1.0664	0.5984	0.4810	0.4228	1.0192	1.0010	0.9920	0.9988	0.6806	0.3915	0.5596	0.4116
0.9	1.6072E-01	1.0681	0.5584	0.4371	0.3794	1.0208	1.0012	0.9924	0.9989	0.6593	0.3920	0.5639	0.4155
1.0	2.1251E-01	1.0698	0.5314	0.4087	0.3518	1.0225	1.0014	0.9928	0.9990	0.6454	0.3927	0.5683	0.4195
1.2	3.0655E-01	1.0736	0.5003	0.3770	0.3216	1.0260	1.0017	0.9936	0.9991	0.6305	0.3944	0.5772	0.4273
1.4	3.7980E-01	1.0779	0.4860	0.3630	0.3085	1.0295	1.0021	0.9942	0.9991	0.6249	0.3965	0.5859	0.4351
1.6	4.3211E-01	1.0824	0.4803	0.3577	0.3037	1.0331	1.0024	0.9948	0.9992	0.6241	0.3988	0.5943	0.4425
1.8	4.6733E-01	1.0871	0.4796	0.3571	0.3033	1.0366	1.0028	0.9953	0.9993	0.6257	0.4012	0.6022	0.4495
2.0	4.8977E-01	1.0920	0.4816	0.3593	0.3055	1.0401	1.0031	0.9957	0.9993	0.6288	0.4036	0.6098	0.4563
2.2	5.0302E-01	1.0971	0.4853	0.3631	0.3092	1.0435	1.0035	0.9960	0.9994	0.6326	0.4060	0.6169	0.4626
2.4	5.0979E-01	1.1022	0.4901	0.3679	0.3139	1.0469	1.0038	0.9964	0.9994	0.6368	0.4083	0.6236	0.4687
2.6	5.1204E-01	1.1074	0.4954	0.3732	0.3191	1.0502	1.0042	0.9966	0.9995	0.6411	0.4105	0.6300	0.4744
2.8	5.1117E-01	1.1127	0.5011	0.3790	0.3246	1.0535	1.0045	0.9969	0.9995	0.6455	0.4125	0.6360	0.4799
3.0	5.0817E-01	1.1180	0.5070	0.3849	0.3303	1.0568	1.0049	0.9971	0.9995	0.6499	0.4144	0.6416	0.4850
3.2	5.0373E-01	1.1234	0.5129	0.3908	0.3361	1.0600	1.0052	0.9973	0.9996	0.6543	0.4163	0.6470	0.4900
3.4	4.9835E-01	1.1288	0.5188	0.3968	0.3419	1.0631	1.0056	0.9974	0.9996	0.6585	0.4179	0.6521	0.4947
3.6	4.9236E-01	1.1342	0.5247	0.4028	0.3476	1.0663	1.0059	0.9976	0.9996	0.6626	0.4195	0.6569	0.4993
3.8	4.8602E-01	1.1396	0.5304	0.4086	0.3533	1.0693	1.0063	0.9977	0.9996	0.6665	0.4210	0.6615	0.5036
4.0	4.7950E-01	1.1450	0.5361	0.4144	0.3589	1.0724	1.0066	0.9978	0.9996	0.6704	0.4223	0.6658	0.5078
4.5	4.6313E-01	1.1587	0.5496	0.4283	0.3724	1.0798	1.0075	0.9981	0.9997	0.6794	0.4253	0.6758	0.5176
5.0	4.4740E-01	1.1723	0.5622	0.4414	0.3853	1.0871	1.0083	0.9983	0.9997	0.6876	0.4277	0.6848	0.5265
5.5	4.3275E-01	1.1860	0.5740	0.4537	0.3973	1.0942	1.0091	0.9985	0.9997	0.6952	0.4296	0.6928	0.5347
6.0	4.1928E-01	1.1996	0.5849	0.4652	0.4087	1.1012	1.0100	0.9986	0.9998	0.7020	0.4311	0.7001	0.5424
6.5	4.0698E-01	1.2133	0.5951	0.4760	0.4194	1.1079	1.0108	0.9987	0.9998	0.7084	0.4322	0.7068	0.5495
7.0	3.9576E-01	1.2270	0.6046	0.4862	0.4295	1.1146	1.0116	0.9988	0.9998	0.7142	0.4331	0.7128	0.5562
7.5	3.8553E-01	1.2406	0.6134	0.4957	0.4389	1.1211	1.0125	0.9989	0.9998	0.7195	0.4337	0.7184	0.5625
8.0	3.7619E-01	1.2543	0.6217	0.5046	0.4479	1.1275	1.0133	0.9990	0.9998	0.7245	0.4340	0.7235	0.5684
9.0	3.5981E-01	1.2816	0.6366	0.5210	0.4643	1.1401	1.0149	0.9991	0.9998	0.7334	0.4341	0.7327	0.5794
10.0	3.4599E-01	1.3089	0.6496	0.5355	0.4789	1.1522	1.0166	0.9992	0.9999	0.7411	0.4335	0.7405	0.5894
11.0	3.3422E-01	1.3362	0.6611	0.5484	0.4920	1.1640	1.0182	0.9993	0.9999	0.7478	0.4324	0.7474	0.5986
12.0	3.2411E-01	1.3636	0.6713	0.5600	0.5038	1.1755	1.0198	0.9994	0.9999	0.7537	0.4309	0.7534	0.6070
13.0	3.1536E-01	1.3911	0.6804	0.5704	0.5145	1.1868	1.0214	0.9994	0.9999	0.7589	0.4290	0.7586	0.6149
14.0	3.0774E-01	1.4186	0.6884	0.5797	0.5241	1.1978	1.0230	0.9995	0.9999	0.7635	0.4267	0.7633	0.6223
16.0	2.9519E-01	1.4741	0.7020	0.5957	0.5406	1.2192	1.0262	0.9996	0.9999	0.7713	0.4216	0.7711	0.6357
18.0	2.8540E-01	1.5301	0.7128	0.6086	0.5542	1.2399	1.0293	0.9996	0.9999	0.7774	0.4157	0.7773	0.6478
20.0	2.7765E-01	1.5866	0.7215	0.6192	0.5653	1.2599	1.0325	0.9997	0.9999	0.7824	0.4093	0.7823	0.6587
25.0	2.6448E-01	1.7313	0.7359	0.6375	0.5849	1.3078	1.0402	0.9997	0.9999	0.7908	0.3916	0.7908	0.6826
30.0	2.5708E-01	1.8814	0.7431	0.6474	0.5959	1.3532	1.0480	0.9998	1.0000	0.7953	0.3723	0.7953	0.7028
35.0	2.5332E-01	2.0377	0.7455	0.6516	0.6010	1.3968	1.0556	0.9998	1.0000	0.7974	0.3522	0.7974	0.7206
40.0	2.5207E-01	2.2009	0.7446	0.6517	0.6017	1.4388	1.0633	0.9999	1.0000	0.7978	0.3315	0.7978	0.7367
45.0	2.5265E-01	2.3717	0.7413	0.6489	0.5993	1.4796	1.0710	0.9999	1.0000	0.7970	0.3104	0.7970	0.7514
50.0	2.5465E-01	2.5505	0.7362	0.6438	0.5945	1.5193	1.0787	0.9999	1.0000	0.7953	0.2890	0.7953	0.7652

POSITRON Z=-100 A= 256

p	F_0L_0	L_0	λ_2	λ_3	λ_4	μ_1	μ_2	Λ_1	Λ_2	ν_{12}	$\hat{\nu}_{12}$	η_{12}	$\hat{\eta}_{12}$
0.1	1.2062E-17	1.0583	14.0500	120.9120	633.2084	1.0122	1.0002	0.9893	0.9985	8.4912	0.3865	0.5326	0.3850
0.2	4.3732E-08	1.0586	3.7777	9.9596	18.8874	1.0127	1.0003	0.9895	0.9985	2.5210	0.3864	0.5338	0.3861
0.3	4.9183E-05	1.0591	1.8773	2.8809	3.6921	1.0135	1.0004	0.9898	0.9986	1.4172	0.3863	0.5357	0.3878
0.4	1.3792E-03	1.0599	1.2143	1.3994	1.4966	1.0145	1.0005	0.9901	0.9986	1.0329	0.3862	0.5382	0.3902
0.5	9.0639E-03	1.0608	0.9097	0.8901	0.8667	1.0157	1.0006	0.9905	0.9986	0.8571	0.3861	0.5413	0.3930
0.6	2.9291E-02	1.0620	0.7464	0.6617	0.6110	1.0171	1.0007	0.9909	0.9987	0.7636	0.3862	0.5449	0.3963
0.7	6.3776E-02	1.0633	0.6500	0.5415	0.4844	1.0187	1.0009	0.9913	0.9988	0.7090	0.3865	0.5488	0.3998
0.8	1.0934E-01	1.0648	0.5895	0.4716	0.4136	1.0203	1.0011	0.9918	0.9988	0.6753	0.3869	0.5530	0.4036
0.9	1.6077E-01	1.0665	0.5498	0.4282	0.3707	1.0220	1.0012	0.9922	0.9989	0.6538	0.3874	0.5573	0.4074
1.0	2.1323E-01	1.0683	0.5230	0.4001	0.3434	1.0238	1.0014	0.9926	0.9989	0.6397	0.3882	0.5618	0.4114
1.2	3.0878E-01	1.0723	0.4922	0.3688	0.3137	1.0274	1.0018	0.9934	0.9990	0.6247	0.3900	0.5707	0.4193
1.4	3.8336E-01	1.0767	0.4781	0.3550	0.3008	1.0311	1.0021	0.9941	0.9991	0.6191	0.3923	0.5795	0.4271
1.6	4.3663E-01	1.0814	0.4726	0.3499	0.2961	1.0348	1.0025	0.9947	0.9992	0.6182	0.3947	0.5880	0.4345
1.8	4.7245E-01	1.0863	0.4719	0.3494	0.2959	1.0385	1.0028	0.9951	0.9993	0.6199	0.3972	0.5961	0.4417
2.0	4.9519E-01	1.0914	0.4741	0.3517	0.2981	1.0421	1.0032	0.9956	0.9993	0.6230	0.3997	0.6037	0.4484
2.2	5.0853E-01	1.0966	0.4779	0.3555	0.3019	1.0457	1.0036	0.9959	0.9994	0.6269	0.4022	0.6110	0.4549
2.4	5.1525E-01	1.1019	0.4827	0.3604	0.3066	1.0492	1.0039	0.9963	0.9994	0.6311	0.4045	0.6178	0.4609
2.6	5.1735E-01	1.1073	0.4882	0.3658	0.3118	1.0526	1.0043	0.9965	0.9995	0.6355	0.4068	0.6242	0.4667
2.8	5.1628E-01	1.1128	0.4939	0.3716	0.3173	1.0561	1.0046	0.9968	0.9995	0.6400	0.4089	0.6303	0.4722
3.0	5.1303E-01	1.1183	0.4999	0.3775	0.3231	1.0594	1.0050	0.9970	0.9995	0.6445	0.4109	0.6361	0.4775
3.2	5.0833E-01	1.1239	0.5059	0.3835	0.3289	1.0628	1.0054	0.9972	0.9996	0.6489	0.4128	0.6415	0.4825
3.4	5.0267E-01	1.1294	0.5119	0.3895	0.3347	1.0661	1.0057	0.9974	0.9996	0.6532	0.4145	0.6467	0.4873
3.6	4.9642E-01	1.1350	0.5178	0.3955	0.3404	1.0693	1.0061	0.9975	0.9996	0.6573	0.4161	0.6516	0.4919
3.8	4.8981E-01	1.1406	0.5237	0.4014	0.3462	1.0725	1.0064	0.9977	0.9996	0.6614	0.4176	0.6562	0.4963
4.0	4.8303E-01	1.1463	0.5294	0.4072	0.3518	1.0757	1.0067	0.9978	0.9996	0.6653	0.4190	0.6606	0.5005
4.5	4.6606E-01	1.1603	0.5431	0.4213	0.3654	1.0834	1.0076	0.9981	0.9997	0.6744	0.4221	0.6708	0.5104
5.0	4.4981E-01	1.1745	0.5559	0.4345	0.3783	1.0910	1.0085	0.9983	0.9997	0.6828	0.4246	0.6799	0.5194
5.5	4.3469E-01	1.1886	0.5679	0.4469	0.3904	1.0983	1.0093	0.9984	0.9997	0.6905	0.4266	0.6881	0.5277
6.0	4.2082E-01	1.2027	0.5790	0.4585	0.4019	1.1055	1.0102	0.9986	0.9998	0.6975	0.4282	0.6956	0.5355
6.5	4.0816E-01	1.2169	0.5893	0.4695	0.4126	1.1125	1.0110	0.9987	0.9998	0.7039	0.4294	0.7023	0.5427
7.0	3.9663E-01	1.2310	0.5989	0.4797	0.4228	1.1194	1.0119	0.9988	0.9998	0.7099	0.4304	0.7085	0.5495
7.5	3.8613E-01	1.2451	0.6079	0.4893	0.4323	1.1262	1.0127	0.9989	0.9998	0.7153	0.4310	0.7142	0.5559
8.0	3.7654E-01	1.2592	0.6162	0.4984	0.4414	1.1328	1.0136	0.9990	0.9998	0.7204	0.4314	0.7194	0.5619
9.0	3.5976E-01	1.2874	0.6314	0.5149	0.4579	1.1458	1.0152	0.9991	0.9998	0.7294	0.4316	0.7287	0.5730
10.0	3.4562E-01	1.3157	0.6447	0.5296	0.4727	1.1583	1.0169	0.9992	0.9999	0.7373	0.4311	0.7367	0.5832
11.0	3.3358E-01	1.3439	0.6564	0.5427	0.4859	1.1705	1.0185	0.9993	0.9999	0.7441	0.4301	0.7437	0.5925
12.0	3.2325E-01	1.3723	0.6668	0.5544	0.4979	1.1824	1.0202	0.9994	0.9999	0.7501	0.4286	0.7498	0.6010
13.0	3.1432E-01	1.4007	0.6760	0.5649	0.5086	1.1941	1.0218	0.9994	0.9999	0.7554	0.4268	0.7552	0.6090
14.0	3.0655E-01	1.4292	0.6842	0.5744	0.5184	1.2054	1.0234	0.9995	0.9999	0.7601	0.4247	0.7599	0.6165
16.0	2.9376E-01	1.4866	0.6980	0.5906	0.5351	1.2275	1.0267	0.9995	0.9999	0.7681	0.4197	0.7679	0.6302
18.0	2.8377E-01	1.5445	0.7091	0.6038	0.5489	1.2488	1.0299	0.9996	0.9999	0.7744	0.4139	0.7743	0.6424
20.0	2.7588E-01	1.6031	0.7179	0.6145	0.5602	1.2694	1.0331	0.9997	0.9999	0.7794	0.4075	0.7793	0.6536
25.0	2.6246E-01	1.7530	0.7326	0.6332	0.5801	1.3187	1.0410	0.9997	0.9999	0.7880	0.3900	0.7880	0.6778
30.0	2.5490E-01	1.9087	0.7401	0.6434	0.5914	1.3655	1.0489	0.9998	1.0000	0.7927	0.3709	0.7927	0.6984
35.0	2.5104E-01	2.0710	0.7426	0.6478	0.5967	1.4102	1.0567	0.9998	1.0000	0.7949	0.3510	0.7949	0.7165
40.0	2.4971E-01	2.2407	0.7418	0.6480	0.5976	1.4534	1.0645	0.9998	1.0000	0.7953	0.3304	0.7953	0.7328
45.0	2.5023E-01	2.4184	0.7385	0.6452	0.5953	1.4953	1.0723	0.9999	1.0000	0.7945	0.3095	0.7946	0.7478
50.0	2.5219E-01	2.6048	0.7335	0.6402	0.5905	1.5360	1.0802	0.9999	1.0000	0.7929	0.2883	0.7930	0.7618

Instr. see p. 22

POSITRON Z=-101 A= 257

p	F_0L_0	L_0	λ_2	λ_3	λ_4	μ_1	μ_2	Λ_1	Λ_2	ν_{12}	$\hat{\nu}_{12}$	η_{12}	$\hat{\eta}_{12}$
0.1	8.2100E-18	1.0559	13.9107	120.7558	639.8596	1.0131	1.0003	0.9891	0.9985	8.5591	0.3810	0.5253	0.3766
0.2	3.7346E-08	1.0563	3.7369	9.9072	18.9141	1.0136	1.0003	0.9892	0.9985	2.5326	0.3810	0.5265	0.3777
0.3	4.5224E-05	1.0568	1.8548	2.8529	3.6667	1.0144	1.0004	0.9895	0.9985	1.4184	0.3809	0.5284	0.3794
0.4	1.3143E-03	1.0576	1.1982	1.3804	1.4770	1.0154	1.0005	0.9899	0.9986	1.0305	0.3809	0.5309	0.3818
0.5	8.8160E-03	1.0586	0.8965	0.8753	0.8514	1.0167	1.0006	0.9903	0.9986	0.8530	0.3809	0.5341	0.3846
0.6	2.8860E-02	1.0598	0.7348	0.6491	0.5983	1.0182	1.0008	0.9907	0.9987	0.7586	0.3611	0.5377	0.3878
0.7	6.3379E-02	1.0611	0.6394	0.5303	0.4733	1.0197	1.0009	0.9911	0.9987	0.7035	0.3814	0.5416	0.3914
0.8	1.0931E-01	1.0627	0.5794	0.4613	0.4034	1.0214	1.0011	0.9916	0.9988	0.6695	0.3819	0.5458	0.3951
0.9	1.6140E-01	1.0644	0.5401	0.4184	0.3612	1.0232	1.0012	0.9920	0.9989	0.6477	0.3826	0.5502	0.3990
1.0	2.1472E-01	1.0663	0.5137	0.3907	0.3343	1.0251	1.0014	0.9925	0.9989	0.6336	0.3834	0.5547	0.4030
1.2	3.1214E-01	1.0705	0.4833	0.3599	0.3051	1.0288	1.0018	0.9933	0.9990	0.6184	0.3854	0.5638	0.4109
1.4	3.8834E-01	1.0750	0.4694	0.3464	0.2925	1.0327	1.0022	0.9939	0.9991	0.6127	0.3877	0.5727	0.4167
1.6	4.4276E-01	1.0798	0.4640	0.3414	0.2880	1.0365	1.0025	0.9945	0.9992	0.6118	0.3903	0.5812	0.4262
1.8	4.7929E-01	1.0849	0.4635	0.3410	0.2878	1.0403	1.0029	0.9950	0.9993	0.6136	0.3929	0.5894	0.4334
2.0	5.0241E-01	1.0902	0.4658	0.3434	0.2901	1.0440	1.0033	0.9955	0.9993	0.6167	0.3955	0.5972	0.4402
2.2	5.1587E-01	1.0955	0.4697	0.3473	0.2939	1.0477	1.0036	0.9959	0.9994	0.6206	0.3981	0.6045	0.4467
2.4	5.2253E-01	1.1010	0.4746	0.3522	0.2986	1.0514	1.0040	0.9962	0.9994	0.6249	0.4005	0.6115	0.4528
2.6	5.2447E-01	1.1066	0.4802	0.3577	0.3039	1.0550	1.0044	0.9965	0.9995	0.6295	0.4029	0.6180	0.4586
2.8	5.2316E-01	1.1122	0.4860	0.3635	0.3094	1.0585	1.0047	0.9967	0.9995	0.6340	0.4051	0.6242	0.4642
3.0	5.1964E-01	1.1179	0.4921	0.3694	0.3152	1.0620	1.0051	0.9970	0.9995	0.6386	0.4071	0.6300	0.4695
3.2	5.1464E-01	1.1236	0.4981	0.3755	0.3210	1.0655	1.0054	0.9972	0.9995	0.6430	0.4091	0.6355	0.4745
3.4	5.0867E-01	1.1294	0.5042	0.3816	0.3268	1.0689	1.0058	0.9973	0.9996	0.6474	0.4109	0.6408	0.4794
3.6	5.0210E-01	1.1351	0.5102	0.3876	0.3326	1.0722	1.0061	0.9975	0.9996	0.6516	0.4126	0.6457	0.4840
3.8	4.9519E-01	1.1409	0.5161	0.3935	0.3383	1.0756	1.0065	0.9976	0.9996	0.6557	0.4141	0.6505	0.4884
4.0	4.8811E-01	1.1467	0.5219	0.3994	0.3440	1.0788	1.0069	0.9978	0.9996	0.6597	0.4156	0.6550	0.4927
4.5	4.7044E-01	1.1612	0.5358	0.4135	0.3576	1.0868	1.0077	0.9980	0.9997	0.6690	0.4188	0.6653	0.5027
5.0	4.5357E-01	1.1757	0.5488	0.4268	0.3706	1.0946	1.0086	0.9982	0.9997	0.6775	0.4214	0.6746	0.5118
5.5	4.3791E-01	1.1903	0.5610	0.4393	0.3828	1.1022	1.0095	0.9984	0.9997	0.6853	0.4235	0.6830	0.5203
6.0	4.2356E-01	1.2048	0.5722	0.4511	0.3943	1.1097	1.0103	0.9986	0.9998	0.6925	0.4252	0.6905	0.5281
6.5	4.1048E-01	1.2194	0.5827	0.4621	0.4051	1.1169	1.0112	0.9987	0.9998	0.6990	0.4266	0.6974	0.5354
7.0	3.9859E-01	1.2339	0.5925	0.4725	0.4154	1.1240	1.0121	0.9988	0.9998	0.7051	0.4276	0.7037	0.5423
7.5	3.8776E-01	1.2484	0.6016	0.4822	0.4250	1.1310	1.0129	0.9989	0.9998	0.7107	0.4283	0.7095	0.5487
8.0	3.7789E-01	1.2629	0.6101	0.4913	0.4341	1.1379	1.0138	0.9990	0.9998	0.7158	0.4288	0.7148	0.5548
9.0	3.6062E-01	1.2920	0.6255	0.5080	0.4508	1.1512	1.0155	0.9991	0.9998	0.7250	0.4291	0.7243	0.5661
10.0	3.4608E-01	1.3210	0.6390	0.5229	0.4657	1.1641	1.0171	0.9992	0.9999	0.7331	0.4288	0.7325	0.5764
11.0	3.3371E-01	1.3501	0.6509	0.5362	0.4791	1.1767	1.0188	0.9993	0.9999	0.7400	0.4279	0.7396	0.5858
12.0	3.2311E-01	1.3793	0.6615	0.5481	0.4912	1.1889	1.0205	0.9994	0.9999	0.7462	0.4265	0.7458	0.5945
13.0	3.1395E-01	1.4085	0.6709	0.5588	0.5021	1.2009	1.0221	0.9994	0.9999	0.7516	0.4248	0.7513	0.6026
14.0	3.0598E-01	1.4378	0.6793	0.5684	0.5120	1.2126	1.0238	0.9995	0.9999	0.7564	0.4227	0.7562	0.6102
16.0	2.9287E-01	1.4969	0.6934	0.5849	0.5290	1.2353	1.0271	0.9995	0.9999	0.7645	0.4179	0.7644	0.6241
18.0	2.8263E-01	1.5565	0.7048	0.5984	0.5430	1.2571	1.0303	0.9996	0.9999	0.7710	0.4123	0.7709	0.6365
20.0	2.7454E-01	1.6168	0.7138	0.6094	0.5546	1.2783	1.0336	0.9996	0.9999	0.7762	0.4061	0.7761	0.6478
25.0	2.6076E-01	1.7713	0.7291	0.6286	0.5750	1.3289	1.0416	0.9997	0.9999	0.7851	0.3890	0.7851	0.6724
30.0	2.5294E-01	1.9318	0.7369	0.6392	0.5867	1.3768	1.0496	0.9998	1.0000	0.7900	0.3702	0.7900	0.6933
35.0	2.4889E-01	2.0993	0.7397	0.6439	0.5923	1.4226	1.0575	0.9998	1.0000	0.7923	0.3505	0.7923	0.7117
40.0	2.4740E-01	2.2746	0.7391	0.6444	0.5935	1.4668	1.0654	0.9998	1.0000	0.7929	0.3302	0.7929	0.7282
45.0	2.4780E-01	2.4583	0.7360	0.6418	0.5914	1.5097	1.0734	0.9999	1.0000	0.7922	0.3095	0.7922	0.7435
50.0	2.4964E-01	2.6510	0.7311	0.6369	0.5868	1.5514	1.0813	0.9999	1.0000	0.7907	0.2886	0.7907	0.7577

Instr. see p. 22

Table III. Ratio of screened to unscreened β-decay Coulomb functions

ELECTRON Z= 2 A= 3

p	$F_0L_0^*/F_0L_0$	λ_2^*/λ_2	μ_1^*/μ_1	μ_2^*/μ_2	Λ_1^*/Λ_1	Λ_2^*/Λ_2	ν_{12}^*/ν_{12}	$\bar{\nu}_{12}^*/\bar{\nu}_{12}$	η_{12}^*/η_{12}	$\bar{\eta}_{12}^*/\bar{\eta}_{12}$
0.1	0.9967	0.9897	1.0001	1.0001	0.9948	0.9949	0.9897	0.9923	9.948E-01	9.948E-01
					Z= 3	A= 6				
0.1	0.9912	0.9659	1.0002	1.0002	0.9826	0.9834	0.9662	0.9612	9.824E-01	9.826E-01
0.2	0.9966	0.9910	1.0002	1.0002	0.9956	0.9957	0.9912	0.9891	9.955E-01	9.956E-01
					Z= 4	A= 8				
0.1	0.9887	0.9446	1.0003	1.0003	0.9713	0.9738	0.9457	0.9318	9.710E-01	9.713E-01
0.2	0.9949	0.9849	1.0003	1.0003	0.9927	0.9929	0.9853	0.9790	9.924E-01	9.927E-01
0.4	0.9981	0.9959	1.0003	1.0003	0.9982	0.9983	0.9962	0.9937	9.980E-01	9.982E-01
					Z= 5	A= 10				
0.1	0.9874	0.9227	1.0004	1.0004	0.9590	0.9643	0.9247	0.9068	9.586E-01	9.590E-01
0.2	0.9933	0.9782	1.0004	1.0004	0.9893	0.9898	0.9788	0.9697	9.889E-01	9.893E-01
0.4	0.9973	0.9940	1.0004	1.0004	0.9974	0.9975	0.9945	0.9914	9.970E-01	9.974E-01
					Z= 6	A= 12				
0.1	0.9874	0.9013	1.0006	1.0005	0.9461	0.9548	0.9037	0.8875	9.456E-01	9.461E-01
0.2	0.9919	0.9710	1.0006	1.0005	0.9857	0.9866	0.9719	0.9613	9.851E-01	9.857E-01
0.4	0.9965	0.9920	1.0006	1.0005	0.9965	0.9966	0.9926	0.9890	9.960E-01	9.965E-01
					Z= 7	A= 14				
0.1	0.9880	0.8802	1.0007	1.0006	0.9322	0.9451	0.8823	0.8723	9.316E-01	9.322E-01
0.2	0.9908	0.9633	1.0008	1.0007	0.9817	0.9831	0.9645	0.9536	9.810E-01	9.817E-01
0.4	0.9957	0.9897	1.0007	1.0007	0.9955	0.9956	0.9905	0.9861	9.948E-01	9.955E-01
					Z= 8	A= 16				
0.1	0.9892	0.8599	1.0009	1.0007	0.9173	0.9348	0.8607	0.8605	9.167E-01	9.174E-01
0.2	0.9899	0.9551	1.0009	1.0009	0.9774	0.9795	0.9566	0.9465	9.766E-01	9.774E-01
0.4	0.9949	0.9872	1.0009	1.0009	0.9944	0.9946	0.9882	0.9835	9.936E-01	9.944E-01
					Z= 9	A= 19				
0.1	0.9906	0.8412	1.0011	1.0009	0.9019	0.9241	0.8393	0.8514	9.011E-01	9.020E-01
0.2	0.9891	0.9468	1.0011	1.0010	0.9728	0.9757	0.9484	0.9405	9.719E-01	9.729E-01
0.4	0.9941	0.9846	1.0011	1.0011	0.9933	0.9935	0.9858	0.9811	9.922E-01	9.933E-01

Instr. see p. 25

ELECTRON Z= 10 A= 20

p	$F_0L_0^*/F_0L_0$	λ_2^*/λ_2	μ_1^*/μ_1	μ_2^*/μ_2	Λ_1^*/Λ_1	Λ_2^*/Λ_2	ν_{12}^*/ν_{12}	$\hat{\nu}_{12}^*/\hat{\nu}_{12}$	η_{12}^*/η_{12}	$\hat{\eta}_{12}^*/\hat{\eta}_{12}$
0.1	0.9920	0.8237	1.0013	1.0010	0.8853	0.9124	0.8177	0.8442	8.844E-01	8.855E-01
0.2	0.9886	0.9382	1.0014	1.0012	0.9680	0.9717	0.9398	0.9350	9.668E-01	9.680E-01
0.4	0.9933	0.9818	1.0013	1.0012	0.9920	0.9924	0.9832	0.9788	9.908E-01	9.920E-01
0.6	0.9955	0.9911	1.0012	1.0012	0.9967	0.9968	0.9923	0.9899	9.955E-01	9.967E-01
Z= 11 A= 23										
0.1	0.9933	0.8078	1.0016	1.0012	0.8679	0.8996	0.7963	0.8387	8.668E-01	8.680E-01
0.2	0.9882	0.9296	1.0016	1.0014	0.9628	0.9674	0.9310	0.9304	9.615E-01	9.629E-01
0.4	0.9925	0.9789	1.0015	1.0014	0.9907	0.9911	0.9805	0.9767	9.892E-01	9.907E-01
0.6	0.9949	0.9896	1.0014	1.0014	0.9961	0.9962	0.9910	0.9889	9.947E-01	9.961E-01
Z= 12 A= 24										
0.1	0.9937	0.7962	1.0018	1.0013	0.8535	0.8896	0.7802	0.8328	8.523E-01	8.536E-01
0.2	0.9876	0.9215	1.0018	1.0015	0.9580	0.9640	0.9231	0.9224	9.566E-01	9.582E-01
0.4	0.9916	0.9759	1.0018	1.0016	0.9893	0.9900	0.9778	0.9714	9.877E-01	9.894E-01
0.6	0.9942	0.9880	1.0016	1.0016	0.9955	0.9957	0.9896	0.9854	9.939E-01	9.955E-01
Z= 13 A= 27										
0.1	0.9947	0.7867	1.0020	1.0015	0.8379	0.8786	0.7648	0.8296	8.366E-01	8.380E-01
0.2	0.9877	0.9142	1.0020	1.0017	0.9532	0.9602	0.9154	0.9192	9.516E-01	9.534E-01
0.4	0.9909	0.9731	1.0020	1.0018	0.9880	0.9888	0.9752	0.9698	9.862E-01	9.881E-01
0.6	0.9936	0.9866	1.0018	1.0018	0.9950	0.9952	0.9884	0.9845	9.932E-01	9.950E-01
Z= 14 A= 29										
0.1	0.9955	0.7790	1.0022	1.0016	0.8224	0.8674	0.7510	0.8274	8.209E-01	8.223E-01
0.2	0.9879	0.9074	1.0023	1.0019	0.9484	0.9565	0.9078	0.9167	9.467E-01	9.486E-01
0.4	0.9904	0.9704	1.0022	1.0020	0.9867	0.9877	0.9727	0.9685	9.847E-01	9.868E-01
0.6	0.9931	0.9851	1.0020	1.0020	0.9944	0.9947	0.9872	0.9838	9.925E-01	9.945E-01
Z= 15 A= 31										
0.1	0.9960	0.7729	1.0025	1.0018	0.8068	0.8560	0.7387	0.8255	8.050E-01	8.066E-01
0.2	0.9881	0.9009	1.0025	1.0021	0.9436	0.9527	0.9004	0.9144	9.417E-01	9.438E-01
0.4	0.9899	0.9677	1.0024	1.0022	0.9854	0.9865	0.9701	0.9672	9.832E-01	9.855E-01
0.6	0.9925	0.9837	1.0023	1.0021	0.9939	0.9941	0.9859	0.9830	9.917E-01	9.939E-01

ELECTRON Z= 16 A= 34

p	$F_0L_0^*/F_0L_0$	λ_2^*/λ_2	μ_1^*/μ_1	μ_2^*/μ_2	Λ_1^*/Λ_1	Λ_2^*/Λ_2	ν_{12}^*/ν_{12}	$\vartheta_{12}^*/\vartheta_{12}$	η_{12}^*/η_{12}	$\hat{\eta}_{12}^*/\hat{\eta}_{12}$
0.1	0.9962	0.7677	1.0027	1.0019	0.7910	0.8443	0.7274	0.8239	7.888E-01	7.904E-01
0.2	0.9883	0.8946	1.0027	1.0022	0.9386	0.9489	0.8931	0.9122	9.365E-01	9.388E-01
0.4	0.9894	0.9649	1.0026	1.0024	0.9841	0.9853	0.9675	0.9658	9.817E-01	9.842E-01
0.6	0.9920	0.9822	1.0025	1.0023	0.9933	0.9936	0.9846	0.9822	9.909E-01	9.933E-01
					Z= 17	A= 37				
0.1	0.9962	0.7637	1.0030	1.0021	0.7749	0.8322	0.7175	0.8226	7.723E-01	7.740E-01
0.2	0.9885	0.8887	1.0030	1.0024	0.9336	0.9451	0.8859	0.9102	9.313E-01	9.338E-01
0.4	0.9889	0.9622	1.0029	1.0026	0.9827	0.9841	0.9648	0.9645	9.801E-01	9.828E-01
0.6	0.9915	C.9807	1.0027	1.0026	0.9927	0.9931	0.9833	0.9814	9.901E-01	9.927E-01
					Z= 18	A= 39				
0.1	0.9961	0.7605	1.0033	1.0023	0.7586	0.8198	0.7087	0.8214	7.554E-01	7.572E-01
0.2	0.9887	0.8831	1.0032	1.0026	C.9285	0.9411	0.8788	0.9084	9.260E-01	9.287E-01
0.4	0.9884	C.9594	1.0031	1.0028	0.9813	0.9829	0.9622	0.9631	9.785E-01	9.814E-01
0.6	0.9909	0.9791	1.0029	1.0028	0.9921	0.9925	0.9820	0.9806	9.893E-01	9.921E-01
0.8	0.9927	0.9868	1.0027	1.0026	0.9959	0.9960	0.9894	0.9881	9.933E-01	9.959E-01
					Z= 19	A= 41				
0.1	0.9958	0.7580	1.0035	1.0025	0.7419	0.8068	0.7009	0.8203	7.380E-01	7.399E-01
0.2	0.9889	0.8777	1.0035	1.0028	0.9232	0.9371	0.8717	0.9067	9.206E-01	9.234E-01
0.4	0.9880	0.9565	1.0034	1.0031	0.9798	0.9817	0.9594	0.9618	9.768E-01	9.800E-01
0.6	0.9904	0.9775	1.0032	1.0030	0.9914	0.9919	0.9806	0.9798	9.885E-01	9.915E-01
0.8	0.9922	0.9857	1.0029	1.0028	0.9956	0.9957	0.9886	0.9876	9.928E-01	9.956E-01
					Z= 20	A= 43				
0.1	0.9956	0.7567	1.0038	1.0027	0.7262	0.7959	0.6958	0.8189	7.214E-01	7.233E-01
0.2	0.9892	0.8731	1.0038	1.0030	0.9182	0.9333	0.8654	0.9052	9.153E-01	9.183E-01
0.4	0.9877	0.9539	1.0036	1.0033	0.9784	0.9805	0.9568	0.9611	9.752E-01	9.786E-01
0.6	0.9900	0.9761	1.0034	1.0032	0.9909	0.9914	0.9793	0.9794	9.877E-01	9.909E-01
0.8	0.9918	0.9847	1.0031	1.0030	0.9953	0.9954	0.9878	0.9874	9.923E-01	9.953E-01
					Z= 21	A= 47				
0.1	0.9954	0.7560	1.0041	1.0028	0.7103	0.7845	0.6917	0.8180	7.047E-01	7.065E-01
0.2	0.9894	0.8688	1.0040	1.0032	0.9131	0.9295	0.8592	0.9040	9.100E-01	9.132E-01
0.4	0.9874	0.9514	1.0039	1.0035	0.9770	0.9793	0.9543	0.9604	9.736E-01	9.772E-01
0.6	0.9895	0.9746	1.0036	1.0034	0.9902	C.9908	0.9780	0.9791	9.869E-01	9.903E-01
0.8	0.9914	0.9838	1.0033	1.0032	0.9949	0.9951	0.9870	0.9873	9.918E-01	9.950E-01

Instr. see p. 25

ELECTRON Z= 22 A= 48

p	$F_0L_0^*/F_0L_0$	λ_2^*/λ_2	μ_1^*/μ_1	μ_2^*/μ_2	Λ_1^*/Λ_1	Λ_2^*/Λ_2	ν_{12}^*/ν_{12}	$\hat{\nu}_{12}^*/\hat{\nu}_{12}$	η_{12}^*/η_{12}	$\hat{\eta}_{12}^*/\hat{\eta}_{12}$
0.1	0.9957	0.7636	1.0042	1.0030	0.6968	0.7708	0.6935	0.8275	6.895E-01	6.914E-01
0.2	0.9905	0.8696	1.0042	1.0034	0.9098	0.9258	0.8562	0.9118	9.063E-01	9.096E-01
0.4	0.9880	0.9510	1.0040	1.0036	0.9764	0.9784	0.9532	0.9659	9.728E-01	9.765E-01
0.6	0.9898	0.9743	1.0037	1.0035	0.9901	0.9905	0.9776	0.9832	9.866E-01	9.901E-01
0.8	0.9915	0.9836	1.0033	1.0033	0.9949	0.9950	0.9868	0.9905	9.916E-01	9.949E-01
					Z= 23	A= 51				
0.1	0.9945	0.7558	1.0046	1.0033	0.6772	0.7581	0.6851	0.8174	6.688E-01	6.706E-01
0.2	0.9896	0.8609	1.0046	1.0036	0.9026	0.9215	0.8470	0.9020	8.989E-01	9.024E-01
0.4	0.9867	0.9461	1.0044	1.0039	0.9740	0.9768	0.9490	0.9586	9.702E-01	9.742E-01
0.6	0.9886	0.9715	1.0041	1.0038	0.9890	0.9896	0.9752	0.9780	9.851E-01	9.891E-01
0.8	0.9905	0.9817	1.0038	1.0036	0.9943	0.9945	0.9853	0.9867	9.907E-01	9.943E-01
					Z= 24	A= 53				
0.1	0.9940	0.7557	1.0050	1.0035	0.6595	0.7440	0.6826	0.8166	6.494E-01	6.512E-01
0.2	0.9895	0.8571	1.0049	1.0038	0.8970	0.9172	0.8408	0.9008	8.930E-01	8.967E-01
0.4	0.9864	0.9434	1.0047	1.0042	0.9725	0.9754	0.9462	0.9574	9.683E-01	9.726E-01
0.6	0.9881	0.9698	1.0044	1.0041	0.9883	0.9890	0.9738	0.9772	9.842E-01	9.884E-01
0.8	0.9900	0.9806	1.0040	1.0038	0.9939	0.9942	0.9844	0.9861	9.901E-01	9.940E-01
					Z= 25	A= 55				
0.1	0.9934	0.7560	1.0053	1.0037	0.6416	0.7293	0.6811	0.8162	6.294E-01	6.311E-01
0.2	0.9894	0.8537	1.0052	1.0041	0.8914	0.9128	0.8349	0.8999	8.870E-01	8.910E-01
0.4	0.9860	0.9408	1.0050	1.0044	0.9709	0.9741	0.9434	0.9565	9.665E-01	9.711E-01
0.6	0.9876	0.9682	1.0047	1.0043	0.9876	0.9884	0.9723	0.9765	9.833E-01	9.877E-01
0.8	0.9895	0.9795	1.0043	1.0041	0.9935	0.9938	0.9835	0.9856	9.895E-01	9.936E-01
1.0	0.9910	0.9852	1.0038	1.0037	0.9962	0.9964	0.9890	0.9904	9.925E-01	9.962E-01
					Z= 26	A= 56				
0.1	0.9928	0.7563	1.0056	1.0039	0.6229	0.7140	0.6801	0.8154	6.083E-01	6.099E-01
0.2	0.9892	0.8503	1.0055	1.0043	0.8856	0.9082	0.8289	0.8990	8.808E-01	8.849E-01
0.4	0.9857	0.9381	1.0053	1.0047	0.9692	0.9728	0.9406	0.9555	9.646E-01	9.694E-01
0.6	0.9871	0.9665	1.0049	1.0046	0.9869	0.9878	0.9708	0.9758	9.823E-01	9.870E-01
0.8	0.9890	0.9783	1.0045	1.0043	0.9931	0.9935	0.9826	0.9851	9.889E-01	9.932E-01
1.0	0.9905	0.9843	1.0041	1.0039	0.9960	0.9961	0.9883	0.9900	9.920E-01	9.960E-01

ELECTRON Z= 27 A= 59

p	$F_0L_0^*/F_0L_0$	λ_2^*/λ_2	μ_1^*/μ_1	μ_2^*/μ_2	Λ_1^*/Λ_1	Λ_2^*/Λ_2	ν_{12}^*/ν_{12}	$\hat{\nu}_{12}^*/\hat{\nu}_{12}$	η_{12}^*/η_{12}	$\hat{\eta}_{12}^*/\hat{\eta}_{12}$
0.1	0.9922	0.7569	1.0059	1.0042	0.6041	0.6985	0.6801	0.8148	5.865E-01	5.880E-01
0.2	0.9889	0.8473	1.0059	1.0045	0.8798	0.9035	0.8232	0.8982	8.745E-01	8.788E-01
0.4	0.9853	0.9355	1.0056	1.0049	0.9676	0.9714	0.9378	0.9546	9.627E-01	9.678E-01
0.6	0.9866	0.9649	1.0052	1.0048	0.9861	0.9871	0.9693	0.9751	9.813E-01	9.863E-01
0.8	0.9885	0.9772	1.0048	1.0045	0.9928	0.9931	0.9817	0.9847	9.882E-01	9.928E-01
1.0	0.9900	0.9835	1.0043	1.0042	0.9958	0.9959	0.9877	0.9897	9.916E-01	9.958E-01

Z= 28 A= 61

p	$F_0L_0^*/F_0L_0$	λ_2^*/λ_2	μ_1^*/μ_1	μ_2^*/μ_2	Λ_1^*/Λ_1	Λ_2^*/Λ_2	ν_{12}^*/ν_{12}	$\hat{\nu}_{12}^*/\hat{\nu}_{12}$	η_{12}^*/η_{12}	$\hat{\eta}_{12}^*/\hat{\eta}_{12}$
0.1	0.9916	0.7576	1.0063	1.0044	0.5847	0.6824	0.6807	0.8141	5.638E-01	5.650E-01
0.2	0.9886	0.8444	1.0062	1.0048	0.8739	0.8987	0.8177	0.8975	8.681E-01	8.725E-01
0.4	0.9850	0.9329	1.0059	1.0052	0.9659	0.9700	0.9349	0.9537	9.608E-01	9.661E-01
0.6	0.9862	0.9632	1.0055	1.0051	0.9854	0.9865	0.9678	0.9745	9.803E-01	9.855E-01
0.8	0.9880	0.9760	1.0050	1.0048	0.9924	0.9928	0.9807	0.9842	9.876E-01	9.925E-01
1.0	0.9896	0.9826	1.0046	1.0044	0.9955	0.9957	0.9870	0.9894	9.911E-01	9.956E-01

Z= 29 A= 65

p	$F_0L_0^*/F_0L_0$	λ_2^*/λ_2	μ_1^*/μ_1	μ_2^*/μ_2	Λ_1^*/Λ_1	Λ_2^*/Λ_2	ν_{12}^*/ν_{12}	$\hat{\nu}_{12}^*/\hat{\nu}_{12}$	η_{12}^*/η_{12}	$\hat{\eta}_{12}^*/\hat{\eta}_{12}$
0.1	0.9910	0.7585	1.0066	1.0047	0.5650	0.6659	0.6821	0.8134	5.401E-01	5.411E-01
0.2	0.9883	0.8418	1.0065	1.0050	0.8677	0.8938	0.8123	0.8968	8.614E-01	8.661E-01
0.4	0.9846	0.9303	1.0062	1.0054	0.9642	0.9686	0.9321	0.9529	9.588E-01	9.644E-01
0.6	0.9857	0.9615	1.0058	1.0053	0.9846	0.9858	0.9662	0.9739	9.793E-01	9.848E-01
0.8	0.9875	0.9749	1.0053	1.0050	0.9920	0.9924	0.9798	0.9837	9.870E-01	9.921E-01
1.0	0.9891	0.9817	1.0048	1.0046	0.9953	0.9955	0.9864	0.9890	9.907E-01	9.953E-01

Z= 30 A= 67

p	$F_0L_0^*/F_0L_0$	λ_2^*/λ_2	μ_1^*/μ_1	μ_2^*/μ_2	Λ_1^*/Λ_1	Λ_2^*/Λ_2	ν_{12}^*/ν_{12}	$\hat{\nu}_{12}^*/\hat{\nu}_{12}$	η_{12}^*/η_{12}	$\hat{\eta}_{12}^*/\hat{\eta}_{12}$
0.1	0.9903	0.7593	1.0070	1.0050	0.5447	0.6489	0.6841	0.8126	5.152E-01	5.159E-01
0.2	0.9879	0.8394	1.0069	1.0053	0.8615	0.8887	0.8071	0.8962	8.546E-01	8.594E-01
0.4	0.9843	0.9278	1.0066	1.0057	0.9624	0.9671	0.9293	0.9522	9.568E-01	9.626E-01
0.6	0.9852	0.9598	1.0061	1.0056	0.9839	0.9852	0.9647	0.9733	9.783E-01	9.840E-01
0.8	0.9870	0.9737	1.0056	1.0053	0.9916	0.9920	0.9788	0.9833	9.863E-01	9.917E-01
1.0	0.9887	0.9808	1.0051	1.0048	0.9951	0.9953	0.9857	0.9887	9.902E-01	9.951E-01

Z= 31 A= 71

p	$F_0L_0^*/F_0L_0$	λ_2^*/λ_2	μ_1^*/μ_1	μ_2^*/μ_2	Λ_1^*/Λ_1	Λ_2^*/Λ_2	ν_{12}^*/ν_{12}	$\hat{\nu}_{12}^*/\hat{\nu}_{12}$	η_{12}^*/η_{12}	$\hat{\eta}_{12}^*/\hat{\eta}_{12}$
0.1	0.9897	0.7602	1.0073	1.0052	0.5239	0.6314	0.6866	0.8117	4.891E-01	4.894E-01
0.2	0.9875	0.8370	1.0072	1.0056	0.8552	0.8834	0.8020	0.8955	8.475E-01	8.525E-01
0.4	0.9839	0.9252	1.0069	1.0060	0.9606	0.9656	0.9264	0.9514	9.547E-01	9.608E-01
0.6	0.9847	0.9581	1.0064	1.0059	0.9831	0.9845	0.9631	0.9726	9.772E-01	9.832E-01
0.8	0.9865	0.9725	1.0059	1.0055	0.9911	0.9916	0.9778	0.9828	9.856E-01	9.912E-01
1.0	0.9882	0.9799	1.0053	1.0051	0.9948	0.9950	0.9850	0.9884	9.897E-01	9.949E-01
1.2	0.9896	0.9843	1.0048	1.0047	0.9967	0.9968	0.9890	0.9917	9.920E-01	9.967E-01

Instr. see p. 25

ELECTRON Z= 32 A= 72

p	$F_0L_0^*/F_0L_0$	λ_2^*/λ_2	μ_1^*/μ_1	μ_2^*/μ_2	Λ_1^*/Λ_1	Λ_2^*/Λ_2	ν_{12}^*/ν_{12}	$\vartheta_{12}^*/\vartheta_{12}$	η_{12}^*/η_{12}	$\hat{\eta}_{12}^*/\hat{\eta}_{12}$
0.1	0.9891	0.7617	1.0077	1.0055	0.5051	0.6164	0.6906	0.8110	4.645E-01	4.643E-01
0.2	0.9871	0.8353	1.0076	1.0058	0.8492	0.8787	0.7979	0.8950	8.408E-01	8.459E-01
0.4	0.9836	0.9230	1.0072	1.0062	0.9590	0.9642	0.9237	0.9509	9.527E-01	9.590E-01
0.6	0.9843	0.9566	1.0067	1.0061	0.9823	0.9838	0.9616	0.9723	9.762E-01	9.825E-01
0.8	0.9861	0.9713	1.0062	1.0058	0.9908	0.9913	0.9769	0.9826	9.850E-01	9.909E-01
1.0	0.9878	0.9791	1.0056	1.0053	0.9946	0.9948	0.9844	0.9882	9.893E-01	9.946E-01
1.2	0.9892	0.9836	1.0050	1.0049	0.9966	0.9967	0.9885	0.9916	9.917E-01	9.966E-01

Z= 33 A= 77

p	$F_0L_0^*/F_0L_0$	λ_2^*/λ_2	μ_1^*/μ_1	μ_2^*/μ_2	Λ_1^*/Λ_1	Λ_2^*/Λ_2	ν_{12}^*/ν_{12}	$\vartheta_{12}^*/\vartheta_{12}$	η_{12}^*/η_{12}	$\hat{\eta}_{12}^*/\hat{\eta}_{12}$
0.1	0.9885	0.7634	1.0080	1.0058	0.4866	0.6013	0.6954	0.8106	4.393E-01	4.387E-01
0.2	0.9867	0.8339	1.0079	1.0061	0.8434	0.8739	0.7941	0.8945	8.340E-01	8.392E-01
0.4	0.9833	0.9209	1.0076	1.0065	0.9573	0.9628	0.9211	0.9504	9.507E-01	9.573E-01
0.6	0.9839	0.9550	1.0070	1.0064	0.9816	0.9832	0.9602	0.9719	9.752E-01	9.817E-01
0.8	0.9857	0.9702	1.0064	1.0060	0.9904	0.9909	0.9759	0.9823	9.844E-01	9.905E-01
1.0	0.9873	0.9782	1.0058	1.0055	0.9944	0.9946	0.9837	0.9880	9.888E-01	9.944E-01
1.2	0.9888	0.9829	1.0053	1.0051	0.9964	0.9966	0.9880	0.9914	9.913E-01	9.964E-01

Z= 34 A= 78

p	$F_0L_0^*/F_0L_0$	λ_2^*/λ_2	μ_1^*/μ_1	μ_2^*/μ_2	Λ_1^*/Λ_1	Λ_2^*/Λ_2	ν_{12}^*/ν_{12}	$\vartheta_{12}^*/\vartheta_{12}$	η_{12}^*/η_{12}	$\hat{\eta}_{12}^*/\hat{\eta}_{12}$
0.1	0.9879	0.7650	1.0084	1.0060	0.4678	0.5860	0.7004	0.8099	4.130E-01	4.118E-01
0.2	0.9862	0.8325	1.0083	1.0064	0.8374	0.8690	0.7904	0.8939	8.271E-01	8.324E-01
0.4	0.9829	0.9187	1.0079	1.0067	0.9556	0.9614	0.9185	0.9497	9.487E-01	9.555E-01
0.6	0.9834	0.9534	1.0073	1.0066	0.9808	0.9825	0.9587	0.9713	9.742E-01	9.810E-01
0.8	0.9852	0.9691	1.0067	1.0063	0.9900	0.9906	0.9750	0.9819	9.837E-01	9.901E-01
1.0	0.9869	0.9773	1.0061	1.0058	0.9941	0.9944	0.9830	0.9877	9.883E-01	9.942E-01
1.2	0.9884	0.9822	1.0055	1.0053	0.9963	0.9964	0.9876	0.9911	9.909E-01	9.963E-01

Z= 35 A= 79

p	$F_0L_0^*/F_0L_0$	λ_2^*/λ_2	μ_1^*/μ_1	μ_2^*/μ_2	Λ_1^*/Λ_1	Λ_2^*/Λ_2	ν_{12}^*/ν_{12}	$\vartheta_{12}^*/\vartheta_{12}$	η_{12}^*/η_{12}	$\hat{\eta}_{12}^*/\hat{\eta}_{12}$
0.1	0.9873	0.7667	1.0087	1.0063	0.4489	0.5706	0.7060	0.8093	3.856E-01	3.837E-01
0.2	0.9857	0.8313	1.0086	1.0066	0.8314	0.8641	0.7871	0.8934	8.200E-01	8.254E-01
0.4	0.9826	0.9166	1.0082	1.0070	0.9538	0.9600	0.9160	0.9491	9.466E-01	9.537E-01
0.6	0.9830	0.9519	1.0077	1.0069	0.9801	0.9818	0.9572	0.9708	9.731E-01	9.802E-01
0.8	0.9847	0.9679	1.0070	1.0065	0.9895	0.9902	0.9740	0.9815	9.830E-01	9.897E-01
1.0	0.9864	0.9764	1.0064	1.0060	0.9939	0.9942	0.9824	0.9874	9.878E-01	9.939E-01
1.2	0.9880	0.9815	1.0058	1.0055	0.9961	0.9963	0.9870	0.9909	9.906E-01	9.961E-01

Instr. see p. 25

ELECTRON Z= 36 A= 84

p	$F_0L_0^*/F_0L_0$	λ_2^*/λ_2	μ_1^*/μ_1	μ_2^*/μ_2	Λ_1^*/Λ_1	Λ_2^*/Λ_2	ν_{12}^*/ν_{12}	$\vartheta_{12}^*/\vartheta_{12}$	η_{12}^*/η_{12}	$\tilde{\eta}_{12}^*/\tilde{\eta}_{12}$
0.1	0.9866	0.7685	1.0091	1.0066	0.4302	0.5550	0.7122	0.8088	3.572E-01	3.544E-01
0.2	0.9852	0.8304	1.0090	1.0069	0.8254	0.8591	0.7841	0.8930	8.128E-01	8.184E-01
0.4	0.9822	0.9146	1.0086	1.0073	0.9521	0.9586	0.9135	0.9486	9.446E-01	9.520E-01
0.6	0.9826	0.9504	1.0080	1.0072	0.9793	0.9812	0.9557	0.9704	9.721E-01	9.794E-01
0.8	0.9843	0.9668	1.0073	1.0068	0.9891	0.9898	0.9731	0.9811	9.824E-01	9.893E-01
1.0	0.9860	0.9756	1.0066	1.0063	0.9936	0.9939	0.9817	0.9871	9.873E-01	9.937E-01
1.2	0.9876	0.9808	1.0060	1.0057	0.9960	0.9961	0.9866	0.9906	9.902E-01	9.960E-01
					Z= 37	A= 88				
0.1	0.9859	0.7702	1.0095	1.0069	0.4112	0.5394	0.7184	0.8082	3.272E-01	3.235E-01
0.2	0.9846	0.8295	1.0094	1.0072	0.8193	0.8541	0.7812	0.8924	8.054E-01	8.110E-01
0.4	0.9818	0.9127	1.0089	1.0076	0.9504	0.9571	0.9109	0.9480	9.425E-01	9.501E-01
0.6	0.9821	0.9488	1.0083	1.0074	0.9785	0.9805	0.9542	0.9698	9.710E-01	9.786E-01
0.8	0.9838	0.9656	1.0076	1.0070	0.9887	0.9895	0.9721	0.9807	9.817E-01	9.888E-01
1.0	0.9856	0.9746	1.0069	1.0065	0.9934	0.9937	0.9810	0.9867	9.869E-01	9.935E-01
1.2	0.9871	0.9801	1.0062	1.0060	0.9958	0.9960	0.9861	0.9904	9.898E-01	9.958E-01
					Z= 38	A= 89				
0.1	0.9825	0.7736	1.0098	1.0072	0.3881	0.5345	0.7310	0.8077	2.908E-01	2.851E-01
0.2	0.9827	0.8304	1.0096	1.0076	0.8135	0.8472	0.7810	0.8933	7.977E-01	8.034E-01
0.4	0.9808	0.9115	1.0092	1.0079	0.9491	0.9552	0.9088	0.9491	9.409E-01	9.488E-01
0.6	0.9813	0.9476	1.0086	1.0077	0.9779	0.9799	0.9530	0.9695	9.703E-01	9.782E-01
0.8	0.9830	0.9647	1.0079	1.0072	0.9884	0.9892	0.9714	0.9795	9.812E-01	9.886E-01
1.0	0.9848	0.9738	1.0072	1.0067	0.9931	0.9935	0.9805	0.9852	9.865E-01	9.933E-01
1.2	0.9864	0.9793	1.0065	1.0061	0.9957	0.9959	0.9856	0.9889	9.895E-01	9.958E-01
1.4	0.9878	0.9830	1.0059	1.0056	0.9971	0.9972	0.9888	0.9913	9.914E-01	9.971E-01
					Z= 39	A= 91				
0.1	0.9819	0.7754	1.0101	1.0075	0.3686	0.5191	0.7379	0.8073	2.571E-01	2.500E-01
0.2	0.9822	0.8298	1.0100	1.0079	0.8072	0.8421	0.7789	0.8929	7.898E-01	7.954E-01
0.4	0.9805	0.9098	1.0095	1.0082	0.9474	0.9537	0.9064	0.9489	9.387E-01	9.469E-01
0.6	0.9809	0.9462	1.0089	1.0080	0.9772	0.9792	0.9516	0.9693	9.692E-01	9.774E-01
0.8	0.9826	0.9636	1.0082	1.0075	0.9880	0.9888	0.9704	0.9793	9.805E-01	9.882E-01
1.0	0.9844	0.9730	1.0075	1.0069	0.9929	0.9933	0.9798	0.9851	9.860E-01	9.931E-01
1.2	0.9860	0.9786	1.0068	1.0064	0.9955	0.9957	0.9851	0.9888	9.891E-01	9.956E-01
1.4	0.9874	0.9824	1.0061	1.0058	0.9970	0.9971	0.9884	0.9912	9.911E-01	9.970E-01

 Instr. see p. 25

ELECTRON Z= 40 A= 93

p	$F_0L_0^*/F_0L_0$	λ_2^*/λ_2	μ_1^*/μ_1	μ_2^*/μ_2	Λ_1^*/Λ_1	Λ_2^*/Λ_2	ν_{12}^*/ν_{12}	$\vartheta_{12}^*/\vartheta_{12}$	η_{12}^*/η_{12}	$\hat{\eta}_{12}^*/\hat{\eta}_{12}$
0.1	0.9813	0.7773	1.0105	1.0078	0.3493	0.5037	0.7451	0.8071	2.217E-01	2.130E-01
0.2	0.9816	0.8294	1.0104	1.0082	0.8011	0.8370	0.7771	0.8926	7.817E-01	7.874E-01
0.4	0.9801	0.9081	1.0099	1.0085	0.9456	0.9522	0.9041	0.9487	9.366E-01	9.450E-01
0.6	0.9805	0.9448	1.0092	1.0082	0.9764	0.9785	0.9502	0.9692	9.681E-01	9.766E-01
0.8	0.9821	0.9625	1.0085	1.0077	0.9876	0.9884	0.9695	0.9793	9.798E-01	9.878E-01
1.0	0.9840	0.9721	1.0077	1.0072	0.9927	0.9931	0.9792	0.9851	9.855E-01	9.929E-01
1.2	0.9856	0.9780	1.0070	1.0066	0.9953	0.9956	0.9846	0.9887	9.887E-01	9.955E-01
1.4	0.9871	0.9818	1.0064	1.0060	0.9969	0.9970	0.9880	0.9912	9.908E-01	9.969E-01

Z= 41 A= 97

0.1	0.9807	0.7791	1.0109	1.0081	0.3297	0.4881	0.7525	0.8067	1.841E-01	1.734E-01
0.2	0.9811	0.8291	1.0107	1.0085	0.7948	0.8318	0.7755	0.8922	7.733E-01	7.790E-01
0.4	0.9798	0.9065	1.0102	1.0088	0.9439	0.9506	0.9017	0.9484	9.344E-01	9.431E-01
0.6	0.9801	0.9435	1.0095	1.0085	0.9756	0.9778	0.9487	0.9690	9.671E-01	9.758E-01
0.8	0.9817	0.9614	1.0087	1.0080	0.9872	0.9881	0.9686	0.9791	9.792E-01	9.874E-01
1.0	0.9836	0.9713	1.0080	1.0074	0.9924	0.9929	0.9785	0.9850	9.850E-01	9.926E-01
1.2	0.9853	0.9773	1.0072	1.0068	0.9952	0.9954	0.9841	0.9887	9.884E-01	9.953E-01
1.4	0.9867	0.9812	1.0066	1.0062	0.9968	0.9969	0.9876	0.9911	9.905E-01	9.968E-01

Z= 42 A= 96

0.1	0.9801	0.7810	1.0113	1.0084	0.3103	0.4724	0.7601	0.8064	1.441E-01	1.313E-01
0.2	0.9805	0.8288	1.0111	1.0088	0.7886	0.8266	0.7742	0.8919	7.647E-01	7.703E-01
0.4	0.9794	0.9050	1.0106	1.0091	0.9421	0.9491	0.8994	0.9482	9.321E-01	9.411E-01
0.6	0.9797	0.9421	1.0098	1.0088	0.9749	0.9771	0.9473	0.9688	9.660E-01	9.750E-01
0.8	0.9813	0.9604	1.0090	1.0083	0.9867	0.9877	0.9677	0.9791	9.785E-01	9.869E-01
1.0	0.9832	0.9704	1.0082	1.0076	0.9922	0.9926	0.9779	0.9849	9.845E-01	9.924E-01
1.2	0.9849	0.9766	1.0075	1.0070	0.9950	0.9953	0.9836	0.9886	9.880E-01	9.951E-01
1.4	0.9864	0.9806	1.0068	1.0064	0.9967	0.9968	0.9872	0.9911	9.902E-01	9.967E-01

Z= 43 A= 99

0.1	0.9795	0.7827	1.0116	1.0087	0.2906	0.4566	0.7679	0.8060	1.013E-01	8.580E-02
0.2	0.9799	0.8286	1.0115	1.0091	0.7822	0.8213	0.7731	0.8915	7.558E-01	7.613E-01
0.4	0.9790	0.9034	1.0109	1.0094	0.9403	0.9475	0.8972	0.9479	9.298E-01	9.391E-01
0.6	0.9793	0.9407	1.0102	1.0091	0.9741	0.9764	0.9459	0.9687	9.649E-01	9.741E-01
0.8	0.9809	0.9593	1.0093	1.0085	0.9863	0.9873	0.9667	0.9789	9.778E-01	9.865E-01
1.0	0.9828	0.9696	1.0085	1.0079	0.9919	0.9924	0.9772	0.9848	9.840E-01	9.921E-01
1.2	0.9845	0.9759	1.0077	1.0072	0.9949	0.9951	0.9831	0.9885	9.876E-01	9.950E-01
1.4	0.9860	0.9800	1.0070	1.0066	0.9966	0.9967	0.9868	0.9911	9.899E-01	9.966E-01

ELECTRON Z= 44 A= 99

p	$F_0L_0^*/F_0L_0$	λ_2^*/λ_2	μ_1^*/μ_1	μ_2^*/μ_2	Λ_1^*/Λ_1	Λ_2^*/Λ_2	ν_{12}^*/ν_{12}	$\hat{\nu}_{12}^*/\hat{\nu}_{12}$	η_{12}^*/η_{12}	$\hat{\eta}_{12}^*/\hat{\eta}_{12}$
0.1	0.9789	0.7845	1.0120	1.0091	0.2710	0.4408	0.7758	0.8057	5.553E-02	3.698E-02
0.2	0.9793	0.8285	1.0119	1.0094	0.7759	0.8161	0.7723	0.8912	7.466E-01	7.520E-01
0.4	0.9786	0.9020	1.0113	1.0097	0.9386	0.9460	0.8950	0.9477	9.275E-01	9.370E-01
0.6	0.9789	0.9394	1.0105	1.0094	0.9733	0.9757	0.9445	0.9685	9.637E-01	9.733E-01
0.8	0.9805	0.9583	1.0096	1.0088	0.9859	0.9869	0.9658	0.9788	9.771E-01	9.861E-01
1.0	0.9824	0.9687	1.0088	1.0081	0.9917	0.9922	0.9766	0.9848	9.835E-01	9.919E-01
1.2	0.9841	0.9752	1.0080	1.0075	0.9947	0.9950	0.9827	0.9885	9.872E-01	9.948E-01
1.4	0.9857	0.9794	1.0072	1.0068	0.9965	0.9966	0.9864	0.9910	9.895E-01	9.965E-01
1.8	0.9881	0.9847	1.0061	1.0058	0.9982	0.9983	0.9907	0.9941	9.923E-01	9.982E-01

Z= 45 A= 103

p	$F_0L_0^*/F_0L_0$	λ_2^*/λ_2	μ_1^*/μ_1	μ_2^*/μ_2	Λ_1^*/Λ_1	Λ_2^*/Λ_2	ν_{12}^*/ν_{12}	$\hat{\nu}_{12}^*/\hat{\nu}_{12}$	η_{12}^*/η_{12}	$\hat{\eta}_{12}^*/\hat{\eta}_{12}$
0.1	0.9783	0.7862	1.0124	1.0094	0.2513	0.4248	0.7839	0.8054	6.136E-03	-1.604E-02
0.2	0.9787	0.8285	1.0122	1.0097	0.7696	0.8108	0.7717	0.8908	7.371E-01	7.423E-01
0.4	0.9782	0.9006	1.0116	1.0100	0.9368	0.9444	0.8928	0.9475	9.252E-01	9.349E-01
0.6	0.9785	0.9381	1.0108	1.0097	0.9725	0.9750	0.9431	0.9684	9.626E-01	9.724E-01
0.8	0.9801	0.9572	1.0099	1.0091	0.9855	0.9865	0.9649	0.9787	9.764E-01	9.856E-01
1.0	0.9820	0.9679	1.0090	1.0084	0.9914	0.9919	0.9759	0.9847	9.830E-01	9.916E-01
1.2	0.9838	0.9745	1.0082	1.0077	0.9946	0.9948	0.9822	0.9884	9.868E-01	9.947E-01
1.4	0.9853	0.9788	1.0075	1.0070	0.9964	0.9965	0.9860	0.9910	9.892E-01	9.964E-01
1.8	0.9878	0.9842	1.0062	1.0060	0.9981	0.9982	0.9904	0.9940	9.921E-01	9.981E-01

Z= 46 A= 106

p	$F_0L_0^*/F_0L_0$	λ_2^*/λ_2	μ_1^*/μ_1	μ_2^*/μ_2	Λ_1^*/Λ_1	Λ_2^*/Λ_2	ν_{12}^*/ν_{12}	$\hat{\nu}_{12}^*/\hat{\nu}_{12}$	η_{12}^*/η_{12}	$\hat{\eta}_{12}^*/\hat{\eta}_{12}$
0.1	0.9777	0.7879	1.0128	1.0097	0.2316	0.4087	0.7921	0.8050	-4.742E-02	-7.391E-02
0.2	0.9781	0.8285	1.0126	1.0100	0.7632	0.8054	0.7714	0.8904	7.273E-01	7.322E-01
0.4	0.9778	0.8993	1.0120	1.0103	0.9349	0.9428	0.8907	0.9472	9.228E-01	9.327E-01
0.6	0.9781	0.9368	1.0112	1.0099	0.9717	0.9743	0.9417	0.9682	9.614E-01	9.715E-01
0.8	0.9797	0.9562	1.0102	1.0093	0.9851	0.9861	0.9640	0.9786	9.756E-01	9.852E-01
1.0	0.9816	0.9670	1.0093	1.0086	0.9912	0.9917	0.9752	0.9846	9.825E-01	9.913E-01
1.2	0.9834	0.9737	1.0085	1.0079	0.9944	0.9947	0.9816	0.9884	9.864E-01	9.945E-01
1.4	0.9850	0.9782	1.0077	1.0073	0.9962	0.9964	0.9856	0.9909	9.889E-01	9.963E-01
1.8	0.9875	0.9837	1.0064	1.0061	0.9981	0.9982	0.9901	0.9940	9.918E-01	9.981E-01

Z= 47 A= 109

p	$F_0L_0^*/F_0L_0$	λ_2^*/λ_2	μ_1^*/μ_1	μ_2^*/μ_2	Λ_1^*/Λ_1	Λ_2^*/Λ_2	ν_{12}^*/ν_{12}	$\hat{\nu}_{12}^*/\hat{\nu}_{12}$	η_{12}^*/η_{12}	$\hat{\eta}_{12}^*/\hat{\eta}_{12}$
0.1	0.9771	0.7896	1.0132	1.0100	0.2120	0.3926	0.8004	0.8046	-1.054E-01	-1.370E-01
0.2	0.9775	0.8286	1.0130	1.0103	0.7569	0.8001	0.7714	0.8901	7.171E-01	7.217E-01
0.4	0.9774	0.8980	1.0124	1.0106	0.9331	0.9412	0.8886	0.9470	9.203E-01	9.305E-01
0.6	0.9777	0.9356	1.0115	1.0102	0.9709	0.9736	0.9403	0.9681	9.603E-01	9.707E-01
0.8	0.9793	0.9551	1.0105	1.0096	0.9847	0.9858	0.9630	0.9785	9.749E-01	9.847E-01
1.0	0.9812	0.9662	1.0096	1.0089	0.9909	0.9915	0.9746	0.9845	9.820E-01	9.911E-01
1.2	0.9830	0.9730	1.0087	1.0082	0.9942	0.9945	0.9811	0.9883	9.860E-01	9.943E-01
1.4	0.9846	0.9776	1.0079	1.0075	0.9961	0.9963	0.9852	0.9908	9.886E-01	9.962E-01
1.8	0.9872	0.9833	1.0066	1.0063	0.9980	0.9981	0.9899	0.9940	9.916E-01	9.980E-01

 Instr. see p. 25

ELECTRON Z= 48 A= 111

p	$F_0L_0^*/F_0L_0$	λ_2^*/λ_2	μ_1^*/μ_1	μ_2^*/μ_2	Λ_1^*/Λ_1	Λ_2^*/Λ_2	ν_{12}^*/ν_{12}	$\hat{\nu}_{12}^*/\hat{\nu}_{12}$	η_{12}^*/η_{12}	$\hat{\eta}_{12}^*/\hat{\eta}_{12}$
0.1	0.9765	0.7912	1.0136	1.0104	0.1920	0.3763	0.8088	0.8042	-1.693E-01	-2.070E-01
0.2	0.9769	0.8287	1.0134	1.0107	0.7504	0.7946	0.7714	0.8897	7.063E-01	7.106E-01
0.4	0.9769	0.8967	1.0127	1.0109	0.9313	0.9396	0.8866	0.9467	9.178E-01	9.282E-01
0.6	0.9773	0.9343	1.0118	1.0105	0.9701	0.9728	0.9389	0.9679	9.591E-01	9.697E-01
0.8	0.9789	0.9541	1.0109	1.0099	0.9842	0.9854	0.9621	0.9784	9.742E-01	9.843E-01
1.0	0.9808	0.9653	1.0099	1.0091	0.9907	0.9912	0.9739	0.9844	9.815E-01	9.908E-01
1.2	0.9827	0.9723	1.0090	1.0084	0.9941	0.9944	0.9806	0.9882	9.856E-01	9.942E-01
1.4	0.9843	0.9770	1.0082	1.0077	0.9960	0.9962	0.9848	0.9908	9.882E-01	9.961E-01
1.8	0.9869	0.9828	1.0068	1.0065	0.9980	0.9981	0.9896	0.9940	9.914E-01	9.980E-01

Z= 49 A= 113

p	$F_0L_0^*/F_0L_0$	λ_2^*/λ_2	μ_1^*/μ_1	μ_2^*/μ_2	Λ_1^*/Λ_1	Λ_2^*/Λ_2	ν_{12}^*/ν_{12}	$\hat{\nu}_{12}^*/\hat{\nu}_{12}$	η_{12}^*/η_{12}	$\hat{\eta}_{12}^*/\hat{\eta}_{12}$
0.1	0.9759	0.7928	1.0140	1.0107	0.1722	0.3600	0.8173	0.8038	-2.393E-01	-2.845E-01
0.2	0.9763	0.8288	1.0138	1.0110	0.7440	0.7892	0.7718	0.8893	6.952E-01	6.990E-01
0.4	0.9765	0.8955	1.0131	1.0112	C.9295	0.9380	0.8846	0.9465	9.152E-01	9.259E-01
0.6	0.9769	0.9331	1.0122	1.0108	0.9693	0.9721	0.9375	0.9677	9.579E-01	9.688E-01
0.8	0.9785	0.9531	1.0112	1.0102	0.9838	0.9850	0.9612	0.9783	9.734E-01	9.838E-01
1.0	0.9804	0.9645	1.0102	1.0094	0.9904	0.9910	0.9732	0.9844	9.809E-01	9.905E-01
1.2	0.9823	0.9716	1.0092	1.0086	0.9939	0.9942	0.9801	0.9882	9.852E-01	9.940E-01
1.4	0.9839	0.9764	1.0084	1.0079	0.9959	0.9961	0.9844	0.9907	9.879E-01	9.960E-01
1.8	0.9866	0.9823	1.0070	1.0067	0.9979	0.9980	0.9893	0.9939	9.911E-01	9.979E-01

Z= 50 A= 117

p	$F_0L_0^*/F_0L_0$	λ_2^*/λ_2	μ_1^*/μ_1	μ_2^*/μ_2	Λ_1^*/Λ_1	Λ_2^*/Λ_2	ν_{12}^*/ν_{12}	$\hat{\nu}_{12}^*/\hat{\nu}_{12}$	η_{12}^*/η_{12}	$\hat{\eta}_{12}^*/\hat{\eta}_{12}$
0.1	0.9753	0.7944	1.0144	1.0110	0.1524	0.3437	0.8259	0.8034	-3.168E-01	-3.712E-01
0.2	0.9757	0.8290	1.0142	1.0113	0.7375	0.7838	0.7724	0.8890	6.836E-01	6.869E-01
0.4	0.9761	0.8944	1.0135	1.0115	0.9276	0.9364	0.8827	0.9462	9.126E-01	9.235E-01
0.6	0.9765	0.9319	1.0125	1.0111	0.9685	0.9714	0.9362	0.9676	9.566E-01	9.679E-01
0.8	0.9781	0.9521	1.0115	1.0104	0.9834	0.9846	0.9602	0.9782	9.727E-01	9.833E-01
1.0	0.9800	0.9637	1.0104	1.0097	0.9902	0.9908	0.9726	0.9843	9.804E-01	9.903E-01
1.2	0.9819	0.9709	1.0095	1.0089	0.9938	0.9941	0.9796	0.9881	9.848E-01	9.938E-01
1.4	0.9836	0.9758	1.0086	1.0081	0.9958	0.9960	0.9840	0.9907	9.876E-01	9.959E-01
1.8	0.9863	0.9819	1.0072	1.0069	0.9979	0.9979	0.9890	0.9939	9.909E-01	9.978E-01

Z= 51 A= 125

p	$F_0L_0^*/F_0L_0$	λ_2^*/λ_2	μ_1^*/μ_1	μ_2^*/μ_2	Λ_1^*/Λ_1	Λ_2^*/Λ_2	ν_{12}^*/ν_{12}	$\hat{\nu}_{12}^*/\hat{\nu}_{12}$	η_{12}^*/η_{12}	$\hat{\eta}_{12}^*/\hat{\eta}_{12}$
0.1	0.9747	0.7959	1.0148	1.0114	0.1325	0.3272	0.8345	0.8029	-4.036E-01	-4.692E-01
0.2	0.9751	0.8292	1.0146	1.0117	0.7310	0.7783	0.7732	0.8886	6.714E-01	6.740E-01
0.4	0.9756	0.8932	1.0139	1.0119	0.9258	0.9347	0.8808	0.9459	9.100E-01	9.210E-01
0.6	0.9761	0.9307	1.0129	1.0114	0.9677	0.9706	0.9348	0.9674	9.554E-01	9.669E-01
0.8	0.9777	0.9510	1.0118	1.0107	0.9829	0.9842	0.9593	0.9781	9.719E-01	9.828E-01
1.0	0.9797	0.9628	1.0107	1.0099	0.9899	0.9905	0.9719	0.9842	9.799E-01	9.900E-01
1.2	0.9816	0.9702	1.0098	1.0091	0.9936	0.9939	0.9791	0.9880	9.844E-01	9.937E-01
1.4	0.9833	0.9752	1.0089	1.0083	0.9957	0.9959	0.9836	0.9906	9.872E-01	9.957E-01
1.8	0.9860	0.9814	1.0074	1.0070	0.9978	0.9979	0.9887	0.9938	9.906E-01	9.978E-01

ELECTRON Z= 52 A= 124

p	$F_0L_0^*/F_0L_0$	λ_2^*/λ_2	μ_1^*/μ_1	μ_2^*/μ_2	Λ_1^*/Λ_1	Λ_2^*/Λ_2	ν_{12}^*/ν_{12}	$\vartheta_{12}^*/\vartheta_{12}$	η_{12}^*/η_{12}	$\hat{\eta}_{12}^*/\hat{\eta}_{12}$
0.1	0.9741	0.7975	1.0152	1.0117	0.1127	0.3106	0.8432	0.8025	-5.012E-01	-5.807E-01
0.2	0.9745	0.8295	1.0150	1.0120	0.7246	0.7728	0.7742	0.8882	6.586E-01	6.605E-01
0.4	0.9751	0.8922	1.0143	1.0122	0.9239	0.9331	0.8790	0.9457	9.072E-01	9.185E-01
0.6	0.9757	0.9296	1.0132	1.0118	0.9669	0.9699	0.9335	0.9672	9.541E-01	9.659E-01
0.8	0.9773	0.9500	1.0121	1.0110	0.9825	0.9838	0.9584	0.9779	9.711E-01	9.823E-01
1.0	0.9793	0.9620	1.0110	1.0102	0.9897	0.9903	0.9712	0.9841	9.793E-01	9.897E-01
1.2	0.9812	0.9695	1.0100	1.0094	0.9934	0.9938	0.9786	0.9880	9.839E-01	9.935E-01
1.4	0.9829	0.9746	1.0091	1.0086	0.9956	0.9958	0.9832	0.9905	9.869E-01	9.956E-01
1.8	0.9857	0.9809	1.0076	1.0072	0.9978	0.9978	0.9884	0.9938	9.904E-01	9.977E-01

Z= 53 A= 127

p	$F_0L_0^*/F_0L_0$	λ_2^*/λ_2	μ_1^*/μ_1	μ_2^*/μ_2	Λ_1^*/Λ_1	Λ_2^*/Λ_2	ν_{12}^*/ν_{12}	$\vartheta_{12}^*/\vartheta_{12}$	η_{12}^*/η_{12}	$\hat{\eta}_{12}^*/\hat{\eta}_{12}$
0.1	0.9735	0.7989	1.0157	1.0121	0.0926	0.2940	0.8520	0.8020	-6.124E-01	-7.096E-01
0.2	0.9739	0.8297	1.0154	1.0124	0.7180	0.7672	0.7754	0.8877	6.450E-01	6.460E-01
0.4	0.9747	0.8911	1.0146	1.0125	0.9220	0.9314	0.8772	0.9454	9.044E-01	9.159E-01
0.6	0.9753	0.9284	1.0136	1.0121	0.9661	0.9691	0.9322	0.9670	9.528E-01	9.649E-01
0.8	0.9769	0.9490	1.0124	1.0113	0.9821	0.9834	0.9575	0.9778	9.703E-01	9.818E-01
1.0	0.9789	0.9611	1.0113	1.0104	0.9894	0.9900	0.9706	0.9840	9.787E-01	9.894E-01
1.2	0.9809	0.9688	1.0103	1.0096	0.9933	0.9936	0.9781	0.9879	9.835E-01	9.933E-01
1.4	0.9826	0.9739	1.0094	1.0088	0.9955	0.9957	0.9828	0.9905	9.865E-01	9.955E-01
1.8	0.9855	0.9804	1.0078	1.0074	0.9977	0.9978	0.9881	0.9937	9.901E-01	9.977E-01

Z= 54 A= 131

p	$F_0L_0^*/F_0L_0$	λ_2^*/λ_2	μ_1^*/μ_1	μ_2^*/μ_2	Λ_1^*/Λ_1	Λ_2^*/Λ_2	ν_{12}^*/ν_{12}	$\vartheta_{12}^*/\vartheta_{12}$	η_{12}^*/η_{12}	$\hat{\eta}_{12}^*/\hat{\eta}_{12}$
0.1	0.9729	0.8003	1.0161	1.0124	0.0727	0.2773	0.8608	0.8016	-7.398E-01	-8.597E-01
0.2	0.9733	0.8300	1.0158	1.0127	0.7115	0.7617	0.7769	0.8873	6.308E-01	6.307E-01
0.4	0.9742	0.8902	1.0150	1.0128	0.9202	0.9298	0.8755	0.9451	9.015E-01	9.132E-01
0.6	0.9749	0.9273	1.0139	1.0124	0.9652	0.9684	0.9308	0.9669	9.515E-01	9.639E-01
0.8	0.9765	0.9480	1.0128	1.0116	0.9816	0.9830	0.9566	0.9777	9.695E-01	9.813E-01
1.0	0.9785	0.9603	1.0116	1.0107	0.9892	0.9898	0.9699	0.9839	9.782E-01	9.891E-01
1.2	0.9805	0.9680	1.0105	1.0098	0.9931	0.9934	0.9776	0.9878	9.831E-01	9.931E-01
1.4	0.9823	0.9733	1.0096	1.0090	0.9954	0.9956	0.9824	0.9904	9.862E-01	9.954E-01
1.8	0.9852	0.9799	1.0080	1.0076	0.9976	0.9977	0.9879	0.9936	9.899E-01	9.976E-01
2.2	0.9874	0.9839	1.0068	1.0065	0.9987	0.9987	0.9908	0.9954	9.919E-01	9.986E-01

ELECTRON Z= 55 A= 135

p	$F_0L_0^*/F_0L_0$	λ_2^*/λ_2	μ_1^*/μ_1	μ_2^*/μ_2	Λ_1^*/Λ_1	Λ_2^*/Λ_2	ν_{12}^*/ν_{12}	$\vartheta_{12}^*/\vartheta_{12}$	η_{12}^*/η_{12}	$\hat{\eta}_{12}^*/\hat{\eta}_{12}$
0.1	0.9723	0.8017	1.0165	1.0128	0:0527	0.2605	0.8696	0.8010	-8.881E-01	-1.038E 00
0.2	0.9728	0.8303	1.0163	1.0130	0.7049	0.7560	0.7785	0.8869	6.157E-01	6.143E-01
0.4	0.9737	0.8892	1.0154	1.0132	0.9183	0.9281	0.8738	0.9449	8.986E-01	9.104E-01
0.6	0.9745	0.9262	1.0143	1.0127	0.9644	0.9676	0.9295	0.9667	9.502E-01	9.628E-01
0.8	0.9762	0.9470	1.0131	1.0119	0.9812	0.9825	0.9556	0.9776	9.687E-01	9.808E-01
1.0	0.9782	0.9594	1.0119	1.0110	0.9889	0.9896	0.9692	0.9838	9.776E-01	9.888E-01
1.2	0.9802	0.9673	1.0108	1.0101	0.9929	0.9933	0.9771	0.9878	9.827E-01	9.929E-01
1.4	0.9820	0.9727	1.0099	1.0092	0.9952	0.9955	0.9820	0.9903	9.858E-01	9.952E-01
1.8	0.9849	0.9794	1.0082	1.0078	0.9976	0.9977	0.9876	0.9935	9.896E-01	9.976E-01
2.2	0.9871	0.9835	1.0070	1.0067	0.9986	0.9987	0.9906	0.9953	9.917E-01	9.985E-01

Z= 56 A= 135

p	$F_0L_0^*/F_0L_0$	λ_2^*/λ_2	μ_1^*/μ_1	μ_2^*/μ_2	Λ_1^*/Λ_1	Λ_2^*/Λ_2	ν_{12}^*/ν_{12}	$\vartheta_{12}^*/\vartheta_{12}$	η_{12}^*/η_{12}	$\hat{\eta}_{12}^*/\hat{\eta}_{12}$
0.1	0.9717	0.8030	1.0169	1.0132	0.0326	0.2437	0.8785	0.8005	-1.063E 00	-1.253E 00
0.2	0.9722	0.8306	1.0167	1.0134	0.6983	0.7504	0.7804	0.8865	5.996E-01	5.967E-01
0.4	0.9732	0.8883	1.0158	1.0135	0.9164	0.9264	0.8722	0.9446	8.955E-01	9.075E-01
0.6	0.9741	0.9251	1.0147	1.0130	0.9636	0.9669	0.9282	0.9665	9.488E-01	9.617E-01
0.8	0.9758	0.9461	1.0134	1.0122	0.9807	0.9821	0.9547	0.9774	9.679E-01	9.803E-01
1.0	0.9778	0.9586	1.0122	1.0112	0.9886	0.9893	0.9686	0.9837	9.770E-01	9.885E-01
1.2	0.9798	0.9666	1.0111	1.0103	0.9928	0.9931	0.9765	0.9877	9.822E-01	9.927E-01
1.4	0.9816	0.9720	1.0101	1.0095	0.9951	0.9953	0.9815	0.9903	9.855E-01	9.951E-01
1.8	0.9846	0.9789	1.0084	1.0080	0.9975	0.9976	0.9873	0.9935	9.893E-01	9.975E-01
2.2	0.9869	0.9831	1.0072	1.0068	0.9986	0.9986	0.9903	0.9953	9.915E-01	9.985E-01

Z= 57 A= 139

p	$F_0L_0^*/F_0L_0$	λ_2^*/λ_2	μ_1^*/μ_1	μ_2^*/μ_2	Λ_1^*/Λ_1	Λ_2^*/Λ_2	ν_{12}^*/ν_{12}	$\vartheta_{12}^*/\vartheta_{12}$	η_{12}^*/η_{12}	$\hat{\eta}_{12}^*/\hat{\eta}_{12}$
0.1	0.9711	0.8044	1.0174	1.0135	0.0128	0.2269	0.8875	0.8000	-1.272E 00	-1.516E 00
0.2	0.9716	0.8310	1.0171	1.0138	0.6918	0.7448	0.7824	0.8861	5.825E-01	5.779E-01
0.4	0.9727	0.8874	1.0162	1.0138	0.9145	0.9248	0.8707	0.9443	8.924E-01	9.046E-01
0.6	0.9737	0.9240	1.0150	1.0133	0.9627	0.9661	0.9270	0.9663	9.475E-01	9.606E-01
0.8	0.9754	0.9451	1.0137	1.0125	0.9803	0.9817	0.9538	0.9773	9.671E-01	9.797E-01
1.0	0.9774	0.9578	1.0125	1.0115	0.9884	0.9891	0.9679	0.9836	9.765E-01	9.882E-01
1.2	0.9795	0.9659	1.0114	1.0106	0.9926	0.9930	0.9760	0.9876	9.818E-01	9.925E-01
1.4	0.9813	0.9714	1.0103	1.0097	0.9950	0.9952	0.9811	0.9903	9.851E-01	9.950E-01
1.8	0.9844	0.9784	1.0087	1.0082	0.9975	0.9976	0.9870	0.9935	9.891E-01	9.974E-01
2.2	0.9866	0.9826	1.0074	1.0070	0.9986	0.9986	0.9901	0.9952	9.913E-01	9.984E-01

ELECTRON Z= 58 A= 140

p	$F_0L_0^*/F_0L_0$	λ_2^*/λ_2	μ_1^*/μ_1	μ_2^*/μ_2	Λ_1^*/Λ_1	Λ_2^*/Λ_2	ν_{12}^*/ν_{12}	$\vartheta_{12}^*/\vartheta_{12}$	η_{12}^*/η_{12}	$\hat{\eta}_{12}^*/\hat{\eta}_{12}$
0.1	0.9705	0.8056	1.0178	1.0139	-0.0075	0.2099	0.8964	0.7994	-1.529E 00	-1.851E 00
0.2	0.9710	0.8313	1.0175	1.0141	0.6852	0.7391	0.7847	0.8856	5.641E-01	5.574E-01
0.4	0.9722	0.8865	1.0166	1.0142	0.9126	0.9231	0.8692	0.9440	8.892E-01	9.014E-01
0.6	0.9733	0.9229	1.0154	1.0136	0.9619	0.9653	0.9257	0.9661	9.460E-01	9.595E-01
0.8	0.9750	0.9441	1.0141	1.0128	0.9799	0.9813	0.9529	0.9772	9.662E-01	9.791E-01
1.0	0.9771	0.9569	1.0128	1.0118	0.9881	0.9888	0.9672	0.9836	9.759E-01	9.879E-01
1.2	0.9791	0.9652	1.0116	1.0108	0.9924	0.9928	0.9755	0.9876	9.813E-01	9.923E-01
1.4	0.9810	0.9708	1.0106	1.0099	0.9949	0.9951	0.9807	0.9902	9.847E-01	9.948E-01
1.8	0.9841	0.9779	1.0089	1.0084	0.9974	0.9975	0.9867	0.9934	9.888E-01	9.973E-01
2.2	0.9864	0.9822	1.0076	1.0072	0.9985	0.9986	0.9899	0.9951	9.911E-01	9.984E-01

Z= 59 A= 141

p	$F_0L_0^*/F_0L_0$	λ_2^*/λ_2	μ_1^*/μ_1	μ_2^*/μ_2	Λ_1^*/Λ_1	Λ_2^*/Λ_2	ν_{12}^*/ν_{12}	$\vartheta_{12}^*/\vartheta_{12}$	η_{12}^*/η_{12}	$\hat{\eta}_{12}^*/\hat{\eta}_{12}$
0.1	0.9700	0.8068	1.0182	1.0143	-0.0275	0.1929	0.9054	0.7989	-1.849E 00	-2.287E 00
0.2	0.9704	0.8316	1.0180	1.0145	0.6786	0.7334	0.7871	0.8851	5.444E-01	5.353E-01
0.4	0.9718	0.8857	1.0170	1.0145	0.9107	0.9214	0.8678	0.9437	8.859E-01	8.982E-01
0.6	0.9729	0.9219	1.0158	1.0140	0.9610	0.9646	0.9244	0.9659	9.446E-01	9.583E-01
0.8	0.9746	0.9432	1.0144	1.0131	0.9794	0.9809	0.9520	0.9771	9.654E-01	9.786E-01
1.0	0.9767	0.9561	1.0131	1.0121	0.9878	0.9886	0.9666	0.9835	9.753E-01	9.875E-01
1.2	0.9788	0.9644	1.0119	1.0111	0.9923	0.9926	0.9750	0.9875	9.809E-01	9.921E-01
1.4	0.9807	0.9702	1.0108	1.0102	0.9948	0.9950	0.9803	0.9902	9.844E-01	9.947E-01
1.8	0.9838	0.9774	1.0091	1.0086	0.9973	0.9974	0.9864	0.9933	9.885E-01	9.973E-01
2.2	0.9862	0.9818	1.0078	1.0073	0.9985	0.9985	0.9896	0.9950	9.909E-01	9.984E-01

Z= 60 A= 142

p	$F_0L_0^*/F_0L_0$	λ_2^*/λ_2	μ_1^*/μ_1	μ_2^*/μ_2	Λ_1^*/Λ_1	Λ_2^*/Λ_2	ν_{12}^*/ν_{12}	$\vartheta_{12}^*/\vartheta_{12}$	η_{12}^*/η_{12}	$\hat{\eta}_{12}^*/\hat{\eta}_{12}$
0.1	0.9694	0.8080	1.0187	1.0146	-0.0478	0.1757	0.9144	0.7982	-2.264E 00	-2.883E 00
0.2	0.9699	0.8319	1.0184	1.0148	0.6719	0.7277	0.7897	0.8846	5.231E-01	5.109E-01
0.4	0.9712	0.8849	1.0174	1.0149	0.9087	0.9197	0.8664	0.9434	8.824E-01	8.948E-01
0.6	0.9725	0.9208	1.0161	1.0143	0.9602	0.9638	0.9232	0.9657	9.431E-01	9.571E-01
0.8	0.9742	0.9422	1.0147	1.0134	0.9790	0.9805	0.9511	0.9769	9.645E-01	9.780E-01
1.0	0.9763	0.9552	1.0134	1.0123	0.9876	0.9883	0.9659	0.9833	9.747E-01	9.872E-01
1.2	0.9784	0.9637	1.0122	1.0113	0.9921	0.9925	0.9745	0.9874	9.804E-01	9.919E-01
1.4	0.9804	0.9695	1.0111	1.0104	0.9947	0.9949	0.9799	0.9901	9.840E-01	9.946E-01
1.8	0.9835	0.9769	1.0093	1.0088	0.9973	0.9974	0.9861	0.9933	9.882E-01	9.972E-01
2.2	0.9859	0.9814	1.0080	1.0075	0.9984	0.9985	0.9894	0.9950	9.907E-01	9.983E-01

Instr. see p. 25

ELECTRON Z= 61 A= 147

p	$F_0L_0^*/F_0L_0$	λ_2^*/λ_2	μ_1^*/μ_1	μ_2^*/μ_2	Λ_1^*/Λ_1	Λ_2^*/Λ_2	ν_{12}^*/ν_{12}	$\hat{\vartheta}_{12}^*/\hat{\vartheta}_{12}$	η_{12}^*/η_{12}	$\hat{\eta}_{12}^*/\hat{\eta}_{12}$
0.1	0.9688	0.8091	1.0191	1.0150	-0.0681	0.1585	0.9235	0.7976	-2.822E 00	-3.744E 00
0.2	0.9693	0.8323	1.0188	1.0152	0.6652	0.7219	0.7925	0.8841	5.000E-01	4.843E-01
0.4	0.9707	0.8841	1.0178	1.0152	0.9068	0.9179	0.8651	0.9431	8.788E-01	8.913E-01
0.6	0.9720	0.9198	1.0165	1.0146	0.9593	0.9630	0.9220	0.9655	9.416E-01	9.558E-01
0.8	0.9739	0.9412	1.0151	1.0137	0.9785	0.9800	0.9502	0.9768	9.636E-01	9.774E-01
1.0	0.9760	0.9544	1.0137	1.0126	0.9873	0.9881	0.9652	0.9832	9.741E-01	9.868E-01
1.2	0.9781	0.9630	1.0125	1.0116	0.9919	0.9923	0.9740	0.9873	9.799E-01	9.917E-01
1.4	0.9801	0.9689	1.0114	1.0106	0.9946	0.9948	0.9795	0.9900	9.836E-01	9.944E-01
1.8	0.9833	0.9764	1.0096	1.0090	0.9972	0.9973	0.9858	0.9932	9.880E-01	9.971E-01
2.2	0.9857	0.9809	1.0082	1.0077	0.9984	0.9985	0.9891	0.9949	9.904E-01	9.983E-01

Z= 62 A= 150

p	$F_0L_0^*/F_0L_0$	λ_2^*/λ_2	μ_1^*/μ_1	μ_2^*/μ_2	Λ_1^*/Λ_1	Λ_2^*/Λ_2	ν_{12}^*/ν_{12}	$\hat{\vartheta}_{12}^*/\hat{\vartheta}_{12}$	η_{12}^*/η_{12}	$\hat{\eta}_{12}^*/\hat{\eta}_{12}$
0.1	0.9682	0.8102	1.0196	1.0154	-0.0883	0.1412	0.9326	0.7969	-3.611E 00	-5.103E 00
0.2	0.9687	0.8326	1.0193	1.0156	0.6585	0.7161	0.7954	0.8836	4.748E-01	4.548E-01
0.4	0.9702	0.8834	1.0183	1.0156	0.9048	0.9162	0.8639	0.9428	8.751E-01	8.876E-01
0.6	0.9716	0.9188	1.0169	1.0150	0.9585	0.9622	0.9208	0.9653	9.401E-01	9.545E-01
0.8	0.9735	0.9403	1.0154	1.0140	0.9781	0.9796	0.9493	0.9766	9.627E-01	9.767E-01
1.0	0.9756	0.9536	1.0140	1.0129	0.9870	0.9878	0.9645	0.9831	9.734E-01	9.865E-01
1.2	0.9778	0.9622	1.0127	1.0118	0.9918	0.9922	0.9734	0.9872	9.794E-01	9.915E-01
1.4	0.9797	0.9682	1.0116	1.0109	0.9945	0.9947	0.9790	0.9900	9.832E-01	9.943E-01
1.8	0.9830	0.9759	1.0098	1.0092	0.9972	0.9973	0.9854	0.9932	9.877E-01	9.970E-01
2.2	0.9855	0.9805	1.0084	1.0079	0.9984	0.9984	0.9889	0.9949	9.902E-01	9.982E-01

Z= 63 A= 152

p	$F_0L_0^*/F_0L_0$	λ_2^*/λ_2	μ_1^*/μ_1	μ_2^*/μ_2	Λ_1^*/Λ_1	Λ_2^*/Λ_2	ν_{12}^*/ν_{12}	$\hat{\vartheta}_{12}^*/\hat{\vartheta}_{12}$	η_{12}^*/η_{12}	$\hat{\eta}_{12}^*/\hat{\eta}_{12}$
0.1	0.9677	0.8113	1.0200	1.0158	-0.1085	0.1240	0.9417	0.7963	-4.817E 00	-7.573E 00
0.2	0.9682	0.8329	1.0197	1.0160	0.6518	0.7103	0.7986	0.8831	4.474E-01	4.221E-01
0.4	0.9697	0.8826	1.0187	1.0159	0.9029	0.9145	0.8627	0.9424	8.712E-01	8.837E-01
0.6	0.9712	0.9178	1.0173	1.0153	0.9576	0.9614	0.9196	0.9651	9.385E-01	9.532E-01
0.8	0.9731	0.9393	1.0158	1.0143	0.9776	0.9792	0.9484	0.9765	9.618E-01	9.761E-01
1.0	0.9753	0.9527	1.0143	1.0132	0.9868	0.9875	0.9638	0.9830	9.728E-01	9.861E-01
1.2	0.9775	0.9615	1.0130	1.0121	0.9916	0.9920	0.9729	0.9871	9.790E-01	9.912E-01
1.4	0.9794	0.9676	1.0119	1.0111	0.9943	0.9946	0.9786	0.9899	9.829E-01	9.941E-01
1.8	0.9828	0.9754	1.0100	1.0094	0.9971	0.9972	0.9851	0.9931	9.874E-01	9.970E-01
2.2	0.9853	0.9801	1.0086	1.0081	0.9983	0.9984	0.9887	0.9948	9.900E-01	9.982E-01

ELECTRON Z= 64 A= 155

p	$F_0L_0^*/F_0L_0$	λ_2^*/λ_2	μ_1^*/μ_1	μ_2^*/μ_2	Λ_1^*/Λ_1	Λ_2^*/Λ_2	ν_{12}^*/ν_{12}	$\vartheta_{12}^*/\vartheta_{12}$	η_{12}^*/η_{12}	$\hat{\eta}_{12}^*/\hat{\eta}_{12}$
0.1	0.9671	0.8123	1.0205	1.0162	-0.1288	0.1067	0.9508	0.7955	-6.895E 00	-1.347E 01
0.2	0.9676	0.8333	1.0202	1.0163	0.6451	0.7045	0.8020	0.8825	4.170E-01	3.855E-01
0.4	0.9692	0.8819	1.0191	1.0163	0.9009	0.9127	0.8616	0.9421	8.672E-01	8.796E-01
0.6	0.9708	0.9168	1.0177	1.0156	0.9567	0.9606	0.9184	0.9649	9.368E-01	9.518E-01
0.8	0.9727	0.9384	1.0161	1.0146	0.9771	0.9788	0.9475	0.9763	9.608E-01	9.754E-01
1.0	0.9749	0.9519	1.0147	1.0135	0.9865	0.9873	0.9632	0.9829	9.721E-01	9.858E-01
1.2	0.9771	0.9608	1.0133	1.0124	0.9914	0.9918	0.9724	0.9871	9.785E-01	9.910E-01
1.4	0.9791	0.9669	1.0121	1.0113	0.9942	0.9945	0.9782	0.9898	9.825E-01	9.940E-01
1.8	0.9825	0.9748	1.0102	1.0096	0.9970	0.9971	0.9848	0.9931	9.871E-01	9.969E-01
2.2	0.9850	0.9797	1.0088	1.0082	0.9983	0.9984	0.9884	0.9948	9.898E-01	9.982E-01
2.6	0.9870	0.9829	1.0077	1.0072	0.9989	0.9990	0.9906	0.9958	9.915E-01	9.988E-01

Z= 65 A= 159

p	$F_0L_0^*/F_0L_0$	λ_2^*/λ_2	μ_1^*/μ_1	μ_2^*/μ_2	Λ_1^*/Λ_1	Λ_2^*/Λ_2	ν_{12}^*/ν_{12}	$\vartheta_{12}^*/\vartheta_{12}$	η_{12}^*/η_{12}	$\hat{\eta}_{12}^*/\hat{\eta}_{12}$
0.1	0.9665	0.8134	1.0209	1.0166	-0.1488	0.0894	0.9600	0.7949	-1.133E 01	-4.608E 01
0.2	0.9670	0.8336	1.0206	1.0167	0.6384	0.6987	0.8055	0.8820	3.836E-01	3.442E-01
0.4	0.9687	0.8813	1.0195	1.0167	0.8990	0.9110	0.8607	0.9418	8.630E-01	8.753E-01
0.6	0.9704	0.9159	1.0181	1.0160	0.9559	0.9599	0.9173	0.9647	9.352E-01	9.504E-01
0.8	0.9724	0.9375	1.0165	1.0149	0.9767	0.9783	0.9466	0.9762	9.599E-01	9.747E-01
1.0	0.9746	0.9511	1.0150	1.0137	0.9862	0.9870	0.9625	0.9828	9.715E-01	9.854E-01
1.2	0.9768	0.9600	1.0136	1.0126	0.9912	0.9917	0.9719	0.9870	9.780E-01	9.908E-01
1.4	0.9788	0.9663	1.0124	1.0116	0.9941	0.9943	0.9778	0.9897	9.821E-01	9.938E-01
1.8	0.9822	0.9743	1.0104	1.0098	0.9970	0.9971	0.9845	0.9931	9.868E-01	9.968E-01
2.2	0.9848	0.9792	1.0090	1.0084	0.9983	0.9983	0.9882	0.9948	9.895E-01	9.981E-01
2.6	0.9868	0.9825	1.0079	1.0074	0.9989	0.9989	0.9904	0.9959	9.913E-01	9.988E-01

Z= 66 A= 161

p	$F_0L_0^*/F_0L_0$	λ_2^*/λ_2	μ_1^*/μ_1	μ_2^*/μ_2	Λ_1^*/Λ_1	Λ_2^*/Λ_2	ν_{12}^*/ν_{12}	$\vartheta_{12}^*/\vartheta_{12}$	η_{12}^*/η_{12}	$\hat{\eta}_{12}^*/\hat{\eta}_{12}$
0.1	0.9660	0.8143	1.0214	1.0170	-0.1693	0.0718	0.9692	0.7941	-2.745E 01	3.789E 01
0.2	0.9665	0.8339	1.0211	1.0171	0.6316	0.6927	0.8092	0.8814	3.460E-01	2.968E-01
0.4	0.9682	0.8806	1.0200	1.0170	0.8970	0.9092	0.8597	0.9414	8.586E-01	8.707E-01
0.6	0.9700	0.9149	1.0185	1.0163	0.9550	0.9591	0.9161	0.9644	9.334E-01	9.489E-01
0.8	0.9720	0.9365	1.0169	1.0152	0.9762	0.9779	0.9457	0.9760	9.589E-01	9.740E-01
1.0	0.9743	0.9502	1.0153	1.0140	0.9860	0.9868	0.9618	0.9827	9.708E-01	9.850E-01
1.2	0.9765	0.9593	1.0139	1.0129	0.9911	0.9915	0.9713	0.9869	9.775E-01	9.905E-01
1.4	0.9785	0.9656	1.0127	1.0118	0.9940	0.9942	0.9773	0.9897	9.817E-01	9.937E-01
1.8	0.9820	0.9738	1.0106	1.0100	0.9969	0.9970	0.9842	0.9930	9.866E-01	9.967E-01
2.2	0.9846	0.9788	1.0091	1.0086	0.9982	0.9983	0.9879	0.9948	9.893E-01	9.981E-01
2.6	0.9866	0.9821	1.0080	1.0075	0.9989	0.9989	0.9902	0.9960	9.911E-01	9.987E-01

ELECTRON Z= 67 A= 165

p	$F_0L_0^*/F_0L_0$	λ_2^*/λ_2	μ_1^*/μ_1	μ_2^*/μ_2	Λ_1^*/Λ_1	Λ_2^*/Λ_2	ν_{12}^*/ν_{12}	$\hat{\nu}_{12}^*/\hat{\nu}_{12}$	η_{12}^*/η_{12}	$\hat{\eta}_{12}^*/\hat{\eta}_{12}$
0.1	0.9654	0.8152	1.0219	1.0174	-0.1895	0.0543	0.9784	0.7933	9.004E 01	1.417E 01
0.2	0.9660	0.8343	1.0215	1.0175	0.6249	0.6869	0.8131	0.8808	3.039E-01	2.425E-01
0.4	0.9677	0.8799	1.0204	1.0174	0.8950	0.9075	0.8588	0.9411	8.541E-01	8.659E-01
0.6	0.9695	0.9140	1.0189	1.0167	0.9541	0.9583	0.9150	0.9642	9.317E-01	9.473E-01
0.8	0.9716	0.9356	1.0172	1.0155	0.9757	0.9775	0.9449	0.9758	9.579E-01	9.733E-01
1.0	0.9739	0.9494	1.0156	1.0143	0.9857	0.9865	0.9612	0.9826	9.701E-01	9.846E-01
1.2	0.9762	0.9586	1.0142	1.0131	0.9909	0.9913	0.9708	0.9868	9.770E-01	9.903E-01
1.4	0.9783	0.9650	1.0129	1.0121	0.9939	0.9941	0.9769	0.9896	9.813E-01	9.935E-01
1.8	0.9817	0.9733	1.0109	1.0102	0.9969	0.9970	0.9839	0.9930	9.863E-01	9.967E-01
2.2	0.9844	0.9783	1.0093	1.0088	0.9982	0.9982	0.9877	0.9948	9.891E-01	9.980E-01
2.6	0.9864	0.9818	1.0082	1.0077	0.9989	0.9989	0.9900	0.9960	9.909E-01	9.988E-01

Z= 68 A= 166

p	$F_0L_0^*/F_0L_0$	λ_2^*/λ_2	μ_1^*/μ_1	μ_2^*/μ_2	Λ_1^*/Λ_1	Λ_2^*/Λ_2	ν_{12}^*/ν_{12}	$\hat{\nu}_{12}^*/\hat{\nu}_{12}$	η_{12}^*/η_{12}	$\hat{\eta}_{12}^*/\hat{\eta}_{12}$
0.1	0.9649	0.8161	1.0223	1.0178	-0.2101	0.0366	0.9877	0.7925	1.821E 01	8.980E 00
0.2	0.9654	0.8345	1.0220	1.0179	0.6180	0.6809	0.8171	0.8802	2.560E-01	1.788E-01
0.4	0.9672	0.8793	1.0208	1.0178	0.8930	0.9057	0.8580	0.9407	8.492E-01	8.607E-01
0.6	0.9691	0.9131	1.0193	1.0170	0.9532	0.9574	0.9139	0.9640	9.298E-01	9.456E-01
0.8	0.9713	0.9347	1.0176	1.0159	0.9753	0.9770	0.9440	0.9757	9.569E-01	9.725E-01
1.0	0.9736	0.9485	1.0160	1.0146	0.9854	0.9863	0.9605	0.9824	9.694E-01	9.841E-01
1.2	0.9759	0.9578	1.0145	1.0134	0.9907	0.9912	0.9703	0.9867	9.765E-01	9.900E-01
1.4	0.9780	0.9643	1.0132	1.0123	0.9938	0.9940	0.9765	0.9895	9.808E-01	9.933E-01
1.8	0.9815	0.9727	1.0111	1.0104	0.9968	0.9969	0.9836	0.9929	9.860E-01	9.966E-01
2.2	0.9842	0.9779	1.0095	1.0090	0.9982	0.9982	0.9874	0.9948	9.888E-01	9.980E-01
2.6	0.9862	0.9814	1.0084	1.0079	0.9989	0.9989	0.9898	0.9960	9.907E-01	9.987E-01

Z= 69 A= 171

p	$F_0L_0^*/F_0L_0$	λ_2^*/λ_2	μ_1^*/μ_1	μ_2^*/μ_2	Λ_1^*/Λ_1	Λ_2^*/Λ_2	ν_{12}^*/ν_{12}	$\hat{\nu}_{12}^*/\hat{\nu}_{12}$	η_{12}^*/η_{12}	$\hat{\eta}_{12}^*/\hat{\eta}_{12}$
0.1	0.9644	0.8170	1.0228	1.0182	-0.2303	0.0190	0.9970	0.7916	1.047E 01	6.704E 00
0.2	0.9649	0.8349	1.0225	1.0183	0.6112	0.6749	0.8213	0.8795	2.012E-01	1.037E-01
0.4	0.9667	0.8787	1.0213	1.0181	0.8910	0.9039	0.8573	0.9403	8.442E-01	8.552E-01
0.6	0.9687	0.9122	1.0197	1.0173	0.9523	0.9566	0.9128	0.9637	9.280E-01	9.439E-01
0.8	0.9709	0.9338	1.0179	1.0162	0.9748	0.9766	0.9431	0.9755	9.558E-01	9.718E-01
1.0	0.9732	0.9477	1.0163	1.0149	0.9851	0.9860	0.9598	0.9823	9.687E-01	9.837E-01
1.2	0.9756	0.9571	1.0148	1.0137	0.9905	0.9910	0.9697	0.9865	9.759E-01	9.898E-01
1.4	0.9777	0.9636	1.0135	1.0125	0.9936	0.9939	0.9760	0.9894	9.804E-01	9.931E-01
1.8	0.9812	0.9722	1.0113	1.0106	0.9967	0.9968	0.9833	0.9929	9.857E-01	9.965E-01
2.2	0.9840	0.9775	1.0097	1.0092	0.9981	0.9982	0.9872	0.9949	9.886E-01	9.980E-01
2.6	0.9861	0.9810	1.0085	1.0081	0.9988	0.9988	0.9896	0.9961	9.905E-01	9.987E-01

ELECTRON Z= 70 A= 170

p	$F_0L_0^*/F_0L_0$	λ_2^*/λ_2	μ_1^*/μ_1	μ_2^*/μ_2	Λ_1^*/Λ_1	Λ_2^*/Λ_2	ν_{12}^*/ν_{12}	$\vartheta_{12}^*/\vartheta_{12}$	η_{12}^*/η_{12}	$\hat{\eta}_{12}^*/\hat{\eta}_{12}$
0.1	0.9638	0.8178	1.0233	1.0186	-0.2507	0.0013	1.0063	0.7908	7.508E 00	5.424E 00
0.2	0.9643	0.8352	1.0229	1.0187	0.6044	0.6690	0.8257	0.8789	1.376E-01	1.305E-02
0.4	0.9663	0.8780	1.0217	1.0185	0.8890	0.9021	0.8566	0.9399	8.388E-01	8.493E-01
0.6	0.9683	0.9113	1.0201	1.0177	0.9514	0.9558	0.9118	0.9635	9.260E-01	9.421E-01
0.8	0.9705	0.9329	1.0183	1.0165	0.9743	0.9762	0.9423	0.9753	9.548E-01	9.710E-01
1.0	0.9729	0.9469	1.0166	1.0152	0.9848	0.9857	0.9592	0.9822	9.680E-01	9.833E-01
1.2	0.9753	0.9563	1.0151	1.0139	0.9904	0.9908	0.9692	0.9865	9.754E-01	9.895E-01
1.4	0.9774	0.9630	1.0137	1.0128	0.9935	0.9938	0.9756	0.9893	9.800E-01	9.929E-01
1.8	0.9810	0.9717	1.0115	1.0109	0.9967	0.9968	0.9830	0.9928	9.853E-01	9.964E-01
2.2	0.9838	0.9770	1.0099	1.0094	0.9981	0.9981	0.9870	0.9948	9.884E-01	9.979E-01
2.6	0.9859	0.9806	1.0087	1.0083	0.9988	0.9988	0.9894	0.9960	9.903E-01	9.987E-01

Z= 71 A= 175

p	$F_0L_0^*/F_0L_0$	λ_2^*/λ_2	μ_1^*/μ_1	μ_2^*/μ_2	Λ_1^*/Λ_1	Λ_2^*/Λ_2	ν_{12}^*/ν_{12}	$\vartheta_{12}^*/\vartheta_{12}$	η_{12}^*/η_{12}	$\hat{\eta}_{12}^*/\hat{\eta}_{12}$
0.1	0.9633	0.8186	1.0238	1.0190	-0.2711	-0.0164	1.0157	0.7899	5.941E 00	4.605E 00
0.2	0.9638	0.8354	1.0234	1.0191	0.5976	0.6630	0.8303	0.8782	6.292E-02	-9.829E-02
0.4	0.9658	0.8774	1.0222	1.0189	0.8870	0.9003	0.8561	0.9395	8.331E-01	8.430E-01
0.6	0.9679	0.9104	1.0205	1.0181	0.9505	0.9550	0.9107	0.9632	9.240E-01	9.403E-01
0.8	0.9702	0.9320	1.0187	1.0168	0.9738	0.9757	0.9414	0.9751	9.537E-01	9.701E-01
1.0	0.9726	0.9460	1.0170	1.0155	0.9846	0.9855	0.9585	0.9820	9.673E-01	9.828E-01
1.2	0.9750	0.9556	1.0154	1.0142	0.9902	0.9907	0.9687	0.9864	9.748E-01	9.892E-01
1.4	0.9771	0.9623	1.0140	1.0130	0.9934	0.9937	0.9752	0.9892	9.796E-01	9.927E-01
1.8	0.9808	0.9711	1.0118	1.0111	0.9966	0.9967	0.9826	0.9927	9.850E-01	9.963E-01
2.2	0.9835	0.9765	1.0101	1.0095	0.9981	0.9981	0.9867	0.9948	9.881E-01	9.978E-01
2.6	0.9857	0.9802	1.0088	1.0084	0.9988	0.9988	0.9892	0.9960	9.901E-01	9.987E-01

Z= 72 A= 177

p	$F_0L_0^*/F_0L_0$	λ_2^*/λ_2	μ_1^*/μ_1	μ_2^*/μ_2	Λ_1^*/Λ_1	Λ_2^*/Λ_2	ν_{12}^*/ν_{12}	$\vartheta_{12}^*/\vartheta_{12}$	η_{12}^*/η_{12}	$\hat{\eta}_{12}^*/\hat{\eta}_{12}$
0.1	0.9628	0.8193	1.0242	1.0194	-0.2915	-0.0343	1.0252	0.7889	4.970E 00	4.033E 00
0.2	0.9633	0.8357	1.0239	1.0195	0.5907	0.6569	0.8351	0.8775	-2.634E-02	-2.388E-01
0.4	0.9653	0.8768	1.0226	1.0193	0.8850	0.8985	0.8556	0.9391	8.271E-01	8.361E-01
0.6	0.9675	0.9095	1.0209	1.0184	0.9496	0.9542	0.9097	0.9629	9.219E-01	9.383E-01
0.8	0.9698	0.9311	1.0191	1.0172	0.9734	0.9753	0.9406	0.9750	9.525E-01	9.692E-01
1.0	0.9723	0.9452	1.0173	1.0158	0.9843	0.9852	0.9578	0.9819	9.665E-01	9.823E-01
1.2	0.9747	0.9548	1.0157	1.0145	0.9900	0.9905	0.9681	0.9862	9.743E-01	9.889E-01
1.4	0.9769	0.9616	1.0143	1.0133	0.9933	0.9935	0.9747	0.9892	9.791E-01	9.926E-01
1.8	0.9805	0.9705	1.0120	1.0113	0.9966	0.9967	0.9823	0.9927	9.847E-01	9.961E-01
2.2	0.9834	0.9761	1.0103	1.0097	0.9980	0.9981	0.9865	0.9947	9.879E-01	9.978E-01
2.6	0.9855	0.9798	1.0090	1.0085	0.9988	0.9988	0.9890	0.9959	9.899E-01	9.986E-01
3.0	0.9873	0.9825	1.0080	1.0076	0.9992	0.9992	0.9907	0.9968	9.913E-01	9.991E-01

ELECTRON Z= 73 A= 181

p	$F_0L_0^*/F_0L_0$	λ_2^*/λ_2	μ_1^*/μ_1	μ_2^*/μ_2	Λ_1^*/Λ_1	Λ_2^*/Λ_2	ν_{12}^*/ν_{12}	$\hat{\nu}_{12}^*/\hat{\nu}_{12}$	η_{12}^*/η_{12}	$\hat{\eta}_{12}^*/\hat{\eta}_{12}$
0.1	0.9623	0.8201	1.0247	1.0198	-0.3119	-0.0521	1.0346	0.7880	4.309E 00	3.613E 00
0.2	0.9628	0.8359	1.0244	1.0199	0.5838	0.6509	0.8400	0.8768	-1.348E-01	-4.216E-01
0.4	0.9648	0.8763	1.0231	1.0197	0.8829	0.8967	0.8552	0.9387	8.207E-01	8.287E-01
0.6	0.9671	0.9086	1.0213	1.0188	0.9487	0.9534	0.9087	0.9627	9.197E-01	9.362E-01
0.8	0.9695	0.9301	1.0195	1.0175	0.9729	0.9748	0.9397	0.9748	9.514E-01	9.683E-01
1.0	0.9720	0.9444	1.0176	1.0161	0.9840	0.9849	0.9571	0.9817	9.658E-01	9.818E-01
1.2	0.9744	0.9540	1.0160	1.0148	0.9898	0.9903	0.9676	0.9861	9.737E-01	9.886E-01
1.4	0.9766	0.9609	1.0146	1.0135	0.9932	0.9934	0.9743	0.9891	9.787E-01	9.923E-01
1.8	0.9803	0.9700	1.0122	1.0115	0.9965	0.9966	0.9820	0.9927	9.844E-01	9.960E-01
2.2	0.9832	0.9756	1.0105	1.0099	0.9980	0.9980	0.9862	0.9946	9.876E-01	9.977E-01
2.6	0.9854	0.9794	1.0092	1.0087	0.9987	0.9988	0.9888	0.9959	9.897E-01	9.985E-01
3.0	0.9871	0.9821	1.0082	1.0078	0.9992	0.9992	0.9905	0.9967	9.911E-01	9.991E-01

Z= 74 A= 182

p	$F_0L_0^*/F_0L_0$	λ_2^*/λ_2	μ_1^*/μ_1	μ_2^*/μ_2	Λ_1^*/Λ_1	Λ_2^*/Λ_2	ν_{12}^*/ν_{12}	$\hat{\nu}_{12}^*/\hat{\nu}_{12}$	η_{12}^*/η_{12}	$\hat{\eta}_{12}^*/\hat{\eta}_{12}$
0.1	0.9618	0.8208	1.0252	1.0203	-0.3319	-0.0699	1.0441	0.7871	3.829E 00	3.289E 00
0.2	0.9623	0.8362	1.0249	1.0203	0.5771	0.6449	0.8451	0.8761	-2.696E-01	-6.694E-01
0.4	0.9643	0.8757	1.0235	1.0200	0.8809	0.8949	0.8550	0.9383	8.139E-01	8.206E-01
0.6	0.9667	0.9077	1.0217	1.0191	0.9478	0.9525	0.9078	0.9624	9.175E-01	9.340E-01
0.8	0.9691	0.9293	1.0198	1.0178	0.9724	0.9744	0.9389	0.9746	9.502E-01	9.673E-01
1.0	0.9717	0.9435	1.0180	1.0164	0.9837	0.9847	0.9565	0.9816	9.650E-01	9.813E-01
1.2	0.9741	0.9533	1.0163	1.0150	0.9896	0.9901	0.9671	0.9860	9.732E-01	9.883E-01
1.4	0.9764	0.9603	1.0148	1.0138	0.9930	0.9933	0.9738	0.9890	9.782E-01	9.921E-01
1.8	0.9801	0.9694	1.0124	1.0117	0.9964	0.9965	0.9817	0.9926	9.841E-01	9.959E-01
2.2	0.9830	0.9751	1.0107	1.0101	0.9980	0.9980	0.9859	0.9946	9.874E-01	9.976E-01
2.6	0.9852	0.9790	1.0093	1.0089	0.9987	0.9987	0.9886	0.9958	9.895E-01	9.985E-01
3.0	0.9870	0.9817	1.0083	1.0079	0.9992	0.9991	0.9903	0.9967	9.909E-01	9.990E-01

Z= 75 A= 187

p	$F_0L_0^*/F_0L_0$	λ_2^*/λ_2	μ_1^*/μ_1	μ_2^*/μ_2	Λ_1^*/Λ_1	Λ_2^*/Λ_2	ν_{12}^*/ν_{12}	$\hat{\nu}_{12}^*/\hat{\nu}_{12}$	η_{12}^*/η_{12}	$\hat{\eta}_{12}^*/\hat{\eta}_{12}$
0.1	0.9613	0.8214	1.0257	1.0207	-0.3525	-0.0879	1.0537	0.7860	3.465E 00	3.033E 00
0.2	0.9618	0.8364	1.0254	1.0207	0.5701	0.6387	0.8504	0.8754	-4.426E-01	-1.026E 00
0.4	0.9639	0.8751	1.0240	1.0204	0.8789	0.8931	0.8547	0.9378	8.066E-01	8.117E-01
0.6	0.9663	0.9069	1.0222	1.0195	0.9469	0.9517	0.9068	0.9621	9.151E-01	9.316E-01
0.8	0.9688	0.9284	1.0202	1.0182	0.9719	0.9739	0.9381	0.9744	9.489E-01	9.663E-01
1.0	0.9714	0.9427	1.0183	1.0167	0.9834	0.9844	0.9558	0.9814	9.642E-01	9.808E-01
1.2	0.9738	0.9525	1.0166	1.0153	0.9895	0.9900	0.9665	0.9859	9.726E-01	9.879E-01
1.4	0.9761	0.9596	1.0151	1.0141	0.9929	0.9932	0.9734	0.9889	9.777E-01	9.919E-01
1.8	0.9799	0.9688	1.0127	1.0119	0.9964	0.9965	0.9813	0.9925	9.837E-01	9.958E-01
2.2	0.9828	0.9746	1.0109	1.0103	0.9979	0.9980	0.9857	0.9946	9.871E-01	9.975E-01
2.6	0.9851	0.9785	1.0095	1.0090	0.9987	0.9987	0.9884	0.9958	9.893E-01	9.984E-01
3.0	0.9869	0.9813	1.0085	1.0081	0.9991	0.9991	0.9901	0.9966	9.907E-01	9.990E-01

ELECTRON Z= 76 A= 188

p	$F_0L_0^*/F_0L_0$	λ_2^*/λ_2	μ_1^*/μ_1	μ_2^*/μ_2	Λ_1^*/Λ_1	Λ_2^*/Λ_2	ν_{12}^*/ν_{12}	$\vartheta_{12}^*/\vartheta_{12}$	η_{12}^*/η_{12}	$\hat{\eta}_{12}^*/\hat{\eta}_{12}$
0.1	0.9608	0.8220	1.0262	1.0211	-0.3730	-0.1060	1.0633	0.7849	3.179E 00	2.825E 00
0.2	0.9613	0.8366	1.0259	1.0212	0.5632	0.6326	0.8559	0.8746	-6.725E-01	-1.584E 00
0.4	0.9634	0.8746	1.0245	1.0208	0.8768	0.8912	0.8546	0.9373	7.987E-01	8.019E-01
0.6	0.9659	0.9060	1.0226	1.0199	0.9460	0.9509	0.9059	0.9618	9.126E-01	9.291E-01
0.8	0.9684	0.9275	1.0206	1.0185	0.9714	0.9735	0.9372	0.9742	9.477E-01	9.653E-01
1.0	0.9710	0.9418	1.0187	1.0170	0.9831	0.9841	0.9552	0.9813	9.634E-01	9.802E-01
1.2	0.9735	0.9518	1.0169	1.0156	0.9893	0.9898	0.9660	0.9858	9.720E-01	9.876E-01
1.4	0.9758	0.9589	1.0154	1.0143	0.9928	0.9931	0.9729	0.9888	9.773E-01	9.917E-01
1.8	0.9797	0.9682	1.0129	1.0121	0.9963	0.9964	0.9810	0.9924	9.834E-01	9.957E-01
2.2	0.9826	0.9741	1.0111	1.0105	0.9979	0.9979	0.9854	0.9945	9.868E-01	9.974E-01
2.6	0.9849	0.9781	1.0096	1.0092	0.9987	0.9987	0.9881	0.9958	9.890E-01	9.984E-01
3.0	0.9867	0.9810	1.0087	1.0082	0.9991	0.9991	0.9899	0.9965	9.905E-01	9.989E-01

Z= 77 A= 191

p	$F_0L_0^*/F_0L_0$	λ_2^*/λ_2	μ_1^*/μ_1	μ_2^*/μ_2	Λ_1^*/Λ_1	Λ_2^*/Λ_2	ν_{12}^*/ν_{12}	$\vartheta_{12}^*/\vartheta_{12}$	η_{12}^*/η_{12}	$\hat{\eta}_{12}^*/\hat{\eta}_{12}$
0.1	0.9603	0.8227	1.0267	1.0216	-0.3932	-0.1241	1.0730	0.7838	2.948E 00	2.652E 00
0.2	0.9608	0.8368	1.0263	1.0216	0.5563	0.6264	0.8615	0.8738	-9.928E-01	-2.579E 00
0.4	0.9629	0.8740	1.0249	1.0212	0.8748	0.8894	0.8546	0.9369	7.902E-01	7.911E-01
0.6	0.9655	0.9051	1.0230	1.0202	0.9451	0.9500	0.9051	0.9615	9.101E-01	9.264E-01
0.8	0.9681	0.9266	1.0210	1.0188	0.9709	0.9730	0.9364	0.9740	9.464E-01	9.641E-01
1.0	0.9707	0.9410	1.0190	1.0173	0.9828	0.9838	0.9545	0.9811	9.625E-01	9.796E-01
1.2	0.9733	0.9510	1.0173	1.0159	0.9891	0.9896	0.9655	0.9857	9.714E-01	9.872E-01
1.4	0.9756	0.9582	1.0157	1.0146	0.9927	0.9930	0.9725	0.9887	9.768E-01	9.914E-01
1.8	0.9795	0.9677	1.0132	1.0123	0.9963	0.9964	0.9807	0.9923	9.831E-01	9.955E-01
2.2	0.9825	0.9736	1.0113	1.0106	0.9979	0.9979	0.9851	0.9944	9.866E-01	9.973E-01
2.6	0.9848	0.9777	1.0098	1.0093	0.9987	0.9987	0.9879	0.9956	9.888E-01	9.983E-01
3.0	0.9866	0.9806	1.0088	1.0084	0.9991	0.9991	0.9897	0.9964	9.903E-01	9.989E-01

Z= 78 A= 194

p	$F_0L_0^*/F_0L_0$	λ_2^*/λ_2	μ_1^*/μ_1	μ_2^*/μ_2	Λ_1^*/Λ_1	Λ_2^*/Λ_2	ν_{12}^*/ν_{12}	$\vartheta_{12}^*/\vartheta_{12}$	η_{12}^*/η_{12}	$\hat{\eta}_{12}^*/\hat{\eta}_{12}$
0.1	0.9598	0.8232	1.0272	1.0220	-0.4132	-0.1420	1.0827	0.7828	2.757E 00	2.506E 00
0.2	0.9604	0.8370	1.0268	1.0220	0.5495	0.6203	0.8674	0.8730	-1.471E 00	-4.866E 00
0.4	0.9625	0.8735	1.0254	1.0216	0.8727	0.8875	0.8547	0.9364	7.810E-01	7.791E-01
0.6	0.9651	0.9043	1.0235	1.0206	0.9442	0.9492	0.9043	0.9612	9.074E-01	9.235E-01
0.8	0.9678	0.9257	1.0214	1.0192	0.9705	0.9726	0.9357	0.9738	9.450E-01	9.630E-01
1.0	0.9705	0.9402	1.0194	1.0176	0.9826	0.9836	0.9539	0.9810	9.617E-01	9.790E-01
1.2	0.9730	0.9502	1.0176	1.0162	0.9889	0.9894	0.9649	0.9855	9.707E-01	9.869E-01
1.4	0.9754	0.9575	1.0160	1.0148	0.9925	0.9928	0.9721	0.9886	9.763E-01	9.912E-01
1.8	0.9793	0.9671	1.0134	1.0126	0.9962	0.9963	0.9804	0.9922	9.827E-01	9.954E-01
2.2	0.9823	0.9731	1.0115	1.0108	0.9978	0.9979	0.9849	0.9943	9.863E-01	9.973E-01
2.6	0.9847	0.9772	1.0101	1.0095	0.9986	0.9986	0.9877	0.9955	9.886E-01	9.983E-01
3.0	0.9865	0.9802	1.0090	1.0085	0.9991	0.9991	0.9895	0.9964	9.902E-01	9.988E-01

ELECTRON Z= 79 A= 197

p	$F_0L_0^*/F_0L_0$	λ_2^*/λ_2	μ_1^*/μ_1	μ_2^*/μ_2	Λ_1^*/Λ_1	Λ_2^*/Λ_2	ν_{12}^*/ν_{12}	$\hat{\nu}_{12}^*/\hat{\nu}_{12}$	η_{12}^*/η_{12}	$\hat{\eta}_{12}^*/\hat{\eta}_{12}$
0.1	0.9593	0.8237	1.0278	1.0224	-0.4336	-0.1602	1.0926	0.7816	2.598E 00	2.381E 00
0.2	0.9599	0.8372	1.0274	1.0224	0.5425	0.6141	0.8734	0.8721	-2.267E 00	-1.566E 01
0.4	0.9621	0.8729	1.0259	1.0220	0.8707	0.8857	0.8549	0.9359	7.709E-01	7.654E-01
0.6	0.9647	0.9035	1.0239	1.0210	0.9433	0.9483	0.9034	0.9609	9.045E-01	9.204E-01
0.8	0.9674	0.9248	1.0218	1.0195	0.9700	0.9721	0.9349	0.9735	9.436E-01	9.617E-01
1.0	0.9702	0.9393	1.0198	1.0180	0.9823	0.9833	0.9532	0.9808	9.608E-01	9.783E-01
1.2	0.9728	0.9494	1.0179	1.0165	0.9887	0.9893	0.9644	0.9854	9.701E-01	9.865E-01
1.4	0.9751	0.9568	1.0163	1.0151	0.9924	0.9927	0.9716	0.9885	9.758E-01	9.909E-01
1.8	0.9791	0.9665	1.0137	1.0128	0.9961	0.9962	0.9800	0.9922	9.824E-01	9.953E-01
2.2	0.9821	0.9726	1.0117	1.0110	0.9978	0.9978	0.9846	0.9943	9.860E-01	9.972E-01
2.6	0.9845	0.9768	1.0103	1.0097	0.9986	0.9986	0.9874	0.9954	9.883E-01	9.982E-01
3.0	0.9864	0.9798	1.0092	1.0086	0.9991	0.9991	0.9893	0.9963	9.900E-01	9.988E-01
4.0	0.9896	0.9848	1.0068	1.0068	0.9996	0.9996	0.9923	0.9978	9.925E-01	9.993E-01

Z= 80 A= 198

p	$F_0L_0^*/F_0L_0$	λ_2^*/λ_2	μ_1^*/μ_1	μ_2^*/μ_2	Λ_1^*/Λ_1	Λ_2^*/Λ_2	ν_{12}^*/ν_{12}	$\hat{\nu}_{12}^*/\hat{\nu}_{12}$	η_{12}^*/η_{12}	$\hat{\eta}_{12}^*/\hat{\eta}_{12}$
0.1	0.9589	0.8242	1.0283	1.0229	-0.4539	-0.1786	1.1025	0.7804	2.461E 00	2.273E 00
0.2	0.9594	0.8373	1.0279	1.0229	0.5356	0.6079	0.8796	0.8713	-3.852E 00	2.021E 01
0.4	0.9616	0.8724	1.0264	1.0224	0.8686	0.8838	0.8552	0.9354	7.599E-01	7.500E-01
0.6	0.9643	0.9026	1.0244	1.0213	0.9423	0.9475	0.9027	0.9606	9.015E-01	9.171E-01
0.8	0.9671	0.9239	1.0222	1.0199	0.9695	0.9716	0.9341	0.9733	9.422E-01	9.604E-01
1.0	0.9699	0.9385	1.0201	1.0183	0.9820	0.9830	0.9526	0.9807	9.599E-01	9.777E-01
1.2	0.9725	0.9486	1.0182	1.0167	0.9885	0.9891	0.9639	0.9853	9.694E-01	9.861E-01
1.4	0.9749	0.9560	1.0166	1.0153	0.9923	0.9926	0.9712	0.9884	9.753E-01	9.907E-01
1.8	0.9789	0.9659	1.0139	1.0130	0.9961	0.9962	0.9797	0.9921	9.820E-01	9.952E-01
2.2	0.9820	0.9721	1.0119	1.0112	0.9978	0.9978	0.9843	0.9941	9.857E-01	9.971E-01
2.6	0.9844	0.9763	1.0105	1.0098	0.9986	0.9986	0.9872	0.9953	9.881E-01	9.981E-01
3.0	0.9863	0.9794	1.0093	1.0088	0.9991	0.9991	0.9891	0.9963	9.898E-01	9.987E-01
4.0	0.9895	0.9845	1.0069	1.0069	0.9996	0.9996	0.9921	0.9978	9.924E-01	9.992E-01

Z= 81 A= 203

p	$F_0L_0^*/F_0L_0$	λ_2^*/λ_2	μ_1^*/μ_1	μ_2^*/μ_2	Λ_1^*/Λ_1	Λ_2^*/Λ_2	ν_{12}^*/ν_{12}	$\hat{\nu}_{12}^*/\hat{\nu}_{12}$	η_{12}^*/η_{12}	$\hat{\eta}_{12}^*/\hat{\eta}_{12}$
0.1	0.9584	0.8247	1.0288	1.0233	-0.4739	-0.1967	1.1124	0.7791	2.344E 00	2.179E 00
0.2	0.9590	0.8375	1.0284	1.0233	0.5287	0.6017	0.8860	0.8704	-8.555E 00	7.029E 00
0.4	0.9612	0.8719	1.0269	1.0228	0.8665	0.8819	0.8557	0.9349	7.477E-01	7.323E-01
0.6	0.9640	0.9018	1.0248	1.0217	0.9414	0.9466	0.9019	0.9602	8.984E-01	9.135E-01
0.8	0.9668	0.9230	1.0226	1.0202	0.9690	0.9712	0.9333	0.9731	9.406E-01	9.589E-01
1.0	0.9696	0.9376	1.0205	1.0186	0.9817	0.9827	0.9519	0.9805	9.590E-01	9.769E-01
1.2	0.9723	0.9479	1.0186	1.0170	0.9883	0.9889	0.9633	0.9851	9.688E-01	9.856E-01
1.4	0.9747	0.9553	1.0169	1.0156	0.9922	0.9925	0.9707	0.9882	9.748E-01	9.904E-01
1.8	0.9787	0.9653	1.0142	1.0132	0.9960	0.9961	0.9794	0.9920	9.817E-01	9.950E-01
2.2	0.9819	0.9716	1.0122	1.0114	0.9977	0.9977	0.9840	0.9941	9.855E-01	9.970E-01
2.6	0.9843	0.9759	1.0107	1.0100	0.9986	0.9986	0.9870	0.9953	9.879E-01	9.981E-01
3.0	0.9861	0.9791	1.0095	1.0089	0.9991	0.9990	0.9889	0.9962	9.896E-01	9.986E-01
4.0	0.9894	0.9842	1.0071	1.0070	0.9996	0.9996	0.9920	0.9976	9.922E-01	9.992E-01

ELECTRON Z= 82 A= 208

p	$F_0L_0^*/F_0L_0$	λ_2^*/λ_2	μ_1^*/μ_1	μ_2^*/μ_2	Λ_1^*/Λ_1	Λ_2^*/Λ_2	ν_{12}^*/ν_{12}	$\hat{\nu}_{12}^*/\hat{\nu}_{12}$	η_{12}^*/η_{12}	$\hat{\eta}_{12}^*/\hat{\eta}_{12}$
0.1	0.9580	0.8252	1.0293	1.0238	-0.4938	-0.2150	1.1224	0.7779	2.241E 00	2.095E 00
0.2	0.9586	0.8376	1.0289	1.0237	0.5218	0.5954	0.8927	0.8695	-6.829E 02	4.551E 00
0.4	0.9608	0.8713	1.0274	1.0233	0.8644	0.8800	0.8562	0.9344	7.342E-01	7.117E-01
0.6	0.9636	0.9010	1.0253	1.0221	0.9405	0.9458	0.9012	0.9599	8.950E-01	9.095E-01
0.8	0.9665	0.9221	1.0230	1.0206	0.9685	0.9707	0.9326	0.9728	9.391E-01	9.574E-01
1.0	0.9694	0.9368	1.0209	1.0189	0.9814	0.9824	0.9513	0.9803	9.580E-01	9.762E-01
1.2	0.9720	0.9471	1.0189	1.0173	0.9882	0.9887	0.9628	0.9850	9.681E-01	9.852E-01
1.4	0.9745	0.9546	1.0172	1.0159	0.9920	0.9923	0.9703	0.9881	9.743E-01	9.901E-01
1.8	0.9786	0.9647	1.0144	1.0134	0.9959	0.9960	0.9790	0.9919	9.813E-01	9.948E-01
2.2	0.9817	0.9710	1.0124	1.0116	0.9977	0.9977	0.9838	0.9940	9.852E-01	9.969E-01
2.6	0.9841	0.9755	1.0109	1.0101	0.9986	0.9986	0.9867	0.9953	9.876E-01	9.980E-01
3.0	0.9860	0.9787	1.0096	1.0091	0.9990	0.9990	0.9888	0.9962	9.894E-01	9.986E-01
4.0	0.9894	0.9839	1.0072	1.0071	0.9996	0.9996	0.9918	0.9976	9.921E-01	9.991E-01

Z= 83 A= 214

p	$F_0L_0^*/F_0L_0$	λ_2^*/λ_2	μ_1^*/μ_1	μ_2^*/μ_2	Λ_1^*/Λ_1	Λ_2^*/Λ_2	ν_{12}^*/ν_{12}	$\hat{\nu}_{12}^*/\hat{\nu}_{12}$	η_{12}^*/η_{12}	$\hat{\eta}_{12}^*/\hat{\eta}_{12}$
0.1	0.9575	0.8256	1.0299	1.0242	-0.5135	-0.2332	1.1325	0.7766	2.151E 00	2.021E 00
0.2	0.9581	0.8377	1.0294	1.0242	0.5149	0.5892	0.8995	0.8685	1.065E 01	3.503E 00
0.4	0.9604	0.8708	1.0279	1.0237	0.8624	0.8781	0.8569	0.9338	7.191E-01	6.876E-01
0.6	0.9633	0.9001	1.0257	1.0225	0.9395	0.9449	0.9006	0.9596	8.915E-01	9.052E-01
0.8	0.9662	0.9212	1.0234	1.0209	0.9680	0.9702	0.9319	0.9726	9.374E-01	9.558E-01
1.0	0.9691	0.9359	1.0212	1.0192	0.9811	0.9822	0.9507	0.9801	9.570E-01	9.753E-01
1.2	0.9718	0.9463	1.0193	1.0176	0.9880	0.9885	0.9623	0.9849	9.674E-01	9.847E-01
1.4	0.9743	0.9539	1.0175	1.0161	0.9919	0.9922	0.9698	0.9880	9.737E-01	9.898E-01
1.8	0.9784	0.9641	1.0147	1.0137	0.9959	0.9960	0.9787	0.9918	9.809E-01	9.947E-01
2.2	0.9816	0.9705	1.0127	1.0118	0.9976	0.9977	0.9835	0.9939	9.849E-01	9.968E-01
2.6	0.9840	0.9750	1.0110	1.0103	0.9985	0.9985	0.9865	0.9952	9.874E-01	9.979E-01
3.0	0.9859	0.9783	1.0098	1.0092	0.9990	0.9990	0.9886	0.9961	9.892E-01	9.985E-01
4.0	0.9893	0.9836	1.0073	1.0072	0.9996	0.9996	0.9917	0.9976	9.919E-01	9.991E-01

Z= 84 A= 210

p	$F_0L_0^*/F_0L_0$	λ_2^*/λ_2	μ_1^*/μ_1	μ_2^*/μ_2	Λ_1^*/Λ_1	Λ_2^*/Λ_2	ν_{12}^*/ν_{12}	$\hat{\nu}_{12}^*/\hat{\nu}_{12}$	η_{12}^*/η_{12}	$\hat{\eta}_{12}^*/\hat{\eta}_{12}$
0.1	0.9571	0.8260	1.0304	1.0247	-0.5338	-0.2517	1.1429	0.7752	2.070E 00	1.955E 00
0.2	0.9577	0.8377	1.0300	1.0246	0.5079	0.5829	0.9065	0.8676	5.734E 00	2.923E 00
0.4	0.9600	0.8703	1.0284	1.0241	0.8603	0.8762	0.8577	0.9333	7.019E-01	6.586E-01
0.6	0.9629	0.8993	1.0262	1.0229	0.9386	0.9440	0.9000	0.9592	8.876E-01	9.003E-01
0.8	0.9659	0.9204	1.0238	1.0213	0.9675	0.9698	0.9311	0.9724	9.357E-01	9.540E-01
1.0	0.9688	0.9351	1.0216	1.0196	0.9808	0.9819	0.9500	0.9799	9.560E-01	9.745E-01
1.2	0.9716	0.9455	1.0196	1.0179	0.9878	0.9884	0.9617	0.9847	9.667E-01	9.842E-01
1.4	0.9741	0.9531	1.0178	1.0164	0.9918	0.9921	0.9694	0.9879	9.732E-01	9.895E-01
1.8	0.9782	0.9635	1.0150	1.0139	0.9958	0.9959	0.9783	0.9917	9.805E-01	9.945E-01
2.2	0.9814	0.9700	1.0129	1.0120	0.9976	0.9976	0.9832	0.9939	9.846E-01	9.967E-01
2.6	0.9839	0.9746	1.0112	1.0105	0.9985	0.9985	0.9863	0.9952	9.872E-01	9.978E-01
3.0	0.9858	0.9779	1.0099	1.0093	0.9990	0.9990	0.9884	0.9961	9.890E-01	9.984E-01
4.0	0.9892	0.9833	1.0075	1.0072	0.9996	0.9996	0.9915	0.9974	9.918E-01	9.990E-01

ELECTRON Z= 85 A= 21

p	$F_0L_0^*/F_0L_0$	λ_2^*/λ_2	μ_1^*/μ_1	μ_2^*/μ_2	Λ_1^*/Λ_1	Λ_2^*/Λ_2	ν_{12}^*/ν_{12}	$\hat{\nu}_{12}^*/\hat{\nu}_{12}$	η_{12}^*/η_{12}	$\hat{\eta}_{12}^*/\hat{\eta}_{12}$
0.1	0.9567	0.8263	1.0310	1.0251	-0.5535	-0.2702	1.1532	0.7738	1.999E 00	1.895E 00
0.2	0.9573	0.8378	1.0305	1.0251	0.5009	0.5765	0.9138	0.8665	4.117E 00	2.556E 00
0.4	0.9596	0.8697	1.0289	1.0245	0.8582	0.8743	0.8587	0.9327	6.824E-01	6.234E-01
0.6	0.9626	0.8985	1.0266	1.0233	0.9377	0.9432	0.8994	0.9588	8.835E-01	8.950E-01
0.8	0.9656	0.9194	1.0243	1.0216	0.9670	0.9693	0.9304	0.9721	9.339E-01	9.521E-01
1.0	0.9686	0.9342	1.0220	1.0199	0.9805	0.9816	0.9494	0.9797	9.549E-01	9.735E-01
1.2	0.9714	0.9447	1.0199	1.0182	0.9876	0.9882	0.9612	0.9846	9.659E-01	9.837E-01
1.4	0.9739	0.9524	1.0181	1.0167	0.9917	0.9920	0.9689	0.9878	9.726E-01	9.891E-01
1.8	0.9781	0.9628	1.0152	1.0141	0.9958	0.9959	0.9780	0.9916	9.801E-01	9.943E-01
2.2	0.9813	0.9695	1.0131	1.0122	0.9976	0.9976	0.9829	0.9938	9.843E-01	9.966E-01
2.6	0.9838	0.9741	1.0114	1.0107	0.9985	0.9985	0.9860	0.9952	9.869E-01	9.977E-01
3.0	0.9857	0.9775	1.0100	1.0095	0.9990	0.9990	0.9882	0.9961	9.888E-01	9.983E-01
4.0	0.9891	0.9830	1.0076	1.0073	0.9996	0.9996	0.9914	0.9973	9.916E-01	9.989E-01

Z= 86 A= 219

p	$F_0L_0^*/F_0L_0$	λ_2^*/λ_2	μ_1^*/μ_1	μ_2^*/μ_2	Λ_1^*/Λ_1	Λ_2^*/Λ_2	ν_{12}^*/ν_{12}	$\hat{\nu}_{12}^*/\hat{\nu}_{12}$	η_{12}^*/η_{12}	$\hat{\eta}_{12}^*/\hat{\eta}_{12}$
0.1	0.9563	0.8266	1.0315	1.0256	-0.5732	-0.2887	1.1637	0.7723	1.934E 00	1.840E 00
0.2	0.9569	0.8378	1.0311	1.0255	0.4940	0.5702	0.9213	0.8655	3.308E 00	2.301E 00
0.4	0.9592	0.8691	1.0294	1.0249	0.8561	0.8724	0.8598	0.9321	6.600E-01	5.796E-01
0.6	0.9622	0.8976	1.0271	1.0237	0.9367	0.9423	0.8989	0.9584	8.791E-01	8.890E-01
0.8	0.9653	0.9185	1.0247	1.0220	0.9665	0.9688	0.9297	0.9718	9.320E-01	9.500E-01
1.0	0.9684	0.9333	1.0224	1.0202	0.9802	0.9813	0.9488	0.9796	9.538E-01	9.725E-01
1.2	0.9712	0.9439	1.0203	1.0185	0.9874	0.9880	0.9607	0.9844	9.652E-01	9.831E-01
1.4	0.9737	0.9517	1.0185	1.0170	0.9915	0.9919	0.9685	0.9876	9.720E-01	9.888E-01
1.8	0.9779	0.9622	1.0155	1.0144	0.9957	0.9958	0.9776	0.9915	9.797E-01	9.942E-01
2.2	0.9811	0.9690	1.0133	1.0124	0.9975	0.9976	0.9827	0.9937	9.840E-01	9.965E-01
2.6	0.9837	0.9736	1.0116	1.0108	0.9985	0.9985	0.9858	0.9951	9.867E-01	9.977E-01
3.0	0.9856	0.9771	1.0102	1.0097	0.9990	0.9990	0.9880	0.9961	9.886E-01	9.982E-01
4.0	0.9890	0.9827	1.0077	1.0073	0.9996	0.9996	0.9912	0.9973	9.914E-01	9.988E-01

Z= 87 A= 221

p	$F_0L_0^*/F_0L_0$	λ_2^*/λ_2	μ_1^*/μ_1	μ_2^*/μ_2	Λ_1^*/Λ_1	Λ_2^*/Λ_2	ν_{12}^*/ν_{12}	$\hat{\nu}_{12}^*/\hat{\nu}_{12}$	η_{12}^*/η_{12}	$\hat{\eta}_{12}^*/\hat{\eta}_{12}$
0.1	0.9559	0.8269	1.0321	1.0261	-0.5925	-0.3072	1.1742	0.7709	1.875E 00	1.791E 00
0.2	0.9565	0.8379	1.0316	1.0260	0.4872	0.5638	0.9290	0.8645	2.823E 00	2.114E 00
0.4	0.9589	0.8686	1.0299	1.0254	0.8540	0.8705	0.8610	0.9315	6.338E-01	5.233E-01
0.6	0.9619	0.8968	1.0276	1.0241	0.9358	0.9414	0.8985	0.9581	8.744E-01	8.823E-01
0.8	0.9651	0.9177	1.0251	1.0223	0.9660	0.9683	0.9291	0.9716	9.300E-01	9.478E-01
1.0	0.9681	0.9325	1.0228	1.0205	0.9799	0.9810	0.9482	0.9794	9.526E-01	9.715E-01
1.2	0.9710	0.9431	1.0206	1.0188	0.9872	0.9878	0.9602	0.9843	9.644E-01	9.825E-01
1.4	0.9735	0.9509	1.0188	1.0172	0.9914	0.9917	0.9680	0.9875	9.714E-01	9.884E-01
1.8	0.9778	0.9616	1.0158	1.0146	0.9956	0.9957	0.9773	0.9915	9.794E-01	9.940E-01
2.2	0.9810	0.9684	1.0135	1.0125	0.9975	0.9975	0.9824	0.9937	9.837E-01	9.963E-01
2.6	0.9836	0.9732	1.0118	1.0110	0.9984	0.9984	0.9856	0.9950	9.864E-01	9.975E-01
3.0	0.9856	0.9766	1.0103	1.0098	0.9990	0.9990	0.9878	0.9960	9.883E-01	9.981E-01
4.0	0.9890	0.9824	1.0079	1.0073	0.9996	0.9996	0.9910	0.9971	9.913E-01	9.986E-01

ELECTRON Z= 88 A= 222

p	$F_0L_0^*/F_0L_0$	λ_2^*/λ_2	μ_1^*/μ_1	μ_2^*/μ_2	Λ_1^*/Λ_1	Λ_2^*/Λ_2	ν_{12}^*/ν_{12}	$\hat{\nu}_{12}^*/\hat{\nu}_{12}$	η_{12}^*/η_{12}	$\hat{\eta}_{12}^*/\hat{\eta}_{12}$
0.1	0.9555	0.8272	1.0326	1.0265	-0.6120	-0.3258	1.1850	0.7693	1.822E 00	1.745E 00
0.2	0.9562	0.8378	1.0321	1.0264	0.4802	0.5574	0.9370	0.8634	2.500E 00	1.971E 00
0.4	0.9585	0.8681	1.0304	1.0258	0.8519	0.8685	0.8624	0.9309	6.027E-01	4.483E-01
0.6	0.9616	0.8960	1.0281	1.0245	0.9348	0.9405	0.8980	0.9577	8.692E-01	8.746E-01
0.8	0.9648	0.9168	1.0256	1.0227	0.9655	0.9679	0.9284	0.9713	9.279E-01	9.453E-01
1.0	0.9679	0.9316	1.0232	1.0209	0.9796	0.9807	0.9476	0.9792	9.514E-01	9.703E-01
1.2	0.9708	0.9423	1.0210	1.0191	0.9870	0.9876	0.9597	0.9841	9.636E-01	9.818E-01
1.4	0.9733	0.9502	1.0191	1.0175	0.9913	0.9916	0.9676	0.9874	9.708E-01	9.880E-01
1.8	0.9776	0.9609	1.0160	1.0148	0.9956	0.9957	0.9769	0.9913	9.789E-01	9.938E-01
2.2	0.9809	0.9679	1.0137	1.0127	0.9975	0.9975	0.9821	0.9936	9.834E-01	9.962E-01
2.6	0.9835	0.9727	1.0119	1.0112	0.9984	0.9984	0.9854	0.9950	9.862E-01	9.974E-01
3.0	0.9855	0.9762	1.0104	1.0100	0.9990	0.9989	0.9876	0.9960	9.881E-01	9.981E-01
4.0	0.9889	0.9820	1.0081	1.0075	0.9996	0.9996	0.9909	0.9970	9.911E-01	9.986E-01

Z= 89 A= 225

p	$F_0L_0^*/F_0L_0$	λ_2^*/λ_2	μ_1^*/μ_1	μ_2^*/μ_2	Λ_1^*/Λ_1	Λ_2^*/Λ_2	ν_{12}^*/ν_{12}	$\hat{\nu}_{12}^*/\hat{\nu}_{12}$	η_{12}^*/η_{12}	$\hat{\eta}_{12}^*/\hat{\eta}_{12}$
0.1	0.9552	0.8274	1.0332	1.0270	-0.6313	-0.3445	1.1959	0.7676	1.773E 00	1.704E 00
0.2	0.9558	0.8378	1.0327	1.0269	0.4733	0.5510	0.9452	0.8622	2.268E 00	1.858E 00
0.4	0.9582	0.8675	1.0309	1.0262	0.8497	0.8666	0.8640	0.9302	5.652E-01	3.432E-01
0.6	0.9613	0.8951	1.0286	1.0249	0.9339	0.9396	0.8977	0.9572	8.635E-01	8.657E-01
0.8	0.9646	0.9158	1.0260	1.0231	0.9649	0.9674	0.9278	0.9710	9.257E-01	9.425E-01
1.0	0.9677	0.9307	1.0236	1.0212	0.9793	0.9804	0.9470	0.9790	9.502E-01	9.690E-01
1.2	0.9706	0.9414	1.0213	1.0194	0.9868	0.9874	0.9591	0.9839	9.627E-01	9.811E-01
1.4	0.9732	0.9494	1.0194	1.0178	0.9912	0.9915	0.9671	0.9872	9.702E-01	9.875E-01
1.8	0.9775	0.9603	1.0163	1.0150	0.9955	0.9956	0.9766	0.9912	9.785E-01	9.936E-01
2.2	0.9808	0.9673	1.0139	1.0129	0.9974	0.9975	0.9818	0.9935	9.831E-01	9.961E-01
2.6	0.9834	0.9722	1.0121	1.0114	0.9984	0.9984	0.9851	0.9950	9.859E-01	9.973E-01
3.0	0.9854	0.9758	1.0106	1.0101	0.9990	0.9989	0.9874	0.9959	9.879E-01	9.980E-01
4.0	0.9889	0.9817	1.0082	1.0076	0.9996	0.9996	0.9907	0.9969	9.910E-01	9.985E-01

Z= 90 A= 228

p	$F_0L_0^*/F_0L_0$	λ_2^*/λ_2	μ_1^*/μ_1	μ_2^*/μ_2	Λ_1^*/Λ_1	Λ_2^*/Λ_2	ν_{12}^*/ν_{12}	$\hat{\nu}_{12}^*/\hat{\nu}_{12}$	η_{12}^*/η_{12}	$\hat{\eta}_{12}^*/\hat{\eta}_{12}$
0.1	0.9548	0.8276	1.0338	1.0275	-0.6499	-0.3630	1.2069	0.7660	1.727E 00	1.665E 00
0.2	0.9555	0.8378	1.0333	1.0274	0.4666	0.5446	0.9537	0.8611	2.094E 00	1.767E 00
0.4	0.9578	0.8669	1.0315	1.0267	0.8477	0.8646	0.8658	0.9296	5.191E-01	1.853E-01
0.6	0.9611	0.8943	1.0290	1.0253	0.9329	0.9388	0.8974	0.9568	8.574E-01	8.554E-01
0.8	0.9643	0.9150	1.0264	1.0234	0.9644	0.9669	0.9272	0.9707	9.233E-01	9.394E-01
1.0	0.9675	0.9298	1.0239	1.0215	0.9790	0.9802	0.9464	0.9787	9.489E-01	9.677E-01
1.2	0.9704	0.9406	1.0217	1.0197	0.9866	0.9872	0.9586	0.9838	9.619E-01	9.804E-01
1.4	0.9730	0.9486	1.0197	1.0180	0.9910	0.9914	0.9667	0.9871	9.695E-01	9.870E-01
1.8	0.9774	0.9596	1.0166	1.0153	0.9954	0.9955	0.9763	0.9911	9.781E-01	9.933E-01
2.2	0.9807	0.9668	1.0142	1.0132	0.9974	0.9974	0.9816	0.9935	9.827E-01	9.959E-01
2.6	0.9833	0.9717	1.0123	1.0115	0.9984	0.9984	0.9849	0.9949	9.856E-01	9.971E-01
3.0	0.9853	0.9753	1.0108	1.0103	0.9989	0.9989	0.9871	0.9958	9.876E-01	9.979E-01
4.0	0.9888	0.9814	1.0084	1.0077	0.9996	0.9995	0.9905	0.9969	9.908E-01	9.984E-01

Instr. see p. 25

ELECTRON Z= 91 A= 234

p	$F_0L_0^*/F_0L_0$	λ_2^*/λ_2	μ_1^*/μ_1	μ_2^*/μ_2	Λ_1^*/Λ_1	Λ_2^*/Λ_2	ν_{12}^*/ν_{12}	$\hat{\nu}_{12}^*/\hat{\nu}_{12}$	η_{12}^*/η_{12}	$\hat{\eta}_{12}^*/\hat{\eta}_{12}$
0.1	0.9545	0.8277	1.0344	1.0280	-0.6687	-0.3819	1.2181	0.7643	1.686E 00	1.629E 00
0.2	0.9551	0.8377	1.0338	1.0278	0.4597	0.5381	0.9625	0.8599	1.958E 00	1.691E 00
0.4	0.9575	0.8663	1.0320	1.0271	0.8455	0.8627	0.8677	0.9289	4.607E-01	-7.934E-02
0.6	0.9608	0.8934	1.0295	1.0257	0.9320	0.9378	0.8972	0.9564	8.505E-01	8.431E-01
0.8	0.9641	0.9140	1.0269	1.0238	0.9639	0.9664	0.9266	0.9704	9.207E-01	9.360E-01
1.0	0.9673	0.9289	1.0244	1.0219	0.9787	0.9799	0.9458	0.9785	9.475E-01	9.661E-01
1.2	0.9702	0.9397	1.0221	1.0200	0.9865	0.9871	0.9581	0.9836	9.609E-01	9.795E-01
1.4	0.9729	0.9479	1.0201	1.0183	0.9909	0.9912	0.9662	0.9870	9.689E-01	9.865E-01
1.8	0.9773	0.9590	1.0168	1.0155	0.9954	0.9955	0.9759	0.9911	9.777E-01	9.930E-01
2.2	0.9806	0.9662	1.0144	1.0134	0.9974	0.9974	0.9813	0.9934	9.824E-01	9.958E-01
2.6	0.9832	0.9712	1.0125	1.0117	0.9984	0.9983	0.9846	0.9948	9.854E-01	9.970E-01
3.0	0.9853	0.9749	1.0110	1.0104	0.9989	0.9989	0.9869	0.9958	9.874E-01	9.977E-01
4.0	0.9888	0.9810	1.0086	1.0078	0.9995	0.9995	0.9904	0.9968	9.906E-01	9.984E-01

Z= 92 A= 233

p	$F_0L_0^*/F_0L_0$	λ_2^*/λ_2	μ_1^*/μ_1	μ_2^*/μ_2	Λ_1^*/Λ_1	Λ_2^*/Λ_2	ν_{12}^*/ν_{12}	$\hat{\nu}_{12}^*/\hat{\nu}_{12}$	η_{12}^*/η_{12}	$\hat{\eta}_{12}^*/\hat{\eta}_{12}$
0.1	0.9541	0.8279	1.0349	1.0284	-0.6871	-0.4005	1.2295	0.7625	1.647E 00	1.596E 00
0.2	0.9548	0.8376	1.0344	1.0283	0.4529	0.5317	0.9716	0.8586	1.849E 00	1.627E 00
0.4	0.9572	0.8658	1.0326	1.0275	0.8434	0.8607	0.8699	0.9282	3.843E-01	-6.172E-01
0.6	0.9605	0.8926	1.0300	1.0261	0.9310	0.9370	0.8971	0.9560	8.428E-01	8.283E-01
0.8	0.9639	0.9131	1.0273	1.0242	0.9634	0.9659	0.9260	0.9701	9.180E-01	9.321E-01
1.0	0.9671	0.9280	1.0248	1.0222	0.9784	0.9796	0.9453	0.9783	9.460E-01	9.645E-01
1.2	0.9701	0.9389	1.0224	1.0203	0.9863	0.9869	0.9576	0.9834	9.600E-01	9.786E-01
1.4	0.9727	0.9471	1.0204	1.0186	0.9908	0.9911	0.9658	0.9868	9.682E-01	9.859E-01
1.8	0.9772	0.9583	1.0171	1.0158	0.9953	0.9954	0.9756	0.9910	9.772E-01	9.928E-01
2.2	0.9805	0.9656	1.0146	1.0136	0.9973	0.9973	0.9810	0.9933	9.821E-01	9.956E-01
2.6	0.9832	0.9707	1.0127	1.0119	0.9984	0.9983	0.9844	0.9947	9.851E-01	9.969E-01
3.0	0.9852	0.9744	1.0112	1.0105	0.9989	0.9989	0.9867	0.9957	9.872E-01	9.976E-01
4.0	0.9887	0.9807	1.0087	1.0080	0.9995	0.9995	0.9902	0.9969	9.904E-01	9.983E-01

Z= 93 A= 237

p	$F_0L_0^*/F_0L_0$	λ_2^*/λ_2	μ_1^*/μ_1	μ_2^*/μ_2	Λ_1^*/Λ_1	Λ_2^*/Λ_2	ν_{12}^*/ν_{12}	$\hat{\nu}_{12}^*/\hat{\nu}_{12}$	η_{12}^*/η_{12}	$\hat{\eta}_{12}^*/\hat{\eta}_{12}$
0.1	0.9538	0.8279	1.0355	1.0289	-0.7055	-0.4196	1.2411	0.7606	1.611E 00	1.565E 00
0.2	0.9545	0.8375	1.0350	1.0288	0.4460	0.5251	0.9809	0.8573	1.759E 00	1.572E 00
0.4	0.9569	0.8651	1.0331	1.0280	0.8413	0.8587	0.8722	0.9274	2.796E-01	-2.297E 00
0.6	0.9603	0.8917	1.0305	1.0265	0.9300	0.9360	0.8970	0.9555	8.341E-01	8.101E-01
0.8	0.9637	0.9122	1.0278	1.0246	0.9629	0.9654	0.9255	0.9698	9.150E-01	9.276E-01
1.0	0.9669	0.9271	1.0252	1.0226	0.9781	0.9793	0.9447	0.9781	9.445E-01	9.626E-01
1.2	0.9699	0.9380	1.0228	1.0206	0.9861	0.9867	0.9571	0.9832	9.590E-01	9.776E-01
1.4	0.9726	0.9463	1.0207	1.0189	0.9907	0.9910	0.9653	0.9867	9.674E-01	9.853E-01
1.8	0.9771	0.9576	1.0174	1.0160	0.9953	0.9953	0.9752	0.9908	9.767E-01	9.925E-01
2.2	0.9805	0.9650	1.0148	1.0138	0.9973	0.9973	0.9807	0.9932	9.817E-01	9.954E-01
2.6	0.9831	0.9702	1.0128	1.0121	0.9983	0.9983	0.9841	0.9947	9.848E-01	9.968E-01
3.0	0.9851	0.9740	1.0114	1.0107	0.9989	0.9989	0.9865	0.9956	9.869E-01	9.975E-01
4.0	0.9887	0.9803	1.0088	1.0080	0.9995	0.9995	0.9900	0.9968	9.902E-01	9.981E-01

ELECTRON Z= 94 A= 238

p	$F_0L_0^*/F_0L_0$	λ_2^*/λ_2	μ_1^*/μ_1	μ_2^*/μ_2	Λ_1^*/Λ_1	Λ_2^*/Λ_2	ν_{12}^*/ν_{12}	$\vartheta_{12}^*/\vartheta_{12}$	η_{12}^*/η_{12}	$\hat{\eta}_{12}^*/\hat{\eta}_{12}$
0.1	0.9536	0.8280	1.0361	1.0294	-0.7231	-0.4382	1.2528	0.7588	1.578E 00	1.536E 00
0.2	0.9543	0.8373	1.0356	1.0293	0.4394	0.5187	0.9906	0.8560	1.684E 00	1.524E 00
0.4	0.9567	0.8645	1.0337	1.0284	0.8392	0.8567	0.8749	0.9267	1.273E-01	5.083E 01
0.6	0.9600	0.8909	1.0311	1.0269	0.9291	0.9351	0.8970	0.9550	8.244E-01	7.871E-01
0.8	0.9635	0.9112	1.0282	1.0249	0.9624	0.9649	0.9250	0.9695	9.119E-01	9.224E-01
1.0	0.9668	0.9262	1.0256	1.0229	0.9778	0.9790	0.9442	0.9778	9.429E-01	9.605E-01
1.2	0.9698	0.9372	1.0232	1.0209	0.9859	0.9865	0.9566	0.9830	9.580E-01	9.765E-01
1.4	0.9725	0.9455	1.0211	1.0192	0.9905	0.9908	0.9649	0.9865	9.667E-01	9.847E-01
1.8	0.9770	0.9569	1.0177	1.0162	0.9952	0.9953	0.9748	0.9907	9.763E-01	9.921E-01
2.2	0.9804	0.9644	1.0150	1.0140	0.9973	0.9973	0.9804	0.9931	9.813E-01	9.952E-01
2.6	0.9831	0.9696	1.0131	1.0122	0.9983	0.9983	0.9839	0.9945	9.845E-01	9.966E-01
3.0	0.9851	0.9735	1.0116	1.0108	0.9989	0.9989	0.9862	0.9955	9.867E-01	9.974E-01
4.0	0.9887	0.9799	1.0090	1.0082	0.9995	0.9995	0.9898	0.9967	9.900E-01	9.980E-01
6.0	0.9924	0.9864	1.0064	1.0056	0.9999	0.9999	0.9931	0.9980	9.932E-01	9.986E-01

Z= 95 A= 243

p	$F_0L_0^*/F_0L_0$	λ_2^*/λ_2	μ_1^*/μ_1	μ_2^*/μ_2	Λ_1^*/Λ_1	Λ_2^*/Λ_2	ν_{12}^*/ν_{12}	$\vartheta_{12}^*/\vartheta_{12}$	η_{12}^*/η_{12}	$\hat{\eta}_{12}^*/\hat{\eta}_{12}$
0.1	0.9533	0.8281	1.0367	1.0299	-0.7402	-0.4571	1.2648	0.7568	1.546E 00	1.509E 00
0.2	0.9540	0.8372	1.0362	1.0297	0.4328	0.5121	1.0007	0.8547	1.620E 00	1.483E 00
0.4	0.9564	0.8639	1.0342	1.0289	0.8371	0.8548	0.8777	0.9260	-1.152E-01	3.843E 00
0.6	0.9598	0.8900	1.0316	1.0273	0.9281	0.9342	0.8971	0.9546	8.131E-01	7.571E-01
0.8	0.9633	0.9103	1.0287	1.0253	0.9619	0.9644	0.9246	0.9692	9.084E-01	9.163E-01
1.0	0.9666	0.9253	1.0260	1.0233	0.9775	0.9787	0.9436	0.9776	9.411E-01	9.581E-01
1.2	0.9697	0.9363	1.0236	1.0213	0.9857	0.9863	0.9561	0.9828	9.569E-01	9.753E-01
1.4	0.9724	0.9447	1.0214	1.0195	0.9904	0.9907	0.9645	0.9864	9.659E-01	9.839E-01
1.8	0.9769	0.9562	1.0179	1.0165	0.9951	0.9952	0.9745	0.9906	9.758E-01	9.917E-01
2.2	0.9804	0.9638	1.0153	1.0142	0.9972	0.9972	0.9801	0.9930	9.810E-01	9.949E-01
2.6	0.9830	0.9691	1.0133	1.0124	0.9983	0.9983	0.9836	0.9944	9.842E-01	9.965E-01
3.0	0.9851	0.9730	1.0118	1.0110	0.9989	0.9988	0.9860	0.9954	9.864E-01	9.973E-01
4.0	0.9886	0.9795	1.0091	1.0083	0.9995	0.9995	0.9896	0.9967	9.898E-01	9.979E-01
6.0	0.9924	0.9861	1.0064	1.0057	0.9999	0.9999	0.9929	0.9981	9.930E-01	9.983E-01

ELECTRON Z= 96 A= 245

p	$F_0L_0^*/F_0L_0$	λ_2^*/λ_2	μ_1^*/μ_1	μ_2^*/μ_2	Λ_1^*/Λ_1	Λ_2^*/Λ_2	ν_{12}^*/ν_{12}	$\hat{\vartheta}_{12}^*/\hat{\vartheta}_{12}$	η_{12}^*/η_{12}	$\hat{\eta}_{12}^*/\hat{\eta}_{12}$
0.1	0.9531	0.8281	1.0373	1.0304	-0.7569	-0.4759	1.2771	0.7548	1.517E 00	1.483E 00
0.2	0.9538	0.8370	1.0368	1.0302	0.4263	0.5056	1.0111	0.8533	1.565E 00	1.446E 00
0.4	0.9562	0.8633	1.0348	1.0293	0.8350	0.8528	0.8809	0.9252	-5.647E-01	2.441E 00
0.6	0.9596	0.8892	1.0321	1.0277	0.9272	0.9333	0.8974	0.9541	8.000E-01	7.160E-01
0.8	0.9631	0.9094	1.0292	1.0257	0.9614	0.9639	0.9242	0.9688	9.046E-01	9.090E-01
1.0	0.9665	0.9243	1.0264	1.0236	0.9772	0.9784	0.9431	0.9773	9.393E-01	9.554E-01
1.2	0.9696	0.9354	1.0239	1.0216	0.9855	0.9861	0.9556	0.9826	9.557E-01	9.739E-01
1.4	0.9723	0.9438	1.0218	1.0197	0.9903	0.9906	0.9640	0.9862	9.651E-01	9.831E-01
1.8	0.9769	0.9555	1.0182	1.0167	0.9951	0.9951	0.9741	0.9905	9.752E-01	9.913E-01
2.2	0.9803	0.9632	1.0155	1.0144	0.9972	0.9972	0.9798	0.9929	9.806E-01	9.947E-01
2.6	0.9830	0.9686	1.0135	1.0126	0.9983	0.9982	0.9833	0.9944	9.839E-01	9.963E-01
3.0	0.9850	0.9726	1.0120	1.0111	0.9989	0.9988	0.9858	0.9954	9.861E-01	9.971E-01
4.0	0.9886	0.9792	1.0093	1.0084	0.9995	0.9995	0.9894	0.9967	9.896E-01	9.978E-01
6.0	0.9924	0.9859	1.0064	1.0058	0.9999	0.9999	0.9928	0.9981	9.929E-01	9.983E-01

Z= 97 A= 249

0.1	0.9528	0.8280	1.0380	1.0309	-0.7739	-0.4953	1.2897	0.7526	1.489E 00	1.459E 00
0.2	0.9535	0.8367	1.0374	1.0307	0.4195	0.4989	1.0220	0.8518	1.517E 00	1.414E 00
0.4	0.9560	0.8626	1.0354	1.0298	0.8329	0.8507	0.8842	0.9243	-1.685E 00	1.956E 00
0.6	0.9594	0.8882	1.0326	1.0282	0.9262	0.9324	0.8977	0.9535	7.845E-01	6.564E-01
0.8	0.9630	0.9084	1.0297	1.0261	0.9609	0.9634	0.9238	0.9684	9.003E-01	9.001E-01
1.0	0.9664	0.9234	1.0269	1.0240	0.9769	0.9781	0.9426	0.9771	9.373E-01	9.522E-01
1.2	0.9695	0.9345	1.0243	1.0219	0.9853	0.9859	0.9551	0.9824	9.545E-01	9.724E-01
1.4	0.9722	0.9430	1.0221	1.0201	0.9902	0.9904	0.9636	0.9860	9.643E-01	9.822E-01
1.8	0.9768	0.9548	1.0185	1.0170	0.9950	0.9951	0.9738	0.9904	9.747E-01	9.909E-01
2.2	0.9803	0.9625	1.0158	1.0146	0.9972	0.9971	0.9795	0.9928	9.802E-01	9.944E-01
2.6	0.9829	0.9680	1.0138	1.0128	0.9982	0.9982	0.9831	0.9943	9.836E-01	9.960E-01
3.0	0.9850	0.9721	1.0122	1.0112	0.9988	0.9988	0.9855	0.9953	9.859E-01	9.969E-01
4.0	0.9886	0.9788	1.0093	1.0086	0.9995	0.9995	0.9892	0.9966	9.894E-01	9.976E-01
6.0	0.9924	0.9856	1.0064	1.0059	0.9999	0.9998	0.9926	0.9982	9.927E-01	9.980E-01

ELECTRON Z= 98 A= 249

p	$F_0L_0^*/F_0L_0$	λ_2^*/λ_2	μ_1^*/μ_1	μ_2^*/μ_2	Λ_1^*/Λ_1	Λ_2^*/Λ_2	ν_{12}^*/ν_{12}	$\hat{\nu}_{12}^*/\hat{\nu}_{12}$	η_{12}^*/η_{12}	$\hat{\eta}_{12}^*/\hat{\eta}_{12}$
0.1	0.9526	0.8280	1.0386	1.0314	-0.7899	-0.5142	1.3026	0.7504	1.463E 00	1.437E 00
0.2	0.9533	0.8365	1.0380	1.0312	0.4131	0.4924	1.0333	0.8503	1.475E 00	1.385E 00
0.4	0.9558	0.8620	1.0360	1.0303	0.8308	0.8487	0.8880	0.9235	-9.511E 00	1.709E 00
0.6	0.9592	0.8873	1.0332	1.0286	0.9253	0.9315	0.8981	0.9530	7.658E-01	5.620E-01
0.8	0.9628	0.9074	1.0301	1.0265	0.9604	0.9629	0.9234	0.9681	8.956E-01	8.890E-01
1.0	0.9663	0.9224	1.0273	1.0243	0.9766	0.9778	0.9422	0.9768	9.351E-01	9.484E-01
1.2	0.9694	0.9336	1.0247	1.0222	0.9851	0.9857	0.9547	0.9822	9.532E-01	9.706E-01
1.4	0.9722	0.9421	1.0224	1.0203	0.9900	0.9903	0.9632	0.9859	9.634E-01	9.811E-01
1.8	0.9767	0.9540	1.0188	1.0172	0.9949	0.9950	0.9734	0.9902	9.742E-01	9.904E-01
2.2	0.9802	0.9619	1.0161	1.0148	0.9971	0.9971	0.9792	0.9927	9.798E-01	9.941E-01
2.6	0.9829	0.9674	1.0140	1.0129	0.9982	0.9982	0.9828	0.9942	9.832E-01	9.958E-01
3.0	0.9850	0.9716	1.0124	1.0114	0.9988	0.9988	0.9853	0.9952	9.856E-01	9.967E-01
4.0	0.9886	0.9784	1.0095	1.0087	0.9995	0.9995	0.9890	0.9967	9.892E-01	9.975E-01
6.0	0.9924	0.9853	1.0066	1.0060	0.9999	0.9998	0.9925	0.9981	9.925E-01	9.979E-01

Z= 99 A= 253

p	$F_0L_0^*/F_0L_0$	λ_2^*/λ_2	μ_1^*/μ_1	μ_2^*/μ_2	Λ_1^*/Λ_1	Λ_2^*/Λ_2	ν_{12}^*/ν_{12}	$\hat{\nu}_{12}^*/\hat{\nu}_{12}$	η_{12}^*/η_{12}	$\hat{\eta}_{12}^*/\hat{\eta}_{12}$
0.1	0.9524	0.8279	1.0392	1.0319	-0.8052	-0.5335	1.3159	0.7481	1.438E 00	1.415E 00
0.2	0.9531	0.8362	1.0386	1.0317	0.4067	0.4857	1.0450	0.8487	1.437E 00	1.358E 00
0.4	0.9556	0.8613	1.0365	1.0307	0.8288	0.8467	0.8921	0.9227	6.180E 00	1.559E 00
0.6	0.9591	0.8864	1.0337	1.0290	0.9243	0.9305	0.8987	0.9525	7.429E-01	3.893E-01
0.8	0.9627	0.9064	1.0306	1.0269	0.9599	0.9624	0.9231	0.9677	8.904E-01	8.747E-01
1.0	0.9662	0.9215	1.0277	1.0247	0.9763	0.9775	0.9417	0.9765	9.328E-01	9.440E-01
1.2	0.9693	0.9327	1.0251	1.0225	0.9849	0.9855	0.9542	0.9820	9.519E-01	9.685E-01
1.4	0.9721	0.9413	1.0228	1.0206	0.9899	0.9902	0.9627	0.9857	9.624E-01	9.800E-01
1.8	0.9767	0.9533	1.0191	1.0174	0.9949	0.9949	0.9730	0.9901	9.736E-01	9.898E-01
2.2	0.9802	0.9613	1.0163	1.0150	0.9971	0.9971	0.9789	0.9926	9.794E-01	9.937E-01
2.6	0.9829	0.9669	1.0143	1.0131	0.9982	0.9982	0.9825	0.9941	9.829E-01	9.956E-01
3.0	0.9850	0.9711	1.0125	1.0115	0.9988	0.9988	0.9850	0.9951	9.853E-01	9.964E-01
4.0	0.9886	0.9780	1.0097	1.0089	0.9995	0.9995	0.9888	0.9966	9.890E-01	9.974E-01
6.0	0.9924	0.9851	1.0068	1.0061	0.9999	0.9998	0.9923	0.9982	9.924E-01	9.978E-01

ELECTRON Z= 100 A= 254

p	$F_0L_0^*/F_0L_0$	λ_2^*/λ_2	μ_1^*/μ_1	μ_2^*/μ_2	Λ_1^*/Λ_1	Λ_2^*/Λ_2	ν_{12}^*/ν_{12}	$\hat{\nu}_{12}^*/\hat{\nu}_{12}$	η_{12}^*/η_{12}	$\hat{\eta}_{12}^*/\hat{\eta}_{12}$
0.1	0.9522	0.8277	1.0399	1.0324	-0.8199	-0.5525	1.3294	0.7457	1.414E 00	1.395E 00
0.2	0.9530	0.8360	1.0393	1.0322	0.4005	0.4791	1.0572	0.8471	1.403E 00	1.334E 00
0.4	0.9555	0.8606	1.0371	1.0312	0.8267	0.8446	0.8966	0.9218	3.021E 00	1.459E 00
0.6	0.9590	0.8855	1.0342	1.0294	0.9234	0.9296	0.8995	0.9519	7.138E-01	-3.160E-02
0.8	0.9626	0.9054	1.0311	1.0273	0.9594	0.9619	0.9229	0.9673	8.844E-01	8.557E-01
1.0	0.9661	0.9205	1.0282	1.0250	0.9760	0.9772	0.9413	0.9762	9.302E-01	9.385E-01
1.2	0.9693	0.9318	1.0255	1.0229	0.9848	0.9853	0.9538	0.9818	9.504E-01	9.661E-01
1.4	0.9721	0.9404	1.0232	1.0209	0.9898	0.9900	0.9623	0.9855	9.615E-01	9.786E-01
1.8	0.9767	0.9526	1.0194	1.0177	0.9948	0.9949	0.9727	0.9900	9.730E-01	9.892E-01
2.2	0.9802	0.9606	1.0166	1.0152	0.9970	0.9970	0.9785	0.9925	9.789E-01	9.933E-01
2.6	0.9829	0.9663	1.0144	1.0133	0.9982	0.9981	0.9822	0.9940	9.826E-01	9.952E-01
3.0	0.9850	0.9706	1.0127	1.0116	0.9988	0.9988	0.9848	0.9950	9.850E-01	9.961E-01
4.0	0.9886	0.9776	1.0098	1.0090	0.9995	0.9995	0.9886	0.9966	9.887E-01	9.971E-01
6.0	0.9924	0.9848	1.0069	1.0062	0.9999	0.9998	0.9921	0.9982	9.922E-01	9.977E-01

Z= 101 A= 254

p	$F_0L_0^*/F_0L_0$	λ_2^*/λ_2	μ_1^*/μ_1	μ_2^*/μ_2	Λ_1^*/Λ_1	Λ_2^*/Λ_2	ν_{12}^*/ν_{12}	$\hat{\nu}_{12}^*/\hat{\nu}_{12}$	η_{12}^*/η_{12}	$\hat{\eta}_{12}^*/\hat{\eta}_{12}$
0.1	0.9521	0.8276	1.0405	1.0329	-0.8342	-0.5717	1.3435	0.7432	1.391E 00	1.375E 00
0.2	0.9528	0.8356	1.0399	1.0327	0.3943	0.4724	1.0701	0.8454	1.373E 00	1.312E 00
0.4	0.9553	0.8599	1.0377	1.0317	0.8246	0.8426	0.9015	0.9209	2.232E 00	1.386E 00
0.6	0.9588	0.8846	1.0348	1.0299	0.9224	0.9287	0.9004	0.9513	6.757E-01	-2.614E 00
0.8	0.9625	0.9044	1.0316	1.0277	0.9589	0.9614	0.9228	0.9669	8.774E-01	8.287E-01
1.0	0.9660	0.9195	1.0286	1.0254	0.9757	0.9769	0.9409	0.9759	9.274E-01	9.317E-01
1.2	0.9692	0.9308	1.0259	1.0232	0.9846	0.9851	0.9533	0.9816	9.488E-01	9.632E-01
1.4	0.9721	0.9395	1.0235	1.0212	0.9896	0.9899	0.9619	0.9853	9.604E-01	9.770E-01
1.8	0.9767	0.9518	1.0197	1.0179	0.9947	0.9948	0.9723	0.9898	9.724E-01	9.885E-01
2.2	0.9802	0.9600	1.0168	1.0154	0.9970	0.9970	0.9782	0.9924	9.785E-01	9.929E-01
2.6	0.9829	0.9657	1.0147	1.0134	0.9982	0.9981	0.9819	0.9939	9.822E-01	9.949E-01
3.0	0.9850	0.9701	1.0129	1.0118	0.9988	0.9988	0.9845	0.9950	9.847E-01	9.958E-01
4.0	0.9886	0.9772	1.0100	1.0091	0.9995	0.9995	0.9884	0.9966	9.885E-01	9.968E-01
6.0	0.9924	0.9845	1.0072	1.0063	0.9998	0.9998	0.9920	0.9981	9.920E-01	9.976E-01

Instr. see p. 25

ELECTRON Z= 102 A= 256

p	$F_0L_0^*/F_0L_0$	λ_2^*/λ_2	μ_1^*/μ_1	μ_2^*/μ_2	Λ_1^*/Λ_1	Λ_2^*/Λ_2	ν_{12}^*/ν_{12}	$\hat{\nu}_{12}^*/\hat{\nu}_{12}$	η_{12}^*/η_{12}	$\hat{\eta}_{12}^*/\hat{\eta}_{12}$
0.1	0.9520	0.8274	1.0412	1.0334	-0.8474	-0.5909	1.3580	0.7406	1.370E 00	1.357E 00
0.2	0.9527	0.8353	1.0405	1.0332	0.3883	0.4657	1.0835	0.8437	1.345E 00	1.292E 00
0.4	0.9552	0.8592	1.0383	1.0321	0.8226	0.8405	0.9068	0.9199	1.873E 00	1.332E 00
0.6	0.9587	0.8836	1.0354	1.0303	0.9215	0.9277	0.9015	0.9507	6.234E-01	3.264E 00
0.8	0.9625	0.9034	1.0321	1.0281	0.9583	0.9609	0.9227	0.9665	8.694E-01	7.877E-01
1.0	0.9660	0.9185	1.0291	1.0257	0.9754	0.9766	0.9405	0.9756	9.243E-01	9.229E-01
1.2	0.9692	0.9299	1.0263	1.0235	0.9844	0.9849	0.9529	0.9813	9.471E-01	9.596E-01
1.4	0.9720	0.9386	1.0239	1.0215	0.9895	0.9898	0.9615	0.9851	9.593E-01	9.751E-01
1.8	0.9767	0.9510	1.0200	1.0182	0.9947	0.9947	0.9720	0.9897	9.717E-01	9.876E-01
2.2	0.9802	0.9593	1.0171	1.0156	0.9970	0.9969	0.9779	0.9922	9.780E-01	9.924E-01
2.6	0.9829	0.9652	1.0148	1.0136	0.9981	0.9981	0.9817	0.9939	9.818E-01	9.945E-01
3.0	0.9850	0.9696	1.0131	1.0119	0.9988	0.9987	0.9843	0.9949	9.844E-01	9.954E-01
4.0	0.9886	0.9768	1.0102	1.0093	0.9995	0.9995	0.9882	0.9965	9.882E-01	9.966E-01
6.0	0.9924	0.9842	1.0073	1.0064	0.9998	0.9998	0.9918	0.9981	9.918E-01	9.976E-01

Instr. see p. 25

POSITRON Z= -2 A= 3

p	$F_0L_0^*/F_0L_0$	λ_2^*/λ_2	μ_1^*/μ_1	μ_2^*/μ_2	Λ_1^*/Λ_1	Λ_2^*/Λ_2	ν_{12}^*/ν_{12}	$\hat{\nu}_{12}^*/\hat{\nu}_{12}$	η_{12}^*/η_{12}	$\hat{\eta}_{12}^*/\hat{\eta}_{12}$
0.1	1.0468E 00	1.0567	0.9997	0.9997	1.0291	1.0261	1.0575	0.9864	1.0294	1.0291
0.2	1.0100E 00	1.0153	0.9997	0.9997	1.0074	1.0072	1.0150	1.0013	1.0077	1.0074
					Z= -3	A= 6				
0.1	1.0825E 00	1.0773	0.9996	0.9997	1.0416	1.0354	1.0811	0.9604	1.0420	1.0416
0.2	1.0162E 00	1.0220	0.9996	0.9996	1.0108	1.0102	1.0218	0.9955	1.0112	1.0108
0.4	1.0039E 00	1.0060	0.9996	0.9996	1.0026	1.0026	1.0056	1.0003	1.0030	1.0026
					Z= -4	A= 8				
0.1	1.1317E 00	1.0949	0.9995	0.9996	1.0544	1.0440	1.1043	0.9364	1.0550	1.0545
0.2	1.0240E 00	1.0291	0.9994	0.9995	1.0145	1.0135	1.0292	0.9895	1.0151	1.0145
0.4	1.0056E 00	1.0080	0.9994	0.9995	1.0035	1.0034	1.0075	0.9988	1.0041	1.0035
					Z= -5	A= 10				
0.1	1.1976E 00	1.1090	0.9994	0.9995	1.0676	1.0527	1.1272	0.9142	1.0683	1.0678
0.2	1.0340E 00	1.0365	0.9993	0.9993	1.0186	1.0169	1.0371	0.9838	1.0193	1.0186
0.4	1.0075E 00	1.0103	0.9993	0.9993	1.0045	1.0044	1.0097	0.9971	1.0052	1.0045
					Z= -6	A= 12				
0.1	1.2828E 00	1.1193	0.9993	0.9994	1.0810	1.0616	1.1494	0.8943	1.0819	1.0813
0.2	1.0463E 00	1.0441	0.9991	0.9992	1.0230	1.0204	1.0456	0.9775	1.0239	1.0230
0.4	1.0098E 00	1.0128	0.9991	0.9991	1.0056	1.0054	1.0120	0.9959	1.0065	1.0056
					Z= -7	A= 14				
0.1	1.3898E 00	1.1257	0.9992	0.9994	1.0944	1.0710	1.1705	0.8769	1.0954	1.0947
0.2	1.0611E 00	1.0515	0.9990	0.9991	1.0276	1.0241	1.0543	0.9713	1.0287	1.0277
0.4	1.0125E 00	1.0154	0.9989	0.9990	1.0068	1.0065	1.0145	0.9946	1.0079	1.0068
					Z= -8	A= 17				
0.1	1.5238E 00	1.1295	0.9991	0.9993	1.1082	1.0814	1.1914	0.8620	1.1094	1.1087
0.2	1.0787E 00	1.0588	0.9988	0.9989	1.0325	1.0279	1.0635	0.9650	1.0338	1.0326
0.4	1.0155E 00	1.0181	0.9987	0.9988	1.0081	1.0077	1.0172	0.9931	1.0094	1.0081

POSITRON Z= -9 A= 19

p	$F_0L_0^*/F_0L_0$	λ_2^*/λ_2	μ_1^*/μ_1	μ_2^*/μ_2	Λ_1^*/Λ_1	Λ_2^*/Λ_2	ν_{12}^*/ν_{12}	$\vartheta_{12}^*/\vartheta_{12}$	η_{12}^*/η_{12}	$\hat{\eta}_{12}^*/\hat{\eta}_{12}$
0.1	1.6893E 00	1.1317	0.9991	0.9992	1.1226	1.0930	1.2126	0.8499	1.1240	1.1232
0.2	1.0993E 00	1.0659	0.9986	0.9988	1.0375	1.0319	1.0729	0.9587	1.0390	1.0377
0.4	1.0189E 00	1.0210	0.9985	0.9986	1.0094	1.0089	1.0201	0.9915	1.0109	1.0095
0.6	1.0085E 00	1.0104	0.9986	0.9986	1.0039	1.0038	1.0092	0.9963	1.0053	1.0039

Z= -10 A= 21

0.1	1.8783E 00	1.1264	0.9991	0.9991	1.1337	1.1028	1.2261	0.8386	1.1352	1.1344
0.2	1.1232E 00	1.0706	0.9985	0.9987	1.0420	1.0350	1.0806	0.9507	1.0437	1.0423
0.4	1.0229E 00	1.0237	0.9983	0.9984	1.0108	1.0100	1.0229	0.9898	1.0125	1.0109
0.6	1.0102E 00	1.0120	0.9984	0.9985	1.0045	1.0043	1.0105	0.9960	1.0061	1.0046

Z= -11 A= 23

0.1	2.1027E 00	1.1219	0.9990	0.9991	1.1463	1.1132	1.2415	0.8301	1.1480	1.1472
0.2	1.1482E 00	1.0753	0.9983	0.9986	1.0466	1.0385	1.0886	0.9447	1.0485	1.0469
0.4	1.0268E 00	1.0263	0.9981	0.9983	1.0121	1.0111	1.0256	0.9879	1.0141	1.0122
0.6	1.0118E 00	1.0134	0.9982	0.9983	1.0051	1.0048	1.0118	0.9952	1.0069	1.0051

Z= -12 A= 24

0.1	2.3684E 00	1.1165	0.9990	0.9990	1.1590	1.1239	1.2558	0.8231	1.1609	1.1601
0.2	1.1759E 00	1.0792	0.9982	0.9985	1.0511	1.0419	1.0964	0.9393	1.0532	1.0516
0.4	1.0310E 00	1.0288	0.9979	0.9981	1.0134	1.0123	1.0283	0.9863	1.0156	1.0136
0.6	1.0134E 00	1.0147	0.9980	0.9981	1.0056	1.0054	1.0131	0.9942	1.0076	1.0057

Z= -13 A= 27

0.1	2.6859E 00	1.1104	0.9990	0.9990	1.1720	1.1348	1.2694	0.8171	1.1741	1.1733
0.2	1.2065E 00	1.0824	0.9981	0.9984	1.0557	1.0454	1.1039	0.9340	1.0580	1.0562
0.4	1.0355E 00	1.0313	0.9977	0.9979	1.0148	1.0134	1.0310	0.9845	1.0171	1.0149
0.6	1.0151E 00	1.0162	0.9978	0.9979	1.0062	1.0059	1.0144	0.9935	1.0084	1.0063

Z= -14 A= 29

0.1	3.0686E 00	1.1041	0.9990	0.9989	1.1855	1.1460	1.2826	0.8116	1.1878	1.1870
0.2	1.2406E 00	1.0850	0.9980	0.9982	1.0602	1.0489	1.1112	0.9290	1.0627	1.0609
0.4	1.0405E 00	1.0338	0.9975	0.9978	1.0162	1.0146	1.0338	0.9828	1.0187	1.0164
0.6	1.0170E 00	1.0176	0.9976	0.9977	1.0068	1.0064	1.0158	0.9927	1.0092	1.0069

 Instr. see p. 25

POSITRON Z= -15 A= 31

p	$F_0L_0^*/F_0L_0$	λ_2^*/λ_2	μ_1^*/μ_1	μ_2^*/μ_2	Λ_1^*/Λ_1	Λ_2^*/Λ_2	ν_{12}^*/ν_{12}	$\hat{\nu}_{12}^*/\hat{\nu}_{12}$	η_{12}^*/η_{12}	$\hat{\eta}_{12}^*/\hat{\eta}_{12}$
0.1	3.5298E 00	1.0977	0.9990	0.9988	1.1995	1.1573	1.2954	0.8066	1.2020	1.2012
0.2	1.2781E 00	1.0870	0.9979	0.9981	1.0648	1.0524	1.1184	0.9244	1.0675	1.0655
0.4	1.0458E 00	1.0362	0.9973	0.9976	1.0176	1.0157	1.0366	0.9810	1.0204	1.0178
0.6	1.0190E 00	1.0191	0.9974	0.9975	1.0074	1.0070	1.0172	0.9920	1.0101	1.0075

Z= -16 A= 33

p	$F_0L_0^*/F_0L_0$	λ_2^*/λ_2	μ_1^*/μ_1	μ_2^*/μ_2	Λ_1^*/Λ_1	Λ_2^*/Λ_2	ν_{12}^*/ν_{12}	$\hat{\nu}_{12}^*/\hat{\nu}_{12}$	η_{12}^*/η_{12}	$\hat{\eta}_{12}^*/\hat{\eta}_{12}$
0.1	4.0889E 00	1.0909	0.9990	0.9988	1.2139	1.1687	1.3074	0.8017	1.2166	1.2159
0.2	1.3194E 00	1.0885	0.9978	0.9980	1.0693	1.0560	1.1253	0.9199	1.0723	1.0702
0.4	1.0516E 00	1.0386	0.9971	0.9974	1.0190	1.0169	1.0394	0.9793	1.0220	1.0193
0.6	1.0211E 00	1.0205	0.9972	0.9974	1.0080	1.0075	1.0186	0.9911	1.0109	1.0082
0.8	1.0124E 00	1.0131	0.9974	0.9975	1.0042	1.0040	1.0108	0.9950	1.0068	1.0042

Z= -17 A= 35

p	$F_0L_0^*/F_0L_0$	λ_2^*/λ_2	μ_1^*/μ_1	μ_2^*/μ_2	Λ_1^*/Λ_1	Λ_2^*/Λ_2	ν_{12}^*/ν_{12}	$\hat{\nu}_{12}^*/\hat{\nu}_{12}$	η_{12}^*/η_{12}	$\hat{\eta}_{12}^*/\hat{\eta}_{12}$
0.1	4.7732E 00	1.0840	0.9990	0.9988	1.2290	1.1802	1.3187	0.7968	1.2318	1.2313
0.2	1.3647E 00	1.0896	0.9977	0.9979	1.0739	1.0597	1.1321	0.9158	1.0771	1.0750
0.4	1.0579E 00	1.0409	0.9969	0.9972	1.0204	1.0181	1.0423	0.9775	1.0237	1.0208
0.6	1.0234E 00	1.0220	0.9970	0.9972	1.0087	1.0081	1.0200	0.9903	1.0118	1.0088
0.8	1.0137E 00	1.0141	0.9972	0.9973	1.0045	1.0043	1.0117	0.9945	1.0074	1.0046

Z= -18 A= 38

p	$F_0L_0^*/F_0L_0$	λ_2^*/λ_2	μ_1^*/μ_1	μ_2^*/μ_2	Λ_1^*/Λ_1	Λ_2^*/Λ_2	ν_{12}^*/ν_{12}	$\hat{\nu}_{12}^*/\hat{\nu}_{12}$	η_{12}^*/η_{12}	$\hat{\eta}_{12}^*/\hat{\eta}_{12}$
0.1	5.5665E 00	1.0743	0.9991	0.9987	1.2420	1.1894	1.3248	0.7915	1.2451	1.2447
0.2	1.4127E 00	1.0899	0.9976	0.9978	1.0784	1.0631	1.1384	0.9118	1.0818	1.0796
0.4	1.0642E 00	1.0431	0.9967	0.9971	1.0219	1.0193	1.0451	0.9759	1.0254	1.0223
0.6	1.0257E 00	1.0234	0.9968	0.9970	1.0093	1.0086	1.0214	0.9895	1.0126	1.0095
0.8	1.0149E 00	1.0151	0.9970	0.9971	1.0048	1.0046	1.0125	0.9940	1.0079	1.0049

Z= -19 A= 41

p	$F_0L_0^*/F_0L_0$	λ_2^*/λ_2	μ_1^*/μ_1	μ_2^*/μ_2	Λ_1^*/Λ_1	Λ_2^*/Λ_2	ν_{12}^*/ν_{12}	$\hat{\nu}_{12}^*/\hat{\nu}_{12}$	η_{12}^*/η_{12}	$\hat{\eta}_{12}^*/\hat{\eta}_{12}$
0.1	6.5263E 00	1.0649	0.9991	0.9987	1.2556	1.1987	1.3303	0.7864	1.2588	1.2586
0.2	1.4645E 00	1.0897	0.9975	0.9977	1.0828	1.0666	1.1443	0.9082	1.0864	1.0842
0.4	1.0709E 00	1.0452	0.9965	0.9969	1.0233	1.0204	1.0479	0.9743	1.0271	1.0238
0.6	1.0280E 00	1.0248	0.9966	0.9968	1.0099	1.0092	1.0228	0.9888	1.0135	1.0101
0.8	1.0161E 00	1.0161	0.9968	0.9969	1.0051	1.0049	1.0134	0.9936	1.0084	1.0053

Z= -20 A= 44

p	$F_0L_0^*/F_0L_0$	λ_2^*/λ_2	μ_1^*/μ_1	μ_2^*/μ_2	Λ_1^*/Λ_1	Λ_2^*/Λ_2	ν_{12}^*/ν_{12}	$\hat{\nu}_{12}^*/\hat{\nu}_{12}$	η_{12}^*/η_{12}	$\hat{\eta}_{12}^*/\hat{\eta}_{12}$
0.1	7.5289E 00	1.0614	0.9991	0.9987	1.2699	1.2094	1.3390	0.7890	1.2732	1.2731
0.2	1.5028E 00	1.0906	0.9974	0.9976	1.0863	1.0703	1.1496	0.9117	1.0901	1.0878
0.4	1.0741E 00	1.0463	0.9964	0.9968	1.0240	1.0213	1.0494	0.9761	1.0279	1.0245
0.6	1.0288E 00	1.0254	0.9965	0.9967	1.0101	1.0095	1.0234	0.9897	1.0138	1.0103
0.8	1.0164E 00	1.0163	0.9967	0.9968	1.0052	1.0050	1.0136	0.9940	1.0086	1.0053

POSITRON Z= -21 A= 45

p	$F_0L_0^*/F_0L_0$	λ_2^*/λ_2	μ_1^*/μ_1	μ_2^*/μ_2	Λ_1^*/Λ_1	Λ_2^*/Λ_2	ν_{12}^*/ν_{12}	$\hat{\nu}_{12}^*/\hat{\nu}_{12}$	η_{12}^*/η_{12}	$\hat{\eta}_{12}^*/\hat{\eta}_{12}$
0.1	9.1877E 00	1.0481	0.9993	0.9987	1.2848	1.2187	1.3413	0.7775	1.2883	1.2886
0.2	1.5831E 00	1.0889	0.9974	0.9975	1.0918	1.0742	1.1560	0.9022	1.0958	1.0935
0.4	1.0861E 00	1.0492	0.9962	0.9966	1.0262	1.0228	1.0534	0.9711	1.0305	1.0269
0.6	1.0333E 00	1.0277	0.9961	0.9964	1.0112	1.0103	1.0258	0.9874	1.0153	1.0115
0.8	1.0189E 00	1.0181	0.9964	0.9966	1.0058	1.0055	1.0151	0.9928	1.0096	1.0060

Z= -22 A= 48

p	$F_0L_0^*/F_0L_0$	λ_2^*/λ_2	μ_1^*/μ_1	μ_2^*/μ_2	Λ_1^*/Λ_1	Λ_2^*/Λ_2	ν_{12}^*/ν_{12}	$\hat{\nu}_{12}^*/\hat{\nu}_{12}$	η_{12}^*/η_{12}	$\hat{\eta}_{12}^*/\hat{\eta}_{12}$
0.1	1.1033E 01	1.0395	0.9994	0.9988	1.2997	1.2288	1.3452	0.7728	1.3034	1.3040
0.2	1.6517E 00	1.0883	0.9973	0.9974	1.0965	1.0783	1.1620	0.8994	1.1008	1.0985
0.4	1.0946E 00	1.0511	0.9960	0.9964	1.0278	1.0240	1.0563	0.9693	1.0323	1.0285
0.6	1.0363E 00	1.0292	0.9959	0.9962	1.0119	1.0109	1.0273	0.9865	1.0162	1.0123
0.8	1.0204E 00	1.0191	0.9962	0.9964	1.0062	1.0058	1.0161	0.9923	1.0102	1.0064
1.0	1.0139E 00	1.0139	0.9965	0.9966	1.0036	1.0035	1.0107	0.9949	1.0072	1.0038

Z= -23 A= 51

p	$F_0L_0^*/F_0L_0$	λ_2^*/λ_2	μ_1^*/μ_1	μ_2^*/μ_2	Λ_1^*/Λ_1	Λ_2^*/Λ_2	ν_{12}^*/ν_{12}	$\hat{\nu}_{12}^*/\hat{\nu}_{12}$	η_{12}^*/η_{12}	$\hat{\eta}_{12}^*/\hat{\eta}_{12}$
0.1	1.3317E 01	1.0311	0.9995	0.9988	1.3147	1.2390	1.3480	0.7685	1.3185	1.3194
0.2	1.7258E 00	1.0877	0.9972	0.9973	1.1014	1.0825	1.1680	0.8970	1.1059	1.1035
0.4	1.1036E 00	1.0529	0.9958	0.9963	1.0293	1.0252	1.0591	0.9677	1.0341	1.0301
0.6	1.0394E 00	1.0306	0.9957	0.9960	1.0126	1.0115	1.0288	0.9856	1.0172	1.0130
0.8	1.0220E 00	1.0202	0.9959	0.9962	1.0066	1.0062	1.0170	0.9919	1.0108	1.0068
1.0	1.0149E 00	1.0147	0.9963	0.9964	1.0038	1.0037	1.0114	0.9946	1.0077	1.0040

Z= -24 A= 52

p	$F_0L_0^*/F_0L_0$	λ_2^*/λ_2	μ_1^*/μ_1	μ_2^*/μ_2	Λ_1^*/Λ_1	Λ_2^*/Λ_2	ν_{12}^*/ν_{12}	$\hat{\nu}_{12}^*/\hat{\nu}_{12}$	η_{12}^*/η_{12}	$\hat{\eta}_{12}^*/\hat{\eta}_{12}$
0.1	1.6188E 01	1.0227	0.9997	0.9988	1.3296	1.2494	1.3498	0.7642	1.3336	1.3349
0.2	1.8074E 00	1.0871	0.9972	0.9972	1.1064	1.0869	1.1741	0.8946	1.1112	1.1088
0.4	1.1133E 00	1.0547	0.9956	0.9961	1.0309	1.0265	1.0620	0.9660	1.0359	1.0318
0.6	1.0427E 00	1.0321	0.9955	0.9958	1.0134	1.0121	1.0304	0.9848	1.0182	1.0138
0.8	1.0237E 00	1.0212	0.9957	0.9959	1.0070	1.0065	1.0180	0.9914	1.0114	1.0072
1.0	1.0160E 00	1.0155	0.9961	0.9962	1.0041	1.0039	1.0120	0.9943	1.0081	1.0043

Z= -25 A= 55

p	$F_0L_0^*/F_0L_0$	λ_2^*/λ_2	μ_1^*/μ_1	μ_2^*/μ_2	Λ_1^*/Λ_1	Λ_2^*/Λ_2	ν_{12}^*/ν_{12}	$\hat{\nu}_{12}^*/\hat{\nu}_{12}$	η_{12}^*/η_{12}	$\hat{\eta}_{12}^*/\hat{\eta}_{12}$
0.1	1.9768E 01	1.0145	0.9999	0.9988	1.3443	1.2598	1.3505	0.7603	1.3485	1.3502
0.2	1.8958E 00	1.0864	0.9971	0.9971	1.1115	1.0914	1.1801	0.8924	1.1166	1.1142
0.4	1.1234E 00	1.0564	0.9954	0.9959	1.0325	1.0278	1.0649	0.9644	1.0378	1.0335
0.6	1.0461E 00	1.0335	0.9952	0.9956	1.0141	1.0127	1.0320	0.9840	1.0192	1.0146
0.8	1.0255E 00	1.0223	0.9955	0.9957	1.0073	1.0069	1.0190	0.9909	1.0121	1.0077
1.0	1.0171E 00	1.0163	0.9958	0.9960	1.0043	1.0041	1.0127	0.9940	1.0086	1.0045

 Instr. see p. 25

POSITRON Z= -26 A= 56

p	$F_0L_0^*/F_0L_0$	λ_2^*/λ_2	μ_1^*/μ_1	μ_2^*/μ_2	Λ_1^*/Λ_1	Λ_2^*/Λ_2	ν_{12}^*/ν_{12}	$\hat{\nu}_{12}^*/\hat{\nu}_{12}$	η_{12}^*/η_{12}	$\hat{\eta}_{12}^*/\hat{\eta}_{12}$
0.1	2.4274E 01	1.0066	1.0000	0.9988	1.3590	1.2705	1.3505	0.7567	1.3634	1.3655
0.2	1.9923E 00	1.0857	0.9970	0.9969	1.1168	1.0960	1.1862	0.8904	1.1222	1.1197
0.4	1.1341E 00	1.0579	0.9952	0.9957	1.0341	1.0291	1.0677	0.9628	1.0397	1.0352
0.6	1.0498E 00	1.0349	0.9950	0.9954	1.0148	1.0134	1.0336	0.9832	1.0202	1.0154
0.8	1.0273E 00	1.0234	0.9952	0.9955	1.0077	1.0072	1.0200	0.9904	1.0127	1.0081
1.0	1.0183E 00	1.0172	0.9956	0.9958	1.0045	1.0043	1.0134	0.9937	1.0091	1.0048

Z= -27 A= 57

p	$F_0L_0^*/F_0L_0$	λ_2^*/λ_2	μ_1^*/μ_1	μ_2^*/μ_2	Λ_1^*/Λ_1	Λ_2^*/Λ_2	ν_{12}^*/ν_{12}	$\hat{\nu}_{12}^*/\hat{\nu}_{12}$	η_{12}^*/η_{12}	$\hat{\eta}_{12}^*/\hat{\eta}_{12}$
0.1	2.9965E 01	0.9992	1.0002	0.9988	1.3738	1.2816	1.3501	0.7534	1.3783	1.3808
0.2	2.0978E 00	1.0849	0.9970	0.9968	1.1223	1.1007	1.1923	0.8885	1.1279	1.1254
0.4	1.1453E 00	1.0595	0.9950	0.9956	1.0357	1.0304	1.0706	0.9613	1.0416	1.0370
0.6	1.0535E 00	1.0364	0.9948	0.9952	1.0156	1.0140	1.0352	0.9823	1.0213	1.0162
0.8	1.0292E 00	1.0245	0.9950	0.9953	1.0081	1.0076	1.0210	0.9899	1.0134	1.0086
1.0	1.0195E 00	1.0180	0.9954	0.9956	1.0048	1.0045	1.0141	0.9934	1.0096	1.0050

Z= -28 A= 60

p	$F_0L_0^*/F_0L_0$	λ_2^*/λ_2	μ_1^*/μ_1	μ_2^*/μ_2	Λ_1^*/Λ_1	Λ_2^*/Λ_2	ν_{12}^*/ν_{12}	$\hat{\nu}_{12}^*/\hat{\nu}_{12}$	η_{12}^*/η_{12}	$\hat{\eta}_{12}^*/\hat{\eta}_{12}$
0.1	3.7195E 01	0.9923	1.0005	0.9989	1.3886	1.2932	1.3495	0.7505	1.3933	1.3964
0.2	2.2133E 00	1.0841	0.9969	0.9967	1.1279	1.1055	1.1983	0.8866	1.1337	1.1312
0.4	1.1572E 00	1.0609	0.9949	0.9954	1.0373	1.0317	1.0734	0.9598	1.0435	1.0387
0.6	1.0575E 00	1.0378	0.9945	0.9950	1.0163	1.0146	1.0369	0.9815	1.0223	1.0171
0.8	1.0312E 00	1.0256	0.9948	0.9951	1.0086	1.0079	1.0220	0.9894	1.0141	1.0090
1.0	1.0207E 00	1.0189	0.9952	0.9954	1.0050	1.0047	1.0148	0.9931	1.0101	1.0053
1.2	1.0154E 00	1.0148	0.9956	0.9957	1.0032	1.0030	1.0108	0.9950	1.0078	1.0034

Z= -29 A= 65

p	$F_0L_0^*/F_0L_0$	λ_2^*/λ_2	μ_1^*/μ_1	μ_2^*/μ_2	Λ_1^*/Λ_1	Λ_2^*/Λ_2	ν_{12}^*/ν_{12}	$\hat{\nu}_{12}^*/\hat{\nu}_{12}$	η_{12}^*/η_{12}	$\hat{\eta}_{12}^*/\hat{\eta}_{12}$
0.1	4.6441E 01	0.9859	1.0007	0.9989	1.4038	1.3053	1.3488	0.7478	1.4086	1.4122
0.2	2.3401E 00	1.0832	0.9969	0.9966	1.1336	1.1104	1.2043	0.8848	1.1398	1.1373
0.4	1.1698E 00	1.0623	0.9947	0.9952	1.0389	1.0331	1.0763	0.9584	1.0454	1.0405
0.6	1.0617E 00	1.0392	0.9943	0.9948	1.0171	1.0153	1.0385	0.9806	1.0234	1.0179
0.8	1.0333E 00	1.0267	0.9945	0.9949	1.0090	1.0083	1.0231	0.9889	1.0148	1.0095
1.0	1.0220E 00	1.0198	0.9949	0.9952	1.0053	1.0050	1.0155	0.9927	1.0106	1.0056
1.2	1.0163E 00	1.0155	0.9954	0.9955	1.0033	1.0032	1.0113	0.9948	1.0082	1.0036

Z= -30 A= 67

p	$F_0L_0^*/F_0L_0$	λ_2^*/λ_2	μ_1^*/μ_1	μ_2^*/μ_2	Λ_1^*/Λ_1	Λ_2^*/Λ_2	ν_{12}^*/ν_{12}	$\hat{\nu}_{12}^*/\hat{\nu}_{12}$	η_{12}^*/η_{12}	$\hat{\eta}_{12}^*/\hat{\eta}_{12}$
0.1	5.7492E 01	0.9786	1.0009	0.9989	1.4170	1.3156	1.3452	0.7450	1.4220	1.4261
0.2	2.4724E 00	1.0816	0.9968	0.9965	1.1390	1.1148	1.2092	0.8829	1.1454	1.1429
0.4	1.1824E 00	1.0635	0.9945	0.9950	1.0405	1.0344	1.0790	0.9571	1.0473	1.0423
0.6	1.0658E 00	1.0405	0.9941	0.9946	1.0179	1.0159	1.0401	0.9799	1.0244	1.0188
0.8	1.0353E 00	1.0277	0.9943	0.9947	1.0094	1.0086	1.0241	0.9885	1.0155	1.0099
1.0	1.0232E 00	1.0206	0.9947	0.9950	1.0055	1.0052	1.0162	0.9924	1.0111	1.0059
1.2	1.0172E 00	1.0162	0.9952	0.9953	1.0035	1.0033	1.0118	0.9945	1.0085	1.0038

POSITRON Z= -31 A= 69

p	$F_0L_0^*/F_0L_0$	λ_2^*/λ_2	μ_1^*/μ_1	μ_2^*/μ_2	Λ_1^*/Λ_1	Λ_2^*/Λ_2	ν_{12}^*/ν_{12}	$\hat{\nu}_{12}^*/\hat{\nu}_{12}$	η_{12}^*/η_{12}	$\hat{\eta}_{12}^*/\hat{\eta}_{12}$
0.1	7.1304E 01	0.9717	1.0012	0.9990	1.4301	1.3260	1.3413	0.7425	1.4351	1.4399
0.2	2.6150E 00	1.0798	0.9968	0.9964	1.1443	1.1191	1.2139	0.8811	1.1510	1.1486
0.4	1.1954E 00	1.0645	0.9944	0.9949	1.0421	1.0357	1.0815	0.9558	1.0491	1.0440
0.6	1.0701E 00	1.0417	0.9939	0.9944	1.0186	1.0165	1.0416	0.9791	1.0254	1.0196
0.8	1.0374E 00	1.0287	0.9941	0.9945	1.0098	1.0090	1.0251	0.9880	1.0161	1.0104
1.0	1.0245E 00	1.0214	0.9945	0.9948	1.0057	1.0054	1.0169	0.9921	1.0115	1.0061
1.2	1.0181E 00	1.0169	0.9949	0.9951	1.0036	1.0035	1.0123	0.9944	1.0089	1.0039

Z= -32 A= 72

p	$F_0L_0^*/F_0L_0$	λ_2^*/λ_2	μ_1^*/μ_1	μ_2^*/μ_2	Λ_1^*/Λ_1	Λ_2^*/Λ_2	ν_{12}^*/ν_{12}	$\hat{\nu}_{12}^*/\hat{\nu}_{12}$	η_{12}^*/η_{12}	$\hat{\eta}_{12}^*/\hat{\eta}_{12}$
0.1	8.8866E 01	0.9650	1.0015	0.9990	1.4432	1.3365	1.3370	0.7400	1.4483	1.4538
0.2	2.7718E 00	1.0780	0.9968	0.9963	1.1498	1.1236	1.2184	0.8791	1.1568	1.1544
0.4	1.2093E 00	1.0655	0.9942	0.9947	1.0437	1.0370	1.0841	0.9544	1.0510	1.0457
0.6	1.0746E 00	1.0429	0.9937	0.9942	1.0194	1.0171	1.0432	0.9783	1.0265	1.0204
0.8	1.0396E 00	1.0297	0.9938	0.9943	1.0102	1.0093	1.0261	0.9875	1.0168	1.0109
1.0	1.0259E 00	1.0222	0.9943	0.9946	1.0060	1.0056	1.0176	0.9918	1.0120	1.0064
1.2	1.0190E 00	1.0175	0.9947	0.9950	1.0038	1.0036	1.0129	0.9941	1.0093	1.0041

Z= -33 A= 73

p	$F_0L_0^*/F_0L_0$	λ_2^*/λ_2	μ_1^*/μ_1	μ_2^*/μ_2	Λ_1^*/Λ_1	Λ_2^*/Λ_2	ν_{12}^*/ν_{12}	$\hat{\nu}_{12}^*/\hat{\nu}_{12}$	η_{12}^*/η_{12}	$\hat{\eta}_{12}^*/\hat{\eta}_{12}$
0.1	1.1104E 02	0.9588	1.0019	0.9991	1.4563	1.3472	1.3328	0.7377	1.4616	1.4678
0.2	2.9420E 00	1.0761	0.9968	0.9962	1.1554	1.1280	1.2228	0.8772	1.1626	1.1604
0.4	1.2237E 00	1.0663	0.9941	0.9945	1.0452	1.0383	1.0867	0.9532	1.0529	1.0474
0.6	1.0793E 00	1.0441	0.9934	0.9940	1.0201	1.0177	1.0447	0.9775	1.0275	1.0213
0.8	1.0419E 00	1.0308	0.9936	0.9941	1.0106	1.0097	1.0271	0.9870	1.0175	1.0113
1.0	1.0273E 00	1.0230	0.9941	0.9944	1.0062	1.0058	1.0183	0.9914	1.0125	1.0067
1.2	1.0200E 00	1.0182	0.9945	0.9948	1.0039	1.0037	1.0134	0.9939	1.0097	1.0043
1.4	1.0158E 00	1.0150	0.9950	0.9952	1.0026	1.0025	1.0104	0.9953	1.0079	1.0029

Z= -34 A= 80

p	$F_0L_0^*/F_0L_0$	λ_2^*/λ_2	μ_1^*/μ_1	μ_2^*/μ_2	Λ_1^*/Λ_1	Λ_2^*/Λ_2	ν_{12}^*/ν_{12}	$\hat{\nu}_{12}^*/\hat{\nu}_{12}$	η_{12}^*/η_{12}	$\hat{\eta}_{12}^*/\hat{\eta}_{12}$
0.1	1.3903E 02	0.9531	1.0022	0.9991	1.4695	1.3581	1.3287	0.7358	1.4748	1.4818
0.2	3.1258E 00	1.0742	0.9968	0.9962	1.1610	1.1324	1.2269	0.8755	1.1685	1.1663
0.4	1.2385E 00	1.0671	0.9939	0.9944	1.0468	1.0396	1.0891	0.9520	1.0547	1.0492
0.6	1.0840E 00	1.0452	0.9932	0.9938	1.0208	1.0184	1.0462	0.9767	1.0285	1.0221
0.8	1.0442E 00	1.0317	0.9934	0.9939	1.0110	1.0100	1.0281	0.9865	1.0182	1.0118
1.0	1.0287E 00	1.0238	0.9938	0.9942	1.0065	1.0060	1.0190	0.9911	1.0130	1.0070
1.2	1.0210E 00	1.0189	0.9943	0.9946	1.0041	1.0039	1.0139	0.9936	1.0101	1.0045
1.4	1.0165E 00	1.0156	0.9948	0.9950	1.0027	1.0026	1.0108	0.9952	1.0082	1.0031

Instr. see p. 25

POSITRON Z= -35 A= 79

p	$F_0L_0^*/F_0L_0$	λ_2^*/λ_2	μ_1^*/μ_1	μ_2^*/μ_2	Λ_1^*/Λ_1	Λ_2^*/Λ_2	ν_{12}^*/ν_{12}	$\hat{\nu}_{12}^*/\hat{\nu}_{12}$	η_{12}^*/η_{12}	$\hat{\eta}_{12}^*/\hat{\eta}_{12}$
0.1	1.7474E 02	0.9477	1.0025	0.9992	1.4828	1.3692	1.3246	0.7339	1.4883	1.4960
0.2	3.3278E 00	1.0720	0.9968	0.9961	1.1666	1.1368	1.2308	0.8735	1.1744	1.1724
0.4	1.2542E 00	1.0678	0.9938	0.9942	1.0484	1.0410	1.0916	0.9508	1.0566	1.0509
0.6	1.0891E 00	1.0463	0.9930	0.9936	1.0216	1.0190	1.0477	0.9758	1.0296	1.0230
0.8	1.0466E 00	1.0327	0.9932	0.9937	1.0114	1.0104	1.0291	0.9859	1.0189	1.0123
1.0	1.0301E 00	1.0246	0.9936	0.9940	1.0067	1.0063	1.0197	0.9908	1.0136	1.0073
1.2	1.0220E 00	1.0196	0.9941	0.9944	1.0043	1.0040	1.0145	0.9933	1.0105	1.0047
1.4	1.0173E 00	1.0162	0.9946	0.9948	1.0029	1.0027	1.0112	0.9950	1.0085	1.0032

Z= -36 A= 82

p	$F_0L_0^*/F_0L_0$	λ_2^*/λ_2	μ_1^*/μ_1	μ_2^*/μ_2	Λ_1^*/Λ_1	Λ_2^*/Λ_2	ν_{12}^*/ν_{12}	$\hat{\nu}_{12}^*/\hat{\nu}_{12}$	η_{12}^*/η_{12}	$\hat{\eta}_{12}^*/\hat{\eta}_{12}$
0.1	2.1653E 02	0.9404	1.0030	0.9993	1.4944	1.3776	1.3185	0.7299	1.4999	1.5087
0.2	3.5710E 00	1.0656	0.9971	0.9962	1.1723	1.1383	1.2311	0.8665	1.1803	1.1788
0.4	1.2701E 00	1.0700	0.9935	0.9939	1.0505	1.0432	1.0958	0.9502	1.0592	1.0532
0.6	1.0942E 00	1.0472	0.9929	0.9934	1.0222	1.0198	1.0491	0.9749	1.0305	1.0237
0.8	1.0493E 00	1.0334	0.9930	0.9935	1.0117	1.0107	1.0298	0.9849	1.0194	1.0127
1.0	1.0319E 00	1.0252	0.9935	0.9938	1.0069	1.0064	1.0203	0.9898	1.0140	1.0076
1.2	1.0233E 00	1.0201	0.9940	0.9942	1.0044	1.0041	1.0149	0.9926	1.0108	1.0049
1.4	1.0183E 00	1.0167	0.9944	0.9947	1.0030	1.0028	1.0116	0.9944	1.0088	1.0034

Z= -37 A= 85

p	$F_0L_0^*/F_0L_0$	λ_2^*/λ_2	μ_1^*/μ_1	μ_2^*/μ_2	Λ_1^*/Λ_1	Λ_2^*/Λ_2	ν_{12}^*/ν_{12}	$\hat{\nu}_{12}^*/\hat{\nu}_{12}$	η_{12}^*/η_{12}	$\hat{\eta}_{12}^*/\hat{\eta}_{12}$
0.1	2.7344E 02	0.9353	1.0034	0.9994	1.5082	1.3880	1.3142	0.7279	1.5137	1.5234
0.2	3.8094E 00	1.0630	0.9972	0.9962	1.1776	1.1425	1.2338	0.8648	1.1858	1.1846
0.4	1.2869E 00	1.0708	0.9933	0.9937	1.0522	1.0447	1.0984	0.9492	1.0612	1.0551
0.6	1.0993E 00	1.0484	0.9927	0.9932	1.0229	1.0204	1.0507	0.9743	1.0315	1.0246
0.8	1.0518E 00	1.0343	0.9928	0.9933	1.0121	1.0111	1.0308	0.9845	1.0201	1.0132
1.0	1.0333E 00	1.0260	0.9933	0.9936	1.0072	1.0066	1.0209	0.9896	1.0145	1.0079
1.2	1.0243E 00	1.0208	0.9938	0.9940	1.0045	1.0043	1.0154	0.9924	1.0112	1.0051
1.4	1.0191E 00	1.0173	0.9943	0.9945	1.0031	1.0029	1.0120	0.9943	1.0092	1.0035

Z= -38 A= 86

p	$F_0L_0^*/F_0L_0$	λ_2^*/λ_2	μ_1^*/μ_1	μ_2^*/μ_2	Λ_1^*/Λ_1	Λ_2^*/Λ_2	ν_{12}^*/ν_{12}	$\hat{\nu}_{12}^*/\hat{\nu}_{12}$	η_{12}^*/η_{12}	$\hat{\eta}_{12}^*/\hat{\eta}_{12}$
0.1	3.4583E 02	0.9305	1.0038	0.9995	1.5218	1.3984	1.3097	0.7261	1.5274	1.5381
0.2	4.0670E 00	1.0604	0.9973	0.9961	1.1829	1.1466	1.2362	0.8631	1.1913	1.1903
0.4	1.3041E 00	1.0715	0.9932	0.9936	1.0539	1.0461	1.1010	0.9483	1.0632	1.0570
0.6	1.1046E 00	1.0494	0.9925	0.9930	1.0237	1.0211	1.0522	0.9737	1.0326	1.0254
0.8	1.0543E 00	1.0353	0.9926	0.9931	1.0125	1.0114	1.0318	0.9841	1.0208	1.0137
1.0	1.0348E 00	1.0268	0.9931	0.9934	1.0074	1.0069	1.0216	0.9893	1.0150	1.0082
1.2	1.0253E 00	1.0214	0.9936	0.9939	1.0047	1.0044	1.0159	0.9922	1.0116	1.0053
1.4	1.0198E 00	1.0178	0.9941	0.9943	1.0032	1.0030	1.0124	0.9941	1.0095	1.0037

POSITRON Z= -39 A= 89

p	$F_0L_0^*/F_0L_0$	λ_2^*/λ_2	μ_1^*/μ_1	μ_2^*/μ_2	Λ_1^*/Λ_1	Λ_2^*/Λ_2	ν_{12}^*/ν_{12}	$\hat{\nu}_{12}^*/\hat{\nu}_{12}$	η_{12}^*/η_{12}	$\hat{\eta}_{12}^*/\hat{\eta}_{12}$
0.1	4.3886E 02	0.9258	1.0042	0.9997	1.5356	1.4088	1.3051	0.7243	1.5412	1.5529
0.2	4.3482E 00	1.0578	0.9974	0.9961	1.1882	1.1509	1.2385	0.8615	1.1969	1.1961
0.4	1.3222E 00	1.0722	0.9931	0.9934	1.0556	1.0475	1.1035	0.9474	1.0652	1.0589
0.6	1.1101E 00	1.0505	0.9922	0.9928	1.0245	1.0218	1.0537	0.9732	1.0337	1.0263
0.8	1.0568E 00	1.0362	0.9924	0.9928	1.0129	1.0118	1.0328	0.9838	1.0215	1.0142
1.0	1.0363E 00	1.0276	0.9928	0.9932	1.0076	1.0071	1.0223	0.9891	1.0155	1.0085
1.2	1.0263E 00	1.0221	0.9934	0.9937	1.0049	1.0046	1.0165	0.9920	1.0120	1.0055
1.4	1.0206E 00	1.0184	0.9939	0.9941	1.0033	1.0031	1.0128	0.9940	1.0098	1.0038
1.8	1.0144E 00	1.0137	0.9948	0.9950	1.0017	1.0016	1.0087	0.9962	1.0072	1.0021

Z= -40 A= 90

p	$F_0L_0^*/F_0L_0$	λ_2^*/λ_2	μ_1^*/μ_1	μ_2^*/μ_2	Λ_1^*/Λ_1	Λ_2^*/Λ_2	ν_{12}^*/ν_{12}	$\hat{\nu}_{12}^*/\hat{\nu}_{12}$	η_{12}^*/η_{12}	$\hat{\eta}_{12}^*/\hat{\eta}_{12}$
0.1	5.5797E 02	0.9212	1.0047	0.9998	1.5492	1.4192	1.3004	0.7226	1.5548	1.5676
0.2	4.6538E 00	1.0552	0.9975	0.9961	1.1936	1.1550	1.2404	0.8599	1.2024	1.2019
0.4	1.3409E 00	1.0728	0.9929	0.9932	1.0573	1.0489	1.1060	0.9465	1.0672	1.0608
0.6	1.1157E 00	1.0515	0.9920	0.9926	1.0252	1.0224	1.0553	0.9727	1.0348	1.0272
0.8	1.0595E 00	1.0371	0.9922	0.9926	1.0134	1.0122	1.0338	0.9834	1.0222	1.0147
1.0	1.0379E 00	1.0283	0.9926	0.9930	1.0079	1.0073	1.0230	0.9888	1.0160	1.0088
1.2	1.0274E 00	1.0227	0.9932	0.9935	1.0050	1.0047	1.0170	0.9919	1.0124	1.0057
1.4	1.0214E 00	1.0189	0.9937	0.9939	1.0034	1.0032	1.0132	0.9938	1.0101	1.0040
1.8	1.0149E 00	1.0142	0.9946	0.9948	1.0017	1.0016	1.0090	0.9961	1.0074	1.0022

Z= -41 A= 91

p	$F_0L_0^*/F_0L_0$	λ_2^*/λ_2	μ_1^*/μ_1	μ_2^*/μ_2	Λ_1^*/Λ_1	Λ_2^*/Λ_2	ν_{12}^*/ν_{12}	$\hat{\nu}_{12}^*/\hat{\nu}_{12}$	η_{12}^*/η_{12}	$\hat{\eta}_{12}^*/\hat{\eta}_{12}$
0.1	7.1153E 02	0.9169	1.0052	0.9999	1.5629	1.4296	1.2956	0.7209	1.5685	1.5825
0.2	4.9875E 00	1.0526	0.9976	0.9960	1.1989	1.1593	1.2422	0.8584	1.2080	1.2078
0.4	1.3605E 00	1.0733	0.9928	0.9931	1.0591	1.0504	1.1085	0.9456	1.0693	1.0628
0.6	1.1216E 00	1.0525	0.9918	0.9923	1.0260	1.0231	1.0568	0.9721	1.0359	1.0281
0.8	1.0622E 00	1.0380	0.9919	0.9924	1.0138	1.0126	1.0348	0.9831	1.0229	1.0152
1.0	1.0395E 00	1.0291	0.9924	0.9928	1.0081	1.0075	1.0237	0.9886	1.0165	1.0091
1.2	1.0284E 00	1.0234	0.9930	0.9933	1.0052	1.0048	1.0175	0.9917	1.0128	1.0059
1.4	1.0222E 00	1.0195	0.9935	0.9938	1.0035	1.0033	1.0137	0.9937	1.0105	1.0041
1.8	1.0154E 00	1.0146	0.9945	0.9947	1.0018	1.0017	1.0093	0.9961	1.0077	1.0022

Z= -42 A= 94

p	$F_0L_0^*/F_0L_0$	λ_2^*/λ_2	μ_1^*/μ_1	μ_2^*/μ_2	Λ_1^*/Λ_1	Λ_2^*/Λ_2	ν_{12}^*/ν_{12}	$\hat{\nu}_{12}^*/\hat{\nu}_{12}$	η_{12}^*/η_{12}	$\hat{\eta}_{12}^*/\hat{\eta}_{12}$
0.1	9.0878E 02	0.9127	1.0057	1.0000	1.5765	1.4399	1.2908	0.7193	1.5821	1.5973
0.2	5.3494E 00	1.0500	0.9977	0.996C	1.2042	1.1635	1.2438	0.8569	1.2136	1.2137
0.4	1.3807E 00	1.0737	0.9926	0.9929	1.0608	1.0518	1.1110	0.9448	1.0714	1.0647
0.6	1.1275E 00	1.0535	0.9916	0.9921	1.0268	1.0238	1.0583	0.9716	1.0370	1.0290
0.8	1.0650E 00	1.0390	0.9917	0.9922	1.0142	1.0129	1.0358	0.9827	1.0236	1.0157
1.0	1.0411E 00	1.0299	0.9922	0.9926	1.0084	1.0078	1.0244	0.9883	1.0170	1.0094
1.2	1.0295E 00	1.0241	0.9928	0.9931	1.0053	1.0050	1.0180	0.9915	1.0132	1.0062
1.4	1.0230E 00	1.0201	0.9933	0.9936	1.0036	1.0034	1.0141	0.9935	1.0108	1.0043
1.8	1.0159E 00	1.0151	0.9943	0.9945	1.0018	1.0017	1.0096	0.9959	1.0079	1.0023

POSITRON Z= -43 A= 97

p	$F_0L_0^*/F_0L_0$	λ_2^*/λ_2	μ_1^*/μ_1	μ_2^*/μ_2	Λ_1^*/Λ_1	Λ_2^*/Λ_2	ν_{12}^*/ν_{12}	$\hat{\nu}_{12}^*/\hat{\nu}_{12}$	η_{12}^*/η_{12}	$\hat{\eta}_{12}^*/\hat{\eta}_{12}$
0.1	1.1638E 03	0.9087	1.0062	1.0002	1.5902	1.4503	1.2860	0.7178	1.5957	1.6122
0.2	5.7447E 00	1.0474	0.9979	0.9960	1.2096	1.1677	1.2452	0.8554	1.2191	1.2195
0.4	1.4017E 00	1.0742	0.9925	0.9928	1.0625	1.0532	1.1134	0.9439	1.0734	1.0667
0.6	1.1337E 00	1.0545	0.9914	0.9919	1.0276	1.0245	1.0599	0.9711	1.0381	1.0300
0.8	1.0678E 00	1.0399	0.9915	0.9920	1.0146	1.0133	1.0368	0.9824	1.0243	1.0162
1.0	1.0428E 00	1.0307	0.9920	0.9924	1.0086	1.0080	1.0252	0.9881	1.0175	1.0098
1.2	1.0307E 00	1.0247	0.9926	0.9929	1.0055	1.0051	1.0186	0.9914	1.0136	1.0064
1.4	1.0238E 00	1.0206	0.9931	0.9934	1.0037	1.0035	1.0145	0.9934	1.0111	1.0044
1.8	1.0165E 00	1.0155	0.9942	0.9943	1.0019	1.0018	1.0099	0.9958	1.0082	1.0024
					Z= -44	A= 98				
0.1	1.4941E 03	0.9048	1.0068	1.0003	1.6038	1.4607	1.2811	0.7163	1.6093	1.6272
0.2	6.1764E 00	1.0449	0.9981	0.9959	1.2149	1.1719	1.2464	0.8540	1.2247	1.2255
0.4	1.4235E 00	1.0745	0.9924	0.9926	1.0643	1.0547	1.1157	0.9431	1.0755	1.0687
0.6	1.1401E 00	1.0554	0.9912	0.9917	1.0284	1.0252	1.0614	0.9706	1.0392	1.0309
0.8	1.0707E 00	1.0408	0.9913	0.9918	1.0150	1.0137	1.0378	0.9821	1.0251	1.0167
1.0	1.0445E 00	1.0314	0.9918	0.9922	1.0089	1.0082	1.0259	0.9879	1.0181	1.0101
1.2	1.0318E 00	1.0254	0.9924	0.9927	1.0057	1.0053	1.0191	0.9912	1.0140	1.0066
1.4	1.0246E 00	1.0212	0.9929	0.9932	1.0038	1.0036	1.0149	0.9933	1.0115	1.0046
1.8	1.0170E 00	1.0160	0.9940	0.9942	1.0019	1.0018	1.0102	0.9958	1.0084	1.0025
					Z= -45	A= 103				
0.1	1.9196E 03	0.9011	1.0073	1.0004	1.6173	1.4710	1.2762	0.7150	1.6228	1.6421
0.2	6.6444E 00	1.0423	0.9982	0.9959	1.2202	1.1761	1.2474	0.8527	1.2302	1.2314
0.4	1.4460E 00	1.0748	0.9923	0.9924	1.0660	1.0561	1.1180	0.9423	1.0776	1.0707
0.6	1.1465E 00	1.0563	0.9910	0.9915	1.0292	1.0258	1.0629	0.9702	1.0403	1.0318
0.8	1.0737E 00	1.0416	0.9911	0.9916	1.0155	1.0141	1.0388	0.9818	1.0258	1.0172
1.0	1.0462E 00	1.0322	0.9916	0.9920	1.0091	1.0084	1.0266	0.9877	1.0186	1.0104
1.2	1.0329E 00	1.0260	0.9922	0.9925	1.0058	1.0054	1.0196	0.9910	1.0145	1.0068
1.4	1.0255E 00	1.0218	0.9928	0.9930	1.0039	1.0037	1.0153	0.9931	1.0118	1.0047
1.8	1.0175E 00	1.0164	0.9938	0.9940	1.0020	1.0019	1.0105	0.9957	1.0087	1.0026
					Z= -46	A= 104				
0.1	2.4757E 03	0.8975	1.0079	1.0006	1.6310	1.4813	1.2713	0.7136	1.6364	1.6572
0.2	7.1600E 00	1.0398	0.9984	0.9959	1.2256	1.1803	1.2483	0.8513	1.2359	1.2374
0.4	1.4695E 00	1.0751	0.9922	0.9923	1.0678	1.0576	1.1203	0.9415	1.0797	1.0727
0.6	1.1533E 00	1.0572	0.9908	0.9913	1.0300	1.0265	1.0644	0.9697	1.0415	1.0328
0.8	1.0768E 00	1.0425	0.9909	0.9914	1.0159	1.0144	1.0398	0.9815	1.0266	1.0178
1.0	1.0480E 00	1.0330	0.9913	0.9918	1.0094	1.0087	1.0273	0.9875	1.0192	1.0108
1.2	1.0341E 00	1.0267	0.9920	0.9923	1.0060	1.0056	1.0202	0.9909	1.0149	1.0070
1.4	1.0263E 00	1.0224	0.9926	0.9928	1.0040	1.0038	1.0158	0.9930	1.0122	1.0049
1.8	1.0181E 00	1.0169	0.9937	0.9939	1.0021	1.0020	1.0108	0.9956	1.0089	1.0027

POSITRON Z= -47 A= 107

p	$F_0L_0^*/F_0L_0$	λ_2^*/λ_2	μ_1^*/μ_1	μ_2^*/μ_2	Λ_1^*/Λ_1	Λ_2^*/Λ_2	ν_{12}^*/ν_{12}	$\hat{\vartheta}_{12}^*/\hat{\vartheta}_{12}$	η_{12}^*/η_{12}	$\hat{\eta}_{12}^*/\hat{\eta}_{12}$
0.1	3.1959E 03	0.8941	1.0086	1.0008	1.6446	1.4916	1.2664	0.7123	1.6498	1.6723
0.2	7.7199E 00	1.0373	0.9986	0.9959	1.2310	1.1846	1.2489	0.8501	1.2414	1.2434
0.4	1.4938E 00	1.0753	0.9921	0.9921	1.0696	1.0590	1.1225	0.9407	1.0818	1.0748
0.6	1.1602E 00	1.0581	0.9906	0.9911	1.0308	1.0272	1.0659	0.9692	1.0426	1.0338
0.8	1.0799E 00	1.0434	0.9906	0.9911	1.0163	1.0148	1.0408	0.9812	1.0273	1.0183
1.0	1.0498E 00	1.0337	0.9911	0.9916	1.0097	1.0089	1.0280	0.9872	1.0197	1.0111
1.2	1.0353E 00	1.0274	0.9917	0.9921	1.0061	1.0057	1.0207	0.9907	1.0153	1.0073
1.4	1.0272E 00	1.0229	0.9924	0.9927	1.0041	1.0039	1.0162	0.9929	1.0125	1.0050
1.8	1.0186E 00	1.0173	0.9935	0.9937	1.0021	1.0020	1.0111	0.9955	1.0092	1.0028

Z= -48 A= 109

p	$F_0L_0^*/F_0L_0$	λ_2^*/λ_2	μ_1^*/μ_1	μ_2^*/μ_2	Λ_1^*/Λ_1	Λ_2^*/Λ_2	ν_{12}^*/ν_{12}	$\hat{\vartheta}_{12}^*/\hat{\vartheta}_{12}$	η_{12}^*/η_{12}	$\hat{\eta}_{12}^*/\hat{\eta}_{12}$
0.1	4.1323E 03	0.8908	1.0092	1.0009	1.6581	1.5019	1.2614	0.7111	1.6632	1.6873
0.2	8.3315E 00	1.0349	0.9989	0.9959	1.2363	1.1888	1.2494	0.8488	1.2470	1.2494
0.4	1.5190E 00	1.0754	0.9920	0.9920	1.0713	1.0605	1.1247	0.9399	1.0839	1.0768
0.6	1.1673E 00	1.0589	0.9904	0.9909	1.0316	1.0279	1.0674	0.9688	1.0437	1.0347
0.8	1.0831E 00	1.0442	0.9904	0.9909	1.0168	1.0152	1.0418	0.9809	1.0280	1.0188
1.0	1.0516E 00	1.0345	0.9909	0.9913	1.0099	1.0091	1.0287	0.9870	1.0202	1.0114
1.2	1.0365E 00	1.0280	0.9915	0.9919	1.0063	1.0059	1.0213	0.9906	1.0157	1.0075
1.4	1.0281E 00	1.0235	0.9922	0.9925	1.0042	1.0040	1.0166	0.9928	1.0129	1.0052
1.8	1.0192E 00	1.0178	0.9933	0.9935	1.0022	1.0021	1.0114	0.9954	1.0094	1.0029

Z= -49 A= 113

p	$F_0L_0^*/F_0L_0$	λ_2^*/λ_2	μ_1^*/μ_1	μ_2^*/μ_2	Λ_1^*/Λ_1	Λ_2^*/Λ_2	ν_{12}^*/ν_{12}	$\hat{\vartheta}_{12}^*/\hat{\vartheta}_{12}$	η_{12}^*/η_{12}	$\hat{\eta}_{12}^*/\hat{\eta}_{12}$
0.1	5.3567E 03	0.8876	1.0099	1.0011	1.6717	1.5121	1.2565	0.7098	1.6767	1.7025
0.2	9.0021E 00	1.0324	0.9991	0.9959	1.2417	1.1930	1.2497	0.8476	1.2525	1.2555
0.4	1.5451E 00	1.0755	0.9919	0.9918	1.0731	1.0620	1.1268	0.9392	1.0860	1.0789
0.6	1.1746E 00	1.0597	0.9902	0.9907	1.0324	1.0286	1.0689	0.9683	1.0449	1.0357
0.8	1.0864E 00	1.0451	0.9902	0.9907	1.0172	1.0156	1.0428	0.9806	1.0288	1.0194
1.0	1.0535E 00	1.0353	0.9907	0.9911	1.0102	1.0094	1.0295	0.9868	1.0208	1.0118
1.2	1.0377E 00	1.0287	0.9913	0.9917	1.0065	1.0060	1.0218	0.9904	1.0162	1.0077
1.4	1.0290E 00	1.0241	0.9920	0.9923	1.0044	1.0041	1.0171	0.9927	1.0132	1.0054
1.8	1.0197E 00	1.0182	0.9932	0.9934	1.0022	1.0021	1.0117	0.9953	1.0097	1.0030

Z= -50 A= 116

p	$F_0L_0^*/F_0L_0$	λ_2^*/λ_2	μ_1^*/μ_1	μ_2^*/μ_2	Λ_1^*/Λ_1	Λ_2^*/Λ_2	ν_{12}^*/ν_{12}	$\hat{\vartheta}_{12}^*/\hat{\vartheta}_{12}$	η_{12}^*/η_{12}	$\hat{\eta}_{12}^*/\hat{\eta}_{12}$
0.1	6.9522E 03	0.8846	1.0106	1.0013	1.6852	1.5224	1.2517	0.7087	1.6901	1.7177
0.2	9.7329E 00	1.0301	0.9993	0.9959	1.2470	1.1972	1.2499	0.8464	1.2581	1.2615
0.4	1.5720E 00	1.0756	0.9918	0.9917	1.0749	1.0634	1.1289	0.9385	1.0881	1.0810
0.6	1.1820E 00	1.0605	0.9901	0.9905	1.0332	1.0293	1.0704	0.9679	1.0461	1.0367
0.8	1.0898E 00	1.0459	0.9900	0.9905	1.0176	1.0160	1.0438	0.9803	1.0295	1.0200
1.0	1.0554E 00	1.0360	0.9905	0.9909	1.0104	1.0096	1.0302	0.9866	1.0213	1.0121
1.2	1.0390E 00	1.0293	0.9911	0.9915	1.0066	1.0062	1.0224	0.9903	1.0166	1.0079
1.4	1.0298E 00	1.0247	0.9918	0.9921	1.0045	1.0042	1.0175	0.9926	1.0136	1.0055
1.8	1.0203E 00	1.0187	0.9930	0.9932	1.0023	1.0022	1.0120	0.9953	1.0099	1.0031
2.2	1.0155E 00	1.0150	0.9939	0.9941	1.0013	1.0013	1.0091	0.9969	1.0079	1.0019

POSITRON Z= -51 A= 121

p	$F_0L_0^*/F_0L_0$	λ_2^*/λ_2	μ_1^*/μ_1	μ_2^*/μ_2	Λ_1^*/Λ_1	Λ_2^*/Λ_2	ν_{12}^*/ν_{12}	$\hat{\nu}_{12}^*/\hat{\nu}_{12}$	η_{12}^*/η_{12}	$\hat{\eta}_{12}^*/\hat{\eta}_{12}$
0.1	9.0477E 03	0.8816	1.0113	1.0015	1.6988	1.5326	1.2468	0.7076	1.7035	1.7330
0.2	1.0538E 01	1.0277	0.9996	0.9959	1.2524	1.2015	1.2499	0.8453	1.2637	1.2677
0.4	1.6002E 00	1.0756	0.9917	0.9915	1.0767	1.0649	1.1310	0.9377	1.0902	1.0831
0.6	1.1897E 00	1.0613	0.9899	0.9903	1.0341	1.0300	1.0719	0.9674	1.0472	1.0377
0.8	1.0932E 00	1.0468	0.9898	0.9903	1.0181	1.0164	1.0448	0.9800	1.0303	1.0205
1.0	1.0573E 00	1.0368	0.9903	0.9907	1.0107	1.0098	1.0309	0.9864	1.0219	1.0125
1.2	1.0403E 00	1.0300	0.9909	0.9913	1.0068	1.0063	1.0229	0.9901	1.0170	1.0082
1.4	1.0308E 00	1.0253	0.9916	0.9919	1.0046	1.0043	1.0180	0.9924	1.0139	1.0057
1.8	1.0209E 00	1.0191	0.9928	0.9930	1.0023	1.0022	1.0123	0.9951	1.0102	1.0032
2.2	1.0159E 00	1.0154	0.9938	0.9940	1.0013	1.0013	1.0093	0.9968	1.0081	1.0020

Z= -52 A= 121

p	$F_0L_0^*/F_0L_0$	λ_2^*/λ_2	μ_1^*/μ_1	μ_2^*/μ_2	Λ_1^*/Λ_1	Λ_2^*/Λ_2	ν_{12}^*/ν_{12}	$\hat{\nu}_{12}^*/\hat{\nu}_{12}$	η_{12}^*/η_{12}	$\hat{\eta}_{12}^*/\hat{\eta}_{12}$
0.1	1.1783E 04	0.8788	1.0121	1.0017	1.7123	1.5428	1.2420	0.7066	1.7168	1.7482
0.2	1.1414E 01	1.0254	0.9999	0.9960	1.2577	1.2057	1.2498	0.8442	1.2692	1.2738
0.4	1.6291E 00	1.0756	0.9916	0.9914	1.0785	1.0664	1.1330	0.9370	1.0924	1.0852
0.6	1.1976E 00	1.0620	0.9897	0.9901	1.0349	1.0307	1.0733	0.9670	1.0484	1.0388
0.8	1.0967E 00	1.0476	0.9896	0.9901	1.0185	1.0168	1.0458	0.9797	1.0311	1.0211
1.0	1.0593E 00	1.0375	0.9901	0.9905	1.0110	1.0101	1.0316	0.9862	1.0225	1.0128
1.2	1.0415E 00	1.0307	0.9907	0.9911	1.0070	1.0065	1.0235	0.9900	1.0175	1.0084
1.4	1.0317E 00	1.0258	0.9914	0.9917	1.0047	1.0044	1.0184	0.9923	1.0143	1.0059
1.8	1.0214E 00	1.0196	0.9927	0.9929	1.0024	1.0023	1.0126	0.9951	1.0105	1.0033
2.2	1.0163E 00	1.0158	0.9936	0.9938	1.0014	1.0013	1.0095	0.9968	1.0083	1.0020

Z= -53 A= 127

p	$F_0L_0^*/F_0L_0$	λ_2^*/λ_2	μ_1^*/μ_1	μ_2^*/μ_2	Λ_1^*/Λ_1	Λ_2^*/Λ_2	ν_{12}^*/ν_{12}	$\hat{\nu}_{12}^*/\hat{\nu}_{12}$	η_{12}^*/η_{12}	$\hat{\eta}_{12}^*/\hat{\eta}_{12}$
0.1	1.5383E 04	0.8761	1.0128	1.0019	1.7259	1.5531	1.2373	0.7056	1.7301	1.7636
0.2	1.2378E 01	1.0232	1.0002	0.9960	1.2631	1.2099	1.2496	0.8431	1.2748	1.2800
0.4	1.6591E 00	1.0756	0.9916	0.9913	1.0803	1.0679	1.1349	0.9363	1.0945	1.0874
0.6	1.2057E 00	1.0627	0.9895	0.9899	1.0357	1.0315	1.0748	0.9666	1.0496	1.0398
0.8	1.1003E 00	1.0484	0.9894	0.9899	1.0190	1.0172	1.0469	0.9794	1.0319	1.0217
1.0	1.0613E 00	1.0383	0.9899	0.9903	1.0112	1.0103	1.0324	0.9860	1.0230	1.0132
1.2	1.0429E 00	1.0313	0.9905	0.9909	1.0072	1.0067	1.0240	0.9898	1.0179	1.0087
1.4	1.0326E 00	1.0264	0.9912	0.9915	1.0048	1.0045	1.0188	0.9922	1.0146	1.0061
1.8	1.0220E 00	1.0201	0.9925	0.9927	1.0025	1.0023	1.0129	0.9951	1.0107	1.0034
2.2	1.0167E 00	1.0162	0.9935	0.9937	1.0014	1.0013	1.0098	0.9967	1.0085	1.0021

POSITRON Z= -54 A= 128

p	$F_0L_0^*/F_0L_0$	λ_2^*/λ_2	μ_1^*/μ_1	μ_2^*/μ_2	Λ_1^*/Λ_1	Λ_2^*/Λ_2	ν_{12}^*/ν_{12}	$\vartheta_{12}^*/\vartheta_{12}$	η_{12}^*/η_{12}	$\hat{\eta}_{12}^*/\hat{\eta}_{12}$
0.1	2.0115E 04	0.8734	1.0137	1.0022	1.7394	1.5633	1.2326	0.7046	1.7435	1.7791
0.2	1.3435E 01	1.0209	1.0005	0.9960	1.2685	1.2141	1.2492	0.8420	1.2803	1.2862
0.4	1.6903E 00	1.0756	0.9915	0.9911	1.0821	1.0694	1.1368	0.9356	1.0967	1.0896
0.6	1.2140E 00	1.0634	0.9893	0.9897	1.0366	1.0322	1.0762	0.9661	1.0508	1.0408
0.8	1.1040E 00	1.0492	0.9892	0.9897	1.0194	1.0175	1.0479	0.9792	1.0326	1.0223
1.0	1.0634E 00	1.0390	0.9896	0.9901	1.0115	1.0106	1.0331	0.9858	1.0236	1.0136
1.2	1.0442E 00	1.0320	0.9903	0.9907	1.0073	1.0068	1.0246	0.9897	1.0184	1.0089
1.4	1.0336E 00	1.0270	0.9910	0.9913	1.0049	1.0046	1.0193	0.9921	1.0150	1.0062
1.8	1.0226E 00	1.0205	0.9923	0.9925	1.0025	1.0024	1.0133	0.9949	1.0110	1.0035
2.2	1.0171E 00	1.0166	0.9933	0.9935	1.0014	1.0014	1.0100	0.9967	1.0087	1.0022

Z= -55 A= 133

p	$F_0L_0^*/F_0L_0$	λ_2^*/λ_2	μ_1^*/μ_1	μ_2^*/μ_2	Λ_1^*/Λ_1	Λ_2^*/Λ_2	ν_{12}^*/ν_{12}	$\vartheta_{12}^*/\vartheta_{12}$	η_{12}^*/η_{12}	$\hat{\eta}_{12}^*/\hat{\eta}_{12}$
0.1	2.6305E 04	0.8709	1.0145	1.0024	1.7528	1.5734	1.2279	0.7038	1.7566	1.7944
0.2	1.4586E 01	1.0188	1.0009	0.9960	1.2738	1.2183	1.2487	0.8411	1.2858	1.2923
0.4	1.7222E 00	1.0755	0.9914	0.9910	1.0839	1.0708	1.1386	0.9350	1.0988	1.0917
0.6	1.2224E 00	1.0641	0.9892	0.9895	1.0374	1.0329	1.0776	0.9657	1.0519	1.0419
0.8	1.1077E 00	1.0500	0.9890	0.9894	1.0199	1.0179	1.0488	0.9789	1.0334	1.0229
1.0	1.0654E 00	1.0398	0.9894	0.9899	1.0118	1.0108	1.0338	0.9856	1.0242	1.0139
1.2	1.0455E 00	1.0327	0.9901	0.9905	1.0075	1.0070	1.0252	0.9896	1.0188	1.0092
1.4	1.0345E 00	1.0276	0.9908	0.9911	1.0050	1.0047	1.0197	0.9920	1.0154	1.0064
1.8	1.0232E 00	1.0210	0.9922	0.9924	1.0026	1.0024	1.0136	0.9949	1.0113	1.0036
2.2	1.0175E 00	1.0169	0.9932	0.9934	1.0015	1.0014	1.0103	0.9966	1.0089	1.0022

Z= -56 A= 134

p	$F_0L_0^*/F_0L_0$	λ_2^*/λ_2	μ_1^*/μ_1	μ_2^*/μ_2	Λ_1^*/Λ_1	Λ_2^*/Λ_2	ν_{12}^*/ν_{12}	$\vartheta_{12}^*/\vartheta_{12}$	η_{12}^*/η_{12}	$\hat{\eta}_{12}^*/\hat{\eta}_{12}$
0.1	3.4514E 04	0.8685	1.0154	1.0026	1.7664	1.5836	1.2232	0.7028	1.7699	1.8100
0.2	1.5861E 01	1.0166	1.0012	0.9961	1.2792	1.2226	1.2481	0.8401	1.2914	1.2986
0.4	1.7556E 00	1.0754	0.9914	0.9908	1.0858	1.0723	1.1404	0.9343	1.1010	1.0939
0.6	1.2312E 00	1.0648	0.9890	0.9893	1.0382	1.0336	1.0790	0.9653	1.0531	1.0429
0.8	1.1115E 00	1.0508	0.9888	0.9892	1.0203	1.0183	1.0499	0.9786	1.0342	1.0235
1.0	1.0676E 00	1.0405	0.9892	0.9897	1.0120	1.0110	1.0346	0.9855	1.0247	1.0143
1.2	1.0469E 00	1.0333	0.9899	0.9903	1.0077	1.0071	1.0257	0.9894	1.0193	1.0094
1.4	1.0355E 00	1.0282	0.9907	0.9910	1.0052	1.0048	1.0202	0.9919	1.0158	1.0066
1.8	1.0238E 00	1.0215	0.9920	0.9922	1.0026	1.0025	1.0139	0.9948	1.0116	1.0037
2.2	1.0180E 00	1.0173	0.9931	0.9932	1.0015	1.0014	1.0105	0.9966	1.0092	1.0023

POSITRON Z= -57 A= 131

p	$F_0L_0^*/F_0L_0$	λ_2^*/λ_2	μ_1^*/μ_1	μ_2^*/μ_2	Λ_1^*/Λ_1	Λ_2^*/Λ_2	ν_{12}^*/ν_{12}	$\vartheta_{12}^*/\vartheta_{12}$	η_{12}^*/η_{12}	$\hat{\eta}_{12}^*/\hat{\eta}_{12}$
0.1	4.5318E 04	0.8662	1.0162	1.0029	1.7799	1.5938	1.2186	0.7020	1.7831	1.8256
0.2	1.7255E 01	1.0146	1.0016	0.9961	1.2846	1.2268	1.2474	0.8392	1.2969	1.3049
0.4	1.7900E 00	1.0752	0.9914	0.9907	1.0876	1.0738	1.1421	0.9337	1.1032	1.0961
0.6	1.2400E 00	1.0654	0.9888	0.9892	1.0391	1.0343	1.0804	0.9650	1.0543	1.0440
0.8	1.1154E 00	1.0515	0.9885	0.9890	1.0208	1.0187	1.0508	0.9784	1.0350	1.0241
1.0	1.0697E 00	1.0412	0.9890	0.9895	1.0123	1.0113	1.0353	0.9853	1.0253	1.0147
1.2	1.0483E 00	1.0340	0.9897	0.9901	1.0078	1.0073	1.0263	0.9893	1.0197	1.0097
1.4	1.0365E 00	1.0288	0.9905	0.9908	1.0053	1.0049	1.0206	0.9918	1.0161	1.0068
1.8	1.0244E 00	1.0220	0.9918	0.9920	1.0027	1.0026	1.0142	0.9948	1.0118	1.0038
2.2	1.0184E 00	1.0177	0.9929	0.9931	1.0016	1.0015	1.0107	0.9965	1.0094	1.0024

Z= -58 A= 136

p	$F_0L_0^*/F_0L_0$	λ_2^*/λ_2	μ_1^*/μ_1	μ_2^*/μ_2	Λ_1^*/Λ_1	Λ_2^*/Λ_2	ν_{12}^*/ν_{12}	$\vartheta_{12}^*/\vartheta_{12}$	η_{12}^*/η_{12}	$\hat{\eta}_{12}^*/\hat{\eta}_{12}$
0.1	5.9653E 04	0.8639	1.0172	1.0031	1.7934	1.6039	1.2141	0.7012	1.7963	1.8414
0.2	1.8796E 01	1.0125	1.0020	0.9962	1.2900	1.2310	1.2466	0.8382	1.3025	1.3113
0.4	1.8259E 00	1.0751	0.9913	0.9906	1.0895	1.0753	1.1439	0.9330	1.1054	1.0984
0.6	1.2493E 00	1.0661	0.9887	0.9890	1.0399	1.0350	1.0818	0.9646	1.0556	1.0451
0.8	1.1194E 00	1.0523	0.9883	0.9888	1.0213	1.0191	1.0519	0.9781	1.0358	1.0247
1.0	1.0720E 00	1.0420	0.9888	0.9892	1.0126	1.0115	1.0360	0.9851	1.0259	1.0151
1.2	1.0497E 00	1.0346	0.9895	0.9899	1.0080	1.0074	1.0269	0.9892	1.0202	1.0100
1.4	1.0375E 00	1.0294	0.9903	0.9906	1.0054	1.0051	1.0211	0.9917	1.0165	1.0070
1.8	1.0250E 00	1.0224	0.9917	0.9919	1.0028	1.0026	1.0145	0.9947	1.0121	1.0039
2.2	1.0188E 00	1.0181	0.9928	0.9929	1.0016	1.0015	1.0110	0.9965	1.0096	1.0025

Z= -59 A= 138

p	$F_0L_0^*/F_0L_0$	λ_2^*/λ_2	μ_1^*/μ_1	μ_2^*/μ_2	Λ_1^*/Λ_1	Λ_2^*/Λ_2	ν_{12}^*/ν_{12}	$\vartheta_{12}^*/\vartheta_{12}$	η_{12}^*/η_{12}	$\hat{\eta}_{12}^*/\hat{\eta}_{12}$
0.1	7.8639E 04	0.8617	1.0181	1.0034	1.8070	1.6142	1.2097	0.7004	1.8096	1.8572
0.2	2.0488E 01	1.0105	1.0024	0.9962	1.2954	1.2353	1.2457	0.8373	1.3081	1.3176
0.4	1.8629E 00	1.0749	0.9913	0.9905	1.0913	1.0768	1.1455	0.9324	1.1075	1.1006
0.6	1.2586E 00	1.0666	0.9885	0.9888	1.0408	1.0358	1.0832	0.9642	1.0568	1.0462
0.8	1.1235E 00	1.0531	0.9881	0.9886	1.0217	1.0195	1.0529	0.9779	1.0366	1.0253
1.0	1.0742E 00	1.0427	0.9886	0.9890	1.0129	1.0118	1.0368	0.9849	1.0265	1.0155
1.2	1.0512E 00	1.0353	0.9893	0.9897	1.0082	1.0076	1.0274	0.9890	1.0207	1.0102
1.4	1.0385E 00	1.0300	0.9901	0.9904	1.0055	1.0052	1.0216	0.9916	1.0169	1.0072
1.8	1.0256E 00	1.0229	0.9915	0.9917	1.0028	1.0027	1.0149	0.9947	1.0124	1.0040
2.2	1.0192E 00	1.0185	0.9926	0.9928	1.0016	1.0015	1.0112	0.9964	1.0098	1.0025

POSITRON Z= -60 A= 141

p	$F_0L_0^*/F_0L_0$	λ_2^*/λ_2	μ_1^*/μ_1	μ_2^*/μ_2	Λ_1^*/Λ_1	Λ_2^*/Λ_2	ν_{12}^*/ν_{12}	$\hat{\nu}_{12}^*/\hat{\nu}_{12}$	η_{12}^*/η_{12}	$\hat{\eta}_{12}^*/\hat{\eta}_{12}$
0.1	1.0376E 05	0.8597	1.0191	1.0036	1.8206	1.6243	1.2053	0.6997	1.8228	1.8731
0.2	2.2347E 01	1.0086	1.0028	0.9963	1.3008	1.2395	1.2447	0.8365	1.3136	1.3240
0.4	1.9011E 00	1.0747	0.9913	0.9903	1.0932	1.0784	1.1471	0.9318	1.1097	1.1029
0.6	1.2682E 00	1.0672	0.9884	0.9886	1.0417	1.0365	1.0846	0.9638	1.0580	1.0473
0.8	1.1276E 00	1.0538	0.9879	0.9884	1.0222	1.0200	1.0538	0.9776	1.0374	1.0259
1.0	1.0765E 00	1.0434	0.9884	0.9888	1.0131	1.0120	1.0375	0.9848	1.0271	1.0159
1.2	1.0526E 00	1.0360	0.9891	0.9895	1.0084	1.0078	1.0280	0.9889	1.0211	1.0105
1.4	1.0395E 00	1.0306	0.9899	0.9902	1.0056	1.0053	1.0220	0.9915	1.0173	1.0074
1.8	1.0262E 00	1.0234	0.9913	0.9915	1.0029	1.0027	1.0152	0.9946	1.0127	1.0042
2.2	1.0197E 00	1.0189	0.9925	0.9926	1.0017	1.0016	1.0115	0.9963	1.0100	1.0026

Z= -61 A= 143

p	$F_0L_0^*/F_0L_0$	λ_2^*/λ_2	μ_1^*/μ_1	μ_2^*/μ_2	Λ_1^*/Λ_1	Λ_2^*/Λ_2	ν_{12}^*/ν_{12}	$\hat{\nu}_{12}^*/\hat{\nu}_{12}$	η_{12}^*/η_{12}	$\hat{\eta}_{12}^*/\hat{\eta}_{12}$
0.1	1.3703E 05	0.8577	1.0201	1.0039	1.8340	1.6344	1.2009	0.6990	1.8359	1.8890
0.2	2.4391E 01	1.0067	1.0033	0.9964	1.3062	1.2437	1.2436	0.8356	1.3191	1.3304
0.4	1.9407E 00	1.0745	0.9913	0.9902	1.0950	1.0799	1.1487	0.9312	1.1119	1.1052
0.6	1.2780E 00	1.0678	0.9882	0.9884	1.0425	1.0372	1.0859	0.9634	1.0592	1.0484
0.8	1.1318E 00	1.0546	0.9877	0.9882	1.0226	1.0204	1.0548	0.9774	1.0382	1.0266
1.0	1.0788E 00	1.0442	0.9882	0.9886	1.0134	1.0123	1.0383	0.9846	1.0277	1.0163
1.2	1.0541E 00	1.0366	0.9889	0.9893	1.0085	1.0079	1.0286	0.9888	1.0216	1.0108
1.4	1.0406E 00	1.0312	0.9897	0.9900	1.0058	1.0054	1.0225	0.9914	1.0177	1.0076
1.8	1.0268E 00	1.0239	0.9911	0.9914	1.0029	1.0028	1.0155	0.9945	1.0130	1.0043
2.2	1.0201E 00	1.0193	0.9923	0.9925	1.0017	1.0016	1.0118	0.9963	1.0103	1.0027
2.6	1.0161E 00	1.0163	0.9932	0.9933	1.0011	1.0010	1.0095	0.9974	1.0085	1.0018

Z= -62 A= 144

p	$F_0L_0^*/F_0L_0$	λ_2^*/λ_2	μ_1^*/μ_1	μ_2^*/μ_2	Λ_1^*/Λ_1	Λ_2^*/Λ_2	ν_{12}^*/ν_{12}	$\hat{\nu}_{12}^*/\hat{\nu}_{12}$	η_{12}^*/η_{12}	$\hat{\eta}_{12}^*/\hat{\eta}_{12}$
0.1	1.8133E 05	0.8557	1.0212	1.0042	1.8475	1.6445	1.1965	0.6984	1.8490	1.9050
0.2	2.6648E 01	1.0048	1.0037	0.9964	1.3116	1.2480	1.2425	0.8348	1.3247	1.3369
0.4	1.9817E 00	1.0742	0.9913	0.9901	1.0969	1.0814	1.1502	0.9306	1.1141	1.1075
0.6	1.2881E 00	1.0684	0.9881	0.9882	1.0434	1.0380	1.0873	0.9631	1.0605	1.0495
0.8	1.1362E 00	1.0553	0.9875	0.9880	1.0231	1.0208	1.0558	0.9772	1.0391	1.0272
1.0	1.0812E 00	1.0449	0.9880	0.9884	1.0137	1.0125	1.0390	0.9844	1.0283	1.0167
1.2	1.0556E 00	1.0373	0.9887	0.9891	1.0087	1.0081	1.0292	0.9887	1.0221	1.0111
1.4	1.0416E 00	1.0318	0.9895	0.9898	1.0059	1.0055	1.0230	0.9914	1.0181	1.0078
1.8	1.0275E 00	1.0244	0.9910	0.9912	1.0030	1.0028	1.0159	0.9945	1.0133	1.0044
2.2	1.0205E 00	1.0198	0.9922	0.9923	1.0017	1.0016	1.0120	0.9962	1.0105	1.0028
2.6	1.0165E 00	1.0166	0.9931	0.9932	1.0011	1.0010	1.0097	0.9973	1.0087	1.0019

 Instr. see p. 25

POSITRON Z= -63 A= 147

p	$F_0L_0^*/F_0L_0$	λ_2^*/λ_2	μ_1^*/μ_1	μ_2^*/μ_2	Λ_1^*/Λ_1	Λ_2^*/Λ_2	ν_{12}^*/ν_{12}	$\hat{\nu}_{12}^*/\hat{\nu}_{12}$	η_{12}^*/η_{12}	$\hat{\eta}_{12}^*/\hat{\eta}_{12}$
0.1	2.3998E 05	0.8539	1.0223	1.0045	1.8609	1.6545	1.1923	0.6978	1.8620	1.9210
0.2	2.9120E 01	1.0030	1.0042	0.9965	1.3169	1.2522	1.2412	0.8341	1.3302	1.3434
0.4	2.0238E 00	1.0739	0.9913	0.9900	1.0987	1.0829	1.1516	0.9301	1.1163	1.1098
0.6	1.2983E 00	1.0589	0.9879	0.9880	1.0442	1.0387	1.0886	0.9627	1.0617	1.0506
0.8	1.1405E 00	1.0560	0.9874	0.9877	1.0236	1.0212	1.0568	0.9769	1.0399	1.0279
1.0	1.0836E 00	1.0456	0.9878	0.9882	1.0140	1.0128	1.0397	0.9843	1.0289	1.0171
1.2	1.0571E 00	1.0380	0.9885	0.9889	1.0089	1.0082	1.0297	0.9886	1.0225	1.0114
1.4	1.0427E 00	1.0324	0.9893	0.9896	1.0060	1.0056	1.0234	0.9913	1.0185	1.0080
1.8	1.0281E 00	1.0249	0.9908	0.9910	1.0031	1.0029	1.0162	0.9944	1.0136	1.0045
2.2	1.0209E 00	1.0202	0.9920	0.9922	1.0018	1.0017	1.0123	0.9962	1.0107	1.0029
2.6	1.0168E 00	1.0170	0.9930	0.9931	1.0011	1.0011	1.0099	0.9972	1.0089	1.0019

Z= -64 A= 154

p	$F_0L_0^*/F_0L_0$	λ_2^*/λ_2	μ_1^*/μ_1	μ_2^*/μ_2	Λ_1^*/Λ_1	Λ_2^*/Λ_2	ν_{12}^*/ν_{12}	$\hat{\nu}_{12}^*/\hat{\nu}_{12}$	η_{12}^*/η_{12}	$\hat{\eta}_{12}^*/\hat{\eta}_{12}$
0.1	3.1871E 05	0.8521	1.0234	1.0048	1.8746	1.6646	1.1881	0.6972	1.8751	1.9372
0.2	3.1873E 01	1.0012	1.0047	0.9966	1.3224	1.2564	1.2399	0.8333	1.3358	1.3500
0.4	2.0679E 00	1.0737	0.9913	0.9899	1.1006	1.0844	1.1530	0.9295	1.1186	1.1122
0.6	1.3089E 00	1.0694	0.9878	0.9878	1.0451	1.0394	1.0899	0.9624	1.0630	1.0518
0.8	1.1450E 00	1.0568	0.9872	0.9875	1.0241	1.0216	1.0578	0.9767	1.0407	1.0285
1.0	1.0860E 00	1.0463	0.9876	0.9880	1.0143	1.0130	1.0405	0.9841	1.0295	1.0176
1.2	1.0587E 00	1.0386	0.9883	0.9886	1.0091	1.0084	1.0303	0.9884	1.0230	1.0117
1.4	1.0438E 00	1.0330	0.9891	0.9894	1.0061	1.0057	1.0239	0.9912	1.0189	1.0082
1.8	1.0287E 00	1.0254	0.9906	0.9908	1.0031	1.0030	1.0165	0.9944	1.0139	1.0046
2.2	1.0214E 00	1.0206	0.9919	0.9920	1.0018	1.0017	1.0125	0.9962	1.0110	1.0030
2.6	1.0171E 00	1.0173	0.9928	0.9929	1.0011	1.0011	1.0101	0.9973	1.0091	1.0020

Z= -65 A= 155

p	$F_0L_0^*/F_0L_0$	λ_2^*/λ_2	μ_1^*/μ_1	μ_2^*/μ_2	Λ_1^*/Λ_1	Λ_2^*/Λ_2	ν_{12}^*/ν_{12}	$\hat{\nu}_{12}^*/\hat{\nu}_{12}$	η_{12}^*/η_{12}	$\hat{\eta}_{12}^*/\hat{\eta}_{12}$
0.1	4.2328E 05	0.8504	1.0246	1.0051	1.8881	1.6747	1.1840	0.6966	1.8882	1.9535
0.2	3.4892E 01	0.9995	1.0053	0.9967	1.3278	1.2607	1.2385	0.8326	1.3413	1.3565
0.4	2.1131E 00	1.0734	0.9914	0.9897	1.1025	1.0859	1.1543	0.9290	1.1208	1.1145
0.6	1.3197E 00	1.0699	0.9876	0.9876	1.0460	1.0402	1.0912	0.9620	1.0642	1.0530
0.8	1.1496E 00	1.0575	0.9870	0.9873	1.0245	1.0220	1.0588	0.9765	1.0415	1.0292
1.0	1.0885E 00	1.0470	0.9874	0.9878	1.0145	1.0133	1.0412	0.9840	1.0301	1.0180
1.2	1.0602E 00	1.0393	0.9881	0.9884	1.0093	1.0086	1.0309	0.9883	1.0235	1.0120
1.4	1.0448E 00	1.0336	0.9889	0.9892	1.0062	1.0058	1.0244	0.9911	1.0193	1.0084
1.8	1.0293E 00	1.0259	0.9905	0.9907	1.0032	1.0030	1.0169	0.9943	1.0142	1.0048
2.2	1.0218E 00	1.0210	0.9917	0.9919	1.0018	1.0017	1.0128	0.9962	1.0112	1.0030
2.6	1.0175E 00	1.0177	0.9927	0.9928	1.0012	1.0011	1.0103	0.9973	1.0093	1.0021

POSITRON Z= -66 A= 160

p	$F_0L_0^*/F_0L_0$	λ_2^*/λ_2	μ_1^*/μ_1	μ_2^*/μ_2	Λ_1^*/Λ_1	Λ_2^*/Λ_2	ν_{12}^*/ν_{12}	$\hat{\nu}_{12}^*/\hat{\nu}_{12}$	η_{12}^*/η_{12}	$\hat{\eta}_{12}^*/\hat{\eta}_{12}$
0.1	5.6339E 05	0.8488	1.0258	1.0054	1.9017	1.6848	1.1799	0.6961	1.9013	1.9700
0.2	3.8246E 01	0.9978	1.0058	0.9968	1.3333	1.2649	1.2371	0.8318	1.3469	1.3632
0.4	2.1604E 00	1.0731	0.9914	0.9896	1.1044	1.0875	1.1556	0.9284	1.1230	1.1169
0.6	1.3308E 00	1.0704	0.9875	0.9874	1.0469	1.0409	1.0925	0.9617	1.0655	1.0542
0.8	1.1542E 00	1.0582	0.9868	0.9871	1.0250	1.0224	1.0597	0.9762	1.0424	1.0299
1.0	1.0910E 00	1.0478	0.9872	0.9875	1.0148	1.0135	1.0420	0.9838	1.0308	1.0184
1.2	1.0618E 00	1.0400	0.9879	0.9882	1.0095	1.0087	1.0315	0.9882	1.0240	1.0123
1.4	1.0460E 00	1.0342	0.9887	0.9890	1.0064	1.0059	1.0249	0.9910	1.0197	1.0087
1.8	1.0300E 00	1.0264	0.9903	0.9905	1.0033	1.0031	1.0172	0.9943	1.0145	1.0049
2.2	1.0223E 00	1.0214	0.9916	0.9917	1.0019	1.0018	1.0131	0.9962	1.0115	1.0031
2.6	1.0178E 00	1.0180	0.9926	0.9927	1.0012	1.0011	1.0105	0.9972	1.0095	1.0021

Z= -67 A= 161

p	$F_0L_0^*/F_0L_0$	λ_2^*/λ_2	μ_1^*/μ_1	μ_2^*/μ_2	Λ_1^*/Λ_1	Λ_2^*/Λ_2	ν_{12}^*/ν_{12}	$\hat{\nu}_{12}^*/\hat{\nu}_{12}$	η_{12}^*/η_{12}	$\hat{\eta}_{12}^*/\hat{\eta}_{12}$
0.1	7.5010E 05	0.8472	1.0270	1.0058	1.9152	1.6949	1.1759	0.6956	1.9143	1.9864
0.2	4.1930E 01	0.9962	1.0064	0.9969	1.3387	1.2692	1.2356	0.8312	1.3524	1.3699
0.4	2.2089E 00	1.0728	0.9915	0.9895	1.1063	1.0890	1.1569	0.9279	1.1253	1.1193
0.6	1.3420E 00	1.0709	0.9874	0.9873	1.0478	1.0416	1.0937	0.9614	1.0667	1.0554
0.8	1.1590E 00	1.0589	0.9866	0.9869	1.0255	1.0228	1.0607	0.9760	1.0432	1.0306
1.0	1.0936E 00	1.0485	0.9870	0.9873	1.0151	1.0138	1.0427	0.9837	1.0314	1.0189
1.2	1.0634E 00	1.0406	0.9877	0.9880	1.0096	1.0089	1.0321	0.9881	1.0245	1.0126
1.4	1.0471E 00	1.0348	0.9885	0.9888	1.0065	1.0061	1.0253	0.9909	1.0201	1.0089
1.8	1.0306E 00	1.0269	0.9901	0.9903	1.0033	1.0031	1.0175	0.9942	1.0148	1.0051
2.2	1.0227E 00	1.0219	0.9914	0.9916	1.0019	1.0018	1.0133	0.9961	1.0117	1.0032
2.6	1.0181E 00	1.0184	0.9924	0.9925	1.0012	1.0011	1.0107	0.9972	1.0097	1.0022

Z= -68 A= 166

p	$F_0L_0^*/F_0L_0$	λ_2^*/λ_2	μ_1^*/μ_1	μ_2^*/μ_2	Λ_1^*/Λ_1	Λ_2^*/Λ_2	ν_{12}^*/ν_{12}	$\hat{\nu}_{12}^*/\hat{\nu}_{12}$	η_{12}^*/η_{12}	$\hat{\eta}_{12}^*/\hat{\eta}_{12}$
0.1	1.0001E 06	0.8457	1.0283	1.0061	1.9288	1.7049	1.1720	0.6951	1.9273	2.0030
0.2	4.6006E 01	0.9546	1.0070	0.9970	1.3441	1.2734	1.2340	0.8305	1.3579	1.3766
0.4	2.2591E 00	1.0725	0.9915	0.9894	1.1082	1.0905	1.1581	0.9274	1.1275	1.1218
0.6	1.3535E 00	1.0714	0.9873	0.9871	1.0487	1.0424	1.0950	0.9610	1.0680	1.0566
0.8	1.1638E 00	1.0596	0.9864	0.9867	1.0260	1.0232	1.0617	0.9758	1.0441	1.0313
1.0	1.0962E 00	1.0492	0.9868	0.9871	1.0154	1.0140	1.0434	0.9835	1.0320	1.0193
1.2	1.0650E 00	1.0413	0.9875	0.9878	1.0098	1.0091	1.0327	0.9880	1.0250	1.0129
1.4	1.0482E 00	1.0354	0.9884	0.9886	1.0066	1.0062	1.0258	0.9909	1.0205	1.0091
1.8	1.0313E 00	1.0274	0.9900	0.9901	1.0034	1.0032	1.0179	0.9942	1.0151	1.0052
2.2	1.0232E 00	1.0223	0.9913	0.9914	1.0020	1.0018	1.0136	0.9961	1.0119	1.0033
2.6	1.0185E 00	1.0188	0.9923	0.9924	1.0012	1.0012	1.0109	0.9972	1.0099	1.0023

 Instr. see p. 25

POSITRON   Z= -69   A= 167

p	$F_0L_0^*/F_0L_0$	λ_2^*/λ_2	μ_1^*/μ_1	μ_2^*/μ_2	Λ_1^*/Λ_1	Λ_2^*/Λ_2	ν_{12}^*/ν_{12}	$\vartheta_{12}^*/\vartheta_{12}$	η_{12}^*/η_{12}	$\hat{\eta}_{12}^*/\hat{\eta}_{12}$
0.1	1.3355E 06	0.8443	1.0296	1.0064	1.9424	1.7150	1.1681	0.6947	1.9403	2.0198
0.2	5.0519E 01	0.9931	1.0077	0.9971	1.3495	1.2776	1.2324	0.8299	1.3634	1.3834
0.4	2.3112E 00	1.0722	0.9916	0.9893	1.1101	1.0920	1.1592	0.9269	1.1298	1.1242
0.6	1.3653E 00	1.0719	0.9872	0.9869	1.0496	1.0431	1.0962	0.9607	1.0693	1.0578
0.8	1.1687E 00	1.0603	0.9863	0.9865	1.0265	1.0237	1.0626	0.9756	1.0449	1.0320
1.0	1.0988E 00	1.0499	0.9866	0.9869	1.0157	1.0143	1.0442	0.9834	1.0327	1.0198
1.2	1.0667E 00	1.0420	0.9873	0.9876	1.0100	1.0092	1.0333	0.9879	1.0255	1.0132
1.4	1.0493E 00	1.0360	0.9882	0.9884	1.0067	1.0063	1.0263	0.9908	1.0209	1.0093
1.8	1.0319E 00	1.0279	0.9898	0.9900	1.0034	1.0033	1.0182	0.9941	1.0154	1.0053
2.2	1.0236E 00	1.0227	0.9911	0.9913	1.0020	1.0019	1.0139	0.9961	1.0122	1.0034
2.6	1.0188E 00	1.0191	0.9922	0.9923	1.0012	1.0012	1.0112	0.9972	1.0101	1.0024

Z= -70   A= 172

p	$F_0L_0^*/F_0L_0$	λ_2^*/λ_2	μ_1^*/μ_1	μ_2^*/μ_2	Λ_1^*/Λ_1	Λ_2^*/Λ_2	ν_{12}^*/ν_{12}	$\vartheta_{12}^*/\vartheta_{12}$	η_{12}^*/η_{12}	$\hat{\eta}_{12}^*/\hat{\eta}_{12}$
0.1	1.7859E 06	0.8429	1.0310	1.0068	1.9560	1.7250	1.1643	0.6943	1.9533	2.0366
0.2	5.5520E 01	0.9916	1.0083	0.9972	1.3550	1.2819	1.2308	0.8293	1.3690	1.3902
0.4	2.3652E 00	1.0719	0.9917	0.9892	1.1120	1.0936	1.1603	0.9264	1.1320	1.1267
0.6	1.3774E 00	1.0723	0.9871	0.9867	1.0505	1.0439	1.0974	0.9604	1.0706	1.0590
0.8	1.1737E 00	1.0610	0.9861	0.9863	1.0270	1.0241	1.0636	0.9754	1.0458	1.0327
1.0	1.1015E 00	1.0506	0.9864	0.9867	1.0160	1.0145	1.0449	0.9832	1.0333	1.0202
1.2	1.0684E 00	1.0426	0.9871	0.9874	1.0102	1.0094	1.0338	0.9878	1.0260	1.0135
1.4	1.0505E 00	1.0366	0.9880	0.9882	1.0069	1.0064	1.0268	0.9907	1.0213	1.0096
1.8	1.0326E 00	1.0284	0.9896	0.9898	1.0035	1.0033	1.0186	0.9941	1.0157	1.0055
2.2	1.0240E 00	1.0232	0.9910	0.9911	1.0020	1.0019	1.0141	0.9961	1.0124	1.0035
2.6	1.0191E 00	1.0195	0.9920	0.9921	1.0013	1.0012	1.0114	0.9971	1.0103	1.0024

Z= -71   A= 170

p	$F_0L_0^*/F_0L_0$	λ_2^*/λ_2	μ_1^*/μ_1	μ_2^*/μ_2	Λ_1^*/Λ_1	Λ_2^*/Λ_2	ν_{12}^*/ν_{12}	$\vartheta_{12}^*/\vartheta_{12}$	η_{12}^*/η_{12}	$\hat{\eta}_{12}^*/\hat{\eta}_{12}$
0.1	2.3899E 06	0.8416	1.0324	1.0072	1.9696	1.7350	1.1604	0.6939	1.9663	2.0536
0.2	6.1050E 01	0.9901	1.0090	0.9973	1.3604	1.2861	1.2290	0.8287	1.3745	1.3971
0.4	2.4210E 00	1.0716	0.9918	0.9891	1.1139	1.0951	1.1613	0.9259	1.1343	1.1292
0.6	1.3898E 00	1.0728	0.9870	0.9865	1.0514	1.0446	1.0986	0.9601	1.0719	1.0603
0.8	1.1788E 00	1.0616	0.9859	0.9861	1.0274	1.0245	1.0646	0.9752	1.0467	1.0334
1.0	1.1042E 00	1.0513	0.9862	0.9865	1.0163	1.0148	1.0457	0.9831	1.0340	1.0207
1.2	1.0701E 00	1.0433	0.9869	0.9872	1.0104	1.0096	1.0344	0.9877	1.0265	1.0139
1.4	1.0516E 00	1.0372	0.9878	0.9880	1.0070	1.0065	1.0273	0.9906	1.0217	1.0098
1.8	1.0333E 00	1.0289	0.9895	0.9896	1.0036	1.0034	1.0189	0.9940	1.0160	1.0056
2.2	1.0245E 00	1.0236	0.9908	0.9910	1.0021	1.0019	1.0144	0.9960	1.0127	1.0036
2.6	1.0195E 00	1.0199	0.9919	0.9920	1.0013	1.0012	1.0116	0.9972	1.0105	1.0025

POSITRON Z= -72 A= 178

p	$F_0L_0^*/F_0L_0$	λ_2^*/λ_2	μ_1^*/μ_1	μ_2^*/μ_2	Λ_1^*/Λ_1	Λ_2^*/Λ_2	ν_{12}^*/ν_{12}	$\hat{\nu}_{12}^*/\hat{\nu}_{12}$	η_{12}^*/η_{12}	$\hat{\eta}_{12}^*/\hat{\eta}_{12}$
0.1	3.1973E 06	0.8403	1.0338	1.0075	1.9831	1.7450	1.1567	0.6936	1.9791	2.0705
0.2	6.7129E 01	0.9887	1.0097	0.9974	1.3658	1.2903	1.2272	0.8282	1.3800	1.4039
0.4	2.4783E 00	1.0713	0.9919	0.9890	1.1158	1.0966	1.1623	0.9255	1.1365	1.1317
0.6	1.4022E 00	1.0732	0.9869	0.9864	1.0523	1.0454	1.0998	0.9598	1.0732	1.0615
0.8	1.1839E 00	1.0623	0.9857	0.9859	1.0279	1.0249	1.0655	0.9751	1.0475	1.0342
1.0	1.1070E 00	1.0520	0.9860	0.9863	1.0166	1.0151	1.0464	0.9830	1.0346	1.0212
1.2	1.0717E 00	1.0440	0.9867	0.9870	1.0106	1.0097	1.0350	0.9877	1.0270	1.0142
1.4	1.0528E 00	1.0378	0.9876	0.9878	1.0071	1.0066	1.0277	0.9906	1.0222	1.0101
1.8	1.0339E 00	1.0294	0.9893	0.9894	1.0036	1.0034	1.0193	0.9940	1.0163	1.0058
2.2	1.0249E 00	1.0240	0.9907	0.9908	1.0021	1.0020	1.0147	0.9960	1.0129	1.0037
2.6	1.0198E 00	1.0203	0.9918	0.9919	1.0013	1.0012	1.0118	0.9971	1.0107	1.0026
3.0	1.0164E 00	1.0176	0.9927	0.9927	1.0009	1.0008	1.0099	0.9977	1.0092	1.0019

Z= -73 A= 179

p	$F_0L_0^*/F_0L_0$	λ_2^*/λ_2	μ_1^*/μ_1	μ_2^*/μ_2	Λ_1^*/Λ_1	Λ_2^*/Λ_2	ν_{12}^*/ν_{12}	$\hat{\nu}_{12}^*/\hat{\nu}_{12}$	η_{12}^*/η_{12}	$\hat{\eta}_{12}^*/\hat{\eta}_{12}$
0.1	4.2923E 06	0.8391	1.0353	1.0079	1.9968	1.7550	1.1531	0.6933	1.9921	2.0878
0.2	7.3937E 01	0.9873	1.0105	0.9976	1.3713	1.2945	1.2254	0.8276	1.3856	1.4109
0.4	2.5383E 00	1.0710	0.9921	0.9890	1.1177	1.0982	1.1633	0.9250	1.1388	1.1343
0.6	1.4152E 00	1.0736	0.9868	0.9862	1.0532	1.0461	1.1010	0.9595	1.0745	1.0628
0.8	1.1892E 00	1.0630	0.9856	0.9857	1.0284	1.0253	1.0665	0.9749	1.0484	1.0349
1.0	1.1098E 00	1.0527	0.9858	0.9861	1.0169	1.0153	1.0472	0.9829	1.0353	1.0217
1.2	1.0735E 00	1.0446	0.9865	0.9868	1.0108	1.0099	1.0356	0.9876	1.0276	1.0145
1.4	1.0540E 00	1.0385	0.9874	0.9876	1.0073	1.0067	1.0282	0.9905	1.0226	1.0103
1.8	1.0346E 00	1.0300	0.9891	0.9893	1.0037	1.0035	1.0196	0.9940	1.0166	1.0059
2.2	1.0254E 00	1.0244	0.9905	0.9907	1.0021	1.0020	1.0149	0.9960	1.0132	1.0038
2.6	1.0201E 00	1.0206	0.9916	0.9917	1.0013	1.0013	1.0121	0.9971	1.0110	1.0027
3.0	1.0167E 00	1.0179	0.9926	0.9925	1.0009	1.0009	1.0101	0.9976	1.0094	1.0020

Z= -74 A= 180

p	$F_0L_0^*/F_0L_0$	λ_2^*/λ_2	μ_1^*/μ_1	μ_2^*/μ_2	Λ_1^*/Λ_1	Λ_2^*/Λ_2	ν_{12}^*/ν_{12}	$\hat{\nu}_{12}^*/\hat{\nu}_{12}$	η_{12}^*/η_{12}	$\hat{\eta}_{12}^*/\hat{\eta}_{12}$
0.1	5.7674E 06	0.8379	1.0368	1.0083	2.0105	1.7651	1.1494	0.6930	2.0051	2.1051
0.2	8.1488E 01	0.9860	1.0112	0.9977	1.3768	1.2988	1.2236	0.8271	1.3911	1.4180
0.4	2.6004E 00	1.0707	0.9922	0.9889	1.1197	1.0997	1.1642	0.9246	1.1411	1.1369
0.6	1.4284E 00	1.0740	0.9867	0.9860	1.0541	1.0469	1.1021	0.9592	1.0758	1.0641
0.8	1.1946E 00	1.0636	0.9854	0.9855	1.0289	1.0258	1.0674	0.9747	1.0493	1.0357
1.0	1.1127E 00	1.0534	0.9856	0.9858	1.0172	1.0156	1.0479	0.9827	1.0359	1.0222
1.2	1.0753E 00	1.0453	0.9863	0.9866	1.0110	1.0101	1.0362	0.9875	1.0281	1.0149
1.4	1.0552E 00	1.0391	0.9872	0.9874	1.0074	1.0069	1.0287	0.9905	1.0230	1.0106
1.8	1.0353E 00	1.0305	0.9890	0.9891	1.0038	1.0036	1.0200	0.9939	1.0170	1.0061
2.2	1.0258E 00	1.0249	0.9904	0.9905	1.0022	1.0021	1.0152	0.9960	1.0135	1.0040
2.6	1.0205E 00	1.0210	0.9915	0.9916	1.0014	1.0013	1.0123	0.9971	1.0112	1.0028
3.0	1.0170E 00	1.0182	0.9924	0.9924	1.0009	1.0009	1.0103	0.9976	1.0096	1.0020

POSITRON Z= -75 A= 183

p	$F_0L_0^*/F_0L_0$	λ_2^*/λ_2	μ_1^*/μ_1	μ_2^*/μ_2	Λ_1^*/Λ_1	Λ_2^*/Λ_2	ν_{12}^*/ν_{12}	$\hat{\nu}_{12}^*/\hat{\nu}_{12}$	η_{12}^*/η_{12}	$\hat{\eta}_{12}^*/\hat{\eta}_{12}$
0.1	7.7494E 06	0.8369	1.0384	1.0087	2.0241	1.7750	1.1459	0.6927	2.0180	2.1226
0.2	8.9812E 01	0.9847	1.0120	0.9979	1.3823	1.3030	1.2217	0.8266	1.3966	1.4250
0.4	2.6642E 00	1.0704	0.9924	0.9888	1.1216	1.1013	1.1650	0.9242	1.1434	1.1395
0.6	1.4418E 00	1.0744	0.9866	0.9858	1.0550	1.0477	1.1032	0.9590	1.0771	1.0654
0.8	1.2001E 00	1.0643	0.9853	0.9853	1.0294	1.0262	1.0683	0.9745	1.0502	1.0364
1.0	1.1155E 00	1.0541	0.9854	0.9856	1.0175	1.0158	1.0487	0.9826	1.0366	1.0227
1.2	1.0770E 00	1.0460	0.9862	0.9864	1.0111	1.0102	1.0368	0.9874	1.0286	1.0152
1.4	1.0564E 00	1.0397	0.9870	0.9872	1.0075	1.0070	1.0292	0.9904	1.0235	1.0108
1.8	1.0360E 00	1.0310	0.9888	0.9889	1.0038	1.0036	1.0204	0.9939	1.0173	1.0063
2.2	1.0263E 00	1.0254	0.9902	0.9903	1.0022	1.0021	1.0155	0.9960	1.0137	1.0041
2.6	1.0208E 00	1.0214	0.9914	0.9915	1.0014	1.0013	1.0125	0.9971	1.0114	1.0028
3.0	1.0172E 00	1.0186	0.9923	0.9923	1.0009	1.0009	1.0105	0.9977	1.0098	1.0021

Z= -76 A= 186

p	$F_0L_0^*/F_0L_0$	λ_2^*/λ_2	μ_1^*/μ_1	μ_2^*/μ_2	Λ_1^*/Λ_1	Λ_2^*/Λ_2	ν_{12}^*/ν_{12}	$\hat{\nu}_{12}^*/\hat{\nu}_{12}$	η_{12}^*/η_{12}	$\hat{\eta}_{12}^*/\hat{\eta}_{12}$
0.1	1.0414E 07	0.8358	1.0400	1.0091	2.0377	1.7849	1.1424	0.6926	2.0308	2.1400
0.2	9.9022E 01	0.9835	1.0128	0.9980	1.3877	1.3072	1.2198	0.8262	1.4021	1.4321
0.4	2.7301E 00	1.0700	0.9925	0.9887	1.1235	1.1028	1.1658	0.9238	1.1457	1.1421
0.6	1.4554E 00	1.0748	0.9866	0.9857	1.0559	1.0484	1.1043	0.9587	1.0784	1.0667
0.8	1.2056E 00	1.0649	0.9851	0.9851	1.0299	1.0266	1.0693	0.9743	1.0511	1.0372
1.0	1.1184E 00	1.0548	0.9853	0.9854	1.0178	1.0161	1.0494	0.9825	1.0373	1.0232
1.2	1.0788E 00	1.0467	0.9860	0.9862	1.0113	1.0104	1.0374	0.9873	1.0291	1.0156
1.4	1.0576E 00	1.0403	0.9869	0.9870	1.0076	1.0071	1.0297	0.9904	1.0239	1.0111
1.8	1.0366E 00	1.0315	0.9886	0.9887	1.0039	1.0037	1.0207	0.9939	1.0176	1.0064
2.2	1.0267E 00	1.0258	0.9901	0.9902	1.0023	1.0021	1.0158	0.9959	1.0140	1.0042
2.6	1.0211E 00	1.0218	0.9912	0.9913	1.0014	1.0013	1.0128	0.9971	1.0116	1.0029
3.0	1.0175E 00	1.0189	0.9922	0.9922	1.0009	1.0009	1.0107	0.9977	1.0099	1.0022

Z= -77 A= 191

p	$F_0L_0^*/F_0L_0$	λ_2^*/λ_2	μ_1^*/μ_1	μ_2^*/μ_2	Λ_1^*/Λ_1	Λ_2^*/Λ_2	ν_{12}^*/ν_{12}	$\hat{\nu}_{12}^*/\hat{\nu}_{12}$	η_{12}^*/η_{12}	$\hat{\eta}_{12}^*/\hat{\eta}_{12}$
0.1	1.4038E 07	0.8348	1.0417	1.0095	2.0514	1.7949	1.1389	0.6923	2.0437	2.1579
0.2	1.0935E 02	0.9822	1.0137	0.9982	1.3933	1.3115	1.2178	0.8257	1.4077	1.4393
0.4	2.7991E 00	1.0697	0.9927	0.9886	1.1255	1.1044	1.1666	0.9234	1.1480	1.1448
0.6	1.4696E 00	1.0752	0.9865	0.9855	1.0569	1.0492	1.1054	0.9584	1.0798	1.0681
0.8	1.2113E 00	1.0656	0.9850	0.9849	1.0304	1.0271	1.0702	0.9742	1.0520	1.0380
1.0	1.1214E 00	1.0555	0.9851	0.9852	1.0181	1.0164	1.0501	0.9824	1.0379	1.0237
1.2	1.0806E 00	1.0473	0.9858	0.9860	1.0115	1.0106	1.0380	0.9872	1.0297	1.0160
1.4	1.0588E 00	1.0410	0.9867	0.9868	1.0078	1.0072	1.0302	0.9903	1.0244	1.0114
1.8	1.0373E 00	1.0321	0.9885	0.9886	1.0040	1.0037	1.0211	0.9938	1.0180	1.0066
2.2	1.0272E 00	1.0263	0.9899	0.9900	1.0023	1.0022	1.0161	0.9959	1.0142	1.0043
2.6	1.0215E 00	1.0222	0.9911	0.9912	1.0014	1.0014	1.0130	0.9971	1.0118	1.0030
3.0	1.0178E 00	1.0193	0.9921	0.9920	1.0010	1.0009	1.0109	0.9976	1.0101	1.0022

POSITRON Z= -78 A= 194

p	$F_0L_0^*/F_0L_0$	λ_2^*/λ_2	μ_1^*/μ_1	μ_2^*/μ_2	Λ_1^*/Λ_1	Λ_2^*/Λ_2	ν_{12}^*/ν_{12}	$\hat{\nu}_{12}^*/\hat{\nu}_{12}$	η_{12}^*/η_{12}	$\hat{\eta}_{12}^*/\hat{\eta}_{12}$
0.1	1.8940E 07	0.8339	1.0435	1.0100	2.0652	1.8050	1.1355	0.6921	2.0567	2.1759
0.2	1.2079E 02	0.9811	1.0146	0.9983	1.3988	1.3157	1.2159	0.8253	1.4132	1.4466
0.4	2.8702E 00	1.0694	0.9929	0.9886	1.1275	1.1059	1.1673	0.9230	1.1503	1.1475
0.6	1.4839E 00	1.0756	0.9864	0.9853	1.0578	1.0500	1.1065	0.9582	1.0811	1.0694
0.8	1.2170E 00	1.0663	0.9848	0.9846	1.0309	1.0275	1.0711	0.9740	1.0529	1.0388
1.0	1.1244E 00	1.0562	0.9849	0.9850	1.0184	1.0166	1.0509	0.9823	1.0386	1.0242
1.2	1.0825E 00	1.0480	0.9856	0.9858	1.0117	1.0108	1.0386	0.9872	1.0302	1.0163
1.4	1.0601E 00	1.0416	0.9865	0.9866	1.0079	1.0073	1.0307	0.9903	1.0248	1.0117
1.8	1.0380E 00	1.0326	0.9883	0.9884	1.0041	1.0038	1.0215	0.9938	1.0183	1.0068
2.2	1.0277E 00	1.0267	0.9898	0.9899	1.0023	1.0022	1.0164	0.9959	1.0145	1.0044
2.6	1.0218E 00	1.0226	0.9910	0.9911	1.0015	1.0014	1.0132	0.9971	1.0120	1.0031
3.0	1.0180E 00	1.0196	0.9920	0.9919	1.0010	1.0009	1.0111	0.9975	1.0103	1.0023

Z= -79 A= 193

p	$F_0L_0^*/F_0L_0$	λ_2^*/λ_2	μ_1^*/μ_1	μ_2^*/μ_2	Λ_1^*/Λ_1	Λ_2^*/Λ_2	ν_{12}^*/ν_{12}	$\hat{\nu}_{12}^*/\hat{\nu}_{12}$	η_{12}^*/η_{12}	$\hat{\eta}_{12}^*/\hat{\eta}_{12}$
0.1	2.5545E 07	0.8330	1.0453	1.0104	2.0789	1.8149	1.1321	0.6920	2.0695	2.1938
0.2	1.3346E 02	0.9800	1.0155	0.9985	1.4043	1.3199	1.2139	0.8250	1.4188	1.4539
0.4	2.9435E 00	1.0691	0.9931	0.9885	1.1294	1.1075	1.1679	0.9226	1.1526	1.1502
0.6	1.4985E 00	1.0760	0.9864	0.9852	1.0587	1.0507	1.1076	0.9580	1.0825	1.0708
0.8	1.2228E 00	1.0669	0.9847	0.9844	1.0315	1.0279	1.0720	0.9739	1.0538	1.0396
1.0	1.1275E 00	1.0569	0.9847	0.9848	1.0187	1.0169	1.0516	0.9822	1.0393	1.0248
1.2	1.0843E 00	1.0487	0.9854	0.9856	1.0119	1.0109	1.0392	0.9871	1.0308	1.0167
1.4	1.0613E 00	1.0423	0.9863	0.9864	1.0080	1.0074	1.0312	0.9902	1.0253	1.0119
1.8	1.0387E 00	1.0332	0.9881	0.9882	1.0041	1.0039	1.0218	0.9938	1.0186	1.0070
2.2	1.0281E 00	1.0272	0.9896	0.9897	1.0024	1.0022	1.0167	0.9959	1.0148	1.0045
2.6	1.0221E 00	1.0230	0.9908	0.9909	1.0015	1.0014	1.0135	0.9970	1.0123	1.0032
3.0	1.0183E 00	1.0200	0.9919	0.9918	1.0010	1.0009	1.0113	0.9976	1.0105	1.0024

Z= -80 A= 200

p	$F_0L_0^*/F_0L_0$	λ_2^*/λ_2	μ_1^*/μ_1	μ_2^*/μ_2	Λ_1^*/Λ_1	Λ_2^*/Λ_2	ν_{12}^*/ν_{12}	$\hat{\nu}_{12}^*/\hat{\nu}_{12}$	η_{12}^*/η_{12}	$\hat{\eta}_{12}^*/\hat{\eta}_{12}$
0.1	3.4511E 07	0.8322	1.0471	1.0108	2.0927	1.8248	1.1289	0.6919	2.0823	2.2120
0.2	1.4758E 02	0.9789	1.0165	0.9987	1.4098	1.3241	1.2119	0.8246	1.4243	1.4612
0.4	3.0196E 00	1.0688	0.9934	0.9884	1.1314	1.1090	1.1685	0.9222	1.1549	1.1529
0.6	1.5134E 00	1.0764	0.9864	0.9850	1.0597	1.0515	1.1086	0.9577	1.0838	1.0722
0.8	1.2287E 00	1.0675	0.9845	0.9842	1.0320	1.0283	1.0729	0.9737	1.0548	1.0405
1.0	1.1306E 00	1.0576	0.9846	0.9846	1.0190	1.0172	1.0524	0.9821	1.0400	1.0253
1.2	1.0862E 00	1.0494	0.9852	0.9853	1.0121	1.0111	1.0398	0.9871	1.0313	1.0171
1.4	1.0626E 00	1.0429	0.9861	0.9862	1.0082	1.0076	1.0317	0.9902	1.0257	1.0122
1.8	1.0394E 00	1.0337	0.9880	0.9880	1.0042	1.0039	1.0222	0.9938	1.0190	1.0071
2.2	1.0286E 00	1.0276	0.9895	0.9896	1.0024	1.0023	1.0170	0.9959	1.0151	1.0047
2.6	1.0225E 00	1.0234	0.9907	0.9908	1.0015	1.0014	1.0137	0.9970	1.0125	1.0033
3.0	1.0185E 00	1.0203	0.9917	0.9916	1.0010	1.0010	1.0115	0.9976	1.0107	1.0025
4.0	1.0130E 00	1.0152	0.9936	0.9935	1.0004	1.0004	1.0083	0.9985	1.0079	1.0015

POSITRON Z= −81 A= 201

p	$F_0L_0^*/F_0L_0$	λ_2^*/λ_2	μ_1^*/μ_1	μ_2^*/μ_2	Λ_1^*/Λ_1	Λ_2^*/Λ_2	ν_{12}^*/ν_{12}	$\vartheta_{12}^*/\vartheta_{12}$	η_{12}^*/η_{12}	$\hat{\eta}_{12}^*/\hat{\eta}_{12}$
0.1	4.6620E 07	0.8314	1.0490	1.0113	2.1063	1.8346	1.1256	0.6918	2.0950	2.2303
0.2	1.6323E 02	0.9779	1.0175	0.9989	1.4153	1.3283	1.2098	0.8243	1.4298	1.4687
0.4	3.0982E 00	1.0686	0.9936	0.9884	1.1333	1.1106	1.1691	0.9219	1.1573	1.1557
0.6	1.5286E 00	1.0768	0.9864	0.9848	1.0606	1.0522	1.1096	0.9575	1.0852	1.0736
0.8	1.2347E 00	1.0682	0.9844	0.9840	1.0325	1.0288	1.0738	0.9736	1.0557	1.0413
1.0	1.1337E 00	1.0583	0.9844	0.9844	1.0193	1.0174	1.0531	0.9820	1.0407	1.0258
1.2	1.0881E 00	1.0501	0.9851	0.9851	1.0123	1.0113	1.0404	0.9870	1.0319	1.0175
1.4	1.0639E 00	1.0435	0.9860	0.9861	1.0083	1.0077	1.0322	0.9901	1.0262	1.0125
1.8	1.0401E 00	1.0343	0.9878	0.9878	1.0043	1.0040	1.0226	0.9938	1.0193	1.0073
2.2	1.0290E 00	1.0281	0.9893	0.9894	1.0025	1.0023	1.0172	0.9959	1.0153	1.0048
2.6	1.0228E 00	1.0238	0.9906	0.9906	1.0015	1.0014	1.0139	0.9971	1.0127	1.0034
3.0	1.0188E 00	1.0207	0.9916	0.9915	1.0010	1.0010	1.0117	0.9976	1.0109	1.0025
4.0	1.0132E 00	1.0155	0.9935	0.9934	1.0004	1.0004	1.0084	0.9985	1.0080	1.0016

Z= −82 A= 203

p	$F_0L_0^*/F_0L_0$	λ_2^*/λ_2	μ_1^*/μ_1	μ_2^*/μ_2	Λ_1^*/Λ_1	Λ_2^*/Λ_2	ν_{12}^*/ν_{12}	$\vartheta_{12}^*/\vartheta_{12}$	η_{12}^*/η_{12}	$\hat{\eta}_{12}^*/\hat{\eta}_{12}$
0.1	6.3157E 07	0.8306	1.0510	1.0118	2.1202	1.8447	1.1224	0.6918	2.1080	2.2490
0.2	1.8076E 02	0.9769	1.0185	0.9991	1.4208	1.3326	1.2077	0.8239	1.4354	1.4762
0.4	3.1800E 00	1.0683	0.9939	0.9883	1.1353	1.1121	1.1696	0.9216	1.1596	1.1585
0.6	1.5442E 00	1.0771	0.9863	0.9847	1.0615	1.0530	1.1106	0.9573	1.0865	1.0750
0.8	1.2409E 00	1.0688	0.9843	0.9838	1.0330	1.0292	1.0747	0.9734	1.0566	1.0421
1.0	1.1369E 00	1.0590	0.9842	0.9842	1.0196	1.0177	1.0538	0.9819	1.0414	1.0264
1.2	1.0900E 00	1.0508	0.9849	0.9849	1.0125	1.0115	1.0410	0.9869	1.0324	1.0179
1.4	1.0652E 00	1.0442	0.9858	0.9859	1.0084	1.0078	1.0327	0.9901	1.0267	1.0128
1.8	1.0408E 00	1.0348	0.9876	0.9877	1.0043	1.0041	1.0230	0.9938	1.0197	1.0075
2.2	1.0295E 00	1.0286	0.9892	0.9892	1.0025	1.0023	1.0176	0.9959	1.0156	1.0049
2.6	1.0231E 00	1.0242	0.9904	0.9905	1.0016	1.0015	1.0142	0.9971	1.0130	1.0035
3.0	1.0190E 00	1.0211	0.9915	0.9914	1.0010	1.0010	1.0119	0.9976	1.0111	1.0026
4.0	1.0134E 00	1.0158	0.9934	0.9933	1.0005	1.0004	1.0086	0.9986	1.0082	1.0016

Z= −83 A= 207

p	$F_0L_0^*/F_0L_0$	λ_2^*/λ_2	μ_1^*/μ_1	μ_2^*/μ_2	Λ_1^*/Λ_1	Λ_2^*/Λ_2	ν_{12}^*/ν_{12}	$\vartheta_{12}^*/\vartheta_{12}$	η_{12}^*/η_{12}	$\hat{\eta}_{12}^*/\hat{\eta}_{12}$
0.1	8.5615E 07	0.8299	1.0531	1.0122	2.1341	1.8546	1.1193	0.6917	2.1209	2.2678
0.2	2.0029E 02	0.9760	1.0196	0.9993	1.4264	1.3368	1.2057	0.8236	1.4409	1.4838
0.4	3.2648E 00	1.0680	0.9942	0.9883	1.1373	1.1137	1.1701	0.9212	1.1620	1.1614
0.6	1.5602E 00	1.0775	0.9863	0.9845	1.0625	1.0538	1.1116	0.9571	1.0879	1.0765
0.8	1.2471E 00	1.0695	0.9842	0.9837	1.0335	1.0297	1.0756	0.9733	1.0576	1.0430
1.0	1.1401E 00	1.0598	0.9841	0.9840	1.0199	1.0180	1.0546	0.9818	1.0421	1.0270
1.2	1.0920E 00	1.0515	0.9847	0.9847	1.0127	1.0116	1.0416	0.9869	1.0330	1.0183
1.4	1.0665E 00	1.0448	0.9856	0.9857	1.0086	1.0079	1.0332	0.9901	1.0271	1.0131
1.8	1.0415E 00	1.0354	0.9875	0.9875	1.0044	1.0041	1.0233	0.9938	1.0200	1.0077
2.2	1.0299E 00	1.0291	0.9890	0.9891	1.0025	1.0024	1.0179	0.9959	1.0159	1.0051
2.6	1.0234E 00	1.0246	0.9903	0.9903	1.0016	1.0015	1.0144	0.9970	1.0132	1.0036
3.0	1.0193E 00	1.0214	0.9914	0.9913	1.0011	1.0010	1.0122	0.9977	1.0113	1.0027
4.0	1.0135E 00	1.0161	0.9933	0.9932	1.0005	1.0004	1.0087	0.9986	1.0083	1.0017

POSITRON Z= -84 A= 208

p	$F_0L_0^*/F_0L_0$	λ_2^*/λ_2	μ_1^*/μ_1	μ_2^*/μ_2	Λ_1^*/Λ_1	Λ_2^*/Λ_2	ν_{12}^*/ν_{12}	$\hat{\nu}_{12}^*/\hat{\nu}_{12}$	η_{12}^*/η_{12}	$\hat{\eta}_{12}^*/\hat{\eta}_{12}$
0.1	1.1610E 08	0.8293	1.0552	1.0127	2.1480	1.8645	1.1161	0.6917	2.1336	2.2867
0.2	2.2200E 02	0.9750	1.0207	0.9995	1.4319	1.3410	1.2036	0.8234	1.4465	1.4915
0.4	3.3525E 00	1.0678	0.9945	0.9882	1.1393	1.1153	1.1705	0.9209	1.1643	1.1643
0.6	1.5765E 00	1.0779	0.9863	0.9844	1.0635	1.0546	1.1126	0.9569	1.0893	1.0780
0.8	1.2534E 00	1.0701	0.9841	0.9835	1.0341	1.0301	1.0765	0.9732	1.0585	1.0439
1.0	1.1434E 00	1.0605	0.9839	0.9837	1.0202	1.0182	1.0553	0.9818	1.0428	1.0276
1.2	1.0940E 00	1.0522	0.9845	0.9845	1.0129	1.0118	1.0423	0.9868	1.0336	1.0187
1.4	1.0678E 00	1.0455	0.9854	0.9854	1.0087	1.0080	1.0337	0.9900	1.0276	1.0134
1.8	1.0422E 00	1.0360	0.9873	0.9873	1.0045	1.0042	1.0237	0.9938	1.0204	1.0079
2.2	1.0304E 00	1.0296	0.9889	0.9889	1.0026	1.0024	1.0182	0.9959	1.0162	1.0052
2.6	1.0238E 00	1.0251	0.9902	0.9902	1.0016	1.0015	1.0147	0.9970	1.0134	1.0037
3.0	1.0196E 00	1.0218	0.9913	0.9911	1.0011	1.0010	1.0124	0.9976	1.0115	1.0028
4.0	1.0137E 00	1.0164	0.9933	0.9931	1.0005	1.0004	1.0089	0.9986	1.0085	1.0017

Z= -85 A= 211

p	$F_0L_0^*/F_0L_0$	λ_2^*/λ_2	μ_1^*/μ_1	μ_2^*/μ_2	Λ_1^*/Λ_1	Λ_2^*/Λ_2	ν_{12}^*/ν_{12}	$\hat{\nu}_{12}^*/\hat{\nu}_{12}$	η_{12}^*/η_{12}	$\hat{\eta}_{12}^*/\hat{\eta}_{12}$
0.1	1.5749E 08	0.8287	1.0573	1.0132	2.1619	1.8745	1.1131	0.6918	2.1465	2.3058
0.2	2.4609E 02	0.9742	1.0219	0.9997	1.4375	1.3453	1.2015	0.8231	1.4520	1.4992
0.4	3.4426E 00	1.0675	0.9948	0.9882	1.1413	1.1168	1.1709	0.9206	1.1667	1.1671
0.6	1.5929E 00	1.0783	0.9864	0.9842	1.0644	1.0553	1.1135	0.9567	1.0907	1.0794
0.8	1.2598E 00	1.0707	0.9840	0.9833	1.0346	1.0305	1.0774	0.9731	1.0595	1.0448
1.0	1.1467E 00	1.0612	0.9838	0.9835	1.0205	1.0185	1.0560	0.9817	1.0435	1.0282
1.2	1.0959E 00	1.0528	0.9844	0.9843	1.0131	1.0120	1.0429	0.9868	1.0342	1.0191
1.4	1.0691E 00	1.0461	0.9853	0.9853	1.0089	1.0082	1.0342	0.9900	1.0281	1.0138
1.8	1.0429E 00	1.0365	0.9871	0.9871	1.0045	1.0042	1.0241	0.9938	1.0207	1.0081
2.2	1.0308E 00	1.0300	0.9887	0.9888	1.0026	1.0025	1.0185	0.9959	1.0165	1.0054
2.6	1.0241E 00	1.0255	0.9901	0.9901	1.0016	1.0015	1.0149	0.9970	1.0137	1.0038
3.0	1.0198E 00	1.0222	0.9911	0.9910	1.0011	1.0010	1.0126	0.9976	1.0117	1.0029
4.0	1.0139E 00	1.0166	0.9931	0.9930	1.0005	1.0005	1.0090	0.9986	1.0086	1.0018

Z= -86 A= 212

p	$F_0L_0^*/F_0L_0$	λ_2^*/λ_2	μ_1^*/μ_1	μ_2^*/μ_2	Λ_1^*/Λ_1	Λ_2^*/Λ_2	ν_{12}^*/ν_{12}	$\hat{\nu}_{12}^*/\hat{\nu}_{12}$	η_{12}^*/η_{12}	$\hat{\eta}_{12}^*/\hat{\eta}_{12}$
0.1	2.1404E 08	0.8282	1.0596	1.0137	2.1758	1.8844	1.1101	0.6918	2.1593	2.3252
0.2	2.7313E 02	0.9734	1.0231	0.9999	1.4431	1.3495	1.1993	0.8229	1.4576	1.5071
0.4	3.5369E 00	1.0673	0.9952	0.9881	1.1433	1.1184	1.1713	0.9204	1.1690	1.1701
0.6	1.6099E 00	1.0787	0.9864	0.9841	1.0654	1.0561	1.1144	0.9565	1.0921	1.0809
0.8	1.2663E 00	1.0714	0.9839	0.9831	1.0351	1.0310	1.0783	0.9730	1.0604	1.0457
1.0	1.1500E 00	1.0619	0.9836	0.9833	1.0209	1.0188	1.0568	0.9816	1.0443	1.0288
1.2	1.0980E 00	1.0536	0.9842	0.9841	1.0133	1.0122	1.0435	0.9867	1.0347	1.0195
1.4	1.0704E 00	1.0468	0.9851	0.9851	1.0090	1.0083	1.0348	0.9900	1.0286	1.0141
1.8	1.0436E 00	1.0371	0.9870	0.9869	1.0046	1.0043	1.0245	0.9938	1.0211	1.0083
2.2	1.0313E 00	1.0305	0.9886	0.9886	1.0027	1.0025	1.0188	0.9959	1.0168	1.0055
2.6	1.0244E 00	1.0259	0.9900	0.9899	1.0017	1.0016	1.0152	0.9970	1.0139	1.0039
3.0	1.0200E 00	1.0225	0.9910	0.9909	1.0011	1.0011	1.0128	0.9977	1.0119	1.0030
4.0	1.0140E 00	1.0169	0.9931	0.9929	1.0005	1.0005	1.0092	0.9986	1.0088	1.0018

POSITRON Z= -87 A= 220

p	$F_0L_0^*/F_0L_0$	λ_2^*/λ_2	μ_1^*/μ_1	μ_2^*/μ_2	Λ_1^*/Λ_1	Λ_2^*/Λ_2	ν_{12}^*/ν_{12}	$\vartheta_{12}^*/\vartheta_{12}$	η_{12}^*/η_{12}	$\hat{\eta}_{12}^*/\hat{\eta}_{12}$
0.1	2.9118E 08	0.8276	1.0619	1.0142	2.1898	1.8943	1.1071	0.6918	2.1722	2.3448
0.2	3.0332E 02	0.9726	1.0243	1.0001	1.4487	1.3537	1.1972	0.8227	1.4632	1.5150
0.4	3.6346E 00	1.0671	0.9956	0.9881	1.1454	1.1200	1.1716	0.9201	1.1714	1.1731
0.6	1.6273E 00	1.0790	0.9864	0.9839	1.0664	1.0569	1.1153	0.9563	1.0935	1.0825
0.8	1.2730E 00	1.0720	0.9838	0.9829	1.0356	1.0314	1.0791	0.9728	1.0614	1.0466
1.0	1.1534E 00	1.0626	0.9835	0.9831	1.0212	1.0191	1.0575	0.9816	1.0450	1.0294
1.2	1.1000E 00	1.0543	0.9840	0.9839	1.0135	1.0123	1.0441	0.9867	1.0353	1.0200
1.4	1.0718E 00	1.0475	0.9849	0.9848	1.0091	1.0084	1.0353	0.9900	1.0291	1.0144
1.8	1.0444E 00	1.0377	0.9868	0.9868	1.0047	1.0044	1.0249	0.9937	1.0215	1.0085
2.2	1.0318E 00	1.0310	0.9884	0.9884	1.0027	1.0025	1.0191	0.9959	1.0171	1.0057
2.6	1.0247E 00	1.0263	0.9898	0.9898	1.0017	1.0016	1.0155	0.9971	1.0142	1.0041
3.0	1.0203E 00	1.0229	0.9909	0.9907	1.0011	1.0011	1.0130	0.9976	1.0121	1.0031
4.0	1.0142E 00	1.0172	0.9930	0.9928	1.0005	1.0005	1.0093	0.9987	1.0090	1.0019

Z= -88 A= 224

p	$F_0L_0^*/F_0L_0$	λ_2^*/λ_2	μ_1^*/μ_1	μ_2^*/μ_2	Λ_1^*/Λ_1	Λ_2^*/Λ_2	ν_{12}^*/ν_{12}	$\vartheta_{12}^*/\vartheta_{12}$	η_{12}^*/η_{12}	$\hat{\eta}_{12}^*/\hat{\eta}_{12}$
0.1	3.9573E 08	0.8272	1.0643	1.0148	2.2037	1.9041	1.1042	0.6920	2.1848	2.3644
0.2	3.3671E 02	0.9718	1.0256	1.0004	1.4542	1.3579	1.1951	0.8225	1.4687	1.5229
0.4	3.7347E 0C	1.0668	0.9960	0.9881	1.1474	1.1215	1.1718	0.9198	1.1738	1.1761
0.6	1.6447E 00	1.0794	0.9865	0.9836	1.0673	1.0577	1.1162	0.9561	1.0949	1.0840
0.8	1.2797E 00	1.0727	0.9837	0.9827	1.0362	1.0319	1.0800	0.9727	1.0624	1.0475
1.0	1.1568E 00	1.0633	0.9833	0.9829	1.0215	1.0193	1.0582	0.9815	1.0457	1.0300
1.2	1.1020E 00	1.0550	0.9839	0.9837	1.0138	1.0125	1.0447	0.9867	1.0359	1.0204
1.4	1.0731E 00	1.0482	0.9848	0.9847	1.0093	1.0085	1.0358	0.9899	1.0295	1.0147
1.8	1.0451E 00	1.0382	0.9867	0.9866	1.0048	1.0044	1.0253	0.9937	1.0219	1.0087
2.2	1.0322E 00	1.0315	0.9883	0.9883	1.0028	1.0026	1.0194	0.9959	1.0174	1.0058
2.6	1.0251E 00	1.0268	0.9897	0.9896	1.0017	1.0016	1.0157	0.9971	1.0144	1.0042
3.0	1.0205E C0	1.0233	0.9908	0.9906	1.0011	1.0011	1.0132	0.9977	1.0124	1.0032
4.0	1.0143E 00	1.0175	0.9929	0.9927	1.0005	1.0005	1.0095	0.9986	1.0091	1.0020

Z= -89 A= 226

p	$F_0L_0^*/F_0L_0$	λ_2^*/λ_2	μ_1^*/μ_1	μ_2^*/μ_2	Λ_1^*/Λ_1	Λ_2^*/Λ_2	ν_{12}^*/ν_{12}	$\vartheta_{12}^*/\vartheta_{12}$	η_{12}^*/η_{12}	$\hat{\eta}_{12}^*/\hat{\eta}_{12}$
0.1	5.3976E 08	0.8268	1.0668	1.0153	2.2179	1.9141	1.1013	0.6921	2.1978	2.3846
0.2	3.7445E 02	0.9711	1.0270	1.0006	1.4599	1.3622	1.1929	0.8223	1.4743	1.5311
0.4	3.8400E 00	1.0667	0.9964	0.988C	1.1494	1.1231	1.1721	0.9196	1.1762	1.1792
0.6	1.6629E 00	1.0798	0.9865	0.9836	1.0683	1.0585	1.1171	0.9560	1.0963	1.0856
0.8	1.2865E 00	1.0733	0.9836	0.9825	1.0367	1.0323	1.0809	0.9726	1.0634	1.0485
1.0	1.1604E 00	1.0640	0.9832	0.9827	1.0218	1.0196	1.0590	0.9814	1.0465	1.0306
1.2	1.1041E 00	1.0557	0.9837	0.9835	1.0140	1.0127	1.0453	0.9866	1.0365	1.0209
1.4	1.0745E 00	1.0488	0.9846	0.9845	1.0094	1.0087	1.0363	0.9899	1.0301	1.0151
1.8	1.0458E 00	1.0388	0.9865	0.9864	1.0048	1.0045	1.0257	0.9937	1.0222	1.0089
2.2	1.0327E 00	1.0321	0.9882	0.9881	1.0028	1.0026	1.0197	0.9959	1.0177	1.0060
2.6	1.0254E 00	1.0272	0.9895	0.9895	1.0017	1.0016	1.0160	0.9971	1.0147	1.0043
3.0	1.0208E 00	1.0237	0.9907	0.9905	1.0012	1.0011	1.0135	0.9977	1.0126	1.0033
4.0	1.0145E 00	1.0178	0.9928	0.9926	1.0005	1.0005	1.0097	0.9986	1.0093	1.0020

Instr. see p. 25

POSITRON Z= -90 A= 228

p	$F_0L_0^*/F_0L_0$	λ_2^*/λ_2	μ_1^*/μ_1	μ_2^*/μ_2	Λ_1^*/Λ_1	Λ_2^*/Λ_2	ν_{12}^*/ν_{12}	$\vartheta_{12}^*/\vartheta_{12}$	η_{12}^*/η_{12}	$\hat{\eta}_{12}^*/\hat{\eta}_{12}$
0.1	7.3498E 08	0.8264	1.0693	1.0158	2.2318	1.9239	1.0985	0.6923	2.2105	2.4048
0.2	4.1618E 02	0.9705	1.0284	1.0009	1.4655	1.3664	1.1908	0.8222	1.4798	1.5392
0.4	3.9476E 00	1.0665	0.9968	0.9880	1.1514	1.1247	1.1723	0.9194	1.1786	1.1822
0.6	1.6811E 00	1.0802	0.9866	0.9835	1.0693	1.0593	1.1179	0.9558	1.0978	1.0872
0.8	1.2934E 00	1.0740	0.9835	0.9823	1.0372	1.0328	1.0817	0.9726	1.0644	1.0495
1.0	1.1638E 00	1.0647	0.9831	0.9825	1.0222	1.0199	1.0597	0.9814	1.0472	1.0313
1.2	1.1062E 00	1.0564	0.9836	0.9833	1.0142	1.0129	1.0459	0.9866	1.0371	1.0214
1.4	1.0759E 00	1.0495	0.9844	0.9843	1.0096	1.0088	1.0369	0.9899	1.0306	1.0155
1.8	1.0465E 00	1.0394	0.9864	0.9862	1.0049	1.0046	1.0261	0.9937	1.0226	1.0092
2.2	1.0331E 00	1.0326	0.9880	0.9880	1.0028	1.0026	1.0200	0.9959	1.0180	1.0061
2.6	1.0257E 00	1.0277	0.9895	0.9894	1.0018	1.0017	1.0162	0.9971	1.0149	1.0044
3.0	1.0210E 00	1.0241	0.9906	0.9904	1.0012	1.0011	1.0137	0.9977	1.0128	1.0033
4.0	1.0146E 00	1.0181	0.9928	0.9925	1.0005	1.0005	1.0098	0.9985	1.0094	1.0021

Z= -91 A= 229

p	$F_0L_0^*/F_0L_0$	λ_2^*/λ_2	μ_1^*/μ_1	μ_2^*/μ_2	Λ_1^*/Λ_1	Λ_2^*/Λ_2	ν_{12}^*/ν_{12}	$\vartheta_{12}^*/\vartheta_{12}$	η_{12}^*/η_{12}	$\hat{\eta}_{12}^*/\hat{\eta}_{12}$
0.1	1.0045E 09	0.8261	1.0720	1.0164	2.2462	1.9339	1.0957	0.6924	2.2236	2.4256
0.2	4.6338E 02	0.9698	1.0299	1.0011	1.4712	1.3707	1.1886	0.8221	1.4855	1.5476
0.4	4.0609E 00	1.0663	0.9973	0.9880	1.1535	1.1263	1.1724	0.9191	1.1810	1.1854
0.6	1.7000E 00	1.0806	0.9867	0.9833	1.0703	1.0601	1.1188	0.9557	1.0992	1.0889
0.8	1.3005E 00	1.0746	0.9835	0.9821	1.0378	1.0332	1.0825	0.9725	1.0654	1.0504
1.0	1.1674E 00	1.0655	0.9829	0.9823	1.0225	1.0202	1.0604	0.9814	1.0480	1.0319
1.2	1.1083E 00	1.0571	0.9834	0.9831	1.0144	1.0131	1.0466	0.9866	1.0377	1.0218
1.4	1.0773E 00	1.0502	0.9843	0.9840	1.0097	1.0089	1.0374	0.9899	1.0311	1.0158
1.8	1.0473E 00	1.0400	0.9862	0.9860	1.0050	1.0046	1.0265	0.9937	1.0230	1.0094
2.2	1.0336E 00	1.0331	0.9879	0.9878	1.0029	1.0027	1.0204	0.9959	1.0183	1.0063
2.6	1.0260E 00	1.0281	0.9893	0.9892	1.0018	1.0017	1.0165	0.9971	1.0152	1.0046
3.0	1.0213E 00	1.0245	0.9905	0.9902	1.0012	1.0011	1.0139	0.9977	1.0130	1.0035
4.0	1.0148E 00	1.0184	0.9926	0.9923	1.0005	1.0005	1.0100	0.9986	1.0096	1.0022

Z= -92 A= 231

p	$F_0L_0^*/F_0L_0$	λ_2^*/λ_2	μ_1^*/μ_1	μ_2^*/μ_2	Λ_1^*/Λ_1	Λ_2^*/Λ_2	ν_{12}^*/ν_{12}	$\vartheta_{12}^*/\vartheta_{12}$	η_{12}^*/η_{12}	$\hat{\eta}_{12}^*/\hat{\eta}_{12}$
0.1	1.3693E 09	0.8258	1.0747	1.0170	2.2601	1.9437	1.0929	0.6927	2.2362	2.4461
0.2	5.1545E 02	0.9693	1.0314	1.0014	1.4768	1.3748	1.1864	0.8221	1.4910	1.5559
0.4	4.1764E 00	1.0662	0.9978	0.9880	1.1555	1.1278	1.1725	0.9190	1.1834	1.1886
0.6	1.7190E 00	1.0810	0.9868	0.9832	1.0713	1.0608	1.1196	0.9556	1.1007	1.0905
0.8	1.3076E 00	1.0753	0.9834	0.9819	1.0383	1.0337	1.0834	0.9724	1.0664	1.0514
1.0	1.1710E 00	1.0662	0.9828	0.9821	1.0228	1.0204	1.0612	0.9813	1.0488	1.0326
1.2	1.1104E 00	1.0579	0.9833	0.9829	1.0146	1.0132	1.0472	0.9866	1.0384	1.0223
1.4	1.0786E 00	1.0509	0.9841	0.9839	1.0099	1.0090	1.0379	0.9899	1.0316	1.0162
1.8	1.0480E 00	1.0406	0.9861	0.9859	1.0051	1.0047	1.0269	0.9938	1.0234	1.0097
2.2	1.0340E 00	1.0336	0.9877	0.9876	1.0029	1.0027	1.0207	0.9960	1.0186	1.0065
2.6	1.0263E 00	1.0286	0.9892	0.9891	1.0018	1.0017	1.0168	0.9971	1.0154	1.0047
3.0	1.0215E 00	1.0249	0.9904	0.9901	1.0012	1.0012	1.0141	0.9977	1.0132	1.0036
4.0	1.0149E 00	1.0187	0.9925	0.9922	1.0005	1.0005	1.0102	0.9986	1.0098	1.0022

 Instr. see p. 25

POSITRON Z= -93 A= 235

p	$F_0L_0^*/F_0L_0$	λ_2^*/λ_2	μ_1^*/μ_1	μ_2^*/μ_2	Λ_1^*/Λ_1	Λ_2^*/Λ_2	ν_{12}^*/ν_{12}	$\vartheta_{12}^*/\vartheta_{12}$	η_{12}^*/η_{12}	$\hat{\eta}_{12}^*/\hat{\eta}_{12}$
0.1	1.8710E 09	0.8256	1.0776	1.0176	2.2744	1.9536	1.0903	0.6929	2.2490	2.4672
0.2	5.7398E 02	0.9687	1.0329	1.0016	1.4825	1.3791	1.1843	0.8220	1.4966	1.5644
0.4	4.2967E 00	1.0660	0.9983	0.9879	1.1576	1.1294	1.1726	0.9188	1.1858	1.1918
0.6	1.7384E 00	1.0814	0.9870	0.9831	1.0723	1.0616	1.1203	0.9555	1.1021	1.0922
0.8	1.3148E 00	1.0759	0.9834	0.9817	1.0389	1.0341	1.0842	0.9723	1.0674	1.0525
1.0	1.1746E 00	1.0669	0.9827	0.9819	1.0232	1.0207	1.0619	0.9813	1.0495	1.0333
1.2	1.1126E 00	1.0586	0.9831	0.9827	1.0148	1.0134	1.0478	0.9866	1.0390	1.0228
1.4	1.0800E 00	1.0516	0.9840	0.9837	1.0100	1.0092	1.0385	0.9899	1.0321	1.0166
1.8	1.0487E 00	1.0412	0.9859	0.9857	1.0051	1.0048	1.0273	0.9938	1.0238	1.0099
2.2	1.0345E 00	1.0341	0.9876	0.9875	1.0030	1.0028	1.0210	0.9960	1.0189	1.0067
2.6	1.0266E 00	1.0290	0.9891	0.9889	1.0019	1.0017	1.0170	0.9971	1.0157	1.0048
3.0	1.0217E 00	1.0253	0.9902	0.9900	1.0012	1.0012	1.0144	0.9978	1.0135	1.0037
4.0	1.0151E 00	1.0190	0.9925	0.9922	1.0005	1.0005	1.0103	0.9986	1.0099	1.0023

Z= -94 A= 238

p	$F_0L_0^*/F_0L_0$	λ_2^*/λ_2	μ_1^*/μ_1	μ_2^*/μ_2	Λ_1^*/Λ_1	Λ_2^*/Λ_2	ν_{12}^*/ν_{12}	$\vartheta_{12}^*/\vartheta_{12}$	η_{12}^*/η_{12}	$\hat{\eta}_{12}^*/\hat{\eta}_{12}$
0.1	2.5553E 09	0.8254	1.0805	1.0181	2.2885	1.9633	1.0876	0.6932	2.2617	2.4883
0.2	6.3921E 02	0.9682	1.0346	1.0019	1.4881	1.3832	1.1821	0.8220	1.5021	1.5729
0.4	4.4210E 00	1.0659	0.9989	0.9879	1.1596	1.1310	1.1726	0.9186	1.1883	1.1951
0.6	1.7582E 00	1.0818	0.9871	0.9829	1.0733	1.0624	1.1211	0.9554	1.1036	1.0939
0.8	1.3221E 00	1.0766	0.9834	0.9815	1.0394	1.0346	1.0850	0.9723	1.0684	1.0535
1.0	1.1783E 00	1.0677	0.9826	0.9817	1.0235	1.0210	1.0626	0.9813	1.0503	1.0340
1.2	1.1147E 00	1.0593	0.9830	0.9825	1.0150	1.0136	1.0484	0.9866	1.0396	1.0233
1.4	1.0814E 00	1.0523	0.9838	0.9835	1.0101	1.0093	1.0390	0.9899	1.0326	1.0169
1.8	1.0494E 00	1.0418	0.9858	0.9855	1.0052	1.0048	1.0277	0.9938	1.0242	1.0101
2.2	1.0349E 00	1.0347	0.9875	0.9873	1.0030	1.0028	1.0213	0.9960	1.0192	1.0068
2.6	1.0269E 00	1.0295	0.9889	0.9888	1.0019	1.0018	1.0173	0.9971	1.0160	1.0050
3.0	1.0220E 00	1.0257	0.9901	0.9899	1.0013	1.0012	1.0146	0.9978	1.0137	1.0038
4.0	1.0152E 00	1.0193	0.9924	0.9920	1.0005	1.0005	1.0105	0.9985	1.0101	1.0024

Z= -95 A= 239

p	$F_0L_0^*/F_0L_0$	λ_2^*/λ_2	μ_1^*/μ_1	μ_2^*/μ_2	Λ_1^*/Λ_1	Λ_2^*/Λ_2	ν_{12}^*/ν_{12}	$\vartheta_{12}^*/\vartheta_{12}$	η_{12}^*/η_{12}	$\hat{\eta}_{12}^*/\hat{\eta}_{12}$
0.1	3.5078E 09	0.8253	1.0836	1.0187	2.3031	1.9734	1.0850	0.6935	2.2749	2.5104
0.2	7.1349E 02	0.9678	1.0363	1.0022	1.4940	1.3876	1.1799	0.8220	1.5079	1.5818
0.4	4.5523E 00	1.0658	0.9995	0.9879	1.1617	1.1326	1.1726	0.9185	1.1908	1.1984
0.6	1.7788E 00	1.0822	0.9872	0.9828	1.0743	1.0632	1.1219	0.9552	1.1051	1.0956
0.8	1.3297E 00	1.0773	0.9833	0.9814	1.0400	1.0350	1.0858	0.9722	1.0694	1.0546
1.0	1.1821E 00	1.0685	0.9825	0.9815	1.0238	1.0213	1.0633	0.9812	1.0511	1.0347
1.2	1.1170E 00	1.0601	0.9829	0.9823	1.0152	1.0138	1.0490	0.9866	1.0402	1.0238
1.4	1.0829E 00	1.0530	0.9837	0.9833	1.0103	1.0094	1.0395	0.9900	1.0332	1.0173
1.8	1.0502E 00	1.0425	0.9856	0.9853	1.0053	1.0049	1.0281	0.9938	1.0246	1.0104
2.2	1.0354E 00	1.0352	0.9873	0.9871	1.0031	1.0028	1.0217	0.9960	1.0195	1.0070
2.6	1.0272E 00	1.0300	0.9888	0.9886	1.0019	1.0018	1.0176	0.9972	1.0162	1.0052
3.0	1.0222E 00	1.0261	0.9900	0.9897	1.0013	1.0012	1.0149	0.9979	1.0139	1.0039
4.0	1.0153E 00	1.0197	0.9923	0.9919	1.0006	1.0005	1.0107	0.9986	1.0103	1.0025

Instr. see p. 25

POSITRON Z= -96 A= 243

p	$F_0L_0^*/F_0L_0$	λ_2^*/λ_2	μ_1^*/μ_1	μ_2^*/μ_2	Λ_1^*/Λ_1	Λ_2^*/Λ_2	ν_{12}^*/ν_{12}	$\hat{\nu}_{12}^*/\hat{\nu}_{12}$	η_{12}^*/η_{12}	$\hat{\eta}_{12}^*/\hat{\eta}_{12}$
0.1	4.7992E 09	0.8252	1.0867	1.0194	2.3174	1.9832	1.0824	0.6938	2.2876	2.5322
0.2	7.9544E 02	0.9674	1.0380	1.0025	1.4997	1.3918	1.1777	0.8220	1.5135	1.5906
0.4	4.6861E 00	1.0658	1.0001	0.9879	1.1638	1.1342	1.1726	0.9183	1.1932	1.2018
0.6	1.7994E 00	1.0827	0.9874	0.9827	1.0753	1.0640	1.1226	0.9552	1.1066	1.0974
0.8	1.3372E 00	1.0780	0.9833	0.9812	1.0406	1.0355	1.0866	0.9722	1.0705	1.0556
1.0	1.1859E 00	1.0692	0.9824	0.9813	1.0242	1.0215	1.0641	0.9812	1.0519	1.0354
1.2	1.1192E 00	1.0608	0.9827	0.9820	1.0155	1.0140	1.0497	0.9866	1.0409	1.0244
1.4	1.0843E 00	1.0538	0.9836	0.9831	1.0104	1.0095	1.0401	0.9900	1.0337	1.0178
1.8	1.0509E 00	1.0431	0.9855	0.9851	1.0054	1.0050	1.0285	0.9939	1.0250	1.0107
2.2	1.0358E 00	1.0358	0.9872	0.9870	1.0031	1.0029	1.0220	0.9960	1.0199	1.0072
2.6	1.0276E 00	1.0305	0.9887	0.9885	1.0019	1.0018	1.0179	0.9972	1.0165	1.0053
3.0	1.0224E 00	1.0265	0.9899	0.9896	1.0013	1.0012	1.0151	0.9978	1.0142	1.0040
4.0	1.0155E 00	1.0200	0.9922	0.9918	1.0006	1.0005	1.0108	0.9986	1.0104	1.0026

Z= -97 A= 245

p	$F_0L_0^*/F_0L_0$	λ_2^*/λ_2	μ_1^*/μ_1	μ_2^*/μ_2	Λ_1^*/Λ_1	Λ_2^*/Λ_2	ν_{12}^*/ν_{12}	$\hat{\nu}_{12}^*/\hat{\nu}_{12}$	η_{12}^*/η_{12}	$\hat{\eta}_{12}^*/\hat{\eta}_{12}$
0.1	6.5854E 09	0.8251	1.0900	1.0200	2.3319	1.9932	1.0799	0.6942	2.3007	2.5547
0.2	8.8789E 02	0.9670	1.0399	1.0028	1.5055	1.3960	1.1756	0.8220	1.5191	1.5996
0.4	4.8257E 00	1.0657	1.0008	0.9879	1.1659	1.1358	1.1725	0.9182	1.1957	1.2052
0.6	1.8206E 00	1.0831	0.9876	0.9825	1.0763	1.0648	1.1233	0.9551	1.1081	1.0992
0.8	1.3449E 00	1.0786	0.9833	0.9810	1.0411	1.0359	1.0874	0.9722	1.0715	1.0567
1.0	1.1897E 00	1.0700	0.9823	0.9811	1.0245	1.0218	1.0648	0.9812	1.0527	1.0362
1.2	1.1214E 00	1.0616	0.9826	0.9818	1.0157	1.0142	1.0503	0.9866	1.0415	1.0249
1.4	1.0857E 00	1.0545	0.9834	0.9829	1.0106	1.0097	1.0406	0.9900	1.0342	1.0182
1.8	1.0516E 00	1.0438	0.9853	0.9849	1.0054	1.0050	1.0289	0.9939	1.0254	1.0110
2.2	1.0363E 00	1.0363	0.9871	0.9868	1.0031	1.0029	1.0223	0.9960	1.0202	1.0074
2.6	1.0279E 00	1.0309	0.9886	0.9884	1.0020	1.0018	1.0182	0.9972	1.0168	1.0055
3.0	1.0226E 00	1.0270	0.9898	0.9895	1.0013	1.0012	1.0153	0.9979	1.0144	1.0041
4.0	1.0156E 00	1.0203	0.9921	0.9917	1.0006	1.0005	1.0110	0.9986	1.0106	1.0027
6.0	1.0099E 00	1.0134	0.9949	0.9943	1.0002	1.0002	1.0071	0.9985	1.0069	1.0016

POSITRON Z= -98 A= 248

p	$F_0L_0^*/F_0L_0$	λ_2^*/λ_2	μ_1^*/μ_1	μ_2^*/μ_2	Λ_1^*/Λ_1	Λ_2^*/Λ_2	ν_{12}^*/ν_{12}	$\vartheta_{12}^*/\vartheta_{12}$	η_{12}^*/η_{12}	$\hat{\eta}_{12}^*/\hat{\eta}_{12}$
0.1	9.0258E 09	0.8251	1.0934	1.0206	2.3463	2.0030	1.0774	0.6946	2.3135	2.5772
0.2	9.9096E 02	0.9667	1.0418	1.0031	1.5112	1.4002	1.1734	0.8221	1.5248	1.6087
0.4	4.9698E 00	1.0657	1.0015	0.9879	1.1680	1.1373	1.1724	0.9181	1.1982	1.2087
0.6	1.8421E 00	1.0835	0.9878	0.9824	1.0773	1.0656	1.1240	0.9550	1.1096	1.1010
0.8	1.3527E 00	1.0793	0.9833	0.9808	1.0417	1.0364	1.0882	0.9721	1.0726	1.0578
1.0	1.1935E 00	1.0707	0.9823	0.9809	1.0248	1.0221	1.0655	0.9812	1.0535	1.0369
1.2	1.1236E 00	1.0624	0.9825	0.9816	1.0159	1.0143	1.0509	0.9866	1.0422	1.0255
1.4	1.0872E 00	1.0552	0.9833	0.9826	1.0107	1.0098	1.0412	0.9900	1.0348	1.0186
1.8	1.0524E 00	1.0444	0.9852	0.9847	1.0055	1.0051	1.0294	0.9939	1.0258	1.0112
2.2	1.0367E 00	1.0369	0.9870	0.9867	1.0032	1.0030	1.0227	0.9961	1.0205	1.0076
2.6	1.0282E 00	1.0314	0.9885	0.9882	1.0020	1.0019	1.0184	0.9973	1.0171	1.0056
3.0	1.0229E 00	1.0274	0.9897	0.9893	1.0013	1.0013	1.0156	0.9979	1.0146	1.0043
4.0	1.0157E 00	1.0206	0.9920	0.9916	1.0006	1.0005	1.0112	0.9986	1.0108	1.0027
6.0	1.0100E 00	1.0136	0.9948	0.9943	1.0002	1.0002	1.0072	0.9986	1.0070	1.0018

Z= -99 A= 251

p	$F_0L_0^*/F_0L_0$	λ_2^*/λ_2	μ_1^*/μ_1	μ_2^*/μ_2	Λ_1^*/Λ_1	Λ_2^*/Λ_2	ν_{12}^*/ν_{12}	$\vartheta_{12}^*/\vartheta_{12}$	η_{12}^*/η_{12}	$\hat{\eta}_{12}^*/\hat{\eta}_{12}$
0.1	1.2388E 10	0.8252	1.0969	1.0213	2.3608	2.0128	1.0749	0.6950	2.3264	2.6001
0.2	1.1068E 03	0.9664	1.0438	1.0034	1.5170	1.4044	1.1712	0.8222	1.5304	1.6179
0.4	5.1194E 00	1.0657	1.0022	0.9880	1.1701	1.1389	1.1722	0.9180	1.2006	1.2123
0.6	1.8641E 00	1.0840	0.9881	0.9823	1.0784	1.0664	1.1247	0.9550	1.1111	1.1029
0.8	1.3605E 00	1.0800	0.9834	0.9806	1.0423	1.0368	1.0890	0.9721	1.0737	1.0590
1.0	1.1974E 00	1.0715	0.9822	0.9807	1.0252	1.0224	1.0662	0.9813	1.0543	1.0377
1.2	1.1259E 00	1.0631	0.9824	0.9814	1.0161	1.0145	1.0516	0.9866	1.0428	1.0260
1.4	1.0886E 00	1.0560	0.9832	0.9825	1.0109	1.0099	1.0417	0.9900	1.0353	1.0190
1.8	1.0531E 00	1.0451	0.9851	0.9846	1.0056	1.0052	1.0298	0.9940	1.0262	1.0115
2.2	1.0371E 00	1.0374	0.9868	0.9865	1.0032	1.0030	1.0230	0.9961	1.0209	1.0078
2.6	1.0284E 00	1.0319	0.9884	0.9880	1.0020	1.0019	1.0187	0.9973	1.0174	1.0058
3.0	1.0231E 00	1.0276	0.9896	0.9892	1.0014	1.0013	1.0158	0.9979	1.0149	1.0044
4.0	1.0159E 00	1.0210	0.9920	0.9915	1.0006	1.0006	1.0114	0.9987	1.0110	1.0029
6.0	1.0101E 00	1.0139	0.9948	0.9941	1.0002	1.0002	1.0073	0.9985	1.0072	1.0018

POSITRON Z=-100 A= 256

p	$F_0L_0^*/F_0L_0$	λ_2^*/λ_2	μ_1^*/μ_1	μ_2^*/μ_2	Λ_1^*/Λ_1	Λ_2^*/Λ_2	ν_{12}^*/ν_{12}	$\hat{\nu}_{12}^*/\hat{\nu}_{12}$	η_{12}^*/η_{12}	$\hat{\eta}_{12}^*/\hat{\eta}_{12}$
0.1	1.7018E 10	0.8253	1.1006	1.0219	2.3754	2.0227	1.0725	0.6955	2.3393	2.6235
0.2	1.2367E 03	0.9662	1.0458	1.0038	1.5228	1.4087	1.1691	0.8223	1.5361	1.6272
0.4	5.2748E 00	1.0657	1.0030	0.9880	1.1722	1.1405	1.1720	0.9180	1.2032	1.2159
0.6	1.8865E 00	1.0845	0.9883	0.9822	1.0794	1.0672	1.1253	0.9549	1.1126	1.1048
0.8	1.3685E 00	1.0807	0.9834	0.9805	1.0428	1.0373	1.0898	0.9721	1.0747	1.0601
1.0	1.2014E 00	1.0723	0.9821	0.9805	1.0255	1.0227	1.0669	0.9813	1.0551	1.0385
1.2	1.1282E 00	1.0639	0.9823	0.9812	1.0164	1.0147	1.0522	0.9867	1.0435	1.0266
1.4	1.0901E 00	1.0567	0.9830	0.9823	1.0110	1.0100	1.0423	0.9901	1.0359	1.0195
1.8	1.0538E 00	1.0457	0.9849	0.9844	1.0057	1.0052	1.0302	0.9940	1.0266	1.0118
2.2	1.0376E 00	1.0380	0.9867	0.9863	1.0033	1.0030	1.0234	0.9961	1.0212	1.0080
2.6	1.0287E 00	1.0324	0.9883	0.9879	1.0021	1.0019	1.0190	0.9973	1.0176	1.0059
3.0	1.0233E 00	1.0283	0.9895	0.9891	1.0014	1.0013	1.0161	0.9979	1.0151	1.0046
4.0	1.0160E 00	1.0213	0.9919	0.9914	1.0006	1.0006	1.0116	0.9987	1.0111	1.0029
6.0	1.0102E 00	1.0141	0.9947	0.9941	1.0002	1.0002	1.0074	0.9986	1.0073	1.0018

Z=-101 A= 257

p	$F_0L_0^*/F_0L_0$	λ_2^*/λ_2	μ_1^*/μ_1	μ_2^*/μ_2	Λ_1^*/Λ_1	Λ_2^*/Λ_2	ν_{12}^*/ν_{12}	$\hat{\nu}_{12}^*/\hat{\nu}_{12}$	η_{12}^*/η_{12}	$\hat{\eta}_{12}^*/\hat{\eta}_{12}$
0.1	2.3396E 10	0.8254	1.1044	1.0226	2.3901	2.0325	1.0701	0.6960	2.3522	2.6472
0.2	1.3827E 03	0.9660	1.0480	1.0041	1.5287	1.4129	1.1669	0.8225	1.5418	1.6368
0.4	5.4360E 00	1.0657	1.0038	0.9880	1.1744	1.1421	1.1718	0.9179	1.2057	1.2195
0.6	1.9094E 00	1.0850	0.9886	0.9821	1.0804	1.0681	1.1259	0.9549	1.1142	1.1067
0.8	1.3766E 00	1.0815	0.9834	0.9803	1.0434	1.0378	1.0905	0.9721	1.0758	1.0613
1.0	1.2053E 00	1.0731	0.9821	0.9803	1.0259	1.0230	1.0677	0.9813	1.0560	1.0393
1.2	1.1305E 00	1.0647	0.9822	0.9810	1.0166	1.0149	1.0528	0.9867	1.0442	1.0272
1.4	1.0916E 00	1.0575	0.9829	0.9821	1.0112	1.0102	1.0428	0.9901	1.0365	1.0199
1.8	1.0546E 00	1.0464	0.9848	0.9842	1.0058	1.0053	1.0307	0.9940	1.0271	1.0121
2.2	1.0380E 00	1.0386	0.9866	0.9862	1.0033	1.0031	1.0237	0.9962	1.0216	1.0083
2.6	1.0290E 00	1.0329	0.9881	0.9877	1.0021	1.0019	1.0193	0.9973	1.0179	1.0061
3.0	1.0235E 00	1.0287	0.9894	0.9889	1.0014	1.0013	1.0163	0.9980	1.0154	1.0047
4.0	1.0161E 00	1.0216	0.9919	0.9913	1.0006	1.0006	1.0117	0.9986	1.0113	1.0030
6.0	1.0102E 00	1.0143	0.9946	0.9940	1.0002	1.0002	1.0075	0.9986	1.0074	1.0018

 Instr. see p. 25

Table IV. Integrated statistical factor

$f_0(E_0)$

E_0 [keV]	0	1	2	3	4	5	6	7	8	9
1	2.278E-07	3.182E-07	4.318E-07	5.719E-07	7.419E-07	9.453E-07	1.186E-06	1.467E-06	1.794E-06	2.169E-06
2	2.598E-06	3.084E-06	3.632E-06	4.247E-06	4.933E-06	5.696E-06	6.539E-06	7.468E-06	8.489E-06	9.606E-06
3	1.082E-05	1.215E-05	1.359E-05	1.515E-05	1.683E-05	1.864E-05	2.059E-05	2.268E-05	2.492E-05	2.731E-05
4	2.987E-05	3.259E-05	3.548E-05	3.856E-05	4.183E-05	4.529E-05	4.895E-05	5.281E-05	5.690E-05	6.120E-05
5	6.574E-05	7.051E-05	7.553E-05	8.081E-05	8.634E-05	9.214E-05	9.821E-05	1.046E-04	1.112E-04	1.182E-04
6	1.254E-04	1.330E-04	1.409E-04	1.491E-04	1.577E-04	1.666E-04	1.759E-04	1.856E-04	1.956E-04	2.060E-04
7	2.168E-04	2.281E-04	2.397E-04	2.518E-04	2.642E-04	2.772E-04	2.905E-04	3.044E-04	3.187E-04	3.335E-04
8	3.488E-04	3.646E-04	3.809E-04	3.977E-04	4.150E-04	4.329E-04	4.513E-04	4.704E-04	4.899E-04	5.101E-04
9	5.308E-04	5.522E-04	5.742E-04	5.968E-04	6.200E-04	6.439E-04	6.685E-04	6.937E-04	7.196E-04	7.463E-04
	00	10	20	30	40	50	60	70	80	90
1	7.736E-04	1.088E-03	1.487E-03	1.983E-03	2.590E-03	3.323E-03	4.196E-03	5.227E-03	6.433E-03	7.831E-03
2	9.441E-03	1.128E-02	1.337E-02	1.574E-02	1.840E-02	2.138E-02	2.471E-02	2.840E-02	3.249E-02	3.700E-02
3	4.195E-02	4.739E-02	5.333E-02	5.981E-02	6.686E-02	7.452E-02	8.281E-02	9.177E-02	1.014E-01	1.119E-01
4	1.231E-01	1.351E-01	1.479E-01	1.617E-01	1.765E-01	1.922E-01	2.089E-01	2.267E-01	2.457E-01	2.658E-01
5	2.871E-01	3.097E-01	3.336E-01	3.589E-01	3.857E-01	4.139E-01	4.436E-01	4.750E-01	5.080E-01	5.427E-01
6	5.792E-01	6.175E-01	6.577E-01	6.999E-01	7.441E-01	7.905E-01	8.389E-01	8.897E-01	9.427E-01	9.981E-01
7	1.056E 00	1.116E 00	1.179E 00	1.245E 00	1.314E 00	1.385E 00	1.459E 00	1.536E 00	1.617E 00	1.700E 00
8	1.787E 00	1.877E 00	1.971E 00	2.068E 00	2.169E 00	2.274E 00	2.382E 00	2.495E 00	2.611E 00	2.732E 00
9	2.856E 00	2.986E 00	3.119E 00	3.257E 00	3.400E 00	3.548E 00	3.700E 00	3.858E 00	4.021E 00	4.189E 00
	000	100	200	300	400	500	600	700	800	900
1	4.362E 00	6.419E 00	9.161E 00	1.274E 01	1.732E 01	2.310E 01	3.028E 01	3.912E 01	4.986E 01	6.279E 01
2	7.823E 01	9.651E 01	1.180E 02	1.431E 02	1.723E 02	2.059E 02	2.446E 02	2.888E 02	3.391E 02	3.961E 02
3	4.604E 02	5.328E 02	6.139E 02	7.045E 02	8.054E 02	9.175E 02	1.042E 03	1.179E 03	1.330E 03	1.496E 03
4	1.678E 03	1.878E 03	2.095E 03	2.333E 03	2.591E 03	2.872E 03	3.177E 03	3.506E 03	3.862E 03	4.247E 03
5	4.662E 03	5.108E 03	5.587E 03	6.102E 03	6.654E 03	7.245E 03	7.877E 03	8.552E 03	9.272E 03	1.004E 04
6	1.086E 04	1.173E 04	1.265E 04	1.364E 04	1.468E 04	1.579E 04	1.696E 04	1.819E 04	1.950E 04	2.089E 04
7	2.235E 04	2.389E 04	2.552E 04	2.723E 04	2.904E 04	3.093E 04	3.293E 04	3.503E 04	3.723E 04	3.954E 04
8	4.197E 04	4.451E 04	4.717E 04	4.996E 04	5.288E 04	5.594E 04	5.913E 04	6.247E 04	6.595E 04	6.959E 04
9	7.339E 04	7.736E 04	8.149E 04	8.580E 04	9.028E 04	9.496E 04	9.982E 04	1.049E 05	1.101E 05	1.156E 05

 Instr. see p. 25

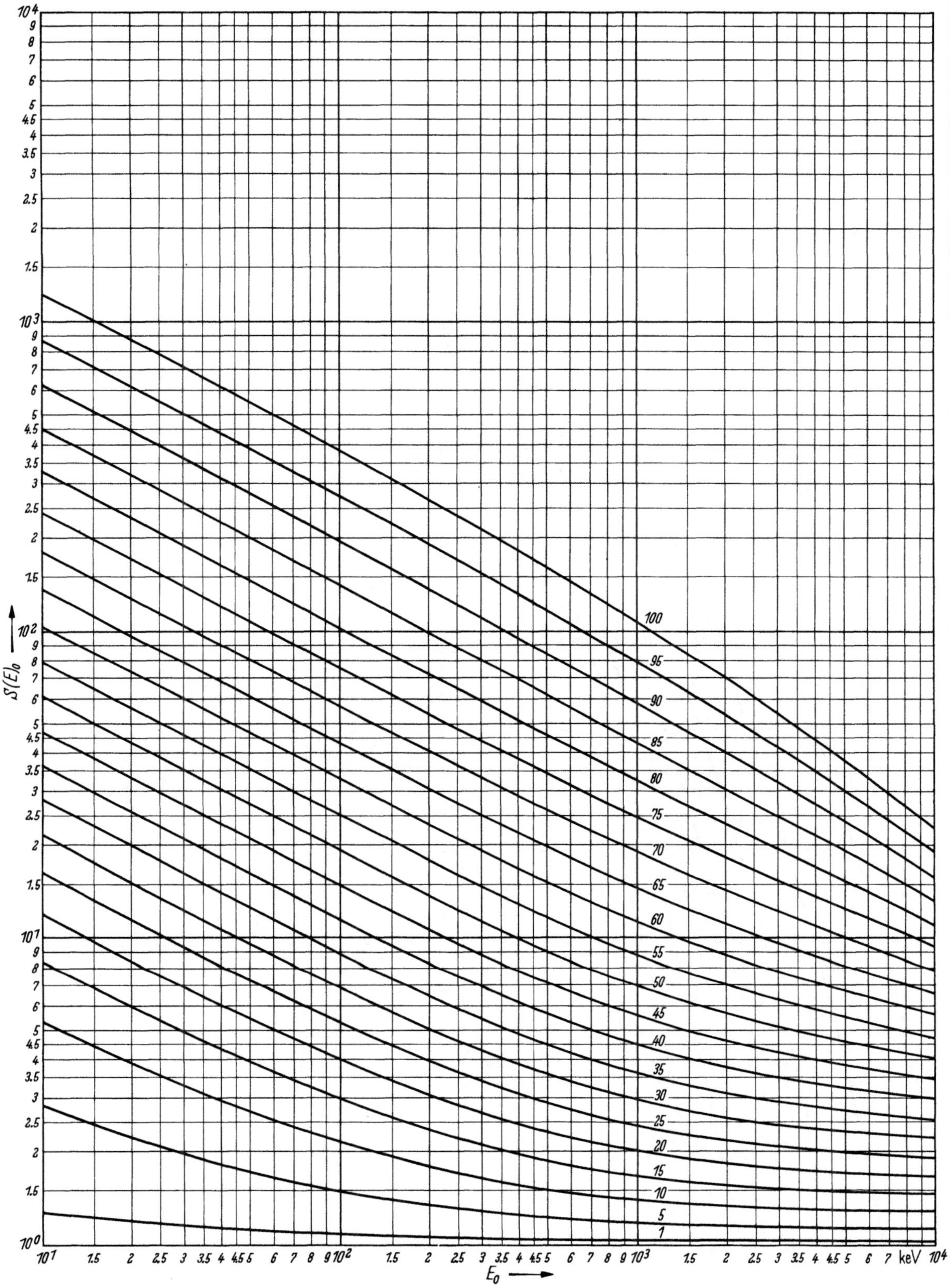

Diagram I. Ratio $S(E_0)$ of the integrated unscreened Fermi function to the integrated statistical factor for electrons as a function of the maximum kinetic energy E_0 with the atomic number Z of the daughter nucleus as parameter.

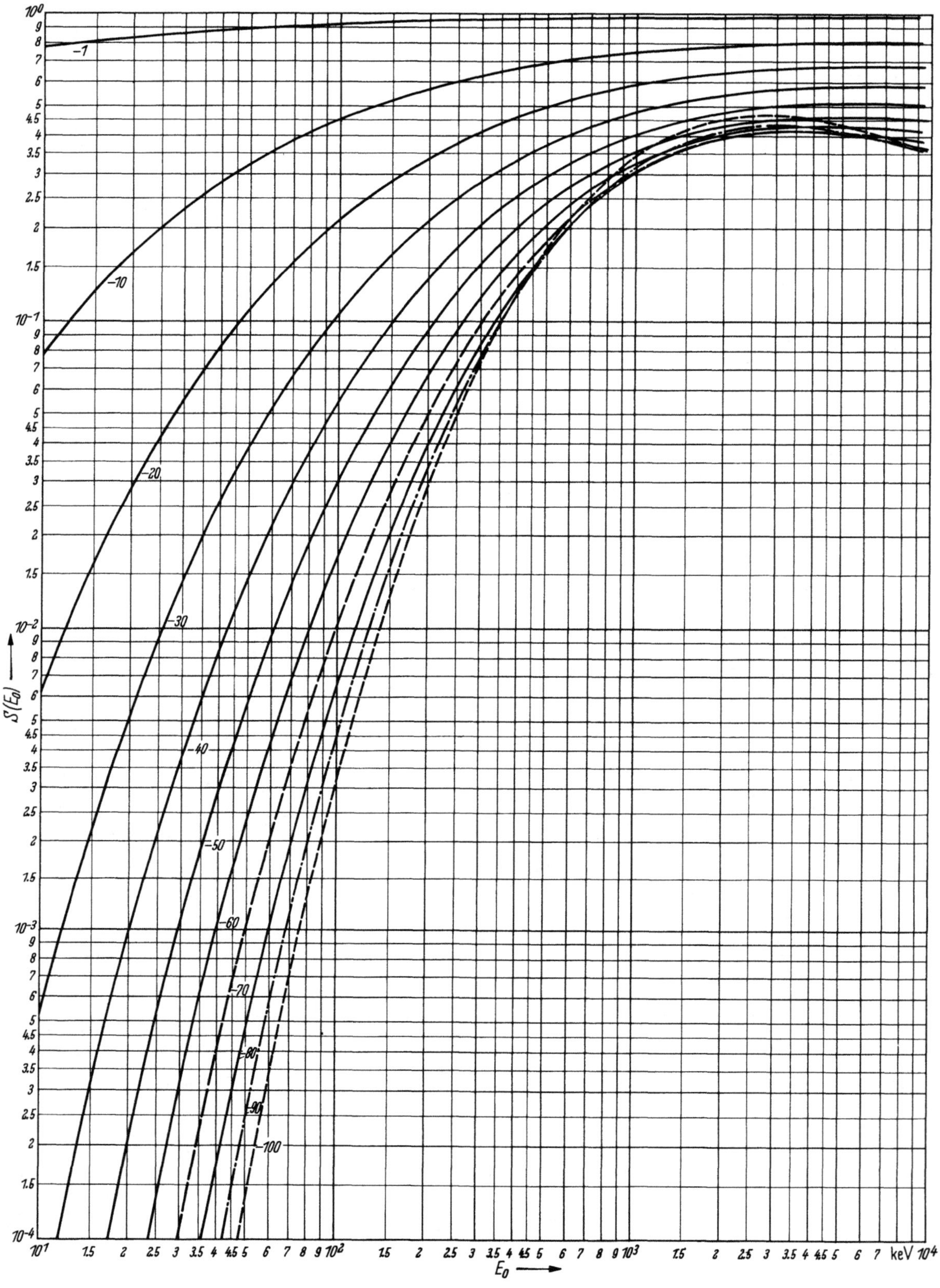

Diagram II a. Ratio $S(E_0)$ of the integrated unscreened Fermi function to the integrated statistical factor for positrons as a function of the maximum kinetic energy E_0 with the atomic number Z of the daughter nucleus as parameter.

 Instr. see p. 25

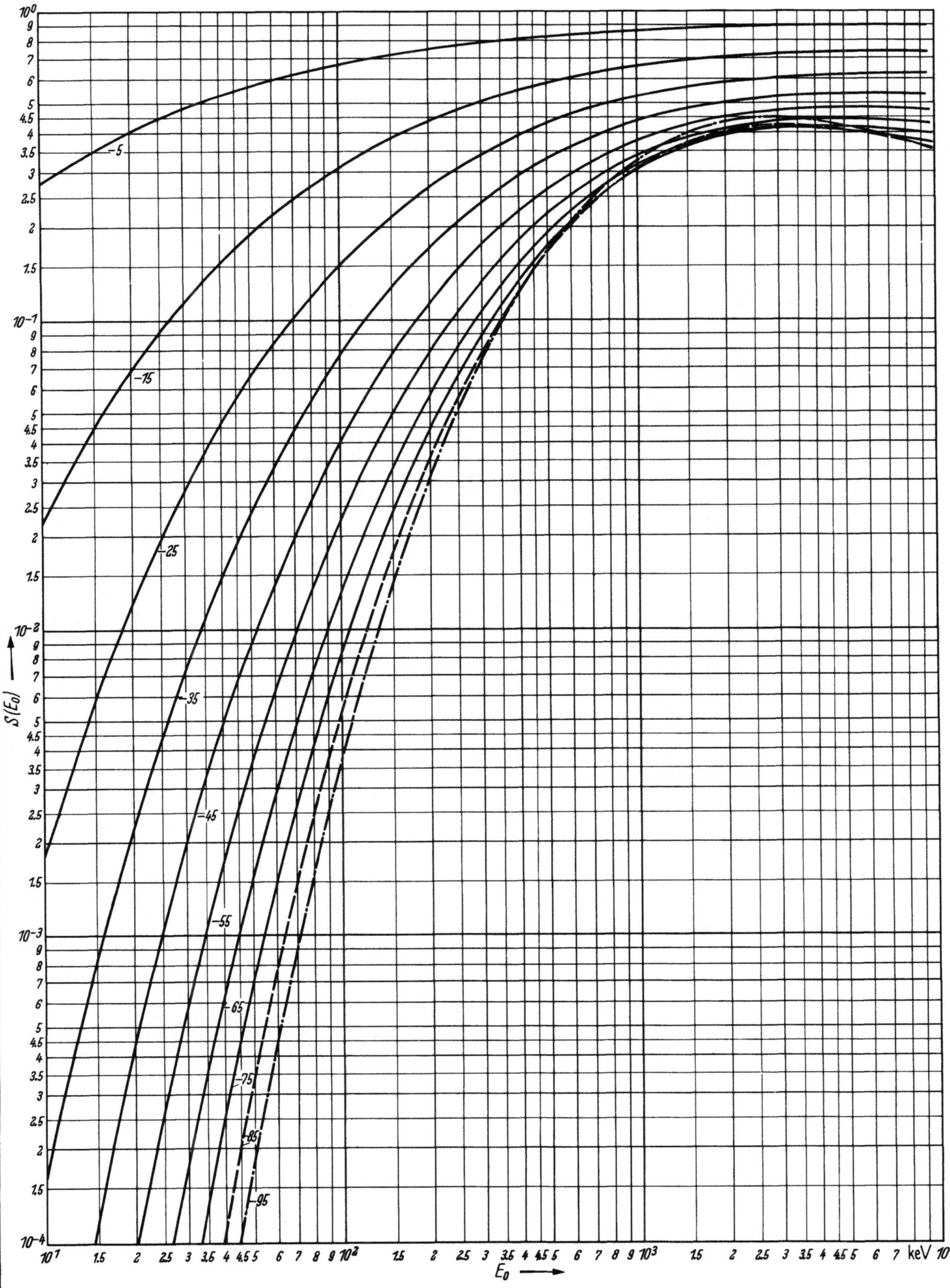

Diagram II b. Ratio $S(E_0)$ of the integrated unscreened Fermi function to the integrated statistical factor for positrons as a function of the maximum kinetic energy E_0 with the atomic number Z of the daughter nucleus as parameter.

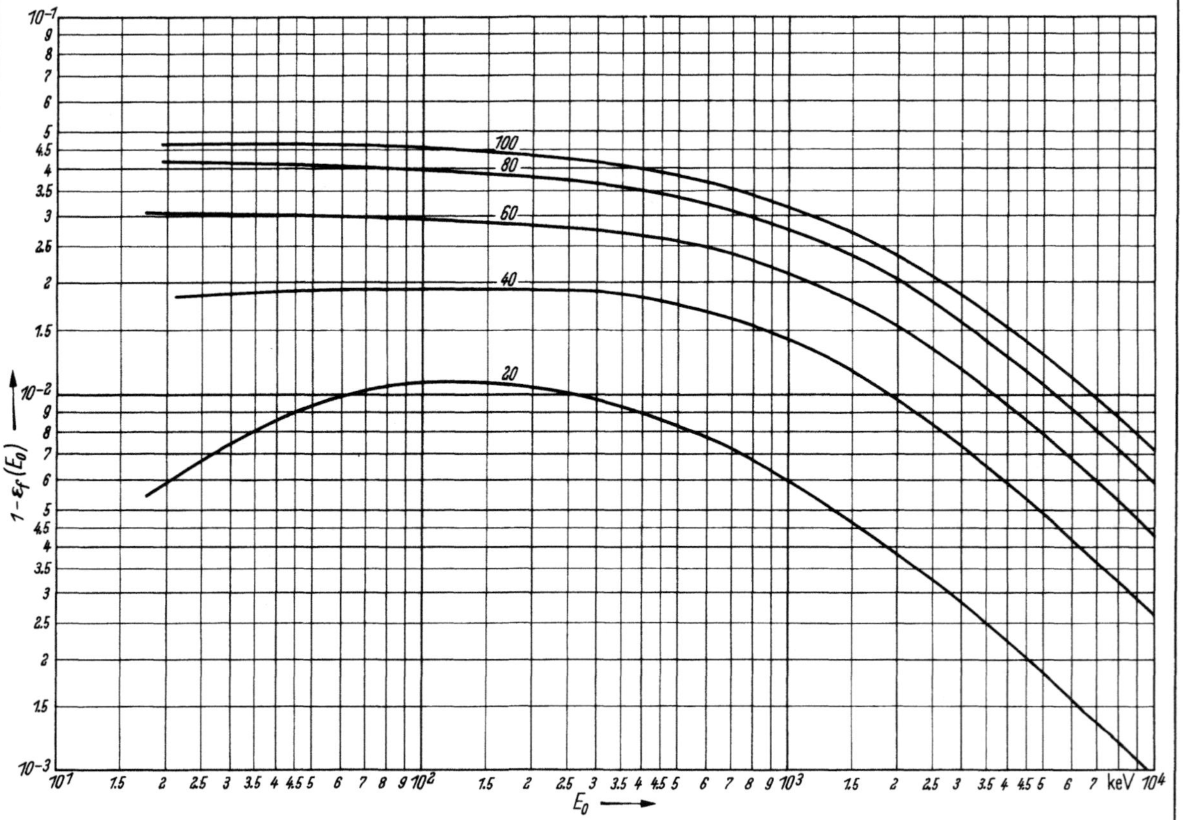

Diagram III. Ratio $\varepsilon_f(E_0)$ of the integrated Fermi function with screening to the same function without screening for electrons as a function of the maximum kinetic E_0 energy with the atomic number Z of the daughter nucleus as parameter.

 Instr. see p. 25

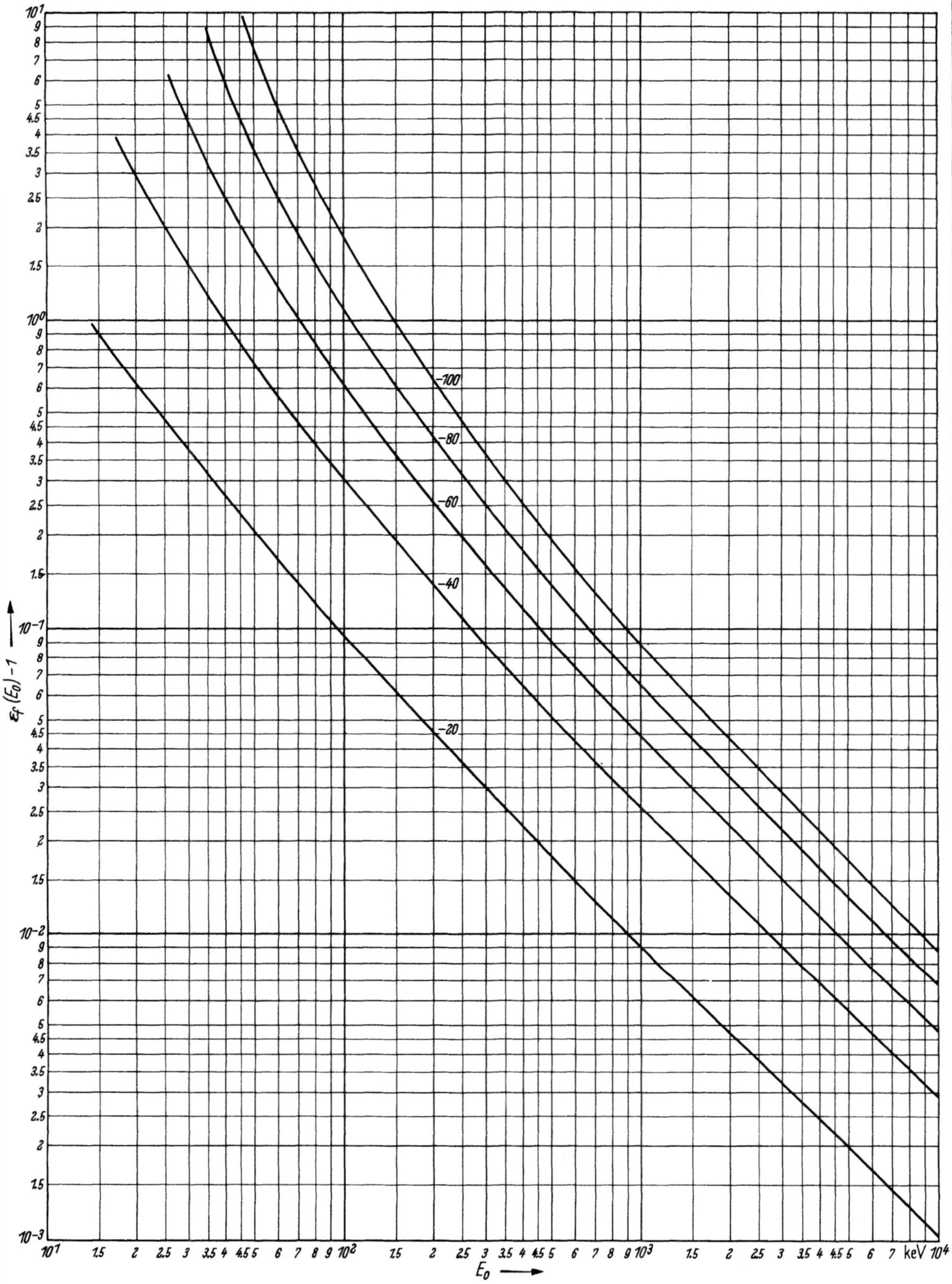

Diagram IV. Ratio $\varepsilon_f(E_0)$ of the integrated Fermi function with screening to the same function without screening for positrons as a function of the maximum kinetic energy E_0 with the atomic number Z of the daughter nucleus as parameter.

Table V. Coulomb functions of the bound orbital electron for electron capture

	Z	A	K
BETA			1.247E-03
F(R)			-3.640E-06
G(R)	1	1	1.247E-03
EB(CO)			1.361E 01
EB(EX)			1.361E 01
BETA			3.125E-03
F(R)			-1.824E-05
G(R)	2	3	3.125E-03
EB(CO)			2.175E 01
EB(EX)			2.500E 01
BETA			6.070E-03
F(R)			-5.315E-05
G(R)	3	3	6.069E-03
EB(CO)			6.191E 01
EB(EX)			5.500E 01
BETA			9.444E-03
F(R)			-1.102E-04
G(R)	4	6	9.441E-03
EB(CO)			1.197E 02
EB(EX)			1.120E 02
BETA			1.333E-02
F(R)			-1.945E-04
G(R)	5	8	1.332E-02
EB(CO)			1.971E 02
EB(EX)			1.880E 02
BETA			1.767E-02
F(R)			-3.093E-04
G(R)	6	10	1.766E-02
EB(CO)			2.934E 02
EB(EX)			2.840E 02
BETA			2.243E-02
F(R)			-4.579E-04
G(R)	7	12	2.241E-02
EB(CO)			4.091E 02
EB(EX)			4.020E 02

	Z	A	K	L_I	L_{II}	L_{III}
BETA			2.758E-02			
F(R)			-6.433E-04			
G(R)	8	14	2.754E-02			
EB(CO)			5.449E 02			
EB(EX)			5.320E 02			
BETA			3.310E-02			
F(R)			-8.684E-04			
G(R)	9	17	3.305E-02			
EB(CO)			6.991E 02			
EB(EX)			6.860E 02			
BETA			3.900E-02			
F(R)			-1.137E-03			
G(R)	10	19	3.893E-02			
EB(CO)			8.720E 02			
EB(EX)			8.670E 02			
BETA			4.529E-02			
F(R)			-1.452E-03			
G(R)	11	21	4.519E-02			
EB(CO)			1.080E 03			
EB(EX)			1.072E 03			
BETA			5.190E-02	1.381E-02	4.209E-04	5.302E-02
F(R)			-1.815E-03	-4.830E-04	4.198E-04	-4.864E-08
G(R)	12	23	5.177E-02	1.378E-02	1.719E-05	2.430E-06
EB(CO)			1.311E 03	9.321E 01	6.236E 01	6.201E 01
EB(EX)			1.305E 03	8.920E 01	5.200E 01	5.200E 01
BETA			5.890E-02	1.591E-02	5.347E-04	5.711E-02
F(R)			-2.230E-03	-6.023E-04	5.332E-04	-6.769E-08
G(R)	13	24	5.873E-02	1.586E-02	2.344E-05	3.121E-06
EB(CO)			1.566E 03	1.219E 02	8.629E 01	8.577E 01
EB(EX)			1.560E 03	1.174E 02	7.300E 01	7.300E 01
BETA			6.625E-02	1.812E-02	6.665E-04	6.188E-02
F(R)			-2.700E-03	-7.389E-04	6.643E-04	-9.420E-08
G(R)	14	27	6.603E-02	1.806E-02	3.131E-05	4.033E-06
EB(CO)			1.845E 03	1.543E 02	1.136E 02	1.128E 02
EB(EX)			1.839E 03	1.490E 02	1.000E 02	9.900E 01

 Instr. see p. 26

	Z	A	K	L_{I}	L_{II}	L_{III}
BETA			7.395E-02	2.048E-02	8.180E-04	6.725E-02
F(R)			-3.229E-03	-8.943E-04	8.148E-04	-1.264E-07
G(R)	15	29	7.367E-02	2.040E-02	4.091E-05	5.051E-06
EB(CO)			2.148E 03	1.900E 02	1.440E 02	1.429E 02
EB(EX)			2.149E 03	1.850E 02	1.360E 02	1.350E 02
BETA			8.200E-02	2.298E-02	9.911E-04	7.309E-02
F(R)			-3.817E-03	-1.070E-03	9.868E-04	-1.664E-07
G(R)	16	31	8.165E-02	2.288E-02	5.258E-05	6.234E-06
EB(CO)			2.475E 03	2.293E 02	1.778E 02	1.765E 02
EB(EX)			2.472E 03	2.250E 02	1.650E 02	1.640E 02
BETA			9.041E-02	2.564E-02	1.188E-03	7.932E-02
F(R)			-4.470E-03	-1.268E-03	1.182E-03	-2.156E-07
G(R)	17	33	8.996E-02	2.551E-02	6.663E-05	7.600E-06
EB(CO)			2.827E 03	2.725E 02	2.156E 02	2.137E 02
EB(EX)			2.822E 03	2.690E 02	2.020E 02	2.000E 02
BETA			9.916E-02	2.843E-02	1.409E-03	8.595E-02
F(R)			-5.189E-03	-1.488E-03	1.401E-03	-2.750E-07
G(R)	18	35	9.862E-02	2.828E-02	8.327E-05	9.153E-06
EB(CO)			3.202E 03	3.189E 02	2.565E 02	2.541E 02
EB(EX)			3.202E 03	3.160E 02	2.470E 02	2.450E 02
BETA			1.075E-01	3.116E-02	1.659E-03	9.244E-02
F(R)			-5.933E-03	-1.721E-03	1.649E-03	-3.497E-07
G(R)	19	38	1.068E-01	3.097E-02	1.031E-04	1.103E-05
EB(CO)			3.590E 03	3.703E 02	3.052E 02	3.021E 02
EB(EX)			3.607E 03	3.670E 02	2.970E 02	2.940E 02
BETA			1.171E-01	3.426E-02	1.933E-03	9.838E-02
F(R)			-6.800E-03	-1.991E-03	1.920E-03	-4.380E-07
G(R)	20	41	1.163E-01	3.403E-02	1.261E-04	1.312E-05
EB(CO)			4.027E 03	4.342E 02	3.632E 02	3.593E 02
EB(EX)			4.038E 03	4.230E 02	3.500E 02	3.460E 02
BETA			1.272E-01	3.786E-02	2.272E-03	1.060E-01
F(R)			-7.757E-03	-2.310E-03	2.255E-03	-5.511E-07
G(R)	21	44	1.263E-01	3.758E-02	1.551E-04	1.572E-05
EB(CO)			4.514E 03	5.039E 02	4.290E 02	4.239E 02
EB(EX)			4.493E 03	4.820E 02	4.070E 02	4.030E 02

	Z	A	K	L_{I}	L_{II}	L_{III}
BETA			1.371E-01	4.102E-02	2.578E-03	1.131E-01
F(R)			-8.754E-03	-2.620E-03	2.558E-03	-6.569E-07
G(R)	22	45	1.360E-01	4.069E-02	1.836E-04	1.789E-05
EB(CO)			4.958E 03	5.630E 02	4.819E 02	4.758E 02
EB(EX)			4.965E 03	5.440E 02	4.600E 02	4.540E 02
BETA			1.477E-01	4.459E-02	2.945E-03	1.210E-01
F(R)			-9.854E-03	-2.976E-03	2.919E-03	-7.976E-07
G(R)	23	48	1.464E-01	4.420E-02	2.186E-04	2.077E-05
EB(CO)			5.456E 03	6.313E 02	5.449E 02	5.374E 02
EB(EX)			5.465E 03	6.110E 02	5.200E 02	5.130E 02
BETA			1.587E-01	4.833E-02	3.348E-03	1.292E-01
F(R)			-1.104E-02	-3.364E-03	3.316E-03	-9.605E-07
G(R)	24	51	1.572E-01	4.787E-02	2.586E-04	2.397E-05
EB(CO)			5.981E 03	7.033E 02	6.118E 02	6.026E 02
EB(EX)			5.989E 03	6.820E 02	5.830E 02	5.740E 02
BETA			1.702E-01	5.222E-02	3.784E-03	1.379E-01
F(R)			-1.233E-02	-3.784E-03	3.745E-03	-1.132E-06
G(R)	25	52	1.684E-01	5.168E-02	3.033E-04	2.712E-05
EB(CO)			6.528E 03	7.781E 02	6.812E 02	6.701E 02
EB(EX)			6.540E 03	7.540E 02	6.520E 02	6.410E 02
BETA			1.820E-01	5.625E-02	4.257E-03	1.467E-01
F(R)			-1.370E-02	-4.238E-03	4.210E-03	-1.343E-06
G(R)	26	55	1.800E-01	5.562E-02	3.540E-04	3.092E-05
EB(CO)			7.100E 03	8.572E 02	7.548E 02	7.416E 02
EB(EX)			7.112E 03	8.420E 02	7.210E 02	7.090E 02
BETA			1.943E-01	6.046E-02	4.771E-03	1.559E-01
F(R)			-1.518E-02	-4.728E-03	4.714E-03	-1.563E-06
G(R)	27	56	1.920E-01	5.974E-02	4.106E-04	3.466E-05
EB(CO)			7.698E 03	9.396E 02	8.318E 02	8.162E 02
EB(EX)			7.709E 03	9.290E 02	7.940E 02	7.790E 02
BETA			2.071E-01	6.484E-02	5.326E-03	1.654E-01
F(R)			-1.677E-02	-5.255E-03	5.258E-03	-1.809E-06
G(R)	28	57	2.044E-01	6.401E-02	4.738E-04	3.868E-05
EB(CO)			8.320E 03	1.026E 03	9.121E 02	8.938E 02
EB(EX)			8.333E 03	1.012E 03	8.720E 02	8.550E 02

Instr. see p. 26

	Z	A	K	L_I	L_{II}	L_{III}	M_I	M_{II}	M_{III}	M_{IV}	M_V
BETA			2.203E-01	6.938E-02	5.925E-03	1.752E-01					
F(R)			-1.847E-02	-5.819E-03	5.844E-03	-2.108E-06					
G(R)	29	60	2.173E-01	6.842E-02	5.446E-04	4.350E-05					
EB(CO)			8.968E 03	1.115E 03	9.961E 02	9.746E 02					
EB(EX)			8.979E 03	1.100E 03	9.520E 02	9.320E 02					
BETA			2.339E-01	7.405E-02	6.566E-03	1.853E-01	2.979E-02	2.647E-03	2.400E-01	9.772E-03	1.832E 00
F(R)			-2.027E-02	-6.422E-03	6.470E-03	-2.467E-06	-2.584E-03	2.608E-03	-9.941E-07	4.216E-07	-7.062E-11
G(R)	30	65	2.305E-01	7.296E-02	6.236E-04	4.921E-05	2.935E-02	2.514E-04	1.983E-05	2.325E-08	2.031E-09
EB(CO)			9.641E 03	1.208E 03	1.083E 03	1.058E 03	1.482E 02	1.059E 02	1.023E 02	2.789E 01	2.738E 01
EB(EX)			9.659E 03	1.196E 03	1.044E 03	1.021E 03	1.390E 02	8.800E 01	8.800E 01	9.000E 00	9.000E 00
BETA			2.483E-01	7.892E-02	7.257E-03	1.947E-01	3.200E-02	2.961E-03	2.461E-01	9.681E-03	1.499E 00
F(R)			-2.222E-02	-7.068E-03	7.144E-03	-2.827E-06	-2.866E-03	2.915E-03	-1.154E-06	5.079E-07	-8.864E-11
G(R)	31	67	2.444E-01	7.768E-02	7.104E-04	5.457E-05	3.149E-02	2.899E-04	2.227E-05	2.888E-08	2.467E-09
EB(CO)			1.036E 04	1.319E 03	1.186E 03	1.157E 03	1.699E 02	1.246E 02	1.204E 02	4.048E 01	3.984E 01
EB(EX)			1.037E 04	1.302E 03	1.142E 03	1.115E 03	1.580E 02	1.060E 02	1.020E 02	1.700E 01	1.700E 01
BETA			2.634E-01	8.397E-02	7.997E-03	2.041E-01	3.437E-02	3.311E-03	2.538E-01	9.901E-03	1.315E 00
F(R)			-2.432E-02	-7.758E-03	7.864E-03	-3.225E-06	-3.175E-03	3.256E-03	-1.336E-06	6.092E-07	-1.106E-10
G(R)	32	69	2.590E-01	8.257E-02	8.061E-04	6.031E-05	3.379E-02	3.337E-04	2.497E-05	3.570E-08	2.982E-09
EB(CO)			1.111E 04	1.436E 03	1.294E 03	1.261E 03	1.933E 02	1.448E 02	1.398E 02	5.466E 01	5.387E 01
EB(EX)			1.110E 04	1.414E 03	1.249E 03	1.218E 03	1.810E 02	1.260E 02	1.220E 02	3.100E 01	3.000E 01
BETA			2.790E-01	8.925E-02	8.785E-03	2.139E-01	3.677E-02	3.677E-03	2.621E-01	1.032E-02	1.215E 00
F(R)			-2.654E-02	-8.497E-03	8.630E-03	-3.681E-06	-3.501E-03	3.612E-03	-1.541E-06	7.237E-07	-1.372E-10
G(R)	33	72	2.741E-01	8.767E-02	9.113E-04	6.672E-05	3.612E-02	3.814E-04	2.794E-05	4.368E-08	3.585E-09
EB(CO)			1.189E 04	1.558E 03	1.407E 03	1.369E 03	2.172E 02	1.654E 02	1.596E 02	6.916E 01	6.819E 01
EB(EX)			1.187E 04	1.530E 03	1.360E 03	1.325E 03	2.060E 02	1.480E 02	1.420E 02	4.300E 01	4.200E 01
BETA			2.950E-01	9.480E-02	9.640E-03	2.241E-01	3.925E-02	4.067E-03	2.709E-01	1.084E-02	1.153E 00
F(R)			-2.889E-02	-9.293E-03	9.461E-03	-4.151E-06	-3.848E-03	3.991E-03	-1.752E-06	8.430E-07	-1.650E-10
G(R)	34	73	2.895E-01	9.302E-02	1.028E-03	7.303E-05	3.851E-02	4.335E-04	3.083E-05	5.231E-08	4.186E-09
EB(CO)			1.268E 04	1.683E 03	1.525E 03	1.482E 03	2.423E 02	1.872E 02	1.805E 02	8.428E 01	8.313E 01
EB(EX)			1.266E 04	1.653E 03	1.477E 03	1.436E 03	2.320E 02	1.690E 02	1.620E 02	5.800E 01	5.700E 01
BETA			3.112E-01	1.005E-01	1.055E-02	2.347E-01	4.180E-02	4.484E-03	2.803E-01	1.145E-02	1.114E 00
F(R)			-3.135E-02	-1.013E-02	1.034E-02	-4.790E-06	-4.215E-03	4.395E-03	-2.037E-06	1.001E-06	-2.075E-10
G(R)	35	80	3.050E-01	9.847E-02	1.157E-03	8.185E-05	4.097E-02	4.917E-04	3.480E-05	6.397E-08	5.114E-09
EB(CO)			1.351E 04	1.814E 03	1.649E 03	1.600E 03	2.685E 02	2.102E 02	2.025E 02	1.003E 02	9.894E 01
EB(EX)			1.347E 04	1.782E 03	1.596E 03	1.550E 03	2.570E 02	1.890E 02	1.830E 02	7.000E 01	7.000E 01

Instr. see p. 26

	Z	A	K	L_{I}	L_{II}	L_{III}	M_{I}	M_{II}	M_{III}	M_{IV}	M_{V}
BETA			3.283E-01	1.065E-01	1.154E-02	2.456E-01	4.450E-02	4.936E-03	2.904E-01	1.215E-02	1.091E 00
F(R)			-3.400E-02	-1.104E-02	1.130E-02	-5.324E-06	-4.613E-03	4.833E-03	-2.280E-06	1.142E-06	-2.419E-10
G(R)	36	79	3.215E-01	1.043E-01	1.297E-03	8.844E-05	4.357E-02	5.550E-04	3.787E-05	7.487E-08	5.795E-09
EB(CO)			1.436E 04	1.949E 03	1.777E 03	1.722E 03	2.960E 02	2.344E 02	2.255E 02	1.172E 02	1.156E 02
EB(EX)			1.432E 04	1.920E 03	1.726E 03	1.675E 03	2.880E 02	2.200E 02	2.110E 02	8.700E 01	8.700E 01
BETA			3.453E-01	1.125E-01	1.257E-02	2.558E-01	4.737E-02	5.417E-03	2.884E-01	1.190E-02	9.226E-01
F(R)			-3.672E-02	-1.197E-02	1.229E-02	-5.990E-06	-5.043E-03	5.298E-03	-2.590E-06	1.306E-06	-2.873E-10
G(R)	37	82	3.377E-01	1.100E-01	1.449E-03	9.680E-05	4.632E-02	6.248E-04	4.185E-05	8.788E-08	6.695E-09
EB(CO)			1.523E 04	2.098E 03	1.920E 03	1.858E 03	3.486E 02	2.830E 02	2.728E 02	1.559E 02	1.540E 02
EB(EX)			1.520E 04	2.065E 03	1.863E 03	1.805E 03	3.220E 02	2.490E 02	2.380E 02	1.120E 02	1.100E 02
BETA			3.634E-01	1.189E-01	1.369E-02	2.673E-01	5.038E-02	5.949E-03	3.005E-01	1.276E-02	9.291E-01
F(R)			-3.966E-02	-1.299E-02	1.337E-02	-6.729E-06	-5.504E-03	5.811E-03	-2.935E-06	1.504E-06	-3.431E-10
G(R)	38	85	3.550E-01	1.161E-01	1.618E-03	1.059E-04	4.921E-02	7.033E-04	4.617E-05	1.039E-07	7.785E-09
EB(CO)			1.613E 04	2.246E 03	2.061E 03	1.990E 03	3.796E 02	3.107E 02	2.990E 02	1.759E 02	1.737E 02
EB(EX)			1.611E 04	2.216E 03	2.007E 03	1.940E 03	3.580E 02	2.790E 02	2.690E 02	1.360E 02	1.340E 02
BETA			3.823E-01	1.256E-01	1.489E-02	2.791E-01	5.356E-02	6.524E-03	3.129E-01	1.367E-02	9.362E-01
F(R)			-4.278E-02	-1.407E-02	1.453E-02	-7.483E-06	-6.001E-03	6.365E-03	-3.292E-06	1.713E-06	-4.017E-10
G(R)	39	86	3.730E-01	1.226E-01	1.802E-03	1.147E-04	5.224E-02	7.897E-04	5.044E-05	1.212E-07	8.879E-09
EB(CO)			1.707E 04	2.400E 03	2.208E 03	2.129E 03	4.127E 02	3.404E 02	3.271E 02	1.977E 02	1.952E 02
EB(EX)			1.704E 04	2.373E 03	2.156E 03	2.080E 03	3.950E 02	3.130E 02	3.010E 02	1.600E 02	1.580E 02
BETA			4.017E-01	1.326E-01	1.617E-02	2.911E-01	5.684E-02	7.136E-03	3.256E-01	1.463E-02	9.463E-01
F(R)			-4.606E-02	-1.522E-02	1.576E-02	-8.361E-06	-6.526E-03	6.954E-03	-3.707E-06	1.957E-06	-4.749E-10
G(R)	40	89	3.914E-01	1.292E-01	2.004E-03	1.249E-04	5.537E-02	8.844E-04	5.537E-05	1.419E-07	1.023E-08
EB(CO)			1.804E 04	2.559E 03	2.361E 03	2.273E 03	4.470E 02	3.714E 02	3.563E 02	2.206E 02	2.177E 02
EB(EX)			1.800E 04	2.532E 03	2.307E 03	2.223E 03	4.310E 02	3.440E 02	3.300E 02	1.830E 02	1.800E 02
BETA			4.219E-01	1.398E-01	1.754E-02	3.035E-01	6.028E-02	7.795E-03	3.386E-01	1.563E-02	9.567E-01
F(R)			-4.955E-02	-1.644E-02	1.707E-02	-9.252E-06	-7.089E-03	7.586E-03	-4.133E-06	2.212E-06	-5.506E-10
G(R)	41	90	4.106E-01	1.361E-01	2.222E-03	1.348E-04	5.865E-02	9.879E-04	6.022E-05	1.641E-07	1.158E-08
EB(CO)			1.903E 04	2.725E 03	2.520E 03	2.422E 03	4.835E 02	4.044E 02	3.873E 02	2.452E 02	2.419E 02
EB(EX)			1.899E 04	2.698E 03	2.465E 03	2.371E 03	4.680E 02	3.780E 02	3.620E 02	2.080E 02	2.050E 02
BETA			4.429E-01	1.474E-01	1.900E-02	3.163E-01	6.387E-02	8.500E-03	3.520E-01	1.669E-02	9.691E-01
F(R)			-5.323E-02	-1.774E-02	1.846E-02	-1.022E-05	-7.688E-03	8.261E-03	-4.595E-06	2.490E-06	-6.357E-10
G(R)	42	91	4.304E-01	1.432E-01	2.460E-03	1.453E-04	6.206E-02	1.101E-03	6.535E-05	1.891E-07	1.305E-08
EB(CO)			2.005E 04	2.897E 03	2.685E 03	2.575E 03	5.215E 02	4.389E 02	4.197E 02	2.711E 02	2.673E 02
EB(EX)			2.000E 04	2.866E 03	2.625E 03	2.520E 03	5.050E 02	4.090E 02	3.920E 02	2.300E 02	2.270E 02

Instr. see p. 26

	Z	A	K	L_I	L_{II}	L_{III}	M_I	M_{II}	M_{III}	M_{IV}	M_V
BETA			4.644E-01	1.552E-01	2.054E-02	3.293E-01	6.759E-02	9.252E-03	3.657E-01	1.780E-02	9.821E-01
F(R)			-5.710E-02	-1.911E-02	1.994E-02	-1.134E-05	-8.323E-03	8.979E-03	-5.135E-06	2.816E-06	-7.420E-10
G(R)	43	94	4.507E-01	1.506E-01	2.719E-03	1.575E-04	6.559E-02	1.225E-03	7.131E-05	2.187E-07	1.487E-08
EB(CO)			2.111E 04	3.075E 03	2.857E 03	2.735E 03	5.616E 02	4.755E 02	4.539E 02	2.987E 02	2.944E 02
EB(EX)			2.105E 04	3.043E 03	2.793E 03	2.677E 03	5.440E 02	4.460E 02	4.260E 02	2.560E 02	2.530E 02
BETA			4.867E-01	1.634E-01	2.219E-02	3.426E-01	7.147E-02	1.005E-02	3.797E-01	1.896E-02	9.963E-01
F(R)			-6.118E-02	-2.056E-02	2.150E-02	-1.255E-05	-8.997E-03	9.744E-03	-5.722E-06	3.173E-06	-8.622E-10
G(R)	44	97	4.717E-01	1.583E-01	3.000E-03	1.703E-04	6.926E-02	1.359E-03	7.765E-05	2.520E-07	1.689E-08
EB(CO)			2.219E 04	3.260E 03	3.035E 03	2.900E 03	6.035E 02	5.138E 02	4.896E 02	3.277E 02	3.229E 02
EB(EX)			2.212E 04	3.224E 03	2.967E 03	2.838E 03	5.850E 02	4.840E 02	4.610E 02	2.830E 02	2.790E 02
BETA			5.100E-01	1.719E-01	2.395E-02	3.564E-01	7.554E-02	1.091E-02	3.942E-01	2.019E-02	1.012E 00
F(R)			-6.550E-02	-2.210E-02	2.318E-02	-1.377E-05	-9.717E-03	1.056E-02	-6.320E-06	3.541E-06	-9.847E-10
G(R)	45	98	4.936E-01	1.663E-01	3.304E-03	1.827E-04	7.310E-02	1.506E-03	8.383E-05	2.874E-07	1.885E-08
EB(CO)			2.331E 04	3.451E 03	3.219E 03	3.070E 03	6.471E 02	5.537E 02	5.267E 02	3.582E 02	3.527E 02
EB(EX)			2.322E 04	3.412E 03	3.146E 03	3.003E 03	6.270E 02	5.210E 02	4.960E 02	3.110E 02	3.070E 02
BETA			5.337E-01	1.806E-01	2.580E-02	3.704E-01	7.974E-02	1.183E-02	4.088E-01	2.147E-02	1.027E 00
F(R)			-7.000E-02	-2.372E-02	2.493E-02	-1.529E-05	-1.047E-02	1.143E-02	-7.057E-06	3.995E-06	-1.151E-09
G(R)	46	103	5.158E-01	1.745E-01	3.634E-03	1.984E-04	7.705E-02	1.666E-03	9.156E-05	3.313E-07	2.156E-08
EB(CO)			2.445E 04	3.650E 03	3.411E 03	3.246E 03	6.932E 02	5.962E 02	5.661E 02	3.908E 02	3.846E 02
EB(EX)			2.435E 04	3.605E 03	3.330E 03	3.173E 03	6.700E 02	5.590E 02	5.310E 02	3.400E 02	3.340E 02
BETA			5.587E-01	1.898E-01	2.779E-02	3.849E-01	8.415E-02	1.280E-02	4.241E-01	2.282E-02	1.045E 00
F(R)			-7.479E-02	-2.545E-02	2.681E-02	-1.672E-05	-1.128E-02	1.236E-02	-7.762E-06	4.435E-06	-1.306E-09
G(R)	47	104	5.391E-01	1.832E-01	3.990E-03	2.123E-04	8.119E-02	1.839E-03	9.854E-05	3.754E-07	2.394E-08
EB(CO)			2.563E 04	3.854E 03	3.607E 03	3.427E 03	7.406E 02	6.399E 02	6.065E 02	4.243E 02	4.174E 02
EB(EX)			2.551E 04	3.806E 03	3.524E 03	3.351E 03	7.170E 02	6.020E 02	5.710E 02	3.710E 02	3.670E 02
BETA			5.843E-01	1.994E-01	2.989E-02	3.996E-01	8.873E-02	1.384E-02	4.396E-01	2.422E-02	1.063E 00
F(R)			-7.981E-02	-2.727E-02	2.880E-02	-1.836E-05	-1.214E-02	1.334E-02	-8.575E-06	4.944E-06	-1.497E-09
G(R)	48	107	5.630E-01	1.921E-01	4.376E-03	2.283E-04	8.548E-02	2.027E-03	1.066E-04	4.271E-07	2.686E-08
EB(CO)			2.683E 04	4.065E 03	3.812E 03	3.614E 03	7.905E 02	6.861E 02	6.490E 02	4.599E 02	4.522E 02
EB(EX)			2.671E 04	4.018E 03	3.727E 03	3.537E 03	7.700E 02	6.510E 02	6.170E 02	4.110E 02	4.040E 02
BETA			6.110E-01	2.093E-01	3.213E-02	4.148E-01	9.353E-02	1.496E-02	4.555E-01	2.569E-02	1.081E 00
F(R)			-8.510E-02	-2.920E-02	3.091E-02	-2.008E-05	-1.305E-02	1.439E-02	-9.427E-06	5.482E-06	-1.700E-09
G(R)	49	109	5.878E-01	2.013E-01	4.793E-03	2.444E-04	8.997E-02	2.232E-03	1.147E-04	4.831E-07	2.987E-08
EB(CO)			2.807E 04	4.284E 03	4.024E 03	3.806E 03	8.427E 02	7.345E 02	6.935E 02	4.974E 02	4.888E 02
EB(EX)			2.794E 04	4.238E 03	3.938E 03	3.730E 03	8.250E 02	7.020E 02	6.640E 02	4.510E 02	4.430E 02

Instr. see p. 26

	Z	A	K	L_{I}	L_{II}	L_{III}	M_{I}	M_{II}	M_{III}	M_{IV}	M_{V}
BETA			6.383E-01	2.196E-01	3.450E-02	4.303E-01	9.848E-02	1.613E-02	4.718E-01	2.723E-02	1.100E 00
F(R)			-9.063E-02	-3.122E-02	3.313E-02	-2.204E-05	-1.401E-02	1.550E-02	-1.040E-05	6.100E-06	-1.947E-09
G(R)	50	113	6.131E-01	2.109E-01	5.243E-03	2.630E-04	9.458E-02	2.452E-03	1.241E-04	5.483E-07	3.354E-08
EB(CO)			2.934E 04	4.509E 03	4.242E 03	4.004E 03	8.967E 02	7.847E 02	7.394E 02	5.363E 02	5.267E 02
EB(EX)			2.920E 04	4.465E 03	4.156E 03	3.929E 03	8.840E 02	7.570E 02	7.150E 02	4.930E 02	4.850E 02
BETA			6.668E-01	2.303E-01	3.701E-02	4.463E-01	1.037E-01	1.739E-02	4.885E-01	2.884E-02	1.120E 00
F(R)			-9.647E-02	-3.337E-02	3.550E-02	-2.409E-05	-1.503E-02	1.668E-02	-1.143E-05	6.755E-06	-2.212E-09
G(R)	51	116	6.394E-01	2.208E-01	5.728E-03	2.817E-04	9.940E-02	2.692E-03	1.336E-04	6.190E-07	3.734E-08
EB(CO)			3.065E 04	4.742E 03	4.468E 03	4.208E 03	9.531E 02	8.372E 02	7.873E 02	5.771E 02	5.665E 02
EB(EX)			3.049E 04	4.698E 03	4.381E 03	4.132E 03	9.440E 02	8.120E 02	7.660E 02	5.370E 02	5.280E 02
BETA			6.960E-01	2.413E-01	3.967E-02	4.626E-01	1.090E-01	1.872E-02	5.056E-01	3.051E-02	1.140E 00
F(R)			-1.025E-01	-3.562E-02	3.798E-02	-2.643E-05	-1.610E-02	1.792E-02	-1.260E-05	7.501E-06	-2.532E-09
G(R)	52	121	6.663E-01	2.310E-01	6.250E-03	3.031E-04	1.044E-01	2.950E-03	1.444E-04	7.008E-07	4.192E-08
EB(CO)			3.198E 04	4.982E 03	4.700E 03	4.417E 03	1.011E 03	8.915E 02	8.367E 02	6.193E 02	6.076E 02
EB(EX)			3.181E 04	4.939E 03	4.612E 03	4.341E 03	1.006E 03	8.700E 02	8.180E 02	5.820E 02	5.720E 02
BETA			7.270E-01	2.531E-01	4.253E-02	4.795E-01	1.147E-01	2.015E-02	5.234E-01	3.228E-02	1.162E 00
F(R)			-1.091E-01	-3.804E-02	4.065E-02	-2.858E-05	-1.725E-02	1.926E-02	-1.369E-05	8.208E-06	-2.814E-09
G(R)	53	121	6.949E-01	2.419E-01	6.815E-03	3.214E-04	1.097E-01	3.230E-03	1.539E-04	7.808E-07	4.571E-08
EB(CO)			3.335E 04	5.229E 03	4.940E 03	4.632E 03	1.072E 03	9.485E 02	8.884E 02	6.638E 02	6.508E 02
EB(EX)			3.317E 04	5.188E 03	4.852E 03	4.557E 03	1.072E 03	9.310E 02	8.750E 02	6.310E 02	6.190E 02
BETA			7.582E-01	2.651E-01	4.551E-02	4.966E-01	1.205E-01	2.165E-02	5.413E-01	3.411E-02	1.184E 00
F(R)			-1.157E-01	-4.054E-02	4.342E-02	-3.134E-05	-1.844E-02	2.066E-02	-1.508E-05	9.103E-06	-3.222E-09
G(R)	54	127	7.234E-01	2.529E-01	7.421E-03	3.458E-04	1.150E-01	3.531E-03	1.664E-04	8.824E-07	5.136E-08
EB(CO)			3.475E 04	5.484E 03	5.188E 03	4.852E 03	1.135E 03	1.007E 03	9.415E 02	7.096E 02	6.953E 02
EB(EX)			3.458E 04	5.452E 03	5.100E 03	4.781E 03	1.143E 03	9.990E 02	9.370E 02	6.840E 02	6.690E 02
BETA			7.913E-01	2.778E-01	4.872E-02	5.144E-01	1.267E-01	2.327E-02	5.600E-01	3.604E-02	1.207E 00
F(R)			-1.229E-01	-4.322E-02	4.641E-02	-3.388E-05	-1.972E-02	2.216E-02	-1.638E-05	9.952E-06	-3.585E-09
G(R)	55	128	7.537E-01	2.645E-01	8.075E-03	3.670E-04	1.207E-01	3.858E-03	1.774E-04	9.817E-07	5.609E-08
EB(CO)			3.619E 04	5.746E 03	5.443E 03	5.079E 03	1.201E 03	1.069E 03	9.967E 02	7.573E 02	7.417E 02
EB(EX)			3.599E 04	5.713E 03	5.359E 03	5.012E 03	1.218E 03	1.071E 03	1.002E 03	7.390E 02	7.210E 02
BETA			8.249E-01	2.908E-01	5.208E-02	5.324E-01	1.331E-01	2.497E-02	5.788E-01	3.803E-02	1.229E 00
F(R)			-1.303E-01	-4.602E-02	4.952E-02	-3.695E-05	-2.106E-02	2.374E-02	-1.794E-05	1.097E-05	-4.063E-09
G(R)	56	133	7.843E-01	2.764E-01	8.776E-03	3.930E-04	1.265E-01	4.209E-03	1.908E-04	1.102E-06	6.243E-08
EB(CO)			3.766E 04	6.017E 03	5.707E 03	5.312E 03	1.269E 03	1.133E 03	1.054E 03	8.076E 02	7.904E 02
EB(EX)			3.744E 04	5.987E 03	5.624E 03	5.247E 03	1.293E 03	1.131E 03	1.061E 03	7.970E 02	7.740E 02

Instr. see p. 26

	Z	A	K	L_I	L_{II}	L_{III}	M_I	M_{II}	M_{III}	M_{IV}	M_V
BETA			8.605E-01	3.046E-01	5.569E-02	5.511E-01	1.398E-01	2.680E-02	5.985E-01	4.013E-02	1.254E 00
F(R)			-1.382E-01	-4.901E-02	5.286E-02	-3.986E-05	-2.250E-02	2.544E-02	-1.944E-05	1.196E-05	-4.502E-09
G(R)	57	134	8.167E-01	2.890E-01	9.533E-03	4.164E-04	1.326E-01	4.588E-03	2.030E-04	1.222E-06	6.796E-08
EB(CO)			3.916E 04	6.295E 03	5.977E 03	5.549E 03	1.339E 03	1.199E 03	1.113E 03	8.590E 02	8.402E 02
EB(EX)			3.893E 04	6.267E 03	5.891E 03	5.483E 03	1.363E 03	1.205E 03	1.125E 03	8.510E 02	8.320E 02
BETA			8.984E-01	3.193E-01	5.959E-02	5.706E-01	1.470E-01	2.878E-02	6.190E-01	4.234E-02	1.279E 00
F(R)			-1.466E-01	-5.222E-02	5.646E-02	-4.255E-05	-2.404E-02	2.726E-02	-2.083E-05	1.289E-05	-4.884E-09
G(R)	58	131	8.511E-01	3.024E-01	1.035E-02	4.367E-04	1.392E-01	5.000E-03	2.138E-04	1.339E-06	7.245E-08
EB(CO)			4.070E 04	6.581E 03	6.256E 03	5.794E 03	1.412E 03	1.268E 03	1.174E 03	9.126E 02	8.921E 02
EB(EX)			4.044E 04	6.549E 03	6.165E 03	5.724E 03	1.434E 03	1.276E 03	1.186E 03	9.020E 02	8.830E 02
BETA			9.358E-01	3.340E-01	6.359E-02	5.901E-01	1.541E-01	3.081E-02	6.395E-01	4.461E-02	1.304E 00
F(R)			-1.552E-01	-5.549E-02	6.014E-02	-4.624E-05	-2.562E-02	2.914E-02	-2.273E-05	1.415E-05	-5.501E-09
G(R)	59	136	8.849E-01	3.157E-01	1.122E-02	4.664E-04	1.457E-01	5.439E-03	2.292E-04	1.494E-06	8.020E-08
EB(CO)			4.227E 04	6.874E 03	6.542E 03	6.043E 03	1.487E 03	1.338E 03	1.237E 03	9.676E 02	9.453E 02
EB(EX)			4.199E 04	6.835E 03	6.441E 03	5.965E 03	1.508E 03	1.337E 03	1.237E 03	9.510E 02	9.320E 02
BETA			9.754E-01	3.495E-01	6.789E-02	6.103E-01	1.618E-01	3.300E-02	6.608E-01	4.700E-02	1.331E 00
F(R)			-1.643E-01	-5.899E-02	6.408E-02	-4.985E-05	-2.731E-02	3.115E-02	-2.460E-05	1.539E-05	-6.095E-09
G(R)	60	138	9.206E-01	3.298E-01	1.216E-02	4.943E-04	1.526E-01	5.913E-03	2.439E-04	1.652E-06	8.736E-08
EB(CO)			4.388E 04	7.176E 03	6.836E 03	6.298E 03	1.564E 03	1.412E 03	1.302E 03	1.024E 03	1.000E 03
EB(EX)			4.357E 04	7.126E 03	6.722E 03	6.208E 03	1.575E 03	1.404E 03	1.294E 03	1.004E 03	9.790E 02
BETA			1.016E 00	3.657E-01	7.243E-02	6.310E-01	1.697E-01	3.532E-02	6.826E-01	4.948E-02	1.358E 00
F(R)			-1.738E-01	-6.267E-02	6.824E-02	-5.380E-05	-2.909E-02	3.327E-02	-2.666E-05	1.677E-05	-6.773E-09
G(R)	61	141	9.572E-01	3.444E-01	1.317E-02	5.247E-04	1.598E-01	6.423E-03	2.599E-04	1.829E-06	9.549E-08
EB(CO)			4.552E 04	7.487E 03	7.140E 03	6.559E 03	1.645E 03	1.488E 03	1.368E 03	1.084E 03	1.057E 03
EB(EX)			4.518E 04	7.428E 03	7.013E 03	6.460E 03	1.651E 03	1.472E 03	1.357E 03	1.056E 03	1.030E 03
BETA			1.059E 00	3.826E-01	7.727E-02	6.523E-01	1.780E-01	3.780E-02	7.049E-01	5.206E-02	1.385E 00
F(R)			-1.838E-01	-6.656E-02	7.265E-02	-5.789E-05	-3.097E-02	3.554E-02	-2.879E-05	1.820E-05	-7.480E-09
G(R)	62	143	9.954E-01	3.596E-01	1.425E-02	5.552E-04	1.673E-01	6.972E-03	2.761E-04	2.017E-06	1.037E-07
EB(CO)			4.720E 04	7.806E 03	7.452E 03	6.827E 03	1.728E 03	1.567E 03	1.438E 03	1.145E 03	1.116E 03
EB(EX)			4.684E 04	7.737E 03	7.312E 03	6.717E 03	1.729E 03	1.542E 03	1.421E 03	1.108E 03	1.082E 03
BETA			1.103E 00	4.003E-01	8.242E-02	6.743E-01	1.867E-01	4.043E-02	7.280E-01	5.477E-02	1.413E 00
F(R)			-1.943E-01	-7.069E-02	7.733E-02	-6.209E-05	-3.297E-02	3.794E-02	-3.099E-05	1.969E-05	-8.210E-09
G(R)	63	144	1.035E 00	3.756E-01	1.541E-02	5.859E-04	1.751E-01	7.564E-03	2.924E-04	2.216E-06	1.120E-07
EB(CO)			4.892E 04	8.134E 03	7.772E 03	7.101E 03	1.814E 03	1.648E 03	1.508E 03	1.208E 03	1.177E 03
EB(EX)			4.852E 04	8.052E 03	7.618E 03	6.977E 03	1.800E 03	1.613E 03	1.481E 03	1.158E 03	1.128E 03
EB(EX)											

Instr. see p. 26

	Z	A	K	L_{I}	L_{II}	L_{III}	M_{I}	M_{II}	M_{III}	M_{IV}	M_{V}
BETA			1.149E 00	4.187E-01	8.783E-02	6.966E-01	1.957E-01	4.322E-02	7.515E-01	5.758E-02	1.442E 00
F(R)			-2.053E-01	-7.500E-02	8.225E-02	-6.683E-05	-3.507E-02	4.047E-02	-3.348E-05	2.137E-05	-9.082E-09
G(R)	64	147	1.076E 00	3.919E-01	1.666E-02	6.206E-04	1.832E-01	8.199E-03	3.108E-04	2.444E-06	1.220E-07
EB(CO)			5.068E 04	8.472E 03	8.103E 03	7.381E 03	1.903E 03	1.733E 03	1.582E 03	1.274E 03	1.240E 03
EB(EX)			5.024E 04	8.376E 03	7.931E 03	7.243E 03	1.882E 03	1.690E 03	1.544E 03	1.221E 03	1.181E 03
BETA			1.195E 00	4.372E-01	9.347E-02	7.193E-01	2.049E-01	4.612E-02	7.754E-01	6.048E-02	1.471E 00
F(R)			-2.165E-01	-7.945E-02	8.734E-02	-7.246E-05	-3.724E-02	4.310E-02	-3.643E-05	2.336E-05	-1.020E-08
G(R)	65	154	1.116E 00	4.085E-01	1.798E-02	6.623E-04	1.914E-01	8.874E-03	3.329E-04	2.713E-06	1.349E-07
EB(CO)			5.247E 04	8.817E 03	8.440E 03	7.666E 03	1.994E 03	1.819E 03	1.657E 03	1.340E 03	1.304E 03
EB(EX)			5.200E 04	8.708E 03	8.252E 03	7.515E 03	1.967E 03	1.768E 03	1.611E 03	1.280E 03	1.240E 03
BETA			1.244E 00	4.574E-01	9.960E-02	7.430E-01	2.148E-01	4.928E-02	8.005E-01	6.355E-02	1.502E 00
F(R)			-2.287E-01	-8.428E-02	9.288E-02	-7.752E-05	-3.959E-02	4.596E-02	-3.912E-05	2.520E-05	-1.115E-08
G(R)	66	155	1.160E 00	4.264E-01	1.941E-02	6.977E-04	2.002E-01	9.609E-03	3.519E-04	2.970E-06	1.452E-07
EB(CO)			5.430E 04	9.172E 03	8.788E 03	7.958E 03	2.088E 03	1.909E 03	1.734E 03	1.409E 03	1.370E 03
EB(EX)			5.379E 04	9.047E 03	8.581E 03	7.790E 03	2.046E 03	1.843E 03	1.677E 03	1.335E 03	1.295E 03
BETA			1.294E 00	4.779E-01	1.060E-01	7.671E-01	2.249E-01	5.258E-02	8.258E-01	6.671E-02	1.533E 00
F(R)			-2.412E-01	-8.926E-02	9.862E-02	-8.353E-05	-4.202E-02	4.892E-02	-4.229E-05	2.736E-05	-1.238E-08
G(R)	67	160	1.204E 00	4.445E-01	2.094E-02	7.404E-04	2.092E-01	1.039E-02	3.747E-04	3.274E-06	1.587E-07
EB(CO)			5.617E 04	9.536E 03	9.143E 03	8.255E 03	2.185E 03	2.001E 03	1.813E 03	1.480E 03	1.438E 03
EB(EX)			5.562E 04	9.395E 03	8.919E 03	8.071E 03	2.127E 03	1.922E 03	1.739E 03	1.390E 03	1.345E 03
BETA			1.348E 00	4.998E-01	1.129E-01	7.922E-01	2.357E-01	5.614E-02	8.523E-01	7.003E-02	1.565E 00
F(R)			-2.545E-01	-9.462E-02	1.048E-01	-8.924E-05	-4.465E-02	5.213E-02	-4.533E-05	2.945E-05	-1.349E-08
G(R)	68	161	1.252E 00	4.639E-01	2.259E-02	7.791E-04	2.188E-01	1.124E-02	3.956E-04	3.576E-06	1.705E-07
EB(CO)			5.808E 04	9.910E 03	9.510E 03	8.559E 03	2.285E 03	2.097E 03	1.894E 03	1.554E 03	1.509E 03
EB(EX)			5.749E 04	9.752E 03	9.265E 03	8.358E 03	2.212E 03	2.006E 03	1.811E 03	1.455E 03	1.410E 03
BETA			1.402E 00	5.221E-01	1.201E-01	8.175E-01	2.468E-01	5.987E-02	8.790E-01	7.347E-02	1.597E 00
F(R)			-2.682E-01	-1.002E-01	1.112E-01	-9.599E-05	-4.736E-02	5.546E-02	-4.893E-05	3.191E-05	-1.493E-08
G(R)	69	166	1.299E 00	4.836E-01	2.434E-02	8.256E-04	2.285E-01	1.214E-02	4.206E-04	3.932E-06	1.859E-07
EB(CO)			6.003E 04	1.029E 04	9.886E 03	8.869E 03	2.389E 03	2.196E 03	1.978E 03	1.629E 03	1.581E 03
EB(EX)			6.939E 04	1.012E 04	9.618E 03	8.648E 03	2.307E 03	2.093E 03	1.888E 03	1.515E 03	1.471E 03
BETA			1.459E 00	5.460E-01	1.278E-01	8.440E-01	2.586E-01	6.389E-02	9.070E-01	7.708E-02	1.631E 00
F(R)			-2.829E-01	-1.061E-01	1.182E-01	-1.024E-04	-5.028E-02	5.905E-02	-5.236E-05	3.429E-05	-1.624E-08
G(R)	70	167	1.349E 00	5.046E-01	2.623E-02	8.679E-04	2.390E-01	1.312E-02	4.436E-04	4.285E-06	1.992E-07
EB(CO)			6.202E 04	1.069E 04	1.027E 04	9.185E 03	2.495E 03	2.297E 03	2.063E 03	1.706E 03	1.654E 03
EB(EX)			6.133E 04	1.049E 04	9.978E 03	8.943E 03	2.398E 03	2.171E 03	1.951E 03	1.578E 03	1.528E 03

	Z	A	K	L_I	L_{II}	L_{III}	M_I	M_{II}	M_{III}	M_{IV}	M_V
BETA			1.517E 00	5.703E-01	1.359E-01	8.707E-01	2.706E-01	6.808E-02	9.353E-01	8.080E-02	1.665E 00
F(R)			-2.979E-01	-1.123E-01	1.253E-01	-1.100E-04	-5.330E-02	6.279E-02	-5.641E-05	3.709E-05	-1.792E-08
G(R)	71	172	1.400E 00	5.258E-01	2.824E-02	9.186E-04	2.495E-01	1.415E-02	4.710E-04	4.701E-06	2.167E-07
EB(CO)			6.405E 04	1.109E 04	1.067E 04	9.507E 03	2.604E 03	2.402E 03	2.151E 03	1.786E 03	1.730E 03
EB(EX)			6.331E 04	1.087E 04	1.035E 04	9.244E 03	2.492E 03	2.265E 03	2.026E 03	1.639E 03	1.589E 03
BETA			1.581E 00	5.969E-01	1.447E-01	8.991E-01	2.838E-01	7.268E-02	9.653E-01	8.475E-02	1.700E 00
F(R)			-3.144E-01	-1.190E-01	1.332E-01	-1.165E-04	-5.660E-02	6.688E-02	-5.997E-05	3.957E-05	-1.923E-08
G(R)	72	170	1.455E 00	5.491E-01	3.043E-02	9.596E-04	2.611E-01	1.529E-02	4.936E-04	5.083E-06	2.293E-07
EB(CO)			6.613E 04	1.150E 04	1.107E 04	9.836E 03	2.717E 03	2.510E 03	2.241E 03	1.867E 03	1.808E 03
EB(EX)			6.535E 04	1.127E 04	1.074E 04	9.561E 03	2.602E 03	2.364E 03	2.110E 03	1.718E 03	1.662E 03
BETA			1.642E 00	6.229E-01	1.537E-01	9.269E-01	2.967E-01	7.735E-02	9.946E-01	8.874E-02	1.735E 00
F(R)			-3.305E-01	-1.257E-01	1.411E-01	-1.256E-04	-5.991E-02	7.101E-02	-6.486E-05	4.296E-05	-2.140E-08
G(R)	73	178	1.508E 00	5.717E-01	3.271E-02	1.020E-03	2.723E-01	1.647E-02	5.264E-04	5.597E-06	2.517E-07
EB(CO)			6.825E 04	1.193E 04	1.149E 04	1.017E 04	2.834E 03	2.622E 03	2.334E 03	1.952E 03	1.888E 03
EB(EX)			6.741E 04	1.168E 04	1.114E 04	9.881E 03	2.703E 03	2.469E 03	2.194E 03	1.793E 03	1.735E 03
BETA			1.710E 00	6.513E-01	1.635E-01	9.564E-01	3.109E-01	8.247E-02	1.026E 00	9.299E-02	1.772E 00
F(R)			-3.482E-01	-1.331E-01	1.497E-01	-1.337E-04	-6.353E-02	7.553E-02	-6.923E-05	4.601E-05	-2.318E-08
G(R)	74	179	1.566E 00	5.964E-01	3.520E-02	1.070E-03	2.846E-01	1.777E-02	5.540E-04	6.075E-06	2.689E-07
EB(CO)			7.040E 04	1.236E 04	1.192E 04	1.051E 04	2.953E 03	2.736E 03	2.428E 03	2.038E 03	1.969E 03
EB(EX)			6.952E 04	1.210E 04	1.154E 04	1.020E 04	2.818E 03	2.575E 03	2.281E 03	1.871E 03	1.809E 03
BETA			1.779E 00	6.811E-01	1.739E-01	9.868E-01	3.257E-01	8.791E-02	1.058E 00	9.741E-02	1.810E 00
F(R)			-3.668E-01	-1.408E-01	1.589E-01	-1.422E-04	-6.736E-02	8.032E-02	-7.384E-05	4.925E-05	-2.508E-08
G(R)	75	180	1.626E 00	6.221E-01	3.787E-02	1.123E-03	2.974E-01	1.915E-02	5.829E-04	6.589E-06	2.870E-07
EB(CO)			7.261E 04	1.281E 04	1.236E 04	1.086E 04	3.075E 03	2.854E 03	2.525E 03	2.125E 03	2.053E 03
EB(EX)			7.168E 04	1.253E 04	1.196E 04	1.053E 04	2.931E 03	2.681E 03	2.367E 03	1.949E 03	1.883E 03
BETA			1.851E 00	7.118E-01	1.849E-01	1.018E 00	3.410E-01	9.365E-02	1.091E 00	1.020E-01	1.848E 00
F(R)			-3.860E-01	-1.489E-01	1.685E-01	-1.516E-04	-7.135E-02	8.535E-02	-7.899E-05	5.287E-05	-2.732E-08
G(R)	76	183	1.688E 00	6.486E-01	4.072E-02	1.181E-03	3.107E-01	2.064E-02	6.152E-04	7.167E-06	3.084E-07
EB(CO)			7.485E 04	1.327E 04	1.281E 04	1.121E 04	3.202E 03	2.976E 03	2.624E 03	2.216E 03	2.139E 03
EB(EX)			7.387E 04	1.296E 04	1.238E 04	1.087E 04	3.050E 03	2.792E 03	2.457E 03	2.031E 03	1.960E 03
BETA			1.925E 00	7.439E-01	1.965E-01	1.049E 00	3.570E-01	9.974E-02	1.125E 00	1.067E-01	1.887E 00
F(R)			-4.062E-01	-1.574E-01	1.787E-01	-1.616E-04	-7.557E-02	9.069E-02	-8.446E-05	5.671E-05	-2.972E-08
G(R)	77	186	1.751E 00	6.762E-01	4.377E-02	1.242E-03	3.244E-01	2.223E-02	6.489E-04	7.789E-06	3.310E-07
EB(CO)			7.715E 04	1.374E 04	1.327E 04	1.158E 04	3.333E 03	3.102E 03	2.726E 03	2.310E 03	2.227E 03
EB(EX)			7.611E 04	1.342E 04	1.282E 04	1.122E 04	3.172E 03	2.909E 03	2.550E 03	2.116E 03	2.041E 03

Instr. see p. 26

	Z	A	K	L_I	L_{II}	L_{III}	M_I	M_{II}	M_{III}	M_{IV}	M_V
BETA			2.001E 00	7.769E-01	2.087E-01	1.082E 00	3.734E-01	1.061E-01	1.159E 00	1.117E-01	1.926E 00
F(R)			-4.270E-01	-1.663E-01	1.893E-01	-1.727E-04	-7.995E-02	9.626E-02	-9.051E-05	6.097E-05	-3.251E-08
G(R)	78	191	1.816E 00	7.044E-01	4.700E-02	1.310E-03	3.385E-01	2.392E-02	6.863E-04	8.484E-06	3.574E-07
EB(CO)			7.949E 04	1.422E 04	1.374E 04	1.194E 04	3.466E 03	3.230E 03	2.830E 03	2.404E 03	2.316E 03
EB(EX)			7.840E 04	1.389E 04	1.327E 04	1.156E 04	3.297E 03	3.026E 03	2.645E 03	2.202E 03	2.122E 03
BETA			2.082E 00	8.120E-01	2.218E-01	1.115E 00	3.909E-01	1.130E-01	1.195E 00	1.168E-01	1.968E 00
F(R)			-4.491E-01	-1.757E-01	2.006E-01	-1.839E-04	-8.464E-02	1.022E-01	-9.664E-05	6.531E-05	-3.529E-08
G(R)	79	194	1.884E 00	7.343E-01	5.049E-02	1.377E-03	3.535E-01	2.574E-02	7.232E-04	9.203E-06	3.830E-07
EB(CO)			8.187E 04	1.471E 04	1.423E 04	1.231E 04	3.603E 03	3.362E 03	2.935E 03	2.501E 03	2.408E 03
EB(EX)			8.072E 04	1.435E 04	1.373E 04	1.192E 04	3.425E 03	3.149E 03	2.743E 03	2.291E 03	2.206E 03
BETA			2.168E 00	8.498E-01	2.360E-01	1.150E 00	4.098E-01	1.205E-01	1.232E 00	1.223E-01	2.010E 00
F(R)			-4.729E-01	-1.859E-01	2.129E-01	-1.944E-04	-8.971E-02	1.087E-01	-1.025E-04	6.947E-05	-3.777E-08
G(R)	80	193	1.957E 00	7.666E-01	5.428E-02	1.437E-03	3.696E-01	2.773E-02	7.571E-04	9.911E-06	4.047E-07
EB(CO)			8.431E 04	1.522E 04	1.473E 04	1.270E 04	3.745E 03	3.498E 03	3.044E 03	2.601E 03	2.502E 03
EB(EX)			8.310E 04	1.485E 04	1.421E 04	1.228E 04	3.562E 03	3.280E 03	2.847E 03	2.385E 03	2.295E 03
BETA			2.252E 00	8.870E-01	2.504E-01	1.185E 00	4.283E-01	1.281E-01	1.269E 00	1.278E-01	2.052E 00
F(R)			-4.965E-01	-1.962E-01	2.254E-01	-2.080E-04	-9.479E-02	1.153E-01	-1.100E-04	7.478E-05	-4.146E-08
G(R)	81	200	2.027E 00	7.981E-01	5.823E-02	1.518E-03	3.853E-01	2.980E-02	8.021E-04	1.080E-05	4.388E-07
EB(CO)			8.679E 04	1.573E 04	1.524E 04	1.308E 04	3.890E 03	3.639E 03	3.155E 03	2.703E 03	2.598E 03
EB(EX)			8.553E 04	1.534E 04	1.470E 04	1.266E 04	3.700E 03	3.416E 03	2.956E 03	2.485E 03	2.389E 03
BETA			2.344E 00	9.278E-01	2.663E-01	1.222E 00	4.470E-01	1.365E-01	1.308E 00	1.336E-01	2.096E 00
F(R)			-5.223E-01	-2.074E-01	2.390E-01	-2.205E-04	-9.997E-02	1.225E-01	-1.169E-04	7.971E-05	-4.460E-08
G(R)	82	201	2.105E 00	8.326E-01	6.255E-02	1.588E-03	4.010E-01	3.207E-02	8.417E-04	1.166E-05	4.661E-07
EB(CO)			8.933E 04	1.626E 04	1.576E 04	1.348E 04	4.041E 03	3.784E 03	3.269E 03	2.808E 03	2.696E 03
EB(EX)			8.801E 04	1.586E 04	1.520E 04	1.304E 04	3.850E 03	3.554E 03	3.066E 03	2.586E 03	2.484E 03
BETA			2.439E 00	9.703E-01	2.830E-01	1.260E 00	4.703E-01	1.453E-01	1.349E 00	1.397E-01	2.141E 00
F(R)			-5.491E-01	-2.192E-01	2.534E-01	-2.339E-04	-1.063E-01	1.301E-01	-1.243E-04	8.504E-05	-4.809E-08
G(R)	83	203	2.184E 00	8.683E-01	6.715E-02	1.664E-03	4.209E-01	3.449E-02	8.843E-04	1.259E-05	4.964E-07
EB(CO)			9.191E 04	1.681E 04	1.629E 04	1.387E 04	4.193E 03	3.931E 03	3.384E 03	2.914E 03	2.796E 03
EB(EX)			9.053E 04	1.639E 04	1.571E 04	1.342E 04	3.999E 03	3.696E 03	3.177E 03	2.687E 03	2.579E 03
BETA			2.536E 00	1.014E 00	3.006E-01	1.298E 00	4.921E-01	1.546E-01	1.390E 00	1.460E-01	2.187E 00
F(R)			-5.769E-01	-2.314E-01	2.684E-01	-2.487E-04	-1.124E-01	1.380E-01	-1.326E-04	9.095E-05	-5.214E-08
G(R)	84	207	2.265E 00	9.050E-01	7.205E-02	1.748E-03	4.392E-01	3.707E-02	9.315E-04	1.363E-05	5.317E-07
EB(CO)			9.455E 04	1.736E 04	1.684E 04	1.428E 04	4.351E 03	4.083E 03	3.501E 03	3.022E 03	2.897E 03
EB(EX)			9.311E 04	1.693E 04	1.624E 04	1.381E 04	4.154E 03	3.845E 03	3.293E 03	2.793E 03	2.679E 03

	Z	A	K	L_I	L_{II}	L_{III}	M_I	M_{II}	M_{III}	M_{IV}	M_V
BETA			2.640E 00	1.061E 00	3.197E-01	1.338E 00	5.156E-01	1.647E-01	1.432E 00	1.526E-01	2.234E 00
F(R)			-6.066E-01	-2.446E-01	2.847E-01	-2.632E-04	-1.190E-01	1.466E-01	-1.407E-04	9.679E-05	-5.596E-08
G(R)	85	208	2.352E 00	9.443E-01	7.736E-02	1.828E-03	4.589E-01	3.987E-02	9.765E-04	1.467E-05	5.638E-07
EB(CO)			9.724E 04	1.794E 04	1.741E 04	1.469E 04	4.513E 03	4.240E 03	3.622E 03	3.134E 03	3.001E 03
EB(EX)			9.574E 04	1.749E 04	1.678E 04	1.421E 04	4.314E 03	3.999E 03	3.411E 03	2.902E 03	2.781E 03
BETA			2.746E 00	1.109E 00	3.397E-01	1.379E 00	5.400E-01	1.753E-01	1.476E 00	1.594E-01	2.282E 00
F(R)			-6.373E-01	-2.583E-01	3.017E-01	-2.793E-04	-1.258E-01	1.556E-01	-1.497E-04	1.033E-04	-6.040E-08
G(R)	86	211	2.440E 00	9.846E-01	8.301E-02	1.916E-03	4.792E-01	4.286E-02	1.027E-03	1.584E-05	6.014E-07
EB(CO)			9.998E 04	1.852E 04	1.799E 04	1.511E 04	4.680E 03	4.402E 03	3.745E 03	3.248E 03	3.108E 03
EB(EX)			9.842E 04	1.806E 04	1.733E 04	1.461E 04	4.478E 03	4.158E 03	3.534E 03	3.014E 03	2.886E 03
BETA			2.859E 00	1.161E 00	3.613E-01	1.421E 00	5.658E-01	1.867E-01	1.522E 00	1.665E-01	2.331E 00
F(R)			-6.700E-01	-2.730E-01	3.199E-01	-2.953E-04	-1.331E-01	1.653E-01	-1.588E-04	1.098E-04	-6.473E-08
G(R)	87	212	2.533E 00	1.027E 00	8.911E-02	2.002E-03	5.008E-01	4.608E-02	1.076E-03	1.703E-05	6.370E-07
EB(CO)			1.028E 05	1.912E 04	1.858E 04	1.554E 04	4.850E 03	4.567E 03	3.870E 03	3.364E 03	3.216E 03
EB(EX)			1.011E 05	1.864E 04	1.790E 04	1.502E 04	4.647E 03	4.321E 03	3.660E 03	3.128E 03	2.994E 03
BETA			2.969E 00	1.212E 00	3.833E-01	1.463E 00	5.914E-01	1.984E-01	1.567E 00	1.738E-01	2.380E 00
F(R)			-7.024E-01	-2.877E-01	3.384E-01	-3.152E-04	-1.405E-01	1.751E-01	-1.699E-04	1.178E-04	-7.079E-08
G(R)	88	220	2.623E 00	1.069E 00	9.545E-02	2.111E-03	5.219E-01	4.943E-02	1.138E-03	1.849E-05	6.885E-07
EB(CO)			1.056E 05	1.974E 04	1.919E 04	1.597E 04	5.026E 03	4.737E 03	3.998E 03	3.482E 03	3.326E 03
EB(EX)			1.039E 05	1.924E 04	1.848E 04	1.544E 04	4.821E 03	4.489E 03	3.798E 03	3.246E 03	3.104E 03
BETA			3.087E 00	1.267E 00	4.072E-01	1.508E 00	6.192E-01	2.111E-01	1.614E 00	1.814E-01	2.431E 00
F(R)			-7.374E-01	-3.037E-01	3.585E-01	-3.344E-04	-1.485E-01	1.858E-01	-1.808E-04	1.257E-04	-7.646E-08
G(R)	89	224	2.720E 00	1.115E 00	1.024E-01	2.214E-03	5.448E-01	5.310E-02	1.196E-03	1.996E-05	7.352E-07
EB(CO)			1.085E 05	2.037E 04	1.981E 04	1.641E 04	5.207E 03	4.913E 03	4.129E 03	3.604E 03	3.439E 03
EB(EX)			1.068E 05	1.983E 04	1.907E 04	1.587E 04	4.996E 03	4.659E 03	3.919E 03	3.365E 03	3.215E 03
BETA			3.213E 00	1.325E 00	4.330E-01	1.554E 00	6.487E-01	2.248E-01	1.663E 00	1.894E-01	2.484E 00
F(R)			-7.747E-01	-3.207E-01	3.801E-01	-3.537E-04	-1.571E-01	1.973E-01	-1.917E-04	1.336E-04	-8.199E-08
G(R)	90	226	2.823E 00	1.163E 00	1.098E-01	2.315E-03	5.691E-01	5.705E-02	1.254E-03	2.145E-05	7.795E-07
EB(CO)			1.115E 05	2.101E 04	2.045E 04	1.686E 04	5.391E 03	5.092E 03	4.261E 03	3.727E 03	3.553E 03
EB(EX)			1.096E 05	2.047E 04	1.969E 04	1.630E 04	5.182E 03	4.831E 03	4.046E 03	3.491E 03	3.332E 03
BETA			3.345E 00	1.387E 00	4.605E-01	1.601E 00	6.798E-01	2.394E-01	1.714E 00	1.978E-01	2.537E 00
F(R)			-8.139E-01	-3.388E-01	4.031E-01	-3.740E-04	-1.661E-01	2.095E-01	-2.032E-04	1.420E-04	-8.792E-08
G(R)	91	228	2.930E 00	1.214E 00	1.178E-01	2.420E-03	5.947E-01	6.130E-02	1.314E-03	2.306E-05	8.264E-07
EB(CO)			1.146E 05	2.168E 04	2.111E 04	1.731E 04	5.583E 03	5.278E 03	4.398E 03	3.854E 03	3.670E 03
EB(EX)			1.126E 05	2.111E 04	2.031E 04	1.673E 04	5.367E 03	5.002E 03	4.175E 03	3.612E 03	3.441E 03

 Instr. see p. 26

	Z	A	K	L_I	L_{II}	L_{III}	M_I	M_{II}	M_{III}	M_{IV}	M_V
BETA			3.483E 00	1.452E 00	4.900E-01	1.650E 00	7.127E-01	2.551E-01	1.767E 00	2.065E-01	2.593E 00
F(R)			-8.553E-01	-3.580E-01	4.276E-01	-3.948E-04	-1.758E-01	2.226E-01	-2.151E-04	1.506E-04	-9.392E-08
G(R)	92	229	3.042E 00	1.267E 00	1.265E-01	2.525E-03	6.216E-01	6.588E-02	1.375E-03	2.473E-05	8.731E-07
EB(CO)			1.176E 05	2.235E 04	2.178E 04	1.777E 04	5.777E 03	5.467E 03	4.535E 03	3.981E 03	3.788E 03
EB(EX)			1.156E 05	2.176E 04	2.095E 04	1.717E 04	5.549E 03	5.181E 03	4.304E 03	3.728E 03	3.552E 03
BETA			3.627E 00	1.520E 00	5.212E-01	1.700E 00	7.470E-01	2.717E-01	1.821E 00	2.155E-01	2.649E 00
F(R)			-8.984E-01	-3.781E-01	4.535E-01	-4.171E-04	-1.859E-01	2.364E-01	-2.279E-04	1.600E-04	-1.006E-07
G(R)	93	231	3.158E 00	1.322E 00	1.357E-01	2.639E-03	6.496E-01	7.079E-02	1.441E-03	2.655E-05	9.248E-07
EB(CO)			1.208E 05	2.305E 04	2.247E 04	1.824E 04	5.979E 03	5.664E 03	4.676E 03	4.113E 03	3.910E 03
EB(EX)			1.187E 05	2.241E 04	2.160E 04	1.761E 04	5.724E 03	5.354E 03	4.422E 03	3.847E 03	3.666E 03
BETA			3.773E 00	1.590E 00	5.541E-01	1.751E 00	7.824E-01	2.893E-01	1.876E 00	2.248E-01	2.706E 00
F(R)			-9.429E-01	-3.990E-01	4.806E-01	-4.417E-04	-1.964E-01	2.509E-01	-2.420E-04	1.703E-04	-1.082E-07
G(R)	94	235	3.275E 00	1.379E 00	1.455E-01	2.763E-03	6.783E-01	7.601E-02	1.513E-03	2.857E-05	9.844E-07
EB(CO)			1.240E 05	2.377E 04	2.318E 04	1.871E 04	6.186E 03	5.865E 03	4.820E 03	4.247E 03	4.033E 03
EB(EX)			1.218E 05	2.310E 04	2.226E 04	1.805E 04	5.914E 03	5.540E 03	4.555E 03	3.969E 03	3.774E 03
BETA			3.927E 00	1.665E 00	5.894E-01	1.804E 00	8.200E-01	3.081E-01	1.933E 00	2.345E-01	2.765E 00
F(R)			-9.899E-01	-4.213E-01	5.097E-01	-4.670E-04	-2.076E-01	2.664E-01	-2.565E-04	1.809E-04	-1.161E-07
G(R)	95	238	3.399E 00	1.439E 00	1.561E-01	2.889E-03	7.086E-01	8.163E-02	1.586E-03	3.068E-05	1.044E-06
EB(CO)			1.273E 05	2.450E 04	2.391E 04	1.919E 04	6.399E 03	6.073E 03	4.967E 03	4.385E 03	4.159E 03
EB(EX)			1.249E 05	2.379E 04	2.294E 04	1.850E 04	6.114E 03	5.730E 03	4.687E 03	4.092E 03	3.886E 03
BETA			4.091E 00	1.744E 00	6.275E-01	1.859E 00	8.600E-01	3.284E-01	1.992E 00	2.448E-01	2.826E 00
F(R)			-1.040E 00	-4.451E-01	5.409E-01	-4.923E-04	-2.197E-01	2.830E-01	-2.711E-04	1.916E-04	-1.238E-07
G(R)	96	239	3.530E 00	1.503E 00	1.675E-01	3.013E-03	7.409E-01	8.773E-02	1.658E-03	3.284E-05	1.101E-06
EB(CO)			1.307E 05	2.525E 04	2.466E 04	1.968E 04	6.614E 03	6.283E 03	5.113E 03	4.521E 03	4.284E 03
EB(EX)			1.282E 05	2.450E 04	2.364E 04	1.896E 04	6.310E 03	5.920E 03	4.820E 03	4.220E 03	4.000E 03
BETA			4.257E 00	1.825E 00	6.674E-01	1.914E 00	9.011E-01	3.497E-01	2.052E 00	2.552E-01	2.887E 00
F(R)			-1.091E 00	-4.698E-01	5.735E-01	-5.208E-04	-2.321E-01	3.004E-01	-2.876E-04	2.037E-04	-1.329E-07
G(R)	97	243	3.662E 00	1.568E 00	1.796E-01	3.153E-03	7.738E-01	9.419E-02	1.740E-03	3.528E-05	1.171E-06
EB(CO)			1.341E 05	2.602E 04	2.542E 04	2.017E 04	6.839E 03	6.503E 03	5.266E 03	4.664E 03	4.415E 03
EB(EX)			1.314E 05	2.523E 04	2.435E 04	1.944E 04	6.520E 03	6.120E 03	4.960E 03	4.350E 03	4.120E 03
BETA			4.435E 00	1.912E 00	7.105E-01	1.972E 00	9.450E-01	3.727E-01	2.115E 00	2.663E-01	2.951E 00
F(R)			-1.146E 00	-4.962E-01	6.086E-01	-5.494E-04	-2.454E-01	3.192E-01	-3.041E-04	2.159E-04	-1.419E-07
G(R)	98	245	3.802E 00	1.637E 00	1.928E-01	3.291E-03	8.089E-01	1.012E-01	1.820E-03	3.779E-05	1.237E-06
EB(CO)			1.376E 05	2.681E 04	2.621E 04	2.067E 04	7.068E 03	6.727E 03	5.419E 03	4.807E 03	4.546E 03
EB(EX)			1.348E 05	2.597E 04	2.508E 04	1.991E 04	6.730E 03	6.330E 03	5.110E 03	4.480E 03	4.240E 03

	Z	A	K	L_I	L_{II}	L_{III}	M_I	M_{II}	M_{III}	M_{IV}	M_V
BETA			4.618E 00	2.002E 00	7.563E-01	2.032E 00	9.908E-01	3.972E-01	2.180E 00	2.777E-01	3.015E 00
F(R)			-1.203E 00	-5.240E-01	6.458E-01	-5.801E-04	-2.594E-01	3.391E-01	-3.220E-04	2.291E-04	-1.519E-07
G(R)	99	248	3.947E 00	1.709E 00	2.068E-01	3.438E-03	8.453E-01	1.087E-01	1.907E-03	4.051E-05	1.310E-06
EB(CO)			1.412E 05	2.763E 04	2.702E 04	2.118E 04	7.305E 03	6.959E 03	5.576E 03	4.954E 03	4.680E 03
EB(EX)			1.382E 05	2.763E 04	2.582E 04	2.040E 04	6.950E 03	6.540E 03	5.250E 03	4.620E 03	4.370E 03
BETA			4.809E 00	2.097E 00	8.052E-01	2.093E 00	1.039E 00	4.234E-01	2.246E 00	2.896E-01	3.081E 00
F(R)			-1.263E 00	-5.532E-01	6.853E-01	-6.123E-04	-2.742E-01	3.602E-01	-3.408E-04	2.430E-04	-1.624E-07
G(R)	100	251	4.097E 00	1.784E 00	2.219E-01	3.591E-03	8.835E-01	1.168E-01	1.997E-03	4.341E-05	1.387E-06
EB(CO)			1.448E 05	2.846E 04	2.786E 04	2.170E 04	7.548E 03	7.197E 03	5.736E 03	5.104E 03	4.817E 03
EB(EX)			1.416E 05	2.750E 04	2.658E 04	2.089E 04	7.170E 03	6.750E 03	5.400E 03	4.760E 03	4.490E 03
BETA			5.004E 00	2.195E 00	8.568E-01	2.155E 00	1.088E 00	4.509E-01	2.313E 00	3.018E-01	3.147E 00
F(R)			-1.325E 00	-5.836E-01	7.268E-01	-6.476E-04	-2.896E-01	3.824E-01	-3.613E-04	2.582E-04	-1.745E-07
G(R)	101	256	4.249E 00	1.861E 00	2.380E-01	3.759E-03	9.226E-01	1.253E-01	2.095E-03	4.660E-05	1.475E-06
EB(CO)			1.485E 05	2.932E 04	2.871E 04	2.222E 04	7.797E 03	7.442E 03	5.898E 03	5.256E 03	4.955E 03
EB(EX)			1.452E 05	2.830E 04	2.736E 04	2.139E 04	7.400E 03	6.970E 03	5.550E 03	4.900E 03	4.620E 03
BETA			5.216E 00	2.302E 00	9.134E-01	2.221E 00	1.143E 00	4.812E-01	2.385E 00	3.148E-01	3.217E 00
F(R)			-1.392E 00	-6.169E-01	7.723E-01	-6.817E-04	-3.064E-01	4.067E-01	-3.813E-04	2.731E-04	-1.855E-07
G(R)	102	257	4.415E 00	1.945E 00	2.556E-01	3.916E-03	9.653E-01	1.348E-01	2.189E-03	4.978E-05	1.552E-06
EB(CO)			1.523E 05	3.020E 04	2.959E 04	2.275E 04	8.053E 03	7.694E 03	6.063E 03	5.411E 03	5.096E 03
EB(EX)			1.487E 05	2.911E 04	2.816E 04	2.190E 04	7.630E 03	7.200E 03	5.700E 03	5.040E 03	4.750E 03

 Instr. see p. 26

5 References

1 Tables for the Analysis of Beta Spectra. National Bureau of Standards Applied Mathematics Series, No. **13** (1952).
2 Rose, M. E., and C. L. Perry: Phys. Rev. **90**, 479 (1953).
— —, and N. M. Dismuke: Oak Ridge National Laboratory Report ORNL — 1459 (1953).
See also Rose, M. E.: In: Beta- and Gamma Ray Spectroscopy. Ed. by. K. Siegbahn. Amsterdam: North Holland Publishing Co. 1955.
3 Dzhelepov, B. S., and L. N. Zyrianova: Influence of atomic electric fields on beta decay. Moscow: Akad. Nauk SSSR 1956.
4 Sliv, L. A., and B. A. Volchok: Tables of Coulomb phases and amplitudes taking into account the finite nuclear size. (Academy of Sciences of the USSR, Moscow-Leningrad 1956.) Translation 1957, AEC-tr-2875.
5 Bhalla, C. P., and M. E. Rose: Phys. Rev. **128**, 774 (1962).
Oak Ridge National Laboratory Report ORNL — 2954 (1960).
ORNL — 3207 (1961).
Bhalla, C. P.: National Bureau of Standards Monograph — NBS — 81 (1964).
6 Bühring, W.: Kernforschungszentrum Karlsruhe. Report No. 559 (1967).
7 Huffäcker, J. N., and C. E. Laird: Nucl. Phys. A **92**, 584 (1967).
8 Rose, M. E.: Phys. Rev. **49**, 727 (1936).
9 Longmire, C., and H. Brown: Phys. Rev. **75**, 1102 (1949).
10 Garret, W. R., and C. P. Bhalla: Z. Physik **198**, 453 (1967).
11 Matese, J. J., and R. Johnson: Phys. Rev. **150**, 846 (1966).
12 Reitz, J. R.: Phys. Rev. **77**, 10 (1950).
Relativistic Electron Wave Functions for a Thomas-Fermi-Dirac-Statistical Atom; University of Chicago, 1949.
13 Durand III, L.: Phys. Rev. **135**, B 310 (1964).
14 Brown, L. S.: Phys. Rev. **135**, B 314 (1964).
15 Eman, B.: Risö Report No. 146 (1966).
16 Bühring, W.: Nucl. Phys. **61**, 110 (1965).
17 Bühring, W.: Nucl. Phys. **40**, 472 (1963)
18 Bühring, W.: Nucl. Phys. **49**, 190 (1963).
19 Bühring,W., and L. Schülke: Nucl. Phys. **65**, 369 (1965).
20 Stech, B., and L. Schülke: Z. Physik **179**, 314 (1964).
21 Schülke, L.: Z. Physik **179**, 331 (1964).
22 Schopper, H.: Weak interactions and nuclear beta decay. Amsterdam: North Holland Publishing Co. 1966.
23 Slater, J. C.: Phys. Rev. **36**, 51 (1930).
24 Wapstra, A. H., G. J. Nijgh, and R. van Lieshout: Nuclear spectroscopy tables. Amsterdam: North Holland Publishing Co. 1959.
25 Hartree, D. K.: Rept. Progr. Phys. **11**, 113 (1946).
26 Brysk, H., and M. E. Rose: Oak Ridge National Laboratory Report ORNL-1830 (1955).
Rev. Mod. Phys. **30**, 1169 (1958).
27 Band, I. M., L. N. Zyrianova, and Tsin Chen-Zhui: Izv. Akad. Nauk SSSR, Ser. Fiz. **20**, 1387 (1956).
28 — —, and Iu. P. Suslov: Izv. Akad. Nauk SSSR, Ser. Fiz. **22**, 952 (1958).
29 Brewer, H. R., D. S. Harmer, and D. H. Hay: Phys. Rev. Lett. **7**, 319 (1961).
Georgia Institute of Technology Report GIT-ES-B 151-1 (1961).
30 Winter, G.: Deutsche Akademie der Wissenschaften zu Berlin Report ZfK-PhA 22 (1966). Nucl. Phys. A 113, 617 (1968).
31 Zyrianova, L N.: Conference on the Electron Capture and Higher Order Processes in Nuclear Decays, Debrecen, Hungary, 1968.
32 Odiot, S., and R. Daudel: J. Phys. Radium **17**, 60 (1956).
33 Bahcall, J. N.: Phys. Rev. Letters **9**, 500 (1962).
Phys. Rev. **129**, 2683 (1963); **131**, 1756 (1963); **132**, 362 (1963).
34 Vatai, E.: Conference on the Electron Capture and Higher Order Processes in Nuclear Decays, Debrecen, Hungary, 1968.
35 Bouchez, R., and P. Depommier: Rept. Progr. Phys. **23**, 395 (1960).
36 Konopinski, E. J.: The theory of Beta radioactivity. Oxford: Oxford University Press 1966.
37 Wu, C. S., and S. A. Moszkowski: Beta decay. New York: John Wiley and Sons 1966.
38 Preston, M. A.: Physics of the nucleus. Reading: Addison-Wesley Publ. Co., Inc. 1962, pp. 383—458.
39 De Benedetti, S.: Nuclear interactions. New York: John Wiley and Sons, Inc. 1964, pp. 527—607.
40 Konopinski, E. J., and M. E. Rose: In: $\alpha \cdot \beta \cdot \gamma$ ray spectroscopy. Ed. by K. Siegbahn. Amsterdam: North Holland Publishing Co. 1965, p. 1327.
41 Weidenmüller, H. A.: Rev. Mod. Phys. **33**, 574 (1961).
42 Tolhoek, H. A.: Selected topics in nuclear theory. Vienna: Intern. Atomic Energy Agency 1963.
43 Blin Stoyle, R. J., and S. C. K. Nair: Advan. Phys. (Phil. Mag. Suppl.) **15**, 493 (1966).
44 Berényi, D.: Rev. Mod. Phys. **40**, 390 (1968).
45 $\alpha \cdot \beta \cdot \gamma$ ray spectroscopy. Ed. by K. Siegbahn. Amsterdam: North Holland Publishing Co. 1965.
46 Appel, H.: Numerical tables for angular correlation computations (3 J-symbols, 6 J-symbols, F-coefficients, and others). Landolt-Börnstein, New Series, Group I, Vol. 3. Berlin: Springer 1968.
47 Fano, U.: Nat. Bur. Standards Rep. 1214 (1951).

48 De Groot, S. R., H. A. Tolhoek, and W. J. Huiskamp: In: $\alpha \cdot \beta \cdot \gamma$ ray spectroscopy. Ed. by K. Siegbahn. Amsterdam: North Holland Publishing Co. 1965, p. 1201.
49 Behrens, H., and W. Bühring: Kernforschungszentrum Karlsruhe. Report, to be published.
50 Rose, M. E.: Relativistic electron theory. New York: John Wiley and Sons 1961, p. 191ff., p. 241ff.
51 Tietz, T.: Nuovo Cimento **36**, 1365 (1965).
52 Strand, T. G., and T. Tietz: Nuovo Cimento **41** B, 89 (1966).
53 Bonham, R. A., and T. G. Strand: J. Chem. Phys. **39**, 2200 (1963).
54 Hagström, S., C. Nordlin, and K. Siegbahn: In: $\alpha \cdot \beta \cdot \gamma$ ray spectroscopy. Ed. by K. Siegbahn. Amsterdam: North Holland Publishing Co. 1965, p. 845.
55 Fahlman, A., K. Hamrin, R. Nordberg, C. Nordling, and K. Siegbahn: Phys. Rev. Letters **14**, 127 (1965).
56 Slater, J.: Quantum theory of atomic structure. New York: McGraw-Hill Book Co. 1960, Vol. 2, p. 329.

GPSR Compliance

The European Union's (EU) General Product Safety Regulation (GPSR) is a set of rules that requires consumer products to be safe and our obligations to ensure this.

If you have any concerns about our products, you can contact us on ProductSafety@springernature.com

In case Publisher is established outside the EU, the EU authorized representative is:

Springer Nature Customer Service Center GmbH
Europaplatz 3
69115 Heidelberg, Germany

Batch number: 10684220

Printed by Printforce, the Netherlands